GRAPHS OF PARENT FUNCTIONS

Linear Function

$f(x) = mx + b$

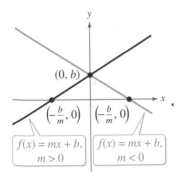

Domain: $(-\infty, \infty)$
Range $(m \neq 0)$: $(-\infty, \infty)$
x-intercept: $(-b/m, 0)$
y-intercept: $(0, b)$
Increasing when $m > 0$
Decreasing when $m < 0$

Absolute Value Function

$f(x) = |x| = \begin{cases} x, & x \geq 0 \\ -x, & x < 0 \end{cases}$

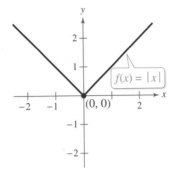

Domain: $(-\infty, \infty)$
Range: $[0, \infty)$
Intercept: $(0, 0)$
Decreasing on $(-\infty, 0)$
Increasing on $(0, \infty)$
Even function
y-axis symmetry

Square Root Function

$f(x) = \sqrt{x}$

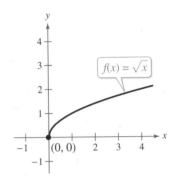

Domain: $[0, \infty)$
Range: $[0, \infty)$
Intercept: $(0, 0)$
Increasing on $(0, \infty)$

Greatest Integer Function

$f(x) = [\![x]\!]$

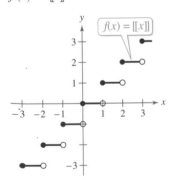

Domain: $(-\infty, \infty)$
Range: the set of integers
x-intercepts: in the interval $[0, 1)$
y-intercept: $(0, 0)$
Constant between each pair of
  consecutive integers
Jumps vertically one unit at
  each integer value

Quadratic (Squaring) Function

$f(x) = ax^2$

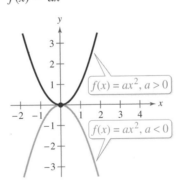

Domain: $(-\infty, \infty)$
Range $(a > 0)$: $[0, \infty)$
Range $(a < 0)$: $(-\infty, 0]$
Intercept: $(0, 0)$
Decreasing on $(-\infty, 0)$ for $a > 0$
Increasing on $(0, \infty)$ for $a > 0$
Increasing on $(-\infty, 0)$ for $a < 0$
Decreasing on $(0, \infty)$ for $a < 0$
Even function
y-axis symmetry
Relative minimum $(a > 0)$,
  relative maximum $(a < 0)$,
  or vertex: $(0, 0)$

Cubic Function

$f(x) = x^3$

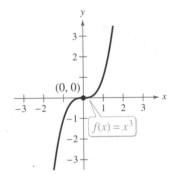

Domain: $(-\infty, \infty)$
Range: $(-\infty, \infty)$
Intercept: $(0, 0)$
Increasing on $(-\infty, \infty)$
Odd function
Origin symmetry

Rational (Reciprocal) Function

$$f(x) = \frac{1}{x}$$

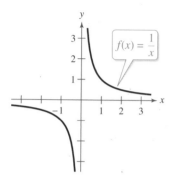

Domain: $(-\infty, 0) \cup (0, \infty)$
Range: $(-\infty, 0) \cup (0, \infty)$
No intercepts
Decreasing on $(-\infty, 0)$ and $(0, \infty)$
Odd function
Origin symmetry
Vertical asymptote: y-axis
Horizontal asymptote: x-axis

Exponential Function

$$f(x) = a^x, \ a > 1$$

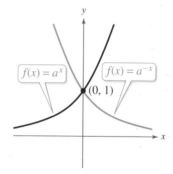

Domain: $(-\infty, \infty)$
Range: $(0, \infty)$
Intercept: $(0, 1)$
Increasing on $(-\infty, \infty)$
 for $f(x) = a^x$
Decreasing on $(-\infty, \infty)$
 for $f(x) = a^{-x}$
Horizontal asymptote: x-axis
Continuous

Logarithmic Function

$$f(x) = \log_a x, \ a > 1$$

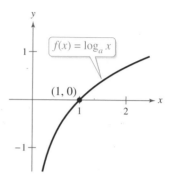

Domain: $(0, \infty)$
Range: $(-\infty, \infty)$
Intercept: $(1, 0)$
Increasing on $(0, \infty)$
Vertical asymptote: y-axis
Continuous
Reflection of graph of $f(x) = a^x$
 in the line $y = x$

Sine Function

$$f(x) = \sin x$$

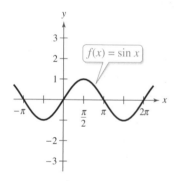

Domain: $(-\infty, \infty)$
Range: $[-1, 1]$
Period: 2π
x-intercepts: $(n\pi, 0)$
y-intercept: $(0, 0)$
Odd function
Origin symmetry

Cosine Function

$$f(x) = \cos x$$

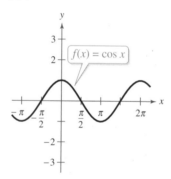

Domain: $(-\infty, \infty)$
Range: $[-1, 1]$
Period: 2π
x-intercepts: $\left(\frac{\pi}{2} + n\pi, 0\right)$
y-intercept: $(0, 1)$
Even function
y-axis symmetry

Tangent Function

$$f(x) = \tan x$$

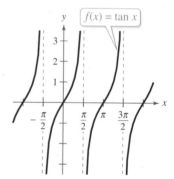

Domain: all $x \neq \dfrac{\pi}{2} + n\pi$
Range: $(-\infty, \infty)$
Period: π
x-intercepts: $(n\pi, 0)$
y-intercept: $(0, 0)$
Vertical asymptotes:
$$x = \frac{\pi}{2} + n\pi$$
Odd function
Origin symmetry

PRECALCULUS
with LIMITS

with

CalcChat® *and* **CalcView®**

4e

Ron Larson
The Pennsylvania State University
The Behrend College

Paul Battaglia
Manasquan High School

CENGAGE
Learning·

Australia • Brazil • Mexico • Singapore • United Kingdom • United States

CENGAGE
Learning

Precalculus with Limits
with CalcChat and CalcView
Fourth Edition

Ron Larson
Paul Battaglia

Product Director: Terry Boyle

Product Manager: Gary Whalen

Product Manager, Advanced and Elective Products Program:
    Maureen McLaughlin

Senior Content Developer: Stacy Green

Senior Content Developer, Advanced and Elective Products
    Program: Philip Lanza

Associate Content Developer: Samantha Lugtu

Product Assistant: Katharine Werring

Media Developer: Lynh Pham

Marketing Manager: Ryan Ahern

Content Project Manager: Jennifer Risden

Manufacturing Planner: Doug Bertke

Production Service: Larson Texts, Inc.

Photo Researcher: Lumina Datamatics

Text Researcher: Lumina Datamatics

Illustrator: Larson Texts, Inc.

Text Designer: Larson Texts, Inc.

Cover Designer: Larson Texts, Inc.

Front Cover Image: Vadim Sadovski/Shutterstock.com

Back Cover Image: © iStockphoto.com/Alejandro Rivers

Compositor: Larson Texts, Inc.

For product information and technology assistance, contact us at
Cengage Learning Customer & Sales Support, 1-888-915-3276.

For permission to use material from this text or product,
submit all requests online at **www.cengage.com/permissions.**
Further permissions questions can be emailed to
permissionrequest@cengage.com.

Library of Congress Control Number: 2016945349

ISBN: 978-1-337-27105-9

Cengage Learning
20 Channel Center Street
Boston, MA 02210
USA

Cengage Learning is a leading provider of customized learning solutions with employees residing in nearly 40 different countries and sales in more than 125 countries around the world. Find your local representative at **www.cengage.com.**

Cengage Learning products are represented in Canada by Nelson Education, Ltd.

To learn more about Cengage Learning Solutions, visit **www.cengage.com.**

To find online supplements and other instructional support, please visit **www.cengagebrain.com.**

QR Code is a registered trademark of Denso Wave Incorporated

Printed in the United States of America
Print Number: 05 Print Year: 2018

Contents

*Available at the text-specific website *www.cengagebrain.com*

Preface

Welcome to *Precalculus with Limits,* Fourth Edition. We are excited to offer you a new edition with even more resources that will help you understand and master precalculus with limits. This textbook includes features and resources that continue to make *Precalculus with Limits* a valuable learning tool for students and a trustworthy teaching tool for instructors.

Precalculus with Limits provides the clear instruction, precise mathematics, and thorough coverage that you expect for your course. Additionally, this new edition provides you with **free** access to three companion websites:

- **CalcView.com**—video solutions to selected exercises

- **CalcChat.com**—worked-out solutions to odd-numbered exercises and access to online tutors

- **LarsonPrecalculus.com**—companion website with resources to supplement your learning

These websites will help enhance and reinforce your understanding of the material presented in this text and prepare you for future mathematics courses. CalcView® and CalcChat® are also available as free mobile apps.

Features

NEW CalcView®

The website *CalcView.com* contains video solutions of selected exercises. Watch teachers progress step-by-step through solutions, providing guidance to help you solve the exercises. The CalcView mobile app is available for free at the Apple® App Store® or Google Play™ store. The app features an embedded QR Code® reader that can be used to scan the on-page codes and go directly to the videos. You can also access the videos at *CalcView.com*.

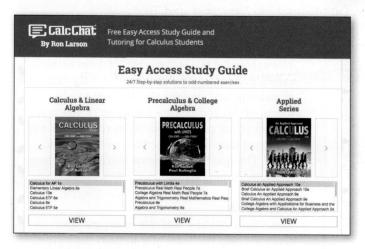

UPDATED CalcChat®

In each exercise set, be sure to notice the reference to *CalcChat.com*. This website provides free step-by-step solutions to all odd-numbered exercises in many of our textbooks. Additionally, you can chat with a tutor, at no charge, during the hours posted at the site. For over 14 years, hundreds of thousands of students have visited this site for help. The CalcChat mobile app is also available as a free download at the Apple® App Store® or Google Play™ store and features an embedded QR Code® reader.

App Store is a service mark of Apple Inc. Google Play is a trademark of Google Inc.
QR Code is a registered trademark of Denso Wave Incorporated.

REVISED LarsonPrecalculus.com

All companion website features have been updated based on this revision, plus we have added a new Collaborative Project feature. Access to these features is free. You can view and listen to worked-out solutions of Checkpoint problems in English or Spanish, explore examples, download data sets, watch lesson videos, and much more.

NEW Collaborative Project

You can find these extended group projects at *LarsonPrecalculus.com*. Check your understanding of the chapter concepts by solving in-depth, real-life problems. These collaborative projects provide an interesting and engaging way for you and other students to work together and investigate ideas.

REVISED Exercise Sets

The exercise sets have been carefully and extensively examined to ensure they are rigorous and relevant, and include topics our users have suggested. The exercises have been reorganized and titled so you can better see the connections between examples and exercises. Multi-step, real-life exercises reinforce problem-solving skills and mastery of concepts by giving you the opportunity to apply the concepts in real-life situations. Error Analysis exercises have been added throughout the text to help you identify common mistakes.

Table of Contents Changes

Based on market research and feedback from users, Section 6.5, The Complex Plane, has been added. In addition, examples on finding the magnitude of a scalar multiple (Section 6.3), multiplying in the complex plane (Section 6.6), using matrices to transform vectors (Section 8.2), and further applications of 2×2 matrices (Section 8.5) have been added.

Chapter Opener

Each Chapter Opener highlights real-life applications used in the examples and exercises.

Section Objectives

A bulleted list of learning objectives provides you the opportunity to preview what will be presented in the upcoming section.

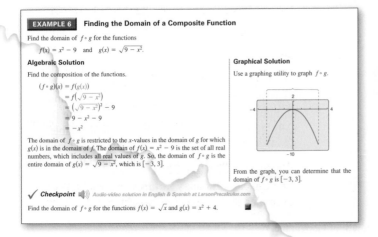

Side-By-Side Examples

Throughout the text, we present solutions to many examples from multiple perspectives—algebraically, graphically, and numerically. The side-by-side format of this pedagogical feature helps you to see that a problem can be solved in more than one way and to see that different methods yield the same result. The side-by-side format also addresses many different learning styles.

Remarks

These hints and tips reinforce or expand upon concepts, help you learn how to study mathematics, caution you about common errors, address special cases, or show alternative or additional steps to a solution of an example.

Checkpoints

Accompanying every example, the Checkpoint problems encourage immediate practice and check your understanding of the concepts presented in the example. View and listen to worked-out solutions of the Checkpoint problems in English or Spanish at *LarsonPrecalculus.com*.

Technology

The technology feature gives suggestions for effectively using tools such as calculators, graphing utilities, and spreadsheet programs to help deepen your understanding of concepts, ease lengthy calculations, and provide alternate solution methods for verifying answers obtained by hand.

Historical Notes

These notes provide helpful information regarding famous mathematicians and their work.

Algebra of Calculus

Throughout the text, special emphasis is given to the algebraic techniques used in calculus. Algebra of Calculus examples and exercises are integrated throughout the text and are identified by the symbol $\int$.

Summarize

The Summarize feature at the end of each section helps you organize the lesson's key concepts into a concise summary, providing you with a valuable study tool.

Vocabulary Exercises

The vocabulary exercises appear at the beginning of the exercise set for each section. These problems help you review previously learned vocabulary terms that you will use in solving the section exercises.

▷ TECHNOLOGY Use a graphing utility to check the result of Example 2. To do this, enter

$$Y1 = -(\sin(X))^3$$

and

$$Y2 = \sin(X)(\cos(X))^2$$
$$- \sin(X).$$

Select the *line* style for Y1 and the *path* style for Y2, then graph both equations in the same viewing window. The two graphs *appear* to coincide, so it is reasonable to assume that their expressions are equivalent. Note that the actual equivalence of the expressions can only be verified algebraically, as in Example 2. This graphical approach is only to check your work.

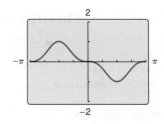

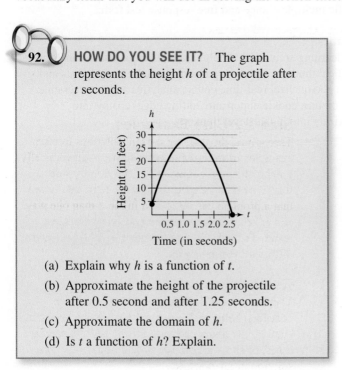

92. HOW DO YOU SEE IT? The graph represents the height *h* of a projectile after *t* seconds.

(a) Explain why *h* is a function of *t*.

(b) Approximate the height of the projectile after 0.5 second and after 1.25 seconds.

(c) Approximate the domain of *h*.

(d) Is *t* a function of *h*? Explain.

How Do You See It?

The How Do You See It? feature in each section presents a real-life exercise that you will solve by visual inspection using the concepts learned in the lesson. This exercise is excellent for classroom discussion or test preparation.

Project

The projects at the end of selected sections involve in-depth applied exercises in which you will work with large, real-life data sets, often creating or analyzing models. These projects are offered online at *LarsonPrecalculus.com*.

Chapter Summary

The Chapter Summary includes explanations and examples of the objectives taught in each chapter.

Teacher's Edition / ISBN-13: 978-1-337-27106-6
The Teacher's edition is the complete student text plus point-of-use annotations for teachers, including Extra Examples, Avoiding Common Errors, Closing the Lesson questions, and Assignment Guides. Answers to all text exercises, Vocabulary Checks, and Checkpoints are also provided.

Complete Solutions Manual / ISBN-13: 978-1-337-28033-4
This manual contains solutions to all exercises from the text, including Chapter Review Exercises, Chapter Tests, and Practice Tests with solutions.

Teacher Resource Guide / ISBN-13: 978-1-337-28041-9
This guide contains suggested time to allot, points to stress, text discussion topics, core materials for lecture, workshop/discussion suggestions, group work exercises in a form suitable for handout, and suggested homework assignments.

Cengage Learning Testing Powered by Cognero (login.cengage.com)
CLT is a flexible online system that allows you to author, edit, and manage test bank content; create multiple test versions in an instant; and deliver tests from your LMS, your classroom, or wherever you want. This is available online via *www.cengage.com/login.*

Instructor Companion Site
Everything you need for your course in one place! This collection of book-specific lecture and class tools is available online via *www.cengage.com/login.* Access and download PowerPoint® presentations, images, teacher's guide, and more.

Test Bank (on instructor companion site)
This contains text-specific multiple-choice and free response test forms.

MindTap for Mathematics
MindTap® is the digital learning solution that helps teachers engage and transform today's students into critical thinkers. Through paths of dynamic assignments and applications that you can personalize, real-time course analytics and an accessible reader, MindTap helps you turn cookie cutter into cutting edge, apathy into engagement, and memorizers into higher-level thinkers.

Student Resources

Student Study and Solutions Manual / ISBN-13: 978-1-337-27985-7
This guide offers step-by-step solutions for all odd-numbered text exercises, Chapter Tests, and Cumulative Tests. It also contains Practice Tests.

Note-Taking Guide / ISBN-13: 978-1-337-27992-5
This is an innovative study aid, in the form of a notebook organizer, that helps students develop a section-by-section summary of key concepts.

CengageBrain.com
To access additional course materials, please visit *www.cengagebrain.com*. At the *CengageBrain.com* home page, search for the ISBN of your title (from the back cover of your book) using the search box at the top of the page. This will take you to the product page where these resources can be found.

MindTap for Mathematics
MindTap® provides you with the tools you need to better manage your limited time—you can complete assignments whenever and wherever you are ready to learn with course material specially customized for you by your teacher and streamlined in one proven, easy-to-use interface. With an array of tools and apps—from note taking to flashcards—you'll get a true understanding of course concepts, helping you to achieve better grades and setting the groundwork for your future courses. This access code entitles you to one term of usage.

Acknowledgments

We would like to thank the many people who have helped us prepare the text and the supplements package. Their encouragement, criticisms, and suggestions have been invaluable.

Thank you to all of the instructors who took the time to review the changes in this edition and to provide suggestions for improving it. Without your help, this book would not be possible.

Reviewers of the Fourth Edition

Gurdial Arora, *Xavier University of Louisiana*
Russell C. Chappell, *Twinsburg High School, Ohio*
Darlene Martin, *Lawson State Community College*
John Fellers, *North Allegheny School District*
Professor Steven Sikes, *Collin College*
Ann Slate, *Surry Community College*
John Elias, *Glenda Dawson High School*
Kathy Wood, *Lansing Catholic High School*
Darin Bauguess, *Surry Community College*
Brianna Kurtz, *Daytona State College*

Reviewers of the Previous Editions

Timothy Andrew Brown, *South Georgia College;* Blair E. Caboot, *Keystone College;* Shannon Cornell, *Amarillo College;* Gayla Dance, *Millsaps College;* Paul Finster, *El Paso Community College;* Paul A. Flasch, *Pima Community College West Campus;* Vadas Gintautas, *Chatham University;* Lorraine A. Hughes, *Mississippi State University;* Shu-Jen Huang, *University of Florida;* Renyetta Johnson, *East Mississippi Community College;* George Keihany, *Fort Valley State University;* Mulatu Lemma, *Savannah State University;* William Mays Jr., *Salem Community College;* Marcella Melby, *University of Minnesota;* Jonathan Prewett, *University of Wyoming;* Denise Reid, *Valdosta State University;* David L. Sonnier, *Lyon College;* David H. Tseng, *Miami Dade College—Kendall Campus;* Kimberly Walters, *Mississippi State University;* Richard Weil, *Brown College;* Solomon Willis, *Cleveland Community College;* Bradley R. Young, *Darton College*

Our thanks to Robert Hostetler, The Behrend College, The Pennsylvania State University, and David Heyd, The Behrend College, The Pennsylvania State University, for their significant contributions to previous editions of this text.

We would also like to thank the staff at Larson Texts, Inc. who assisted with proofreading the manuscript, preparing and proofreading the art package, and checking and typesetting the supplements.

On a personal level, we are grateful to our spouses, Deanna Gilbert Larson and Janice Battaglia, for their love, patience, and support. Also, a special thanks goes to R. Scott O'Neil. If you have suggestions for improving this text, please feel free to write to us. Over the past two decades, we have received many useful comments from both teachers and students, and we value these comments very highly.

Ron Larson
Paul Battaglia

PRECALCULUS
with LIMITS

with

CalcChat® *and* **CalcView®**

4e

1 Functions and Their Graphs

Snowstorm *(Exercise 47, page 66)*

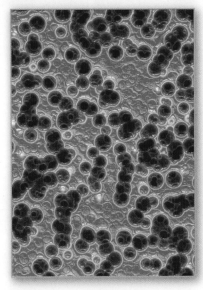

Bacteria *(Example 8, page 80)*

Average Speed *(Example 7, page 54)*

Alternative-Fuel Stations
(Example 10, page 42)

Americans with Disabilities Act *(page 28)*

1.1 Rectangular Coordinates

The Cartesian plane can help you visualize relationships between two variables. For example, in Exercise 37 on page 9, given how far north and west one city is from another, plotting points to represent the cities can help you visualize these distances and determine the flying distance between the cities.

- Plot points in the Cartesian plane.
- Use the Distance Formula to find the distance between two points.
- Use the Midpoint Formula to find the midpoint of a line segment.
- Use a coordinate plane to model and solve real-life problems.

The Cartesian Plane

Just as you can represent real numbers by points on a real number line, you can represent ordered pairs of real numbers by points in a plane called the **rectangular coordinate system,** or the **Cartesian plane,** named after the French mathematician René Descartes (1596–1650).

Two real number lines intersecting at right angles form the Cartesian plane, as shown in Figure 1.1. The horizontal real number line is usually called the **x-axis,** and the vertical real number line is usually called the **y-axis.** The point of intersection of these two axes is the **origin,** and the two axes divide the plane into four **quadrants.**

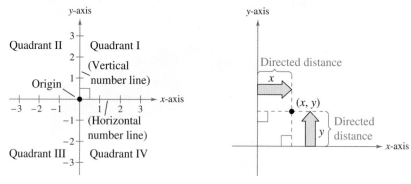

Figure 1.1

Figure 1.2

Each point in the plane corresponds to an **ordered pair** (x, y) of real numbers x and y, called **coordinates** of the point. The **x-coordinate** represents the directed distance from the y-axis to the point, and the **y-coordinate** represents the directed distance from the x-axis to the point, as shown in Figure 1.2.

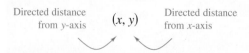

The notation (x, y) denotes both a point in the plane and an open interval on the real number line. The context will tell you which meaning is intended.

EXAMPLE 1   Plotting Points in the Cartesian Plane

Plot the points $(-1, 2)$, $(3, 4)$, $(0, 0)$, $(3, 0)$, and $(-2, -3)$.

Solution   To plot the point $(-1, 2)$, imagine a vertical line through -1 on the x-axis and a horizontal line through 2 on the y-axis. The intersection of these two lines is the point $(-1, 2)$. Plot the other four points in a similar way, as shown in Figure 1.3.

✓ **Checkpoint** 🔊))) *Audio-video solution in English & Spanish at LarsonPrecalculus.com*

Plot the points $(-3, 2)$, $(4, -2)$, $(3, 1)$, $(0, -2)$, and $(-1, -2)$.

Figure 1.3

The beauty of a rectangular coordinate system is that it allows you to *see* relationships between two variables. It would be difficult to overestimate the importance of Descartes's introduction of coordinates in the plane. Today, his ideas are in common use in virtually every scientific and business-related field.

EXAMPLE 2 **Sketching a Scatter Plot**

The table shows the numbers N (in millions) of subscribers to a cellular telecommunication service in the United States from 2005 through 2014, where t represents the year. Sketch a scatter plot of the data. *(Source: CTIA-The Wireless Association)*

Solution To sketch a *scatter plot* of the data shown in the table, represent each pair of values by an ordered pair (t, N) and plot the resulting points. For example, let $(2005, 207.9)$ represent the first pair of values. Note that in the scatter plot below, the break in the t-axis indicates omission of the years before 2005, and the break in the N-axis indicates omission of the numbers less than 150 million.

Year, t	Subscribers, N
2005	207.9
2006	233.0
2007	255.4
2008	270.3
2009	285.6
2010	296.3
2011	316.0
2012	326.5
2013	335.7
2014	355.4

✓ Checkpoint Audio-video solution in English & Spanish at LarsonPrecalculus.com

The table shows the numbers N (in thousands) of cellular telecommunication service employees in the United States from 2005 through 2014, where t represents the year. Sketch a scatter plot of the data. *(Source: CTIA-The Wireless Association)*

t	N
2005	233.1
2006	253.8
2007	266.8
2008	268.5
2009	249.2
2010	250.4
2011	238.1
2012	230.1
2013	230.4
2014	232.2

▷ **TECHNOLOGY** The scatter plot in Example 2 is only one way to represent the data graphically. You could also represent the data using a bar graph or a line graph. Use a graphing utility to represent the data given in Example 2 graphically.

In Example 2, you could let $t = 1$ represent the year 2005. In that case, there would not be a break in the horizontal axis, and the labels 1 through 10 (instead of 2005 through 2014) would be on the tick marks.

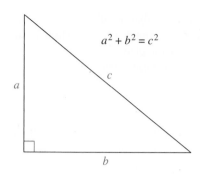

$$a^2 + b^2 = c^2$$

Figure 1.4

The Pythagorean Theorem and The Distance Formula

The Pythagorean Theorem is used extensively throughout this course.

> ## Pythagorean Theorem
>
> For a right triangle with hypotenuse length c and sides lengths a and b, you have $a^2 + b^2 = c^2$, as shown in Figure 1.4. (The converse is also true. That is, if $a^2 + b^2 = c^2$, then the triangle is a right triangle.)

Using the points (x_1, y_1) and (x_2, y_2), you can form a right triangle, as shown in Figure 1.5. The length of the hypotenuse of the right triangle is the distance d between the two points. The length of the vertical side of the triangle is $|y_2 - y_1|$ and the length of the horizontal side is $|x_2 - x_1|$. By the Pythagorean Theorem,

$$d^2 = |x_2 - x_1|^2 + |y_2 - y_1|^2$$
$$d = \sqrt{|x_2 - x_1|^2 + |y_2 - y_1|^2}$$
$$= \sqrt{(x_2 - x_1)^2 + (y_2 - y_1)^2}.$$

This result is the **Distance Formula.**

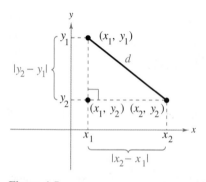

Figure 1.5

> ## The Distance Formula
>
> The distance d between the points (x_1, y_1) and (x_2, y_2) in the plane is
>
> $$d = \sqrt{(x_2 - x_1)^2 + (y_2 - y_1)^2}.$$

EXAMPLE 3   **Finding a Distance**

Find the distance between the points $(-2, 1)$ and $(3, 4)$.

Algebraic Solution

Let $(x_1, y_1) = (-2, 1)$ and $(x_2, y_2) = (3, 4)$. Then apply the Distance Formula.

$$d = \sqrt{(x_2 - x_1)^2 + (y_2 - y_1)^2} \quad \text{Distance Formula}$$
$$= \sqrt{[3 - (-2)]^2 + (4 - 1)^2} \quad \text{Substitute for } x_1, y_1, x_2, \text{ and } y_2.$$
$$= \sqrt{(5)^2 + (3)^2} \quad \text{Simplify.}$$
$$= \sqrt{34} \quad \text{Simplify.}$$
$$\approx 5.83 \quad \text{Use a calculator.}$$

So, the distance between the points is about 5.83 units.

Check

$$d^2 \stackrel{?}{=} 5^2 + 3^2 \quad \text{Pythagorean Theorem}$$
$$\left(\sqrt{34}\right)^2 \stackrel{?}{=} 5^2 + 3^2 \quad \text{Substitute for } d.$$
$$34 = 34 \quad \text{Distance checks. } \checkmark$$

Graphical Solution

Use centimeter graph paper to plot the points $A(-2, 1)$ and $B(3, 4)$. Carefully sketch the line segment from A to B. Then use a centimeter ruler to measure the length of the segment.

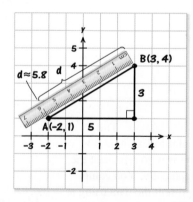

The line segment measures about 5.8 centimeters. So, the distance between the points is about 5.8 units.

✓ *Checkpoint* ◀))) *Audio-video solution in English & Spanish at LarsonPrecalculus.com*

Find the distance between the points $(3, 1)$ and $(-3, 0)$.

EXAMPLE 4 **Verifying a Right Triangle**

Show that the points

$$(2, 1), \quad (4, 0), \quad \text{and} \quad (5, 7)$$

are vertices of a right triangle.

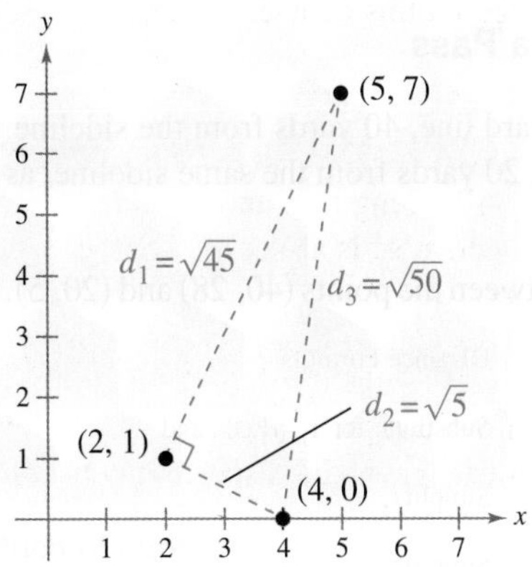

Solution The three points are plotted in Figure 1.6. Using the Distance Formula, the lengths of the three sides are

$$d_1 = \sqrt{(5-2)^2 + (7-1)^2} = \sqrt{9+36} = \sqrt{45},$$

$$d_2 = \sqrt{(4-2)^2 + (0-1)^2} = \sqrt{4+1} = \sqrt{5}, \text{ and}$$

$$d_3 = \sqrt{(5-4)^2 + (7-0)^2} = \sqrt{1+49} = \sqrt{50}.$$

Because $(d_1)^2 + (d_2)^2 = 45 + 5 = 50 = (d_3)^2$, you can conclude by the converse of the Pythagorean Theorem that the triangle is a right triangle.

✓ **Checkpoint** ◀))) Audio-video solution in English & Spanish at LarsonPrecalculus.com

Show that the points $(2, -1)$, $(5, 5)$, and $(6, -3)$ are vertices of a right triangle. ■

Figure 1.6

▷ **ALGEBRA HELP** To review the techniques for evaluating a radical, see Appendix A.2.

The Midpoint Formula

To find the **midpoint** of the line segment that joins two points in a coordinate plane, find the average values of the respective coordinates of the two endpoints using the **Midpoint Formula**.

The Midpoint Formula

The midpoint of the line segment joining the points (x_1, y_1) and (x_2, y_2) is

$$\text{Midpoint} = \left(\frac{x_1 + x_2}{2}, \frac{y_1 + y_2}{2}\right).$$

For a proof of the Midpoint Formula, see Proofs in Mathematics on page 110.

EXAMPLE 5 **Finding the Midpoint of a Line Segment**

Find the midpoint of the line segment joining the points

$$(-5, -3) \quad \text{and} \quad (9, 3).$$

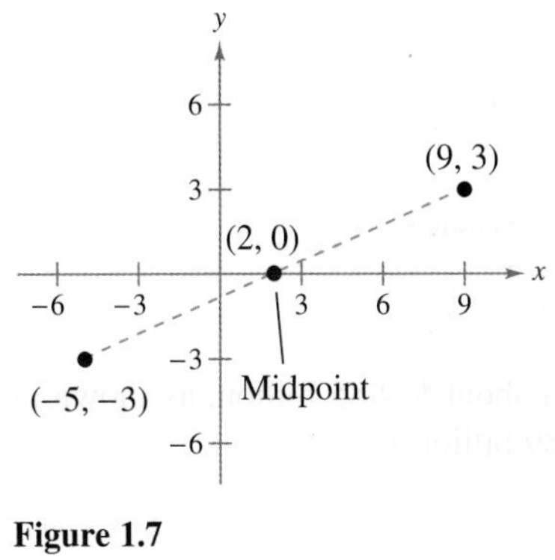

Solution Let $(x_1, y_1) = (-5, -3)$ and $(x_2, y_2) = (9, 3)$.

$$\text{Midpoint} = \left(\frac{x_1 + x_2}{2}, \frac{y_1 + y_2}{2}\right) \qquad \text{Midpoint Formula}$$

$$= \left(\frac{-5 + 9}{2}, \frac{-3 + 3}{2}\right) \qquad \text{Substitute for } x_1, y_1, x_2, \text{ and } y_2.$$

$$= (2, 0) \qquad \text{Simplify.}$$

The midpoint of the line segment is $(2, 0)$, as shown in Figure 1.7.

Figure 1.7

✓ **Checkpoint** ◀))) Audio-video solution in English & Spanish at LarsonPrecalculus.com

Find the midpoint of the line segment joining the points

$$(-2, 8) \quad \text{and} \quad (4, -10).$$ ■

1.1 Exercises

See **CalcChat.com** for tutorial help and worked-out solutions to odd-numbered exercises.

Vocabulary: Fill in the blanks.

1. An ordered pair of real numbers can be represented in a plane called the rectangular coordinate system or the _____ plane.
2. The x- and y-axes divide the coordinate plane into four _____.
3. The _____ _____ is derived from the Pythagorean Theorem.
4. Finding the average values of the respective coordinates of the two endpoints of a line segment in a coordinate plane is also known as using the _____ _____.

Skills and Applications

Plotting Points in the Cartesian Plane In Exercises 5 and 6, plot the points.

5. $(2, 4), (3, -1), (-6, 2), (-4, 0), (-1, -8), (1.5, -3.5)$
6. $(1, -5), (-2, -7), (3, 3), (-2, 4), (0, 5), \left(\frac{2}{3}, \frac{5}{2}\right)$

Finding the Coordinates of a Point In Exercises 7 and 8, find the coordinates of the point.

7. The point is three units to the left of the y-axis and four units above the x-axis.
8. The point is on the x-axis and 12 units to the left of the y-axis.

Determining Quadrant(s) for a Point In Exercises 9–14, determine the quadrant(s) in which (x, y) could be located.

9. $x > 0$ and $y < 0$
10. $x < 0$ and $y < 0$
11. $x = -4$ and $y > 0$
12. $x < 0$ and $y = 7$
13. $x + y = 0, x \neq 0, y \neq 0$
14. $xy > 0$

Sketching a Scatter Plot In Exercises 15 and 16, sketch a scatter plot of the data shown in the table.

15. The table shows the number y of Wal-Mart stores for each year x from 2008 through 2014. *(Source: Wal-Mart Stores, Inc.)*

Year, x	Number of Stores, y
2008	7720
2009	8416
2010	8970
2011	10,130
2012	10,773
2013	10,942
2014	11,453

16. The table shows the lowest temperature on record y (in degrees Fahrenheit) in Duluth, Minnesota, for each month x, where $x = 1$ represents January. *(Source: NOAA)*

Month, x	Temperature, y
1	-39
2	-39
3	-29
4	-5
5	17
6	27
7	35
8	32
9	22
10	8
11	-23
12	-34

Finding a Distance In Exercises 17–22, find the distance between the points.

17. $(-2, 6), (3, -6)$
18. $(8, 5), (0, 20)$
19. $(1, 4), (-5, -1)$
20. $(1, 3), (3, -2)$
21. $\left(\frac{1}{2}, \frac{4}{3}\right), (2, -1)$
22. $(9.5, -2.6), (-3.9, 8.2)$

Verifying a Right Triangle In Exercises 23 and 24, (a) find the length of each side of the right triangle, and (b) show that these lengths satisfy the Pythagorean Theorem.

23.
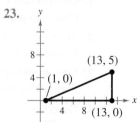

24.
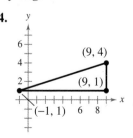

The symbol ▦ and a red exercise number indicates that a video solution can be seen at *CalcView.com*.

Verifying a Polygon In Exercises 25–28, show that the points form the vertices of the polygon.

25. Right triangle: $(4, 0)$, $(2, 1)$, $(-1, -5)$
26. Right triangle: $(-1, 3)$, $(3, 5)$, $(5, 1)$
27. Isosceles triangle: $(1, -3)$, $(3, 2)$, $(-2, 4)$
28. Isosceles triangle: $(2, 3)$, $(4, 9)$, $(-2, 7)$

Plotting, Distance, and Midpoint In Exercises 29–36, (a) plot the points, (b) find the distance between the points, and (c) find the midpoint of the line segment joining the points.

29. $(6, -3)$, $(6, 5)$
30. $(1, 4)$, $(8, 4)$
31. $(1, 1)$, $(9, 7)$
32. $(1, 12)$, $(6, 0)$
33. $(-1, 2)$, $(5, 4)$
34. $(2, 10)$, $(10, 2)$
35. $(-16.8, 12.3)$, $(5.6, 4.9)$
36. $\left(\frac{1}{2}, 1\right)$, $\left(-\frac{5}{2}, \frac{4}{3}\right)$

• • **37. Flying Distance** • • • • • • • • • • • • • • • • •

An airplane flies from Naples, Italy, in a straight line to Rome, Italy, which is 120 kilometers north and 150 kilometers west of Naples. How far does the plane fly?

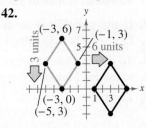

38. **Sports** A soccer player passes the ball from a point that is 18 yards from the endline and 12 yards from the sideline. A teammate who is 42 yards from the same endline and 50 yards from the same sideline receives the pass. (See figure.) How long is the pass?

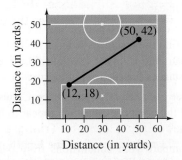

39. **Sales** The Coca-Cola Company had sales of $35,123 million in 2010 and $45,998 million in 2014. Use the Midpoint Formula to estimate the sales in 2012. Assume that the sales followed a linear pattern. *(Source: The Coca-Cola Company)*

40. **Revenue per Share** The revenue per share for Twitter, Inc. was $1.17 in 2013 and $3.25 in 2015. Use the Midpoint Formula to estimate the revenue per share in 2014. Assume that the revenue per share followed a linear pattern. *(Source: Twitter, Inc.)*

Translating Points in the Plane In Exercises 41–44, find the coordinates of the vertices of the polygon after the given translation to a new position in the plane.

41.

42.

43. Original coordinates of vertices: $(-7, -2)$, $(-2, 2)$, $(-2, -4)$, $(-7, -4)$
Shift: eight units up, four units to the right

44. Original coordinates of vertices: $(5, 8)$, $(3, 6)$, $(7, 6)$
Shift: 6 units down, 10 units to the left

45. **Minimum Wage** Use the graph below, which shows the minimum wages in the United States (in dollars) from 1950 through 2015. *(Source: U.S. Department of Labor)*

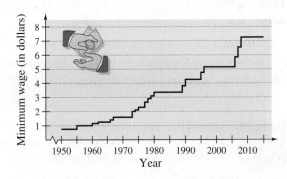

(a) Which decade shows the greatest increase in the minimum wage?

(b) Approximate the percent increases in the minimum wage from 1985 to 2000 and from 2000 to 2015.

(c) Use the percent increase from 2000 to 2015 to predict the minimum wage in 2030.

(d) Do you believe that your prediction in part (c) is reasonable? Explain.

46. **Exam Scores** The table shows the mathematics entrance test scores x and the final examination scores y in an algebra course for a sample of 10 students.

x	22	29	35	40	44	48	53	58	65	76
y	53	74	57	66	79	90	76	93	83	99

(a) Sketch a scatter plot of the data.

(b) Find the entrance test score of any student with a final exam score in the 80s.

(c) Does a higher entrance test score imply a higher final exam score? Explain.

Exploration

True or False?   In Exercises 47–50, determine whether the statement is true or false. Justify your answer.

47. If the point (x, y) is in Quadrant II, then the point $(2x, -3y)$ is in Quadrant III.

48. To divide a line segment into 16 equal parts, you have to use the Midpoint Formula 16 times.

49. The points $(-8, 4)$, $(2, 11)$, and $(-5, 1)$ represent the vertices of an isosceles triangle.

50. If four points represent the vertices of a polygon, and the four side lengths are equal, then the polygon must be a square.

51. Think About It   When plotting points on the rectangular coordinate system, when should you use different scales for the x- and y-axes? Explain.

52. Think About It   What is the y-coordinate of any point on the x-axis? What is the x-coordinate of any point on the y-axis?

53. Using the Midpoint Formula   A line segment has (x_1, y_1) as one endpoint and (x_m, y_m) as its midpoint. Find the other endpoint (x_2, y_2) of the line segment in terms of $x_1, y_1, x_m,$ and y_m.

54. Using the Midpoint Formula   Use the result of Exercise 53 to find the endpoint (x_2, y_2) of each line segment with the given endpoint (x_1, y_1) and midpoint (x_m, y_m).

(a) $(x_1, y_1) = (1, -2)$
     $(x_m, y_m) = (4, -1)$
(b) $(x_1, y_1) = (-5, 11)$
     $(x_m, y_m) = (2, 4)$

55. Using the Midpoint Formula   Use the Midpoint Formula three times to find the three points that divide the line segment joining (x_1, y_1) and (x_2, y_2) into four equal parts.

56. Using the Midpoint Formula   Use the result of Exercise 55 to find the points that divide each line segment joining the given points into four equal parts.

(a) $(x_1, y_1) = (1, -2)$
     $(x_2, y_2) = (4, -1)$
(b) $(x_1, y_1) = (-2, -3)$
     $(x_2, y_2) = (0, 0)$

57. Proof   Prove that the diagonals of the parallelogram in the figure intersect at their midpoints.

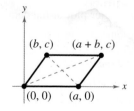

58. **HOW DO YOU SEE IT?**   Use the plot of the point (x_0, y_0) in the figure. Match the transformation of the point with the correct plot. Explain. [The plots are labeled (i), (ii), (iii), and (iv).]

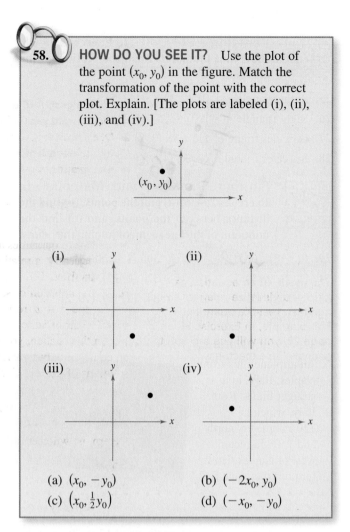

(a) $(x_0, -y_0)$ (b) $(-2x_0, y_0)$
(c) $\left(x_0, \frac{1}{2}y_0\right)$ (d) $(-x_0, -y_0)$

59. Collinear Points   Three or more points are collinear when they all lie on the same line. Use the steps below to determine whether the set of points $\{A(2, 3), B(2, 6), C(6, 3)\}$ and the set of points $\{A(8, 3), B(5, 2), C(2, 1)\}$ are collinear.

(a) For each set of points, use the Distance Formula to find the distances from A to B, from B to C, and from A to C. What relationship exists among these distances for each set of points?

(b) Plot each set of points in the Cartesian plane. Do all the points of either set appear to lie on the same line?

(c) Compare your conclusions from part (a) with the conclusions you made from the graphs in part (b). Make a general statement about how to use the Distance Formula to determine collinearity.

60. Make a Conjecture

(a) Use the result of Exercise 58(a) to make a conjecture about the new location of a point when the sign of the y-coordinate is changed.

(b) Use the result of Exercise 58(d) to make a conjecture about the new location of a point when the signs of both x- and y-coordinates are changed.

1.2 Graphs of Equations

The graph of an equation can help you visualize relationships between real-life quantities. For example, in Exercise 85 on page 21, you will use a graph to analyze life expectancy.

- ■ Sketch graphs of equations.
- ■ Find *x*- and *y*-intercepts of graphs of equations.
- ■ Use symmetry to sketch graphs of equations.
- ■ Write equations of circles.
- ■ Use graphs of equations to solve real-life problems.

The Graph of an Equation

In Section 1.1, you used a coordinate system to graphically represent the relationship between two quantities as points in a coordinate plane.

Frequently, a relationship between two quantities is expressed as an **equation in two variables.** For example, $y = 7 - 3x$ is an equation in x and y. An ordered pair (a, b) is a **solution** or **solution point** of an equation in x and y when the substitutions $x = a$ and $y = b$ result in a true statement. For example, $(1, 4)$ is a solution of $y = 7 - 3x$ because $4 = 7 - 3(1)$ is a true statement.

In this section, you will review some basic procedures for sketching the graph of an equation in two variables. The **graph of an equation** is the set of all points that are solutions of the equation.

EXAMPLE 1 Determining Solution Points

Determine whether (a) $(2, 13)$ and (b) $(-1, -3)$ lie on the graph of $y = 10x - 7$.

Solution

a. $y = 10x - 7$ Write original equation.

$13 \overset{?}{=} 10(2) - 7$ Substitute 2 for *x* and 13 for *y*.

$13 = 13$ $(2, 13)$ is a solution. ✓

The point $(2, 13)$ *does* lie on the graph of $y = 10x - 7$ because it is a solution point of the equation.

b. $y = 10x - 7$ Write original equation.

$-3 \overset{?}{=} 10(-1) - 7$ Substitute −1 for *x* and −3 for *y*.

$-3 \neq -17$ $(-1, -3)$ is not a solution.

The point $(-1, -3)$ *does not* lie on the graph of $y = 10x - 7$ because it is *not* a solution point of the equation.

✓ **Checkpoint** ◀))) Audio-video solution in English & Spanish at LarsonPrecalculus.com

Determine whether (a) $(3, -5)$ and (b) $(-2, 26)$ lie on the graph of $y = 14 - 6x$. ■

The basic technique used for sketching the graph of an equation is the **point-plotting method.**

> ▷ **ALGEBRA HELP** When evaluating an expression or an equation, remember to follow the Basic Rules of Algebra. To review these rules, see Appendix A.1.

The Point-Plotting Method of Graphing

1. When possible, isolate one of the variables.

2. Construct a table of values showing several solution points.

3. Plot these points in a rectangular coordinate system.

4. Connect the points with a smooth curve or line.

EXAMPLE 5 **Testing for Symmetry**

Test $y = 2x^3$ for symmetry with respect to both axes and the origin.

Solution

x-Axis: $y = 2x^3$ Write original equation.

$-y = 2x^3$ Replace y with $-y$. Result is *not* an equivalent equation.

y-Axis: $y = 2x^3$ Write original equation.

$y = 2(-x)^3$ Replace x with $-x$.

$y = -2x^3$ Simplify. Result is *not* an equivalent equation.

Origin: $y = 2x^3$ Write original equation.

$-y = 2(-x)^3$ Replace y with $-y$ and x with $-x$.

$-y = -2x^3$ Simplify.

$y = 2x^3$ Simplify. Result is an equivalent equation.

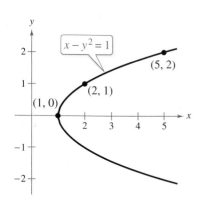

Figure 1.15

Of the three tests for symmetry, the test for origin symmetry is the only one satisfied. So, the graph of $y = 2x^3$ is symmetric with respect to the origin (see Figure 1.15).

✓ *Checkpoint*))) *Audio-video solution in English & Spanish at LarsonPrecalculus.com*

Test $y^2 = 6 - x$ for symmetry with respect to both axes and the origin.

EXAMPLE 6 **Using Symmetry as a Sketching Aid**

Use symmetry to sketch the graph of $x - y^2 = 1$.

Solution   Of the three tests for symmetry, the test for x-axis symmetry is the only one satisfied, because $x - (-y)^2 = 1$ is equivalent to $x - y^2 = 1$. So, the graph is symmetric with respect to the x-axis. Find solution points above (or below) the x-axis and then use symmetry to obtain the graph, as shown in Figure 1.16.

✓ *Checkpoint*))) *Audio-video solution in English & Spanish at LarsonPrecalculus.com*

Use symmetry to sketch the graph of $y = x^2 - 4$.

Figure 1.16

EXAMPLE 7 **Sketching the Graph of an Equation**

Sketch the graph of $y = |x - 1|$.

Solution   This equation fails all three tests for symmetry, so its graph is not symmetric with respect to either axis or to the origin. The absolute value bars tell you that y is always nonnegative. Construct a table of values. Then plot and connect the points, as shown in Figure 1.17. Notice from the table that $x = 0$ when $y = 1$. So, the y-intercept is $(0, 1)$. Similarly, $y = 0$ when $x = 1$. So, the x-intercept is $(1, 0)$.

x	-2	-1	0	1	2	3	4
$y = |x - 1|$	3	2	1	0	1	2	3
(x, y)	$(-2, 3)$	$(-1, 2)$	$(0, 1)$	$(1, 0)$	$(2, 1)$	$(3, 2)$	$(4, 3)$

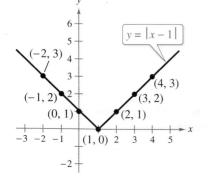

Figure 1.17

✓ *Checkpoint*))) *Audio-video solution in English & Spanish at LarsonPrecalculus.com*

Sketch the graph of $y = |x - 2|$.

Circles

A **circle** is a set of points (x, y) in a plane that are the same distance r from a point called the center, (h, k), as shown at the right. By the Distance Formula,

$$\sqrt{(x - h)^2 + (y - k)^2} = r.$$

By squaring each side of this equation, you obtain the **standard form of the equation of a circle.** For example, for a circle with its center at $(h, k) = (1, 3)$ and radius $r = 4$,

$$\sqrt{(x - 1)^2 + (y - 3)^2} = 4 \qquad \text{Substitute for } h, k, \text{ and } r.$$
$$(x - 1)^2 + (y - 3)^2 = 16. \qquad \text{Square each side.}$$

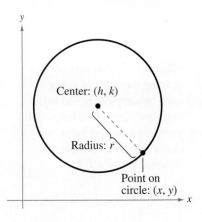

Standard Form of the Equation of a Circle

A point (x, y) lies on the circle of **radius** r and **center** (h, k) if and only if

$$(x - h)^2 + (y - k)^2 = r^2.$$

From this result, the standard form of the equation of a circle with radius r *and center at the origin,* $(h, k) = (0, 0)$, is

$$x^2 + y^2 = r^2. \qquad \text{Circle with radius } r \text{ and center at origin}$$

EXAMPLE 8   Writing the Equation of a Circle

The point $(3, 4)$ lies on a circle whose center is at $(-1, 2)$, as shown in Figure 1.18. Write the standard form of the equation of this circle.

Solution

The radius of the circle is the distance between $(-1, 2)$ and $(3, 4)$.

$$r = \sqrt{(x - h)^2 + (y - k)^2} \qquad \text{Distance Formula}$$
$$= \sqrt{[3 - (-1)]^2 + (4 - 2)^2} \qquad \text{Substitute for } x, y, h, \text{ and } k.$$
$$= \sqrt{4^2 + 2^2} \qquad \text{Simplify.}$$
$$= \sqrt{16 + 4} \qquad \text{Simplify.}$$
$$= \sqrt{20} \qquad \text{Radius}$$

Using $(h, k) = (-1, 2)$ and $r = \sqrt{20}$, the equation of the circle is

$$(x - h)^2 + (y - k)^2 = r^2 \qquad \text{Equation of circle}$$
$$[x - (-1)]^2 + (y - 2)^2 = \left(\sqrt{20}\right)^2 \qquad \text{Substitute for } h, k, \text{ and } r.$$
$$(x + 1)^2 + (y - 2)^2 = 20. \qquad \text{Standard form}$$

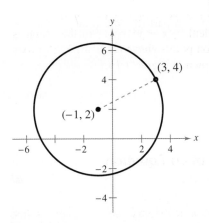

Figure 1.18

✓ **Checkpoint** 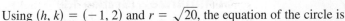 *Audio-video solution in English & Spanish at LarsonPrecalculus.com*

The point $(1, -2)$ lies on a circle whose center is at $(-3, -5)$. Write the standard form of the equation of this circle.

To find h and k from the standard form of the equation of a circle, you may want to rewrite one or both of the quantities in parentheses. For example, $x + 1 = x - (-1)$.

Application

In this course, you will learn that there are many ways to approach a problem. Example 9 illustrates three common approaches.

A numerical approach: Construct and use a table.

A graphical approach: Draw and use a graph.

An algebraic approach: Use the rules of algebra.

EXAMPLE 9 **Maximum Weight**

The maximum weight y (in pounds) for a man in the United States Marine Corps can be approximated by the mathematical model

$$y = 0.040x^2 - 0.11x + 3.9, \quad 58 \le x \le 80$$

where x is the man's height (in inches). *(Source: U.S. Department of Defense)*

a. Construct a table of values that shows the maximum weights for men with heights of 62, 64, 66, 68, 70, 72, 74, and 76 inches.

b. Use the table of values to sketch a graph of the model. Then use the graph to estimate *graphically* the maximum weight for a man whose height is 71 inches.

c. Use the model to confirm *algebraically* the estimate you found in part (b).

Solution

a. Use a calculator to construct a table, as shown at the left.

b. Use the table of values to sketch the graph of the equation, as shown in Figure 1.19. From the graph, you can estimate that a height of 71 inches corresponds to a weight of about 198 pounds.

c. To confirm algebraically the estimate you found in part (b), substitute 71 for x in the model.

$$y = 0.040(71)^2 - 0.11(71) + 3.9$$
$$\approx 197.7$$

So, the graphical estimate of 198 pounds is fairly good.

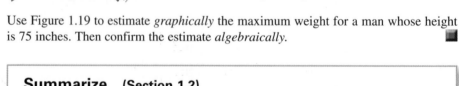 ✓ *Checkpoint* ◀))) *Audio-video solution in English & Spanish at LarsonPrecalculus.com*

Use Figure 1.19 to estimate *graphically* the maximum weight for a man whose height is 75 inches. Then confirm the estimate *algebraically*.

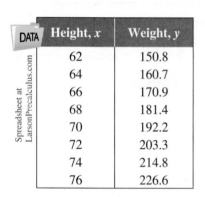

DATA	Height, x	Weight, y
	62	150.8
	64	160.7
	66	170.9
	68	181.4
	70	192.2
	72	203.3
	74	214.8
	76	226.6

Spreadsheet at LarsonPrecalculus.com

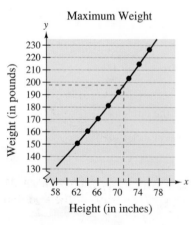

Figure 1.19

Summarize (Section 1.2)

1. Explain how to sketch the graph of an equation *(page 11)*. For examples of sketching graphs of equations, see Examples 2 and 3.

2. Explain how to find the x- and y-intercepts of a graph *(page 14)*. For an example of finding x- and y-intercepts, see Example 4.

3. Explain how to use symmetry to graph an equation *(page 15)*. For an example of using symmetry to graph an equation, see Example 6.

4. State the standard form of the equation of a circle *(page 17)*. For an example of writing the standard form of the equation of a circle, see Example 8.

5. Describe an example of how to use the graph of an equation to solve a real-life problem *(page 18, Example 9)*.

1.2 Exercises

See **CalcChat.com** for tutorial help and worked-out solutions to odd-numbered exercises.

Vocabulary: Fill in the blanks.

1. An ordered pair (a, b) is a _____ of an equation in x and y when the substitutions $x = a$ and $y = b$ result in a true statement.
2. The set of all solution points of an equation is the _____ of the equation.
3. The points at which a graph intersects or touches an axis are the _____ of the graph.
4. A graph is symmetric with respect to the _____ if, whenever (x, y) is on the graph, $(-x, y)$ is also on the graph.
5. The equation $(x - h)^2 + (y - k)^2 = r^2$ is the standard form of the equation of a _____ with center _____ and radius _____.
6. When you construct and use a table to solve a problem, you are using a _____ approach.

Skills and Applications

 Determining Solution Points In Exercises 7–14, determine whether each point lies on the graph of the equation.

Equation	Points	
7. $y = \sqrt{x + 4}$	(a) $(0, 2)$	(b) $(5, 3)$
8. $y = \sqrt{5 - x}$	(a) $(1, 2)$	(b) $(5, 0)$
9. $y = x^2 - 3x + 2$	(a) $(2, 0)$	(b) $(-2, 8)$
10. $y = 3 - 2x^2$	(a) $(-1, 1)$	(b) $(-2, 11)$
11. $y = 4 - |x - 2|$	(a) $(1, 5)$	(b) $(6, 0)$
12. $y = |x - 1| + 2$	(a) $(2, 3)$	(b) $(-1, 0)$
13. $x^2 + y^2 = 20$	(a) $(3, -2)$	(b) $(-4, 2)$
14. $2x^2 + 5y^2 = 8$	(a) $(6, 0)$	(b) $(0, 4)$

 Sketching the Graph of an Equation In Exercises 15–18, complete the table. Use the resulting solution points to sketch the graph of the equation.

15. $y = -2x + 5$

x	-1	0	1	2	$\frac{5}{2}$
y					
(x, y)					

16. $y + 1 = \frac{3}{4}x$

x	-2	0	1	$\frac{4}{3}$	2
y					
(x, y)					

17. $y + 3x = x^2$

x	-1	0	1	2	3
y					
(x, y)					

18. $y = 5 - x^2$

x	-2	-1	0	1	2
y					
(x, y)					

 Identifying x- and y-Intercepts In Exercises 19–22, identify the x- and y-intercepts of the graph. Verify your results algebraically.

19. $y = (x - 3)^2$

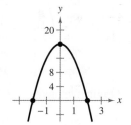

20. $y = 16 - 4x^2$

21. $y = |x + 2|$

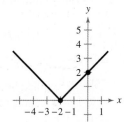

22. $y^2 = 4 - x$

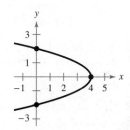

Finding x- and y-Intercepts In Exercises 23–32, find the x- and y-intercepts of the graph of the equation.

23. $y = 5x - 6$

24. $y = 8 - 3x$

25. $y = \sqrt{x + 4}$

26. $y = \sqrt{2x - 1}$

27. $y = |3x - 7|$

28. $y = -|x + 10|$

29. $y = 2x^3 - 4x^2$

30. $y = x^4 - 25$

31. $y^2 = 6 - x$

32. $y^2 = x + 1$

get me yalled up.

Testing for Symmetry In Exercises 33–40, use the algebraic tests to check for symmetry with respect to both axes and the origin.

33. $x^2 - y = 0$

34. $x - y^2 = 0$

35. $y = x^3$

36. $y = x^4 - x^2 + 3$

37. $y = \dfrac{x}{x^2 + 1}$

38. $y = \dfrac{1}{x^2 + 1}$

39. $xy^2 + 10 = 0$

40. $xy = 4$

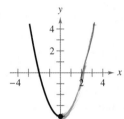

Using Symmetry as a Sketching Aid In Exercises 41–44, assume that the graph has the given type of symmetry. Complete the graph of the equation. To print an enlarged copy of the graph, go to *MathGraphs.com*.

41.

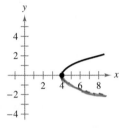

y-Axis symmetry

42.

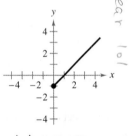

x-Axis symmetry

43.

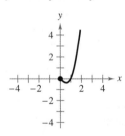

Origin symmetry

44.

y-Axis symmetry

I used this all year lol I year lol

Sketching the Graph of an Equation In Exercises 45–56, find any intercepts and test for symmetry. Then sketch the graph of the equation.

45. $y = -3x + 1$

46. $y = 2x - 3$

47. $y = x^2 - 2x$

48. $y = -x^2 - 2x$

49. $y = x^3 + 3$

50. $y = x^3 - 1$

51. $y = \sqrt{x - 3}$

52. $y = \sqrt{1 - x}$

53. $y = |x - 6|$

54. $y = 1 - |x|$

55. $x = y^2 - 1$

56. $x = y^2 - 5$

Using Technology In Exercises 57–66, use a graphing utility to graph the equation. Use a standard setting. Approximate any intercepts.

57. $y = 3 - \frac{1}{2}x$

58. $y = \frac{2}{3}x - 1$

59. $y = x^2 - 4x + 3$

60. $y = x^2 + x - 2$

61. $y = \dfrac{2x}{x - 1}$

62. $y = \dfrac{4}{x^2 + 1}$

63. $y = \sqrt[3]{x + 1}$

64. $y = x\sqrt{x + 6}$

65. $y = |x + 3|$

66. $y = 2 - |x|$

Writing the Equation of a Circle In Exercises 67–74, write the standard form of the equation of the circle with the given characteristics.

67. Center: $(0, 0)$; Radius: 3

68. Center: $(0, 0)$; Radius: 7

69. Center: $(-4, 5)$; Radius: 2

70. Center: $(1, -3)$; Radius: $\sqrt{11}$

71. Center: $(3, 8)$; Solution point: $(-9, 13)$

72. Center: $(-2, -6)$; Solution point: $(1, -10)$

73. Endpoints of a diameter: $(3, 2), (-9, -8)$

74. Endpoints of a diameter: $(11, -5), (3, 15)$

Sketching a Circle In Exercises 75–80, find the center and radius of the circle with the given equation. Then sketch the circle.

75. $x^2 + y^2 = 25$

76. $x^2 + y^2 = 16$

77. $(x - 1)^2 + (y + 3)^2 = 9$

78. $x^2 + (y - 1)^2 = 1$

79. $\left(x - \frac{1}{2}\right)^2 + \left(y - \frac{1}{2}\right)^2 = \frac{9}{4}$

80. $(x - 2)^2 + (y + 3)^2 = \frac{16}{9}$

81. Depreciation A hospital purchases a new magnetic resonance imaging (MRI) machine for \$1.2 million. The depreciated value y (reduced value) after t years is given by $y = 1,200,000 - 80,000t$, $0 \le t \le 10$. Sketch the graph of the equation.

82. Depreciation You purchase an all-terrain vehicle (ATV) for \$9500. The depreciated value y (reduced value) after t years is given by $y = 9500 - 1000t$, $0 \le t \le 6$. Sketch the graph of the equation.

83. Geometry A regulation NFL playing field of length x and width y has a perimeter of $346\frac{2}{3}$ or $\frac{1040}{3}$ yards.

(a) Draw a rectangle that gives a visual representation of the problem. Use the specified variables to label the sides of the rectangle.

(b) Show that the width of the rectangle is $y = \frac{520}{3} - x$ and its area is $A = x\left(\frac{520}{3} - x\right)$.

(c) Use a graphing utility to graph the area equation. Be sure to adjust your window settings.

(d) From the graph in part (c), estimate the dimensions of the rectangle that yield a maximum area.

(e) Use your school's library, the Internet, or some other reference source to find the actual dimensions and area of a regulation NFL playing field and compare your findings with the results of part (d).

$x = 86\frac{2}{3}$ $x = \frac{260}{3}$ yd $y = \frac{260}{3}$

The symbol indicates an exercise or a part of an exercise in which you are instructed to use a graphing utility.

84. Architecture The arch support of a bridge is modeled by $y = -0.0012x^2 + 300$, where x and y are measured in feet and the x-axis represents the ground.

(a) Use a graphing utility to graph the equation.

(b) Find one x-intercept of the graph. Explain how to use the intercept and the symmetry of the graph to find the width of the arch support.

85. Population Statistics

The table shows the life expectancies of a child (at birth) in the United States for selected years from 1940 through 2010. *(Source: U.S. National Center for Health Statistics)*

DATA	Year	Life Expectancy, y
	1940	62.9
	1950	68.2
	1960	69.7
	1970	70.8
	1980	73.7
	1990	75.4
	2000	76.8
	2010	78.7

Spreadsheet at LarsonPrecalculus.com

A model for the life expectancy during this period is

$$y = \frac{63.6 + 0.97t}{1 + 0.01t}, \quad 0 \le t \le 70$$

where y represents the life expectancy and t is the time in years, with $t = 0$ corresponding to 1940.

(a) Use a graphing utility to graph the data from the table and the model in the same viewing window. How well does the model fit the data? Explain.

(b) Determine the life expectancy in 1990 both graphically and algebraically.

(c) Use the graph to determine the year when life expectancy was approximately 70.1. Verify your answer algebraically.

(d) Find the y-intercept of the graph of the model. What does it represent in the context of the problem?

(e) Do you think this model can be used to predict the life expectancy of a child 50 years from now? Explain.

86. Electronics The resistance y (in ohms) of 1000 feet of solid copper wire at 68 degrees Fahrenheit is

$$y = \frac{10{,}370}{x^2}$$

where x is the diameter of the wire in mils (0.001 inch).

(a) Complete the table.

x	5	10	20	30	40	50
y						

x	60	70	80	90	100
y					

(b) Use the table of values in part (a) to sketch a graph of the model. Then use your graph to estimate the resistance when $x = 85.5$.

(c) Use the model to confirm algebraically the estimate you found in part (b).

(d) What can you conclude about the relationship between the diameter of the copper wire and the resistance?

Exploration

True or False? In Exercises 87–89, determine whether the statement is true or false. Justify your answer.

87. The graph of a linear equation cannot be symmetric with respect to the origin.

88. The graph of a linear equation can have either no x-intercepts or only one x-intercept.

89. A circle can have a total of zero, one, two, three, or four x- and y-intercepts.

90. 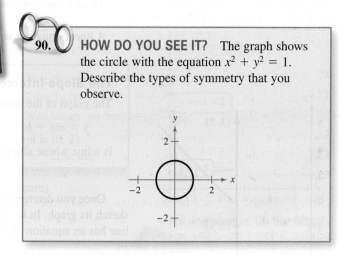 **HOW DO YOU SEE IT?** The graph shows the circle with the equation $x^2 + y^2 = 1$. Describe the types of symmetry that you observe.

91. Think About It Find a and b when the graph of $y = ax^2 + bx^3$ is symmetric with respect to (a) the y-axis and (b) the origin. (There are many correct answers.)

Finding the Slope of a Line

Given an equation of a line, you can find its slope by writing the equation in slope-intercept form. When you are not given an equation, you can still find the slope by using two points on the line. For example, consider the line passing through the points (x_1, y_1) and (x_2, y_2) in the figure below.

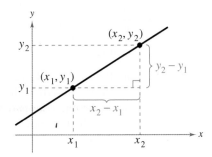

As you move from left to right along this line, a change of $(y_2 - y_1)$ units in the vertical direction corresponds to a change of $(x_2 - x_1)$ units in the horizontal direction.

$$y_2 - y_1 = \text{change in } y = \text{rise}$$

and

$$x_2 - x_1 = \text{change in } x = \text{run}$$

The ratio of $(y_2 - y_1)$ to $(x_2 - x_1)$ represents the slope of the line that passes through the points (x_1, y_1) and (x_2, y_2).

$$\text{Slope} = \frac{\text{change in } y}{\text{change in } x} = \frac{\text{rise}}{\text{run}} = \frac{y_2 - y_1}{x_2 - x_1}$$

The Slope of a Line Passing Through Two Points

The **slope** m of the nonvertical line through (x_1, y_1) and (x_2, y_2) is

$$m = \frac{y_2 - y_1}{x_2 - x_1}$$

where $x_1 \neq x_2$.

When using the formula for slope, the *order of subtraction* is important. Given two points on a line, you are free to label either one of them as (x_1, y_1) and the other as (x_2, y_2). However, once you do this, you must form the numerator and denominator using the same order of subtraction.

$$m = \frac{y_2 - y_1}{x_2 - x_1} \qquad m = \frac{y_1 - y_2}{x_1 - x_2} \qquad m = \frac{y_2 - y_1}{x_1 - x_2}$$

Correct   Correct   Incorrect

For example, the slope of the line passing through the points $(3, 4)$ and $(5, 7)$ can be calculated as

$$m = \frac{7 - 4}{5 - 3} = \frac{3}{2}$$

or as

$$m = \frac{4 - 7}{3 - 5} = \frac{-3}{-2} = \frac{3}{2}.$$

EXAMPLE 2   **Finding the Slope of a Line Through Two Points**

Find the slope of the line passing through each pair of points.

a. $(-2, 0)$ and $(3, 1)$ **b.** $(-1, 2)$ and $(2, 2)$

c. $(0, 4)$ and $(1, -1)$ **d.** $(3, 4)$ and $(3, 1)$

Solution

a. Letting $(x_1, y_1) = (-2, 0)$ and $(x_2, y_2) = (3, 1)$, you find that the slope is

$$m = \frac{y_2 - y_1}{x_2 - x_1} = \frac{1 - 0}{3 - (-2)} = \frac{1}{5}.$$ See Figure 1.21.

b. The slope of the line passing through $(-1, 2)$ and $(2, 2)$ is

$$m = \frac{2 - 2}{2 - (-1)} = \frac{0}{3} = 0.$$ See Figure 1.22.

c. The slope of the line passing through $(0, 4)$ and $(1, -1)$ is

$$m = \frac{-1 - 4}{1 - 0} = \frac{-5}{1} = -5.$$ See Figure 1.23.

d. The slope of the line passing through $(3, 4)$ and $(3, 1)$ is

$$m = \frac{1 - 4}{3 - 3} = \frac{-3}{0}.$$ See Figure 1.24.

Division by 0 is undefined, so the slope is undefined and the line is vertical.

• • REMARK   In Figures 1.21 through 1.24, note the relationships between slope and the orientation of the line.

a. Positive slope: line rises from left to right

b. Zero slope: line is horizontal

c. Negative slope: line falls from left to right

d. Undefined slope: line is vertical

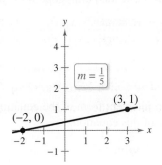

Figure 1.21

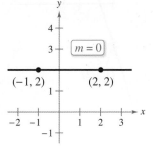

Figure 1.22

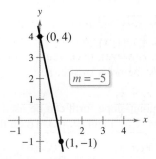

Figure 1.23

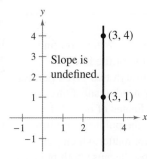

Figure 1.24

✓ **Checkpoint** ◀))) *Audio-video solution in English & Spanish at LarsonPrecalculus.com*

Find the slope of the line passing through each pair of points.

a. $(-5, -6)$ and $(2, 8)$ **b.** $(4, 2)$ and $(2, 5)$

c. $(0, 0)$ and $(0, -6)$ **d.** $(0, -1)$ and $(3, -1)$

Writing Linear Equations in Two Variables

If (x_1, y_1) is a point on a line of slope m and (x, y) is *any other* point on the line, then

$$\frac{y - y_1}{x - x_1} = m.$$

This equation in the variables x and y can be rewritten in the **point-slope form** of the equation of a line

$$y - y_1 = m(x - x_1).$$

Point-Slope Form of the Equation of a Line

The equation of the line with slope m passing through the point (x_1, y_1) is

$$y - y_1 = m(x - x_1).$$

The point-slope form is useful for *finding* the equation of a line. You should remember this form.

EXAMPLE 3   **Using the Point-Slope Form**

Find the slope-intercept form of the equation of the line that has a slope of 3 and passes through the point $(1, -2)$.

Solution   Use the point-slope form with $m = 3$ and $(x_1, y_1) = (1, -2)$.

$$y - y_1 = m(x - x_1) \qquad \text{Point-slope form}$$
$$y - (-2) = 3(x - 1) \qquad \text{Substitute for } m, x_1, \text{ and } y_1.$$
$$y + 2 = 3x - 3 \qquad \text{Simplify.}$$
$$y = 3x - 5 \qquad \text{Write in slope-intercept form.}$$

The slope-intercept form of the equation of the line is $y = 3x - 5$. Figure 1.25 shows the graph of this equation.

✓ *Checkpoint* 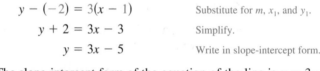 *Audio-video solution in English & Spanish at LarsonPrecalculus.com*

Find the slope-intercept form of the equation of the line that has the given slope and passes through the given point.

a. $m = 2$, $(3, -7)$

b. $m = -\frac{2}{3}$, $(1, 1)$

c. $m = 0$, $(1, 1)$

The point-slope form can be used to find an equation of the line passing through two points (x_1, y_1) and (x_2, y_2). To do this, first find the slope of the line.

$$m = \frac{y_2 - y_1}{x_2 - x_1}, \quad x_1 \neq x_2$$

Then use the point-slope form to obtain the equation.

$$y - y_1 = \frac{y_2 - y_1}{x_2 - x_1}(x - x_1) \qquad \text{Two-point form}$$

This is sometimes called the **two-point form** of the equation of a line.

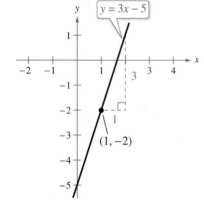

Figure 1.25

• **REMARK**   When you find an equation of the line that passes through two given points, you only need to substitute the coordinates of one of the points in the point-slope form. It does not matter which point you choose because both points will yield the same result.

Parallel and Perpendicular Lines

Slope can tell you whether two nonvertical lines in a plane are parallel, perpendicular, or neither.

> ### Parallel and Perpendicular Lines
>
> 1. Two distinct nonvertical lines are **parallel** if and only if their slopes are equal. That is,
>
> $$m_1 = m_2.$$
>
> 2. Two nonvertical lines are **perpendicular** if and only if their slopes are negative reciprocals of each other. That is,
>
> $$m_1 = \frac{-1}{m_2}.$$

EXAMPLE 4 **Finding Parallel and Perpendicular Lines**

Find the slope-intercept form of the equations of the lines that pass through the point $(2, -1)$ and are (a) parallel to and (b) perpendicular to the line $2x - 3y = 5$.

Solution Write the equation of the given line in slope-intercept form.

$2x - 3y = 5$	Write original equation.
$-3y = -2x + 5$	Subtract $2x$ from each side.
$y = \frac{2}{3}x - \frac{5}{3}$	Write in slope-intercept form.

Notice that the line has a slope of $m = \frac{2}{3}$.

a. Any line parallel to the given line must also have a slope of $\frac{2}{3}$. Use the point-slope form with $m = \frac{2}{3}$ and $(x_1, y_1) = (2, -1)$.

$y - (-1) = \frac{2}{3}(x - 2)$	Write in point-slope form.
$3(y + 1) = 2(x - 2)$	Multiply each side by 3.
$3y + 3 = 2x - 4$	Distributive Property
$y = \frac{2}{3}x - \frac{7}{3}$	Write in slope-intercept form.

Notice the similarity between the slope-intercept form of this equation and the slope-intercept form of the given equation.

b. Any line perpendicular to the given line must have a slope of $-\frac{3}{2}$ (because $-\frac{3}{2}$ is the negative reciprocal of $\frac{2}{3}$). Use the point-slope form with $m = -\frac{3}{2}$ and $(x_1, y_1) = (2, -1)$.

$y - (-1) = -\frac{3}{2}(x - 2)$	Write in point-slope form.
$2(y + 1) = -3(x - 2)$	Multiply each side by 2.
$2y + 2 = -3x + 6$	Distributive Property
$y = -\frac{3}{2}x + 2$	Write in slope-intercept form.

The graphs of all three equations are shown in Figure 1.26.

✓ *Checkpoint*  *Audio-video solution in English & Spanish at LarsonPrecalculus.com*

Find the slope-intercept form of the equations of the lines that pass through the point $(-4, 1)$ and are (a) parallel to and (b) perpendicular to the line $5x - 3y = 8$. ■

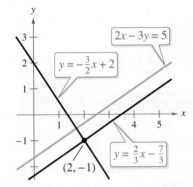

Figure 1.26

▷ **TECHNOLOGY** On a graphing utility, lines will not appear to have the correct slope unless you use a viewing window that has a square setting. For instance, graph the lines in Example 4 using the standard setting $-10 \le x \le 10$ and $-10 \le y \le 10$. Then reset the viewing window with the square setting $-9 \le x \le 9$ and $-6 \le y \le 6$. On which setting do the lines $y = \frac{2}{3}x - \frac{5}{3}$ and $y = -\frac{3}{2}x + 2$ appear to be perpendicular?

Applications

In real-life problems, the slope of a line can be interpreted as either a *ratio* or a *rate*. When the *x*-axis and *y*-axis have the same unit of measure, the slope has no units and is a **ratio.** When the *x*-axis and *y*-axis have different units of measure, the slope is a **rate** or **rate of change.**

EXAMPLE 5   **Using Slope as a Ratio**

The maximum recommended slope of a wheelchair ramp is $\frac{1}{12}$. A business installs a wheelchair ramp that rises 22 inches over a horizontal length of 24 feet. Is the ramp steeper than recommended?   *(Source: ADA Standards for Accessible Design)*

Solution   The horizontal length of the ramp is 24 feet or 12(24) = 288 inches (see figure). So, the slope of the ramp is

$$\text{Slope} = \frac{\text{vertical change}}{\text{horizontal change}} = \frac{22 \text{ in.}}{288 \text{ in.}} \approx 0.076.$$

Because $\frac{1}{12} \approx 0.083$, the slope of the ramp is not steeper than recommended.

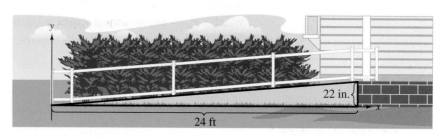

The Americans with Disabilities Act (ADA) became law on July 26, 1990. It is the most comprehensive formulation of rights for persons with disabilities in U.S. (and world) history.

✓ *Checkpoint* ◀))) *Audio-video solution in English & Spanish at LarsonPrecalculus.com*

The business in Example 5 installs a second ramp that rises 36 inches over a horizontal length of 32 feet. Is the ramp steeper than recommended?

EXAMPLE 6   **Using Slope as a Rate of Change**

A kitchen appliance manufacturing company determines that the total cost *C* (in dollars) of producing *x* units of a blender is given by

$$C = 25x + 3500. \qquad \text{Cost equation}$$

Interpret the *y*-intercept and slope of this line.

Solution   The *y*-intercept (0, 3500) tells you that the cost of producing 0 units is $3500. This is the *fixed cost* of production—it includes costs that must be paid regardless of the number of units produced. The slope of $m = 25$ tells you that the cost of producing each unit is $25, as shown in Figure 1.27. Economists call the cost per unit the *marginal cost.* When the production increases by one unit, the "margin," or extra amount of cost, is $25. So, the cost increases at a rate of $25 per unit.

✓ *Checkpoint* ◀))) *Audio-video solution in English & Spanish at LarsonPrecalculus.com*

An accounting firm determines that the value *V* (in dollars) of a copier *t* years after its purchase is given by

$$V = -300t + 1500.$$

Interpret the *y*-intercept and slope of this line.

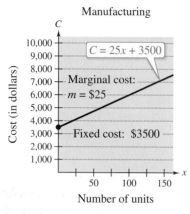

Manufacturing

$C = 25x + 3500$

Marginal cost: $m = \$25$

Fixed cost: $3500

Number of units

Production cost
Figure 1.27

Businesses can deduct most of their expenses in the same year they occur. One exception is the cost of property that has a useful life of more than 1 year. Such costs must be *depreciated* (decreased in value) over the useful life of the property. Depreciating the *same amount* each year is called *linear* or *straight-line depreciation*. The *book value* is the difference between the original value and the total amount of depreciation accumulated to date.

EXAMPLE 7   Straight-Line Depreciation

A college purchased exercise equipment worth $12,000 for the new campus fitness center. The equipment has a useful life of 8 years. The salvage value at the end of 8 years is $2000. Write a linear equation that describes the book value of the equipment each year.

Solution   Let V represent the value of the equipment at the end of year t. Represent the initial value of the equipment by the data point $(0, 12,000)$ and the salvage value of the equipment by the data point $(8, 2000)$. The slope of the line is

$$m = \frac{2000 - 12,000}{8 - 0} = -\$1250$$

which represents the annual depreciation in *dollars per year*. Using the point-slope form, write an equation of the line.

$$V - 12,000 = -1250(t - 0) \qquad \text{Write in point-slope form.}$$
$$V = -1250t + 12,000 \qquad \text{Write in slope-intercept form.}$$

The table shows the book value at the end of each year, and Figure 1.28 shows the graph of the equation.

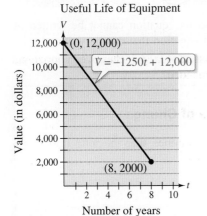

Useful Life of Equipment

$V = -1250t + 12,000$

(0, 12,000)

(8, 2000)

Value (in dollars)

Number of years

Straight-line depreciation
Figure 1.28

Year, t	Value, V
0	12,000
1	10,750
2	9500
3	8250
4	7000
5	5750
6	4500
7	3250
8	2000

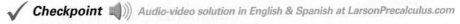

✓ *Checkpoint* ◀))) *Audio-video solution in English & Spanish at LarsonPrecalculus.com*

A manufacturing firm purchases a machine worth $24,750. The machine has a useful life of 6 years. After 6 years, the machine will have to be discarded and replaced, because it will have no salvage value. Write a linear equation that describes the book value of the machine each year.

In many real-life applications, the two data points that determine the line are often given in a disguised form. Note how the data points are described in Example 7.

NIKE

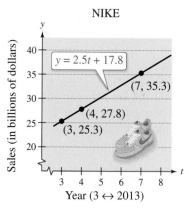

Figure 1.29

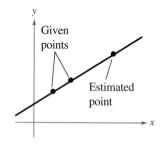

Linear extrapolation
Figure 1.30

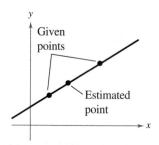

Linear interpolation
Figure 1.31

EXAMPLE 8 Predicting Sales

The sales for NIKE were approximately \$25.3 billion in 2013 and \$27.8 billion in 2014. Using only this information, write a linear equation that gives the sales in terms of the year. Then predict the sales in 2017. *(Source: NIKE Inc.)*

Solution Let $t = 3$ represent 2013. Then the two given values are represented by the data points $(3, 25.3)$ and $(4, 27.8)$ The slope of the line through these points is

$$m = \frac{27.8 - 25.3}{4 - 3} = 2.5.$$

Use the point-slope form to write an equation that relates the sales y and the year t.

$$y - 25.3 = 2.5(t - 3) \qquad \text{Write in point-slope form.}$$
$$y = 2.5t + 17.8 \qquad \text{Write in slope-intercept form.}$$

According to this equation, the sales in 2017 will be

$$y = 2.5(7) + 17.8 = 17.5 + 17.8 = \$35.3 \text{ billion. (See Figure 1.29.)}$$

✓ *Checkpoint* 🔊))) *Audio-video solution in English & Spanish at LarsonPrecalculus.com*

The sales for Foot Locker were approximately \$6.5 billion in 2013 and \$7.2 billion in 2014. Repeat Example 8 using this information. *(Source: Foot Locker)* ◼

The prediction method illustrated in Example 8 is called **linear extrapolation.** Note in Figure 1.30 that an extrapolated point does not lie between the given points. When the estimated point lies between two given points, as shown in Figure 1.31, the procedure is called **linear interpolation.**

The slope of a vertical line is undefined, so its equation cannot be written in slope-intercept form. However, every line has an equation that can be written in the **general form** $Ax + By + C = 0$, where A and B are not both zero.

Summary of Equations of Lines

1. General form: $\qquad Ax + By + C = 0$
2. Vertical line: $\qquad x = a$
3. Horizontal line: $\qquad y = b$
4. Slope-intercept form: $\quad y = mx + b$
5. Point-slope form: $\qquad y - y_1 = m(x - x_1)$
6. Two-point form: $\qquad y - y_1 = \dfrac{y_2 - y_1}{x_2 - x_1}(x - x_1)$

Summarize (Section 1.3)

1. Explain how to use slope to graph a linear equation in two variables *(page 22)* and how to find the slope of a line passing through two points *(page 24)*. For examples of using and finding slopes, see Examples 1 and 2.

2. State the point-slope form of the equation of a line *(page 26)*. For an example of using point-slope form, see Example 3.

3. Explain how to use slope to identify parallel and perpendicular lines *(page 27)*. For an example of finding parallel and perpendicular lines, see Example 4.

4. Describe examples of how to use slope and linear equations in two variables to model and solve real-life problems *(pages 28–30, Examples 5–8)*.

1.3 Exercises

See **CalcChat.com** for tutorial help and worked-out solutions to odd-numbered exercises.

Vocabulary: Fill in the blanks.

1. The simplest mathematical model for relating two variables is the _____ equation in two variables $y = mx + b$.

2. For a line, the ratio of the change in y to the change in x is the _____ of the line.

3. The _____-_____ form of the equation of a line with slope m passing through the point (x_1, y_1) is $y - y_1 = m(x - x_1)$.

4. Two distinct nonvertical lines are _____ if and only if their slopes are equal.

5. Two nonvertical lines are _____ if and only if their slopes are negative reciprocals of each other.

6. When the x-axis and y-axis have different units of measure, the slope can be interpreted as a _____.

7. _____ _____ is the prediction method used to estimate a point on a line when the point does not lie between the given points.

8. Every line has an equation that can be written in _____ form.

Skills and Applications

Identifying Lines In Exercises 9 and 10, identify the line that has each slope.

9. (a) $m = \frac{2}{3}$
   (b) m is undefined.
   (c) $m = -2$

10. (a) $m = 0$
    (b) $m = -\frac{3}{4}$
    (c) $m = 1$

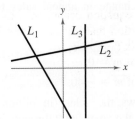

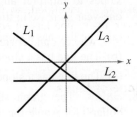

Sketching Lines In Exercises 11 and 12, sketch the lines through the point with the given slopes on the same set of coordinate axes.

Point	Slopes
11. $(2, 3)$	(a) 0 (b) 1 (c) 2 (d) -3
12. $(-4, 1)$	(a) 3 (b) -3 (c) $\frac{1}{2}$ (d) Undefined

Estimating the Slope of a Line In Exercises 13 and 14, estimate the slope of the line.

13.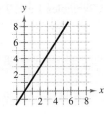
14.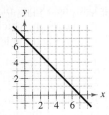

 Graphing a Linear Equation In Exercises 15–24, find the slope and y-intercept (if possible) of the line. Sketch the line.

15. $y = 5x + 3$
16. $y = -x - 10$
17. $y = -\frac{3}{4}x - 1$
18. $y = \frac{2}{3}x + 2$
19. $y - 5 = 0$
20. $x + 4 = 0$
21. $5x - 2 = 0$
22. $3y + 5 = 0$
23. $7x - 6y = 30$
24. $2x + 3y = 9$

 Finding the Slope of a Line Through Two Points In Exercises 25–34, find the slope of the line passing through the pair of points.

25. $(0, 9), (6, 0)$
26. $(10, 0), (0, -5)$
27. $(-3, -2), (1, 6)$
28. $(2, -1), (-2, 1)$
29. $(5, -7), (8, -7)$
30. $(-2, 1), (-4, -5)$
31. $(-6, -1), (-6, 4)$
32. $(0, -10), (-4, 0)$
33. $(4.8, 3.1), (-5.2, 1.6)$
34. $\left(\frac{11}{2}, -\frac{4}{3}\right), \left(-\frac{3}{2}, -\frac{1}{3}\right)$

Using the Slope and a Point In Exercises 35–42, use the slope of the line and the point on the line to find three additional points through which the line passes. (There are many correct answers.)

35. $m = 0$, $(5, 7)$
36. $m = 0$, $(3, -2)$
37. $m = 2$, $(-5, 4)$
38. $m = -2$, $(0, -9)$
39. $m = -\frac{1}{3}$, $(4, 5)$
40. $m = \frac{1}{4}$, $(3, -4)$
41. m is undefined, $(-4, 3)$
42. m is undefined, $(2, 14)$

Using the Point-Slope Form In Exercises 43–54, find the slope-intercept form of the equation of the line that has the given slope and passes through the given point. Sketch the line.

43. $m = 3$, $(0, -2)$
44. $m = -1$, $(0, 10)$
45. $m = -2$, $(-3, 6)$
46. $m = 4$, $(0, 0)$
47. $m = -\frac{1}{3}$, $(4, 0)$
48. $m = \frac{1}{4}$, $(8, 2)$
49. $m = -\frac{1}{2}$, $(2, -3)$
50. $m = \frac{3}{4}$, $(-2, -5)$
51. $m = 0$, $\left(4, \frac{5}{2}\right)$
52. $m = 6$, $\left(2, \frac{3}{2}\right)$
53. $m = 5$, $(-5.1, 1.8)$
54. $m = 0$, $(-2.5, 3.25)$

Finding an Equation of a Line In Exercises 55–64, find an equation of the line passing through the pair of points. Sketch the line.

55. $(5, -1)$, $(-5, 5)$
56. $(4, 3)$, $(-4, -4)$
57. $(-7, 2)$, $(-7, 5)$
58. $(-6, -3)$, $(2, -3)$
59. $\left(2, \frac{1}{2}\right)$, $\left(\frac{1}{2}, \frac{5}{4}\right)$
60. $(1, 1)$, $\left(6, -\frac{2}{3}\right)$
61. $(1, 0.6)$, $(-2, -0.6)$
62. $(-8, 0.6)$, $(2, -2.4)$
63. $(2, -1)$, $\left(\frac{1}{3}, -1\right)$
64. $\left(\frac{7}{3}, -8\right)$, $\left(\frac{7}{3}, 1\right)$

Parallel and Perpendicular Lines In Exercises 65–68, determine whether the lines are parallel, perpendicular, or neither.

65. L_1: $y = -\frac{2}{3}x - 3$
    L_2: $y = -\frac{2}{3}x + 4$
66. L_1: $y = \frac{1}{4}x - 1$
    L_2: $y = 4x + 7$
67. L_1: $y = \frac{1}{2}x - 3$
    L_2: $y = -\frac{1}{2}x + 1$
68. L_1: $y = -\frac{4}{5}x - 5$
    L_2: $y = \frac{5}{4}x + 1$

Parallel and Perpendicular Lines In Exercises 69–72, determine whether the lines L_1 and L_2 passing through the pairs of points are parallel, perpendicular, or neither.

69. L_1: $(0, -1)$, $(5, 9)$
    L_2: $(0, 3)$, $(4, 1)$
70. L_1: $(-2, -1)$, $(1, 5)$
    L_2: $(1, 3)$, $(5, -5)$
71. L_1: $(-6, -3)$, $(2, -3)$
    L_2: $\left(3, -\frac{1}{2}\right)$, $\left(6, -\frac{1}{2}\right)$
72. L_1: $(4, 8)$, $(-4, 2)$
    L_2: $(3, -5)$, $\left(-1, \frac{1}{3}\right)$

Finding Parallel and Perpendicular Lines In Exercises 73–80, find equations of the lines that pass through the given point and are (a) parallel to and (b) perpendicular to the given line.

73. $4x - 2y = 3$, $(2, 1)$
74. $x + y = 7$, $(-3, 2)$
75. $3x + 4y = 7$, $\left(-\frac{2}{3}, \frac{7}{8}\right)$
76. $5x + 3y = 0$, $\left(\frac{7}{8}, \frac{3}{4}\right)$
77. $y + 5 = 0$, $(-2, 4)$
78. $x - 4 = 0$, $(3, -2)$
79. $x - y = 4$, $(2.5, 6.8)$
80. $6x + 2y = 9$, $(-3.9, -1.4)$

Using Intercept Form In Exercises 81–86, use the *intercept form* to find the general form of the equation of the line with the given intercepts. The intercept form of the equation of a line with intercepts $(a, 0)$ and $(0, b)$ is

$$\frac{x}{a} + \frac{y}{b} = 1, \quad a \neq 0, \quad b \neq 0.$$

81. x-intercept: $(3, 0)$
    y-intercept: $(0, 5)$
82. x-intercept: $(-3, 0)$
    y-intercept: $(0, 4)$
83. x-intercept: $\left(-\frac{1}{6}, 0\right)$
    y-intercept: $\left(0, -\frac{2}{3}\right)$
84. x-intercept: $\left(\frac{2}{3}, 0\right)$
    y-intercept: $(0, -2)$
85. Point on line: $(1, 2)$
    x-intercept: $(c, 0)$, $c \neq 0$
    y-intercept: $(0, c)$, $c \neq 0$
86. Point on line: $(-3, 4)$
    x-intercept: $(d, 0)$, $d \neq 0$
    y-intercept: $(0, d)$, $d \neq 0$

87. **Sales** The slopes of lines representing annual sales y in terms of time x in years are given below. Use the slopes to interpret any change in annual sales for a one-year increase in time.

    (a) The line has a slope of $m = 135$.

    (b) The line has a slope of $m = 0$.

    (c) The line has a slope of $m = -40$.

88. **Sales** The graph shows the sales (in billions of dollars) for Apple Inc. in the years 2009 through 2015. (*Source: Apple Inc.*)

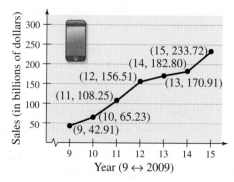

    (a) Use the slopes of the line segments to determine the years in which the sales showed the greatest increase and the least increase.

    (b) Find the slope of the line segment connecting the points for the years 2009 and 2015.

    (c) Interpret the meaning of the slope in part (b) in the context of the problem.

89. Road Grade   You are driving on a road that has a 6% uphill grade. This means that the slope of the road is $\frac{6}{100}$. Approximate the amount of vertical change in your position when you drive 200 feet.

90. Road Grade

From the top of a mountain road, a surveyor takes several horizontal measurements x and several vertical measurements y, as shown in the table (x and y are measured in feet).

x	300	600	900	1200
y	-25	-50	-75	-100

x	1500	1800	2100
y	-125	-150	-175

(a) Sketch a scatter plot of the data.

(b) Use a straightedge to sketch the line that you think best fits the data.

(c) Find an equation for the line you sketched in part (b).

(d) Interpret the meaning of the slope of the line in part (c) in the context of the problem.

(e) The surveyor needs to put up a road sign that indicates the steepness of the road. For example, a surveyor would put up a sign that states "8% grade" on a road with a downhill grade that has a slope of $-\frac{8}{100}$. What should the sign state for the road in this problem?

Rate of Change   **In Exercises 91 and 92, you are given the dollar value of a product in 2016 and the rate at which the value of the product is expected to change during the next 5 years. Use this information to write a linear equation that gives the dollar value V of the product in terms of the year t. (Let $t = 16$ represent 2016.)**

	2016 Value	Rate
91.	$3000	$150 decrease per year
92.	$200	$6.50 increase per year

93. Cost   The cost C of producing n computer laptop bags is given by

$$C = 1.25n + 15{,}750, \quad n > 0.$$

Explain what the C-intercept and the slope represent.

94. Monthly Salary   A pharmaceutical salesperson receives a monthly salary of $5000 plus a commission of 7% of sales. Write a linear equation for the salesperson's monthly wage W in terms of monthly sales S.

95. Depreciation   A sandwich shop purchases a used pizza oven for $875. After 5 years, the oven will have to be discarded and replaced. Write a linear equation giving the value V of the equipment during the 5 years it will be in use.

96. Depreciation   A school district purchases a high-volume printer, copier, and scanner for $24,000. After 10 years, the equipment will have to be replaced. Its value at that time is expected to be $2000. Write a linear equation giving the value V of the equipment during the 10 years it will be in use.

97. Temperature Conversion   Write a linear equation that expresses the relationship between the temperature in degrees Celsius C and degrees Fahrenheit F. Use the fact that water freezes at $0°C$ ($32°F$) and boils at $100°C$ ($212°F$).   *F changes 180, C → 100°*

98. Neurology   The average weight of a male child's brain is 970 grams at age 1 and 1270 grams at age 3. (*Source: American Neurological Association*)

(a) Assuming that the relationship between brain weight y and age t is linear, write a linear model for the data.

(b) What is the slope and what does it tell you about brain weight?

(c) Use your model to estimate the average brain weight at age 2.

(d) Use your school's library, the Internet, or some other reference source to find the actual average brain weight at age 2. How close was your estimate?

(e) Do you think your model could be used to determine the average brain weight of an adult? Explain.

99. Cost, Revenue, and Profit   A roofing contractor purchases a shingle delivery truck with a shingle elevator for $42,000. The vehicle requires an average expenditure of $9.50 per hour for fuel and maintenance, and the operator is paid $11.50 per hour.

(a) Write a linear equation giving the total cost C of operating this equipment for t hours. (Include the purchase cost of the equipment.)

(b) Assuming that customers are charged $45 per hour of machine use, write an equation for the revenue R obtained from t hours of use.

(c) Use the formula for profit $P = R - C$ to write an equation for the profit obtained from t hours of use.

(d) Use the result of part (c) to find the break-even point—that is, the number of hours this equipment must be used to yield a profit of 0 dollars.

100. Geometry   The length and width of a rectangular garden are 15 meters and 10 meters, respectively. A walkway of width x surrounds the garden.

(a) Draw a diagram that gives a visual representation of the problem.

(b) Write the equation for the perimeter y of the walkway in terms of x.

(c) Use a graphing utility to graph the equation for the perimeter.

(d) Determine the slope of the graph in part (c). For each additional one-meter increase in the width of the walkway, determine the increase in its perimeter.

Exploration

True or False?   **In Exercises 101 and 102, determine whether the statement is true or false. Justify your answer.**

101. A line with a slope of $-\frac{5}{7}$ is steeper than a line with a slope of $-\frac{6}{7}$.

102. The line through $(-8, 2)$ and $(-1, 4)$ and the line through $(0, -4)$ and $(-7, 7)$ are parallel.

103. Right Triangle   Explain how you can use slope to show that the points $A(-1, 5)$, $B(3, 7)$, and $C(5, 3)$ are the vertices of a right triangle.

104. Vertical Line   Explain why the slope of a vertical line is undefined.

105. Error Analysis   Describe the error.

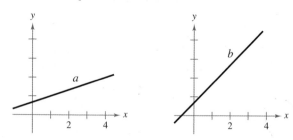

Line b has a greater slope than line a.   ✗

106. Perpendicular Segments   Find d_1 and d_2 in terms of m_1 and m_2, respectively (see figure). Then use the Pythagorean Theorem to find a relationship between m_1 and m_2.

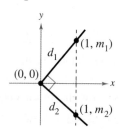

107. Think About It   Is it possible for two lines with positive slopes to be perpendicular? Explain.

108. Slope and Steepness   The slopes of two lines are -4 and $\frac{5}{2}$. Which is steeper? Explain.

109. Comparing Slopes   Use a graphing utility to compare the slopes of the lines $y = mx$, where $m = 0.5$, 1, 2, and 4. Which line rises most quickly? Now, let $m = -0.5$, -1, -2, and -4. Which line falls most quickly? Use a square setting to obtain a true geometric perspective. What can you conclude about the slope and the "rate" at which the line rises or falls?

110.   **HOW DO YOU SEE IT?**   Match the description of the situation with its graph. Also determine the slope and y-intercept of each graph and interpret the slope and y-intercept in the context of the situation. [The graphs are labeled (i), (ii), (iii), and (iv).]

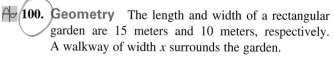

(a) A person is paying \$20 per week to a friend to repay a \$200 loan.

(b) An employee receives \$12.50 per hour plus \$2 for each unit produced per hour.

(c) A sales representative receives \$30 per day for food plus \$0.32 for each mile traveled.

(d) A computer that was purchased for \$750 depreciates \$100 per year.

Finding a Relationship for Equidistance   **In Exercises 111–114, find a relationship between x and y such that (x, y) is equidistant (the same distance) from the two points.**

111. $(4, -1)$, $(-2, 3)$ **112.** $(6, 5)$, $(1, -8)$

113. $\left(3, \frac{5}{2}\right)$, $(-7, 1)$ **114.** $\left(-\frac{1}{2}, -4\right)$, $\left(\frac{7}{2}, \frac{5}{4}\right)$

Project: Bachelor's Degrees   To work an extended application analyzing the numbers of bachelor's degrees earned by women in the United States from 2002 through 2013, visit this text's website at *LarsonPrecalculus.com*. (*Source: National Center for Education Statistics*)

1.4 Functions

Functions are used to model and solve real-life problems. For example, in Exercise 70 on page 47, you will use a function that models the force of water against the face of a dam.

■ Determine whether relations between two variables are functions, and use function notation.
■ Find the domains of functions.
■ Use functions to model and solve real-life problems.
■ Evaluate difference quotients.

Introduction to Functions and Function Notation

Many everyday phenomena involve two quantities that are related to each other by some rule of correspondence. The mathematical term for such a rule of correspondence is a **relation.** In mathematics, equations and formulas often represent relations. For example, the simple interest I earned on \$1000 for 1 year is related to the annual interest rate r by the formula $I = 1000r$.

The formula $I = 1000r$ represents a special kind of relation that matches each item from one set with *exactly one* item from a different set. Such a relation is a **function.**

Definition of Function

A **function** f from a set A to a set B is a relation that assigns to each element x in the set A exactly one element y in the set B. The set A is the **domain** (or set of inputs) of the function f, and the set B contains the **range** (or set of outputs).

To help understand this definition, look at the function below, which relates the time of day to the temperature.

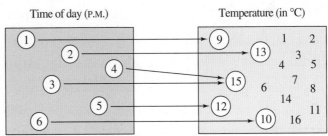

Set A is the domain.
Inputs: 1, 2, 3, 4, 5, 6

Set B contains the range.
Outputs: 9, 10, 12, 13, 15

The ordered pairs below can represent this function. The first coordinate (x-value) is the input and the second coordinate (y-value) is the output.

$$\{(1, 9), (2, 13), (3, 15), (4, 15), (5, 12), (6, 10)\}$$

Characteristics of a Function from Set *A* to Set *B*

1. Each element in A must be matched with an element in B.

2. Some elements in B may not be matched with any element in A.

3. Two or more elements in A may be matched with the same element in B.

4. An element in A (the domain) cannot be matched with two different elements in B.

Here are four common ways to represent functions.

> ### Four Ways to Represent a Function
>
> 1. *Verbally* by a sentence that describes how the input variable is related to the output variable
>
> 2. *Numerically* by a table or a list of ordered pairs that matches input values with output values
>
> 3. *Graphically* by points in a coordinate plane in which the horizontal positions represent the input values and the vertical positions represent the output values
>
> 4. *Algebraically* by an equation in two variables

To determine whether a relation is a function, you must decide whether each input value is matched with exactly one output value. When any input value is matched with two or more output values, the relation is not a function.

EXAMPLE 1   Testing for Functions

Determine whether the relation represents y as a function of x.

a. The input value x is the number of representatives from a state, and the output value y is the number of senators.

b.

Input, x	Output, y
2	11
2	10
3	8
4	5
5	1

c.

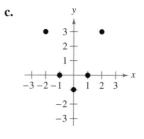

Solution

a. This verbal description *does* describe y as a function of x. Regardless of the value of x, the value of y is always 2. This is an example of a *constant function*.

b. This table *does not* describe y as a function of x. The input value 2 is matched with two different y-values.

c. The graph *does* describe y as a function of x. Each input value is matched with exactly one output value.

✓ *Checkpoint*  *Audio-video solution in English & Spanish at LarsonPrecalculus.com*

Determine whether the relation represents y as a function of x.

a. *Domain, x Range, y*

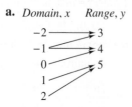

b.

Input, x	0	1	2	3	4
Output, y	-4	-2	0	2	4

Representing functions by sets of ordered pairs is common in *discrete mathematics*. In algebra, however, it is more common to represent functions by equations or formulas involving two variables. For example, the equation

$$y = x^2 \qquad \text{\textit{y} is a function of \textit{x}.}$$

represents the variable y as a function of the variable x. In this equation, x is the **independent variable** and y is the **dependent variable.** The domain of the function is the set of all values taken on by the independent variable x, and the range of the function is the set of all values taken on by the dependent variable y.

HISTORICAL NOTE

Many consider Leonhard Euler (1707–1783), a Swiss mathematician, to be the most prolific and productive mathematician in history. One of his greatest influences on mathematics was his use of symbols, or notation. Euler introduced the function notation $y = f(x)$.

EXAMPLE 2 **Testing for Functions Represented Algebraically**

See LarsonPrecalculus.com for an interactive version of this type of example.

Determine whether each equation represents y as a function of x.

a. $x^2 + y = 1$

b. $-x + y^2 = 1$

Solution To determine whether y is a function of x, solve for y in terms of x.

a. Solving for y yields

$x^2 + y = 1$	Write original equation.
$y = 1 - x^2.$	Solve for y.

To each value of x there corresponds exactly one value of y. So, y is a function of x.

b. Solving for y yields

$-x + y^2 = 1$	Write original equation.
$y^2 = 1 + x$	Add x to each side.
$y = \pm\sqrt{1 + x}.$	Solve for y.

The $\pm$ indicates that to a given value of x there correspond two values of y. So, y is not a function of x.

✓ **Checkpoint** 🔊))) *Audio-video solution in English & Spanish at LarsonPrecalculus.com*

Determine whether each equation represents y as a function of x.

a. $x^2 + y^2 = 8$ **b.** $y - 4x^2 = 36$

When using an equation to represent a function, it is convenient to name the function for easy reference. For example, the equation $y = 1 - x^2$ describes y as a function of x. By renaming this function "f," you can write the input, output, and equation using **function notation.**

Input	Output	Equation
x	$f(x)$	$f(x) = 1 - x^2$

The symbol $f(x)$ is read as *the value of f at x* or simply *f of x*. The symbol $f(x)$ corresponds to the y-value for a given x. So, $y = f(x)$. Keep in mind that f is the *name* of the function, whereas $f(x)$ is the *value* of the function at x. For example, the function $f(x) = 3 - 2x$ has *function values* denoted by $f(-1)$, $f(0)$, $f(2)$, and so on. To find these values, substitute the specified input values into the given equation.

For $x = -1$,	$f(-1) = 3 - 2(-1) = 3 + 2 = 5.$
For $x = 0$,	$f(0) = 3 - 2(0) = 3 - 0 = 3.$
For $x = 2$,	$f(2) = 3 - 2(2) = 3 - 4 = -1.$

Although it is often convenient to use f as a function name and x as the independent variable, other letters may be used as well. For example,

$$f(x) = x^2 - 4x + 7, \quad f(t) = t^2 - 4t + 7, \quad \text{and} \quad g(s) = s^2 - 4s + 7$$

all define the same function. In fact, the role of the independent variable is that of a "placeholder." Consequently, the function can be described by

$$f(\boxed{}) = (\boxed{})^2 - 4(\boxed{}) + 7.$$

EXAMPLE 3   Evaluating a Function

Let $g(x) = -x^2 + 4x + 1$. Find each function value.

a. $g(2)$ **b.** $g(t)$ **c.** $g(x + 2)$

Solution

a. Replace x with 2 in $g(x) = -x^2 + 4x + 1$.

$$g(2) = -(2)^2 + 4(2) + 1$$
$$= -4 + 8 + 1$$
$$= 5$$

b. Replace x with t.

$$g(t) = -(t)^2 + 4(t) + 1$$
$$= -t^2 + 4t + 1$$

c. Replace x with $x + 2$.

$$g(x + 2) = -(x + 2)^2 + 4(x + 2) + 1$$
$$= -(x^2 + 4x + 4) + 4x + 8 + 1$$
$$= -x^2 - 4x - 4 + 4x + 8 + 1$$
$$= -x^2 + 5$$

REMARK In Example 3(c), note that $g(x + 2)$ is not equal to $g(x) + g(2)$. In general, $g(u + v) \neq g(u) + g(v)$.

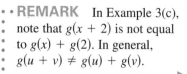

✓ **Checkpoint** Audio-video solution in English & Spanish at LarsonPrecalculus.com

Let $f(x) = 10 - 3x^2$. Find each function value.

a. $f(2)$ **b.** $f(-4)$ **c.** $f(x - 1)$

A function defined by two or more equations over a specified domain is called a **piecewise-defined function**.

EXAMPLE 4   A Piecewise-Defined Function

Evaluate the function when $x = -1, 0,$ and 1.

$$f(x) = \begin{cases} x^2 + 1, & x < 0 \\ x - 1, & x \geq 0 \end{cases}$$

Solution Because $x = -1$ is less than 0, use $f(x) = x^2 + 1$ to obtain $f(-1) = (-1)^2 + 1 = 2$. For $x = 0$, use $f(x) = x - 1$ to obtain $f(0) = (0) - 1 = -1$. For $x = 1$, use $f(x) = x - 1$ to obtain $f(1) = (1) - 1 = 0$.

✓ **Checkpoint** Audio-video solution in English & Spanish at LarsonPrecalculus.com

Evaluate the function given in Example 4 when $x = -2, 2,$ and 3.

EXAMPLE 5 Finding Values for Which $f(x) = 0$

Find all real values of x for which $f(x) = 0$.

a. $f(x) = -2x + 10$ **b.** $f(x) = x^2 - 5x + 6$

Solution For each function, set $f(x) = 0$ and solve for x.

a. $-2x + 10 = 0$ — Set $f(x)$ equal to 0.

$\qquad -2x = -10$ — Subtract 10 from each side.

$\qquad\quad x = 5$ — Divide each side by -2.

So, $f(x) = 0$ when $x = 5$.

b. $\quad x^2 - 5x + 6 = 0$ — Set $f(x)$ equal to 0.

$\quad (x - 2)(x - 3) = 0$ — Factor.

$\qquad\quad x - 2 = 0 \implies x = 2$ — Set 1st factor equal to 0 and solve.

$\qquad\quad x - 3 = 0 \implies x = 3$ — Set 2nd factor equal to 0 and solve.

So, $f(x) = 0$ when $x = 2$ or $x = 3$.

✓ **Checkpoint** ◄))) *Audio-video solution in English & Spanish at LarsonPrecalculus.com*

Find all real values of x for which $f(x) = 0$, where $f(x) = x^2 - 16$.

EXAMPLE 6 Finding Values for Which $f(x) = g(x)$

Find the values of x for which $f(x) = g(x)$.

a. $f(x) = x^2 + 1$ and $g(x) = 3x - x^2$

b. $f(x) = x^2 - 1$ and $g(x) = -x^2 + x + 2$

Solution

a. $\qquad x^2 + 1 = 3x - x^2$ — Set $f(x)$ equal to $g(x)$.

$\qquad 2x^2 - 3x + 1 = 0$ — Write in general form.

$\quad (2x - 1)(x - 1) = 0$ — Factor.

$\qquad\quad 2x - 1 = 0 \implies x = \frac{1}{2}$ — Set 1st factor equal to 0 and solve.

$\qquad\quad\; x - 1 = 0 \implies x = 1$ — Set 2nd factor equal to 0 and solve.

So, $f(x) = g(x)$ when $x = \dfrac{1}{2}$ or $x = 1$.

b. $\qquad x^2 - 1 = -x^2 + x + 2$ — Set $f(x)$ equal to $g(x)$.

$\qquad 2x^2 - x - 3 = 0$ — Write in general form.

$\quad (2x - 3)(x + 1) = 0$ — Factor.

$\qquad\quad 2x - 3 = 0 \implies x = \frac{3}{2}$ — Set 1st factor equal to 0 and solve.

$\qquad\quad\; x + 1 = 0 \implies x = -1$ — Set 2nd factor equal to 0 and solve.

So, $f(x) = g(x)$ when $x = \dfrac{3}{2}$ or $x = -1$.

✓ **Checkpoint** ◄))) *Audio-video solution in English & Spanish at LarsonPrecalculus.com*

Find the values of x for which $f(x) = g(x)$, where $f(x) = x^2 + 6x - 24$ and $g(x) = 4x - x^2$.

The Domain of a Function

▷ TECHNOLOGY Use a
graphing utility to graph the
functions $y = \sqrt{4 - x^2}$ and
$y = \sqrt{x^2 - 4}$. What is the
domain of each function?
Do the domains of these two
functions overlap? If so, for
what values do the domains
overlap?

The domain of a function can be described explicitly or it can be *implied* by the expression used to define the function. The **implied domain** is the set of all real numbers for which the expression is defined. For example, the function

$$f(x) = \frac{1}{x^2 - 4}$$ Domain excludes x-values that result in division by zero.

has an implied domain consisting of all real x other than $x = \pm 2$. These two values are excluded from the domain because division by zero is undefined. Another common type of implied domain is that used to avoid even roots of negative numbers. For example, the function

$$f(x) = \sqrt{x}$$ Domain excludes x-values that result in even roots of negative numbers.

is defined only for $x \geq 0$. So, its implied domain is the interval $[0, \infty)$. In general, the domain of a function *excludes* values that cause division by zero *or* that result in the even root of a negative number.

EXAMPLE 7   **Finding the Domains of Functions**

Find the domain of each function.

a. f: $\{(-3, 0), (-1, 4), (0, 2), (2, 2), (4, -1)\}$ **b.** $g(x) = \dfrac{1}{x + 5}$

c. Volume of a sphere: $V = \frac{4}{3}\pi r^3$ **d.** $h(x) = \sqrt{4 - 3x}$

Solution

a. The domain of f consists of all first coordinates in the set of ordered pairs.

Domain $= \{-3, -1, 0, 2, 4\}$

b. Excluding x-values that yield zero in the denominator, the domain of g is the set of all real numbers x except $x = -5$.

c. This function represents the volume of a sphere, so the values of the radius r must be positive. The domain is the set of all real numbers r such that $r > 0$.

d. This function is defined only for x-values for which

$$4 - 3x \geq 0.$$

By solving this inequality, you can conclude that $x \leq \frac{4}{3}$. So, the domain is the interval $\left(-\infty, \frac{4}{3}\right]$.

✓ *Checkpoint*  Audio-video solution in English & Spanish at LarsonPrecalculus.com

Find the domain of each function.

a. f: $\{(-2, 2), (-1, 1), (0, 3), (1, 1), (2, 2)\}$ **b.** $g(x) = \dfrac{1}{3 - x}$

c. Circumference of a circle: $C = 2\pi r$ **d.** $h(x) = \sqrt{x - 16}$ ■

In Example 7(c), note that the domain of a function may be implied by the physical context. For example, from the equation

$$V = \frac{4}{3}\pi r^3$$

you have no reason to restrict r to positive values, but the physical context implies that a sphere cannot have a negative or zero radius.

Applications

EXAMPLE 8 **The Dimensions of a Container**

You work in the marketing department of a soft-drink company and are experimenting with a new can for iced tea that is slightly narrower and taller than a standard can. For your experimental can, the ratio of the height to the radius is 4.

a. Write the volume of the can as a function of the radius r.

b. Write the volume of the can as a function of the height h.

Solution

a. $V(r) = \pi r^2 h = \pi r^2 (4r) = 4\pi r^3$
 Write V as a function of r.

b. $V(h) = \pi r^2 h = \pi \left(\dfrac{h}{4}\right)^2 h = \dfrac{\pi h^3}{16}$
 Write V as a function of h.

✓ **Checkpoint** ◀))) *Audio-video solution in English & Spanish at LarsonPrecalculus.com*

For the experimental can described in Example 8, write the *surface area* as a function of (a) the radius r and (b) the height h.

EXAMPLE 9 **The Path of a Baseball**

A batter hits a baseball at a point 3 feet above ground at a velocity of 100 feet per second and an angle of 45°. The path of the baseball is given by the function

$$f(x) = -0.0032x^2 + x + 3$$

where $f(x)$ is the height of the baseball (in feet) and x is the horizontal distance from home plate (in feet). Will the baseball clear a 10-foot fence located 300 feet from home plate?

Algebraic Solution

Find the height of the baseball when $x = 300$.

$f(x) = -0.0032x^2 + x + 3$
 Write original function.

$f(300) = -0.0032(300)^2 + 300 + 3$
 Substitute 300 for x.

$\quad\quad = 15$
 Simplify.

When $x = 300$, the height of the baseball is 15 feet. So, the baseball will clear a 10-foot fence.

Graphical Solution

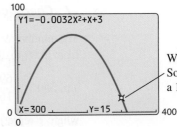

When $x = 300$, $y = 15$. So, the ball will clear a 10-foot fence.

✓ **Checkpoint** ◀))) *Audio-video solution in English & Spanish at LarsonPrecalculus.com*

A second baseman throws a baseball toward the first baseman 60 feet away. The path of the baseball is given by the function

$$f(x) = -0.004x^2 + 0.3x + 6$$

where $f(x)$ is the height of the baseball (in feet) and x is the horizontal distance from the second baseman (in feet). The first baseman can reach 8 feet high. Can the first baseman catch the baseball without jumping?

Flexible-fuel vehicles are designed to operate on gasoline, E85, or a mixture of the two fuels. The concentration of ethanol in E85 fuel ranges from 51% to 83%, depending on where and when the E85 is produced.

EXAMPLE 10 **Alternative-Fuel Stations**

The number S of fuel stations that sold E85 (a gasoline-ethanol blend) in the United States increased in a linear pattern from 2008 through 2011, and then increased in a different linear pattern from 2012 through 2015, as shown in the bar graph. These two patterns can be approximated by the function

$$S(t) = \begin{cases} 260.8t - 439, & 8 \le t \le 11 \\ 151.2t + 714, & 12 \le t \le 15 \end{cases}$$

where t represents the year, with $t = 8$ corresponding to 2008. Use this function to approximate the number of stations that sold E85 each year from 2008 to 2015. *(Source: Alternative Fuels Data Center)*

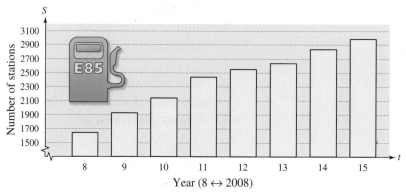

Number of Stations Selling E85 in the U.S.

Solution From 2008 through 2011, use $S(t) = 260.8t - 439$.

1647	1908	2169	2430
2008	2009	2010	2011

From 2012 to 2015, use $S(t) = 151.2t + 714$.

2528	2680	2831	2982
2012	2013	2014	2015

✓ **Checkpoint** ◀))) *Audio-video solution in English & Spanish at LarsonPrecalculus.com*

The number S of fuel stations that sold compressed natural gas in the United States from 2009 to 2015 can be approximated by the function

$$S(t) = \begin{cases} 69t + 151, & 9 \le t \le 11 \\ 160t - 803, & 12 \le t \le 15 \end{cases}$$

where t represents the year, with $t = 9$ corresponding to 2009. Use this function to approximate the number of stations that sold compressed natural gas each year from 2009 through 2015. *(Source: Alternative Fuels Data Center)* ◼

Difference Quotients

One of the basic definitions in calculus uses the ratio

$$\frac{f(x + h) - f(x)}{h}, \quad h \ne 0.$$

This ratio is a **difference quotient,** as illustrated in Example 11.

REMARK You may find it easier to calculate the difference quotient in Example 11 by first finding $f(x + h)$, and then substituting the resulting expression into the difference quotient

$$\frac{f(x + h) - f(x)}{h}.$$

EXAMPLE 11 **Evaluating a Difference Quotient**

For $f(x) = x^2 - 4x + 7$, find $\dfrac{f(x + h) - f(x)}{h}$.

Solution

$$\frac{f(x + h) - f(x)}{h} = \frac{[(x + h)^2 - 4(x + h) + 7] - (x^2 - 4x + 7)}{h}$$

$$= \frac{x^2 + 2xh + h^2 - 4x - 4h + 7 - x^2 + 4x - 7}{h}$$

$$= \frac{2xh + h^2 - 4h}{h} = \frac{h(2x + h - 4)}{h} = 2x + h - 4, \quad h \neq 0$$

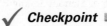 **Checkpoint** ◀))) Audio-video solution in English & Spanish at LarsonPrecalculus.com

For $f(x) = x^2 + 2x - 3$, find $\dfrac{f(x + h) - f(x)}{h}$.

Summary of Function Terminology

Function: A **function** is a relationship between two variables such that to each value of the independent variable there corresponds exactly one value of the dependent variable.

Function notation: $y = f(x)$

$\quad$ f is the *name* of the function.

$\quad$ y is the **dependent variable.**

$\quad$ x is the **independent variable.**

$\quad$ $f(x)$ is the *value of the function at x.*

Domain: The **domain** of a function is the set of all values (inputs) of the independent variable for which the function is defined. If x is in the domain of f, then f is *defined* at x. If x is not in the domain of f, then f is *undefined* at x.

Range: The **range** of a function is the set of all values (outputs) taken on by the dependent variable (that is, the set of all function values).

Implied domain: If f is defined by an algebraic expression and the domain is not specified, then the **implied domain** consists of all real numbers for which the expression is defined.

Summarize (Section 1.4)

1. State the definition of a function and describe function notation *(pages 35–39)*. For examples of determining functions and using function notation, see Examples 1–6.

2. State the definition of the implied domain of a function *(page 40)*. For an example of finding the domains of functions, see Example 7.

3. Describe examples of how functions can model real-life problems *(pages 41 and 42, Examples 8–10)*.

4. State the definition of a difference quotient *(page 42)*. For an example of evaluating a difference quotient, see Example 11.

1.4 Exercises

See **CalcChat.com** for tutorial help and worked-out solutions to odd-numbered exercises.

Vocabulary: Fill in the blanks.

1. A relation that assigns to each element x from a set of inputs, or _____, exactly one element y in a set of outputs, or _____, is a _____.

2. For an equation that represents y as a function of x, the set of all values taken on by the _____ variable x is the domain, and the set of all values taken on by the _____ variable y is the range.

3. If the domain of the function f is not given, then the set of values of the independent variable for which the expression is defined is the _____ _____.

4. One of the basic definitions in calculus uses the ratio $\dfrac{f(x + h) - f(x)}{h}$, $h \neq 0$. This ratio is a _____ _____.

Skills and Applications

 Testing for Functions In Exercises 5–8, determine whether the relation represents y as a function of x.

5. Domain, x   Range, y

6. Domain, x   Range, y

7.

Input, x	10	7	4	7	10
Output, y	3	6	9	12	15

8.

Input, x	-2	0	2	4	6
Output, y	1	1	1	1	1

Testing for Functions In Exercises 9 and 10, which sets of ordered pairs represent functions from A to B? Explain.

9. $A = \{0, 1, 2, 3\}$ and $B = \{-2, -1, 0, 1, 2\}$
   (a) $\{(0, 1), (1, -2), (2, 0), (3, 2)\}$
   (b) $\{(0, -1), (2, 2), (1, -2), (3, 0), (1, 1)\}$
   (c) $\{(0, 0), (1, 0), (2, 0), (3, 0)\}$
   (d) $\{(0, 2), (3, 0), (1, 1)\}$

10. $A = \{a, b, c\}$ and $B = \{0, 1, 2, 3\}$
    (a) $\{(a, 1), (c, 2), (c, 3), (b, 3)\}$
    (b) $\{(a, 1), (b, 2), (c, 3)\}$
    (c) $\{(1, a), (0, a), (2, c), (3, b)\}$
    (d) $\{(c, 0), (b, 0), (a, 3)\}$

 Testing for Functions Represented Algebraically In Exercises 11–18, determine whether the equation represents y as a function of x.

11. $x^2 + y^2 = 4$

12. $x^2 - y = 9$

13. $y = \sqrt{16 - x^2}$

14. $y = \sqrt{x + 5}$

15. $y = 4 - |x|$

16. $|y| = 4 - x$

17. $y = -75$

18. $x - 1 = 0$

 Evaluating a Function In Exercises 19–30, find each function value, if possible.

19. $f(x) = 3x - 5$
    (a) $f(1)$ (b) $f(-3)$ (c) $f(x + 2)$

20. $V(r) = \frac{4}{3}\pi r^3$
    (a) $V(3)$ (b) $V\left(\frac{3}{2}\right)$ (c) $V(2r)$

21. $g(t) = 4t^2 - 3t + 5$
    (a) $g(2)$ (b) $g(t - 2)$ (c) $g(t) - g(2)$

22. $h(t) = -t^2 + t + 1$
    (a) $h(2)$ (b) $h(-1)$ (c) $h(x + 1)$

23. $f(y) = 3 - \sqrt{y}$
    (a) $f(4)$ (b) $f(0.25)$ (c) $f(4x^2)$

24. $f(x) = \sqrt{x + 8} + 2$
    (a) $f(-8)$ (b) $f(1)$ (c) $f(x - 8)$

25. $q(x) = 1/(x^2 - 9)$
    (a) $q(0)$ (b) $q(3)$ (c) $q(y + 3)$

26. $q(t) = (2t^2 + 3)/t^2$
    (a) $q(2)$ (b) $q(0)$ (c) $q(-x)$

27. $f(x) = |x|/x$
    (a) $f(2)$ (b) $f(-2)$ (c) $f(x - 1)$

28. $f(x) = |x| + 4$
    (a) $f(2)$ (b) $f(-2)$ (c) $f(x^2)$

29. $f(x) = \begin{cases} 2x + 1, & x < 0 \\ 2x + 2, & x \geq 0 \end{cases}$
    (a) $f(-1)$ (b) $f(0)$ (c) $f(2)$

30. $f(x) = \begin{cases} -3x - 3, & x < -1 \\ x^2 + 2x - 1, & x \geq -1 \end{cases}$
    (a) $f(-2)$ (b) $f(-1)$ (c) $f(1)$

Evaluating a Function In Exercises 31–34, complete the table.

31. $f(x) = -x^2 + 5$

x	-2	-1	0	1	2
$f(x)$					

32. $h(t) = \frac{1}{2}|t + 3|$

t	-5	-4	-3	-2	-1
$h(t)$					

33. $f(x) = \begin{cases} -\frac{1}{2}x + 4, & x \le 0 \\ (x - 2)^2, & x > 0 \end{cases}$

x	-2	-1	0	1	2
$f(x)$					

34. $f(x) = \begin{cases} 9 - x^2, & x < 3 \\ x - 3, & x \ge 3 \end{cases}$

x	1	2	3	4	5
$f(x)$					

 Finding Values for Which $f(x) = 0$ In Exercises 35–42, find all real values of x for which $f(x) = 0$.

35. $f(x) = 15 - 3x$ **36.** $f(x) = 4x + 6$

37. $f(x) = \dfrac{3x - 4}{5}$ **38.** $f(x) = \dfrac{12 - x^2}{8}$

39. $f(x) = x^2 - 81$ **40.** $f(x) = x^2 - 6x - 16$

41. $f(x) = x^3 - x$

42. $f(x) = x^3 - x^2 - 3x + 3$

 Finding Values for Which $f(x) = g(x)$ In Exercises 43–46, find the value(s) of x for which $f(x) = g(x)$.

43. $f(x) = x^2$, $g(x) = x + 2$

44. $f(x) = x^2 + 2x + 1$, $g(x) = 5x + 19$

45. $f(x) = x^4 - 2x^2$, $g(x) = 2x^2$

46. $f(x) = \sqrt{x} - 4$, $g(x) = 2 - x$

 Finding the Domain of a Function In Exercises 47–56, find the domain of the function.

47. $f(x) = 5x^2 + 2x - 1$

48. $g(x) = 1 - 2x^2$

49. $g(y) = \sqrt{y + 6}$

50. $f(t) = \sqrt[3]{t + 4}$

51. $g(x) = \dfrac{1}{x} - \dfrac{3}{x + 2}$ **52.** $h(x) = \dfrac{6}{x^2 - 4x}$

53. $f(s) = \dfrac{\sqrt{s - 1}}{s - 4}$ **54.** $f(x) = \dfrac{\sqrt{x + 6}}{6 + x}$

55. $f(x) = \dfrac{x - 4}{\sqrt{x}}$

56. $f(x) = \dfrac{x + 2}{\sqrt{x - 10}}$

57. Maximum Volume An open box of maximum volume is made from a square piece of material 24 centimeters on a side by cutting equal squares from the corners and turning up the sides (see figure).

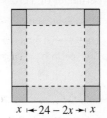 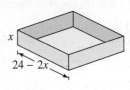

(a) The table shows the volumes V (in cubic centimeters) of the box for various heights x (in centimeters). Use the table to estimate the maximum volume.

Height, x	1	2	3	4	5	6
Volume, V	484	800	972	1024	980	864

(b) Plot the points (x, V) from the table in part (a). Does the relation defined by the ordered pairs represent V as a function of x?

(c) Given that V is a function of x, write the function and determine its domain.

58. Maximum Profit The cost per unit in the production of an MP3 player is $60. The manufacturer charges $90 per unit for orders of 100 or less. To encourage large orders, the manufacturer reduces the charge by $0.15 per MP3 player for each unit ordered in excess of 100 (for example, the charge is reduced to $87 per MP3 player for an order size of 120).

(a) The table shows the profits P (in dollars) for various numbers of units ordered, x. Use the table to estimate the maximum profit.

Units, x	130	140	150	160	170
Profit, P	3315	3360	3375	3360	3315

(b) Plot the points (x, P) from the table in part (a). Does the relation defined by the ordered pairs represent P as a function of x?

(c) Given that P is a function of x, write the function and determine its domain. (*Note:* $P = R - C$, where R is revenue and C is cost.)

59. Geometry   Write the area A of a square as a function of its perimeter P.

60. Geometry   Write the area A of a circle as a function of its circumference C.

61. Path of a Ball   You throw a baseball to a child 25 feet away. The height y (in feet) of the baseball is given by

$$y = -\tfrac{1}{10}x^2 + 3x + 6$$

where x is the horizontal distance (in feet) from where you threw the ball. Can the child catch the baseball while holding a baseball glove at a height of 5 feet?

62. Postal Regulations   A rectangular package has a combined length and girth (perimeter of a cross section) of 108 inches (see figure).

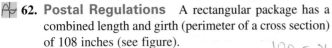

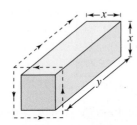

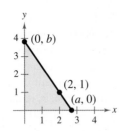

(a) Write the volume V of the package as a function of x. What is the domain of the function?

(b) Use a graphing utility to graph the function. Be sure to use an appropriate window setting.

(c) What dimensions will maximize the volume of the package? Explain.

63. Geometry   A right triangle is formed in the first quadrant by the x- and y-axes and a line through the point $(2, 1)$ (see figure). Write the area A of the triangle as a function of x, and determine the domain of the function.

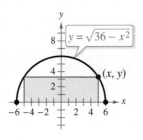

64. Geometry   A rectangle is bounded by the x-axis and the semicircle $y = \sqrt{36 - x^2}$ (see figure). Write the area A of the rectangle as a function of x, and graphically determine the domain of the function.

65. Pharmacology   The percent p of prescriptions filled with generic drugs at CVS Pharmacies from 2008 through 2014 (see figure) can be approximated by the model

$$p(t) = \begin{cases} 2.77t + 45.2, & 8 \le t \le 11 \\ 1.95t + 55.9, & 12 \le t \le 14 \end{cases}$$

where t represents the year, with $t = 8$ corresponding to 2008. Use this model to find the percent of prescriptions filled with generic drugs in each year from 2008 through 2014.   (*Source: CVS Health*)

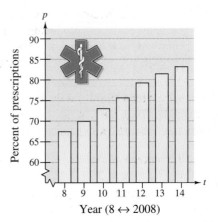

Year (8 ↔ 2008)

66. Median Sale Price   The median sale price p (in thousands of dollars) of an existing one-family home in the United States from 2002 through 2014 (see figure) can be approximated by the model

$$p(t) = \begin{cases} -0.757t^2 + 20.80t + 127.2, & 2 \le t \le 6 \\ 3.879t^2 - 82.50t + 605.8, & 7 \le t \le 11 \\ -4.171t^2 + 124.34t - 714.2, & 12 \le t \le 14 \end{cases}$$

where t represents the year, with $t = 2$ corresponding to 2002. Use this model to find the median sale price of an existing one-family home in each year from 2002 through 2014.   (*Source: National Association of Realtors*)

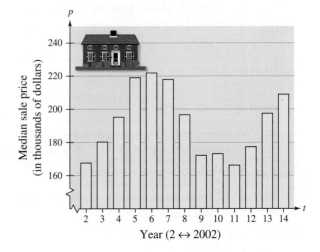

Year (2 ↔ 2002)

67. Cost, Revenue, and Profit A company produces a product for which the variable cost is $12.30 per unit and the fixed costs are $98,000. The product sells for $17.98. Let x be the number of units produced and sold.

(a) The total cost for a business is the sum of the variable cost and the fixed costs. Write the total cost C as a function of the number of units produced.

(b) Write the revenue R as a function of the number of units sold.

(c) Write the profit P as a function of the number of units sold. (*Note: $P = R - C$*)

68. Average Cost The inventor of a new game believes that the variable cost for producing the game is $0.95 per unit and the fixed costs are $6000. The inventor sells each game for $1.69. Let x be the number of games produced.

(a) The total cost for a business is the sum of the variable cost and the fixed costs. Write the total cost C as a function of the number of games produced.

(b) Write the average cost per unit $\overline{C} = \dfrac{C}{x}$ as a function of x.

69. Height of a Balloon A balloon carrying a transmitter ascends vertically from a point 3000 feet from the receiving station.

(a) Draw a diagram that gives a visual representation of the problem. Let h represent the height of the balloon and let d represent the distance between the balloon and the receiving station.

(b) Write the height of the balloon as a function of d. What is the domain of the function?

70. Physics

The function $F(y) = 149.76\sqrt{10}\,y^{5/2}$ estimates the force F (in tons) of water against the face of a dam, where y is the depth of the water (in feet).

(a) Complete the table. What can you conclude from the table?

y	5	10	20	30	40
$F(y)$					

(b) Use the table to approximate the depth at which the force against the dam is 1,000,000 tons.

(c) Find the depth at which the force against the dam is 1,000,000 tons algebraically.

71. Transportation For groups of 80 or more people, a charter bus company determines the rate per person according to the formula

$$\text{Rate} = 8 - 0.05(n - 80), \quad n \geq 80$$

where the rate is given in dollars and n is the number of people.

(a) Write the revenue R for the bus company as a function of n.

(b) Use the function in part (a) to complete the table. What can you conclude?

n	90	100	110	120	130	140	150
$R(n)$							

72. E-Filing The table shows the numbers of tax returns (in millions) made through e-file from 2007 through 2014. Let $f(t)$ represent the number of tax returns made through e-file in the year t. (*Source: eFile*)

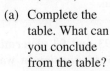

DATA Year	Number of Tax Returns Made Through E-File
2007	80.0
2008	89.9
2009	95.0
2010	98.7
2011	112.2
2012	112.1
2013	114.4
2014	125.8

(a) Find $\dfrac{f(2014) - f(2007)}{2014 - 2007}$ and interpret the result in the context of the problem.

(b) Make a scatter plot of the data.

(c) Find a linear model for the data algebraically. Let N represent the number of tax returns made through e-file and let $t = 7$ correspond to 2007.

(d) Use the model found in part (c) to complete the table.

t	7	8	9	10	11	12	13	14
N								

(e) Compare your results from part (d) with the actual data.

(f) Use a graphing utility to find a linear model for the data. Let $x = 7$ correspond to 2007. How does the model you found in part (c) compare with the model given by the graphing utility?

Evaluating a Difference Quotient In Exercises 73–80, find the difference quotient and simplify your answer.

73. $f(x) = x^2 - 2x + 4,\quad \dfrac{f(2 + h) - f(2)}{h},\quad h \neq 0$

74. $f(x) = 5x - x^2,\quad \dfrac{f(5 + h) - f(5)}{h},\quad h \neq 0$

75. $f(x) = x^3 + 3x,\quad \dfrac{f(x + h) - f(x)}{h},\quad h \neq 0$

76. $f(x) = 4x^3 - 2x,\quad \dfrac{f(x + h) - f(x)}{h},\quad h \neq 0$

77. $g(x) = \dfrac{1}{x^2},\quad \dfrac{g(x) - g(3)}{x - 3},\quad x \neq 3$

78. $f(t) = \dfrac{1}{t - 2},\quad \dfrac{f(t) - f(1)}{t - 1},\quad t \neq 1$

79. $f(x) = \sqrt{5x},\quad \dfrac{f(x) - f(5)}{x - 5},\quad x \neq 5$

80. $f(x) = x^{2/3} + 1,\quad \dfrac{f(x) - f(8)}{x - 8},\quad x \neq 8$

Modeling Data In Exercises 81–84, determine which of the following functions

$$f(x) = cx,\ g(x) = cx^2,\ h(x) = c\sqrt{|x|},\ \text{and}\ r(x) = \dfrac{c}{x}$$

can be used to model the data and determine the value of the constant c that will make the function fit the data in the table.

81.

x	-4	-1	0	1	4
y	-32	-2	0	-2	-32

82.

x	-4	-1	0	1	4
y	-1	$-\frac{1}{4}$	0	$\frac{1}{4}$	1

83.

x	-4	-1	0	1	4
y	-8	-32	Undefined	32	8

84.

x	-4	-1	0	1	4
y	6	3	0	3	6

Exploration

True or False? In Exercises 85–88, determine whether the statement is true or false. Justify your answer.

85. Every relation is a function.

86. Every function is a relation.

87. For the function
$$f(x) = x^4 - 1$$
the domain is $(-\infty, \infty)$ and the range is $(0, \infty)$.

88. The set of ordered pairs $\{(-8, -2), (-6, 0), (-4, 0), (-2, 2), (0, 4), (2, -2)\}$ represents a function.

89. Error Analysis Describe the error.

The functions
$$f(x) = \sqrt{x - 1}\quad \text{and}\quad g(x) = \dfrac{1}{\sqrt{x - 1}}$$
have the same domain, which is the set of all real numbers x such that $x \geq 1$.

90. Think About It Consider
$$f(x) = \sqrt{x - 2}\quad \text{and}\quad g(x) = \sqrt[3]{x - 2}.$$
Why are the domains of f and g different?

91. Think About It Given $f(x) = x^2$, is f the independent variable? Why or why not?

92. 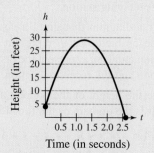 **HOW DO YOU SEE IT?** The graph represents the height h of a projectile after t seconds.

(a) Explain why h is a function of t.

(b) Approximate the height of the projectile after 0.5 second and after 1.25 seconds.

(c) Approximate the domain of h.

(d) Is t a function of h? Explain.

Think About It In Exercises 93 and 94, determine whether the statements use the word *function* in ways that are mathematically correct. **Explain.**

93. (a) The sales tax on a purchased item is a function of the selling price.

(b) Your score on the next algebra exam is a function of the number of hours you study the night before the exam.

94. (a) The amount in your savings account is a function of your salary.

(b) The speed at which a free-falling baseball strikes the ground is a function of the height from which it was dropped.

1.5 Analyzing Graphs of Functions

- Use the Vertical Line Test for functions.
- Find the zeros of functions.
- Determine intervals on which functions are increasing or decreasing.
- Determine relative minimum and relative maximum values of functions.
- Determine the average rate of change of a function.
- Identify even and odd functions.

Graphs of functions can help you visualize relationships between variables in real life. For example, in Exercise 90 on page 59, you will use the graph of a function to visually represent the temperature in a city over a 24-hour period.

The Graph of a Function

In Section 1.4, you studied functions from an algebraic point of view. In this section, you will study functions from a graphical perspective.

The **graph of a function** f is the collection of ordered pairs $(x, f(x))$ such that x is in the domain of f. As you study this section, remember that

$x =$ the directed distance from the y-axis

$y = f(x) =$ the directed distance from the x-axis

as shown in the figure at the right.

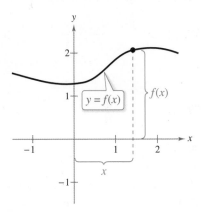

EXAMPLE 1   **Finding the Domain and Range of a Function**

Use the graph of the function f, shown in Figure 1.32, to find (a) the domain of f, (b) the function values $f(-1)$ and $f(2)$, and (c) the range of f.

Solution

a. The closed dot at $(-1, 1)$ indicates that $x = -1$ is in the domain of f, whereas the open dot at $(5, 2)$ indicates that $x = 5$ is not in the domain. So, the domain of f is all x in the interval $[-1, 5)$.

b. One point on the graph of f is $(-1, 1)$, so $f(-1) = 1$. Another point on the graph of f is $(2, -3)$, so $f(2) = -3$.

c. The graph does not extend below $f(2) = -3$ or above $f(0) = 3$, so the range of f is the interval $[-3, 3]$.

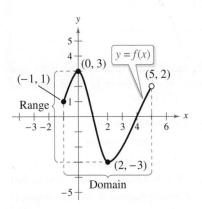

Figure 1.32

✓ *Checkpoint* ◀))) *Audio-video solution in English & Spanish at LarsonPrecalculus.com*

Use the graph of the function f to find (a) the domain of f, (b) the function values $f(0)$ and $f(3)$, and (c) the range of f.

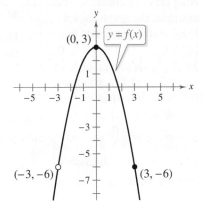

• • **REMARK** The use of dots (open or closed) at the extreme left and right points of a graph indicates that the graph does not extend beyond these points. If such dots are not on the graph, then assume that the graph extends beyond these points.

By the definition of a function, at most one y-value corresponds to a given x-value. So, no two points on the graph of a function have the same x-coordinate, or lie on the same vertical line. It follows, then, that a vertical line can intersect the graph of a function at most once. This observation provides a convenient visual test called the **Vertical Line Test** for functions.

Vertical Line Test for Functions

A set of points in a coordinate plane is the graph of y as a function of x if and only if no *vertical* line intersects the graph at more than one point.

EXAMPLE 2 **Vertical Line Test for Functions**

Use the Vertical Line Test to determine whether each graph represents y as a function of x.

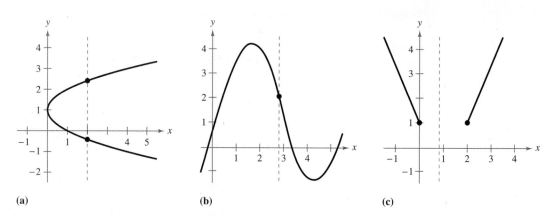

(a) (b) (c)

Solution

a. This *is not* a graph of y as a function of x, because there are vertical lines that intersect the graph twice. That is, for a particular input x, there is more than one output y.

b. This *is* a graph of y as a function of x, because every vertical line intersects the graph at most once. That is, for a particular input x, there is at most one output y.

c. This *is* a graph of y as a function of x, because every vertical line intersects the graph at most once. That is, for a particular input x, there is at most one output y. (Note that when a vertical line does not intersect the graph, it simply means that the function is undefined for that particular value of x.)

✓ *Checkpoint*  *Audio-video solution in English & Spanish at LarsonPrecalculus.com*

Use the Vertical Line Test to determine whether the graph represents y as a function of x.

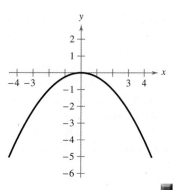

▷ TECHNOLOGY Most graphing utilities graph functions of x more easily than other types of equations. For example, the graph shown in (a) above represents the equation $x - (y - 1)^2 = 0$. To duplicate this graph using a graphing utility, you must first solve the equation for y to obtain $y = 1 \pm \sqrt{x}$, and then graph the two equations $y_1 = 1 + \sqrt{x}$ and $y_2 = 1 - \sqrt{x}$ in the same viewing window.

▷ **ALGEBRA HELP** The solution to Example 3 involves solving equations. To review the techniques for solving equations, see Appendix A.5.

Zeros of a Function

If the graph of a function of x has an x-intercept at $(a, 0)$, then a is a **zero** of the function.

> **Zeros of a Function**
>
> The **zeros of a function** $y = f(x)$ are the x-values for which $f(x) = 0$.

EXAMPLE 3 **Finding the Zeros of Functions**

Find the zeros of each function algebraically.

a. $f(x) = 3x^2 + x - 10$

b. $g(x) = \sqrt{10 - x^2}$

c. $h(t) = \dfrac{2t - 3}{t + 5}$

Solution To find the zeros of a function, set the function equal to zero and solve for the independent variable.

a.

$$3x^2 + x - 10 = 0 \qquad \text{Set } f(x) \text{ equal to 0.}$$
$$(3x - 5)(x + 2) = 0 \qquad \text{Factor.}$$
$$3x - 5 = 0 \implies x = \tfrac{5}{3} \qquad \text{Set 1st factor equal to 0 and solve.}$$
$$x + 2 = 0 \implies x = -2 \qquad \text{Set 2nd factor equal to 0 and solve.}$$

The zeros of f are $x = \tfrac{5}{3}$ and $x = -2$. In Figure 1.33, note that the graph of f has $\left(\tfrac{5}{3}, 0\right)$ and $(-2, 0)$ as its x-intercepts.

b.

$$\sqrt{10 - x^2} = 0 \qquad \text{Set } g(x) \text{ equal to 0.}$$
$$10 - x^2 = 0 \qquad \text{Square each side.}$$
$$10 = x^2 \qquad \text{Add } x^2 \text{ to each side.}$$
$$\pm\sqrt{10} = x \qquad \text{Extract square roots.}$$

The zeros of g are $x = -\sqrt{10}$ and $x = \sqrt{10}$. In Figure 1.34, note that the graph of g has $\left(-\sqrt{10}, 0\right)$ and $\left(\sqrt{10}, 0\right)$ as its x-intercepts.

c.

$$\frac{2t - 3}{t + 5} = 0 \qquad \text{Set } h(t) \text{ equal to 0.}$$
$$2t - 3 = 0 \qquad \text{Multiply each side by } t + 5.$$
$$2t = 3 \qquad \text{Add 3 to each side.}$$
$$t = \frac{3}{2} \qquad \text{Divide each side by 2.}$$

The zero of h is $t = \tfrac{3}{2}$. In Figure 1.35, note that the graph of h has $\left(\tfrac{3}{2}, 0\right)$ as its t-intercept.

✓ **Checkpoint** ◀))) Audio-video solution in English & Spanish at LarsonPrecalculus.com

Find the zeros of each function.

a. $f(x) = 2x^2 + 13x - 24$ **b.** $g(t) = \sqrt{t - 25}$ **c.** $h(x) = \dfrac{x^2 - 2}{x - 1}$

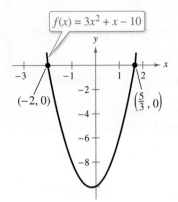

Zeros of f: $x = -2$, $x = \tfrac{5}{3}$
Figure 1.33

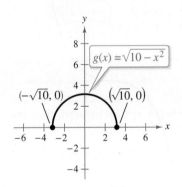

Zeros of g: $x = \pm\sqrt{10}$
Figure 1.34

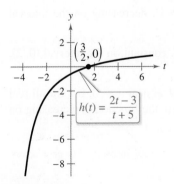

Zero of h: $t = \tfrac{3}{2}$
Figure 1.35

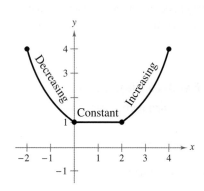

Figure 1.36

Increasing and Decreasing Functions

The more you know about the graph of a function, the more you know about the function itself. Consider the graph shown in Figure 1.36. As you move from *left to right*, this graph falls from $x = -2$ to $x = 0$, is constant from $x = 0$ to $x = 2$, and rises from $x = 2$ to $x = 4$.

Increasing, Decreasing, and Constant Functions

A function f is **increasing** on an interval when, for any x_1 and x_2 in the interval,

$$x_1 < x_2 \quad \text{implies} \quad f(x_1) < f(x_2).$$

A function f is **decreasing** on an interval when, for any x_1 and x_2 in the interval,

$$x_1 < x_2 \quad \text{implies} \quad f(x_1) > f(x_2).$$

A function f is **constant** on an interval when, for any x_1 and x_2 in the interval,

$$f(x_1) = f(x_2).$$

EXAMPLE 4   **Describing Function Behavior**

Determine the open intervals on which each function is increasing, decreasing, or constant.

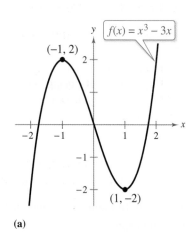

(a)

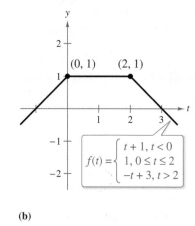

(b)

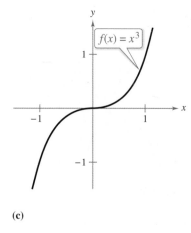

(c)

Solution

a. This function is increasing on the interval $(-\infty, -1)$, decreasing on the interval $(-1, 1)$, and increasing on the interval $(1, \infty)$.

b. This function is increasing on the interval $(-\infty, 0)$, constant on the interval $(0, 2)$, and decreasing on the interval $(2, \infty)$.

c. This function may appear to be constant on an interval near $x = 0$, but for all real values of x_1 and x_2, if $x_1 < x_2$, then $(x_1)^3 < (x_2)^3$. So, the function is increasing on the interval $(-\infty, \infty)$.

✓ **Checkpoint**  *Audio-video solution in English & Spanish at LarsonPrecalculus.com*

Graph the function

$$f(x) = x^3 + 3x^2 - 1.$$

Then determine the open intervals on which the function is increasing, decreasing, or constant. ∎

Relative Minimum and Relative Maximum Values

The points at which a function changes its increasing, decreasing, or constant behavior are helpful in determining the **relative minimum** or **relative maximum** values of the function.

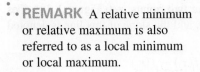

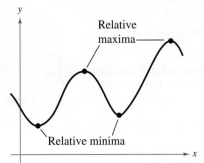

Figure 1.37

• **REMARK** A relative minimum or relative maximum is also referred to as a local minimum or local maximum.

Definitions of Relative Minimum and Relative Maximum

A function value $f(a)$ is a **relative minimum** of f when there exists an interval (x_1, x_2) that contains a such that

$$x_1 < x < x_2 \quad \text{implies} \quad f(a) \le f(x).$$

A function value $f(a)$ is a **relative maximum** of f when there exists an interval (x_1, x_2) that contains a such that

$$x_1 < x < x_2 \quad \text{implies} \quad f(a) \ge f(x).$$

Figure 1.37 shows several different examples of relative minima and relative maxima. In Section 2.1, you will study a technique for finding the *exact point* at which a second-degree polynomial function has a relative minimum or relative maximum. For the time being, however, you can use a graphing utility to find reasonable approximations of these points.

EXAMPLE 5 **Approximating a Relative Minimum**

Use a graphing utility to approximate the relative minimum of the function

$$f(x) = 3x^2 - 4x - 2.$$

Solution The graph of f is shown in Figure 1.38. By using the *zoom* and *trace* features or the *minimum* feature of a graphing utility, you can approximate that the relative minimum of the function occurs at the point

$$(0.67, -3.33).$$

So, the relative minimum is approximately -3.33. Later, in Section 2.1, you will learn how to determine that the exact point at which the relative minimum occurs is $\left(\frac{2}{3}, -\frac{10}{3}\right)$ and the exact relative minimum is $-\frac{10}{3}$.

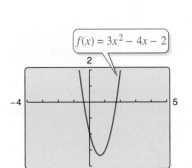

Figure 1.38

✓ Checkpoint 🔊))) *Audio-video solution in English & Spanish at LarsonPrecalculus.com*

Use a graphing utility to approximate the relative maximum of the function

$$f(x) = -4x^2 - 7x + 3.$$

You can also use the *table* feature of a graphing utility to numerically approximate the relative minimum of the function in Example 5. Using a table that begins at 0.6 and increments the value of x by 0.01, you can approximate that the minimum of

$$f(x) = 3x^2 - 4x - 2$$

occurs at the point $(0.67, -3.33)$.

▷ **TECHNOLOGY** When you use a graphing utility to approximate the x- and y-values of the point where a relative minimum or relative maximum occurs, the *zoom* feature will often produce graphs that are nearly flat. To overcome this problem, manually change the vertical setting of the viewing window. The graph will stretch vertically when the values of Ymin and Ymax are closer together.

Average Rate of Change

In Section 1.3, you learned that the slope of a line can be interpreted as a *rate of change*. For a nonlinear graph, the **average rate of change** between any two points $(x_1, f(x_1))$ and $(x_2, f(x_2))$ is the slope of the line through the two points (see Figure 1.39). The line through the two points is called a **secant line,** and the slope of this line is denoted as m_{sec}.

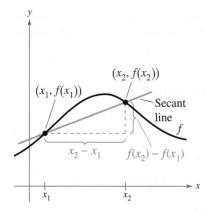

$$\text{Average rate of change of } f \text{ from } x_1 \text{ to } x_2 = \frac{f(x_2) - f(x_1)}{x_2 - x_1}$$

$$= \frac{\text{change in } y}{\text{change in } x}$$

$$= m_{\text{sec}}$$

Figure 1.39

EXAMPLE 6 Average Rate of Change of a Function

Find the average rates of change of $f(x) = x^3 - 3x$ (a) from $x_1 = -2$ to $x_2 = -1$ and (b) from $x_1 = 0$ to $x_2 = 1$ (see Figure 1.40).

Solution

a. The average rate of change of f from $x_1 = -2$ to $x_2 = -1$ is

$$\frac{f(x_2) - f(x_1)}{x_2 - x_1} = \frac{f(-1) - f(-2)}{-1 - (-2)} = \frac{2 - (-2)}{1} = 4.$$ Secant line has positive slope.

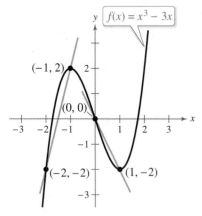

Figure 1.40

b. The average rate of change of f from $x_1 = 0$ to $x_2 = 1$ is

$$\frac{f(x_2) - f(x_1)}{x_2 - x_1} = \frac{f(1) - f(0)}{1 - 0} = \frac{-2 - 0}{1} = -2.$$ Secant line has negative slope.

✓ **Checkpoint**)))) *Audio-video solution in English & Spanish at LarsonPrecalculus.com*

Find the average rates of change of $f(x) = x^2 + 2x$ (a) from $x_1 = -3$ to $x_2 = -2$ and (b) from $x_1 = -2$ to $x_2 = 0$.

EXAMPLE 7 Finding Average Speed

The distance s (in feet) a moving car is from a stoplight is given by the function

$$s(t) = 20t^{3/2}$$

where t is the time (in seconds). Find the average speed of the car (a) from $t_1 = 0$ to $t_2 = 4$ seconds and (b) from $t_1 = 4$ to $t_2 = 9$ seconds.

Solution

a. The average speed of the car from $t_1 = 0$ to $t_2 = 4$ seconds is

$$\frac{s(t_2) - s(t_1)}{t_2 - t_1} = \frac{s(4) - s(0)}{4 - 0} = \frac{160 - 0}{4} = 40 \text{ feet per second.}$$

b. The average speed of the car from $t_1 = 4$ to $t_2 = 9$ seconds is

$$\frac{s(t_2) - s(t_1)}{t_2 - t_1} = \frac{s(9) - s(4)}{9 - 4} = \frac{540 - 160}{5} = 76 \text{ feet per second.}$$

Average speed is an average rate of change.

✓ **Checkpoint**)))) *Audio-video solution in English & Spanish at LarsonPrecalculus.com*

In Example 7, find the average speed of the car (a) from $t_1 = 0$ to $t_2 = 1$ second and (b) from $t_1 = 1$ second to $t_2 = 4$ seconds. ∎

Even and Odd Functions

In Section 1.2, you studied different types of symmetry of a graph. In the terminology of functions, a function is said to be **even** when its graph is symmetric with respect to the y-axis and **odd** when its graph is symmetric with respect to the origin. The symmetry tests in Section 1.2 yield the tests for even and odd functions below.

Tests for Even and Odd Functions

A function $y = f(x)$ is **even** when, for each x in the domain of f, $f(-x) = f(x)$.

A function $y = f(x)$ is **odd** when, for each x in the domain of f, $f(-x) = -f(x)$.

EXAMPLE 8   **Even and Odd Functions**

See LarsonPrecalculus.com for an interactive version of this type of example.

a. The function $g(x) = x^3 - x$ is odd because $g(-x) = -g(x)$, as follows.

$$g(-x) = (-x)^3 - (-x) \qquad \text{Substitute } -x \text{ for } x.$$
$$= -x^3 + x \qquad \text{Simplify.}$$
$$= -(x^3 - x) \qquad \text{Distributive Property}$$
$$= -g(x) \qquad \text{Test for odd function}$$

b. The function $h(x) = x^2 + 1$ is even because $h(-x) = h(x)$, as follows.

$$h(-x) = (-x)^2 + 1 = x^2 + 1 = h(x) \qquad \text{Test for even function}$$

Figure 1.41 shows the graphs and symmetry of these two functions.

✓ **Checkpoint** 🔊))) *Audio-video solution in English & Spanish at LarsonPrecalculus.com*

Determine whether each function is even, odd, or neither. Then describe the symmetry.

a. $f(x) = 5 - 3x$ **b.** $g(x) = x^4 - x^2 - 1$ **c.** $h(x) = 2x^3 + 3x$ ◼

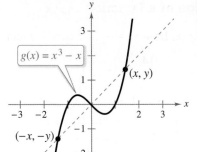

(a) Symmetric to origin: Odd Function

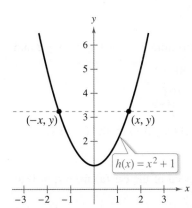

(b) Symmetric to y-axis: Even Function

Figure 1.41

Summarize   (Section 1.5)

1. State the Vertical Line Test for functions *(page 50)*. For an example of using the Vertical Line Test, see Example 2.

2. Explain how to find the zeros of a function *(page 51)*. For an example of finding the zeros of functions, see Example 3.

3. Explain how to determine intervals on which functions are increasing or decreasing *(page 52)*. For an example of describing function behavior, see Example 4.

4. Explain how to determine relative minimum and relative maximum values of functions *(page 53)*. For an example of approximating a relative minimum, see Example 5.

5. Explain how to determine the average rate of change of a function *(page 54)*. For examples of determining average rates of change, see Examples 6 and 7.

6. State the definitions of an even function and an odd function *(page 55)*. For an example of identifying even and odd functions, see Example 8.

1.5 Exercises

See CalcChat.com for tutorial help and worked-out solutions to odd-numbered exercises.

Vocabulary: Fill in the blanks.

1. The _____ _____ _____ is used to determine whether a graph represents y as a function of x.
2. The _____ of a function $y = f(x)$ are the values of x for which $f(x) = 0$.
3. A function f is _____ on an interval when, for any x_1 and x_2 in the interval, $x_1 < x_2$ implies $f(x_1) > f(x_2)$.
4. A function value $f(a)$ is a relative _____ of f when there exists an interval (x_1, x_2) containing a such that $x_1 < x < x_2$ implies $f(a) \geq f(x)$.
5. The _____ _____ _____ _____ between any two points $(x_1, f(x_1))$ and $(x_2, f(x_2))$ is the slope of the line through the two points, and this line is called the _____ line.
6. A function f is _____ when, for each x in the domain of f, $f(-x) = -f(x)$.

Skills and Applications

 Domain, Range, and Values of a Function In Exercises 7–10, use the graph of the function to find the domain and range of f and each function value.

7. (a) $f(-1)$ (b) $f(0)$ 8. (a) $f(-1)$ (b) $f(0)$
    (c) $f(1)$ (d) $f(2)$ (c) $f(1)$ (d) $f(3)$

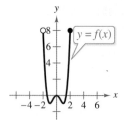 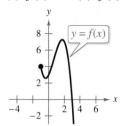

9. (a) $f(2)$ (b) $f(1)$ 10. (a) $f(-2)$ (b) $f(1)$
    (c) $f(3)$ (d) $f(-1)$ (c) $f(0)$ (d) $f(2)$

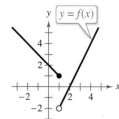 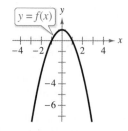

 Vertical Line Test for Functions In Exercises 11–14, use the Vertical Line Test to determine whether the graph represents y as a function of x. To print an enlarged copy of the graph, go to *MathGraphs.com*.

11.

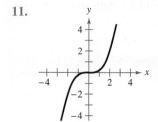

12.

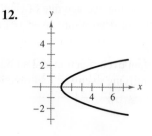

13.

14.

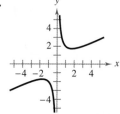

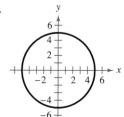 **Finding the Zeros of a Function** In Exercises 15–26, find the zeros of the function algebraically.

15. $f(x) = 3x + 18$
16. $f(x) = 15 - 2x$
17. $f(x) = 2x^2 - 7x - 30$
18. $f(x) = 3x^2 + 22x - 16$
19. $f(x) = \dfrac{x + 3}{2x^2 - 6}$
20. $f(x) = \dfrac{x^2 - 9x + 14}{4x}$
21. $f(x) = \frac{1}{3}x^3 - 2x$
22. $f(x) = -25x^4 + 9x^2$
23. $f(x) = x^3 - 4x^2 - 9x + 36$
24. $f(x) = 4x^3 - 24x^2 - x + 6$
25. $f(x) = \sqrt{2x} - 1$
26. $f(x) = \sqrt{3x + 2}$

Graphing and Finding Zeros In Exercises 27–32, (a) use a graphing utility to graph the function and find the zeros of the function and (b) verify your results from part (a) algebraically.

27. $f(x) = x^2 - 6x$ 28. $f(x) = 2x^2 - 13x - 7$
29. $f(x) = \sqrt{2x + 11}$ 30. $f(x) = \sqrt{3x - 14} - 8$
31. $f(x) = \dfrac{3x - 1}{x - 6}$ 32. $f(x) = \dfrac{2x^2 - 9}{3 - x}$

Pg. 56-57   1-46; 49-64   Joseph Atkinson

1.) Vertical line test

2.) zeroes

7.) $-2 < x \le 2$ Domain; range: $-1 \le x \le 8$; $a = -1$, $b = 0$, $c = -1$, $d = 8$

2nd catalog (0)   19.) $0 = \frac{x+3}{2x^2-6}(2x^2-6)$ $0 = x+3$ $\boxed{x = -3}$ $x+3 = 0$ $\boxed{x = -3}$

Asml prgrm   21.) $f(x) = \frac{1}{3}x^3 - 2x \rightarrow 0 = x^3 - 6x \rightarrow 0 = x(x^2-6)$ $\boxed{x = 0 \quad x = \pm\sqrt{6}}$

#65, #68   HOMEWORK
   (65A.)

68)
A. $s_0 = 6.5$; $V_0 = 72$ feet/sec
    $S = -16t^2 + 72t + 6.5$

Describing Function Behavior In Exercises 33–40, determine the open intervals on which the function is increasing, decreasing, or constant.

33. $f(x) = -\frac{1}{2}x^3$

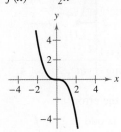

34. $f(x) = x^2 - 4x$

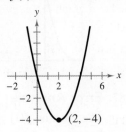

35. $f(x) = \sqrt{x^2 - 1}$

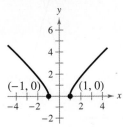

36. $f(x) = x^3 - 3x^2 + 2$

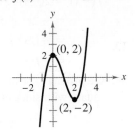

37. $f(x) = |x + 1| + |x - 1|$

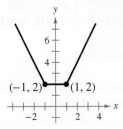

38. $f(x) = \dfrac{x^2 + x + 1}{x + 1}$

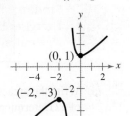

39. $f(x) = \begin{cases} 2x + 1, & x \le -1 \\ x^2 - 2, & x > -1 \end{cases}$

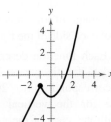

40. $f(x) = \begin{cases} x + 3, & x \le 0 \\ 3, & 0 < x \le 2 \\ 2x + 1, & x > 2 \end{cases}$

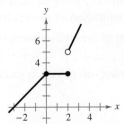

Describing Function Behavior In Exercises 41–48, use a graphing utility to graph the function and visually determine the open intervals on which the function is increasing, decreasing, or constant. Use a table of values to verify your results.

41. $f(x) = 3$ **42.** $g(x) = x$

43. $g(x) = \frac{1}{2}x^2 - 3$ **44.** $f(x) = 3x^4 - 6x^2$

45. $f(x) = \sqrt{1 - x}$ **46.** $f(x) = x\sqrt{x + 3}$

47. $f(x) = x^{3/2}$ **48.** $f(x) = x^{2/3}$

 Approximating Relative Minima or Maxima In Exercises 49–54, use a graphing utility to approximate (to two decimal places) any relative minima or maxima of the function.

49. $f(x) = x(x + 3)$

50. $f(x) = -x^2 + 3x - 2$

51. $h(x) = x^3 - 6x^2 + 15$

52. $f(x) = x^3 - 3x^2 - x + 1$

53. $h(x) = (x - 1)\sqrt{x}$

54. $g(x) = x\sqrt{4 - x}$

 Graphical Reasoning In Exercises 55–60, graph the function and determine the interval(s) for which $f(x) \ge 0$.

55. $f(x) = 4 - x$ **56.** $f(x) = 4x + 2$

57. $f(x) = 9 - x^2$ **58.** $f(x) = x^2 - 4x$

59. $f(x) = \sqrt{x - 1}$ **60.** $f(x) = |x + 5|$

 Average Rate of Change of a Function In Exercises 61–64, find the average rate of change of the function from x_1 to x_2.

Function	x-Values
61. $f(x) = -2x + 15$	$x_1 = 0, x_2 = 3$
62. $f(x) = x^2 - 2x + 8$	$x_1 = 1, x_2 = 5$
63. $f(x) = x^3 - 3x^2 - x$	$x_1 = -1, x_2 = 2$
64. $f(x) = -x^3 + 6x^2 + x$	$x_1 = 1, x_2 = 6$

65. Research and Development The amounts (in billions of dollars) the U.S. federal government spent on research and development for defense from 2010 through 2014 can be approximated by the model

$$y = 0.5079t^2 - 8.168t + 95.08$$

where t represents the year, with $t = 0$ corresponding to 2010. *(Source: American Association for the Advancement of Science)*

(a) Use a graphing utility to graph the model.

(b) Find the average rate of change of the model from 2010 to 2014. Interpret your answer in the context of the problem.

66. Finding Average Speed Use the information in Example 7 to find the average speed of the car from $t_1 = 0$ to $t_2 = 9$ seconds. Explain why the result is less than the value obtained in part (b) of Example 7.

 Physics In Exercises 67–70, (a) use the position equation $s = -16t^2 + v_0t + s_0$ to write a function that represents the situation, (b) use a graphing utility to graph the function, (c) find the average rate of change of the function from t_1 to t_2, (d) describe the slope of the secant line through t_1 and t_2, (e) find the equation of the secant line through t_1 and t_2, and (f) graph the secant line in the same viewing window as your position function.

67. An object is thrown upward from a height of 6 feet at a velocity of 64 feet per second.

$t_1 = 0, t_2 = 3$

68. An object is thrown upward from a height of 6.5 feet at a velocity of 72 feet per second.

$t_1 = 0, t_2 = 4$

69. An object is thrown upward from ground level at a velocity of 120 feet per second.

$t_1 = 3, t_2 = 5$

70. An object is dropped from a height of 80 feet.

$t_1 = 1, t_2 = 2$

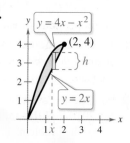 **Even, Odd, or Neither?** In Exercises 71–76, determine whether the function is even, odd, or neither. Then describe the symmetry.

71. $f(x) = x^6 - 2x^2 + 3$ **72.** $g(x) = x^3 - 5x$
73. $h(x) = x\sqrt{x + 5}$ **74.** $f(x) = x\sqrt{1 - x^2}$
75. $f(s) = 4s^{3/2}$ **76.** $g(s) = 4s^{2/3}$

Even, Odd, or Neither? In Exercises 77–82, sketch a graph of the function and determine whether it is even, odd, or neither. Verify your answer algebraically.

77. $f(x) = -9$ **78.** $f(x) = 5 - 3x$
79. $f(x) = -|x - 5|$ **80.** $h(x) = x^2 - 4$
81. $f(x) = \sqrt[3]{4x}$ **82.** $f(x) = \sqrt[3]{x - 4}$

Height of a Rectangle In Exercises 83 and 84, write the height h of the rectangle as a function of x.

83.
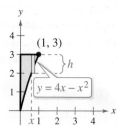

84.
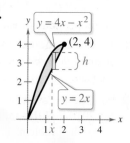

Length of a Rectangle In Exercises 85 and 86, write the length L of the rectangle as a function of y.

85.
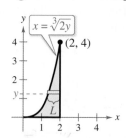

86.
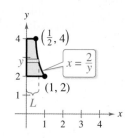

87. Error Analysis Describe the error.

The function $f(x) = 2x^3 - 5$ is odd because $f(-x) = -f(x)$, as follows.

$$f(-x) = 2(-x)^3 - 5$$
$$= -2x^3 - 5$$
$$= -(2x^3 - 5)$$
$$= -f(x)$$

88. Geometry Corners of equal size are cut from a square with sides of length 8 meters (see figure).

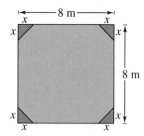

(a) Write the area A of the resulting figure as a function of x. Determine the domain of the function.

(b) Use a graphing utility to graph the area function over its domain. Use the graph to find the range of the function.

(c) Identify the figure that results when x is the maximum value in the domain of the function. What would be the length of each side of the figure?

89. Coordinate Axis Scale Each function described below models the specified data for the years 2006 through 2016, with $t = 6$ corresponding to 2006. Estimate a reasonable scale for the vertical axis (e.g., hundreds, thousands, millions, etc.) of the graph and justify your answer. (There are many correct answers.)

(a) $f(t)$ represents the average salary of college professors.

(b) $f(t)$ represents the U.S. population.

(c) $f(t)$ represents the percent of the civilian workforce that is unemployed.

(d) $f(t)$ represents the number of games a college football team wins.

90. Temperature

The table shows the temperatures y (in degrees Fahrenheit) in a city over a 24-hour period. Let x represent the time of day, where $x = 0$ corresponds to 6 A.M.

Time, x	Temperature, y
0	34
2	50
4	60
6	64
8	63
10	59
12	53
14	46
16	40
18	36
20	34
22	37
24	45

Spreadsheet at LarsonPrecalculus.com

These data can be approximated by the model

$$y = 0.026x^3 - 1.03x^2 + 10.2x + 34, \quad 0 \le x \le 24.$$

(a) Use a graphing utility to create a scatter plot of the data. Then graph the model in the same viewing window.

(b) How well does the model fit the data?

(c) Use the graph to approximate the times when the temperature was increasing and decreasing.

(d) Use the graph to approximate the maximum and minimum temperatures during this 24-hour period.

(e) Could this model predict the temperatures in the city during the next 24-hour period? Why or why not?

Exploration

True or False? In Exercises 91–93, determine whether the statement is true or false. Justify your answer.

91. A function with a square root cannot have a domain that is the set of real numbers.

92. It is possible for an odd function to have the interval $[0, \infty)$ as its domain.

93. It is impossible for an even function to be increasing on its entire domain.

94. **HOW DO YOU SEE IT?** Use the graph of the function to answer parts (a)–(e).

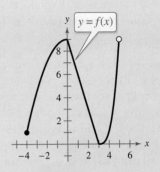

$y = f(x)$

(a) Find the domain and range of f.

(b) Find the zero(s) of f.

(c) Determine the open intervals on which f is increasing, decreasing, or constant.

(d) Approximate any relative minimum or relative maximum values of f.

(e) Is f even, odd, or neither?

Think About It In Exercises 95 and 96, find the coordinates of a second point on the graph of a function f when the given point is on the graph and the function is (a) even and (b) odd.

95. $\left(-\frac{5}{3}, -7\right)$ **96.** $(2a, 2c)$

97. Writing Use a graphing utility to graph each function. Write a paragraph describing any similarities and differences you observe among the graphs.

(a) $y = x$ (b) $y = x^2$ (c) $y = x^3$
(d) $y = x^4$ (e) $y = x^5$ (f) $y = x^6$

98. Graphical Reasoning Graph each of the functions with a graphing utility. Determine whether each function is even, odd, or neither.

$$f(x) = x^2 - x^4 \qquad g(x) = 2x^3 + 1$$
$$h(x) = x^5 - 2x^3 + x \qquad j(x) = 2 - x^6 - x^8$$
$$k(x) = x^5 - 2x^4 + x - 2 \qquad p(x) = x^9 + 3x^5 - x^3 + x$$

What do you notice about the equations of functions that are odd? What do you notice about the equations of functions that are even? Can you describe a way to identify a function as odd or even by inspecting the equation? Can you describe a way to identify a function as neither odd nor even by inspecting the equation?

99. Even, Odd, or Neither? Determine whether g is even, odd, or neither when f is an even function. Explain.

(a) $g(x) = -f(x)$ (b) $g(x) = f(-x)$
(c) $g(x) = f(x) - 2$ (d) $g(x) = f(x - 2)$

1.6 A Library of Parent Functions

Piecewise-defined functions model many real-life situations. For example, in Exercise 47 on page 66, you will write a piecewise-defined function to model the depth of snow during a snowstorm.

- ■ Identify and graph linear and squaring functions.
- ■ Identify and graph cubic, square root, and reciprocal functions.
- ■ Identify and graph step and other piecewise-defined functions.
- ■ Recognize graphs of parent functions.

Linear and Squaring Functions

One of the goals of this text is to enable you to recognize the basic shapes of the graphs of different types of functions. For example, you know that the graph of the **linear function** $f(x) = ax + b$ is a line with slope $m = a$ and y-intercept at $(0, b)$. The graph of a linear function has the characteristics below.

- • The domain of the function is the set of all real numbers.
- • When $m \neq 0$, the range of the function is the set of all real numbers.
- • The graph has an x-intercept at $(-b/m, 0)$ and a y-intercept at $(0, b)$.
- • The graph is increasing when $m > 0$, decreasing when $m < 0$, and constant when $m = 0$.

EXAMPLE 1   **Writing a Linear Function**

Write the linear function f for which $f(1) = 3$ and $f(4) = 0$.

Solution   To find the equation of the line that passes through $(x_1, y_1) = (1, 3)$ and $(x_2, y_2) = (4, 0)$, first find the slope of the line.

$$m = \frac{y_2 - y_1}{x_2 - x_1} = \frac{0 - 3}{4 - 1} = \frac{-3}{3} = -1$$

Next, use the point-slope form of the equation of a line.

$y - y_1 = m(x - x_1)$	Point-slope form
$y - 3 = -1(x - 1)$	Substitute for x_1, y_1, and m.
$y = -x + 4$	Simplify.
$f(x) = -x + 4$	Function notation

The figure below shows the graph of this function.

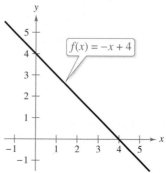

$f(x) = -x + 4$

✓ **Checkpoint** ◀))) *Audio-video solution in English & Spanish at LarsonPrecalculus.com*

Write the linear function f for which $f(-2) = 6$ and $f(4) = -9$.

There are two special types of linear functions, the **constant function** and the **identity function.** A constant function has the form

$$f(x) = c$$

and has a domain of all real numbers with a range consisting of a single real number c. The graph of a constant function is a horizontal line, as shown in Figure 1.42. The identity function has the form

$$f(x) = x.$$

Its domain and range are the set of all real numbers. The identity function has a slope of $m = 1$ and a y-intercept at $(0, 0)$. The graph of the identity function is a line for which each x-coordinate equals the corresponding y-coordinate. The graph is always increasing, as shown in Figure 1.43.

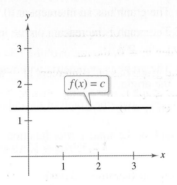

Figure 1.42

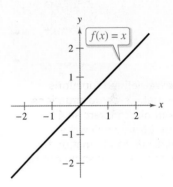

Figure 1.43

The graph of the **squaring function**

$$f(x) = x^2$$

is a U-shaped curve with the characteristics below.

- The domain of the function is the set of all real numbers.
- The range of the function is the set of all nonnegative real numbers.
- The function is even.
- The graph has an intercept at $(0, 0)$.
- The graph is decreasing on the interval $(-\infty, 0)$ and increasing on the interval $(0, \infty)$.
- The graph is symmetric with respect to the y-axis.
- The graph has a relative minimum at $(0, 0)$.

The figure below shows the graph of the squaring function.

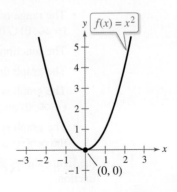

Cubic, Square Root, and Reciprocal Functions

Here are the basic characteristics of the graphs of the **cubic, square root,** and **reciprocal functions.**

1. The graph of the *cubic* function

$$f(x) = x^3$$

has the characteristics below.

- The domain of the function is the set of all real numbers.
- The range of the function is the set of all real numbers.
- The function is odd.
- The graph has an intercept at $(0, 0)$.
- The graph is increasing on the interval $(-\infty, \infty)$.
- The graph is symmetric with respect to the origin.

The figure shows the graph of the cubic function.

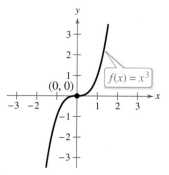

Cubic function

2. The graph of the *square root* function

$$f(x) = \sqrt{x}$$

has the characteristics below.

- The domain of the function is the set of all nonnegative real numbers.
- The range of the function is the set of all nonnegative real numbers.
- The graph has an intercept at $(0, 0)$.
- The graph is increasing on the interval $(0, \infty)$.

The figure shows the graph of the square root function.

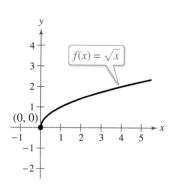

Square root function

3. The graph of the *reciprocal* function

$$f(x) = \frac{1}{x}$$

has the characteristics below.

- The domain of the function is $(-\infty, 0) \cup (0, \infty)$.
- The range of the function is $(-\infty, 0) \cup (0, \infty)$.
- The function is odd.
- The graph does not have any intercepts.
- The graph is decreasing on the intervals $(-\infty, 0)$ and $(0, \infty)$.
- The graph is symmetric with respect to the origin.

The figure shows the graph of the reciprocal function.

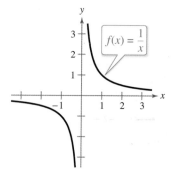

Reciprocal function

Step and Piecewise-Defined Functions

Functions whose graphs resemble sets of stairsteps are known as **step functions**. One common type of step function is the **greatest integer function,** denoted by $[\![x]\!]$ and defined as

$$f(x) = [\![x]\!] = \textit{the greatest integer less than or equal to } x.$$

Here are several examples of evaluating the greatest integer function.

$$[\![-1]\!] = (\text{greatest integer} \le -1) = -1$$
$$\left[\!\left[-\tfrac{1}{2}\right]\!\right] = \left(\text{greatest integer} \le -\tfrac{1}{2}\right) = -1$$
$$\left[\!\left[\tfrac{1}{10}\right]\!\right] = \left(\text{greatest integer} \le \tfrac{1}{10}\right) = 0$$
$$[\![1.5]\!] = (\text{greatest integer} \le 1.5) = 1$$
$$[\![1.9]\!] = (\text{greatest integer} \le 1.9) = 1$$

The graph of the greatest integer function

$$f(x) = [\![x]\!]$$

has the characteristics below, as shown in Figure 1.44.

- The domain of the function is the set of all real numbers.
- The range of the function is the set of all integers.
- The graph has a y-intercept at $(0, 0)$ and x-intercepts in the interval $[0, 1)$.
- The graph is constant between each pair of consecutive integer values of x.
- The graph jumps vertically one unit at each integer value of x.

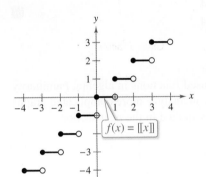

Figure 1.44

▷ TECHNOLOGY Most graphing utilities display graphs in *connected* mode, which works well for graphs that do not have breaks. For graphs that do have breaks, such as the graph of the greatest integer function, it may be better to use *dot* mode. Graph the greatest integer function [often called Int(x)] in *connected* and *dot* modes, and compare the two results.

EXAMPLE 2 **Evaluating a Step Function**

Evaluate the function $f(x) = [\![x]\!] + 1$ when $x = -1, 2,$ and $\tfrac{3}{2}$.

Solution For $x = -1$, the greatest integer ≤ -1 is -1, so

$$f(-1) = [\![-1]\!] + 1 = -1 + 1 = 0.$$

For $x = 2$, the greatest integer ≤ 2 is 2, so

$$f(2) = [\![2]\!] + 1 = 2 + 1 = 3.$$

For $x = \tfrac{3}{2}$, the greatest integer $\le \tfrac{3}{2}$ is 1, so

$$f\!\left(\tfrac{3}{2}\right) = \left[\!\left[\tfrac{3}{2}\right]\!\right] + 1 = 1 + 1 = 2.$$

Verify your answers by examining the graph of $f(x) = [\![x]\!] + 1$ shown in Figure 1.45.

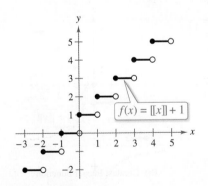

Figure 1.45

✓ *Checkpoint* ◀))) *Audio-video solution in English & Spanish at LarsonPrecalculus.com*

Evaluate the function $f(x) = [\![x + 2]\!]$ when $x = -\tfrac{3}{2}, 1,$ and $-\tfrac{5}{2}$. ■

Recall from Section 1.4 that a piecewise-defined function is defined by two or more equations over a specified domain. To graph a piecewise-defined function, graph each equation separately over the specified domain, as shown in Example 3.

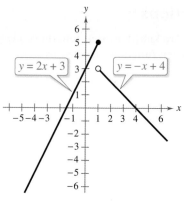

Figure 1.46

EXAMPLE 3 **Graphing a Piecewise-Defined Function**

See LarsonPrecalculus.com for an interactive version of this type of example.

Sketch the graph of $f(x) = \begin{cases} 2x + 3, & x \le 1 \\ -x + 4, & x > 1 \end{cases}$.

Solution This piecewise-defined function consists of two linear functions. At $x = 1$ and to the left of $x = 1$, the graph is the line $y = 2x + 3$, and to the right of $x = 1$, the graph is the line $y = -x + 4$, as shown in Figure 1.46. Notice that the point $(1, 5)$ is a solid dot and the point $(1, 3)$ is an open dot. This is because $f(1) = 2(1) + 3 = 5$.

✓ *Checkpoint* ◀))) Audio-video solution in English & Spanish at LarsonPrecalculus.com

Sketch the graph of $f(x) = \begin{cases} -\frac{1}{2}x - 6, & x \le -4 \\ x + 5, & x > -4 \end{cases}$.

Parent Functions

The graphs below represent the most commonly used functions in algebra. Familiarity with the characteristics of these graphs will help you analyze more complicated graphs obtained from these graphs by the transformations studied in the next section.

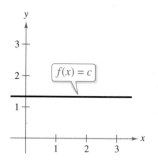

(a) Constant Function

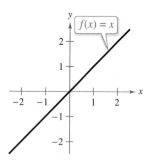

(b) Identity Function

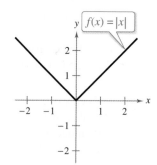

(c) Absolute Value Function

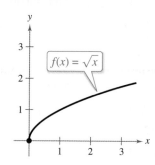

(d) Square Root Function

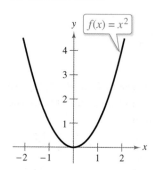

(e) Squaring Function

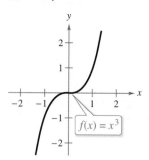

(f) Cubic Function

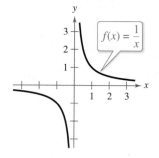

(g) Reciprocal Function

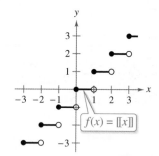

(h) Greatest Integer Function

Summarize (Section 1.6)

1. Explain how to identify and graph linear and squaring functions (*pages 60 and 61*). For an example involving a linear function, see Example 1.

2. Explain how to identify and graph cubic, square root, and reciprocal functions (*page 62*).

3. Explain how to identify and graph step and other piecewise-defined functions (*page 63*). For examples involving these functions, see Examples 2 and 3.

4. Identify and sketch the graphs of parent functions (*page 64*).

1.6 Exercises

See **CalcChat.com** for tutorial help and worked-out solutions to odd-numbered exercises.

Vocabulary

In Exercises 1–9, write the most specific name of the function.

1. $f(x) = [\![x]\!]$

2. $f(x) = x$

3. $f(x) = 1/x$

4. $f(x) = x^2$

5. $f(x) = \sqrt{x}$

6. $f(x) = c$

7. $f(x) = |x|$

8. $f(x) = x^3$

9. $f(x) = ax + b$

10. Fill in the blank: The constant function and the identity function are two special types of _____ functions.

Skills and Applications

 Writing a Linear Function In Exercises 11–14, (a) write the linear function f that has the given function values and (b) sketch the graph of the function.

11. $f(1) = 4$, $f(0) = 6$

12. $f(-3) = -8$, $f(1) = 2$

13. $f\left(\frac{1}{2}\right) = -\frac{5}{3}$, $f(6) = 2$

14. $f\left(\frac{3}{5}\right) = \frac{1}{2}$, $f(4) = 9$

Graphing a Function In Exercises 15–26, use a graphing utility to graph the function. Be sure to choose an appropriate viewing window.

15. $f(x) = 2.5x - 4.25$

16. $f(x) = \frac{5}{6} - \frac{2}{3}x$

17. $g(x) = x^2 + 3$

18. $f(x) = -2x^2 - 1$

19. $f(x) = x^3 - 1$

20. $f(x) = (x - 1)^3 + 2$

21. $f(x) = \sqrt{x} + 4$

22. $h(x) = \sqrt{x + 2} + 3$

23. $f(x) = \dfrac{1}{x - 2}$

24. $k(x) = 3 + \dfrac{1}{x + 3}$

25. $g(x) = |x| - 5$

26. $f(x) = |x - 1|$

 Evaluating a Step Function In Exercises 27–30, evaluate the function for the given values.

27. $f(x) = [\![x]\!]$

 (a) $f(2.1)$ (b) $f(2.9)$ (c) $f(-3.1)$ (d) $f\left(\frac{7}{2}\right)$

28. $h(x) = [\![x + 3]\!]$

 (a) $h(-2)$ (b) $h\left(\frac{1}{2}\right)$ (c) $h(4.2)$ (d) $h(-21.6)$

29. $k(x) = [\![2x + 1]\!]$

 (a) $k\left(\frac{1}{3}\right)$ (b) $k(-2.1)$ (c) $k(1.1)$ (d) $k\left(\frac{2}{3}\right)$

30. $g(x) = -7[\![x + 4]\!] + 6$

 (a) $g\left(\frac{1}{8}\right)$ (b) $g(9)$ (c) $g(-4)$ (d) $g\left(\frac{3}{2}\right)$

Graphing a Step Function In Exercises 31–34, sketch the graph of the function.

31. $g(x) = -[\![x]\!]$

32. $g(x) = 4[\![x]\!]$

33. $g(x) = [\![x]\!] - 1$

34. $g(x) = [\![x - 3]\!]$

Graphing a Piecewise-Defined Function In Exercises 35–40, sketch the graph of the function.

35.  $g(x) = \begin{cases} x + 6, & x \le -4 \\ \frac{1}{2}x - 4, & x > -4 \end{cases}$

36. $f(x) = \begin{cases} 4 + x, & x \le 2 \\ x^2 + 2, & x > 2 \end{cases}$

37. $f(x) = \begin{cases} 1 - (x - 1)^2, & x \le 2 \\ \sqrt{x - 2}, & x > 2 \end{cases}$

38. $f(x) = \begin{cases} \sqrt{4 + x}, & x < 0 \\ \sqrt{4 - x}, & x \ge 0 \end{cases}$

39. $h(x) = \begin{cases} 4 - x^2, & x < -2 \\ 3 + x, & -2 \le x < 0 \\ x^2 + 1, & x \ge 0 \end{cases}$

40. $k(x) = \begin{cases} 2x + 1, & x \le -1 \\ 2x^2 - 1, & -1 < x \le 1 \\ 1 - x^2, & x > 1 \end{cases}$

Graphing a Function In Exercises 41 and 42, (a) use a graphing utility to graph the function and (b) state the domain and range of the function.

41. $s(x) = 2\left(\frac{1}{4}x - [\![\frac{1}{4}x]\!]\right)$

42. $k(x) = 4\left(\frac{1}{2}x - [\![\frac{1}{2}x]\!]\right)^2$

43. Wages A mechanic's pay is $14 per hour for regular time and time-and-a-half for overtime. The weekly wage function is

$$W(h) = \begin{cases} 14h, & 0 < h \le 40 \\ 21(h - 40) + 560, & h > 40 \end{cases}$$

where h is the number of hours worked in a week.

(a) Evaluate $W(30)$, $W(40)$, $W(45)$, and $W(50)$.

(b) The company decreases the regular work week to 36 hours. What is the new weekly wage function?

(c) The company increases the mechanic's pay to $16 per hour. What is the new weekly wage function? Use a regular work week of 40 hours.

44. Revenue The table shows the monthly revenue y (in thousands of dollars) of a landscaping business for each month of the year 2016, with $x = 1$ representing January.

DATA	Month, x	Revenue, y
	1	5.2
	2	5.6
	3	6.6
	4	8.3
	5	11.5
	6	15.8
	7	12.8
	8	10.1
	9	8.6
	10	6.9
	11	4.5
	12	2.7

Spreadsheet at LarsonPrecalculus.com

A mathematical model that represents these data is

$$f(x) = \begin{cases} -1.97x + 26.3 \\ 0.505x^2 - 1.47x + 6.3 \end{cases}.$$

(a) Use a graphing utility to graph the model. What is the domain of each part of the piecewise-defined function? How can you tell?

(b) Find $f(5)$ and $f(11)$ and interpret your results in the context of the problem.

(c) How do the values obtained from the model in part (b) compare with the actual data values?

45. Fluid Flow The intake pipe of a 100-gallon tank has a flow rate of 10 gallons per minute, and two drainpipes have flow rates of 5 gallons per minute each. The figure shows the volume V of fluid in the tank as a function of time t. Determine whether the input pipe and each drainpipe are open or closed in specific subintervals of the 1 hour of time shown in the graph. (There are many correct answers.)

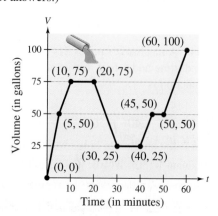

46. Delivery Charges The cost of mailing a package weighing up to, but not including, 1 pound is \$2.72. Each additional pound or portion of a pound costs \$0.50.

(a) Use the greatest integer function to create a model for the cost C of mailing a package weighing x pounds, where $x > 0$.

(b) Sketch the graph of the function.

• • 47. Snowstorm • • • • • • • • • • • • • • •

During a nine-hour snowstorm, it snows at a rate of 1 inch per hour for the first 2 hours, at a rate of 2 inches per hour for the next 6 hours, and at a rate of 0.5 inch per hour for the final hour. Write and graph a piecewise-defined function that gives the depth of the snow during the snowstorm. How many inches of snow accumulated from the storm?

48. HOW DO YOU SEE IT? For each graph of f shown below, answer parts (a)–(d).

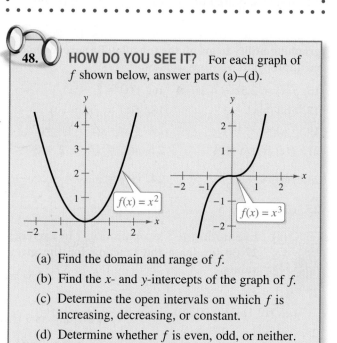

(a) Find the domain and range of f.

(b) Find the x- and y-intercepts of the graph of f.

(c) Determine the open intervals on which f is increasing, decreasing, or constant.

(d) Determine whether f is even, odd, or neither. Then describe the symmetry.

Exploration

True or False? In Exercises 49 and 50, determine whether the statement is true or false. Justify your answer.

49. A piecewise-defined function will always have at least one x-intercept or at least one y-intercept.

50. A linear equation will always have an x-intercept and a y-intercept.

1.7 Transformations of Functions

Transformations of functions model many real-life applications. For example, in Exercise 61 on page 74, you will use a transformation of a function to model the number of horsepower required to overcome wind drag on an automobile.

■ **Use vertical and horizontal shifts to sketch graphs of functions.**
■ **Use reflections to sketch graphs of functions.**
■ **Use nonrigid transformations to sketch graphs of functions.**

Shifting Graphs

Many functions have graphs that are transformations of the parent graphs summarized in Section 1.6. For example, you obtain the graph of

$$h(x) = x^2 + 2$$

by shifting the graph of $f(x) = x^2$ *up* two units, as shown in Figure 1.47. In function notation, h and f are related as follows.

$$h(x) = x^2 + 2 = f(x) + 2 \qquad \text{Upward shift of two units}$$

Similarly, you obtain the graph of

$$g(x) = (x - 2)^2$$

by shifting the graph of $f(x) = x^2$ to the *right* two units, as shown in Figure 1.48. In this case, the functions g and f have the following relationship.

$$g(x) = (x - 2)^2 = f(x - 2) \qquad \text{Right shift of two units}$$

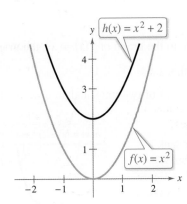

Figure 1.47

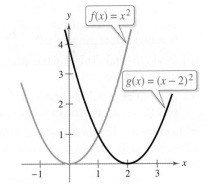

Figure 1.48

The list below summarizes this discussion about horizontal and vertical shifts.

• • REMARK In items 3 and 4, be sure you see that $h(x) = f(x - c)$ corresponds to a *right* shift and $h(x) = f(x + c)$ corresponds to a *left* shift for $c > 0$.

Vertical and Horizontal Shifts

Let c be a positive real number. **Vertical and horizontal shifts** in the graph of $y = f(x)$ are represented as follows.

1. Vertical shift c units *up:* $h(x) = f(x) + c$

2. Vertical shift c units *down:* $h(x) = f(x) - c$

3. Horizontal shift c units to the *right:* $h(x) = f(x - c)$

4. Horizontal shift c units to the *left:* $h(x) = f(x + c)$

Some graphs are obtained from combinations of vertical and horizontal shifts, as demonstrated in Example 1(b). Vertical and horizontal shifts generate a *family of functions*, each with the same shape but at a different location in the plane.

> **EXAMPLE 1** **Shifting the Graph of a Function**

Use the graph of $f(x) = x^3$ to sketch the graph of each function.

a. $g(x) = x^3 - 1$

b. $h(x) = (x + 2)^3 + 1$

Solution

a. Relative to the graph of $f(x) = x^3$, the graph of

$$g(x) = x^3 - 1$$

is a downward shift of one unit, as shown below.

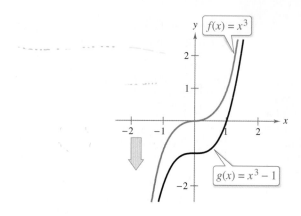

b. Relative to the graph of $f(x) = x^3$, the graph of

$$h(x) = (x + 2)^3 + 1$$

is a left shift of two units and an upward shift of one unit, as shown below.

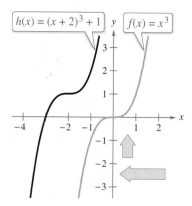

✓ **Checkpoint** 🔊))) *Audio-video solution in English & Spanish at LarsonPrecalculus.com*

Use the graph of $f(x) = x^3$ to sketch the graph of each function.

a. $h(x) = x^3 + 5$

b. $g(x) = (x - 3)^3 + 2$

In Example 1(a), note that $g(x) = f(x) - 1$ and in Example 1(b), $h(x) = f(x + 2) + 1$. In Example 1(b), you obtain the same result whether the vertical shift precedes the horizontal shift or the horizontal shift precedes the vertical shift.

Reflecting Graphs

Another common type of transformation is a **reflection.** For example, if you consider the x-axis to be a mirror, then the graph of $h(x) = -x^2$ is the mirror image (or reflection) of the graph of $f(x) = x^2$, as shown in Figure 1.49.

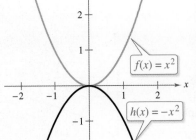

$f(x) = x^2$

$h(x) = -x^2$

Figure 1.49

Reflections in the Coordinate Axes

Reflections in the coordinate axes of the graph of $y = f(x)$ are represented as follows.

1. Reflection in the x-axis: $h(x) = -f(x)$

2. Reflection in the y-axis: $h(x) = f(-x)$

EXAMPLE 2   **Writing Equations from Graphs**

The graph of the function

$$f(x) = x^4$$

is shown in Figure 1.50. Each graph below is a transformation of the graph of f. Write an equation for the function represented by each graph.

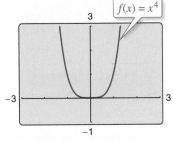

$f(x) = x^4$

Figure 1.50

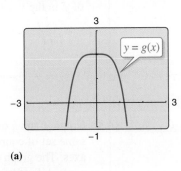

$y = g(x)$

(a)

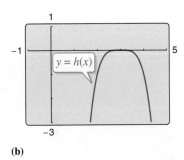

$y = h(x)$

(b)

Solution

a. The graph of g is a reflection in the x-axis *followed by* an upward shift of two units of the graph of $f(x) = x^4$. So, an equation for g is

$$g(x) = -x^4 + 2.$$

b. The graph of h is a right shift of three units *followed by* a reflection in the x-axis of the graph of $f(x) = x^4$. So, an equation for h is

$$h(x) = -(x - 3)^4.$$

✓ **Checkpoint** 🔊))) *Audio-video solution in English & Spanish at LarsonPrecalculus.com*

The graph is a transformation of the graph of $f(x) = x^4$. Write an equation for the function represented by the graph.

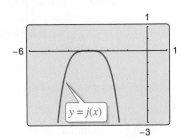

$y = j(x)$

EXAMPLE 3   **Reflections and Shifts**

Compare the graph of each function with the graph of $f(x) = \sqrt{x}$.

a. $g(x) = -\sqrt{x}$ **b.** $h(x) = \sqrt{-x}$ **c.** $k(x) = -\sqrt{x+2}$

Algebraic Solution

a. The graph of g is a reflection of the graph of f in the x-axis because

$$g(x) = -\sqrt{x}$$
$$= -f(x).$$

b. The graph of h is a reflection of the graph of f in the y-axis because

$$h(x) = \sqrt{-x}$$
$$= f(-x).$$

c. The graph of k is a left shift of two units followed by a reflection in the x-axis because

$$k(x) = -\sqrt{x+2}$$
$$= -f(x+2).$$

Graphical Solution

a. Graph f and g on the same set of coordinate axes. The graph of g is a reflection of the graph of f in the x-axis.

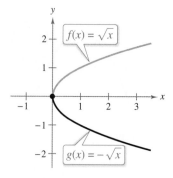

b. Graph f and h on the same set of coordinate axes. The graph of h is a reflection of the graph of f in the y-axis.

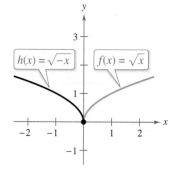

c. Graph f and k on the same set of coordinate axes. The graph of k is a left shift of two units followed by a reflection in the x-axis of the graph of f.

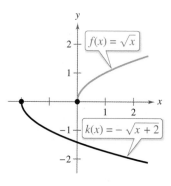

✓ *Checkpoint* *Audio-video solution in English & Spanish at LarsonPrecalculus.com*

Compare the graph of each function with the graph of

$$f(x) = \sqrt{x-1}.$$

a. $g(x) = -\sqrt{x-1}$ **b.** $h(x) = \sqrt{-x-1}$

When sketching the graphs of functions involving square roots, remember that you must restrict the domain to exclude negative numbers inside the radical. For instance, here are the domains of the functions in Example 3.

Domain of $g(x) = -\sqrt{x}$: $x \geq 0$

Domain of $h(x) = \sqrt{-x}$: $x \leq 0$

Domain of $k(x) = -\sqrt{x+2}$: $x \geq -2$

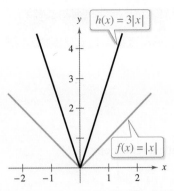

Figure 1.51

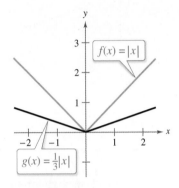

Figure 1.52

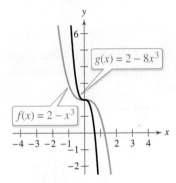

Figure 1.53

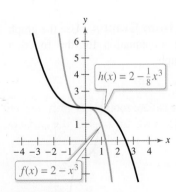

Figure 1.54

Nonrigid Transformations

Horizontal shifts, vertical shifts, and reflections are **rigid transformations** because the basic shape of the graph is unchanged. These transformations change only the *position* of the graph in the coordinate plane. **Nonrigid transformations** are those that cause a *distortion*—a change in the shape of the original graph. For example, a nonrigid transformation of the graph of $y = f(x)$ is represented by $g(x) = cf(x)$, where the transformation is a **vertical stretch** when $c > 1$ and a **vertical shrink** when $0 < c < 1$. Another nonrigid transformation of the graph of $y = f(x)$ is represented by $h(x) = f(cx)$, where the transformation is a **horizontal shrink** when $c > 1$ and a **horizontal stretch** when $0 < c < 1$.

EXAMPLE 4 **Nonrigid Transformations**

Compare the graph of each function with the graph of $f(x) = |x|$.

  a. $h(x) = 3|x|$ **b.** $g(x) = \frac{1}{3}|x|$

Solution

  a. Relative to the graph of $f(x) = |x|$, the graph of $h(x) = 3|x| = 3f(x)$ is a vertical stretch (each y-value is multiplied by 3). (See Figure 1.51.)
  b. Similarly, the graph of $g(x) = \frac{1}{3}|x| = \frac{1}{3}f(x)$ is a vertical shrink $\left(\text{each } y\text{-value is multiplied by } \frac{1}{3}\right)$ of the graph of f. (See Figure 1.52.)

✓ ***Checkpoint*** 🔊))) *Audio-video solution in English & Spanish at LarsonPrecalculus.com*

Compare the graph of each function with the graph of $f(x) = x^2$.

  a. $g(x) = 4x^2$ **b.** $h(x) = \frac{1}{4}x^2$

EXAMPLE 5 **Nonrigid Transformations**

See LarsonPrecalculus.com for an interactive version of this type of example.

Compare the graph of each function with the graph of $f(x) = 2 - x^3$.

  a. $g(x) = f(2x)$ **b.** $h(x) = f\left(\frac{1}{2}x\right)$

Solution

  a. Relative to the graph of $f(x) = 2 - x^3$, the graph of $g(x) = f(2x) = 2 - (2x)^3 = 2 - 8x^3$ is a horizontal shrink ($c > 1$). (See Figure 1.53.)
  b. Similarly, the graph of $h(x) = f\left(\frac{1}{2}x\right) = 2 - \left(\frac{1}{2}x\right)^3 = 2 - \frac{1}{8}x^3$ is a horizontal stretch ($0 < c < 1$) of the graph of f. (See Figure 1.54.)

✓ ***Checkpoint*** 🔊))) *Audio-video solution in English & Spanish at LarsonPrecalculus.com*

Compare the graph of each function with the graph of $f(x) = x^2 + 3$.

  a. $g(x) = f(2x)$ **b.** $h(x) = f\left(\frac{1}{2}x\right)$

Summarize   (Section 1.7)

1. Explain how to shift the graph of a function vertically and horizontally *(page 67)*. For an example of shifting the graph of a function, see Example 1.
2. Explain how to reflect the graph of a function in the x-axis and in the y-axis *(page 69)*. For examples of reflecting graphs of functions, see Examples 2 and 3.
3. Describe nonrigid transformations of the graph of a function *(page 71)*. For examples of nonrigid transformations, see Examples 4 and 5.

1.7 Exercises

See **CalcChat.com** for tutorial help and worked-out solutions to odd-numbered exercises.

Vocabulary

In Exercises 1–3, fill in the blanks.

1. Horizontal shifts, vertical shifts, and reflections are _____ transformations.

2. A reflection in the x-axis of the graph of $y = f(x)$ is represented by $h(x) = $ _____, while a reflection in the y-axis of the graph of $y = f(x)$ is represented by $h(x) = $ _____.

3. A nonrigid transformation of the graph of $y = f(x)$ represented by $g(x) = cf(x)$ is a _____ _____ when $c > 1$ and a _____ _____ when $0 < c < 1$.

4. Match each function h with the transformation it represents, where $c > 0$.
   (a) $h(x) = f(x) + c$ (i) A horizontal shift of f, c units to the right
   (b) $h(x) = f(x) - c$ (ii) A vertical shift of f, c units down
   (c) $h(x) = f(x + c)$ (iii) A horizontal shift of f, c units to the left
   (d) $h(x) = f(x - c)$ (iv) A vertical shift of f, c units up

Skills and Applications

5. **Shifting the Graph of a Function** For each function, sketch the graphs of the function when $c = -2$, -1, 1, and 2 on the same set of coordinate axes.
   (a) $f(x) = |x| + c$ (b) $f(x) = |x - c|$

6. **Shifting the Graph of a Function** For each function, sketch the graphs of the function when $c = -3$, -2, 2, and 3 on the same set of coordinate axes.
   (a) $f(x) = \sqrt{x} + c$ (b) $f(x) = \sqrt{x - c}$

7. **Shifting the Graph of a Function** For each function, sketch the graphs of the function when $c = -4$, -1, 2, and 5 on the same set of coordinate axes.
   (a) $f(x) = [\![x]\!] + c$ (b) $f(x) = [\![x + c]\!]$

8. **Shifting the Graph of a Function** For each function, sketch the graphs of the function when $c = -3$, -2, 1, and 2 on the same set of coordinate axes.
   (a) $f(x) = \begin{cases} x^2 + c, & x < 0 \\ -x^2 + c, & x \geq 0 \end{cases}$

   (b) $f(x) = \begin{cases} (x + c)^2, & x < 0 \\ -(x + c)^2, & x \geq 0 \end{cases}$

Sketching Transformations In Exercises 9 and 10, use the graph of f to sketch each graph. To print an enlarged copy of the graph, go to *MathGraphs.com*.

9. (a) $y = f(-x)$
   (b) $y = f(x) + 4$
   (c) $y = 2f(x)$
   (d) $y = -f(x - 4)$
   (e) $y = f(x) - 3$
   (f) $y = -f(x) - 1$
   (g) $y = f(2x)$

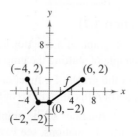

10. (a) $y = f(x - 5)$
    (b) $y = -f(x) + 3$
    (c) $y = \frac{1}{3}f(x)$
    (d) $y = -f(x + 1)$
    (e) $y = f(-x)$
    (f) $y = f(x) - 10$
    (g) $y = f\left(\frac{1}{3}x\right)$

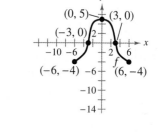

11. **Writing Equations from Graphs** Use the graph of $f(x) = x^2$ to write an equation for the function represented by each graph.
    (a)   (b)

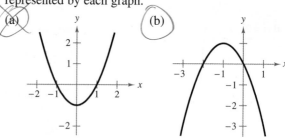

12. **Writing Equations from Graphs** Use the graph of $f(x) = x^3$ to write an equation for the function represented by each graph.
    (a)   (b)

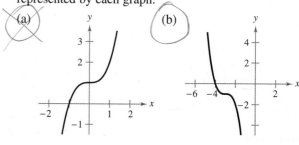

13. Writing Equations from Graphs Use the graph of $f(x) = |x|$ to write an equation for the function represented by each graph.

(a)   (b)

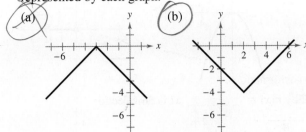

14. Writing Equations from Graphs Use the graph of $f(x) = \sqrt{x}$ to write an equation for the function represented by each graph.

(a)   (b)

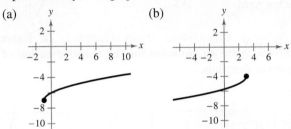

 Writing Equations from Graphs In Exercises 15–20, identify the parent function and the transformation represented by the graph. Write an equation for the function represented by the graph.

15. **16.**

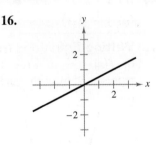

17. **18.**

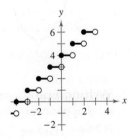

19. **20.**

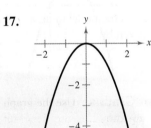 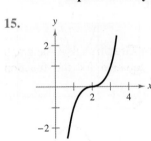

 Describing Transformations In Exercises 21–38, g is related to one of the parent functions described in Section 1.6. (a) Identify the parent function f. (b) Describe the sequence of transformations from f to g. (c) Sketch the graph of g. (d) Use function notation to write g in terms of f.

21. $g(x) = x^2 + 6$ **22.** $g(x) = x^2 - 2$

23. $g(x) = -(x - 2)^3$ **24.** $g(x) = -(x + 1)^3$

25. $g(x) = -3 - (x + 1)^2$

26. $g(x) = 4 - (x - 2)^2$

27. $g(x) = |x - 1| + 2$ **28.** $g(x) = |x + 3| - 2$

29. $g(x) = 2\sqrt{x}$ **30.** $g(x) = \frac{1}{2}\sqrt{x}$

31. $g(x) = 2[\![x]\!] - 1$ **32.** $g(x) = -[\![x]\!] + 1$

33. $g(x) = |2x|$ **34.** $g(x) = \left|\frac{1}{2}x\right|$

35. $g(x) = -2x^2 + 1$ **36.** $g(x) = \frac{1}{2}x^2 - 2$

37. $g(x) = 3|x - 1| + 2$

38. $g(x) = -2|x + 1| - 3$

 Writing an Equation from a Description In Exercises 39–46, write an equation for the function whose graph is described.

39. The shape of $f(x) = x^2$, but shifted three units to the right and seven units down

40. The shape of $f(x) = x^2$, but shifted two units to the left, nine units up, and then reflected in the x-axis

41. The shape of $f(x) = x^3$, but shifted 13 units to the right

42. The shape of $f(x) = x^3$, but shifted six units to the left, six units down, and then reflected in the y-axis

43. The shape of $f(x) = |x|$, but shifted 12 units up and then reflected in the x-axis

44. The shape of $f(x) = |x|$, but shifted four units to the left and eight units down

45. The shape of $f(x) = \sqrt{x}$, but shifted six units to the left and then reflected in both the x-axis and the y-axis

46. The shape of $f(x) = \sqrt{x}$, but shifted nine units down and then reflected in both the x-axis and the y-axis

47. Writing Equations from Graphs Use the graph of $f(x) = x^2$ to write an equation for the function represented by each graph.

(a)   (b)

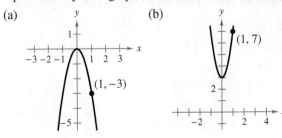

48. Writing Equations from Graphs Use the graph of

$f(x) = x^3$

to write an equation for the function represented by each graph.

(a)
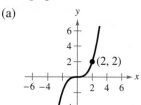
(2, 2)

(b)
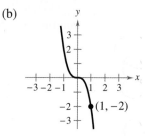
(1, −2)

49. Writing Equations from Graphs Use the graph of

$f(x) = |x|$

to write an equation for the function represented by each graph.

(a)
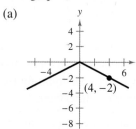
(4, −2)

(b)
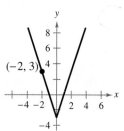
(−2, 3)

50. Writing Equations from Graphs Use the graph of

$f(x) = \sqrt{x}$

to write an equation for the function represented by each graph.

(a)
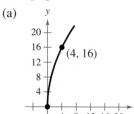
(4, 16)

(b)
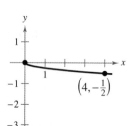
$\left(4, -\frac{1}{2}\right)$

Writing Equations from Graphs In Exercises 51–56, identify the parent function and the transformation represented by the graph. Write an equation for the function represented by the graph. Then use a graphing utility to verify your answer.

51.
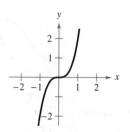

52.
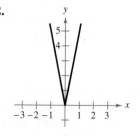

53.
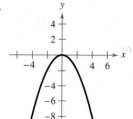

54.
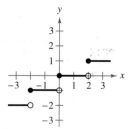

55.
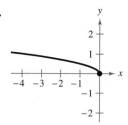

56.
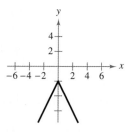

Writing Equations from Graphs In Exercises 57–60, write an equation for the transformation of the parent function.

57.
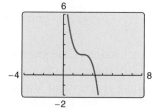

58.
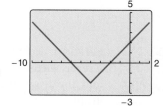

59.
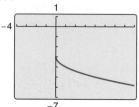

60.
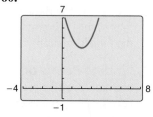

61. Automobile Aerodynamics

The horsepower H required to overcome wind drag on a particular automobile is given by

$H(x) = 0.00004636x^3$

where x is the speed of the car (in miles per hour).

(a) Use a graphing utility to graph the function.

(b) Rewrite the horsepower function so that x represents the speed in kilometers per hour. [Find $H(x/1.6)$.] Identify the type of transformation applied to the graph of the horsepower function.

62. Households The number N (in millions) of households in the United States from 2000 through 2014 can be approximated by

$$N(x) = -0.023(x - 33.12)^2 + 131, \quad 0 \le t \le 14$$

where t represents the year, with $t = 0$ corresponding to 2000. *(Source: U.S. Census Bureau)*

(a) Describe the transformation of the parent function $f(x) = x^2$. Then use a graphing utility to graph the function over the specified domain.

(b) Find the average rate of change of the function from 2000 to 2014. Interpret your answer in the context of the problem.

(c) Use the model to predict the number of households in the United States in 2022. Does your answer seem reasonable? Explain.

Exploration

True or False? In Exercises 63–66, determine whether the statement is true or false. Justify your answer.

63. The graph of $y = f(-x)$ is a reflection of the graph of $y = f(x)$ in the x-axis.

64. The graph of $y = -f(x)$ is a reflection of the graph of $y = f(x)$ in the y-axis.

65. The graphs of $f(x) = |x| + 6$ and $f(x) = |-x| + 6$ are identical.

66. If the graph of the parent function $f(x) = x^2$ is shifted six units to the right, three units up, and reflected in the x-axis, then the point $(-2, 19)$ will lie on the graph of the transformation.

67. Finding Points on a Graph The graph of $y = f(x)$ passes through the points $(0, 1)$, $(1, 2)$, and $(2, 3)$. Find the corresponding points on the graph of $y = f(x + 2) - 1$.

68. Think About It Two methods of graphing a function are plotting points and translating a parent function as shown in this section. Which method of graphing do you prefer to use for each function? Explain.

(a) $f(x) = 3x^2 - 4x + 1$ (b) $f(x) = 2(x - 1)^2 - 6$

69. Error Analysis Describe the error.

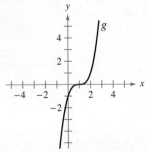

The graph of g is a right shift of one unit of the graph of $f(x) = x^3$. So, an equation for g is $g(x) = (x + 1)^3$.

70. **HOW DO YOU SEE IT?** Use the graph of $y = f(x)$ to find the open intervals on which the graph of each transformation is increasing and decreasing. If not possible, state the reason.

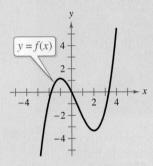

(a) $y = f(-x)$ (b) $y = -f(x)$ (c) $y = \frac{1}{2}f(x)$

(d) $y = -f(x - 1)$ (e) $y = f(x - 2) + 1$

71. Describing Profits Management originally predicted that the profits from the sales of a new product could be approximated by the graph of the function f shown. The actual profits are represented by the graph of the function g along with a verbal description. Use the concepts of transformations of graphs to write g in terms of f.

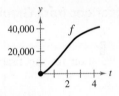

(a) The profits were only three-fourths as large as expected.

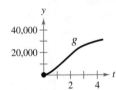

(b) The profits were consistently $10,000 greater than predicted.

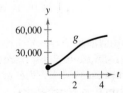

(c) There was a two-year delay in the introduction of the product. After sales began, profits grew as expected.

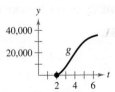

72. Reversing the Order of Transformations Reverse the order of transformations in Example 2(a). Do you obtain the same graph? Do the same for Example 2(b). Do you obtain the same graph? Explain.

1.8 Combinations of Functions: Composite Functions

Arithmetic combinations of functions are used to model and solve real-life problems. For example, in Exercise 60 on page 82, you will use arithmetic combinations of functions to analyze numbers of pets in the United States.

■ Add, subtract, multiply, and divide functions.
■ Find the composition of one function with another function.
■ Use combinations and compositions of functions to model and solve real-life problems.

Arithmetic Combinations of Functions

Just as two real numbers can be combined by the operations of addition, subtraction, multiplication, and division to form other real numbers, two *functions* can be combined to create new functions. For example, the functions $f(x) = 2x - 3$ and $g(x) = x^2 - 1$ can be combined to form the sum, difference, product, and quotient of f and g.

$$f(x) + g(x) = (2x - 3) + (x^2 - 1) = x^2 + 2x - 4 \qquad \text{Sum}$$

$$f(x) - g(x) = (2x - 3) - (x^2 - 1) = -x^2 + 2x - 2 \qquad \text{Difference}$$

$$f(x)g(x) = (2x - 3)(x^2 - 1) = 2x^3 - 3x^2 - 2x + 3 \qquad \text{Product}$$

$$\frac{f(x)}{g(x)} = \frac{2x - 3}{x^2 - 1}, \quad x \neq \pm 1 \qquad \text{Quotient}$$

The domain of an **arithmetic combination** of functions f and g consists of all real numbers that are common to the domains of f and g. In the case of the quotient $f(x)/g(x)$, there is the further restriction that $g(x) \neq 0$.

Sum, Difference, Product, and Quotient of Functions

Let f and g be two functions with overlapping domains. Then, for all x common to both domains, the *sum, difference, product,* and *quotient* of f and g are defined as follows.

1. Sum: $\qquad (f + g)(x) = f(x) + g(x)$

2. Difference: $(f - g)(x) = f(x) - g(x)$

3. Product: $\qquad (fg)(x) = f(x) \cdot g(x)$

4. Quotient: $\qquad \left(\dfrac{f}{g}\right)(x) = \dfrac{f(x)}{g(x)}, \quad g(x) \neq 0$

EXAMPLE 1 **Finding the Sum of Two Functions**

Given $f(x) = 2x + 1$ and $g(x) = x^2 + 2x - 1$, find $(f + g)(x)$. Then evaluate the sum when $x = 3$.

Solution The sum of f and g is

$$(f + g)(x) = f(x) + g(x) = (2x + 1) + (x^2 + 2x - 1) = x^2 + 4x.$$

When $x = 3$, the value of this sum is

$$(f + g)(3) = 3^2 + 4(3) = 21.$$

✓ **Checkpoint** ◀))) *Audio-video solution in English & Spanish at LarsonPrecalculus.com*

Given $f(x) = x^2$ and $g(x) = 1 - x$, find $(f + g)(x)$. Then evaluate the sum when $x = 2$. ∎

EXAMPLE 2   **Finding the Difference of Two Functions**

Given $f(x) = 2x + 1$ and $g(x) = x^2 + 2x - 1$, find $(f - g)(x)$. Then evaluate the difference when $x = 2$.

Solution   The difference of f and g is

$$(f - g)(x) = f(x) - g(x) = (2x + 1) - (x^2 + 2x - 1) = -x^2 + 2.$$

When $x = 2$, the value of this difference is

$$(f - g)(2) = -(2)^2 + 2 = -2.$$

✓ *Checkpoint* ◀)))) *Audio-video solution in English & Spanish at LarsonPrecalculus.com*

Given $f(x) = x^2$ and $g(x) = 1 - x$, find $(f - g)(x)$. Then evaluate the difference when $x = 3$.

EXAMPLE 3   **Finding the Product of Two Functions**

Given $f(x) = x^2$ and $g(x) = x - 3$, find $(fg)(x)$. Then evaluate the product when $x = 4$.

Solution   The product of f and g is

$$(fg)(x) = f(x)g(x) = (x^2)(x - 3) = x^3 - 3x^2.$$

When $x = 4$, the value of this product is

$$(fg)(4) = 4^3 - 3(4)^2 = 16.$$

✓ *Checkpoint* ◀)))) *Audio-video solution in English & Spanish at LarsonPrecalculus.com*

Given $f(x) = x^2$ and $g(x) = 1 - x$, find $(fg)(x)$. Then evaluate the product when $x = 3$. ∎

In Examples 1–3, both f and g have domains that consist of all real numbers. So, the domains of $f + g$, $f - g$, and fg are also the set of all real numbers. Remember to consider any restrictions on the domains of f and g when forming the sum, difference, product, or quotient of f and g.

EXAMPLE 4   **Finding the Quotients of Two Functions**

Find $(f/g)(x)$ and $(g/f)(x)$ for the functions $f(x) = \sqrt{x}$ and $g(x) = \sqrt{4 - x^2}$. Then find the domains of f/g and g/f.

Solution   The quotient of f and g is

$$\left(\frac{f}{g}\right)(x) = \frac{f(x)}{g(x)} = \frac{\sqrt{x}}{\sqrt{4 - x^2}}$$

and the quotient of g and f is

$$\left(\frac{g}{f}\right)(x) = \frac{g(x)}{f(x)} = \frac{\sqrt{4 - x^2}}{\sqrt{x}}.$$

• • **REMARK**   Note that the domain of f/g includes $x = 0$, but not $x = 2$, because $x = 2$ yields a zero in the denominator, whereas the domain of g/f includes $x = 2$, but not $x = 0$, because $x = 0$ yields a zero in the denominator.

The domain of f is $[0, \infty)$ and the domain of g is $[-2, 2]$. The intersection of these domains is $[0, 2]$. So, the domains of f/g and g/f are as follows.

Domain of f/g: $[0, 2)$ Domain of g/f: $(0, 2]$

✓ *Checkpoint* ◀)))) *Audio-video solution in English & Spanish at LarsonPrecalculus.com*

Find $(f/g)(x)$ and $(g/f)(x)$ for the functions $f(x) = \sqrt{x - 3}$ and $g(x) = \sqrt{16 - x^2}$. Then find the domains of f/g and g/f. ∎

Composition of Functions

Another way of combining two functions is to form the **composition** of one with the other. For example, if $f(x) = x^2$ and $g(x) = x + 1$, then the composition of f with g is

$$f(g(x)) = f(x + 1)$$
$$= (x + 1)^2.$$

This composition is denoted as $f \circ g$ and reads as "f composed with g."

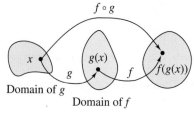

Domain of g

Domain of f

Figure 1.55

Definition of Composition of Two Functions

The **composition** of the function f with the function g is

$$(f \circ g)(x) = f(g(x)).$$

The domain of $f \circ g$ is the set of all x in the domain of g such that $g(x)$ is in the domain of f. (See Figure 1.55.)

EXAMPLE 5   Compositions of Functions

See LarsonPrecalculus.com for an interactive version of this type of example.

Given $f(x) = x + 2$ and $g(x) = 4 - x^2$, find the following.

a. $(f \circ g)(x)$ **b.** $(g \circ f)(x)$ **c.** $(g \circ f)(-2)$

Solution

a. The composition of f with g is as shown.

$$(f \circ g)(x) = f(g(x)) \qquad \text{Definition of } f \circ g$$
$$= f(4 - x^2) \qquad \text{Definition of } g(x)$$
$$= (4 - x^2) + 2 \qquad \text{Definition of } f(x)$$
$$= -x^2 + 6 \qquad \text{Simplify.}$$

b. The composition of g with f is as shown.

$$(g \circ f)(x) = g(f(x)) \qquad \text{Definition of } g \circ f$$
$$= g(x + 2) \qquad \text{Definition of } f(x)$$
$$= 4 - (x + 2)^2 \qquad \text{Definition of } g(x)$$
$$= 4 - (x^2 + 4x + 4) \qquad \text{Expand.}$$
$$= -x^2 - 4x \qquad \text{Simplify.}$$

Note that, in this case, $(f \circ g)(x) \neq (g \circ f)(x)$.

c. Evaluate the result of part (b) when $x = -2$.

$$(g \circ f)(-2) = -(-2)^2 - 4(-2) \qquad \text{Substitute.}$$
$$= -4 + 8 \qquad \text{Simplify.}$$
$$= 4 \qquad \text{Simplify.}$$

REMARK The tables of values below help illustrate the composition $(f \circ g)(x)$ in Example 5(a).

x	0	1	2	3
$g(x)$	4	3	0	-5

$g(x)$	4	3	0	-5
$f(g(x))$	6	5	2	-3

x	0	1	2	3
$f(g(x))$	6	5	2	-3

Note that the first two tables are combined (or "composed") to produce the values in the third table.

✓ **Checkpoint** ◀))) Audio-video solution in English & Spanish at LarsonPrecalculus.com

Given $f(x) = 2x + 5$ and $g(x) = 4x^2 + 1$, find the following.

a. $(f \circ g)(x)$ **b.** $(g \circ f)(x)$ **c.** $(f \circ g)\left(-\frac{1}{2}\right)$

EXAMPLE 6 **Finding the Domain of a Composite Function**

Find the domain of $f \circ g$ for the functions

$$f(x) = x^2 - 9 \quad \text{and} \quad g(x) = \sqrt{9 - x^2}.$$

Algebraic Solution

Find the composition of the functions.

$$
\begin{aligned}
(f \circ g)(x) &= f(g(x)) \\
&= f\left(\sqrt{9 - x^2}\right) \\
&= \left(\sqrt{9 - x^2}\right)^2 - 9 \\
&= 9 - x^2 - 9 \\
&= -x^2
\end{aligned}
$$

The domain of $f \circ g$ is restricted to the x-values in the domain of g for which $g(x)$ is in the domain of f. The domain of $f(x) = x^2 - 9$ is the set of all real numbers, which includes all real values of g. So, the domain of $f \circ g$ is the entire domain of $g(x) = \sqrt{9 - x^2}$, which is $[-3, 3]$.

Graphical Solution

Use a graphing utility to graph $f \circ g$.

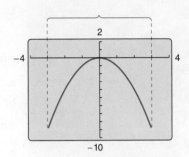

From the graph, you can determine that the domain of $f \circ g$ is $[-3, 3]$.

✓ *Checkpoint*  *Audio-video solution in English & Spanish at LarsonPrecalculus.com*

Find the domain of $f \circ g$ for the functions $f(x) = \sqrt{x}$ and $g(x) = x^2 + 4$.

In Examples 5 and 6, you formed the composition of two given functions. In calculus, it is also important to be able to identify two functions that make up a given composite function. For example, the function $h(x) = (3x - 5)^3$ is the composition of $f(x) = x^3$ and $g(x) = 3x - 5$. That is,

$$h(x) = (3x - 5)^3 = [g(x)]^3 = f(g(x)).$$

Basically, to "decompose" a composite function, look for an "inner" function and an "outer" function. In the function h above, $g(x) = 3x - 5$ is the inner function and $f(x) = x^3$ is the outer function.

EXAMPLE 7 **Decomposing a Composite Function**

Write the function $h(x) = \dfrac{1}{(x - 2)^2}$ as a composition of two functions.

Solution Consider $g(x) = x - 2$ as the inner function and $f(x) = \dfrac{1}{x^2} = x^{-2}$ as the outer function. Then write

$$
\begin{aligned}
h(x) &= \frac{1}{(x - 2)^2} \\
&= (x - 2)^{-2} \\
&= f(x - 2) \\
&= f(g(x)).
\end{aligned}
$$

✓ *Checkpoint*  *Audio-video solution in English & Spanish at LarsonPrecalculus.com*

Write the function $h(x) = \dfrac{\sqrt[3]{8 - x}}{5}$ as a composition of two functions.

Application

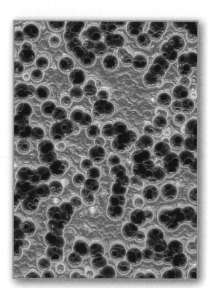

Refrigerated foods can have two types of bacteria: pathogenic bacteria, which can cause foodborne illness, and spoilage bacteria, which give foods an unpleasant look, smell, taste, or texture.

EXAMPLE 8 Bacteria Count

The number N of bacteria in a refrigerated food is given by

$$N(T) = 20T^2 - 80T + 500, \quad 2 \le T \le 14$$

where T is the temperature of the food in degrees Celsius. When the food is removed from refrigeration, the temperature of the food is given by

$$T(t) = 4t + 2, \quad 0 \le t \le 3$$

where t is the time in hours.

a. Find and interpret $(N \circ T)(t)$.

b. Find the time when the bacteria count reaches 2000.

Solution

a. $(N \circ T)(t) = N(T(t))$

$$= 20(4t + 2)^2 - 80(4t + 2) + 500$$

$$= 20(16t^2 + 16t + 4) - 320t - 160 + 500$$

$$= 320t^2 + 320t + 80 - 320t - 160 + 500$$

$$= 320t^2 + 420$$

The composite function $N \circ T$ represents the number of bacteria in the food as a function of the amount of time the food has been out of refrigeration.

b. The bacteria count reaches 2000 when $320t^2 + 420 = 2000$. By solving this equation algebraically, you find that the count reaches 2000 when $t \approx 2.2$ hours. Note that the negative solution $t \approx -2.2$ hours is rejected because it is not in the domain of the composite function.

✓ **Checkpoint**  *Audio-video solution in English & Spanish at LarsonPrecalculus.com*

The number N of bacteria in a refrigerated food is given by

$$N(T) = 8T^2 - 14T + 200, \quad 2 \le T \le 12$$

where T is the temperature of the food in degrees Celsius. When the food is removed from refrigeration, the temperature of the food is given by

$$T(t) = 2t + 2, \quad 0 \le t \le 5$$

where t is the time in hours.

a. Find $(N \circ T)(t)$.

b. Find the time when the bacteria count reaches 1000. ∎

Summarize   (Section 1.8)

1. Explain how to add, subtract, multiply, and divide functions *(page 76)*. For examples of finding arithmetic combinations of functions, see Examples 1–4.

2. Explain how to find the composition of one function with another function *(page 78)*. For examples that use compositions of functions, see Examples 5–7.

3. Describe a real-life example that uses a composition of functions *(page 80, Example 8)*.

1.8 Exercises

See CalcChat.com for tutorial help and worked-out solutions to odd-numbered exercises.

Vocabulary: Fill in the blanks.

1. Two functions f and g can be combined by the arithmetic operations of _____, _____, _____, and _____ to create new functions.

2. The _____ of the function f with the function g is $(f \circ g)(x) = f(g(x))$.

Skills and Applications

Graphing the Sum of Two Functions In Exercises 3 and 4, use the graphs of f and g to graph $h(x) = (f + g)(x)$. To print an enlarged copy of the graph, go to *MathGraphs.com*.

3.
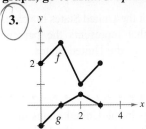

4.
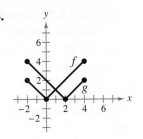

Finding Arithmetic Combinations of Functions In Exercises 5–12, find (a) $(f + g)(x)$, (b) $(f - g)(x)$, (c) $(fg)(x)$, and (d) $(f/g)(x)$. What is the domain of f/g?

5. $f(x) = x + 2$, $g(x) = x - 2$

6. $f(x) = 2x - 5$, $g(x) = 2 - x$

7. $f(x) = x^2$, $g(x) = 4x - 5$

8. $f(x) = 3x + 1$, $g(x) = x^2 - 16$

9. $f(x) = x^2 + 6$, $g(x) = \sqrt{1 - x}$

10. $f(x) = \sqrt{x^2 - 4}$, $g(x) = \dfrac{x^2}{x^2 + 1}$

11. $f(x) = \dfrac{x}{x + 1}$, $g(x) = x^3$

12. $f(x) = \dfrac{2}{x}$, $g(x) = \dfrac{1}{x^2 - 1}$

 Evaluating an Arithmetic Combination of Functions In Exercises 13–24, evaluate the function for $f(x) = x + 3$ and $g(x) = x^2 - 2$.

13. $(f + g)(2)$

14. $(f + g)(-1)$

15. $(f - g)(0)$

16. $(f - g)(1)$

17. $(f - g)(3t)$

18. $(f + g)(t - 2)$

19. $(fg)(6)$

20. $(fg)(-6)$

21. $(f/g)(5)$

22. $(f/g)(0)$

23. $(f/g)(-1) - g(3)$

24. $(fg)(5) + f(4)$

Graphical Reasoning In Exercises 25–28, use a graphing utility to graph f, g, and $f + g$ in the same viewing window. Which function contributes most to the magnitude of the sum when $0 \le x \le 2$? Which function contributes most to the magnitude of the sum when $x > 6$?

25. $f(x) = 3x$, $g(x) = -\dfrac{x^3}{10}$

26. $f(x) = \dfrac{x}{2}$, $g(x) = \sqrt{x}$

27. $f(x) = 3x + 2$, $g(x) = -\sqrt{x + 5}$

28. $f(x) = x^2 - \dfrac{1}{2}$, $g(x) = -3x^2 - 1$

 Finding Compositions of Functions In Exercises 29–34, find (a) $f \circ g$, (b) $g \circ f$, and (c) $g \circ g$.

29. $f(x) = x + 8$, $g(x) = x - 3$

30. $f(x) = -4x$, $g(x) = x + 7$

31. $f(x) = x^2$, $g(x) = x - 1$

32. $f(x) = 3x$, $g(x) = x^4$

33. $f(x) = \sqrt[3]{x - 1}$, $g(x) = x^3 + 1$

34. $f(x) = x^3$, $g(x) = \dfrac{1}{x}$

 Finding Domains of Functions and Composite Functions In Exercises 35–42, find (a) $f \circ g$ and (b) $g \circ f$. Find the domain of each function and of each composite function.

35. $f(x) = \sqrt{x + 4}$, $g(x) = x^2$

36. $f(x) = \sqrt[3]{x - 5}$, $g(x) = x^3 + 1$

37. $f(x) = x^3$, $g(x) = x^{2/3}$

38. $f(x) = x^5$, $g(x) = \sqrt[4]{x}$

39. $f(x) = |x|$, $g(x) = x + 6$

40. $f(x) = |x - 4|$, $g(x) = 3 - x$

41. $f(x) = \dfrac{1}{x}$, $g(x) = x + 3$

42. $f(x) = \dfrac{3}{x^2 - 1}$, $g(x) = x + 1$

Graphing Combinations of Functions In Exercises 43 and 44, on the same set of coordinate axes, (a) graph the functions f, g, and $f + g$ and (b) graph the functions f, g, and $f \circ g$.

43. $f(x) = \frac{1}{2}x$, $g(x) = x - 4$

44. $f(x) = x + 3$, $g(x) = x^2$

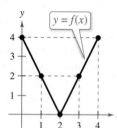 **Evaluating Combinations of Functions** In Exercises 45–48, use the graphs of f and g to evaluate the functions.

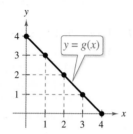

45. (a) $(f + g)(3)$ (b) $(f/g)(2)$

46. (a) $(f - g)(1)$ (b) $(fg)(4)$

47. (a) $(f \circ g)(2)$ (b) $(g \circ f)(2)$

48. (a) $(f \circ g)(1)$ (b) $(g \circ f)(3)$

 Decomposing a Composite Function In Exercises 49–56, find two functions f and g such that $(f \circ g)(x) = h(x)$. (There are many correct answers.)

49. $h(x) = (2x + 1)^2$

50. $h(x) = (1 - x)^3$

51. $h(x) = \sqrt[3]{x^2 - 4}$

52. $h(x) = \sqrt{9 - x}$

53. $h(x) = \dfrac{1}{x + 2}$

54. $h(x) = \dfrac{4}{(5x + 2)^2}$

55. $h(x) = \dfrac{-x^2 + 3}{4 - x^2}$

56. $h(x) = \dfrac{27x^3 + 6x}{10 - 27x^3}$

57. Stopping Distance The research and development department of an automobile manufacturer determines that when a driver is required to stop quickly to avoid an accident, the distance (in feet) the car travels during the driver's reaction time is given by $R(x) = \frac{3}{4}x$, where x is the speed of the car in miles per hour. The distance (in feet) the car travels while the driver is braking is given by $B(x) = \frac{1}{15}x^2$.

(a) Find the function that represents the total stopping distance T.

(b) Graph the functions R, B, and T on the same set of coordinate axes for $0 \le x \le 60$.

(c) Which function contributes most to the magnitude of the sum at higher speeds? Explain.

58. Business The annual cost C (in thousands of dollars) and revenue R (in thousands of dollars) for a company each year from 2010 through 2016 can be approximated by the models

$$C = 254 - 9t + 1.1t^2 \quad \text{and} \quad R = 341 + 3.2t$$

where t is the year, with $t = 10$ corresponding to 2010.

(a) Write a function P that represents the annual profit of the company.

(b) Use a graphing utility to graph C, R, and P in the same viewing window.

59. Vital Statistics Let $b(t)$ be the number of births in the United States in year t, and let $d(t)$ represent the number of deaths in the United States in year t, where $t = 10$ corresponds to 2010.

(a) If $p(t)$ is the population of the United States in year t, find the function $c(t)$ that represents the percent change in the population of the United States.

(b) Interpret $c(16)$.

• **60. Pets** • • • • • • • • • • • • • • • • • • •

Let $d(t)$ be the number of dogs in the United States in year t, and let $c(t)$ be the number of cats in the United States in year t, where $t = 10$ corresponds to 2010.

(a) Find the function $p(t)$ that represents the total number of dogs and cats in the United States.

(b) Interpret $p(16)$.

(c) Let $n(t)$ represent the population of the United States in year t, where $t = 10$ corresponds to 2010. Find and interpret $h(t) = p(t)/n(t)$.

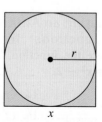

61. Geometry A square concrete foundation is a base for a cylindrical tank (see figure).

(a) Write the radius r of the tank as a function of the length x of the sides of the square.

(b) Write the area A of the circular base of the tank as a function of the radius r.

(c) Find and interpret $(A \circ r)(x)$.

62. Biology The number N of bacteria in a refrigerated food is given by

$$N(T) = 10T^2 - 20T + 600, \quad 2 \le T \le 20$$

where T is the temperature of the food in degrees Celsius. When the food is removed from refrigeration, the temperature of the food is given by

$$T(t) = 3t + 2, \quad 0 \le t \le 6$$

where t is the time in hours.

(a) Find and interpret $(N \circ T)(t)$.

(b) Find the bacteria count after 0.5 hour.

(c) Find the time when the bacteria count reaches 1500.

63. Salary You are a sales representative for a clothing manufacturer. You are paid an annual salary, plus a bonus of 3% of your sales over $500,000. Consider the two functions $f(x) = x - 500,000$ and $g(x) = 0.03x$. When x is greater than $500,000, which of the following represents your bonus? Explain.

(a) $f(g(x))$

(b) $g(f(x))$

64. Consumer Awareness The suggested retail price of a new hybrid car is p dollars. The dealership advertises a factory rebate of $2000 and a 10% discount.

(a) Write a function R in terms of p giving the cost of the hybrid car after receiving the rebate from the factory.

(b) Write a function S in terms of p giving the cost of the hybrid car after receiving the dealership discount.

(c) Find and interpret $(R \circ S)(p)$ and $(S \circ R)(p)$.

(d) Find $(R \circ S)(25,795)$ and $(S \circ R)(25,795)$. Which yields the lower cost for the hybrid car? Explain.

Exploration

True or False? In Exercises 65 and 66, determine whether the statement is true or false. Justify your answer.

65. If $f(x) = x + 1$ and $g(x) = 6x$, then

$$(f \circ g)(x) = (g \circ f)(x).$$

66. When you are given two functions f and g and a constant c, you can find $(f \circ g)(c)$ if and only if $g(c)$ is in the domain of f.

Siblings In Exercises 67 and 68, three siblings are three different ages. The oldest is twice the age of the middle sibling, and the middle sibling is six years older than one-half the age of the youngest.

67. (a) Write a composite function that gives the oldest sibling's age in terms of the youngest. Explain how you arrived at your answer.

(b) If the oldest sibling is 16 years old, find the ages of the other two siblings.

68. (a) Write a composite function that gives the youngest sibling's age in terms of the oldest. Explain how you arrived at your answer.

(b) If the youngest sibling is 2 years old, find the ages of the other two siblings.

69. Proof Prove that the product of two odd functions is an even function, and that the product of two even functions is an even function.

70. Conjecture Use examples to hypothesize whether the product of an odd function and an even function is even or odd. Then prove your hypothesis.

71. Writing Functions Write two unique functions f and g such that $(f \circ g)(x) = (g \circ f)(x)$ and f and g are (a) linear functions and (b) polynomial functions with degrees greater than one.

72. HOW DO YOU SEE IT? The graphs labeled L_1, L_2, L_3, and L_4 represent four different pricing discounts, where p is the original price (in dollars) and S is the sale price (in dollars). Match each function with its graph. Describe the situations in parts (c) and (d).

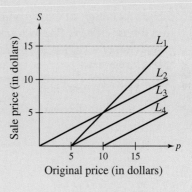

(a) $f(p)$: A 50% discount is applied.

(b) $g(p)$: A $5 discount is applied.

(c) $(g \circ f)(p)$

(d) $(f \circ g)(p)$

73. Proof

(a) Given a function f, prove that g is even and h is odd, where $g(x) = \frac{1}{2}[f(x) + f(-x)]$ and

$$h(x) = \frac{1}{2}[f(x) - f(-x)].$$

(b) Use the result of part (a) to prove that any function can be written as a sum of even and odd functions. [*Hint:* Add the two equations in part (a).]

(c) Use the result of part (b) to write each function as a sum of even and odd functions.

$$f(x) = x^2 - 2x + 1, \quad k(x) = \frac{1}{x + 1}$$

1.9 Inverse Functions

Inverse functions can help you model and solve real-life problems. For example, in Exercise 90 on page 92, you will write an inverse function and use it to determine the percent load interval for a diesel engine.

■ Find inverse functions informally and verify that two functions are inverse functions of each other.
■ Use graphs to verify that two functions are inverse functions of each other.
■ Use the Horizontal Line Test to determine whether functions are one-to-one.
■ Find inverse functions algebraically.

Inverse Functions

Recall from Section 1.4 that a set of ordered pairs can represent a function. For example, the function $f(x) = x + 4$ from the set $A = \{1, 2, 3, 4\}$ to the set $B = \{5, 6, 7, 8\}$ can be written as

$$f(x) = x + 4\colon \{(1, 5), (2, 6), (3, 7), (4, 8)\}.$$

In this case, by interchanging the first and second coordinates of each of the ordered pairs, you form the **inverse function** of f, which is denoted by f^{-1}. It is a function from the set B to the set A, and can be written as

$$f^{-1}(x) = x - 4\colon \{(5, 1), (6, 2), (7, 3), (8, 4)\}.$$

Note that the domain of f is equal to the range of f^{-1}, and vice versa, as shown in the figure below. Also note that the functions f and f^{-1} have the effect of "undoing" each other. In other words, when you form the composition of f with f^{-1} or the composition of f^{-1} with f, you obtain the identity function.

$$f(f^{-1}(x)) = f(x - 4) = (x - 4) + 4 = x$$
$$f^{-1}(f(x)) = f^{-1}(x + 4) = (x + 4) - 4 = x$$

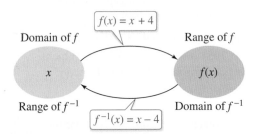

EXAMPLE 1   **Finding an Inverse Function Informally**

Find the inverse function of $f(x) = 4x$. Then verify that both $f(f^{-1}(x))$ and $f^{-1}(f(x))$ are equal to the identity function.

Solution   The function f *multiplies* each input by 4. To "undo" this function, you need to *divide* each input by 4. So, the inverse function of $f(x) = 4x$ is

$$f^{-1}(x) = \frac{x}{4}.$$

Verify that $f(f^{-1}(x)) = x$ and $f^{-1}(f(x)) = x$.

$$f(f^{-1}(x)) = f\left(\frac{x}{4}\right) = 4\left(\frac{x}{4}\right) = x \qquad f^{-1}(f(x)) = f^{-1}(4x) = \frac{4x}{4} = x$$

✓ **Checkpoint** ◉))) *Audio-video solution in English & Spanish at LarsonPrecalculus.com*

Find the inverse function of $f(x) = \frac{1}{5}x$. Then verify that both $f(f^{-1}(x))$ and $f^{-1}(f(x))$ are equal to the identity function. ■

Definition of Inverse Function

Let f and g be two functions such that

$$f(g(x)) = x \qquad \text{for every } x \text{ in the domain of } g$$

and

$$g(f(x)) = x \qquad \text{for every } x \text{ in the domain of } f.$$

Under these conditions, the function g is the **inverse function** of the function f. The function g is denoted by f^{-1} (read "f-inverse"). So,

$$f(f^{-1}(x)) = x \quad \text{and} \quad f^{-1}(f(x)) = x.$$

The domain of f must be equal to the range of f^{-1}, and the range of f must be equal to the domain of f^{-1}.

Do not be confused by the use of -1 to denote the inverse function f^{-1}. In this text, whenever f^{-1} is written, it *always* refers to the inverse function of the function f and *not* to the reciprocal of $f(x)$.

If the function g is the inverse function of the function f, then it must also be true that the function f is the inverse function of the function g. So, it is correct to say that the functions f and g are *inverse functions of each other*.

EXAMPLE 2 **Verifying Inverse Functions**

Which of the functions is the inverse function of $f(x) = \dfrac{5}{x-2}$?

$$g(x) = \frac{x-2}{5} \qquad h(x) = \frac{5}{x} + 2$$

Solution By forming the composition of f with g, you have

$$f(g(x)) = f\left(\frac{x-2}{5}\right) = \frac{5}{\left(\dfrac{x-2}{5}\right)-2} = \frac{25}{x-12} \neq x.$$

This composition is not equal to the identity function x, so g *is not* the inverse function of f. By forming the composition of f with h, you have

$$f(h(x)) = f\left(\frac{5}{x} + 2\right) = \frac{5}{\left(\dfrac{5}{x}+2\right)-2} = \frac{5}{\left(\dfrac{5}{x}\right)} = x.$$

So, it appears that h *is* the inverse function of f. Confirm this by showing that the composition of h with f is also equal to the identity function.

$$h(f(x)) = h\left(\frac{5}{x-2}\right) = \frac{5}{\left(\dfrac{5}{x-2}\right)} + 2 = x - 2 + 2 = x$$

Check to see that the domain of f is the same as the range of h and vice versa.

✓ **Checkpoint** ◀))) *Audio-video solution in English & Spanish at LarsonPrecalculus.com*

Which of the functions is the inverse function of $f(x) = \dfrac{x-4}{7}$?

$$g(x) = 7x + 4 \qquad h(x) = \frac{7}{x-4}$$

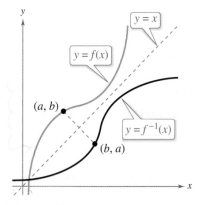

Figure 1.56

The Graph of an Inverse Function

The graphs of a function f and its inverse function f^{-1} are related to each other in this way: If the point (a, b) lies on the graph of f, then the point (b, a) must lie on the graph of f^{-1}, and vice versa. This means that the graph of f^{-1} is a *reflection* of the graph of f in the line $y = x$, as shown in Figure 1.56.

EXAMPLE 3 **Verifying Inverse Functions Graphically**

Verify graphically that the functions $f(x) = 2x - 3$ and $g(x) = \frac{1}{2}(x + 3)$ are inverse functions of each other.

Solution Sketch the graphs of f and g on the same rectangular coordinate system, as shown in Figure 1.57. It appears that the graphs are reflections of each other in the line $y = x$. Further verify this reflective property by testing a few points on each graph. Note that for each point (a, b) on the graph of f, the point (b, a) is on the graph of g.

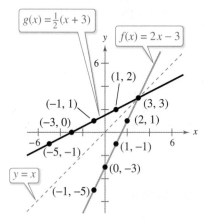

Figure 1.57

Graph of $f(x) = 2x - 3$	Graph of $g(x) = \frac{1}{2}(x + 3)$
$(-1, -5)$	$(-5, -1)$
$(0, -3)$	$(-3, 0)$
$(1, -1)$	$(-1, 1)$
$(2, 1)$	$(1, 2)$
$(3, 3)$	$(3, 3)$

The graphs of f and g are reflections of each other in the line $y = x$. So, f and g are inverse functions of each other.

✓ *Checkpoint*))) Audio-video solution in English & Spanish at LarsonPrecalculus.com

Verify graphically that the functions $f(x) = 4x - 1$ and $g(x) = \frac{1}{4}(x + 1)$ are inverse functions of each other.

EXAMPLE 4 **Verifying Inverse Functions Graphically**

Verify graphically that the functions $f(x) = x^2$ $(x \geq 0)$ and $g(x) = \sqrt{x}$ are inverse functions of each other.

Solution Sketch the graphs of f and g on the same rectangular coordinate system, as shown in Figure 1.58. It appears that the graphs are reflections of each other in the line $y = x$. Test a few points on each graph.

Graph of $f(x) = x^2,\ x \geq 0$	Graph of $g(x) = \sqrt{x}$
$(0, 0)$	$(0, 0)$
$(1, 1)$	$(1, 1)$
$(2, 4)$	$(4, 2)$
$(3, 9)$	$(9, 3)$

The graphs of f and g are reflections of each other in the line $y = x$. So, f and g are inverse functions of each other.

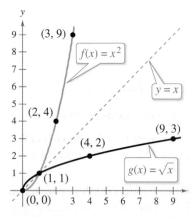

Figure 1.58

✓ *Checkpoint*))) Audio-video solution in English & Spanish at LarsonPrecalculus.com

Verify graphically that the functions $f(x) = x^2 + 1$ $(x \geq 0)$ and $g(x) = \sqrt{x - 1}$ are inverse functions of each other.

One-to-One Functions

The reflective property of the graphs of inverse functions gives you a graphical test for determining whether a function has an inverse function. This test is the **Horizontal Line Test** for inverse functions.

Horizontal Line Test for Inverse Functions

A function f has an inverse function if and only if no *horizontal* line intersects the graph of f at more than one point.

If no horizontal line intersects the graph of f at more than one point, then no y-value corresponds to more than one x-value. This is the essential characteristic of **one-to-one functions.**

One-to-One Functions

A function f is **one-to-one** when each value of the dependent variable corresponds to exactly one value of the independent variable. A function f has an inverse function if and only if f is one-to-one.

Consider the table of values for the function $f(x) = x^2$ on the left. The output $f(x) = 4$ corresponds to two inputs, $x = -2$ and $x = 2$, so f is not one-to-one. In the table on the right, x and y are interchanged. Here $x = 4$ corresponds to both $y = -2$ and $y = 2$, so this table does not represent a function. So, $f(x) = x^2$ is not one-to-one and does not have an inverse function.

x	$f(x) = x^2$
-2	4
-1	1
0	0
1	1
2	4
3	9

x	y
4	-2
1	-1
0	0
1	1
4	2
9	3

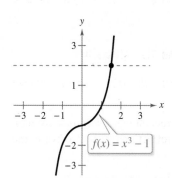

Figure 1.59

EXAMPLE 5 Applying the Horizontal Line Test

See LarsonPrecalculus.com for an interactive version of this type of example.

a. The graph of the function $f(x) = x^3 - 1$ is shown in Figure 1.59. No horizontal line intersects the graph of f at more than one point, so f *is* a one-to-one function and *does* have an inverse function.

b. The graph of the function $f(x) = x^2 - 1$ is shown in Figure 1.60. It is possible to find a horizontal line that intersects the graph of f at more than one point, so f *is not* a one-to-one function and *does not* have an inverse function.

✓ **Checkpoint** ◀))) *Audio-video solution in English & Spanish at LarsonPrecalculus.com*

Use the graph of f to determine whether the function has an inverse function.

a. $f(x) = \frac{1}{2}(3 - x)$　　**b.** $f(x) = |x|$

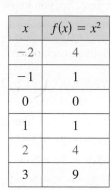

Figure 1.60

Finding Inverse Functions Algebraically

For relatively simple functions (such as the one in Example 1), you can find inverse functions by inspection. For more complicated functions, however, it is best to use the guidelines below. The key step in these guidelines is Step 3—interchanging the roles of x and y. This step corresponds to the fact that inverse functions have ordered pairs with the coordinates reversed.

REMARK Note what happens when you try to find the inverse function of a function that is not one-to-one.

$$f(x) = x^2 + 1 \qquad \text{Original function}$$

$$y = x^2 + 1 \qquad \text{Replace } f(x) \text{ with } y.$$

$$x = y^2 + 1 \qquad \text{Interchange } x \text{ and } y.$$

$$x - 1 = y^2 \qquad \text{Isolate } y\text{-term.}$$

$$y = \pm\sqrt{x - 1} \qquad \text{Solve for } y.$$

You obtain two y-values for each x.

Finding an Inverse Function

1. Use the Horizontal Line Test to decide whether f has an inverse function.

2. In the equation for $f(x)$, replace $f(x)$ with y.

3. Interchange the roles of x and y, and solve for y.

4. Replace y with $f^{-1}(x)$ in the new equation.

5. Verify that f and f^{-1} are inverse functions of each other by showing that the domain of f is equal to the range of f^{-1}, the range of f is equal to the domain of f^{-1}, and $f(f^{-1}(x)) = x$ and $f^{-1}(f(x)) = x$.

EXAMPLE 6 **Finding an Inverse Function Algebraically**

Find the inverse function of

$$f(x) = \frac{5 - x}{3x + 2}.$$

$$f(x) = \frac{5 - x}{3x + 2}$$

Figure 1.61

Solution The graph of f is shown in Figure 1.61. This graph passes the Horizontal Line Test. So, you know that f is one-to-one and has an inverse function.

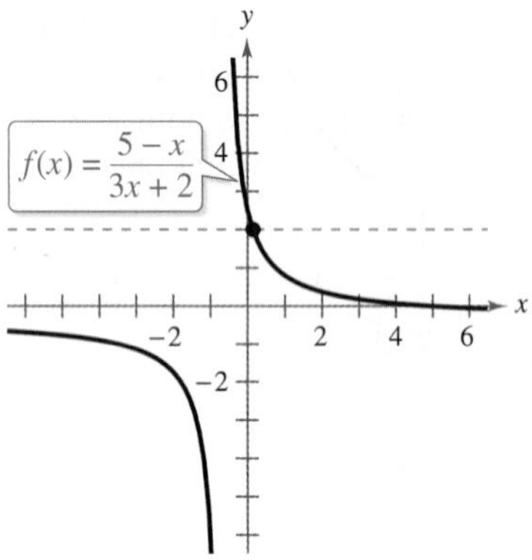

$$f(x) = \frac{5 - x}{3x + 2} \qquad \text{Write original function.}$$

$$y = \frac{5 - x}{3x + 2} \qquad \text{Replace } f(x) \text{ with } y.$$

$$x = \frac{5 - y}{3y + 2} \qquad \text{Interchange } x \text{ and } y.$$

$$x(3y + 2) = 5 - y \qquad \text{Multiply each side by } 3y + 2.$$

$$3xy + 2x = 5 - y \qquad \text{Distributive Property}$$

$$3xy + y = 5 - 2x \qquad \text{Collect terms with } y.$$

$$y(3x + 1) = 5 - 2x \qquad \text{Factor.}$$

$$y = \frac{5 - 2x}{3x + 1} \qquad \text{Solve for } y.$$

$$f^{-1}(x) = \frac{5 - 2x}{3x + 1} \qquad \text{Replace } y \text{ with } f^{-1}(x).$$

Check that $f(f^{-1}(x)) = x$ and $f^{-1}(f(x)) = x$.

✓ **Checkpoint** ◀))) *Audio-video solution in English & Spanish at LarsonPrecalculus.com*

Find the inverse function of

$$f(x) = \frac{5 - 3x}{x + 2}.$$

EXAMPLE 7   **Finding an Inverse Function Algebraically**

Find the inverse function of

$$f(x) = \sqrt{2x - 3}.$$

Solution The graph of f is shown in the figure below. This graph passes the Horizontal Line Test. So, you know that f is one-to-one and has an inverse function.

$f(x) = \sqrt{2x - 3}$	Write original function.
$y = \sqrt{2x - 3}$	Replace $f(x)$ with y.
$x = \sqrt{2y - 3}$	Interchange x and y.
$x^2 = 2y - 3$	Square each side.
$2y = x^2 + 3$	Isolate y-term.
$y = \dfrac{x^2 + 3}{2}$	Solve for y.
$f^{-1}(x) = \dfrac{x^2 + 3}{2},\ x \geq 0$	Replace y with $f^{-1}(x)$.

The graph of f^{-1} in the figure is the reflection of the graph of f in the line $y = x$. Note that the range of f is the interval $[0, \infty)$, which implies that the domain of f^{-1} is the interval $[0, \infty)$. Moreover, the domain of f is the interval $\left[\frac{3}{2}, \infty\right)$, which implies that the range of f^{-1} is the interval $\left[\frac{3}{2}, \infty\right)$. Verify that $f(f^{-1}(x)) = x$ and $f^{-1}(f(x)) = x$.

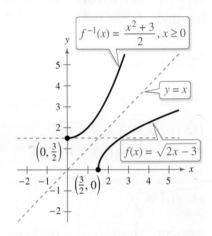

✓ *Checkpoint* *Audio-video solution in English & Spanish at LarsonPrecalculus.com*

Find the inverse function of

$$f(x) = \sqrt[3]{10 + x}.$$

Summarize   (Section 1.9)

1. State the definition of an inverse function *(page 85)*. For examples of finding inverse functions informally and verifying inverse functions, see Examples 1 and 2.

2. Explain how to use graphs to verify that two functions are inverse functions of each other *(page 86)*. For examples of verifying inverse functions graphically, see Examples 3 and 4.

3. Explain how to use the Horizontal Line Test to determine whether a function is one-to-one *(page 87)*. For an example of applying the Horizontal Line Test, see Example 5.

4. Explain how to find an inverse function algebraically *(page 88)*. For examples of finding inverse functions algebraically, see Examples 6 and 7.

1.9 Exercises

See **CalcChat.com** for tutorial help and worked-out solutions to odd-numbered exercises.

Vocabulary: Fill in the blanks.

1. If $f(g(x))$ and $g(f(x))$ both equal x, then the function g is the _____ function of the function f.
2. The inverse function of f is denoted by _____.
3. The domain of f is the _____ of f^{-1}, and the _____ of f^{-1} is the range of f.
4. The graphs of f and f^{-1} are reflections of each other in the line _____.
5. A function f is _____ when each value of the dependent variable corresponds to exactly one value of the independent variable.
6. A graphical test for the existence of an inverse function of f is the _____ Line Test.

Skills and Applications

Finding an Inverse Function Informally In Exercises 7–14, find the inverse function of f informally. Verify that $f(f^{-1}(x)) = x$ and $f^{-1}(f(x)) = x$.

7. $f(x) = 6x$
8. $f(x) = \dfrac{1}{3}x$

9. $f(x) = 3x + 1$
10. $f(x) = \dfrac{x-3}{2}$

11. $f(x) = x^2 - 4, \ x \geq 0$
12. $f(x) = x^2 + 2, \ x \geq 0$
13. $f(x) = x^3 + 1$

14. $f(x) = \dfrac{x^5}{4}$

Verifying Inverse Functions In Exercises 15–18, verify that f and g are inverse functions algebraically.

15. $f(x) = \dfrac{x-9}{4}, \quad g(x) = 4x + 9$

16. $f(x) = -\dfrac{3}{2}x - 4, \quad g(x) = -\dfrac{2x+8}{3}$

17. $f(x) = \dfrac{x^3}{4}, \quad g(x) = \sqrt[3]{4x}$

18. $f(x) = x^3 + 5, \quad g(x) = \sqrt[3]{x-5}$

Sketching the Graph of an Inverse Function In Exercises 19 and 20, use the graph of the function to sketch the graph of its inverse function $y = f^{-1}(x)$.

19.
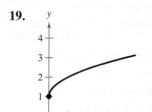

20.
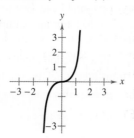

Verifying Inverse Functions In Exercises 21–32, verify that f and g are inverse functions (a) algebraically and (b) graphically.

21. $f(x) = x - 5, \quad g(x) = x + 5$

22. $f(x) = 2x, \quad g(x) = \dfrac{x}{2}$

23. $f(x) = 7x + 1, \quad g(x) = \dfrac{x-1}{7}$

24. $f(x) = 3 - 4x, \quad g(x) = \dfrac{3-x}{4}$

25. $f(x) = x^3, \quad g(x) = \sqrt[3]{x}$

26. $f(x) = \dfrac{x^3}{3}, \quad g(x) = \sqrt[3]{3x}$

27. $f(x) = \sqrt{x + 5}, \quad g(x) = x^2 - 5, \quad x \geq 0$
28. $f(x) = 1 - x^3, \quad g(x) = \sqrt[3]{1-x}$

29. $f(x) = \dfrac{1}{x}, \quad g(x) = \dfrac{1}{x}$

30. $f(x) = \dfrac{1}{1+x}, \quad x \geq 0, \quad g(x) = \dfrac{1-x}{x}, \quad 0 < x \leq 1$

31. $f(x) = \dfrac{x-1}{x+5}, \quad g(x) = -\dfrac{5x+1}{x-1}$

32. $f(x) = \dfrac{x+3}{x-2}, \quad g(x) = \dfrac{2x+3}{x-1}$

Using a Table to Determine an Inverse Function In Exercises 33 and 34, does the function have an inverse function?

33.

x	-1	0	1	2	3	4
$f(x)$	-2	1	2	1	-2	-6

34.

x	-3	-2	-1	0	2	3
$f(x)$	10	6	4	1	-3	-10

Using a Table to Find an Inverse Function In Exercises 35 and 36, use the table of values for $y = f(x)$ to complete a table for $y = f^{-1}(x)$.

35.

x	-1	0	1	2	3	4
$f(x)$	3	5	7	9	11	13

36.

x	-3	-2	-1	0	1	2
$f(x)$	10	5	0	-5	-10	-15

Applying the Horizontal Line Test In Exercises 37–40, does the function have an inverse function?

37.
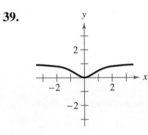

38.
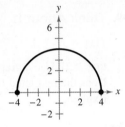

39.

40.
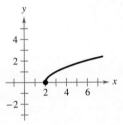

 Applying the Horizontal Line Test In Exercises 41–44, use a graphing utility to graph the function, and use the Horizontal Line Test to determine whether the function has an inverse function.

41. $g(x) = (x + 3)^2 + 2$ **42.** $f(x) = \frac{1}{5}(x + 2)^3$
43. $f(x) = x\sqrt{9 - x^2}$ **44.** $h(x) = |x| - |x - 4|$

 Finding and Analyzing Inverse Functions In Exercises 45–54, (a) find the inverse function of f, (b) graph both f and f^{-1} on the same set of coordinate axes, (c) describe the relationship between the graphs of f and f^{-1}, and (d) state the domains and ranges of f and f^{-1}.

45. $f(x) = x^5 - 2$ **46.** $f(x) = x^3 + 8$
47. $f(x) = \sqrt{4 - x^2}, \quad 0 \le x \le 2$
48. $f(x) = x^2 - 2, \quad x \le 0$
49. $f(x) = \dfrac{4}{x}$ **50.** $f(x) = -\dfrac{2}{x}$
51. $f(x) = \dfrac{x + 1}{x - 2}$ **52.** $f(x) = \dfrac{x - 2}{3x + 5}$
53. $f(x) = \sqrt[3]{x - 1}$ **54.** $f(x) = x^{3/5}$

Finding an Inverse Function In Exercises 55–70, determine whether the function has an inverse function. If it does, find the inverse function.

55. $f(x) = x^4$ **56.** $f(x) = \dfrac{1}{x^2}$
57. $g(x) = \dfrac{x + 1}{6}$ **58.** $f(x) = 3x + 5$
59. $p(x) = -4$ **60.** $f(x) = 0$
61. $f(x) = (x + 3)^2, \quad x \ge -3$
62. $q(x) = (x - 5)^2$
63. $f(x) = \begin{cases} x + 3, & x < 0 \\ 6 - x, & x \ge 0 \end{cases}$
64. $f(x) = \begin{cases} -x, & x \le 0 \\ x^2 - 3x, & x > 0 \end{cases}$
65. $h(x) = |x + 1| - 1$
66. $f(x) = |x - 2|, \quad x \le 2$
67. $f(x) = \sqrt{2x + 3}$
68. $f(x) = \sqrt{x - 2}$
69. $f(x) = \dfrac{6x + 4}{4x + 5}$
70. $f(x) = \dfrac{5x - 3}{2x + 5}$

Restricting the Domain In Exercises 71–78, restrict the domain of the function f so that the function is one-to-one and has an inverse function. Then find the inverse function f^{-1}. State the domains and ranges of f and f^{-1}. Explain your results. (There are many correct answers.)

71. $f(x) = |x + 2|$ **72.** $f(x) = |x - 5|$
73. $f(x) = (x + 6)^2$ **74.** $f(x) = (x - 4)^2$
75. $f(x) = -2x^2 + 5$
76. $f(x) = \frac{1}{2}x^2 - 1$
77. $f(x) = |x - 4| + 1$
78. $f(x) = -|x - 1| - 2$

Composition with Inverses In Exercises 79–84, use the functions $f(x) = \frac{1}{8}x - 3$ and $g(x) = x^3$ to find the value or function.

79. $(f^{-1} \circ g^{-1})(1)$ **80.** $(g^{-1} \circ f^{-1})(-3)$
81. $(f^{-1} \circ f^{-1})(4)$ **82.** $(g^{-1} \circ g^{-1})(-1)$
83. $(f \circ g)^{-1}$ **84.** $g^{-1} \circ f^{-1}$

Composition with Inverses In Exercises 85–88, use the functions $f(x) = x + 4$ and $g(x) = 2x - 5$ to find the function.

85. $g^{-1} \circ f^{-1}$ **86.** $f^{-1} \circ g^{-1}$
87. $(f \circ g)^{-1}$ **88.** $(g \circ f)^{-1}$

89. Hourly Wage   Your wage is $10.00 per hour plus $0.75 for each unit produced per hour. So, your hourly wage y in terms of the number of units produced x is $y = 10 + 0.75x$.

   (a) Find the inverse function. What does each variable represent in the inverse function?

   (b) Determine the number of units produced when your hourly wage is $24.25.

90. Diesel Mechanics

The function

$$y = 0.03x^2 + 245.50, \quad 0 < x < 100$$

approximates the exhaust temperature y in degrees Fahrenheit, where x is the percent load for a diesel engine.

   (a) Find the inverse function. What does each variable represent in the inverse function?

   (b) Use a graphing utility to graph the inverse function.

   (c) The exhaust temperature of the engine must not exceed 500 degrees Fahrenheit. What is the percent load interval?

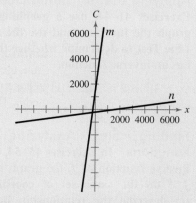

Exploration

True or False?   **In Exercises 91 and 92, determine whether the statement is true or false. Justify your answer.**

91. If f is an even function, then f^{-1} exists.

92. If the inverse function of f exists and the graph of f has a y-intercept, then the y-intercept of f is an x-intercept of f^{-1}.

Creating a Table   **In Exercises 93 and 94, use the graph of the function f to create a table of values for the given points. Then create a second table that can be used to find f^{-1}, and sketch the graph of f^{-1}, if possible.**

93.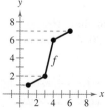

94.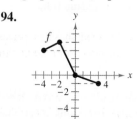

95. Proof   Prove that if f and g are one-to-one functions, then $(f \circ g)^{-1}(x) = (g^{-1} \circ f^{-1})(x)$.

96. Proof   Prove that if f is a one-to-one odd function, then f^{-1} is an odd function.

97. Think About It   The function $f(x) = k(2 - x - x^3)$ has an inverse function, and $f^{-1}(3) = -2$. Find k.

98. Think About It   Consider the functions $f(x) = x + 2$ and $f^{-1}(x) = x - 2$. Evaluate $f(f^{-1}(x))$ and $f^{-1}(f(x))$ for the given values of x. What can you conclude about the functions?

x	-10	0	7	45
$f(f^{-1}(x))$				
$f^{-1}(f(x))$				

99. Think About It   Restrict the domain of

$$f(x) = x^2 + 1$$

to $x \geq 0$. Use a graphing utility to graph the function. Does the restricted function have an inverse function? Explain.

100.   **HOW DO YOU SEE IT?**   The cost C for a business to make personalized T-shirts is given by

$$C(x) = 7.50x + 1500$$

where x represents the number of T-shirts.

   (a) The graphs of C and C^{-1} are shown below. Match each function with its graph.

   (b) Explain what $C(x)$ and $C^{-1}(x)$ represent in the context of the problem.

One-to-One Function Representation   **In Exercises 101 and 102, determine whether the situation can be represented by a one-to-one function. If so, write a statement that best describes the inverse function.**

101. The number of miles n a marathon runner has completed in terms of the time t in hours

102. The depth of the tide d at a beach in terms of the time t over a 24-hour period

1.10 Mathematical Modeling and Variation

Mathematical models have a wide variety of real-life applications. For example, in Exercise 71 on page 103, you will use variation to model ocean temperatures at various depths.

- Use mathematical models to approximate sets of data points.
- Use the *regression* feature of a graphing utility to find equations of least squares regression lines.
- Write mathematical models for direct variation.
- Write mathematical models for direct variation as an *n*th power.
- Write mathematical models for inverse variation.
- Write mathematical models for combined variation.
- Write mathematical models for joint variation.

Introduction

In this section, you will study two techniques for fitting models to data: *least squares regression* and *direct and inverse variation*.

EXAMPLE 1 **Using a Mathematical Model**

The table shows the populations y (in millions) of the United States from 2008 through 2015. *(Source: U.S. Census Bureau)*

Year	2008	2009	2010	2011	2012	2013	2014	2015
Population, y	304.1	306.8	309.3	311.7	314.1	316.5	318.9	321.2

Spreadsheet at LarsonPrecalculus.com

A linear model that approximates the data is

$$y = 2.43t + 284.9, \quad 8 \le t \le 15$$

where t represents the year, with $t = 8$ corresponding to 2008. Plot the actual data *and* the model on the same graph. How closely does the model represent the data?

Solution Figure 1.62 shows the actual data and the model plotted on the same graph. From the graph, it appears that the model is a "good fit" for the actual data. To see how well the model fits, compare the actual values of y with the values of y found using the model. The values found using the model are labeled y^* in the table below.

t	8	9	10	11	12	13	14	15
y	304.1	306.8	309.3	311.7	314.1	316.5	318.9	321.2
y^*	304.3	306.8	309.2	311.6	314.1	316.5	318.9	321.4

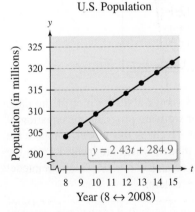

U.S. Population

$y = 2.43t + 284.9$

Year (8 ↔ 2008)

Figure 1.62

✓ **Checkpoint** 🔊))) *Audio-video solution in English & Spanish at LarsonPrecalculus.com*

The ordered pairs below give the median sales prices y (in thousands of dollars) of new homes sold in a neighborhood from 2009 through 2016. *(Spreadsheet at LarsonPrecalculus.com)*

(2009, 179.4)   (2011, 191.0)   (2013, 202.6)   (2015, 214.9)
(2010, 185.4)   (2012, 196.7)   (2014, 208.7)   (2016, 221.4)

A linear model that approximates the data is $y = 5.96t + 125.5, 9 \le t \le 16$, where t represents the year, with $t = 9$ corresponding to 2009. Plot the actual data *and* the model on the same graph. How closely does the model represent the data?

Least Squares Regression and Graphing Utilities

So far in this text, you have worked with many different types of mathematical models that approximate real-life data. In some instances the model was given (as in Example 1), whereas in other instances you found the model using algebraic techniques or a graphing utility.

To find a model that approximates a set of data most accurately, statisticians use a measure called the **sum of the squared differences,** which is the sum of the squares of the differences between actual data values and model values. The "best-fitting" linear model, called the **least squares regression line,** is the one with the least sum of the squared differences.

Recall that you can approximate this line visually by plotting the data points and drawing the line that appears to best fit the data—or you can enter the data points into a graphing utility or software program and use the *linear regression* feature.

When you use the *regression* feature of a graphing utility or software program, an "*r*-value" may be output. This is the **correlation coefficient** of the data and gives a measure of how well the model fits the data. The closer the value of $|r|$ is to 1, the better the fit.

EXAMPLE 2   Finding a Least Squares Regression Line

See LarsonPrecalculus.com for an interactive version of this type of example.

The table shows the numbers E (in millions) of Medicare private health plan enrollees from 2008 through 2015. Construct a scatter plot that represents the data and find the equation of the least squares regression line for the data.   *(Source: U.S. Centers for Medicare and Medicaid Services)*

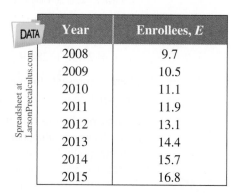

Year	Enrollees, E
2008	9.7
2009	10.5
2010	11.1
2011	11.9
2012	13.1
2013	14.4
2014	15.7
2015	16.8

Medicare Private
Health Plan Enrollees

$E = 1.03t + 1.0$

Year (8 ↔ 2008)

Figure 1.63

Solution   Let $t = 8$ represent 2008. Figure 1.63 shows a scatter plot of the data. Using the *regression* feature of a graphing utility or software program, the equation of the least squares regression line is $E = 1.03t + 1.0$. To check this model, compare the actual E-values with the E-values found using the model, which are labeled E^* in the table at the left. The correlation coefficient for this model is $r \approx 0.992$, so the model is a good fit.

t	E	E^*
8	9.7	9.2
9	10.5	10.3
10	11.1	11.3
11	11.9	12.3
12	13.1	13.4
13	14.4	14.4
14	15.7	15.4
15	16.8	16.5

✓ *Checkpoint* 🔊))) *Audio-video solution in English & Spanish at LarsonPrecalculus.com*

The ordered pairs below give the numbers E (in millions) of Medicare Advantage enrollees in health maintenance organization plans from 2008 through 2015. *(Spreadsheet at LarsonPrecalculus.com)* Construct a scatter plot that represents the data and find the equation of the least squares regression line for the data.   *(Source: U.S. Centers for Medicare and Medicaid Services)*

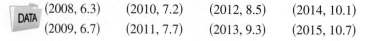

DATA   (2008, 6.3)   (2010, 7.2)   (2012, 8.5)   (2014, 10.1)
       (2009, 6.7)   (2011, 7.7)   (2013, 9.3)   (2015, 10.7)

Direct Variation

There are two basic types of linear models. The more general model has a nonzero y-intercept.

$$y = mx + b, \quad b \neq 0$$

The simpler model

$$y = kx$$

has a y-intercept of zero. In the simpler model, y **varies directly** as x, or is **directly proportional** to x.

Direct Variation

The statements below are equivalent.

1. y **varies directly** as x.
2. y is **directly proportional** to x.
3. $y = kx$ for some nonzero constant k.

k is the **constant of variation** or the **constant of proportionality.**

EXAMPLE 3　Direct Variation

In Pennsylvania, the state income tax is directly proportional to *gross income*. You work in Pennsylvania and your state income tax deduction is $46.05 for a gross monthly income of $1500. Find a mathematical model that gives the Pennsylvania state income tax in terms of gross income.

Solution

Verbal model:

| State income tax | = | k | · | Gross income |

Labels:　State income tax = y　(dollars)
　　　　　Gross income = x　(dollars)
　　　　　Income tax rate = k　(percent in decimal form)

Equation:　$y = kx$

To find the state income tax rate k, substitute the given information into the equation $y = kx$ and solve.

$$y = kx \qquad \text{Write direct variation model.}$$

$$46.05 = k(1500) \qquad \text{Substitute 46.05 for } y \text{ and 1500 for } x.$$

$$0.0307 = k \qquad \text{Divide each side by 1500.}$$

So, the equation (or model) for state income tax in Pennsylvania is

$$y = 0.0307x.$$

In other words, Pennsylvania has a state income tax rate of 3.07% of gross income. Figure 1.64 shows the graph of this equation.

✓ *Checkpoint* *Audio-video solution in English & Spanish at LarsonPrecalculus.com*

The simple interest on an investment is directly proportional to the amount of the investment. For example, an investment of $2500 earns $187.50 after 1 year. Find a mathematical model that gives the interest I after 1 year in terms of the amount invested P.

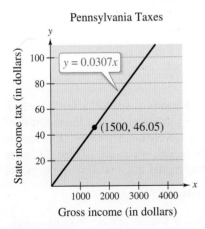

Pennsylvania Taxes

$y = 0.0307x$

(1500, 46.05)

State income tax (in dollars)

Gross income (in dollars)

Figure 1.64

Direct Variation as an *n*th Power

Another type of direct variation relates one variable to a *power* of another variable. For example, in the formula for the area of a circle

$$A = \pi r^2$$

the area A is directly proportional to the square of the radius r. Note that for this formula, π is the constant of proportionality.

•• **REMARK** Note that the direct variation model $y = kx$ is a special case of $y = kx^n$ with $n = 1$.

$\cdots\cdots\cdots\cdots\cdots\cdots\triangleright$

Direct Variation as an *n*th Power

The statements below are equivalent.

1. y **varies directly as the *n*th power** of x.
2. y is **directly proportional to the *n*th power** of x.
3. $y = kx^n$ for some nonzero constant k.

EXAMPLE 4 **Direct Variation as an *n*th Power**

The distance a ball rolls down an inclined plane is directly proportional to the square of the time it rolls. During the first second, the ball rolls 8 feet. (See Figure 1.65.)

a. Write an equation relating the distance traveled to the time.

b. How far does the ball roll during the first 3 seconds?

Solution

a. Letting d be the distance (in feet) the ball rolls and letting t be the time (in seconds), you have

$$d = kt^2.$$

Now, $d = 8$ when $t = 1$, so you have

$d = kt^2$	Write direct variation model.
$8 = k(\ )^2$	Substitute 8 for d and 1 for t.
$8 = k$	Simplify.

and, the equation relating distance to time is

$$d = 8t^2.$$

b. When $t = 3$, the distance traveled is

$d = 8(3)^2$	Substitute 3 for t.
$= 8(9)$	Simplify.
$= 72$ feet.	Simplify.

So, the ball rolls 72 feet during the first 3 seconds.

✓ **Checkpoint** 🔊 *Audio-video solution in English & Spanish at LarsonPrecalculus.com*

Neglecting air resistance, the distance s an object falls varies directly as the square of the duration t of the fall. An object falls a distance of 144 feet in 3 seconds. How far does it fall in 6 seconds?

In Examples 3 and 4, the direct variations are such that an *increase* in one variable corresponds to an *increase* in the other variable. You should not, however, assume that this always occurs with direct variation. For example, for the model $y = -3x$, an increase in x results in a *decrease* in y, and yet y is said to vary directly as x.

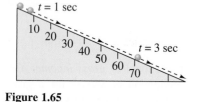

$t = 0$ sec
$t = 1$ sec
10 20 30 40 50 60 70
$t = 3$ sec

Figure 1.65

Inverse Variation

> ### Inverse Variation
>
> The statements below are equivalent.
>
> 1. y **varies inversely** as x.
>
> 2. y is **inversely proportional** to x.
>
> 3. $y = \dfrac{k}{x}$ for some nonzero constant k.

If x and y are related by an equation of the form $y = k/x^n$, then y varies inversely as the nth power of x (or y is inversely proportional to the nth power of x).

EXAMPLE 5 Inverse Variation

A company has found that the demand for one of its products varies inversely as the price of the product. When the price is \$6.25, the demand is 400 units. Approximate the demand when the price is \$5.75.

Solution

Let p be the price and let x be the demand. The demand varies inversely as the price, so you have

$$x = \frac{k}{p}.$$

Now, $x = 400$ when $p = 6.25$, so you have

$x = \dfrac{k}{p}$	Write inverse variation model.
$400 = \dfrac{k}{6.25}$	Substitute 400 for x and 6.25 for p.
$(400)(6.25) = k$	Multiply each side by 6.25.
$2500 = k$	Simplify.

and the equation relating price and demand is

$$x = \frac{2500}{p}.$$

When $p = 5.75$, the demand is

$x = \dfrac{2500}{p}$	Write inverse variation model.
$= \dfrac{2500}{5.75}$	Substitute 5.75 for p.
≈ 435 units.	Simplify.

So, the demand for the product is about 435 units when the price is \$5.75.

Supply and demand are fundamental concepts in economics. The law of demand states that, all other factors remaining equal, the lower the price of the product, the higher the quantity demanded. The law of supply states that the higher the price of the product, the higher the quantity supplied. *Equilibrium* occurs when the demand and the supply are the same.

✓ *Checkpoint* *Audio-video solution in English & Spanish at LarsonPrecalculus.com*

The company in Example 5 has found that the demand for another of its products also varies inversely as the price of the product. When the price is \$2.75, the demand is 600 units. Approximate the demand when the price is \$3.25.

Combined Variation

Some applications of variation involve problems with *both* direct and inverse variations in the same model. These types of models have **combined variation.**

EXAMPLE 6   **Combined Variation**

A gas law states that the volume of an enclosed gas varies inversely as the pressure (Figure 1.66) *and* directly as the temperature. The pressure of a gas is 0.75 kilogram per square centimeter when the temperature is 294 K and the volume is 8000 cubic centimeters.

a. Write an equation relating pressure, temperature, and volume.

b. Find the pressure when the temperature is 300 K and the volume is 7000 cubic centimeters.

Solution

a. Volume V varies directly as temperature T and inversely as pressure P, so you have

$$V = \frac{kT}{P}.$$

Now, $P = 0.75$ when $T = 294$ and $V = 8000$, so you have

$$V = \frac{kT}{P} \qquad \text{Write combined variation model.}$$

$$8000 = \frac{k(294)}{0.75} \qquad \text{Substitute 8000 for } V, 294 \text{ for } T, \text{ and 0.75 for } P.$$

$$\frac{6000}{294} = k \qquad \text{Simplify.}$$

$$\frac{1000}{49} = k \qquad \text{Simplify.}$$

and the equation relating pressure, temperature, and volume is

$$V = \frac{1000}{49}\left(\frac{T}{P}\right).$$

b. Isolate P on one side of the equation by multiplying each side by P and dividing each side by V to obtain $P = \frac{1000}{49}\left(\frac{T}{V}\right)$. When $T = 300$ and $V = 7000$, the pressure is

$$P = \frac{1000}{49}\left(\frac{T}{V}\right) \qquad \text{Combined variation model solved for } P.$$

$$= \frac{1000}{49}\left(\frac{300}{7000}\right) \qquad \text{Substitute 300 for } T \text{ and 7000 for } V.$$

$$= \frac{300}{343} \qquad \text{Simplify.}$$

$$\approx 0.87 \text{ kilogram per square centimeter.} \qquad \text{Simplify.}$$

So, the pressure is about 0.87 kilogram per square centimeter when the temperature is 300 K and the volume is 7000 cubic centimeters.

✓ **Checkpoint** 🔊))) *Audio-video solution in English & Spanish at LarsonPrecalculus.com*

The resistance of a copper wire carrying an electrical current is directly proportional to its length and inversely proportional to its cross-sectional area. A copper wire with a diameter of 0.0126 inch has a resistance of 64.9 ohms per thousand feet. What length of 0.0201-inch-diameter copper wire will produce a resistance of 33.5 ohms? ◼

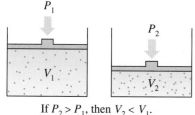

P_1

P_2

V_1

V_2

If $P_2 > P_1$, then $V_2 < V_1$.

If the temperature is held constant and pressure increases, then the volume *decreases.*

Figure 1.66

Joint Variation

> ### Joint Variation
>
> The statements below are equivalent.
>
> **1.** z **varies jointly** as x and y.
>
> **2.** z is **jointly proportional** to x and y.
>
> **3.** $z = kxy$ for some nonzero constant k.

If x, y, and z are related by an equation of the form $z = kx^n y^m$, then z varies jointly as the nth power of x and the mth power of y.

EXAMPLE 7   Joint Variation

The *simple* interest for an investment is jointly proportional to the time and the principal. After one quarter (3 months), the interest on a principal of \$5000 is \$43.75. (a) Write an equation relating the interest, principal, and time. (b) Find the interest after three quarters.

Solution

a. Interest I (in dollars) is jointly proportional to principal P (in dollars) and time t (in years), so you have

$$I = kPt.$$

For $I = 43.75$, $P = 5000$, and $t = \frac{3}{12} = \frac{1}{4}$, you have $43.75 = k(5000)\left(\frac{1}{4}\right)$, which implies that $k = 4(43.75)/5000 = 0.035$. So, the equation relating interest, principal, and time is

$$I = 0.035Pt$$

which is the familiar equation for simple interest where the constant of proportionality, 0.035, represents an annual interest rate of 3.5%.

b. When $P = \$5000$ and $t = \frac{3}{4}$, the interest is $I = (0.035)(5000)\left(\frac{3}{4}\right) = \$131.25.$

✓ **Checkpoint** 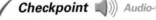 *Audio-video solution in English & Spanish at LarsonPrecalculus.com*

The kinetic energy E of an object varies jointly with the object's mass m and the square of the object's velocity v. An object with a mass of 50 kilograms traveling at 16 meters per second has a kinetic energy of 6400 joules. What is the kinetic energy of an object with a mass of 70 kilograms traveling at 20 meters per second? ■

> ## Summarize   (Section 1.10)
>
> **1.** Explain how to use a mathematical model to approximate a set of data points *(page 93)*. For an example of using a mathematical model to approximate a set of data points, see Example 1.
>
> **2.** Explain how to use the *regression* feature of a graphing utility to find the equation of a least squares regression line *(page 94)*. For an example of finding the equation of a least squares regression line, see Example 2.
>
> **3.** Explain how to write mathematical models for direct variation, direct variation as an nth power, inverse variation, combined variation, and joint variation *(pages 95–99)*. For examples of these types of variation, see Examples 3–7.

1.10 Exercises

See **CalcChat.com** for tutorial help and worked-out solutions to odd-numbered exercises.

Vocabulary: Fill in the blanks.

1. Two techniques for fitting models to data are direct and inverse _____ and least squares _____.

2. Statisticians use a measure called the _____ of the _____ _____ to find a model that approximates a set of data most accurately.

3. The linear model with the least sum of the squared differences is called the _____ _____ _____ line.

4. An r-value, or _____ _____, of a set of data gives a measure of how well a model fits the data.

5. The direct variation model $y = kx^n$ can be described as "y varies directly as the nth power of x," or "y is _____ _____ to the nth power of x."

6. The mathematical model $y = \dfrac{2}{x}$ is an example of _____ variation.

7. Mathematical models that involve both direct and inverse variation have _____ variation.

8. The joint variation model $z = kxy$ can be described as "z varies jointly as x and y," or "z is _____ _____ to x and y."

Skills and Applications

Mathematical Models In Exercises 9 and 10, (a) plot the actual data and the model of the same graph and (b) describe how closely the model represents the data. If the model does not closely represent the data, suggest another type of model that may be a better fit.

9. The ordered pairs below give the civilian noninstitutional U.S. populations y (in millions of people) 16 years of age and over not in the civilian labor force from 2006 through 2014. (*Spreadsheet at LarsonPrecalculus.com*)

(2006, 77.4)	(2011, 86.0)
(2007, 78.7)	(2012, 88.3)
(2008, 79.5)	(2013, 90.3)
(2009, 81.7)	(2014, 92.0)
(2010, 83.9)	

   A model for the data is $y = 1.92t + 65.0$, $6 \le t \le 14$, where t represents the years, with $t = 6$ corresponding to 2006. (*Source: U.S. Bureau of Labor Statistics*)

10. The ordered pairs below give the revenues y (in billions of dollars) for Activision Blizzard, Inc., from 2008 through 2014. (*Spreadsheet at LarsonPrecalculus.com*)

    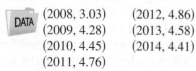

(2008, 3.03)	(2012, 4.86)
(2009, 4.28)	(2013, 4.58)
(2010, 4.45)	(2014, 4.41)
(2011, 4.76)	

    A model for the data is $y = 0.184t + 2.32$, $8 \le t \le 14$, where t represents the year, with $t = 8$ corresponding to 2008. (*Source: Activision Blizzard, Inc.*)

Sketching a Line In Exercises 11–16, sketch the line that you think best approximates the data in the scatter plot. Then find an equation of the line. To print an enlarged copy of the graph, go to *MathGraphs.com*.

11.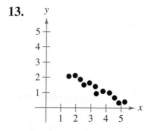

12.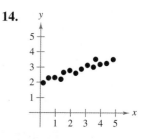

13.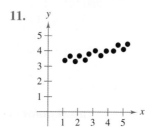

14.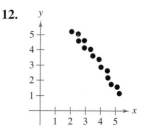

15.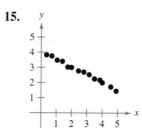

16.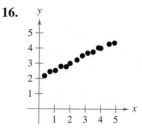

17. Sports The ordered pairs below give the winning times (in seconds) of the women's 100-meter freestyle in the Olympics from 1984 through 2012. *(Spreadsheet at LarsonPrecalculus.com)* *(Source: International Olympic Committee)*

 (1984, 55.92)   (2000, 53.83)
(1988, 54.93)   (2004, 53.84)
(1992, 54.64)   (2008, 53.12)
(1996, 54.50)   (2012, 53.00)

(a) Sketch a scatter plot of the data. Let y represent the winning time (in seconds) and let $t = 84$ represent 1984.

(b) Sketch the line that you think best approximates the data and find an equation of the line.

(c) Use the *regression* feature of a graphing utility to find the equation of the least squares regression line that fits the data.

(d) Compare the linear model you found in part (b) with the linear model you found in part (c).

18. Broadway The ordered pairs below give the starting year and gross ticket sales S (in millions of dollars) for each Broadway season in New York City from 1997 through 2014. *(Spreadsheet at LarsonPrecalculus.com)* *(Source: The Broadway League)*

(1997, 558)   (2003, 771)   (2009, 1020)
(1998, 588)   (2004, 769)   (2010, 1081)
(1999, 603)   (2005, 862)   (2011, 1139)
(2000, 666)   (2006, 939)   (2012, 1139)
(2001, 643)   (2007, 938)   (2013, 1269)
(2002, 721)   (2008, 943)   (2014, 1365)

(a) Use a graphing utility to create a scatter plot of the data. Let $t = 7$ represent 1997.

(b) Use the *regression* feature of the graphing utility to find the equation of the least squares regression line that fits the data.

(c) Use the graphing utility to graph the scatter plot you created in part (a) and the model you found in part (b) in the same viewing window. How closely does the model represent the data?

(d) Use the model to predict the gross ticket sales during the season starting in 2021.

(e) Interpret the meaning of the slope of the linear model in the context of the problem.

Direct Variation In Exercises 19–24, find a direct variation model that relates y and x.

19. $x = 2, y = 14$ 20. $x = 5, y = 12$
21. $x = 5, y = 1$ 22. $x = -24, y = 3$
23. $x = 4, y = 8\pi$ 24. $x = \pi, y = -1$

 Direct Variation as an *n*th Power In Exercises 25–28, use the given values of k and n to complete the table for the direct variation model $y = kx^n$. Plot the points in a rectangular coordinate system.

x	2	4	6	8	10
$y = kx^n$					

25. $k = 1, n = 2$ 26. $k = 2, n = 2$
27. $k = \frac{1}{2}, n = 3$ 28. $k = \frac{1}{4}, n = 3$

Inverse Variation as an *n*th Power In Exercises 29–32, use the given values of k and n to complete the table for the inverse variation model $y = k/x^n$. Plot the points in a rectangular coordinate system.

x	2	4	6	8	10
$y = k/x^n$					

29. $k = 2, n = 1$ 30. $k = 5, n = 1$
31. $k = 10, n = 2$ 32. $k = 20, n = 2$

Think About It In Exercises 33 and 34, use the graph to determine whether y varies directly as some power of x or inversely as some power of x. Explain.

33.
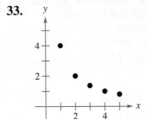

34.
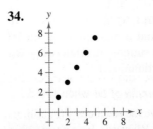

 Determining Variation In Exercises 35–38, determine whether the variation model represented by the ordered pairs (x, y) is of the form $y = kx$ or $y = k/x$, and find k. Then write a model that relates y and x.

35. $(5, 1), \left(10, \frac{1}{2}\right), \left(15, \frac{1}{3}\right), \left(20, \frac{1}{4}\right), \left(25, \frac{1}{5}\right)$

36. $(5, 2), (10, 4), (15, 6), (20, 8), (25, 10)$

37. $(5, -3.5), (10, -7), (15, -10.5), (20, -14), (25, -17.5)$

38. $(5, 24), (10, 12), (15, 8), (20, 6), \left(25, \frac{24}{5}\right)$

Finding a Mathematical Model In Exercises 39–48, find a mathematical model for the verbal statement.

39. A varies directly as the square of r.

40. V varies directly as the cube of l.

41. y varies inversely as the square of x.

42. h varies inversely as the square root of s.

43. F varies directly as g and inversely as r^2.

44. z varies jointly as the square of x and the cube of y.

45. *Newton's Law of Cooling:* The rate of change R of the temperature of an object is directly proportional to the difference between the temperature T of the object and the temperature T_e of the environment.

46. *Boyle's Law:* For a constant temperature, the pressure P of a gas is inversely proportional to the volume V of the gas.

47. *Direct Current:* The electric power P of a direct current circuit is jointly proportional to the voltage V and the electric current I.

48. *Newton's Law of Universal Gravitation:* The gravitational attraction F between two objects of masses m_1 and m_2 is jointly proportional to the masses and inversely proportional to the square of the distance r between the objects.

Describing a Formula In Exercises 49–52, use variation terminology to describe the formula.

49. $y = 2x^2$

50. $t = \dfrac{72}{r}$

51. $A = \frac{1}{2}bh$

52. $K = \frac{1}{2}mv^2$

Finding a Mathematical Model In Exercises 53–60, find a mathematical model that represents the statement. (Determine the constant of proportionality.)

53. y is directly proportional to x. ($y = 54$ when $x = 3$.)

54. A varies directly as r^2. ($A = 9\pi$ when $r = 3$.)

55. y varies inversely as x. ($y = 3$ when $x = 25$.)

56. y is inversely proportional to x^3. ($y = 7$ when $x = 2$.)

57. z varies jointly as x and y. ($z = 64$ when $x = 4$ and $y = 8$.)

58. F is jointly proportional to r and the third power of s. ($F = 4158$ when $r = 11$ and $s = 3$.)

59. P varies directly as x and inversely as the square of y. ($P = \frac{28}{3}$ when $x = 42$ and $y = 9$.)

60. z varies directly as the square of x and inversely as y. ($z = 6$ when $x = 6$ and $y = 4$.)

61. Simple Interest The simple interest on an investment is directly proportional to the amount of the investment. An investment of \$3250 earns \$113.75 after 1 year. Find a mathematical model that gives the interest I after 1 year in terms of the amount invested P.

62. Simple Interest The simple interest on an investment is directly proportional to the amount of the investment. An investment of \$6500 earns \$211.25 after 1 year. Find a mathematical model that gives the interest I after 1 year in terms of the amount invested P.

63. Measurement Use the fact that 13 inches is approximately the same length as 33 centimeters to find a mathematical model that relates centimeters y to inches x. Then use the model to find the numbers of centimeters in 10 inches and 20 inches.

64. Measurement Use the fact that 14 gallons is approximately the same amount as 53 liters to find a mathematical model that relates liters y to gallons x. Then use the model to find the numbers of liters in 5 gallons and 25 gallons.

Hooke's Law In Exercises 65–68, use Hooke's Law, which states that the distance a spring stretches (or compresses) from its natural, or equilibrium, length varies directly as the applied force on the spring.

65. A force of 220 newtons stretches a spring 0.12 meter. What force stretches the spring 0.16 meter?

66. A force of 265 newtons stretches a spring 0.15 meter.

 (a) What force stretches the spring 0.1 meter?

 (b) How far does a force of 90 newtons stretch the spring?

67. The coiled spring of a toy supports the weight of a child. The weight of a 25-pound child compresses the spring a distance of 1.9 inches. The toy does not work properly when a weight compresses the spring more than 3 inches. What is the maximum weight for which the toy works properly?

68. An overhead garage door has two springs, one on each side of the door. A force of 15 pounds is required to stretch each spring 1 foot. Because of a pulley system, the springs stretch only one-half the distance the door travels. The door moves a total of 8 feet, and the springs are at their natural lengths when the door is open. Find the combined lifting force applied to the door by the springs when the door is closed.

69. Ecology The diameter of the largest particle that a stream can move is approximately directly proportional to the square of the velocity of the stream. When the velocity is $\frac{1}{4}$ mile per hour, the stream can move coarse sand particles about 0.02 inch in diameter. Approximate the velocity required to carry particles 0.12 inch in diameter.

70. Work The work W required to lift an object varies jointly with the object's mass m and the height h that the object is lifted. The work required to lift a 120-kilogram object 1.8 meters is 2116.8 joules. Find the amount of work required to lift a 100-kilogram object 1.5 meters.

•• **71. Ocean Temperatures** ••••••••••

The ordered pairs below give the average water temperatures C (in degrees Celsius) at several depths d (in meters) in the Indian Ocean. *(Spreadsheet at LarsonPrecalculus.com)* *(Source: NOAA)*

DATA (1000, 4.85) (2500, 1.888)
(1500, 3.525) (3000, 1.583)
(2000, 2.468) (3500, 1.422)

(a) Sketch a scatter plot of the data.

(b) Determine whether a direct variation model or an inverse variation model better fits the data.

(c) Find k for each pair of coordinates. Then find the mean value of k to find the constant of proportionality for the model you chose in part (b).

(d) Use your model to approximate the depth at which the water temperature is 3°C.

72. Light Intensity The ordered pairs below give the intensities y (in microwatts per square centimeter) of the light measured by a light probe located x centimeters from a light source. *(Spreadsheet at LarsonPrecalculus.com)*

DATA (30, 0.1881) (38, 0.1172) (46, 0.0775)
(34, 0.1543) (42, 0.0998) (50, 0.0645)

A model that approximates the data is $y = 171.33/x^2$.

(a) Use a graphing utility to plot the data points and the model in the same viewing window.

(b) Use the model to approximate the light intensity 25 centimeters from the light source.

73. Music The fundamental frequency (in hertz) of a piano string is directly proportional to the square root of its tension and inversely proportional to its length and the square root of its mass density. A string has a frequency of 100 hertz. Find the frequency of a string with each property.

(a) Four times the tension

(b) Twice the length

(c) Four times the tension and twice the length

74. Beam Load The maximum load that a horizontal beam can safely support varies jointly as the width of the beam and the square of its depth and inversely as the length of the beam. Determine how each change affects the beam's maximum load.

(a) Doubling the width

(b) Doubling the depth

(c) Halving the length

(d) Halving the width and doubling the length

Exploration

True or False? In Exercises 75 and 76, decide whether the statement is true or false. Justify your answer.

75. If y is directly proportional to x and x is directly proportional to z, then y is directly proportional to z.

76. If y is inversely proportional to x and x is inversely proportional to z, then y is inversely proportional to z.

77. Error Analysis Describe the error.

In the equation for the surface area of a sphere, $S = 4\pi r^2$, the surface area S varies jointly with π and the square of the radius r.

78. HOW DO YOU SEE IT? Discuss how well a linear model approximates the data shown in each scatter plot.

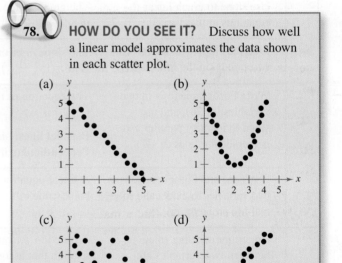

79. Think About It Let $y = 2x + 2$ and $t = x + 1$. What kind of variation do y and t have? Explain.

Project: Fraud and Identity Theft To work an extended application analyzing the numbers of fraud complaints and identity theft victims in the United States in 2014, visit this text's website at *LarsonPrecalculus.com*. *(Source: U.S. Federal Trade Commission)*

Chapter Summary

	What Did You Learn?	**Explanation/Examples**	**Review Exercises**
Section 1.1	Plot points in the Cartesian plane *(p. 2)*, use the Distance Formula *(p. 4)* and the Midpoint Formula *(p. 5)*, and use a coordinate plane to model and solve real-life problems *(p. 6)*.	For an ordered pair (x, y), the x-coordinate is the directed distance from the y-axis to the point, and the y-coordinate is the directed distance from the x-axis to the point. The coordinate plane can be used to estimate the annual sales of a company. (See Example 7.)	1–6
Section 1.2	Sketch graphs of equations *(p. 11)*, find x- and y-intercepts *(p. 14)*, and use symmetry to sketch graphs of equations *(p. 15)*.	To find x-intercepts, let y be zero and solve for x. To find y-intercepts, let x be zero and solve for y. Graphs can have symmetry with respect to one of the coordinate axes or with respect to the origin.	7–22
	Write equations of circles *(p. 17)*.	A point (x, y) lies on the circle of radius r and center (h, k) if and only if $(x - h)^2 + (y - k)^2 = r^2$.	23–27
	Use graphs of equations to solve real-life problems *(p. 18)*.	The graph of an equation can be used to estimate the maximum weight for a man in the U.S. Marine Corps. (See Example 9.)	28
Section 1.3	Use slope to graph linear equations in two variables *(p. 22)*.	The graph of the equation $y = mx + b$ is a line whose slope is m and whose y-intercept is $(0, b)$.	29–32
	Find the slope of a line given two points on the line *(p. 24)*.	The slope m of the nonvertical line through (x_1, y_1) and (x_2, y_2) is $m = (y_2 - y_1)/(x_2 - x_1)$, where $x_1 \neq x_2$.	33, 34
	Write linear equations in two variables *(p. 26)*, and use slope to identify parallel and perpendicular lines *(p. 27)*.	The equation of the line with slope m passing through the point (x_1, y_1) is $y - y_1 = m(x - x_1)$. **Parallel lines:** $m_1 = m_2$ **Perpendicular lines:** $m_1 = -1/m_2$	35–40
	Use slope and linear equations in two variables to model and solve real-life problems *(p. 28)*.	A linear equation in two variables can help you describe the book value of exercise equipment each year. (See Example 7.)	41, 42
Section 1.4	Determine whether relations between two variables are functions and use function notation *(p. 35)*, and find the domains of functions *(p. 40)*.	A function f from a set A (domain) to a set B (range) is a relation that assigns to each element x in the set A exactly one element y in the set B. **Equation:** $f(x) = 5 - x^2$ **$f(2)$:** $f(2) = 5 - 2^2 = 1$ **Domain of $f(x) = 5 - x^2$:** All real numbers	43–50
	Use functions to model and solve real-life problems *(p. 41)*.	A function can model the path of a baseball. (See Example 9.)	51, 52
	Evaluate difference quotients *(p. 42)*.	**Difference quotient:** $\dfrac{f(x + h) - f(x)}{h}, \quad h \neq 0$	53, 54
Section 1.5	Use the Vertical Line Test for functions *(p. 50)*.	A set of points in a coordinate plane is the graph of y as a function of x if and only if no *vertical* line intersects the graph at more than one point.	55, 56
	Find the zeros of functions *(p. 51)*.	**Zeros of $y = f(x)$:** x-values for which $f(x) = 0$	57, 58

	What Did You Learn?	Explanation/Examples	Review Exercises
Section 1.5	Determine intervals on which functions are increasing or decreasing *(p. 52)*, relative minimum and maximum values of functions *(p. 53)*, and the average rate of change of a function *(p. 54)*.	To determine whether a function is increasing, decreasing, or constant on an interval, determine whether the graph of the function rises, falls, or is constant from left to right. The points at which the behavior of a function changes can help determine relative minimum or relative maximum values. The average rate of change between any two points is the slope of the line (secant line) through the two points.	59–64
	Identify even and odd functions *(p. 55)*.	**Even:** For each x in the domain of f, $f(-x) = f(x)$. **Odd:** For each x in the domain of f, $f(-x) = -f(x)$.	65, 66
Section 1.6	Identify and graph different types of functions *(pp. 60, 62–64)*, and recognize graphs of parent functions *(p. 64)*.	**Linear:** $f(x) = ax + b$; **Squaring:** $f(x) = x^2$; **Cubic:** $f(x) = x^3$; **Square Root:** $f(x) = \sqrt{x}$; **Reciprocal:** $f(x) = 1/x$ Eight of the most commonly used functions in algebra are shown on page 64.	67–70
Section 1.7	Use vertical and horizontal shifts *(p. 67)*, reflections *(p. 69)*, and nonrigid transformations *(p. 71)* to sketch graphs of functions.	**Vertical shifts:** $h(x) = f(x) + c$ or $h(x) = f(x) - c$ **Horizontal shifts:** $h(x) = f(x - c)$ or $h(x) = f(x + c)$ **Reflection in x-axis:** $h(x) = -f(x)$ **Reflection in y-axis:** $h(x) = f(-x)$ **Nonrigid transformations:** $h(x) = cf(x)$ or $h(x) = f(cx)$	71–80
Section 1.8	Add, subtract, multiply, and divide functions *(p. 76)*, find compositions of functions *(p. 78)*, and use combinations and compositions of functions to model and solve real-life problems *(p. 80)*.	$(f + g)(x) = f(x) + g(x)$ $(f - g)(x) = f(x) - g(x)$ $(fg)(x) = f(x) \cdot g(x)$ $(f/g)(x) = f(x)/g(x), g(x) \neq 0$ The composition of the function f with the function g is $(f \circ g)(x) = f(g(x))$. A composite function can be used to represent the number of bacteria in food as a function of the amount of time the food has been out of refrigeration. (See Example 8.)	81–86
Section 1.9	Find inverse functions informally and verify that two functions are inverse functions of each other *(p. 84)*.	Let f and g be two functions such that $f(g(x)) = x$ for every x in the domain of g and $g(f(x)) = x$ for every x in the domain of f. Under these conditions, the function g is the inverse function of the function f.	87, 88
	Use graphs to verify inverse functions *(p. 86)*, use the Horizontal Line Test *(p. 87)*, and find inverse functions algebraically *(p. 88)*.	If the point (a, b) lies on the graph of f, then the point (b, a) must lie on the graph of f^{-1}, and vice versa. In short, the graph of f^{-1} is a reflection of the graph of f in the line $y = x$. To find an inverse function, replace $f(x)$ with y, interchange the roles of x and y, solve for y, and then replace y with $f^{-1}(x)$.	89–94
Section 1.10	Use mathematical models to approximate sets of data points *(p. 93)*, and use the *regression* feature of a graphing utility to find equations of least squares regression lines *(p. 94)*.	To see how well a model fits a set of data, compare the actual values of y with the model values. (See Example 1.) The sum of the squared differences is the sum of the squares of the differences between actual data values and model values. The least squares regression line is the linear model with the least sum of the squared differences.	95
	Write mathematical models for direct variation, direct variation as an nth power, inverse variation, combined variation, and joint variation *(pp. 95–99)*.	**Direct variation:** $y = kx$ for some nonzero constant k. **Direct variation as an nth power:** $y = kx^n$ for some nonzero constant k. **Inverse variation:** $y = k/x$ for some nonzero constant k. **Joint variation:** $z = kxy$ for some nonzero constant k.	96, 97

Review Exercises

See CalcChat.com for tutorial help and worked-out solutions to odd-numbered exercises.

1.1 Plotting Points in the Cartesian Plane In Exercises 1 and 2, plot the points.

1. $(5, 5), (-2, 0), (-3, 6), (-1, -7)$

2. $(0, 6), (8, 1), (5, -4), (-3, -3)$

Determining Quadrant(s) for a Point In Exercises 3 and 4, determine the quadrant(s) in which (x, y) could be located.

3. $x > 0$ and $y = -2$

4. $xy = 4$

5. **Plotting, Distance, and Midpoint** Plot the points $(-2, 6)$ and $(4, -3)$. Then find the distance between the points and the midpoint of the line segment joining the points.

6. **Sales** Barnes & Noble had annual sales of $6.8 billion in 2013 and $6.1 billion in 2015. Use the Midpoint Formula to estimate the sales in 2014. Assume that the annual sales follow a linear pattern. *(Source: Barnes & Noble, Inc.)*

1.2 Sketching the Graph of an Equation In Exercises 7–10, construct a table of values that consists of several points of the equation. Use the resulting solution points to sketch the graph of the equation.

7. $y = 3x - 5$

8. $y = -\frac{1}{2}x + 2$

9. $y = x^2 - 3x$

10. $y = 2x^2 - x - 9$

Finding x- and y-Intercepts In Exercises 11–14, find the x- and y-intercepts of the graph of the equation.

11. $y = 2x + 7$

12. $y = |x + 1| - 3$

13. $y = (x - 3)^2 - 4$

14. $y = x\sqrt{4 - x^2}$

Intercepts, Symmetry, and Graphing In Exercises 15–22, find any intercepts and test for symmetry. Then sketch the graph of the equation.

15. $y = -4x + 1$

16. $y = 5x - 6$

17. $y = 6 - x^2$

18. $y = x^2 - 12$

19. $y = x^3 + 5$

20. $y = -6 - x^3$

21. $y = \sqrt{x + 5}$

22. $y = |x| + 9$

Sketching a Circle In Exercises 23–26, find the center and radius of the circle with the given equation. Then sketch the circle.

23. $x^2 + y^2 = 9$

24. $x^2 + y^2 = 4$

25. $(x + 2)^2 + y^2 = 16$

26. $x^2 + (y - 8)^2 = 81$

27. **Writing the Equation of a Circle** Write the standard form of the equation of the circle for which the endpoints of a diameter are $(0, 0)$ and $(4, -6)$.

28. **Physics** The force F (in pounds) required to stretch a spring x inches from its natural length (see figure) is

$$F = \frac{5}{4}x, \quad 0 \le x \le 20.$$

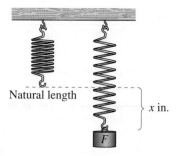

(a) Use the model to complete the table.

x	0	4	8	12	16	20
Force, F						

(b) Sketch a graph of the model.

(c) Use the graph to estimate the force necessary to stretch the spring 10 inches.

1.3 Graphing a Linear Equation In Exercises 29–32, find the slope and y-intercept (if possible) of the line. Sketch the line.

29. $y = -\frac{1}{2}x + 1$

30. $2x - 3y = 6$

31. $y = 1$

32. $x = -6$

Finding the Slope of a Line Through Two Points In Exercises 33 and 34, find the slope of the line passing through the pair of points.

33. $(5, -2), (-1, 4)$

34. $(-1, 6), (3, -2)$

Using the Point-Slope Form In Exercises 35 and 36, find the slope-intercept form of the equation of the line that has the given slope and passes through the given point. Sketch the line.

35. $m = \frac{1}{3}, \quad (6, -5)$

36. $m = -\frac{3}{4}, \quad (-4, -2)$

Finding an Equation of a Line In Exercises 37 and 38, find an equation of the line passing through the pair of points. Sketch the line.

37. $(-6, 4), (4, 9)$

38. $(-9, -3), (-3, -5)$

Finding Parallel and Perpendicular Lines In Exercises 39 and 40, find equations of the lines that pass through the given point and are (a) parallel to and (b) perpendicular to the given line.

39. $5x - 4y = 8$, $(3, -2)$

40. $2x + 3y = 5$, $(-8, 3)$

41. Sales A discount outlet offers a 20% discount on all items. Write a linear equation giving the sale price S for an item with a list price L.

42. Hourly Wage A manuscript translator charges a starting fee of \$50 plus \$2.50 per page translated. Write a linear equation for the amount A earned for translating p pages.

1.4 Testing for Functions Represented Algebraically In Exercises 43–46, determine whether the equation represents y as a function of x.

43. $16x - y^4 = 0$

44. $2x - y - 3 = 0$

45. $y = \sqrt{1 - x}$

46. $|y| = x + 2$

Evaluating a Function In Exercises 47 and 48, find each function value.

47. $f(x) = x^2 + 1$

  (a) $f(2)$

  (b) $f(-4)$

  (c) $f(t^2)$

  (d) $f(t + 1)$

48. $h(x) = |x - 2|$

  (a) $h(-4)$

  (b) $h(-2)$

  (c) $h(0)$

  (d) $h(-x + 2)$

Finding the Domain of a Function In Exercises 49 and 50, find the domain of the function.

49. $f(x) = \sqrt{25 - x^2}$

50. $h(x) = \dfrac{x}{x^2 - x - 6}$

Physics In Exercises 51 and 52, the velocity of a ball projected upward from ground level is given by $v(t) = -32t + 48$, where t is the time in seconds and v is the velocity in feet per second.

51. Find the velocity when $t = 1$.

52. Find the time when the ball reaches its maximum height. [*Hint:* Find the time when $v(t) = 0$.]

𝑓 Evaluating a Difference Quotient In Exercises 53 and 54, find the difference quotient and simplify your answer.

53. $f(x) = 2x^2 + 3x - 1$, $\dfrac{f(x + h) - f(x)}{h}$, $h \neq 0$

54. $f(x) = x^3 - 5x^2 + x$, $\dfrac{f(x + h) - f(x)}{h}$, $h \neq 0$

1.5 Vertical Line Test for Functions In Exercises 55 and 56, use the Vertical Line Test to determine whether the graph represents y as a function of x. To print an enlarged copy of the graph, go to *MathGraphs.com*.

55.

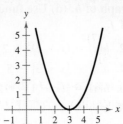

56.

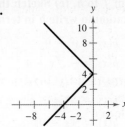

Finding the Zeros of a Function In Exercises 57 and 58, find the zeros of the function algebraically.

57. $f(x) = 3x^2 - 16x + 21$

58. $f(x) = 5x^2 + 4x - 1$

Describing Function Behavior In Exercises 59 and 60, use a graphing utility to graph the function and visually determine the open intervals on which the function is increasing, decreasing, or constant.

59. $f(x) = |x| + |x + 1|$ **60.** $f(x) = (x^2 - 4)^2$

Approximating Relative Minima or Maxima In Exercises 61 and 62, use a graphing utility to approximate (to two decimal places) any relative minima or maxima of the function.

61. $f(x) = -x^2 + 2x + 1$

62. $f(x) = x^3 - 4x^2 - 1$

𝑓 Average Rate of Change of a Function In Exercises 63 and 64, find the average rate of change of the function from x_1 to x_2.

63. $f(x) = -x^2 + 8x - 4$, $x_1 = 0, x_2 = 4$

64. $f(x) = x^3 + 2x + 1$, $x_1 = 1, x_2 = 3$

Even, Odd, or Neither? In Exercises 65 and 66, determine whether the function is even, odd, or neither. Then describe the symmetry.

65. $f(x) = x^4 - 20x^2$ **66.** $f(x) = 2x\sqrt{x^2 + 3}$

1.6 Writing a Linear Function In Exercises 67 and 68, (a) write the linear function f that has the given function values and (b) sketch the graph of the function.

67. $f(2) = -6$, $f(-1) = 3$

68. $f(0) = -5$, $f(4) = -8$

Graphing a Function In Exercises 69 and 70, sketch the graph of the function.

69. $g(x) = [\![x]\!] - 2$

70. $f(x) = \begin{cases} 5x - 3, & x \geq -1 \\ -4x + 5, & x < -1 \end{cases}$

1.7 **Describing Transformations** In Exercises 71–80, h is related to one of the parent functions described in this chapter. (a) Identify the parent function f. (b) Describe the sequence of transformations from f to h. (c) Sketch the graph of h. (d) Use function notation to write h in terms of f.

71. $h(x) = x^2 - 9$ **72.** $h(x) = (x - 2)^3 + 2$

73. $h(x) = -\sqrt{x} + 4$ **74.** $h(x) = |x + 3| - 5$

75. $h(x) = -(x + 2)^2 + 3$ **76.** $h(x) = \frac{1}{2}(x - 1)^2 - 2$

77. $h(x) = -[\![x]\!] + 6$ **78.** $h(x) = -\sqrt{x + 1} + 9$

79. $h(x) = 5[\![x - 9]\!]$ **80.** $h(x) = -\frac{1}{3}x^3$

1.8 **Finding Arithmetic Combinations of Functions** In Exercises 81 and 82, find (a) $(f + g)(x)$, (b) $(f - g)(x)$, (c) $(fg)(x)$, and (d) $(f/g)(x)$. What is the domain of f/g?

81. $f(x) = x^2 + 3$, $g(x) = 2x - 1$
82. $f(x) = x^2 - 4$, $g(x) = \sqrt{3 - x}$

Finding Domains of Functions and Composite Functions In Exercises 83 and 84, find (a) $f \circ g$ and (b) $g \circ f$. Find the domain of each function and of each composite function.

83. $f(x) = \frac{1}{3}x - 3$, $g(x) = 3x + 1$
84. $f(x) = x^3 - 4$, $g(x) = \sqrt[3]{x + 7}$

Retail In Exercises 85 and 86, the price of a washing machine is x dollars. The function

$$f(x) = x - 100$$

gives the price of the washing machine after a \$100 rebate. The function

$$g(x) = 0.95x$$

gives the price of the washing machine after a 5% discount.

85. Find and interpret $(f \circ g)(x)$.
86. Find and interpret $(g \circ f)(x)$.

1.9 **Finding an Inverse Function Informally** In Exercises 87 and 88, find the inverse function of f informally. Verify that $f(f^{-1}(x)) = x$ and $f^{-1}(f(x)) = x$.

87. $f(x) = 3x + 8$ **88.** $f(x) = \dfrac{x - 4}{5}$

Applying the Horizontal Line Test In Exercises 89 and 90, use a graphing utility to graph the function, and use the Horizontal Line Test to determine whether the function has an inverse function.

89. $f(x) = (x - 1)^2$

90. $h(t) = \dfrac{2}{t - 3}$

Finding and Analyzing Inverse Functions In Exercises 91 and 92, (a) find the inverse function of f, (b) graph both f and f^{-1} on the same set of coordinate axes, (c) describe the relationship between the graphs of f and f^{-1}, and (d) state the domains and ranges of f and f^{-1}.

91. $f(x) = \frac{1}{2}x - 3$ **92.** $f(x) = \sqrt{x + 1}$

Restricting the Domain In Exercises 93 and 94, restrict the domain of the function f to an interval on which the function is increasing, and find f^{-1} on that interval.

93. $f(x) = 2(x - 4)^2$ **94.** $f(x) = |x - 2|$

1.10

95. Agriculture The ordered pairs below give the amount B (in millions of pounds) of beef produced on private farms each year from 2007 through 2014. *(Spreadsheet at LarsonPrecalculus.com)* *(Source: United States Department of Agriculture)*

DATA

(2007, 102.7) (2010, 84.2) (2013, 70.4)
(2008, 95.9) (2011, 75.0) (2014, 67.9)
(2009, 90.2) (2012, 76.3)

(a) Use a graphing utility to create a scatter plot of the data. Let t represent the year, with $t = 7$ corresponding to 2007.

(b) Use the *regression* feature of the graphing utility to find the equation of the least squares regression line that fits the data. Then graph the model and the scatter plot you found in part (a) in the same viewing window. How closely does the model represent the data?

96. Travel Time The travel time between two cities is inversely proportional to the average speed. A train travels between the cities in 3 hours at an average speed of 65 miles per hour. How long does it take to travel between the cities at an average speed of 80 miles per hour?

97. Cost The cost of constructing a wooden box with a square base varies jointly as the height of the box and the square of the width of the box. Constructing a box of height 16 inches and of width 6 inches costs \$28.80. How much does it cost to construct a box of height 14 inches and of width 8 inches?

Exploration

True or False? In Exercises 98 and 99, determine whether the statement is true or false. Justify your answer.

98. Relative to the graph of $f(x) = \sqrt{x}$, the graph of the function $h(x) = -\sqrt{x + 9} - 13$ is shifted 9 units to the left and 13 units down, then reflected in the x-axis.

99. If f and g are two inverse functions, then the domain of g is equal to the range of f.

Chapter Test

See **CalcChat.com** for tutorial help and worked-out solutions to odd-numbered exercises.

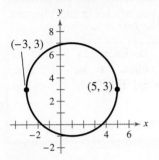

Figure for 6

Take this test as you would take a test in class. When you are finished, check your work against the answers given in the back of the book.

1. Plot the points $(-2, 5)$ and $(6, 0)$. Then find the distance between the points and the midpoint of the line segment joining the points.

2. A cylindrical can has a radius of 4 centimeters. Write the volume V of the can as a function of the height h.

In Exercises 3–5, find any intercepts and test for symmetry. Then sketch the graph of the equation.

3. $y = 3 - 5x$
4. $y = 4 - |x|$
5. $y = x^2 - 1$

6. Write the standard form of the equation of the circle shown at the left.

In Exercises 7 and 8, find an equation of the line passing through the pair of points. Sketch the line.

7. $(-2, 5), (1, -7)$
8. $(-4, -7), \left(1, \frac{4}{3}\right)$

9. Find equations of the lines that pass through the point $(0, 4)$ and are (a) parallel to and (b) perpendicular to the line $5x + 2y = 3$.

10. Let $f(x) = \dfrac{\sqrt{x+9}}{x^2 - 81}$. Find (a) $f(7)$, (b) $f(-5)$, and (c) $f(x - 9)$.

11. Find the domain of $f(x) = 10 - \sqrt{3 - x}$.

In Exercises 12–14, (a) find the zeros of the function, (b) use a graphing utility to graph the function, (c) approximate the open intervals on which the function is increasing, decreasing, or constant, and (d) determine whether the function is even, odd, or neither.

12. $f(x) = |x + 5|$
13. $f(x) = 4x\sqrt{3 - x}$
14. $f(x) = 2x^6 + 5x^4 - x^2$

15. Sketch the graph of $f(x) = \begin{cases} 3x + 7, & x \le -3 \\ 4x^2 - 1, & x > -3 \end{cases}$.

In Exercises 16–18, (a) identify the parent function f in the transformation, (b) describe the sequence of transformations from f to h, and (c) sketch the graph of h.

16. $h(x) = 4[\![x]\!]$
17. $h(x) = \sqrt{x + 5} + 8$
18. $h(x) = -2(x - 5)^3 + 3$

In Exercises 19 and 20, find (a) $(f + g)(x)$, (b) $(f - g)(x)$, (c) $(fg)(x)$, (d) $(f/g)(x)$, (e) $(f \circ g)(x)$, and (f) $(g \circ f)(x)$.

19. $f(x) = 3x^2 - 7, \quad g(x) = -x^2 - 4x + 5$
20. $f(x) = 1/x, \quad g(x) = 2\sqrt{x}$

In Exercises 21–23, determine whether the function has an inverse function. If it does, find the inverse function.

21. $f(x) = x^3 + 8$
22. $f(x) = |x^2 - 3| + 6$
23. $f(x) = 3x\sqrt{x}$

In Exercises 24–26, find the mathematical model that represents the statement. (Determine the constant of proportionality.)

24. v varies directly as the square root of s. ($v = 24$ when $s = 16$.)

25. A varies jointly as x and y. ($A = 500$ when $x = 15$ and $y = 8$.)

26. b varies inversely as a. ($b = 32$ when $a = 1.5$.)

Proofs in Mathematics ■ ■ ■ ■ ■ ■ ■ ■ ■ ■ ■ ■ ■ ■ ■

What does the word *proof* mean to you? In mathematics, the word *proof* means a valid argument. When you prove a statement or theorem, you must use facts, definitions, and accepted properties in a logical order. You can also use previously proved theorems in your proof. For example, the proof of the Midpoint Formula below uses the Distance Formula. There are several different proof methods, which you will see in later chapters.

> **The Midpoint Formula**   *(p.5)*
>
> The midpoint of the line segment joining the points (x_1, y_1) and (x_2, y_2) is
>
> $$\text{Midpoint} = \left(\frac{x_1 + x_2}{2}, \frac{y_1 + y_2}{2} \right).$$

Proof

Using the figure, you must show that $d_1 = d_2$ and $d_1 + d_2 = d_3$.

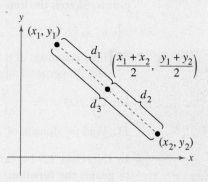

By the Distance Formula, you obtain

$$d_1 = \sqrt{\left(\frac{x_1 + x_2}{2} - x_1 \right)^2 + \left(\frac{y_1 + y_2}{2} - y_1 \right)^2}$$

$$= \sqrt{\left(\frac{x_2 - x_1}{2} \right)^2 + \left(\frac{y_2 - y_1}{2} \right)^2}$$

$$= \frac{1}{2}\sqrt{(x_2 - x_1)^2 + (y_2 - y_1)^2},$$

$$d_2 = \sqrt{\left(x_2 - \frac{x_1 + x_2}{2} \right)^2 + \left(y_2 - \frac{y_1 + y_2}{2} \right)^2}$$

$$= \sqrt{\left(\frac{x_2 - x_1}{2} \right)^2 + \left(\frac{y_2 - y_1}{2} \right)^2}$$

$$= \frac{1}{2}\sqrt{(x_2 - x_1)^2 + (y_2 - y_1)^2},$$

and

$$d_3 = \sqrt{(x_2 - x_1)^2 + (y_2 - y_1)^2}.$$

So, it follows that $d_1 = d_2$ and $d_1 + d_2 = d_3$.   ■

P.S. Problem Solving

1. **Monthly Wages** As a salesperson, you receive a monthly salary of $2000, plus a commission of 7% of sales. You receive an offer for a new job at $2300 per month, plus a commission of 5% of sales.

   (a) Write a linear equation for your current monthly wage W_1 in terms of your monthly sales S.

   (b) Write a linear equation for the monthly wage W_2 of your new job offer in terms of the monthly sales S.

   (c) Use a graphing utility to graph both equations in the same viewing window. Find the point of intersection. What does the point of intersection represent?

   (d) You expect sales of $20,000 per month. Should you change jobs? Explain.

2. **Cellphone Keypad** For the numbers 2 through 9 on a cellphone keypad (see figure), consider two relations: one mapping numbers onto letters, and the other mapping letters onto numbers. Are both relations functions? Explain.

3. **Sums and Differences of Functions** What can be said about the sum and difference of each pair of functions?

   (a) Two even functions

   (b) Two odd functions

   (c) An odd function and an even function

4. **Inverse Functions** The functions

   $$f(x) = x \quad \text{and} \quad g(x) = -x$$

   are their own inverse functions. Graph each function and explain why this is true. Graph other linear functions that are their own inverse functions. Find a formula for a family of linear functions that are their own inverse functions.

5. **Proof** Prove that a function of the form

   $$y = a_{2n}x^{2n} + a_{2n-2}x^{2n-2} + \cdots + a_2 x^2 + a_0$$

   is an even function.

6. **Miniature Golf** A golfer is trying to make a hole-in-one on the miniature golf green shown. The golf ball is at the point $(2.5, 2)$ and the hole is at the point $(9.5, 2)$. The golfer wants to bank the ball off the side wall of the green at the point (x, y). Find the coordinates of the point (x, y). Then write an equation for the path of the ball.

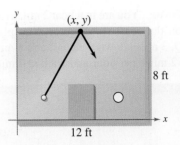

Figure for 6

7. **Titanic** At 2:00 P.M. on April 11, 1912, the *Titanic* left Cobh, Ireland, on her voyage to New York City. At 11:40 P.M. on April 14, the *Titanic* struck an iceberg and sank, having covered only about 2100 miles of the approximately 3400-mile trip.

   (a) What was the total duration of the voyage in hours?

   (b) What was the average speed in miles per hour?

   (c) Write a function relating the distance of the *Titanic* from New York City and the number of hours traveled. Find the domain and range of the function.

   (d) Graph the function in part (c).

8. **Average Rate of Change** Consider the function $f(x) = -x^2 + 4x - 3$. Find the average rate of change of the function from x_1 to x_2.

   (a) $x_1 = 1, x_2 = 2$

   (b) $x_1 = 1, x_2 = 1.5$

   (c) $x_1 = 1, x_2 = 1.25$

   (d) $x_1 = 1, x_2 = 1.125$

   (e) $x_1 = 1, x_2 = 1.0625$

   (f) Does the average rate of change seem to be approaching one value? If so, state the value.

   (g) Find the equations of the secant lines through the points $(x_1, f(x_1))$ and $(x_2, f(x_2))$ for parts (a)–(e).

   (h) Find the equation of the line through the point $(1, f(1))$ using your answer from part (f) as the slope of the line.

9. **Inverse of a Composition** Consider the functions $f(x) = 4x$ and $g(x) = x + 6$.

   (a) Find $(f \circ g)(x)$.

   (b) Find $(f \circ g)^{-1}(x)$.

   (c) Find $f^{-1}(x)$ and $g^{-1}(x)$.

   (d) Find $(g^{-1} \circ f^{-1})(x)$ and compare the result with that of part (b).

   (e) Repeat parts (a) through (d) for $f(x) = x^3 + 1$ and $g(x) = 2x$.

   (f) Write two one-to-one functions f and g, and repeat parts (a) through (d) for these functions.

   (g) Make a conjecture about $(f \circ g)^{-1}(x)$ and $(g^{-1} \circ f^{-1})(x)$.

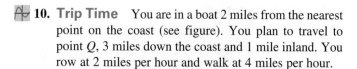

10. Trip Time You are in a boat 2 miles from the nearest point on the coast (see figure). You plan to travel to point Q, 3 miles down the coast and 1 mile inland. You row at 2 miles per hour and walk at 4 miles per hour.

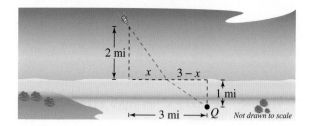

(a) Write the total time T (in hours) of the trip as a function of the distance x (in miles).

(b) Determine the domain of the function.

(c) Use a graphing utility to graph the function. Be sure to choose an appropriate viewing window.

(d) Find the value of x that minimizes T.

(e) Write a brief paragraph interpreting these values.

11. Heaviside Function The **Heaviside function**

$$H(x) = \begin{cases} 1, & x \geq 0 \\ 0, & x < 0 \end{cases}$$

is widely used in engineering applications. (See figure.) To print an enlarged copy of the graph, go to *MathGraphs.com*.

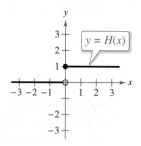

Sketch the graph of each function by hand.

(a) $H(x) - 2$

(b) $H(x - 2)$

(c) $-H(x)$

(d) $H(-x)$

(e) $\frac{1}{2}H(x)$

(f) $-H(x - 2) + 2$

12. Repeated Composition Let $f(x) = \dfrac{1}{1 - x}$.

(a) Find the domain and range of f.

(b) Find $f(f(x))$. What is the domain of this function?

(c) Find $f(f(f(x)))$. Is the graph a line? Why or why not?

13. Associative Property with Compositions Show that the Associative Property holds for compositions of functions—that is,

$$(f \circ (g \circ h))(x) = ((f \circ g) \circ h)(x).$$

14. Graphical Reasoning Use the graph of the function f to sketch the graph of each function. To print an enlarged copy of the graph, go to *MathGraphs.com*.

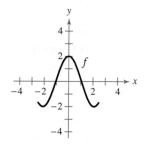

(a) $f(x + 1)$

(b) $f(x) + 1$

(c) $2f(x)$

(d) $f(-x)$

(e) $-f(x)$

(f) $|f(x)|$

(g) $f(|x|)$

15. Graphical Reasoning Use the graphs of f and f^{-1} to complete each table of function values.

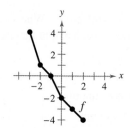 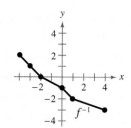

(a)

x	-4	-2	0	4
$(f(f^{-1}(x)))$				

(b)

x	-3	-2	0	1
$(f + f^{-1})(x)$				

(c)

x	-3	-2	0	1
$(f \cdot f^{-1})(x)$				

(d)

x	-4	-3	0	4
$|f^{-1}(x)|$				

2 Polynomial and Rational Functions

Candle Making Kits *(Example 12, page 161)*

Electrical Circuit
(Example 87, page 151)

Lyme Disease *(Exercise 82, page 144)*

Tree Growth
(Exercise 98, page 135)

Path of a Diver *(Exercise 67, page 121)*

2.1 Quadratic Functions and Models

Quadratic functions have many real-life applications. For example, in Exercise 67 on page 121, you will use a quadratic function that models the path of a diver.

- Analyze graphs of quadratic functions.
- Write quadratic functions in standard form and use the results to sketch their graphs.
- Find minimum and maximum values of quadratic functions in real-life applications.

The Graph of a Quadratic Function

In this and the next section, you will study graphs of polynomial functions. Section 1.6 introduced basic functions such as linear, constant, and squaring functions.

$f(x) = ax + b$ Linear function

$f(x) = c$ Constant function

$f(x) = x^2$ Squaring function

These are examples of **polynomial functions.**

Definition of a Polynomial Function

Let n be a nonnegative integer and let $a_n, a_{n-1}, \ldots, a_2, a_1, a_0$ be real numbers with $a_n \neq 0$. The function

$$f(x) = a_n x^n + a_{n-1} x^{n-1} + \cdots + a_2 x^2 + a_1 x + a_0$$

is a **polynomial function of x with degree n.**

Polynomial functions are classified by degree. For example, a constant function $f(x) = c$ with $c \neq 0$ has degree 0, and a linear function $f(x) = ax + b$ with $a \neq 0$ has degree 1. In this section, you will study **quadratic functions,** which are second-degree polynomial functions.

For example, each function listed below is a quadratic function.

$f(x) = x^2 + 6x + 2$

$g(x) = 2(x + 1)^2 - 3$

$h(x) = 9 + \frac{1}{4}x^2$

$k(x) = (x - 2)(x + 1)$

Note that the squaring function is a simple quadratic function.

Definition of a Quadratic Function

Let a, b, and c be real numbers with $a \neq 0$. The function

$$f(x) = ax^2 + bx + c$$ Quadratic function

is a **quadratic function.**

Time, t	Height, h
0	6
4	774
8	1030
12	774
16	6

Often, quadratic functions can model real-life data. For example, the table at the left shows the heights h (in feet) of a projectile fired from an initial height of 6 feet with an initial velocity of 256 feet per second at selected values of time t (in seconds). A quadratic model for the data in the table is

$$h(t) = -16t^2 + 256t + 6, \quad 0 \leq t \leq 16.$$

The graph of a quadratic function is a "U"-shaped curve called a **parabola.** Parabolas occur in many real-life applications—including those that involve reflective properties of satellite dishes and flashlight reflectors. You will study these properties in Section 10.2.

All parabolas are symmetric with respect to a line called the **axis of symmetry,** or simply the **axis** of the parabola. The point where the axis intersects the parabola is the **vertex** of the parabola. When the leading coefficient is positive, the graph of

$$f(x) = ax^2 + bx + c$$

is a parabola that opens upward. When the leading coefficient is negative, the graph is a parabola that opens downward. The next two figures show the axes and vertices of parabolas for cases where $a > 0$ and $a < 0$.

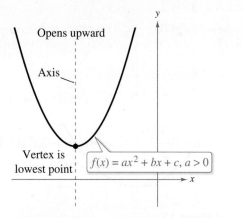

Leading coefficient is positive.

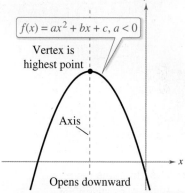

Leading coefficient is negative.

The simplest type of quadratic function is one in which $b = c = 0$. In this case, the function has the form $f(x) = ax^2$. Its graph is a parabola whose vertex is $(0, 0)$. When $a > 0$, the vertex is the point with the *minimum y*-value on the graph, and when $a < 0$, the vertex is the point with the *maximum y*-value on the graph, as shown in the figures below.

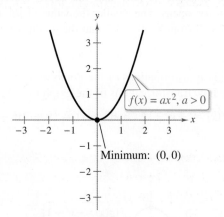

Leading coefficient is positive.

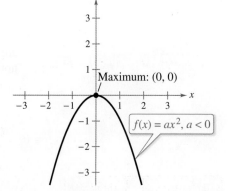

Leading coefficient is negative.

When sketching the graph of $f(x) = ax^2$, it is helpful to use the graph of $y = x^2$ as a reference, as suggested in Section 1.7. There you learned that when $a > 1$, the graph of $y = af(x)$ is a vertical stretch of the graph of $y = f(x)$. When $0 < a < 1$, the graph of $y = af(x)$ is a vertical shrink of the graph of $y = f(x)$. Example 1 demonstrates this again.

EXAMPLE 1　**Sketching Graphs of Quadratic Functions**

See LarsonPrecalculus.com for an interactive version of this type of example.

Sketch the graph of each quadratic function and compare it with the graph of $y = x^2$.

a. $f(x) = \frac{1}{3}x^2$　　**b.** $g(x) = 2x^2$

▷ **ALGEBRA HELP**　To review techniques for shifting, reflecting, stretching, and shrinking graphs, see Section 1.7.

Solution

a. Compared with $y = x^2$, each output of $f(x) = \frac{1}{3}x^2$ "shrinks" by a factor of $\frac{1}{3}$, producing the broader parabola shown in Figure 2.1.

b. Compared with $y = x^2$, each output of $g(x) = 2x^2$ "stretches" by a factor of 2, producing the narrower parabola shown in Figure 2.2.

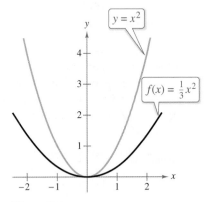

Figure 2.1　　　　　　　　　**Figure 2.2**

✓ *Checkpoint* ◀))) *Audio-video solution in English & Spanish at LarsonPrecalculus.com*

Sketch the graph of each quadratic function and compare it with the graph of $y = x^2$.

a. $f(x) = \frac{1}{4}x^2$　　**b.** $g(x) = -\frac{1}{6}x^2$　　**c.** $h(x) = \frac{5}{2}x^2$　　**d.** $k(x) = -4x^2$　　■

In Example 1, note that the coefficient a determines how wide the parabola $f(x) = ax^2$ opens. The smaller the value of $|a|$, the wider the parabola opens. Recall from Section 1.7 that the graphs of

$$y = f(x \pm c), \quad y = f(x) \pm c, \quad y = f(-x), \quad \text{and} \quad y = -f(x)$$

are rigid transformations of the graph of $y = f(x)$. For example, in the figures below, notice how transformations of the graph of $y = x^2$ can produce the graphs of

$$f(x) = -x^2 + 1 \quad \text{and} \quad g(x) = (x + 2)^2 - 3.$$

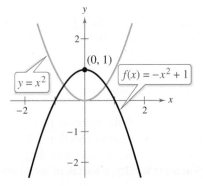

Reflection in x-axis followed by an upward shift of one unit

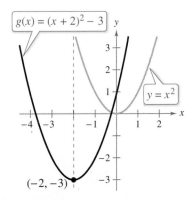

Left shift of two units followed by a downward shift of three units

The Standard Form of a Quadratic Function

The **standard form** of a quadratic function is $f(x) = a(x - h)^2 + k$. This form is especially convenient for sketching a parabola because it identifies the vertex of the parabola as (h, k).

Standard Form of a Quadratic Function

The quadratic function

$$f(x) = a(x - h)^2 + k, \quad a \neq 0$$

is in **standard form.** The graph of f is a parabola whose axis is the vertical line $x = h$ and whose vertex is the point (h, k). When $a > 0$, the parabola opens upward, and when $a < 0$, the parabola opens downward.

To graph a parabola, it is helpful to begin by writing the quadratic function in standard form using the process of completing the square, as illustrated in Example 2. In this example, notice that when completing the square, you *add and subtract* the square of half the coefficient of x within the parentheses instead of adding the value to each side of the equation as is done in Appendix A.5.

▷ **ALGEBRA HELP** To review techniques for completing the square, see Appendix A.5.

EXAMPLE 2   **Using Standard Form to Graph a Parabola**

Sketch the graph of $f(x) = 2x^2 + 8x + 7$. Identify the vertex and the axis of the parabola.

Solution   Begin by writing the quadratic function in standard form. Notice that the first step in completing the square is to factor out any coefficient of x^2 that is not 1.

$$\begin{aligned}
f(x) &= 2x^2 + 8x + 7 && \text{Write original function.} \\
&= 2(x^2 + 4x) + 7 && \text{Factor 2 out of } x\text{-terms.} \\
&= 2(x^2 + 4x + 4 - 4) + 7 && \text{Add and subtract 4 within parentheses.}
\end{aligned}$$

$$\underset{(4/2)^2}{\underline{\qquad\uparrow\qquad}}$$

$$\begin{aligned}
&= 2(x^2 + 4x + 4) - 2(4) + 7 && \text{Distributive Property} \\
&= 2(x^2 + 4x + 4) - 8 + 7 && \text{Simplify.} \\
&= 2(x + 2)^2 - 1 && \text{Write in standard form.}
\end{aligned}$$

The graph of f is a parabola that opens upward and has its vertex at $(-2, -1)$. This corresponds to a left shift of two units and a downward shift of one unit relative to the graph of $y = 2x^2$, as shown in the figure. The axis of the parabola is the vertical line through the vertex, $x = -2$, also shown in the figure.

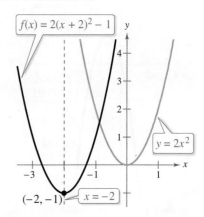

$f(x) = 2(x + 2)^2 - 1$

$y = 2x^2$

$(-2, -1)$ $x = -2$

✓ **Checkpoint** ◀))) *Audio-video solution in English & Spanish at LarsonPrecalculus.com*

Sketch the graph of $f(x) = 3x^2 - 6x + 4$. Identify the vertex and the axis of the parabola. ∎

▷ **ALGEBRA HELP** To review techniques for solving quadratic equations, see Appendix A.5.

To find the x-intercepts of the graph of $f(x) = ax^2 + bx + c$, you must solve the equation $ax^2 + bx + c = 0$. When $ax^2 + bx + c$ does not factor, use completing the square or the Quadratic Formula to find the x-intercepts. Remember, however, that a parabola may not have x-intercepts.

EXAMPLE 3 **Finding the Vertex and x-Intercepts of a Parabola**

Sketch the graph of $f(x) = -x^2 + 6x - 8$. Identify the vertex and x-intercepts.

Solution

$$f(x) = -x^2 + 6x - 8 \qquad \text{Write original function.}$$
$$= -(x^2 - 6x) - 8 \qquad \text{Factor } -1 \text{ out of } x\text{-terms.}$$
$$= -(x^2 - 6x + 9 - 9) - 8 \qquad \text{Add and subtract 9 within parentheses.}$$

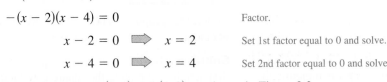

$$\overbrace{}^{(-6/2)^2}$$

$$= -(x^2 - 6x + 9) - (-9) - 8 \qquad \text{Distributive Property}$$
$$= -(x - 3)^2 + 1 \qquad \text{Write in standard form.}$$

The graph of f is a parabola that opens downward with vertex $(3, 1)$. Next, find the x-intercepts of the graph.

$$-(x^2 - 6x + 8) = 0 \qquad \text{Factor out } -1.$$
$$-(x - 2)(x - 4) = 0 \qquad \text{Factor.}$$
$$x - 2 = 0 \implies x = 2 \qquad \text{Set 1st factor equal to 0 and solve.}$$
$$x - 4 = 0 \implies x = 4 \qquad \text{Set 2nd factor equal to 0 and solve.}$$

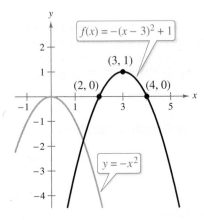

Figure 2.3

So, the x-intercepts are $(2, 0)$ and $(4, 0)$, as shown in Figure 2.3.

✓ *Checkpoint* ◀))) *Audio-video solution in English & Spanish at LarsonPrecalculus.com*

Sketch the graph of $f(x) = x^2 - 4x + 3$. Identify the vertex and x-intercepts.

EXAMPLE 4 **Writing a Quadratic Function**

Write the standard form of the quadratic function whose graph is a parabola with vertex $(1, 2)$ and that passes through the point $(3, -6)$.

Solution The vertex is $(h, k) = (1, 2)$, so the equation has the form

$$f(x) = a(x - 1)^2 + 2. \qquad \text{Substitute for } h \text{ and } k \text{ in standard form.}$$

The parabola passes through the point $(3, -6)$, so it follows that $f(3) = -6$. So,

$$f(x) = a(x - 1)^2 + 2 \qquad \text{Write in standard form.}$$
$$-6 = a(3 - 1)^2 + 2 \qquad \text{Substitute 3 for } x \text{ and } -6 \text{ for } f(x).$$
$$-6 = 4a + 2 \qquad \text{Simplify.}$$
$$-8 = 4a \qquad \text{Subtract 2 from each side.}$$
$$-2 = a. \qquad \text{Divide each side by 4.}$$

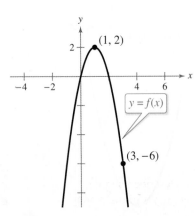

Figure 2.4

The function in standard form is $f(x) = -2(x - 2)^2 + 2$. Figure 2.4 shows the graph of f.

✓ *Checkpoint* ◀))) *Audio-video solution in English & Spanish at LarsonPrecalculus.com*

Write the standard form of the quadratic function whose graph is a parabola with vertex $(-4, 11)$ and that passes through the point $(-6, 15)$. ■

Finding Minimum and Maximum Values

Many applications involve finding the maximum or minimum value of a quadratic function. By completing the square within the quadratic function $f(x) = ax^2 + bx + c$, you can rewrite the function in standard form (see Exercise 79).

$$f(x) = a\left(x + \frac{b}{2a}\right)^2 + \left(c - \frac{b^2}{4a}\right) \qquad \text{Standard form}$$

So, the vertex of the graph of f is $\left(-\dfrac{b}{2a},\ f\left(-\dfrac{b}{2a}\right)\right)$.

Minimum and Maximum Values of Quadratic Functions

Consider the function $f(x) = ax^2 + bx + c$ with vertex $\left(-\dfrac{b}{2a}, f\left(-\dfrac{b}{2a}\right)\right)$.

1. When $a > 0$, f has a *minimum* at $x = -\dfrac{b}{2a}$. The minimum value is $f\left(-\dfrac{b}{2a}\right)$.

2. When $a < 0$, f has a *maximum* at $x = -\dfrac{b}{2a}$. The maximum value is $f\left(-\dfrac{b}{2a}\right)$.

EXAMPLE 5   Maximum Height of a Baseball

The path of a baseball after being hit is modeled by $f(x) = -0.0032x^2 + x + 3$, where $f(x)$ is the height of the baseball (in feet) and x is the horizontal distance from home plate (in feet). What is the maximum height of the baseball?

Algebraic Solution

For this quadratic function, you have

$$f(x) = ax^2 + bx + c = -0.0032x^2 + x + 3$$

which shows that $a = -0.0032$ and $b = 1$. Because $a < 0$, the function has a maximum at $x = -b/(2a)$. So, the baseball reaches its maximum height when it is

$$x = -\frac{b}{2a} = -\frac{1}{2(-0.0032)} = 156.25 \text{ feet}$$

from home plate. At this distance, the maximum height is

$$f(156.25) = -0.0032(156.25)^2 + 156.25 + 3 = 81.125 \text{ feet.}$$

Graphical Solution

The maximum height is $y = 81.125$ feet at $x = 156.25$ feet.

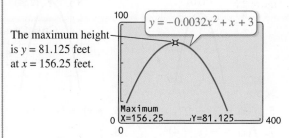

✔ *Checkpoint* ◀))) *Audio-video solution in English & Spanish at LarsonPrecalculus.com*

Rework Example 5 when the path of the baseball is modeled by

$$f(x) = -0.007x^2 + x + 4.$$

Summarize   (Section 2.1)

1. State the definition of a quadratic function and describe its graph *(pages 114–116)*. For an example of sketching graphs of quadratic functions, see Example 1.

2. State the standard form of a quadratic function *(page 117)*. For examples that use the standard form of a quadratic function, see Examples 2–4.

3. Explain how to find the minimum or maximum value of a quadratic function *(page 119)*. For a real-life application, see Example 5.

2.1 Exercises

See **CalcChat.com** for tutorial help and worked-out solutions to odd-numbered exercises.

Vocabulary: Fill in the blanks.

1. Linear, constant, and squaring functions are examples of _____ functions.
2. A polynomial function of x with degree n has the form $f(x) = a_n x^n + a_{n-1} x^{n-1} + \cdots + a_1 x + a_0$ $(a_n \neq 0)$, where n is a _____ _____ and $a_n, a_{n-1}, \ldots, a_1, a_0$ are _____ numbers.
3. A _____ function is a second-degree polynomial function, and its graph is called a _____.
4. When the graph of a quadratic function opens downward, its leading coefficient is _____ and the vertex of the graph is a _____.

Skills and Applications

Matching In Exercises 5–8, match the quadratic function with its graph. [The graphs are labeled (a), (b), (c), and (d).]

(a)
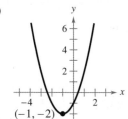

(b)
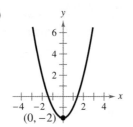

(c)
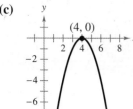

(d)
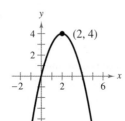

5. $f(x) = x^2 - 2$
6. $f(x) = (x + 1)^2 - 2$
7. $f(x) = -(x - 4)^2$
8. $f(x) = 4 - (x - 2)^2$

 Sketching Graphs of Quadratic Functions In Exercises 9–12, sketch the graph of each quadratic function and compare it with the graph of $y = x^2$.

9. (a) $f(x) = \frac{1}{2}x^2$ (b) $g(x) = -\frac{1}{8}x^2$
   (c) $h(x) = \frac{3}{2}x^2$ (d) $k(x) = -3x^2$
10. (a) $f(x) = x^2 + 1$ (b) $g(x) = x^2 - 1$
    (c) $h(x) = x^2 + 3$ (d) $k(x) = x^2 - 3$
11. (a) $f(x) = (x - 1)^2$ (b) $g(x) = (3x)^2 + 1$
    (c) $h(x) = \left(\frac{1}{3}x\right)^2 - 3$ (d) $k(x) = (x + 3)^2$
12. (a) $f(x) = -\frac{1}{2}(x - 2)^2 + 1$
    (b) $g(x) = \left[\frac{1}{2}(x - 1)\right]^2 - 3$
    (c) $h(x) = -\frac{1}{2}(x + 2)^2 - 1$
    (d) $k(x) = [2(x + 1)]^2 + 4$

 Using Standard Form to Graph a Parabola In Exercises 13–26, write the quadratic function in standard form and sketch its graph. Identify the vertex, axis of symmetry, and x-intercept(s).

13. $f(x) = x^2 - 6x$ 14. $g(x) = x^2 - 8x$
15. $h(x) = x^2 - 8x + 16$ 16. $g(x) = x^2 + 2x + 1$
17. $f(x) = x^2 - 6x + 2$ 18. $f(x) = x^2 + 16x + 61$
19. $f(x) = x^2 - 8x + 21$ 20. $f(x) = x^2 + 12x + 40$
21. $f(x) = x^2 - x + \frac{5}{4}$ 22. $f(x) = x^2 + 3x + \frac{1}{4}$
23. $f(x) = -x^2 + 2x + 5$ 24. $f(x) = -x^2 - 4x + 1$
25. $h(x) = 4x^2 - 4x + 21$ 26. $f(x) = 2x^2 - x + 1$

Using Technology In Exercises 27–34, use a graphing utility to graph the quadratic function. Identify the vertex, axis of symmetry, and x-intercept(s). Then check your results algebraically by writing the quadratic function in standard form.

27. $f(x) = -(x^2 + 2x - 3)$ 28. $f(x) = -(x^2 + x - 30)$
29. $g(x) = x^2 + 8x + 11$ 30. $f(x) = x^2 + 10x + 14$
31. $f(x) = -2x^2 + 12x - 18$
32. $f(x) = -4x^2 + 24x - 41$
33. $g(x) = \frac{1}{2}(x^2 + 4x - 2)$
34. $f(x) = \frac{3}{5}(x^2 + 6x - 5)$

Writing a Quadratic Function In Exercises 35 and 36, write the standard form of the quadratic function whose graph is the parabola shown.

35.
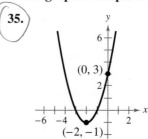

36.
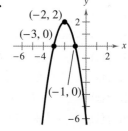

Writing a Quadratic Function In Exercises 37–46, write the standard form of the quadratic function whose graph is a parabola with the given vertex and that passes through the given point.

37. Vertex: $(-2, 5)$; point: $(0, 9)$

38. Vertex: $(-3, -10)$; point: $(0, 8)$

39. Vertex: $(1, -2)$; point: $(-1, 14)$

40. Vertex: $(2, 3)$; point: $(0, 2)$

41. Vertex: $(5, 12)$; point: $(7, 15)$

42. Vertex: $(-2, -2)$; point: $(-1, 0)$

43. Vertex: $\left(-\frac{1}{4}, \frac{3}{2}\right)$; point: $(-2, 0)$

44. Vertex: $\left(\frac{5}{2}, -\frac{3}{4}\right)$; point: $(-2, 4)$

45. Vertex: $\left(-\frac{5}{2}, 0\right)$; point: $\left(-\frac{7}{2}, -\frac{16}{3}\right)$

46. Vertex: $(6, 6)$; point: $\left(\frac{61}{10}, \frac{3}{2}\right)$

Graphical Reasoning In Exercises 47–50, determine the x-intercept(s) of the graph visually. Then find the x-intercept(s) algebraically to confirm your results.

47. $y = x^2 - 2x - 3$ **48.** $y = x^2 - 4x - 5$

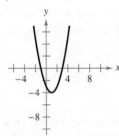
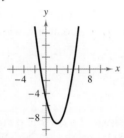

49. $y = 2x^2 + 5x - 3$ **50.** $y = -2x^2 + 5x + 3$

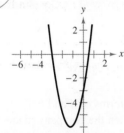
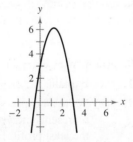

 **Using Technology** In Exercises 51–56, use a graphing utility to graph the quadratic function. Find the x-intercept(s) of the graph and compare them with the solutions of the corresponding quadratic equation when $f(x) = 0$.

51. $f(x) = x^2 - 4x$

52. $f(x) = -2x^2 + 10x$

53. $f(x) = x^2 - 9x + 18$

54. $f(x) = x^2 - 8x - 20$

55. $f(x) = 2x^2 - 7x - 30$

56. $f(x) = \frac{7}{10}(x^2 + 12x - 45)$

Finding Quadratic Functions In Exercises 57–62, find two quadratic functions, one that opens upward and one that opens downward, whose graphs have the given x-intercepts. (There are many correct answers.)

57. $(-3, 0), (3, 0)$ **58.** $(-5, 0), (5, 0)$

59. $(-1, 0), (4, 0)$ **60.** $(-2, 0), (3, 0)$

61. $(-3, 0), \left(-\frac{1}{2}, 0\right)$ **62.** $\left(-\frac{3}{2}, 0\right), (-5, 0)$

Number Problems In Exercises 63–66, find two positive real numbers whose product is a maximum.

63. The sum is 110.

64. The sum is S.

65. The sum of the first and twice the second is 24.

66. The sum of the first and three times the second is 42.

67. Path of a Diver

The path of a diver is modeled by

$$f(x) = -\frac{4}{9}x^2 + \frac{24}{9}x + 12$$

where $f(x)$ is the height (in feet) and x is the horizontal distance (in feet) from the end of the diving board. What is the maximum height of the diver?

68. Height of a Ball The path of a punted football is modeled by

$$f(x) = -\frac{16}{2025}x^2 + \frac{9}{5}x + 1.5$$

where $f(x)$ is the height (in feet) and x is the horizontal distance (in feet) from the point at which the ball is punted.

(a) How high is the ball when it is punted?

(b) What is the maximum height of the punt?

(c) How long is the punt?

69. Minimum Cost A manufacturer of lighting fixtures has daily production costs of $C = 800 - 10x + 0.25x^2$, where C is the total cost (in dollars) and x is the number of units produced. What daily production number yields a minimum cost?

70. Maximum Profit The profit P (in hundreds of dollars) that a company makes depends on the amount x (in hundreds of dollars) the company spends on advertising according to the model $P = 230 + 20x - 0.5x^2$. What expenditure for advertising yields a maximum profit?

71. Maximum Revenue The total revenue R earned (in thousands of dollars) from manufacturing handheld video games is given by $R(p) = -25p^2 + 1200p$, where p is the price per unit (in dollars).

(a) Find the revenues when the prices per unit are $20, $25, and $30.

(b) Find the unit price that yields a maximum revenue. What is the maximum revenue? Explain.

72. Maximum Revenue The total revenue R earned per day (in dollars) from a pet-sitting service is given by $R(p) = -12p^2 + 150p$, where p is the price charged per pet (in dollars).

(a) Find the revenues when the prices per pet are $4, $6, and $8.

(b) Find the unit price that yields a maximum revenue. What is the maximum revenue? Explain.

73. Maximum Area A rancher has 200 feet of fencing to enclose two adjacent rectangular corrals (see figure).

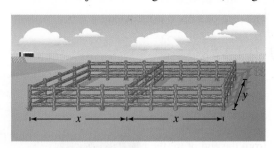

(a) Write the area A of the corrals as a function of x.

(b) What dimensions produce a maximum enclosed area?

74. Maximum Area A Norman window is constructed by adjoining a semicircle to the top of an ordinary rectangular window (see figure). The perimeter of the window is 16 feet.

(a) Write the area A of the window as a function of x.

(b) What dimensions produce a window of maximum area?

Exploration

True or False? **In Exercises 75 and 76, determine whether the statement is true or false. Justify your answer.**

75. The graph of $f(x) = -12x^2 - 1$ has no x-intercepts.

76. The graphs of $f(x) = -4x^2 - 10x + 7$ and $g(x) = 12x^2 + 30x + 1$ have the same axis of symmetry.

Think About It **In Exercises 77 and 78, find the values of b such that the function has the given maximum or minimum value.**

77. $f(x) = -x^2 + bx - 75$; Maximum value: 25

78. $f(x) = x^2 + bx - 25$; Minimum value: -50

79. Verifying the Vertex Write the quadratic function

$$f(x) = ax^2 + bx + c$$

in standard form to verify that the vertex occurs at

$$\left(-\frac{b}{2a}, f\left(-\frac{b}{2a} \right) \right).$$

80. **HOW DO YOU SEE IT?** The graph shows a quadratic function of the form

$$P(t) = at^2 + bt + c$$

which represents the yearly profit for a company, where $P(t)$ is the profit in year t.

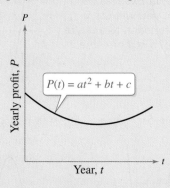

(a) Is the value of a positive, negative, or zero? Explain.

(b) Write an expression in terms of a and b that represents the year t when the company made the least profit.

(c) The company made the same yearly profits in 2008 and 2016. Estimate the year in which the company made the least profit.

81. Proof Assume that the function

$$f(x) = ax^2 + bx + c, \quad a \neq 0$$

has two real zeros. Prove that the x-coordinate of the vertex of the graph is the average of the zeros of f. (*Hint:* Use the Quadratic Formula.)

Project: Height of a Basketball To work an extended application analyzing the height of a dropped basketball, visit this text's website at *LarsonPrecalculus.com.*

2.2 Polynomial Functions of Higher Degree

Polynomial functions have many real-life applications. For example, in Exercise 98 on page 135, you will use a polynomial function to analyze the growth of a red oak tree.

■ Use transformations to sketch graphs of polynomial functions.
■ Use the Leading Coefficient Test to determine the end behaviors of graphs of polynomial functions.
■ Find real zeros of polynomial functions and use them as sketching aids.
■ Use the Intermediate Value Theorem to help locate real zeros of polynomial functions.

Graphs of Polynomial Functions

In this section, you will study basic features of the graphs of polynomial functions. One feature is that the graph of a polynomial function is **continuous.** Essentially, this means that the graph of a polynomial function has no breaks, holes, or gaps, as shown in Figure 2.5(a). The graph shown in Figure 2.5(b) is an example of a piecewise-defined function that is not continuous.

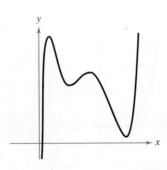

(a) Polynomial functions have continuous graphs.

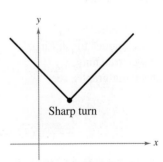

(b) Functions with graphs that are not continuous are not polynomial functions.

Figure 2.5

Another feature of the graph of a polynomial function is that it has only smooth, rounded turns, as shown in Figure 2.6(a). The graph of a polynomial function cannot have a sharp turn, such as the one shown in Figure 2.6(b).

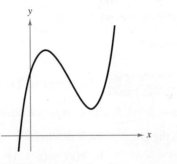

(a) Polynomial functions have graphs with smooth, rounded turns.

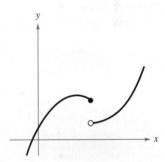

Sharp turn

(b) Functions with graphs that have sharp turns are not polynomial functions.

Figure 2.6

Sketching graphs of polynomial functions of degree greater than 2 is often more involved than sketching graphs of polynomial functions of degree 0, 1, or 2. However, using the features presented in this section, along with your knowledge of point plotting, intercepts, and symmetry, you should be able to make reasonably accurate sketches by hand.

. . . . ▷

REMARK For functions of the form $f(x) = x^n$, if n is even, then the graph of the function is symmetric with respect to the y-axis, and if n is odd, then the graph of the function is symmetric with respect to the origin.

The polynomial functions that have the simplest graphs are monomial functions of the form $f(x) = x^n$, where n is an integer greater than zero. When n is *even*, the graph is similar to the graph of $f(x) = x^2$, and when n is *odd*, the graph is similar to the graph of $f(x) = x^3$, as shown in Figure 2.7. Moreover, the greater the value of n, the flatter the graph near the origin. Polynomial functions of the form $f(x) = x^n$ are often referred to as **power functions.**

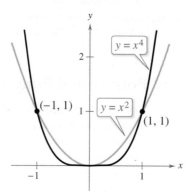

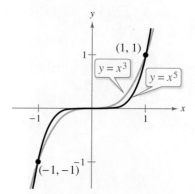

(a) When n is even, the graph of $y = x^n$ touches the x-axis at the x-intercept.

(b) When n is odd, the graph of $y = x^n$ crosses the x-axis at the x-intercept.

Figure 2.7

EXAMPLE 1   **Sketching Transformations of Monomial Functions**

See LarsonPrecalculus.com for an interactive version of this type of example.

Sketch the graph of each function.

a. $f(x) = -x^5$ **b.** $h(x) = (x + 1)^4$

Solution

a. The degree of $f(x) = -x^5$ is odd, so its graph is similar to the graph of $y = x^3$. In Figure 2.8, note that the negative coefficient has the effect of reflecting the graph in the x-axis.

b. The degree of $h(x) = (x + 1)^4$ is even, so its graph is similar to the graph of $y = x^2$. In Figure 2.9, note that the graph of h is a left shift by one unit of the graph of $y = x^4$.

▷ **ALGEBRA HELP** To review techniques for shifting, reflecting, stretching, and shrinking graphs, see Section 1.7.

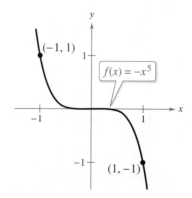

Figure 2.8

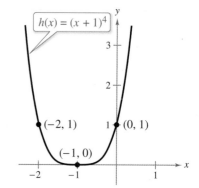

Figure 2.9

✓ *Checkpoint* ◀))) Audio-video solution in English & Spanish at LarsonPrecalculus.com

Sketch the graph of each function.

a. $f(x) = (x + 5)^4$ **b.** $g(x) = x^4 - 7$

c. $h(x) = 7 - x^4$ **d.** $k(x) = \frac{1}{4}(x - 3)^4$

The Leading Coefficient Test

In Example 1, note that both graphs eventually rise or fall without bound as x moves to the left or to the right. A polynomial function's degree (even or odd) and its leading coefficient (positive or negative) determine whether the graph of the function eventually rises or falls, as described in the **Leading Coefficient Test.**

Leading Coefficient Test

As x moves without bound to the left or to the right, the graph of the polynomial function

$$f(x) = a_n x^n + \cdots + a_1 x + a_0, \quad a_n \neq 0$$

eventually rises or falls in the manner described below.

1. When n is *odd*:

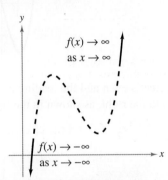

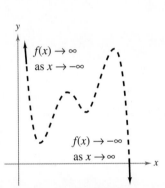

If the leading coefficient is positive $(a_n > 0)$, then the graph falls to the left and rises to the right.

If the leading coefficient is negative $(a_n < 0)$, then the graph rises to the left and falls to the right.

2. When n is *even*:

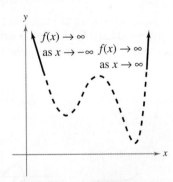

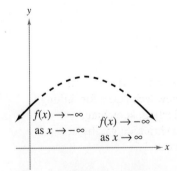

If the leading coefficient is positive $(a_n > 0)$, then the graph rises to the left and to the right.

If the leading coefficient is negative $(a_n < 0)$, then the graph falls to the left and to the right.

The dashed portions of the graphs indicate that the test determines *only* the right-hand and left-hand behavior of the graph.

▷

REMARK The notation "$f(x) \to -\infty$ as $x \to -\infty$" means that the graph falls to the left. The notation "$f(x) \to \infty$ as $x \to \infty$" means that the graph rises to the right. Identify and interpret similar notation for the other two possible types of end behavior given in the Leading Coefficient Test.

As you continue to study polynomial functions and their graphs, you will notice that the degree of a polynomial plays an important role in determining other characteristics of the polynomial function and its graph.

EXAMPLE 2   **Applying the Leading Coefficient Test**

Describe the left-hand and right-hand behavior of the graph of each function.

a. $f(x) = -x^3 + 4x$ **b.** $f(x) = x^4 - 5x^2 + 4$ **c.** $f(x) = x^5 - x$

Solution

a. The degree is odd and the leading coefficient is negative, so the graph rises to the left and falls to the right, as shown in the figure below.

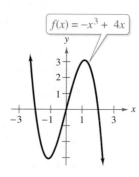

b. The degree is even and the leading coefficient is positive, so the graph rises to the left and to the right, as shown in the figure below.

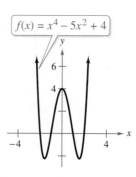

c. The degree is odd and the leading coefficient is positive, so the graph falls to the left and rises to the right, as shown in the figure below.

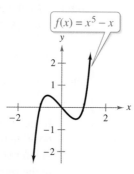

✓ **Checkpoint** 🔊 *Audio-video solution in English & Spanish at LarsonPrecalculus.com*

Describe the left-hand and right-hand behavior of the graph of each function.

a. $f(x) = \frac{1}{4}x^3 - 2x$ **b.** $f(x) = -3.6x^5 + 5x^3 - 1$ ■

In Example 2, note that the Leading Coefficient Test tells you only whether the graph *eventually* rises or falls to the left or to the right. You must use other tests to determine other characteristics of the graph, such as intercepts and minimum and maximum points.

Real Zeros of Polynomial Functions

It is possible to show that for a polynomial function f of degree n, the two statements below are true.

• • • • • • • • • • • • • • ▷

··REMARK Remember that the *zeros* of a function of x are the x-values for which the function is zero.

1. The function f has, at most, n real zeros. (You will study this result in detail in the discussion of the Fundamental Theorem of Algebra in Section 2.5.)

2. The graph of f has, at most, $n - 1$ turning points. (Turning points, also called relative minima or relative maxima, are points at which the graph changes from increasing to decreasing or vice versa.)

Finding the zeros of a polynomial function is an important problem in algebra. There is a strong interplay between graphical and algebraic approaches to this problem.

Real Zeros of Polynomial Functions

When f is a polynomial function and a is a real number, the statements listed below are equivalent.

1. $x = a$ is a *zero* of the function f.

2. $x = a$ is a *solution* of the polynomial equation $f(x) = 0$.

3. $(x - a)$ is a *factor* of the polynomial $f(x)$.

4. $(a, 0)$ is an *x-intercept* of the graph of f.

EXAMPLE 3 **Finding Real Zeros of a Polynomial Function**

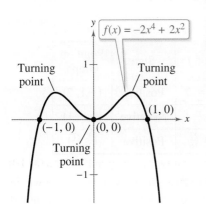

Figure 2.10

Find all real zeros of $f(x) = -2x^4 + 2x^2$. Then determine the maximum possible number of turning points of the graph of the function.

Solution To find the real zeros of the function, set $f(x)$ equal to zero and then solve for x.

$$-2x^4 + 2x^2 = 0 \qquad \text{Set } f(x) \text{ equal to 0.}$$
$$-2x^2(x^2 - 1) = 0 \qquad \text{Remove common monomial factor.}$$
$$-2x^2(x - 1)(x + 1) = 0 \qquad \text{Factor completely.}$$

So, the real zeros are $x = 0$, $x = 1$, and $x = -1$, and the corresponding x-intercepts occur at $(0, 0)$, $(1, 0)$, and $(-1, 0)$. The function is a fourth-degree polynomial, so the graph of f can have at most $4 - 1 = 3$ turning points. In this case, the graph of f has three turning points. Figure 2.10 shows the graph of f.

✓ *Checkpoint*  *Audio-video solution in English & Spanish at LarsonPrecalculus.com*

Find all real zeros of $f(x) = x^3 - 12x^2 + 36x$. Then determine the maximum possible number of turning points of the graph of the function. ∎

▷ **ALGEBRA HELP** The solution to Example 3 uses polynomial factoring. To review the techniques for factoring polynomials, see Appendix A.3.

In Example 3, note that the factor $-2x^2$ yields the *repeated* zero $x = 0$. The exponent is even, so the graph touches the x-axis at $x = 0$.

Repeated Zeros

A factor $(x - a)^k$, $k > 1$, yields a **repeated zero** $x = a$ of **multiplicity** k.

1. When k is odd, the graph *crosses* the x-axis at $x = a$.

2. When k is even, the graph *touches* the x-axis (but does not cross the x-axis) at $x = a$.

To graph polynomial functions, use the fact that a polynomial function can change signs only at its zeros. Between two consecutive zeros, a polynomial must be entirely positive or entirely negative. (This follows from the Intermediate Value Theorem, which you will study later in this section.) This means that when you put the real zeros of a polynomial function in order, they divide the real number line into intervals in which the function has no sign changes. These resulting intervals are **test intervals** in which you choose a representative x-value to determine whether the value of the polynomial function is positive (the graph lies above the x-axis) or negative (the graph lies below the x-axis).

▷ TECHNOLOGY Example 4 uses an *algebraic approach* to describe the graph of the function. A graphing utility can complement this approach. Remember to find a viewing window that shows all significant features of the graph. For instance, viewing window (a) illustrates all of the significant features of the function in Example 4, but viewing window (b) does not.

(a)

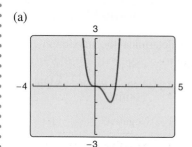

(b)

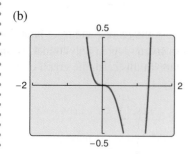

| EXAMPLE 4 | **Sketching the Graph of a Polynomial Function** |

Sketch the graph of $f(x) = 3x^4 - 4x^3$.

Solution

1. *Apply the Leading Coefficient Test.* The leading coefficient is positive and the degree is even, so you know that the graph eventually rises to the left and to the right (see Figure 2.11).

2. *Find the Real Zeros of the Function.* Factoring $f(x) = 3x^4 - 4x^3$ as $f(x) = x^3(3x - 4)$ shows that the real zeros of f are $x = 0$ and $x = \frac{4}{3}$ (both of odd multiplicity). So, the x-intercepts occur at $(0, 0)$ and $\left(\frac{4}{3}, 0\right)$. Add these points to your graph, as shown in Figure 2.11.

3. *Plot a Few Additional Points.* Use the zeros of the polynomial to find the test intervals. In each test interval, choose a representative x-value and evaluate the polynomial function, as shown in the table

Test Interval	Representative x-Value	Value of f	Sign	Point on Graph
$(-\infty, 0)$	-1	$f(-1) = 7$	Positive	$(-1, 7)$
$\left(0, \frac{4}{3}\right)$	1	$f(1) = -1$	Negative	$(1, -1)$
$\left(\frac{4}{3}, \infty\right)$	$\frac{3}{2}$	$f\left(\frac{3}{2}\right) = \frac{27}{16}$	Positive	$\left(\frac{3}{2}, \frac{27}{16}\right)$

4. *Draw the Graph.* Draw a continuous curve through the points, as shown in Figure 2.12. Both zeros are of odd multiplicity, so you know that the graph should cross the x-axis at $x = 0$ and $x = \frac{4}{3}$.

• • • • • • • • • • • • • • • • ▷

•• REMARK   If you are unsure of the shape of a portion of the graph of a polynomial function, then plot some additional points. For instance, in Example 4, it is helpful to plot the additional point $\left(\frac{1}{2}, -\frac{5}{16}\right)$, as shown in Figure 2.12.

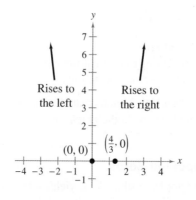

Figure 2.11

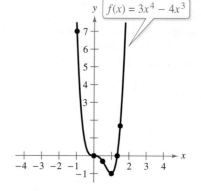

Figure 2.12

✓ *Checkpoint* ◀))) *Audio-video solution in English & Spanish at LarsonPrecalculus.com*

Sketch the graph of $f(x) = 2x^3 - 6x^2$.

A polynomial function is in **standard form** when its terms are in descending order of exponents from left to right. To avoid making a mistake when applying the Leading Coefficient Test, write the polynomial function in standard form first, if necessary.

EXAMPLE 5   **Sketching the Graph of a Polynomial Function**

Sketch the graph of $f(x) = -\frac{9}{2}x + 6x^2 - 2x^3$.

Solution

1. *Write in Standard Form and Apply the Leading Coefficient Test.*   In standard form, the polynomial function is $f(x) = -2x^3 + 6x^2 - \frac{9}{2}x$. The leading coefficient is negative and the degree is odd, so you know that the graph eventually rises to the left and falls to the right (see Figure 2.13).

2. *Find the Real Zeros of the Function.*   Factoring

$$f(x) = -2x^3 + 6x^2 - \frac{9}{2}x$$
$$= -\frac{1}{2}x(4x^2 - 12x + 9)$$
$$= -\frac{1}{2}x(2x - 3)^2$$

shows that the real zeros of f are $x = 0$ (odd multiplicity) and $x = \frac{3}{2}$ (even multiplicity). So, the x-intercepts occur at $(0, 0)$ and $\left(\frac{3}{2}, 0\right)$. Add these points to your graph, as shown in Figure 2.13.

3. *Plot a Few Additional Points.*   Use the zeros of the polynomial to find the test intervals. In each test interval, choose a representative x-value and evaluate the polynomial function, as shown in the table.

Test Interval	Representative x-Value	Value of f	Sign	Point on Graph
$(-\infty, 0)$	$-\frac{1}{2}$	$f\left(-\frac{1}{2}\right) = 4$	Positive	$\left(-\frac{1}{2}, 4\right)$
$\left(0, \frac{3}{2}\right)$	$\frac{1}{2}$	$f\left(\frac{1}{2}\right) = -1$	Negative	$\left(\frac{1}{2}, -1\right)$
$\left(\frac{3}{2}, \infty\right)$	2	$f(2) = -1$	Negative	$(2, -1)$

REMARK Observe in Example 5 that the sign of $f(x)$ is positive to the left of and negative to the right of the zero $x = 0$. Similarly, the sign of $f(x)$ is negative to the left and to the right of the zero $x = \frac{3}{2}$. This illustrates that (1) if the zero of a polynomial function is of *odd* multiplicity, then the graph crosses the x-axis at that zero, and (2) if the zero is of *even* multiplicity, then the graph touches the x-axis at that zero.

4. *Draw the Graph.*   Draw a continuous curve through the points, as shown in Figure 2.14. From the multiplicities of the zeros, you know that the graph crosses the x-axis at $(0, 0)$ but does not cross the x-axis at $\left(\frac{3}{2}, 0\right)$.

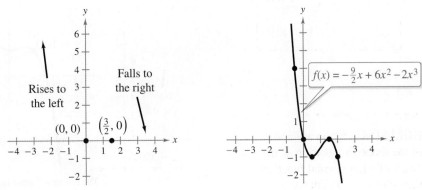

Figure 2.13   **Figure 2.14**

✓ *Checkpoint*))) *Audio-video solution in English & Spanish at LarsonPrecalculus.com*

Sketch the graph of $f(x) = -\frac{1}{4}x^4 + \frac{3}{2}x^3 - \frac{9}{4}x^2$.

The Intermediate Value Theorem

The **Intermediate Value Theorem** implies that if

$$(a, f(a)) \quad \text{and} \quad (b, f(b))$$

are two points on the graph of a polynomial function such that $f(a) \neq f(b)$, then for any number d between $f(a)$ and $f(b)$ there must be a number c between a and b such that $f(c) = d$. (See figure below.)

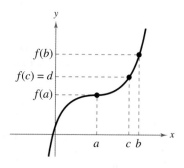

> ### Intermediate Value Theorem
>
> Let a and b be real numbers such that $a < b$. If f is a polynomial function such that $f(a) \neq f(b)$, then, in the interval $[a, b]$, f takes on every value between $f(a)$ and $f(b)$.

REMARK Note that $f(a)$ and $f(b)$ must be of opposite signs in order to guarantee that a zero exists between them. If $f(a)$ and $f(b)$ are of the same sign, then it is inconclusive whether a zero exists between them.

One application of the Intermediate Value Theorem is in helping you locate real zeros of a polynomial function. If there exists a value $x = a$ at which a polynomial function is negative, and another value $x = b$ at which it is positive (or if it is positive when $x = a$ and negative when $x = b$), then the function has at least one real zero between these two values. For example, the function

$$f(x) = x^3 + x^2 + 1$$

is negative when $x = -2$ and positive when $x = -1$. So, it follows from the Intermediate Value Theorem that f must have a real zero somewhere between -2 and -1, as shown in the figure below.

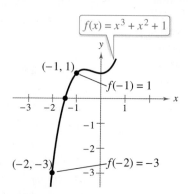

The function f must have a real zero somewhere between -2 and -1.

By continuing this line of reasoning, it is possible to approximate real zeros of a polynomial function to any desired accuracy. Example 6 further demonstrates this concept.

the *table* feature of a graphing utility can help you approximate real zeros of polynomial functions. For instance, in Example 6, construct a table that shows function values for integer values of x. Scrolling through the table, notice that $f(-1)$ and $f(0)$ differ in sign.

X	Y1	
-2	-11	
-1	-1	
0	1	
1	1	
2	5	
3	19	
4	49	
X=0		

So, by the Intermediate Value Theorem, the function has a real zero between -1 and 0. Adjust your table to show function values for $-1 \le x \le 0$ using increments of 0.1. Scrolling through this table, notice that $f(-0.8)$ and $f(-0.7)$ differ in sign.

X	Y1	
-1	-1	
-.9	-.539	
-.8	-.152	
-.7	.167	
-.6	.424	
-.5	.625	
-.4	.776	
X=-.7		

So, the function has a real zero between -0.8 and -0.7. Repeating this process with smaller increments, you should obtain $x \approx -0.755$ as the real zero of the function to three decimal places, as stated in Example 6. Use the *zero* or *root* feature of the graphing utility to confirm this result.

EXAMPLE 6 **Using the Intermediate Value Theorem**

Use the Intermediate Value Theorem to approximate the real zero of

$$f(x) = x^3 - x^2 + 1.$$

Solution Begin by computing a few function values.

x	-2	-1	0	1
$f(x)$	-11	-1	1	1

The value $f(-1)$ is negative and $f(0)$ is positive, so by the Intermediate Value Theorem, the function has a real zero between -1 and 0. To pinpoint this zero more closely, divide the interval $[-1, 0]$ into tenths and evaluate the function at each point. When you do this, you will find that

$$f(-0.8) = -0.152$$

and

$$f(-0.7) = 0.167.$$

So, f must have a real zero between -0.8 and -0.7, as shown in the figure. For a more accurate approximation, compute function values between $f(-0.8)$ and $f(-0.7)$ and apply the Intermediate Value Theorem again. Continue this process to verify that

$$x \approx -0.755$$

is an approximation (to the nearest thousandth) of the real zero of f.

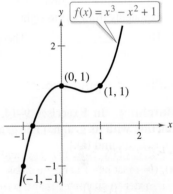

The function f has a real zero between -0.8 and -0.7.

✓ *Checkpoint* Audio-video solution in English & Spanish at LarsonPrecalculus.com

Use the Intermediate Value Theorem to approximate the real zero of

$$f(x) = x^3 - 3x^2 - 2.$$

Summarize (Section 2.2)

1. Explain how to use transformations to sketch graphs of polynomial functions *(page 124)*. For an example of sketching transformations of monomial functions, see Example 1.

2. Explain how to apply the Leading Coefficient Test *(page 125)*. For an example of applying the Leading Coefficient Test, see Example 2.

3. Explain how to find real zeros of polynomial functions and use them as sketching aids *(page 127)*. For examples involving finding real zeros of polynomial functions, see Examples 3–5.

4. Explain how to use the Intermediate Value Theorem to help locate real zeros of polynomial functions *(page 130)*. For an example of using the Intermediate Value Theorem, see Example 6.

2.2 Exercises

See **CalcChat.com** for tutorial help and worked-out solutions to odd-numbered exercises.

Vocabulary: **Fill in the blanks.**

1. The graph of a polynomial function is _____, which means that the graph has no breaks, holes, or gaps.

2. The _____ _____ _____ is used to determine the left-hand and right-hand behavior of the graph of a polynomial function.

3. A polynomial function of degree n has at most _____ real zeros and at most _____ turning points.

4. When $x = a$ is a zero of a polynomial function f, the three statements below are true.
   (a) $x = a$ is a _____ of the polynomial equation $f(x) = 0$.
   (b) _____ is a factor of the polynomial $f(x)$.
   (c) $(a, 0)$ is an _____ of the graph of f.

5. When a real zero $x = a$ of a polynomial function f is of even multiplicity, the graph of f _____ the x-axis at $x = a$, and when it is of odd multiplicity, the graph of f _____ the x-axis at $x = a$.

6. A factor $(x - a)^k$, $k > 1$, yields a _____ _____ $x = a$ of _____ k.

7. A polynomial function is written in _____ form when its terms are written in descending order of exponents from left to right.

8. The _____ _____ Theorem states that if f is a polynomial function such that $f(a) \neq f(b)$, then, in the interval $[a, b]$, f takes on every value between $f(a)$ and $f(b)$.

Skills and Applications

Matching **In Exercises 9–14, match the polynomial function with its graph. [The graphs are labeled (a), (b), (c), (d), (e), and (f).]**

9. $f(x) = -2x^2 - 5x$

10. $f(x) = 2x^3 - 3x + 1$

11. $f(x) = -\frac{1}{4}x^4 + 3x^2$

12. $f(x) = -\frac{1}{3}x^3 + x^2 - \frac{4}{3}$

13. $f(x) = x^4 + 2x^3$

14. $f(x) = \frac{1}{5}x^5 - 2x^3 + \frac{9}{5}x$

(a)

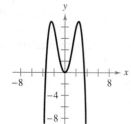

(b)

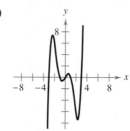

(c)

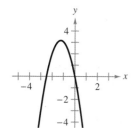

(d)

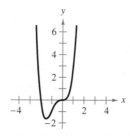

(e)

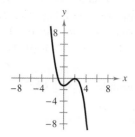

(f)

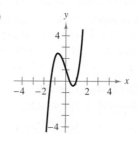

 Sketching   Transformations   of Monomial Functions **In Exercises 15–18, sketch the graph of $y = x^n$ and each transformation.**

15. $y = x^3$
   (a) $f(x) = (x - 4)^3$
   (b) $f(x) = x^3 - 4$
   (c) $f(x) = -\frac{1}{4}x^3$
   (d) $f(x) = (x - 4)^3 - 4$

16. $y = x^5$
   (a) $f(x) = (x + 1)^5$
   (b) $f(x) = x^5 + 1$
   (c) $f(x) = 1 - \frac{1}{2}x^5$
   (d) $f(x) = -\frac{1}{2}(x + 1)^5$

17. $y = x^4$
   (a) $f(x) = (x + 3)^4$
   (b) $f(x) = x^4 - 3$
   (c) $f(x) = 4 - x^4$
   (d) $f(x) = \frac{1}{2}(x - 1)^4$
   (e) $f(x) = (2x)^4 + 1$
   (f) $f(x) = \left(\frac{1}{2}x\right)^4 - 2$

18. $y = x^6$
   (a) $f(x) = (x - 5)^6$
   (b) $f(x) = \frac{1}{8}x^6$
   (c) $f(x) = (x + 3)^6 - 4$
   (d) $f(x) = -\frac{1}{4}x^6 + 1$
   (e) $f(x) = \left(\frac{1}{4}x\right)^6 - 2$
   (f) $f(x) = (2x)^6 - 1$

Applying the Leading Coefficient Test
In Exercises 19–28, describe the left-hand
and right-hand behavior of the graph of the
polynomial function.

19. $f(x) = 12x^3 + 4x$ **20.** $f(x) = 2x^2 - 3x + 1$

21. $g(x) = 5 - \frac{7}{2}x - 3x^2$ **22.** $h(x) = 1 - x^6$

23. $h(x) = 6x - 9x^3 + x^2$ **24.** $g(x) = 8 + \frac{1}{4}x^5 - x^4$

25. $f(x) = 9.8x^6 - 1.2x^3$

26. $h(x) = 1 - 0.5x^5 - 2.7x^3$

27. $f(s) = -\frac{7}{8}(s^3 + 5s^2 - 7s + 1)$

28. $h(t) = -\frac{4}{3}(t - 6t^3 + 2t^4 + 9)$

Using Technology In Exercises 29–32, use a
graphing utility to graph the functions f and g in the
same viewing window. Zoom out sufficiently far to show
that the left-hand and right-hand behaviors of f and g
appear identical.

29. $f(x) = 3x^3 - 9x + 1,$ $g(x) = 3x^3$

30. $f(x) = -\frac{1}{3}(x^3 - 3x + 2),$ $g(x) = -\frac{1}{3}x^3$

31. $f(x) = -(x^4 - 4x^3 + 16x),$ $g(x) = -x^4$

32. $f(x) = 3x^4 - 6x^2,$ $g(x) = 3x^4$

**Finding Real Zeros of a Polynomial
Function** In Exercises 33–48, (a) find all
real zeros of the polynomial function,
(b) determine whether the multiplicity of
each zero is even or odd, (c) determine the
maximum possible number of turning points
of the graph of the function, and (d) use a
graphing utility to graph the function and
verify your answers.

33. $f(x) = x^2 - 36$ **34.** $f(x) = 81 - x^2$

35. $h(t) = t^2 - 6t + 9$ **36.** $f(x) = x^2 + 10x + 25$

37. $f(x) = \frac{1}{3}x^2 + \frac{1}{3}x - \frac{2}{3}$ **38.** $f(x) = \frac{1}{2}x^2 + \frac{5}{2}x - \frac{3}{2}$

39. $g(x) = 5x(x^2 - 2x - 1)$ **40.** $f(t) = t^2(3t^2 - 10t + 7)$

41. $f(x) = 3x^3 - 12x^2 + 3x$

42. $f(x) = x^4 - x^3 - 30x^2$

43. $g(t) = t^5 - 6t^3 + 9t$ **44.** $f(x) = x^5 + x^3 - 6x$

45. $f(x) = 3x^4 + 9x^2 + 6$ **46.** $f(t) = 2t^4 - 2t^2 - 40$

47. $g(x) = x^3 + 3x^2 - 4x - 12$

48. $f(x) = x^3 - 4x^2 - 25x + 100$

Using Technology In Exercises 49–52, (a) use a
graphing utility to graph the function, (b) use the
graph to approximate any x-intercepts of the graph,
(c) find any real zeros of the function algebraically, and
(d) compare the results of part (c) with those of part (b).

49. $y = 4x^3 - 20x^2 + 25x$

50. $y = 4x^3 + 4x^2 - 8x - 8$

51. $y = x^5 - 5x^3 + 4x$ **52.** $y = \frac{1}{5}x^5 - \frac{9}{5}x^3$

Finding a Polynomial Function In
Exercises 53–62, find a polynomial function
that has the given zeros. (There are many
correct answers.)

53. $0, 7$ **54.** $-2, 5$

55. $0, -2, -4$ **56.** $0, 1, 6$

57. $4, -3, 3, 0$ **58.** $-2, -1, 0, 1, 2$

59. $1 + \sqrt{2}, 1 - \sqrt{2}$ **60.** $4 + \sqrt{3}, 4 - \sqrt{3}$

61. $2, 2 + \sqrt{5}, 2 - \sqrt{5}$ **62.** $3, 2 + \sqrt{7}, 2 - \sqrt{7}$

Finding a Polynomial Function In
Exercises 63–70, find a polynomial of degree
n that has the given zero(s). (There are many
correct answers.)

Zero(s)	Degree
63. $x = -3$	$n = 2$
64. $x = -\sqrt{2}, \sqrt{2}$	$n = 2$
65. $x = -5, 0, 1$	$n = 3$
66. $x = -2, 6$	$n = 3$
67. $x = -5, 1, 2$	$n = 4$
68. $x = -4, -1$	$n = 4$
69. $x = 0, -\sqrt{3}, \sqrt{3}$	$n = 5$
70. $x = -1, 4, 7, 8$	$n = 5$

**Sketching the Graph of a Polynomial
Function** In Exercises 71–84, sketch the
graph of the function by (a) applying the
Leading Coefficient Test, (b) finding the real
zeros of the polynomial, (c) plotting sufficient
solution points, and (d) drawing a continuous
curve through the points.

71. $f(t) = \frac{1}{4}(t^2 - 2t + 15)$ **72.** $g(x) = -x^2 + 10x - 16$

73. $f(x) = x^3 - 25x$ **74.** $g(x) = -9x^2 + x^4$

75. $f(x) = -8 + \frac{1}{2}x^4$ **76.** $f(x) = 8 - x^3$

77. $f(x) = 3x^3 - 15x^2 + 18x$

78. $f(x) = -4x^3 + 4x^2 + 15x$

79. $f(x) = -5x^2 - x^3$ **80.** $f(x) = -48x^2 + 3x^4$

81. $f(x) = 9x^2(x + 2)^3$ **82.** $h(x) = \frac{1}{3}x^3(x - 4)^2$

83. $g(t) = -\frac{1}{4}(t - 2)^2(t + 2)^2$

84. $g(x) = \frac{1}{10}(x + 1)^2(x - 3)^3$

Using Technology In Exercises 85–88, use a
graphing utility to graph the function. Use the *zero*
or *root* feature to approximate the real zeros of the
function. Then determine whether the multiplicity of
each zero is even or odd.

85. $f(x) = x^3 - 16x$ **86.** $f(x) = \frac{1}{4}x^4 - 2x^2$

87. $g(x) = \frac{1}{5}(x + 1)^2(x - 3)(2x - 9)$

88. $h(x) = \frac{1}{5}(x + 2)^2(3x - 5)^2$

Using the Intermediate Value Theorem
**In Exercises 89–92, (a) use the Intermediate
Value Theorem and the *table* feature of a
graphing utility to find intervals one unit in
length in which the polynomial function is
guaranteed to have a zero. (b) Adjust the
table to approximate the zeros of the function
to the nearest thousandth.**

89. $f(x) = x^3 - 3x^2 + 3$

90. $f(x) = 0.11x^3 - 2.07x^2 + 9.81x - 6.88$

91. $g(x) = 3x^4 + 4x^3 - 3$ **92.** $h(x) = x^4 - 10x^2 + 3$

93. Maximum Volume You construct an open box from a square piece of material, 36 inches on a side, by cutting equal squares with sides of length x from the corners and turning up the sides (see figure).

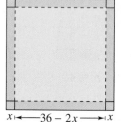
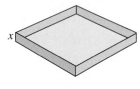

(a) Write a function V that represents the volume of the box.

(b) Determine the domain of the function V.

(c) Use a graphing utility to construct a table that shows the box heights x and the corresponding volumes $V(x)$. Use the table to estimate the dimensions that produce a maximum volume.

(d) Use the graphing utility to graph V and use the graph to estimate the value of x for which $V(x)$ is a maximum. Compare your result with that of part (c).

94. Maximum Volume You construct an open box with locking tabs from a square piece of material, 24 inches on a side, by cutting equal sections from the corners and folding along the dashed lines (see figure).

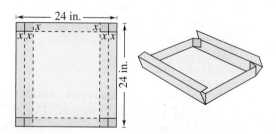

(a) Write a function V that represents the volume of the box.

(b) Determine the domain of the function V.

(c) Sketch a graph of the function and estimate the value of x for which $V(x)$ is a maximum.

95. Revenue The revenue R (in millions of dollars) for a software company from 2003 through 2016 can be modeled by

$R = 6.212t^3 - 152.87t^2 + 990.2t - 414, \; 3 \le t \le 16$

where t represents the year, with $t = 3$ corresponding to 2003.

(a) Use a graphing utility to approximate any relative minima or maxima of the model over its domain.

(b) Use the graphing utility to approximate the intervals on which the revenue for the company is increasing and decreasing over its domain.

(c) Use the results of parts (a) and (b) to describe the company's revenue during this time period.

96. Revenue The revenue R (in millions of dollars) for a construction company from 2003 through 2010 can be modeled by

$R = 0.1104t^4 - 5.152t^3 + 88.20t^2 - 654.8t + 1907, \; 7 \le t \le 16$

where t represents the year, with $t = 7$ corresponding to 2007.

(a) Use a graphing utility to approximate any relative minima or maxima of the model over its domain.

(b) Use the graphing utility to approximate the intervals on which the revenue for the company is increasing and decreasing over its domain.

(c) Use the results of parts (a) and (b) to describe the company's revenue during this time period.

97. Revenue The revenue R (in millions of dollars) for a beverage company is related to its advertising expense by the function

$R = \dfrac{1}{100,000}(-x^3 + 600x^2), \quad 0 \le x \le 400$

where x is the amount spent on advertising (in tens of thousands of dollars). Use the graph of this function to estimate the point on the graph at which the function is increasing most rapidly. This point is called the *point of diminishing returns* because any expense above this amount will yield less return per dollar invested in advertising.

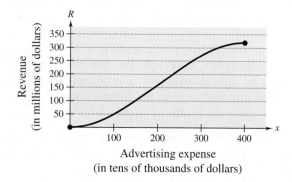

98. Arboriculture

The growth of a red oak tree is approximated by the function

$G = -0.003t^3 + 0.137t^2 + 0.458t - 0.839,$

$2 \leq t \leq 34$

where G is the height of the tree (in feet) and t is its age (in years).

(a) Use a graphing utility to graph the function.

(b) Estimate the age of the tree when it is growing most rapidly. This point is called the *point of diminishing returns* because the increase in size will be less with each additional year.

(c) Using calculus, the point of diminishing returns can be found by finding the vertex of the parabola

$y = -0.009t^2 + 0.274t + 0.458.$

Find the vertex of this parabola.

(d) Compare your results from parts (b) and (c).

Exploration

True or False? In Exercises 99–102, determine whether the statement is true or false. Justify your answer.

99. If the graph of a polynomial function falls to the right, then its leading coefficient is negative.

100. A fifth-degree polynomial function can have five turning points in its graph.

101. It is possible for a polynomial with an even degree to have a range of $(-\infty, \infty)$.

102. If f is a polynomial function of x such that $f(2) = -6$ and $f(6) = 6$, then f has at most one real zero between $x = 2$ and $x = 6$.

103. Modeling Polynomials Sketch the graph of a fourth-degree polynomial function that has a zero of multiplicity 2 and a negative leading coefficient. Sketch the graph of another polynomial function with the same characteristics except that the leading coefficient is positive.

104. Modeling Polynomials Sketch the graph of a fifth-degree polynomial function that has a zero of multiplicity 2 and a negative leading coefficient. Sketch the graph of another polynomial function with the same characteristics except that the leading coefficient is positive.

105. Graphical Reasoning Sketch the graph of the function $f(x) = x^4$. Explain how the graph of each function g differs (if it does) from the graph of f. Determine whether g is even, odd, or neither.

(a) $g(x) = f(x) + 2$ (b) $g(x) = f(x + 2)$

(c) $g(x) = f(-x)$ (d) $g(x) = -f(x)$

(e) $g(x) = f\left(\frac{1}{2}x\right)$ (f) $g(x) = \frac{1}{2}f(x)$

(g) $g(x) = f(x^{3/4})$ (h) $g(x) = (f \circ f)(x)$

106. HOW DO YOU SEE IT? For each graph, describe a polynomial function that could represent the graph. (Indicate the degree of the function and the sign of its leading coefficient.)

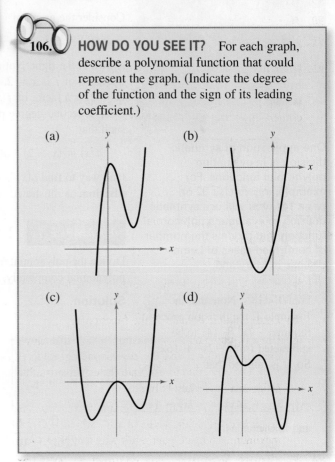

107. Think About It Use a graphing utility to graph the functions

$y_1 = -\frac{1}{3}(x - 2)^5 + 1$ and $y_2 = \frac{3}{5}(x + 2)^5 - 3.$

(a) Determine whether the graphs of y_1 and y_2 are increasing or decreasing. Explain.

(b) Will the graph of

$g(x) = a(x - h)^5 + k$

always be strictly increasing or strictly decreasing? If so, is this behavior determined by a, h, or k? Explain.

(c) Use a graphing utility to graph

$f(x) = x^5 - 3x^2 + 2x + 1.$

Use a graph and the result of part (b) to determine whether f can be written in the form $f(x) = a(x - h)^5 + k$. Explain.

2.3 Polynomial and Synthetic Division

One application of synthetic division is in evaluating polynomial functions. For example, in Exercise 82 on page 144, you will use synthetic division to evaluate a polynomial function that models the number of confirmed cases of Lyme disease in Maryland.

■ Use long division to divide polynomials by other polynomials.
■ Use synthetic division to divide polynomials by binomials of the form $(x - k)$.
■ Use the Remainder Theorem and the Factor Theorem.

Long Division of Polynomials

Consider the graph of

$$f(x) = 6x^3 - 19x^2 + 16x - 4$$

shown at the right. Notice that one of the zeros of f is $x = 2$. This means that $(x - 2)$ is a factor of $f(x)$, and there exists a second-degree polynomial $q(x)$ such that

$$f(x) = (x - 2) \cdot q(x).$$

One way to find $q(x)$ is to use **long division,** as illustrated in Example 1.

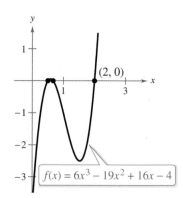

EXAMPLE 1   Long Division of Polynomials

Divide the polynomial $6x^3 - 19x^2 + 16x - 4$ by $x - 2$, and use the result to factor the polynomial completely.

Solution

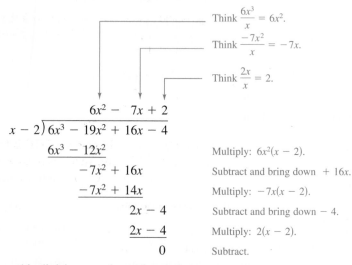

> • • **REMARK** Note that in Example 1, the division process requires $-7x^2 + 14x$ to be subtracted from $-7x^2 + 16x$. So, it is implied that
>
> $$\frac{-7x^2 + 16x}{-(-7x^2 + 14x)} = \frac{-7x^2 + 16x}{7x^2 - 14x}$$
>
> and is written as
>
> $$\begin{array}{r} -7x^2 + 16x \\ -7x^2 + 14x \\ \hline 2x. \end{array}$$

Think $\dfrac{6x^3}{x} = 6x^2$.

Think $\dfrac{-7x^2}{x} = -7x$.

Think $\dfrac{2x}{x} = 2$.

$$\begin{array}{r} 6x^2 - 7x + 2 \\ x - 2\overline{)\,6x^3 - 19x^2 + 16x - 4} \\ \underline{6x^3 - 12x^2} \\ -7x^2 + 16x \\ \underline{-7x^2 + 14x} \\ 2x - 4 \\ \underline{2x - 4} \\ 0 \end{array}$$

Multiply: $6x^2(x - 2)$.

Subtract and bring down $+ 16x$.

Multiply: $-7x(x - 2)$.

Subtract and bring down $- 4$.

Multiply: $2(x - 2)$.

Subtract.

From this division, you have shown that

$$6x^3 - 19x^2 + 16x - 4 = (x - 2)(6x^2 - 7x + 2)$$

and by factoring the quadratic $6x^2 - 7x + 2$, you have

$$6x^3 - 19x^2 + 16x - 4 = (x - 2)(2x - 1)(3x - 2).$$

> • • **REMARK** Note that the factorization found in Example 1 agrees with the graph of f above. The three x-intercepts occur at $(2, 0)$, $\left(\frac{1}{2}, 0\right)$, and $\left(\frac{2}{3}, 0\right)$.

✓ **Checkpoint** 🔊 *Audio-video solution in English & Spanish at LarsonPrecalculus.com*

Divide the polynomial $9x^3 + 36x^2 - 49x - 196$ by $x + 4$, and use the result to factor the polynomial completely. ■

In Example 1, $x - 2$ is a factor of the polynomial

$$6x^3 - 19x^2 + 16x - 4$$

and the long division process produces a remainder of zero. Often, long division will produce a nonzero remainder. For example, when you divide $x^2 + 3x + 5$ by $x + 1$, you obtain a remainder of 3.

$$
\begin{array}{r}
x + 2 \quad \longleftarrow \text{Quotient} \\
\text{Divisor} \longrightarrow x + 1 \overline{) x^2 + 3x + 5} \quad \longleftarrow \text{Dividend} \\
\underline{x^2 + x} \\
2x + 5 \\
\underline{2x + 2} \\
3 \quad \longleftarrow \text{Remainder}
\end{array}
$$

In fractional form, you can write this result as

$$
\overbrace{\frac{x^3 + 3x + 5}{\underbrace{x + 1}_{\text{Divisor}}}}^{\text{Dividend}} = \overbrace{x + 2}^{\text{Quotient}} + \frac{\overset{\text{Remainder}}{\downarrow}3}{\underbrace{x + 1}_{\text{Divisor}}}.
$$

This implies that

$$x^2 + 3x + 5 = (x + 1)(x + 2) + 3 \qquad \text{Multiply each side by } (x + 1).$$

which illustrates a theorem called the **Division Algorithm.**

The Division Algorithm

If $f(x)$ and $d(x)$ are polynomials such that $d(x) \neq 0$, and the degree of $d(x)$ is less than or equal to the degree of $f(x)$, then there exist unique polynomials $q(x)$ and $r(x)$ such that

$$f(x) = d(x)q(x) + r(x)$$

$$
\begin{array}{cccc}
\uparrow & \uparrow & \uparrow & \uparrow \\
\text{Dividend} & \text{Quotient} & & \text{Remainder} \\
& \text{Divisor} & &
\end{array}
$$

where $r(x) = 0$ or the degree of $r(x)$ is less than the degree of $d(x)$. If the remainder $r(x)$ is zero, then $d(x)$ *divides evenly* into $f(x)$.

Another way to write the Division Algorithm is

$$\frac{f(x)}{d(x)} = q(x) + \frac{r(x)}{d(x)}.$$

In the Division Algorithm, the rational expression $f(x)/d(x)$ is **improper** because the degree of $f(x)$ is greater than or equal to the degree of $d(x)$. On the other hand, the rational expression $r(x)/d(x)$ is **proper** because the degree of $r(x)$ is less than the degree of $d(x)$.

If necessary, follow these steps before you apply the Division Algorithm.

1. Write the terms of the dividend and divisor in descending powers of the variable.

2. Insert placeholders with zero coefficients for missing powers of the variable.

Note how Examples 2 and 3 apply these steps.

EXAMPLE 2   **Long Division of Polynomials**

Divide $x^3 - 1$ by $x - 1$. Check the result.

Solution   There is no x^2-term or x-term in the dividend $x^3 - 1$, so you need to rewrite the dividend as $x^3 + 0x^2 + 0x - 1$ before you apply the Division Algorithm.

$$
\begin{array}{r}
x^2 + x + 1 \\
x - 1 \overline{\smash{)}\, x^3 + 0x^2 + 0x - 1} \\
\underline{x^3 - x^2} \\
x^2 + 0x \\
\underline{x^2 - x} \\
x - 1 \\
\underline{x - 1} \\
0
\end{array}
$$

Multiply: $x^2(x - 1)$.

Subtract and bring down $0x$.

Multiply: $x(x - 1)$.

Subtract and bring down -1.

Multiply: $1(x - 1)$.

Subtract.

So, $x - 1$ divides evenly into $x^3 - 1$, and you can write

$$\frac{x^3 - 1}{x - 1} = x^2 + x + 1, \quad x \neq 1.$$

Check the result by multiplying.

$$(x - 1)(x^2 + x + 1) = x^3 + x^2 + x - x^2 - x - 1$$
$$= x^3 - 1$$

✓ *Checkpoint*   ))) *Audio-video solution in English & Spanish at LarsonPrecalculus.com*

Divide $x^3 - 2x^2 - 9$ by $x - 3$. Check the result.

EXAMPLE 3   **Long Division of Polynomials**

See LarsonPrecalculus.com for an interactive version of this type of example.

Divide $-5x^2 - 2 + 3x + 2x^4 + 4x^3$ by $2x - 3 + x^2$. Check the result.

Solution   Write the terms of the dividend and divisor in descending powers of x.

$$
\begin{array}{r}
2x^2 \quad\quad + 1 \\
x^2 + 2x - 3 \overline{\smash{)}\, 2x^4 + 4x^3 - 5x^2 + 3x - 2} \\
\underline{2x^4 + 4x^3 - 6x^2} \\
x^2 + 3x - 2 \\
\underline{x^2 + 2x - 3} \\
x + 1
\end{array}
$$

Multiply: $2x^2(x^2 + 2x - 3)$.

Subtract and bring down $3x - 2$.

Multiply: $1(x^2 + 2x - 3)$.

Subtract.

Note that the first subtraction eliminated two terms from the dividend. When this happens, the quotient skips a term. You can write the result as

$$\frac{2x^4 + 4x^3 - 5x^2 + 3x - 2}{x^2 + 2x - 3} = 2x^2 + 1 + \frac{x + 1}{x^2 + 2x - 3}.$$

Check the result by multiplying.

$$(x^2 + 2x - 3)(2x^2 + 1) + x + 1 = 2x^4 + x^2 + 4x^3 + 2x - 6x^2 - 3 + x + 1$$
$$= 2x^4 + 4x^3 - 5x^2 + 3x - 2$$

✓ *Checkpoint*   ))) *Audio-video solution in English & Spanish at LarsonPrecalculus.com*

Divide $-x^3 + 9x + 6x^4 - x^2 - 3$ by $1 + 3x$. Check the result. ∎

Synthetic Division

For long division of polynomials by divisors of the form $x - k$, there is a shortcut called **synthetic division.** The pattern for synthetic division of a cubic polynomial is summarized below. (The pattern for higher-degree polynomials is similar.)

Synthetic Division (for a Cubic Polynomial)

To divide $ax^3 + bx^2 + cx + d$ by $x - k$, use this pattern.

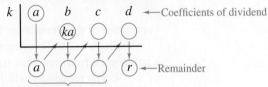

Vertical pattern: Add terms in columns.
Diagonal pattern: Multiply results by k.

This algorithm for synthetic division works only for divisors of the form $x - k$. Remember that $x + k = x - (-k)$.

EXAMPLE 4   **Using Synthetic Division**

Use synthetic division to divide

$$x^4 - 10x^2 - 2x + 4 \quad \text{by} \quad x + 3.$$

Solution   Begin by setting up an array. Include a zero for the missing x^3-term in the dividend.

$$-3 \ \lfloor 1 \quad 0 \quad -10 \quad -2 \quad 4$$

Then, use the synthetic division pattern by adding terms in columns and multiplying the results by -3.

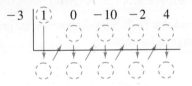

Divisor: $x + 3$ Dividend: $x^4 - 10x^2 - 2x + 4$

$$
\begin{array}{r|rrrr r}
-3 & 1 & 0 & -10 & -2 & 4 \\
   &   & -3 & 9 & 3 & -3 \\
\hline
   & 1 & -3 & -1 & 1 & \;1 \quad \leftarrow \text{Remainder: 1}
\end{array}
$$

Quotient: $x^3 - 3x^2 - x + 1$

So, you have

$$\frac{x^4 - 10x^2 - 2x + 4}{x + 3} = x^3 - 3x^2 - x + 1 + \frac{1}{x + 3}.$$

✓ *Checkpoint* 🔊))) *Audio-video solution in English & Spanish at LarsonPrecalculus.com*

Use synthetic division to divide $5x^3 + 8x^2 - x + 6$ by $x + 2$.

The Remainder and Factor Theorems

The remainder obtained in the synthetic division process has an important interpretation, as described in the **Remainder Theorem.**

The Remainder Theorem

If a polynomial $f(x)$ is divided by $x - k$, then the remainder is

$$r = f(k).$$

For a proof of the Remainder Theorem, see Proofs in Mathematics on page 193.

The Remainder Theorem tells you that synthetic division can be used to evaluate a polynomial function. That is, to evaluate a polynomial $f(x)$ when $x = k$, divide $f(x)$ by $x - k$. The remainder will be $f(k)$, as illustrated in Example 5.

EXAMPLE 5   **Using the Remainder Theorem**

Use the Remainder Theorem to evaluate

$$f(x) = 3x^3 + 8x^2 + 5x - 7$$

when $x = -2$. Check your answer.

Solution   Using synthetic division gives the result below.

$$
\begin{array}{r|rrrr}
-2 & 3 & 8 & 5 & -7 \\
   &   & -6 & -4 & -2 \\
\hline
   & 3 & 2 & 1 & -9
\end{array}
$$

The remainder is $r = -9$, so

$$f(-2) = -9. \qquad {\scriptstyle r = f(k)}$$

This means that $(-2, -9)$ is a point on the graph of f. Check this by substituting $x = -2$ in the original function.

Check

$$
\begin{aligned}
f(-2) &= 3(-2)^3 + 8(-2)^2 + 5(-2) - 7 \\
&= 3(-8) + 8(4) - 10 - 7 \\
&= -24 + 32 - 10 - 7 \\
&= -9
\end{aligned}
$$

✓ *Checkpoint* 🔊))) *Audio-video solution in English & Spanish at LarsonPrecalculus.com*

Use the Remainder Theorem to find each function value given

$$f(x) = 4x^3 + 10x^2 - 3x - 8.$$

Check your answer.

a. $f(-1)$ **b.** $f(4)$

c. $f\left(\tfrac{1}{2}\right)$ **d.** $f(-3)$

▷ TECHNOLOGY One way to evaluate a function with your graphing utility is to enter the function in the equation editor and use the *table* feature in *ask* mode. When you enter values in the X column of a table in *ask* mode, the corresponding function values are displayed in the function column.

Another important theorem is the **Factor Theorem,** stated below.

The Factor Theorem

A polynomial $f(x)$ has a factor $(x - k)$ if and only if $f(k) = 0$.

For a proof of the Factor Theorem, see Proofs in Mathematics on page 193.

Using the Factor Theorem, you can test whether a polynomial has $(x - k)$ as a factor by evaluating the polynomial at $x = k$. If the result is 0, then $(x - k)$ is a factor.

EXAMPLE 6　Factoring a Polynomial: Repeated Division

Show that $(x - 2)$ and $(x + 3)$ are factors of

$$f(x) = 2x^4 + 7x^3 - 4x^2 - 27x - 18.$$

Then find the remaining factors of $f(x)$.

Algebraic Solution

Using synthetic division with the factor $(x - 2)$ gives the result below.

$$
\begin{array}{r|rrrrr}
2 & 2 & 7 & -4 & -27 & -18 \\
  &   & 4 & 22 & 36 & 18 \\
\hline
  & 2 & 11 & 18 & 9 & 0
\end{array}
$$

→ 0 remainder, so $f(2) = 0$ and $(x - 2)$ is a factor.

Take the result of this division and perform synthetic division again using the factor $(x + 3)$.

$$
\begin{array}{r|rrrr}
-3 & 2 & 11 & 18 & 9 \\
   &   & -6 & -15 & -9 \\
\hline
   & 2 & 5 & 3 & 0
\end{array}
$$

$\underbrace{\qquad\qquad\qquad}_{2x^2 + 5x + 3}$

→ 0 remainder, so $f(-3) = 0$ and $(x + 3)$ is a factor.

The resulting quadratic expression factors as

$$2x^2 + 5x + 3 = (2x + 3)(x + 1)$$

so the complete factorization of $f(x)$ is

$$f(x) = (x - 2)(x + 3)(2x + 3)(x + 1).$$

Graphical Solution

The graph of $f(x) = 2x^4 + 7x^3 - 4x^2 - 27x - 18$ has four x-intercepts (see figure). These occur at $x = -3$, $x = -\frac{3}{2}$, $x = -1$, and $x = 2$. (Check this algebraically.) This implies that $(x + 3)$, $\left(x + \frac{3}{2}\right)$, $(x + 1)$, and $(x - 2)$ are factors of $f(x)$. [Note that $\left(x + \frac{3}{2}\right)$ and $(2x + 3)$ are equivalent factors because they both yield the same zero, $x = -\frac{3}{2}$.]

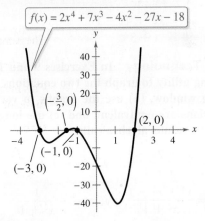

$f(x) = 2x^4 + 7x^3 - 4x^2 - 27x - 18$

✓ **Checkpoint** ◄))) *Audio-video solution in English & Spanish at LarsonPrecalculus.com*

Show that $(x + 3)$ is a factor of $f(x) = x^3 - 19x - 30$. Then find the remaining factors of $f(x)$.

Summarize　(Section 2.3)

1. Explain how to use long division to divide two polynomials *(pages 136 and 137)*. For examples of long division of polynomials, see Examples 1–3.

2. Describe the algorithm for synthetic division *(page 139)*. For an example of synthetic division, see Example 4.

3. State the Remainder Theorem and the Factor Theorem *(pages 140 and 141)*. For an example of using the Remainder Theorem, see Example 5. For an example of using the Factor Theorem, see Example 6.

2.3 Exercises

Vocabulary

1. Two forms of the Division Algorithm are shown below. Identify and label each term or function.

$$f(x) = d(x)q(x) + r(x) \qquad \frac{f(x)}{d(x)} = q(x) + \frac{r(x)}{d(x)}$$

In Exercises 2–6, fill in the blanks.

2. In the Division Algorithm, the rational expression $r(x)/d(x)$ is _____ because the degree of $r(x)$ is less than the degree of $d(x)$.

3. In the Division Algorithm, the rational expression $f(x)/d(x)$ is _____ because the degree of $f(x)$ is greater than or equal to the degree of $d(x)$.

4. A shortcut for long division of polynomials is _____ _____, in which the divisor must be of the form $x - k$.

5. The _____ Theorem states that a polynomial $f(x)$ has a factor $(x - k)$ if and only if $f(k) = 0$.

6. The _____ Theorem states that if a polynomial $f(x)$ is divided by $x - k$, then the remainder is $r = f(k)$.

Skills and Applications

Using the Division Algorithm   In Exercises 7 and 8, use long division to verify that $y_1 = y_2$.

7. $y_1 = \dfrac{x^2}{x + 2}$, $y_2 = x - 2 + \dfrac{4}{x + 2}$

8. $y_1 = \dfrac{x^3 - 3x^2 + 4x - 1}{x + 3}$, $y_2 = x^2 - 6x + 22 - \dfrac{67}{x + 3}$

Using Technology   In Exercises 9 and 10, (a) use a graphing utility to graph the two equations in the same viewing window, (b) use the graphs to verify that the expressions are equivalent, and (c) use long division to verify the results algebraically.

9. $y_1 = \dfrac{x^2 + 2x - 1}{x + 3}$, $y_2 = x - 1 + \dfrac{2}{x + 3}$

10. $y_1 = \dfrac{x^4 + x^2 - 1}{x^2 + 1}$, $y_2 = x^2 - \dfrac{1}{x^2 + 1}$

Long Division of Polynomials   In Exercises 11–24, use long division to divide.

11. $(2x^2 + 10x + 12) \div (x + 3)$

12. $(5x^2 - 17x - 12) \div (x - 4)$

13. $(4x^3 - 7x^2 - 11x + 5) \div (4x + 5)$

14. $(6x^3 - 16x^2 + 17x - 6) \div (3x - 2)$

15. $(x^4 + 5x^3 + 6x^2 - x - 2) \div (x + 2)$

16. $(x^3 + 4x^2 - 3x - 12) \div (x - 3)$

17. $(6x + 5) \div (x + 1)$ **18.** $(9x - 4) \div (3x + 2)$

19. $(x^3 - 9) \div (x^2 + 1)$ **20.** $(x^5 + 7) \div (x^4 - 1)$

21. $(3x + 2x^3 - 9 - 8x^2) \div (x^2 + 1)$

22. $(5x^3 - 16 - 20x + x^4) \div (x^2 - x - 3)$

23. $\dfrac{x^4}{(x - 1)^3}$

24. $\dfrac{2x^3 - 4x^2 - 15x + 5}{(x - 1)^2}$

Using Synthetic Division   In Exercises 25–44, use synthetic division to divide.

25. $(2x^3 - 10x^2 + 14x - 24) \div (x - 4)$

26. $(5x^3 + 18x^2 + 7x - 6) \div (x + 3)$

27. $(6x^3 + 7x^2 - x + 26) \div (x - 3)$

28. $(2x^3 + 12x^2 + 14x - 3) \div (x + 4)$

29. $(4x^3 - 9x + 8x^2 - 18) \div (x + 2)$

30. $(9x^3 - 16x - 18x^2 + 32) \div (x - 2)$

31. $(-x^3 + 75x - 250) \div (x + 10)$

32. $(3x^3 - 16x^2 - 72) \div (x - 6)$

33. $(x^3 - 3x^2 + 5) \div (x - 4)$

34. $(5x^3 + 6x + 8) \div (x + 2)$

35. $\dfrac{10x^4 - 50x^3 - 800}{x - 6}$

36. $\dfrac{x^5 - 13x^4 - 120x + 80}{x + 3}$

37. $\dfrac{x^3 + 512}{x + 8}$

38. $\dfrac{x^3 - 729}{x - 9}$

39. $\dfrac{-3x^4}{x - 2}$

40. $\dfrac{-2x^5}{x + 2}$

41. $\dfrac{180x - x^4}{x - 6}$

42. $\dfrac{5 - 3x + 2x^2 - x^3}{x + 1}$

43. $\dfrac{4x^3 + 16x^2 - 23x - 15}{x + \frac{1}{2}}$

44. $\dfrac{3x^3 - 4x^2 + 5}{x - \frac{3}{2}}$

Using the Remainder Theorem In Exercises 45–50, write the function in the form $f(x) = (x - k)q(x) + r$ for the given value of k, and demonstrate that $f(k) = r$.

45. $f(x) = x^3 - x^2 - 10x + 7, \quad k = 3$

46. $f(x) = x^3 - 4x^2 - 10x + 8, \quad k = -2$

47. $f(x) = 15x^4 + 10x^3 - 6x^2 + 14, \quad k = -\frac{2}{3}$

48. $f(x) = 10x^3 - 22x^2 - 3x + 4, \quad k = \frac{1}{5}$

49. $f(x) = -4x^3 + 6x^2 + 12x + 4, \quad k = 1 - \sqrt{3}$

50. $f(x) = -3x^3 + 8x^2 + 10x - 8, \quad k = 2 + \sqrt{2}$

 Using the Remainder Theorem In Exercises 51–54, use the Remainder Theorem and synthetic division to find each function value. Verify your answers using another method.

51. $f(x) = 2x^3 - 7x + 3$

   (a) $f(1)$ (b) $f(-2)$ (c) $f(3)$ (d) $f(2)$

52. $g(x) = 2x^6 + 3x^4 - x^2 + 3$

   (a) $g(2)$ (b) $g(1)$ (c) $g(3)$ (d) $g(-1)$

53. $h(x) = x^3 - 5x^2 - 7x + 4$

   (a) $h(3)$ (b) $h(\frac{1}{2})$ (c) $h(-2)$ (d) $h(-5)$

54. $f(x) = 4x^4 - 16x^3 + 7x^2 + 20$

   (a) $f(1)$ (b) $f(-2)$ (c) $f(5)$ (d) $f(-10)$

Using the Factor Theorem In Exercises 55–62, use synthetic division to show that x is a solution of the third-degree polynomial equation, and use the result to factor the polynomial completely. List all real solutions of the equation.

55. $x^3 + 6x^2 + 11x + 6 = 0, \quad x = -3$

56. $x^3 - 52x - 96 = 0, \quad x = -6$

57. $2x^3 - 15x^2 + 27x - 10 = 0, \quad x = \frac{1}{2}$

58. $48x^3 - 80x^2 + 41x - 6 = 0, \quad x = \frac{2}{3}$

59. $x^3 + 2x^2 - 3x - 6 = 0, \quad x = \sqrt{3}$

60. $x^3 + 2x^2 - 2x - 4 = 0, \quad x = \sqrt{2}$

61. $x^3 - 3x^2 + 2 = 0, \quad x = 1 + \sqrt{3}$

62. $x^3 - x^2 - 13x - 3 = 0, \quad x = 2 - \sqrt{5}$

 Factoring a Polynomial In Exercises 63–70, (a) verify the given factors of $f(x)$, (b) find the remaining factor(s) of $f(x)$, (c) use your results to write the complete factorization of $f(x)$, (d) list all real zeros of f, and (e) confirm your results by using a graphing utility to graph the function.

Function	Factors
63. $f(x) = 2x^3 + x^2 - 5x + 2$	$(x + 2), (x - 1)$
64. $f(x) = 3x^3 - x^2 - 8x - 4$	$(x + 1), (x - 2)$
65. $f(x) = x^4 - 8x^3 + 9x^2$ $+ 38x - 40$	$(x - 5), (x + 2)$
66. $f(x) = 8x^4 - 14x^3 - 71x^2$ $- 10x + 24$	$(x + 2), (x - 4)$
67. $f(x) = 6x^3 + 41x^2 - 9x - 14$	$(2x + 1), (3x - 2)$
68. $f(x) = 10x^3 - 11x^2 - 72x + 45$	$(2x + 5), (5x - 3)$
69. $f(x) = 2x^3 - x^2 - 10x + 5$	$(2x - 1), (x + \sqrt{5})$
70. $f(x) = x^3 + 3x^2 - 48x - 144$	$(x + 4\sqrt{3}), (x + 3)$

Approximating Zeros In Exercises 71–76, (a) use the *zero* or *root* feature of a graphing utility to approximate the zeros of the function accurate to three decimal places, (b) determine the exact value of one of the zeros, and (c) use synthetic division to verify your result from part (b), and then factor the polynomial completely.

71. $f(x) = x^3 - 2x^2 - 5x + 10$

72. $g(x) = x^3 + 3x^2 - 2x - 6$

73. $h(t) = t^3 - 2t^2 - 7t + 2$

74. $f(s) = s^3 - 12s^2 + 40s - 24$

75. $h(x) = x^5 - 7x^4 + 10x^3 + 14x^2 - 24x$

76. $g(x) = 6x^4 - 11x^3 - 51x^2 + 99x - 27$

Simplifying Rational Expressions In Exercises 77–80, simplify the rational expression by using long division or synthetic division.

77. $\dfrac{x^3 + x^2 - 64x - 64}{x + 8}$

78. $\dfrac{4x^3 - 8x^2 + x + 3}{2x - 3}$

79. $\dfrac{x^4 + 6x^3 + 11x^2 + 6x}{x^2 + 3x + 2}$

80. $\dfrac{x^4 + 9x^3 - 5x^2 - 36x + 4}{x^2 - 4}$

81. Profit A company that produces calculators estimates that the profit P (in dollars) from selling a specific model of calculator is given by

$$P = -152x^3 + 7545x^2 - 169{,}625, \quad 0 \le x \le 45$$

where x is the advertising expense (in tens of thousands of dollars). For this model of calculator, an advertising expense of \$400,000 $(x = 40)$ results in a profit of \$2,174,375.

(a) Use a graphing utility to graph the profit function.

(b) Use the graph from part (a) to estimate another amount the company can spend on advertising that results in the same profit.

(c) Use synthetic division to confirm the result of part (b) algebraically.

• **82. Lyme Disease** • • • • • • • • • • • • • •

The numbers N of confirmed cases
of Lyme disease in
Maryland from 2007
through 2014 are
shown in the table,
where t represents
the year, with $t = 7$
corresponding to
2007. *(Source:
Centers for Disease Control and Prevention)*

DATA	Year, t	Number, N
	7	2576
	8	1746
	9	1466
	10	1163
	11	938
	12	1113
	13	801
	14	957

Spreadsheet at
LarsonPrecalculus.com

(a) Use a graphing utility to create a scatter plot of
the data.

(b) Use the *regression* feature of the graphing utility
to find a *quartic* model for the data. (A quartic
model has the form $at^4 + bt^3 + ct^2 + dt + e$,
where a, b, c, d, and e are constant and t is
variable.) Graph the model in the same viewing
window as the scatter plot.

(c) Use the model to create a table of estimated values
of N. Compare the model with the original data.

(d) Use synthetic division to confirm algebraically
your estimated value for the year 2014.

Exploration

True or False?　In Exercises 83–86, determine whether
the statement is true or false. Justify your answer.

83. If $(7x + 4)$ is a factor of some polynomial function
$f(x)$, then $\frac{4}{7}$ is a zero of f.

84. $(2x - 1)$ is a factor of the polynomial

$$6x^6 + x^5 - 92x^4 + 45x^3 + 184x^2 + 4x - 48.$$

85. The rational expression $\dfrac{x^3 + 2x^2 - 7x + 4}{x^2 - 4x - 12}$ is improper.

86. The equation

$$\frac{x^3 - 3x^2 + 4}{x + 1} = x^2 - 4x + 4$$

is true for all values of x.

Think About It　In Exercises 87 and 88, perform the
division. Assume that n is a positive integer.

87. $\dfrac{x^{3n} + 9x^{2n} + 27x^n + 27}{x^n + 3}$　**88.** $\dfrac{x^{3n} - 3x^{2n} + 5x^n - 6}{x^n - 2}$

89. Error Analysis　Describe the error.

Use synthetic division to find the remainder when
$x^2 + 3x - 5$ is divided by $x + 1$.

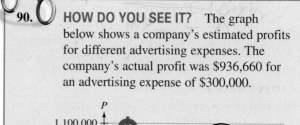

90.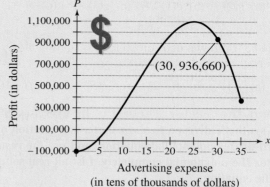

HOW DO YOU SEE IT?　The graph
below shows a company's estimated profits
for different advertising expenses. The
company's actual profit was \$936,660 for
an advertising expense of \$300,000.

(a) From the graph, it appears that the company
could have obtained the same profit for a lesser
advertising expense. Use the graph to estimate
this expense.

(b) The company's model is

$$P = -140.75x^3 + 5348.3x^2 - 76,560,$$
$$0 \le x \le 35$$

where P is the profit (in dollars) and x is the
advertising expense (in tens of thousands of
dollars). Explain how you could verify the
lesser expense from part (a) algebraically.

Exploration　In Exercises 91 and 92, find the constant
c such that the denominator will divide evenly into the
numerator.

91. $\dfrac{x^3 + 4x^2 - 3x + c}{x - 5}$　**92.** $\dfrac{x^5 - 2x^2 + x + c}{x + 2}$

93. Think About It　Find the value of k such that $x - 4$
is a factor of $x^3 - kx^2 + 2kx - 8$.

2.4 Complex Numbers

■ Use the imaginary unit *i* to write complex numbers.
■ Add, subtract, and multiply complex numbers.
■ Use complex conjugates to write the quotient of two complex numbers in standard form.
■ Find complex solutions of quadratic equations.

The Imaginary Unit *i*

You have learned that some quadratic equations have no real solutions. For example, the quadratic equation

$$x^2 + 1 = 0$$

has no real solution because there is no real number x that can be squared to produce -1. To overcome this deficiency, mathematicians created an expanded system of numbers using the **imaginary unit *i*,** defined as

$$i = \sqrt{-1} \qquad \text{Imaginary unit}$$

where $i^2 = -1$. By adding real numbers to real multiples of this imaginary unit, you obtain the set of **complex numbers.** Each complex number can be written in the **standard form $a + bi$.** For example, the standard form of the complex number $-5 + \sqrt{-9}$ is $-5 + 3i$ because

$$-5 + \sqrt{-9} = -5 + \sqrt{3^2(-1)} = -5 + 3\sqrt{-1} = -5 + 3i.$$

Complex numbers are often used in electrical engineering. For example, in Exercise 87 on page 151, you will use complex numbers to find the impedance of an electrical circuit.

Definition of a Complex Number

Let a and b be real numbers. The number $a + bi$ is a **complex number** written in **standard form.** The real number a is the **real part** and the number bi (where b is a real number) is the **imaginary part** of the complex number.

When $b = 0$, the number $a + bi$ is a real number. When $b \neq 0$, the number $a + bi$ is an **imaginary number.** A number of the form bi, where $b \neq 0$, is a **pure imaginary number.**

Every real number a can be written as a complex number using $b = 0$. That is, for every real number a, $a = a + 0i$. So, the set of real numbers is a subset of the set of complex numbers, as shown in the figure below.

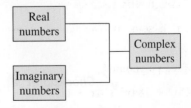

Equality of Complex Numbers

Two complex numbers $a + bi$ and $c + di$, written in standard form, are equal to each other

$$a + bi = c + di \qquad \text{Equality of two complex numbers}$$

if and only if $a = c$ and $b = d$.

Operations with Complex Numbers

To add (or subtract) two complex numbers, add (or subtract) the real and imaginary parts of the numbers separately.

Addition and Subtraction of Complex Numbers

For two complex numbers $a + bi$ and $c + di$ written in standard form, the sum and difference are

Sum: $(a + bi) + (c + di) = (a + c) + (b + d)i$

Difference: $(a + bi) - (c + di) = (a - c) + (b - d)i$

The **additive identity** in the complex number system is zero (the same as in the real number system). Furthermore, the **additive inverse** of the complex number $a + bi$ is

$-(a + bi) = -a - bi.$ Additive inverse

So, you have $(a + bi) + (-a - bi) = 0 + 0i = 0.$

EXAMPLE 1 Adding and Subtracting Complex Numbers

a. $(4 + 7i) + (1 - 6i) = 4 + 7i + 1 - 6i$ Remove parentheses.

$= (4 + 1) + (7 - 6)i$ Group like terms.

$= 5 + i$ Write in standard form.

▷ **b.** $(1 + 2i) + (3 - 2i) = 1 + 2i + 3 - 2i$ Remove parentheses.

$= (1 + 3) + (2 - 2)i$ Group like terms.

$= 4 + 0i$ Simplify.

$= 4$ Write in standard form.

• • REMARK Note that the sum of two complex numbers can be a real number.

c. $3i - (-2 + 3i) - (2 + 5i) = 3i + 2 - 3i - 2 - 5i$

$= (2 - 2) + (3 - 3 - 5)i$

$= 0 - 5i$

$= -5i$

d. $(3 + 2i) + (4 - i) - (7 + i) = 3 + 2i + 4 - i - 7 - i$

$= (3 + 4 - 7) + (2 - 1 - 1)i$

$= 0 + 0i$

$= 0$

✓ Checkpoint ◀))) *Audio-video solution in English & Spanish at LarsonPrecalculus.com*

Perform each operation and write the result in standard form.

a. $(7 + 3i) + (5 - 4i)$

b. $(3 + 4i) - (5 - 3i)$

c. $2i + (-3 - 4i) - (-3 - 3i)$

d. $(5 - 3i) + (3 + 5i) - (8 + 2i)$

Many of the properties of real numbers are valid for complex numbers as well. Here are some examples.

Associative Properties of Addition and Multiplication

Commutative Properties of Addition and Multiplication

Distributive Property of Multiplication Over Addition

Note the use of these properties when multiplying two complex numbers.

$$(a + bi)(c + di) = a(c + di) + bi(c + di)$$ Distributive Property

$$= ac + (ad)i + (bc)i + (bd)i^2$$ Distributive Property

$$= ac + (ad)i + (bc)i + (bd)(-1)$$ $i^2 = -1$

$$= ac - bd + (ad)i + (bc)i$$ Commutative Property

$$= (ac - bd) + (ad + bc)i$$ Associative Property

▷ **ALGEBRA HELP** To review the FOIL method, see Appendix A.3.

The procedure shown above is similar to multiplying two binomials and combining like terms, as in the FOIL method. So, you do not need to memorize this procedure.

EXAMPLE 2 **Multiplying Complex Numbers**

See LarsonPrecalculus.com for an interactive version of this type of example.

a. $4(-2 + 3i) = 4(-2) + 4(3i)$ Distributive Property

$$= -8 + 12i$$ Simplify.

b. $(2 - i)(4 + 3i) = 8 + 6i - 4i - 3i^2$ FOIL Method

$$= 8 + 6i - 4i - 3(-1)$$ $i^2 = -1$

$$= (8 + 3) + (6 - 4)i$$ Group like terms.

$$= 11 + 2i$$ Write in standard form.

c. $(3 + 2i)(3 - 2i) = 9 - 6i + 6i - 4i^2$ FOIL Method

$$= 9 - 6i + 6i - 4(-1)$$ $i^2 = -1$

$$= 9 + 4$$ Simplify.

$$= 13$$ Write in standard form.

d. $(3 + 2i)^2 = (3 + 2i)(3 + 2i)$ Square of a binomial

$$= 9 + 6i + 6i + 4i^2$$ FOIL Method

$$= 9 + 6i + 6i + 4(-1)$$ $i^2 = -1$

$$= 9 + 12i - 4$$ Simplify.

$$= 5 + 12i$$ Write in standard form.

✓ **Checkpoint** ◀))) *Audio-video solution in English & Spanish at LarsonPrecalculus.com*

Perform each operation and write the result in standard form.

a. $-5(3 - 2i)$

b. $(2 - 4i)(3 + 3i)$

c. $(4 + 5i)(4 - 5i)$

d. $(4 + 2i)^2$

Complex Conjugates

Notice in Example 2(c) that the product of two complex numbers can be a real number. This occurs with pairs of complex numbers of the form $a + bi$ and $a - bi$, called **complex conjugates.**

$$(a + bi)(a - bi) = a^2 - abi + abi - b^2i^2$$
$$= a^2 - b^2(-1)$$
$$= a^2 + b^2$$

> **REMARK** Recall that the product of $a - b\sqrt{m}$ or $a + b\sqrt{m}$ and its conjugate is rational. Similarly, the product of a complex number and its conjugate is real.

EXAMPLE 3   Multiplying Conjugates

Multiply each complex number by its complex conjugate.

a. $1 + i$ **b.** $4 - 3i$

Solution

a. The complex conjugate of $1 + i$ is $1 - i$.

$$(1 + i)(1 - i) = 1^2 - i^2 = 1 - (-1) = 2$$

b. The complex conjugate of $4 - 3i$ is $4 + 3i$.

$$(4 - 3i)(4 + 3i) = 4^2 - (3i)^2 = 16 - 9i^2 = 16 - 9(-1) = 25$$

✓ **Checkpoint** ◀))) *Audio-video solution in English & Spanish at LarsonPrecalculus.com*

Multiply each complex number by its complex conjugate.

a. $3 + 6i$ **b.** $2 - 5i$

To write the quotient of $a + bi$ and $c + di$ in standard form, where c and d are not both zero, multiply the numerator and denominator by the complex conjugate of the *denominator* to obtain

$$\frac{a + bi}{c + di} = \frac{a + bi}{c + di}\left(\frac{c - di}{c - di}\right) = \frac{(ac + bd) + (bc - ad)i}{c^2 + d^2} = \frac{ac + bd}{c^2 + d^2} + \left(\frac{bc - ad}{c^2 + d^2}\right)i.$$

> **REMARK** Note that when you multiply a quotient of complex numbers by
>
> $$\frac{c - di}{c - di}$$
>
> you are multiplying the quotient by a form of 1. So, you are not changing the original expression, you are only writing an equivalent expression.

EXAMPLE 4   A Quotient of Complex Numbers in Standard Form

$$\frac{2 + 3i}{4 - 2i} = \frac{2 + 3i}{4 - 2i}\left(\frac{4 + 2i}{4 + 2i}\right)$$ Multiply numerator and denominator by complex conjugate of denominator.

$$= \frac{8 + 4i + 12i + 6i^2}{16 - 4i^2}$$ Expand.

$$= \frac{8 - 6 + 16i}{16 + 4}$$ $i^2 = -1$

$$= \frac{2 + 16i}{20}$$ Simplify.

$$= \frac{1}{10} + \frac{4}{5}i$$ Write in standard form.

✓ **Checkpoint** ◀))) *Audio-video solution in English & Spanish at LarsonPrecalculus.com*

Write $\dfrac{2 + i}{2 - i}$ in standard form.

Complex Solutions of Quadratic Equations

You can write a number such as $\sqrt{-3}$ in standard form by factoring out $i = \sqrt{-1}$.

$$\sqrt{-3} = \sqrt{3(-1)} = \sqrt{3}\sqrt{-1} = \sqrt{3}i$$

The number $\sqrt{3}i$ is the *principal square root* of -3.

> **·· REMARK** The definition of
> principal square root uses the
> rule
> $$\sqrt{ab} = \sqrt{a}\sqrt{b}$$
> for $a > 0$ and $b < 0$. This rule
> is not valid when *both a and b*
> are negative. For example,
> $$\sqrt{-5}\sqrt{-5} = \sqrt{5(-1)}\sqrt{5(-1)}$$
> $$= \sqrt{5}i\sqrt{5}i$$
> $$= \sqrt{25}i^2$$
> $$= 5i^2$$
> $$= -5$$
> whereas
> $$\sqrt{(-5)(-5)} = \sqrt{25} = 5.$$
> Be sure to convert complex
> numbers to standard form *before*
> performing any operations.

▷ **ALGEBRA HELP** To review
the Quadratic Formula, see
Appendix A.5.

Principal Square Root of a Negative Number

When a is a positive real number, the **principal square root** of $-a$ is defined as
$$\sqrt{-a} = \sqrt{a}i.$$

EXAMPLE 5   **Writing Complex Numbers in Standard Form**

a. $\sqrt{-3}\sqrt{-12} = \sqrt{3}i\sqrt{12}i = \sqrt{36}i^2 = 6(-1) = -6$

b. $\sqrt{-48} - \sqrt{-27} = \sqrt{48}i - \sqrt{27}i = 4\sqrt{3}i - 3\sqrt{3}i = \sqrt{3}i$

c. $\left(-1 + \sqrt{-3}\right)^2 = \left(-1 + \sqrt{3}i\right)^2 = 1 - 2\sqrt{3}i + 3(-1)$
$$= -2 - 2\sqrt{3}i$$

✓ **Checkpoint** ◀))) Audio-video solution in English & Spanish at LarsonPrecalculus.com

Write $\sqrt{-14}\sqrt{-2}$ in standard form.

EXAMPLE 6   **Complex Solutions of a Quadratic Equation**

Solve $3x^2 - 2x + 5 = 0$.

Solution

$$x = \frac{-(-2) \pm \sqrt{(-2)^2 - 4(3)(5)}}{2(3)} \qquad \text{Quadratic Formula}$$

$$= \frac{2 \pm \sqrt{-56}}{6} \qquad \text{Simplify.}$$

$$= \frac{2 \pm 2\sqrt{14}i}{6} \qquad \text{Write } \sqrt{-56} \text{ in standard form.}$$

$$= \frac{1}{3} \pm \frac{\sqrt{14}}{3}i \qquad \text{Write solution in standard form.}$$

✓ **Checkpoint** ◀))) Audio-video solution in English & Spanish at LarsonPrecalculus.com

Solve $8x^2 + 14x + 9 = 0$. ■

Summarize (Section 2.4)

1. Explain how to write complex numbers using the imaginary unit i *(page 145)*.

2. Explain how to add, subtract, and multiply complex numbers *(pages 146 and 147, Examples 1 and 2)*.

3. Explain how to use complex conjugates to write the quotient of two complex numbers in standard form *(page 148, Example 4)*.

4. Explain how to find complex solutions of a quadratic equation *(page 149, Example 6)*.

2.4 Exercises

See CalcChat.com for tutorial help and worked-out solutions to odd-numbered exercises.

Vocabulary: Fill in the blanks.

1. A _____ number has the form $a + bi$, where $a \neq 0$, $b = 0$.
2. An _____ number has the form $a + bi$, where $a \neq 0$, $b \neq 0$.
3. A _____ _____ number has the form $a + bi$, where $a = 0$, $b \neq 0$.
4. The imaginary unit i is defined as $i =$ _____, where $i^2 =$ _____.
5. When a is a positive real number, the _____ _____ root of $-a$ is defined as $\sqrt{-a} = \sqrt{a}\,i$.
6. The numbers $a + bi$ and $a - bi$ are called _____ _____, and their product is a real number $a^2 + b^2$.

Skills and Applications

Equality of Complex Numbers In Exercises 7–10, find real numbers a and b such that the equation is true.

7. $a + bi = 9 + 8i$
8. $a + bi = 10 - 5i$
9. $(a - 2) + (b + 1)i = 6 + 5i$
10. $(a + 2) + (b - 3)i = 4 + 7i$

 Writing a Complex Number in Standard Form In Exercises 11–22, write the complex number in standard form.

11. $2 + \sqrt{-25}$
12. $4 + \sqrt{-49}$
13. $1 - \sqrt{-12}$
14. $2 - \sqrt{-18}$
15. $\sqrt{-40}$
16. $\sqrt{-27}$
17. 23
18. 50
19. $-6i + i^2$
20. $-2i^2 + 4i$
21. $\sqrt{-0.04}$
22. $\sqrt{-0.0025}$

Adding or Subtracting Complex Numbers In Exercises 23–30, perform the operation and write the result in standard form.

23. $(5 + i) + (2 + 3i)$
24. $(13 - 2i) + (-5 + 6i)$
25. $(9 - i) - (8 - i)$
26. $(3 + 2i) - (6 + 13i)$
27. $\left(-2 + \sqrt{-8}\right) + \left(5 - \sqrt{-50}\right)$
28. $\left(8 + \sqrt{-18}\right) - \left(4 + 3\sqrt{2}i\right)$
29. $13i - (14 - 7i)$
30. $25 + (-10 + 11i) + 15i$

Multiplying Complex Numbers In Exercises 31–38, perform the operation and write the result in standard form.

31. $(1 + i)(3 - 2i)$
32. $(7 - 2i)(3 - 5i)$
33. $12i(1 - 9i)$
34. $-8i(9 + 4i)$
35. $\left(\sqrt{2} + 3i\right)\left(\sqrt{2} - 3i\right)$
36. $\left(4 + \sqrt{7}i\right)\left(4 - \sqrt{7}i\right)$
37. $(6 + 7i)^2$
38. $(5 - 4i)^2$

Multiplying Conjugates In Exercises 39–46, write the complex conjugate of the complex number. Then multiply the number by its complex conjugate.

39. $9 + 2i$
40. $8 - 10i$
41. $-1 - \sqrt{5}i$
42. $-3 + \sqrt{2}i$
43. $\sqrt{-20}$
44. $\sqrt{-15}$
45. $\sqrt{6}$
46. $1 + \sqrt{8}$

 A Quotient of Complex Numbers in Standard Form In Exercises 47–54, write the quotient in standard form.

47. $\dfrac{2}{4 - 5i}$
48. $\dfrac{13}{1 - i}$
49. $\dfrac{5 + i}{5 - i}$
50. $\dfrac{6 - 7i}{1 - 2i}$
51. $\dfrac{9 - 4i}{i}$
52. $\dfrac{8 + 16i}{2i}$
53. $\dfrac{3i}{(4 - 5i)^2}$
54. $\dfrac{5i}{(2 + 3i)^2}$

 Performing Operations with Complex Numbers In Exercises 55–58, perform the operation and write the result in standard form.

55. $\dfrac{2}{1 + i} - \dfrac{3}{1 - i}$
56. $\dfrac{2i}{2 + i} + \dfrac{5}{2 - i}$
57. $\dfrac{i}{3 - 2i} + \dfrac{2i}{3 + 8i}$
58. $\dfrac{1 + i}{i} - \dfrac{3}{4 - i}$

Writing a Complex Number in Standard Form In Exercises 59–66, write the complex number in standard form.

59. $\sqrt{-6}\sqrt{-2}$
60. $\sqrt{-5}\sqrt{-10}$
61. $\left(\sqrt{-15}\right)^2$
62. $\left(\sqrt{-75}\right)^2$
63. $\sqrt{-8} + \sqrt{-50}$
64. $\sqrt{-45} - \sqrt{-5}$
65. $\left(3 + \sqrt{-5}\right)\left(7 - \sqrt{-10}\right)$
66. $\left(2 - \sqrt{-6}\right)^2$

Complex Solutions of a Quadratic Equation In Exercises 67–76, use the Quadratic Formula to solve the quadratic equation.

67. $x^2 - 2x + 2 = 0$ **68.** $x^2 + 6x + 10 = 0$

69. $4x^2 + 16x + 17 = 0$ **70.** $9x^2 - 6x + 37 = 0$

71. $4x^2 + 16x + 21 = 0$ **72.** $16t^2 - 4t + 3 = 0$

73. $\frac{3}{2}x^2 - 6x + 9 = 0$ **74.** $\frac{7}{8}x^2 - \frac{3}{4}x + \frac{5}{16} = 0$

75. $1.4x^2 - 2x + 10 = 0$ **76.** $4.5x^2 - 3x + 12 = 0$

Simplifying a Complex Number In Exercises 77–86, simplify the complex number and write it in standard form.

77. $-6i^3 + i^2$ **78.** $4i^2 - 2i^3$

79. $-14i^5$ **80.** $(-i)^3$

81. $\left(\sqrt{-72}\right)^3$ **82.** $\left(\sqrt{-2}\right)^6$

83. $\dfrac{1}{i^3}$ **84.** $\dfrac{1}{(2i)^3}$

85. $(3i)^4$ **86.** $(-i)^6$

• • 87. Impedance of a Circuit • • • • • • • • • • •

The opposition to current in an electrical circuit is called its impedance. The impedance z in a parallel circuit with two pathways satisfies the equation

$$\frac{1}{z} = \frac{1}{z_1} + \frac{1}{z_2}$$

where z_1 is the impedance (in ohms) of pathway 1 and z_2 is the impedance (in ohms) of pathway 2.

(a) The impedance of each pathway in a parallel circuit is found by adding the impedances of all components in the pathway. Use the table to find z_1 and z_2

	Resistor	Inductor	Capacitor
Symbol	$-\!\!\bigwedge\!\!-$ $a\,\Omega$	$-\!\text{000}\!-$ $b\,\Omega$	$-\!\vert\!\vert\!-$ $c\,\Omega$
Impedance	a	bi	$-ci$

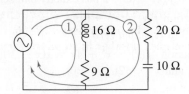

(b) Find the impedance z.

88. Cube of a Complex Number Cube each complex number.

(a) $-1 + \sqrt{3}i$ (b) $-1 - \sqrt{3}i$

Exploration

True or False? In Exercises 89–92, determine whether the statement is true or false. Justify your answer.

89. The sum of two complex numbers is always a real number.

90. There is no complex number that is equal to its complex conjugate.

91. $-i\sqrt{6}$ is a solution of $x^4 - x^2 + 14 = 56$.

92. $i^{44} + i^{150} - i^{74} - i^{109} + i^{61} = -1$

93. Pattern Recognition Find the missing values.

$i^1 = i$	$i^2 = -1$	$i^3 = -i$	$i^4 = 1$
$i^5 = \rule{1cm}{0.3cm}$	$i^6 = \rule{1cm}{0.3cm}$	$i^7 = \rule{1cm}{0.3cm}$	$i^8 = \rule{1cm}{0.3cm}$
$i^9 = \rule{1cm}{0.3cm}$	$i^{10} = \rule{1cm}{0.3cm}$	$i^{11} = \rule{1cm}{0.3cm}$	$i^{12} = \rule{1cm}{0.3cm}$

What pattern do you see? Write a brief description of how you would find i raised to any positive integer power.

94. **HOW DO YOU SEE IT?** The coordinate system shown below is called the complex plane. In the complex plane, the point (a, b) corresponds to the complex number $a + bi$.

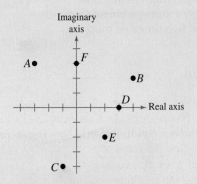

Match each complex number with its corresponding point.

(i) 3 (ii) $3i$ (iii) $4 + 2i$

(iv) $2 - 2i$ (v) $-3 + 3i$ (vi) $-1 - 4i$

95. Error Analysis Describe the error.

$$\sqrt{-6}\sqrt{-6} = \sqrt{(-6)(-6)} = \sqrt{36} = 6 \quad \text{✗}$$

96. Proof Prove that the complex conjugate of the product of two complex numbers $a_1 + b_1i$ and $a_2 + b_2i$ is the product of their complex conjugates.

97. Proof Prove that the complex conjugate of the sum of two complex numbers $a_1 + b_1i$ and $a_2 + b_2i$ is the sum of their complex conjugates.

2.5 Zeros of Polynomial Functions

Finding zeros of polynomial functions is an important part of solving many real-life problems. For example, in Exercise 105 on page 164, you will use the zeros of a polynomial function to redesign a storage bin so that it can hold five times as much food.

- Use the Fundamental Theorem of Algebra to determine numbers of zeros of polynomial functions.
- Find rational zeros of polynomial functions.
- Find complex zeros using conjugate pairs.
- Find zeros of polynomials by factoring.
- Use Descartes's Rule of Signs and the Upper and Lower Bound Rules to find zeros of polynomials.
- Find zeros of polynomials in real-life applications.

The Fundamental Theorem of Algebra

In the complex number system, every nth-degree polynomial function has *precisely n* zeros. This important result is derived from the **Fundamental Theorem of Algebra,** first proved by German mathematician Carl Friedrich Gauss (1777–1855).

> ### The Fundamental Theorem of Algebra
>
> If $f(x)$ is a polynomial of degree n, where $n > 0$, then f has at least one zero in the complex number system.

Using the Fundamental Theorem of Algebra and the equivalence of zeros and factors, you obtain the **Linear Factorization Theorem.**

> ### Linear Factorization Theorem
>
> If $f(x)$ is a polynomial of degree n, where $n > 0$, then $f(x)$ has precisely n linear factors
>
> $$f(x) = a_n(x - c_1)(x - c_2) \cdots (x - c_n)$$
>
> where $c_1, c_2, \ldots, c_n$ are complex numbers.

For a proof of the Linear Factorization Theorem, see Proofs in Mathematics on page 194.

Note that the Fundamental Theorem of Algebra and the Linear Factorization Theorem tell you only that the zeros or factors of a polynomial exist, not how to find them. Such theorems are called **existence theorems.**

•• **REMARK** Recall that in order to find the zeros of a function f, set $f(x)$ equal to 0 and solve the resulting equation for x. For instance, the function in Example 1(a) has a zero at $x = 2$ because

$$x - 2 = 0$$

$$x = 2.$$

EXAMPLE 1 **Zeros of Polynomial Functions**

See LarsonPrecalculus.com for an interactive version of this type of example.

a. The first-degree polynomial function $f(x) = x - 2$ has exactly *one* zero: $x = 2$.

b. The second-degree polynomial function $f(x) = x^2 - 6x + 9 = (x - 3)(x - 3)$ has exactly *two* zeros: $x = 3$ and $x = 3$ (a *repeated zero*).

c. The third-degree polynomial function $f(x) = x^3 + 4x = x(x - 2i)(x + 2i)$ has exactly *three* zeros: $x = 0$, $x = 2i$, and $x = -2i$.

✓ *Checkpoint*))) *Audio-video solution in English & Spanish at LarsonPrecalculus.com*

Determine the number of zeros of the polynomial function $f(x) = x^4 - 1$. ∎

The Rational Zero Test

The **Rational Zero Test** relates the possible rational zeros of a polynomial (having integer coefficients) to the leading coefficient and to the constant term of the polynomial.

> **The Rational Zero Test**
>
> If the polynomial
>
> $$f(x) = a_n x^n + a_{n-1} x^{n-1} + \cdots + a_2 x^2 + a_1 x + a_0$$
>
> has *integer* coefficients, then every rational zero of f has the form
>
> $$\text{Rational zero} = \frac{p}{q}$$
>
> where p and q have no common factors other than 1, and
>
> $p = $ a factor of the constant term a_0
>
> $q = $ a factor of the leading coefficient a_n.

Although they were not contemporaries, French mathematician Jean Le Rond d'Alembert (1717–1783) worked independently of Carl Friedrich Gauss in trying to prove the Fundamental Theorem of Algebra. His efforts were such that, in France, the Fundamental Theorem of Algebra is frequently known as d'Alembert's Theorem.

To use the Rational Zero Test, you should first list all rational numbers whose numerators are factors of the constant term and whose denominators are factors of the leading coefficient.

$$\text{Possible rational zeros:} \quad \frac{\text{Factors of constant term}}{\text{Factors of leading coefficient}}$$

Having formed this list of *possible rational zeros,* use a trial-and-error method to determine which, if any, are actual zeros of the polynomial. Note that when the leading coefficient is 1, the possible rational zeros are simply the factors of the constant term.

EXAMPLE 2 **Rational Zero Test with Leading Coefficient of 1**

Find (if possible) the rational zeros of

$$f(x) = x^3 + x + 1.$$

Solution   The leading coefficient is 1, so the possible rational zeros are the factors of the constant term.

Possible rational zeros: 1 and -1

Testing these possible zeros shows that neither works.

$$f(1) = (1)^3 + 1 + 1$$
$$= 3$$
$$f(-1) = (-1)^3 + (-1) + 1$$
$$= -1$$

So, the given polynomial has *no* rational zeros. Note from the graph of f in Figure 2.15 that f does have one real zero between -1 and 0. However, by the Rational Zero Test, you know that this real zero is *not* a rational number.

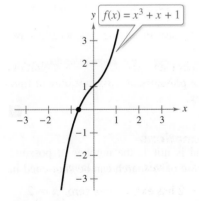

Figure 2.15

The graph is labeled $f(x) = x^3 + x + 1$.

✓ **Checkpoint** 🔊))) Audio-video solution in English & Spanish at LarsonPrecalculus.com

Find (if possible) the rational zeros of

$$f(x) = x^3 + 2x^2 + 6x - 4.$$

EXAMPLE 3 **Rational Zero Test with Leading Coefficient of 1**

Find the rational zeros of

$$f(x) = x^4 - x^3 + x^2 - 3x - 6.$$

Solution The leading coefficient is 1, so the possible rational zeros are the factors of the constant term.

Possible rational zeros: $\pm 1, \pm 2, \pm 3, \pm 6$

By applying synthetic division successively, you find that $x = -1$ and $x = 2$ are the only two rational zeros.

$$
\begin{array}{r|rrrrr}
-1 & 1 & -1 & 1 & -3 & -6 \\
   &   & -1 & 2 & -3 & 6 \\
\hline
   & 1 & -2 & 3 & -6 & 0
\end{array}
\longrightarrow \quad \text{0 remainder, so } x = -1 \text{ is a zero.}
$$

$$
\begin{array}{r|rrrr}
2 & 1 & -2 & 3 & -6 \\
  &   & 2 & 0 & 6 \\
\hline
  & 1 & 0 & 3 & 0
\end{array}
\longrightarrow \quad \text{0 remainder, so } x = 2 \text{ is a zero.}
$$

So, $f(x)$ factors as

$$f(x) = (x + 1)(x - 2)(x^2 + 3).$$

The factor $(x^2 + 3)$ produces no real zeros, so $x = -1$ and $x = 2$ are the only *real* zeros of f. The figure below verifies this.

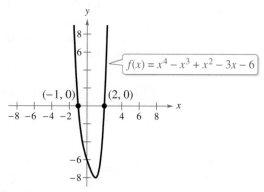

✓ **Checkpoint** 🔊))) *Audio-video solution in English & Spanish at LarsonPrecalculus.com*

Find the rational zeros of

$$f(x) = x^3 - 15x^2 + 75x - 125.$$

When the leading coefficient of a polynomial is not 1, the number of possible rational zeros can increase dramatically. In such cases, the search can be shortened in several ways.

1. A graphing utility can help to speed up the calculations.

2. A graph can give good estimates of the locations of the zeros.

3. The Intermediate Value Theorem, along with a table of values, can give approximations of the zeros.

4. Synthetic division can be used to test the possible rational zeros.

After finding the first zero, the search becomes simpler by working with the lower-degree polynomial obtained in synthetic division, as shown in Example 3.

• • **REMARK** When there are few possible rational zeros, as in Example 2, it may be quicker to test the zeros by evaluating the function. When there are more possible rational zeros, as in Example 3, it may be quicker to use a different approach to test the zeros, such as using synthetic division or sketching a graph.

EXAMPLE 4   **Using the Rational Zero Test**

Find the rational zeros of $f(x) = 2x^3 + 3x^2 - 8x + 3$.

Solution   The leading coefficient is 2 and the constant term is 3.

$$\textit{Possible rational zeros: } \frac{\text{Factors of } 3}{\text{Factors of } 2} = \frac{\pm 1, \pm 3}{\pm 1, \pm 2} = \pm 1, \pm 3, \pm\frac{1}{2}, \pm\frac{3}{2}$$

▷

By synthetic division, $x = 1$ is a rational zero.

$$
\begin{array}{r|rrrr}
1 & 2 & 3 & -8 & 3 \\
  &   & 2 & 5 & -3 \\
\hline
  & 2 & 5 & -3 & 0 \\
\end{array}
$$

· · REMARK   Remember that when you find the rational zeros of a polynomial function with many possible rational zeros, as in Example 4, you must use trial and error. There is no quick algebraic method to determine which of the possibilities is an actual zero; however, sketching a graph may be helpful.

So, $f(x)$ factors as

$$f(x) = (x - 1)(2x^2 + 5x - 3)$$
$$= (x - 1)(2x - 1)(x + 3)$$

which shows that the rational zeros of f are $x = 1$, $x = \frac{1}{2}$, and $x = -3$.

✓ **Checkpoint** ◀))) *Audio-video solution in English & Spanish at LarsonPrecalculus.com*

Find the rational zeros of

$$f(x) = 2x^3 + x^2 - 13x + 6.$$

■

Recall from Section 2.2 that if $x = a$ is a zero of the polynomial function f, then $x = a$ is a solution of the polynomial equation $f(x) = 0$.

EXAMPLE 5   **Solving a Polynomial Equation**

Find all real solutions of $-10x^3 + 15x^2 + 16x - 12 = 0$.

Solution   The leading coefficient is -10 and the constant term is -12.

$$\textit{Possible rational solutions: } \frac{\text{Factors of } -12}{\text{Factors of } -10} = \frac{\pm 1, \pm 2, \pm 3, \pm 4, \pm 6, \pm 12}{\pm 1, \pm 2, \pm 5, \pm 10}$$

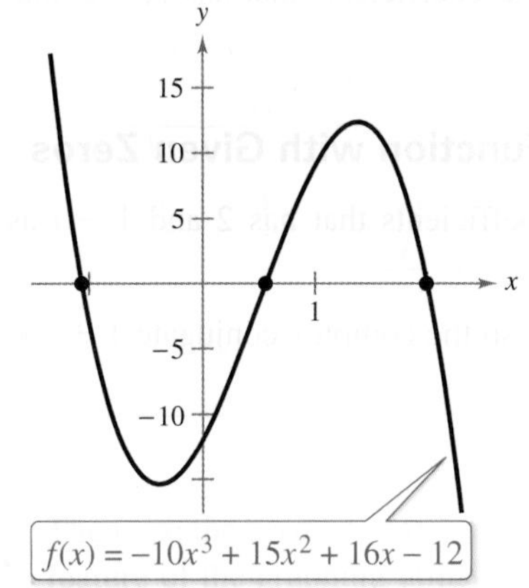

With so many possibilities (32, in fact), it is worth your time to sketch a graph. In Figure 2.16, three reasonable solutions appear to be $x = -\frac{6}{5}$, $x = \frac{1}{2}$, and $x = 2$. Testing these by synthetic division shows that $x = 2$ is the only rational solution. So, you have

$$(x - 2)(-10x^2 - 5x + 6) = 0.$$

Using the Quadratic Formula to solve $-10x^2 - 5x + 6 = 0$, you find that the two additional solutions are irrational numbers.

$$x = \frac{5 + \sqrt{265}}{-20} \approx -1.0639$$

and

$$x = \frac{5 - \sqrt{265}}{-20} \approx 0.5639$$

$$f(x) = -10x^3 + 15x^2 + 16x - 12$$

Figure 2.16

▷ **ALGEBRA HELP**   To review the Quadratic Formula, see Appendix A.5.

✓ **Checkpoint** ◀))) *Audio-video solution in English & Spanish at LarsonPrecalculus.com*

Find all real solutions of

$$-2x^3 - 5x^2 + 15x + 18 = 0.$$

■

2.5 Exercises

See **CalcChat.com** for tutorial help and worked-out solutions to odd-numbered exercises.

Vocabulary: Fill in the blanks.

1. The _____ _____ of _____ states that if $f(x)$ is a polynomial of degree n $(n > 0)$, then f has at least one zero in the complex number system.

2. The _____ _____ _____ states that if $f(x)$ is a polynomial of degree n $(n > 0)$, then $f(x)$ has precisely n linear factors, $f(x) = a_n(x - c_1)(x - c_2) \cdots (x - c_n)$, where $c_1, c_2, \ldots, c_n$ are complex numbers.

3. The test that gives a list of the possible rational zeros of a polynomial function is the _____ _____ Test.

4. If $a + bi$, where $b \neq 0$, is a complex zero of a polynomial with real coefficients, then so is its _____ _____, $a - bi$.

5. Every polynomial of degree $n > 0$ with real coefficients can be written as the product of _____ and _____ factors with real coefficients, where the _____ factors have no real zeros.

6. A quadratic factor that cannot be factored further as a product of linear factors containing real numbers is _____ over the _____.

7. The theorem that can be used to determine the possible numbers of positive and negative real zeros of a function is called _____ _____ of _____.

8. A real number c is a _____ bound for the real zeros of f when no real zeros are less than c, and is a _____ bound when no real zeros are greater than c.

Skills and Applications

Zeros of Polynomial Functions In Exercises 9–14, determine the number of zeros of the polynomial function.

9. $f(x) = x^3 + 2x^2 + 1$ 10. $f(x) = x^4 - 3x$

11. $g(x) = x^4 - x^5$ 12. $f(x) = x^3 - x^6$

13. $f(x) = (x + 5)^2$

14. $h(t) = (t - 1)^2 - (t + 1)^2$

Using the Rational Zero Test In Exercises 15–18, use the Rational Zero Test to list the possible rational zeros of f. Verify that the zeros of f shown in the graph are contained in the list.

15. $f(x) = x^3 + 2x^2 - x - 2$

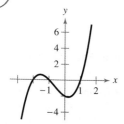

16. $f(x) = x^3 - 4x^2 - 4x + 16$

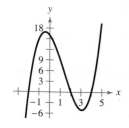

17. $f(x) = 2x^4 - 17x^3 + 35x^2 + 9x - 45$

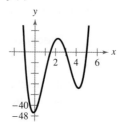

18. $f(x) = 4x^5 - 8x^4 - 5x^3 + 10x^2 + x - 2$

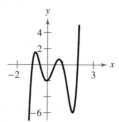

Using the Rational Zero Test In Exercises 19–28, find (if possible) the rational zeros of the function.

19. $f(x) = x^3 - 7x - 6$ 20. $f(x) = x^3 - 13x + 12$

21. $g(t) = t^3 - 4t^2 + 4$ 22. $h(x) = x^3 - 19x + 30$

23. $h(t) = t^3 + 8t^2 + 13t + 6$

24. $g(x) = x^3 + 8x^2 + 12x + 18$

25. $C(x) = 2x^3 + 3x^2 - 1$

26. $f(x) = 3x^3 - 19x^2 + 33x - 9$

27. $g(x) = 9x^4 - 9x^3 - 58x^2 + 4x + 24$

28. $f(x) = 2x^4 - 15x^3 + 23x^2 + 15x - 25$

Solving a Polynomial Equation In Exercises 29–32, find all real solutions of the polynomial equation.

29. $-5x^3 + 11x^2 - 4x - 2 = 0$
30. $8x^3 + 10x^2 - 15x - 6 = 0$
31. $x^4 + 6x^3 + 3x^2 - 16x + 6 = 0$
32. $x^4 + 8x^3 + 14x^2 - 17x - 42 = 0$

Using the Rational Zero Test In Exercises 33–36, (a) list the possible rational zeros of f, (b) sketch the graph of f so that some of the possible zeros in part (a) can be disregarded, and then (c) determine all real zeros of f.

33. $f(x) = x^3 + x^2 - 4x - 4$
34. $f(x) = -3x^3 + 20x^2 - 36x + 16$
35. $f(x) = -4x^3 + 15x^2 - 8x - 3$
36. $f(x) = 4x^3 - 12x^2 - x + 15$

Using the Rational Zero Test In Exercises 37–40, (a) list the possible rational zeros of f, (b) use a graphing utility to graph f so that some of the possible zeros in part (a) can be disregarded, and then (c) determine all real zeros of f.

37. $f(x) = -2x^4 + 13x^3 - 21x^2 + 2x + 8$
38. $f(x) = 4x^4 - 17x^2 + 4$
39. $f(x) = 32x^3 - 52x^2 + 17x + 3$
40. $f(x) = 4x^3 + 7x^2 - 11x - 18$

Finding a Polynomial Function with Given Zeros In Exercises 41–46, find a polynomial function with real coefficients that has the given zeros. (There are many correct answers.)

41. $1, 5i$
42. $4, -3i$
43. $2, 2, 1 + i$
44. $-1, 5, 3 - 2i$
45. $\frac{2}{3}, -1, 3 + \sqrt{2}i$
46. $-\frac{5}{2}, -5, 1 + \sqrt{3}i$

Finding a Polynomial Function with Given Zeros In Exercises 47–50, find the polynomial function f with real coefficients that has the given degree, zeros, and solution point.

Degree	Zeros	Solution Point
47. 4	$-2, 1, i$	$f(0) = -4$
48. 4	$-1, 2, \sqrt{2}i$	$f(1) = 12$
49. 3	$-3, 1 + \sqrt{3}i$	$f(-2) = 12$
50. 3	$-2, 1 - \sqrt{2}i$	$f(-1) = -12$

Factoring a Polynomial In Exercises 51–54, write the polynomial (a) as the product of factors that are irreducible over the *rationals*, (b) as the product of linear and quadratic factors that are irreducible over the *reals*, and (c) in completely factored form.

51. $f(x) = x^4 + 2x^2 - 8$
52. $f(x) = x^4 + 6x^2 - 27$
53. $f(x) = x^4 - 2x^3 - 3x^2 + 12x - 18$
    (*Hint:* One factor is $x^2 - 6$.)
54. $f(x) = x^4 - 3x^3 - x^2 - 12x - 20$
    (*Hint:* One factor is $x^2 + 4$.)

Finding the Zeros of a Polynomial Function In Exercises 55–60, use the given zero to find all the zeros of the function.

Function	Zero
55. $f(x) = x^3 - x^2 + 4x - 4$	$2i$
56. $f(x) = 2x^3 + 3x^2 + 18x + 27$	$3i$
57. $g(x) = x^3 - 8x^2 + 25x - 26$	$3 + 2i$
58. $g(x) = x^3 + 9x^2 + 25x + 17$	$-4 + i$
59. $h(x) = x^4 - 6x^3 + 14x^2 - 18x + 9$	$1 - \sqrt{2}i$
60. $h(x) = x^4 + x^3 - 3x^2 - 13x + 14$	$-2 + \sqrt{3}i$

Finding the Zeros of a Polynomial Function In Exercises 61–72, write the polynomial as the product of linear factors and list all the zeros of the function.

61. $f(x) = x^2 + 36$ 62. $f(x) = x^2 + 49$
63. $h(x) = x^2 - 2x + 17$ 64. $g(x) = x^2 + 10x + 17$
65. $f(x) = x^4 - 16$ 66. $f(y) = y^4 - 256$
67. $f(z) = z^2 - 2z + 2$
68. $h(x) = x^3 - 3x^2 + 4x - 2$
69. $g(x) = x^3 - 3x^2 + x + 5$
70. $f(x) = x^3 - x^2 + x + 39$
71. $g(x) = x^4 - 4x^3 + 8x^2 - 16x + 16$
72. $h(x) = x^4 + 6x^3 + 10x^2 + 6x + 9$

Finding the Zeros of a Polynomial Function In Exercises 73–78, find all the zeros of the function. When there is an extended list of possible rational zeros, use a graphing utility to graph the function in order to disregard any of the possible rational zeros that are obviously not zeros of the function.

73. $f(x) = x^3 + 24x^2 + 214x + 740$
74. $f(s) = 2s^3 - 5s^2 + 12s - 5$
75. $f(x) = 16x^3 - 20x^2 - 4x + 15$
76. $f(x) = 9x^3 - 15x^2 + 11x - 5$
77. $f(x) = 2x^4 + 5x^3 + 4x^2 + 5x + 2$
78. $g(x) = x^5 - 8x^4 + 28x^3 - 56x^2 + 64x - 32$

Using Descartes's Rule of Signs In Exercises 79–86, use Descartes's Rule of Signs to determine the possible numbers of positive and negative real zeros of the function.

79. $g(x) = 2x^3 - 3x^2 - 3$
80. $h(x) = 4x^2 - 8x + 3$
81. $h(x) = 2x^3 + 3x^2 + 1$
82. $h(x) = 2x^4 - 3x - 2$
83. $g(x) = 6x^4 + 2x^3 - 3x^2 + 2$
84. $f(x) = 4x^3 - 3x^2 - 2x - 1$
85. $f(x) = 5x^3 + x^2 - x + 5$
86. $f(x) = 3x^3 - 2x^2 - x + 3$

Verifying Upper and Lower Bounds In Exercises 87–90, use synthetic division to verify the upper and lower bounds of the real zeros of f.

87. $f(x) = x^3 + 3x^2 - 2x + 1$
   (a) Upper: $x = 1$ (b) Lower: $x = -4$
88. $f(x) = x^3 - 4x^2 + 1$
   (a) Upper: $x = 4$ (b) Lower: $x = -1$
89. $f(x) = x^4 - 4x^3 + 16x - 16$
   (a) Upper: $x = 5$ (b) Lower: $x = -3$
90. $f(x) = 2x^4 - 8x + 3$
   (a) Upper: $x = 3$ (b) Lower: $x = -4$

Finding Real Zeros of a Polynomial Function In Exercises 91–94, find all real zeros of the function.

91. $f(x) = 16x^3 - 12x^2 - 4x + 3$
92. $f(z) = 12z^3 - 4z^2 - 27z + 9$
93. $f(y) = 4y^3 + 3y^2 + 8y + 6$
94. $g(x) = 3x^3 - 2x^2 + 15x - 10$

Finding the Rational Zeros of a Polynomial In Exercises 95–98, find the rational zeros of the polynomial function.

95. $P(x) = x^4 - \frac{25}{4}x^2 + 9 = \frac{1}{4}(4x^4 - 25x^2 + 36)$
96. $f(x) = x^3 - \frac{3}{2}x^2 - \frac{23}{2}x + 6$
   $= \frac{1}{2}(2x^3 - 3x^2 - 23x + 12)$
97. $f(x) = x^3 - \frac{1}{4}x^2 - x + \frac{1}{4} = \frac{1}{4}(4x^3 - x^2 - 4x + 1)$
98. $f(z) = z^3 + \frac{11}{6}z^2 - \frac{1}{2}z - \frac{1}{3} = \frac{1}{6}(6z^3 + 11z^2 - 3z - 2)$

Rational and Irrational Zeros In Exercises 99–102, match the cubic function with the numbers of rational and irrational zeros.

(a) **Rational zeros: 0; irrational zeros: 1**
(b) **Rational zeros: 3; irrational zeros: 0**
(c) **Rational zeros: 1; irrational zeros: 2**
(d) **Rational zeros: 1; irrational zeros: 0**

99. $f(x) = x^3 - 1$
100. $f(x) = x^3 - 2$
101. $f(x) = x^3 - x$
102. $f(x) = x^3 - 2x$

103. Geometry You want to make an open box from a rectangular piece of material, 15 centimeters by 9 centimeters, by cutting equal squares from the corners and turning up the sides.

   (a) Let x represent the side length of each of the squares removed. Draw a diagram showing the squares removed from the original piece of material and the resulting dimensions of the open box.

   (b) Use the diagram to write the volume V of the box as a function of x. Determine the domain of the function.

   (c) Sketch the graph of the function and approximate the dimensions of the box that yield a maximum volume.

   (d) Find values of x such that $V = 56$. Which of these values is a physical impossibility in the construction of the box? Explain.

104. Geometry A rectangular package to be sent by a delivery service (see figure) has a combined length and girth (perimeter of a cross section) of 120 inches.

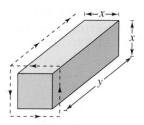

   (a) Use the diagram to write the volume V of the package as a function of x.

   (b) Use a graphing utility to graph the function and approximate the dimensions of the package that yield a maximum volume.

   (c) Find values of x such that $V = 13,500$. Which of these values is a physical impossibility in the construction of the package? Explain.

105. Geometry

A bulk food storage bin with dimensions 2 feet by 3 feet by 4 feet needs to be increased in size to hold five times as much food as the current bin.

   (a) Assume each dimension is increased by the same amount. Write a function that represents the volume V of the new bin.

   (b) Find the dimensions of the new bin.

106. Cost The ordering and transportation cost C (in thousands of dollars) for machine parts is given by

$$C(x) = 100\left(\frac{200}{x^2} + \frac{x}{x+30}\right), \quad x \geq 1$$

where x is the order size (in hundreds). In calculus, it can be shown that the cost is a minimum when

$$3x^3 - 40x^2 - 2400x - 36{,}000 = 0.$$

Use a graphing utility to approximate the optimal order size to the nearest hundred units.

Exploration

True or False? In Exercises 107 and 108, decide whether the statement is true or false. Justify your answer.

107. It is possible for a third-degree polynomial function with integer coefficients to have no real zeros.

108. If $x = -i$ is a zero of the function

$$f(x) = x^3 + ix^2 + ix - 1$$

then $x = i$ must also be a zero of f.

Think About It In Exercises 109–114, determine (if possible) the zeros of the function g when the function f has zeros at $x = r_1$, $x = r_2$, and $x = r_3$.

109. $g(x) = -f(x)$

110. $g(x) = 3f(x)$

111. $g(x) = f(x - 5)$

112. $g(x) = f(2x)$

113. $g(x) = 3 + f(x)$

114. $g(x) = f(-x)$

115. Think About It A cubic polynomial function f has real zeros -2, $\frac{1}{2}$, and 3, and its leading coefficient is negative. Write an equation for f and sketch its graph. How many different polynomial functions are possible for f?

116. Think About It Sketch the graph of a fifth-degree polynomial function whose leading coefficient is positive and that has a zero at $x = 3$ of multiplicity 2.

Writing an Equation In Exercises 117 and 118, the graph of a cubic polynomial function $y = f(x)$ is shown. One of the zeros is $1 + i$. Write an equation for f.

117.

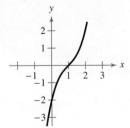

118.

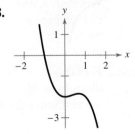

119. Error Analysis Describe the error.

The graph of a quartic (fourth-degree) polynomial $y = f(x)$ is shown. One of the zeros is i.

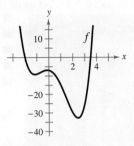

The function is $f(x) = (x + 2)(x - 3.5)(x - i)$. ✗

120. HOW DO YOU SEE IT? Use the information in the table to answer each question.

Interval	Value of $f(x)$
$(-\infty, -2)$	Positive
$(-2, 1)$	Negative
$(1, 4)$	Negative
$(4, \infty)$	Positive

(a) What are the three real zeros of the polynomial function f?

(b) What can be said about the behavior of the graph of f at $x = 1$?

(c) What is the least possible degree of f? Explain. Can the degree of f ever be odd? Explain.

(d) Is the leading coefficient of f positive or negative? Explain.

(e) Sketch a graph of a function that exhibits the behavior described in the table.

121. Think About It Let $y = f(x)$ be a quartic (fourth-degree) polynomial with leading coefficient $a = 1$ and

$$f(i) = f(2i) = 0.$$

Write an equation for f.

122. Think About It Let $y = f(x)$ be a cubic polynomial with leading coefficient $a = -1$ and

$$f(2) = f(i) = 0.$$

Write an equation for f.

123. Writing an Equation Write the equation for a quadratic function f (with integer coefficients) that has the given zeros. Assume that b is a positive integer.

(a) $\pm\sqrt{b}i$ (b) $a \pm bi$

2.6 Rational Functions

Rational functions have many real-life applications. For example, in Exercise 69 on page 176, you will use a rational function to determine the cost of supplying recycling bins to the population of a rural township.

- Find domains of rational functions.
- Find vertical and horizontal asymptotes of graphs of rational functions.
- Sketch graphs of rational functions.
- Sketch graphs of rational functions that have slant asymptotes.
- Use rational functions to model and solve real-life problems.

Introduction

A **rational function** is a quotient of polynomial functions. It can be written in the form

$$f(x) = \frac{N(x)}{D(x)}$$

where $N(x)$ and $D(x)$ are polynomials and $D(x)$ is not the zero polynomial.

The *domain* of a rational function of x includes all real numbers except x-values that make the denominator zero. Much of the discussion of rational functions will focus on the behavior of their graphs near x-values excluded from the domain.

EXAMPLE 1 **Finding the Domain of a Rational Function**

See LarsonPrecalculus.com for an interactive version of this type of example.

Find the domain of $f(x) = \dfrac{1}{x}$ and discuss the behavior of f near any excluded x-values.

Solution The denominator is zero when $x = 0$, so the domain of f is all real numbers except $x = 0$. To determine the behavior of f near this excluded value, evaluate $f(x)$ to the left and right of $x = 0$, as shown in the tables below.

x	-1	-0.5	-0.1	-0.01	-0.001	$\rightarrow 0$
$f(x)$	-1	-2	-10	-100	-1000	$\rightarrow -\infty$

x	$0 \leftarrow$	0.001	0.01	0.1	0.5	1
$f(x)$	$\infty \leftarrow$	1000	100	10	2	1

Note that as x approaches 0 *from the left,* $f(x)$ decreases without bound. In contrast, as x approaches 0 *from the right,* $f(x)$ increases without bound. The graph of f is shown below.

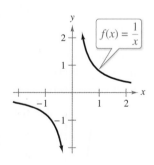

- - - - - - - - - - - - - - - - ▷

REMARK Recall from Section 1.6 that the rational function

$$f(x) = \frac{1}{x}$$

is also referred to as the *reciprocal function.*

✓ **Checkpoint** 🔊))) *Audio-video solution in English & Spanish at LarsonPrecalculus.com*

Find the domain of $f(x) = \dfrac{3x}{x - 1}$ and discuss the behavior of f near any excluded x-values.

Vertical and Horizontal Asymptotes

In Example 1, the behavior of f near $x = 0$ is as denoted below.

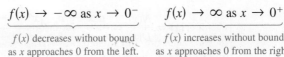

$$f(x) \to -\infty \text{ as } x \to 0^- \qquad f(x) \to \infty \text{ as } x \to 0^+$$

$f(x)$ decreases without bound as x approaches 0 from the left. $f(x)$ increases without bound as x approaches 0 from the right.

The line $x = 0$ is a **vertical asymptote** of the graph of f, as shown in Figure 2.19. Notice that the graph of f also has a **horizontal asymptote**—the line $y = 0$. The behavior of f near $y = 0$ is as denoted below.

$$f(x) \to 0 \text{ as } x \to -\infty \qquad f(x) \to 0 \text{ as } x \to \infty$$

$f(x)$ approaches 0 as x decreases without bound. $f(x)$ approaches 0 as x increases without bound.

Vertical asymptote: $x = 0$

$f(x) = \dfrac{1}{x}$

Horizontal asymptote: $y = 0$

Figure 2.19

Definitions of Vertical and Horizontal Asymptotes

1. The line $x = a$ is a **vertical asymptote** of the graph of f when

$$f(x) \to \infty \quad \text{or} \quad f(x) \to -\infty$$

as $x \to a$, either from the right or from the left.

2. The line $y = b$ is a **horizontal asymptote** of the graph of f when

$$f(x) \to b$$

as $x \to \infty$ or $x \to -\infty$.

Eventually (as $x \to \infty$ or $x \to -\infty$), the distance between the horizontal asymptote and the points on the graph must approach zero. Figure 2.20 shows the vertical and horizontal asymptotes of the graphs of three rational functions.

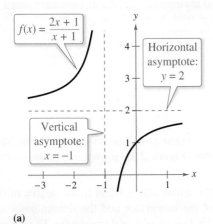

$f(x) = \dfrac{2x + 1}{x + 1}$

Horizontal asymptote: $y = 2$

Vertical asymptote: $x = -1$

(a)

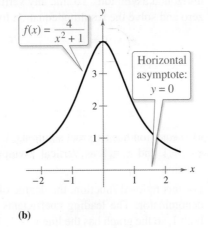

$f(x) = \dfrac{4}{x^2 + 1}$

Horizontal asymptote: $y = 0$

(b)

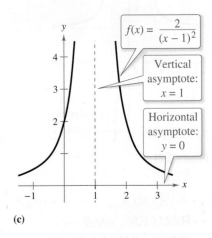

$f(x) = \dfrac{2}{(x - 1)^2}$

Vertical asymptote: $x = 1$

Horizontal asymptote: $y = 0$

(c)

Figure 2.20

Verify numerically the horizontal asymptotes shown in Figure 2.20. For example, to show that the line $y = 2$ is the horizontal asymptote of the graph of

$$f(x) = \frac{2x + 1}{x + 1}$$

create a table that shows the value of $f(x)$ as x increases and decreases without bound. The graphs of $f(x) = \dfrac{1}{x}$ in Figure 2.19 and $f(x) = \dfrac{2x + 1}{x + 1}$ in Figure 2.20(a) are **hyperbolas.** You will study hyperbolas in Chapter 10.

Vertical and Horizontal Asymptotes

Let f be the rational function

$$f(x) = \frac{N(x)}{D(x)} = \frac{a_n x^n + a_{n-1}x^{n-1} + \cdots + a_1 x + a_0}{b_m x^m + b_{m-1}x^{m-1} + \cdots + b_1 x + b_0}$$

where $N(x)$ and $D(x)$ have no common factors.

1. The graph of f has *vertical* asymptotes at the zeros of $D(x)$.

2. The graph of f has at most one *horizontal* asymptote determined by comparing the degrees of $N(x)$ and $D(x)$.

   a. When $n < m$, the graph of f has the line $y = 0$ (the x-axis) as a horizontal asymptote.

   b. When $n = m$, the graph of f has the line $y = \dfrac{a_n}{b_m}$ (ratio of the leading coefficients) as a horizontal asymptote.

   c. When $n > m$, the graph of f has no horizontal asymptote.

EXAMPLE 2   Finding Vertical and Horizontal Asymptotes

Find all vertical and horizontal asymptotes of the graph of each rational function.

a. $f(x) = \dfrac{2x^2}{x^2 - 1}$ **b.** $f(x) = \dfrac{x^2 + x - 2}{x^2 - x - 6}$

Solution

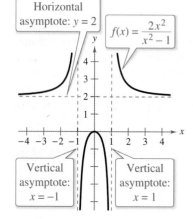

Horizontal asymptote: $y = 2$

$f(x) = \dfrac{2x^2}{x^2 - 1}$

Vertical asymptote: $x = -1$

Vertical asymptote: $x = 1$

Figure 2.21

a. For this rational function, the degree of the numerator is *equal* to the degree of the denominator. The leading coefficient of the numerator is 2 and the leading coefficient of the denominator is 1, so the graph has the line $y = 2/1 = 2$ as a horizontal asymptote. To find any vertical asymptotes, set the denominator equal to zero and solve the resulting equation for x.

$$x^2 - 1 = 0 \qquad \text{Set denominator equal to zero.}$$
$$(x + 1)(x - 1) = 0 \qquad \text{Factor.}$$
$$x + 1 = 0 \implies x = -1 \qquad \text{Set 1st factor equal to 0.}$$
$$x - 1 = 0 \implies x = 1 \qquad \text{Set 2nd factor equal to 0.}$$

This equation has two real solutions, $x = -1$ and $x = 1$, so the graph has the lines $x = -1$ and $x = 1$ as vertical asymptotes. Figure 2.21 shows the graph of this function.

b. For this rational function, the degree of the numerator is equal to the degree of the denominator. The leading coefficients of the numerator and the denominator are both 1, so the graph has the line $y = 1/1 = 1$ as a horizontal asymptote. To find any vertical asymptotes, first factor the numerator and denominator as follows.

$$f(x) = \frac{x^2 + x - 2}{x^2 - x - 6} = \frac{(x - 1)(x + 2)}{(x + 2)(x - 3)} = \frac{x - 1}{x - 3}, \quad x \neq -2$$

Setting the denominator $x - 3$ (of the simplified function) equal to zero, you find that the graph has the line $x = 3$ as a vertical asymptote.

·· REMARK There is a *hole* in the graph of f at $x = -2$. In Example 6, you will sketch the graph of a rational function that has a hole.

✓ **Checkpoint** ◀))) *Audio-video solution in English & Spanish at LarsonPrecalculus.com*

Find all vertical and horizontal asymptotes of the graph of $f(x) = \dfrac{3x^2 + 7x - 6}{x^2 + 4x + 3}$. ∎

Sketching the Graph of a Rational Function

To sketch the graph of a rational function, use the following guidelines.

Guidelines for Graphing Rational Functions

Let $f(x) = \dfrac{N(x)}{D(x)}$, where $N(x)$ and $D(x)$ are polynomials and $D(x)$ is not the zero polynomial.

1. Simplify f, if possible. List any restrictions on the domain of f that are not implied by the simplified function.

2. Find and plot the y-intercept (if any) by evaluating $f(0)$.

3. Find the zeros of the numerator (if any). Then plot the corresponding x-intercepts.

4. Find the zeros of the denominator (if any). Then sketch the corresponding vertical asymptotes.

5. Find and sketch the horizontal asymptote (if any) by using the rule for finding the horizontal asymptote of a rational function on page 168.

6. Plot at least one point *between* and one point *beyond* each x-intercept and vertical asymptote.

7. Use smooth curves to complete the graph between and beyond the vertical asymptotes.

The concept of *test intervals* from Section 2.2 can be extended to graphing rational functions. Be aware, however, that although a polynomial function can change signs only at its zeros, a rational function can change signs both at its zeros and at its undefined values (the x-values for which its denominator is zero). So, to form the test intervals in which a rational function has no sign changes, arrange the x-values representing the zeros of both the numerator and the denominator of the rational function in increasing order.

You may also want to test for symmetry when graphing rational functions, especially for simple rational functions. Recall from Section 1.6 that the graph of the reciprocal function $f(x) = \dfrac{1}{x}$ is symmetric with respect to the origin.

▷ **TECHNOLOGY** Some graphing utilities have difficulty graphing rational functions with vertical asymptotes. In connected mode, the graphing utility may connect portions of the graph that are not supposed to be connected. For example, the graph on the left should consist of two unconnected portions—one to the left of $x = 2$ and the other to the right of $x = 2$. Changing the mode of the graphing utility to *dot* mode eliminates this problem. In *dot* mode, however, the graph is represented as a collection of dots (as shown in the graph on the right) rather than as a smooth curve.

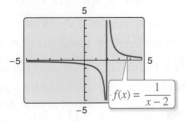

$f(x) = \dfrac{1}{x - 2}$

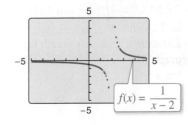

$f(x) = \dfrac{1}{x - 2}$

EXAMPLE 3   **Sketching the Graph of a Rational Function**

• • **REMARK** You can use transformations to help you sketch graphs of rational functions. For instance, the graph of g in Example 3 is a vertical stretch and a right shift of the graph of $f(x) = 1/x$ because

$$g(x) = \frac{3}{x - 2}$$

$$= 3\left(\frac{1}{x - 2}\right)$$

$$= 3f(x - 2).$$

▷

Sketch the graph of $g(x) = \dfrac{3}{x - 2}$ and state its domain.

Solution

y-intercept: $\left(0, -\frac{3}{2}\right)$, because $g(0) = -\frac{3}{2}$

x-intercept: none, because there are no zeros of the numerator

Vertical asymptote: $x = 2$, zero of denominator

Horizontal asymptote: $y = 0$, because degree of $N(x)$ < degree of $D(x)$

Additional points:

Test Interval	Representative x-Value	Value of g	Sign	Point on Graph
$(-\infty, 2)$	-4	$g(-4) = -\frac{1}{2}$	Negative	$\left(-4, -\frac{1}{2}\right)$
$(2, \infty)$	3	$g(3) = 3$	Positive	$(3, 3)$

By plotting the intercept, asymptotes, and a few additional points, you obtain the graph shown in Figure 2.22. The domain of g is all real numbers except $x = 2$.

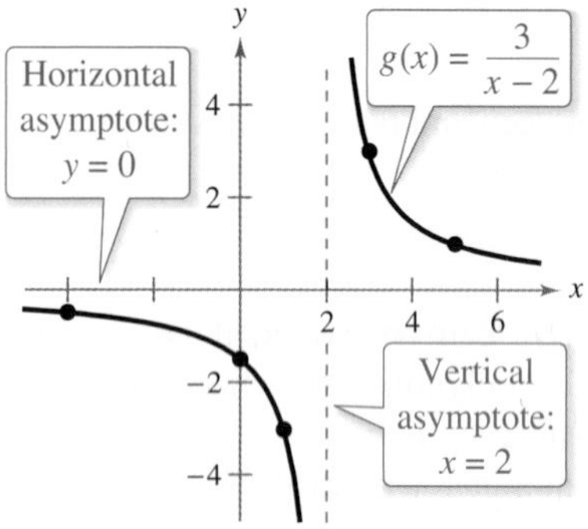

✓ **Checkpoint** 🔊))) *Audio-video solution in English & Spanish at LarsonPrecalculus.com*

Sketch the graph of $f(x) = \dfrac{1}{x + 3}$ and state its domain.

Figure 2.22

EXAMPLE 4   **Sketching the Graph of a Rational Function**

Sketch the graph of $f(x) = (2x - 1)/x$ and state its domain.

Solution

y-intercept: none, because $x = 0$ is not in the domain

x-intercept: $\left(\frac{1}{2}, 0\right)$, because $2x - 1 = 0$ when $x = \frac{1}{2}$

Vertical asymptote: $x = 0$, zero of denominator

Horizontal asymptote: $y = 2$, because degree of $N(x)$ = degree of $D(x)$

Additional points:

Test Interval	Representative x-Value	Value of f	Sign	Point on Graph
$(-\infty, 0)$	-1	$f(-1) = 3$	Positive	$(-1, 3)$
$\left(0, \frac{1}{2}\right)$	$\frac{1}{4}$	$f\left(\frac{1}{4}\right) = -2$	Negative	$\left(\frac{1}{4}, -2\right)$
$\left(\frac{1}{2}, \infty\right)$	4	$f(4) = \frac{7}{4}$	Positive	$\left(4, \frac{7}{4}\right)$

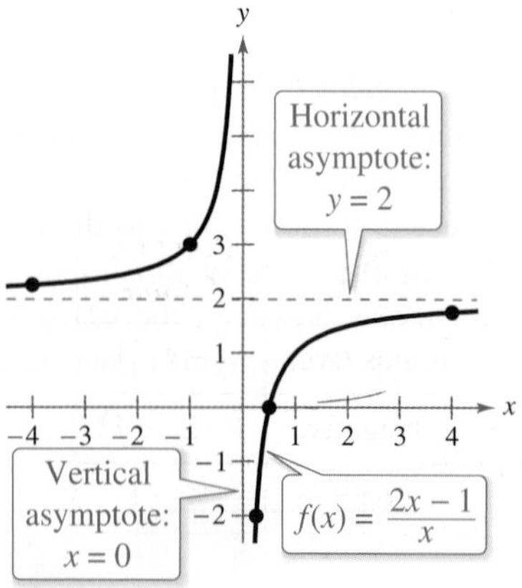

By plotting the intercept, asymptotes, and a few additional points, you obtain the graph shown in Figure 2.23. The domain of f is all real numbers except $x = 0$.

Figure 2.23

✓ **Checkpoint** 🔊))) *Audio-video solution in English & Spanish at LarsonPrecalculus.com*

Sketch the graph of $g(x) = (3 + 2x)/(1 + x)$ and state its domain. ■

| EXAMPLE 5 | **Sketching the Graph of a Rational Function** |

Sketch the graph of $f(x) = x/(x^2 - x - 2)$.

Solution Factoring the denominator, you have $f(x) = x/[(x + 1)(x - 2)]$.

Intercept: $\qquad\qquad$ $(0, 0)$, because $f(0) = 0$

Vertical asymptotes: $\quad x = -1, x = 2$, zeros of denominator

Horizontal asymptote: $y = 0$, because degree of $N(x) <$ degree of $D(x)$

Additional points:

Test Interval	Representative x-Value	Value of f	Sign	Point on Graph
$(-\infty, -1)$	-3	$f(-3) = \frac{3}{10}$	Negative	$\left(-3, -\frac{3}{10}\right)$
$(-1, 0)$	$-\frac{1}{2}$	$f\left(-\frac{1}{2}\right) = \frac{2}{5}$	Positive	$\left(-\frac{1}{2}, \frac{2}{5}\right)$
$(0, 2)$	1	$f(1) = -\frac{1}{2}$	Negative	$\left(1, -\frac{1}{2}\right)$
$(2, \infty)$	3	$f(3) = \frac{3}{4}$	Positive	$\left(3, \frac{3}{4}\right)$

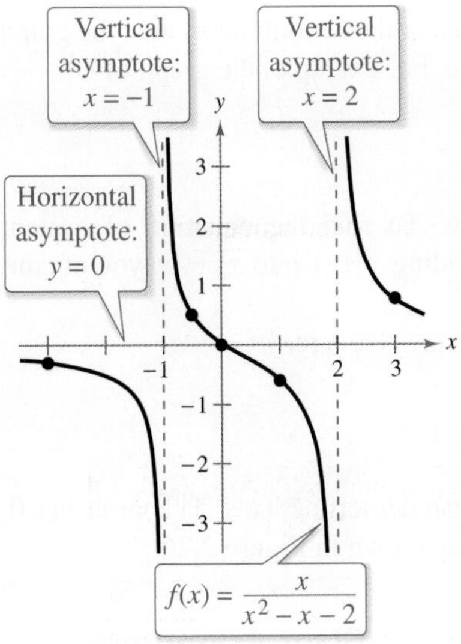

Figure 2.24 shows the graph of this function.

✓ **Checkpoint** 🔊))) *Audio-video solution in English & Spanish at LarsonPrecalculus.com*

Sketch the graph of $f(x) = 3x/(x^2 + x - 2)$.

| EXAMPLE 6 | **A Rational Function with Common Factors** |

Sketch the graph of $f(x) = (x^2 - 9)/(x^2 - 2x - 3)$.

Solution By factoring the numerator and denominator, you have

$$f(x) = \frac{x^2 - 9}{x^2 - 2x - 3} = \frac{(x - 3)(x + 3)}{(x - 3)(x + 1)} = \frac{x + 3}{x + 1}, \quad x \neq 3.$$

y-intercept: $\qquad\qquad$ $(0, 3)$, because $f(0) = 3$

x-intercept: $\qquad\qquad$ $(-3, 0)$, because $x + 3 = 0$ when $x = -3$

Vertical asymptote: $\quad x = -1$, zero of (simplified) denominator

Horizontal asymptote: $y = 1$, because degree of $N(x) =$ degree of $D(x)$

Additional points:

Test Interval	Representative x-Value	Value of f	Sign	Point on Graph
$(-\infty, -3)$	-4	$f(-4) = \frac{1}{3}$	Positive	$\left(-4, \frac{1}{3}\right)$
$(-3, -1)$	-2	$f(-2) = -1$	Negative	$(-2, -1)$
$(-1, \infty)$	2	$f(2) = \frac{5}{3}$	Positive	$\left(2, \frac{5}{3}\right)$

Figure 2.25 shows the graph of this function. Notice that there is a hole in the graph at $x = 3$, because the numerator and denominator have a common factor of $x - 3$.

✓ **Checkpoint** 🔊))) *Audio-video solution in English & Spanish at LarsonPrecalculus.com*

Sketch the graph of $f(x) = (x^2 - 4)/(x^2 - x - 6)$. ◼

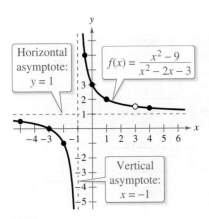

Vertical asymptote: $x = -1$

Vertical asymptote: $x = 2$

Horizontal asymptote: $y = 0$

$f(x) = \dfrac{x}{x^2 - x - 2}$

Figure 2.24

•• **REMARK** If you are unsure of the shape of a portion of the graph of a rational function, then plot some additional points. Also note that when the numerator and the denominator of a rational function have a common factor, the graph of the function has a *hole* at the zero of the common factor. (See Example 6.)

Horizontal asymptote: $y = 1$

$f(x) = \dfrac{x^2 - 9}{x^2 - 2x - 3}$

Vertical asymptote: $x = -1$

Hole at $x = 3$

Figure 2.25

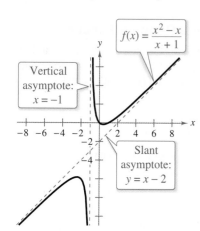

$f(x) = \dfrac{x^2 - x}{x + 1}$

Vertical asymptote: $x = -1$

Slant asymptote: $y = x - 2$

Figure 2.26

Slant Asymptotes

Consider a rational function whose denominator is of degree 1 or greater. If the degree of the numerator is exactly *one more* than the degree of the denominator, then the graph of the function has a **slant** (or **oblique**) **asymptote.** For example, the graph of

$$f(x) = \frac{x^2 - x}{x + 1}$$

has a slant asymptote, as shown in Figure 2.26. To find the equation of a slant asymptote, use long division. For example, by dividing $x + 1$ into $x^2 - x$, you obtain

$$f(x) = \frac{x^2 - x}{x + 1} = \underbrace{x - 2}_{\text{Slant asymptote}} + \frac{2}{x + 1}.$$

Slant asymptote
$(y = x - 2)$

As x increases or decreases without bound, the remainder term $2/(x + 1)$ approaches 0, so the graph of f approaches the line $y = x - 2$, as shown in Figure 2.26.

EXAMPLE 7 **A Rational Function with a Slant Asymptote**

Sketch the graph of $f(x) = \dfrac{x^2 - x - 2}{x - 1}$.

Solution Factoring the numerator as $(x - 2)(x + 1)$ enables you to recognize the *x*-intercepts. Using long division

$$f(x) = \frac{x^2 - x - 2}{x - 1} = x - \frac{2}{x - 1}$$

enables you to recognize that the line $y = x$ is a slant asymptote of the graph.

y-intercept: $(0, 2)$, because $f(0) = 2$

x-intercepts: $(2, 0)$ and $(-1, 0)$, because $x - 2 = 0$ when $x = 2$ and $x + 1 = 0$ when $x = -1$

Vertical asymptote: $x = 1$, zero of denominator

Slant asymptote: $y = x$

Additional points:

Test Interval	Representative x-Value	Value of f	Sign	Point on Graph
$(-\infty, -1)$	-2	$f(-2) = -\frac{4}{3}$	Negative	$\left(-2, -\frac{4}{3}\right)$
$(-1, 1)$	$\frac{1}{2}$	$f\left(\frac{1}{2}\right) = \frac{9}{2}$	Positive	$\left(\frac{1}{2}, \frac{9}{2}\right)$
$(1, 2)$	$\frac{3}{2}$	$f\left(\frac{3}{2}\right) = -\frac{5}{2}$	Negative	$\left(\frac{3}{2}, -\frac{5}{2}\right)$
$(2, \infty)$	3	$f(3) = 2$	Positive	$(3, 2)$

Figure 2.27 shows the graph of the function.

✓ **Checkpoint** ◄))) *Audio-video solution in English & Spanish at LarsonPrecalculus.com*

Sketch the graph of $f(x) = \dfrac{3x^2 + 1}{x}$.

Slant asymptote: $y = x$

Vertical asymptote: $x = 1$

$f(x) = \dfrac{x^2 - x - 2}{x - 1}$

Figure 2.27

Application

There are many examples of asymptotic behavior in real life. For instance, Example 8 shows how a vertical asymptote can help you to analyze the cost of removing pollutants from smokestack emissions.

EXAMPLE 8 **Cost-Benefit Model**

A utility company burns coal to generate electricity. The cost C (in dollars) of removing $p\%$ of the smokestack pollutants is given by

$$C = \frac{80,000p}{100 - p}, \quad 0 \le p \le 100.$$

You are a member of a state legislature considering a law that would require utility companies to remove 90% of the pollutants from their smokestack emissions. The current law requires 85% removal. How much additional cost would the utility company incur as a result of the new law?

Algebraic Solution

The current law requires 85% removal, so the current cost to the utility company is

$$C = \frac{80,000(85)}{100 - 85} \qquad \text{Evaluate } C \text{ when } p = 85.$$

$$\approx \$453,333.$$

The cost to remove 90% of the pollutants would be

$$C = \frac{80,000(90)}{100 - 90} \qquad \text{Evaluate } C \text{ when } p = 90.$$

$$= \$720,000.$$

So, the new law would require the utility company to spend an additional

$$720,000 - 453,333 = \$266,667. \qquad \begin{array}{l}\text{Subtract 85\% removal cost}\\ \text{from 90\% removal cost.}\end{array}$$

Graphical Solution

Use a graphing utility to graph the function

$$y_1 = \frac{80,000x}{100 - x}$$

and use the *value* feature to approximate the values of y_1 when $x = 85$ and $x = 90$, as shown below. Note that the graph has a vertical asymptote at

$$x = 100.$$

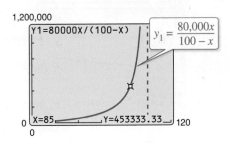

When $x = 85$, $y_1 \approx 453,333$.

When $x = 90$, $y_1 = 720,000$.

So, the new law would require the utility company to spend an additional

$$720,000 - 453,333 = \$266,667.$$

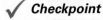 ✓ **Checkpoint** 🔊))) *Audio-video solution in English & Spanish at LarsonPrecalculus.com*

The cost C (in millions of dollars) of removing $p\%$ of the industrial and municipal pollutants discharged into a river is given by

$$C = \frac{255p}{100 - p}, \quad 0 \le p < 100.$$

a. Find the costs of removing 20%, 45%, and 80% of the pollutants.

b. According to the model, is it possible to remove 100% of the pollutants? Explain.

EXAMPLE 9 **Finding a Minimum Area**

A rectangular page contains 48 square inches of print. The margins at the top and bottom of the page are each 1 inch deep. The margins on each side are $1\frac{1}{2}$ inches wide. What should the dimensions of the page be to use the least amount of paper?

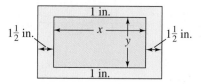

Figure 2.28

Graphical Solution

Let A be the area to be minimized. From Figure 2.28, you can write $A = (x + 3)(y + 2)$. The printed area inside the margins is given by $xy = 48$ or $y = 48/x$. To find the minimum area, rewrite the equation for A in terms of just one variable by substituting $48/x$ for y.

$$A = (x + 3)\left(\frac{48}{x} + 2\right) = \frac{(x + 3)(48 + 2x)}{x}, \quad x > 0$$

The graph of this rational function is shown below. Because x represents the width of the printed area, you need to consider only the portion of the graph for which x is positive. Use the *minimum* feature of a graphing utility to estimate that the minimum value of A occurs when $x \approx 8.5$ inches. The corresponding value of y is $48/8.5 \approx 5.6$ inches. So, the dimensions should be $x + 3 \approx 11.5$ inches by $y + 2 \approx 7.6$ inches.

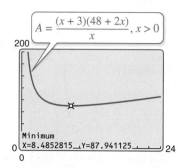

Numerical Solution

Let A be the area to be minimized. From Figure 2.28, you can write $A = (x + 3)(y + 2)$. The printed area inside the margins is given by $xy = 48$ or $y = 48/x$. To find the minimum area, rewrite the equation for A in terms of just one variable by substituting $48/x$ for y.

$$A = (x + 3)\left(\frac{48}{x} + 2\right) = \frac{(x + 3)(48 + 2x)}{x}, \quad x > 0$$

Use the *table* feature of a graphing utility to create a table of values for the function $y_1 = [(x + 3)(48 + 2x)]/x$ beginning at $x = 1$ and increasing by 1. The minimum value of y_1 occurs when x is somewhere between 8 and 9, as shown in Figure 2.29. To approximate the minimum value of y_1 to one decimal place, change the table to begin at $x = 8$ and increase by 0.1. The minimum value of y_1 occurs when $x \approx 8.5$, as shown in Figure 2.30. The corresponding value of y is $48/8.5 \approx 5.6$ inches. So, the dimensions should be $x + 3 \approx 11.5$ inches by $y + 2 \approx 7.6$ inches.

X	Y₁			X	Y₁	
6	90			8.2	87.961	
7	88.571			8.3	87.949	
8	88			8.4	87.943	
9	88			8.5	87.941	
10	88.4			8.6	87.944	
11	89.091			8.7	87.952	
12	90			8.8	87.964	
X=8				X=8.5		

Figure 2.29   **Figure 2.30**

✓ *Checkpoint* ◀))) *Audio-video solution in English & Spanish at LarsonPrecalculus.com*

Rework Example 9 when the margins on each side are 2 inches wide and the page contains 40 square inches of print.

Summarize (Section 2.6)

1. State the definition of a rational function and describe the domain *(page 166)*. For an example of finding the domain of a rational function, see Example 1.

2. Explain how to find the vertical and horizontal asymptotes of the graph of a rational function *(page 168)*. For an example of finding vertical and horizontal asymptotes of graphs of rational functions, see Example 2.

3. Explain how to sketch the graph of a rational function *(page 169)*. For examples of sketching the graphs of rational functions, see Examples 3–6.

4. Explain how to determine whether the graph of a rational function has a slant asymptote *(page 172)*. For an example of sketching the graph of a rational function that has a slant asymptote, see Example 7.

5. Describe examples of how to use rational functions to model and solve real-life problems *(pages 173 and 174, Examples 8 and 9)*.

2.6 Exercises

See **CalcChat.com** for tutorial help and worked-out solutions to odd-numbered exercises.

Vocabulary: Fill in the blanks.

1. Functions of the form $f(x) = N(x)/D(x)$, where $N(x)$ and $D(x)$ are polynomials and $D(x)$ is not the zero polynomial, are called _____ _____.

2. When $f(x) \to \pm\infty$ as $x \to a$ from the left or the right, $x = a$ is a _____ _____ of the graph of f.

3. When $f(x) \to b$ as $x \to \pm\infty$, $y = b$ is a _____ _____ of the graph of f.

4. For the rational function $f(x) = N(x)/D(x)$, if the degree of $N(x)$ is exactly one more than the degree of $D(x)$, then the graph of f has a _____ (or oblique) _____.

Skills and Applications

Finding the Domain of a Rational Function In Exercises 5–8, find the domain of the function and discuss the behavior of f near any excluded x-values.

5. $f(x) = \dfrac{1}{x-1}$

6. $f(x) = \dfrac{5x}{x+2}$

7. $f(x) = \dfrac{3x^2}{x^2-1}$

8. $f(x) = \dfrac{2x}{x^2-4}$

Finding Vertical and Horizontal Asymptotes In Exercises 9–16, find all vertical and horizontal asymptotes of the graph of the function.

9. $f(x) = \dfrac{4}{x^2}$

10. $f(x) = \dfrac{1}{(x-2)^3}$

11. $f(x) = \dfrac{5+x}{5-x}$

12. $f(x) = \dfrac{3-7x}{3+2x}$

13. $f(x) = \dfrac{x^3}{x^2-x}$

14. $f(x) = \dfrac{4x^2}{x+2}$

15. $f(x) = \dfrac{x^2-3x-4}{2x^2+x-1}$

16. $f(x) = \dfrac{-4x^2+1}{x^2+x+3}$

Sketching the Graph of a Rational Function In Exercises 17–38, (a) state the domain of the function, (b) identify all intercepts, (c) find any vertical or horizontal asymptotes, and (d) plot additional solution points as needed to sketch the graph of the rational function.

17. $f(x) = \dfrac{1}{x+1}$

18. $f(x) = \dfrac{1}{x-3}$

19. $h(x) = \dfrac{-1}{x+4}$

20. $g(x) = \dfrac{1}{6-x}$

21. $C(x) = \dfrac{2x+3}{x+2}$

22. $P(x) = \dfrac{1-3x}{1-x}$

23. $f(x) = \dfrac{x^2}{x^2+9}$

24. $f(t) = \dfrac{1-2t}{t}$

25. $g(s) = \dfrac{4s}{s^2+4}$

26. $f(x) = -\dfrac{x}{(x-2)^2}$

27. $h(x) = \dfrac{2x}{x^2-3x-4}$

28. $g(x) = \dfrac{3x}{x^2+2x-3}$

29. $f(x) = \dfrac{x-4}{x^2-16}$

30. $f(x) = \dfrac{x+1}{x^2-1}$

31. $f(t) = \dfrac{t^2-1}{t-1}$

32. $f(x) = \dfrac{x^2-36}{x+6}$

33. $f(x) = \dfrac{x^2-25}{x^2-4x-5}$

34. $f(x) = \dfrac{x^2-4}{x^2-3x+2}$

35. $f(x) = \dfrac{x^2+3x}{x^2+x-6}$

36. $f(x) = \dfrac{5(x+4)}{x^2+x-12}$

37. $f(x) = \dfrac{2x^2-5x-3}{x^3-2x^2-x+2}$

38. $f(x) = \dfrac{x^2-x-2}{x^3-2x^2-5x+6}$

Matching In Exercises 39–42, match the rational function with its graph. [The graphs are labeled (a)–(d).]

(a)

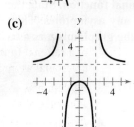

(b)

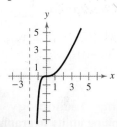

(c)

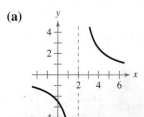

(d)

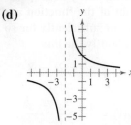

39. $f(x) = \dfrac{4}{x+2}$

40. $f(x) = \dfrac{5}{x-2}$

41. $f(x) = \dfrac{2x^2}{x^2-4}$

42. $f(x) = \dfrac{3x^3}{(x+2)^2}$

Comparing Graphs of Functions In Exercises 43–46, (a) state the domains of f and g, (b) use a graphing utility to graph f and g in the same viewing window, and (c) explain why the graphing utility may not show the difference in the domains of f and g.

43. $f(x) = \dfrac{x^2 - 1}{x + 1}$, $g(x) = x - 1$

44. $f(x) = \dfrac{x^2(x - 2)}{x^2 - 2x}$, $g(x) = x$

45. $f(x) = \dfrac{x - 2}{x^2 - 2x}$, $g(x) = \dfrac{1}{x}$

46. $f(x) = \dfrac{2x - 6}{x^2 - 7x + 12}$, $g(x) = \dfrac{2}{x - 4}$

A Rational Function with a Slant Asymptote In Exercises 47–60, (a) state the domain of the function, (b) identify all intercepts, (c) find any vertical or slant asymptotes, and (d) plot additional solution points as needed to sketch the graph of the rational function.

47. $h(x) = \dfrac{x^2 - 4}{x}$

48. $g(x) = \dfrac{x^2 + 5}{x}$

49. $f(x) = \dfrac{2x^2 + 1}{x}$

50. $f(x) = \dfrac{-x^2 - 2}{x}$

51. $g(x) = \dfrac{x^2 + 1}{x}$

52. $h(x) = \dfrac{x^2}{x - 1}$

53. $f(t) = -\dfrac{t^2 + 1}{t + 5}$

54. $f(x) = \dfrac{x^2 + 1}{x + 1}$

55. $f(x) = \dfrac{x^3}{x^2 - 4}$

56. $g(x) = \dfrac{x^3}{2x^2 - 8}$

57. $f(x) = \dfrac{x^2 - x + 1}{x - 1}$

58. $f(x) = \dfrac{2x^2 - 5x + 5}{x - 2}$

59. $f(x) = \dfrac{2x^3 - x^2 - 2x + 1}{x^2 + 3x + 2}$

60. $f(x) = \dfrac{2x^3 + x^2 - 8x - 4}{x^2 - 3x + 2}$

 Using Technology In Exercises 61–64, use a graphing utility to graph the rational function. State the domain of the function and find any asymptotes. Then zoom out sufficiently far so that the graph appears as a line. Identify the line.

61. $f(x) = \dfrac{x^2 + 2x - 8}{x + 2}$

62. $f(x) = \dfrac{2x^2 + x}{x + 1}$

63. $g(x) = \dfrac{1 + 3x^2 - x^3}{x^2}$

64. $h(x) = \dfrac{12 - 2x - x^2}{2(4 + x)}$

Graphical Reasoning In Exercises 65–68, (a) use the graph to determine any x-intercepts of the graph of the rational function and (b) set $y = 0$ and solve the resulting equation to confirm your result in part (a).

65. $y = \dfrac{x + 1}{x - 3}$

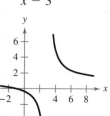

66. $y = \dfrac{2x}{x - 3}$

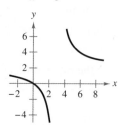

67. $y = \dfrac{1}{x} - x$

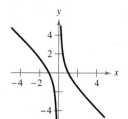

68. $y = x - 3 + \dfrac{2}{x}$

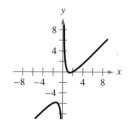

69. Recycling

The cost C (in dollars) of supplying recycling bins to $p\%$ of the population of a rural township is given by

$$C = \dfrac{25{,}000p}{100 - p}, \quad 0 \le p < 100.$$

(a) Use a graphing utility to graph the cost function.

(b) Find the costs of supplying bins to 15%, 50%, and 90% of the population.

(c) According to the model, is it possible to supply bins to 100% of the population? Explain.

70. Population Growth The game commission introduces 100 deer into newly acquired state game lands. The population N of the herd is modeled by

$$N = \dfrac{20(5 + 3t)}{1 + 0.04t}, \quad t \ge 0$$

where t is the time in years.

 (a) Use a graphing utility to graph this model.

(b) Find the populations when $t = 5$, $t = 10$, and $t = 25$.

(c) What is the limiting size of the herd as time increases?

71. Page Design A rectangular page contains 64 square inches of print. The margins at the top and bottom of the page are each 1 inch deep. The margins on each side are $1\frac{1}{2}$ inches wide. What should the dimensions of the page be to use the least amount of paper?

72. Page Design A page that is x inches wide and y inches high contains 30 square inches of print. The top and bottom margins are each 1 inch deep, and the margins on each side are 2 inches wide (see figure).

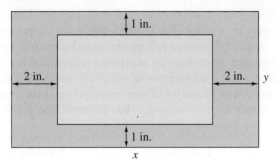

(a) Write a functions for the total area A of the page in terms of x.

(b) Determine the domain of the function based on the physical constraints of the problem.

(c) Use a graphing utility to graph the area function and approximate the dimensions of the page that use the least amount of paper.

73. Average Speed A driver's average speed is 50 miles per hour on a round trip between two cities 100 miles apart. The average speeds for going and returning were x and y miles per hour, respectively.

(a) Show that $y = (25x)/(x - 25)$.

(b) Determine the vertical and horizontal asymptotes of the graph of the function.

(c) Use a graphing utility to graph the function.

(d) Complete the table.

x	30	35	40	45	50	55	60
y							

(e) Are the results in the table what you expected? Explain.

(f) Is it possible to average 20 miles per hour in one direction and still average 50 miles per hour on the round trip? Explain.

74. Medicine The concentration C of a chemical in the bloodstream t hours after injection into muscle tissue is given by

$$C = \frac{3t^2 + t}{t^3 + 50}, \quad t > 0.$$

Use a graphing utility to graph the function. Determine the horizontal asymptote of the graph of the function and interpret its meaning in the context of the problem.

Exploration

True or False? In Exercises 75–77, determine whether the statement is true or false. Justify your answer.

75. The graph of a polynomial function can have infinitely many vertical asymptotes.

76. The graph of a rational function can never cross one of its asymptotes.

77. The graph of a rational function can have a vertical asymptote, a horizontal asymptote, and a slant asymptote.

78. HOW DO YOU SEE IT? The graph of a rational function

$$f(x) = \frac{N(x)}{D(x)}$$

is shown below. Determine which of the statements about the function is false. Justify your answer.

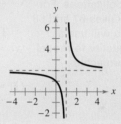

(a) $D(1) = 0$.

(b) The degree of $N(x)$ and $D(x)$ are equal.

(c) The ratio of the leading coefficients of $N(x)$ and $D(x)$ is 1.

79. Writing Is every rational function a polynomial function? Is every polynomial function a rational function? Explain.

Writing a Rational Function In Exercises 80–82, write a rational function f whose graph has the specified characteristics. (There are many correct answers.)

80. Vertical asymptote: None

Horizontal asymptote: $y = 2$

81. Vertical asymptotes: $x = -2, x = 1$

Horizontal asymptote: None

82. Vertical asymptote: $x = 2$

Slant asymptote: $y = x + 1$

Zero of the function: $x = -2$

Project: Department of Defense To work an extended application analyzing the total numbers of military personnel on active duty from 1984 through 2014, visit this text's website at *LarsonPrecalculus.com*. (*Source: U.S. Department of Defense*)

2.7 Nonlinear Inequalities

■ Solve polynomial inequalities.
■ Solve rational inequalities.
■ Use nonlinear inequalities to model and solve real-life problems.

Polynomial Inequalities

To solve a polynomial inequality such as $x^2 - 2x - 3 < 0$, use the fact that a polynomial can change signs only at its *zeros* (the x-values that make the polynomial equal to zero). Between two consecutive zeros, a polynomial must be entirely positive or entirely negative. This means that when the real zeros of a polynomial are put in order, they divide the real number line into intervals in which the polynomial has no sign changes. These zeros are the **key numbers** of the inequality, and the resulting open intervals are the *test intervals* for the inequality. For example, the polynomial $x^2 - 2x - 3$ factors as

$$x^2 - 2x - 3 = (x + 1)(x - 3)$$

so it has two zeros,

$$x = -1 \quad \text{and} \quad x = 3.$$

These zeros divide the real number line into three test intervals:

$$(-\infty, -1), \quad (-1, 3), \quad \text{and} \quad (3, \infty). \quad \text{(See figure below.)}$$

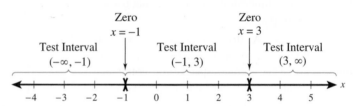

Three test intervals for $x^2 - 2x - 3$

To solve the inequality $x^2 - 2x - 3 < 0$, you need to test only one value from each of these test intervals. When a value from a test interval satisfies the original inequality, you can conclude that the interval is a solution of the inequality.

Use the same basic approach, generalized below, to find the solution set of any polynomial inequality.

Nonlinear inequalities have many real-life applications. For example, in Exercises 67 and 68 on page 186, you will use a polynomial inequality to model the height of a projectile.

• • **REMARK** The solution set of

$$x^2 - 2x - 3 < 0$$

discussed above, is the open interval $(-1, 3)$. Use Step 3 to verify this. By choosing the representative x-values $x = -2$, $x = 0$, and $x = 4$, you will find that the value of the polynomial is negative only in $(-1, 3)$.

Test Intervals for a Polynomial Inequality

To determine the intervals on which the values of a polynomial are entirely negative or entirely positive, use the steps below.

1. Find all real zeros of the polynomial, and arrange the zeros in increasing order. These zeros are the key numbers of the inequality.

2. Use the key numbers of the inequality to determine the test intervals.

3. Choose one representative x-value in each test interval and evaluate the polynomial at that value. When the value of the polynomial is negative, the polynomial has negative values for every x-value in the interval. When the value of the polynomial is positive, the polynomial has positive values for every x-value in the interval.

EXAMPLE 1 **Solving a Polynomial Inequality**

Solve $x^2 - x - 6 < 0$. Then graph the solution set.

▷ **ALGEBRA HELP** To review the techniques for factoring polynomials, see Appendix A.5.

Solution Factoring the polynomial

$$x^2 - x - 6 = (x + 2)(x - 3)$$

shows that the key numbers are $x = -2$ and $x = 3$. So, the inequality's test intervals are

$$(-\infty, -2), \quad (-2, 3), \quad \text{and} \quad (3, \infty) \qquad \text{Test intervals}$$

In each test interval, choose a representative x-value and evaluate the polynomial.

Test Interval	x-Value	Polynomial Value	Conclusion
$(-\infty, -2)$	$x = -3$	$(-3)^2 - (-3) - 6 = 6$	Positive
$(-2, 3)$	$x = 0$	$(0)^2 - (0) - 6 = -6$	Negative
$(3, \infty)$	$x = 4$	$(4)^2 - (4) - 6 = 6$	Positive

The inequality is satisfied for all x-values in $(-2, 3)$. This implies that the solution set of the inequality

$$x^2 - x - 6 < 0$$

is the interval $(-2, 3)$, as shown on the number line below. Note that the original inequality contains a "less than" symbol. This means that the solution set does not contain the endpoints of the test interval $(-2, 3)$.

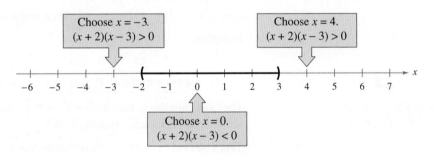

✓ *Checkpoint* 🔊))) *Audio-video solution in English & Spanish at LarsonPrecalculus.com*

Solve $x^2 - x - 20 < 0$. Then graph the solution set. ■

As with linear inequalities, you can check the reasonableness of a solution by substituting x-values into the original inequality. For instance, to check the solution found in Example 1, substitute several x-values from the interval $(-2, 3)$ into the inequality

$$x^2 - x - 6 < 0.$$

Regardless of which x-values you choose, the inequality should be satisfied.

You can also use a graph to check the result of Example 1. Sketch the graph of

$$y = x^2 - x - 6$$

as shown in Figure 2.31. Notice that the graph is below the x-axis on the interval $(-2, 3)$.

In Example 1, the polynomial inequality is in general form (with the polynomial on one side and zero on the other). Whenever this is not the case, you should begin by writing the inequality in general form.

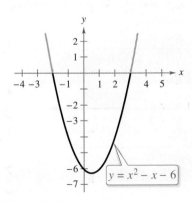

Figure 2.31

EXAMPLE 2 **Solving a Polynomial Inequality**

See LarsonPrecalculus.com for an interactive version of this type of example.

Solve $4x^2 - 5x > 6$.

Algebraic Solution

$$4x^2 - 5x - 6 > 0 \qquad \text{Write in general form.}$$
$$(x - 2)(4x + 3) > 0 \qquad \text{Factor.}$$

Key numbers: $x = -\frac{3}{4}$, $x = 2$

Test intervals: $\left(-\infty, -\frac{3}{4}\right), \left(-\frac{3}{4}, 2\right), (2, \infty)$

Test: Is $(x - 2)(4x + 3) > 0$?

Testing these intervals shows that the polynomial $4x^2 - 5x - 6$ is positive on the open intervals $\left(-\infty, -\frac{3}{4}\right)$ and $(2, \infty)$. So, the solution set of the inequality is $\left(-\infty, -\frac{3}{4}\right) \cup (2, \infty)$.

Graphical Solution

First write the polynomial inequality $4x^2 - 5x > 6$ as $4x^2 - 5x - 6 > 0$. Then use a graphing utility to graph $y = 4x^2 - 5x - 6$.

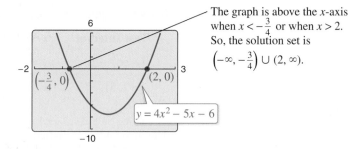

The graph is above the x-axis when $x < -\frac{3}{4}$ or when $x > 2$. So, the solution set is $\left(-\infty, -\frac{3}{4}\right) \cup (2, \infty)$.

✓ *Checkpoint* ◄))) Audio-video solution in English & Spanish at LarsonPrecalculus.com

Solve $2x^2 + 3x < 5$ (a) algebraically and (b) graphically.

EXAMPLE 3 **Solving a Polynomial Inequality**

Solve $2x^3 - 3x^2 - 32x > -48$. Then graph the solution set.

Solution

$$2x^3 - 3x^2 - 32x + 48 > 0 \qquad \text{Write in general form.}$$
$$(x - 4)(x + 4)(2x - 3) > 0 \qquad \text{Factor by grouping.}$$

The key numbers are $x = -4$, $x = \frac{3}{2}$, and $x = 4$, and the test intervals are $(-\infty, -4), \left(-4, \frac{3}{2}\right), \left(\frac{3}{2}, 4\right)$, and $(4, \infty)$.

Test Interval	x-Value	Polynomial Value	Conclusion
$(-\infty, -4)$	$x = -5$	$2(-5)^3 - 3(-5)^2 - 32(-5) + 48 = -117$	Negative
$\left(-4, \frac{3}{2}\right)$	$x = 0$	$2(0)^3 - 3(0)^2 - 32(0) + 48 = 48$	Positive
$\left(\frac{3}{2}, 4\right)$	$x = 2$	$2(2)^3 - 3(2)^2 - 32(2) + 48 = -12$	Negative
$(4, \infty)$	$x = 5$	$2(5)^3 - 3(5)^2 - 32(5) + 48 = 63$	Positive

The inequality is satisfied on the open intervals $\left(-4, \frac{3}{2}\right)$ and $(4, \infty)$. So, the solution set is $\left(-4, \frac{3}{2}\right) \cup (4, \infty)$, as shown on the number line below.

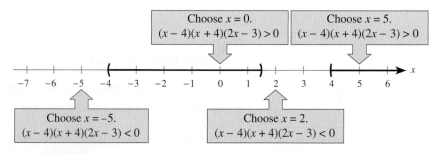

✓ *Checkpoint* ◄))) Audio-video solution in English & Spanish at LarsonPrecalculus.com

Solve $3x^3 - x^2 - 12x > -4$. Then graph the solution set.

You may find it easier to determine the sign of a polynomial from its *factored* form. For instance, in Example 2, when you substitute the test value $x = 1$ into the factored form

$$(x - 2)(4x + 3)$$

the sign pattern of the factors is

$$(-)(+)$$

which yields a negative result. Use factored forms to determine the signs of the polynomials in other examples in this section.

When solving a polynomial inequality, be sure to account for the inequality symbol. For instance, in Example 2, note that the original inequality symbol is "greater than" and the solution consists of two open intervals. If the original inequality had been

$$4x^2 - 5x \geq 6$$

then the solution set would have been

$$\left(-\infty, -\tfrac{3}{4}\right] \cup [2, \infty).$$

Each of the polynomial inequalities in Examples 1, 2, and 3 has a solution set that consists of a single interval or the union of two intervals. When solving the exercises for this section, watch for unusual solution sets, as illustrated in Example 4.

EXAMPLE 4   Unusual Solution Sets

a. The solution set of

$$x^2 + 2x + 4 > 0$$

consists of the entire set of real numbers, $(-\infty, \infty)$. In other words, the value of the quadratic polynomial $x^2 + 2x + 4$ is positive for every real value of x.

b. The solution set of

$$x^2 + 2x + 1 \leq 0$$

consists of the single real number $\{-1\}$, because the inequality has only one key number, $x = -1$, and it is the only value that satisfies the inequality.

c. The solution set of

$$x^2 + 3x + 5 < 0$$

is empty. In other words, $x^2 + 3x + 5$ is not less than zero for any value of x.

d. The solution set of

$$x^2 - 4x + 4 > 0$$

consists of all real numbers except $x = 2$. This solution set can be written in interval notation as

$$(-\infty, 2) \cup (2, \infty).$$

✓ **Checkpoint**))) *Audio-video solution in English & Spanish at LarsonPrecalculus.com*

What is unusual about the solution set of each inequality?

a. $x^2 + 6x + 9 < 0$

b. $x^2 + 4x + 4 \leq 0$

c. $x^2 - 6x + 9 > 0$

d. $x^2 - 2x + 1 \geq 0$

Linear Factorization Theorem   *(p. 152)*

If $f(x)$ is a polynomial of degree n, where $n > 0$, then $f(x)$ has precisely n linear factors

$$f(x) = a_n(x - c_1)(x - c_2) \cdots (x - c_n)$$

where $c_1, c_2, \ldots, c_n$ are complex numbers.

Proof

Using the Fundamental Theorem of Algebra, you know that f must have at least one zero, c_1. Consequently, $(x - c_1)$ is a factor of $f(x)$, and you have

$$f(x) = (x - c_1)f_1(x).$$

If the degree of $f_1(x)$ is greater than zero, then you again apply the Fundamental Theorem of Algebra to conclude that f_1 must have a zero c_2, which implies that

$$f(x) = (x - c_1)(x - c_2)f_2(x).$$

It is clear that the degree of $f_1(x)$ is $n - 1$, that the degree of $f_2(x)$ is $n - 2$, and that you can repeatedly apply the Fundamental Theorem of Algebra n times until you obtain

$$f(x) = a_n(x - c_1)(x - c_2) \cdots (x - c_n)$$

where a_n is the leading coefficient of the polynomial $f(x)$. ∎

Factors of a Polynomial   *(p. 157)*

Every polynomial of degree $n > 0$ with real coefficients can be written as the product of linear and quadratic factors with real coefficients, where the quadratic factors have no real zeros.

Proof

To begin, use the Linear Factorization Theorem to conclude that $f(x)$ can be *completely* factored in the form

$$f(x) = d(x - c_1)(x - c_2)(x - c_3) \cdots (x - c_n).$$

If each c_i is real, then there is nothing more to prove. If any c_i is imaginary $(c_i = a + bi, b \neq 0)$, then you know that the conjugate $c_j = a - bi$ is also a zero, because the coefficients of $f(x)$ are real. By multiplying the corresponding factors, you obtain

$$(x - c_i)(x - c_j) = [x - (a + bi)][x - (a - bi)]$$
$$= [(x - a) - bi][(x - a) + bi]$$
$$= (x - a)^2 + b^2$$
$$= x^2 - 2ax + (a^2 + b^2)$$

where each coefficient is real. ∎

P.S. Problem Solving ▪ ▪ ▪ ▪ ▪ ▪ ▪ ▪ ▪ ▪ ▪ ▪ ▪ ▪

1. Verifying the Remainder Theorem Show that if $f(x) = ax^3 + bx^2 + cx + d$, then $f(k) = r$, where $r = ak^3 + bk^2 + ck + d$, using long division. In other words, verify the Remainder Theorem for a third-degree polynomial function.

2. Babylonian Mathematics In 2000 B.C., the Babylonians solved polynomial equations by referring to tables of values. One such table gave the values of $y^3 + y^2$. To be able to use this table, the Babylonians sometimes used the method below to manipulate the equation.

$$ax^3 + bx^2 = c \qquad \text{Original equation}$$

$$\frac{a^3x^3}{b^3} + \frac{a^2x^2}{b^2} = \frac{a^2c}{b^3} \qquad \text{Multiply each side by } \frac{a^2}{b^3}.$$

$$\left(\frac{ax}{b}\right)^3 + \left(\frac{ax}{b}\right)^2 = \frac{a^2c}{b^3} \qquad \text{Rewrite.}$$

Then they would find $(a^2c)/b^3$ in the $y^3 + y^2$ column of the table. They knew that the corresponding y-value was equal to $(ax)/b$, so they could conclude that $x = (by)/a$.

(a) Calculate $y^3 + y^2$ for $y = 1, 2, 3, \ldots, 10$. Record the values in a table.

(b) Use the table from part (a) and the method above to solve each equation.

    (i) $x^3 + x^2 = 252$

    (ii) $x^3 + 2x^2 = 288$

    (iii) $3x^3 + x^2 = 90$

    (iv) $2x^3 + 5x^2 = 2500$

    (v) $7x^3 + 6x^2 = 1728$

    (vi) $10x^3 + 3x^2 = 297$

(c) Using the methods from this chapter, verify your solution of each equation.

3. Finding Dimensions At a glassware factory, molten cobalt glass is poured into molds to make paperweights. Each mold is a rectangular prism whose height is 3 inches greater than the length of each side of the square base. A machine pours 20 cubic inches of liquid glass into each mold. What are the dimensions of the mold?

4. True or False? Determine whether the statement is true or false. If false, provide one or more reasons why the statement is false and correct the statement. Let $f(x) = ax^3 + bx^2 + cx + d$, $a \neq 0$, and let $f(2) = -1$. Then

$$\frac{f(x)}{x+1} = q(x) + \frac{2}{x+1}$$

where $q(x)$ is a second-degree polynomial.

5. Finding the Equation of a Parabola The parabola shown in the figure has an equation of the form $y = ax^2 + bx + c$. Find the equation of this parabola using each method.

(a) Find the equation analytically.

(b) Use the regression feature of a graphing utility to find the equation.

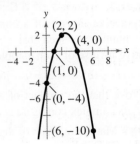

6. Finding the Slope of a Tangent Line One of the fundamental themes of calculus is to find the slope of the tangent line to a curve at a point. To see how this can be done, consider the point $(2, 4)$ on the graph of the quadratic function $f(x) = x^2$, as shown in the figure.

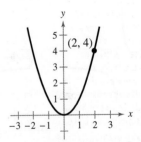

(a) Find the slope m_1 of the line joining $(2, 4)$ and $(3, 9)$. Is the slope of the tangent line at $(2, 4)$ greater than or less than the slope of the line through $(2, 4)$ and $(3, 9)$?

(b) Find the slope m_2 of the line joining $(2, 4)$ and $(1, 1)$. Is the slope of the tangent line at $(2, 4)$ greater than or less than the slope of the line through $(2, 4)$ and $(1, 1)$?

(c) Find the slope m_3 of the line joining $(2, 4)$ and $(2.1, 4.41)$. Is the slope of the tangent line at $(2, 4)$ greater than or less than the slope of the line through $(2, 4)$ and $(2.1, 4.41)$?

(d) Find the slope m_h of the line joining $(2, 4)$ and $(2 + h, f(2 + h))$ in terms of the nonzero number h.

(e) Evaluate the slope formula from part (d) for $h = -1$, 1, and 0.1. Compare these values with those in parts (a)–(c).

(f) What can you conclude the slope m_{tan} of the tangent line at $(2, 4)$ to be? Explain.

7. **Writing Cubic Functions** For each part, write a cubic function of the form $f(x) = (x - k)q(x) + r$ whose graph has the specified characteristics. (There are many correct answers.)

   (a) Passes through the point $(2, 5)$ and rises to the right

   (b) Passes through the point $(-3, 1)$ and falls to the right

8. **Multiplicative Inverse of a Complex Number** The multiplicative inverse of a complex number z is a complex number z_m such that $z \cdot z_m = 1$. Find the multiplicative inverse of each complex number.

   (a) $z = 1 + i$ (b) $z = 3 - i$ (c) $z = -2 + 8i$

9. **Proof** Prove that the product of a complex number $a + bi$ and its complex conjugate is a real number.

10. **Matching** Match the graph of the rational function

$$f(x) = \frac{ax + b}{cx + d}$$

with the given conditions.

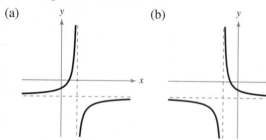

(a)

(b)

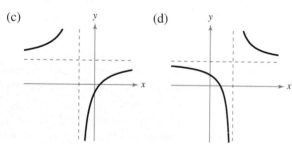

(c)

(d)

   (i) $a > 0$ (ii) $a > 0$ (iii) $a < 0$ (iv) $a > 0$
   $b < 0$ $b > 0$ $b > 0$ $b < 0$
   $c > 0$ $c < 0$ $c > 0$ $c > 0$
   $d < 0$ $d < 0$ $d < 0$ $d > 0$

11. **Effects of Values on a Graph** Consider the function

$$f(x) = \frac{ax}{(x - b)^2}.$$

   (a) Determine the effect on the graph of f when $b \neq 0$ and a is varied. Consider cases in which a is positive and a is negative.

   (b) Determine the effect on the graph of f when $a \neq 0$ and b is varied.

12. **Distinct Vision** The endpoints of the interval over which distinct vision is possible are called the *near point* and *far point* of the eye (see figure). With increasing age, these points normally change. The table shows the approximate near points y (in inches) for various ages x (in years).

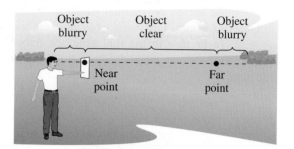

Age, x	Near Point, y
16	3.0
32	4.7
44	9.8
50	19.7
60	39.4

   (a) Use the *regression* feature of a graphing utility to find a quadratic model for the data. Use the graphing utility to plot the data and graph the model in the same viewing window.

   (b) Find a rational model for the data. Take the reciprocals of the near points to generate the points $(x, 1/y)$. Use the *regression* feature of the graphing utility to find a linear model for the data. The resulting line has the form

$$\frac{1}{y} = ax + b.$$

   Solve for y. Use the graphing utility to plot the data and graph the model in the same viewing window.

   (c) Use the *table* feature of the graphing utility to construct a table showing the predicted near point based on each model for each of the ages in the original table. How well do the models fit the original data?

   (d) Use both models to estimate the near point for a person who is 25 years old. Which model is a better fit?

   (e) Do you think either model can be used to predict the near point for a person who is 70 years old? Explain.

13. **Zeros of a Cubic Function** Can a cubic function with real coefficients have two real zeros and one complex zero? Explain.

3 Exponential and Logarithmic Functions

Earthquakes *(Example 6, page 242)*

Beaver Population *(Exercise 83, page 234)*

Sound Intensity *(Exercises 79–82, page 224)*

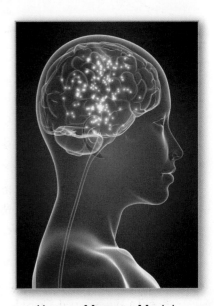

Human Memory Model *(Exercise 83, page 218)*

Nuclear Reactor Accident *(Example 9, page 205)*

Clockwise from top left, Alexander Kuguchin/Shutterstock.com; Somjin Klong-ugkara/Shutterstock.com; Sebastian Kaulitzki/Shutterstock.com; Fotokon/Shutterstock.com; Titima Ongkantong/Shutterstock.com

3.1 Exponential Functions and Their Graphs

- Recognize and evaluate exponential functions with base *a*.
- Graph exponential functions and use the One-to-One Property.
- Recognize, evaluate, and graph exponential functions with base e.
- Use exponential functions to model and solve real-life problems.

Exponential Functions

So far, this text has dealt mainly with **algebraic functions,** which include polynomial functions and rational functions. In this chapter, you will study two types of nonalgebraic functions—*exponential functions* and *logarithmic functions*. These functions are examples of **transcendental functions.** This section will focus on exponential functions.

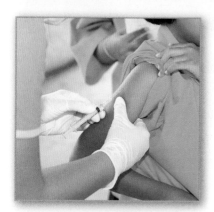

Exponential functions can help you model and solve real-life problems. For example, in Exercise 66 on page 208, you will use an exponential function to model the concentration of a drug in the bloodstream.

> ### Definition of Exponential Function
> The **exponential function** *f* **with base** *a* is denoted by
>
> $$f(x) = a^x$$
>
> where $a > 0$, $a \neq 1$, and x is any real number.

The base a of an exponential function cannot be 1 because $a = 1$ yields $f(x) = 1^x = 1$. This is a constant function, not an exponential function.

You have evaluated a^x for integer and rational values of x. For example, you know that $4^3 = 64$ and $4^{1/2} = 2$. However, to evaluate 4^x for any real number x, you need to interpret forms with *irrational* exponents. For the purposes of this text, it is sufficient to think of $a^{\sqrt{2}}$ (where $\sqrt{2} \approx 1.41421356$) as the number that has the successively closer approximations

$$a^{1.4}, a^{1.41}, a^{1.414}, a^{1.4142}, a^{1.41421}, \ldots.$$

EXAMPLE 1   Evaluating Exponential Functions

Use a calculator to evaluate each function at the given value of x.

Function	Value
a. $f(x) = 2^x$	$x = -3.1$
b. $f(x) = 2^{-x}$	$x = \pi$
c. $f(x) = 0.6^x$	$x = \frac{3}{2}$

Solution

Function Value	Calculator Keystrokes	Display
a. $f(-3.1) = 2^{-3.1}$	2 ∧ (−) 3.1 ENTER	0.1166291
b. $f(\pi) = 2^{-\pi}$	2 ∧ (−) π ENTER	0.1133147
c. $f\left(\frac{3}{2}\right) = (0.6)^{3/2}$	.6 ∧ (3 ÷ 2) ENTER	0.4647580

✓ **Checkpoint** ◀))) *Audio-video solution in English & Spanish at LarsonPrecalculus.com*

Use a calculator to evaluate $f(x) = 8^{-x}$ at $x = \sqrt{2}$.

When evaluating exponential functions with a calculator, it may be necessary to enclose fractional exponents in parentheses. Some calculators do not correctly interpret an exponent that consists of an expression unless parentheses are used.

Graphs of Exponential Functions

The graphs of all exponential functions have similar characteristics, as shown in Examples 2, 3, and 5.

▷ **ALGEBRA HELP** To review the techniques for sketching the graph of an equation, see Section 1.2.

EXAMPLE 2   Graphs of $y = a^x$

In the same coordinate plane, sketch the graph of each function.

a. $f(x) = 2^x$ **b.** $g(x) = 4^x$

Solution Begin by constructing a table of values.

x	-3	-2	-1	0	1	2
2^x	$\frac{1}{8}$	$\frac{1}{4}$	$\frac{1}{2}$	1	2	4
4^x	$\frac{1}{64}$	$\frac{1}{16}$	$\frac{1}{4}$	1	4	16

To sketch the graph of each function, plot the points from the table and connect them with a smooth curve, as shown in Figure 3.1. Note that both graphs are increasing. Moreover, the graph of $g(x) = 4^x$ is increasing more rapidly than the graph of $f(x) = 2^x$.

✓ *Checkpoint* ◀))) *Audio-video solution in English & Spanish at LarsonPrecalculus.com*

In the same coordinate plane, sketch the graph of each function.

a. $f(x) = 3^x$ **b.** $g(x) = 9^x$

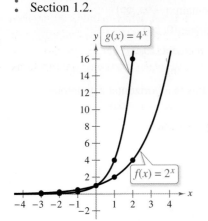

Figure 3.1

The table in Example 2 was evaluated by hand for integer values of x. You can also evaluate $f(x)$ and $g(x)$ for noninteger values of x by using a calculator.

EXAMPLE 3   Graphs of $y = a^{-x}$

In the same coordinate plane, sketch the graph of each function.

a. $F(x) = 2^{-x}$ **b.** $G(x) = 4^{-x}$

Solution Begin by constructing a table of values.

x	-2	-1	0	1	2	3
2^{-x}	4	2	1	$\frac{1}{2}$	$\frac{1}{4}$	$\frac{1}{8}$
4^{-x}	16	4	1	$\frac{1}{4}$	$\frac{1}{16}$	$\frac{1}{64}$

To sketch the graph of each function, plot the points from the table and connect them with a smooth curve, as shown in Figure 3.2. Note that both graphs are decreasing. Moreover, the graph of $G(x) = 4^{-x}$ is decreasing more rapidly than the graph of $F(x) = 2^{-x}$.

✓ *Checkpoint* ◀))) *Audio-video solution in English & Spanish at LarsonPrecalculus.com*

In the same coordinate plane, sketch the graph of each function.

a. $f(x) = 3^{-x}$ **b.** $g(x) = 9^{-x}$

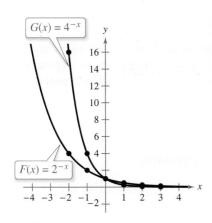

Figure 3.2

Note that it is possible to use one of the properties of exponents to rewrite the functions in Example 3 with positive exponents.

$$F(x) = 2^{-x} = \frac{1}{2^x} = \left(\frac{1}{2}\right)^x \quad \text{and} \quad G(x) = 4^{-x} = \frac{1}{4^x} = \left(\frac{1}{4}\right)^x$$

Comparing the functions in Examples 2 and 3, observe that

$$F(x) = 2^{-x} = f(-x) \quad \text{and} \quad G(x) = 4^{-x} = g(-x).$$

Consequently, the graph of F is a reflection (in the y-axis) of the graph of f. The graphs of G and g have the same relationship. The graphs in Figures 3.1 and 3.2 are typical of the exponential functions $y = a^x$ and $y = a^{-x}$. They have one y-intercept and one horizontal asymptote (the x-axis), and they are continuous. Here is a summary of the basic characteristics of the graphs of these exponential functions.

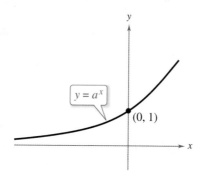

Graph of $y = a^x$, $a > 1$
- Domain: $(-\infty, \infty)$
- Range: $(0, \infty)$
- y-intercept: $(0, 1)$
- Increasing
- x-axis is a horizontal asymptote $(a^x \to 0 \text{ as } x \to -\infty)$.
- Continuous

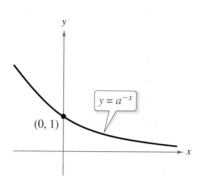

Graph of $y = a^{-x}$, $a > 1$
- Domain: $(-\infty, \infty)$
- Range: $(0, \infty)$
- y-intercept: $(0, 1)$
- Decreasing
- x-axis is a horizontal asymptote $(a^{-x} \to 0 \text{ as } x \to \infty)$.
- Continuous

Notice that the graph of an exponential function is always increasing or always decreasing, so the graph passes the Horizontal Line Test. Therefore, an exponential function is a one-to-one function. You can use the following **One-to-One Property** to solve simple exponential equations.

For $a > 0$ and $a \neq 1$, $a^x = a^y$ if and only if $x = y$.   One-to-One Property

EXAMPLE 4   Using the One-to-One Property

a.

$9 = 3^{x+1}$	Original equation
$3^2 = 3^{x+1}$	$9 = 3^2$
$2 = x + 1$	One-to-One Property
$1 = x$	Solve for x.

b.

$\left(\frac{1}{2}\right)^x = 8$	Original equation
$2^{-x} = 2^3$	$\left(\frac{1}{2}\right)^x = 2^{-x}, 8 = 2^3$
$x = -3$	One-to-One Property

✓ **Checkpoint** ◀))) *Audio-video solution in English & Spanish at LarsonPrecalculus.com*

Use the One-to-One Property to solve the equation for x.

a. $8 = 2^{2x-1}$ **b.** $\left(\frac{1}{3}\right)^{-x} = 27$

In Example 5, notice how the graph of $y = a^x$ can be used to sketch the graphs of functions of the form $f(x) = b \pm a^{x+c}$.

▷ **ALGEBRA HELP** To review the techniques for transforming the graph of a function, see Section 1.7.

EXAMPLE 5 **Transformations of Graphs of Exponential Functions**

See LarsonPrecalculus.com for an interactive version of this type of example.

Describe the transformation of the graph of $f(x) = 3^x$ that yields each graph.

a.
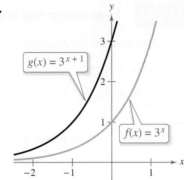

b.
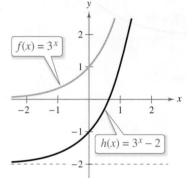

c.
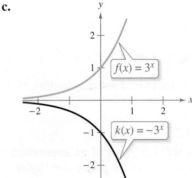

d.
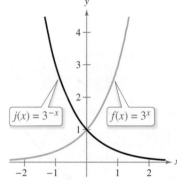

Solution

a. Because $g(x) = 3^{x+1} = f(x + 1)$, the graph of g is obtained by shifting the graph of f one unit to the *left*.

b. Because $h(x) = 3^x - 2 = f(x) - 2$, the graph of h is obtained by shifting the graph of f *down* two units.

c. Because $k(x) = -3^x = -f(x)$, the graph of k is obtained by *reflecting* the graph of f in the x-axis.

d. Because $j(x) = 3^{-x} = f(-x)$, the graph of j is obtained by *reflecting* the graph of f in the y-axis.

✓ *Checkpoint* ◀))) Audio-video solution in English & Spanish at LarsonPrecalculus.com

Describe the transformation of the graph of $f(x) = 4^x$ that yields the graph of each function.

a. $g(x) = 4^{x-2}$ **b.** $h(x) = 4^x + 3$ **c.** $k(x) = 4^{-x} - 3$

Note how each transformation in Example 5 affects the y-intercept and the horizontal asymptote.

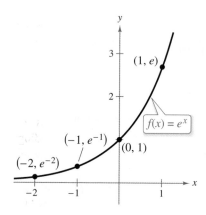

Figure 3.3

The Natural Base e

In many applications, the most convenient choice for a base is the irrational number

$$e \approx 2.718281828 \ldots .$$

This number is called the **natural base.** The function $f(x) = e^x$ is called the **natural exponential function.** Figure 3.3 shows its graph. Be sure you see that for the exponential function $f(x) = e^x$, e is the constant $2.718281828 \ldots$, whereas x is the variable.

EXAMPLE 6　**Evaluating the Natural Exponential Function**

Use a calculator to evaluate the function $f(x) = e^x$ at each value of x.

a. $x = -2$　　　**b.** $x = -1$

c. $x = 0.25$　　**d.** $x = -0.3$

Solution

Function Value	Calculator Keystrokes	Display
a. $f(-2) = e^{-2}$	(eˣ)(−) 2 (ENTER)	0.1353353
b. $f(-1) = e^{-1}$	(eˣ)(−) 1 (ENTER)	0.3678794
c. $f(0.25) = e^{0.25}$	(eˣ) 0.25 (ENTER)	1.2840254
d. $f(-0.3) = e^{-0.3}$	(eˣ)(−) 0.3 (ENTER)	0.7408182

✓ *Checkpoint*  Audio-video solution in English & Spanish at LarsonPrecalculus.com

Use a calculator to evaluate the function $f(x) = e^x$ at each value of x.

a. $x = 0.3$

b. $x = -1.2$

c. $x = 6.2$

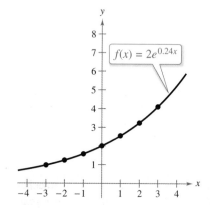

Figure 3.4

EXAMPLE 7　**Graphing Natural Exponential Functions**

Sketch the graph of each natural exponential function.

a. $f(x) = 2e^{0.24x}$

b. $g(x) = \frac{1}{2}e^{-0.58x}$

Solution　Begin by using a graphing utility to construct a table of values.

x	-3	-2	-1	0	1	2	3
$f(x)$	0.974	1.238	1.573	2.000	2.542	3.232	4.109
$g(x)$	2.849	1.595	0.893	0.500	0.280	0.157	0.088

To graph each function, plot the points from the table and connect them with a smooth curve, as shown in Figures 3.4 and 3.5. Note that the graph in Figure 3.4 is increasing, whereas the graph in Figure 3.5 is decreasing.

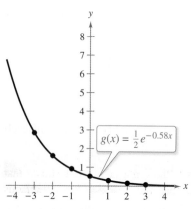

Figure 3.5

✓ *Checkpoint* 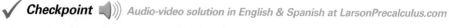 Audio-video solution in English & Spanish at LarsonPrecalculus.com

Sketch the graph of $f(x) = 5e^{0.17x}$.

Applications

One of the most familiar examples of exponential growth is an investment earning *continuously compounded interest*. The formula for *interest compounded n times per year* is

$$A = P\left(1 + \frac{r}{n}\right)^{nt}.$$

In this formula, A is the balance in the account, P is the principal (or original deposit), r is the annual interest rate (in decimal form), n is the number of compoundings per year, and t is the time in years. Exponential functions can be used to *develop* this formula and show how it leads to continuous compounding.

Consider a principal P invested at an annual interest rate r, compounded once per year. When the interest is added to the principal at the end of the first year, the new balance P_1 is

$$P_1 = P + Pr$$
$$= P(1 + r).$$

This pattern of multiplying the balance by $1 + r$ repeats each successive year, as shown here.

Year	Balance After Each Compounding
0	$P = P$
1	$P_1 = P(1 + r)$
2	$P_2 = P_1(1 + r) = P(1 + r)(1 + r) = P(1 + r)^2$
3	$P_3 = P_2(1 + r) = P(1 + r)^2(1 + r) = P(1 + r)^3$
$\vdots$	$\vdots$
t	$P_t = P(1 + r)^t$

To accommodate more frequent (quarterly, monthly, or daily) compounding of interest, let n be the number of compoundings per year and let t be the number of years. Then the rate per compounding is r/n, and the account balance after t years is

$$A = P\left(1 + \frac{r}{n}\right)^{nt}. \qquad \text{Amount (balance) with } n \text{ compoundings per year}$$

When the number of compoundings n increases without bound, the process approaches what is called **continuous compounding.** In the formula for n compoundings per year, let $m = n/r$. This yields a new expression.

$$A = P\left(1 + \frac{r}{n}\right)^{nt} \qquad \text{Amount with } n \text{ compoundings per year}$$

$$= P\left(1 + \frac{r}{mr}\right)^{mrt} \qquad \text{Substitute } mr \text{ for } n.$$

$$= P\left(1 + \frac{1}{m}\right)^{mrt} \qquad \text{Simplify.}$$

$$= P\left[\left(1 + \frac{1}{m}\right)^m\right]^{rt} \qquad \text{Property of exponents}$$

m	$\left(1 + \dfrac{1}{m}\right)^m$
1	2
10	2.59374246
100	2.704813829
1,000	2.716923932
10,000	2.718145927
100,000	2.718268237
1,000,000	2.718280469
10,000,000	2.718281693
$\downarrow$	$\downarrow$
∞	e

As m increases without bound (that is, as $m \to \infty$), the table at the left shows that $[1 + (1/m)]^m \to e$. This allows you to conclude that the formula for continuous compounding is

$$A = Pe^{rt}. \qquad \text{Substitute } e \text{ for } [1 + (1/m)]^m.$$

•• REMARK   Be sure you see
that, when using the formulas
for compound interest, you must
write the annual interest rate in
decimal form. For example, you
must write 6% as 0.06.
⊳

Formulas for Compound Interest

After t years, the balance A in an account with principal P and annual interest rate r (in decimal form) is given by one of these two formulas.

1. For n compoundings per year: $A = P\left(1 + \dfrac{r}{n}\right)^{nt}$

2. For continuous compounding: $A = Pe^{rt}$

EXAMPLE 8   **Compound Interest**

You invest $12,000 at an annual rate of 3%. Find the balance after 5 years for each type of compounding.

a. Quarterly

b. Monthly

c. Continuous

Solution

a. For quarterly compounding, use $n = 4$ to find the balance after 5 years.

$$A = P\left(1 + \frac{r}{n}\right)^{nt} \qquad \text{Formula for compound interest}$$

$$= 12{,}000\left(1 + \frac{0.03}{4}\right)^{4(5)} \qquad \text{Substitute for } P, r, n, \text{ and } t.$$

$$\approx 13{,}934.21 \qquad \text{Use a calculator.}$$

b. For monthly compounding, use $n = 12$ to find the balance after 5 years.

$$A = P\left(1 + \frac{r}{n}\right)^{nt} \qquad \text{Formula for compound interest}$$

$$= 12{,}000\left(1 + \frac{0.03}{12}\right)^{12(5)} \qquad \text{Substitute for } P, r, n, \text{ and } t.$$

$$\approx \$13{,}939.40 \qquad \text{Use a calculator.}$$

c. Use the formula for continuous compounding to find the balance after 5 years.

$$A = Pe^{rt} \qquad \text{Formula for continuous compounding}$$

$$= 12{,}000e^{0.03(5)} \qquad \text{Substitute for } P, r, \text{ and } t.$$

$$\approx \$13{,}942.01 \qquad \text{Use a calculator.}$$

✓ *Checkpoint* ◀))) *Audio-video solution in English & Spanish at LarsonPrecalculus.com*

You invest $6000 at an annual rate of 4%. Find the balance after 7 years for each type of compounding.

a. Quarterly   **b.** Monthly   **c.** Continuous   ◼

In Example 8, note that continuous compounding yields more than quarterly and monthly compounding. This is typical of the two types of compounding. That is, for a given principal, interest rate, and time, continuous compounding will always yield a larger balance than compounding n times per year.

EXAMPLE 9 **Radioactive Decay**

In 1986, a nuclear reactor accident occurred in Chernobyl in what was then the Soviet Union. The explosion spread highly toxic radioactive chemicals, such as plutonium (^{239}Pu), over hundreds of square miles, and the government evacuated the city and the surrounding area. To see why the city is now uninhabited, consider the model

$$P = 10\left(\frac{1}{2}\right)^{t/24,100}$$

which represents the amount of plutonium P that remains (from an initial amount of 10 pounds) after t years. Sketch the graph of this function over the interval from $t = 0$ to $t = 100{,}000$, where $t = 0$ represents 1986. How much of the 10 pounds will remain in the year 2020? How much of the 10 pounds will remain after 100,000 years?

The International Atomic Energy Authority ranks nuclear incidents and accidents by severity using a scale from 1 to 7 called the International Nuclear and Radiological Event Scale (INES). A level 7 ranking is the most severe. To date, the Chernobyl accident and an accident at Japan's Fukushima Daiichi power plant in 2011 are the only two disasters in history to be given an INES level 7 ranking.

Solution The graph of this function is shown in the figure at the right. Note from this graph that plutonium has a *half-life* of about 24,100 years. That is, after 24,100 years, *half* of the original amount will remain. After another 24,100 years, one-quarter of the original amount will remain, and so on. In the year 2020 ($t = 34$), there will still be

$$P = 10\left(\frac{1}{2}\right)^{34/24,100}$$
$$\approx 10\left(\frac{1}{2}\right)^{0.0014108}$$
$$\approx 9.990 \text{ pounds}$$

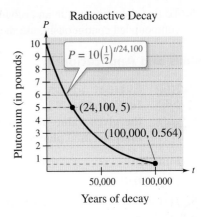

Radioactive Decay

$$P = 10\left(\frac{1}{2}\right)^{t/24,100}$$

(24,100, 5)

(100,000, 0.564)

Plutonium (in pounds)

Years of decay

of plutonium remaining. After 100,000 years, there will still be

$$P = 10\left(\frac{1}{2}\right)^{100,000/24,100}$$
$$\approx 0.564 \text{ pound}$$

of plutonium remaining.

✓ *Checkpoint* 🔊))) *Audio-video solution in English & Spanish at LarsonPrecalculus.com*

In Example 9, how much of the 10 pounds will remain in the year 2089? How much of the 10 pounds will remain after 125,000 years?

Summarize (Section 3.1)

1. State the definition of the exponential function f with base a *(page 198)*. For an example of evaluating exponential functions, see Example 1.

2. Describe the basic characteristics of the graphs of the exponential functions $y = a^x$ and $y = a^{-x}$, $a > 1$ *(page 200)*. For examples of graphing exponential functions, see Examples 2, 3, and 5.

3. State the definitions of the natural base and the natural exponential function *(page 202)*. For examples of evaluating and graphing natural exponential functions, see Examples 6 and 7.

4. Describe real-life applications involving exponential functions *(pages 204 and 205, Examples 8 and 9)*.

3.1 Exercises

See **CalcChat.com** for tutorial help and worked-out solutions to odd-numbered exercises.

Vocabulary: Fill in the blanks.

1. Polynomial and rational functions are examples of _____ functions.
2. Exponential and logarithmic functions are examples of nonalgebraic functions, also called _____ functions.
3. The _____ Property can be used to solve simple exponential equations.
4. The exponential function $f(x) = e^x$ is called the _____ _____ function, and the base e is called the _____ base.
5. To find the amount A in an account after t years with principal P and an annual interest rate r (in decimal form) compounded n times per year, use the formula _____.
6. To find the amount A in an account after t years with principal P and an annual interest rate r (in decimal form) compounded continuously, use the formula _____.

Skills and Applications

 Evaluating an Exponential Function In Exercises 7–12, evaluate the function at the given value of x. Round your result to three decimal places.

Function	Value
7. $f(x) = 0.9^x$	$x = 1.4$
8. $f(x) = 4.7^x$	$x = -\pi$
9. $f(x) = 3^x$	$x = \frac{2}{5}$
10. $f(x) = \left(\frac{2}{3}\right)^{5x}$	$x = \frac{3}{10}$
11. $f(x) = 5000(2^x)$	$x = -1.5$
12. $f(x) = 200(1.2)^{12x}$	$x = 24$

Matching an Exponential Function with Its Graph In Exercises 13–16, match the exponential function with its graph. [The graphs are labeled (a), (b), (c), and (d).]

(a)
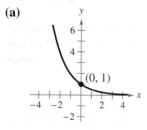

(b)
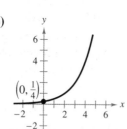

(c)
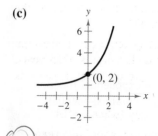

(d)
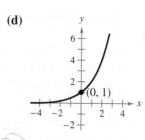

 13. $f(x) = 2^x$

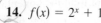 14. $f(x) = 2^x + 1$

15. $f(x) = 2^{-x}$

16. $f(x) = 2^{x-2}$

 Graphing an Exponential Function In Exercises 17–24, use a graphing utility to construct a table of values for the function. Then sketch the graph of the function.

17. $f(x) = 7^x$	18. $f(x) = 7^{-x}$
19. $f(x) = \left(\frac{1}{4}\right)^{-x}$	20. $f(x) = \left(\frac{1}{4}\right)^x$
21. $f(x) = 4^{x-1}$	22. $f(x) = 4^{x+1}$
23. $f(x) = 2^{x+1} + 3$	24. $f(x) = 3^{x-2} + 1$

 Using the One-to-One Property In Exercises 25–28, use the One-to-One Property to solve the equation for x.

25. $3^{x+1} = 27$	26. $2^{x-2} = 64$
27. $\left(\frac{1}{2}\right)^x = 32$	28. $5^{x-2} = \frac{1}{125}$

 Transformations of the Graph of an Exponential Function In Exercises 29–32, describe the transformation(s) of the graph of f that yield(s) the graph of g.

29. $f(x) = 3^x$, $g(x) = 3^x + 1$
30. $f(x) = \left(\frac{7}{2}\right)^x$, $g(x) = -\left(\frac{7}{2}\right)^{-x}$
31. $f(x) = 10^x$, $g(x) = 10^{-x+3}$
32. $f(x) = 0.3^x$, $g(x) = -0.3^x + 5$

 Evaluating a Natural Exponential Function In Exercises 33–36, evaluate the function at the given value of x. Round your result to three decimal places.

Function	Value
33. $f(x) = e^x$	$x = 1.9$
34. $f(x) = 1.5e^{x/2}$	$x = 240$
35. $f(x) = 5000e^{0.06x}$	$x = 6$
36. $f(x) = 250e^{0.05x}$	$x = 20$

Graphing a Natural Exponential Function In Exercises 37–40, use a graphing utility to construct a table of values for the function. Then sketch the graph of the function.

37. $f(x) = 3e^{x+4}$

38. $f(x) = 2e^{-1.5x}$

39. $f(x) = 2e^{x-2} + 4$

40. $f(x) = 2 + e^{x-5}$

Graphing a Natural Exponential Function In Exercises 41–44, use a graphing utility to graph the exponential function.

41. $s(t) = 2e^{0.5t}$

42. $s(t) = 3e^{-0.2t}$

43. $g(x) = 1 + e^{-x}$

44. $h(x) = e^{x-2}$

Using the One-to-One Property In Exercises 45–48, use the One-to-One Property to solve the equation for x.

45. $e^{3x+2} = e^3$

46. $e^{2x-1} = e^4$

47. $e^{x^2-3} = e^{2x}$

48. $e^{x^2+6} = e^{5x}$

 Compound Interest In Exercises 49–52, complete the table by finding the balance A when P dollars is invested at rate r for t years and compounded n times per year.

n	1	2	4	12	365	Continuous
A						

49. $P = \$1500, r = 2\%, t = 10$ years

50. $P = \$2500, r = 3.5\%, t = 10$ years

51. $P = \$2500, r = 4\%, t = 20$ years

52. $P = \$1000, r = 6\%, t = 40$ years

Compound Interest In Exercises 53–56, complete the table by finding the balance A when $\$12,000$ is invested at rate r for t years, compounded continuously.

t	10	20	30	40	50
A					

53. $r = 4\%$

54. $r = 6\%$

55. $r = 6.5\%$

56. $r = 3.5\%$

57. **Trust Fund** On the day of a child's birth, a parent deposits $30,000 in a trust fund that pays 5% interest, compounded continuously. Determine the balance in this account on the child's 25th birthday.

58. **Trust Fund** A philanthropist deposits $5000 in a trust fund that pays 7.5% interest, compounded continuously. The balance will be given to the college from which the philanthropist graduated after the money has earned interest for 50 years. How much will the college receive?

59. **Inflation** Assuming that the annual rate of inflation averages 4% over the next 10 years, the approximate costs C of goods or services during any year in that decade can be modeled by $C(t) = P(1.04)^t$, where t is the time in years and P is the present cost. The price of an oil change for your car is presently $29.88. Estimate the price 10 years from now.

60. **Computer Virus** The number V of computers infected by a virus increases according to the model $V(t) = 100e^{4.6052t}$, where t is the time in hours. Find the number of computers infected after (a) 1 hour, (b) 1.5 hours, and (c) 2 hours.

61. **Population Growth** The projected population of the United States for the years 2025 through 2055 can be modeled by $P = 307.58e^{0.0052t}$, where P is the population (in millions) and t is the time (in years), with $t = 25$ corresponding to 2025. (*Source: U.S. Census Bureau*)

   (a) Use a graphing utility to graph the function for the years 2025 through 2055.

   (b) Use the *table* feature of the graphing utility to create a table of values for the same time period as in part (a).

   (c) According to the model, during what year will the population of the United States exceed 430 million?

62. **Population** The population P (in millions) of Italy from 2003 through 2015 can be approximated by the model $P = 57.59e^{0.0051t}$, where t represents the year, with $t = 3$ corresponding to 2003. (*Source: U.S. Census Bureau*)

   (a) According to the model, is the population of Italy increasing or decreasing? Explain.

   (b) Find the populations of Italy in 2003 and 2015.

   (c) Use the model to predict the populations of Italy in 2020 and 2025.

63. **Radioactive Decay** Let Q represent a mass (in grams) of radioactive plutonium (^{239}Pu), whose half-life is 24,100 years. The quantity of plutonium present after t years is $Q = 16\left(\frac{1}{2}\right)^{t/24,100}$.

   (a) Determine the initial quantity (when $t = 0$).

   (b) Determine the quantity present after 75,000 years.

   (c) Use a graphing utility to graph the function over the interval $t = 0$ to $t = 150,000$.

64. **Radioactive Decay** Let Q represent a mass (in grams) of carbon (^{14}C), whose half-life is 5715 years. The quantity of carbon 14 present after t years is $Q = 10\left(\frac{1}{2}\right)^{t/5715}$.

   (a) Determine the initial quantity (when $t = 0$).

   (b) Determine the quantity present after 2000 years.

   (c) Sketch the graph of the function over the interval $t = 0$ to $t = 10,000$.

65. Depreciation The value of a wheelchair conversion van that originally cost $49,810 depreciates so that each year it is worth $\frac{7}{8}$ of its value for the previous year.

(a) Find a model for $V(t)$, the value of the van after t years.

(b) Determine the value of the van 4 years after it was purchased.

66. Chemistry

Immediately following an injection, the concentration of a drug in the bloodstream is 300 milligrams per milliliter. After t hours, the concentration is 75% of the level of the previous hour.

(a) Find a model for $C(t)$, the concentration of the drug after t hours.

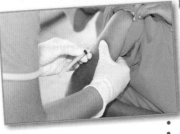

(b) Determine the concentration of the drug after 8 hours.

Exploration

True or False? In Exercises 67 and 68, determine whether the statement is true or false. Justify your answer.

67. The line $y = -2$ is an asymptote for the graph of $f(x) = 10^x - 2$.

68. $e = \dfrac{271,801}{99,990}$

Think About It In Exercises 69–72, use properties of exponents to determine which functions (if any) are the same.

69. $f(x) = 3^{x-2}$
$g(x) = 3^x - 9$
$h(x) = \frac{1}{9}(3^x)$

70. $f(x) = 4^x + 12$
$g(x) = 2^{2x+6}$
$h(x) = 64(4^x)$

71. $f(x) = 16(4^{-x})$
$g(x) = \left(\frac{1}{4}\right)^{x-2}$
$h(x) = 16(2^{-2x})$

72. $f(x) = e^{-x} + 3$
$g(x) = e^{3-x}$
$h(x) = -e^{x-3}$

73. Solving Inequalities Graph the functions $y = 3^x$ and $y = 4^x$ and use the graphs to solve each inequality.

(a) $4^x < 3^x$ (b) $4^x > 3^x$

74. Using Technology Use a graphing utility to graph each function. Use the graph to find where the function is increasing and decreasing, and approximate any relative maximum or minimum values.

(a) $f(x) = x^2 e^{-x}$

(b) $g(x) = x 2^{3-x}$

75. Graphical Reasoning Use a graphing utility to graph $y_1 = [1 + (1/x)]^x$ and $y_2 = e$ in the same viewing window. Using the *trace* feature, explain what happens to the graph of y_1 as x increases.

76. Graphical Reasoning Use a graphing utility to graph

$$f(x) = \left(1 + \frac{0.5}{x}\right)^x \quad \text{and} \quad g(x) = e^{0.5}$$

in the same viewing window. What is the relationship between f and g as x increases and decreases without bound?

77. Comparing Graphs Use a graphing utility to graph each pair of functions in the same viewing window. Describe any similarities and differences in the graphs.

(a) $y_1 = 2^x, y_2 = x^2$

(b) $y_1 = 3^x, y_2 = x^3$

78. **HOW DO YOU SEE IT?** The figure shows the graphs of $y = 2^x$, $y = e^x$, $y = 10^x$, $y = 2^{-x}$, $y = e^{-x}$, and $y = 10^{-x}$. Match each function with its graph. [The graphs are labeled (a) through (f).] Explain your reasoning.

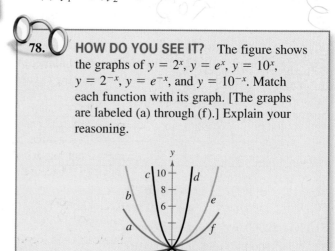

79. Think About It Which functions are exponential?

(a) $f(x) = 3x$ (b) $g(x) = 3x^2$

(c) $h(x) = 3^x$ (d) $k(x) = 2^{-x}$

80. Compound Interest Use the formula

$$A = P\left(1 + \frac{r}{n}\right)^{nt}$$

to calculate the balance A of an investment when $P = \$3000$, $r = 6\%$, and $t = 10$ years, and compounding is done (a) by the day, (b) by the hour, (c) by the minute, and (d) by the second. Does increasing the number of compoundings per year result in unlimited growth of the balance? Explain.

Project: Population per Square Mile To work an extended application analyzing the population per square mile of the United States, visit this text's website at *LarsonPrecalculus.com*. (*Source: U.S. Census Bureau*)

3.2 Logarithmic Functions and Their Graphs

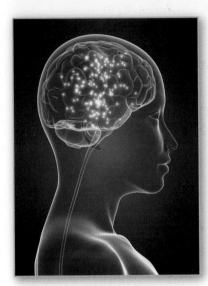

Logarithmic functions can often model scientific observations. For example, in Exercise 83 on page 218, you will use a logarithmic function that models human memory.

■ Recognize and evaluate logarithmic functions with base *a*.
■ Graph logarithmic functions.
■ Recognize, evaluate, and graph natural logarithmic functions.
■ Use logarithmic functions to model and solve real-life problems.

Logarithmic Functions

In Section 3.1, you learned that the exponential function $f(x) = a^x$ is one-to-one. It follows that $f(x) = a^x$ must have an inverse function. This inverse function is the **logarithmic function with base *a*.**

Definition of Logarithmic Function with Base *a*

For $x > 0$, $a > 0$, and $a \neq 1$,

$\quad y = \log_a x$ if and only if $x = a^y$.

The function

$\quad f(x) = \log_a x \qquad$ Read as "log base *a* of *x*."

is the **logarithmic function with base *a*.**

The equations $y = \log_a x$ and $x = a^y$ are equivalent. For example, $2 = \log_3 9$ is equivalent to $9 = 3^2$, and $5^3 = 125$ is equivalent to $\log_5 125 = 3$.

When evaluating logarithms, remember that *a logarithm is an exponent*. This means that $\log_a x$ is the exponent to which *a* must be raised to obtain *x*. For example, $\log_2 8 = 3$ because 2 raised to the third power is 8.

EXAMPLE 1 **Evaluating Logarithms**

Evaluate each logarithm at the given value of *x*.

a. $f(x) = \log_2 x$, $x = 32$ $\qquad$ **b.** $f(x) = \log_3 x$, $x = 1$

c. $f(x) = \log_4 x$, $x = 2$ $\qquad$ **d.** $f(x) = \log_{10} x$, $x = \frac{1}{100}$

Solution

a. $f(32) = \log_2 32 = 5 \qquad$ because $\quad 2^5 = 32$.

b. $f(1) = \log_3 1 = 0 \qquad$ because $\quad 3^0 = 1$.

c. $f(2) = \log_4 2 = \frac{1}{2} \qquad$ because $\quad 4^{1/2} = \sqrt{4} = 2$.

d. $f\left(\dfrac{1}{100}\right) = \log_{10} \dfrac{1}{100} = -2 \quad$ because $\quad 10^{-2} = \dfrac{1}{10^2} = \dfrac{1}{100}$.

✓ **Checkpoint** ◀))) *Audio-video solution in English & Spanish at LarsonPrecalculus.com*

Evaluate each logarithm at the given value of *x*.

a. $f(x) = \log_6 x, x = 1$ $\quad$ **b.** $f(x) = \log_5 x, x = \frac{1}{125}$ $\quad$ **c.** $f(x) = \log_7 x, x = 343$

The logarithmic function with base 10 is called the **common logarithmic function.** It is denoted by $\log_{10}$ or simply log. On most calculators, it is denoted by $\boxed{\text{LOG}}$. Example 2 shows how to use a calculator to evaluate common logarithmic functions. You will learn how to use a calculator to calculate logarithms with any base in Section 3.3.

EXAMPLE 2 Evaluating Common Logarithms on a Calculator

Use a calculator to evaluate the function $f(x) = \log x$ at each value of x.

a. $x = 10$ **b.** $x = \frac{1}{3}$ **c.** $x = -2$

Solution

Function Value	Calculator Keystrokes	Display
a. $f(10) = \log 10$	LOG 10 ENTER	1
b. $f\left(\frac{1}{3}\right) = \log \frac{1}{3}$	LOG (1 ÷ 3) ENTER	−0.4771213
c. $f(-2) = \log(-2)$	LOG (−) 2 ENTER	ERROR

Note that the calculator displays an error message (or a complex number) when you try to evaluate $\log(-2)$. This occurs because there is no real number power to which 10 can be raised to obtain -2.

✓ Checkpoint *Audio-video solution in English & Spanish at LarsonPrecalculus.com*

Use a calculator to evaluate the function $f(x) = \log x$ at each value of x.

a. $x = 275$ **b.** $x = -\frac{1}{2}$ **c.** $x = \frac{1}{2}$

The definition of the logarithmic function with base a leads to several properties.

Properties of Logarithms

1. $\log_a 1 = 0$ because $a^0 = 1$.
2. $\log_a a = 1$ because $a^1 = a$.
3. $\log_a a^x = x$ and $a^{\log_a x} = x$ — Inverse Properties
4. If $\log_a x = \log_a y$, then $x = y$. — One-to-One Property

EXAMPLE 3 Using Properties of Logarithms

a. Simplify $\log_4 1$. **b.** Simplify $\log_{\sqrt{7}} \sqrt{7}$. **c.** Simplify $6^{\log_6 20}$.

Solution

a. $\log_4 1 = 0$ — Property 1

b. $\log_{\sqrt{7}} \sqrt{7} = 1$ — Property 2

c. $6^{\log_6 20} = 20$ — Property 3 (Inverse Property)

✓ Checkpoint *Audio-video solution in English & Spanish at LarsonPrecalculus.com*

a. Simplify $\log_9 9$. **b.** Simplify $20^{\log_{20} 3}$. **c.** Simplify $\log_{\sqrt{3}} 1$.

EXAMPLE 4 Using the One-to-One Property

a. $\log_3 x = \log_3 12$ — Original equation

$x = 12$ — One-to-One Property

b. $\log(2x + 1) = \log 3x \Rightarrow 2x + 1 = 3x \Rightarrow 1 = x$

c. $\log_4(x^2 - 6) = \log_4 10 \Rightarrow x^2 - 6 = 10 \Rightarrow x^2 = 16 \Rightarrow x = \pm 4$

✓ Checkpoint *Audio-video solution in English & Spanish at LarsonPrecalculus.com*

Solve $\log_5 (x^2 + 3) = \log_5 12$ for x.

Graphs of Logarithmic Functions

To sketch the graph of $y = \log_a x$, use the fact that the graphs of inverse functions are reflections of each other in the line $y = x$.

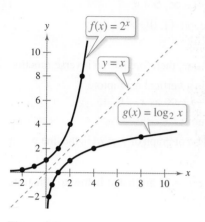

Figure 3.6

EXAMPLE 5   Graphing Exponential and Logarithmic Functions

In the same coordinate plane, sketch the graph of each function.

a. $f(x) = 2^x$ **b.** $g(x) = \log_2 x$

Solution

a. For $f(x) = 2^x$, construct a table of values. By plotting these points and connecting them with a smooth curve, you obtain the graph shown in Figure 3.6.

x	-2	-1	0	1	2	3
$f(x) = 2^x$	$\frac{1}{4}$	$\frac{1}{2}$	1	2	4	8

b. Because $g(x) = \log_2 x$ is the inverse function of $f(x) = 2^x$, the graph of g is obtained by plotting the points $(f(x), x)$ and connecting them with a smooth curve. The graph of g is a reflection of the graph of f in the line $y = x$, as shown in Figure 3.6.

✓ **Checkpoint** 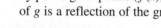 *Audio-video solution in English & Spanish at LarsonPrecalculus.com*

In the same coordinate plane, sketch the graphs of (a) $f(x) = 8^x$ and (b) $g(x) = \log_8 x$.

EXAMPLE 6   Sketching the Graph of a Logarithmic Function

Sketch the graph of $f(x) = \log x$. Identify the vertical asymptote.

Solution   Begin by constructing a table of values. Note that some of the values can be obtained without a calculator by using the properties of logarithms. Others require a calculator.

	Without calculator				With calculator		
x	$\frac{1}{100}$	$\frac{1}{10}$	1	10	2	5	8
$f(x) = \log x$	-2	-1	0	1	0.301	0.699	0.903

Next, plot the points and connect them with a smooth curve, as shown in the figure below. The vertical asymptote is $x = 0$ (y-axis).

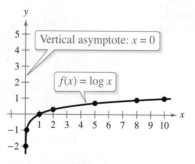

✓ **Checkpoint** *Audio-video solution in English & Spanish at LarsonPrecalculus.com*

Sketch the graph of $f(x) = \log_3 x$ by constructing a table of values without using a calculator. Identify the vertical asymptote.

The graph in Example 6 is typical for functions of the form $f(x) = \log_a x$, $a > 1$. They have one x-intercept and one vertical asymptote. Notice how slowly the graph rises for $x > 1$. Here are the basic characteristics of logarithmic graphs.

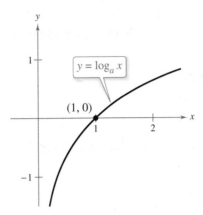

Graph of $y = \log_a x$, $a > 1$

- Domain: $(0, \infty)$
- Range: $(-\infty, \infty)$
- x-intercept: $(1, 0)$
- Increasing
- One-to-one, therefore has an inverse function
- y-axis is a vertical asymptote
  $(\log_a x \to -\infty$ as $x \to 0^+)$.
- Continuous
- Reflection of graph of $y = a^x$ in the line $y = x$

Some basic characteristics of the graph of $f(x) = a^x$ are listed below to illustrate the inverse relation between $f(x) = a^x$ and $g(x) = \log_a x$.

- Domain: $(-\infty, \infty)$ • Range: $(0, \infty)$
- y-intercept: $(0, 1)$ • x-axis is a horizontal asymptote $(a^x \to 0$ as $x \to -\infty)$.

The next example uses the graph of $y = \log_a x$ to sketch the graphs of functions of the form $f(x) = b \pm \log_a(x + c)$.

EXAMPLE 7   **Shifting Graphs of Logarithmic Functions**

See LarsonPrecalculus.com for an interactive version of this type of example.

Use the graph of $f(x) = \log x$ to sketch the graph of each function.

a. $g(x) = \log(x - 1)$ **b.** $h(x) = 2 + \log x$

Solution

a. Because $g(x) = \log(x - 1) = f(x - 1)$, the graph of g can be obtained by shifting the graph of f one unit to the right, as shown in Figure 3.7.

b. Because $h(x) = 2 + \log x = 2 + f(x)$, the graph of h can be obtained by shifting the graph of f two units up, as shown in Figure 3.8.

•• **REMARK**   Notice that
the vertical transformation in
Figure 3.8 keeps the y-axis as
the vertical asymptote, but the
horizontal transformation in
Figure 3.7 yields a new vertical
asymptote of $x = 1$.

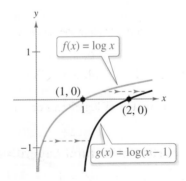

Figure 3.7

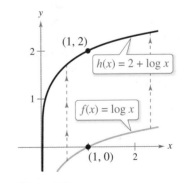

Figure 3.8

▷ **ALGEBRA HELP**   To review
the techniques for shifting,
reflecting, and stretching
graphs, see Section 1.7.

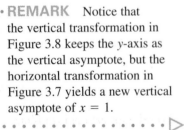

✓ **Checkpoint**))) *Audio-video solution in English & Spanish at LarsonPrecalculus.com*

Use the graph of $f(x) = \log_3 x$ to sketch the graph of each function.

a. $g(x) = -1 + \log_3 x$ **b.** $h(x) = \log_3(x + 3)$

The Natural Logarithmic Function

By looking back at the graph of the natural exponential function introduced on page 202 in Section 3.1, you will see that $f(x) = e^x$ is one-to-one and so has an inverse function. This inverse function is called the **natural logarithmic function** and is denoted by the special symbol $\ln x$, read as "the natural log of x" or "el en of x."

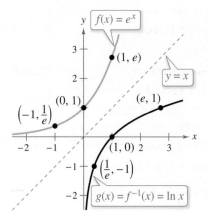

Reflection of graph of $f(x) = e^x$ in the line $y = x$
Figure 3.9

The Natural Logarithmic Function

The function

$$f(x) = \log_e x = \ln x, \quad x > 0$$

is called the **natural logarithmic function.**

The equations $y = \ln x$ and $x = e^y$ are equivalent. Note that the natural logarithm $\ln x$ is written without a base. The base is understood to be e.

Because the functions $f(x) = e^x$ and $g(x) = \ln x$ are inverse functions of each other, their graphs are reflections of each other in the line $y = x$, as shown in Figure 3.9.

EXAMPLE 8 **Evaluating the Natural Logarithmic Function**

Use a calculator to evaluate the function $f(x) = \ln x$ at each value of x.

a. $x = 2$

b. $x = 0.3$

c. $x = -1$

d. $x = 1 + \sqrt{2}$

Solution

Function Value	Calculator Keystrokes	Display
a. $f(2) = \ln 2$	LN 2 ENTER	0.6931472
b. $f(0.3) = \ln 0.3$	LN .3 ENTER	−1.2039728
c. $f(-1) = \ln(-1)$	LN (−) 1 ENTER	ERROR
d. $f\left(1 + \sqrt{2}\right) = \ln\left(1 + \sqrt{2}\right)$	LN (1 + √ 2) ENTER	0.8813736

✓ **Checkpoint** 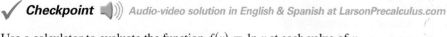 Audio-video solution in English & Spanish at LarsonPrecalculus.com

Use a calculator to evaluate the function $f(x) = \ln x$ at each value of x.

a. $x = 0.01$ **b.** $x = 4$

c. $x = \sqrt{3} + 2$ **d.** $x = \sqrt{3} - 2$

The properties of logarithms on page 210 are also valid for natural logarithms.

> **TECHNOLOGY** On most calculators, the natural logarithm is denoted by LN as illustrated in Example 8.

> **REMARK** In Example 8(c), be sure you see that $\ln(-1)$ gives an error message on most calculators. This occurs because the domain of $\ln x$ is the set of *positive real numbers* (see Figure 3.9). So, $\ln(-1)$ is undefined.

Properties of Natural Logarithms

1. $\ln 1 = 0$ because $e^0 = 1$.

2. $\ln e = 1$ because $e^1 = e$.

3. $\ln e^x = x$ and $e^{\ln x} = x$ Inverse Properties

4. If $\ln x = \ln y$, then $x = y$.   One-to-One Property

EXAMPLE 9   **Using Properties of Natural Logarithms**

Use the properties of natural logarithms to simplify each expression.

a. $\ln \dfrac{1}{e}$ **b.** $e^{\ln 5}$ **c.** $\dfrac{\ln 1}{3}$ **d.** $2 \ln e$

Solution

a. $\ln \dfrac{1}{e} = \ln e^{-1} = -1$ Property 3 (Inverse Property)

b. $e^{\ln 5} = 5$ Property 3 (Inverse Property)

c. $\dfrac{\ln 1}{3} = \dfrac{0}{3} = 0$ Property 1

d. $2 \ln e = 2(1) = 2$ Property 2

✓ **Checkpoint**  *Audio-video solution in English & Spanish at LarsonPrecalculus.com*

Use the properties of natural logarithms to simplify each expression.

a. $\ln e^{1/3}$ **b.** $5 \ln 1$ **c.** $\frac{3}{4} \ln e$ **d.** $e^{\ln 7}$

EXAMPLE 10   **Finding the Domains of Logarithmic Functions**

Find the domain of each function.

a. $f(x) = \ln(x - 2)$ **b.** $g(x) = \ln(2 - x)$ **c.** $h(x) = \ln x^2$

Solution

a. Because $\ln(x - 2)$ is defined only when

$$x - 2 > 0$$

it follows that the domain of f is $(2, \infty)$, as shown in Figure 3.10.

b. Because $\ln(2 - x)$ is defined only when

$$2 - x > 0$$

it follows that the domain of g is $(-\infty, 2)$, as shown in Figure 3.11.

c. Because $\ln x^2$ is defined only when

$$x^2 > 0$$

it follows that the domain of h is all real numbers except $x = 0$, as shown in Figure 3.12.

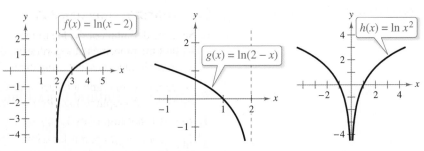

Figure 3.10   Figure 3.11   Figure 3.12

✓ **Checkpoint**  *Audio-video solution in English & Spanish at LarsonPrecalculus.com*

Find the domain of $f(x) = \ln(x + 3)$. ■

Application

EXAMPLE 11   **Human Memory Model**

Students participating in a psychology experiment attended several lectures on a subject and took an exam. Every month for a year after the exam, the students took a retest to see how much of the material they remembered. The average scores for the group are given by the *human memory model* $f(t) = 75 - 6 \ln(t + 1)$, $0 \le t \le 12$, where t is the time in months.

a. What was the average score on the original exam ($t = 0$)?

b. What was the average score at the end of $t = 2$ months?

c. What was the average score at the end of $t = 6$ months?

Algebraic Solution

a. The original average score was

$$f(0) = 75 - 6 \ln(0 + 1) \qquad \text{Substitute 0 for } t.$$

$$= 75 - 6 \ln 1 \qquad \text{Simplify.}$$

$$= 75 - 6(0) \qquad \text{Property of natural logarithms}$$

$$= 75. \qquad \text{Solution}$$

b. After 2 months, the average score was

$$f(2) = 75 - 6 \ln(2 + 1) \qquad \text{Substitute 2 for } t.$$

$$= 75 - 6 \ln 3 \qquad \text{Simplify.}$$

$$\approx 75 - 6(1.0986) \qquad \text{Use a calculator.}$$

$$\approx 68.41. \qquad \text{Solution}$$

c. After 6 months, the average score was

$$f(6) = 75 - 6 \ln(6 + 1) \qquad \text{Substitute 6 for } t.$$

$$= 75 - 6 \ln 7 \qquad \text{Simplify.}$$

$$\approx 75 - 6(1.9459) \qquad \text{Use a calculator.}$$

$$\approx 63.32. \qquad \text{Solution}$$

Graphical Solution

a.

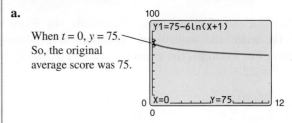

When $t = 0$, $y = 75$. So, the original average score was 75.

b.

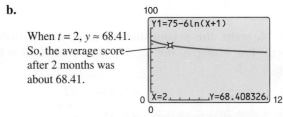

When $t = 2$, $y \approx 68.41$. So, the average score after 2 months was about 68.41.

c.

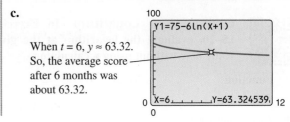

When $t = 6$, $y \approx 63.32$. So, the average score after 6 months was about 63.32.

 ✔ *Checkpoint* ◀))) *Audio-video solution in English & Spanish at LarsonPrecalculus.com*

In Example 11, find the average score at the end of (a) $t = 1$ month, (b) $t = 9$ months, and (c) $t = 12$ months.

Summarize   (Section 3.2)

1. State the definition of the logarithmic function with base *a* *(page 209)* and make a list of the properties of logarithms *(page 210)*. For examples of evaluating logarithmic functions and using the properties of logarithms, see Examples 1–4.

2. Explain how to graph a logarithmic function *(pages 211 and 212)*. For examples of graphing logarithmic functions, see Examples 5–7.

3. State the definition of the natural logarithmic function and make a list of the properties of natural logarithms *(page 213)*. For examples of evaluating natural logarithmic functions and using the properties of natural logarithms, see Examples 8 and 9.

4. Describe a real-life application that uses a logarithmic function to model and solve a problem *(page 215, Example 11)*.

3.2 Exercises

See **CalcChat.com** for tutorial help and worked-out solutions to odd-numbered exercises.

Vocabulary: Fill in the blanks.

1. The inverse function of the exponential function $f(x) = a^x$ is the _____ function with base a.
2. The common logarithmic function has base _____.
3. The logarithmic function $f(x) = \ln x$ is the _____ logarithmic function and has base _____.
4. The Inverse Properties of logarithms state that $\log_a a^x = x$ and _____.
5. The One-to-One Property of natural logarithms states that if $\ln x = \ln y$, then _____.
6. The domain of the natural logarithmic function is the set of _____ _____ _____.

Skills and Applications

Writing an Exponential Equation In Exercises 7–10, write the logarithmic equation in exponential form. For example, the exponential form of $\log_5 25 = 2$ is $5^2 = 25$.

7. $\log_4 16 = 2$
8. $\log_9 \frac{1}{81} = -2$
9. $\log_{12} 12 = 1$
10. $\log_{32} 4 = \frac{2}{5}$

Writing a Logarithmic Equation In Exercises 11–14, write the exponential equation in logarithmic form. For example, the logarithmic form of $2^3 = 8$ is $\log_2 8 = 3$.

11. $5^3 = 125$
12. $9^{3/2} = 27$
13. $4^{-3} = \frac{1}{64}$
14. $24^0 = 1$

 Evaluating a Logarithm In Exercises 15–20, evaluate the logarithm at the given value of x without using a calculator.

Function	Value
15. $f(x) = \log_2 x$	$x = 64$
16. $f(x) = \log_{25} x$	$x = 5$
17. $f(x) = \log_8 x$	$x = 1$
18. $f(x) = \log x$	$x = 10$
19. $g(x) = \log_a x$	$x = a^{-2}$
20. $g(x) = \log_b x$	$x = \sqrt{b}$

Evaluating a Common Logarithm on a Calculator In Exercises 21–24, use a calculator to evaluate $f(x) = \log x$ at the given value of x. Round your result to three decimal places.

21. $x = \frac{7}{8}$
22. $x = \frac{1}{500}$
23. $x = 12.5$
24. $x = 96.75$

 Using Properties of Logarithms In Exercises 25–28, use the properties of logarithms to simplify the expression.

25. $\log_8 8$
26. $\log_\pi \pi^2$
27. $\log_{7.5} 1$
28. $5^{\log_5 3}$

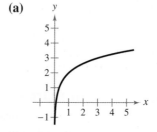 **Using the One-to-One Property** In Exercises 29–32, use the One-to-One Property to solve the equation for x.

29. $\log_5(x + 1) = \log_5 6$
30. $\log_2(x - 3) = \log_2 9$
31. $\log 11 = \log(x^2 + 7)$
32. $\log(x^2 + 6x) = \log 27$

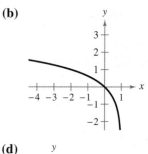 **Graphing Exponential and Logarithmic Functions** In Exercises 33–36, sketch the graphs of f and g in the same coordinate plane.

33. $f(x) = 7^x$, $g(x) = \log_7 x$
34. $f(x) = 5^x$, $g(x) = \log_5 x$
35. $f(x) = 6^x$, $g(x) = \log_6 x$
36. $f(x) = 10^x$, $g(x) = \log x$

Matching a Logarithmic Function with Its Graph In Exercises 37–40, use the graph of $g(x) = \log_3 x$ to match the given function with its graph. Then describe the relationship between the graphs of f and g. [The graphs are labeled (a), (b), (c), and (d).]

(a)

(b)

(c)

(d)

37. $f(x) = \log_3 x + 2$
38. $f(x) = \log_3(x - 1)$
39. $f(x) = \log_3(1 - x)$
40. $f(x) = -\log_3 x$

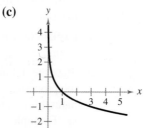
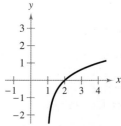

Sketching the Graph of a Logarithmic Function In Exercises 41–48, find the domain, x-intercept, and vertical asymptote of the logarithmic function and sketch its graph.

41. $f(x) = \log_4 x$ **42.** $g(x) = \log_6 x$

43. $y = \log_3 x + 1$

44. $h(x) = \log_4(x - 3)$

45. $f(x) = -\log_6(x + 2)$

46. $y = \log_5(x - 1) + 4$

47. $y = \log \dfrac{x}{7}$

48. $y = \log(-2x)$

Writing a Natural Exponential Equation In Exercises 49–52, write the logarithmic equation in exponential form.

49. $\ln \frac{1}{2} = -0.693 \ldots$ **50.** $\ln 7 = 1.945 \ldots$

51. $\ln 250 = 5.521 \ldots$ **52.** $\ln 1 = 0$

Writing a Natural Logarithmic Equation In Exercises 53–56, write the exponential equation in logarithmic form.

53. $e^2 = 7.3890 \ldots$ **54.** $e^{-3/4} = 0.4723 \ldots$

55. $e^{-4x} = \frac{1}{2}$ **56.** $e^{2x} = 3$

 Evaluating a Logarithmic Function In Exercises 57–60, use a calculator to evaluate the function at the given value of x. Round your result to three decimal places.

Function	Value
57. $f(x) = \ln x$	$x = 18.42$
58. $f(x) = 3 \ln x$	$x = 0.74$
59. $g(x) = 8 \ln x$	$x = \sqrt{5}$
60. $g(x) = -\ln x$	$x = \frac{1}{2}$

 Using Properties of Natural Logarithms In Exercises 61–66, use the properties of natural logarithms to simplify the expression.

61. $e^{\ln 4}$ **62.** $\ln \dfrac{1}{e^2}$

63. $2.5 \ln 1$ **64.** $\dfrac{\ln e}{\pi}$

65. $\ln e^{\ln e}$ **66.** $e^{\ln(1/e)}$

Graphing a Natural Logarithmic Function In Exercises 67–70, find the domain, x-intercept, and vertical asymptote of the logarithmic function and sketch its graph.

67. $f(x) = \ln(x - 4)$ **68.** $h(x) = \ln(x + 5)$

69. $g(x) = \ln(-x)$ **70.** $f(x) = \ln(3 - x)$

Graphing a Natural Logarithmic Function In Exercises 71–74, use a graphing utility to graph the function. Be sure to use an appropriate viewing window.

71. $f(x) = \ln(x - 1)$ **72.** $f(x) = \ln(x + 2)$

73. $f(x) = -\ln x + 8$ **74.** $f(x) = 3 \ln x - 1$

Using the One-to-One Property In Exercises 75–78, use the One-to-One Property to solve the equation for x.

75. $\ln(x + 4) = \ln 12$ **76.** $\ln(x - 7) = \ln 7$

77. $\ln(x^2 - x) = \ln 6$ **78.** $\ln(x^2 - 2) = \ln 23$

79. Monthly Payment The model

$$t = 16.625 \ln \frac{x}{x - 750}, \quad x > 750$$

approximates the length of a home mortgage of \$150,000 at 6% in terms of the monthly payment. In the model, t is the length of the mortgage in years and x is the monthly payment in dollars.

(a) Approximate the lengths of a \$150,000 mortgage at 6% when the monthly payment is \$897.72 and when the monthly payment is \$1659.24.

(b) Approximate the total amounts paid over the term of the mortgage with a monthly payment of \$897.72 and with a monthly payment of \$1659.24. What amount of the total is interest costs in each case?

(c) What is the vertical asymptote for the model? Interpret its meaning in the context of the problem.

80. Telephone Service The percent P of households in the United States with wireless-only telephone service from 2005 through 2014 can be approximated by the model

$$P = -3.42 + 1.297t \ln t, \quad 5 \le t \le 14$$

where t represents the year, with $t = 5$ corresponding to 2005. (*Source: National Center for Health Statistics*)

(a) Approximate the percents of households with wireless-only telephone service in 2008 and 2012.

(b) Use a graphing utility to graph the function.

(c) Can the model be used to predict the percent of households with wireless-only telephone service in 2020? in 2030? Explain.

81. Population The time t (in years) for the world population to double when it is increasing at a continuous rate r (in decimal form) is given by $t = (\ln 2)/r$.

(a) Complete the table and interpret your results.

r	0.005	0.010	0.015	0.020	0.025	0.030
t						

(b) Use a graphing utility to graph the function.

82. Compound Interest A principal P, invested at $5\frac{1}{2}\%$ and compounded continuously, increases to an amount K times the original principal after t years, where $t = (\ln K)/0.055$.

(a) Complete the table and interpret your results.

K	1	2	4	6	8	10	12
t							

(b) Sketch a graph of the function.

83. Human Memory Model

Students in a mathematics class took an exam and then took a retest monthly with an equivalent exam. The average scores for the class are given by the human memory model

$f(t) = 80 - 17 \log(t + 1), \quad 0 \leq t \leq 12$

where t is the time in months.

(a) Use a graphing utility to graph the model over the specified domain.

(b) What was the average score on the original exam ($t = 0$)?

(c) What was the average score after 4 months?

(d) What was the average score after 10 months?

84. Sound Intensity The relationship between the number of decibels β and the intensity of a sound I (in watts per square meter) is

$\beta = 10 \log \dfrac{I}{10^{-12}}.$

(a) Determine the number of decibels of a sound with an intensity of 1 watt per square meter.

(b) Determine the number of decibels of a sound with an intensity of 10^{-2} watt per square meter.

(c) The intensity of the sound in part (a) is 100 times as great as that in part (b). Is the number of decibels 100 times as great? Explain.

Exploration

True or False? **In Exercises 85 and 86, determine whether the statement is true or false. Justify your answer.**

85. The graph of $f(x) = \log_6 x$ is a reflection of the graph of $g(x) = 6^x$ in the x-axis.

86. The graph of $f(x) = \ln(-x)$ is a reflection of the graph of $h(x) = e^{-x}$ in the line $y = -x$.

87. Graphical Reasoning Use a graphing utility to graph f and g in the same viewing window and determine which is increasing at the greater rate as x approaches $+\infty$. What can you conclude about the rate of growth of the natural logarithmic function?

(a) $f(x) = \ln x, \quad g(x) = \sqrt{x}$

(b) $f(x) = \ln x, \quad g(x) = \sqrt[4]{x}$

88. **HOW DO YOU SEE IT?** The figure shows the graphs of $f(x) = 3^x$ and $g(x) = \log_3 x$. [The graphs are labeled m and n.]

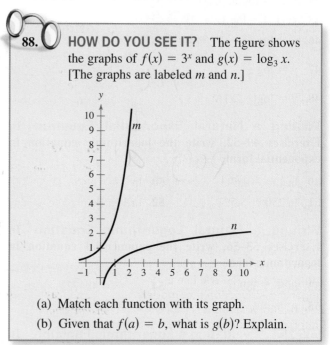

(a) Match each function with its graph.

(b) Given that $f(a) = b$, what is $g(b)$? Explain.

Error Analysis **In Exercises 89 and 90, describe the error.**

89.

x	1	2	8
y	0	1	3

From the table, you can conclude that y is an exponential function of x.

90.

x	1	2	5
y	2	4	32

From the table, you can conclude that y is a logarithmic function of x.

91. Numerical Analysis

(a) Complete the table for the function $f(x) = (\ln x)/x$.

x	1	5	10	10^2	10^4	10^6
$f(x)$						

(b) Use the table in part (a) to determine what value $f(x)$ approaches as x increases without bound.

(c) Use a graphing utility to confirm the result of part (b).

92. Writing Explain why $\log_a x$ is defined only for $0 < a < 1$ and $a > 1$.

3.3 Properties of Logarithms

- Use the change-of-base formula to rewrite and evaluate logarithmic expressions.
- Use properties of logarithms to evaluate or rewrite logarithmic expressions.
- Use properties of logarithms to expand or condense logarithmic expressions.
- Use logarithmic functions to model and solve real-life problems.

Change of Base

Most calculators have only two types of log keys, LOG for common logarithms (base 10) and LN for natural logarithms (base e). Although common logarithms and natural logarithms are the most frequently used, you may occasionally need to evaluate logarithms with other bases. To do this, use the **change-of-base formula.**

Logarithmic functions have many real-life applications. For example, in Exercises 79–82 on page 224, you will use a logarithmic function that models the relationship between the number of decibels and the intensity of a sound.

> **Change-of-Base Formula**
>
> Let a, b, and x be positive real numbers such that $a \neq 1$ and $b \neq 1$. Then $\log_a x$ can be converted to a different base as follows.
>
> | Base b | Base 10 | Base e |
> |----------|---------|----------|
> | $\log_a x = \dfrac{\log_b x}{\log_b a}$ | $\log_a x = \dfrac{\log x}{\log a}$ | $\log_a x = \dfrac{\ln x}{\ln a}$ |

One way to look at the change-of-base formula is that logarithms with base a are *constant multiples* of logarithms with base b. The constant multiplier is

$$\frac{1}{\log_b a}.$$

EXAMPLE 1 **Changing Bases Using Common Logarithms**

$$\log_4 25 = \frac{\log 25}{\log 4} \qquad \log_a x = \frac{\log x}{\log a}$$

$$\approx \frac{1.39794}{0.60206} \qquad \text{Use a calculator.}$$

$$\approx 2.3219 \qquad \text{Simplify.}$$

✓ **Checkpoint** ◀))) *Audio-video solution in English & Spanish at LarsonPrecalculus.com*

Evaluate $\log_2 12$ using the change-of-base formula and common logarithms.

EXAMPLE 2 **Changing Bases Using Natural Logarithms**

$$\log_4 25 = \frac{\ln 25}{\ln 4} \qquad \log_a x = \frac{\ln x}{\ln a}$$

$$\approx \frac{3.21888}{1.38629} \qquad \text{Use a calculator.}$$

$$\approx 2.3219 \qquad \text{Simplify.}$$

✓ **Checkpoint** ◀))) *Audio-video solution in English & Spanish at LarsonPrecalculus.com*

Evaluate $\log_2 12$ using the change-of-base formula and natural logarithms.

Properties of Logarithms

You know from the preceding section that the logarithmic function with base a is the *inverse function* of the exponential function with base a. So, it makes sense that the properties of exponents have corresponding properties involving logarithms. For example, the exponential property $a^m a^n = a^{m+n}$ has the corresponding logarithmic property $\log_a(uv) = \log_a u + \log_a v$.

> ·· **REMARK**   There is no
> property that can be used
> to rewrite $\log_a(u \pm v)$.
> Specifically, $\log_a(u + v)$ is *not*
> equal to $\log_a u + \log_a v$.

Properties of Logarithms

Let a be a positive number such that $a \neq 1$, let n be a real number, and let u and v be positive real numbers.

	Logarithm with Base a	**Natural Logarithm**
1. Product Property:	$\log_a(uv) = \log_a u + \log_a v$	$\ln(uv) = \ln u + \ln v$
2. Quotient Property:	$\log_a \dfrac{u}{v} = \log_a u - \log_a v$	$\ln \dfrac{u}{v} = \ln u - \ln v$
3. Power Property:	$\log_a u^n = n \log_a u$	$\ln u^n = n \ln u$

For proofs of the properties listed above, see Proofs in Mathematics on page 256.

EXAMPLE 3   Using Properties of Logarithms

Write each logarithm in terms of $\ln 2$ and $\ln 3$.

a. $\ln 6$ **b.** $\ln \dfrac{2}{27}$

Solution

a. $\ln 6 = \ln(2 \cdot 3)$ Rewrite 6 as $2 \cdot 3$.

$ = \ln 2 + \ln 3$ Product Property

b. $\ln \dfrac{2}{27} = \ln 2 - \ln 27$ Quotient Property

$\phantom{\ln \dfrac{2}{27}} = \ln 2 - \ln 3^3$ Rewrite 27 as 3^3.

$\phantom{\ln \dfrac{2}{27}} = \ln 2 - 3 \ln 3$ Power Property

✓ **Checkpoint** 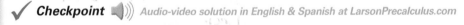 Audio-video solution in English & Spanish at LarsonPrecalculus.com

Write each logarithm in terms of $\log 3$ and $\log 5$.

a. $\log 75$ **b.** $\log \dfrac{9}{125}$

EXAMPLE 4   Using Properties of Logarithms

Find the exact value of $\log_5 \sqrt[3]{5}$ without using a calculator.

Solution

$\log_5 \sqrt[3]{5} = \log_5 5^{1/3} = \frac{1}{3} \log_5 5 = \frac{1}{3}(1) = \frac{1}{3}$

✓ **Checkpoint** 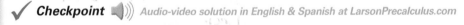 Audio-video solution in English & Spanish at LarsonPrecalculus.com

Find the exact value of $\ln e^6 - \ln e^2$ without using a calculator.

Rewriting Logarithmic Expressions

The properties of logarithms are useful for rewriting logarithmic expressions in forms that simplify the operations of algebra. This is true because these properties convert complicated products, quotients, and exponential forms into simpler sums, differences, and products, respectively.

EXAMPLE 5 **Expanding Logarithmic Expressions**

Expand each logarithmic expression.

a. $\log_4 5x^3y$ **b.** $\ln \dfrac{\sqrt{3x-5}}{7}$

Solution

a. $\log_4 5x^3y = \log_4 5 + \log_4 x^3 + \log_4 y$ Product Property

$\qquad\qquad = \log_4 5 + 3\log_4 x + \log_4 y$ Power Property

b. $\ln \dfrac{\sqrt{3x-5}}{7} = \ln \dfrac{(3x-5)^{1/2}}{7}$ Rewrite using rational exponent.

$\qquad\qquad = \ln(3x-5)^{1/2} - \ln 7$ Quotient Property

$\qquad\qquad = \dfrac{1}{2}\ln(3x-5) - \ln 7$ Power Property

▷ **ALGEBRA HELP** To review rewriting radicals and rational exponents, see Appendix A.2.

✓ *Checkpoint* ◀))) *Audio-video solution in English & Spanish at LarsonPrecalculus.com*

Expand the expression $\log_3 \dfrac{4x^2}{\sqrt{y}}$.

Example 5 uses the properties of logarithms to *expand* logarithmic expressions. Example 6 reverses this procedure and uses the properties of logarithms to *condense* logarithmic expressions.

EXAMPLE 6 **Condensing Logarithmic Expressions**

See LarsonPrecalculus.com for an interactive version of this type of example.

Condense each logarithmic expression.

a. $\frac{1}{2}\log x + 3\log(x+1)$ **b.** $2\ln(x+2) - \ln x$ **c.** $\frac{1}{3}[\log_2 x + \log_2(x+1)]$

Solution

a. $\frac{1}{2}\log x + 3\log(x+1) = \log x^{1/2} + \log(x+1)^3$ Power Property

$\qquad\qquad = \log[\sqrt{x}\,(x+1)^3]$ Product Property

b. $2\ln(x+2) - \ln x = \ln(x+2)^2 - \ln x$ Power Property

$\qquad\qquad = \ln\dfrac{(x+2)^2}{x}$ Quotient Property

c. $\frac{1}{3}[\log_2 x + \log_2(x+1)] = \frac{1}{3}\log_2[x(x+1)]$ Product Property

$\qquad\qquad = \log_2[x(x+1)]^{1/3}$ Power Property

$\qquad\qquad = \log_2 \sqrt[3]{x(x+1)}$ Rewrite with a radical.

✓ *Checkpoint* ◀))) *Audio-video solution in English & Spanish at LarsonPrecalculus.com*

Condense the expression $2[\log(x+3) - 2\log(x-2)]$.

Application

One way to determine a possible relationship between the x- and y-values of a set of nonlinear data is to take the natural logarithm of each x-value and each y-value. If the plotted points $(\ln x, \ln y)$ lie on a line, then x and y are related by the equation $\ln y = m \ln x$, where m is the slope of the line.

EXAMPLE 7 **Finding a Mathematical Model**

The table shows the mean distance x from the sun and the period y (the time it takes a planet to orbit the sun, in years) for each of the six planets that are closest to the sun. In the table, the mean distance is given in astronomical units (where one astronomical unit is defined as Earth's mean distance from the sun). The points from the table are plotted in Figure 3.13. Find an equation that relates y and x.

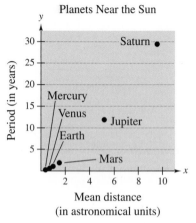

Planets Near the Sun

Mean distance (in astronomical units)

Figure 3.13

DATA	Planet	Mean Distance, x	Period, y
	Mercury	0.387	0.241
	Venus	0.723	0.615
	Earth	1.000	1.000
	Mars	1.524	1.881
	Jupiter	5.203	11.862
	Saturn	9.537	29.457

Spreadsheet at LarsonPrecalculus.com

Planet	$\ln x$	$\ln y$
Mercury	-0.949	-1.423
Venus	-0.324	-0.486
Earth	0.000	0.000
Mars	0.421	0.632
Jupiter	1.649	2.473
Saturn	2.255	3.383

Solution From Figure 3.13, it is not clear how to find an equation that relates y and x. To solve this problem, make a table of values giving the natural logarithms of all x- and y-values of the data (see the table at the left). Plot each point $(\ln x, \ln y)$. These points appear to lie on a line (see Figure 3.14). Choose two points to determine the slope of the line. Using the points $(0.421, 0.632)$ and $(0, 0)$, the slope of the line is

$$m = \frac{0.632 - 0}{0.421 - 0} \approx 1.5 = \frac{3}{2}.$$

By the point-slope form, the equation of the line is $Y = \frac{3}{2}X$, where $Y = \ln y$ and $X = \ln x$. So, an equation that relates y and x is $\ln y = \frac{3}{2} \ln x$.

✓ **Checkpoint** ◀))) *Audio-video solution in English & Spanish at LarsonPrecalculus.com*

Find a logarithmic equation that relates y and x for the following ordered pairs.

$(0.37, 0.51), (1.00, 1.00), (2.72, 1.95), (7.39, 3.79), (20.09, 7.39)$

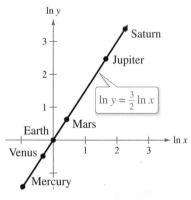

Figure 3.14

$\ln y = \frac{3}{2} \ln x$

Summarize (Section 3.3)

1. State the change-of-base formula *(page 219)*. For examples of using the change-of-base formula to rewrite and evaluate logarithmic expressions, see Examples 1 and 2.

2. Make a list of the properties of logarithms *(page 220)*. For examples of using the properties of logarithms to evaluate or rewrite logarithmic expressions, see Examples 3 and 4.

3. Explain how to use the properties of logarithms to expand or condense logarithmic expressions *(page 221)*. For examples of expanding and condensing logarithmic expressions, see Examples 5 and 6.

4. Describe an example of how to use a logarithmic function to model and solve a real-life problem *(page 222, Example 7)*.

89. Comparing M

temperature of 7
temperature of 2
measured every
The results are r
(t, T), where t i
temperature (in d

$(0, 78.0°)$, $(5, 66.$
$(20, 46.3°)$, $(25,$

(a) Subtract the
temperatures
utility to plo

(b) An exponen
$T - 21 = 5$
the model. (
original data

(c) Use the g
$(t, \ln(T - 2$
be linear. Us
utility to fi
line has the
equivalent
verify that
part (b).

(d) Fit a rationa
of the y-co
generate the

$$\left(t, \frac{1}{T - 21},\right.$$

Use the gra
observe th
regression
to these dat

$$\frac{1}{T - 21} =$$

Solve for T
rational fur

90. Writing Wr
the transforma
necessary to c
logarithms of
plot? Why did
lead to a linea

Exploration

True or False?
the statement is
Justify your answ

91. $f(0) = 0$
92. $f(ax) = f(a)$

3.3 Exercises

See **CalcChat.com** for tutorial help and worked-out solutions to odd-numbered exercises.

Vocabulary

In Exercises 1–3, fill in the blanks.

1. To evaluate a logarithm to any base, use the _____ formula.
2. The change-of-base formula for base e is $\log_a x = $ _____.
3. When you consider $\log_a x$ to be a constant multiple of $\log_b x$, the constant multiplier is _____.
4. Name the property of logarithms illustrated by each statement.

(a) $\ln(uv) = \ln u + \ln v$ (b) $\log_a u^n = n \log_a u$ (c) $\ln \dfrac{u}{v} = \ln u - \ln v$

Skills and Applications

 Changing Bases In Exercises 5–8, rewrite the logarithm as a ratio of (a) common logarithms and (b) natural logarithms.

5. $\log_5 16$
6. $\log_{1/5} 4$
7. $\log_x \frac{3}{10}$
8. $\log_{2.6} x$

 Using the Change-of-Base Formula In Exercises 9–12, evaluate the logarithm using the change-of-base formula. Round your result to three decimal places.

9. $\log_3 17$
10. $\log_{0.4} 12$
11. $\log_\pi 0.5$
12. $\log_{2/3} 0.125$

 Using Properties of Logarithms In Exercises 13–18, use the properties of logarithms to write the logarithm in terms of $\log_3 5$ and $\log_3 7$.

13. $\log_3 35$
14. $\log_3 \frac{5}{7}$
15. $\log_3 \frac{7}{25}$
16. $\log_3 175$
17. $\log_3 \frac{21}{5}$
18. $\log_3 \frac{45}{49}$

 Using Properties of Logarithms In Exercises 19–32, find the exact value of the logarithmic expression without using a calculator. (If this is not possible, state the reason.)

19. $\log_3 9$
20. $\log_5 \frac{1}{125}$
21. $\log_6 \sqrt[3]{\frac{1}{6}}$
22. $\log_2 \sqrt[4]{8}$
23. $\log_2(-2)$
24. $\log_3(-27)$
25. $\ln \sqrt[4]{e^3}$
26. $\ln(1/\sqrt{e})$
27. $\ln e^2 + \ln e^5$
28. $2 \ln e^6 - \ln e^5$ $2(6) - 5$
29. $\log_5 75 - \log_5 3$
30. $\log_4 2 + \log_4 32$
31. $\log_4 8$
32. $\log_8 16$

Using Properties of Logarithms In Exercises 33–40, approximate the logarithm using the properties of logarithms, given $\log_b 2 \approx 0.3562$, $\log_b 3 \approx 0.5646$, and $\log_b 5 \approx 0.8271$.

33. $\log_b 10$
34. $\log_b \frac{2}{3}$
35. $\log_b 0.04$
36. $\log_b \sqrt{2}$
37. $\log_b 45$
38. $\log_b(3b^2)$
39. $\log_b(2b)^{-2}$
40. $\log_b \sqrt[3]{3b}$

 Expanding a Logarithmic Expression In Exercises 41–60, use the properties of logarithms to expand the expression as a sum, difference, and/or constant multiple of logarithms. (Assume all variables are positive.)

41. $\ln 7x$
42. $\log_3 13z$
43. $\log_8 x^4$
44. $\ln(xy)^3$
45. $\log_5 \frac{5}{x}$
46. $\log_6 \frac{w^2}{v}$
47. $\ln \sqrt{z}$
48. $\ln \sqrt[3]{t}$
49. $\ln xyz^2$
50. $\log_4 11b^2c$
51. $\ln z(z-1)^2$, $z > 1$
52. $\ln \dfrac{x^2 - 1}{x^3}$, $x > 1$
53. $\log_2 \dfrac{\sqrt{a^2 - 4}}{7}$, $a > 2$
54. $\ln \dfrac{3}{\sqrt{x^2 + 1}}$
55. $\log_5 \dfrac{x^2}{y^2 z^3}$
56. $\log_{10} \dfrac{xy^4}{z^5}$
57. $\ln \sqrt[3]{\dfrac{yz}{x^2}}$
58. $\log_2 x^4 \sqrt{\dfrac{y}{z^3}}$
59. $\ln \sqrt[4]{x^3(x^2 + 3)}$
60. $\ln \sqrt{x^2(x + 2)}$

Condensing

In Exercises (
to the logarith

61. ln 3 + ln x

63. $\frac{2}{3}\log_7(z-2)$

65. $\log_3 5x - 4\log_3 x$

67. $\log x + 2\log(x+1)$

69. $\log x - 2\log y + 3$ l

70. $3\log_3 x + \frac{1}{4}\log_3 y -$

71. $\ln x - [\ln(x+1) + 1$

72. $4[\ln z + \ln(z+5)] -$

73. $\frac{1}{2}[2\ln(x+3) + \ln x$

74. $2[3\ln x - \ln(x+1)$

75. $\frac{1}{3}[\log_8 y + 2\log_8(y +$

76. $\frac{1}{2}[\log_4(x+1) + 2\log$

Comparing Logarith
77 and 78, determine w
expressions are equal. J

77. $\dfrac{\log_2 32}{\log_2 4}$, $\log_2\dfrac{32}{4}$,

78. $\log_7\sqrt{70}$, $\log_7 35$,

• • **Sound Intensity** •

• **In Exercises 79–82, u**
• **The relationship**
• **between the number**
• **of decibels** β **and the**
• **intensity of a sound**
• I **(in watts per**
• **square meter) is**

$\beta = 10\log\dfrac{I}{10^{-12}}.$

• **79.** Use the properties
•   formula in a simp
•   number of decibe
•   of 10^{-6} watt per

• **80.** Find the differenc
•   office with an int
•   per square meter
•   intensity of 3.16

• **81.** Find the differenc
•   vacuum cleaner v
•   square meter and
•   of 10^{-11} watt per

• **82.** You and your roc
•   at the same time
•   much louder is th
•   playing compared

Logarithmic Models

On the Richter scale, the magnitude R of an earthquake of intensity I is given by

$$R = \log\frac{I}{I_0}$$

where $I_0 = 1$ is the minimum intensity used for comparison. (Intensity is a measure of the wave energy of an earthquake.)

On April 25, 2015, an earthquake of magnitude 7.8 struck in Nepal. The city of Kathmandu took extensive damage, including the collapse of the 203-foot Dharahara Tower, built by Nepal's first prime minister in 1832.

EXAMPLE 6 **Magnitudes of Earthquakes**

Find the intensity of each earthquake.

a. Piedmont, California, in 2015: $R = 4.0$ **b.** Nepal in 2015: $R = 7.8$

Solution

a. Because $I_0 = 1$ and $R = 4.0$, you have

$$4.0 = \log\frac{I}{1}$$ Substitute 1 for I_0 and 4.0 for R.

$$10^{4.0} = 10^{\log I}$$ Exponentiate each side.

$$10^{4.0} = I$$ Inverse Property

$$10{,}000 = I.$$ Simplify.

b. For $R = 7.8$, you have

$$7.8 = \log\frac{I}{1}$$ Substitute 1 for I_0 and 7.8 for R.

$$10^{7.8} = 10^{\log I}$$ Exponentiate each side.

$$10^{7.8} = I$$ Inverse Property

$$63{,}000{,}000 \approx I.$$ Use a calculator.

Note that an increase of 3.8 units on the Richter scale (from 4.0 to 7.8) represents an increase in intensity by a factor of $10^{7.8}/10^4 \approx 63{,}000{,}000/10{,}000 = 6300$. In other words, the intensity of the earthquake in Nepal was about 6300 times as great as that of the earthquake in Piedmont, California.

✓ **Checkpoint**))) *Audio-video solution in English & Spanish at LarsonPrecalculus.com*

Find the intensities of earthquakes whose magnitudes are (a) $R = 6.0$ and (b) $R = 7.9$.

Summarize *(Section 3.5)*

1. State the five most common types of models involving exponential and logarithmic functions *(page 236)*.

2. Describe examples of real-life applications that use exponential growth and decay functions *(pages 237–239, Examples 1–3)*.

3. Describe an example of a real-life application that uses a Gaussian function *(page 240, Example 4)*.

4. Describe an example of a real-life application that uses a logistic growth function *(page 241, Example 5)*.

5. Describe an example of a real-life application that uses a logarithmic function *(page 242, Example 6)*.

3.5 Exercises

See **CalcChat.com** for tutorial help and worked-out solutions to odd-numbered exercises.

Vocabulary: Fill in the blanks.

1. An exponential growth model has the form _____, and an exponential decay model has the form _____.
2. A logarithmic model has the form _____ or _____.
3. In probability and statistics, Gaussian models commonly represent populations that are _____ _____.
4. A logistic growth model has the form _____.

Skills and Applications

Solving for a Variable In Exercises 5 and 6, (a) solve for P and (b) solve for t.

5. $A = Pe^{rt}$

6. $A = P\left(1 + \dfrac{r}{n}\right)^{nt}$

 Compound Interest In Exercises 7–12, find the missing values assuming continuously compounded interest.

Initial Investment	Annual % Rate	Time to Double	Amount After 10 Years
7. $1000	3.5%		
8. $750	$10\frac{1}{2}$%		
9. $750		$7\frac{3}{4}$ yr	
10. $500			$1505.00
11.	4.5%		$10,000.00
12.		12 yr	$2000.00

Compound Interest In Exercises 13 and 14, determine the principal P that must be invested at rate r, compounded monthly, so that $500,000 will be available for retirement in t years.

13. $r = 5\%, t = 10$

14. $r = 3\frac{1}{2}\%, t = 15$

Compound Interest In Exercises 15 and 16, determine the time necessary for P dollars to double when it is invested at interest rate r compounded (a) annually, (b) monthly, (c) daily, and (d) continuously.

15. $r = 10\%$

16. $r = 6.5\%$

17. **Compound Interest** Complete the table for the time t (in years) necessary for P dollars to triple when it is invested at an interest rate r compounded (a) continuously and (b) annually.

r	2%	4%	6%	8%	10%	12%
t						

18. **Modeling Data** Draw scatter plots of the data in Exercise 17. Use the *regression* feature of a graphing utility to find models for the data.

19. **Comparing Models** If $1 is invested over a 10-year period, then the balance A after t years is given by either $A = 1 + 0.075[\![t]\!]$ or $A = e^{0.07t}$ depending on whether the interest is simple interest at $7\frac{1}{2}\%$ or continuous compound interest at 7%. Graph each function on the same set of axes. Which grows at a greater rate? (Remember that $[\![t]\!]$ is the greatest integer function discussed in Section 1.6.)

20. **Comparing Models** If $1 is invested over a 10-year period, then the balance A after t years is given by either $A = 1 + 0.06[\![t]\!]$ or $A = [1 + (0.055/365)]^{[\![365t]\!]}$ depending on whether the interest is simple interest at 6% or compound interest at $5\frac{1}{2}\%$ compounded daily. Use a graphing utility to graph each function in the same viewing window. Which grows at a greater rate?

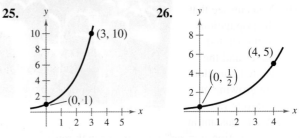 **Radioactive Decay** In Exercises 21–24, find the missing value for the radioactive isotope.

Isotope	Half-life (years)	Initial Quantity	Amount After 1000 Years
21. ^{226}Ra	1599	10 g	
22. ^{14}C	5715	6.5 g	
23. ^{14}C	5715		2 g
24. ^{239}Pu	24,100		0.4 g

Finding an Exponential Model In Exercises 25–28, find the exponential model that fits the points shown in the graph or table.

25.

26.

27.

x	0	4
y	5	1

28.

x	0	3
y	1	$\frac{1}{4}$

Geology In Exercises 45 and 46, use the Richter scale

$$R = \log \frac{I}{I_0}$$

for measuring the magnitude R of an earthquake.

45. Find the intensity I of an earthquake measuring R on the Richter scale (let $I_0 = 1$).

  (a) Peru in 2015: $R = 7.6$

  (b) Pakistan in 2015: $R = 5.6$

  (c) Indonesia in 2015: $R = 6.6$

46. Find the magnitude R of each earthquake of intensity I (let $I_0 = 1$).

  (a) $I = 199{,}500{,}000$

  (b) $I = 48{,}275{,}000$

  (c) $I = 17{,}000$

Intensity of Sound In Exercises 47–50, use the following information for determining sound intensity. The number of decibels β of a sound with an intensity of I watts per square meter is given by $\beta = 10 \log(I/I_0)$, where I_0 is an intensity of 10^{-12} watt per square meter, corresponding roughly to the faintest sound that can be heard by the human ear. In Exercises 47 and 48, find the number of decibels β of the sound.

47. (a) $I = 10^{-10}$ watt per m^2 (quiet room)

  (b) $I = 10^{-5}$ watt per m^2 (busy street corner)

  (c) $I = 10^{-8}$ watt per m^2 (quiet radio)

  (d) $I = 10^{-3}$ watt per m^2 (loud car horn)

48. (a) $I = 10^{-11}$ watt per m^2 (rustle of leaves)

  (b) $I = 10^2$ watt per m^2 (jet at 30 meters)

  (c) $I = 10^{-4}$ watt per m^2 (door slamming)

  (d) $I = 10^{-6}$ watt per m^2 (normal conversation)

49. Due to the installation of noise suppression materials, the noise level in an auditorium decreased from 93 to 80 decibels. Find the percent decrease in the intensity of the noise as a result of the installation of these materials.

50. Due to the installation of a muffler, the noise level of an engine decreased from 88 to 72 decibels. Find the percent decrease in the intensity of the noise as a result of the installation of the muffler.

pH Levels In Exercises 51–56, use the acidity model $\text{pH} = -\log[\text{H}^+]$, where acidity (pH) is a measure of the hydrogen ion concentration $[\text{H}^+]$ (measured in moles of hydrogen per liter) of a solution.

51. Find the pH when $[\text{H}^+] = 2.3 \times 10^{-5}$.

52. Find the pH when $[\text{H}^+] = 1.13 \times 10^{-5}$.

53. Compute $[\text{H}^+]$ for a solution in which pH $= 5.8$.

54. Compute $[\text{H}^+]$ for a solution in which pH $= 3.2$.

55. Apple juice has a pH of 2.9 and drinking water has a pH of 8.0. The hydrogen ion concentration of the apple juice is how many times the concentration of drinking water?

56. The pH of a solution decreases by one unit. By what factor does the hydrogen ion concentration increase?

57. **Forensics** At 8:30 A.M., a coroner went to the home of a person who had died during the night. In order to estimate the time of death, the coroner took the person's temperature twice. At 9:00 A.M. the temperature was 85.7°F, and at 11:00 A.M. the temperature was 82.8°F. From these two temperatures, the coroner was able to determine that the time elapsed since death and the body temperature were related by the formula

$$t = -10 \ln \frac{T - 70}{98.6 - 70}$$

where t is the time in hours elapsed since the person died and T is the temperature (in degrees Fahrenheit) of the person's body. (This formula comes from a general cooling principle called *Newton's Law of Cooling*. It uses the assumptions that the person had a normal body temperature of 98.6°F at death and that the room temperature was a constant 70°F.) Use the formula to estimate the time of death of the person.

58. **Home Mortgage** A $120,000 home mortgage for 30 years at $7\frac{1}{2}\%$ has a monthly payment of $839.06. Part of the monthly payment covers the interest charge on the unpaid balance, and the remainder of the payment reduces the principal. The amount paid toward the interest is

$$u = M - \left(M - \frac{Pr}{12}\right)\left(1 + \frac{r}{12}\right)^{12t}$$

and the amount paid toward the reduction of the principal is

$$v = \left(M - \frac{Pr}{12}\right)\left(1 + \frac{r}{12}\right)^{12t}.$$

In these formulas, P is the amount of the mortgage, r is the interest rate (in decimal form), M is the monthly payment, and t is the time in years.

  (a) Use a graphing utility to graph each function in the same viewing window. (The viewing window should show all 30 years of mortgage payments.)

  (b) In the early years of the mortgage, is the greater part of the monthly payment paid toward the interest or the principal? Approximate the time when the monthly payment is evenly divided between interest and principal reduction.

  (c) Repeat parts (a) and (b) for a repayment period of 20 years ($M = 966.71). What can you conclude?

59. Home Mortgage The total interest u paid on a home mortgage of P dollars at interest rate r (in decimal form) for t years is

$$u = P\left[\frac{rt}{1 - \left(\dfrac{1}{1 + r/12}\right)^{12t}} - 1\right].$$

Consider a $120,000 home mortgage at $7\frac{1}{2}\%$.

(a) Use a graphing utility to graph the total interest function.

(b) Approximate the length of the mortgage for which the total interest paid is the same as the size of the mortgage. Is it possible that some people are paying twice as much in interest charges as the size of the mortgage?

60. Car Speed The table shows the time t (in seconds) required for a car to attain a speed of s miles per hour from a standing start.

DATA	Speed, s	Time, t
	30	3.4
	40	5.0
	50	7.0
	60	9.3
	70	12.0
	80	15.8
	90	20.0

Spreadsheet at LarsonPrecalculus.com

Two models for these data are given below.

$t_1 = 40.757 + 0.556s - 15.817 \ln s$

$t_2 = 1.2259 + 0.0023s^2$

(a) Use the *regression* feature of a graphing utility to find a linear model t_3 and an exponential model t_4 for the data.

(b) Use the graphing utility to graph the data and each model in the same viewing window.

(c) Create a table comparing the data with estimates obtained from each model.

(d) Use the results of part (c) to find the sum of the absolute values of the differences between the data and the estimated values found using each model. Based on the four sums, which model do you think best fits the data? Explain.

Exploration

True or False? In Exercises 61–64, determine whether the statement is true or false. Justify your answer.

61. The domain of a logistic growth function cannot be the set of real numbers.

62. A logistic growth function will always have an x-intercept.

63. The graph of $f(x) = \dfrac{4}{1 + 6e^{-2x}} + 5$ is the graph of $g(x) = \dfrac{4}{1 + 6e^{-2x}}$ shifted to the right five units.

64. The graph of a Gaussian model will never have an x-intercept.

65. Writing Use your school's library, the Internet, or some other reference source to write a paper describing John Napier's work with logarithms.

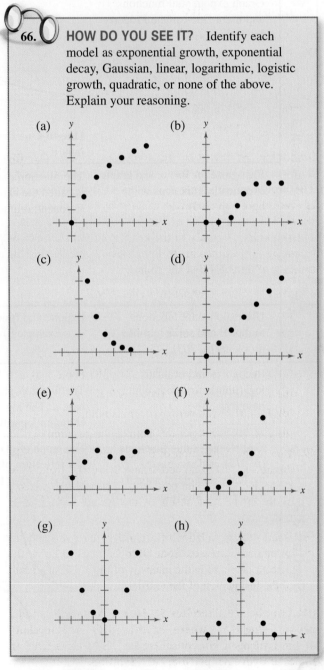

66. **HOW DO YOU SEE IT?** Identify each model as exponential growth, exponential decay, Gaussian, linear, logarithmic, logistic growth, quadratic, or none of the above. Explain your reasoning.

(a) (b) (c) (d) (e) (f) (g) (h)

Project: Sales per Share To work an extended application analyzing the sales per share for Kohl's Corporation from 1999 through 2014, visit this text's website at *LarsonPrecalculus.com*. *(Source: Kohl's Corporation)*

Chapter Summary

What Did You Learn?	Explanation/Examples	Review Exercises
Section 3.1 Recognize and evaluate exponential functions with base a *(p. 198)*.	The exponential function f with base a is denoted by $f(x) = a^x$, where $a > 0$, $a \neq 1$, and x is any real number.	1–6
Graph exponential functions and use a One-to-One Property *(p. 199)*.	**One-to-One Property:** For $a > 0$ and $a \neq 1$, $a^x = a^y$ if and only if $x = y$.	7–20
Recognize, evaluate, and graph exponential functions with base e *(p. 202)*.	The function $f(x) = e^x$ is called the natural exponential function.	21–28
Use exponential functions to model and solve real-life problems *(p. 203)*.	Exponential functions are used in compound interest formulas (see Example 8) and in radioactive decay models (see Example 9).	29–32
Section 3.2 Recognize and evaluate logarithmic functions with base a *(p. 209)*.	For $x > 0$, $a > 0$, and $a \neq 1$, $y = \log_a x$ if and only if $x = a^y$. The function $f(x) = \log_a x$ is the logarithmic function with base a. The logarithmic function with base 10 is called the common logarithmic function. It is denoted by $\log_{10}$ or $\log$.	33–44
Graph logarithmic functions *(p. 211)*, and recognize, evaluate, and graph natural logarithmic functions *(p. 213)*.	The graph of $g(x) = \log_a x$ is a reflection of the graph of $f(x) = a^x$ in the line $y = x$. The function $g(x) = \ln x$, $x > 0$, is called the natural logarithmic function. Its graph is a reflection of the graph of $f(x) = e^x$ in the line $y = x$.	45–56
Use logarithmic functions to model and solve real-life problems *(p. 215)*.	A logarithmic function can model human memory. (See Example 11.)	57, 58

	What Did You Learn?	**Explanation/Examples**	**Review Exercises**
Section 3.3	Use the change-of-base formula to rewrite and evaluate logarithmic expressions *(p. 219)*.	Let a, b, and x be positive real numbers such that $a \neq 1$ and $b \neq 1$. Then $\log_a x$ can be converted to a different base as follows. **Base b** $\log_a x = \dfrac{\log_b x}{\log_b a}$ **Base 10** $\log_a x = \dfrac{\log x}{\log a}$ **Base e** $\log_a x = \dfrac{\ln x}{\ln a}$	59–62

	Use properties of logarithms to evaluate, rewrite, expand, or condense logarithmic expressions *(pp. 220–221)*.	Let a be a positive number such that $a \neq 1$, let n be a real number, and let u and v be positive real numbers.
		
		1. Product Property: $\log_a(uv) = \log_a u + \log_a v$
		    $\ln(uv) = \ln u + \ln v$
		
		2. Quotient Property: $\log_a(u/v) = \log_a u - \log_a v$
		    $\ln(u/v) = \ln u - \ln v$
		
		3. Power Property: $\log_a u^n = n \log_a u,\ \ln u^n = n \ln u$
			63–78

	Use logarithmic functions to model and solve real-life problems *(p. 222)*.	Logarithmic functions can help you find an equation that relates the periods of several planets and their distances from the sun. (See Example 7.)	79, 80
Section 3.4	Solve simple exponential and logarithmic equations *(p. 226)*.	One-to-One Properties and Inverse Properties of exponential or logarithmic functions are used to solve exponential or logarithmic equations.	81–86
	Solve more complicated exponential equations *(p. 227)* and logarithmic equations *(p. 229)*.	To solve more complicated equations, rewrite the equations to allow the use of the One-to-One Properties or Inverse Properties of exponential or logarithmic functions. (See Examples 2–9.)	87–102
	Use exponential and logarithmic equations to model and solve real-life problems *(p. 231)*.	Exponential and logarithmic equations can help you determine how long it will take to double an investment (see Example 10) and find the year in which an industry had a given amount of sales (see Example 11).	103, 104
Section 3.5	Recognize the five most common types of models involving exponential and logarithmic functions *(p. 236)*.	**1. Exponential growth model:** $y = ae^{bx}$, $b > 0$ **2. Exponential decay model:** $y = ae^{-bx}$, $b > 0$ **3. Gaussian model:** $y = ae^{-(x-b)^2/c}$ **4. Logistic growth model:** $y = \dfrac{a}{1 + be^{-rx}}$ **5. Logarithmic models:** $y = a + b \ln x$, $y = a + b \log x$	105–110
	Use exponential growth and decay functions to model and solve real-life problems *(p. 237)*.	An exponential growth function can help you model a population of fruit flies (see Example 2), and an exponential decay function can help you estimate the age of a fossil (see Example 3).	111, 112
	Use Gaussian functions *(p. 240)*, logistic growth functions *(p. 241)*, and logarithmic functions *(p. 242)* to model and solve real-life problems.	A Gaussian function can help you model SAT mathematics scores for college-bound seniors. (See Example 4.) A logistic growth function can help you model the spread of a flu virus. (See Example 5.) A logarithmic function can help you find the intensity of an earthquake given its magnitude. (See Example 6.)	113–115

Review Exercises

See CalcChat.com for tutorial help and worked-out solutions to odd-numbered exercises.

3.1 **Evaluating an Exponential Function** In Exercises 1–6, evaluate the function at the given value of x. Round your result to three decimal places.

1. $f(x) = 0.3^x, \quad x = 1.5$ **2.** $f(x) = 30^x, \quad x = \sqrt{3}$

3. $f(x) = 2^x, \quad x = \frac{2}{3}$ **4.** $f(x) = \left(\frac{1}{2}\right)^{2x}, \quad x = \pi$

5. $f(x) = 7(0.2^x), \quad x = -\sqrt{11}$

6. $f(x) = -14(5^x), \quad x = -0.8$

Graphing an Exponential Function In Exercises 7–12, use a graphing utility to construct a table of values for the function. Then sketch the graph of the function.

7. $f(x) = 4^{-x} + 4$ **8.** $f(x) = 2.65^{x-1}$

9. $f(x) = 5^{x-2} + 4$ **10.** $f(x) = 2^{x-6} - 5$

11. $f(x) = \left(\frac{1}{2}\right)^{-x} + 3$ **12.** $f(x) = \left(\frac{1}{8}\right)^{x+2} - 5$

Using a One-to-One Property In Exercises 13–16, use a One-to-One Property to solve the equation for x.

13. $\left(\frac{1}{3}\right)^{x-3} = 9$ **14.** $3^{x+3} = \frac{1}{81}$

15. $e^{3x-5} = e^7$ **16.** $e^{8-2x} = e^{-3}$

Transforming the Graph of an Exponential Function In Exercises 17–20, describe the transformation of the graph of f that yields the graph of g.

17. $f(x) = 5^x, \quad g(x) = 5^x + 1$

18. $f(x) = 6^x, \quad g(x) = 6^{x+1}$

19. $f(x) = 3^x, \quad g(x) = 1 - 3^x$

20. $f(x) = \left(\frac{1}{2}\right)^x, \quad g(x) = -\left(\frac{1}{2}\right)^{x+2}$

Evaluating the Natural Exponential Function In Exercises 21–24, evaluate $f(x) = e^x$ at the given value of x. Round your result to three decimal places.

21. $x = 3.4$ **22.** $x = -2.5$

23. $x = \frac{3}{5}$ **24.** $x = \frac{2}{7}$

Graphing a Natural Exponential Function In Exercises 25–28, use a graphing utility to construct a table of values for the function. Then sketch the graph of the function.

25. $h(x) = e^{-x/2}$ **26.** $h(x) = 2 - e^{-x/2}$

27. $f(x) = e^{x+2}$ **28.** $s(t) = 4e^{t-1}$

29. Waiting Times The average time between new posts on a message board is 3 minutes. The probability F of waiting less than t minutes until the next post is approximated by the model $F(t) = 1 - e^{-t/3}$. A message has just been posted. Find the probability that the next post will be within (a) 1 minute, (b) 2 minutes, and (c) 5 minutes.

30. Depreciation After t years, the value V of a car that originally cost $23,970 is given by $V(t) = 23,970\left(\frac{3}{4}\right)^t$.

(a) Use a graphing utility to graph the function.

(b) Find the value of the car 2 years after it was purchased.

(c) According to the model, when does the car depreciate most rapidly? Is this realistic? Explain.

(d) According to the model, when will the car have no value?

Compound Interest In Exercises 31 and 32, complete the table by finding the balance A when P dollars is invested at rate r for t years and compounded n times per year.

n	1	2	4	12	365	Continuous
A						

31. $P = \$5000, \quad r = 3\%, \quad t = 10$ years

32. $P = \$4500, \quad r = 2.5\%, \quad t = 30$ years

3.2 **Writing a Logarithmic Equation** In Exercises 33–36, write the exponential equation in logarithmic form. For example, the logarithmic form of $2^3 = 8$ is $\log_2 8 = 3$.

33. $3^3 = 27$ **34.** $25^{3/2} = 125$

35. $e^{0.8} = 2.2255\ldots$ **36.** $e^0 = 1$

Evaluating a Logarithm In Exercises 37–40, evaluate the logarithm at the given value of x without using a calculator.

37. $f(x) = \log x, \quad x = 1000$ **38.** $g(x) = \log_9 x, \quad x = 3$

39. $g(x) = \log_2 x, \quad x = \frac{1}{4}$ **40.** $f(x) = \log_3 x, \quad x = \frac{1}{81}$

Using a One-to-One Property In Exercises 41–44, use a One-to-One Property to solve the equation for x.

41. $\log_4(x + 7) = \log_4 14$ **42.** $\log_8(3x - 10) = \log_8 5$

43. $\ln(x + 9) = \ln 4$ **44.** $\log(3x - 2) = \log 7$

Sketching the Graph of a Logarithmic Function In Exercises 45–48, find the domain, x-intercept, and vertical asymptote of the logarithmic function and sketch its graph.

45. $g(x) = \log_7 x$

46. $f(x) = \log \dfrac{x}{3}$

47. $f(x) = 4 - \log(x + 5)$

48. $f(x) = \log(x - 3) + 1$

Evaluating a Logarithmic Function In Exercises 49–52, use a calculator to evaluate the function at the given value of x. Round your result to three decimal places, if necessary.

49. $f(x) = \ln x, \ x = 22.6$ 50. $f(x) = \ln x, \ x = e^{-12}$

51. $f(x) = \frac{1}{2} \ln x, \ x = \sqrt{e}$

52. $f(x) = 5 \ln x, \ x = 0.98$

Graphing a Natural Logarithmic Function In Exercises 53–56, find the domain, x-intercept, and vertical asymptote of the logarithmic function and sketch its graph.

53. $f(x) = \ln x + 6$ 54. $f(x) = \ln x - 5$

55. $h(x) = \ln(x - 6)$ 56. $f(x) = \ln(x + 4)$

57. **Astronomy** The formula $M = m - 5 \log(d/10)$ gives the distance d (in parsecs) from Earth to a star with apparent magnitude m and absolute magnitude M. The star Rasalhague has an apparent magnitude of 2.08 and an absolute magnitude of 1.3. Find the distance from Earth to Rasalhague.

58. **Snow Removal** The number of miles s of roads cleared of snow is approximated by the model

$$s = 25 - \frac{13 \ln(h/12)}{\ln 3}, \quad 2 \leq h \leq 15$$

where h is the depth (in inches) of the snow. Use this model to find s when $h = 10$ inches.

3.3 **Using the Change-of-Base Formula** In Exercises 59–62, evaluate the logarithm using the change-of-base formula (a) with common logarithms and (b) with natural logarithms. Round your results to three decimal places.

59. $\log_2 6$ 60. $\log_{12} 200$

61. $\log_{1/2} 5$ 62. $\log_4 0.75$

Using Properties of Logarithms In Exercises 63–66, use the properties of logarithms to write the logarithm in terms of $\log_2 3$ and $\log_2 5$.

63. $\log_2 \frac{5}{3}$ 64. $\log_2 45$

65. $\log_2 \frac{9}{5}$ 66. $\log_2 \frac{20}{9}$

Expanding a Logarithmic Expression In Exercises 67–72, use the properties of logarithms to expand the expression as a sum, difference, and/or constant multiple of logarithms. (Assume all variables are positive.)

67. $\log 7x^2$ 68. $\log 11x^3$

69. $\log_3 \dfrac{9}{\sqrt{x}}$ 70. $\log_7 \dfrac{\sqrt[3]{x}}{19}$

71. $\ln x^2 y^2 z$ 72. $\ln\left(\dfrac{y-1}{3}\right)^2, \quad y > 1$

Condensing a Logarithmic Expression In Exercises 73–78, condense the expression to the logarithm of a single quantity.

73. $\ln 7 + \ln x$

74. $\log_2 y - \log_2 3$

75. $\log x - \frac{1}{2} \log y$

76. $3 \ln x + 2 \ln(x + 1)$

77. $\frac{1}{2} \log_3 x - 2 \log_3(y + 8)$

78. $5 \ln(x - 2) - \ln(x + 2) - 3 \ln x$

79. **Climb Rate** The time t (in minutes) for a small plane to climb to an altitude of h feet is modeled by

$$t = 50 \log[18{,}000/(18{,}000 - h)]$$

where 18,000 feet is the plane's absolute ceiling.

   (a) Determine the domain of the function in the context of the problem.

   (b) Use a graphing utility to graph the function and identify any asymptotes.

   (c) As the plane approaches its absolute ceiling, what can be said about the time required to increase its altitude?

   (d) Find the time it takes for the plane to climb to an altitude of 4000 feet.

80. **Human Memory Model** Students in a learning theory study took an exam and then retested monthly for 6 months with an equivalent exam. The data obtained in the study are given by the ordered pairs (t, s), where t is the time (in months) after the initial exam and s is the average score for the class. Use the data to find a logarithmic equation that relates t and s.

   $(1, 84.2), (2, 78.4), (3, 72.1),$
   $(4, 68.5), (5, 67.1), (6, 65.3)$

3.4 **Solving a Simple Equation** In Exercises 81–86, solve for x.

81. $5^x = 125$

82. $6^x = \frac{1}{216}$

83. $e^x = 3$

84. $\log x - \log 5 = 0$

85. $\ln x = 4$

86. $\ln x = -1.6$

Solving an Exponential Equation In Exercises 87–90, solve the exponential equation algebraically. Approximate the result to three decimal places.

87. $e^{4x} = e^{x^2+3}$

88. $e^{3x} = 25$

89. $2^x - 3 = 29$

90. $e^{2x} - 6e^x + 8 = 0$

Solving a Logarithmic Equation In Exercises 91–98, solve the logarithmic equation algebraically. Approximate the result to three decimal places.

91. $\ln 3x = 8.2$ **92.** $4 \ln 3x = 15$

93. $\ln x + \ln(x - 3) = 1$

94. $\ln(x + 2) - \ln x = 2$

95. $\log_8(x - 1) = \log_8(x - 2) - \log_8(x + 2)$

96. $\log_6(x + 2) - \log_6 x = \log_6(x + 5)$

97. $\log(1 - x) = -1$

98. $\log(-x - 4) = 2$

Using Technology In Exercises 99–102, use a graphing utility to graphically solve the equation. Approximate the result to three decimal places. Verify your result algebraically.

99. $25e^{-0.3x} = 12$

100. $2 = 5 - e^{x+7}$

101. $2 \ln(x + 3) - 3 = 0$

102. $2 \ln x - \ln(3x - 1) = 0$

103. Compound Interest You deposit $8500 in an account that pays 1.5% interest, compounded continuously. How long will it take for the money to triple?

104. Meteorology The speed of the wind S (in miles per hour) near the center of a tornado and the distance d (in miles) the tornado travels are related by the model $S = 93 \log d + 65$. On March 18, 1925, a large tornado struck portions of Missouri, Illinois, and Indiana with a wind speed at the center of about 283 miles per hour. Approximate the distance traveled by this tornado.

3.5 Matching a Function with Its Graph In Exercises 105–110, match the function with its graph. [The graphs are labeled (a), (b), (c), (d), (e), and (f).]

(a)

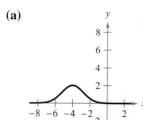

(b)

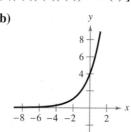

(c)

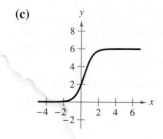

(d)

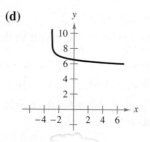

(e)

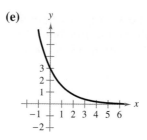

(f)

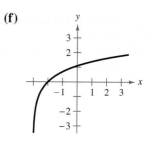

105. $y = 3e^{-2x/3}$ **106.** $y = 4e^{2x/3}$

107. $y = \ln(x + 3)$ **108.** $y = 7 - \log(x + 3)$

109. $y = 2e^{-(x+4)^2/3}$ **110.** $y = \dfrac{6}{1 + 2e^{-2x}}$

111. Finding an Exponential Model Find the exponential model $y = ae^{bx}$ that fits the points $(0, 2)$ and $(4, 3)$.

112. Wildlife Population A species of bat is in danger of becoming extinct. Five years ago, the total population of the species was 2000. Two years ago, the total population of the species was 1400. What was the total population of the species one year ago?

113. Test Scores The test scores for a biology test follow the normal distribution

$$y = 0.0499e^{-(x - 71)^2/128}, \quad 40 \le x \le 100$$

where x is the test score. Use a graphing utility to graph the equation and estimate the average test score.

114. Typing Speed In a typing class, the average number N of words per minute typed after t weeks of lessons is

$$N = 157/(1 + 5.4e^{-0.12t}).$$

Find the time necessary to type (a) 50 words per minute and (b) 75 words per minute.

115. Sound Intensity The relationship between the number of decibels β and the intensity of a sound I (in watts per square meter) is

$$\beta = 10 \log(I/10^{-12}).$$

Find the intensity I for each decibel level β.
(a) $\beta = 60$ (b) $\beta = 135$ (c) $\beta = 1$

Exploration

116. Graph of an Exponential Function Consider the graph of $y = e^{kt}$. Describe the characteristics of the graph when k is positive and when k is negative.

True or False? In Exercises 117 and 118, determine whether the equation is true or false. Justify your answer.

117. $\log_b b^{2x} = 2x$

118. $\ln(x + y) = \ln x + \ln y$

Chapter Test

See CalcChat.com for tutorial help and worked-out solutions to odd-numbered exercises.

Take this test as you would take a test in class. When you are finished, check your work against the answers given in the back of the book.

In Exercises 1–4, evaluate the expression. Round your result to three decimal places.

1. $0.7^{2.5}$ **2.** $3^{-\pi}$ **3.** $e^{-7/10}$ **4.** $e^{3.1}$

In Exercises 5–7, use a graphing utility to construct a table of values for the function. Then sketch the graph of the function.

5. $f(x) = 10^{-x}$ **6.** $f(x) = -6^{x-2}$ **7.** $f(x) = 1 - e^{2x}$

8. Evaluate (a) $\log_7 7^{-0.89}$ and (b) $4.6 \ln e^2$.

In Exercises 9–11, find the domain, x-intercept, and vertical asymptote of the logarithmic function and sketch its graph.

9. $f(x) = 4 + \log x$ **10.** $f(x) = \ln(x - 4)$ **11.** $f(x) = 1 + \ln(x + 6)$

In Exercises 12–14, evaluate the logarithm using the change-of-base formula. Round your result to three decimal places.

12. $\log_5 35$ **13.** $\log_{16} 0.63$ **14.** $\log_{3/4} 24$

In Exercises 15–17, use the properties of logarithms to expand the expression as a sum, difference, and/or constant multiple of logarithms. (Assume all variables are positive.)

15. $\log_2 3a^4$ **16.** $\ln \dfrac{\sqrt{x}}{7}$ **17.** $\log \dfrac{10x^2}{y^3}$

In Exercises 18–20, condense the expression to the logarithm of a single quantity.

18. $\log_3 13 + \log_3 y$ **19.** $4 \ln x - 4 \ln y$

20. $3 \ln x - \ln(x + 3) + 2 \ln y$

In Exercises 21–26, solve the equation algebraically. Approximate the result to three decimal places, if necessary.

21. $5^x = \dfrac{1}{25}$ **22.** $3e^{-5x} = 132$

23. $\dfrac{1025}{8 + e^{4x}} = 5$ **24.** $\ln x = \dfrac{1}{2}$

25. $18 + 4 \ln x = 7$ **26.** $\log x + \log(x - 15) = 2$

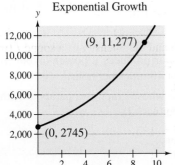

Exponential Growth

Figure for 27

27. Find the exponential growth model that fits the points shown in the graph.

28. The half-life of radioactive actinium (^{227}Ac) is 21.77 years. What percent of a present amount of radioactive actinium will remain after 19 years?

29. A model that can predict a child's height H (in centimeters) based on the child's age is $H = 70.228 + 5.104x + 9.222 \ln x$, $\frac{1}{4} \le x \le 6$, where x is the child's age in years.   (*Source: Snapshots of Applications in Mathematics*)

(a) Construct a table of values for the model. Then sketch the graph of the model.

(b) Use the graph from part (a) to predict the height of a four-year-old child. Then confirm your prediction algebraically.

Cumulative Test for Chapters 1–3

See CalcChat.com for tutorial help and worked-out solutions to odd-numbered exercises.

Take this test as you would take a test in class. When you are finished, check your work against the answers given in the back of the book.

1. Plot the points $(-2, 5)$ and $(3, -1)$. Find the midpoint of the line segment joining the points and the distance between the points.

In Exercises 2–4, sketch the graph of the equation.

2. $x - 3y + 12 = 0$ **3.** $y = x^2 - 9$ **4.** $y = \sqrt{4 - x}$

5. Find the slope-intercept form of the equation of the line passing through $\left(-\frac{1}{2}, 1\right)$ and $(3, 8)$.

6. Explain why the graph at the left does not represent y as a function of x.

7. Let $f(x) = \dfrac{x}{x - 2}$. Find each function value, if possible.

    (a) $f(6)$ (b) $f(2)$ (c) $f(s + 2)$

8. Compare the graph of each function with the graph of $y = \sqrt[3]{x}$. (*Note:* It is not necessary to sketch the graphs.)

    (a) $r(x) = \frac{1}{2}\sqrt[3]{x}$ (b) $h(x) = \sqrt[3]{x} + 2$ (c) $g(x) = \sqrt[3]{x + 2}$

In Exercises 9 and 10, find (a) $(f + g)(x)$, (b) $(f - g)(x)$, (c) $(fg)(x)$, and (d) $(f/g)(x)$. What is the domain of f/g?

9. $f(x) = x - 4$, $g(x) = 3x + 1$

10. $f(x) = \sqrt{x - 1}$, $g(x) = x^2 + 1$

In Exercises 11 and 12, find (a) $f \circ g$ and (b) $g \circ f$. Find the domain of each composite function.

11. $f(x) = 2x^2$, $g(x) = \sqrt{x + 6}$

12. $f(x) = x - 2$, $g(x) = |x|$

13. Determine whether $h(x) = 3x - 4$ has an inverse function. If is does, find the inverse function.

14. The power P produced by a wind turbine varies directly as the cube of the wind speed S. A wind speed of 27 miles per hour produces a power output of 750 kilowatts. Find the output for a wind speed of 40 miles per hour.

15. Write the standard form of the quadratic function whose graph is a parabola with vertex $(-8, 5)$ and that passes through the point $(-4, -7)$.

In Exercises 16–18, sketch the graph of the function.

16. $h(x) = -x^2 + 10x - 21$

17. $f(t) = -\dfrac{1}{2}(t - 1)^2(t + 2)^2$

18. $g(s) = s^3 - 3s^2$

In Exercises 19–21, find all the zeros of the function.

19. $f(x) = x^3 + 2x^2 + 4x + 8$

20. $f(x) = x^4 + 4x^3 - 21x^2$

21. $f(x) = 2x^4 - 11x^3 + 30x^2 - 62x - 40$

Figure for 6

22. Use long division to divide: $\dfrac{6x^3 - 4x^2}{2x^2 + 1}$.

23. Use synthetic division to divide $3x^4 + 2x^2 - 5x + 3$ by $x - 2$.

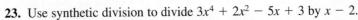 **24.** Use the Intermediate Value Theorem and the *table* feature of a graphing utility to find an interval one unit in length in which the function $g(x) = x^3 + 3x^2 - 6$ is guaranteed to have a zero. Then adjust the table to approximate the real zero to the nearest thousandth.

In Exercises 25–27, sketch the graph of the rational function. Identify all intercepts and find any asymptotes.

25. $f(x) = \dfrac{2x}{x^2 + 2x - 3}$

26. $f(x) = \dfrac{x^2 - 4}{x^2 + x - 2}$

27. $f(x) = \dfrac{x^3 - 2x^2 - 9x + 18}{x^2 + 4x + 3}$

In Exercises 28 and 29, solve the inequality. Then graph the solution set.

28. $2x^3 - 18x \le 0$

29. $\dfrac{1}{x + 1} \ge \dfrac{1}{x + 5}$

In Exercises 30 and 31, describe the transformations of the graph of f that yield the graph of g.

30. $f(x) = \left(\frac{2}{5}\right)^x, \quad g(x) = -\left(\frac{2}{5}\right)^{-x+3}$

31. $f(x) = 2.2^x, \quad g(x) = -2.2^x + 4$

In Exercises 32–35, use a calculator to evaluate the expression. Round your result to three decimal places.

32. $\log 98$

33. $\log \frac{6}{7}$

34. $\ln \sqrt{31}$

35. $\ln\left(\sqrt{30} - 4\right)$

36. Use the properties of logarithms to expand $\ln\left(\dfrac{x^2 - 25}{x^4}\right)$, where $x > 5$.

37. Condense $2 \ln x - \frac{1}{2} \ln(x + 5)$ to the logarithm of a single quantity.

In Exercises 38–40, solve the equation algebraically. Approximate the result to three decimal places.

38. $6e^{2x} = 72$

39. $e^{2x} - 13e^x + 42 = 0$

40. $\ln \sqrt{x + 2} = 3$

41. On the day a grandchild is born, a grandparent deposits $2500 in a fund earning 7.5% interest, compounded continuously. Determine the balance in the account on the grandchild's 25th birthday.

42. The number N of bacteria in a culture is given by the model $N = 175e^{kt}$, where t is the time in hours. If $N = 420$ when $t = 8$, then estimate the time required for the population to double in size.

43. The population P (in millions) of Texas from 2001 through 2014 can be approximated by the model $P = 20.913e^{0.0184t}$, where t represents the year, with $t = 1$ corresponding to 2001. According to this model, when will the population reach 32 million? *(Source: U.S. Census Bureau)*

Proofs in Mathematics ■ ■ ■ ■ ■ ■ ■ ■ ■ ■ ■ ■ ■ ■ ■

Each of the three properties of logarithms listed below can be proved by using properties of exponential functions.

SLIDE RULES

William Oughtred (1574–1660) is credited with inventing the slide rule. The slide rule is a computational device with a sliding portion and a fixed portion. A slide rule enables you to perform multiplication by using the Product Property of Logarithms. There are other slide rules that allow for the calculation of roots and trigonometric functions. Mathematicians and engineers used slide rules until the hand-held calculator came into widespread use in the 1970s.

Properties of Logarithms (p. 220)

Let a be a positive number such that $a \neq 1$, let n be a real number, and let u and v be positive real numbers.

	Logarithm with Base a	Natural Logarithm
1. Product Property:	$\log_a(uv) = \log_a u + \log_a v$	$\ln(uv) = \ln u + \ln v$
2. Quotient Property:	$\log_a \dfrac{u}{v} = \log_a u - \log_a v$	$\ln \dfrac{u}{v} = \ln u - \ln v$
3. Power Property:	$\log_a u^n = n \log_a u$	$\ln u^n = n \ln u$

Proof

Let

$$x = \log_a u \quad \text{and} \quad y = \log_a v.$$

The corresponding exponential forms of these two equations are

$$a^x = u \quad \text{and} \quad a^y = v.$$

To prove the Product Property, multiply u and v to obtain

$$uv = a^x a^y$$
$$= a^{x+y}.$$

The corresponding logarithmic form of $uv = a^{x+y}$ is $\log_a(uv) = x + y$. So,

$$\log_a(uv) = \log_a u + \log_a v.$$

To prove the Quotient Property, divide u by v to obtain

$$\frac{u}{v} = \frac{a^x}{a^y}$$
$$= a^{x-y}.$$

The corresponding logarithmic form of $\dfrac{u}{v} = a^{x-y}$ is $\log_a \dfrac{u}{v} = x - y$. So,

$$\log_a \frac{u}{v} = \log_a u - \log_a v.$$

To prove the Power Property, substitute a^x for u in the expression $\log_a u^n$.

$\log_a u^n = \log_a(a^x)^n$	Substitute a^x for u.
$= \log_a a^{nx}$	Property of Exponents
$= nx$	Inverse Property
$= n \log_a u$	Substitute $\log_a u$ for x.

So, $\log_a u^n = n \log_a u$. ■

P.S. Problem Solving ▪ ▪ ▪ ▪ ▪ ▪ ▪ ▪ ▪ ▪ ▪ ▪ ▪

1. **Graphical Reasoning** Graph the exponential function $y = a^x$ for $a = 0.5$, 1.2, and 2.0. Which of these curves intersects the line $y = x$? Determine all positive numbers a for which the curve $y = a^x$ intersects the line $y = x$.

2. **Graphical Reasoning** Use a graphing utility to graph each of the functions $y_1 = e^x$, $y_2 = x^2$, $y_3 = x^3$, $y_4 = \sqrt{x}$, and $y_5 = |x|$. Which function increases at the greatest rate as x approaches ∞?

3. **Conjecture** Use the result of Exercise 2 to make a conjecture about the rate of growth of $y_1 = e^x$ and $y = x^n$, where n is a natural number and x approaches ∞.

4. **Implication of "Growing Exponentially"** Use the results of Exercises 2 and 3 to describe what is implied when it is stated that a quantity is growing exponentially.

5. **Exponential Function** Given the exponential function

   $$f(x) = a^x$$

   show that

   (a) $f(u + v) = f(u) \cdot f(v)$ and (b) $f(2x) = [f(x)]^2$.

6. **Hyperbolic Functions** Given that

   $$f(x) = \frac{e^x + e^{-x}}{2} \quad \text{and} \quad g(x) = \frac{e^x - e^{-x}}{2}$$

   show that

   $$[f(x)]^2 - [g(x)]^2 = 1.$$

7. **Graphical Reasoning** Use a graphing utility to compare the graph of the function $y = e^x$ with the graph of each function. [$n!$ (read "n factorial") is defined as $n! = 1 \cdot 2 \cdot 3 \cdots (n - 1) \cdot n$.]

   (a) $y_1 = 1 + \dfrac{x}{1!}$

   (b) $y_2 = 1 + \dfrac{x}{1!} + \dfrac{x^2}{2!}$

   (c) $y_3 = 1 + \dfrac{x}{1!} + \dfrac{x^2}{2!} + \dfrac{x^3}{3!}$

8. **Identifying a Pattern** Identify the pattern of successive polynomials given in Exercise 7. Extend the pattern one more term and compare the graph of the resulting polynomial function with the graph of $y = e^x$. What do you think this pattern implies?

9. **Finding an Inverse Function** Graph the function

   $$f(x) = e^x - e^{-x}.$$

   From the graph, the function appears to be one-to-one. Assume that f has an inverse function and find $f^{-1}(x)$.

10. **Finding a Pattern for an Inverse Function** Find a pattern for $f^{-1}(x)$ when

    $$f(x) = \frac{a^x + 1}{a^x - 1}$$

    where $a > 0$, $a \neq 1$.

11. **Determining the Equation of a Graph** Determine whether the graph represents equation (a), (b), or (c). Explain your reasoning.

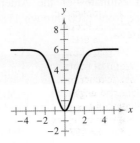

    (a) $y = 6e^{-x^2/2}$

    (b) $y = \dfrac{6}{1 + e^{-x/2}}$

    (c) $y = 6(1 - e^{-x^2/2})$

12. **Simple and Compound Interest** You have two options for investing $500. The first earns 7% interest compounded annually, and the second earns 7% simple interest. The figure shows the growth of each investment over a 30-year period.

    (a) Determine which graph represents each type of investment. Explain your reasoning.

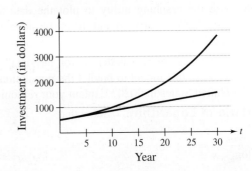

    (b) Verify your answer in part (a) by finding the equations that model the investment growth and by graphing the models.

    (c) Which option would you choose? Explain.

13. **Radioactive Decay** Two different samples of radioactive isotopes are decaying. The isotopes have initial amounts of c_1 and c_2 and half-lives of k_1 and k_2, respectively. Find an expression for the time t required for the samples to decay to equal amounts.

14. Bacteria Decay A lab culture initially contains 500 bacteria. Two hours later, the number of bacteria decreases to 200. Find the exponential decay model of the form

$$B = B_0 a^{kt}$$

that approximates the number of bacteria B in the culture after t hours.

15. Colonial Population The table shows the colonial population estimates of the American colonies for each decade from 1700 through 1780. *(Source: U.S. Census Bureau)*

Year	Population
1700	250,900
1710	331,700
1720	466,200
1730	629,400
1740	905,600
1750	1,170,800
1760	1,593,600
1770	2,148,100
1780	2,780,400

DATA — Spreadsheet at LarsonPrecalculus.com

Let y represent the population in the year t, with $t = 0$ corresponding to 1700.

(a) Use the *regression* feature of a graphing utility to find an exponential model for the data.

(b) Use the *regression* feature of the graphing utility to find a quadratic model for the data.

(c) Use the graphing utility to plot the data and the models from parts (a) and (b) in the same viewing window.

(d) Which model is a better fit for the data? Would you use this model to predict the population of the United States in 2020? Explain your reasoning.

16. Ratio of Logarithms Show that

$$\frac{\log_a x}{\log_{a/b} x} = 1 + \log_a \frac{1}{b}.$$

17. Solving a Logarithmic Equation Solve

$(\ln x)^2 = \ln x^2.$

18. Graphical Reasoning Use a graphing utility to compare the graph of each function with the graph of $y = \ln x$.

(a) $y_1 = x - 1$

(b) $y_2 = (x - 1) - \frac{1}{2}(x - 1)^2$

(c) $y_3 = (x - 1) - \frac{1}{2}(x - 1)^2 + \frac{1}{3}(x - 1)^3$

19. Identifying a Pattern Identify the pattern of successive polynomials given in Exercise 18. Extend the pattern one more term and compare the graph of the resulting polynomial function with the graph of $y = \ln x$. What do you think the pattern implies?

20. Finding Slope and y-Intercept Take the natural log of each side of each equation below.

$$y = ab^x, \quad y = ax^b$$

(a) What are the slope and y-intercept of the line relating x and $\ln y$ for $y = ab^x$?

(b) What are the slope and y-intercept of the line relating $\ln x$ and $\ln y$ for $y = ax^b$?

Ventilation Rate In Exercises 21 and 22, use the model

$$y = 80.4 - 11 \ln x, \quad 100 \le x \le 1500$$

which approximates the minimum required ventilation rate in terms of the air space per child in a public school classroom. In the model, x is the air space (in cubic feet) per child and y is the ventilation rate (in cubic feet per minute) per child.

21. Use a graphing utility to graph the model and approximate the required ventilation rate when there are 300 cubic feet of air space per child.

22. In a classroom designed for 30 students, the air conditioning system can move 450 cubic feet of air per minute.

(a) Determine the ventilation rate per child in a full classroom.

(b) Estimate the air space required per child.

(c) Determine the minimum number of square feet of floor space required for the room when the ceiling height is 30 feet.

Using Technology In Exercises 23–26, (a) use a graphing utility to create a scatter plot of the data, (b) decide whether the data could best be modeled by a linear model, an exponential model, or a logarithmic model, (c) explain why you chose the model you did in part (b), (d) use the *regression* feature of the graphing utility to find the model you chose in part (b) for the data and graph the model with the scatter plot, and (e) determine how well the model you chose fits the data.

23. $(1, 2.0), (1.5, 3.5), (2, 4.0), (4, 5.8), (6, 7.0), (8, 7.8)$

24. $(1, 4.4), (1.5, 4.7), (2, 5.5), (4, 9.9), (6, 18.1), (8, 33.0)$

25. $(1, 7.5), (1.5, 7.0), (2, 6.8), (4, 5.0), (6, 3.5), (8, 2.0)$

26. $(1, 5.0), (1.5, 6.0), (2, 6.4), (4, 7.8), (6, 8.6), (8, 9.0)$

4 Trigonometry

Television Coverage *(Exercise 85, page 317)*

Waterslide Design
(Exercise 30, page 335)

Respiratory Cycle *(Exercise 80, page 306)*

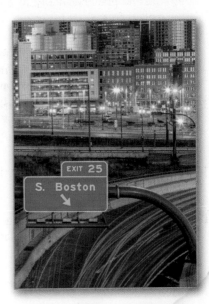

Temperature of a City
(Exercise 99, page 296)

Skateboard Ramp *(Example 10, page 283)*

259

4.1 Radian and Degree Measure

- ■ Describe angles.
- ■ Use radian measure.
- ■ Use degree measure.
- ■ Use angles and their measure to model and solve real-life problems.

Angles and their measure have a wide variety of real-life applications. For example, in Exercise 68 on page 269, you will use angles and their measure to model the distance a cyclist travels.

Angles

As derived from the Greek language, the word **trigonometry** means "measurement of triangles." Originally, trigonometry dealt with relationships among the sides and angles of triangles and was instrumental in the development of astronomy, navigation, and surveying. With the development of calculus and the physical sciences in the 17th century, a different perspective arose—one that viewed the classic trigonometric relationships as *functions* with the set of real numbers as their domains. Consequently, the applications of trigonometry expanded to include a vast number of physical phenomena, such as sound waves, planetary orbits, vibrating strings, pendulums, and orbits of atomic particles. This text incorporates *both* perspectives, starting with angles and their measure.

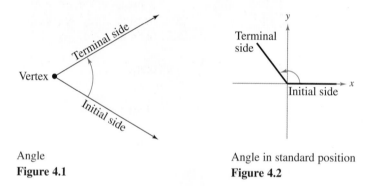

Angle
Figure 4.1

Angle in standard position
Figure 4.2

Rotating a ray (half-line) about its endpoint determines an **angle.** The starting position of the ray is the **initial side** of the angle, and the position after rotation is the **terminal side,** as shown in Figure 4.1. The endpoint of the ray is the **vertex** of the angle. This perception of an angle fits a coordinate system in which the origin is the vertex and the initial side coincides with the positive *x*-axis. Such an angle is in **standard position,** as shown in Figure 4.2. Counterclockwise rotation generates **positive angles** and clockwise rotation generates **negative angles,** as shown in Figure 4.3. Labels for angles can be Greek letters such as α (alpha), β (beta), and θ (theta) or uppercase letters such as A, B, and C. In Figure 4.4, note that angles α and β have the same initial and terminal sides. Such angles are **coterminal.**

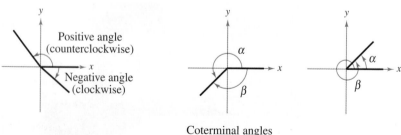

Figure 4.3

Coterminal angles
Figure 4.4

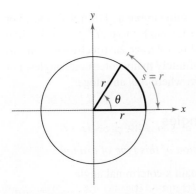

Arc length = radius when θ = 1 radian.
Figure 4.5

Radian Measure

The amount of rotation from the initial side to the terminal side determines the **measure of an angle.** One way to measure angles is in *radians.* This type of measure is especially useful in calculus. To define a radian, use a **central angle** of a circle, which is an angle whose vertex is the center of the circle, as shown in Figure 4.5.

Definition of a Radian

One **radian** (rad) is the measure of a central angle θ that intercepts an arc s equal in length to the radius r of the circle. (See Figure 4.5.) Algebraically, this means that

$$\theta = \frac{s}{r}$$

where θ is measured in radians. (Note that $\theta = 1$ when $s = r$.)

The circumference of a circle is $2\pi r$ units, so it follows that a central angle of one full revolution (counterclockwise) corresponds to an arc length of $s = 2\pi r$. Moreover, $2\pi \approx 6.28$, so there are just over six radius lengths in a full circle, as shown in Figure 4.6. The units of measure for s and r are the same, so the ratio s/r has no units—it is a real number.

The measure of an angle of one full revolution is $s/r = 2\pi r/r = 2\pi$ radians, so you can obtain the following.

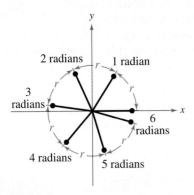

Figure 4.6

$$\frac{1}{2} \text{ revolution} = \frac{2\pi}{2} = \pi \text{ radians} \qquad \frac{1}{4} \text{ revolution} = \frac{2\pi}{4} = \frac{\pi}{2} \text{ radians}$$

$$\frac{1}{6} \text{ revolution} = \frac{2\pi}{6} = \frac{\pi}{3} \text{ radians}$$

These and other common angles are shown below.

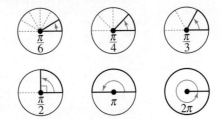

·· REMARK The phrase "θ lies in a quadrant" is an abbreviation for the phrase "the terminal side of θ lies in a quadrant." The terminal sides of the "quadrantal angles" 0, $\pi/2$, π, and $3\pi/2$ do not lie within quadrants.

Recall that the four quadrants in a coordinate system are numbered I, II, III, and IV. The figure below shows which angles between 0 and 2π lie in each of the four quadrants. Note that angles between 0 and $\pi/2$ are **acute** angles and angles between $\pi/2$ and π are **obtuse** angles.

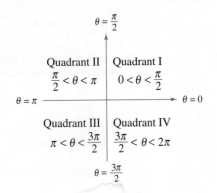

Two angles are coterminal when they have the same initial and terminal sides. For example, the angles 0 and 2π are coterminal, as are the angles $\pi/6$ and $13\pi/6$. To find an angle that is coterminal to a given angle θ, add or subtract 2π (one revolution), as demonstrated in Example 1. A given angle θ has infinitely many coterminal angles. For example, $\theta = \pi/6$ is coterminal with $(\pi/6) + 2n\pi$, where n is an integer.

EXAMPLE 1   Finding Coterminal Angles

See LarsonPrecalculus.com for an interactive version of this type of example.

▷ **ALGEBRA HELP**   To review operations involving fractions, see Appendix A.1.

a. For the positive angle $13\pi/6$, subtract 2π to obtain a coterminal angle.

$$\frac{13\pi}{6} - 2\pi = \frac{\pi}{6} \qquad \text{See Figure 4.7.}$$

b. For the negative angle $-2\pi/3$, add 2π to obtain a coterminal angle.

$$-\frac{2\pi}{3} + 2\pi = \frac{4\pi}{3} \qquad \text{See Figure 4.8.}$$

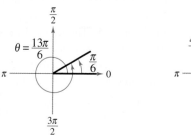

Figure 4.7   **Figure 4.8**

✓ **Checkpoint** 🔊))) *Audio-video solution in English & Spanish at LarsonPrecalculus.com*

Determine two coterminal angles (one positive and one negative) for each angle.

a. $\theta = \dfrac{9\pi}{4}$ **b.** $\theta = -\dfrac{\pi}{3}$

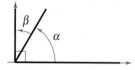

Complementary angles

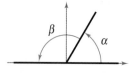

Supplementary angles
Figure 4.9

Two positive angles α and β are **complementary** (complements of each other) when their sum is $\pi/2$. Two positive angles are **supplementary** (supplements of each other) when their sum is π. (See Figure 4.9.)

EXAMPLE 2   Complementary and Supplementary Angles

a. The complement of $\dfrac{2\pi}{5}$ is $\dfrac{\pi}{2} - \dfrac{2\pi}{5} = \dfrac{5\pi}{10} - \dfrac{4\pi}{10} = \dfrac{\pi}{10}.$

The supplement of $\dfrac{2\pi}{5}$ is $\pi - \dfrac{2\pi}{5} = \dfrac{5\pi}{5} - \dfrac{2\pi}{5} = \dfrac{3\pi}{5}.$

b. There is no complement of $4\pi/5$ because $4\pi/5$ is greater than $\pi/2$. (Remember that complements are *positive* angles.) The supplement of $4\pi/5$ is

$$\pi - \frac{4\pi}{5} = \frac{5\pi}{5} - \frac{4\pi}{5} = \frac{\pi}{5}.$$

✓ **Checkpoint** 🔊))) *Audio-video solution in English & Spanish at LarsonPrecalculus.com*

Find (if possible) the complement and supplement of (a) $\pi/6$ and (b) $5\pi/6$.

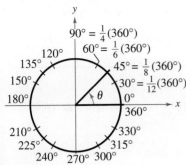

Figure 4.10

$\frac{\pi}{6}$

30°

$\frac{\pi}{4}$

45°

$\frac{\pi}{3}$

60°

$\frac{\pi}{2}$

90°

π

180°

2π

360°

Figure 4.11

▷ **TECHNOLOGY** With calculators, it is convenient to use *decimal* degrees to denote fractional parts of degrees. Historically, however, fractional parts of degrees were expressed in *minutes* and *seconds*, using the prime (′) and double prime (″) notations, respectively. That is,

$1' = $ one minute $= \frac{1}{60}(1°)$

$1'' = $ one second $= \frac{1}{3600}(1°)$.

For example, you would write an angle θ of 64 degrees, 32 minutes, and 47 seconds as $\theta = 64° \, 32' \, 47''$.

Many calculators have special keys for converting an angle in degrees, minutes, and seconds (D° M′ S″) to decimal degree form and vice versa.

Degree Measure

Another way to measure angles is in **degrees**, denoted by the symbol °. A measure of one degree (1°) is equivalent to a rotation of $\frac{1}{360}$ of a complete revolution about the vertex. To measure angles, it is convenient to mark degrees on the circumference of a circle, as shown in Figure 4.10. So, a full revolution (counterclockwise) corresponds to 360°, a half revolution corresponds to 180°, a quarter revolution corresponds to 90°, and so on.

One complete revolution corresponds to 2π radians, so degrees and radians are related by the equations

$$360° = 2\pi \text{ rad} \quad \text{and} \quad 180° = \pi \text{ rad}.$$

From these equations, you obtain

$$1° = \frac{\pi}{180} \text{ rad} \quad \text{and} \quad 1 \text{ rad} = \left(\frac{180}{\pi}\right)°$$

which lead to the conversion rules below.

Conversions Between Degrees and Radians

1. To convert degrees to radians, multiply degrees by $\dfrac{\pi \text{ rad}}{180°}$.

2. To convert radians to degrees, multiply radians by $\dfrac{180°}{\pi \text{ rad}}$.

To apply these two conversion rules, use the basic relationship π rad $= 180°$. (See Figure 4.11.)

When no units of angle measure are specified, *radian measure is implied*. For example, $\theta = 2$ implies that $\theta = 2$ radians.

EXAMPLE 3 **Converting from Degrees to Radians**

a. $135° = (135 \text{ deg})\left(\dfrac{\pi \text{ rad}}{180 \text{ deg}}\right) = \dfrac{3\pi}{4}$ radians   Multiply by $\frac{\pi \text{ rad}}{180°}$.

b. $540° = (540 \text{ deg})\left(\dfrac{\pi \text{ rad}}{180 \text{ deg}}\right) = 3\pi$ radians   Multiply by $\frac{\pi \text{ rad}}{180°}$.

✓ *Checkpoint* ◀))) *Audio-video solution in English & Spanish at LarsonPrecalculus.com*

Convert each degree measure to radian measure as a multiple of π. Do not use a calculator.

a. 60°   **b.** 320°

EXAMPLE 4 **Converting from Radians to Degrees**

a. $-\dfrac{\pi}{2}$ rad $= \left(-\dfrac{\pi}{2} \text{ rad}\right)\left(\dfrac{180 \text{ deg}}{\pi \text{ rad}}\right) = -90°$ Multiply by $\frac{180°}{\pi \text{ rad}}$.

b. 2 rad $= (2 \text{ rad})\left(\dfrac{180 \text{ deg}}{\pi \text{ rad}}\right) = \dfrac{360°}{\pi} \approx 114.59°$ Multiply by $\frac{180°}{\pi \text{ rad}}$.

✓ *Checkpoint* ◀))) *Audio-video solution in English & Spanish at LarsonPrecalculus.com*

Convert each radian measure to degree measure. Do not use a calculator.

a. $\pi/6$ **b.** $5\pi/3$

Applications

To measure arc length along a circle, use the radian measure formula, $\theta = s/r$.

Arc Length

For a circle of radius r, a central angle θ intercepts an arc of length s given by

$$s = r\theta \qquad \text{Length of circular arc}$$

where θ is measured in radians. Note that if $r = 1$, then $s = \theta$, and the radian measure of θ equals the arc length.

EXAMPLE 5 **Finding Arc Length**

A circle has a radius of 4 inches. Find the length of the arc intercepted by a central angle of 240°, as shown in Figure 4.12.

Solution To use the formula $s = r\theta$, first convert 240° to radian measure.

$$240° = (240 \text{ deg})\left(\frac{\pi \text{ rad}}{180 \text{ deg}}\right)$$

$$= \frac{4\pi}{3} \text{ radians}$$

Then, using a radius of $r = 4$ inches, find the arc length.

$$s = r\theta \qquad \text{Length of circular arc}$$

$$= 4\left(\frac{4\pi}{3}\right) \qquad \text{Substitute for } r \text{ and } \theta.$$

$$\approx 16.76 \text{ inches} \qquad \text{Use a calculator.}$$

Note that the units for r determine the units for $r\theta$ because θ is in radian measure, which has no units.

✓ **Checkpoint** 🔊)) *Audio-video solution in English & Spanish at LarsonPrecalculus.com*

A circle has a radius of 27 inches. Find the length of the arc intercepted by a central angle of 160°. ■

The formula for the length of a circular arc can be used to analyze the motion of a particle moving at a *constant speed* along a circular path.

$\theta = 240°$

$r = 4$

Figure 4.12

•• **REMARK**
Linear speed measures how fast the particle moves, and angular speed measures how fast the angle changes. To establish a relationship between linear speed v and angular speed ω, divide each side of the formula for arc length by t, as shown.

$$s = r\theta$$

$$\frac{s}{t} = \frac{r\theta}{t}$$

$$v = r\omega$$

Linear and Angular Speeds

Consider a particle moving at a constant speed along a circular arc of radius r. If s is the length of the arc traveled in time t, then the **linear speed** v of the particle is

$$\text{Linear speed } v = \frac{\text{arc length}}{\text{time}} = \frac{s}{t}.$$

Moreover, if θ is the angle (in radian measure) corresponding to the arc length s, then the **angular speed** ω (the lowercase Greek letter omega) of the particle is

$$\text{Angular speed } \omega = \frac{\text{central angle}}{\text{time}} = \frac{\theta}{t}.$$

EXAMPLE 6   **Finding Linear Speed**

The second hand of a clock is 10.2 centimeters long, as shown at the right. Find the linear speed of the tip of the second hand as it passes around the clock face.

Solution   In one revolution, the arc length traveled is

$$s = 2\pi r$$
$$= 2\pi(10.2) \qquad \text{Substitute for } r.$$
$$= 20.4\pi \text{ centimeters.}$$

The time required for the second hand to travel this distance is

$$t = 1 \text{ minute} = 60 \text{ seconds.}$$

So, the linear speed of the tip of the second hand is

$$v = \frac{s}{t}$$
$$= \frac{20.4\pi \text{ centimeters}}{60 \text{ seconds}}$$
$$\approx 1.07 \text{ centimeters per second.}$$

 Checkpoint *Audio-video solution in English & Spanish at LarsonPrecalculus.com*

The second hand of a clock is 8 centimeters long. Find the linear speed of the tip of the second hand as it passes around the clock face.

EXAMPLE 7   **Finding Angular and Linear Speeds**

The blades of a wind turbine are 116 feet long (see Figure 4.13). The propeller rotates at 15 revolutions per minute.

a. Find the angular speed of the propeller in radians per minute.

b. Find the linear speed of the tips of the blades.

Solution

a. Each revolution corresponds to 2π radians, so the propeller turns $15(2\pi) = 30\pi$ radians per minute. In other words, the angular speed is

$$\omega = \frac{\theta}{t} = \frac{30\pi \text{ radians}}{1 \text{ minute}} = 30\pi \text{ radians per minute.}$$

b. The linear speed is

$$v = \frac{s}{t} = \frac{r\theta}{t} = \frac{116(30\pi) \text{ feet}}{1 \text{ minute}} \approx 10,933 \text{ feet per minute.}$$

 Checkpoint *Audio-video solution in English & Spanish at LarsonPrecalculus.com*

The circular blade on a saw has a radius of 4 inches and it rotates at 2400 revolutions per minute.

a. Find the angular speed of the blade in radians per minute.

b. Find the linear speed of the edge of the blade.

Figure 4.13

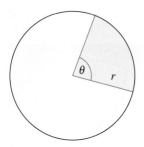

Figure 4.14

A **sector** of a circle is the region bounded by two radii of the circle and their intercepted arc (see Figure 4.14).

Area of a Sector of a Circle

For a circle of radius r, the area A of a sector of the circle with central angle θ is

$$A = \frac{1}{2}r^2\theta$$

where θ is measured in radians.

EXAMPLE 8   **Area of a Sector of a Circle**

A sprinkler on a golf course fairway sprays water over a distance of 70 feet and rotates through an angle of 120° (see Figure 4.15). Find the area of the fairway watered by the sprinkler.

Solution

First convert 120° to radian measure.

$$\theta = 120°$$

$$= (120 \text{ deg})\left(\frac{\pi \text{ rad}}{180 \text{ deg}}\right) \qquad \text{Multiply by } \frac{\pi \text{ rad}}{180°}.$$

$$= \frac{2\pi}{3} \text{ radians}$$

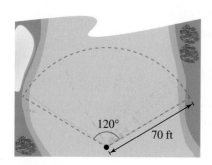

Figure 4.15

Then, using $\theta = 2\pi/3$ and $r = 70$, the area is

$$A = \frac{1}{2}r^2\theta \qquad\qquad \text{Formula for the area of a sector of a circle}$$

$$= \frac{1}{2}(70)^2\left(\frac{2\pi}{3}\right) \qquad \text{Substitute for } r \text{ and } \theta.$$

$$= \frac{4900\pi}{3} \qquad\qquad \text{Multiply.}$$

$$\approx 5131 \text{ square feet.} \qquad \text{Use a calculator.}$$

✔ **Checkpoint** 🔊))) *Audio-video solution in English & Spanish at LarsonPrecalculus.com*

A sprinkler sprays water over a distance of 40 feet and rotates through an angle of 80°. Find the area watered by the sprinkler. ■

Summarize   *(Section 4.1)*

1. Describe an angle *(page 260)*.
2. Explain how to use radian measure *(page 261)*. For examples involving radian measure, see Examples 1 and 2.
3. Explain how to use degree measure *(page 263)*. For examples involving degree measure, see Examples 3 and 4.
4. Describe real-life applications involving angles and their measure *(pages 264–266, Examples 5–8)*.

4.1 Exercises

Vocabulary: Fill in the blanks.

1. Two angles that have the same initial and terminal sides are _____.

2. One _____ is the measure of a central angle that intercepts an arc equal in length to the radius of the circle.

3. Two positive angles that have a sum of $\pi/2$ are _____ angles, and two positive angles that have a sum of π are _____ angles.

4. The angle measure that is equivalent to a rotation of $\frac{1}{360}$ of a complete revolution about an angle's vertex is one _____.

5. The _____ speed of a particle is the ratio of the arc length traveled to the elapsed time, and the _____ speed of a particle is the ratio of the change in the central angle to the elapsed time.

6. The area A of a sector of a circle with radius r and central angle θ, where θ is measured in radians, is given by the formula _____.

Skills and Applications

Estimating an Angle In Exercises 7–10, estimate the angle to the nearest one-half radian.

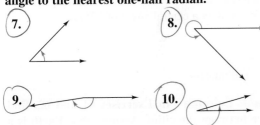

7. 8.

9. 10.

Determining Quadrants In Exercises 11 and 12, determine the quadrant in which each angle lies.

11. (a) $\dfrac{\pi}{4}$ (b) $-\dfrac{5\pi}{4}$ 12. (a) $-\dfrac{\pi}{6}$ (b) $\dfrac{11\pi}{9}$

Sketching Angles In Exercises 13 and 14, sketch each angle in standard position.

13. (a) $\dfrac{\pi}{3}$ (b) $-\dfrac{2\pi}{3}$ 14. (a) $\dfrac{5\pi}{2}$ (b) 4

 Finding Coterminal Angles In Exercises 15 and 16, determine two coterminal angles (one positive and one negative) for each angle. Give your answers in radians.

15. (a) $\dfrac{\pi}{6}$ (b) $-\dfrac{5\pi}{6}$ 16. (a) $\dfrac{2\pi}{3}$ (b) $-\dfrac{9\pi}{4}$

 Complementary and Supplementary Angles In Exercises 17–20, find (if possible) the complement and supplement of each angle.

17. (a) $\dfrac{\pi}{12}$ (b) $\dfrac{11\pi}{12}$ 18. (a) $\dfrac{\pi}{3}$ (b) $\dfrac{\pi}{4}$

19. (a) 1 (b) 2 20. (a) 3 (b) 1.5

Estimating an Angle In Exercises 21–24, estimate the number of degrees in the angle.

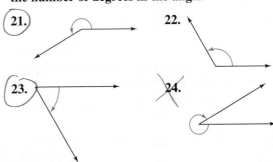

21. 22.

23. 24.

Determining Quadrants In Exercises 25 and 26, determine the quadrant in which each angle lies.

25. (a) $130°$ (b) $-8.3°$

26. (a) $-132°\,50'$ (b) $3.4°$

Sketching Angles In Exercises 27 and 28, sketch each angle in standard position.

27. (a) $270°$ (b) $-120°$ 28. (a) $135°$ (b) $-750°$

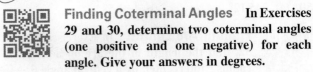 **Finding Coterminal Angles** In Exercises 29 and 30, determine two coterminal angles (one positive and one negative) for each angle. Give your answers in degrees.

29. (a) $120°$ (b) $-210°$ 30. (a) $45°$ (b) $-420°$

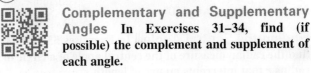 **Complementary and Supplementary Angles** In Exercises 31–34, find (if possible) the complement and supplement of each angle.

31. (a) $18°$ (b) $85°$ 32. (a) $46°$ (b) $93°$

33. (a) $24°$ (b) $126°$ 34. (a) $130°$ (b) $170°$

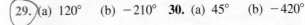

Converting from Degrees to Radians In Exercises 35 and 36, convert each degree measure to radian measure as a multiple of π. Do not use a calculator.

35. (a) 120°   (b) −20°

36. (a) −60°   (b) 144°

Converting from Radians to Degrees In Exercises 37 and 38, convert each radian measure to degree measure. Do not use a calculator.

37. (a) $\dfrac{3\pi}{2}$ (b) $-\dfrac{7\pi}{6}$

38. (a) $-\dfrac{7\pi}{12}$ (b) $\dfrac{5\pi}{4}$

Converting from Degrees to Radians In Exercises 39–42, convert the degree measure to radian measure. Round to three decimal places.

39. 45°   40. −48.27°

41. −0.54°   42. 345°

Converting from Radians to Degrees In Exercises 43–46, convert the radian measure to degree measure. Round to three decimal places, if necessary.

43. $\dfrac{5\pi}{11}$ 44. $\dfrac{15\pi}{8}$

45. −4.2π   46. −0.57

Converting to Decimal Degree Form In Exercises 47 and 48, convert each angle measure to decimal degree form.

47. (a) 54° 45′   (b) −128° 30′

48. (a) 135° 10′ 36″   (b) −408° 16′ 20″

Converting to D° M′ S″ Form In Exercises 49 and 50, convert each angle measure to D° M′ S″ form.

49. (a) 240.6°   (b) −145.8°

50. (a) 345.12°   (b) −3.58°

Finding Arc Length In Exercises 51 and 52, find the length of the arc on a circle of radius r intercepted by a central angle θ.

51. r = 15 inches, θ = 120°

52. r = 3 meters, θ = 150°

Finding the Central Angle In Exercises 53 and 54, find the radian measure of the central angle of a circle of radius r that intercepts an arc of length s.

53. r = 80 kilometers, s = 150 kilometers

54. r = 14 feet, s = 8 feet

Finding the Central Angle In Exercises 55 and 56, find the radian measure of the central angle.

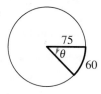

55.   28   56.

Area of a Sector of a Circle In Exercises 57 and 58, find the area of the sector of a circle of radius r and central angle θ.

57. r = 6 inches, θ = $\dfrac{\pi}{3}$ 58. r = 2.5 feet, θ = 225°

Error Analysis In Exercises 59 and 60, describe the error.

59. $20° = (20\ \text{deg})\left(\dfrac{180\ \text{rad}}{\pi\ \text{deg}}\right) = \dfrac{3600}{\pi}\ \text{rad}$ ✗

60. A circle has a radius of 6 millimeters. The length of the arc intercepted by a central angle of 72° is

$$s = r\theta$$
$$= 6(72)$$
$$= 432 \text{ millimeters.}$$ ✗

Earth-Space Science In Exercises 61 and 62, find the distance between the cities. Assume that Earth is a sphere of radius 4000 miles and that the cities are on the same longitude (one city is due north of the other).

City	Latitude
61. Dallas, Texas	32° 47′ 9″ N
Omaha, Nebraska	41° 15′ 50″ N
62. San Francisco, California	37° 47′ 36″ N
Seattle, Washington	47° 37′ 18″ N

63. **Instrumentation** The pointer on a voltmeter is 6 centimeters in length (see figure). Find the number of degrees through which the pointer rotates when it moves 2.5 centimeters on the scale.

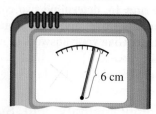

6 cm

64. **Linear and Angular Speed** A $7\frac{1}{4}$-inch circular power saw blade rotates at 5200 revolutions per minute.

(a) Find the angular speed of the saw blade in radians per minute.

(b) Find the linear speed (in feet per minute) of the saw teeth as they contact the wood being cut.

65. Linear and Angular Speed A carousel with a 50-foot diameter makes 4 revolutions per minute.

    (a) Find the angular speed of the carousel in radians per minute.

    (b) Find the linear speed (in feet per minute) of the platform rim of the carousel.

66. Linear and Angular Speed A Blu-ray disc is approximately 12 centimeters in diameter. The drive motor of a Blu-ray player is able to rotate up to 10,000 revolutions per minute.

    (a) Find the maximum angular speed (in radians per second) of a Blu-ray disc as it rotates.

    (b) Find the maximum linear speed (in meters per second) of a point on the outermost track as the disc rotates.

67. Linear and Angular Speed A computerized spin balance machine rotates a 25-inch-diameter tire at 480 revolutions per minute.

    (a) Find the road speed (in miles per hour) at which the tire is being balanced.

    (b) At what rate should the spin balance machine be set so that the tire is being tested for 55 miles per hour?

• •68. Speed of a Bicycle • • • • • • • • • • • • •

The radii of the pedal sprocket, the wheel sprocket, and the wheel of the bicycle in the figure are 4 inches, 2 inches, and 14 inches, respectively. A cyclist pedals at a rate of 1 revolution per second.

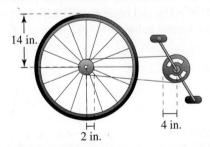

14 in.   2 in.   4 in.

    (a) Find the speed of the bicycle in feet per second and miles per hour.

    (b) Use your result from part (a) to write a function for the distance d (in miles) a cyclist travels in terms of the number n of revolutions of the pedal sprocket.

    (c) Write a function for the distance d (in miles) a cyclist travels in terms of the time t (in seconds). Compare this function with the function from part (b).

69. Area A sprinkler on a golf green is set to spray water over a distance of 15 meters and to rotate through an angle of 150°. Draw a diagram that shows the region that can be irrigated with the sprinkler. Find the area of the region.

70. Area A car's rear windshield wiper rotates 125°. The total length of the wiper mechanism is 25 inches and the length of the wiper blade is 14 inches. Find the area wiped by the wiper blade.

Exploration

True or False? In Exercises 71–74, determine whether the statement is true or false. Justify your answer.

71. An angle measure containing π must be in radian measure.

72. A measurement of 4 radians corresponds to two complete revolutions from the initial side to the terminal side of an angle.

73. The difference between the measures of two coterminal angles is always a multiple of 360° when expressed in degrees and is always a multiple of 2π radians when expressed in radians.

74. An angle that measures $-1260°$ lies in Quadrant III.

75. Writing When the radius of a circle increases and the magnitude of a central angle is held constant, how does the length of the intercepted arc change? Explain.

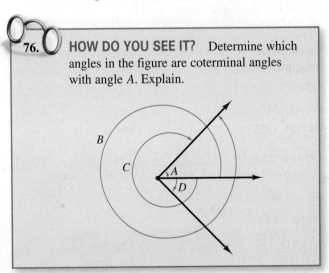

76. HOW DO YOU SEE IT? Determine which angles in the figure are coterminal angles with angle A. Explain.

77. Think About It A fan motor turns at a given angular speed. How does the speed of the tips of the blades change when a fan of greater diameter is installed on the motor? Explain.

78. Think About It Is a degree or a radian the larger unit of measure? Explain.

79. Proof Prove that the area of a circular sector of radius r with central angle θ is $A = \frac{1}{2}\theta r^2$, where θ is measured in radians.

4.2 Trigonometric Functions: The Unit Circle

- Identify a unit circle and describe its relationship to real numbers.
- Evaluate trigonometric functions using the unit circle.
- Use domain and period to evaluate sine and cosine functions, and use a calculator to evaluate trigonometric functions.

The Unit Circle

The two historical perspectives of trigonometry incorporate different methods for introducing the trigonometric functions. One such perspective is based on the unit circle.

Consider the **unit circle** given by

$$x^2 + y^2 = 1 \qquad \text{Unit circle}$$

as shown in the figure below.

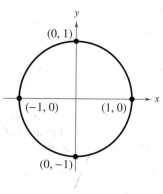

Imagine wrapping the real number line around this circle, with positive numbers corresponding to a counterclockwise wrapping and negative numbers corresponding to a clockwise wrapping, as shown in the figures below.

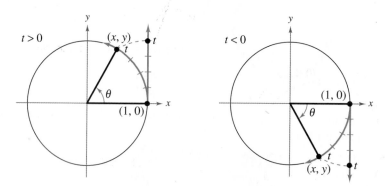

As the real number line wraps around the unit circle, each real number t corresponds to a point (x, y) on the circle. For example, the real number 0 corresponds to the point $(1, 0)$. Moreover, the unit circle has a circumference of 2π, so the real number 2π also corresponds to the point $(1, 0)$.

Each real number t also corresponds to a central angle θ (in standard position) whose radian measure is t. With this interpretation of t, the arc length formula

$$s = r\theta \quad (\text{with } r = 1)$$

indicates that the real number t is the (directional) length of the arc intercepted by the angle θ, given in radians.

Trigonometric functions can help you analyze the movement of an oscillating weight. For example, in Exercise 50 on page 276, you will analyze the displacement of an oscillating weight suspended by a spring using a model that is the product of a trigonometric function and an exponential function.

The Trigonometric Functions

From the preceding discussion, the coordinates x and y are two functions of the real variable t. These coordinates are used to define the six trigonometric functions of a real number t.

sine   cosecant   cosine   secant   tangent   cotangent

Abbreviations for these six functions are sin, csc, cos, sec, tan, and cot, respectively.

> **• • REMARK** Note that the functions in the second row are the *reciprocals* of the corresponding functions in the first row.

Definitions of Trigonometric Functions

Let t be a real number and let (x, y) be the point on the unit circle corresponding to t.

$$\sin t = y \qquad\qquad \cos t = x \qquad\qquad \tan t = \frac{y}{x}, \quad x \neq 0$$

$$\csc t = \frac{1}{y}, \quad y \neq 0 \qquad \sec t = \frac{1}{x}, \quad x \neq 0 \qquad \cot t = \frac{x}{y}, \quad y \neq 0$$

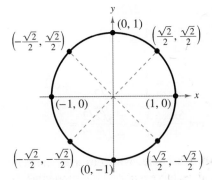

Figure 4.16

In the definitions of the trigonometric functions, note that the tangent and secant are not defined when $x = 0$. For example, $t = \pi/2$ corresponds to $(x, y) = (0, 1)$, so $\tan(\pi/2)$ and $\sec(\pi/2)$ are *undefined*. Similarly, the cotangent and cosecant are not defined when $y = 0$. For example, $t = 0$ corresponds to $(x, y) = (1, 0)$, so $\cot 0$ and $\csc 0$ are *undefined*.

In Figure 4.16, the unit circle is divided into eight equal arcs, corresponding to t-values of

$$0, \frac{\pi}{4}, \frac{\pi}{2}, \frac{3\pi}{4}, \pi, \frac{5\pi}{4}, \frac{3\pi}{2}, \frac{7\pi}{4}, \text{ and } 2\pi.$$

Similarly, in Figure 4.17, the unit circle is divided into 12 equal arcs, corresponding to t-values of

$$0, \frac{\pi}{6}, \frac{\pi}{3}, \frac{\pi}{2}, \frac{2\pi}{3}, \frac{5\pi}{6}, \pi, \frac{7\pi}{6}, \frac{4\pi}{3}, \frac{3\pi}{2}, \frac{5\pi}{3}, \frac{11\pi}{6}, \text{ and } 2\pi.$$

To verify the points on the unit circle in Figure 4.16, note that

$$\left(\frac{\sqrt{2}}{2}, \frac{\sqrt{2}}{2}\right)$$

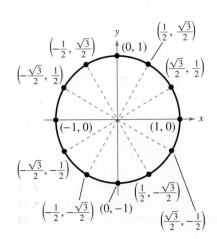

Figure 4.17

lies on the line $y = x$. So, substituting x for y in the equation of the unit circle produces the following

$$x^2 + x^2 = 1 \quad\Longrightarrow\quad 2x^2 = 1 \quad\Longrightarrow\quad x^2 = \frac{1}{2} \quad\Longrightarrow\quad x = \pm\frac{\sqrt{2}}{2}$$

Because the point is in the first quadrant and $y = x$, you have

$$x = \frac{\sqrt{2}}{2} \quad \text{and} \quad y = \frac{\sqrt{2}}{2}.$$

Similar reasoning can be used to verify the rest of the points in Figure 4.16 and the points in Figure 4.17.

Using the (x, y) coordinates in Figures 4.16 and 4.17, you can evaluate the trigonometric functions for these common t-values. Examples 1 and 2 demonstrate this procedure. You should study and learn these exact function values for common t-values because they will help you perform calculations in later sections.

> ▷ **ALGEBRA HELP** To review dividing fractions and rationalizing denominators, see Appendix A.1 and Appendix A.2, respectively.

EXAMPLE 1 **Evaluating Trigonometric Functions**

See LarsonPrecalculus.com for an interactive version of this type of example.

Evaluate the six trigonometric functions at each real number.

a. $t = \dfrac{\pi}{6}$ **b.** $t = \dfrac{5\pi}{4}$ **c.** $t = \pi$ **d.** $t = -\dfrac{\pi}{3}$

Solution For each t-value, begin by finding the corresponding point (x, y) on the unit circle. Then use the definitions of trigonometric functions listed on page 271.

a. $t = \pi/6$ corresponds to the point $(x, y) = \left(\sqrt{3}/2,\, 1/2\right)$.

$$\sin\frac{\pi}{6} = y = \frac{1}{2} \qquad\qquad \csc\frac{\pi}{6} = \frac{1}{y} = \frac{1}{1/2} = 2$$

$$\cos\frac{\pi}{6} = x = \frac{\sqrt{3}}{2} \qquad\qquad \sec\frac{\pi}{6} = \frac{1}{x} = \frac{2}{\sqrt{3}} = \frac{2\sqrt{3}}{3}$$

$$\tan\frac{\pi}{6} = \frac{y}{x} = \frac{1/2}{\sqrt{3}/2} = \frac{1}{\sqrt{3}} = \frac{\sqrt{3}}{3} \qquad\qquad \cot\frac{\pi}{6} = \frac{x}{y} = \frac{\sqrt{3}/2}{1/2} = \sqrt{3}$$

b. $t = 5\pi/4$ corresponds to the point $(x, y) = \left(-\sqrt{2}/2,\, -\sqrt{2}/2\right)$.

$$\sin\frac{5\pi}{4} = y = -\frac{\sqrt{2}}{2} \qquad\qquad \csc\frac{5\pi}{4} = \frac{1}{y} = -\frac{2}{\sqrt{2}} = -\sqrt{2}$$

$$\cos\frac{5\pi}{4} = x = -\frac{\sqrt{2}}{2} \qquad\qquad \sec\frac{5\pi}{4} = \frac{1}{x} = -\frac{2}{\sqrt{2}} = -\sqrt{2}$$

$$\tan\frac{5\pi}{4} = \frac{y}{x} = \frac{-\sqrt{2}/2}{-\sqrt{2}/2} = 1 \qquad\qquad \cot\frac{5\pi}{4} = \frac{x}{y} = \frac{-\sqrt{2}/2}{-\sqrt{2}/2} = 1$$

c. $t = \pi$ corresponds to the point $(x, y) = (-1, 0)$.

$$\sin\pi = y = 0 \qquad\qquad \csc\pi = \frac{1}{y} \text{ is undefined.}$$

$$\cos\pi = x = -1 \qquad\qquad \sec\pi = \frac{1}{x} = \frac{1}{-1} = -1$$

$$\tan\pi = \frac{y}{x} = \frac{0}{-1} = 0 \qquad\qquad \cot\pi = \frac{x}{y} \text{ is undefined.}$$

d. Moving *clockwise* around the unit circle, $t = -\pi/3$ corresponds to the point $(x, y) = \left(1/2,\, -\sqrt{3}/2\right)$.

$$\sin\left(-\frac{\pi}{3}\right) = y = -\frac{\sqrt{3}}{2} \qquad\qquad \csc\left(-\frac{\pi}{3}\right) = \frac{1}{y} = -\frac{2}{\sqrt{3}} = -\frac{2\sqrt{3}}{3}$$

$$\cos\left(-\frac{\pi}{3}\right) = x = \frac{1}{2} \qquad\qquad \sec\left(-\frac{\pi}{3}\right) = \frac{1}{x} = \frac{1}{1/2} = 2$$

$$\tan\left(-\frac{\pi}{3}\right) = \frac{y}{x} = \frac{-\sqrt{3}/2}{1/2} = -\sqrt{3}$$

$$\cot\left(-\frac{\pi}{3}\right) = \frac{x}{y} = \frac{1/2}{-\sqrt{3}/2} = -\frac{1}{\sqrt{3}} = -\frac{\sqrt{3}}{3}$$

✓ *Checkpoint* ◀))) *Audio-video solution in English & Spanish at LarsonPrecalculus.com*

Evaluate the six trigonometric functions at each real number.

a. $t = \pi/2$ **b.** $t = 0$ **c.** $t = -5\pi/6$ **d.** $t = -3\pi/4$

Domain and Period of Sine and Cosine

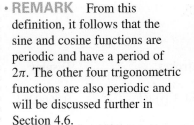

Figure 4.18

The *domain* of the sine and cosine functions is the set of all real numbers. To determine the *range* of these two functions, consider the unit circle shown in Figure 4.18. You know that $\sin t = y$ and $\cos t = x$. Moreover, (x, y) is on the unit circle, so you also know that $-1 \le y \le 1$ and $-1 \le x \le 1$. This means that the values of sine and cosine also range between -1 and 1.

$$-1 \le\ y\ \le 1 \quad \text{and} \quad -1 \le\ x\ \le 1$$
$$-1 \le \sin t \le 1 \qquad\qquad -1 \le \cos t \le 1$$

Adding 2π to each value of t in the interval $[0, 2\pi]$ results in a revolution around the unit circle, as shown in the figure below.

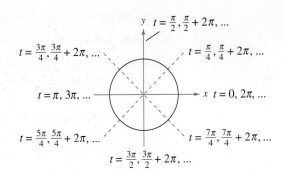

The values of $\sin(t + 2\pi)$ and $\cos(t + 2\pi)$ correspond to those of $\sin t$ and $\cos t$. Repeated revolutions (positive or negative) on the unit circle yield similar results. This leads to the general result

$$\sin(t + 2\pi n) = \sin t \quad \text{and} \quad \cos(t + 2\pi n) = \cos t$$

for any integer n and real number t. Functions that behave in such a repetitive (or cyclic) manner are **periodic.**

•• **REMARK** From this definition, it follows that the sine and cosine functions are periodic and have a period of 2π. The other four trigonometric functions are also periodic and will be discussed further in Section 4.6.

Definition of Periodic Function

A function f is **periodic** when there exists a positive real number c such that

$$f(t + c) = f(t)$$

for all t in the domain of f. The smallest number c for which f is periodic is the **period** of f.

Recall from Section 1.5 that a function f is *even* when $f(-t) = f(t)$ and is *odd* when $f(-t) = -f(t)$.

Even and Odd Trigonometric Functions

The cosine and secant functions are *even.*

$$\cos(-t) = \cos t \qquad\qquad \sec(-t) = \sec t$$

The sine, cosecant, tangent, and cotangent functions are *odd.*

$$\sin(-t) = -\sin t \qquad\qquad \csc(-t) = -\csc t$$

$$\tan(-t) = -\tan t \qquad\qquad \cot(-t) = -\cot t$$

EXAMPLE 2 **Evaluating Sine and Cosine**

a. Because $\frac{13\pi}{6} = 2\pi + \frac{\pi}{6}$, you have $\sin\frac{13\pi}{6} = \sin\left(2\pi + \frac{\pi}{6}\right) = \sin\frac{\pi}{6} = \frac{1}{2}$.

b. Because $-\frac{7\pi}{2} = -4\pi + \frac{\pi}{2}$, you have

$$\cos\left(-\frac{7\pi}{2}\right) = \cos\left(-4\pi + \frac{\pi}{2}\right) = \cos\frac{\pi}{2} = 0.$$

c. For $\sin t = \frac{4}{5}$, $\sin(-t) = -\frac{4}{5}$ because the sine function is odd.

✓ *Checkpoint* 🔊))) *Audio-video solution in English & Spanish at LarsonPrecalculus.com*

a. Use the period of the cosine function to evaluate $\cos(9\pi/2)$.

b. Use the period of the sine function to evaluate $\sin(-7\pi/3)$.

c. Evaluate $\cos t$ given that $\cos(-t) = 0.3$. ■

When evaluating a trigonometric function with a calculator, set the calculator to the desired *mode* of measurement (*degree* or *radian*). Most calculators do not have keys for the cosecant, secant, and cotangent functions. To evaluate these functions, you can use the $\boxed{x^{-1}}$ key with their respective reciprocal functions: sine, cosine, and tangent. For example, to evaluate $\csc(\pi/8)$, use the fact that

$$\csc\frac{\pi}{8} = \frac{1}{\sin(\pi/8)}$$

and enter the keystroke sequence below in *radian* mode.

$\boxed{(}\ \boxed{\text{SIN}}\ \boxed{(}\ \boxed{\pi}\ \boxed{\div}\ 8\ \boxed{)}\ \boxed{)}\ \boxed{x^{-1}}\ \boxed{\text{ENTER}}$ Display 2.6131259

▷ **TECHNOLOGY** When evaluating trigonometric functions with a calculator, remember to enclose all fractional angle measures in parentheses. For example, to evaluate $\sin t$ for $t = \pi/6$, enter

$\boxed{\text{SIN}}\ \boxed{(}\ \boxed{\pi}\ \boxed{\div}\ 6\ \boxed{)}\ \boxed{\text{ENTER}}$.

These keystrokes yield the correct value of 0.5. Note that some calculators automatically place a left parenthesis after trigonometric functions.

EXAMPLE 3 **Using a Calculator**

Function	Mode	Calculator Keystrokes	Display
a. $\sin\frac{2\pi}{3}$	Radian	$\boxed{\text{SIN}}\ \boxed{(}\ 2\ \boxed{\pi}\ \boxed{\div}\ 3\ \boxed{)}\ \boxed{\text{ENTER}}$	0.8660254
b. $\cot 1.5$	Radian	$\boxed{(}\ \boxed{\text{TAN}}\ \boxed{(}\ 1.5\ \boxed{)}\ \boxed{)}\ \boxed{x^{-1}}\ \boxed{\text{ENTER}}$	0.0709148

✓ *Checkpoint* 🔊))) *Audio-video solution in English & Spanish at LarsonPrecalculus.com*

Use a calculator to evaluate (a) $\sin(5\pi/7)$ and (b) $\csc 2.0$. ■

Summarize (Section 4.2)

1. Explain how to identify a unit circle and describe its relationship to real numbers *(page 270)*.

2. State the unit circle definitions of trigonometric functions *(page 271)*. For an example of evaluating trigonometric functions using the unit circle, see Example 1.

3. Explain how to use domain and period to evaluate sine and cosine functions *(page 273)*, and describe how to use a calculator to evaluate trigonometric functions *(page 274)*. For an example of using domain and period to evaluate sine and cosine functions, see Example 2. For an example of using a calculator to evaluate trigonometric functions, see Example 3.

4.2 Exercises

See **CalcChat.com** for tutorial help and worked-out solutions to odd-numbered exercises.

Vocabulary: Fill in the blanks.

1. Each real number t corresponds to a point (x, y) on the _____ _____.
2. A function f is _____ when there exists a positive real number c such that $f(t + c) = f(t)$ for all t in the domain of f.
3. The smallest number c for which a function f is periodic is the _____ of f.
4. A function f is _____ when $f(-t) = -f(t)$ and _____ when $f(-t) = f(t)$.

Skills and Applications

Evaluating Trigonometric Functions In Exercises 5–8, find the exact values of the six trigonometric functions of the real number t.

5.

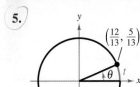

6.

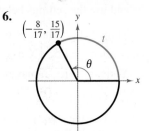

7.

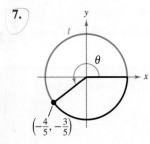

8.

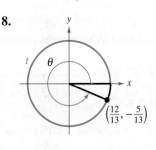

Finding a Point on the Unit Circle In Exercises 9–12, find the point (x, y) on the unit circle that corresponds to the real number t.

9. $t = \pi/2$ 10. $t = \pi/4$

11. $t = 5\pi/6$ 12. $t = 4\pi/3$

Evaluating Sine, Cosine, and Tangent In Exercises 13–22, evaluate (if possible) the sine, cosine, and tangent at the real number.

13. $t = \dfrac{\pi}{4}$ 14. $t = \dfrac{\pi}{3}$

15. $t = -\dfrac{\pi}{6}$ 16. $t = -\dfrac{\pi}{4}$

17. $t = -\dfrac{7\pi}{4}$ 18. $t = -\dfrac{4\pi}{3}$

19. $t = \dfrac{11\pi}{6}$ 20. $t = \dfrac{5\pi}{3}$

21. $t = -\dfrac{3\pi}{2}$ 22. $t = -2\pi$

Evaluating Trigonometric Functions In Exercises 23–30, evaluate (if possible) the six trigonometric functions at the real number.

23. $t = 2\pi/3$ 24. $t = 5\pi/6$

25. $t = 4\pi/3$ 26. $t = 7\pi/4$

27. $t = -5\pi/3$ 28. $t = -3\pi/2$

29. $t = -\pi/2$ 30. $t = -\pi$

Using Period to Evaluate Sine and Cosine In Exercises 31–36, evaluate the trigonometric function using its period as an aid.

31. $\sin 4\pi$ 32. $\cos 3\pi$

33. $\cos(7\pi/3)$ 34. $\sin(9\pi/4)$

35. $\sin(19\pi/6)$ 36. $\sin(-8\pi/3)$

Using the Value of a Function In Exercises 37–42, use the given value to evaluate each function.

37. $\sin t = \frac{1}{2}$
    (a) $\sin(-t)$
    (b) $\csc(-t)$

38. $\sin(-t) = \frac{3}{8}$
    (a) $\sin t$
    (b) $\csc t$

39. $\cos(-t) = -\frac{1}{5}$
    (a) $\cos t$
    (b) $\sec(-t)$

40. $\cos t = -\frac{3}{4}$
    (a) $\cos(-t)$
    (b) $\sec(-t)$

41. $\sin t = \frac{4}{5}$
    (a) $\sin(\pi - t)$
    (b) $\sin(t + \pi)$

42. $\cos t = \frac{4}{5}$
    (a) $\cos(\pi - t)$
    (b) $\cos(t + \pi)$

Using a Calculator In Exercises 43–48, use a calculator to evaluate the trigonometric function. Round your answer to four decimal places. (Be sure the calculator is in the correct mode.)

43. $\sin 0.6$ 44. $\cos(-2.8)$

45. $\tan(\pi/8)$ 46. $\tan(5\pi/7)$

47. $\sec 3.1$ 48. $\cot(-1.1)$

49. Harmonic Motion The displacement from equilibrium of an oscillating weight suspended by a spring is given by

$$y(t) = \frac{1}{2}\cos 6t$$

where y is the displacement in feet and t is the time in seconds. Find the displacement when (a) $t = 0$, (b) $t = \frac{1}{4}$, and (c) $t = \frac{1}{2}$.

• • 50. Harmonic Motion • • • • • • • • • • • • • •

The displacement from equilibrium of an oscillating weight suspended by a spring and subject to the damping effect of friction is given by

$$y(t) = \frac{1}{2}e^{-t}\cos 6t$$

where y is the displacement in feet and t is the time in seconds.

(a) Complete the table

t	0	$\frac{1}{4}$	$\frac{1}{2}$	$\frac{3}{4}$	1
y					

(b) Use the *table* feature of a graphing utility to approximate the time when the weight reaches equilibrium.

(c) What appears to happen to the displacement as t increases?

Exploration

True of False? In Exercises 51–54, determine whether the statement is true or false. Justify your answer.

51. Because $\sin(-t) = -\sin t$, the sine of a negative angle is a negative number.

52. The real number 0 corresponds to the point $(0, 1)$ on the unit circle.

53. $\tan a = \tan(a - 6\pi)$

54. $\cos\left(-\dfrac{7\pi}{2}\right) = \cos\left(\pi + \dfrac{\pi}{2}\right)$

55. Conjecture Let (x_1, y_1) and (x_2, y_2) be points on the unit circle corresponding to $t = t_1$ and $t = \pi - t_1$, respectively.

(a) Identify the symmetry of the points (x_1, y_1) and (x_2, y_2).

(b) Make a conjecture about any relationship between $\sin t_1$ and $\sin(\pi - t_1)$.

(c) Make a conjecture about any relationship between $\cos t_1$ and $\cos(\pi - t_1)$.

56. Using the Unit Circle Use the unit circle to verify that the cosine and secant functions are even and that the sine, cosecant, tangent, and cotangent functions are odd.

57. Error Analysis Describe the error.

Your classmate uses a calculator to evaluate $\tan(\pi/2)$ and gets a result of 0.0274224385.

58. Verifying Expressions Are Not Equal Verify that

$$\sin(t_1 + t_2) \neq \sin t_1 + \sin t_2$$

by approximating sin 0.25, sin 0.75, and sin 1.

59. Using Technology With a graphing utility in *radian* and *parametric* modes, enter the equations

$$X_{1T} = \cos T \quad \text{and} \quad Y_{1T} = \sin T$$

and use the settings below.

Tmin = 0,   Tmax = 6.3,   Tstep = 0.1
Xmin = −1.5,   Xmax = 1.5,   Xscl = 1
Ymin = −1,   Ymax = 1,   Yscl = 1

(a) Graph the entered equations and describe the graph.

(b) Use the *trace* feature to move the cursor around the graph. What do the t-values represent? What do the x- and y-values represent?

(c) What are the least and greatest values of x and y?

60. 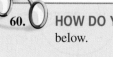 **HOW DO YOU SEE IT?** Use the figure below.

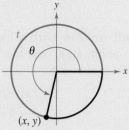

(a) Are all of the trigonometric functions of t defined? Explain.

(b) For those trigonometric functions that are defined, determine whether the sign of the trigonometric function is positive or negative. Explain.

61. Think About It Because $f(t) = \sin t$ is an odd function and $g(t) = \cos t$ is an even function, what can be said about the function $h(t) = f(t)g(t)$?

62. Think About It Because $f(t) = \sin t$ and $g(t) = \tan t$ are odd functions, what can be said about the function $h(t) = f(t)g(t)$?

4.3 Right Triangle Trigonometry

Right triangle trigonometry has many real-life applications. For example, in Exercise 72 on page 287, you will use right triangle trigonometry to analyze the height of a helium-filled balloon.

▪ Evaluate trigonometric functions of acute angles.
▪ Use fundamental trigonometric identities.
▪ Use trigonometric functions to model and solve real-life problems.

The Six Trigonometric Functions

This section introduces the trigonometric functions from a *right triangle* perspective. Consider the right triangle shown below, in which one acute angle is labeled θ. Relative to the angle θ, the three sides of the triangle are the **hypotenuse,** the **opposite side** (the side opposite the angle θ), and the **adjacent side** (the side adjacent to the angle θ).

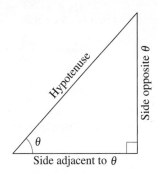

Using the lengths of these three sides, you can form six ratios that define the six trigonometric functions of the acute angle θ.

sine   cosecant   cosine   secant   tangent   cotangent

In the definitions below,

$$0° < \theta < 90°$$

(θ lies in the first quadrant). For such angles, the value of each trigonometric function is *positive*.

Right Triangle Definitions of Trigonometric Functions

Let θ be an *acute* angle of a right triangle. The six trigonometric functions of the angle θ are defined below. (Note that the functions in the second row are the *reciprocals* of the corresponding functions in the first row.)

$$\sin \theta = \frac{\text{opp}}{\text{hyp}} \qquad \cos \theta = \frac{\text{adj}}{\text{hyp}} \qquad \tan \theta = \frac{\text{opp}}{\text{adj}}$$

$$\csc \theta = \frac{\text{hyp}}{\text{opp}} \qquad \sec \theta = \frac{\text{hyp}}{\text{adj}} \qquad \cot \theta = \frac{\text{adj}}{\text{opp}}$$

The abbreviations

opp,   adj,   and   *hyp*

represent the lengths of the three sides of a right triangle.

opp = the length of the side *opposite* θ

adj = the length of the side *adjacent to* θ

hyp = the length of the *hypotenuse*

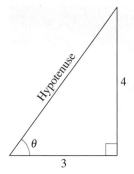

Figure 4.19

| EXAMPLE 1 | **Evaluating Trigonometric Functions** |

See LarsonPrecalculus.com for an interactive version of this type of example.

Use the triangle in Figure 4.19 to find the values of the six trigonometric functions of θ.

Solution   By the Pythagorean Theorem, $(\text{hyp})^2 = (\text{opp})^2 + (\text{adj})^2$, it follows that

$$\text{hyp} = \sqrt{4^2 + 3^2}$$
$$= \sqrt{25}$$
$$= 5.$$

So, the six trigonometric functions of θ are

$$\sin \theta = \frac{\text{opp}}{\text{hyp}} = \frac{4}{5} \qquad \csc \theta = \frac{\text{hyp}}{\text{opp}} = \frac{5}{4}$$

$$\cos \theta = \frac{\text{adj}}{\text{hyp}} = \frac{3}{5} \qquad \sec \theta = \frac{\text{hyp}}{\text{adj}} = \frac{5}{3}$$

$$\tan \theta = \frac{\text{opp}}{\text{adj}} = \frac{4}{3} \qquad \cot \theta = \frac{\text{adj}}{\text{opp}} = \frac{3}{4}.$$

✓ *Checkpoint* 🔊))) *Audio-video solution in English & Spanish at LarsonPrecalculus.com*

Use the triangle below to find the values of the six trigonometric functions of θ.

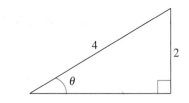

In Example 1, you were given the lengths of two sides of the right triangle, but not the angle θ. Often, you will be asked to find the trigonometric functions of a *given* acute angle θ. To do this, construct a right triangle having θ as one of its angles.

| EXAMPLE 2 | **Evaluating Trigonometric Functions of 45°** |

Find the values of sin 45°, cos 45°, and tan 45°.

Solution   Construct a right triangle having 45° as one of its acute angles, as shown in Figure 4.20. Choose 1 as the length of the adjacent side. From geometry, you know that the other acute angle is also 45°. So, the triangle is isosceles and the length of the opposite side is also 1. By the Pythagorean Theorem, the length of the hypotenuse is $\sqrt{2}$.

$$\sin 45° = \frac{\text{opp}}{\text{hyp}} = \frac{1}{\sqrt{2}} = \frac{\sqrt{2}}{2}$$

$$\cos 45° = \frac{\text{adj}}{\text{hyp}} = \frac{1}{\sqrt{2}} = \frac{\sqrt{2}}{2}$$

$$\tan 45° = \frac{\text{opp}}{\text{adj}} = \frac{1}{1} = 1$$

✓ *Checkpoint* 🔊))) *Audio-video solution in English & Spanish at LarsonPrecalculus.com*

Find the values of cot 45°, sec 45°, and csc 45°.

Figure 4.20

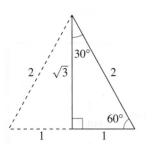

Figure 4.21

• • REMARK The angles 30°, 45°, and 60° ($\pi/6$, $\pi/4$, and $\pi/3$ radians, respectively) occur frequently in trigonometry, so you should learn to construct the triangles shown in Figures 4.20 and 4.21.

EXAMPLE 3 **Evaluating Trigonometric Functions of 30° and 60°**

Use the equilateral triangle shown in Figure 4.21 to find the values of sin 60°, cos 60°, sin 30°, and cos 30°.

Solution For $\theta = 60°$, you have adj = 1, opp = $\sqrt{3}$, and hyp = 2. So,

$$\sin 60° = \frac{\text{opp}}{\text{hyp}} = \frac{\sqrt{3}}{2} \quad \text{and} \quad \cos 60° = \frac{\text{adj}}{\text{hyp}} = \frac{1}{2}.$$

For $\theta = 30°$, adj = $\sqrt{3}$, opp = 1, and hyp = 2. So,

$$\sin 30° = \frac{\text{opp}}{\text{hyp}} = \frac{1}{2} \quad \text{and} \quad \cos 30° = \frac{\text{adj}}{\text{hyp}} = \frac{\sqrt{3}}{2}.$$

✓ *Checkpoint*))) *Audio-video solution in English & Spanish at LarsonPrecalculus.com*

Use the equilateral triangle shown in Figure 4.21 to find the values of tan 60° and tan 30°.

Sines, Cosines, and Tangents of Special Angles

$$\sin 30° = \sin \frac{\pi}{6} = \frac{1}{2} \qquad \cos 30° = \cos \frac{\pi}{6} = \frac{\sqrt{3}}{2} \qquad \tan 30° = \tan \frac{\pi}{6} = \frac{\sqrt{3}}{3}$$

$$\sin 45° = \sin \frac{\pi}{4} = \frac{\sqrt{2}}{2} \qquad \cos 45° = \cos \frac{\pi}{4} = \frac{\sqrt{2}}{2} \qquad \tan 45° = \tan \frac{\pi}{4} = 1$$

$$\sin 60° = \sin \frac{\pi}{3} = \frac{\sqrt{3}}{2} \qquad \cos 60° = \cos \frac{\pi}{3} = \frac{1}{2} \qquad \tan 60° = \tan \frac{\pi}{3} = \sqrt{3}$$

Note that $\sin 30° = \frac{1}{2} = \cos 60°$. This occurs because 30° and 60° are complementary angles. In general, it can be shown from the right triangle definitions that *cofunctions of complementary angles are equal*. That is, if θ is an acute angle, then the relationships below are true.

$$\sin(90° - \theta) = \cos \theta \qquad \cos(90° - \theta) = \sin \theta \qquad \tan(90° - \theta) = \cot \theta$$

$$\cot(90° - \theta) = \tan \theta \qquad \sec(90° - \theta) = \csc \theta \qquad \csc(90° - \theta) = \sec \theta$$

To use a calculator to evaluate trigonometric functions of angles measured in degrees, remember to set the calculator to *degree* mode.

EXAMPLE 4 **Using a Calculator**

Use a calculator to evaluate sec 5° 40′ 12″.

Solution Begin by converting to decimal degree form. $\left[\text{Recall that } 1' = \frac{1}{60}(1°) \text{ and } 1'' = \frac{1}{3600}(1°).\right]$

$$5° \, 40' \, 12'' = 5° + \left(\frac{40}{60}\right)° + \left(\frac{12}{3600}\right)° = 5.67°$$

Then, use a calculator to evaluate sec 5.67°.

Function	Calculator Keystrokes	Display
sec 5° 40′ 12″ = sec 5.67°	(𝖨 [COS] (𝖨 5.67)𝖨)𝖨 [x⁻¹] [ENTER]	1.0049166

✓ *Checkpoint*))) *Audio-video solution in English & Spanish at LarsonPrecalculus.com*

Use a calculator to evaluate csc 34° 30′ 36″.

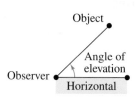

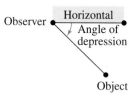

Figure 4.23

Applications Involving Right Triangles

Many applications of trigonometry involve **solving right triangles.** In this type of application, you are usually given one side of a right triangle and one of the acute angles and are asked to find one of the other sides, *or* you are given two sides and are asked to find one of the acute angles.

In Example 8, you are given the **angle of elevation,** which represents the angle from the horizontal upward to an object. In other applications you may be given the **angle of depression,** which represents the angle from the horizontal downward to an object. (See Figure 4.23.)

EXAMPLE 8 **Solving a Right Triangle**

A surveyor stands 115 feet from the base of the Washington Monument, as shown in Figure 4.24. The surveyor measures the angle of elevation to the top of the monument to be 78.3°. How tall is the Washington Monument?

Solution From Figure 4.24,

$$\tan 78.3° = \frac{\text{opp}}{\text{adj}} = \frac{y}{115}$$

where y is the height of the monument. So, the height of the Washington Monument is

$$y = 115 \tan 78.3°$$
$$\approx 115(4.8288)$$
$$\approx 555 \text{ feet.}$$

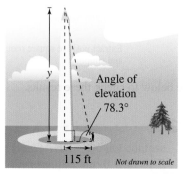

Figure 4.24

✓ *Checkpoint* ◄))) *Audio-video solution in English & Spanish at LarsonPrecalculus.com*

The angle of elevation to the top of a flagpole at a distance of 19 feet from its base is 64.6°. How tall is the flagpole?

EXAMPLE 9 **Solving a Right Triangle**

A lighthouse is 200 yards from a bike path along the edge of a lake. A walkway to the lighthouse is 400 yards long. (See Figure 4.25.) Find the acute angle θ between the bike path and the walkway.

Solution From Figure 4.25, the sine of the angle θ is

$$\sin \theta = \frac{\text{opp}}{\text{hyp}} = \frac{200}{400} = \frac{1}{2}.$$

You should recognize that $\theta = 30°$.

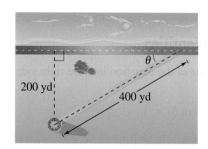

Figure 4.25

✓ *Checkpoint* ◄))) *Audio-video solution in English & Spanish at LarsonPrecalculus.com*

Find the acute angle θ between the two paths shown below.

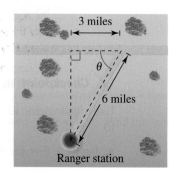

In Example 9, you were able to recognize that the special angle $\theta = 30°$ satisfies the equation $\sin \theta = \frac{1}{2}$. However, when θ is not a special angle, you can *estimate* its value. For example, to estimate the acute angle θ in the equation $\sin \theta = 0.6$, you could reason that $\sin 30° = \frac{1}{2} = 0.5000$ and $\sin 45° = 1/\sqrt{2} \approx 0.7071$, so θ lies somewhere between 30° and 45°. In a later section, you will study a method of determining a more precise value of θ.

EXAMPLE 10 Solving a Right Triangle

Find the length c and the height b of the skateboard ramp below.

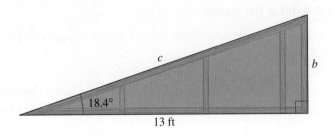

Solution   From the figure,

$$\cos 18.4° = \frac{\text{adj}}{\text{hyp}} = \frac{13}{c}.$$

So, the length of the skateboard ramp is

$$c = \frac{13}{\cos 18.4°} \approx \frac{13}{0.9489} \approx 13.7 \text{ feet.}$$

Also from the figure,

$$\tan 18.4° = \frac{\text{opp}}{\text{adj}} = \frac{b}{13}.$$

So, the height is

$$b = 13 \tan 18.4° \approx 13(0.3327) \approx 4.3 \text{ feet.}$$

✓ **Checkpoint** 🔊))) *Audio-video solution in English & Spanish at LarsonPrecalculus.com*

Find the length c and the horizontal length a of the loading ramp below.

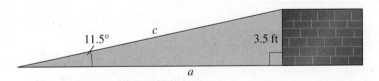

Skateboarders can go to a skatepark, which is a recreational environment built with many different types of ramps and rails.

Summarize   (Section 4.3)

1. State the right triangle definitions of the six trigonometric functions *(page 277)*. For examples of evaluating trigonometric functions of acute angles, see Examples 1–4.

2. List the reciprocal, quotient, and Pythagorean identities *(page 280)*. For examples of using these identities, see Examples 5–7.

3. Describe real-life applications of trigonometric functions *(pages 282 and 283, Examples 8–10)*.

4.3 Exercises

See **CalcChat.com** for tutorial help and worked-out solutions to odd-numbered exercises.

Vocabulary

1. Match each trigonometric function with its right triangle definition.

(a) sine   (b) cosine   (c) tangent   (d) cosecant   (e) secant   (f) cotangent

(i) $\dfrac{\text{hypotenuse}}{\text{adjacent}}$ (ii) $\dfrac{\text{adjacent}}{\text{opposite}}$ (iii) $\dfrac{\text{hypotenuse}}{\text{opposite}}$ (iv) $\dfrac{\text{adjacent}}{\text{hypotenuse}}$ (v) $\dfrac{\text{opposite}}{\text{hypotenuse}}$ (vi) $\dfrac{\text{opposite}}{\text{adjacent}}$

In Exercises 2–4, fill in the blanks.

2. Relative to the acute angle θ, the three sides of a right triangle are the _____ side, the _____ side, and the _____.

3. Cofunctions of _____ angles are equal.

4. An angle of _____ represents the angle from the horizontal upward to an object, whereas an angle of _____ represents the angle from the horizontal downward to an object.

Skills and Applications

 Evaluating Trigonometric Functions In Exercises 5–10, find the exact values of the six trigonometric functions of the angle θ.

5.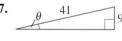

6.

7.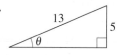

8.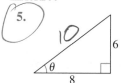

9.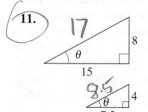

10.

Evaluating Trigonometric Functions In Exercises 11–14, find the exact values of the six trigonometric functions of the angle θ for each of the two triangles. Explain why the function values are the same.

11.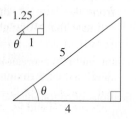

12.

13.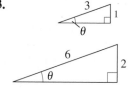

14.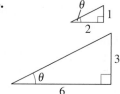

Evaluating Trigonometric Functions In Exercises 15–22, sketch a right triangle corresponding to the trigonometric function of the acute angle θ. Then find the exact values of the other five trigonometric functions of θ.

15. $\cos \theta = \frac{15}{17}$ 16. $\sin \theta = \frac{3}{5}$

17. $\sec \theta = \frac{6}{5}$ 18. $\tan \theta = \frac{4}{5}$

19. $\sin \theta = \frac{1}{5}$ 20. $\sec \theta = \frac{17}{7}$

21. $\cot \theta = 3$ 22. $\csc \theta = 9$

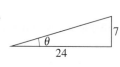 **Evaluating Trigonometric Functions of 30°, 45°, and 60°** In Exercises 23–28, construct an appropriate triangle to find the missing values. ($0° \leq \theta \leq 90°, 0 \leq \theta \leq \pi/2$)

Function	θ (deg)	θ (rad)	Function Value
23. tan	30°		
24. cos	45°		
25. sin		$\frac{\pi}{4}$	
26. tan		$\frac{\pi}{3}$	
27. sec		$\frac{\pi}{4}$	
28. csc		$\frac{\pi}{6}$	

Using a Calculator In Exercises 29–36, use a calculator to evaluate each function. Round your answers to four decimal places. (Be sure the calculator is in the correct mode.)

29. (a) $\sin 20°$ (b) $\cos 70°$

30. (a) $\tan 23.5°$ (b) $\cot 66.5°$

31. (a) $\sin 14.21°$ (b) $\csc 14.21°$

32. (a) $\cot 79.56°$ (b) $\sec 79.56°$

33. (a) $\cos 4° \, 50' \, 15''$ (b) $\sec 4° \, 50' \, 15''$

34. (a) $\sec 42° \, 12'$ (b) $\csc 48° \, 7'$

35. (a) $\cot 17° \, 15'$ (b) $\tan 17° \, 15'$

36. (a) $\sec 56° \, 8' \, 10''$ (b) $\cos 56° \, 8' \, 10''$

 Applying Trigonometric Identities In Exercises 37–42, use the given function value(s) and the trigonometric identities to find the exact value of each indicated trigonometric function.

37. $\sin 60° = \dfrac{\sqrt{3}}{2}, \quad \cos 60° = \dfrac{1}{2}$

   (a) $\sin 30°$ (b) $\cos 30°$

   (c) $\tan 60°$ (d) $\cot 60°$

38. $\sin 30° = \dfrac{1}{2}, \quad \tan 30° = \dfrac{\sqrt{3}}{3}$

   (a) $\csc 30°$ (b) $\cot 60°$

   (c) $\cos 30°$ (d) $\cot 30°$

39. $\cos \theta = \frac{1}{3}$

   (a) $\sin \theta$ (b) $\tan \theta$

   (c) $\sec \theta$ (d) $\csc(90° - \theta)$

40. $\sec \theta = 5$

   (a) $\cos \theta$ (b) $\cot \theta$

   (c) $\cot(90° - \theta)$ (d) $\sin \theta$

41. $\cot \alpha = 3$

   (a) $\tan \alpha$ (b) $\csc \alpha$

   (c) $\cot(90° - \alpha)$ (d) $\sin \alpha$

42. $\cos \beta = \dfrac{\sqrt{7}}{4}$

   (a) $\sec \beta$ (b) $\sin \beta$

   (c) $\cot \beta$ (d) $\sin(90° - \beta)$

 Using Trigonometric Identities In Exercises 43–52, use trigonometric identities to transform the left side of the equation into the right side $(0 < \theta < \pi/2)$.

43. $\tan \theta \cot \theta = 1$

44. $\cos \theta \sec \theta = 1$

45. $\tan \alpha \cos \alpha = \sin \alpha$

46. $\cot \alpha \sin \alpha = \cos \alpha$

47. $(1 + \sin \theta)(1 - \sin \theta) = \cos^2 \theta$

48. $(1 + \cos \theta)(1 - \cos \theta) = \sin^2 \theta$

49. $(\sec \theta + \tan \theta)(\sec \theta - \tan \theta) = 1$

50. $\sin^2 \theta - \cos^2 \theta = 2 \sin^2 \theta - 1$

51. $\dfrac{\sin \theta}{\cos \theta} + \dfrac{\cos \theta}{\sin \theta} = \csc \theta \sec \theta$

52. $\dfrac{\tan \beta + \cot \beta}{\tan \beta} = \csc^2 \beta$

Finding Special Angles of a Triangle In Exercises 53–58, find each value of θ in degrees $(0° < \theta < 90°)$ and radians $(0 < \theta < \pi/2)$ without using a calculator.

53. (a) $\sin \theta = \frac{1}{2}$ (b) $\csc \theta = 2$

54. (a) $\cos \theta = \dfrac{\sqrt{2}}{2}$ (b) $\tan \theta = 1$

55. (a) $\sec \theta = 2$ (b) $\cot \theta = 1$

56. (a) $\tan \theta = \sqrt{3}$ (b) $\csc \theta = \sqrt{2}$

57. (a) $\csc \theta = \dfrac{2\sqrt{3}}{3}$ (b) $\sin \theta = \dfrac{\sqrt{2}}{2}$

58. (a) $\cot \theta = \dfrac{\sqrt{3}}{3}$ (b) $\sec \theta = \sqrt{2}$

Finding Side Lengths of a Triangle In Exercises 59–62, find the exact values of the indicated variables.

59. Find x and y.   60. Find x and r.

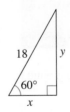 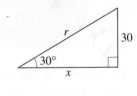

61. Find x and r.   62. Find x and r.

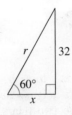 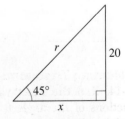

63. **Empire State Building** You are standing 45 meters from the base of the Empire State Building. You estimate that the angle of elevation to the top of the 86th floor (the observatory) is 82°. The total height of the building is another 123 meters above the 86th floor. What is the approximate height of the building? One of your friends is on the 86th floor. What is the distance between you and your friend?

64. Height of a Tower A six-foot person walks from the base of a broadcasting tower directly toward the tip of the shadow cast by the tower. When the person is 132 feet from the tower and 3 feet from the tip of the shadow, the person's shadow starts to appear beyond the tower's shadow.

(a) Draw a right triangle that gives a visual representation of the problem. Label the known quantities of the triangle and use a variable to represent the height of the tower.

(b) Use a trigonometric function to write an equation involving the unknown quantity.

(c) What is the height of the tower?

65. Angle of Elevation You are skiing down a mountain with a vertical height of 1250 feet. The distance from the top of the mountain to the base is 2500 feet. What is the angle of elevation from the base to the top of the mountain?

66. Biology A biologist wants to know the width w of a river to properly set instruments for an experiment. From point A, the biologist walks downstream 100 feet and sights to point C (see figure). From this sighting, it is determined that $\theta = 54°$. How wide is the river?

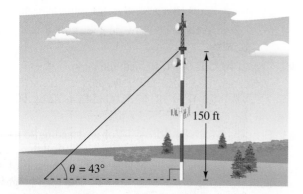

67. Guy Wire A guy wire runs from the ground to a cell tower. The wire is attached to the cell tower 150 feet above the ground. The angle formed between the wire and the ground is 43° (see figure).

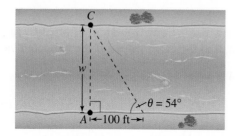

(a) How long is the guy wire?

(b) How far from the base of the tower is the guy wire anchored to the ground?

68. Height of a Mountain In traveling across flat land, you see a mountain directly in front of you. Its angle of elevation (to the peak) is 3.5°. After you drive 13 miles closer to the mountain, the angle of elevation is 9° (see figure). Approximate the height of the mountain.

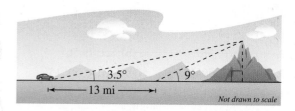

Not drawn to scale

69. Machine Shop Calculations A steel plate has the form of one-fourth of a circle with a radius of 60 centimeters. Two two-centimeter holes are drilled in the plate, positioned as shown in the figure. Find the coordinates of the center of each hole.

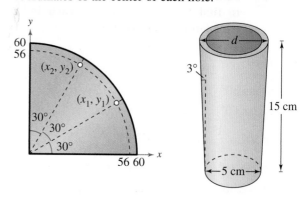

Figure for 69   Figure for 70

70. Machine Shop Calculations A tapered shaft has a diameter of 5 centimeters at the small end and is 15 centimeters long (see figure). The taper is 3°. Find the diameter d of the large end of the shaft.

71. Geometry Use a compass to sketch a quarter of a circle of radius 10 centimeters. Using a protractor, construct an angle of 20° in standard position (see figure). Drop a perpendicular line from the point of intersection of the terminal side of the angle and the arc of the circle. By actual measurement, calculate the coordinates (x, y) of the point of intersection and use these measurements to approximate the six trigonometric functions of a 20° angle.

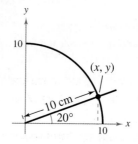

72. Helium-Filled Balloon

A 20-meter line is used to tether a helium-filled balloon. The line makes an angle of approximately 85° with the ground because of a breeze.

(a) Draw a right triangle that gives a visual representation of the problem. Label the known quantities of the triangle and use a variable to represent the height of the balloon.

(b) Use a trigonometric function to write and solve an equation for the height of the balloon.

(c) The breeze becomes stronger and the angle the line makes with the ground decreases. How does this affect the triangle you drew in part (a)?

(d) Complete the table, which shows the heights (in meters) of the balloon for decreasing angle measures θ.

Angle, θ	80°	70°	60°	50°
Height				

Angle, θ	40°	30°	20°	10°
Height				

(e) As θ approaches 0°, how does this affect the height of the balloon? Draw a right triangle to explain your reasoning.

73. Johnstown Inclined Plane The Johnstown Inclined Plane in Pennsylvania is one of the longest and steepest hoists in the world. The railway cars travel a distance of 896.5 feet at an angle of approximately 35.4°, rising to a height of 1693.5 feet above sea level.

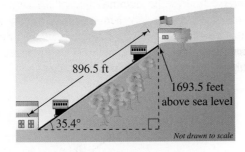

896.5 ft

1693.5 feet above sea level

35.4°

Not drawn to scale

(a) Find the vertical rise of the inclined plane.

(b) Find the elevation of the lower end of the inclined plane.

(c) The cars move up the mountain at a rate of 300 feet per minute. Find the rate at which they rise vertically.

74. Error Analysis Describe the error.

$$\cos 60° = \frac{\text{opp}}{\text{hyp}} = \frac{1}{2} \quad \times$$

Exploration

True or False? In Exercises 75–80, determine whether the statement is true or false. Justify your answer.

75. $\sin 60° \csc 60° = 1$ **76.** $\sec 30° = \csc 30°$

77. $\sin 45° + \cos 45° = 1$ **78.** $\cos 60° - \sin 30° = 0$

79. $\dfrac{\sin 60°}{\sin 30°} = \sin 2°$ **80.** $\tan[(5°)^2] = \tan^2 5°$

81. Think About It You are given the value of $\tan \theta$. Is it possible to find the value of $\sec \theta$ without finding the measure of θ? Explain.

82. **HOW DO YOU SEE IT?** Use the figure below.

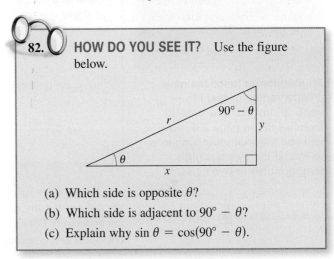

(a) Which side is opposite θ?

(b) Which side is adjacent to $90° - \theta$?

(c) Explain why $\sin \theta = \cos(90° - \theta)$.

83. Think About It Complete the table.

θ	0.1	0.2	0.3	0.4	0.5
$\sin \theta$					

(a) Is θ or $\sin \theta$ greater for θ in the interval $(0, 0.5]$?

(b) As θ approaches 0, how do θ and $\sin \theta$ compare? Explain.

84. Think About It Complete the table.

θ	0°	18°	36°	54°	72°	90°
$\sin \theta$						
$\cos \theta$						

(a) Discuss the behavior of the sine function for $0° \le \theta \le 90°$.

(b) Discuss the behavior of the cosine function for $0° \le \theta \le 90°$.

(c) Use the definitions of the sine and cosine functions to explain the results of parts (a) and (b).

4.4 Trigonometric Functions of Any Angle

- ■ Evaluate trigonometric functions of any angle.
- ■ Find reference angles.
- ■ Evaluate trigonometric functions of real numbers.

Introduction

In Section 4.3, the definitions of trigonometric functions were restricted to acute angles. In this section, the definitions are extended to cover *any* angle. When θ is an *acute* angle, the definitions here coincide with those in the preceding section.

Trigonometric functions have a wide variety of real-life applications. For example, in Exercise 99 on page 296, you will use trigonometric functions to model the average high temperatures in two cities.

Definitions of Trigonometric Functions of Any Angle

Let θ be an angle in standard position with (x, y) a point on the terminal side of θ and $r = \sqrt{x^2 + y^2} \neq 0$.

$$\sin \theta = \frac{y}{r} \qquad\qquad \cos \theta = \frac{x}{r}$$

$$\tan \theta = \frac{y}{x}, \quad x \neq 0 \qquad \cot \theta = \frac{x}{y}, \quad y \neq 0$$

$$\sec \theta = \frac{r}{x}, \quad x \neq 0 \qquad \csc \theta = \frac{r}{y}, \quad y \neq 0$$

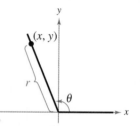

Because $r = \sqrt{x^2 + y^2}$ *cannot* be zero, it follows that the sine and cosine functions are defined for any real value of θ. However, when $x = 0$, the tangent and secant of θ are undefined. For example, the tangent of $90°$ is undefined. Similarly, when $y = 0$, the cotangent and cosecant of θ are undefined.

EXAMPLE 1 **Evaluating Trigonometric Functions**

Let $(-3, 4)$ be a point on the terminal side of θ. Find the sine, cosine, and tangent of θ.

Solution Referring to Figure 4.26, $x = -3$, $y = 4$, and

$$\begin{aligned} r &= \sqrt{x^2 + y^2} \\ &= \sqrt{(-3)^2 + 4^2} \\ &= 5. \end{aligned}$$

So, you have

$$\sin \theta = \frac{y}{r} = \frac{4}{5}$$

$$\cos \theta = \frac{x}{r} = -\frac{3}{5}$$

and

$$\tan \theta = \frac{y}{x} = -\frac{4}{3}.$$

Figure 4.26

▷ **ALGEBRA HELP** The formula $r = \sqrt{x^2 + y^2}$ is an application of the Distance Formula. To review the Distance Formula, see Section 1.1.

✓ *Checkpoint* 🔊))) *Audio-video solution in English & Spanish at LarsonPrecalculus.com*

Let $(-2, 3)$ be a point on the terminal side of θ. Find the sine, cosine, and tangent of θ.

The *signs* of the trigonometric functions in the four quadrants can be determined from the definitions of the functions. For example, $\cos \theta = x/r$, so $\cos \theta$ is positive wherever $x > 0$, which is in Quadrants I and IV. (Remember, r is always positive.) Figure 4.27 shows this and other results. Use similar reasoning to verify the other results.

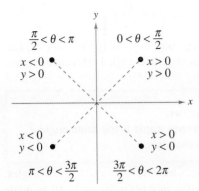

$\frac{\pi}{2} < \theta < \pi$ $0 < \theta < \frac{\pi}{2}$

$x < 0$ $x > 0$
$y > 0$ $y > 0$

$x < 0$ $x > 0$
$y < 0$ $y < 0$

$\pi < \theta < \frac{3\pi}{2}$ $\frac{3\pi}{2} < \theta < 2\pi$

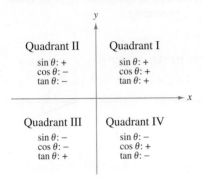

Quadrant II
$\sin \theta$: +
$\cos \theta$: −
$\tan \theta$: −

Quadrant I
$\sin \theta$: +
$\cos \theta$: +
$\tan \theta$: +

Quadrant III
$\sin \theta$: −
$\cos \theta$: −
$\tan \theta$: +

Quadrant IV
$\sin \theta$: −
$\cos \theta$: +
$\tan \theta$: −

Figure 4.27

EXAMPLE 2 Evaluating Trigonometric Functions

Given $\tan \theta = -\frac{5}{4}$ and $\cos \theta > 0$, find $\sin \theta$ and $\sec \theta$.

Solution Note that θ lies in Quadrant IV because that is the only quadrant in which the tangent is negative and the cosine is positive. Moreover, using

$$\tan \theta = \frac{y}{x} = -\frac{5}{4}$$

and the fact that y is negative in Quadrant IV, let $y = -5$ and $x = 4$. So, $r = \sqrt{16 + 25} = \sqrt{41}$ and you have the results below.

$$\sin \theta = \frac{y}{r}$$

$$= \frac{-5}{\sqrt{41}} \qquad \text{Exact value}$$

$$\approx -0.7809 \qquad \text{Approximate value}$$

$$\sec \theta = \frac{r}{x}$$

$$= \frac{\sqrt{41}}{4} \qquad \text{Exact value}$$

$$\approx 1.6008 \qquad \text{Approximate value}$$

✓ **Checkpoint** ◀))) *Audio-video solution in English & Spanish at LarsonPrecalculus.com*

Given $\sin \theta = \frac{4}{5}$ and $\tan \theta < 0$, find $\cos \theta$ and $\tan \theta$.

EXAMPLE 3 Trigonometric Functions of Quadrantal Angles

Evaluate the cosine and tangent functions at the quadrantal angles 0, $\frac{\pi}{2}$, π, and $\frac{3\pi}{2}$.

Solution To begin, choose a point on the terminal side of each angle, as shown in Figure 4.28. For each of the four points, $r = 1$ and you have the results below.

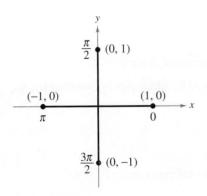

$\frac{\pi}{2}$ • $(0, 1)$

$(-1, 0)$ $(1, 0)$
π 0

$\frac{3\pi}{2}$ • $(0, -1)$

Figure 4.28

$$\cos 0 = \frac{x}{r} = \frac{1}{1} = 1 \qquad \tan 0 = \frac{y}{x} = \frac{0}{1} = 0 \qquad (x, y) = (1, 0)$$

$$\cos \frac{\pi}{2} = \frac{x}{r} = \frac{0}{1} = 0 \qquad \tan \frac{\pi}{2} = \frac{y}{x} = \frac{1}{0} \implies \text{undefined} \qquad (x, y) = (0, 1)$$

$$\cos \pi = \frac{x}{r} = \frac{-1}{1} = -1 \qquad \tan \pi = \frac{y}{x} = \frac{0}{-1} = 0 \qquad (x, y) = (-1, 0)$$

$$\cos \frac{3\pi}{2} = \frac{x}{r} = \frac{0}{1} = 0 \qquad \tan \frac{3\pi}{2} = \frac{y}{x} = \frac{-1}{0} \implies \text{undefined} \qquad (x, y) = (0, -1)$$

✓ **Checkpoint** ◀))) *Audio-video solution in English & Spanish at LarsonPrecalculus.com*

Evaluate the sine and cotangent functions at the quadrantal angle $\frac{3\pi}{2}$.

Reference Angles

The values of the trigonometric functions of angles greater than 90° (or less than 0°) can be determined from their values at corresponding acute angles called **reference angles.**

Definition of a Reference Angle

Let θ be an angle in standard position. Its **reference angle** is the acute angle θ' formed by the terminal side of θ and the horizontal axis.

The three figures below show the reference angles for θ in Quadrants II, III, and IV.

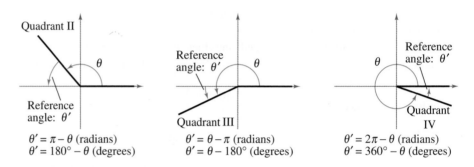

$\theta' = \pi - \theta$ (radians)
$\theta' = 180° - \theta$ (degrees)

$\theta' = \theta - \pi$ (radians)
$\theta' = \theta - 180°$ (degrees)

$\theta' = 2\pi - \theta$ (radians)
$\theta' = 360° - \theta$ (degrees)

EXAMPLE 4 **Finding Reference Angles**

Find the reference angle θ'.

a. $\theta = 300°$ **b.** $\theta = 2.3$ **c.** $\theta = -135°$

Solution

a. Because 300° lies in Quadrant IV, the angle it makes with the x-axis is

$$\theta' = 360° - 300°$$
$$= 60°. \qquad \text{Degrees}$$

Figure 4.29 shows the angle $\theta = 300°$ and its reference angle $\theta' = 60°$.

b. Because 2.3 lies between $\pi/2 \approx 1.5708$ and $\pi \approx 3.1416$, it follows that it is in Quadrant II and its reference angle is

$$\theta' = \pi - 2.3$$
$$\approx 0.8416. \qquad \text{Radians}$$

Figure 4.30 shows the angle $\theta = 2.3$ and its reference angle $\theta' = \pi - 2.3$.

c. First, determine that $-135°$ is coterminal with 225°, which lies in Quadrant III. So, the reference angle is

$$\theta' = 225° - 180°$$
$$= 45°. \qquad \text{Degrees}$$

Figure 4.31 shows the angle $\theta = -135°$ and its reference angle $\theta' = 45°$.

✓ **Checkpoint** 🔊))) *Audio-video solution in English & Spanish at LarsonPrecalculus.com*

Find the reference angle θ'.

a. $\theta = 213°$ **b.** $\theta = \dfrac{14\pi}{9}$ **c.** $\theta = \dfrac{4\pi}{5}$

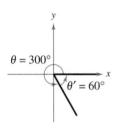

Figure 4.29

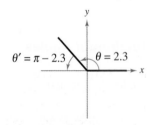

Figure 4.30

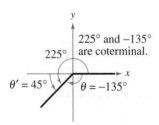

Figure 4.31

Trigonometric Functions of Real Numbers

To see how to use a reference angle to evaluate a trigonometric function, consider the point (x, y) on the terminal side of the angle θ, as shown at the right. You know that

$$\sin \theta = \frac{y}{r}$$

and

$$\tan \theta = \frac{y}{x}.$$

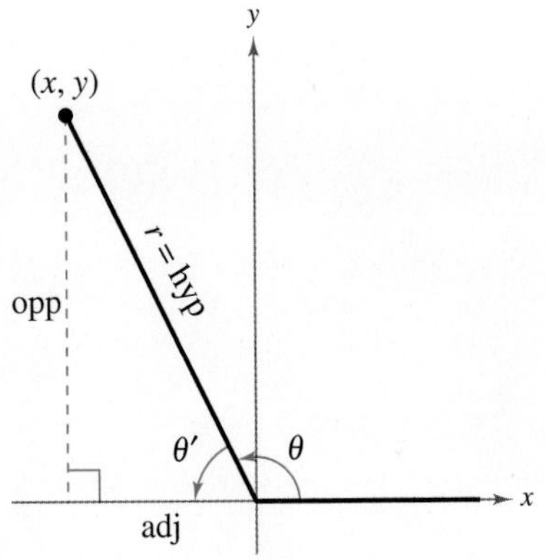

For the right triangle with acute angle θ' and sides of lengths $|x|$ and $|y|$, you have

$$\sin \theta' = \frac{\text{opp}}{\text{hyp}} = \frac{|y|}{r}$$

and

$$\tan \theta' = \frac{\text{opp}}{\text{adj}} = \frac{|y|}{|x|}.$$

opp = $|y|$, adj = $|x|$

So, it follows that $\sin \theta$ and $\sin \theta'$ are equal, *except possibly in sign*. The same is true for $\tan \theta$ and $\tan \theta'$ and for the other four trigonometric functions. In all cases, the quadrant in which θ lies determines the sign of the function value.

Evaluating Trigonometric Functions of Any Angle

To find the value of a trigonometric function of any angle θ:

1. Determine the function value of the associated reference angle θ'.
2. Depending on the quadrant in which θ lies, affix the appropriate sign to the function value.

- **REMARK** Learning the table of values at the right is worth the effort because doing so will increase both your efficiency and your confidence when working in trigonometry. Below is a pattern for the sine function that may help you remember the values.

θ	0°	30°	45°	60°	90°
$\sin \theta$	$\dfrac{\sqrt{0}}{2}$	$\dfrac{\sqrt{1}}{2}$	$\dfrac{\sqrt{2}}{2}$	$\dfrac{\sqrt{3}}{2}$	$\dfrac{\sqrt{4}}{2}$

Reverse the order to get cosine values of the same angles.

Using reference angles and the special angles discussed in the preceding section enables you to greatly extend the scope of *exact* trigonometric function values. For example, knowing the function values of 30° means that you know the function values of all angles for which 30° is a reference angle. For convenience, the table below shows the exact values of the sine, cosine, and tangent functions of special angles and quadrantal angles.

Trigonometric Values of Common Angles

θ (degrees)	0°	30°	45°	60°	90°	180°	270°
θ (radians)	0	$\dfrac{\pi}{6}$	$\dfrac{\pi}{4}$	$\dfrac{\pi}{3}$	$\dfrac{\pi}{2}$	π	$\dfrac{3\pi}{2}$
$\sin \theta$	0	$\dfrac{1}{2}$	$\dfrac{\sqrt{2}}{2}$	$\dfrac{\sqrt{3}}{2}$	1	0	-1
$\cos \theta$	1	$\dfrac{\sqrt{3}}{2}$	$\dfrac{\sqrt{2}}{2}$	$\dfrac{1}{2}$	0	-1	0
$\tan \theta$	0	$\dfrac{\sqrt{3}}{3}$	1	$\sqrt{3}$	Undef.	0	Undef.

| EXAMPLE 5 | **Using Reference Angles** |

See LarsonPrecalculus.com for an interactive version of this type of example.

Evaluate each trigonometric function.

a. $\cos\dfrac{4\pi}{3}$ **b.** $\tan(-210°)$ **c.** $\csc\dfrac{11\pi}{4}$

Solution

a. Because $\theta = 4\pi/3$ lies in Quadrant III, the reference angle is

$$\theta' = \frac{4\pi}{3} - \pi = \frac{\pi}{3}$$

as shown at the right. The cosine is negative in Quadrant III, so

$$\cos\frac{4\pi}{3} = (-)\cos\frac{\pi}{3}$$

$$= -\frac{1}{2}.$$

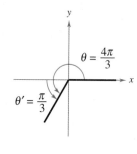

b. Because $-210° + 360° = 150°$, it follows that $-210°$ is coterminal with the second-quadrant angle $150°$. So, the reference angle is

$$\theta' = 180° - 150°$$

$$= 30°$$

as shown at the right. The tangent is negative in Quadrant II, so

$$\tan(-210°) = (-)\tan 30°$$

$$= -\frac{\sqrt{3}}{3}.$$

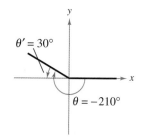

c. Because $(11\pi/4) - 2\pi = 3\pi/4$, it follows that $11\pi/4$ is coterminal with the second-quadrant angle $3\pi/4$. So, the reference angle is

$$\theta' = \pi - \frac{3\pi}{4} = \frac{\pi}{4}$$

as shown at the right. The cosecant is positive in Quadrant II, so

$$\csc\frac{11\pi}{4} = (+)\csc\frac{\pi}{4}$$

$$= \frac{1}{\sin(\pi/4)}$$

$$= \sqrt{2}.$$

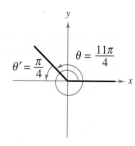

✓ *Checkpoint* 🔊))) *Audio-video solution in English & Spanish at LarsonPrecalculus.com*

Evaluate each trigonometric function.

a. $\sin\dfrac{7\pi}{4}$ **b.** $\cos(-120°)$ **c.** $\tan\dfrac{11\pi}{6}$

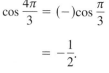

EXAMPLE 6 **Using Trigonometric Identities**

Let θ be an angle in Quadrant II such that $\sin \theta = \frac{1}{3}$. Find (a) $\cos \theta$ and (b) $\tan \theta$ by using trigonometric identities.

Solution

▷ **REMARK** The fundamental trigonometric identities listed in the preceding section (for an acute angle θ) are also valid when θ is any angle in the domain of the function.

a. Using the Pythagorean identity $\sin^2 \theta + \cos^2 \theta = 1$, you obtain

$$\left(\frac{1}{3}\right)^2 + \cos^2 \theta = 1 \implies \cos^2 \theta = 1 - \frac{1}{9} = \frac{8}{9}.$$

You know that $\cos \theta < 0$ in Quadrant II, so use the negative root to obtain

$$\cos \theta = -\frac{\sqrt{8}}{\sqrt{9}} = -\frac{2\sqrt{2}}{3}.$$

b. Using the trigonometric identity $\tan \theta = \dfrac{\sin \theta}{\cos \theta}$, you obtain

$$\tan \theta = \frac{1/3}{-2\sqrt{2}/3} = -\frac{1}{2\sqrt{2}} = -\frac{\sqrt{2}}{4}.$$

✓ *Checkpoint*  *Audio-video solution in English & Spanish at LarsonPrecalculus.com*

Let θ be an angle in Quadrant III such that $\sin \theta = -\frac{4}{5}$. Find (a) $\cos \theta$ and (b) $\tan \theta$ by using trigonometric identities.

EXAMPLE 7 **Using a Calculator**

Use a calculator to evaluate each trigonometric function.

a. $\cot 410°$ b. $\sin(-7)$ c. $\sec \dfrac{\pi}{9}$

Solution

Function	Mode	Calculator Keystrokes	Display
a. $\cot 410°$	Degree	(⟨TAN⟩ (410)) ⟨x⁻¹⟩ ⟨ENTER⟩	0.8390996
b. $\sin(-7)$	Radian	⟨SIN⟩ (⟨(−)⟩ 7) ⟨ENTER⟩	−0.6569866
c. $\sec(\pi/9)$	Radian	(⟨COS⟩ (π ÷ 9)) ⟨x⁻¹⟩ ⟨ENTER⟩	1.0641778

✓ *Checkpoint*  *Audio-video solution in English & Spanish at LarsonPrecalculus.com*

Use a calculator to evaluate each trigonometric function.

a. $\tan 119°$ b. $\csc 5$ c. $\cos \dfrac{\pi}{5}$

Summarize (Section 4.4)

1. State the definitions of the trigonometric functions of any angle *(page 288)*. For examples of evaluating trigonometric functions, see Examples 1–3.

2. Explain how to use a reference angle *(page 290)*. For an example of finding reference angles, see Example 4.

3. Explain how to evaluate a trigonometric function of a real number *(page 291)*. For examples of evaluating trigonometric functions of real numbers, see Examples 5–7.

4.4 Exercises

See **CalcChat.com** for tutorial help and worked-out solutions to odd-numbered exercises.

Vocabulary: Fill in the blanks.

In Exercises 1–6, let θ be an angle in standard position with (x, y) a point on the terminal side of θ and $r = \sqrt{x^2 + y^2} \neq 0$.

1. $\sin \theta = $ _____
2. $\dfrac{r}{y} = $ _____
3. $\tan \theta = $ _____

4. $\sec \theta = $ _____
5. $\dfrac{x}{r} = $ _____
6. $\dfrac{x}{y} = $ _____

7. Because $r = \sqrt{x^2 + y^2}$ cannot be _____, the sine and cosine functions are _____ for any real value of θ.

8. The acute angle formed by the terminal side of an angle θ in standard position and the horizontal axis is the _____ angle of θ and is denoted by θ'.

Skills and Applications

 Evaluating Trigonometric Functions In Exercises 9–12, find the exact values of the six trigonometric functions of each angle θ.

9. (a) 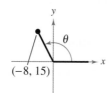 (b)

10. (a) 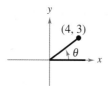 (b)

11. (a) 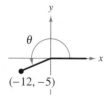 (b)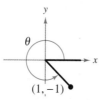

12. (a) 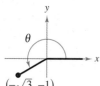 (b)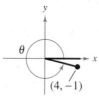

Evaluating Trigonometric Functions In Exercises 13–18, the point is on the terminal side of an angle in standard position. Find the exact values of the six trigonometric functions of the angle.

13. $(5, 12)$
14. $(8, 15)$
15. $(-5, -2)$
16. $(-4, 10)$
17. $(-5.4, 7.2)$
18. $\left(3\frac{1}{2}, -2\sqrt{15}\right)$

Determining a Quadrant In Exercises 19–22, determine the quadrant in which θ lies.

19. $\sin \theta > 0$, $\cos \theta > 0$
20. $\sin \theta < 0$, $\cos \theta < 0$
21. $\csc \theta > 0$, $\tan \theta < 0$
22. $\sec \theta > 0$, $\cot \theta < 0$

 Evaluating Trigonometric Functions In Exercises 23–32, find the exact values of the remaining trigonometric functions of θ satisfying the given conditions.

23. $\tan \theta = \frac{15}{8}$, $\sin \theta > 0$
24. $\cos \theta = \frac{8}{17}$, $\tan \theta < 0$
25. $\sin \theta = 0.6$, θ lies in Quadrant II.
26. $\cos \theta = -0.8$, θ lies in Quadrant III.
27. $\cot \theta = -3$, $\cos \theta > 0$
28. $\csc \theta = 4$, $\cot \theta < 0$
29. $\cos \theta = 0$, $\csc \theta = 1$
30. $\sin \theta = 0$, $\sec \theta = -1$
31. $\cot \theta$ is undefined, $\dfrac{\pi}{2} \leq \theta \leq \dfrac{3\pi}{2}$
32. $\tan \theta$ is undefined, $\pi \leq \theta \leq 2\pi$

 An Angle Formed by a Line Through the Origin In Exercises 33–36, the terminal side of θ lies on the given line in the specified quadrant. Find the exact values of the six trigonometric functions of θ by finding a point on the line.

Line	Quadrant
33. $y = -x$	II
34. $y = \frac{1}{3}x$	III
35. $2x - y = 0$	I
36. $4x + 3y = 0$	IV

Trigonometric Function of a Quadrantal Angle In Exercises 37–46, evaluate the trigonometric function of the quadrantal angle, if possible.

37. $\sin 0$

38. $\csc \dfrac{3\pi}{2}$

39. $\sec \dfrac{3\pi}{2}$

40. $\sec \pi$

41. $\sin \dfrac{\pi}{2}$

42. $\cot 0$

43. $\csc \pi$

44. $\cot \dfrac{\pi}{2}$

45. $\cos \dfrac{9\pi}{2}$

46. $\tan\left(-\dfrac{\pi}{2}\right)$

Finding a Reference Angle In Exercises 47–54, find the reference angle θ'. Sketch θ in standard position and label θ'.

47. $\theta = 160°$

48. $\theta = 309°$

49. $\theta = -125°$

50. $\theta = -215°$

51. $\theta = \dfrac{2\pi}{3}$

52. $\theta = \dfrac{7\pi}{6}$

53. $\theta = 4.8$

54. $\theta = 12.9$

Using a Reference Angle In Exercises 55–68, evaluate the sine, cosine, and tangent of the angle without using a calculator.

55. $225°$

56. $300°$

57. $750°$

58. $675°$

59. $-120°$

60. $-570°$

61. $\dfrac{2\pi}{3}$

62. $\dfrac{3\pi}{4}$

63. $-\dfrac{\pi}{6}$

64. $-\dfrac{2\pi}{3}$

65. $\dfrac{11\pi}{4}$

66. $\dfrac{13\pi}{6}$

67. $-\dfrac{17\pi}{6}$

68. $-\dfrac{23\pi}{4}$

Using a Trigonometric Identity In Exercises 69–74, use the function value to find the indicated trigonometric value in the specified quadrant.

Function Value	Quadrant	Trigonometric Value
69. $\sin \theta = -\frac{3}{5}$	IV	$\cos \theta$
70. $\cot \theta = -3$	II	$\csc \theta$
71. $\tan \theta = \frac{3}{2}$	III	$\sec \theta$
72. $\csc \theta = -2$	IV	$\cot \theta$
73. $\cos \theta = \frac{5}{8}$	I	$\csc \theta$
74. $\sec \theta = -\frac{9}{4}$	III	$\cot \theta$

Using a Calculator In Exercises 75–90, use a calculator to evaluate the trigonometric function. Round your answer to four decimal places. (Be sure the calculator is in the correct mode.)

75. $\sin 10°$

76. $\tan 304°$

77. $\cos(-110°)$

78. $\sin(-330°)$

79. $\cot 178°$

80. $\sec 72°$

81. $\csc 405°$

82. $\cot(-560°)$

83. $\tan \dfrac{\pi}{9}$

84. $\cos \dfrac{2\pi}{7}$

85. $\sec \dfrac{11\pi}{8}$

86. $\csc \dfrac{15\pi}{4}$

87. $\sin(-0.65)$

88. $\cos 1.35$

89. $\csc(-10)$

90. $\sec(-4.6)$

Solving for θ In Exercises 91–96, find two solutions of each equation. Give your answers in degrees $(0° \le \theta < 360°)$ and in radians $(0 \le \theta < 2\pi)$. Do not use a calculator.

91. (a) $\sin \theta = \dfrac{1}{2}$ **92.** (a) $\cos \theta = \dfrac{\sqrt{2}}{2}$

(b) $\sin \theta = -\dfrac{1}{2}$ (b) $\cos \theta = -\dfrac{\sqrt{2}}{2}$

93. (a) $\cos \theta = \dfrac{1}{2}$ **94.** (a) $\sin \theta = \dfrac{\sqrt{3}}{2}$

(b) $\sec \theta = 2$ (b) $\csc \theta = \dfrac{2\sqrt{3}}{3}$

95. (a) $\tan \theta = 1$ **96.** (a) $\cot \theta = 0$

(b) $\cot \theta = -\sqrt{3}$ (b) $\sec \theta = -\sqrt{2}$

97. Distance An airplane, flying at an altitude of 6 miles, is on a flight path that passes directly over an observer (see figure). Let θ be the angle of elevation from the observer to the plane. Find the distance d from the observer to the plane when (a) $\theta = 30°$, (b) $\theta = 90°$, and (c) $\theta = 120°$.

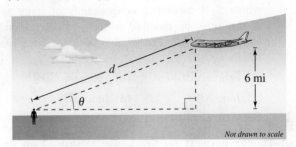

Not drawn to scale

98. Harmonic Motion The displacement from equilibrium of an oscillating weight suspended by a spring is given by $y(t) = 2 \cos 6t$, where y is the displacement in centimeters and t is the time in seconds. Find the displacement when (a) $t = 0$, (b) $t = \frac{1}{4}$, and (c) $t = \frac{1}{2}$.

• • 99. Temperature • • • • • • • • • • • • • • • •

The table shows the average high temperatures (in degrees Fahrenheit) in Boston, Massachusetts (*B*), and Fairbanks, Alaska (*F*), for selected months in 2015. *(Source: U.S. Climate Data)*

Spreadsheet at LarsonPrecalculus.com

DATA Month	Boston, *B*	Fairbanks, *F*
January	33	1
March	41	31
June	72	71
August	83	62
November	56	17

(a) Use the *regression* feature of a graphing utility to find a model of the form

$$y = a \sin(bt + c) + d$$

for each city. Let *t* represent the month, with *t* = 1 corresponding to January.

(b) Use the models from part (a) to estimate the monthly average high temperatures for the two cities in February, April, May, July, September, October, and December.

(c) Use a graphing utility to graph both models in the same viewing window. Compare the temperatures for the two cities.

100. Sales A company that produces snowboards forecasts monthly sales over the next 2 years to be

$$S = 23.1 + 0.442t + 4.3 \cos \frac{\pi t}{6}$$

where *S* is measured in thousands of units and *t* is the time in months, with *t* = 1 corresponding to January 2017. Predict the sales for each of the following months.

(a) February 2017   (b) February 2018

(c) June 2017   (d) June 2018

101. Electric Circuits The current *I* (in amperes) when 100 volts is applied to a circuit is given by

$$I = 5e^{-2t} \sin t$$

where *t* is the time (in seconds) after the voltage is applied. Approximate the current at *t* = 0.7 second after the voltage is applied.

102.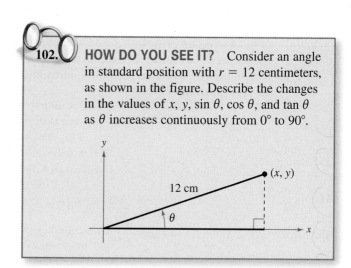

HOW DO YOU SEE IT? Consider an angle in standard position with *r* = 12 centimeters, as shown in the figure. Describe the changes in the values of *x*, *y*, sin *θ*, cos *θ*, and tan *θ* as *θ* increases continuously from 0° to 90°.

Exploration

True or False? **In Exercises 103 and 104, determine whether the statement is true or false. Justify your answer.**

103. In each of the four quadrants, the signs of the secant function and the sine function are the same.

104. The reference angle for an angle *θ* (in degrees) is the angle $\theta' = 360°n - \theta$, where *n* is an integer and $0° \le \theta' \le 360°$.

105. Writing Write a short essay explaining to a classmate how to evaluate the six trigonometric functions of any angle *θ* in standard position. Include an explanation of reference angles and how to use them, the signs of the functions in each of the four quadrants, and the trigonometric values of common angles. Include figures or diagrams in your essay.

106. Think About It The figure shows point *P*(*x*, *y*) on a unit circle and right triangle *OAP*.

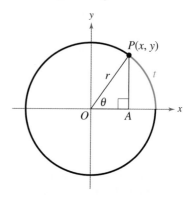

(a) Find sin *t* and cos *t* using the unit circle definitions of sine and cosine (from Section 4.2).

(b) What is the value of *r*? Explain.

(c) Use the definitions of sine and cosine given in this section to find sin *θ* and cos *θ*. Write your answers in terms of *x* and *y*.

(d) Based on your answers to parts (a) and (c), what can you conclude?

4.5 Graphs of Sine and Cosine Functions

Graphs of sine and cosine functions have many scientific applications. For example, in Exercise 80 on page 306, you will use the graph of a sine function to analyze airflow during a respiratory cycle.

- Sketch the graphs of basic sine and cosine functions.
- Use amplitude and period to help sketch the graphs of sine and cosine functions.
- Sketch translations of the graphs of sine and cosine functions.
- Use sine and cosine functions to model real-life data.

Basic Sine and Cosine Curves

In this section, you will study techniques for sketching the graphs of the sine and cosine functions. The graph of the sine function, shown in Figure 4.32, is a **sine curve.** In the figure, the black portion of the graph represents one period of the function and is **one cycle** of the sine curve. The gray portion of the graph indicates that the basic sine curve repeats indefinitely to the left and right. Figure 4.33 shows the graph of the cosine function.

Recall from Section 4.4 that the domain of the sine and cosine functions is the set of all real numbers. Moreover, the range of each function is the interval $[-1, 1]$, and each function has a period of 2π. This information is consistent with the basic graphs shown in Figures 4.32 and 4.33.

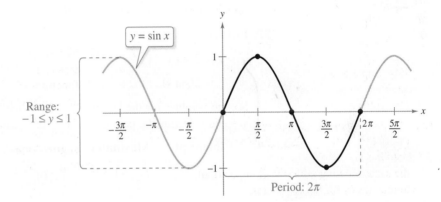

Figure 4.32

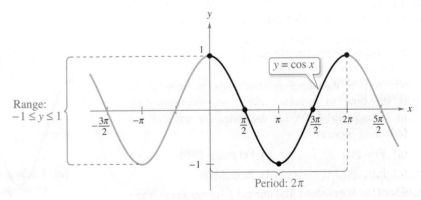

Figure 4.33

Note in Figures 4.32 and 4.33 that the sine curve is symmetric with respect to the *origin*, whereas the cosine curve is symmetric with respect to the *y-axis*. These properties of symmetry follow from the fact that the sine function is odd and the cosine function is even.

To sketch the graphs of the basic sine and cosine functions, it helps to note five **key points** in one period of each graph: the *intercepts, maximum points,* and *minimum points* (see graphs below).

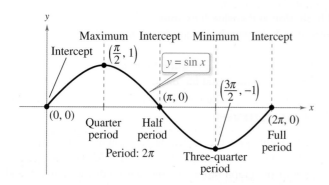

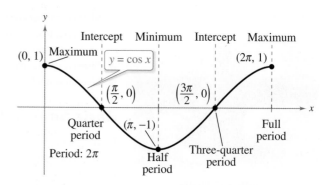

EXAMPLE 1   **Using Key Points to Sketch a Sine Curve**

See LarsonPrecalculus.com for an interactive version of this type of example.

Sketch the graph of

$$y = 2 \sin x$$

on the interval $[-\pi, 4\pi]$.

Solution   Note that

$$y = 2 \sin x$$
$$= 2(\sin x).$$

So, the y-values for the key points have twice the magnitude of those on the graph of $y = \sin x$. Divide the period 2π into four equal parts to obtain the key points

Intercept	Maximum	Intercept	Minimum	Intercept
$(0, 0)$,	$\left(\dfrac{\pi}{2}, 2\right)$,	$(\pi, 0)$,	$\left(\dfrac{3\pi}{2}, -2\right)$, and	$(2\pi, 0)$.

By connecting these key points with a smooth curve and extending the curve in both directions over the interval $[-\pi, 4\pi]$, you obtain the graph below.

▷ **TECHNOLOGY** When using a graphing utility to graph trigonometric functions, pay special attention to the viewing window you use. For example, graph

$$y = \dfrac{\sin 10x}{10}$$

in the standard viewing window in *radian* mode. What do you observe? Use the *zoom* feature to find a viewing window that displays a good view of the graph.

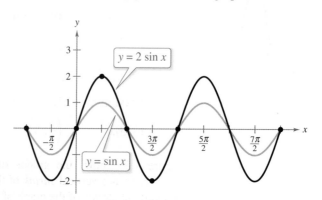

✓ **Checkpoint** 🔊))) *Audio-video solution in English & Spanish at LarsonPrecalculus.com*

Sketch the graph of

$$y = 2 \cos x$$

on the interval $\left[-\dfrac{\pi}{2}, \dfrac{9\pi}{2}\right]$.

Amplitude and Period

In the rest of this section, you will study the effect of each of the constants a, b, c, and d on the graphs of equations of the forms

$$y = d + a \sin(bx - c)$$

and

$$y = d + a \cos(bx - c).$$

A quick review of the transformations you studied in Section 1.7 will help in this investigation.

The constant factor a in $y = a \sin x$ and $y = a \cos x$ acts as a *scaling factor*—a *vertical stretch* or *vertical shrink* of the basic curve. When $|a| > 1$, the basic curve is stretched, and when $0 < |a| < 1$, the basic curve is shrunk. The result is that the graphs of $y = a \sin x$ and $y = a \cos x$ range between $-a$ and a instead of between -1 and 1. The absolute value of a is the **amplitude** of the function. The range of the function for $a > 0$ is $-a \leq y \leq a$.

Definition of the Amplitude of Sine and Cosine Curves

The **amplitude** of $y = a \sin x$ and $y = a \cos x$ represents half the distance between the maximum and minimum values of the function and is given by

$$\text{Amplitude} = |a|.$$

EXAMPLE 2 **Scaling: Vertical Shrinking and Stretching**

In the same coordinate plane, sketch the graph of each function.

a. $y = \dfrac{1}{2} \cos x$

b. $y = 3 \cos x$

Solution

a. The amplitude of $y = \frac{1}{2} \cos x$ is $\frac{1}{2}$, so the maximum value is $\frac{1}{2}$ and the minimum value is $-\frac{1}{2}$. Divide one cycle, $0 \leq x \leq 2\pi$, into four equal parts to obtain the key points

Maximum	Intercept	Minimum	Intercept	Maximum
$\left(0, \dfrac{1}{2}\right)$,	$\left(\dfrac{\pi}{2}, 0\right)$,	$\left(\pi, -\dfrac{1}{2}\right)$,	$\left(\dfrac{3\pi}{2}, 0\right)$, and	$\left(2\pi, \dfrac{1}{2}\right)$.

b. A similar analysis shows that the amplitude of $y = 3 \cos x$ is 3, and the key points are

Maximum	Intercept	Minimum	Intercept	Maximum
$(0, 3)$,	$\left(\dfrac{\pi}{2}, 0\right)$,	$(\pi, -3)$,	$\left(\dfrac{3\pi}{2}, 0\right)$, and	$(2\pi, 3)$.

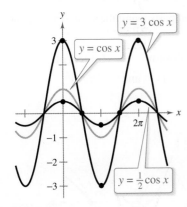

Figure 4.34

Figure 4.34 shows the graphs of these two functions. Notice that the graph of $y = \frac{1}{2} \cos x$ is a vertical *shrink* of the graph of $y = \cos x$ and the graph of $y = 3 \cos x$ is a vertical *stretch* of the graph of $y = \cos x$.

✓ *Checkpoint*  *Audio-video solution in English & Spanish at LarsonPrecalculus.com*

In the same coordinate plane, sketch the graph of each function.

a. $y = \dfrac{1}{3} \sin x$

b. $y = 3 \sin x$

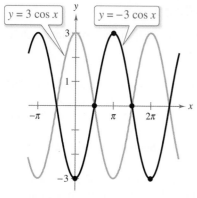

Figure 4.35

You know from Section 1.7 that the graph of $y = -f(x)$ is a **reflection** in the x-axis of the graph of $y = f(x)$. For example, the graph of $y = -3 \cos x$ is a reflection of the graph of $y = 3 \cos x$, as shown in Figure 4.35.

Next, consider the effect of the positive real number b on the graphs of $y = a \sin bx$ and $y = a \cos bx$. For example, compare the graphs of $y = a \sin x$ and $y = a \sin bx$. The graph of $y = a \sin x$ completes one cycle from $x = 0$ to $x = 2\pi$, so it follows that the graph of $y = a \sin bx$ completes one cycle from $x = 0$ to $x = 2\pi/b$.

Period of Sine and Cosine Functions

Let b be a positive real number. The **period** of $y = a \sin bx$ and $y = a \cos bx$ is given by

$$\text{Period} = \frac{2\pi}{b}.$$

Note that when $0 < b < 1$, the period of $y = a \sin bx$ is greater than 2π and represents a *horizontal stretch* of the basic curve. Similarly, when $b > 1$, the period of $y = a \sin bx$ is less than 2π and represents a *horizontal shrink* of the basic curve. These two statements are also true for $y = a \cos bx$. When b is negative, rewrite the function using the identity $\sin(-x) = -\sin x$ or $\cos(-x) = \cos x$.

EXAMPLE 3   **Scaling: Horizontal Stretching**

Sketch the graph of

$$y = \sin \frac{x}{2}.$$

Solution   The amplitude is 1. Moreover, $b = \frac{1}{2}$, so the period is

$$\frac{2\pi}{b} = \frac{2\pi}{\frac{1}{2}} = 4\pi. \qquad \text{Substitute for } b.$$

Now, divide the period-interval $[0, 4\pi]$ into four equal parts using the values π, 2π, and 3π to obtain the key points

Intercept	Maximum	Intercept	Minimum	Intercept
$(0, 0)$,	$(\pi, 1)$,	$(2\pi, 0)$,	$(3\pi, -1)$, and	$(4\pi, 0)$.

The graph is shown below.

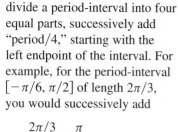

Period: 4π

> **REMARK** In general, to divide a period-interval into four equal parts, successively add "period/4," starting with the left endpoint of the interval. For example, for the period-interval $[-\pi/6, \pi/2]$ of length $2\pi/3$, you would successively add
>
> $$\frac{2\pi/3}{4} = \frac{\pi}{6}$$
>
> to obtain $-\pi/6, 0, \pi/6, \pi/3$, and $\pi/2$ as the x-values for the key points on the graph.

✓ **Checkpoint** ◀))) *Audio-video solution in English & Spanish at LarsonPrecalculus.com*

Sketch the graph of

$$y = \cos \frac{x}{3}.$$

Translations of Sine and Cosine Curves

The constant c in the equations

$$y = a \sin(bx - c) \quad \text{and} \quad y = a \cos(bx - c)$$

results in *horizontal translations* (shifts) of the basic curves. For example, compare the graphs of $y = a \sin bx$ and $y = a \sin(bx - c)$. The graph of $y = a \sin(bx - c)$ completes one cycle from $bx - c = 0$ to $bx - c = 2\pi$. Solve for x to find that the interval for one cycle is

$$\overbrace{\frac{c}{b}}^{\text{Left endpoint}} \leq x \leq \overbrace{\frac{c}{b} + \frac{2\pi}{b}}^{\text{Right endpoint}}.$$

$$\underbrace{\phantom{\frac{c}{b} + \frac{2\pi}{b}}}_{\text{Period}}$$

This implies that the period of $y = a \sin(bx - c)$ is $2\pi/b$, and the graph of $y = a \sin bx$ is shifted by an amount c/b. The number c/b is the **phase shift**.

Graphs of Sine and Cosine Functions

The graphs of $y = a \sin(bx - c)$ and $y = a \cos(bx - c)$ have the characteristics below. (Assume $b > 0$.)

$$\text{Amplitude} = |a| \qquad \text{Period} = \frac{2\pi}{b}$$

The left and right endpoints of a one-cycle interval can be determined by solving the equations $bx - c = 0$ and $bx - c = 2\pi$.

EXAMPLE 4 **Horizontal Translation**

Analyze the graph of $y = \dfrac{1}{2} \sin\left(x - \dfrac{\pi}{3}\right)$.

Algebraic Solution

The amplitude is $\frac{1}{2}$ and the period is $2\pi/1 = 2\pi$. Solving the equations

$$x - \frac{\pi}{3} = 0 \implies x = \frac{\pi}{3}$$

and

$$x - \frac{\pi}{3} = 2\pi \implies x = \frac{7\pi}{3}$$

shows that the interval $[\pi/3, 7\pi/3]$ corresponds to one cycle of the graph. Dividing this interval into four equal parts produces the key points

Intercept	Maximum	Intercept	Minimum	Intercept
$\left(\dfrac{\pi}{3}, 0\right),$	$\left(\dfrac{5\pi}{6}, \dfrac{1}{2}\right),$	$\left(\dfrac{4\pi}{3}, 0\right),$	$\left(\dfrac{11\pi}{6}, -\dfrac{1}{2}\right),$ and	$\left(\dfrac{7\pi}{3}, 0\right).$

Graphical Solution

Use a graphing utility set in *radian* mode to graph $y = (1/2) \sin[x - (\pi/3)]$, as shown in the figure below. Use the *minimum*, *maximum*, and *zero* or *root* features of the graphing utility to approximate the key points $(1.05, 0)$, $(2.62, 0.5)$, $(4.19, 0)$, $(5.76, -0.5)$, and $(7.33, 0)$.

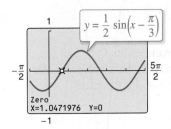

✓ **Checkpoint**))) Audio-video solution in English & Spanish at LarsonPrecalculus.com

Analyze the graph of $y = 2 \cos\left(x - \dfrac{\pi}{2}\right)$.

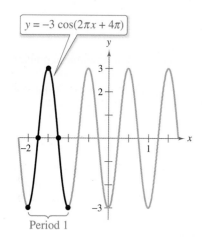

Figure 4.36

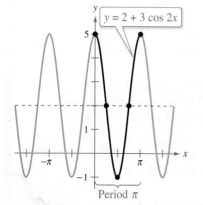

Figure 4.37

EXAMPLE 5 Horizontal Translation

Sketch the graph of

$$y = -3 \cos(2\pi x + 4\pi).$$

Solution The amplitude is 3 and the period is $2\pi/2\pi = 1$. Solving the equations

$$2\pi x + 4\pi = 0$$
$$2\pi x = -4\pi$$
$$x = -2$$

and

$$2\pi x + 4\pi = 2\pi$$
$$2\pi x = -2\pi$$
$$x = -1$$

shows that the interval $[-2, -1]$ corresponds to one cycle of the graph. Dividing this interval into four equal parts produces the key points

Minimum	Intercept	Maximum	Intercept	Minimum
$(-2, -3)$,	$\left(-\dfrac{7}{4}, 0\right)$,	$\left(-\dfrac{3}{2}, 3\right)$,	$\left(-\dfrac{5}{4}, 0\right)$, and	$(-1, -3)$.

Figure 4.36 shows the graph.

✓ **Checkpoint**  Audio-video solution in English & Spanish at LarsonPrecalculus.com

Sketch the graph of

$$y = -\frac{1}{2} \sin(\pi x + \pi).$$

The constant d in the equations

$$y = d + a \sin(bx - c) \quad \text{and} \quad y = d + a \cos(bx - c)$$

results in *vertical translations* of the basic curves. The shift is d units up for $d > 0$ and d units down for $d < 0$. In other words, the graph oscillates about the horizontal line $y = d$ instead of about the x-axis.

EXAMPLE 6 Vertical Translation

Sketch the graph of

$$y = 2 + 3 \cos 2x.$$

Solution The amplitude is 3 and the period is $2\pi/2 = \pi$. The key points over the interval $[0, \pi]$ are

$$(0, 5), \quad \left(\frac{\pi}{4}, 2\right), \quad \left(\frac{\pi}{2}, -1\right), \quad \left(\frac{3\pi}{2}, 2\right), \quad \text{and} \quad (\pi, 5).$$

Figure 4.37 shows the graph. Compared with the graph of $f(x) = 3 \cos 2x$, the graph of $y = 2 + 3 \cos 2x$ is shifted up two units.

✓ **Checkpoint**  Audio-video solution in English & Spanish at LarsonPrecalculus.com

Sketch the graph of

$$y = 2 \cos x - 5.$$

Mathematical Modeling

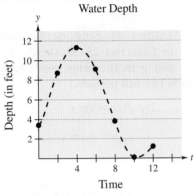

DATA	Time, t	Depth, y
	0	3.4
	2	8.7
	4	11.3
	6	9.1
	8	3.8
	10	0.1
	12	1.2

Spreadsheet at
LarsonPrecalculus.com

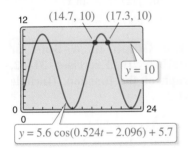

Figure 4.38

EXAMPLE 7   Finding a Trigonometric Model

The table shows the depths (in feet) of the water at the end of a dock every two hours from midnight to noon, where $t = 0$ corresponds to midnight. (a) Use a trigonometric function to model the data. (b) Find the depths at 9 A.M. and 3 P.M. (c) A boat needs at least 10 feet of water to moor at the dock. During what times in the afternoon can it safely dock?

Solution

a. Begin by graphing the data, as shown in Figure 4.38. Use either a sine or cosine model. For example, a cosine model has the form $y = a \cos(bt - c) + d$. The difference between the maximum value and the minimum value is twice the amplitude of the function. So, the amplitude is

$$a = \tfrac{1}{2}[(\text{maximum depth}) - (\text{minimum depth})] = \tfrac{1}{2}(11.3 - 0.1) = 5.6.$$

The cosine function completes one half of a cycle between the times at which the maximum and minimum depths occur. So, the period p is

$$p = 2[(\text{time of min. depth}) - (\text{time of max. depth})] = 2(10 - 4) = 12$$

which implies that $b = 2\pi/p \approx 0.524$. The maximum depth occurs 4 hours after midnight, so consider the left endpoint to be $c/b = 4$, which means that $c \approx 4(0.524) = 2.096$. Moreover, the average depth is $\tfrac{1}{2}(11.3 + 0.1) = 5.7$, so it follows that $d = 5.7$. Substituting the values of a, b, c, and d into the cosine model yields $y = 5.6 \cos(0.524t - 2.096) + 5.7$.

b. The depths at 9 A.M. and 3 P.M. are

$$y = 5.6 \cos(0.524 \cdot 9 - 2.096) + 5.7 \approx 0.84 \text{ foot} \qquad \text{9 A.M.}$$

and

$$y = 5.6 \cos(0.524 \cdot 15 - 2.096) + 5.7 \approx 10.56 \text{ feet.} \qquad \text{3 P.M.}$$

c. Using a graphing utility, graph the model with the line $y = 10$. Using the *intersect* feature, determine that the depth is at least 10 feet between 2:42 P.M. ($t \approx 14.7$) and 5:18 P.M. ($t \approx 17.3$), as shown in Figure 4.39.

(14.7, 10)   (17.3, 10)
$y = 10$
$y = 5.6 \cos(0.524t - 2.096) + 5.7$

Figure 4.39

✓ **Checkpoint** 🔊))) *Audio-video solution in English & Spanish at LarsonPrecalculus.com*

Find a sine model for the data in Example 7.

Summarize   (Section 4.5)

1. Explain how to sketch the graphs of basic sine and cosine functions (*page 297*). For an example of sketching the graph of a sine function, see Example 1.

2. Explain how to use amplitude and period to help sketch the graphs of sine and cosine functions (*pages 299 and 300*). For examples of using amplitude and period to sketch graphs of sine and cosine functions, see Examples 2 and 3.

3. Explain how to sketch translations of the graphs of sine and cosine functions (*page 301*). For examples of translating the graphs of sine and cosine functions, see Examples 4–6.

4. Give an example of using a sine or cosine function to model real-life data (*page 303, Example 7*).

4.5 Exercises

See **CalcChat.com** for tutorial help and worked-out solutions to odd-numbered exercises.

Vocabulary: Fill in the blanks.

1. One period of a sine or cosine function is one _____ of the sine or cosine curve.

2. The _____ of a sine or cosine curve represents half the distance between the maximum and minimum values of the function.

3. For the function $y = a \sin(bx - c)$, $\dfrac{c}{b}$ represents the _____ _____ of one cycle of the graph of the function.

4. For the function $y = d + a \cos(bx - c)$, d represents a _____ _____ of the basic curve.

Skills and Applications

Finding the Period and Amplitude In Exercises 5–12, find the period and amplitude.

5. $y = 2 \sin 5x$

6. $y = 3 \cos 2x$

7. $y = \dfrac{3}{4} \cos \dfrac{\pi x}{2}$

8. $y = -5 \sin \dfrac{\pi x}{3}$

9. $y = -\dfrac{1}{2} \sin \dfrac{5x}{4}$

10. $y = \dfrac{1}{4} \sin \dfrac{x}{6}$

11. $y = -\dfrac{5}{3} \cos \dfrac{\pi x}{12}$

12. $y = -\dfrac{2}{5} \cos 10\pi x$

Describing the Relationship Between Graphs In Exercises 13–24, describe the relationship between the graphs of f and g. Consider amplitude, period, and shifts.

13. $f(x) = \cos x$
    $g(x) = \cos 5x$

14. $f(x) = \sin x$
    $g(x) = 2 \sin x$

15. $f(x) = \cos 2x$
    $g(x) = -\cos 2x$

16. $f(x) = \sin 3x$
    $g(x) = \sin(-3x)$

17. $f(x) = \sin x$
    $g(x) = \sin(x - \pi)$

18. $f(x) = \cos x$
    $g(x) = \cos(x + \pi)$

19. $f(x) = \sin 2x$
    $g(x) = 3 + \sin 2x$

20. $f(x) = \cos 4x$
    $g(x) = -2 + \cos 4x$

21.
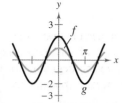

22.
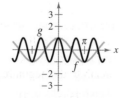

23.
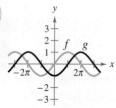

24.
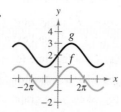

Sketching Graphs of Sine or Cosine Functions In Exercises 25–30, sketch the graphs of f and g in the same coordinate plane. (Include two full periods.)

25. $f(x) = \sin x$
    $g(x) = \sin \dfrac{x}{3}$

26. $f(x) = \sin x$
    $g(x) = 4 \sin x$

27. $f(x) = \cos x$
    $g(x) = 2 + \cos x$

28. $f(x) = \cos x$
    $g(x) = \cos\left(x + \dfrac{\pi}{2}\right)$

29. $f(x) = -\cos x$
    $g(x) = -\cos(x - \pi)$

30. $f(x) = -\sin x$
    $g(x) = -3 \sin x$

Sketching the Graph of a Sine or Cosine Function In Exercises 31–52, sketch the graph of the function. (Include two full periods.)

31. $y = 5 \sin x$

32. $y = \frac{1}{4} \sin x$

33. $y = \frac{1}{3} \cos x$

34. $y = 4 \cos x$

35. $y = \cos \dfrac{x}{2}$

36. $y = \sin 4x$

37. $y = \cos 2\pi x$

38. $y = \sin \dfrac{\pi x}{4}$

39. $y = -\sin \dfrac{2\pi x}{3}$

40. $y = 10 \cos \dfrac{\pi x}{6}$

41. $y = \cos\left(x - \dfrac{\pi}{2}\right)$

42. $y = \sin(x - 2\pi)$

43. $y = 3 \sin(x + \pi)$

44. $y = -4 \cos\left(x + \dfrac{\pi}{4}\right)$

45. $y = 2 - \sin \dfrac{2\pi x}{3}$

46. $y = -3 + 5 \cos \dfrac{\pi t}{12}$

47. $y = 2 + 5 \cos 6\pi x$

48. $y = 2 \sin 3x + 5$

49. $y = 3 \sin(x + \pi) - 3$

50. $y = -3 \sin(6x + \pi)$

51. $y = \dfrac{2}{3} \cos\left(\dfrac{x}{2} - \dfrac{\pi}{4}\right)$

52. $y = 4 \cos\left(\pi x + \dfrac{\pi}{2}\right) - 1$

Describing a Transformation In Exercises 53–58, g **is related to a parent function** $f(x) = \sin(x)$ **or** $f(x) = \cos(x)$. **(a) Describe the sequence of transformations from** f **to** g. **(b) Sketch the graph of** g. **(c) Use function notation to write** g **in terms of** f.

53. $g(x) = \sin(4x - \pi)$

54. $g(x) = \sin(2x + \pi)$

55. $g(x) = \cos\left(x - \dfrac{\pi}{2}\right) + 2$

56. $g(x) = 1 + \cos(x + \pi)$

57. $g(x) = 2\sin(4x - \pi) - 3$

58. $g(x) = 4 - \sin\left(2x + \dfrac{\pi}{2}\right)$

 Graphing a Sine or Cosine Function In Exercises 59–64, use a graphing utility to graph the function. (Include two full periods.) Be sure to choose an appropriate viewing window.

59. $y = -2\sin(4x + \pi)$

60. $y = -4\sin\left(\dfrac{2}{3}x - \dfrac{\pi}{3}\right)$

61. $y = \cos\left(2\pi x - \dfrac{\pi}{2}\right) + 1$

62. $y = 3\cos\left(\dfrac{\pi x}{2} + \dfrac{\pi}{2}\right) - 2$

63. $y = -0.1\sin\left(\dfrac{\pi x}{10} + \pi\right)$

64. $y = \dfrac{1}{100}\cos 120\pi t$

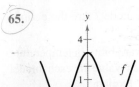 **Graphical Reasoning In Exercises 65–68, find** a **and** d **for the function** $f(x) = a\cos x + d$ **such that the graph of** f **matches the figure.**

65.
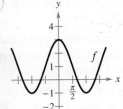

66.
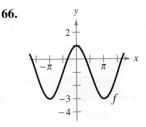

67.
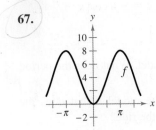

68.
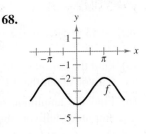

 Graphical Reasoning In Exercises 69–72, find a, b, **and** c **for the function** $f(x) = a\sin(bx - c)$ **such that the graph of** f **matches the figure.**

69.
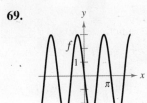

70.
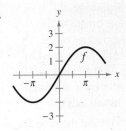

71.
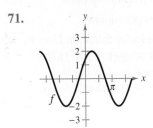

72.
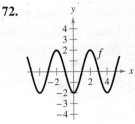

 Using Technology In Exercises 73 and 74, use a graphing utility to graph y_1 **and** y_2 **in the interval** $[-2\pi, 2\pi]$. **Use the graphs to find real numbers** x **such that** $y_1 = y_2$.

73. $y_1 = \sin x$, $y_2 = -\dfrac{1}{2}$ **74.** $y_1 = \cos x$, $y_2 = -1$

Writing an Equation In Exercises 75–78, write an equation for a function with the given characteristics.

75. A sine curve with a period of π, an amplitude of 2, a right phase shift of $\pi/2$, and a vertical translation up 1 unit

76. A sine curve with a period of 4π, an amplitude of 3, a left phase shift of $\pi/4$, and a vertical translation down 1 unit

77. A cosine curve with a period of π, an amplitude of 1, a left phase shift of π, and a vertical translation down $\dfrac{3}{2}$ units

78. A cosine curve with a period of 4π, an amplitude of 3, a right phase shift of $\pi/2$, and a vertical translation up 2 units

79. Respiratory Cycle For a person exercising, the velocity v (in liters per second) of airflow during a respiratory cycle (the time from the beginning of one breath to the beginning of the next) is modeled by

$$v = 1.75\sin(\pi t/2)$$

where t is the time (in seconds). (Inhalation occurs when $v > 0$, and exhalation occurs when $v < 0$.)

(a) Find the time for one full respiratory cycle.

(b) Find the number of cycles per minute.

(c) Sketch the graph of the velocity function.

80. Respiratory Cycle

For a person at rest, the velocity v (in liters per second) of airflow during a respiratory cycle (the time from the beginning of one breath to the beginning of the next) is modeled by $v = 0.85 \sin(\pi t/3)$, where t is the time (in seconds).

(a) Find the time for one full respiratory cycle.

(b) Find the number of cycles per minute.

(c) Sketch the graph of the velocity function. Use the graph to confirm your answer in part (a) by finding two times when new breaths begin. (Inhalation occurs when $v > 0$, and exhalation occurs when $v < 0$.)

81. Biology The function $P = 100 - 20\cos(5\pi t/3)$ approximates the blood pressure P (in millimeters of mercury) at time t (in seconds) for a person at rest.

(a) Find the period of the function.

(b) Find the number of heartbeats per minute.

82. Piano Tuning When tuning a piano, a technician strikes a tuning fork for the A above middle C and sets up a wave motion that can be approximated by $y = 0.001 \sin 880\pi t$, where t is the time (in seconds).

(a) What is the period of the function?

(b) The frequency f is given by $f = 1/p$. What is the frequency of the note?

83. Astronomy The table shows the percent y (in decimal form) of the moon's face illuminated on day x in the year 2018, where $x = 1$ corresponds to January 1. *(Source: U.S. Naval Observatory)*

DATA	x	y
	1	1.0
	8	0.5
	16	0.0
	24	0.5
	31	1.0
	38	0.5

Spreadsheet at LarsonPrecalculus.com

(a) Create a scatter plot of the data.

(b) Find a trigonometric model for the data.

(c) Add the graph of your model in part (b) to the scatter plot. How well does the model fit the data?

(d) What is the period of the model?

(e) Estimate the percent of the moon's face illuminated on March 12, 2018.

84. Meteorology The table shows the maximum daily high temperatures (in degrees Fahrenheit) in Las Vegas L and International Falls I for month t, where $t = 1$ corresponds to January. *(Source: National Climatic Data Center)*

DATA	Month, t	Las Vegas, L	International Falls, I
	1	57.1	13.8
	2	63.0	22.4
	3	69.5	34.9
	4	78.1	51.5
	5	87.8	66.6
	6	98.9	74.2
	7	104.1	78.6
	8	101.8	76.3
	9	93.8	64.7
	10	80.8	51.7
	11	66.0	32.5
	12	57.3	18.1

Spreadsheet at LarsonPrecalculus.com

(a) A model for the temperatures in Las Vegas is

$$L(t) = 80.60 + 23.50 \cos\left(\frac{\pi t}{6} - 3.67\right).$$

Find a trigonometric model for the temperatures in International Falls.

(b) Use a graphing utility to graph the data points and the model for the temperatures in Las Vegas. How well does the model fit the data?

(c) Use the graphing utility to graph the data points and the model for the temperatures in International Falls. How well does the model fit the data?

(d) Use the models to estimate the average maximum temperature in each city. Which value in each model did you use? Explain.

(e) What is the period of each model? Are the periods what you expected? Explain.

(f) Which city has the greater variability in temperature throughout the year? Which value in each model determines this variability? Explain.

85. Ferris Wheel The height h (in feet) above ground of a seat on a Ferris wheel at time t (in seconds) is modeled by

$$h(t) = 53 + 50 \sin\left(\frac{\pi}{10}t - \frac{\pi}{2}\right).$$

(a) Find the period of the model. What does the period tell you about the ride?

(b) Find the amplitude of the model. What does the amplitude tell you about the ride?

(c) Use a graphing utility to graph one cycle of the model.

86. Fuel Consumption The daily consumption C (in gallons) of diesel fuel on a farm is modeled by

$$C = 30.3 + 21.6 \sin\left(\frac{2\pi t}{365} + 10.9\right)$$

where t is the time (in days), with $t = 1$ corresponding to January 1.

(a) What is the period of the model? Is it what you expected? Explain.

(b) What is the average daily fuel consumption? Which value in the model did you use? Explain.

(c) Use a graphing utility to graph the model. Use the graph to approximate the time of the year when consumption exceeds 40 gallons per day.

Exploration

True or False? In Exercises 87–89, determine whether the statement is true or false. Justify your answer.

87. The graph of $g(x) = \sin(x + 2\pi)$ is a translation of the graph of $f(x) = \sin x$ exactly one period to the right, and the two graphs look identical.

88. The function $y = \frac{1}{2}\cos 2x$ has an amplitude that is twice that of the function $y = \cos x$.

89. The graph of $y = -\cos x$ is a reflection of the graph of $y = \sin[x + (\pi/2)]$ in the x-axis.

90. HOW DO YOU SEE IT? The figure below shows the graph of $y = \sin(x - c)$ for

$$c = -\frac{\pi}{4}, \quad 0, \quad \text{and} \quad \frac{\pi}{4}.$$

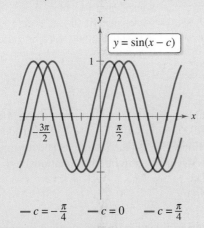

$$-c = -\frac{\pi}{4} \qquad -c = 0 \qquad -c = \frac{\pi}{4}$$

(a) How does the value of c affect the graph?

(b) Which graph is equivalent to that of

$$y = -\cos\left(x + \frac{\pi}{4}\right)?$$

Conjecture In Exercises 91 and 92, graph f and g in the same coordinate plane. (Include two full periods.) Make a conjecture about the functions.

91. $f(x) = \sin x, \quad g(x) = \cos\left(x - \frac{\pi}{2}\right)$

92. $f(x) = \sin x, \quad g(x) = -\cos\left(x + \frac{\pi}{2}\right)$

93. Writing Sketch the graph of $y = \cos bx$ for $b = \frac{1}{2}, 2,$ and 3. How does the value of b affect the graph? How many complete cycles of the graph occur between 0 and 2π for each value of b?

94. Polynomial Approximations Using calculus, it can be shown that the sine and cosine functions can be approximated by the polynomials

$$\sin x \approx x - \frac{x^3}{3!} + \frac{x^5}{5!}$$

and

$$\cos x \approx 1 - \frac{x^2}{2!} + \frac{x^4}{4!}$$

where x is in radians.

(a) Use a graphing utility to graph the sine function and its polynomial approximation in the same viewing window. How do the graphs compare?

(b) Use the graphing utility to graph the cosine function and its polynomial approximation in the same viewing window. How do the graphs compare?

(c) Study the patterns in the polynomial approximations of the sine and cosine functions and predict the next term in each. Then repeat parts (a) and (b). How does the accuracy of the approximations change when an additional term is added?

95. Polynomial Approximations Use the polynomial approximations of the sine and cosine functions in Exercise 94 to approximate each function value. Compare the results with those given by a calculator. Is the error in the approximation the same in each case? Explain.

(a) $\sin \frac{1}{2}$ (b) $\sin 1$

(c) $\sin \frac{\pi}{6}$ (d) $\cos(-0.5)$

(e) $\cos 1$ (f) $\cos \frac{\pi}{4}$

Project: Meteorology To work an extended application analyzing the mean monthly temperature and mean monthly precipitation for Honolulu, Hawaii, visit this text's website at *LarsonPrecalculus.com*. (*Source: National Climatic Data Center*)

4.6 Graphs of Other Trigonometric Functions

Graphs of trigonometric functions have many real-life applications, such as in modeling the distance from a television camera to a unit in a parade, as in Exercise 85 on page 317.

■ Sketch the graphs of tangent functions.
■ Sketch the graphs of cotangent functions.
■ Sketch the graphs of secant and cosecant functions.
■ Sketch the graphs of damped trigonometric functions.

Graph of the Tangent Function

Recall that the tangent function is odd. That is, $\tan(-x) = -\tan x$. Consequently, the graph of $y = \tan x$ is symmetric with respect to the origin. You also know from the identity $\tan x = (\sin x)/(\cos x)$ that the tangent function is undefined for values at which $\cos x = 0$. Two such values are $x = \pm\pi/2 \approx \pm 1.5708$. As shown in the table below, $\tan x$ increases without bound as x approaches $\pi/2$ from the left and decreases without bound as x approaches $-\pi/2$ from the right.

x	$-\dfrac{\pi}{2}$	-1.57	-1.5	$-\dfrac{\pi}{4}$	0	$\dfrac{\pi}{4}$	1.5	1.57	$\dfrac{\pi}{2}$
$\tan x$	Undef.	-1255.8	-14.1	-1	0	1	14.1	1255.8	Undef.

So, the graph of $y = \tan x$ (shown below) has *vertical asymptotes* at $x = \pi/2$ and $x = -\pi/2$. Moreover, the period of the tangent function is π, so vertical asymptotes also occur at $x = (\pi/2) + n\pi$, where n is an integer. The domain of the tangent function is the set of all real numbers other than $x = (\pi/2) + n\pi$, and the range is the set of all real numbers.

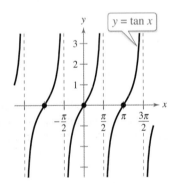

Period: π

Domain: all $x \ne \dfrac{\pi}{2} + n\pi$

Range: $(-\infty, \infty)$

Vertical asymptotes: $x = \dfrac{\pi}{2} + n\pi$

x-intercepts: $(n\pi, 0)$
y-intercept: $(0, 0)$
Symmetry: origin
Odd function

▷ **ALGEBRA HELP**
- To review odd and even functions, see Section 1.5.
- To review symmetry of a graph, see Section 1.2.
- To review fundamental trigonometric identities, see Section 4.3.
- To review asymptotes, see Section 2.6.
- To review domain and range of a function, see Section 1.4.
- To review intercepts of a graph, see Section 1.2.

Sketching the graph of $y = a\tan(bx - c)$ is similar to sketching the graph of $y = a\sin(bx - c)$ in that you locate key points of the graph. When sketching the graph of $y = a\tan(bx - c)$, the key points identify the intercepts and asymptotes. Two consecutive vertical asymptotes can be found by solving the equations

$$bx - c = -\frac{\pi}{2} \quad \text{and} \quad bx - c = \frac{\pi}{2}.$$

On the x-axis, the point halfway between two consecutive vertical asymptotes is an x-intercept of the graph. The period of the function $y = a\tan(bx - c)$ is the distance between two consecutive vertical asymptotes. The amplitude of a tangent function is not defined. After plotting two consecutive asymptotes and the x-intercept between them, plot additional points between the asymptotes and sketch one cycle. Finally, sketch one or two additional cycles to the left and right.

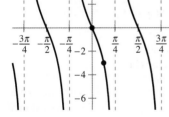

Figure 4.40

EXAMPLE 1   Sketching the Graph of a Tangent Function

Sketch the graph of $y = \tan \dfrac{x}{2}$.

Solution

Solving the equations

$$\frac{x}{2} = -\frac{\pi}{2} \quad \text{and} \quad \frac{x}{2} = \frac{\pi}{2}$$

shows that two consecutive vertical asymptotes occur at $x = -\pi$ and $x = \pi$. Between these two asymptotes, find a few points, including the x-intercept, as shown in the table. Figure 4.40 shows three cycles of the graph.

x	$-\pi$	$-\dfrac{\pi}{2}$	0	$\dfrac{\pi}{2}$	π
$\tan \dfrac{x}{2}$	Undef.	-1	0	1	Undef.

✓ **Checkpoint**))) *Audio-video solution in English & Spanish at LarsonPrecalculus.com*

Sketch the graph of $y = \tan \dfrac{x}{4}$.

EXAMPLE 2   Sketching the Graph of a Tangent Function

Sketch the graph of $y = -3 \tan 2x$.

Solution

Solving the equations

$$2x = -\frac{\pi}{2} \quad \text{and} \quad 2x = \frac{\pi}{2}$$

shows that two consecutive vertical asymptotes occur at $x = -\pi/4$ and $x = \pi/4$. Between these two asymptotes, find a few points, including the x-intercept, as shown in the table. Figure 4.41 shows three cycles of the graph.

Figure 4.41

x	$-\dfrac{\pi}{4}$	$-\dfrac{\pi}{8}$	0	$\dfrac{\pi}{8}$	$\dfrac{\pi}{4}$
$-3 \tan 2x$	Undef.	3	0	-3	Undef.

✓ **Checkpoint**))) *Audio-video solution in English & Spanish at LarsonPrecalculus.com*

Sketch the graph of $y = \tan 2x$.

Compare the graphs in Examples 1 and 2. The graph of $y = a \tan(bx - c)$ increases between consecutive vertical asymptotes when $a > 0$ and decreases between consecutive vertical asymptotes when $a < 0$. In other words, the graph for $a < 0$ is a reflection in the x-axis of the graph for $a > 0$. Also, the period is greater when $0 < b < 1$ than when $b > 1$. In other words, compared with the case where $b = 1$, the period represents a horizontal stretch when $0 < b < 1$ and a horizontal shrink when $b > 1$.

Graph of the Cotangent Function

The graph of the cotangent function is similar to the graph of the tangent function. It also has a period of π. However, the identity

$$y = \cot x = \frac{\cos x}{\sin x}$$

shows that the cotangent function has vertical asymptotes when $\sin x$ is zero, which occurs at $x = n\pi$, where n is an integer. The graph of the cotangent function is shown below. Note that two consecutive vertical asymptotes of the graph of $y = a \cot(bx - c)$ can be found by solving the equations

$$bx - c = 0 \quad \text{and} \quad bx - c = \pi.$$

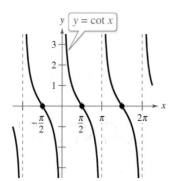

Period: π
Domain: all $x \neq n\pi$
Range: $(-\infty, \infty)$
Vertical asymptotes: $x = n\pi$
x-intercepts: $\left(\dfrac{\pi}{2} + n\pi, 0\right)$
Symmetry: origin
Odd function

EXAMPLE 3   **Sketching the Graph of a Cotangent Function**

Sketch the graph of

$$y = 2 \cot \frac{x}{3}.$$

Solution

Solving the equations

$$\frac{x}{3} = 0 \quad \text{and} \quad \frac{x}{3} = \pi$$

shows that two consecutive vertical asymptotes occur at $x = 0$ and $x = 3\pi$. Between these two asymptotes, find a few points, including the x-intercept, as shown in the table. Figure 4.42 shows three cycles of the graph. Note that the period is 3π, the distance between consecutive asymptotes.

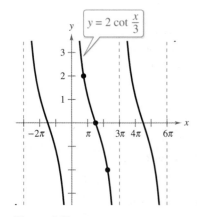

Figure 4.42

x	0	$\dfrac{3\pi}{4}$	$\dfrac{3\pi}{2}$	$\dfrac{9\pi}{4}$	3π
$2 \cot \dfrac{x}{3}$	Undef.	2	0	-2	Undef.

✓ **Checkpoint** ◀))) *Audio-video solution in English & Spanish at LarsonPrecalculus.com*

Sketch the graph of

$$y = \cot \frac{x}{4}.$$

Graphs of the Reciprocal Functions

You can obtain the graphs of the cosecant and secant functions from the graphs of the sine and cosine functions, respectively, using the reciprocal identities

$$\csc x = \frac{1}{\sin x} \quad \text{and} \quad \sec x = \frac{1}{\cos x}.$$

For example, at a given value of x, the y-coordinate of $\sec x$ is the reciprocal of the y-coordinate of $\cos x$. Of course, when $\cos x = 0$, the reciprocal does not exist. Near such values of x, the behavior of the secant function is similar to that of the tangent function. In other words, the graphs of

$$\tan x = \frac{\sin x}{\cos x} \quad \text{and} \quad \sec x = \frac{1}{\cos x}$$

have vertical asymptotes where $\cos x = 0$, that is, at $x = (\pi/2) + n\pi$, where n is an integer. Similarly,

$$\cot x = \frac{\cos x}{\sin x} \quad \text{and} \quad \csc x = \frac{1}{\sin x}$$

have vertical asymptotes where $\sin x = 0$, that is, at $x = n\pi$, where n is an integer.

To sketch the graph of a secant or cosecant function, first make a sketch of its reciprocal function. For example, to sketch the graph of $y = \csc x$, first sketch the graph of $y = \sin x$. Then find reciprocals of the y-coordinates to obtain points on the graph of $y = \csc x$. You can use this procedure to obtain the graphs below.

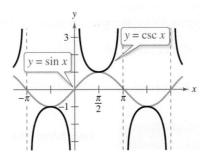

Period: 2π
Domain: all $x \neq n\pi$
Range: $(-\infty, -1] \cup [1, \infty)$
Vertical asymptotes: $x = n\pi$
No intercepts
Symmetry: origin
Odd function

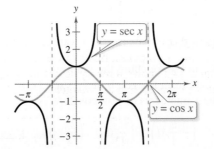

Period: 2π
Domain: all $x \neq \dfrac{\pi}{2} + n\pi$
Range: $(-\infty, -1] \cup [1, \infty)$
Vertical asymptotes: $x = \dfrac{\pi}{2} + n\pi$
y-intercept: $(0, 1)$
Symmetry: y-axis
Even function

In comparing the graphs of the cosecant and secant functions with those of the sine and cosine functions, respectively, note that the "hills" and "valleys" are interchanged. For example, a hill (or maximum point) on the sine curve corresponds to a valley (a relative minimum) on the cosecant curve, and a valley (or minimum point) on the sine curve corresponds to a hill (a relative maximum) on the cosecant curve, as shown in Figure 4.43. Additionally, x-intercepts of the sine and cosine functions become vertical asymptotes of the cosecant and secant functions, respectively (see Figure 4.43).

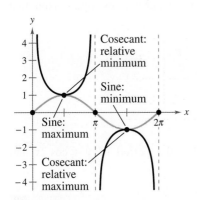

Figure 4.43

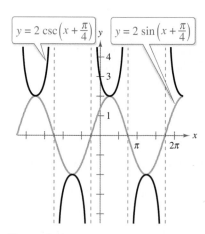

Figure 4.44

EXAMPLE 4   **Sketching the Graph of a Cosecant Function**

Sketch the graph of $y = 2 \csc\left(x + \dfrac{\pi}{4}\right)$.

Solution

Begin by sketching the graph of

$$y = 2 \sin\left(x + \dfrac{\pi}{4}\right).$$

For this function, the amplitude is 2 and the period is 2π. Solving the equations

$$x + \frac{\pi}{4} = 0 \quad \text{and} \quad x + \frac{\pi}{4} = 2\pi$$

shows that one cycle of the sine function corresponds to the interval from $x = -\pi/4$ to $x = 7\pi/4$. The gray curve in Figure 4.44 represents the graph of the sine function. At the midpoint and endpoints of this interval, the sine function is zero. So, the corresponding cosecant function

$$
\begin{aligned}
y &= 2 \csc\left(x + \frac{\pi}{4}\right) \\
&= 2\left(\frac{1}{\sin[x + (\pi/4)]}\right)
\end{aligned}
$$

has vertical asymptotes at $x = -\pi/4$, $x = 3\pi/4$, $x = 7\pi/4$, and so on. The black curve in Figure 4.44 represents the graph of the cosecant function.

 ✓ **Checkpoint** 🔊))) Audio-video solution in English & Spanish at LarsonPrecalculus.com

Sketch the graph of $y = 2 \csc\left(x + \dfrac{\pi}{2}\right)$.

EXAMPLE 5   **Sketching the Graph of a Secant Function**

See LarsonPrecalculus.com for an interactive version of this type of example.

Sketch the graph of $y = \sec 2x$.

Solution

Begin by sketching the graph of $y = \cos 2x$, shown as the gray curve in Figure 4.45. Then, form the graph of $y = \sec 2x$, shown as the black curve in the figure. Note that the x-intercepts of $y = \cos 2x$

$$\left(-\frac{\pi}{4}, 0\right), \quad \left(\frac{\pi}{4}, 0\right), \quad \left(\frac{3\pi}{4}, 0\right), \ldots$$

correspond to the vertical asymptotes

$$x = -\frac{\pi}{4}, \quad x = \frac{\pi}{4}, \quad x = \frac{3\pi}{4}, \ldots$$

of the graph of $y = \sec 2x$. Moreover, notice that the period of $y = \cos 2x$ and $y = \sec 2x$ is π.

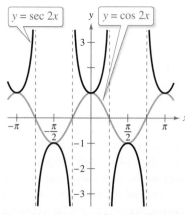

Figure 4.45

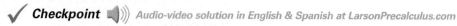 ✓ **Checkpoint** 🔊))) Audio-video solution in English & Spanish at LarsonPrecalculus.com

Sketch the graph of $y = \sec \dfrac{x}{2}$.

Damped Trigonometric Graphs

You can graph a *product* of two functions using properties of the individual functions. For example, consider the function

$$f(x) = x \sin x$$

as the product of the functions $y = x$ and $y = \sin x$. Using properties of absolute value and the fact that $|\sin x| \leq 1$, you have

$$0 \leq |x||\sin x| \leq |x|.$$

Consequently,

$$-|x| \leq x \sin x \leq |x|.$$

which means that the graph of $f(x) = x \sin x$ lies between the lines $y = -x$ and $y = x$. Furthermore,

$$f(x) = x \sin x = \pm x \quad \text{at} \quad x = \frac{\pi}{2} + n\pi$$

and

$$f(x) = x \sin x = 0 \quad \text{at} \quad x = n\pi$$

where n is an integer, so the graph of f touches the line $y = x$ or the line $y = -x$ at $x = (\pi/2) + n\pi$ and has x-intercepts at $x = n\pi$. A sketch of f is shown at the right. In the function $f(x) = x \sin x$, the factor x is called the **damping factor.**

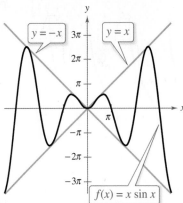

> **REMARK** Do you see why the graph of $f(x) = x \sin x$ touches the lines $y = \pm x$ at $x = (\pi/2) + n\pi$ and why the graph has x-intercepts at $x = n\pi$? Recall that the sine function is equal to ± 1 at odd multiples of $\pi/2$ and is equal to 0 at multiples of π.

EXAMPLE 6   Damped Sine Curve

Sketch the graph of $f(x) = e^{-x} \sin 3x$.

Solution

Consider f as the product of the two functions $y = e^{-x}$ and $y = \sin 3x$, each of which has the set of real numbers as its domain. For any real number x, you know that $e^{-x} > 0$ and $|\sin 3x| \leq 1$. So,

$$e^{-x}|\sin 3x| \leq e^{-x}$$

which means that

$$-e^{-x} \leq e^{-x} \sin 3x \leq e^{-x}.$$

Furthermore,

$$f(x) = e^{-x} \sin 3x = \pm e^{-x} \quad \text{at} \quad x = \frac{\pi}{6} + \frac{n\pi}{3}$$

and

$$f(x) = e^{-x} \sin 3x = 0 \quad \text{at} \quad x = \frac{n\pi}{3}$$

so the graph of f touches the curve $y = e^{-x}$ or the curve $y = -e^{-x}$ at $x = (\pi/6) + (n\pi/3)$ and has intercepts at $x = n\pi/3$. Figure 4.46 shows a sketch of f.

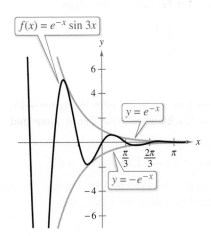

Figure 4.46

✓ **Checkpoint** 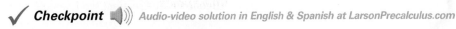 *Audio-video solution in English & Spanish at LarsonPrecalculus.com*

Sketch the graph of $f(x) = e^x \sin 4x$.

Below is a summary of the characteristics of the six basic trigonometric functions.

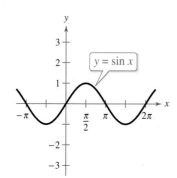

Domain: $(-\infty, \infty)$
Range: $[-1, 1]$
Period: 2π

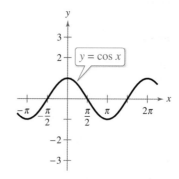

Domain: $(-\infty, \infty)$
Range: $[-1, 1]$
Period: 2π

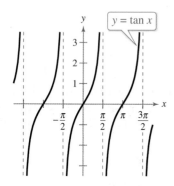

Domain: all $x \neq \dfrac{\pi}{2} + n\pi$

Range: $(-\infty, \infty)$
Period: π

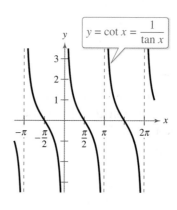

Domain: all $x \neq n\pi$
Range: $(-\infty, \infty)$
Period: π

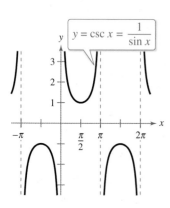

Domain: all $x \neq n\pi$
Range: $(-\infty, -1] \cup [1, \infty)$
Period: 2π

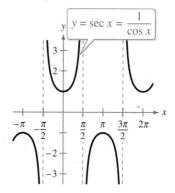

Domain: all $x \neq \dfrac{\pi}{2} + n\pi$

Range: $(-\infty, -1] \cup [1, \infty)$
Period: 2π

Summarize   (Section 4.6)

1. Explain how to sketch the graph of $y = a \tan(bx - c)$ *(page 308)*. For examples of sketching graphs of tangent functions, see Examples 1 and 2.

2. Explain how to sketch the graph of $y = a \cot(bx - c)$ *(page 310)*. For an example of sketching the graph of a cotangent function, see Example 3.

3. Explain how to sketch the graphs of $y = a \csc(bx - c)$ and $y = a \sec(bx - c)$ *(page 311)*. For examples of sketching graphs of cosecant and secant functions, see Examples 4 and 5.

4. Explain how to sketch the graph of a damped trigonometric function *(page 313)*. For an example of sketching the graph of a damped trigonometric function, see Example 6.

4.6 Exercises

See **CalcChat.com** for tutorial help and worked-out solutions to odd-numbered exercises.

Vocabulary: Fill in the blanks.

1. The tangent, cotangent, and cosecant functions are _____, so the graphs of these functions have symmetry with respect to the _____.

2. The graphs of the tangent, cotangent, secant, and cosecant functions have _____ asymptotes.

3. To sketch the graph of a secant or cosecant function, first make a sketch of its _____ function.

4. For the function $f(x) = g(x) \cdot \sin x$, $g(x)$ is called the _____ factor.

5. The period of $y = \tan x$ is _____.

6. The domain of $y = \cot x$ is all real numbers such that _____.

7. The range of $y = \sec x$ is _____.

8. The period of $y = \csc x$ is _____.

Skills and Applications

Matching In Exercises 9–14, match the function with its graph. State the period of the function. [The graphs are labeled (a), (b), (c), (d), (e), and (f).]

(a)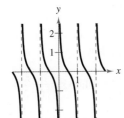

(b)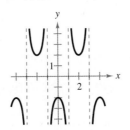

(c)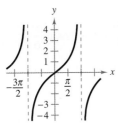

(d)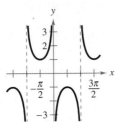

(e)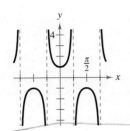

(f)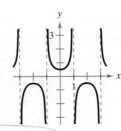

9. $y = \sec 2x$

10. $y = \tan \dfrac{x}{2}$

11. $y = \dfrac{1}{2} \cot \pi x$

12. $y = -\csc x$

13. $y = \dfrac{1}{2} \sec \dfrac{\pi x}{2}$

14. $y = -2 \sec \dfrac{\pi x}{2}$

Sketching the Graph of a Trigonometric Function In Exercises 15–38, sketch the graph of the function. (Include two full periods.)

15. $y = \dfrac{1}{3} \tan x$

16. $y = -\dfrac{1}{2} \tan x$

17. $y = -\dfrac{1}{2} \sec x$

18. $y = \dfrac{1}{4} \sec x$

19. $y = -2 \tan 3x$

20. $y = -3 \tan \pi x$

21. $y = \csc \pi x$

22. $y = 3 \csc 4x$

23. $y = \dfrac{1}{2} \sec \pi x$

24. $y = 2 \sec 3x$

25. $y = \csc \dfrac{x}{2}$

26. $y = \csc \dfrac{x}{3}$

27. $y = 3 \cot 2x$

28. $y = 3 \cot \dfrac{\pi x}{2}$

29. $y = \tan \dfrac{\pi x}{4}$

30. $y = \tan 4x$

31. $y = 2 \csc(x - \pi)$

32. $y = \csc(2x - \pi)$

33. $y = 2 \sec(x + \pi)$

34. $y = \tan(x + \pi)$

35. $y = -\sec \pi x + 1$

36. $y = -2 \sec 4x + 2$

37. $y = \dfrac{1}{4} \csc\left(x + \dfrac{\pi}{4}\right)$

38. $y = 2 \cot\left(x + \dfrac{\pi}{2}\right)$

Graphing a Trigonometric Function In Exercises 39–48, use a graphing utility to graph the function. (Include two full periods.)

39. $y = \tan \dfrac{x}{3}$

40. $y = -\tan 2x$

41. $y = -2 \sec 4x$

42. $y = \sec \pi x$

43. $y = \tan\left(x - \dfrac{\pi}{4}\right)$

44. $y = \dfrac{1}{4} \cot\left(x - \dfrac{\pi}{2}\right)$

45. $y = -\csc(4x - \pi)$

46. $y = 2 \sec(2x - \pi)$

47. $y = 0.1 \tan\left(\dfrac{\pi x}{4} + \dfrac{\pi}{4}\right)$

48. $y = \dfrac{1}{3} \sec\left(\dfrac{\pi x}{2} + \dfrac{\pi}{2}\right)$

Solving a Trigonometric Equation In Exercises 49–56, find the solutions of the equation in the interval $[-2\pi, 2\pi]$. Use a graphing utility to verify your results.

49. $\tan x = 1$

50. $\tan x = \sqrt{3}$

51. $\cot x = -\sqrt{3}$

52. $\cot x = 1$

53. $\sec x = -2$

54. $\sec x = 2$

55. $\csc x = \sqrt{2}$

56. $\csc x = -2$

Even and Odd Trigonometric Functions In Exercises 57–64, use the graph of the function to determine whether the function is even, odd, or neither. Verify your answer algebraically.

57. $f(x) = \sec x$

58. $f(x) = \tan x$

59. $g(x) = \cot x$

60. $g(x) = \csc x$

61. $f(x) = x + \tan x$

62. $f(x) = x^2 - \sec x$

63. $g(x) = x \csc x$

64. $g(x) = x^2 \cot x$

Identifying Damped Trigonometric Functions In Exercises 65–68, match the function with its graph. Describe the behavior of the function as x approaches zero. [The graphs are labeled (a), (b), (c), and (d).]

(a)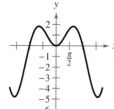

(b)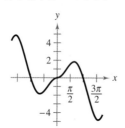

(c)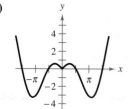

(d)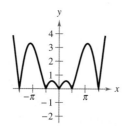

65. $f(x) = |x \cos x|$

66. $f(x) = x \sin x$

67. $g(x) = |x| \sin x$

68. $g(x) = |x| \cos x$

Conjecture In Exercises 69–72, graph the functions f and g. Use the graphs to make a conjecture about the relationship between the functions.

69. $f(x) = \sin x + \cos\left(x + \dfrac{\pi}{2}\right), \quad g(x) = 0$

70. $f(x) = \sin x - \cos\left(x + \dfrac{\pi}{2}\right), \quad g(x) = 2\sin x$

71. $f(x) = \sin^2 x, \quad g(x) = \dfrac{1}{2}(1 - \cos 2x)$

72. $f(x) = \cos^2 \dfrac{\pi x}{2}, \quad g(x) = \dfrac{1}{2}(1 + \cos \pi x)$

Analyzing a Damped Trigonometric Graph In Exercises 73–76, use a graphing utility to graph the function and the damping factor of the function in the same viewing window. Describe the behavior of the function as x increases without bound.

73. $g(x) = e^{-x^2/2} \sin x$

74. $f(x) = e^{-x} \cos x$

75. $f(x) = 2^{-x/4} \cos \pi x$

76. $h(x) = 2^{-x^2/4} \sin x$

Analyzing a Trigonometric Graph In Exercises 77–82, use a graphing utility to graph the function. Describe the behavior of the function as x approaches zero.

77. $y = \dfrac{6}{x} + \cos x, \quad x > 0$

78. $y = \dfrac{4}{x} + \sin 2x, \quad x > 0$

79. $g(x) = \dfrac{\sin x}{x}$

80. $f(x) = \dfrac{1 - \cos x}{x}$

81. $f(x) = \sin \dfrac{1}{x}$

82. $h(x) = x \sin \dfrac{1}{x}$

83. **Meteorology** The normal monthly high temperatures H (in degrees Fahrenheit) in Erie, Pennsylvania, are approximated by

$$H(t) = 57.54 - 18.53 \cos \frac{\pi t}{6} - 14.03 \sin \frac{\pi t}{6}$$

and the normal monthly low temperatures L are approximated by

$$L(t) = 42.03 - 15.99 \cos \frac{\pi t}{6} - 14.32 \sin \frac{\pi t}{6}$$

where t is the time (in months), with $t = 1$ corresponding to January (see figure). (*Source: NOAA*)

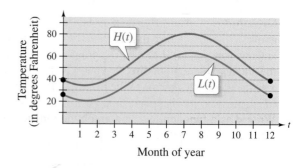

Month of year

(a) What is the period of each function?

(b) During what part of the year is the difference between the normal high and normal low temperatures greatest? When is it least?

(c) The sun is northernmost in the sky around June 21, but the graph shows the warmest temperatures at a later date. Approximate the lag time of the temperatures relative to the position of the sun.

84. Sales The projected monthly sales S (in thousands of units) of lawn mowers are modeled by

$$S = 74 + 3t - 40 \cos \frac{\pi t}{6}$$

where t is the time (in months), with $t = 1$ corresponding to January.

(a) Graph the sales function over 1 year.

(b) What are the projected sales for June?

85. Television Coverage

A television camera is on a reviewing platform 27 meters from the street on which a parade passes from left to right (see figure). Write the distance d from the camera to a unit in the parade as a function of the angle x, and graph the function over the interval $-\pi/2 < x < \pi/2$. (Consider x as negative when a unit in the parade approaches from the left.)

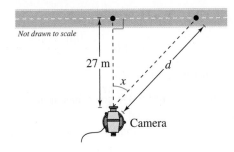

Not drawn to scale

27 m

d

x

Camera

86. Distance A plane flying at an altitude of 7 miles above a radar antenna passes directly over the radar antenna (see figure). Let d be the ground distance from the antenna to the point directly under the plane and let x be the angle of elevation to the plane from the antenna. (d is positive as the plane approaches the antenna.) Write d as a function of x and graph the function over the interval $0 < x < \pi$.

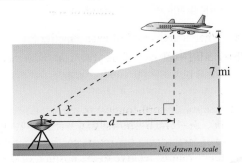

7 mi

x

d

Not drawn to scale

Exploration

True or False? **In Exercises 87 and 88, determine whether the statement is true or false. Justify your answer.**

87. You can obtain the graph of $y = \csc x$ on a calculator by graphing the reciprocal of $y = \sin x$.

88. You can obtain the graph of $y = \sec x$ on a calculator by graphing a translation of the reciprocal of $y = \sin x$.

89. Think About It Consider the function $f(x) = x - \cos x$.

(a) Use a graphing utility to graph the function and verify that there exists a zero between 0 and 1. Use the graph to approximate the zero.

(b) Starting with $x_0 = 1$, generate a sequence x_1, x_2, $x_3, \ldots$, where $x_n = \cos(x_{n-1})$. For example, $x_0 = 1, x_1 = \cos(x_0), x_2 = \cos(x_1), x_3 = \cos(x_2), \ldots$. What value does the sequence approach?

90. HOW DO YOU SEE IT? Determine which function each graph represents. Do not use a calculator. Explain.

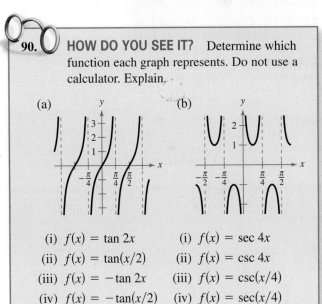

(a)

(b)

(i) $f(x) = \tan 2x$ (i) $f(x) = \sec 4x$

(ii) $f(x) = \tan(x/2)$ (ii) $f(x) = \csc 4x$

(iii) $f(x) = -\tan 2x$ (iii) $f(x) = \csc(x/4)$

(iv) $f(x) = -\tan(x/2)$ (iv) $f(x) = \sec(x/4)$

Graphical Reasoning **In Exercises 91 and 92, use a graphing utility to graph the function. Use the graph to determine the behavior of the function as $x \to c$. (Note: The notation $x \to c^+$ indicates that x approaches c from the right and $x \to c^-$ indicates that x approaches c from the left.)**

(a) $x \to 0^+$ (b) $x \to 0^-$ (c) $x \to \pi^+$ (d) $x \to \pi^-$

91. $f(x) = \cot x$ **92.** $f(x) = \csc x$

Graphical Reasoning **In Exercises 93 and 94, use a graphing utility to graph the function. Use the graph to determine the behavior of the function as $x \to c$.**

(a) $x \to (\pi/2)^+$ (b) $x \to (\pi/2)^-$

(c) $x \to (-\pi/2)^+$ (d) $x \to (-\pi/2)^-$

93. $f(x) = \tan x$ **94.** $f(x) = \sec x$

4.7 Inverse Trigonometric Functions

Inverse trigonometric functions have many applications in real life. For example, in Exercise 100 on page 326, you will use an inverse trigonometric function to model the angle of elevation from a television camera to a space shuttle.

■ Evaluate and graph the inverse sine function.
■ Evaluate and graph other inverse trigonometric functions.
■ Evaluate compositions with inverse trigonometric functions.

Inverse Sine Function

Recall from Section 1.9 that for a function to have an inverse function, it must be one-to-one—that is, it must pass the Horizontal Line Test. Notice in Figure 4.47 that $y = \sin x$ does not pass the test because different values of x yield the same y-value.

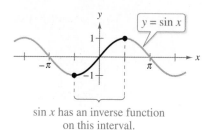

sin x has an inverse function
on this interval.

Figure 4.47

However, when you restrict the domain to the interval $-\pi/2 \le x \le \pi/2$ (corresponding to the black portion of the graph in Figure 4.47), the properties listed below hold.

1. On the interval $[-\pi/2, \pi/2]$, the function $y = \sin x$ is increasing.
2. On the interval $[-\pi/2, \pi/2]$, $y = \sin x$ takes on its full range of values, $-1 \le \sin x \le 1$.
3. On the interval $[-\pi/2, \pi/2]$, $y = \sin x$ is one-to-one.

So, on the restricted domain $-\pi/2 \le x \le \pi/2$, $y = \sin x$ has a unique inverse function called the **inverse sine function.** It is denoted by

$$y = \arcsin x \quad \text{or} \quad y = \sin^{-1} x.$$

The notation $\sin^{-1} x$ is consistent with the inverse function notation $f^{-1}(x)$. The arcsin x notation (read as "the arcsine of x") comes from the association of a central angle with its intercepted *arc length* on a unit circle. So, arcsin x means the angle (or arc) whose sine is x. Both notations, arcsin x and $\sin^{-1} x$ are commonly used in mathematics. You must remember that $\sin^{-1} x$ denotes the *inverse* sine function, *not* $1/\sin x$. The values of arcsin x lie in the interval

$$-\frac{\pi}{2} \le \arcsin x \le \frac{\pi}{2}.$$

Figure 4.48 on the next page shows the graph of $y = \arcsin x$.

⋯⋯⋯⋯⋯⋯⋯⋯▷
REMARK When evaluating the inverse sine function, it helps to remember the phrase "the arcsine of x is the angle (or number) whose sine is x."

Definition of Inverse Sine Function

The **inverse sine function** is defined by

$$y = \arcsin x \quad \text{if and only if} \quad \sin y = x$$

where $-1 \le x \le 1$ and $-\pi/2 \le y \le \pi/2$. The domain of $y = \arcsin x$ is $[-1, 1]$, and the range is $[-\pi/2, \pi/2]$.

•• REMARK As with trigonometric functions, some of the work with inverse trigonometric functions can be done by *exact* calculations rather than by calculator approximations. Exact calculations help to increase your understanding of inverse functions by relating them to the right triangle definitions of trigonometric functions.

EXAMPLE 1   **Evaluating the Inverse Sine Function**

If possible, find the exact value of each expression.

a. $\arcsin\left(-\dfrac{1}{2}\right)$ **b.** $\sin^{-1}\dfrac{\sqrt{3}}{2}$ **c.** $\sin^{-1}2$

Solution

a. You know that $\sin\left(-\dfrac{\pi}{6}\right) = -\dfrac{1}{2}$ and $-\dfrac{\pi}{6}$ lies in $\left[-\dfrac{\pi}{2}, \dfrac{\pi}{2}\right]$, so

$$\arcsin\left(-\dfrac{1}{2}\right) = -\dfrac{\pi}{6}. \qquad \text{Angle whose sine is } -\tfrac{1}{2}$$

b. You know that $\sin\dfrac{\pi}{3} = \dfrac{\sqrt{3}}{2}$ and $\dfrac{\pi}{3}$ lies in $\left[-\dfrac{\pi}{2}, \dfrac{\pi}{2}\right]$, so

$$\sin^{-1}\dfrac{\sqrt{3}}{2} = \dfrac{\pi}{3}. \qquad \text{Angle whose sine is } \sqrt{3}/2$$

c. It is not possible to evaluate $y = \sin^{-1}x$ when $x = 2$ because there is no angle whose sine is 2. Remember that the domain of the inverse sine function is $[-1, 1]$.

✓ **Checkpoint**))) *Audio-video solution in English & Spanish at LarsonPrecalculus.com*

If possible, find the exact value of each expression.

a. $\arcsin 1$ **b.** $\sin^{-1}(-2)$

EXAMPLE 2   **Graphing the Arcsine Function**

See LarsonPrecalculus.com for an interactive version of this type of example.

Sketch the graph of $y = \arcsin x$.

Solution

By definition, the equations $y = \arcsin x$ and $\sin y = x$ are equivalent for $-\pi/2 \le y \le \pi/2$. So, their graphs are the same. From the interval $[-\pi/2, \pi/2]$, assign values to y in the equation $\sin y = x$ to make a table of values.

y	$-\dfrac{\pi}{2}$	$-\dfrac{\pi}{4}$	$-\dfrac{\pi}{6}$	0	$\dfrac{\pi}{6}$	$\dfrac{\pi}{4}$	$\dfrac{\pi}{2}$
$x = \sin y$	-1	$-\dfrac{\sqrt{2}}{2}$	$-\dfrac{1}{2}$	0	$\dfrac{1}{2}$	$\dfrac{\sqrt{2}}{2}$	1

Then plot the points and connect them with a smooth curve. Figure 4.48 shows the graph of $y = \arcsin x$. Note that it is the reflection (in the line $y = x$) of the black portion of the graph in Figure 4.47. Be sure you see that Figure 4.48 shows the *entire* graph of the inverse sine function. Remember that the domain of $y = \arcsin x$ is the closed interval $[-1, 1]$ and the range is the closed interval $[-\pi/2, \pi/2]$.

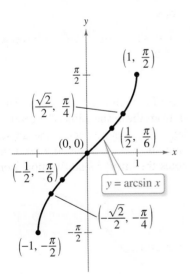

Figure 4.48

✓ **Checkpoint**))) *Audio-video solution in English & Spanish at LarsonPrecalculus.com*

Use a graphing utility to graph $f(x) = \sin x$, $g(x) = \arcsin x$, and $y = x$ in the same viewing window to verify geometrically that g is the inverse function of f. (Be sure to restrict the domain of f properly.)

Other Inverse Trigonometric Functions

The cosine function is decreasing and one-to-one on the interval $0 \le x \le \pi$, as shown in the graph below.

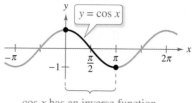

cos x has an inverse function
on this interval.

Consequently, on this interval the cosine function has an inverse function—the **inverse cosine function**—denoted by

$$y = \arccos x \quad \text{or} \quad y = \cos^{-1} x.$$

Similarly, to define an **inverse tangent function,** restrict the domain of $y = \tan x$ to the interval $(-\pi/2, \pi/2)$. The inverse tangent function is denoted by

$$y = \arctan x \quad \text{or} \quad y = \tan^{-1} x.$$

The list below summarizes the definitions of the three most common inverse trigonometric functions. Definitions of the remaining three are explored in Exercises 111–113.

Definitions of the Inverse Trigonometric Functions

Function	Domain	Range
$y = \arcsin x$ if and only if $\sin y = x$	$-1 \le x \le 1$	$-\dfrac{\pi}{2} \le y \le \dfrac{\pi}{2}$
$y = \arccos x$ if and only if $\cos y = x$	$-1 \le x \le 1$	$0 \le y \le \pi$
$y = \arctan x$ if and only if $\tan y = x$	$-\infty < x < \infty$	$-\dfrac{\pi}{2} < y < \dfrac{\pi}{2}$

The graphs of these three inverse trigonometric functions are shown below.

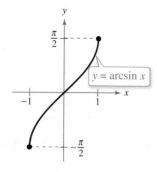

Domain: $[-1, 1]$
Range: $\left[-\dfrac{\pi}{2}, \dfrac{\pi}{2}\right]$
Intercept: $(0, 0)$
Symmetry: origin
Odd function

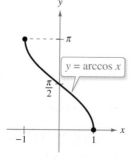

Domain: $[-1, 1]$
Range: $[0, \pi]$
y-intercept: $\left(0, \dfrac{\pi}{2}\right)$

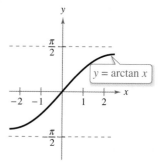

Domain: $(-\infty, \infty)$
Range: $\left(-\dfrac{\pi}{2}, \dfrac{\pi}{2}\right)$
Horizontal asymptotes: $y = \pm\dfrac{\pi}{2}$
Intercept: $(0, 0)$
Symmetry: origin
Odd function

EXAMPLE 3   **Evaluating Inverse Trigonometric Functions**

Find the exact value of each expression.

a. $\arccos \dfrac{\sqrt{2}}{2}$

b. $\arctan 0$

c. $\tan^{-1}(-1)$

Solution

a. You know that $\cos(\pi/4) = \sqrt{2}/2$ and $\pi/4$ lies in $[0, \pi]$, so

$$\arccos \frac{\sqrt{2}}{2} = \frac{\pi}{4}. \qquad \text{Angle whose cosine is } \sqrt{2}/2$$

b. You know that $\tan 0 = 0$ and 0 lies in $(-\pi/2, \pi/2)$, so

$$\arctan 0 = 0. \qquad \text{Angle whose tangent is } 0$$

c. You know that $\tan(-\pi/4) = -1$ and $-\pi/4$ lies in $(-\pi/2, \pi/2)$, so

$$\tan^{-1}(-1) = -\frac{\pi}{4}. \qquad \text{Angle whose tangent is } -1$$

✓ *Checkpoint* *Audio-video solution in English & Spanish at LarsonPrecalculus.com*

Find the exact value of $\cos^{-1}(-1)$.

EXAMPLE 4   **Calculators and Inverse Trigonometric Functions**

Use a calculator to approximate the value of each expression, if possible.

a. $\arctan(-8.45)$

b. $\sin^{-1} 0.2447$

c. $\arccos 2$

Solution

Function	Mode	Calculator Keystrokes
a. $\arctan(-8.45)$	Radian	(TAN⁻¹) (() ((−)) 8.45 ()) (ENTER)

From the display, it follows that $\arctan(-8.45) \approx -1.4530010$.

b. $\sin^{-1} 0.2447$	Radian	(SIN⁻¹) (() 0.2447 ()) (ENTER)

From the display, it follows that $\sin^{-1} 0.2447 \approx 0.2472103$.

c. $\arccos 2$	Radian	(COS⁻¹) (() 2 ()) (ENTER)

The calculator should display an *error message* because the domain of the inverse cosine function is $[-1, 1]$.

✓ *Checkpoint*  *Audio-video solution in English & Spanish at LarsonPrecalculus.com*

Use a calculator to approximate the value of each expression, if possible.

a. $\arctan 4.84$

b. $\arcsin(-1.1)$

c. $\arccos(-0.349)$

In Example 4, had you set the calculator to *degree* mode, the displays would have been in degrees rather than in radians. This convention is peculiar to calculators. By definition, the values of inverse trigonometric functions are *always in radians*.

Compositions with Inverse Trigonometric Functions

▷ **ALGEBRA HELP** To review compositions of functions, see Section 1.8.

Recall from Section 1.9 that for all x in the domains of f and f^{-1}, inverse functions have the properties

$$f(f^{-1}(x)) = x \quad \text{and} \quad f^{-1}(f(x)) = x.$$

Inverse Properties of Trigonometric Functions

If $-1 \le x \le 1$ and $-\pi/2 \le y \le \pi/2$, then

$$\sin(\arcsin x) = x \quad \text{and} \quad \arcsin(\sin y) = y.$$

If $-1 \le x \le 1$ and $0 \le y \le \pi$, then

$$\cos(\arccos x) = x \quad \text{and} \quad \arccos(\cos y) = y.$$

If x is a real number and $-\pi/2 < y < \pi/2$, then

$$\tan(\arctan x) = x \quad \text{and} \quad \arctan(\tan y) = y.$$

Keep in mind that these inverse properties do not apply for arbitrary values of x and y. For example,

$$\arcsin\left(\sin \frac{3\pi}{2}\right) = \arcsin(-1) = -\frac{\pi}{2} \neq \frac{3\pi}{2}.$$

In other words, the property $\arcsin(\sin y) = y$ is not valid for values of y outside the interval $[-\pi/2, \pi/2]$.

EXAMPLE 5 **Using Inverse Properties**

If possible, find the exact value of each expression.

a. $\tan[\arctan(-5)]$ **b.** $\arcsin\left(\sin \dfrac{5\pi}{3}\right)$ **c.** $\cos(\cos^{-1} \pi)$

Solution

a. You know that -5 lies in the domain of the arctangent function, so the inverse property applies, and you have

$$\tan[\arctan(-5)] = -5.$$

b. In this case, $5\pi/3$ does not lie in the range of the arcsine function, $-\pi/2 \le y \le \pi/2$. However, $5\pi/3$ is coterminal with

$$\frac{5\pi}{3} - 2\pi = -\frac{\pi}{3}$$

which does lie in the range of the arcsine function, and you have

$$\arcsin\left(\sin \frac{5\pi}{3}\right) = \arcsin\left[\sin\left(-\frac{\pi}{3}\right)\right] = -\frac{\pi}{3}.$$

c. The expression $\cos(\cos^{-1} \pi)$ is not defined because $\cos^{-1} \pi$ is not defined. Remember that the domain of the inverse cosine function is $[-1, 1]$.

✓ **Checkpoint** ◀))) *Audio-video solution in English & Spanish at LarsonPrecalculus.com*

If possible, find the exact value of each expression.

a. $\tan[\tan^{-1}(-14)]$ **b.** $\sin^{-1}\left(\sin \dfrac{7\pi}{4}\right)$ **c.** $\cos(\arccos 0.54)$

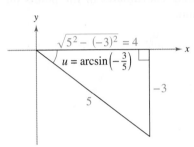

Angle whose cosine is $\frac{2}{3}$
Figure 4.49

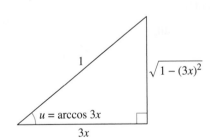

Angle whose sine is $-\frac{3}{5}$
Figure 4.50

EXAMPLE 6 **Evaluating Compositions of Functions**

Find the exact value of each expression.

a. $\tan\left(\arccos \frac{2}{3}\right)$ **b.** $\cos\left[\arcsin\left(-\frac{3}{5}\right)\right]$

Solution

a. If you let $u = \arccos \frac{2}{3}$, then $\cos u = \frac{2}{3}$. The range of the inverse cosine function is $[0, \pi]$ and $\cos u$ is positive, so u is a *first*-quadrant angle. Sketch and label a right triangle with acute angle u, as shown in Figure 4.49. Consequently,

$$\tan\left(\arccos \frac{2}{3}\right) = \tan u = \frac{\text{opp}}{\text{adj}} = \frac{\sqrt{5}}{2}.$$

b. If you let $u = \arcsin\left(-\frac{3}{5}\right)$, then $\sin u = -\frac{3}{5}$. The range of the inverse sine function is $[-\pi/2, \pi/2]$ and $\sin u$ is negative, so u is a *fourth*-quadrant angle. Sketch and label a right triangle with acute angle u, as shown in Figure 4.50. Consequently,

$$\cos\left[\arcsin\left(-\frac{3}{5}\right)\right] = \cos u = \frac{\text{adj}}{\text{hyp}} = \frac{4}{5}.$$

✓ **Checkpoint**))) *Audio-video solution in English & Spanish at LarsonPrecalculus.com*

Find the exact value of $\cos\left[\arctan\left(-\frac{3}{4}\right)\right]$.

EXAMPLE 7 **Some Problems from Calculus**

Write an algebraic expression that is equivalent to each expression.

a. $\sin(\arccos 3x), \quad 0 \le x \le \frac{1}{3}$ **b.** $\cot(\arccos 3x), \quad 0 \le x < \frac{1}{3}$

Solution

If you let $u = \arccos 3x$, then $\cos u = 3x$, where $-1 \le 3x \le 1$. Write

$$\cos u = \frac{\text{adj}}{\text{hyp}} = \frac{3x}{1}$$

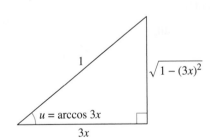

and sketch a right triangle with acute angle u, as shown in Figure 4.51. From this triangle, convert each expression to algebraic form.

a. $\sin(\arccos 3x) = \sin u = \frac{\text{opp}}{\text{hyp}} = \sqrt{1 - 9x^2}, \quad 0 \le x \le \frac{1}{3}$

b. $\cot(\arccos 3x) = \cot u = \frac{\text{adj}}{\text{opp}} = \frac{3x}{\sqrt{1 - 9x^2}}, \quad 0 \le x < \frac{1}{3}$

Angle whose cosine is $3x$
Figure 4.51

✓ **Checkpoint**))) *Audio-video solution in English & Spanish at LarsonPrecalculus.com*

Write an algebraic expression that is equivalent to $\sec(\arctan x)$. ∎

Summarize **(Section 4.7)**

1. State the definition of the inverse sine function *(page 318)*. For examples of evaluating and graphing the inverse sine function, see Examples 1 and 2.

2. State the definitions of the inverse cosine and inverse tangent functions *(page 320)*. For examples of evaluating inverse trigonometric functions, see Examples 3 and 4.

3. State the inverse properties of trigonometric functions *(page 322)*. For examples of finding compositions with inverse trigonometric functions, see Examples 5–7.

100. Videography A television camera at ground level films the lift-off of a space shuttle at a point 750 meters from the launch pad (see figure). Let θ be the angle of elevation to the shuttle and let s be the height of the shuttle.

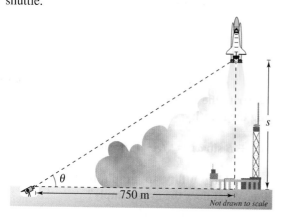

750 m

Not drawn to scale

(a) Write θ as a function of s.

(b) Find θ when $s = 300$ meters and $s = 1200$ meters.

101. Granular Angle of Repose Different types of granular substances naturally settle at different angles when stored in cone-shaped piles. This angle θ is called the *angle of repose* (see figure). When rock salt is stored in a cone-shaped pile 5.5 meters high, the diameter of the pile's base is about 17 meters.

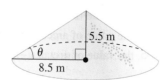

5.5 m

8.5 m

(a) Find the angle of repose for rock salt.

(b) How tall is a pile of rock salt that has a base diameter of 20 meters?

102. Granular Angle of Repose When shelled corn is stored in a cone-shaped pile 20 feet high, the diameter of the pile's base is about 94 feet.

(a) Draw a diagram that gives a visual representation of the problem. Label the known quantities.

(b) Find the angle of repose (see Exercise 101) for shelled corn.

(c) How tall is a pile of shelled corn that has a base diameter of 60 feet?

103. Photography A photographer takes a picture of a three-foot-tall painting hanging in an art gallery. The camera lens is 1 foot below the lower edge of the painting (see figure). The angle β subtended by the camera lens x feet from the painting is given by

$$\beta = \arctan \frac{3x}{x^2 + 4}, \quad x > 0.$$

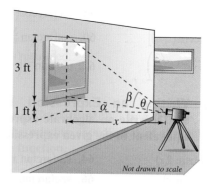

3 ft

1 ft

Not drawn to scale

(a) Use a graphing utility to graph β as a function of x.

(b) Use the graph to approximate the distance from the picture when β is maximum.

(c) Identify the asymptote of the graph and interpret its meaning in the context of the problem.

104. Angle of Elevation An airplane flies at an altitude of 6 miles toward a point directly over an observer. Consider θ and x as shown in the figure.

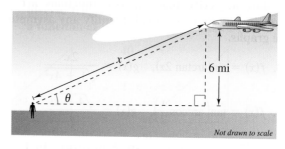

6 mi

Not drawn to scale

(a) Write θ as a function of x.

(b) Find θ when $x = 12$ miles and $x = 7$ miles.

105. Police Patrol A police car with its spotlight on is parked 20 meters from a warehouse. Consider θ and x as shown in the figure.

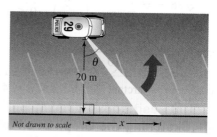

20 m

Not drawn to scale

(a) Write θ as a function of x.

(b) Find θ when $x = 5$ meters and $x = 12$ meters.

Exploration

True or False? In Exercises 106–109, determine whether the statement is true or false. Justify your answer.

106. $\sin \dfrac{5\pi}{6} = \dfrac{1}{2} \implies \arcsin \dfrac{1}{2} = \dfrac{5\pi}{6}$

107. $\tan\left(-\dfrac{\pi}{4}\right) = -1 \implies \arctan(-1) = -\dfrac{\pi}{4}$

108. $\arctan x = \dfrac{\arcsin x}{\arccos x}$ **109.** $\sin^{-1} x = \dfrac{1}{\sin x}$

110. HOW DO YOU SEE IT? Use the figure below to determine the value(s) of x for which each statement is true.

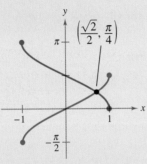

$\left(\dfrac{\sqrt{2}}{2}, \dfrac{\pi}{4}\right)$

— $y = \arcsin x$ — $y = \arccos x$

(a) $\arcsin x < \arccos x$

(b) $\arcsin x = \arccos x$

(c) $\arcsin x > \arccos x$

111. Inverse Cotangent Function Define the inverse cotangent function by restricting the domain of the cotangent function to the interval $(0, \pi)$, and sketch the graph of the inverse trigonometric function.

112. Inverse Secant Function Define the inverse secant function by restricting the domain of the secant function to the intervals $[0, \pi/2)$ and $(\pi/2, \pi]$, and sketch the graph of the inverse trigonometric function.

113. Inverse Cosecant Function Define the inverse cosecant function by restricting the domain of the cosecant function to the intervals $[-\pi/2, 0)$ and $(0, \pi/2]$, and sketch the graph of the inverse trigonometric function.

114. Writing Use the results of Exercises 111–113 to explain how to graph (a) the inverse cotangent function, (b) the inverse secant function, and (c) the inverse cosecant function on a graphing utility.

Evaluating an Inverse Trigonometric Function In Exercises 115–120, use the results of Exercises 111–113 to find the exact value of the expression.

115. $\operatorname{arcsec} \sqrt{2}$ **116.** $\operatorname{arcsec} 1$

117. $\operatorname{arccot}(-1)$ **118.** $\operatorname{arccot}\left(-\sqrt{3}\right)$

119. $\operatorname{arccsc}(-1)$ **120.** $\operatorname{arccsc} \dfrac{2\sqrt{3}}{3}$

Calculators and Inverse Trigonometric Functions In Exercises 121–126, use the results of Exercises 111–113 and a calculator to approximate the value of the expression. Round your result to two decimal places.

121. $\operatorname{arcsec} 2.54$ **122.** $\operatorname{arcsec}(-1.52)$

123. $\operatorname{arccsc}\left(-\dfrac{25}{3}\right)$ **124.** $\operatorname{arccsc}(-12)$

125. $\operatorname{arccot} 5.25$ **126.** $\operatorname{arccot}\left(-\dfrac{16}{7}\right)$

127. Area In calculus, it is shown that the area of the region bounded by the graphs of $y = 0$, $y = 1/(x^2 + 1)$, $x = a$, and $x = b$ (see figure) is given by

$$\text{Area} = \arctan b - \arctan a.$$

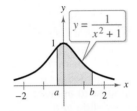

Find the area for each value of a and b.

(a) $a = 0, b = 1$ (b) $a = -1, b = 1$

(c) $a = 0, b = 3$ (d) $a = -1, b = 3$

128. Think About It Use a graphing utility to graph the functions $f(x) = \sqrt{x}$ and $g(x) = 6 \arctan x$. For $x > 0$, it appears that $g > f$. Explain how you know that there exists a positive real number a such that $g < f$ for $x > a$. Approximate the number a.

129. Think About It Consider the functions

$$f(x) = \sin x \quad \text{and} \quad f^{-1}(x) = \arcsin x.$$

(a) Use a graphing utility to graph the composite functions $f \circ f^{-1}$ and $f^{-1} \circ f$.

(b) Explain why the graphs in part (a) are not the graph of the line $y = x$. Why do the graphs of $f \circ f^{-1}$ and $f^{-1} \circ f$ differ?

130. Proof Prove each identity.

(a) $\arcsin(-x) = -\arcsin x$

(b) $\arctan(-x) = -\arctan x$

(c) $\arctan x + \arctan \dfrac{1}{x} = \dfrac{\pi}{2}, \quad x > 0$

(d) $\arcsin x + \arccos x = \dfrac{\pi}{2}$

(e) $\arcsin x = \arctan \dfrac{x}{\sqrt{1 - x^2}}$

4.8 Applications and Models

Right triangles often occur in real-life situations. For example, in Exercise 30 on page 335, you will use right triangles to analyze the design of a new slide at a water park.

- Solve real-life problems involving right triangles.
- Solve real-life problems involving directional bearings.
- Solve real-life problems involving harmonic motion.

Applications Involving Right Triangles

In this section, the three angles of a right triangle are denoted by A, B, and C (where C is the right angle), and the lengths of the sides opposite these angles are denoted by a, b, and c, respectively (where c is the hypotenuse).

EXAMPLE 1 Solving a Right Triangle

See LarsonPrecalculus.com for an interactive version of this type of example.

Solve the right triangle shown at the right for all unknown sides and angles.

Solution Because $C = 90°$, it follows that

$$A + B = 90° \quad \text{and} \quad B = 90° - 34.2° = 55.8°.$$

To solve for a, use the fact that

$$\tan A = \frac{\text{opp}}{\text{adj}} = \frac{a}{b} \quad \Longrightarrow \quad a = b \tan A.$$

So, $a = 19.4 \tan 34.2° \approx 13.2$. Similarly, to solve for c, use the fact that

$$\cos A = \frac{\text{adj}}{\text{hyp}} = \frac{b}{c} \quad \Longrightarrow \quad c = \frac{b}{\cos A}.$$

So, $c = \dfrac{19.4}{\cos 34.2°} \approx 23.5$.

✓ **Checkpoint** ◀))) *Audio-video solution in English & Spanish at LarsonPrecalculus.com*

Solve the right triangle shown at the right for all unknown sides and angles.

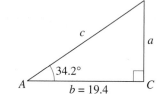

EXAMPLE 2 Finding a Side of a Right Triangle

The height of a mountain is 5000 feet. The distance between its peak and that of an adjacent mountain is 25,000 feet. The angle of elevation between the two peaks is 27°. (See Figure 4.52.) What is the height of the adjacent mountain?

Solution From the figure, $\sin A = a/c$, so

$$a = c \sin A = 25{,}000 \sin 27° \approx 11{,}350.$$

The height of the adjacent mountain is about $11{,}350 + 5000 = 16{,}350$ feet.

✓ **Checkpoint** ◀))) *Audio-video solution in English & Spanish at LarsonPrecalculus.com*

A ladder that is 16 feet long leans against the side of a house. The angle of elevation of the ladder is 80°. Find the height from the top of the ladder to the ground. ∎

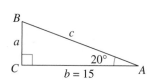

Figure 4.52

EXAMPLE 3 Finding a Side of a Right Triangle

At a point 200 feet from the base of a building, the angle of elevation to the *bottom* of a smokestack is 35°, whereas the angle of elevation to the *top* is 53°, as shown in Figure 4.53. Find the height s of the smokestack alone.

Solution

This problem involves two right triangles. For the smaller right triangle, use the fact that

$$\tan 35° = \frac{a}{200}$$

to find that the height of the building is

$$a = 200 \tan 35°.$$

For the larger right triangle, use the equation

$$\tan 53° = \frac{a + s}{200}$$

to find that

$$a + s = 200 \tan 53°.$$

So, the height of the smokestack is

$$s = 200 \tan 53° - a$$
$$= 200 \tan 53° - 200 \tan 35°$$
$$\approx 125.4 \text{ feet.}$$

✓ Checkpoint 🔊))) *Audio-video solution in English & Spanish at LarsonPrecalculus.com*

At a point 65 feet from the base of a church, the angles of elevation to the bottom of the steeple and the top of the steeple are 35° and 43°, respectively. Find the height of the steeple.

EXAMPLE 4 Finding an Angle of Depression

A swimming pool is 20 meters long and 12 meters wide. The bottom of the pool is slanted so that the water depth is 1.3 meters at the shallow end and 4 meters at the deep end, as shown in Figure 4.54. Find the angle of depression (in degrees) of the bottom of the pool.

Solution Using the tangent function,

$$\tan A = \frac{\text{opp}}{\text{adj}}$$
$$= \frac{2.7}{20}$$
$$= 0.135.$$

So, the angle of depression is

$$A = \arctan 0.135 \approx 0.13419 \text{ radian} \approx 7.69°.$$

✓ Checkpoint 🔊))) *Audio-video solution in English & Spanish at LarsonPrecalculus.com*

From the time a small airplane is 100 feet high and 1600 ground feet from its landing runway, the plane descends in a straight line to the runway. Determine the angle of descent (in degrees) of the plane.

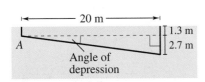

Figure 4.53

Figure 4.54

Trigonometry and Bearings

. . . . ▷
•• **REMARK** In *air navigation*, bearings are measured in degrees *clockwise* from north. The figures below illustrate examples of air navigation bearings

In surveying and navigation, directions can be given in terms of **bearings.** A bearing measures the acute angle that a path or line of sight makes with a fixed north-south line. For example, in the figures below, the bearing S 35° E means 35 degrees east of south, N 80° W means 80 degrees west of north, and N 45° E means 45 degrees east of north.

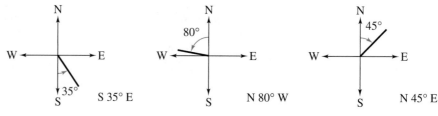

EXAMPLE 5   Finding Directions in Terms of Bearings

A ship leaves port at noon and heads due west at 20 knots, or 20 nautical miles (nmi) per hour. At 2 P.M. the ship changes course to N 54° W, as shown in the figure below. Find the ship's bearing and distance from port at 3 P.M.

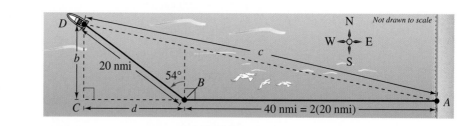

Solution

For triangle *BCD*, you have

$$B = 90° - 54° = 36°.$$

The two sides of this triangle are

$$b = 20 \sin 36° \quad \text{and} \quad d = 20 \cos 36°.$$

For triangle *ACD*, find angle *A*.

$$\tan A = \frac{b}{d + 40} = \frac{20 \sin 36°}{20 \cos 36° + 40} \approx 0.209$$

$$A \approx \arctan 0.209 \approx 0.20603 \text{ radian} \approx 11.80°$$

The angle with the north-south line is 90° − 11.80° = 78.20°. So, the bearing of the ship is N 78.20° W. Finally, from triangle *ACD*, you have

$$\sin A = \frac{b}{c}$$

which yields

$$c = \frac{b}{\sin A} = \frac{20 \sin 36°}{\sin 11.80°} \approx 57.5 \text{ nautical miles.} \qquad \text{Distance from port}$$

✓ *Checkpoint* ◀))) *Audio-video solution in English & Spanish at LarsonPrecalculus.com*

A sailboat leaves a pier heading due west at 8 knots. After 15 minutes, the sailboat changes course to N 16° W at 10 knots. Find the sailboat's bearing and distance from the pier after 12 minutes on this course.

Harmonic Motion

The periodic nature of the trigonometric functions is useful for describing the motion of a point on an object that vibrates, oscillates, rotates, or is moved by wave motion.

For example, consider a ball that is bobbing up and down on the end of a spring. Assume that the maximum distance the ball moves vertically upward or downward from its equilibrium (at rest) position is 10 centimeters (see figure). Assume further that the time it takes for the ball to move from its maximum displacement above zero to its maximum displacement below zero and back again is $t = 4$ seconds. With the ideal conditions of perfect elasticity and no friction or air resistance, the ball would continue to move up and down in a uniform and regular manner.

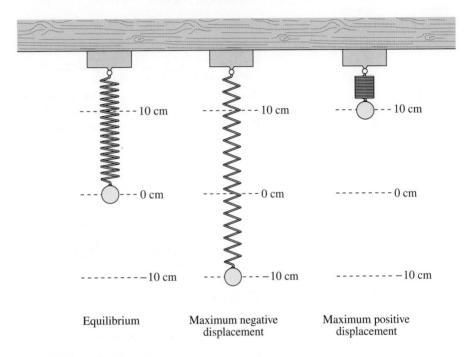

| Equilibrium | Maximum negative displacement | Maximum positive displacement |

The period (time for one complete cycle) of the motion is

Period = 4 seconds

the amplitude (maximum displacement from equilibrium) is

Amplitude = 10 centimeters

and the **frequency** (number of cycles per second) is

Frequency = $\dfrac{1}{4}$ cycle per second.

Motion of this nature can be described by a sine or cosine function and is called **simple harmonic motion.**

Definition of Simple Harmonic Motion

A point that moves on a coordinate line is in **simple harmonic motion** when its distance d from the origin at time t is given by either

$$d = a \sin \omega t \quad \text{or} \quad d = a \cos \omega t$$

where a and ω are real numbers such that $\omega > 0$. The motion has amplitude $|a|$, period $\dfrac{2\pi}{\omega}$, and frequency $\dfrac{\omega}{2\pi}$.

EXAMPLE 6   **Simple Harmonic Motion**

Write an equation for the simple harmonic motion of the ball described on the preceding page.

Solution

The spring is at equilibrium $(d = 0)$ when $t = 0$, so use the equation

$$d = a \sin \omega t.$$

Moreover, the maximum displacement from zero is 10 and the period is 4. Using this information, you have

$$\text{Amplitude} = |a|$$

$$= 10$$

$$\text{Period} = \frac{2\pi}{\omega} = 4 \quad \Longrightarrow \quad \omega = \frac{\pi}{2}.$$

Consequently, an equation of motion is

$$d = 10 \sin \frac{\pi}{2} t.$$

Note that the choice of

$$a = 10 \quad \text{or} \quad a = -10$$

depends on whether the ball initially moves up or down.

✓ **Checkpoint** Audio-video solution in English & Spanish at LarsonPrecalculus.com

Write an equation for simple harmonic motion for which $d = 0$ when $t = 0$, the amplitude is 6 centimeters, and the period is 3 seconds. ■

One illustration of the relationship between sine waves and harmonic motion is in the wave motion that results when you drop a stone into a calm pool of water. The waves move outward in roughly the shape of sine (or cosine) waves, as shown at the right. Now suppose you are fishing in the same pool of water and your fishing bobber does not move horizontally. As the waves move outward from the dropped stone, the fishing bobber moves up and down in simple harmonic motion, as shown below.

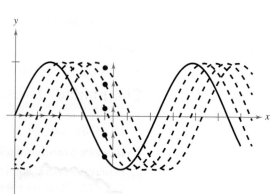

EXAMPLE 7 **Simple Harmonic Motion**

Consider the equation for simple harmonic motion $d = 6 \cos \frac{3\pi}{4} t$. Find (a) the maximum displacement, (b) the frequency, (c) the value of d when $t = 4$, and (d) the least positive value of t for which $d = 0$.

Algebraic Solution

The equation has the form $d = a \cos \omega t$, with $a = 6$ and $\omega = 3\pi/4$.

a. The maximum displacement (from the point of equilibrium) is the amplitude. So, the maximum displacement is 6.

b. Frequency $= \dfrac{\omega}{2\pi}$

$= \dfrac{3\pi/4}{2\pi}$

$= \dfrac{3}{8}$ cycle per unit of time

c. $d = 6 \cos \left[\dfrac{3\pi}{4}(4) \right] = 6 \cos 3\pi = 6(-1) = -6$

d. To find the least positive value of t for which $d = 0$, solve

$$6 \cos \frac{3\pi}{4} t = 0.$$

First divide each side by 6 to obtain

$$\cos \frac{3\pi}{4} t = 0.$$

This equation is satisfied when

$$\frac{3\pi}{4} t = \frac{\pi}{2}, \frac{3\pi}{2}, \frac{5\pi}{2}, \cdots.$$

Multiply these values by $4/(3\pi)$ to obtain

$$t = \frac{2}{3}, 2, \frac{10}{3}, \cdots.$$

So, the least positive value of t is $t = \frac{2}{3}$.

Graphical Solution

Use a graphing utility set in *radian* mode.

a.

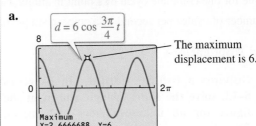

The maximum displacement is 6.

b.

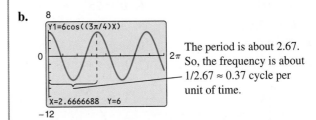

The period is about 2.67. So, the frequency is about $1/2.67 \approx 0.37$ cycle per unit of time.

c.

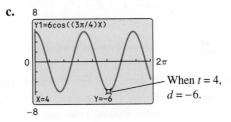

When $t = 4$, $d = -6$.

d. The least positive value of t for which $d = 0$ is $t \approx 0.67$.

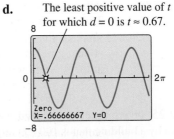

✓ *Checkpoint* ))) *Audio-video solution in English & Spanish at LarsonPrecalculus.com*

Rework Example 7 for the equation $d = 4 \cos 6\pi t$.

Summarize (Section 4.8)

1. Describe real-life applications of right triangles (*pages 328 and 329, Examples 1–4*).

2. Describe a real-life application of a directional bearing (*page 330, Example 5*).

3. Describe real-life applications of simple harmonic motion (*pages 332 and 333, Examples 6 and 7*).

4.8 Exercises

See **CalcChat.com** for tutorial help and worked-out solutions to odd-numbered exercises.

Vocabulary: Fill in the blanks.

1. A _____ measures the acute angle that a path or line of sight makes with a fixed north-south line.

2. A point that moves on a coordinate line is in simple _____ _____ when its distance d from the origin at time t is given by either $d = a \sin \omega t$ or $d = a \cos \omega t$.

3. The time for one complete cycle of a point in simple harmonic motion is its _____.

4. The number of cycles per second of a point in simple harmonic motion is its _____.

Skills and Applications

Solving a Right Triangle In Exercises 5–12, solve the right triangle shown in the figure for all unknown sides and angles. Round your answers to two decimal places.

5. $A = 60°$, $c = 12$
6. $B = 25°$, $b = 4$
7. $B = 72.8°$, $a = 4.4$
8. $A = 8.4°$, $a = 40.5$
9. $a = 3$, $b = 4$
10. $a = 25$, $c = 35$
11. $b = 15.70$, $c = 55.16$
12. $b = 1.32$, $c = 9.45$

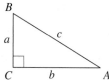

Figure for 5–12   Figure for 13–16

Finding an Altitude In Exercises 13–16, find the altitude of the isosceles triangle shown in the figure. Round your answers to two decimal places.

13. $\theta = 45°$, $b = 6$
14. $\theta = 22°$, $b = 14$
15. $\theta = 32°$, $b = 8$
16. $\theta = 27°$, $b = 11$

17. **Length** The sun is 25° above the horizon. Find the length of a shadow cast by a building that is 100 feet tall (see figure).

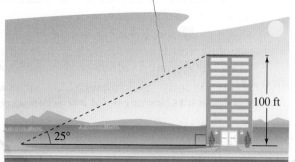

18. **Length** The sun is 20° above the horizon. Find the length of a shadow cast by a park statue that is 12 feet tall.

19. **Height** A ladder that is 20 feet long leans against the side of a house. The angle of elevation of the ladder is 80°. Find the height from the top of the ladder to the ground.

20. **Height** The length of a shadow of a tree is 125 feet when the angle of elevation of the sun is 33°. Approximate the height of the tree.

21. **Height** At a point 50 feet from the base of a church, the angles of elevation to the bottom of the steeple and the top of the steeple are 35° and 48°, respectively. Find the height of the steeple.

22. **Distance** An observer in a lighthouse 350 feet above sea level observes two ships directly offshore. The angles of depression to the ships are 4° and 6.5° (see figure). How far apart are the ships?

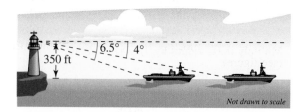

23. **Distance** A passenger in an airplane at an altitude of 10 kilometers sees two towns directly to the east of the plane. The angles of depression to the towns are 28° and 55° (see figure). How far apart are the towns?

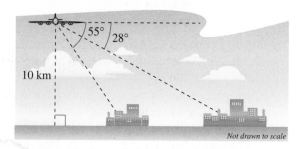

24. Angle of Elevation The height of an outdoor basketball backboard is $12\frac{1}{2}$ feet, and the backboard casts a shadow 17 feet long.

(a) Draw a right triangle that gives a visual representation of the problem. Label the known and unknown quantities.

(b) Use a trigonometric function to write an equation involving the unknown angle of elevation.

(c) Find the angle of elevation.

25. Angle of Elevation An engineer designs a 75-foot cellular telephone tower. Find the angle of elevation to the top of the tower at a point on level ground 50 feet from its base.

26. Angle of Depression A cellular telephone tower that is 120 feet tall is placed on top of a mountain that is 1200 feet above sea level. What is the angle of depression from the top of the tower to a cell phone user who is 5 horizontal miles away and 400 feet above sea level?

27. Angle of Depression A Global Positioning System satellite orbits 12,500 miles above Earth's surface (see figure). Find the angle of depression from the satellite to the horizon. Assume the radius of Earth is 4000 miles.

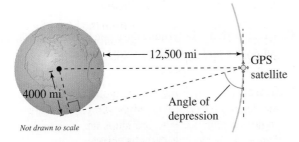

Not drawn to scale

28. Height You are holding one of the tethers attached to the top of a giant character balloon that is floating approximately 20 feet above ground level. You are standing approximately 100 feet ahead of the balloon (see figure).

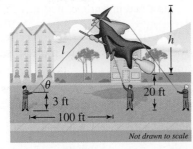

Not drawn to scale

(a) Find an equation for the length l of the tether you are holding in terms of h, the height of the balloon from top to bottom.

(b) Find an equation for the angle of elevation θ from you to the top of the balloon.

(c) The angle of elevation to the top of the balloon is 35°. Find the height h of the balloon.

29. Altitude You observe a plane approaching overhead and assume that its speed is 550 miles per hour. The angle of elevation of the plane is 16° at one time and 57° one minute later. Approximate the altitude of the plane.

30. Waterslide Design

The designers of a water park have sketched a preliminary drawing of a new slide (see figure).

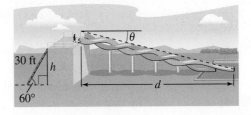

(a) Find the height h of the slide.

(b) Find the angle of depression θ from the top of the slide to the end of the slide at the ground in terms of the horizontal distance d a rider travels.

(c) Safety restrictions require the angle of depression to be no less than 25° and no more than 30°. Find an interval for how far a rider travels horizontally.

31. Speed Enforcement A police department has set up a speed enforcement zone on a straight length of highway. A patrol car is parked parallel to the zone, 200 feet from one end and 150 feet from the other end (see figure).

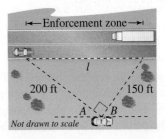

Not drawn to scale

(a) Find the length l of the zone and the measures of angles A and B (in degrees).

(b) Find the minimum amount of time (in seconds) it takes for a vehicle to pass through the zone without exceeding the posted speed limit of 35 miles per hour.

32. Airplane Ascent During takeoff, an airplane's angle of ascent is 18° and its speed is 260 feet per second.

(a) Find the plane's altitude after 1 minute.

(b) How long will it take for the plane to climb to an altitude of 10,000 feet?

33. Air Navigation An airplane flying at 550 miles per hour has a bearing of 52°. After flying for 1.5 hours, how far north and how far east will the plane have traveled from its point of departure?

34. Air Navigation A jet leaves Reno, Nevada, and heads toward Miami, Florida, at a bearing of 100°. The distance between the two cities is approximately 2472 miles.

(a) How far north and how far west is Reno relative to Miami?

(b) The jet is to return directly to Reno from Miami. At what bearing should it travel?

35. Navigation A ship leaves port at noon and has a bearing of S 29° W. The ship sails at 20 knots.

(a) How many nautical miles south and how many nautical miles west will the ship have traveled by 6:00 P.M.?

(b) At 6:00 P.M., the ship changes course to due west. Find the ship's bearing and distance from port at 7:00 P.M.

36. Navigation A privately owned yacht leaves a dock in Myrtle Beach, South Carolina, and heads toward Freeport in the Bahamas at a bearing of S 1.4° E. The yacht averages a speed of 20 knots over the 428-nautical-mile trip.

(a) How long will it take the yacht to make the trip?

(b) How far east and south is the yacht after 12 hours?

(c) A plane leaves Myrtle Beach to fly to Freeport. At what bearing should it travel?

37. Navigation A ship is 45 miles east and 30 miles south of port. The captain wants to sail directly to port. What bearing should the captain take?

38. Air Navigation An airplane is 160 miles north and 85 miles east of an airport. The pilot wants to fly directly to the airport. What bearing should the pilot take?

39. Surveying A surveyor wants to find the distance across a pond (see figure). The bearing from A to B is N 32° W. The surveyor walks 50 meters from A to C, and at the point C the bearing to B is N 68° W.

(a) Find the bearing from A to C.

(b) Find the distance from A to B.

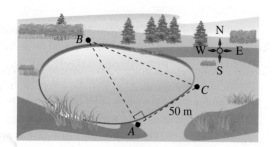

40. Location of a Fire Fire tower A is 30 kilometers due west of fire tower B. A fire is spotted from the towers, and the bearings from A and B are N 76° E and N 56° W, respectively (see figure). Find the distance d of the fire from the line segment AB.

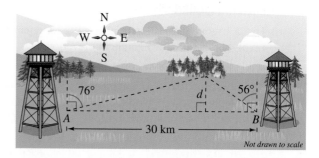

Not drawn to scale

41. Geometry Determine the angle between the diagonal of a cube and the diagonal of its base, as shown in the figure.

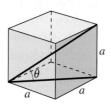

42. Geometry Determine the angle between the diagonal of a cube and its edge, as shown in the figure.

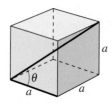

43. Geometry Find the length of the sides of a regular pentagon inscribed in a circle of radius 25 inches.

44. Geometry Find the length of the sides of a regular hexagon inscribed in a circle of radius 25 inches.

Simple Harmonic Motion In Exercises 45–48, find a model for simple harmonic motion satisfying the specified conditions.

Displacement ($t = 0$)	Amplitude	Period
45. 0	4 centimeters	2 seconds
46. 0	3 meters	6 seconds
47. 3 inches	3 inches	1.5 seconds
48. 2 feet	2 feet	10 seconds

49. Tuning Fork A point on the end of a tuning fork moves in simple harmonic motion described by $d = a \sin \omega t$. Find ω given that the tuning fork for middle C has a frequency of 262 vibrations per second.

50. Wave Motion A buoy oscillates in simple harmonic motion as waves go past. The buoy moves a total of 3.5 feet from its low point to its high point (see figure), and it returns to its high point every 10 seconds. Write an equation that describes the motion of the buoy where the high point corresponds to the time $t = 0$.

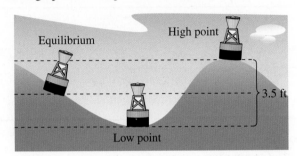

Simple Harmonic Motion In Exercises 51–54, for the simple harmonic motion described by the trigonometric function, find (a) the maximum displacement, (b) the frequency, (c) the value of d when $t = 5$, and (d) the least positive value of t for which $d = 0$. Use a graphing utility to verify your results.

51. $d = 9 \cos \dfrac{6\pi}{5} t$

52. $d = \dfrac{1}{2} \cos 20\pi t$

53. $d = \dfrac{1}{4} \sin 6\pi t$

54. $d = \dfrac{1}{64} \sin 792\pi t$

55. Oscillation of a Spring A ball that is bobbing up and down on the end of a spring has a maximum displacement of 3 inches. Its motion (in ideal conditions) is modeled by $y = \frac{1}{4} \cos 16t, \, t > 0$, where y is measured in feet and t is the time in seconds.

  (a) Graph the function.

  (b) What is the period of the oscillations?

  (c) Determine the first time the weight passes the point of equilibrium $(y = 0)$.

56. Hours of Daylight The numbers of hours H of daylight in Denver, Colorado, on the 15th of each month starting with January are: 9.68, 10.72, 11.92, 13.25, 14.35, 14.97, 14.72, 13.73, 12.47, 11.18, 10.00, and 9.37. A model for the data is

$$H(t) = 12.13 + 2.77 \sin\left(\frac{\pi t}{6} - 1.60\right)$$

where t represents the month, with $t = 1$ corresponding to January. *(Source: United States Navy)*

  (a) Use a graphing utility to graph the data and the model in the same viewing window.

  (b) What is the period of the model? Is it what you expected? Explain.

  (c) What is the amplitude of the model? What does it represent in the context of the problem?

57. Sales The table shows the average sales S (in millions of dollars) of an outerwear manufacturer for each month t, where $t = 1$ corresponds to January.

Time, t	1	2	3	4
Sales, S	13.46	11.15	8.00	4.85

Time, t	5	6	7	8
Sales, S	2.54	1.70	2.54	4.85

Time, t	9	10	11	12
Sales, S	8.00	11.15	13.46	14.30

  (a) Create a scatter plot of the data.

  (b) Find a trigonometric model that fits the data. Graph the model with your scatter plot. How well does the model fit the data?

  (c) What is the period of the model? Do you think it is reasonable given the context? Explain.

  (d) Interpret the meaning of the model's amplitude in the context of the problem.

Exploration

58. **HOW DO YOU SEE IT?** The graph below shows the displacement of an object in simple harmonic motion.

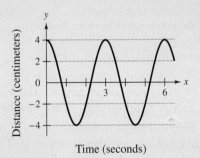

  (a) What is the amplitude?

  (b) What is the period?

  (c) Is the equation of the simple harmonic motion of the form $d = a \sin \omega t$ or $d = a \cos \omega t$?

True or False? In Exercises 59 and 60, determine whether the statement is true or false. Justify your answer.

59. The Leaning Tower of Pisa is not vertical, but when you know the angle of elevation θ to the top of the tower as you stand d feet away from it, its height h can be found using the formula $h = d \tan \theta$.

60. The bearing N 24° E means 24 degrees north of east.

Chapter Summary

	What Did You Learn?	Explanation/Examples	Review Exercises
Section 4.1	Describe angles *(p. 260)*.		1–4
	Use radian measure *(p. 261)* and degree measure *(p. 263)*.	To convert degrees to radians, multiply degrees by $\frac{\pi \text{ rad}}{180°}$. To convert radians to degrees, multiply radians by $\frac{180°}{\pi \text{ rad}}$.	5–14
	Use angles and their measure to model and solve real-life problems *(p. 264)*.	Angles and their measure can be used to find arc length and the area of a sector of a circle. (See Examples 5 and 8.)	15–18
Section 4.2	Identify a unit circle and describe its relationship to real numbers *(p. 270)*.		19–22
	Evaluate trigonometric functions using the unit circle *(p. 271)*.	For the point (x, y) on the unit circle corresponding to a real number t: $\sin t = y$; $\cos t = x$; $\tan t = \frac{y}{x}$, $x \neq 0$; $\csc t = \frac{1}{y}$, $y \neq 0$; $\sec t = \frac{1}{x}$, $x \neq 0$; and $\cot t = \frac{x}{y}$, $y \neq 0$.	23, 24
	Use domain and period to evaluate sine and cosine functions *(p. 273)*, and use a calculator to evaluate trigonometric functions *(p. 274)*.	Because $\frac{13\pi}{6} = 2\pi + \frac{\pi}{6}$, $\sin \frac{13\pi}{6} = \sin \frac{\pi}{6} = \frac{1}{2}$. $\sin \frac{3\pi}{8} \approx 0.9239$, $\cot(-1.2) \approx -0.3888$	25–32
Section 4.3	Evaluate trigonometric functions of acute angles *(p. 277)*.	$\sin \theta = \frac{\text{opp}}{\text{hyp}}$, $\cos \theta = \frac{\text{adj}}{\text{hyp}}$, $\tan \theta = \frac{\text{opp}}{\text{adj}}$ $\csc \theta = \frac{\text{hyp}}{\text{opp}}$, $\sec \theta = \frac{\text{hyp}}{\text{adj}}$, $\cot \theta = \frac{\text{adj}}{\text{opp}}$ $\csc 29° \, 15' = 1/\sin 29.25° \approx 2.0466$	33–38
	Use fundamental trigonometric identities *(p. 280)*.	$\sin \theta = \frac{1}{\csc \theta}$, $\tan \theta = \frac{\sin \theta}{\cos \theta}$, $\sin^2 \theta + \cos^2 \theta = 1$	39, 40
	Use trigonometric functions to model and solve real-life problems *(p. 282)*.	Trigonometric functions can be used to find the height of a monument, the angle between two paths, and the length and height of a ramp. (See Examples 8–10.)	41, 42

	What Did You Learn?	**Explanation/Examples**	**Review Exercises**
Section 4.4	Evaluate trigonometric functions of any angle *(p. 288)*.	Let $(3, 4)$ be a point on the terminal side of θ. Then $\sin\theta = \frac{4}{5}$, $\cos\theta = \frac{3}{5}$, and $\tan\theta = \frac{4}{3}$.	43–50
	Find reference angles *(p. 290)*.	Let θ be an angle in standard position. Its reference angle is the acute angle θ' formed by the terminal side of θ and the horizontal axis.	51–54
	Evaluate trigonometric functions of real numbers *(p. 291)*.	$\cos\dfrac{7\pi}{3} = \dfrac{1}{2}$ because $\theta' = \dfrac{7\pi}{3} - 2\pi = \dfrac{\pi}{3}$ and $\cos\dfrac{\pi}{3} = \dfrac{1}{2}$.	55–62
Section 4.5	Sketch the graphs of sine and cosine functions using amplitude and period *(p. 299)*.		63, 64
	Sketch translations of the graphs of sine and cosine functions *(p. 301)*.	For $y = d + a\sin(bx - c)$ and $y = d + a\cos(bx - c)$, the constant c results in horizontal translations and the constant d results in vertical translations. (See Examples 4–6.)	65–68
	Use sine and cosine functions to model real-life data *(p. 303)*.	A cosine function can be used to model the depth of the water at the end of a dock. (See Example 7.)	69, 70
Section 4.6	Sketch the graphs of tangent *(p. 308)*, cotangent *(p. 310)*, secant *(p. 311)*, and cosecant functions *(p. 311)*.		71–74
	Sketch the graphs of damped trigonometric functions *(p. 313)*.	In $f(x) = x\cos 2x$, the factor x is called the damping factor.	75, 76
Section 4.7	Evaluate and graph inverse trigonometric functions *(p. 318)*.	$\arcsin\left(\dfrac{1}{2}\right) = \dfrac{\pi}{6}$, $\cos^{-1}\left(-\dfrac{\sqrt{2}}{2}\right) = \dfrac{3\pi}{4}$, $\tan^{-1}\sqrt{3} = \dfrac{\pi}{3}$	77–86
	Evaluate compositions with inverse trigonometric functions *(p. 322)*.	$\sin(\sin^{-1} 0.4) = 0.4$, $\cos\left(\arctan\dfrac{5}{12}\right) = \dfrac{12}{13}$	87–92
Section 4.8	Solve real-life problems involving right triangles *(p. 328)*.	A trigonometric function can be used to find the height of a smokestack on top of a building. (See Example 3.)	93, 94
	Solve real-life problems involving directional bearings *(p. 330)*.	Trigonometric functions can be used to find a ship's bearing and distance from a port at a given time. (See Example 5.)	95
	Solve real-life problems involving harmonic motion *(p. 331)*.	Trigonometric functions can be used to describe the motion of a point on an object that vibrates, oscillates, rotates, or is moved by wave motion. (See Examples 6 and 7.)	96

Review Exercises   See CalcChat.com for tutorial help and worked-out solutions to odd-numbered exercises.

4.1 Using Radian or Degree Measure In Exercises 1–4, (a) sketch the angle in standard position, (b) determine the quadrant in which the angle lies, and (c) determine two coterminal angles (one positive and one negative).

1. $\dfrac{15\pi}{4}$

2. $-\dfrac{4\pi}{3}$

3. $-110°$

4. $280°$

Converting from Degrees to Radians In Exercises 5–8, convert the degree measure to radian measure. Round to three decimal places.

5. $450°$

6. $190°$

7. $-16°$

8. $-112°$

Converting from Radians to Degrees In Exercises 9–12, convert the radian measure to degree measure. Round to three decimal places, if necessary.

9. $\dfrac{3\pi}{10}$

10. $-\dfrac{11\pi}{6}$

11. -3.5

12. 5.7

Converting to D° M′ S″ Form In Exercises 13 and 14, convert the angle measure to D° M′ S″ form.

13. $198.4°$

14. $-5.96°$

15. **Arc Length** Find the length of the arc on a circle of radius 20 inches intercepted by a central angle of $138°$.

16. **Phonograph** Phonograph records are vinyl discs that rotate on a turntable. A typical record album is 12 inches in diameter and plays at $33\frac{1}{3}$ revolutions per minute.
    (a) Find the angular speed of a record album.
    (b) Find the linear speed (in inches per minute) of the outer edge of a record album.

Area of a Sector of a Circle In Exercises 17 and 18, find the area of the sector of a circle of radius r and central angle θ.

Radius r	Central Angle θ
17. 20 inches	$150°$
18. 7.5 millimeters	$2\pi/3$ radians

4.2 Finding a Point on the Unit Circle In Exercises 19–22, find the point (x, y) on the unit circle that corresponds to the real number t.

19. $t = 2\pi/3$

20. $t = 7\pi/4$

21. $t = 7\pi/6$

22. $t = -4\pi/3$

Evaluating Trigonometric Functions In Exercises 23 and 24, evaluate (if possible) the six trigonometric functions at the real number.

23. $t = \dfrac{3\pi}{4}$

24. $t = -\dfrac{2\pi}{3}$

Using Period to Evaluate Sine and Cosine In Exercises 25–28, evaluate the trigonometric function using its period as an aid.

25. $\sin \dfrac{11\pi}{4}$

26. $\cos 4\pi$

27. $\cos\left(-\dfrac{17\pi}{6}\right)$

28. $\sin\left(-\dfrac{13\pi}{3}\right)$

Using a Calculator In Exercises 29–32, use a calculator to evaluate the trigonometric function. Round your answer to four decimal places. (Be sure the calculator is in the correct mode.)

29. $\sec \dfrac{12\pi}{5}$

30. $\sin\left(-\dfrac{\pi}{9}\right)$

31. $\tan 33$

32. $\csc 10.5$

4.3 Evaluating Trigonometric Functions In Exercises 33 and 34, find the exact values of the six trigonometric functions of the angle θ.

33.
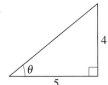

34.
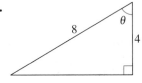

Using a Calculator In Exercises 35–38, use a calculator to evaluate the trigonometric function. Round your answer to four decimal places. (Be sure the calculator is in the correct mode.)

35. $\tan 33°$

36. $\sec 79.3°$

37. $\cot 15° \, 14′$

38. $\cos 78° \, 11′ \, 58″$

Applying Trigonometric Identities In Exercises 39 and 40, use the given function value and the trigonometric identities to find the exact value of each indicated trigonometric function.

39. $\sin \theta = \frac{1}{3}$
    (a) $\csc \theta$
    (b) $\cos \theta$
    (c) $\sec \theta$
    (d) $\tan \theta$

40. $\csc \theta = 5$
    (a) $\sin \theta$
    (b) $\cot \theta$
    (c) $\tan \theta$
    (d) $\sec(90° - \theta)$

41. Railroad Grade A train travels 3.5 kilometers on a straight track with a grade of 1.2° (see figure). What is the vertical rise of the train in that distance?

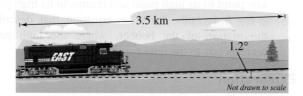

— 3.5 km —

1.2°

Not drawn to scale

42. Guy Wire A guy wire runs from the ground to the top of a 25-foot telephone pole. The angle formed between the wire and the ground is 52°. How far from the base of the pole is the guy wire anchored to the ground? Assume the pole is perpendicular to the ground.

4.4 Evaluating Trigonometric Functions In Exercises 43–46, the point is on the terminal side of an angle in standard position. Find the exact values of the six trigonometric functions of the angle.

43. $(12, 16)$ **44.** $(3, -4)$

45. $(0.3, 0.4)$ **46.** $\left(-\frac{10}{3}, -\frac{2}{3}\right)$

Evaluating Trigonometric Functions In Exercises 47–50, find the exact values of the remaining five trigonometric functions of θ satisfying the given conditions.

47. $\sec \theta = \frac{6}{5}$, $\tan \theta < 0$

48. $\csc \theta = \frac{3}{2}$, $\cos \theta < 0$

49. $\cos \theta = -\frac{2}{5}$, $\sin \theta > 0$

50. $\sin \theta = -\frac{1}{2}$, $\cos \theta > 0$

Finding a Reference Angle In Exercises 51–54, find the reference angle θ'. Sketch θ in standard position and label θ'.

51. $\theta = 264°$ **52.** $\theta = 635°$

53. $\theta = -6\pi/5$ **54.** $\theta = 17\pi/3$

Using a Reference Angle In Exercises 55–58, evaluate the sine, cosine, and tangent of the angle without using a calculator.

55. $-150°$ **56.** $495°$

57. $\pi/3$ **58.** $-5\pi/4$

Using a Calculator In Exercises 59–62, use a calculator to evaluate the trigonometric function. Round your answer to four decimal places. (Be sure the calculator is in the correct mode.)

59. $\sin 106°$

60. $\tan 37°$

61. $\tan(-17\pi/15)$

62. $\cos(-25\pi/7)$

4.5 Sketching the Graph of a Sine or Cosine Function In Exercises 63–68, sketch the graph of the function. (Include two full periods.)

63. $y = \sin 6x$

64. $f(x) = -\cos 3x$

65. $y = 5 + \sin \pi x$

66. $y = -4 - \cos \pi x$

67. $g(t) = \frac{5}{2} \sin(t - \pi)$

68. $g(t) = 3 \cos(t + \pi)$

69. Sound Waves Sound waves can be modeled using sine functions of the form $y = a \sin bx$, where x is measured in seconds.

(a) Write an equation of a sound wave whose amplitude is 2 and whose period is $\frac{1}{264}$ second.

(b) What is the frequency of the sound wave described in part (a)?

70. Meteorology The times S of sunset (Greenwich Mean Time) at 40° north latitude on the 15th of each month starting with January are: 16:59, 17:35, 18:06, 18:38, 19:08, 19:30, 19:28, 18:57, 18:10, 17:21, 16:44, and 16:36. A model (in which minutes have been converted to the decimal parts of an hour) for the data is

$$S(t) = 18.10 - 1.41 \sin\left(\frac{\pi t}{6} + 1.55\right)$$

where t represents the month, with $t = 1$ corresponding to January. *(Source: NOAA)*

(a) Use a graphing utility to graph the data and the model in the same viewing window.

(b) What is the period of the model? Is it what you expected? Explain.

(c) What is the amplitude of the model? What does it represent in the context of the problem?

4.6 Sketching the Graph of a Trigonometric Function In Exercises 71–74, sketch the graph of the function. (Include two full periods.)

71. $f(t) = \tan\left(t + \frac{\pi}{2}\right)$ **72.** $f(x) = \frac{1}{2} \cot x$

73. $f(x) = \frac{1}{2} \csc \frac{x}{2}$ **74.** $h(t) = \sec\left(t - \frac{\pi}{4}\right)$

Analyzing a Damped Trigonometric Graph In Exercises 75 and 76, use a graphing utility to graph the function and the damping factor of the function in the same viewing window. Describe the behavior of the function as x increases without bound.

75. $f(x) = x \cos x$

76. $g(x) = e^x \cos x$

Proofs in Mathematics ■ ■ ■ ■ ■ ■ ■ ■ ■ ■ ■ ■ ■ ■ ■

The Pythagorean Theorem

The Pythagorean Theorem is one of the most famous theorems in mathematics. More than 350 different proofs now exist. James A. Garfield, the twentieth president of the United States, developed a proof of the Pythagorean Theorem in 1876. His proof, shown below, involves the fact that two congruent right triangles and an isosceles right triangle can form a trapezoid.

The Pythagorean Theorem

In a right triangle, the sum of the squares of the lengths of the legs is equal to the square of the length of the hypotenuse, where a and b are the lengths of the legs and c is the length of the hypotenuse.

$$a^2 + b^2 = c^2$$

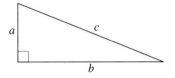

Proof

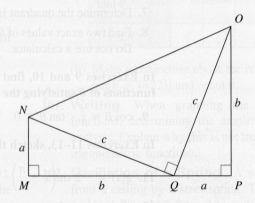

$$\frac{\text{Area of}}{\text{trapezoid } MNOP} = \frac{\text{Area of}}{\triangle MNQ} + \frac{\text{Area of}}{\triangle PQO} + \frac{\text{Area of}}{\triangle NOQ}$$

$$\frac{1}{2}(a + b)(a + b) = \frac{1}{2}ab + \frac{1}{2}ab + \frac{1}{2}c^2$$

$$\frac{1}{2}(a + b)(a + b) = ab + \frac{1}{2}c^2$$

$$(a + b)(a + b) = 2ab + c^2$$

$$a^2 + 2ab + b^2 = 2ab + c^2$$

$$a^2 + b^2 = c^2$$

P.S. Problem Solving ▪ ▪ ▪ ▪ ▪ ▪ ▪ ▪ ▪ ▪ ▪ ▪ ▪

1. Angle of Rotation The restaurant at the top of the Space Needle in Seattle, Washington, is circular and has a radius of 47.25 feet. The dining part of the restaurant revolves, making about one complete revolution every 48 minutes. A dinner party, seated at the edge of the revolving restaurant at 6:45 P.M., finishes at 8:57 P.M.

(a) Find the angle through which the dinner party rotated.

(b) Find the distance the party traveled during dinner.

2. Bicycle Gears A bicycle's gear ratio is the number of times the freewheel turns for every one turn of the chainwheel (see figure). The table shows the numbers of teeth in the freewheel and chainwheel for the first five gears of an 18-speed touring bicycle. The chainwheel completes one rotation for each gear. Find the angle through which the freewheel turns for each gear. Give your answers in both degrees and radians.

DATA Gear Number	Number of Teeth in Freewheel	Number of Teeth in Chainwheel
1	32	24
2	26	24
3	22	24
4	32	40
5	19	24

Spreadsheet at LarsonPrecalculus.com

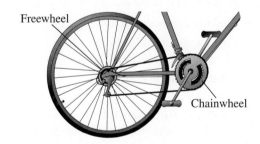

Freewheel

Chainwheel

3. Height of a Ferris Wheel Car A model for the height h (in feet) of a Ferris wheel car is

$$h = 50 + 50 \sin 8\pi t$$

where t is the time (in minutes). (The Ferris wheel has a radius of 50 feet.) This model yields a height of 50 feet when $t = 0$. Alter the model so that the height of the car is 1 foot when $t = 0$.

4. Periodic Function The function f is periodic, with period c. So, $f(t + c) = f(t)$. Determine whether each statement is true or false. Explain.

(a) $f(t - 2c) = f(t)$
(b) $f\left(t + \frac{1}{2}c\right) = f\left(\frac{1}{2}t\right)$
(c) $f\left(\frac{1}{2}[t + c]\right) = f\left(\frac{1}{2}t\right)$
(d) $f\left(\frac{1}{2}[t + 4c]\right) = f\left(\frac{1}{2}t\right)$

5. Surveying A surveyor in a helicopter is determining the width of an island, as shown in the figure.

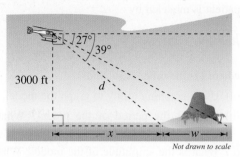

27°
39°
3000 ft
d
x
w
Not drawn to scale

(a) What is the shortest distance d the helicopter must travel to land on the island?

(b) What is the horizontal distance x the helicopter must travel before it is directly over the nearer end of the island?

(c) Find the width w of the island. Explain how you found your answer.

6. Similar Triangles and Trigonometric Functions Use the figure below.

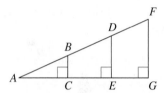

(a) Explain why $\triangle ABC$, $\triangle ADE$, and $\triangle AFG$ are similar triangles.

(b) What does similarity imply about the ratios

$$\frac{BC}{AB}, \quad \frac{DE}{AD}, \quad \text{and} \quad \frac{FG}{AF}?$$

(c) Does the value of $\sin A$ depend on which triangle from part (a) is used to calculate it? Does the value of $\sin A$ change when you use a different right triangle similar to the three given triangles?

(d) Do your conclusions from part (c) apply to the other five trigonometric functions? Explain.

7. Using Technology Use a graphing utility to graph h, and use the graph to determine whether h is even, odd, or neither.

(a) $h(x) = \cos^2 x$
(b) $h(x) = \sin^2 x$

8. Squares of Even and Odd Functions Given that f is an even function and g is an odd function, use the results of Exercise 7 to make a conjecture about each function h.

(a) $h(x) = [f(x)]^2$
(b) $h(x) = [g(x)]^2$

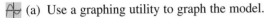

9. Blood Pressure The pressure P (in millimeters of mercury) against the walls of the blood vessels of a patient is modeled by

$$P = 100 - 20 \cos \frac{8\pi t}{3}$$

where t is the time (in seconds).

(a) Use a graphing utility to graph the model.

(b) What is the period of the model? What does it represent in the context of the problem?

(c) What is the amplitude of the model? What does it represent in the context of the problem?

(d) If one cycle of this model is equivalent to one heartbeat, what is the pulse of the patient?

(e) A physician wants the patient's pulse rate to be 64 beats per minute or less. What should the period be? What should the coefficient of t be?

10. Biorhythms A popular theory that attempts to explain the ups and downs of everyday life states that each person has three cycles, called biorhythms, which begin at birth. These three cycles can be modeled by the sine functions below, where t is the number of days since birth.

Physical (23 days): $P = \sin \dfrac{2\pi t}{23}, \quad t \geq 0$

Emotional (28 days): $E = \sin \dfrac{2\pi t}{28}, \quad t \geq 0$

Intellectual (33 days): $I = \sin \dfrac{2\pi t}{33}, \quad t \geq 0$

Consider a person who was born on July 20, 1995.

(a) Use a graphing utility to graph the three models in the same viewing window for $7300 \leq t \leq 7380$.

(b) Describe the person's biorhythms during the month of September 2015.

(c) Calculate the person's three energy levels on September 22, 2015.

11. Graphical Reasoning

(a) Use a graphing utility to graph the functions

$$f(x) = 2 \cos 2x + 3 \sin 3x$$

and

$$g(x) = 2 \cos 2x + 3 \sin 4x.$$

(b) Use the graphs from part (a) to find the period of each function.

(c) Is the function $h(x) = A \cos \alpha x + B \sin \beta x$, where α and β are positive integers, periodic? Explain.

12. Analyzing Trigonometric Functions Two trigonometric functions f and g have periods of 2, and their graphs intersect at $x = 5.35$.

(a) Give one positive value of x less than 5.35 and one value of x greater than 5.35 at which the functions have the same value.

(b) Determine one negative value of x at which the graphs intersect.

(c) Is it true that $f(13.35) = g(-4.65)$? Explain.

13. Refraction When you stand in shallow water and look at an object below the surface of the water, the object will look farther away from you than it really is. This is because when light rays pass between air and water, the water refracts, or bends, the light rays. The index of refraction for water is 1.333. This is the ratio of the sine of θ_1 and the sine of θ_2 (see figure).

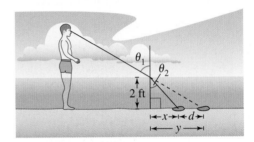

(a) While standing in water that is 2 feet deep, you look at a rock at angle $\theta_1 = 60°$ (measured from a line perpendicular to the surface of the water). Find θ_2.

(b) Find the distances x and y.

(c) Find the distance d between where the rock is and where it appears to be.

(d) What happens to d as you move closer to the rock? Explain.

14. Polynomial Approximation Using calculus, it can be shown that the arctangent function can be approximated by the polynomial

$$\arctan x \approx x - \frac{x^3}{3} + \frac{x^5}{5} - \frac{x^7}{7}$$

where x is in radians.

(a) Use a graphing utility to graph the arctangent function and its polynomial approximation in the same viewing window. How do the graphs compare?

(b) Study the pattern in the polynomial approximation of the arctangent function and predict the next term. Then repeat part (a). How does the accuracy of the approximation change when an additional term is added?

5 Analytic Trigonometry

Projectile Motion
(Example 10, page 387)

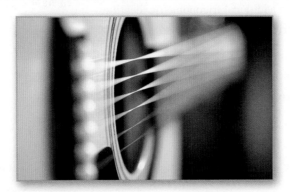

Standing Waves *(Exercise 80, page 379)*

Ferris Wheel *(Exercise 94, page 373)*

Shadow Length
(Exercise 62, page 361)

Friction *(Exercise 65, page 354)*

5.1 Using Fundamental Identities

Fundamental trigonometric identities are useful in simplifying trigonometric expressions. For example, in Exercise 65 on page 354, you will use trigonometric identities to simplify an expression for the coefficient of friction.

■ Recognize and write the fundamental trigonometric identities.
■ Use the fundamental trigonometric identities to evaluate trigonometric functions, simplify trigonometric expressions, and rewrite trigonometric expressions.

Introduction

In Chapter 4, you studied the basic definitions, properties, graphs, and applications of the individual trigonometric functions. In this chapter, you will learn how to use the fundamental identities to perform the four tasks listed below.

1. Evaluate trigonometric functions.
2. Simplify trigonometric expressions.
3. Develop additional trigonometric identities.
4. Solve trigonometric equations.

Fundamental Trigonometric Identities

Reciprocal Identities

$$\sin u = \frac{1}{\csc u} \qquad \cos u = \frac{1}{\sec u} \qquad \tan u = \frac{1}{\cot u}$$

$$\csc u = \frac{1}{\sin u} \qquad \sec u = \frac{1}{\cos u} \qquad \cot u = \frac{1}{\tan u}$$

Quotient Identities

$$\tan u = \frac{\sin u}{\cos u} \qquad \cot u = \frac{\cos u}{\sin u}$$

Pythagorean Identities

$$\sin^2 u + \cos^2 u = 1 \qquad 1 + \tan^2 u = \sec^2 u \qquad 1 + \cot^2 u = \csc^2 u$$

Cofunction Identities

$$\sin\left(\frac{\pi}{2} - u\right) = \cos u \qquad \cos\left(\frac{\pi}{2} - u\right) = \sin u$$

$$\tan\left(\frac{\pi}{2} - u\right) = \cot u \qquad \cot\left(\frac{\pi}{2} - u\right) = \tan u$$

$$\sec\left(\frac{\pi}{2} - u\right) = \csc u \qquad \csc\left(\frac{\pi}{2} - u\right) = \sec u$$

Even/Odd Identities

$$\sin(-u) = -\sin u \qquad \cos(-u) = \cos u \qquad \tan(-u) = -\tan u$$

$$\csc(-u) = -\csc u \qquad \sec(-u) = \sec u \qquad \cot(-u) = -\cot u$$

•• **REMARK** You should learn the fundamental trigonometric identities well, because you will use them frequently in trigonometry and they will also appear in calculus. Note that u can be an angle, a real number, or a variable.

Pythagorean identities are sometimes used in radical form such as

$$\sin u = \pm\sqrt{1 - \cos^2 u}$$

or

$$\tan u = \pm\sqrt{\sec^2 u - 1}$$

where the sign depends on the choice of u.

Using the Fundamental Identities

One common application of trigonometric identities is to use given information about trigonometric functions to evaluate other trigonometric functions.

EXAMPLE 1 **Using Identities to Evaluate a Function**

Use the conditions $\sec u = -\frac{3}{2}$ and $\tan u > 0$ to find the values of all six trigonometric functions.

Solution Using a reciprocal identity, you have

$$\cos u = \frac{1}{\sec u} = \frac{1}{-3/2} = -\frac{2}{3}.$$

Using a Pythagorean identity, you have

$$\sin^2 u = 1 - \cos^2 u \qquad\qquad \text{Pythagorean identity}$$
$$= 1 - \left(-\tfrac{2}{3}\right)^2 \qquad\qquad \text{Substitute } -\tfrac{2}{3} \text{ for } \cos u.$$
$$= \tfrac{5}{9}. \qquad\qquad \text{Simplify.}$$

Because $\sec u < 0$ and $\tan u > 0$, it follows that u lies in Quadrant III. Moreover, $\sin u$ is negative when u is in Quadrant III, so choose the negative root and obtain $\sin u = -\sqrt{5}/3$. Knowing the values of the sine and cosine enables you to find the values of the remaining trigonometric functions.

$$\sin u = -\frac{\sqrt{5}}{3} \qquad\qquad \csc u = \frac{1}{\sin u} = -\frac{3}{\sqrt{5}} = -\frac{3\sqrt{5}}{5}$$

$$\cos u = -\frac{2}{3} \qquad\qquad \sec u = -\frac{3}{2}$$

$$\tan u = \frac{\sin u}{\cos u} = \frac{-\sqrt{5}/3}{-2/3} = \frac{\sqrt{5}}{2} \qquad \cot u = \frac{1}{\tan u} = \frac{2}{\sqrt{5}} = \frac{2\sqrt{5}}{5}$$

 ✓ **Checkpoint** ◀))) Audio-video solution in English & Spanish at LarsonPrecalculus.com

Use the conditions $\tan x = \frac{1}{3}$ and $\cos x < 0$ to find the values of all six trigonometric functions.

EXAMPLE 2 **Simplifying a Trigonometric Expression**

Simplify the expression.

$$\sin x \cos^2 x - \sin x$$

Solution First factor out the common monomial factor $\sin x$ and then use a Pythagorean identity.

$$\sin x \cos^2 x - \sin x = \sin x(\cos^2 x - 1) \qquad \text{Factor out common monomial factor.}$$
$$= -\sin x(1 - \cos^2 x) \qquad \text{Factor out } -1.$$
$$= -\sin x(\sin^2 x) \qquad \text{Pythagorean identity}$$
$$= -\sin^3 x \qquad \text{Multiply.}$$

✓ **Checkpoint** ◀))) Audio-video solution in English & Spanish at LarsonPrecalculus.com

Simplify the expression.

$$\cos^2 x \csc x - \csc x$$

▷ **TECHNOLOGY** Use a graphing utility to check the result of Example 2. To do this, enter

$$Y1 = -(\sin(X))^3$$

and

$$Y2 = \sin(X)(\cos(X))^2$$
$$- \sin(X).$$

Select the *line* style for Y1 and the *path* style for Y2, then graph both equations in the same viewing window. The two graphs *appear* to coincide, so it is reasonable to assume that their expressions are equivalent. Note that the actual equivalence of the expressions can only be verified algebraically, as in Example 2. This graphical approach is only to check your work.

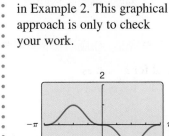

When factoring trigonometric expressions, it is helpful to find a polynomial form that fits the expression, as shown in Example 3.

EXAMPLE 3 Factoring Trigonometric Expressions

Factor each expression.

a. $\sec^2 \theta - 1$ **b.** $4 \tan^2 \theta + \tan \theta - 3$

Solution

a. This expression has the polynomial form $u^2 - v^2$, which is the difference of two squares. It factors as

$$\sec^2 \theta - 1 = (\sec \theta + 1)(\sec \theta - 1).$$

b. This expression has the polynomial form $ax^2 + bx + c$, and it factors as

$$4 \tan^2 \theta + \tan \theta - 3 = (4 \tan \theta - 3)(\tan \theta + 1).$$

✓ **Checkpoint**  Audio-video solution in English & Spanish at LarsonPrecalculus.com

Factor each expression.

a. $1 - \cos^2 \theta$ **b.** $2 \csc^2 \theta - 7 \csc \theta + 6$

In some cases, when factoring or simplifying a trigonometric expression, it is helpful to first rewrite the expression in terms of just *one* trigonometric function or in terms of *sine and cosine only*. These strategies are demonstrated in Examples 4 and 5.

EXAMPLE 4 Factoring a Trigonometric Expression

Factor $\csc^2 x - \cot x - 3$.

Solution Use the identity $\csc^2 x = 1 + \cot^2 x$ to rewrite the expression.

$$\csc^2 x - \cot x - 3 = (1 + \cot^2 x) - \cot x - 3 \qquad \text{Pythagorean identity}$$
$$= \cot^2 x - \cot x - 2 \qquad \text{Combine like terms.}$$
$$= (\cot x - 2)(\cot x + 1) \qquad \text{Factor.}$$

✓ **Checkpoint** Audio-video solution in English & Spanish at LarsonPrecalculus.com

Factor $\sec^2 x + 3 \tan x + 1$.

EXAMPLE 5 Simplifying a Trigonometric Expression

See LarsonPrecalculus.com for an interactive version of this type of example.

$$\sin t + \cot t \cos t = \sin t + \left(\frac{\cos t}{\sin t}\right) \cos t \qquad \text{Quotient identity}$$
$$= \frac{\sin^2 t + \cos^2 t}{\sin t} \qquad \text{Add fractions.}$$
$$= \frac{1}{\sin t} \qquad \text{Pythagorean identity}$$
$$= \csc t \qquad \text{Reciprocal identity}$$

✓ **Checkpoint**  Audio-video solution in English & Spanish at LarsonPrecalculus.com

Simplify $\csc x - \cos x \cot x$.

▷ **ALGEBRA HELP** In Example 3, you factor the difference of two squares and you factor a trinomial. To review the techniques for factoring polynomials, see Appendix A.3.

•• **REMARK** Remember that when adding rational expressions, you must first find the least common denominator (LCD). In Example 5, the LCD is $\sin t$.

EXAMPLE 6   **Adding Trigonometric Expressions**

Perform the addition and simplify: $\dfrac{\sin \theta}{1 + \cos \theta} + \dfrac{\cos \theta}{\sin \theta}$.

Solution

$$\dfrac{\sin \theta}{1 + \cos \theta} + \dfrac{\cos \theta}{\sin \theta} = \dfrac{(\sin \theta)(\sin \theta) + (\cos \theta)(1 + \cos \theta)}{(1 + \cos \theta)(\sin \theta)}$$

$$= \dfrac{\sin^2 \theta + \cos^2 \theta + \cos \theta}{(1 + \cos \theta)(\sin \theta)} \qquad \text{Multiply.}$$

$$= \dfrac{1 + \cos \theta}{(1 + \cos \theta)(\sin \theta)} \qquad \text{Pythagorean identity}$$

$$= \dfrac{1}{\sin \theta} \qquad \text{Divide out common factor.}$$

$$= \csc \theta \qquad \text{Reciprocal identity}$$

✓ **Checkpoint** ◀))) *Audio-video solution in English & Spanish at LarsonPrecalculus.com*

Perform the addition and simplify: $\dfrac{1}{1 + \sin \theta} + \dfrac{1}{1 - \sin \theta}$.

The next two examples involve techniques for rewriting expressions in forms that are used in calculus.

EXAMPLE 7   **Rewriting a Trigonometric Expression**

Rewrite $\dfrac{1}{1 + \sin x}$ so that it is *not* in fractional form.

Solution   From the Pythagorean identity

$$\cos^2 x = 1 - \sin^2 x = (1 - \sin x)(1 + \sin x)$$

multiplying both the numerator and the denominator by $(1 - \sin x)$ will produce a monomial denominator.

$$\dfrac{1}{1 + \sin x} = \dfrac{1}{1 + \sin x} \cdot \dfrac{1 - \sin x}{1 - \sin x} \qquad \begin{array}{l}\text{Multiply numerator and}\\ \text{denominator by } (1 - \sin x).\end{array}$$

$$= \dfrac{1 - \sin x}{1 - \sin^2 x} \qquad \text{Multiply.}$$

$$= \dfrac{1 - \sin x}{\cos^2 x} \qquad \text{Pythagorean identity}$$

$$= \dfrac{1}{\cos^2 x} - \dfrac{\sin x}{\cos^2 x} \qquad \text{Write as separate fractions.}$$

$$= \dfrac{1}{\cos^2 x} - \dfrac{\sin x}{\cos x} \cdot \dfrac{1}{\cos x} \qquad \text{Product of fractions}$$

$$= \sec^2 x - \tan x \sec x \qquad \text{Reciprocal and quotient identities}$$

✓ **Checkpoint** ◀))) *Audio-video solution in English & Spanish at LarsonPrecalculus.com*

Rewrite $\dfrac{\cos^2 \theta}{1 - \sin \theta}$ so that it is *not* in fractional form.

EXAMPLE 8   **Trigonometric Substitution**

Use the substitution $x = 2 \tan \theta$, $0 < \theta < \pi/2$, to write $\sqrt{4 + x^2}$ as a trigonometric function of θ.

Solution   Begin by letting $x = 2 \tan \theta$. Then, you obtain

$$\sqrt{4 + x^2} = \sqrt{4 + (2 \tan \theta)^2} \qquad \text{Substitute } 2 \tan \theta \text{ for } x.$$

$$= \sqrt{4 + 4 \tan^2 \theta} \qquad \text{Property of exponents}$$

$$= \sqrt{4(1 + \tan^2 \theta)} \qquad \text{Factor.}$$

$$= \sqrt{4 \sec^2 \theta} \qquad \text{Pythagorean identity}$$

$$= 2 \sec \theta. \qquad \sec \theta > 0 \text{ for } 0 < \theta < \frac{\pi}{2}$$

✓ *Checkpoint* *Audio-video solution in English & Spanish at LarsonPrecalculus.com*

Use the substitution $x = 3 \sin \theta$, $0 < \theta < \pi/2$, to write $\sqrt{9 - x^2}$ as a trigonometric function of θ.

Figure 5.1 shows the right triangle illustration of the trigonometric substitution $x = 2 \tan \theta$ in Example 8. You can use this triangle to check the solution to Example 8. For $0 < \theta < \pi/2$, you have

$$\text{opp} = x, \quad \text{adj} = 2, \quad \text{and} \quad \text{hyp} = \sqrt{4 + x^2}.$$

Using these expressions,

$$\sec \theta = \frac{\text{hyp}}{\text{adj}} = \frac{\sqrt{4 + x^2}}{2}.$$

So, $2 \sec \theta = \sqrt{4 + x^2}$, and the solution checks.

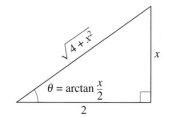

$2 \tan \theta = x \implies \tan \theta = \dfrac{x}{2}$

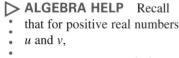

Figure 5.1

EXAMPLE 9   **Rewriting a Logarithmic Expression**

Rewrite $\ln|\csc \theta| + \ln|\tan \theta|$ as a single logarithm and simplify the result.

Solution

$$\ln|\csc \theta| + \ln|\tan \theta| = \ln|\csc \theta \tan \theta| \qquad \text{Product Property of Logarithms}$$

$$= \ln\left|\frac{1}{\sin \theta} \cdot \frac{\sin \theta}{\cos \theta}\right| \qquad \text{Reciprocal and quotient identities}$$

$$= \ln\left|\frac{1}{\cos \theta}\right| \qquad \text{Simplify.}$$

$$= \ln|\sec \theta| \qquad \text{Reciprocal identity}$$

▷ **ALGEBRA HELP**   Recall that for positive real numbers u and v,

$$\ln u + \ln v = \ln(uv).$$

To review the properties of logarithms, see Section 3.3.

✓ *Checkpoint* *Audio-video solution in English & Spanish at LarsonPrecalculus.com*

Rewrite $\ln|\sec x| + \ln|\sin x|$ as a single logarithm and simplify the result.

Summarize   **(Section 5.1)**

1. State the fundamental trigonometric identities *(page 348)*.

2. Explain how to use the fundamental trigonometric identities to evaluate trigonometric functions, simplify trigonometric expressions, and rewrite trigonometric expressions *(pages 349–352)*. For examples of these concepts, see Examples 1–9.

5.1 Exercises

See **CalcChat.com** for tutorial help and worked-out solutions to odd-numbered exercises.

Vocabulary: Fill in the blank to complete the trigonometric identity.

1. $\dfrac{\sin u}{\cos u} = $ _____

2. $\dfrac{1}{\sin u} = $ _____

3. $\dfrac{1}{\tan u} = $ _____

4. $\sec\left(\dfrac{\pi}{2} - u\right) = $ _____

5. $\sin^2 u + \cos^2 u = $ _____

6. $\sin(-u) = $ _____

Skills and Applications

 Using Identities to Evaluate a Function In Exercises 7–12, use the given conditions to find the values of all six trigonometric functions.

7. $\sec x = -\dfrac{5}{2}, \ \tan x < 0$

8. $\csc x = -\dfrac{7}{6}, \ \tan x > 0$

9. $\sin \theta = -\dfrac{3}{4}, \ \cos \theta > 0$

10. $\cos \theta = \dfrac{2}{3}, \ \sin \theta < 0$

11. $\tan x = \dfrac{2}{3}, \ \cos x > 0$

12. $\cot x = \dfrac{7}{4}, \ \sin x < 0$

Matching Trigonometric Expressions In Exercises 13–18, match the trigonometric expression with its simplified form.

(a) $\csc x$ (b) -1 (c) 1

(d) $\sin x \tan x$ (e) $\sec^2 x$ (f) $\sec x$

13. $\sec x \cos x$

14. $\cot^2 x - \csc^2 x$

15. $\cos x(1 + \tan^2 x)$

16. $\cot x \sec x$

17. $\dfrac{\sec^2 x - 1}{\sin^2 x}$

18. $\dfrac{\cos^2[(\pi/2) - x]}{\cos x}$

 Simplifying a Trigonometric Expression In Exercises 19–22, use the fundamental identities to simplify the expression. (There is more than one correct form of each answer).

19. $\dfrac{\tan \theta \cot \theta}{\sec \theta}$

20. $\cos\left(\dfrac{\pi}{2} - x\right) \sec x$

21. $\tan^2 x - \tan^2 x \sin^2 x$

22. $\sin^2 x \sec^2 x - \sin^2 x$

 Factoring a Trigonometric Expression In Exercises 23–32, factor the expression. Use the fundamental identities to simplify, if necessary. (There is more than one correct form of each answer.)

23. $\dfrac{\sec^2 x - 1}{\sec x - 1}$

24. $\dfrac{\cos x - 2}{\cos^2 x - 4}$

25. $1 - 2\cos^2 x + \cos^4 x$

26. $\sec^4 x - \tan^4 x$

27. $\cot^3 x + \cot^2 x + \cot x + 1$

28. $\sec^3 x - \sec^2 x - \sec x + 1$

29. $3\sin^2 x - 5\sin x - 2$

30. $6\cos^2 x + 5\cos x - 6$

31. $\cot^2 x + \csc x - 1$

32. $\sin^2 x + 3\cos x + 3$

Simplifying a Trigonometric Expression In Exercises 33–40, use the fundamental identities to simplify the expression. (There is more than one correct form of each answer.)

33. $\tan \theta \csc \theta$

34. $\tan(-x) \cos x$

35. $\sin \phi(\csc \phi - \sin \phi)$

36. $\cos x(\sec x - \cos x)$

37. $\sin \beta \tan \beta + \cos \beta$

38. $\cot u \sin u + \tan u \cos u$

39. $\dfrac{1 - \sin^2 x}{\csc^2 x - 1}$

40. $\dfrac{\cos^2 y}{1 - \sin y}$

Multiplying Trigonometric Expressions In Exercises 41 and 42, perform the multiplication and use the fundamental identities to simplify. (There is more than one correct form of each answer.)

41. $(\sin x + \cos x)^2$

42. $(2\csc x + 2)(2\csc x - 2)$

 Adding or Subtracting Trigonometric Expressions In Exercises 43–48, perform the addition or subtraction and use the fundamental identities to simplify. (There is more than one correct form of each answer.)

43. $\dfrac{1}{1 + \cos x} + \dfrac{1}{1 - \cos x}$

44. $\dfrac{1}{\sec x + 1} - \dfrac{1}{\sec x - 1}$

45. $\dfrac{\cos x}{1 + \sin x} - \dfrac{\cos x}{1 - \sin x}$

46. $\dfrac{\sin x}{1 + \cos x} + \dfrac{\sin x}{1 - \cos x}$

47. $\tan x - \dfrac{\sec^2 x}{\tan x}$

48. $\dfrac{\cos x}{1 + \sin x} + \dfrac{1 + \sin x}{\cos x}$

Rewriting a Trigonometric Expression In Exercises 49 and 50, rewrite the expression so that it is *not* in fractional form. (There is more than one correct form of each answer.)

49. $\dfrac{\sin^2 y}{1 - \cos y}$

50. $\dfrac{5}{\tan x + \sec x}$

Trigonometric Functions and Expressions In Exercises 51 and 52, use a graphing utility to determine which of the six trigonometric functions is equal to the expression. Verify your answer algebraically.

51. $\dfrac{\tan x + 1}{\sec x + \csc x}$ **52.** $\dfrac{1}{\sin x}\left(\dfrac{1}{\cos x} - \cos x\right)$

 Trigonometric Substitution In Exercises 53–56, use the trigonometric substitution to write the algebraic expression as a trigonometric function of θ, where $0 < \theta < \pi/2$.

53. $\sqrt{9 - x^2}$, $\quad x = 3\cos\theta$
54. $\sqrt{49 - x^2}$, $\quad x = 7\sin\theta$
55. $\sqrt{x^2 - 4}$, $\quad x = 2\sec\theta$
56. $\sqrt{9x^2 + 25}$, $\quad 3x = 5\tan\theta$

Trigonometric Substitution In Exercises 57 and 58, use the trigonometric substitution to write the algebraic equation as a trigonometric equation of θ, where $-\pi/2 < \theta < \pi/2$. Then find $\sin\theta$ and $\cos\theta$.

57. $\sqrt{2} = \sqrt{4 - x^2}$, $\quad x = 2\sin\theta$
58. $5\sqrt{3} = \sqrt{100 - x^2}$, $\quad x = 10\cos\theta$

Solving a Trigonometric Equation In Exercises 59 and 60, use a graphing utility to solve the equation for θ, where $0 \le \theta < 2\pi$.

59. $\sin\theta = \sqrt{1 - \cos^2\theta}$ **60.** $\sec\theta = \sqrt{1 + \tan^2\theta}$

Rewriting a Logarithmic Expression In Exercises 61–64, rewrite the expression as a single logarithm and simplify the result.

61. $\ln|\sin x| + \ln|\cot x|$ **62.** $\ln|\cos x| - \ln|\sin x|$
63. $\ln|\tan t| - \ln(1 - \cos^2 t)$
64. $\ln(\cos^2 t) + \ln(1 + \tan^2 t)$

65. Friction

The forces acting on an object weighing W units on an inclined plane positioned at an angle of θ with the horizontal (see figure) are modeled by $\mu W\cos\theta = W\sin\theta$, where μ is the coefficient of friction. Solve the equation for μ and simplify the result.

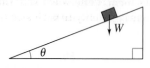

66. Rate of Change The rate of change of the function $f(x) = \sec x + \cos x$ is given by the expression $\sec x \tan x - \sin x$. Show that this expression can also be written as $\sin x \tan^2 x$.

Exploration

True or False? In Exercises 67 and 68, determine whether the statement is true or false. Justify your answer.

67. The quotient identities and reciprocal identities can be used to write any trigonometric function in terms of sine and cosine.

68. A cofunction identity can transform a tangent function into a cosecant function.

Analyzing Trigonometric Functions In Exercises 69 and 70, fill in the blanks. (*Note:* The notation $x \to c^+$ indicates that x approaches c from the right and $x \to c^-$ indicates that x approaches c from the left.)

69. As $x \to \left(\dfrac{\pi}{2}\right)^-$, $\tan x \to$ ▨ and $\cot x \to$ ▨ .

70. As $x \to \pi^+$, $\sin x \to$ ▨ and $\csc x \to$ ▨ .

71. Error Analysis Describe the error.
$$\dfrac{\sin\theta}{\cos(-\theta)} = \dfrac{\sin\theta}{-\cos\theta}$$
$$= -\tan\theta \qquad ✗$$

72. Trigonometric Substitution Use the trigonometric substitution $u = a\tan\theta$, where $-\pi/2 < \theta < \pi/2$ and $a > 0$, to simplify the expression $\sqrt{a^2 + u^2}$.

73. Writing Trigonometric Functions in Terms of Sine Write each of the other trigonometric functions of θ in terms of $\sin\theta$.

74. **HOW DO YOU SEE IT?**
Explain how to use the figure to derive the Pythagorean identities

$$\sin^2\theta + \cos^2\theta = 1,$$
$$1 + \tan^2\theta = \sec^2\theta,$$
and $1 + \cot^2\theta = \csc^2\theta.$

Discuss how to remember these identities and other fundamental trigonometric identities.

75. Rewriting a Trigonometric Expression Rewrite the expression below in terms of $\sin\theta$ and $\cos\theta$.

$$\dfrac{\sec\theta(1 + \tan\theta)}{\sec\theta + \csc\theta}$$

5.2 Verifying Trigonometric Identities

Trigonometric identities enable you to rewrite trigonometric equations that model real-life situations. For example, in Exercise 62 on page 361, trigonometric identities can help you simplify an equation that models the length of a shadow cast by a gnomon (a device used to tell time).

■ **Verify trigonometric identities.**

Verifying Trigonometric Identities

In this section, you will study techniques for verifying trigonometric identities. In the next section, you will study techniques for solving trigonometric equations. The key to both verifying identities *and* solving equations is your ability to use the fundamental identities and the rules of algebra to rewrite trigonometric expressions.

Remember that a *conditional equation* is an equation that is true for only some of the values in the domain of the variable. For example, the conditional equation

$$\sin x = 0 \qquad \text{Conditional equation}$$

is true only for

$$x = n\pi$$

where n is an integer. When you are finding the values of the variable for which the equation is true, you are *solving* the equation.

On the other hand, an equation that is true for all real values in the domain of the variable is an *identity*. For example, the familiar equation

$$\sin^2 x = 1 - \cos^2 x \qquad \text{Identity}$$

is true for all real numbers x. So, it is an identity.

Although there are similarities, verifying that a trigonometric equation is an identity is quite different from solving an equation. There is no well-defined set of rules to follow in verifying trigonometric identities, the process is best learned through practice.

Guidelines for Verifying Trigonometric Identities

1. Work with one side of the equation at a time. It is often better to work with the more complicated side first.

2. Look for opportunities to factor an expression, add fractions, square a binomial, or create a monomial denominator.

3. Look for opportunities to use the fundamental identities. Note which functions are in the final expression you want. Sines and cosines pair up well, as do secants and tangents, and cosecants and cotangents.

4. When the preceding guidelines do not help, try converting all terms to sines and cosines.

5. Always try *something*. Even making an attempt that leads to a dead end can provide insight.

Verifying trigonometric identities is a useful process when you need to convert a trigonometric expression into a form that is more useful algebraically. When you verify an identity, you cannot *assume* that the two sides of the equation are equal because you are trying to verify that they *are* equal. As a result, when verifying identities, you cannot use operations such as adding the same quantity to each side of the equation or cross multiplication.

5.3 Solving Trigonometric Equations

Trigonometric equations have many applications in circular motion. For example, in Exercise 94 on page 373, you will solve a trigonometric equation to determine when a person riding a Ferris wheel will be at certain heights above the ground.

■ Use standard algebraic techniques to solve trigonometric equations.
■ Solve trigonometric equations of quadratic type.
■ Solve trigonometric equations involving multiple angles.
■ Use inverse trigonometric functions to solve trigonometric equations.

Introduction

To solve a trigonometric equation, use standard algebraic techniques (when possible) such as collecting like terms, extracting square roots, and factoring. Your preliminary goal in solving a trigonometric equation is to *isolate* the trigonometric function on one side of the equation. For example, to solve the equation $2 \sin x = 1$, divide each side by 2 to obtain

$$\sin x = \frac{1}{2}.$$

To solve for x, note in the graph of $y = \sin x$ below that the equation $\sin x = \frac{1}{2}$ has solutions $x = \pi/6$ and $x = 5\pi/6$ in the interval $[0, 2\pi)$. Moreover, because $\sin x$ has a period of 2π, there are infinitely many other solutions, which can be written as

$$x = \frac{\pi}{6} + 2n\pi \quad \text{and} \quad x = \frac{5\pi}{6} + 2n\pi \qquad \text{General solution}$$

where n is an integer. Notice the solutions for $n = \pm 1$ in the graph of $y = \sin x$.

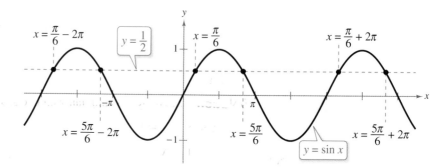

The figure below illustrates another way to show that the equation $\sin x = \frac{1}{2}$ has infinitely many solutions. Any angles that are coterminal with $\pi/6$ or $5\pi/6$ are also solutions of the equation.

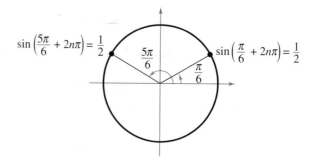

When solving trigonometric equations, write your answer(s) using exact values (when possible) rather than decimal approximations.

EXAMPLE 1   **Collecting Like Terms**

Solve

$$\sin x + \sqrt{2} = -\sin x.$$

Solution   Begin by isolating $\sin x$ on one side of the equation.

$$\sin x + \sqrt{2} = -\sin x \qquad\qquad \text{Write original equation.}$$

$$\sin x + \sin x + \sqrt{2} = 0 \qquad\qquad \text{Add } \sin x \text{ to each side.}$$

$$\sin x + \sin x = -\sqrt{2} \qquad\qquad \text{Subtract } \sqrt{2} \text{ from each side.}$$

$$2\sin x = -\sqrt{2} \qquad\qquad \text{Combine like terms.}$$

$$\sin x = -\frac{\sqrt{2}}{2} \qquad\qquad \text{Divide each side by 2.}$$

The period of $\sin x$ is 2π, so first find all solutions in the interval $[0, 2\pi)$. These solutions are $x = 5\pi/4$ and $x = 7\pi/4$. Finally, add multiples of 2π to each of these solutions to obtain the general form

$$x = \frac{5\pi}{4} + 2n\pi \quad \text{and} \quad x = \frac{7\pi}{4} + 2n\pi \qquad \text{General solution}$$

where n is an integer.

✓ **Checkpoint**  *Audio-video solution in English & Spanish at LarsonPrecalculus.com*

Solve $\sin x - \sqrt{2} = -\sin x$.

EXAMPLE 2   **Extracting Square Roots**

Solve

$$3\tan^2 x - 1 = 0.$$

Solution   Begin by isolating $\tan x$ on one side of the equation.

$$3\tan^2 x - 1 = 0 \qquad\qquad \text{Write original equation.}$$

$$3\tan^2 x = 1 \qquad\qquad \text{Add 1 to each side.}$$

$$\tan^2 x = \frac{1}{3} \qquad\qquad \text{Divide each side by 3.}$$

$$\tan x = \pm\frac{1}{\sqrt{3}} \qquad\qquad \text{Extract square roots.}$$

$$\tan x = \pm\frac{\sqrt{3}}{3} \qquad\qquad \text{Rationalize the denominator.}$$

> •• **REMARK** When you extract square roots, make sure you account for both the positive and negative solutions.

The period of $\tan x$ is π, so first find all solutions in the interval $[0, \pi)$. These solutions are $x = \pi/6$ and $x = 5\pi/6$. Finally, add multiples of π to each of these solutions to obtain the general form

$$x = \frac{\pi}{6} + n\pi \quad \text{and} \quad x = \frac{5\pi}{6} + n\pi \qquad \text{General solution}$$

where n is an integer.

✓ **Checkpoint**  *Audio-video solution in English & Spanish at LarsonPrecalculus.com*

Solve $4\sin^2 x - 3 = 0$.

The equations in Examples 1 and 2 involved only one trigonometric function. When two or more functions occur in the same equation, collect all terms on one side and try to separate the functions by factoring or by using appropriate identities. This may produce factors that yield no solutions, as illustrated in Example 3.

EXAMPLE 3 **Factoring**

Solve $\cot x \cos^2 x = 2 \cot x$.

Solution Begin by collecting all terms on one side of the equation and factoring.

$$\cot x \cos^2 x = 2 \cot x \qquad \text{Write original equation.}$$

$$\cot x \cos^2 x - 2 \cot x = 0 \qquad \text{Subtract } 2 \cot x \text{ from each side.}$$

$$\cot x(\cos^2 x - 2) = 0 \qquad \text{Factor.}$$

Set each factor equal to zero and isolate the trigonometric function, if necessary.

$$\cot x = 0 \quad \text{or} \quad \cos^2 x - 2 = 0$$

$$\cos^2 x = 2$$

$$\cos x = \pm\sqrt{2}$$

In the interval $(0, \pi)$, the equation $\cot x = 0$ has the solution

$$x = \frac{\pi}{2}.$$

No solution exists for $\cos x = \pm\sqrt{2}$ because $\pm\sqrt{2}$ are outside the range of the cosine function. The period of $\cot x$ is π, so add multiples of π to $x = \pi/2$ to get the general form

$$x = \frac{\pi}{2} + n\pi \qquad \text{General solution}$$

where n is an integer. Confirm this graphically by sketching the graph of $y = \cot x \cos^2 x - 2 \cot x$.

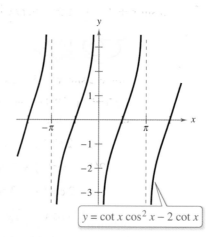

$$y = \cot x \cos^2 x - 2 \cot x$$

Notice that the x-intercepts occur at

$$-\frac{3\pi}{2}, \quad -\frac{\pi}{2}, \quad \frac{\pi}{2}, \quad \frac{3\pi}{2}$$

and so on. These x-intercepts correspond to the solutions of $\cot x \cos^2 x = 2 \cot x$.

✓ **Checkpoint** ◀)) *Audio-video solution in English & Spanish at LarsonPrecalculus.com*

Solve $\sin^2 x = 2 \sin x$.

▷ **ALGEBRA HELP** To review the techniques for solving quadratic equations, see Appendix A.5.

Equations of Quadratic Type

Below are two examples of trigonometric equations of quadratic type

$$ax^2 + bx + c = 0.$$

To solve equations of this type, use factoring (when possible) or use the Quadratic Formula.

Quadratic in sin x	Quadratic in sec x
$2\sin^2 x - \sin x - 1 = 0$	$\sec^2 x - 3\sec x - 2 = 0$
$2(\sin x)^2 - (\sin x) - 1 = 0$	$(\sec x)^2 - 3(\sec x) - 2 = 0$

EXAMPLE 4   **Solving an Equation of Quadratic Type**

Find all solutions of $2\sin^2 x - \sin x - 1 = 0$ in the interval $[0, 2\pi)$.

Algebraic Solution

Treat the equation as quadratic in sin x and factor.

$$2\sin^2 x - \sin x - 1 = 0 \qquad \text{Write original equation.}$$

$$(2\sin x + 1)(\sin x - 1) = 0 \qquad \text{Factor.}$$

Setting each factor equal to zero, you obtain the following solutions in the interval $[0, 2\pi)$.

$$2\sin x + 1 = 0 \qquad \text{or} \quad \sin x - 1 = 0$$

$$\sin x = -\frac{1}{2} \qquad\qquad \sin x = 1$$

$$x = \frac{7\pi}{6}, \frac{11\pi}{6} \qquad\qquad x = \frac{\pi}{2}$$

Graphical Solution

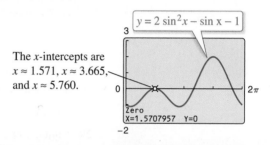

The x-intercepts are $x \approx 1.571$, $x \approx 3.665$, and $x \approx 5.760$.

Use the x-intercepts to conclude that the approximate solutions of $2\sin^2 x - \sin x - 1 = 0$ in the interval $[0, 2\pi)$ are

$$x \approx 1.571 \approx \frac{\pi}{2}, \quad x \approx 3.665 \approx \frac{7\pi}{6}, \quad \text{and} \quad x \approx 5.760 \approx \frac{11\pi}{6}.$$

✓ **Checkpoint** 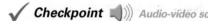 *Audio-video solution in English & Spanish at LarsonPrecalculus.com*

Find all solutions of $2\sin^2 x - 3\sin x + 1 = 0$ in the interval $[0, 2\pi)$.

EXAMPLE 5   **Rewriting with a Single Trigonometric Function**

Solve $2\sin^2 x + 3\cos x - 3 = 0$.

Solution This equation contains both sine and cosine functions. Rewrite the equation so that it has only cosine functions by using the identity $\sin^2 x = 1 - \cos^2 x$.

$$2\sin^2 x + 3\cos x - 3 = 0 \qquad \text{Write original equation.}$$

$$2(1 - \cos^2 x) + 3\cos x - 3 = 0 \qquad \text{Pythagorean identity}$$

$$2\cos^2 x - 3\cos x + 1 = 0 \qquad \text{Multiply each side by } -1.$$

$$(2\cos x - 1)(\cos x - 1) = 0 \qquad \text{Factor.}$$

Setting each factor equal to zero, you obtain the solutions $x = 0$, $x = \pi/3$, and $x = 5\pi/3$ in the interval $[0, 2\pi)$. Because cos x has a period of 2π, the general solution is

$$x = 2n\pi, \quad x = \frac{\pi}{3} + 2n\pi, \quad \text{and} \quad x = \frac{5\pi}{3} + 2n\pi \qquad \text{General solution}$$

where n is an integer.

✓ **Checkpoint** 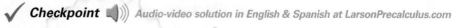 *Audio-video solution in English & Spanish at LarsonPrecalculus.com*

Solve $3\sec^2 x - 2\tan^2 x - 4 = 0$.

Sometimes you square each side of an equation to obtain an equation of quadratic type, as demonstrated in the next example. This procedure can introduce extraneous solutions, so check any solutions in the original equation to determine whether they are valid or extraneous.

·· REMARK You square each side of the equation in Example 6 because the squares of the sine and cosine functions are related by a Pythagorean identity. The same is true for the squares of the secant and tangent functions and for the squares of the cosecant and cotangent functions.

EXAMPLE 6 **Squaring and Converting to Quadratic Type**

See LarsonPrecalculus.com for an interactive version of this type of example.

Find all solutions of $\cos x + 1 = \sin x$ in the interval $[0, 2\pi)$.

Solution It is not clear how to rewrite this equation in terms of a single trigonometric function. Notice what happens when you square each side of the equation.

$$\cos x + 1 = \sin x \qquad \text{Write original equation.}$$

$$\cos^2 x + 2\cos x + 1 = \sin^2 x \qquad \text{Square each side.}$$

$$\cos^2 x + 2\cos x + 1 = 1 - \cos^2 x \qquad \text{Pythagorean identity}$$

$$\cos^2 x + \cos^2 x + 2\cos x + 1 - 1 = 0 \qquad \text{Rewrite equation.}$$

$$2\cos^2 x + 2\cos x = 0 \qquad \text{Combine like terms.}$$

$$2\cos x(\cos x + 1) = 0 \qquad \text{Factor.}$$

Set each factor equal to zero and solve for x.

$$2\cos x = 0 \qquad \text{or} \quad \cos x + 1 = 0$$

$$\cos x = 0 \qquad\qquad \cos x = -1$$

$$x = \frac{\pi}{2}, \frac{3\pi}{2} \qquad\qquad x = \pi$$

Because you squared the original equation, check for extraneous solutions.

Check $x = \dfrac{\pi}{2}$

$$\cos \frac{\pi}{2} + 1 \stackrel{?}{=} \sin \frac{\pi}{2} \qquad \text{Substitute } \frac{\pi}{2} \text{ for } x.$$

$$0 + 1 = 1 \qquad \text{Solution checks.} ✓$$

Check $x = \dfrac{3\pi}{2}$

$$\cos \frac{3\pi}{2} + 1 \stackrel{?}{=} \sin \frac{3\pi}{2} \qquad \text{Substitute } \frac{3\pi}{2} \text{ for } x.$$

$$0 + 1 \neq -1 \qquad \text{Solution does not check.}$$

Check $x = \pi$

$$\cos \pi + 1 \stackrel{?}{=} \sin \pi \qquad \text{Substitute } \pi \text{ for } x.$$

$$-1 + 1 = 0 \qquad \text{Solution checks.} ✓$$

Of the three possible solutions, $x = 3\pi/2$ is extraneous. So, in the interval $[0, 2\pi)$, the only two solutions are

$$x = \frac{\pi}{2} \quad \text{and} \quad x = \pi.$$

✓ *Checkpoint* ◀))) *Audio-video solution in English & Spanish at LarsonPrecalculus.com*

Find all solutions of $\sin x + 1 = \cos x$ in the interval $[0, 2\pi)$.

Functions Involving Multiple Angles

The next two examples involve trigonometric functions of multiple angles of the forms $\cos ku$ and $\tan ku$. To solve equations involving these forms, first solve the equation for ku, and then divide your result by k.

EXAMPLE 7   Solving a Multiple-Angle Equation

Solve $2 \cos 3t - 1 = 0$.

Solution

$2 \cos 3t - 1 = 0$	Write original equation.
$2 \cos 3t = 1$	Add 1 to each side.
$\cos 3t = \dfrac{1}{2}$	Divide each side by 2.

In the interval $[0, 2\pi)$, you know that $3t = \pi/3$ and $3t = 5\pi/3$ are the only solutions, so, in general, you have

$$3t = \frac{\pi}{3} + 2n\pi \quad \text{and} \quad 3t = \frac{5\pi}{3} + 2n\pi.$$

Dividing these results by 3, you obtain the general solution

$$t = \frac{\pi}{9} + \frac{2n\pi}{3} \quad \text{and} \quad t = \frac{5\pi}{9} + \frac{2n\pi}{3} \qquad \text{General solution}$$

where n is an integer.

✓ **Checkpoint** ◀))) *Audio-video solution in English & Spanish at LarsonPrecalculus.com*

Solve $2 \sin 2t - \sqrt{3} = 0$.

EXAMPLE 8   Solving a Multiple-Angle Equation

$3 \tan \dfrac{x}{2} + 3 = 0$	Original equation
$3 \tan \dfrac{x}{2} = -3$	Subtract 3 from each side.
$\tan \dfrac{x}{2} = -1$	Divide each side by 3.

In the interval $[0, \pi)$, you know that $x/2 = 3\pi/4$ is the only solution, so, in general, you have

$$\frac{x}{2} = \frac{3\pi}{4} + n\pi.$$

Multiplying this result by 2, you obtain the general solution

$$x = \frac{3\pi}{2} + 2n\pi \qquad \text{General solution}$$

where n is an integer.

✓ **Checkpoint** ◀))) *Audio-video solution in English & Spanish at LarsonPrecalculus.com*

Solve $2 \tan \dfrac{x}{2} - 2 = 0$.

Using Inverse Functions

EXAMPLE 9 Using Inverse Functions

$$\sec^2 x - 2 \tan x = 4 \qquad \text{Original equation}$$

$$1 + \tan^2 x - 2 \tan x - 4 = 0 \qquad \text{Pythagorean identity}$$

$$\tan^2 x - 2 \tan x - 3 = 0 \qquad \text{Combine like terms.}$$

$$(\tan x - 3)(\tan x + 1) = 0 \qquad \text{Factor.}$$

Setting each factor equal to zero, you obtain two solutions in the interval $(-\pi/2, \pi/2)$. [Recall that the range of the inverse tangent function is $(-\pi/2, \pi/2)$.]

$$x = \arctan 3 \quad \text{and} \quad x = \arctan(-1) = -\pi/4$$

Finally, $\tan x$ has a period of π, so add multiples of π to obtain

$$x = \arctan 3 + n\pi \quad \text{and} \quad x = (-\pi/4) + n\pi \qquad \text{General solution}$$

where n is an integer. You can use a calculator to approximate the value of arctan 3.

✓ **Checkpoint** ◀))) Audio-video solution in English & Spanish at LarsonPrecalculus.com

Solve $4 \tan^2 x + 5 \tan x - 6 = 0$.

EXAMPLE 10 Using the Quadratic Formula

Find all solutions of $\sin^2 x - 3 \sin x - 2 = 0$ in the interval $[0, 2\pi)$.

Solution

The expression $\sin^2 x - 3 \sin x - 2$ cannot be factored, so use the Quadratic Formula.

$$\sin^2 x - 3 \sin x - 2 = 0 \qquad \text{Write original equation.}$$

$$\sin x = \frac{-(-3) \pm \sqrt{(-3)^2 - 4(1)(-2)}}{2(1)} \qquad \text{Quadratic Formula}$$

$$\sin x = \frac{3 \pm \sqrt{17}}{2} \qquad \text{Simplify.}$$

So, $\sin x = \dfrac{3 + \sqrt{17}}{2} \approx 3.5616$ or $\sin x = \dfrac{3 - \sqrt{17}}{2} \approx -0.5616$. The range of the sine function is $[-1, 1]$, so $\sin x = \dfrac{3 + \sqrt{17}}{2}$ has no solution for x. Use a calculator to approximate a solution of $\sin x = \dfrac{3 - \sqrt{17}}{2}$.

$$x = \arcsin\left(\frac{3 - \sqrt{17}}{2}\right) \approx -0.5963$$

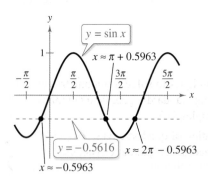

Figure 5.2

Note that this solution is not in the interval $[0, 2\pi)$. To find the solutions in $[0, 2\pi)$, sketch the graphs of $y = \sin x$ and $y = -0.5616$, as shown in Figure 5.2. From the graph, it appears that $\sin x \approx -0.5616$ on the interval $[0, 2\pi)$ when

$$x \approx \pi + 0.5963 \approx 3.7379 \quad \text{and} \quad x \approx 2\pi - 0.5963 \approx 5.6869.$$

So, the solutions of $\sin^2 x - 3 \sin x - 2 = 0$ in $[0, 2\pi)$ are $x \approx 3.7379$ and $x \approx 5.6869$.

✓ **Checkpoint** ◀))) Audio-video solution in English & Spanish at LarsonPrecalculus.com

Find all solutions of $\sin^2 x + 2 \sin x - 1 = 0$ in the interval $[0, 2\pi)$. ∎

EXAMPLE 11 **Surface Area of a Honeycomb Cell**

The surface area S (in square inches) of a honeycomb cell is given by

$$S = 6hs + 1.5s^2\left(\frac{\sqrt{3} - \cos\theta}{\sin\theta}\right), \quad 0° < \theta \le 90°$$

where $h = 2.4$ inches, $s = 0.75$ inch, and θ is the angle shown in the figure at the right. What value of θ gives the minimum surface area?

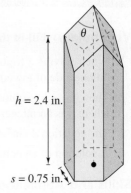

$h = 2.4$ in.

$s = 0.75$ in.

Solution

Letting $h = 2.4$ and $s = 0.75$, you obtain

$$S = 10.8 + 0.84375\left(\frac{\sqrt{3} - \cos\theta}{\sin\theta}\right).$$

Graph this function using a graphing utility set in *degree* mode. Use the *minimum* feature to approximate the minimum point on the graph, as shown in the figure below.

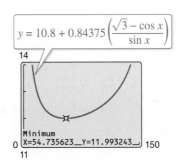

$$y = 10.8 + 0.84375\left(\frac{\sqrt{3} - \cos x}{\sin x}\right)$$

14

Minimum
X=54.735623 Y=11.993243

0   150

11

So, the minimum surface area occurs when

$$\theta \approx 54.7356°.$$

✓ Checkpoint))) *Audio-video solution in English & Spanish at LarsonPrecalculus.com*

Use the equation for the surface area of a honeycomb cell given in Example 11 with $h = 3.2$ inches and $s = 0.75$ inch. What value of θ gives the minimum surface area?

REMARK By using calculus, it can be shown that the *exact* minimum surface area occurs when

$$\theta = \arccos\left(\frac{1}{\sqrt{3}}\right).$$

Summarize (Section 5.3)

1. Explain how to use standard algebraic techniques to solve trigonometric equations *(page 362)*. For examples of using standard algebraic techniques to solve trigonometric equations, see Examples 1–3.

2. Explain how to solve a trigonometric equation of quadratic type *(page 365)*. For examples of solving trigonometric equations of quadratic type, see Examples 4–6.

3. Explain how to solve a trigonometric equation involving multiple angles *(page 367)*. For examples of solving trigonometric equations involving multiple angles, see Examples 7 and 8.

4. Explain how to use inverse trigonometric functions to solve trigonometric equations *(page 368)*. For examples of using inverse trigonometric functions to solve trigonometric equations, see Examples 9–11.

5.3 Exercises

See **CalcChat.com** for tutorial help and worked-out solutions to odd-numbered exercises.

Vocabulary: Fill in the blanks.

1. When solving a trigonometric equation, the preliminary goal is to _____ the trigonometric function on one side of the equation.

2. The _____ solution of the equation $2 \sin \theta + 1 = 0$ is $\theta = \frac{7\pi}{6} + 2n\pi$ and $\theta = \frac{11\pi}{6} + 2n\pi$, where n is an integer.

3. The equation $2 \tan^2 x - 3 \tan x + 1 = 0$ is a trigonometric equation of _____ type.

4. A solution of an equation that does not satisfy the original equation is an _____ solution.

Skills and Applications

Verifying Solutions In Exercises 5–10, verify that each x-value is a solution of the equation.

5. $\tan x - \sqrt{3} = 0$

   (a) $x = \dfrac{\pi}{3}$

   (b) $x = \dfrac{4\pi}{3}$

6. $\sec x - 2 = 0$

   (a) $x = \dfrac{\pi}{3}$

   (b) $x = \dfrac{5\pi}{3}$

7. $3 \tan^2 2x - 1 = 0$

   (a) $x = \dfrac{\pi}{12}$

   (b) $x = \dfrac{5\pi}{12}$

8. $2 \cos^2 4x - 1 = 0$

   (a) $x = \dfrac{\pi}{16}$

   (b) $x = \dfrac{3\pi}{16}$

9. $2 \sin^2 x - \sin x - 1 = 0$

   (a) $x = \dfrac{\pi}{2}$

   (b) $x = \dfrac{7\pi}{6}$

10. $\csc^4 x - 4 \csc^2 x = 0$

   (a) $x = \dfrac{\pi}{6}$

   (b) $x = \dfrac{5\pi}{6}$

 Solving a Trigonometric Equation In Exercises 11–28, solve the equation.

11. $\sqrt{3} \csc x - 2 = 0$

12. $\tan x + \sqrt{3} = 0$

13. $\cos x + 1 = -\cos x$

14. $3 \sin x + 1 = \sin x$

15. $3 \sec^2 x - 4 = 0$

16. $3 \cot^2 x - 1 = 0$

17. $4 \cos^2 x - 1 = 0$

18. $2 - 4 \sin^2 x = 0$

19. $\sin x(\sin x + 1) = 0$

20. $(2 \sin^2 x - 1)(\tan^2 x - 3) = 0$

21. $\cos^3 x - \cos x = 0$

22. $\sec^2 x - 1 = 0$

23. $3 \tan^3 x = \tan x$

24. $\sec x \csc x = 2 \csc x$

25. $2 \cos^2 x + \cos x - 1 = 0$

26. $2 \sin^2 x + 3 \sin x + 1 = 0$

27. $\sec^2 x - \sec x = 2$

28. $\csc^2 x + \csc x = 2$

 Solving a Trigonometric Equation In Exercises 29–38, find all solutions of the equation in the interval $[0, 2\pi)$.

29. $\sin x - 2 = \cos x - 2$

30. $\cos x + \sin x \tan x = 2$

31. $2 \sin^2 x = 2 + \cos x$

32. $\tan^2 x = \sec x - 1$

33. $\sin^2 x = 3 \cos^2 x$

34. $2 \sec^2 x + \tan^2 x - 3 = 0$

35. $2 \sin x + \csc x = 0$

36. $3 \sec x - 4 \cos x = 0$

37. $\csc x + \cot x = 1$

38. $\sec x + \tan x = 1$

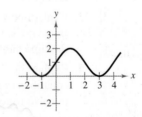 **Solving a Multiple-Angle Equation** In Exercises 39–46, solve the multiple-angle equation.

39. $2 \cos 2x - 1 = 0$

40. $2 \sin 2x + \sqrt{3} = 0$

41. $\tan 3x - 1 = 0$

42. $\sec 4x - 2 = 0$

43. $2 \cos \dfrac{x}{2} - \sqrt{2} = 0$

44. $2 \sin \dfrac{x}{2} + \sqrt{3} = 0$

45. $3 \tan \dfrac{x}{2} - \sqrt{3} = 0$

46. $\tan \dfrac{x}{2} + \sqrt{3} = 0$

Finding x-Intercepts In Exercises 47 and 48, find the x-intercepts of the graph.

47. $y = \sin \dfrac{\pi x}{2} + 1$

48. $y = \sin \pi x + \cos \pi x$

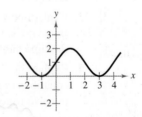

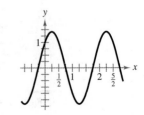

Approximating Solutions In Exercises 49–58, use a graphing utility to approximate (to three decimal places) the solutions of the equation in the interval $[0, 2\pi)$.

49. $5 \sin x + 2 = 0$

50. $2 \tan x + 7 = 0$

51. $\sin x - 3 \cos x = 0$

52. $\sin x + 4 \cos x = 0$

53. $\cos x = x$

54. $\tan x = \csc x$

55. $\sec^2 x - 3 = 0$

56. $\csc^2 x - 5 = 0$

57. $2 \tan^2 x = 15$

58. $6 \sin^2 x = 5$

 Using Inverse Functions In Exercises 59–70, solve the equation.

59. $\tan^2 x + \tan x - 12 = 0$

60. $\tan^2 x - \tan x - 2 = 0$

61. $\sec^2 x - 6 \tan x = -4$

62. $\sec^2 x + \tan x = 3$

63. $2 \sin^2 x + 5 \cos x = 4$

64. $2 \cos^2 x + 7 \sin x = 5$

65. $\cot^2 x - 9 = 0$

66. $\cot^2 x - 6 \cot x + 5 = 0$

67. $\sec^2 x - 4 \sec x = 0$

68. $\sec^2 x + 2 \sec x - 8 = 0$

69. $\csc^2 x + 3 \csc x - 4 = 0$

70. $\csc^2 x - 5 \csc x = 0$

Using the Quadratic Formula In Exercises 71–74, use the Quadratic Formula to find all solutions of the equation in the interval $[0, 2\pi)$. Round your result to four decimal places.

71. $12 \sin^2 x - 13 \sin x + 3 = 0$

72. $3 \tan^2 x + 4 \tan x - 4 = 0$

73. $\tan^2 x + 3 \tan x + 1 = 0$

74. $4 \cos^2 x - 4 \cos x - 1 = 0$

Approximating Solutions In Exercises 75–78, use a graphing utility to approximate (to three decimal places) the solutions of the equation in the given interval.

75. $3 \tan^2 x + 5 \tan x - 4 = 0$, $\left[-\dfrac{\pi}{2}, \dfrac{\pi}{2} \right]$

76. $\cos^2 x - 2 \cos x - 1 = 0$, $[0, \pi]$

77. $4 \cos^2 x - 2 \sin x + 1 = 0$, $\left[-\dfrac{\pi}{2}, \dfrac{\pi}{2} \right]$

78. $2 \sec^2 x + \tan x - 6 = 0$, $\left[-\dfrac{\pi}{2}, \dfrac{\pi}{2} \right]$

Approximating Maximum and Minimum Points In Exercises 79–84, (a) use a graphing utility to graph the function and approximate the maximum and minimum points on the graph in the interval $[0, 2\pi)$, and (b) solve the trigonometric equation and verify that its solutions are the x-coordinates of the maximum and minimum points of f. (Calculus is required to find the trigonometric equation.)

Function	Trigonometric Equation
79. $f(x) = \sin^2 x + \cos x$	$2 \sin x \cos x - \sin x = 0$
80. $f(x) = \cos^2 x - \sin x$	$-2 \sin x \cos x - \cos x = 0$
81. $f(x) = \sin x + \cos x$	$\cos x - \sin x = 0$
82. $f(x) = 2 \sin x + \cos 2x$	$2 \cos x - 4 \sin x \cos x = 0$
83. $f(x) = \sin x \cos x$	$-\sin^2 x + \cos^2 x = 0$
84. $f(x) = \sec x + \tan x - x$	$\sec x \tan x + \sec^2 x = 1$

Number of Points of Intersection In Exercises 85 and 86, use the graph to approximate the number of points of intersection of the graphs of y_1 and y_2.

85. $y_1 = 2 \sin x$
$y_2 = 3x + 1$

86. $y_1 = 2 \sin x$
$y_2 = \frac{1}{2}x + 1$

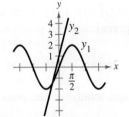 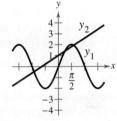

87. Graphical Reasoning Consider the function

$$f(x) = \frac{\sin x}{x}$$

and its graph, shown in the figure below.

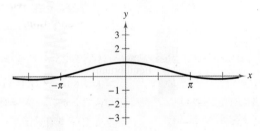

(a) What is the domain of the function?

(b) Identify any symmetry and any asymptotes of the graph.

(c) Describe the behavior of the function as $x \to 0$.

(d) How many solutions does the equation

$$\frac{\sin x}{x} = 0$$

have in the interval $[-8, 8]$? Find the solutions.

88. Graphical Reasoning Consider the function

$$f(x) = \cos \frac{1}{x}$$

and its graph, shown in the figure below.

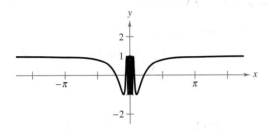

(a) What is the domain of the function?

(b) Identify any symmetry and any asymptotes of the graph.

(c) Describe the behavior of the function as $x \to 0$.

(d) How many solutions does the equation

$$\cos \frac{1}{x} = 0$$

have in the interval $[-1, 1]$? Find the solutions.

(e) Does the equation $\cos(1/x) = 0$ have a greatest solution? If so, then approximate the solution. If not, then explain why.

89. Harmonic Motion A weight is oscillating on the end of a spring (see figure). The displacement from equilibrium of the weight relative to the point of equilibrium is given by

$$y = \tfrac{1}{12}(\cos 8t - 3 \sin 8t)$$

where y is the displacement (in meters) and t is the time (in seconds). Find the times when the weight is at the point of equilibrium ($y = 0$) for $0 \le t \le 1$.

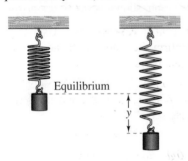

90. Damped Harmonic Motion The displacement from equilibrium of a weight oscillating on the end of a spring is given by

$$y = 1.56e^{-0.22t}\cos 4.9t$$

where y is the displacement (in feet) and t is the time (in seconds). Use a graphing utility to graph the displacement function for $0 \le t \le 10$. Find the time beyond which the distance between the weight and equilibrium does not exceed 1 foot.

91. Equipment Sales The monthly sales S (in hundreds of units) of skiing equipment at a sports store are approximated by

$$S = 58.3 + 32.5 \cos \frac{\pi t}{6}$$

where t is the time (in months), with $t = 1$ corresponding to January. Determine the months in which sales exceed 7500 units.

92. Projectile Motion A baseball is hit at an angle of θ with the horizontal and with an initial velocity of $v_0 = 100$ feet per second. An outfielder catches the ball 300 feet from home plate (see figure). Find θ when the range r of a projectile is given by

$$r = \frac{1}{32}v_0^2 \sin 2\theta.$$

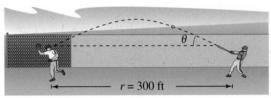

Not drawn to scale

93. Meteorology The table shows the normal daily high temperatures C in Chicago (in degrees Fahrenheit) for month t, with $t = 1$ corresponding to January. *(Source: NOAA)*

Month, t	Chicago, C
1	31.0
2	35.3
3	46.6
4	59.0
5	70.0
6	79.7
7	84.1
8	81.9
9	74.8
10	62.3
11	48.2
12	34.8

Spreadsheet at LarsonPrecalculus.com

(a) Use a graphing utility to create a scatter plot of the data.

(b) Find a cosine model for the temperatures.

(c) Graph the model and the scatter plot in the same viewing window. How well does the model fit the data?

(d) What is the overall normal daily high temperature?

(e) Use the graphing utility to determine the months during which the normal daily high temperature is above 72°F and below 72°F.

94. Ferris Wheel

The height h (in feet) above ground of a seat on a Ferris wheel at time t (in minutes) can be modeled by

$$h(t) = 53 + 50 \sin\left(\frac{\pi}{16}t - \frac{\pi}{2}\right).$$

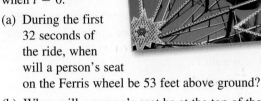

The wheel makes one revolution every 32 seconds. The ride begins when $t = 0$.

(a) During the first 32 seconds of the ride, when will a person's seat on the Ferris wheel be 53 feet above ground?

(b) When will a person's seat be at the top of the Ferris wheel for the first time during the ride? For a ride that lasts 160 seconds, how many times will a person's seat be at the top of the ride, and at what times?

95. Geometry The area of a rectangle inscribed in one arc of the graph of $y = \cos x$ (see figure) is given by

$$A = 2x \cos x, \quad 0 < x < \pi/2.$$

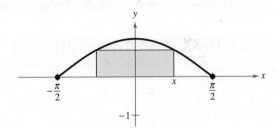

(a) Use a graphing utility to graph the area function, and approximate the area of the largest inscribed rectangle.

(b) Determine the values of x for which $A \geq 1$.

96. Quadratic Approximation Consider the function

$$f(x) = 3 \sin(0.6x - 2).$$

(a) Approximate the zero of the function in the interval $[0, 6]$.

(b) A quadratic approximation agreeing with f at $x = 5$ is

$$g(x) = -0.45x^2 + 5.52x - 13.70.$$

Use a graphing utility to graph f and g in the same viewing window. Describe the result.

(c) Use the Quadratic Formula to find the zeros of g. Compare the zero of g in the interval $[0, 6]$ with the result of part (a).

Fixed Point In Exercises 97 and 98, find the least positive fixed point of the function f. [A *fixed point* of a function f is a real number c such that $f(c) = c$.]

97. $f(x) = \tan(\pi x/4)$

98. $f(x) = \cos x$

Exploration

True or False? In Exercises 99 and 100, determine whether the statement is true or false. Justify your answer.

99. The equation $2 \sin 4t - 1 = 0$ has four times the number of solutions in the interval $[0, 2\pi)$ as the equation $2 \sin t - 1 = 0$.

100. The trigonometric equation $\sin x = 3.4$ can be solved using an inverse trigonometric function.

101. Think About It Explain what happens when you divide each side of the equation $\cot x \cos^2 x = 2 \cot x$ by $\cot x$. Is this a correct method to use when solving equations?

102. HOW DO YOU SEE IT? Explain how to use the figure to solve the equation $2 \cos x - 1 = 0$.

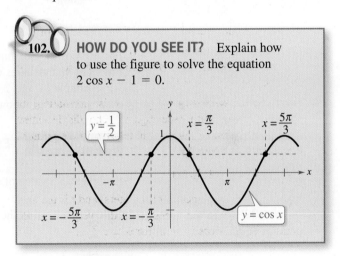

103. Graphical Reasoning Use a graphing utility to confirm the solutions found in Example 6 in two different ways.

(a) Graph both sides of the equation and find the x-coordinates of the points at which the graphs intersect.

Left side: $y = \cos x + 1$

Right side: $y = \sin x$

(b) Graph the equation $y = \cos x + 1 - \sin x$ and find the x-intercepts of the graph.

(c) Do both methods produce the same x-values? Which method do you prefer? Explain.

Project: Meteorology To work an extended application analyzing the normal daily high temperatures in Phoenix, Arizona, and in Seattle, Washington, visit this text's website at *LarsonPrecalculus.com*. (*Source: NOAA*)

5.4 Sum and Difference Formulas

Sum and difference formulas are used to model standing waves, such as those produced in a guitar string. For example, in Exercise 80 on page 379, you will use a sum formula to write the equation of a standing wave.

■ Use sum and difference formulas to evaluate trigonometric functions, verify identities, and solve trigonometric equations.

Using Sum and Difference Formulas

In this section and the next, you will study the uses of several trigonometric identities and formulas.

Sum and Difference Formulas

$\sin(u + v) = \sin u \cos v + \cos u \sin v$

$\sin(u - v) = \sin u \cos v - \cos u \sin v$

$\cos(u + v) = \cos u \cos v - \sin u \sin v$

$\cos(u - v) = \cos u \cos v + \sin u \sin v$

$\tan(u + v) = \dfrac{\tan u + \tan v}{1 - \tan u \tan v}$ $\tan(u - v) = \dfrac{\tan u - \tan v}{1 + \tan u \tan v}$

For a proof of the sum and difference formulas for $\cos(u \pm v)$ and $\tan(u \pm v)$, see Proofs in Mathematics on page 395.

Examples 1 and 2 show how **sum and difference formulas** enable you to find exact values of trigonometric functions involving sums or differences of special angles.

EXAMPLE 1   **Evaluating a Trigonometric Function**

Find the exact value of $\sin \dfrac{\pi}{12}$.

Solution   To find the *exact* value of $\sin(\pi/12)$, use the fact that

$$\frac{\pi}{12} = \frac{\pi}{3} - \frac{\pi}{4}$$

with the formula for $\sin(u - v)$.

$$\sin \frac{\pi}{12} = \sin\left(\frac{\pi}{3} - \frac{\pi}{4}\right)$$

$$= \sin \frac{\pi}{3} \cos \frac{\pi}{4} - \cos \frac{\pi}{3} \sin \frac{\pi}{4}$$

$$= \frac{\sqrt{3}}{2}\left(\frac{\sqrt{2}}{2}\right) - \frac{1}{2}\left(\frac{\sqrt{2}}{2}\right)$$

$$= \frac{\sqrt{6} - \sqrt{2}}{4}$$

Check this result on a calculator by comparing its value to $\sin(\pi/12) \approx 0.2588$.

✔ **Checkpoint** 🔊)) Audio-video solution in English & Spanish at LarsonPrecalculus.com

Find the exact value of $\cos \dfrac{\pi}{12}$.

•• REMARK Another way to solve Example 2 is to use the fact that $75° = 120° - 45°$ with the formula for $\cos(u - v)$.

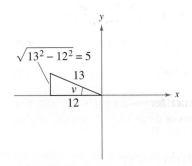

Figure 5.3

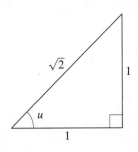

Figure 5.4

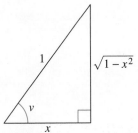

Figure 5.5

EXAMPLE 2 **Evaluating a Trigonometric Function**

Find the exact value of $\cos 75°$.

Solution Use the fact that $75° = 30° + 45°$ with the formula for $\cos(u + v)$.

$$\cos 75° = \cos(30° + 45°)$$

$$= \cos 30° \cos 45° - \sin 30° \sin 45°$$

$$= \frac{\sqrt{3}}{2}\left(\frac{\sqrt{2}}{2}\right) - \frac{1}{2}\left(\frac{\sqrt{2}}{2}\right)$$

$$= \frac{\sqrt{6} - \sqrt{2}}{4}$$

✓ *Checkpoint* ◀))) Audio-video solution in English & Spanish at LarsonPrecalculus.com

Find the exact value of $\sin 75°$.

EXAMPLE 3 **Evaluating a Trigonometric Expression**

Find the exact value of $\sin(u + v)$ given $\sin u = 4/5$, where $0 < u < \pi/2$, and $\cos v = -12/13$, where $\pi/2 < v < \pi$.

Solution Because $\sin u = 4/5$ and u is in Quadrant I, $\cos u = 3/5$, as shown in Figure 5.3. Because $\cos v = -12/13$ and v is in Quadrant II, $\sin v = 5/13$, as shown in Figure 5.4. Use these values in the formula for $\sin(u + v)$.

$$\sin(u + v) = \sin u \cos v + \cos u \sin v$$

$$= \frac{4}{5}\left(-\frac{12}{13}\right) + \frac{3}{5}\left(\frac{5}{13}\right)$$

$$= -\frac{33}{65}$$

✓ *Checkpoint* ◀))) Audio-video solution in English & Spanish at LarsonPrecalculus.com

Find the exact value of $\cos(u + v)$ given $\sin u = 12/13$, where $0 < u < \pi/2$, and $\cos v = -3/5$, where $\pi/2 < v < \pi$.

EXAMPLE 4 **An Application of a Sum Formula**

Write $\cos(\arctan 1 + \arccos x)$ as an algebraic expression.

Solution This expression fits the formula for $\cos(u + v)$. Figure 5.5 shows angles $u = \arctan 1$ and $v = \arccos x$.

$$\cos(u + v) = \cos(\arctan 1) \cos(\arccos x) - \sin(\arctan 1) \sin(\arccos x)$$

$$= \frac{1}{\sqrt{2}} \cdot x - \frac{1}{\sqrt{2}} \cdot \sqrt{1 - x^2}$$

$$= \frac{x - \sqrt{1 - x^2}}{\sqrt{2}}$$

$$= \frac{\sqrt{2}x - \sqrt{2 - 2x^2}}{2}$$

✓ *Checkpoint* ◀))) Audio-video solution in English & Spanish at LarsonPrecalculus.com

Write $\sin(\arctan 1 + \arccos x)$ as an algebraic expression.

Hipparchus, considered the most important of the Greek astronomers, was born about 190 B.C. in Nicaea. He is credited with the invention of trigonometry, and his work contributed to the derivation of the sum and difference formulas for $\sin(A \pm B)$ and $\cos(A \pm B)$.

EXAMPLE 5   Verifying a Cofunction Identity

See LarsonPrecalculus.com for an interactive version of this type of example.

Verify the cofunction identity $\cos\left(\dfrac{\pi}{2} - x\right) = \sin x$.

Solution   Use the formula for $\cos(u - v)$.

$$\cos\left(\frac{\pi}{2} - x\right) = \cos\frac{\pi}{2}\cos x + \sin\frac{\pi}{2}\sin x$$
$$= (0)(\cos x) + (1)(\sin x)$$
$$= \sin x$$

✓ **Checkpoint** 🔊)) *Audio-video solution in English & Spanish at LarsonPrecalculus.com*

Verify the cofunction identity $\sin\left(x - \dfrac{\pi}{2}\right) = -\cos x$.

Sum and difference formulas can be used to derive **reduction formulas** for rewriting expressions such as

$$\sin\left(\theta + \frac{n\pi}{2}\right) \quad \text{and} \quad \cos\left(\theta + \frac{n\pi}{2}\right), \quad \text{where } n \text{ is an integer}$$

as trigonometric functions of only θ.

EXAMPLE 6   Deriving Reduction Formulas

Write each expression as a trigonometric function of only θ.

a. $\cos\left(\theta - \dfrac{3\pi}{2}\right)$

b. $\tan(\theta + 3\pi)$

Solution

a. Use the formula for $\cos(u - v)$.

$$\cos\left(\theta - \frac{3\pi}{2}\right) = \cos\theta\cos\frac{3\pi}{2} + \sin\theta\sin\frac{3\pi}{2}$$
$$= (\cos\theta)(0) + (\sin\theta)(-1)$$
$$= -\sin\theta$$

b. Use the formula for $\tan(u + v)$.

$$\tan(\theta + 3\pi) = \frac{\tan\theta + \tan 3\pi}{1 - \tan\theta\tan 3\pi}$$
$$= \frac{\tan\theta + 0}{1 - (\tan\theta)(0)}$$
$$= \tan\theta$$

✓ **Checkpoint** 🔊)) *Audio-video solution in English & Spanish at LarsonPrecalculus.com*

Write each expression as a trigonometric function of only θ.

a. $\sin\left(\dfrac{3\pi}{2} - \theta\right)$ **b.** $\tan\left(\theta - \dfrac{\pi}{4}\right)$

EXAMPLE 7 **Solving a Trigonometric Equation**

Find all solutions of $\sin[x + (\pi/4)] + \sin[x - (\pi/4)] = -1$ in the interval $[0, 2\pi)$.

Algebraic Solution

Use sum and difference formulas to rewrite the equation.

$$\sin x \cos \frac{\pi}{4} + \cos x \sin \frac{\pi}{4} + \sin x \cos \frac{\pi}{4} - \cos x \sin \frac{\pi}{4} = -1$$

$$2 \sin x \cos \frac{\pi}{4} = -1$$

$$2(\sin x)\left(\frac{\sqrt{2}}{2}\right) = -1$$

$$\sin x = -\frac{1}{\sqrt{2}}$$

$$\sin x = -\frac{\sqrt{2}}{2}$$

So, the solutions in the interval $[0, 2\pi)$ are $x = \frac{5\pi}{4}$ and $x = \frac{7\pi}{4}$.

Graphical Solution

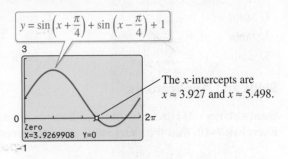

$$y = \sin\left(x + \frac{\pi}{4}\right) + \sin\left(x - \frac{\pi}{4}\right) + 1$$

The x-intercepts are $x \approx 3.927$ and $x \approx 5.498$.

Zero
X=3.9269908   Y=0

Use the x-intercepts of

$$y = \sin[x + (\pi/4)] + \sin[x - (\pi/4)] + 1$$

to conclude that the approximate solutions in the interval $[0, 2\pi)$ are

$$x \approx 3.927 \approx \frac{5\pi}{4} \quad \text{and} \quad x \approx 5.498 \approx \frac{7\pi}{4}.$$

✓ **Checkpoint**))) *Audio-video solution in English & Spanish at LarsonPrecalculus.com*

Find all solutions of $\sin[x + (\pi/2)] + \sin[x - (3\pi/2)] = 1$ in the interval $[0, 2\pi)$. ∎

The next example is an application from calculus.

EXAMPLE 8 **An Application from Calculus** ∫

Verify that $\dfrac{\sin(x + h) - \sin x}{h} = (\cos x)\left(\dfrac{\sin h}{h}\right) - (\sin x)\left(\dfrac{1 - \cos h}{h}\right)$, where $h \neq 0$.

Solution Use the formula for $\sin(u + v)$.

$$\frac{\sin(x + h) - \sin x}{h} = \frac{\sin x \cos h + \cos x \sin h - \sin x}{h}$$

$$= \frac{\cos x \sin h - \sin x(1 - \cos h)}{h}$$

$$= (\cos x)\left(\frac{\sin h}{h}\right) - (\sin x)\left(\frac{1 - \cos h}{h}\right)$$

✓ **Checkpoint**))) *Audio-video solution in English & Spanish at LarsonPrecalculus.com*

Verify that $\dfrac{\cos(x + h) - \cos x}{h} = (\cos x)\left(\dfrac{\cos h - 1}{h}\right) - (\sin x)\left(\dfrac{\sin h}{h}\right)$, where $h \neq 0$.

Summarize (Section 5.4)

1. State the sum and difference formulas for sine, cosine, and tangent *(page 374)*. For examples of using the sum and difference formulas to evaluate trigonometric functions, verify identities, and solve trigonometric equations, see Examples 1–8.

5.4 Exercises

See **CalcChat.com** for tutorial help and worked-out solutions to odd-numbered exercises.

Vocabulary: Fill in the blank.

1. $\sin(u - v) =$ _____
2. $\cos(u + v) =$ _____
3. $\tan(u + v) =$ _____
4. $\sin(u + v) =$ _____
5. $\cos(u - v) =$ _____
6. $\tan(u - v) =$ _____

Skills and Applications

Evaluating Trigonometric Expressions In Exercises 7–10, find the exact value of each expression.

7. (a) $\cos\left(\dfrac{\pi}{4} + \dfrac{\pi}{3}\right)$ (b) $\cos\dfrac{\pi}{4} + \cos\dfrac{\pi}{3}$

8. (a) $\sin\left(\dfrac{7\pi}{6} - \dfrac{\pi}{3}\right)$ (b) $\sin\dfrac{7\pi}{6} - \sin\dfrac{\pi}{3}$

9. (a) $\sin(135° - 30°)$ (b) $\sin 135° - \cos 30°$

10. (a) $\cos(120° + 45°)$ (b) $\cos 120° + \cos 45°$

 Evaluating Trigonometric Functions In Exercises 11–26, find the exact values of the sine, cosine, and tangent of the angle.

11. $\dfrac{11\pi}{12} = \dfrac{3\pi}{4} + \dfrac{\pi}{6}$ 12. $\dfrac{7\pi}{12} = \dfrac{\pi}{3} + \dfrac{\pi}{4}$

13. $\dfrac{17\pi}{12} = \dfrac{9\pi}{4} - \dfrac{5\pi}{6}$ 14. $-\dfrac{\pi}{12} = \dfrac{\pi}{6} - \dfrac{\pi}{4}$

15. $105° = 60° + 45°$ 16. $165° = 135° + 30°$

17. $-195° = 30° - 225°$ 18. $255° = 300° - 45°$

19. $\dfrac{13\pi}{12}$ 20. $\dfrac{19\pi}{12}$

21. $-\dfrac{5\pi}{12}$ 22. $-\dfrac{7\pi}{12}$

23. $285°$ 24. $15°$

25. $-165°$ 26. $-105°$

Rewriting a Trigonometric Expression In Exercises 27–34, write the expression as the sine, cosine, or tangent of an angle.

27. $\sin 3 \cos 1.2 - \cos 3 \sin 1.2$

28. $\cos\dfrac{\pi}{7}\cos\dfrac{\pi}{5} - \sin\dfrac{\pi}{7}\sin\dfrac{\pi}{5}$

29. $\sin 60° \cos 15° + \cos 60° \sin 15°$

30. $\cos 130° \cos 40° - \sin 130° \sin 40°$

31. $\dfrac{\tan(\pi/15) + \tan(2\pi/5)}{1 - \tan(\pi/15)\tan(2\pi/5)}$

32. $\dfrac{\tan 1.1 - \tan 4.6}{1 + \tan 1.1 \tan 4.6}$

33. $\cos 3x \cos 2y + \sin 3x \sin 2y$

34. $\sin x \cos 2x + \cos x \sin 2x$

 Evaluating a Trigonometric Expression In Exercises 35–40, find the exact value of the expression.

35. $\sin\dfrac{\pi}{12}\cos\dfrac{\pi}{4} + \cos\dfrac{\pi}{12}\sin\dfrac{\pi}{4}$

36. $\cos\dfrac{\pi}{16}\cos\dfrac{3\pi}{16} - \sin\dfrac{\pi}{16}\sin\dfrac{3\pi}{16}$

37. $\cos 130° \cos 10° + \sin 130° \sin 10°$

38. $\sin 100° \cos 40° - \cos 100° \sin 40°$

39. $\dfrac{\tan(9\pi/8) - \tan(\pi/8)}{1 + \tan(9\pi/8)\tan(\pi/8)}$

40. $\dfrac{\tan 25° + \tan 110°}{1 - \tan 25° \tan 110°}$

 Evaluating a Trigonometric Expression In Exercises 41–46, find the exact value of the trigonometric expression given that $\sin u = -\dfrac{3}{5}$, where $3\pi/2 < u < 2\pi$, and $\cos v = \dfrac{15}{17}$, where $0 < v < \pi/2$.

41. $\sin(u + v)$ 42. $\cos(u - v)$

43. $\tan(u + v)$ 44. $\csc(u - v)$

45. $\sec(v - u)$ 46. $\cot(u + v)$

Evaluating a Trigonometric Expression In Exercises 47–52, find the exact value of the trigonometric expression given that $\sin u = -\dfrac{7}{25}$ and $\cos v = -\dfrac{4}{5}$. (Both u and v are in Quadrant III.)

47. $\cos(u + v)$ 48. $\sin(u + v)$

49. $\tan(u - v)$ 50. $\cot(v - u)$

51. $\csc(u - v)$ 52. $\sec(v - u)$

An Application of a Sum or Difference Formula In Exercises 53–56, write the trigonometric expression as an algebraic expression.

53. $\sin(\arcsin x + \arccos x)$

54. $\sin(\arctan 2x - \arccos x)$

55. $\cos(\arccos x + \arcsin x)$

56. $\cos(\arccos x - \arctan x)$

$$\sin(300+45) = \sin 345$$

$$\sin 300 \cos 45 + \cos 300 \sin 45$$

$$\sin \frac{\pi}{6} \cos \frac{\pi}{4} + \cos \frac{\pi}{6} \sin \frac{\pi}{4}$$

$$-\frac{1}{2}\left(\frac{\sqrt{2}}{2}\right) + \frac{\sqrt{3}}{2}\left(\frac{\sqrt{2}}{2}\right)$$

$$\boxed{\frac{-\sqrt{2}+\sqrt{6}}{4}}$$

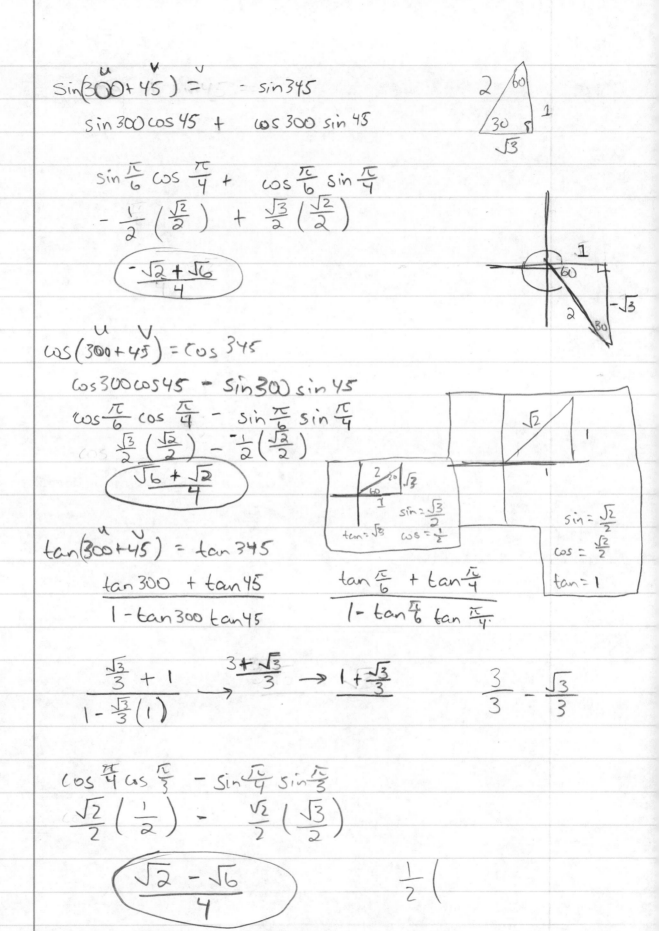

$$\cos(300+45) = \cos 345$$

$$\cos 300 \cos 45 - \sin 300 \sin 45$$

$$\cos \frac{\pi}{6} \cos \frac{\pi}{4} - \sin \frac{\pi}{6} \sin \frac{\pi}{4}$$

$$\frac{\sqrt{3}}{2}\left(\frac{\sqrt{2}}{2}\right) - -\frac{1}{2}\left(\frac{\sqrt{2}}{2}\right)$$

$$\boxed{\frac{\sqrt{6}+\sqrt{2}}{4}}$$

$$\tan(300+45) = \tan 345$$

$$\frac{\tan 300 + \tan 45}{1 - \tan 300 \tan 45} \qquad \frac{\tan \frac{\pi}{6} + \tan \frac{\pi}{4}}{1 - \tan \frac{\pi}{6} \tan \frac{\pi}{4}}$$

$$\frac{\frac{\sqrt{3}}{3}+1}{1-\frac{\sqrt{3}}{3}(1)} \to \frac{3+\sqrt{3}}{3} \to \frac{1+\frac{\sqrt{3}}{3}} \qquad \frac{3}{3} - \frac{\sqrt{3}}{3}$$

$$\cos \frac{\pi}{4} \cos \frac{\pi}{3} - \sin \frac{\pi}{4} \sin \frac{\pi}{3}$$

$$\frac{\sqrt{2}}{2}\left(\frac{1}{2}\right) - \frac{\sqrt{2}}{2}\left(\frac{\sqrt{3}}{2}\right)$$

$$\boxed{\frac{\sqrt{2}-\sqrt{6}}{4}} \qquad \frac{1}{2}($$

Verifying a Trigonometric Identity **In Exercises 57–64, verify the identity.**

57. $\sin\left(\dfrac{\pi}{2} - x\right) = \cos x$ **58.** $\sin\left(\dfrac{\pi}{2} + x\right) = \cos x$

59. $\sin\left(\dfrac{\pi}{6} + x\right) = \dfrac{1}{2}\left(\cos x + \sqrt{3}\sin x\right)$

60. $\cos\left(\dfrac{5\pi}{4} - x\right) = -\dfrac{\sqrt{2}}{2}\left(\cos x + \sin x\right)$

61. $\tan(\theta + \pi) = \tan\theta$ **62.** $\tan\left(\dfrac{\pi}{4} - \theta\right) = \dfrac{1 - \tan\theta}{1 + \tan\theta}$

63. $\cos(\pi - \theta) + \sin\left(\dfrac{\pi}{2} + \theta\right) = 0$

64. $\cos(x + y)\cos(x - y) = \cos^2 x - \sin^2 y$

Deriving a Reduction Formula **In Exercises 65–68, write the expression as a trigonometric function of only θ, and use a graphing utility to confirm your answer graphically.**

65. $\cos\left(\dfrac{3\pi}{2} - \theta\right)$ **66.** $\sin(\pi + \theta)$

67. $\csc\left(\dfrac{3\pi}{2} + \theta\right)$ **68.** $\cot(\theta - \pi)$

Solving a Trigonometric Equation **In Exercises 69–74, find all solutions of the equation in the interval $[0, 2\pi)$.**

69. $\sin(x + \pi) - \sin x + 1 = 0$

70. $\cos(x + \pi) - \cos x - 1 = 0$

71. $\cos\left(x + \dfrac{\pi}{4}\right) - \cos\left(x - \dfrac{\pi}{4}\right) = 1$

72. $\sin\left(x + \dfrac{\pi}{6}\right) - \sin\left(x - \dfrac{7\pi}{6}\right) = \dfrac{\sqrt{3}}{2}$

73. $\tan(x + \pi) + 2\sin(x + \pi) = 0$

74. $\sin\left(x + \dfrac{\pi}{2}\right) - \cos^2 x = 0$

Approximating Solutions **In Exercises 75–78, use a graphing utility to approximate the solutions of the equation in the interval $[0, 2\pi)$.**

75. $\cos\left(x + \dfrac{\pi}{4}\right) + \cos\left(x - \dfrac{\pi}{4}\right) = 1$

76. $\tan(x + \pi) - \cos\left(x + \dfrac{\pi}{2}\right) = 0$

77. $\sin\left(x + \dfrac{\pi}{2}\right) + \cos^2 x = 0$

78. $\cos\left(x - \dfrac{\pi}{2}\right) - \sin^2 x = 0$

79. Harmonic Motion A weight is attached to a spring suspended vertically from a ceiling. When a driving force is applied to the system, the weight moves vertically from its equilibrium position, and this motion is modeled by

$$y = \dfrac{1}{3}\sin 2t + \dfrac{1}{4}\cos 2t$$

where y is the displacement (in feet) from equilibrium of the weight and t is the time (in seconds).

(a) Use the identity

$$a\sin B\theta + b\cos B\theta = \sqrt{a^2 + b^2}\,\sin(B\theta + C)$$

where $C = \arctan(b/a)$, $a > 0$, to write the model in the form

$$y = \sqrt{a^2 + b^2}\,\sin(Bt + C).$$

(b) Find the amplitude of the oscillations of the weight.

(c) Find the frequency of the oscillations of the weight.

80. Standing Waves

The equation of a standing wave is obtained by adding the displacements of two waves traveling in opposite directions (see figure). Assume that each of the waves has amplitude A, period T, and wavelength λ. The models for two such waves are

$$y_1 = A\cos 2\pi\left(\dfrac{t}{T} - \dfrac{x}{\lambda}\right) \quad \text{and} \quad y_2 = A\cos 2\pi\left(\dfrac{t}{T} + \dfrac{x}{\lambda}\right).$$

Show that

$$y_1 + y_2 = 2A\cos\dfrac{2\pi t}{T}\cos\dfrac{2\pi x}{\lambda}.$$

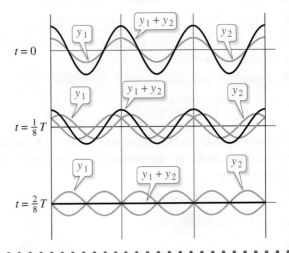

Exploration

81. $\sin(u \pm v) = \sin u \cos v \pm \cos u \sin v$

82. $\cos(u \pm v) = \cos u \cos v \pm \sin u \sin v$

83. When α and β are supplementary,

$\sin \alpha \cos \beta = \cos \alpha \sin \beta$.

84. When A, B, and C form $\triangle ABC$, $\cos(A + B) = -\cos C$.

85. Error Analysis Describe the error.

$$\tan\left(x - \frac{\pi}{4}\right) = \frac{\tan x - \tan(\pi/4)}{1 - \tan x \tan(\pi/4)}$$

$$= \frac{\tan x - 1}{1 - \tan x}$$

$$= -1$$

86. **HOW DO YOU SEE IT?** Explain how to use the figure to justify each statement.

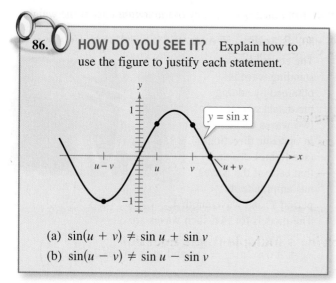

(a) $\sin(u + v) \neq \sin u + \sin v$

(b) $\sin(u - v) \neq \sin u - \sin v$

Verifying an Identity In Exercises 87–90, verify the identity.

87. $\cos(n\pi + \theta) = (-1)^n \cos \theta$, n is an integer

88. $\sin(n\pi + \theta) = (-1)^n \sin \theta$, n is an integer

89. $a \sin B\theta + b \cos B\theta = \sqrt{a^2 + b^2} \sin(B\theta + C)$,

where $C = \arctan(b/a)$ and $a > 0$

90. $a \sin B\theta + b \cos B\theta = \sqrt{a^2 + b^2} \cos(B\theta - C)$,

where $C = \arctan(a/b)$ and $b > 0$

Rewriting a Trigonometric Expression In Exercises 91–94, use the formulas given in Exercises 89 and 90 to write the trigonometric expression in the following forms.

(a) $\sqrt{a^2 + b^2} \sin(B\theta + C)$

(b) $\sqrt{a^2 + b^2} \cos(B\theta - C)$

91. $\sin \theta + \cos \theta$

92. $3 \sin 2\theta + 4 \cos 2\theta$

93. $12 \sin 3\theta + 5 \cos 3\theta$

94. $\sin 2\theta + \cos 2\theta$

Rewriting a Trigonometric Expression In Exercises 95 and 96, use the formulas given in Exercises 89 and 90 to write the trigonometric expression in the form $a \sin B\theta + b \cos B\theta$.

95. $2 \sin[\theta + (\pi/4)]$ **96.** $5 \cos[\theta - (\pi/4)]$

Angle Between Two Lines In Exercises 97 and 98, use the figure, which shows two lines whose equations are $y_1 = m_1 x + b_1$ and $y_2 = m_2 x + b_2$. Assume that both lines have positive slopes. Derive a formula for the angle between the two lines. Then use your formula to find the angle between the given pair of lines.

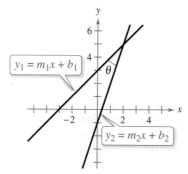

97. $y = x$ and $y = \sqrt{3}x$ **98.** $y = x$ and $y = x/\sqrt{3}$

Graphical Reasoning In Exercises 99 and 100, use a graphing utility to graph y_1 and y_2 in the same viewing window. Use the graphs to determine whether $y_1 = y_2$. Explain your reasoning.

99. $y_1 = \cos(x + 2)$, $y_2 = \cos x + \cos 2$

100. $y_1 = \sin(x + 4)$, $y_2 = \sin x + \sin 4$

101. Proof Write a proof of the formula for $\sin(u + v)$. Write a proof of the formula for $\sin(u - v)$.

102. An Application from Calculus Let $x = \pi/3$ in the identity in Example 8 and define the functions f and g as follows.

$$f(h) = \frac{\sin[(\pi/3) + h] - \sin(\pi/3)}{h}$$

$$g(h) = \cos \frac{\pi}{3}\left(\frac{\sin h}{h}\right) - \sin \frac{\pi}{3}\left(\frac{1 - \cos h}{h}\right)$$

(a) What are the domains of the functions f and g?

(b) Use a graphing utility to complete the table.

h	0.5	0.2	0.1	0.05	0.02	0.01
$f(h)$						
$g(h)$						

(c) Use the graphing utility to graph the functions f and g.

(d) Use the table and the graphs to make a conjecture about the values of the functions f and g as $h \to 0^+$.

5.5 Multiple-Angle and Product-to-Sum Formulas

- ■ Use multiple-angle formulas to rewrite and evaluate trigonometric functions.
- ■ Use power-reducing formulas to rewrite trigonometric expressions.
- ■ Use half-angle formulas to rewrite and evaluate trigonometric functions.
- ■ Use product-to-sum and sum-to-product formulas to rewrite and evaluate trigonometric expressions.
- ■ Use trigonometric formulas to rewrite real-life models.

Multiple-Angle Formulas

In this section, you will study four other categories of trigonometric identities.

1. The first category involves *functions of multiple angles* such as $\sin ku$ and $\cos ku$.
2. The second category involves *squares of trigonometric functions* such as $\sin^2 u$.
3. The third category involves *functions of half-angles* such as $\sin(u/2)$.
4. The fourth category involves *products of trigonometric functions* such as $\sin u \cos v$.

You should learn the **double-angle formulas** because they are used often in trigonometry and calculus. For proofs of these formulas, see Proofs in Mathematics on page 395.

A variety of trigonometric formulas enable you to rewrite trigonometric equations in more convenient forms. For example, in Exercise 71 on page 389, you will use a half-angle formula to rewrite an equation relating the Mach number of a supersonic airplane to the apex angle of the cone formed by the sound waves behind the airplane.

Double-Angle Formulas

$$\sin 2u = 2 \sin u \cos u \qquad \cos 2u = \cos^2 u - \sin^2 u$$

$$\tan 2u = \frac{2 \tan u}{1 - \tan^2 u} \qquad\qquad = 2 \cos^2 u - 1$$

$$= 1 - 2 \sin^2 u$$

EXAMPLE 1   Solving a Multiple-Angle Equation

Solve $2 \cos x + \sin 2x = 0$.

Solution   Begin by rewriting the equation so that it involves trigonometric functions of only x. Then factor and solve.

$$2 \cos x + \sin 2x = 0 \qquad \text{Write original equation.}$$

$$2 \cos x + 2 \sin x \cos x = 0 \qquad \text{Double-angle formula}$$

$$2 \cos x(1 + \sin x) = 0 \qquad \text{Factor.}$$

$$2 \cos x = 0 \quad \text{and} \quad 1 + \sin x = 0 \qquad \text{Set factors equal to zero.}$$

$$x = \frac{\pi}{2}, \frac{3\pi}{2} \qquad\qquad x = \frac{3\pi}{2} \qquad \text{Solutions in } [0, 2\pi)$$

So, the general solution is

$$x = \frac{\pi}{2} + 2n\pi \quad \text{and} \quad x = \frac{3\pi}{2} + 2n\pi$$

where n is an integer. Verify these solutions graphically.

✓ **Checkpoint** 🔊))) *Audio-video solution in English & Spanish at LarsonPrecalculus.com*

Solve $\cos 2x + \cos x = 0$.

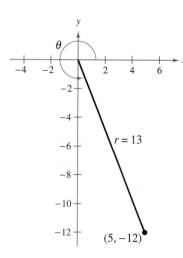

Figure 5.6

> **EXAMPLE 2** **Evaluating Functions Involving Double Angles**

Use the conditions below to find $\sin 2\theta$, $\cos 2\theta$, and $\tan 2\theta$.

$$\cos \theta = \frac{5}{13}, \quad \frac{3\pi}{2} < \theta < 2\pi$$

Solution   From Figure 5.6,

$$\sin \theta = \frac{y}{r} = -\frac{12}{13} \quad \text{and} \quad \tan \theta = \frac{y}{x} = -\frac{12}{5}.$$

Use these values with each of the double-angle formulas.

$$\sin 2\theta = 2 \sin \theta \cos \theta = 2\left(-\frac{12}{13}\right)\left(\frac{15}{13}\right) = -\frac{120}{169}$$

$$\cos 2\theta = 2 \cos^2 \theta - 1 = 2\left(\frac{25}{169}\right) - 1 = -\frac{119}{169}$$

$$\tan 2\theta = \frac{2 \tan \theta}{1 - \tan^2 \theta} = \frac{2\left(-\dfrac{12}{5}\right)}{1 - \left(-\dfrac{12}{5}\right)^2} = \frac{120}{119}$$

✓ **Checkpoint** ◀))) *Audio-video solution in English & Spanish at LarsonPrecalculus.com*

Use the conditions below to find $\sin 2\theta$, $\cos 2\theta$, and $\tan 2\theta$.

$$\sin \theta = \frac{3}{5}, \quad 0 < \theta < \frac{\pi}{2}$$

The double-angle formulas are not restricted to the angles 2θ and θ. Other *double* combinations, such as 4θ and 2θ or 6θ and 3θ, are also valid. Here are two examples.

$$\sin 4\theta = 2 \sin 2\theta \cos 2\theta \quad \text{and} \quad \cos 6\theta = \cos^2 3\theta - \sin^2 3\theta$$

By using double-angle formulas together with the sum formulas given in the preceding section, you can derive other multiple-angle formulas.

> **EXAMPLE 3** **Deriving a Triple-Angle Formula**

Rewrite $\sin 3x$ in terms of $\sin x$.

Solution

$$
\begin{aligned}
\sin 3x &= \sin(2x + x) && \text{Rewrite the angle as a sum.}\\
&= \sin 2x \cos x + \cos 2x \sin x && \text{Sum formula}\\
&= 2 \sin x \cos x \cos x + (1 - 2 \sin^2 x) \sin x && \text{Double-angle formulas}\\
&= 2 \sin x \cos^2 x + \sin x - 2 \sin^3 x && \text{Distributive Property}\\
&= 2 \sin x(1 - \sin^2 x) + \sin x - 2 \sin^3 x && \text{Pythagorean identity}\\
&= 2 \sin x - 2 \sin^3 x + \sin x - 2 \sin^3 x && \text{Distributive Property}\\
&= 3 \sin x - 4 \sin^3 x && \text{Simplify.}
\end{aligned}
$$

✓ **Checkpoint** ◀))) *Audio-video solution in English & Spanish at LarsonPrecalculus.com*

Rewrite $\cos 3x$ in terms of $\cos x$.

Power-Reducing Formulas

The double-angle formulas can be used to obtain the **power-reducing formulas.**

Power-Reducing Formulas

$$\sin^2 u = \frac{1 - \cos 2u}{2}$$

$$\cos^2 u = \frac{1 + \cos 2u}{2}$$

$$\tan^2 u = \frac{1 - \cos 2u}{1 + \cos 2u}$$

For a proof of the power-reducing formulas, see Proofs in Mathematics on page 396. Example 4 shows a typical power reduction used in calculus.

EXAMPLE 4   **Reducing a Power**

Rewrite $\sin^4 x$ in terms of first powers of the cosines of multiple angles.

Solution   Note the repeated use of power-reducing formulas.

$$
\begin{aligned}
\sin^4 x &= (\sin^2 x)^2 && \text{Property of exponents} \\[4pt]
&= \left(\frac{1 - \cos 2x}{2} \right)^2 && \text{Power-reducing formula} \\[4pt]
&= \frac{1}{4}(1 - 2\cos 2x + \cos^2 2x) && \text{Expand.} \\[4pt]
&= \frac{1}{4}\left(1 - 2\cos 2x + \frac{1 + \cos 4x}{2} \right) && \text{Power-reducing formula} \\[4pt]
&= \frac{1}{4} - \frac{1}{2}\cos 2x + \frac{1}{8} + \frac{1}{8}\cos 4x && \text{Distributive Property} \\[4pt]
&= \frac{3}{8} - \frac{1}{2}\cos 2x + \frac{1}{8}\cos 4x && \text{Simplify.} \\[4pt]
&= \frac{1}{8}(3 - 4\cos 2x + \cos 4x) && \text{Factor out common factor.}
\end{aligned}
$$

Use a graphing utility to check this result, as shown below. Notice that the graphs coincide.

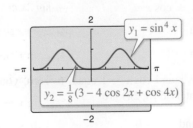

   *Checkpoint* 📢)))   *Audio-video solution in English & Spanish at LarsonPrecalculus.com*

Rewrite $\tan^4 x$ in terms of first powers of the cosines of multiple angles.   ■

Half-Angle Formulas

You can derive some useful alternative forms of the power-reducing formulas by replacing u with $u/2$. The results are called **half-angle formulas.**

> ### Half-Angle Formulas
>
> $$\sin \frac{u}{2} = \pm \sqrt{\frac{1 - \cos u}{2}} \qquad \cos \frac{u}{2} = \pm \sqrt{\frac{1 + \cos u}{2}}$$
>
> $$\tan \frac{u}{2} = \frac{1 - \cos u}{\sin u} = \frac{\sin u}{1 + \cos u}$$
>
> The signs of $\sin \dfrac{u}{2}$ and $\cos \dfrac{u}{2}$ depend on the quadrant in which $\dfrac{u}{2}$ lies.

EXAMPLE 5 **Using a Half-Angle Formula**

Find the exact value of $\sin 105°$.

Solution Begin by noting that $105°$ is half of $210°$. Then, use the half-angle formula for $\sin(u/2)$ and the fact that $105°$ lies in Quadrant II.

$$\sin 105° = \sqrt{\frac{1 - \cos 210°}{2}} = \sqrt{\frac{1 + (\sqrt{3}/2)}{2}} = \frac{\sqrt{2 + \sqrt{3}}}{2}$$

The positive square root is chosen because $\sin \theta$ is positive in Quadrant II.

✓ **Checkpoint** 🔊)) *Audio-video solution in English & Spanish at LarsonPrecalculus.com*

Find the exact value of $\cos 105°$.

EXAMPLE 6 **Solving a Trigonometric Equation**

Find all solutions of $1 + \cos^2 x = 2 \cos^2 \dfrac{x}{2}$ in the interval $[0, 2\pi)$.

Algebraic Solution

$1 + \cos^2 x = 2 \cos^2 \dfrac{x}{2}$	Write original equation.
$1 + \cos^2 x = 2 \left(\pm \sqrt{\dfrac{1 + \cos x}{2}} \right)^2$	Half-angle formula
$1 + \cos^2 x = 1 + \cos x$	Simplify.
$\cos^2 x - \cos x = 0$	Simplify.
$\cos x(\cos x - 1) = 0$	Factor.

By setting the factors $\cos x$ and $\cos x - 1$ equal to zero, you find that the solutions in the interval $[0, 2\pi)$ are

$$x = \frac{\pi}{2}, \quad x = \frac{3\pi}{2}, \quad \text{and} \quad x = 0.$$

Graphical Solution

The x-intercepts are $x \approx 0$, $x \approx 1.571$, and $x \approx 4.712$.

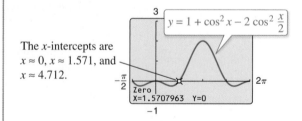

Use the x-intercepts of $y = 1 + \cos^2 x - 2 \cos^2(x/2)$ to conclude that the approximate solutions of $1 + \cos^2 x = 2 \cos^2(x/2)$ in the interval $[0, 2\pi)$ are

$$x = 0, \quad x \approx 1.571 \approx \frac{\pi}{2}, \quad \text{and} \quad x \approx 4.712 \approx \frac{3\pi}{2}.$$

✓ **Checkpoint** 🔊)) *Audio-video solution in English & Spanish at LarsonPrecalculus.com*

Find all solutions of $\cos^2 x = \sin^2(x/2)$ in the interval $[0, 2\pi)$. ◼

Product-to-Sum and Sum-to-Product Formulas

Each of the **product-to-sum formulas** can be proved using the sum and difference formulas discussed in the preceding section.

Product-to-Sum Formulas

$$\sin u \sin v = \frac{1}{2}[\cos(u - v) - \cos(u + v)]$$

$$\cos u \cos v = \frac{1}{2}[\cos(u - v) + \cos(u + v)]$$

$$\sin u \cos v = \frac{1}{2}[\sin(u + v) + \sin(u - v)]$$

$$\cos u \sin v = \frac{1}{2}[\sin(u + v) - \sin(u - v)]$$

Product-to-sum formulas are used in calculus to solve problems involving the products of sines and cosines of two different angles.

EXAMPLE 7   **Writing Products as Sums**

Rewrite the product $\cos 5x \sin 4x$ as a sum or difference.

Solution   Using the appropriate product-to-sum formula, you obtain

$$\cos 5x \sin 4x = \frac{1}{2}[\sin(5x + 4x) - \sin(5x - 4x)]$$

$$= \frac{1}{2}\sin 9x - \frac{1}{2}\sin x.$$

✓ *Checkpoint* ◉))) *Audio-video solution in English & Spanish at LarsonPrecalculus.com*

Rewrite the product $\sin 5x \cos 3x$ as a sum or difference.

Occasionally, it is useful to reverse the procedure and write a sum of trigonometric functions as a product. This can be accomplished with the **sum-to-product formulas.**

Sum-to-Product Formulas

$$\sin u + \sin v = 2\sin\left(\frac{u + v}{2}\right)\cos\left(\frac{u - v}{2}\right)$$

$$\sin u - \sin v = 2\cos\left(\frac{u + v}{2}\right)\sin\left(\frac{u - v}{2}\right)$$

$$\cos u + \cos v = 2\cos\left(\frac{u + v}{2}\right)\cos\left(\frac{u - v}{2}\right)$$

$$\cos u - \cos v = -2\sin\left(\frac{u + v}{2}\right)\sin\left(\frac{u - v}{2}\right)$$

For a proof of the sum-to-product formulas, see Proofs in Mathematics on page 396.

EXAMPLE 8 **Using a Sum-to-Product Formula**

Find the exact value of $\cos 195° + \cos 105°$.

Solution Use the appropriate sum-to-product formula.

$$\cos 195° + \cos 105° = 2 \cos\left(\frac{195° + 105°}{2}\right) \cos\left(\frac{195° - 105°}{2}\right)$$

$$= 2 \cos 150° \cos 45°$$

$$= 2\left(-\frac{\sqrt{3}}{2}\right)\left(\frac{\sqrt{2}}{2}\right)$$

$$= -\frac{\sqrt{6}}{2}$$

✓ **Checkpoint** ◀))) *Audio-video solution in English & Spanish at LarsonPrecalculus.com*

Find the exact value of $\sin 195° + \sin 105°$.

EXAMPLE 9 **Solving a Trigonometric Equation**

See LarsonPrecalculus.com for an interactive version of this type of example.

Solve $\sin 5x + \sin 3x = 0$.

Solution

$$\sin 5x + \sin 3x = 0 \qquad \text{Write original equation.}$$

$$2 \sin\left(\frac{5x + 3x}{2}\right) \cos\left(\frac{5x - 3x}{2}\right) = 0 \qquad \text{Sum-to-product formula}$$

$$2 \sin 4x \cos x = 0 \qquad \text{Simplify.}$$

Set the factor $2 \sin 4x$ equal to zero. The solutions in the interval $[0, 2\pi)$ are

$$x = 0, \frac{\pi}{4}, \frac{\pi}{2}, \frac{3\pi}{4}, \pi, \frac{5\pi}{4}, \frac{3\pi}{2}, \frac{7\pi}{4}.$$

The equation $\cos x = 0$ yields no additional solutions, so the solutions are of the form $x = n\pi/4$, where n is an integer. To confirm this graphically, sketch the graph of $y = \sin 5x + \sin 3x$, as shown below.

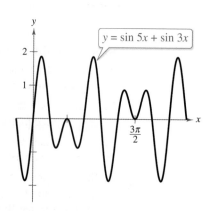

Notice from the graph that the x-intercepts occur at multiples of $\pi/4$.

✓ **Checkpoint** ◀))) *Audio-video solution in English & Spanish at LarsonPrecalculus.com*

Solve $\sin 4x - \sin 2x = 0$.

Application

Kicking a football with an initial velocity of 80 feet per second at an angle of 45° with the horizontal results in a distance traveled of 200 feet.

> **EXAMPLE 10** Projectile Motion

Ignoring air resistance, the range of a projectile fired at an angle θ with the horizontal and with an initial velocity of v_0 feet per second is given by

$$r = \frac{1}{16} v_0^2 \sin \theta \cos \theta$$

where r is the horizontal distance (in feet) that the projectile travels. A football player can kick a football from ground level with an initial velocity of 80 feet per second.

a. Rewrite the projectile motion model in terms of the first power of the sine of a multiple angle.

b. At what angle must the player kick the football so that the football travels 200 feet?

Solution

a. Use a double-angle formula to rewrite the projectile motion model as

$$r = \frac{1}{32} v_0^2 (2 \sin \theta \cos \theta) \qquad \text{Write original model.}$$

$$= \frac{1}{32} v_0^2 \sin 2\theta. \qquad \text{Double-angle formula}$$

b.
$$r = \frac{1}{32} v_0^2 \sin 2\theta \qquad \text{Write projectile motion model.}$$

$$200 = \frac{1}{32}(80)^2 \sin 2\theta \qquad \text{Substitute 200 for } r \text{ and 80 for } v_0.$$

$$200 = 200 \sin 2\theta \qquad \text{Simplify.}$$

$$1 = \sin 2\theta \qquad \text{Divide each side by 200.}$$

You know that $2\theta = \pi/2$. Dividing this result by 2 produces $\theta = \pi/4$, or 45°. So, the player must kick the football at an angle of 45° so that the football travels 200 feet.

✓ **Checkpoint** ◀))) *Audio-video solution in English & Spanish at LarsonPrecalculus.com*

In Example 10, for what angle is the horizontal distance the football travels a maximum?

Summarize (Section 5.5)

1. State the double-angle formulas *(page 381)*. For examples of using multiple-angle formulas to rewrite and evaluate trigonometric functions, see Examples 1–3.
2. State the power-reducing formulas *(page 383)*. For an example of using power-reducing formulas to rewrite a trigonometric expression, see Example 4.
3. State the half-angle formulas *(page 384)*. For examples of using half-angle formulas to rewrite and evaluate trigonometric functions, see Examples 5 and 6.
4. State the product-to-sum and sum-to-product formulas *(page 385)*. For an example of using a product-to-sum formula to rewrite a trigonometric expression, see Example 7. For examples of using sum-to-product formulas to rewrite and evaluate trigonometric functions, see Examples 8 and 9.
5. Describe an example of how to use a trigonometric formula to rewrite a real-life model *(page 387, Example 10)*.

Chapter Summary

What Did You Learn?	Explanation/Examples	Review Exercises
Section 5.1 Recognize and write the fundamental trigonometric identities *(p. 348)*.	**Reciprocal Identities** $\sin u = 1/\csc u \qquad \cos u = 1/\sec u \qquad \tan u = 1/\cot u$ $\csc u = 1/\sin u \qquad \sec u = 1/\cos u \qquad \cot u = 1/\tan u$ **Quotient Identities:** $\tan u = \dfrac{\sin u}{\cos u}, \quad \cot u = \dfrac{\cos u}{\sin u}$ **Pythagorean Identities:** $\sin^2 u + \cos^2 u = 1$, $1 + \tan^2 u = \sec^2 u, \quad 1 + \cot^2 u = \csc^2 u$ **Cofunction Identities** $\sin[(\pi/2) - u] = \cos u \qquad \cos[(\pi/2) - u] = \sin u$ $\tan[(\pi/2) - u] = \cot u \qquad \cot[(\pi/2) - u] = \tan u$ $\sec[(\pi/2) - u] = \csc u \qquad \csc[(\pi/2) - u] = \sec u$ **Even/Odd Identities** $\sin(-u) = -\sin u \qquad \cos(-u) = \cos u \qquad \tan(-u) = -\tan u$ $\csc(-u) = -\csc u \qquad \sec(-u) = \sec u \qquad \cot(-u) = -\cot u$	1–4
Use the fundamental trigonometric identities to evaluate trigonometric functions, simplify trigonometric expressions, and rewrite trigonometric expressions *(p. 349)*.	In some cases, when factoring or simplifying a trigonometric expression, it is helpful to rewrite the expression in terms of just *one* trigonometric function or in terms of *sine and cosine only*.	5–18
Section 5.2 Verify trigonometric identities *(p. 355)*.	**Guidelines for Verifying Trigonometric Identities** 1. Work with one side of the equation at a time. 2. Look to factor an expression, add fractions, square a binomial, or create a monomial denominator. 3. Look to use the fundamental identities. Note which functions are in the final expression you want. Sines and cosines pair up well, as do secants and tangents, and cosecants and cotangents. 4. When the preceding guidelines do not help, try converting all terms to sines and cosines. 5. Always try *something*.	19–26
Section 5.3 Use standard algebraic techniques to solve trigonometric equations *(p. 362)*.	Use standard algebraic techniques (when possible) such as collecting like terms, extracting square roots, and factoring to solve trigonometric equations.	27–32
Solve trigonometric equations of quadratic type *(p. 365)*.	To solve trigonometric equations of quadratic type $ax^2 + bx + c = 0$, use factoring (when possible) or use the Quadratic Formula.	33–36
Solve trigonometric equations involving multiple angles *(p. 367)*.	To solve equations that contain forms such as $\sin ku$ or $\cos ku$, first solve the equation for ku, and then divide your result by k.	37–42
Use inverse trigonometric functions to solve trigonometric equations *(p. 368)*.	After factoring an equation, you may get an equation such as $(\tan x - 3)(\tan x + 1) = 0$. In such cases, use inverse trigonometric functions to solve. (See Example 9.)	43–46

What Did You Learn?	**Explanation/Examples**	**Review Exercises**
Section 5.4 Use sum and difference formulas to evaluate trigonometric functions, verify identities, and solve trigonometric equations (*p. 374*).	**Sum and Difference Formulas** $$\sin(u + v) = \sin u \cos v + \cos u \sin v$$ $$\sin(u - v) = \sin u \cos v - \cos u \sin v$$ $$\cos(u + v) = \cos u \cos v - \sin u \sin v$$ $$\cos(u - v) = \cos u \cos v + \sin u \sin v$$ $$\tan(u + v) = \frac{\tan u + \tan v}{1 - \tan u \tan v}$$ $$\tan(u - v) = \frac{\tan u - \tan v}{1 + \tan u \tan v}$$	47–62
Section 5.5 Use multiple-angle formulas to rewrite and evaluate trigonometric functions (*p. 381*).	**Double-Angle Formulas** $$\sin 2u = 2 \sin u \cos u \qquad \cos 2u = \cos^2 u - \sin^2 u$$ $$\tan 2u = \frac{2 \tan u}{1 - \tan^2 u} \qquad\qquad = 2 \cos^2 u - 1$$ $$\qquad\qquad\qquad\qquad\qquad = 1 - 2 \sin^2 u$$	63–66
Use power-reducing formulas to rewrite trigonometric expressions (*p. 383*).	**Power-Reducing Formulas** $$\sin^2 u = \frac{1 - \cos 2u}{2}, \quad \cos^2 u = \frac{1 + \cos 2u}{2}$$ $$\tan^2 u = \frac{1 - \cos 2u}{1 + \cos 2u}$$	67, 68
Use half-angle formulas to rewrite and evaluate trigonometric functions (*p. 384*).	**Half-Angle Formulas** $$\sin \frac{u}{2} = \pm \sqrt{\frac{1 - \cos u}{2}}, \quad \cos \frac{u}{2} = \pm \sqrt{\frac{1 + \cos u}{2}}$$ $$\tan \frac{u}{2} = \frac{1 - \cos u}{\sin u} = \frac{\sin u}{1 + \cos u}$$ The signs of $\sin \frac{u}{2}$ and $\cos \frac{u}{2}$ depend on the quadrant in which $u/2$ lies.	69–74
Use product-to-sum and sum-to-product formulas to rewrite and evaluate trigonometric expressions (*p. 385*).	**Product-to-Sum Formulas** $$\sin u \sin v = (1/2)[\cos(u - v) - \cos(u + v)]$$ $$\cos u \cos v = (1/2)[\cos(u - v) + \cos(u + v)]$$ $$\sin u \cos v = (1/2)[\sin(u + v) + \sin(u - v)]$$ $$\cos u \sin v = (1/2)[\sin(u + v) - \sin(u - v)]$$ **Sum-to-Product Formulas** $$\sin u + \sin v = 2 \sin\left(\frac{u + v}{2}\right) \cos\left(\frac{u - v}{2}\right)$$ $$\sin u - \sin v = 2 \cos\left(\frac{u + v}{2}\right) \sin\left(\frac{u - v}{2}\right)$$ $$\cos u + \cos v = 2 \cos\left(\frac{u + v}{2}\right) \cos\left(\frac{u - v}{2}\right)$$ $$\cos u - \cos v = -2 \sin\left(\frac{u + v}{2}\right) \sin\left(\frac{u - v}{2}\right)$$	75–78
Use trigonometric formulas to rewrite real-life models (*p. 387*).	A trigonometric formula can be used to rewrite the projectile motion model $r = (1/16) v_0^2 \sin \theta \cos \theta$. (See Example 10.)	79, 80

Review Exercises

See **CalcChat.com** for tutorial help and worked-out solutions to odd-numbered exercises.

5.1 Recognizing a Fundamental Identity In Exercises 1–4, name the trigonometric function that is equivalent to the expression.

1. $\dfrac{\cos x}{\sin x}$

2. $\dfrac{1}{\cos x}$

3. $\sin\left(\dfrac{\pi}{2} - x\right)$

4. $\sqrt{\cot^2 x + 1}$

Using Identities to Evaluate a Function In Exercises 5 and 6, use the given conditions and fundamental trigonometric identities to find the values of all six trigonometric functions.

5. $\cos\theta = -\frac{2}{5}$, $\quad \tan\theta > 0$ **6.** $\cot x = -\frac{2}{3}$, $\quad \cos x < 0$

Simplifying a Trigonometric Expression In Exercises 7–16, use the fundamental trigonometric identities to simplify the expression. (There is more than one correct form of each answer.)

7. $\dfrac{1}{\cot^2 x + 1}$

8. $\dfrac{\tan\theta}{1 - \cos^2\theta}$

9. $\tan^2 x(\csc^2 x - 1)$

10. $\cot^2 x(\sin^2 x)$

11. $\dfrac{\cot\left(\dfrac{\pi}{2} - u\right)}{\cos u}$

12. $\dfrac{\sec^2(-\theta)}{\csc^2\theta}$

13. $\cos^2 x + \cos^2 x \cot^2 x$

14. $(\tan x + 1)^2 \cos x$

15. $\dfrac{1}{\csc\theta + 1} - \dfrac{1}{\csc\theta - 1}$

16. $\dfrac{\tan^2 x}{1 + \sec x}$

Trigonometric Substitution In Exercises 17 and 18, use the trigonometric substitution to write the algebraic expression as a trigonometric function of θ, where $0 < \theta < \pi/2$.

17. $\sqrt{25 - x^2}$, $x = 5\sin\theta$ **18.** $\sqrt{x^2 - 16}$, $x = 4\sec\theta$

5.2 Verifying a Trigonometric Identity In Exercises 19–26, verify the identity.

19. $\cos x(\tan^2 x + 1) = \sec x$

20. $\sec^2 x \cot x - \cot x = \tan x$

21. $\sin\left(\dfrac{\pi}{2} - \theta\right)\tan\theta = \sin\theta$

22. $\cot\left(\dfrac{\pi}{2} - x\right)\csc x = \sec x$

23. $\dfrac{1}{\tan\theta \csc\theta} = \cos\theta$ **24.** $\dfrac{1}{\tan x \csc x \sin x} = \cot x$

25. $\sin^5 x \cos^2 x = (\cos^2 x - 2\cos^4 x + \cos^6 x)\sin x$

26. $\cos^3 x \sin^2 x = (\sin^2 x - \sin^4 x)\cos x$

5.3 Solving a Trigonometric Equation In Exercises 27–32, solve the equation.

27. $\sin x = \sqrt{3} - \sin x$

28. $4\cos\theta = 1 + 2\cos\theta$

29. $3\sqrt{3}\tan u = 3$

30. $\frac{1}{2}\sec x - 1 = 0$

31. $3\csc^2 x = 4$

32. $4\tan^2 u - 1 = \tan^2 u$

Solving a Trigonometric Equation In Exercises 33–42, find all solutions of the equation in the interval $[0, 2\pi)$.

33. $\sin^3 x = \sin x$

34. $2\cos^2 x + 3\cos x = 0$

35. $\cos^2 x + \sin x = 1$

36. $\sin^2 x + 2\cos x = 2$

37. $2\sin 2x - \sqrt{2} = 0$

38. $2\cos\dfrac{x}{2} + 1 = 0$

39. $3\tan^2\left(\dfrac{x}{3}\right) - 1 = 0$

40. $\sqrt{3}\tan 3x = 0$

41. $\cos 4x(\cos x - 1) = 0$

42. $3\csc^2 5x = -4$

Using Inverse Functions In Exercises 43–46, solve the equation.

43. $\tan^2 x - 2\tan x = 0$

44. $2\tan^2 x - 3\tan x = -1$

45. $\tan^2\theta + \tan\theta - 6 = 0$

46. $\sec^2 x + 6\tan x + 4 = 0$

5.4 Evaluating Trigonometric Functions In Exercises 47–50, find the exact values of the sine, cosine, and tangent of the angle.

47. $75° = 120° - 45°$

48. $375° = 135° + 240°$

49. $\dfrac{25\pi}{12} = \dfrac{11\pi}{6} + \dfrac{\pi}{4}$

50. $\dfrac{19\pi}{12} = \dfrac{11\pi}{6} - \dfrac{\pi}{4}$

Rewriting a Trigonometric Expression In Exercises 51 and 52, write the expression as the sine, cosine, or tangent of an angle.

51. $\sin 60° \cos 45° - \cos 60° \sin 45°$

52. $\dfrac{\tan 68° - \tan 115°}{1 + \tan 68° \tan 115°}$

Evaluating a Trigonometric Expression In Exercises 53–56, find the exact value of the trigonometric expression given that $\tan u = \frac{3}{4}$ and $\cos v = -\frac{4}{5}$. (u is in Quadrant I and v is in Quadrant III.)

53. $\sin(u + v)$

54. $\tan(u + v)$

55. $\cos(u - v)$

56. $\sin(u - v)$

Verifying a Trigonometric Identity In Exercises 57–60, verify the identity.

57. $\cos\left(x + \dfrac{\pi}{2}\right) = -\sin x$ **58.** $\tan\left(x - \dfrac{\pi}{2}\right) = -\cot x$

59. $\tan(\pi - x) = -\tan x$ **60.** $\sin(x - \pi) = -\sin x$

Solving a Trigonometric Equation In Exercises 61 and 62, find all solutions of the equation in the interval $[0, 2\pi)$.

61. $\sin\left(x + \dfrac{\pi}{4}\right) - \sin\left(x - \dfrac{\pi}{4}\right) = 1$

62. $\cos\left(x + \dfrac{\pi}{6}\right) - \cos\left(x - \dfrac{\pi}{6}\right) = 1$

5.5 Evaluating Functions Involving Double Angles In Exercises 63 and 64, use the given conditions to find the exact values of $\sin 2u$, $\cos 2u$, and $\tan 2u$ using the double-angle formulas.

63. $\sin u = \frac{4}{5}$, $0 < u < \pi/2$

64. $\cos u = -2/\sqrt{5}$, $\pi/2 < u < \pi$

Verifying a Trigonometric Identity In Exercises 65 and 66, use the double-angle formulas to verify the identity algebraically and use a graphing utility to confirm your result graphically.

65. $\sin 4x = 8\cos^3 x \sin x - 4\cos x \sin x$

66. $\tan^2 x = \dfrac{1 - \cos 2x}{1 + \cos 2x}$

Reducing Powers In Exercises 67 and 68, use the power-reducing formulas to rewrite the expression in terms of first powers of the cosines of multiple angles.

67. $\tan^2 3x$ **68.** $\sin^2 x \cos^2 x$

Using Half-Angle Formulas In Exercises 69 and 70, use the half-angle formulas to determine the exact values of the sine, cosine, and tangent of the angle.

69. $-75°$ **70.** $5\pi/12$

Using Half-Angle Formulas In Exercises 71–74, use the given conditions to (a) determine the quadrant in which $u/2$ lies, and (b) find the exact values of $\sin(u/2)$, $\cos(u/2)$, and $\tan(u/2)$ using the half-angle formulas.

71. $\tan u = \dfrac{4}{3}$, $\pi < u < \dfrac{3\pi}{2}$

72. $\sin u = \dfrac{3}{5}$, $0 < u < \dfrac{\pi}{2}$

73. $\cos u = -\dfrac{2}{7}$, $\dfrac{\pi}{2} < u < \pi$

74. $\tan u = -\dfrac{\sqrt{21}}{2}$, $\dfrac{3\pi}{2} < u < 2\pi$

Using Product-to-Sum Formulas In Exercises 75 and 76, use the product-to-sum formulas to rewrite the product as a sum or difference.

75. $\cos 4\theta \sin 6\theta$

76. $2\sin 7\theta \cos 3\theta$

Using Sum-to-Product Formulas In Exercises 77 and 78, use the sum-to-product formulas to rewrite the sum or difference as a product.

77. $\cos 6\theta + \cos 5\theta$

78. $\sin 3x - \sin x$

79. Projectile Motion A baseball leaves the hand of a player at first base at an angle of θ with the horizontal and at an initial velocity of $v_0 = 80$ feet per second. A player at second base 100 feet away catches the ball. Find θ when the range r of a projectile is

$$r = \dfrac{1}{32}v_0^2 \sin 2\theta.$$

80. Geometry A trough for feeding cattle is 4 meters long and its cross sections are isosceles triangles with the two equal sides being $\frac{1}{2}$ meter (see figure). The angle between the two sides is θ.

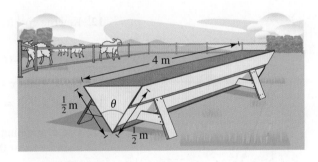

(a) Write the volume of the trough as a function of $\theta/2$.

(b) Write the volume of the trough as a function of θ and determine the value of θ such that the volume is maximized.

Exploration

True or False? In Exercises 81–84, determine whether the statement is true or false. Justify your answer.

81. If $\dfrac{\pi}{2} < \theta < \pi$, then $\cos\dfrac{\theta}{2} < 0$.

82. $\cot x \sin^2 x = \cos x \sin x$

83. $4\sin(-x)\cos(-x) = -2\sin 2x$

84. $4\sin 45° \cos 15° = 1 + \sqrt{3}$

85. Think About It Is it possible for a trigonometric equation that is not an identity to have an infinite number of solutions? Explain.

Chapter Test

See CalcChat.com for tutorial help and worked-out solutions to odd-numbered exercises.

Take this test as you would take a test in class. When you are finished, check your work against the answers given in the back of the book.

1. Use the conditions $\csc \theta = \frac{5}{2}$ and $\tan \theta < 0$ to find the values of all six trigonometric functions.

2. Use the fundamental identities to simplify $\csc^2 \beta (1 - \cos^2 \beta)$.

3. Factor and simplify $\dfrac{\sec^4 x - \tan^4 x}{\sec^2 x + \tan^2 x}$.

4. Add and simplify $\dfrac{\cos \theta}{\sin \theta} + \dfrac{\sin \theta}{\cos \theta}$.

In Exercises 5–10, verify the identity.

5. $\sin \theta \sec \theta = \tan \theta$

6. $\sec^2 x \tan^2 x + \sec^2 x = \sec^4 x$

7. $\dfrac{\csc \alpha + \sec \alpha}{\sin \alpha + \cos \alpha} = \cot \alpha + \tan \alpha$

8. $\tan\left(x + \dfrac{\pi}{2}\right) = -\cot x$

9. $1 + \cos 10y = 2 \cos^2 5y$

10. $\sin \dfrac{\alpha}{3} \cos \dfrac{\alpha}{3} = \dfrac{1}{2} \sin \dfrac{2\alpha}{3}$

11. Rewrite $4 \sin 3\theta \cos 2\theta$ as a sum or difference.

12. Rewrite $\cos\left(\theta + \dfrac{\pi}{2}\right) - \cos\left(\theta - \dfrac{\pi}{2}\right)$ as a product.

In Exercises 13–16, find all solutions of the equation in the interval $[0, 2\pi)$.

13. $\tan^2 x + \tan x = 0$

14. $\sin 2\alpha - \cos \alpha = 0$

15. $4 \cos^2 x - 3 = 0$

16. $\csc^2 x - \csc x - 2 = 0$

17. Use a graphing utility to approximate (to three decimal places) the solutions of $5 \sin x - x = 0$ in the interval $[0, 2\pi)$.

18. Find the exact value of $\cos 105°$ using the fact that $105° = 135° - 30°$.

19. Use the figure to find the exact values of $\sin 2u$, $\cos 2u$, and $\tan 2u$.

20. Cheyenne, Wyoming, has a latitude of 41°N. At this latitude, the number of hours of daylight D can be modeled by

$$D = 2.914 \sin(0.017t - 1.321) + 12.134$$

where t represents the day, with $t = 1$ corresponding to January 1. Use a graphing utility to determine the days on which there are more than 10 hours of daylight. *(Source: U.S. Naval Observatory)*

21. The heights h_1 and h_2 (in feet) above ground of two people in different seats on a Ferris wheel can be modeled by

$$h_1 = 28 \cos 10t + 38$$

and

$$h_2 = 28 \cos\left[10\left(t - \dfrac{\pi}{6}\right)\right] + 38, \quad 0 \le t \le 2$$

where t represents the time (in minutes). When are the two people at the same height?

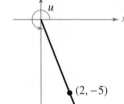

Figure for 19

(2, −5)

Proofs in Mathematics

Sum and Difference Formulas *(p. 374)*

$$\sin(u + v) = \sin u \cos v + \cos u \sin v \qquad \tan(u + v) = \frac{\tan u + \tan v}{1 - \tan u \tan v}$$

$$\sin(u - v) = \sin u \cos v - \cos u \sin v$$

$$\cos(u + v) = \cos u \cos v - \sin u \sin v \qquad \tan(u - v) = \frac{\tan u - \tan v}{1 + \tan u \tan v}$$

$$\cos(u - v) = \cos u \cos v + \sin u \sin v$$

Proof

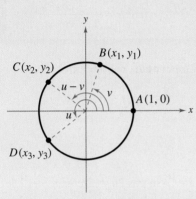

In the proofs of the formulas for $\cos(u \pm v)$, assume that $0 < v < u < 2\pi$. The top figure at the left uses u and v to locate the points $B(x_1, y_1)$, $C(x_2, y_2)$, and $D(x_3, y_3)$ on the unit circle. So, $x_i^2 + y_i^2 = 1$ for $i = 1, 2$, and 3. In the bottom figure, arc lengths AC and BD are equal, so segment lengths AC and BD are also equal. This leads to the following.

$$\sqrt{(x_2 - 1)^2 + (y_2 - 0)^2} = \sqrt{(x_3 - x_1)^2 + (y_3 - y_1)^2}$$

$$x_2^2 - 2x_2 + 1 + y_2^2 = x_3^2 - 2x_1x_3 + x_1^2 + y_3^2 - 2y_1y_3 + y_1^2$$

$$(x_2^2 + y_2^2) + 1 - 2x_2 = (x_3^2 + y_3^2) + (x_1^2 + y_1^2) - 2x_1x_3 - 2y_1y_3$$

$$1 + 1 - 2x_2 = 1 + 1 - 2x_1x_3 - 2y_1y_3$$

$$x_2 = x_3x_1 + y_3y_1$$

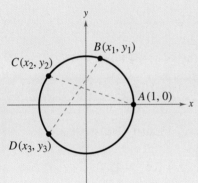

Substitute the values $x_2 = \cos(u - v)$, $x_3 = \cos u$, $x_1 = \cos v$, $y_3 = \sin u$, and $y_1 = \sin v$ to obtain $\cos(u - v) = \cos u \cos v + \sin u \sin v$. To establish the formula for $\cos(u + v)$, consider $u + v = u - (-v)$ and use the formula just derived to obtain

$$\cos(u + v) = \cos[u - (-v)]$$

$$= \cos u \cos(-v) + \sin u \sin(-v)$$

$$= \cos u \cos v - \sin u \sin v.$$

You can use the sum and difference formulas for sine and cosine to prove the formulas for $\tan(u \pm v)$.

$$\tan(u \pm v) = \frac{\sin(u \pm v)}{\cos(u \pm v)} = \frac{\sin u \cos v \pm \cos u \sin v}{\cos u \cos v \mp \sin u \sin v}$$

$$= \frac{\dfrac{\sin u \cos v \pm \cos u \sin v}{\cos u \cos v}}{\dfrac{\cos u \cos v \mp \sin u \sin v}{\cos u \cos v}} = \frac{\dfrac{\sin u \cos v}{\cos u \cos v} \pm \dfrac{\cos u \sin v}{\cos u \cos v}}{\dfrac{\cos u \cos v}{\cos u \cos v} \mp \dfrac{\sin u \sin v}{\cos u \cos v}}$$

$$= \frac{\dfrac{\sin u}{\cos u} \pm \dfrac{\sin v}{\cos v}}{1 \mp \dfrac{\sin u}{\cos u} \cdot \dfrac{\sin v}{\cos v}} = \frac{\tan u \pm \tan v}{1 \mp \tan u \tan v}$$

Double-Angle Formulas *(p. 381)*

$$\sin 2u = 2 \sin u \cos u \qquad\qquad \cos 2u = \cos^2 u - \sin^2 u$$

$$\tan 2u = \frac{2 \tan u}{1 - \tan^2 u} \qquad\qquad\qquad = 2 \cos^2 u - 1$$

$$= 1 - 2 \sin^2 u$$

Proof Prove each Double-Angle Formula by letting $v = u$ in the corresponding sum formula.

$$\sin 2u = \sin(u + u) = \sin u \cos u + \cos u \sin u = 2 \sin u \cos u$$

$$\cos 2u = \cos(u + u) = \cos u \cos u - \sin u \sin u = \cos^2 u - \sin^2 u$$

$$\tan 2u = \tan(u + u) = \frac{\tan u + \tan u}{1 - \tan u \tan u} = \frac{2 \tan u}{1 - \tan^2 u}$$

Power-Reducing Formulas *(p. 383)*

$$\sin^2 u = \frac{1 - \cos 2u}{2} \qquad \cos^2 u = \frac{1 + \cos 2u}{2} \qquad \tan^2 u = \frac{1 - \cos 2u}{1 + \cos 2u}$$

Proof Prove the first formula by solving for $\sin^2 u$ in $\cos 2u = 1 - 2 \sin^2 u$.

$$\cos 2u = 1 - 2 \sin^2 u \qquad \text{Write double-angle formula.}$$

$$2 \sin^2 u = 1 - \cos 2u \qquad \text{Subtract } \cos 2u \text{ from, and add } 2 \sin^2 u \text{ to, each side.}$$

$$\sin^2 u = \frac{1 - \cos 2u}{2} \qquad \text{Divide each side by 2.}$$

Similarly, to prove the second formula, solve for $\cos^2 u$ in $\cos 2u = 2 \cos^2 u - 1$. To prove the third formula, use a quotient identity.

$$\tan^2 u = \frac{\sin^2 u}{\cos^2 u} = \frac{\dfrac{1 - \cos 2u}{2}}{\dfrac{1 + \cos 2u}{2}} = \frac{1 - \cos 2u}{1 + \cos 2u}$$

Sum-to-Product Formulas *(p. 385)*

$$\sin u + \sin v = 2 \sin\left(\frac{u + v}{2}\right) \cos\left(\frac{u - v}{2}\right)$$

$$\sin u - \sin v = 2 \cos\left(\frac{u + v}{2}\right) \sin\left(\frac{u - v}{2}\right)$$

$$\cos u + \cos v = 2 \cos\left(\frac{u + v}{2}\right) \cos\left(\frac{u - v}{2}\right)$$

$$\cos u - \cos v = -2 \sin\left(\frac{u + v}{2}\right) \sin\left(\frac{u - v}{2}\right)$$

Proof To prove the first formula, let $x = u + v$ and $y = u - v$. Then substitute $u = (x + y)/2$ and $v = (x - y)/2$ in the product-to-sum formula.

$$\sin u \cos v = \frac{1}{2}[\sin(u + v) + \sin(u - v)]$$

$$\sin\left(\frac{x + y}{2}\right) \cos\left(\frac{x - y}{2}\right) = \frac{1}{2}(\sin x + \sin y)$$

$$2 \sin\left(\frac{x + y}{2}\right) \cos\left(\frac{x - y}{2}\right) = \sin x + \sin y$$

The other sum-to-product formulas can be proved in a similar manner.

P.S. Problem Solving ▪ ▪ ▪ ▪ ▪ ▪ ▪ ▪ ▪ ▪ ▪ ▪ ▪ ▪

1. **Writing Trigonometric Functions in Terms of Cosine** Write each of the other trigonometric functions of θ in terms of $\cos \theta$.

2. **Verifying a Trigonometric Identity** Verify that for all integers n,
$$\cos\left[\frac{(2n + 1)\pi}{2}\right] = 0.$$

3. **Verifying a Trigonometric Identity** Verify that for all integers n,
$$\sin\left[\frac{(12n + 1)\pi}{6}\right] = \frac{1}{2}.$$

4. **Sound Wave** A sound wave is modeled by
$$p(t) = \frac{1}{4\pi}[p_1(t) + 30p_2(t) + p_3(t) + p_5(t) + 30p_6(t)]$$

where $p_n(t) = \frac{1}{n}\sin(524n\pi t)$, and t represents the time (in seconds).

(a) Find the sine components $p_n(t)$ and use a graphing utility to graph the components. Then verify the graph of p shown below.

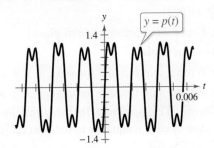

(b) Find the period of each sine component of p. Is p periodic? If so, then what is its period?

(c) Use the graphing utility to find the t-intercepts of the graph of p over one cycle.

(d) Use the graphing utility to approximate the absolute maximum and absolute minimum values of p over one cycle.

5. **Geometry** Three squares of side length s are placed side by side (see figure). Make a conjecture about the relationship between the sum $u + v$ and w. Prove your conjecture by using the identity for the tangent of the sum of two angles.

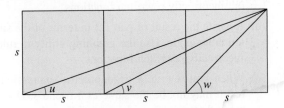

6. **Projectile Motion** The path traveled by an object (neglecting air resistance) that is projected at an initial height of h_0 feet, an initial velocity of v_0 feet per second, and an initial angle θ is given by
$$y = -\frac{16}{v_0^2 \cos^2 \theta}x^2 + (\tan \theta)x + h_0$$

where the horizontal distance x and the vertical distance y are measured in feet. Find a formula for the maximum height of an object projected from ground level at velocity v_0 and angle θ. To do this, find half of the horizontal distance
$$\frac{1}{32}v_0^2 \sin 2\theta$$

and then substitute it for x in the model for the path of a projectile (where $h_0 = 0$).

7. **Geometry** The length of each of the two equal sides of an isosceles triangle is 10 meters (see figure). The angle between the two sides is θ.

(a) Write the area of the triangle as a function of $\theta/2$.

(b) Write the area of the triangle as a function of θ. Determine the value of θ such that the area is a maximum.

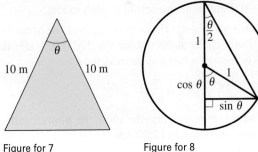

Figure for 7   Figure for 8

8. **Geometry** Use the figure to derive the formulas for
$$\sin\frac{\theta}{2}, \quad \cos\frac{\theta}{2}, \quad \text{and} \quad \tan\frac{\theta}{2}$$

where θ is an acute angle.

9. **Force** The force F (in pounds) on a person's back when he or she bends over at an angle θ from an upright position is modeled by
$$F = \frac{0.6W \sin(\theta + 90°)}{\sin 12°}$$

where W represents the person's weight (in pounds).

(a) Simplify the model.

(b) Use a graphing utility to graph the model, where $W = 185$ and $0° < \theta < 90°$.

(c) At what angle is the force maximized? At what angle is the force minimized?

10. Hours of Daylight The number of hours of daylight that occur at any location on Earth depends on the time of year and the latitude of the location. The equations below model the numbers of hours of daylight in Seward, Alaska (60° latitude), and New Orleans, Louisiana (30° latitude).

$$D = 12.2 - 6.4 \cos\left[\frac{\pi(t + 0.2)}{182.6}\right] \quad \text{Seward}$$

$$D = 12.2 - 1.9 \cos\left[\frac{\pi(t + 0.2)}{182.6}\right] \quad \text{New Orleans}$$

In these models, D represents the number of hours of daylight and t represents the day, with $t = 0$ corresponding to January 1.

(a) Use a graphing utility to graph both models in the same viewing window. Use a viewing window of $0 \le t \le 365$.

(b) Find the days of the year on which both cities receive the same amount of daylight.

(c) Which city has the greater variation in the number of hours of daylight? Which constant in each model would you use to determine the difference between the greatest and least numbers of hours of daylight?

(d) Determine the period of each model.

11. Ocean Tide The tide, or depth of the ocean near the shore, changes throughout the day. The water depth d (in feet) of a bay can be modeled by

$$d = 35 - 28 \cos \frac{\pi}{6.2} t$$

where t represents the time in hours, with $t = 0$ corresponding to 12:00 A.M.

(a) Algebraically find the times at which the high and low tides occur.

(b) If possible, algebraically find the time(s) at which the water depth is 3.5 feet.

(c) Use a graphing utility to verify your results from parts (a) and (b).

12. Piston Heights The heights h (in inches) of pistons 1 and 2 in an automobile engine can be modeled by

$$h_1 = 3.75 \sin 733t + 7.5$$

and

$$h_2 = 3.75 \sin 733\left(t + \frac{4\pi}{3}\right) + 7.5$$

respectively, where t is measured in seconds.

(a) Use a graphing utility to graph the heights of these pistons in the same viewing window for $0 \le t \le 1$.

(b) How often are the pistons at the same height?

13. Index of Refraction The index of refraction n of a transparent material is the ratio of the speed of light in a vacuum to the speed of light in the material. Some common materials and their indices of refraction are air (1.00), water (1.33), and glass (1.50). Triangular prisms are often used to measure the index of refraction based on the formula

$$n = \frac{\sin\left(\dfrac{\theta}{2} + \dfrac{\alpha}{2}\right)}{\sin \dfrac{\theta}{2}}.$$

For the prism shown in the figure, $\alpha = 60°$.

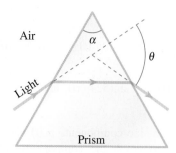

(a) Write the index of refraction as a function of $\cot(\theta/2)$.

(b) Find θ for a prism made of glass.

14. Sum Formulas

(a) Write a sum formula for $\sin(u + v + w)$.

(b) Write a sum formula for $\tan(u + v + w)$.

15. Solving Trigonometric Inequalities Find the solution of each inequality in the interval $[0, 2\pi)$.

(a) $\sin x \ge 0.5$ (b) $\cos x \le -0.5$

(c) $\tan x < \sin x$ (d) $\cos x \ge \sin x$

16. Sum of Fourth Powers Consider the function $f(x) = \sin^4 x + \cos^4 x$.

(a) Use the power-reducing formulas to write the function in terms of cosine to the first power.

(b) Determine another way of rewriting the original function. Use a graphing utility to rule out incorrectly rewritten functions.

(c) Add a trigonometric term to the original function so that it becomes a perfect square trinomial. Rewrite the function as a perfect square trinomial minus the term that you added. Use the graphing utility to rule out incorrectly rewritten functions.

(d) Rewrite the result of part (c) in terms of the sine of a double angle. Use the graphing utility to rule out incorrectly rewritten functions.

(e) When you rewrite a trigonometric expression, the result may not be the same as a friend's. Does this mean that one of you is wrong? Explain.

6 Additional Topics in Trigonometry

Ohm's Law
(Exercise 95, page 453)

Work *(page 434)*

Air Navigation *(Example 11, page 424)*

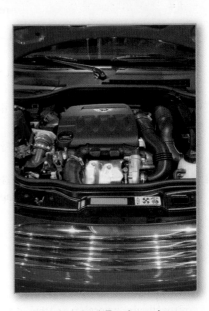

Mechanical Engineering
(Exercise 56, page 415)

Surveying *(page 401)*

6.1 Law of Sines

- Use the Law of Sines to solve oblique triangles (AAS or ASA).
- Use the Law of Sines to solve oblique triangles (SSA).
- Find the areas of oblique triangles.
- Use the Law of Sines to model and solve real-life problems.

Introduction

In Chapter 4, you studied techniques for solving right triangles. In this section and the next, you will solve **oblique triangles**—triangles that have no right angles. As standard notation, the angles of a triangle are labeled A, B, and C, and their opposite sides are labeled a, b, and c, as shown in the figure.

The Law of Sines is a useful tool for solving real-life problems involving oblique triangles. For example, in Exercise 46 on page 407, you will use the Law of Sines to determine the distance from a boat to a shoreline.

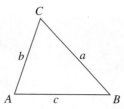

To solve an oblique triangle, you need to know the measure of at least one side and any two other measures of the triangle—the other two sides, two angles, or one angle and one other side. So, there are four cases.

1. Two angles and any side (AAS or ASA)
2. Two sides and an angle opposite one of them (SSA)
3. Three sides (SSS)
4. Two sides and their included angle (SAS)

The first two cases can be solved using the **Law of Sines**, whereas the last two cases require the Law of Cosines (see Section 6.2).

Law of Sines

If ABC is a triangle with sides a, b, and c, then

$$\frac{a}{\sin A} = \frac{b}{\sin B} = \frac{c}{\sin C}.$$

A is acute.

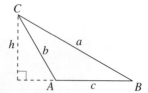

A is obtuse.

The Law of Sines can also be written in the reciprocal form

$$\frac{\sin A}{a} = \frac{\sin B}{b} = \frac{\sin C}{c}.$$

For a proof of the Law of Sines, see Proofs in Mathematics on page 462.

$b = 28$ ft

C

$102°$

a

$29°$

A c B

Figure 6.1

EXAMPLE 1   **Given Two Angles and One Side—AAS**

For the triangle in Figure 6.1, $C = 102°$, $B = 29°$, and $b = 28$ feet. Find the remaining angle and sides.

Solution   The third angle of the triangle is

$$A = 180° - B - C = 180° - 29° - 102° = 49°.$$

By the Law of Sines, you have

$$\frac{a}{\sin A} = \frac{b}{\sin B} = \frac{c}{\sin C}.$$

Using $b = 28$ produces

$$a = \frac{b}{\sin B}(\sin A) = \frac{28}{\sin 29°}(\sin 49°) \approx 43.59 \text{ feet}$$

and

$$c = \frac{b}{\sin B}(\sin C) = \frac{28}{\sin 29°}(\sin 102°) \approx 56.49 \text{ feet.}$$

✓ *Checkpoint* 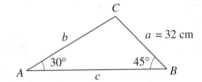 *Audio-video solution in English & Spanish at LarsonPrecalculus.com*

For the triangle shown, $A = 30°$, $B = 45°$, and $a = 32$ centimeters. Find the remaining angle and sides.

C

b $a = 32$ cm

$30°$ $45°$

A c B

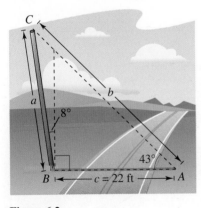

In the 1850s, surveyors used the Law of Sines to calculate the height of Mount Everest. Their calculation was within 30 feet of the currently accepted value.

EXAMPLE 2   **Given Two Angles and One Side—ASA**

A pole tilts toward the sun at an 8° angle from the vertical, and it casts a 22-foot shadow. (See Figure 6.2.) The angle of elevation from the tip of the shadow to the top of the pole is 43°. How tall is the pole?

Solution   In Figure 6.2, $A = 43°$ and

$$B = 90° + 8° = 98°.$$

So, the third angle is

$$C = 180° - A - B = 180° - 43° - 98° = 39°.$$

By the Law of Sines, you have

$$\frac{a}{\sin A} = \frac{c}{\sin C}.$$

The shadow length c is $c = 22$ feet, so the height of the pole is

$$a = \frac{c}{\sin C}(\sin A) = \frac{22}{\sin 39°}(\sin 43°) \approx 23.84 \text{ feet.}$$

✓ *Checkpoint* 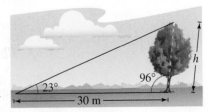 *Audio-video solution in English & Spanish at LarsonPrecalculus.com*

Find the height of the tree shown in the figure.

Figure 6.2

C

a

$8°$

b

$43°$

$B \longleftarrow c = 22$ ft $\longrightarrow A$

$23°$ $96°$ h

$\longmapsto$ 30 m $\longmapsto$

The Ambiguous Case (SSA)

In Examples 1 and 2, you saw that two angles and one side determine a unique triangle. However, if two sides and one opposite angle are given, then three possible situations can occur: (1) no such triangle exists, (2) one such triangle exists, or (3) two distinct triangles exist that satisfy the conditions.

The Ambiguous Case (SSA)

Consider a triangle in which a, b, and A are given. $\quad(h = b \sin A)$

	A is acute.	A is acute.	A is acute.	A is acute.	A is obtuse.	A is obtuse.
Sketch						
Necessary condition	$a < h$	$a = h$	$a \geq b$	$h < a < b$	$a \leq b$	$a > b$
Triangles possible	None	One	One	Two	None	One

EXAMPLE 3 **Single-Solution Case—SSA**

See LarsonPrecalculus.com for an interactive version of this type of example.

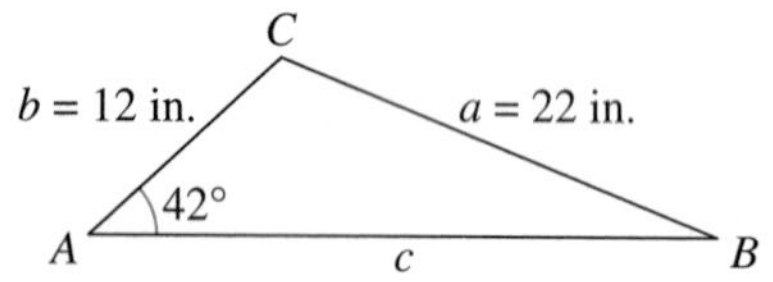

For the triangle in Figure 6.3, $a = 22$ inches, $b = 12$ inches, and $A = 42°$. Find the remaining side and angles.

Solution By the Law of Sines, you have

$$\frac{\sin B}{b} = \frac{\sin A}{a} \qquad \text{Reciprocal form}$$

$$\sin B = b\left(\frac{\sin A}{a}\right) \qquad \text{Multiply each side by } b.$$

$$\sin B = 12\left(\frac{\sin 42°}{22}\right) \qquad \text{Substitute for } A, a, \text{ and } b.$$

$$B \approx 21.41°. \qquad \text{Solve for acute angle } B.$$

Next, subtract to determine that $C \approx 180° - 42° - 21.41° = 116.59°$. Then find the remaining side.

$$\frac{c}{\sin C} = \frac{a}{\sin A} \qquad \text{Law of Sines}$$

$$c = \frac{a}{\sin A}(\sin C) \qquad \text{Multiply each side by } \sin C.$$

$$c \approx \frac{22}{\sin 42°}(\sin 116.59°) \qquad \text{Substitute for } a, A, \text{ and } C.$$

$$c \approx 29.40 \text{ inches} \qquad \text{Simplify.}$$

✓ **Checkpoint** ◀))) *Audio-video solution in English & Spanish at LarsonPrecalculus.com*

Given $A = 31°$, $a = 12$ inches, and $b = 5$ inches, find the remaining side and angles of the triangle.

$b = 12$ in. $\qquad a = 22$ in.

$42°$

One solution: $a \geq b$
Figure 6.3

$a = 15$ ft

$b = 25$ ft

h

$85°$

A

No solution: $a < h$

Figure 6.4

EXAMPLE 4   No-Solution Case—SSA

Show that there is no triangle for which $a = 15$ feet, $b = 25$ feet, and $A = 85°$.

Solution   Begin by making the sketch shown in Figure 6.4. From this figure, it appears that no triangle is possible. Verify this using the Law of Sines.

$$\frac{\sin B}{b} = \frac{\sin A}{a} \qquad \text{Reciprocal form}$$

$$\sin B = b\left(\frac{\sin A}{a}\right) \qquad \text{Multiply each side by } b.$$

$$\sin B = 25\left(\frac{\sin 85°}{15}\right) \approx 1.6603 > 1$$

This contradicts the fact that $|\sin B| \leq 1$. So, no triangle can be formed with sides $a = 15$ feet and $b = 25$ feet and angle $A = 85°$.

✓ **Checkpoint**))) *Audio-video solution in English & Spanish at LarsonPrecalculus.com*

Show that there is no triangle for which $a = 4$ feet, $b = 14$ feet, and $A = 60°$.

EXAMPLE 5   Two-Solution Case—SSA

Find two triangles for which $a = 12$ meters, $b = 31$ meters, and $A = 20.50°$.

Solution   Because $h = b \sin A = 31(\sin 20.50°) \approx 10.86$ meters and $h < a < b$, there are two possible triangles. By the Law of Sines, you have

$$\frac{\sin B}{b} = \frac{\sin A}{a} \qquad \text{Reciprocal form}$$

$$\sin B = b\left(\frac{\sin A}{a}\right) = 31\left(\frac{\sin 20.50°}{12}\right) \approx 0.9047.$$

There are two angles between $0°$ and $180°$ whose sine is approximately 0.9047, $B_1 \approx 64.78°$ and $B_2 \approx 180° - 64.78° = 115.22°$. For $B_1 \approx 64.78°$, you obtain

$$C \approx 180° - 20.50° - 64.78° = 94.72°$$

$$c = \frac{a}{\sin A}(\sin C) \approx \frac{12}{\sin 20.50°}(\sin 94.72°) \approx 34.15 \text{ meters.}$$

For $B_2 \approx 115.22°$, you obtain

$$C \approx 180° - 20.50° - 115.22° = 44.28°$$

$$c = \frac{a}{\sin A}(\sin C) \approx \frac{12}{\sin 20.50°}(\sin 44.28°) \approx 23.92 \text{ meters.}$$

The resulting triangles are shown below.

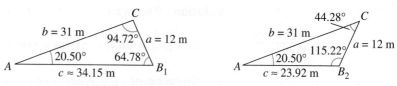

C
$b = 31$ m $94.72°$ $a = 12$ m
$20.50°$ $64.78°$
A $c \approx 34.15$ m B_1

$44.28°$ C
$b = 31$ m $a = 12$ m
$20.50°$ $115.22°$
A $c \approx 23.92$ m B_2

Two solutions: $h < a < b$

✓ **Checkpoint**))) *Audio-video solution in English & Spanish at LarsonPrecalculus.com*

Find two triangles for which $a = 4.5$ feet, $b = 5$ feet, and $A = 58°$.

6.1 Exercises

See **CalcChat.com** for tutorial help and worked-out solutions to odd-numbered exercises.

Vocabulary: **Fill in the blanks.**

1. An _____ triangle is a triangle that has no right angle.

2. For triangle ABC, the Law of Sines is $\dfrac{a}{\sin A}$ = _____ = $\dfrac{c}{\sin C}$.

3. Two _____ and one _____ determine a unique triangle.

4. The area of an oblique triangle ABC is $\frac{1}{2}bc \sin A = \frac{1}{2}ab \sin C$ = _____.

Skills and Applications

 Using the Law of Sines In Exercises 5–22, use the Law of Sines to solve the triangle. Round your answers to two decimal places.

5.
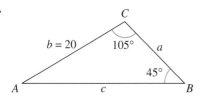

6.
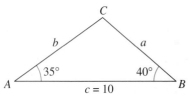

7.
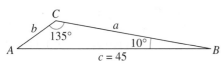

8.
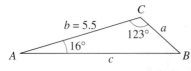

9. $A = 102.4°$, $C = 16.7°$, $a = 21.6$
10. $A = 24.3°$, $C = 54.6°$, $c = 2.68$
11. $A = 83° \, 20'$, $C = 54.6°$, $c = 18.1$
12. $A = 5° \, 40'$, $B = 8° \, 15'$, $b = 4.8$
13. $A = 35°$, $B = 65°$, $c = 10$
14. $A = 120°$, $B = 45°$, $c = 16$
15. $A = 55°$, $B = 42°$, $c = \frac{3}{4}$
16. $B = 28°$, $C = 104°$, $a = 3\frac{5}{8}$
17. $A = 36°$, $a = 8$, $b = 5$
18. $A = 60°$, $a = 9$, $c = 7$
19. $A = 145°$, $a = 14$, $b = 4$
20. $A = 100°$, $a = 125$, $c = 10$
21. $B = 15° \, 30'$, $a = 4.5$, $b = 6.8$
22. $B = 2° \, 45'$, $b = 6.2$, $c = 5.8$

 Using the Law of Sines In Exercises 23–32, use the Law of Sines to solve (if possible) the triangle. If two solutions exist, find both. Round your answers to two decimal places.

23. $A = 110°$, $a = 125$, $b = 100$
24. $A = 110°$, $a = 125$, $b = 200$
25. $A = 76°$, $a = 18$, $b = 20$
26. $A = 76°$, $a = 34$, $b = 21$
27. $A = 58°$, $a = 11.4$, $b = 12.8$
28. $A = 58°$, $a = 4.5$, $b = 12.8$
29. $A = 120°$, $a = b = 25$
30. $A = 120°$, $a = 25$, $b = 24$
31. $A = 45°$, $a = b = 1$
32. $A = 25° \, 4'$, $a = 9.5$, $b = 22$

 Using the Law of Sines In Exercises 33–36, find values for b such that the triangle has (a) one solution, (b) two solutions (if possible), and (c) no solution.

33. $A = 36°$, $a = 5$
34. $A = 60°$, $a = 10$
35. $A = 105°$, $a = 80$
36. $A = 132°$, $a = 215$

 Finding the Area of a Triangle In Exercises 37–44, find the area of the triangle. Round your answers to one decimal place.

37. $A = 125°$, $b = 9$, $c = 6$
38. $C = 150°$, $a = 17$, $b = 10$
39. $B = 39°$, $a = 25$, $c = 12$
40. $A = 72°$, $b = 31$, $c = 44$
41. $C = 103° \, 15'$, $a = 16$, $b = 28$
42. $B = 54° \, 30'$, $a = 62$, $c = 35$
43. $A = 67°$, $B = 43°$, $a = 8$
44. $B = 118°$, $C = 29°$, $a = 52$

45. Height A tree grows at an angle of $4°$ from the vertical due to prevailing winds. At a point 40 meters from the base of the tree, the angle of elevation to the top of the tree is $30°$ (see figure).

(a) Write an equation that you can use to find the height h of the tree.

(b) Find the height of the tree.

46. Distance A boat is traveling due east parallel to the shoreline at a speed of 10 miles per hour. At a given time, the bearing to a lighthouse is S $70°$ E, and 15 minutes later the bearing is S $63°$ E (see figure). The lighthouse is located at the shoreline. What is the distance from the boat to the shoreline?

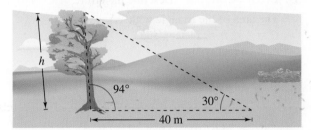

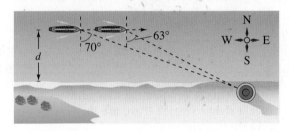

47. Environmental Science The bearing from the Pine Knob fire tower to the Colt Station fire tower is N $65°$ E, and the two towers are 30 kilometers apart. A fire spotted by rangers in each tower has a bearing of N $80°$ E from Pine Knob and S $70°$ E from Colt Station (see figure). Find the distance of the fire from each tower.

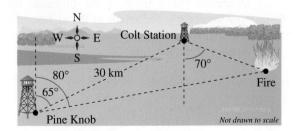

48. Bridge Design A bridge is built across a small lake from a gazebo to a dock (see figure). The bearing from the gazebo to the dock is S $41°$ W. From a tree 100 meters from the gazebo, the bearings to the gazebo and the dock are S $74°$ E and S $28°$ E, respectively. Find the distance from the gazebo to the dock.

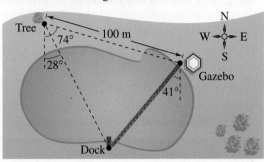

49. Angle of Elevation A 10-meter utility pole casts a 17-meter shadow directly down a slope when the angle of elevation of the sun is $42°$ (see figure). Find θ, the angle of elevation of the ground.

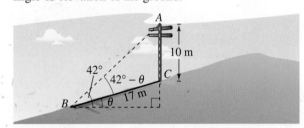

50. Flight Path A plane flies 500 kilometers with a bearing of $316°$ from Naples to Elgin (see figure). The plane then flies 720 kilometers from Elgin to Canton (Canton is due west of Naples). Find the bearing of the flight from Elgin to Canton.

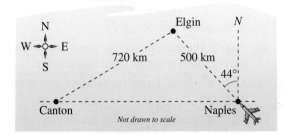

51. Altitude The angles of elevation to an airplane from two points A and B on level ground are $55°$ and $72°$, respectively. The points A and B are 2.2 miles apart, and the airplane is east of both points in the same vertical plane.

(a) Draw a diagram that represents the problem. Show the known quantities on the diagram.

(b) Find the distance between the plane and point B.

(c) Find the altitude of the plane.

(d) Find the distance the plane must travel before it is directly above point A.

52. Height   A flagpole at a right angle to the horizontal is located on a slope that makes an angle of 12° with the horizontal. The flagpole's shadow is 16 meters long and points directly up the slope. The angle of elevation from the tip of the shadow to the sun is 20°.

(a) Draw a diagram that represents the problem. Show the known quantities on the diagram and use a variable to indicate the height of the flagpole.

(b) Write an equation that you can use to find the height of the flagpole.

(c) Find the height of the flagpole.

53. Distance   Air traffic controllers continuously monitor the angles of elevation θ and ϕ to an airplane from an airport control tower and from an observation post 2 miles away (see figure). Write an equation giving the distance d between the plane and the observation post in terms of θ and ϕ.

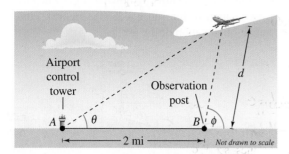

54. Numerical Analysis   In the figure, α and β are positive angles.

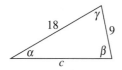

(a) Write α as a function of β.

(b) Use a graphing utility to graph the function in part (a). Determine its domain and range.

(c) Use the result of part (a) to write c as a function of β.

(d) Use the graphing utility to graph the function in part (c). Determine its domain and range.

(e) Complete the table. What can you infer?

β	0.4	0.8	1.2	1.6	2.0	2.4	2.8
α							
c							

Exploration

True or False?   **In Exercises 55–58, determine whether the statement is true or false. Justify your answer.**

55. If a triangle contains an obtuse angle, then it must be oblique.

56. Two angles and one side of a triangle do not necessarily determine a unique triangle.

57. When you know the three angles of an oblique triangle, you can solve the triangle.

58. The ratio of any two sides of a triangle is equal to the ratio of the sines of the opposite angles of the two sides.

59. Error Analysis   Describe the error.

The area of the triangle with $C = 58°$, $b = 11$ feet, and $c = 16$ feet is

$$\text{Area} = \frac{1}{2}(11)(16)(\sin 58°)$$

$$= 88(\sin 58°)$$

$$\approx 74.63 \text{ square feet.}$$

60. **HOW DO YOU SEE IT?**   In the figure, a triangle is to be formed by drawing a line segment of length a from $(4, 3)$ to the positive x-axis. For what value(s) of a can you form (a) one triangle, (b) two triangles, and (c) no triangles? Explain.

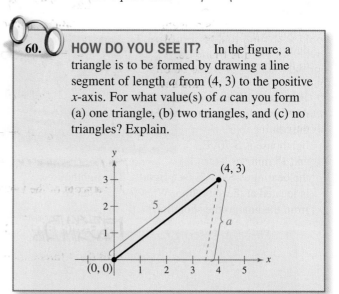

61. Think About It   Can the Law of Sines be used to solve a right triangle? If so, use the Law of Sines to solve the triangle with

$$B = 50°, \quad C = 90°, \quad \text{and} \quad a = 10.$$

Is there another way to solve the triangle? Explain.

62. Using Technology

(a) Write the area A of the shaded region in the figure as a function of θ.

(b) Use a graphing utility to graph the function.

(c) Determine the domain of the function. Explain how decreasing the length of the eight-centimeter line segment affects the area of the region and the domain of the function.

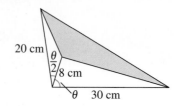

6.2 Law of Cosines

The Law of Cosines is a useful tool for solving real-life problems involving oblique triangles. For example, in Exercise 56 on page 415, you will use the Law of Cosines to determine the total distance a piston moves in an engine.

- Use the Law of Cosines to solve oblique triangles (SSS or SAS).
- Use the Law of Cosines to model and solve real-life problems.
- Use Heron's Area Formula to find areas of triangles.

Introduction

Two cases remain in the list of conditions needed to solve an oblique triangle—SSS and SAS. When you are given three sides (SSS), or two sides and their included angle (SAS), you cannot solve the triangle using the Law of Sines alone. In such cases, use the **Law of Cosines.**

Law of Cosines

Standard Form	**Alternative Form**
$a^2 = b^2 + c^2 - 2bc \cos A$	$\cos A = \dfrac{b^2 + c^2 - a^2}{2bc}$
$b^2 = a^2 + c^2 - 2ac \cos B$	$\cos B = \dfrac{a^2 + c^2 - b^2}{2ac}$
$c^2 = a^2 + b^2 - 2ab \cos C$	$\cos C = \dfrac{a^2 + b^2 - c^2}{2ab}$

For a proof of the Law of Cosines, see Proofs in Mathematics on page 462.

EXAMPLE 1 **Given Three Sides—SSS**

Find the three angles of the triangle shown below.

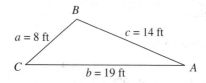

$a = 8$ ft $c = 14$ ft $b = 19$ ft

Solution It is a good idea to find the angle opposite the longest side first—side b in this case. Using the alternative form of the Law of Cosines,

$$\cos B = \frac{a^2 + c^2 - b^2}{2ac} = \frac{8^2 + 14^2 - 19^2}{2(8)(14)} \approx -0.4509.$$

Because $\cos B$ is negative, B is an *obtuse* angle given by $B \approx 116.80°$. At this point, use the Law of Sines to determine A.

$$\sin A = a\left(\frac{\sin B}{b}\right) \approx 8\left(\frac{\sin 116.80°}{19}\right) \approx 0.3758$$

The angle B is obtuse and a triangle can have at most one obtuse angle, so you know that A must be acute. So, $A \approx 22.07°$ and $C \approx 180° - 22.07° - 116.80° = 41.13°$.

✓ **Checkpoint** ◀))) *Audio-video solution in English & Spanish at LarsonPrecalculus.com*

Find the three angles of the triangle whose sides have lengths $a = 6$ centimeters, $b = 8$ centimeters, and $c = 12$ centimeters.

Do you see why it was wise to find the largest angle *first* in Example 1? Knowing the cosine of an angle, you can determine whether the angle is acute or obtuse. That is,

$$\cos \theta > 0 \quad \text{for} \quad 0° < \theta < 90° \qquad \text{Acute}$$

$$\cos \theta < 0 \quad \text{for} \quad 90° < \theta < 180°. \qquad \text{Obtuse}$$

So, in Example 1, after you find that angle B is obtuse, you know that angles A and C must both be acute. Furthermore, if the largest angle is acute, then the remaining two angles must also be acute.

EXAMPLE 2   Given Two Sides and Their Included Angle—SAS

See LarsonPrecalculus.com for an interactive version of this type of example.

Find the remaining angles and side of the triangle shown below.

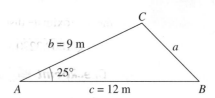

·· REMARK   When solving an oblique triangle given three sides, use the alternative form of the Law of Cosines to solve for an angle. When solving an oblique triangle given two sides and their included angle, use the standard form of the Law of Cosines to solve for the remaining side.

▷ **Solution**   Use the standard form of the Law of Cosines to find side a.

$$a^2 = b^2 + c^2 - 2bc \cos A$$

$$a^2 = 9^2 + 12^2 - 2(9)(12) \cos 25°$$

$$a^2 \approx 29.2375$$

$$a \approx 5.4072 \text{ meters}$$

Next, use the ratio $(\sin A)/a$, the given value of b, and the reciprocal form of the Law of Sines to find B.

$$\frac{\sin B}{b} = \frac{\sin A}{a} \qquad \text{Reciprocal form}$$

$$\sin B = b\left(\frac{\sin A}{a}\right) \qquad \text{Multiply each side by } b.$$

$$\sin B \approx 9\left(\frac{\sin 25°}{5.4072}\right) \qquad \text{Substitute for } A, a, \text{ and } b.$$

$$\sin B \approx 0.7034 \qquad \text{Use a calculator.}$$

There are two angles between $0°$ and $180°$ whose sine is approximately 0.7034, $B_1 \approx 44.70°$ and $B_2 \approx 180° - 44.70° = 135.30°$.

For $B_1 \approx 44.70°$,

$$C_1 \approx 180° - 25° - 44.70° = 110.30°.$$

For $B_2 \approx 135.30°$,

$$C_2 \approx 180° - 25° - 135.30° = 19.70°.$$

Side c is the longest side of the triangle, which means that angle C is the largest angle of the triangle. So, $C \approx 110.30°$ and $B \approx 44.70°$.

✓ *Checkpoint* ◀)))) *Audio-video solution in English & Spanish at LarsonPrecalculus.com*

Given $A = 80°$, $b = 16$ meters, and $c = 12$ meters, find the remaining angles and side of the triangle.

Applications

EXAMPLE 3 An Application of the Law of Cosines

The pitcher's mound on a women's softball field is 43 feet from home plate and the distance between the bases is 60 feet, as shown in Figure 6.8. (The pitcher's mound is *not* halfway between home plate and second base.) How far is the pitcher's mound from first base?

Solution In triangle *HPF*, $H = 45°$ (line segment *HP* bisects the right angle at *H*), $f = 43$, and $p = 60$. Using the standard form of the Law of Cosines for this SAS case,

$$h^2 = f^2 + p^2 - 2fp \cos H$$
$$= 43^2 + 60^2 - 2(43)(60) \cos 45°$$
$$\approx 1800.3290.$$

So, the approximate distance from the pitcher's mound to first base is

$$h \approx \sqrt{1800.3290} \approx 42.43 \text{ feet.}$$

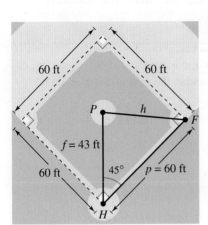

Figure 6.8

✓ **Checkpoint** ◀))) *Audio-video solution in English & Spanish at LarsonPrecalculus.com*

In a softball game, a batter hits a ball to dead center field, a distance of 240 feet from home plate. The center fielder then throws the ball to third base and gets a runner out. The distance between the bases is 60 feet. How far is the center fielder from third base?

EXAMPLE 4 An Application of the Law of Cosines

A ship travels 60 miles due north and then adjusts its course, as shown in Figure 6.9. After traveling 80 miles in this new direction, the ship is 139 miles from its point of departure. Describe the bearing from point *B* to point *C*.

Solution You have $a = 80$, $b = 139$, and $c = 60$. So, using the alternative form of the Law of Cosines,

$$\cos B = \frac{a^2 + c^2 - b^2}{2ac}$$
$$= \frac{80^2 + 60^2 - 139^2}{2(80)(60)}$$
$$\approx -0.9709.$$

So, $B \approx 166.14°$, and the bearing measured from due north from point *B* to point *C* is approximately $180° - 166.14° = 13.86°$, or N 13.86° W.

✓ **Checkpoint** ◀))) *Audio-video solution in English & Spanish at LarsonPrecalculus.com*

A ship travels 40 miles due east and then changes direction, as shown at the right. After traveling 30 miles in this new direction, the ship is 56 miles from its point of departure. Describe the bearing from point *B* to point *C*.

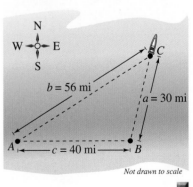

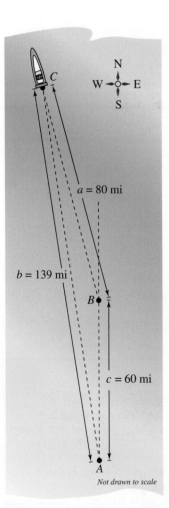

Not drawn to scale

Figure 6.9

Heron's Area Formula

The Law of Cosines can be used to establish a formula for the area of a triangle. This formula is called **Heron's Area Formula** after the Greek mathematician Heron (ca. 10–75 A.D.).

Heron's Area Formula

Given any triangle with sides of lengths a, b, and c, the area of the triangle is

$$\text{Area} = \sqrt{s(s-a)(s-b)(s-c)}$$

where

$$s = \frac{a+b+c}{2}.$$

For a proof of Heron's Area Formula, see Proofs in Mathematics on page 463.

EXAMPLE 5 **Using Heron's Area Formula**

Use Heron's Area Formula to find the area of a triangle with sides of lengths $a = 43$ meters, $b = 53$ meters, and $c = 72$ meters.

Solution First, determine that $s = (a+b+c)/2 = 168/2 = 84$. Then Heron's Area Formula yields

$$\begin{aligned} \text{Area} &= \sqrt{s(s-a)(s-b)(s-c)} \\ &= \sqrt{84(84-43)(84-53)(84-72)} \\ &= \sqrt{84(41)(31)(12)} \\ &\approx 1131.89 \text{ square meters.} \end{aligned}$$

✓ *Checkpoint*  *Audio-video solution in English & Spanish at LarsonPrecalculus.com*

Use Heron's Area Formula to find the area of a triangle with sides of lengths $a = 5$ inches, $b = 9$ inches, and $c = 8$ inches. ∎

You have now studied three different formulas for the area of a triangle.

Standard Formula: $\text{Area} = \dfrac{1}{2}bh$

Oblique Triangle: $\text{Area} = \dfrac{1}{2}bc \sin A = \dfrac{1}{2}ab \sin C = \dfrac{1}{2}ac \sin B$

Heron's Area Formula: $\text{Area} = \sqrt{s(s-a)(s-b)(s-c)}$

Summarize (Section 6.2)

1. State the Law of Cosines *(page 409)*. For examples of using the Law of Cosines to solve oblique triangles (SSS or SAS), see Examples 1 and 2.

2. Describe real-life applications of the Law of Cosines *(page 411, Examples 3 and 4)*.

3. State Heron's Area Formula *(page 412)*. For an example of using Heron's Area Formula to find the area of a triangle, see Example 5.

6.2 Exercises

See CalcChat.com for tutorial help and worked-out solutions to odd-numbered exercises.

Vocabulary: Fill in the blanks.

1. The standard form of the Law of Cosines for $\cos B = \dfrac{a^2 + c^2 - b^2}{2ac}$ is _____.

2. When solving an oblique triangle given three sides, use the _____ form of the Law of Cosines to solve for an angle.

3. When solving an oblique triangle given two sides and their included angle, use the _____ form of the Law of Cosines to solve for the remaining side.

4. The Law of Cosines can be used to establish a formula for the area of a triangle called _____ _____ Formula.

Skills and Applications

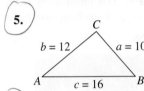 **Using the Law of Cosines** In Exercises 5–24, use the Law of Cosines to solve the triangle. Round your answers to two decimal places.

5.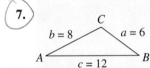

6.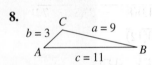

7.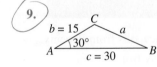

8.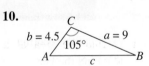

9.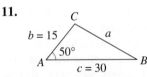

10.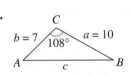

11.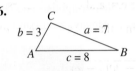

12.

13. $a = 11$, $b = 15$, $c = 21$
14. $a = 55$, $b = 25$, $c = 72$
15. $a = 2.5$, $b = 1.8$, $c = 0.9$
16. $a = 75.4$, $b = 52.5$, $c = 52.5$
17. $A = 120°$, $b = 6$, $c = 7$
18. $A = 48°$, $b = 3$, $c = 14$
19. $B = 10° 35'$, $a = 40$, $c = 30$
20. $B = 75° 20'$, $a = 9$, $c = 6$
21. $B = 125° 40'$, $a = 37$, $c = 37$
22. $C = 15° 15'$, $a = 7.45$, $b = 2.15$
23. $C = 43°$, $a = \frac{4}{9}$, $b = \frac{7}{9}$
24. $C = 101°$, $a = \frac{3}{8}$, $b = \frac{3}{4}$

Finding Measures in a Parallelogram In Exercises 25–30, find the missing values by solving the parallelogram shown in the figure. (The lengths of the diagonals are given by c and d.)

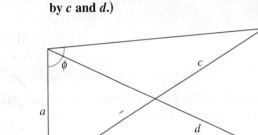

	a	b	c	d	θ	ϕ
25.	5	8			45°	
26.	25	35				120°
27.	10	14	20			
28.	40	60		80		
29.	15		25	20		
30.		25	50	35		

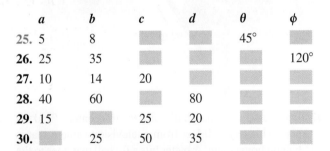 **Solving a Triangle** In Exercises 31–36, determine whether the Law of Cosines is needed to solve the triangle. Then solve (if possible) the triangle. If two solutions exist, find both. Round your answers to two decimal places.

31. $a = 8$, $c = 5$, $B = 40°$
32. $a = 10$, $b = 12$, $C = 70°$
33. $A = 24°$, $a = 4$, $b = 18$
34. $a = 11$, $b = 13$, $c = 7$
35. $A = 42°$, $B = 35°$, $c = 1.2$
36. $B = 12°$, $a = 160$, $b = 63$

Using Heron's Area Formula In Exercises 37–44, use Heron's Area Formula to find the area of the triangle.

37. $a = 6$, $b = 12$, $c = 17$

38. $a = 33$, $b = 36$, $c = 21$

39. $a = 2.5$, $b = 10.2$, $c = 8$

40. $a = 12.32$, $b = 8.46$, $c = 15.9$

41. $a = 1$, $b = \frac{1}{2}$, $c = \frac{5}{4}$

42. $a = \frac{3}{5}$, $b = \frac{4}{3}$, $c = \frac{7}{8}$

43. $A = 80°$, $b = 75$, $c = 41$

44. $C = 109°$, $a = 16$, $b = 3.5$

45. **Surveying** To approximate the length of a marsh, a surveyor walks 250 meters from point A to point B, then turns 75° and walks 220 meters to point C (see figure). Approximate the length AC of the marsh.

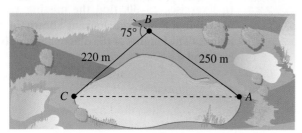

46. **Streetlight Design** Determine the angle θ in the design of the streetlight shown in the figure.

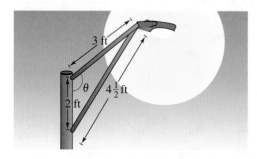

47. **Baseball** A baseball player in center field is approximately 330 feet from a television camera that is behind home plate. A batter hits a fly ball that goes to the wall 420 feet from the camera (see figure). The camera turns 8° to follow the play. Approximately how far does the center fielder have to run to make the catch?

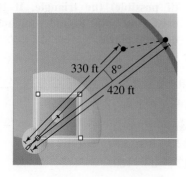

48. **Baseball** On a baseball diamond with 90-foot sides, the pitcher's mound is 60.5 feet from home plate. How far is the pitcher's mound from third base?

49. **Length** A 100-foot vertical tower is built on the side of a hill that makes a 6° angle with the horizontal (see figure). Find the length of each of the two guy wires that are anchored 75 feet uphill and downhill from the base of the tower.

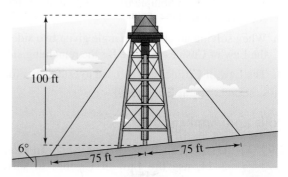

50. **Navigation** On a map, Minneapolis is 165 millimeters due west of Albany, Phoenix is 216 millimeters from Minneapolis, and Phoenix is 368 millimeters from Albany (see figure).

(a) Find the bearing of Minneapolis from Phoenix.

(b) Find the bearing of Albany from Phoenix.

51. **Navigation** A boat race runs along a triangular course marked by buoys A, B, and C. The race starts with the boats headed west for 3700 meters. The other two sides of the course lie to the north of the first side, and their lengths are 1700 meters and 3000 meters. Draw a diagram that gives a visual representation of the problem. Then find the bearings for the last two legs of the race.

52. **Air Navigation** A plane flies 810 miles from Franklin to Centerville with a bearing of 75°. Then it flies 648 miles from Centerville to Rosemount with a bearing of 32°. Draw a diagram that gives a visual representation of the problem. Then find the straight-line distance and bearing from Franklin to Rosemount.

53. **Surveying** A triangular parcel of land has 115 meters of frontage, and the other boundaries have lengths of 76 meters and 92 meters. What angles does the frontage make with the two other boundaries?

54. Surveying A triangular parcel of ground has sides of lengths 725 feet, 650 feet, and 575 feet. Find the measure of the largest angle.

55. Distance Two ships leave a port at 9 A.M. One travels at a bearing of N 53° W at 12 miles per hour, and the other travels at a bearing of S 67° W at s miles per hour.

(a) Use the Law of Cosines to write an equation that relates s and the distance d between the two ships at noon.

(b) Find the speed s that the second ship must travel so that the ships are 43 miles apart at noon.

• • **56. Mechanical Engineering** • • • • • • • • • • •

An engine has a seven-inch connecting rod fastened to a crank (see figure).

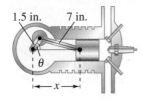

1.5 in.   7 in.

θ

x

(a) Use the Law of Cosines to write an equation giving the relationship between x and θ.

(b) Write x as a function of θ. (Select the sign that yields positive values of x.)

(c) Use a graphing utility to graph the function in part (b).

(d) Use the graph in part (c) to determine the total distance the piston moves in one cycle.

57. Geometry A triangular parcel of land has sides of lengths 200 feet, 500 feet, and 600 feet. Find the area of the parcel.

58. Geometry A parking lot has the shape of a parallelogram (see figure). The lengths of two adjacent sides are 70 meters and 100 meters. The angle between the two sides is 70°. What is the area of the parking lot?

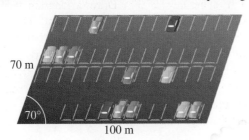

70 m

70°

100 m

59. Geometry You want to buy a triangular lot measuring 510 yards by 840 yards by 1120 yards. The price of the land is $2000 per acre. How much does the land cost? (*Hint:* 1 acre = 4840 square yards)

60. Geometry You want to buy a triangular lot measuring 1350 feet by 1860 feet by 2490 feet. The price of the land is $2200 per acre. How much does the land cost? (*Hint:* 1 acre = 43,560 square feet)

Exploration

True or False? In Exercises 61 and 62, determine whether the statement is true or false. Justify your answer.

61. In Heron's Area Formula, s is the average of the lengths of the three sides of the triangle.

62. In addition to SSS and SAS, the Law of Cosines can be used to solve triangles with AAS conditions.

63. Think About It What familiar formula do you obtain when you use the standard form of the Law of Cosines, $c^2 = a^2 + b^2 - 2ab \cos C$, and you let $C = 90°$? What is the relationship between the Law of Cosines and this formula?

64. Writing Describe how the Law of Cosines can be used to solve the ambiguous case of the oblique triangle ABC, where $a = 12$ feet, $b = 30$ feet, and $A = 20°$. Is the result the same as when the Law of Sines is used to solve the triangle? Describe the advantages and the disadvantages of each method.

65. Writing In Exercise 64, the Law of Cosines was used to solve a triangle in the two-solution case of SSA. Can the Law of Cosines be used to solve the no-solution and single-solution cases of SSA? Explain.

66. **HOW DO YOU SEE IT?** To solve the triangle, would you begin by using the Law of Sines or the Law of Cosines? Explain.

(a)

$b = 16$ C $a = 12$

A $c = 18$ B

(b)

A

$35°$ c

b

$55°$

C $a = 18$ B

67. Proof Use the Law of Cosines to prove each identity.

(a) $\dfrac{1}{2}bc(1 + \cos A) = \dfrac{a + b + c}{2} \cdot \dfrac{-a + b + c}{2}$

(b) $\dfrac{1}{2}bc(1 - \cos A) = \dfrac{a - b + c}{2} \cdot \dfrac{a + b - c}{2}$

6.3 Vectors in the Plane

- Represent vectors as directed line segments.
- Write component forms of vectors.
- Perform basic vector operations and represent vector operations graphically.
- Write vectors as linear combinations of unit vectors.
- Find direction angles of vectors.
- Use vectors to model and solve real-life problems.

Vectors are useful tools for modeling and solving real-life problems involving magnitude and direction. For instance, in Exercise 94 on page 428, you will use vectors to determine the speed and true direction of a commercial jet.

Introduction

Quantities such as force and velocity involve both *magnitude* and *direction* and cannot be completely characterized by a single real number. To represent such a quantity, you can use a **directed line segment,** as shown in Figure 6.10. The directed line segment $\overrightarrow{PQ}$ has **initial point** P and **terminal point** Q. Its **magnitude** (or **length**) is denoted by $\|\overrightarrow{PQ}\|$ and can be found using the Distance Formula.

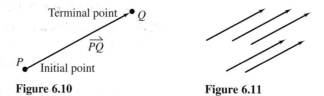

Figure 6.10   **Figure 6.11**

Two directed line segments that have the same magnitude and direction are *equivalent*. For example, the directed line segments in Figure 6.11 are all equivalent. The set of all directed line segments that are equivalent to the directed line segment $\overrightarrow{PQ}$ is a **vector v in the plane,** written $\mathbf{v} = \overrightarrow{PQ}$. Vectors are denoted by lowercase, boldface letters such as **u**, **v**, and **w**.

EXAMPLE 1   **Showing That Two Vectors Are Equivalent**

Show that **u** and **v** in Figure 6.12 are equivalent.

Solution   From the Distance Formula, $\overrightarrow{PQ}$ and $\overrightarrow{RS}$ have the *same magnitude*.

$$\|\overrightarrow{PQ}\| = \sqrt{(3-0)^2 + (2-0)^2} = \sqrt{13}$$
$$\|\overrightarrow{RS}\| = \sqrt{(4-1)^2 + (4-2)^2} = \sqrt{13}$$

Moreover, both line segments have the *same direction* because they are both directed toward the upper right on lines with a slope of

$$\frac{4-2}{4-1} = \frac{2-0}{3-0} = \frac{2}{3}.$$

Because $\overrightarrow{PQ}$ and $\overrightarrow{RS}$ have the same magnitude and direction, **u** and **v** are equivalent.

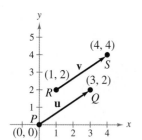

Figure 6.12

✓ *Checkpoint*   ))) *Audio-video solution in English & Spanish at LarsonPrecalculus.com*

Show that **u** and **v** in the figure at the right are equivalent.

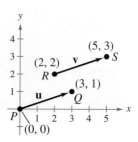

iStockphoto.com/Vladimir Maravic

Component Form of a Vector

The directed line segment whose initial point is the origin is often the most convenient representative of a set of equivalent directed line segments. This representative of the vector **v** is in **standard position.**

A vector whose initial point is the origin $(0, 0)$ can be uniquely represented by the coordinates of its terminal point (v_1, v_2). This is the **component form of a vector v,** written as $\mathbf{v} = \langle v_1, v_2 \rangle$. The coordinates v_1 and v_2 are the *components* of **v**. If both the initial point and the terminal point lie at the origin, then **v** is the **zero vector** and is denoted by $\mathbf{0} = \langle 0, 0 \rangle$.

▷ **TECHNOLOGY** Consult the user's guide for your graphing utility for specific instructions on how to use your graphing utility to graph vectors.

Component Form of a Vector

The component form of the vector with initial point $P(p_1, p_2)$ and terminal point $Q(q_1, q_2)$ is given by

$$\overrightarrow{PQ} = \langle q_1 - p_1, q_2 - p_2 \rangle = \langle v_1, v_2 \rangle = \mathbf{v}.$$

The **magnitude** (or **length**) of **v** is given by

$$\|\mathbf{v}\| = \sqrt{(q_1 - p_1)^2 + (q_2 - p_2)^2} = \sqrt{v_1^2 + v_2^2}.$$

If $\|\mathbf{v}\| = 1$, then **v** is a **unit vector.** Moreover, $\|\mathbf{v}\| = 0$ if and only if **v** is the zero vector **0**.

Two vectors $\mathbf{u} = \langle u_1, u_2 \rangle$ and $\mathbf{v} = \langle v_1, v_2 \rangle$ are *equal* if and only if $u_1 = v_1$ and $u_2 = v_2$. For instance, in Example 1, the vector **u** from $P(0, 0)$ to $Q(3, 2)$ is $\mathbf{u} = \overrightarrow{PQ} = \langle 3 - 0, 2 - 0 \rangle = \langle 3, 2 \rangle$, and the vector **v** from $R(1, 2)$ to $S(4, 4)$ is $\mathbf{v} = \overrightarrow{RS} = \langle 4 - 1, 4 - 2 \rangle = \langle 3, 2 \rangle$. So, the vectors **u** and **v** in Example 1 are equal.

EXAMPLE 2   **Finding the Component Form of a Vector**

Find the component form and magnitude of the vector **v** that has initial point $(4, -7)$ and terminal point $(-1, 5)$.

Algebraic Solution

Let

$$P(4, -7) = (p_1, p_2)$$

and

$$Q(-1, 5) = (q_1, q_2).$$

Then, the components of $\mathbf{v} = \langle v_1, v_2 \rangle$ are

$$v_1 = q_1 - p_1 = -1 - 4 = -5$$

$$v_2 = q_2 - p_2 = 5 - (-7) = 12.$$

So, $\mathbf{v} = \langle -5, 12 \rangle$ and the magnitude of **v** is

$$\|\mathbf{v}\| = \sqrt{(-5)^2 + 12^2}$$

$$= \sqrt{169}$$

$$= 13.$$

Graphical Solution

Use centimeter graph paper to plot the points $P(4, -7)$ and $Q(-1, 5)$. Carefully sketch the vector **v**. Use the sketch to find the components of $\mathbf{v} = \langle v_1, v_2 \rangle$. Then use a centimeter ruler to find the magnitude of **v**. The figure at the right shows that the components of **v** are $v_1 = -5$ and $v_2 = 12$, so $\mathbf{v} = \langle -5, 12 \rangle$. The figure also shows that the magnitude of **v** is $\|\mathbf{v}\| = 13$.

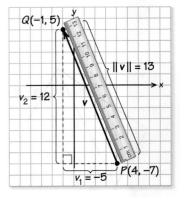

✓ **Checkpoint** 🔊))) *Audio-video solution in English & Spanish at LarsonPrecalculus.com*

Find the component form and magnitude of the vector **v** that has initial point $(-2, 3)$ and terminal point $(-7, 9)$.

•• REMARK   Recall from
Section 4.8 that in air navigation,
bearings are measured in
degrees clockwise from north.

v

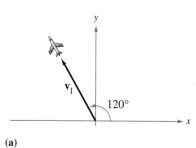

$\mathbf{v}_1$

120°

Figure 6.1

(a)

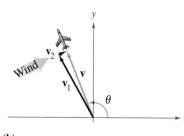

Wind $\mathbf{v}_2$ $\mathbf{v}$

$\mathbf{v}_1$

θ

(b)

Figure 6.24

Pilots can take advantage of
fast-moving air currents called jet
streams to decrease travel time.

$\mathbf{u} + (-\mathbf{v})$

$\mathbf{u} - \mathbf{v} =$

Figure 6.

EXAMPLE 11   **Using Vectors to Find Speed and Direction**

An airplane travels at a speed of 500 miles per hour with a bearing of 330° at a fixed
altitude with a negligible wind velocity, as shown in Figure 6.24(a). (Note that a bearing
of 330° corresponds to a direction angle of 120°.) The airplane encounters a wind with
a velocity of 70 miles per hour in the direction N 45° E, as shown in Figure 6.24(b).
What are the resultant speed and true direction of the airplane?

Solution   Using Figure 6.24, the velocity of the airplane (alone) is

$$\mathbf{v}_1 = 500\langle\cos 120°, \sin 120°\rangle = \langle-250, 250\sqrt{3}\rangle$$

and the velocity of the wind is

$$\mathbf{v}_2 = 70\langle\cos 45°, \sin 45°\rangle = \langle35\sqrt{2}, 35\sqrt{2}\rangle.$$

So, the velocity of the airplane (in the wind) is

$$\mathbf{v} = \mathbf{v}_1 + \mathbf{v}_2$$
$$= \langle-250 + 35\sqrt{2}, 250\sqrt{3} + 35\sqrt{2}\rangle$$
$$\approx \langle-200.5, 482.5\rangle$$

and the resultant speed of the airplane is

$$\|\mathbf{v}\| \approx \sqrt{(-200.5)^2 + (482.5)^2} \approx 522.5 \text{ miles per hour.}$$

To find the direction angle θ of the flight path, you have

$$\tan\theta \approx \frac{482.5}{-200.5} \approx -2.4065.$$

The flight path lies in Quadrant II, so θ lies in Quadrant II, and its reference angle is

$$\theta' \approx |\arctan(-2.4065)| \approx |-1.1770 \text{ radians}| \approx |-67.44°| = 67.44°.$$

So, the direction angle is $\theta \approx 180° - 67.44° = 112.56°$, and the true direction of the
airplane is approximately $270° + (180° - 112.56°) = 337.44°$.

✓ *Checkpoint* 🔊))) *Audio-video solution in English & Spanish at LarsonPrecalculus.com*

Repeat Example 11 for an airplane traveling at a speed of 450 miles per hour with
a bearing of 300° that encounters a wind with a velocity of 40 miles per hour in the
direction N 30° E. ∎

Summarize   **(Section 6.3)**

1. Explain how to represent a vector as a directed line segment *(page 416)*.
   For an example involving vectors represented as directed line segments, see
   Example 1.

2. Explain how to find the component form of a vector *(page 417)*. For an
   example of finding the component form of a vector, see Example 2.

3. Explain how to perform basic vector operations *(page 418)*. For an example
   of performing basic vector operations, see Example 3.

4. Explain how to write a vector as a linear combination of unit vectors
   (page 420). For examples involving unit vectors, see Examples 5–7.

5. Explain how to find the direction angle of a vector *(page 422)*. For an
   example of finding direction angles of vectors, see Example 8.

6. Describe real-life applications of vectors *(pages 423 and 424, Examples 9–11)*.

Component Form of a Vector

The directed line segment whose initial point is the origin is often the most convenient representative of a set of equivalent directed line segments. This representative of the vector **v** is in **standard position.**

A vector whose initial point is the origin $(0, 0)$ can be uniquely represented by the coordinates of its terminal point (v_1, v_2). This is the **component form of a vector v,** written as $\mathbf{v} = \langle v_1, v_2 \rangle$. The coordinates v_1 and v_2 are the *components* of **v**. If both the initial point and the terminal point lie at the origin, then **v** is the **zero vector** and is denoted by $\mathbf{0} = \langle 0, 0 \rangle$.

▷ **TECHNOLOGY** Consult the user's guide for your graphing utility for specific instructions on how to use your graphing utility to graph vectors.

Component Form of a Vector

The component form of the vector with initial point $P(p_1, p_2)$ and terminal point $Q(q_1, q_2)$ is given by

$$\overrightarrow{PQ} = \langle q_1 - p_1, q_2 - p_2 \rangle = \langle v_1, v_2 \rangle = \mathbf{v}.$$

The **magnitude** (or **length**) of **v** is given by

$$\|\mathbf{v}\| = \sqrt{(q_1 - p_1)^2 + (q_2 - p_2)^2} = \sqrt{v_1^2 + v_2^2}.$$

If $\|\mathbf{v}\| = 1$, then **v** is a **unit vector.** Moreover, $\|\mathbf{v}\| = 0$ if and only if **v** is the zero vector **0**.

Two vectors $\mathbf{u} = \langle u_1, u_2 \rangle$ and $\mathbf{v} = \langle v_1, v_2 \rangle$ are *equal* if and only if $u_1 = v_1$ and $u_2 = v_2$. For instance, in Example 1, the vector **u** from $P(0, 0)$ to $Q(3, 2)$ is $\mathbf{u} = \overrightarrow{PQ} = \langle 3 - 0, 2 - 0 \rangle = \langle 3, 2 \rangle$, and the vector **v** from $R(1, 2)$ to $S(4, 4)$ is $\mathbf{v} = \overrightarrow{RS} = \langle 4 - 1, 4 - 2 \rangle = \langle 3, 2 \rangle$. So, the vectors **u** and **v** in Example 1 are equal.

EXAMPLE 2   **Finding the Component Form of a Vector**

Find the component form and magnitude of the vector **v** that has initial point $(4, -7)$ and terminal point $(-1, 5)$.

Algebraic Solution

Let

$$P(4, -7) = (p_1, p_2)$$

and

$$Q(-1, 5) = (q_1, q_2).$$

Then, the components of $\mathbf{v} = \langle v_1, v_2 \rangle$ are

$$v_1 = q_1 - p_1 = -1 - 4 = -5$$

$$v_2 = q_2 - p_2 = 5 - (-7) = 12.$$

So, $\mathbf{v} = \langle -5, 12 \rangle$ and the magnitude of **v** is

$$\|\mathbf{v}\| = \sqrt{(-5)^2 + 12^2}$$

$$= \sqrt{169}$$

$$= 13.$$

Graphical Solution

Use centimeter graph paper to plot the points $P(4, -7)$ and $Q(-1, 5)$. Carefully sketch the vector **v**. Use the sketch to find the components of $\mathbf{v} = \langle v_1, v_2 \rangle$. Then use a centimeter ruler to find the magnitude of **v**. The figure at the right shows that the components of **v** are $v_1 = -5$ and $v_2 = 12$, so $\mathbf{v} = \langle -5, 12 \rangle$. The figure also shows that the magnitude of **v** is $\|\mathbf{v}\| = 13$.

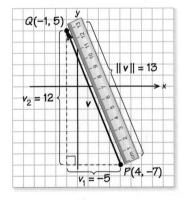

✓ *Checkpoint* ◀))) *Audio-video solution in English & Spanish at LarsonPrecalculus.com*

Find the component form and magnitude of the vector **v** that has initial point $(-2, 3)$ and terminal point $(-7, 9)$.

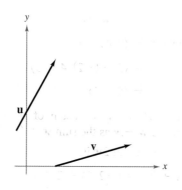

Figure 6.13

Vector Operations

The two basic vector operations are **scalar multiplication** and **vector addition.** In operations with vectors, numbers are usually referred to as **scalars.** In this text, scalars will always be real numbers. Geometrically, the product of a vector **v** and a scalar k is the vector that is $|k|$ times as long as **v.** When k is positive, k**v** has the same direction as **v**, and when k is negative, k**v** has the direction opposite that of **v**, as shown in Figure 6.13.

To add two vectors **u** and **v** geometrically, first position them (without changing their lengths or directions) so that the initial point of the second vector **v** coincides with the terminal point of the first vector **u.** The sum **u** + **v** is the vector formed by joining the initial point of the first vector **u** with the terminal point of the second vector **v**, as shown in the next two figures. This technique is called the **parallelogram law** for vector addition because the vector **u** + **v**, often called the **resultant** of vector addition, is the diagonal of a parallelogram with adjacent sides **u** and **v.**

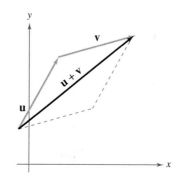

Definitions of Vector Addition and Scalar Multiplication

Let $\mathbf{u} = \langle u_1, u_2 \rangle$ and $\mathbf{v} = \langle v_1, v_2 \rangle$ be vectors and let k be a scalar (a real number). Then the **sum** of **u** and **v** is the vector

$$\mathbf{u} + \mathbf{v} = \langle u_1 + v_1, u_2 + v_2 \rangle \qquad \text{Sum}$$

and the **scalar multiple** of k times **u** is the vector

$$k\mathbf{u} = k\langle u_1, u_2 \rangle = \langle ku_1, ku_2 \rangle. \qquad \text{Scalar multiple}$$

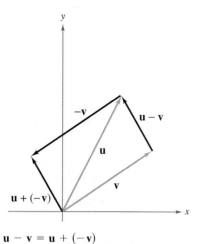

$$\mathbf{u} - \mathbf{v} = \mathbf{u} + (-\mathbf{v})$$
Figure 6.14

The **negative** of $\mathbf{v} = \langle v_1, v_2 \rangle$ is

$$-\mathbf{v} = (-1)\mathbf{v}$$
$$= \langle -v_1, -v_2 \rangle \qquad \text{Negative}$$

and the **difference** of **u** and **v** is

$$\mathbf{u} - \mathbf{v} = \mathbf{u} + (-\mathbf{v}) \qquad \text{Add } (-\mathbf{v}). \text{ See Figure 6.14.}$$
$$= \langle u_1 - v_1, u_2 - v_2 \rangle. \qquad \text{Difference}$$

To represent **u** − **v** geometrically, use directed line segments with the *same* initial point. The difference **u** − **v** is the vector from the terminal point of **v** to the terminal point of **u**, which is equal to

$$\mathbf{u} + (-\mathbf{v})$$

as shown in Figure 6.14.

Example 3 illustrates the component definitions of vector addition and scalar multiplication. In this example, note the geometrical interpretations of each of the vector operations.

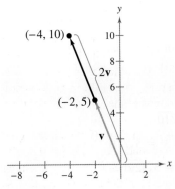

Figure 6.15

EXAMPLE 3 Vector Operations

See LarsonPrecalculus.com for an interactive version of this type of example.

Let $\mathbf{v} = \langle -2, 5 \rangle$ and $\mathbf{w} = \langle 3, 4 \rangle$. Find each vector.

a. $2\mathbf{v}$ **b.** $\mathbf{w} - \mathbf{v}$ **c.** $\mathbf{v} + 2\mathbf{w}$

Solution

a. Multiplying $\mathbf{v} = \langle -2, 5 \rangle$ by the scalar 2, you have

$$2\mathbf{v} = 2\langle -2, 5 \rangle$$
$$= \langle 2(-2), 2(5) \rangle$$
$$= \langle -4, 10 \rangle.$$

Figure 6.15 shows a sketch of $2\mathbf{v}$.

b. The difference of $\mathbf{w}$ and $\mathbf{v}$ is

$$\mathbf{w} - \mathbf{v} = \langle 3, 4 \rangle - \langle -2, 5 \rangle$$
$$= \langle 3 - (-2), 4 - 5 \rangle$$
$$= \langle 5, -1 \rangle.$$

Figure 6.16 shows a sketch of $\mathbf{w} - \mathbf{v}$. Note that the figure shows the vector difference $\mathbf{w} - \mathbf{v}$ as the sum $\mathbf{w} + (-\mathbf{v})$.

c. The sum of $\mathbf{v}$ and $2\mathbf{w}$ is

$$\mathbf{v} + 2\mathbf{w} = \langle -2, 5 \rangle + 2\langle 3, 4 \rangle$$
$$= \langle -2, 5 \rangle + \langle 2(3), 2(4) \rangle$$
$$= \langle -2, 5 \rangle + \langle 6, 8 \rangle$$
$$= \langle -2 + 6, 5 + 8 \rangle$$
$$= \langle 4, 13 \rangle.$$

Figure 6.17 shows a sketch of $\mathbf{v} + 2\mathbf{w}$.

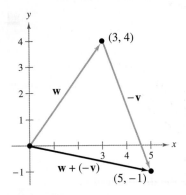

Figure 6.16

✓ *Checkpoint* 🔊))) *Audio-video solution in English & Spanish at LarsonPrecalculus.com*

Let $\mathbf{u} = \langle 1, 4 \rangle$ and $\mathbf{v} = \langle 3, 2 \rangle$. Find each vector.

a. $\mathbf{u} + \mathbf{v}$ **b.** $\mathbf{u} - \mathbf{v}$ **c.** $2\mathbf{u} - 3\mathbf{v}$

Vector addition and scalar multiplication share many of the properties of ordinary arithmetic.

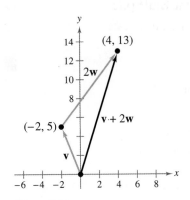

Figure 6.17

Properties of Vector Addition and Scalar Multiplication

Let $\mathbf{u}$, $\mathbf{v}$, and $\mathbf{w}$ be vectors and let c and d be scalars. Then the properties listed below are true.

1. $\mathbf{u} + \mathbf{v} = \mathbf{v} + \mathbf{u}$
2. $(\mathbf{u} + \mathbf{v}) + \mathbf{w} = \mathbf{u} + (\mathbf{v} + \mathbf{w})$
3. $\mathbf{u} + \mathbf{0} = \mathbf{u}$
4. $\mathbf{u} + (-\mathbf{u}) = \mathbf{0}$
5. $c(d\mathbf{u}) = (cd)\mathbf{u}$
6. $(c + d)\mathbf{u} = c\mathbf{u} + d\mathbf{u}$
7. $c(\mathbf{u} + \mathbf{v}) = c\mathbf{u} + c\mathbf{v}$
8. $1(\mathbf{u}) = \mathbf{u}$, $0(\mathbf{u}) = \mathbf{0}$
9. $\|c\mathbf{v}\| = |c|\|\mathbf{v}\|$

• • **REMARK** Property 9 can be stated as: The magnitude of a scalar multiple $c\mathbf{v}$ is the absolute value of c times the magnitude of $\mathbf{v}$.

William Rowan Hamilton (1805–1865), an Irish mathematician, did some of the earliest work with vectors. Hamilton spent many years developing a system of vector-like quantities called quaternions. Although Hamilton was convinced of the benefits of quaternions, the operations he defined did not produce good models for physical phenomena. It was not until the latter half of the nineteenth century that the Scottish physicist James Maxwell (1831–1879) restructured Hamilton's quaternions in a form that is useful for representing physical quantities such as force, velocity, and acceleration.

EXAMPLE 4 Finding the Magnitude of a Scalar Multiple

Let $\mathbf{u} = \langle 1, 3 \rangle$ and $\mathbf{v} = \langle -2, 5 \rangle$. Find the magnitude of each scalar multiple.

a. $\|2\mathbf{u}\|$ **b.** $\|-5\mathbf{u}\|$ **c.** $\|3\mathbf{v}\|$

Solution

a. $\|2\mathbf{u}\| = |2|\|\mathbf{u}\| = |2|\|\langle 1, 3 \rangle\| = |2|\sqrt{1^2 + 3^2} = 2\sqrt{10}$

b. $\|-5\mathbf{u}\| = |-5|\|\mathbf{u}\| = |-5|\|\langle 1, 3 \rangle\| = |-5|\sqrt{1^2 + 3^2} = 5\sqrt{10}$

c. $\|3\mathbf{v}\| = |3|\|\mathbf{v}\| = |3|\|\langle -2, 5 \rangle\| = |3|\sqrt{(-2)^2 + 5^2} = 3\sqrt{29}$

✓ *Checkpoint* 🔊))) *Audio-video solution in English & Spanish at LarsonPrecalculus.com*

Let $\mathbf{u} = \langle 4, -1 \rangle$ and $\mathbf{v} = \langle 3, 2 \rangle$. Find the magnitude of each scalar multiple.

a. $\|3\mathbf{u}\|$ **b.** $\|-2\mathbf{v}\|$ **c.** $\|5\mathbf{v}\|$

Unit Vectors

In many applications of vectors, it is useful to find a unit vector that has the same direction as a given nonzero vector $\mathbf{v}$. To do this, divide $\mathbf{v}$ by its magnitude to obtain

$$\mathbf{u} = \text{unit vector} = \frac{\mathbf{v}}{\|\mathbf{v}\|} = \left(\frac{1}{\|\mathbf{v}\|}\right)\mathbf{v}. \qquad \text{Unit vector in direction of } \mathbf{v}$$

Note that $\mathbf{u}$ is a scalar multiple of $\mathbf{v}$. The vector $\mathbf{u}$ has a magnitude of 1 and the same direction as $\mathbf{v}$. The vector $\mathbf{u}$ is called a **unit vector in the direction of v.**

EXAMPLE 5 Finding a Unit Vector

Find a unit vector $\mathbf{u}$ in the direction of $\mathbf{v} = \langle -2, 5 \rangle$. Verify that $\|\mathbf{u}\| = 1$.

Solution The unit vector $\mathbf{u}$ in the direction of $\mathbf{v}$ is

$$\frac{\mathbf{v}}{\|\mathbf{v}\|} = \frac{\langle -2, 5 \rangle}{\sqrt{(-2)^2 + 5^2}} = \frac{1}{\sqrt{29}}\langle -2, 5 \rangle = \left\langle \frac{-2}{\sqrt{29}}, \frac{5}{\sqrt{29}} \right\rangle.$$

This vector has a magnitude of 1 because

$$\sqrt{\left(\frac{-2}{\sqrt{29}}\right)^2 + \left(\frac{5}{\sqrt{29}}\right)^2} = \sqrt{\frac{4}{29} + \frac{25}{29}} = \sqrt{\frac{29}{29}} = 1.$$

✓ *Checkpoint* 🔊))) *Audio-video solution in English & Spanish at LarsonPrecalculus.com*

Find a unit vector $\mathbf{u}$ in the direction of $\mathbf{v} = \langle 6, -1 \rangle$. Verify that $\|\mathbf{u}\| = 1$.

The unit vectors $\langle 1, 0 \rangle$ and $\langle 0, 1 \rangle$ are the **standard unit vectors** and are denoted by

$$\mathbf{i} = \langle 1, 0 \rangle \quad \text{and} \quad \mathbf{j} = \langle 0, 1 \rangle$$

as shown in Figure 6.18. (Note that the lowercase letter $\mathbf{i}$ is in boldface and not italicized to distinguish it from the imaginary unit $i = \sqrt{-1}$.) These vectors can be used to represent any vector $\mathbf{v} = \langle v_1, v_2 \rangle$, because

$$\mathbf{v} = \langle v_1, v_2 \rangle = v_1\langle 1, 0 \rangle + v_2\langle 0, 1 \rangle = v_1\mathbf{i} + v_2\mathbf{j}.$$

The scalars v_1 and v_2 are the **horizontal** and **vertical components of v,** respectively. The vector sum $v_1\mathbf{i} + v_2\mathbf{j}$ is a **linear combination** of the vectors $\mathbf{i}$ and $\mathbf{j}$. Any vector in the plane can be written as a linear combination of the standard unit vectors $\mathbf{i}$ and $\mathbf{j}$.

Figure 6.18

$\mathbf{j} = \langle 0, 1 \rangle$

$\mathbf{i} = \langle 1, 0 \rangle$

EXAMPLE 6 **Writing a Linear Combination of Unit Vectors**

Let **u** be the vector with initial point $(2, -5)$ and terminal point $(-1, 3)$. Write **u** as a linear combination of the standard unit vectors **i** and **j**.

Solution Begin by writing the component form of the vector **u**. Then write the component form in terms of **i** and **j**.

$$\mathbf{u} = \langle -1 - 2, 3 - (-5) \rangle = \langle -3, 8 \rangle = -3\mathbf{i} + 8\mathbf{j}$$

This result is shown graphically below.

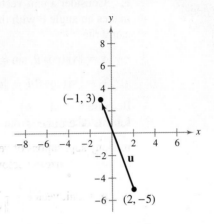

✓ Checkpoint ◀))) *Audio-video solution in English & Spanish at LarsonPrecalculus.com*

Let **u** be the vector with initial point $(-2, 6)$ and terminal point $(-8, 3)$. Write **u** as a linear combination of the standard unit vectors **i** and **j**.

EXAMPLE 7 **Vector Operations**

Let $\mathbf{u} = -3\mathbf{i} + 8\mathbf{j}$ and $\mathbf{v} = 2\mathbf{i} - \mathbf{j}$. Find $2\mathbf{u} - 3\mathbf{v}$.

Solution It is not necessary to convert **u** and **v** to component form to solve this problem. Just perform the operations with the vectors in unit vector form.

$$2\mathbf{u} - 3\mathbf{v} = 2(-3\mathbf{i} + 8\mathbf{j}) - 3(2\mathbf{i} - \mathbf{j})$$
$$= -6\mathbf{i} + 16\mathbf{j} - 6\mathbf{i} + 3\mathbf{j}$$
$$= -12\mathbf{i} + 19\mathbf{j}$$

✓ Checkpoint ◀))) *Audio-video solution in English & Spanish at LarsonPrecalculus.com*

Let $\mathbf{u} = \mathbf{i} - 2\mathbf{j}$ and $\mathbf{v} = -3\mathbf{i} + 2\mathbf{j}$. Find $5\mathbf{u} - 2\mathbf{v}$.

In Example 7, you could perform the operations in component form by writing

$$\mathbf{u} = -3\mathbf{i} + 8\mathbf{j} = \langle -3, 8 \rangle \quad \text{and} \quad \mathbf{v} = 2\mathbf{i} - \mathbf{j} = \langle 2, -1 \rangle.$$

The difference of 2**u** and 3**v** is

$$2\mathbf{u} - 3\mathbf{v} = 2\langle -3, 8 \rangle - 3\langle 2, -1 \rangle$$
$$= \langle -6, 16 \rangle - \langle 6, -3 \rangle$$
$$= \langle -6 - 6, 16 - (-3) \rangle$$
$$= \langle -12, 19 \rangle.$$

Compare this result with the solution to Example 7.

Direction Angles

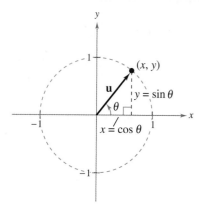

$\|\mathbf{u}\| = 1$
Figure 6.19

If **u** is a unit vector such that θ is the angle (measured counterclockwise) from the positive x-axis to **u**, then the terminal point of **u** lies on the unit circle and you have

$$\mathbf{u} = \langle x, y \rangle$$
$$= \langle \cos \theta, \sin \theta \rangle$$
$$= (\cos \theta)\mathbf{i} + (\sin \theta)\mathbf{j}$$

as shown in Figure 6.19. The angle θ is the **direction angle** of the vector **u**.

Consider a unit vector **u** with direction angle θ. If $\mathbf{v} = a\mathbf{i} + b\mathbf{j}$ is any vector that makes an angle θ with the positive x-axis, then it has the same direction as **u** and you can write

$$\mathbf{v} = \|\mathbf{v}\|\langle \cos \theta, \sin \theta \rangle$$
$$= \|\mathbf{v}\|(\cos \theta)\mathbf{i} + \|\mathbf{v}\|(\sin \theta)\mathbf{j}.$$

Because $\mathbf{v} = a\mathbf{i} + b\mathbf{j} = \|\mathbf{v}\|(\cos \theta)\mathbf{i} + \|\mathbf{v}\|(\sin \theta)\mathbf{j}$, it follows that the direction angle θ for **v** is determined from

$$\tan \theta = \frac{\sin \theta}{\cos \theta} \qquad \text{Quotient identity}$$

$$= \frac{\|\mathbf{v}\| \sin \theta}{\|\mathbf{v}\| \cos \theta} \qquad \text{Multiply numerator and denominator by } \|\mathbf{v}\|.$$

$$= \frac{b}{a}. \qquad \text{Simplify.}$$

EXAMPLE 8   **Finding Direction Angles of Vectors**

Find the direction angle of each vector.

a. $\mathbf{u} = 3\mathbf{i} + 3\mathbf{j}$ **b.** $\mathbf{v} = 3\mathbf{i} - 4\mathbf{j}$

Solution

a. The direction angle is determined from

$$\tan \theta = \frac{b}{a} = \frac{3}{3} = 1.$$

So, $\theta = 45°$, as shown in Figure 6.20.

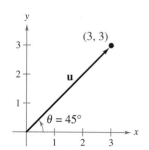

Figure 6.20

b. The direction angle is determined from

$$\tan \theta = \frac{b}{a} = \frac{-4}{3}.$$

Moreover, $\mathbf{v} = 3\mathbf{i} - 4\mathbf{j}$ lies in Quadrant IV, so θ lies in Quadrant IV, and its reference angle is

$$\theta' = \left| \arctan\left(-\frac{4}{3} \right) \right| \approx |-0.9273 \text{ radian}| \approx |-53.13°| = 53.13°.$$

It follows that $\theta \approx 360° - 53.13° = 306.87°$, as shown in Figure 6.21.

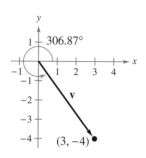

Figure 6.21

✓ *Checkpoint* 🔊)) *Audio-video solution in English & Spanish at LarsonPrecalculus.com*

Find the direction angle of each vector.

a. $\mathbf{v} = -6\mathbf{i} + 6\mathbf{j}$ **b.** $\mathbf{v} = -7\mathbf{i} - 4\mathbf{j}$

Applications

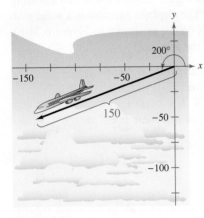

Figure 6.22

EXAMPLE 9 **Finding the Component Form of a Vector**

Find the component form of the vector that represents the velocity of an airplane descending at a speed of 150 miles per hour at an angle 20° below the horizontal, as shown in Figure 6.22.

Solution The velocity vector **v** has a magnitude of 150 and a direction angle of $\theta = 200°$.

$$\mathbf{v} = \|\mathbf{v}\|(\cos\theta)\mathbf{i} + \|\mathbf{v}\|(\sin\theta)\mathbf{j}$$
$$= 150(\cos 200°)\mathbf{i} + 150(\sin 200°)\mathbf{j}$$
$$\approx 150(-0.9397)\mathbf{i} + 150(-0.3420)\mathbf{j}$$
$$\approx -140.96\mathbf{i} - 52.30\mathbf{j}$$
$$= \langle -140.96, -52.30 \rangle$$

Check that **v** has a magnitude of 150.

$$\|\mathbf{v}\| \approx \sqrt{(-140.96)^2 + (-51.30)^2} \approx \sqrt{22,501.41} \approx 150 \qquad \text{Solution checks.}$$

✓ **Checkpoint** 🔊))) *Audio-video solution in English & Spanish at LarsonPrecalculus.com*

Find the component form of the vector that represents the velocity of an airplane descending at a speed of 100 miles per hour at an angle 15° below the horizontal ($\theta = 195°$).

EXAMPLE 10 **Using Vectors to Determine Weight**

A force of 600 pounds is required to pull a boat and trailer up a ramp inclined at 15° from the horizontal. Find the combined weight of the boat and trailer.

Solution Use Figure 6.23 to make the observations below.

$\|\overrightarrow{BA}\|$ = force of gravity = combined weight of boat and trailer

$\|\overrightarrow{BC}\|$ = force against ramp

$\|\overrightarrow{AC}\|$ = force required to move boat up ramp = 600 pounds

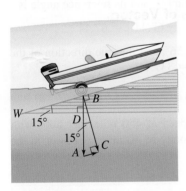

Figure 6.23

Note that $\overrightarrow{AC}$ is parallel to the ramp. So, by construction, triangles BWD and ABC are similar and angle ABC is 15°. In triangle ABC, you have

$$\sin 15° = \frac{\|\overrightarrow{AC}\|}{\|\overrightarrow{BA}\|}$$

$$\sin 15° = \frac{600}{\|\overrightarrow{BA}\|}$$

$$\|\overrightarrow{BA}\| = \frac{600}{\sin 15°}$$

$$\|\overrightarrow{BA}\| \approx 2318.$$

So, the combined weight is approximately 2318 pounds.

✓ **Checkpoint** 🔊))) *Audio-video solution in English & Spanish at LarsonPrecalculus.com*

A force of 500 pounds is required to pull a boat and trailer up a ramp inclined at 12° from the horizontal. Find the combined weight of the boat and trailer. ◼

•• **REMARK** Recall from
Section 4.8 that in air navigation,
bearings are measured in
degrees clockwise from north.

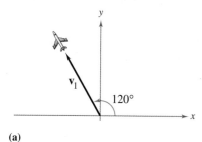

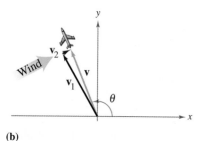

(a)

(b)

Figure 6.24

Pilots can take advantage of
fast-moving air currents called jet
streams to decrease travel time.

EXAMPLE 11 **Using Vectors to Find Speed and Direction**

An airplane travels at a speed of 500 miles per hour with a bearing of 330° at a fixed altitude with a negligible wind velocity, as shown in Figure 6.24(a). (Note that a bearing of 330° corresponds to a direction angle of 120°.) The airplane encounters a wind with a velocity of 70 miles per hour in the direction N 45° E, as shown in Figure 6.24(b). What are the resultant speed and true direction of the airplane?

Solution Using Figure 6.24, the velocity of the airplane (alone) is

$$\mathbf{v}_1 = 500\langle \cos 120°, \sin 120° \rangle = \langle -250, 250\sqrt{3} \rangle$$

and the velocity of the wind is

$$\mathbf{v}_2 = 70\langle \cos 45°, \sin 45° \rangle = \langle 35\sqrt{2}, 35\sqrt{2} \rangle.$$

So, the velocity of the airplane (in the wind) is

$$\mathbf{v} = \mathbf{v}_1 + \mathbf{v}_2$$
$$= \langle -250 + 35\sqrt{2}, 250\sqrt{3} + 35\sqrt{2} \rangle$$
$$\approx \langle -200.5, 482.5 \rangle$$

and the resultant speed of the airplane is

$$\|\mathbf{v}\| \approx \sqrt{(-200.5)^2 + (482.5)^2} \approx 522.5 \text{ miles per hour.}$$

To find the direction angle θ of the flight path, you have

$$\tan \theta \approx \frac{482.5}{-200.5} \approx -2.4065.$$

The flight path lies in Quadrant II, so θ lies in Quadrant II, and its reference angle is

$$\theta' \approx |\arctan(-2.4065)| \approx |-1.1770 \text{ radians}| \approx |-67.44°| = 67.44°.$$

So, the direction angle is $\theta \approx 180° - 67.44° = 112.56°$, and the true direction of the airplane is approximately $270° + (180° - 112.56°) = 337.44°$.

✓ *Checkpoint*  *Audio-video solution in English & Spanish at LarsonPrecalculus.com*

Repeat Example 11 for an airplane traveling at a speed of 450 miles per hour with a bearing of 300° that encounters a wind with a velocity of 40 miles per hour in the direction N 30° E.

Summarize (Section 6.3)

1. Explain how to represent a vector as a directed line segment (*page 416*). For an example involving vectors represented as directed line segments, see Example 1.

2. Explain how to find the component form of a vector (*page 417*). For an example of finding the component form of a vector, see Example 2.

3. Explain how to perform basic vector operations (*page 418*). For an example of performing basic vector operations, see Example 3.

4. Explain how to write a vector as a linear combination of unit vectors (*page 420*). For examples involving unit vectors, see Examples 5–7.

5. Explain how to find the direction angle of a vector (*page 422*). For an example of finding direction angles of vectors, see Example 8.

6. Describe real-life applications of vectors (*pages 423 and 424, Examples 9–11*).

6.3 Exercises

Vocabulary: Fill in the blanks.

1. You can use a _____ _____ _____ to represent a quantity that involves both magnitude and direction.

2. The directed line segment $\overrightarrow{PQ}$ has _____ point P and _____ point Q.

3. The set of all directed line segments that are equivalent to a given directed line segment $\overrightarrow{PQ}$ is a _____ **v** in the plane.

4. Two vectors are equivalent when they have the same _____ and the same _____.

5. The directed line segment whose initial point is the origin is in _____ _____.

6. A vector that has a magnitude of 1 is a _____ _____.

7. The two basic vector operations are scalar _____ and vector _____.

8. The vector sum $v_1\mathbf{i} + v_2\mathbf{j}$ is a _____ _____ of the vectors **i** and **j**, and the scalars v_1 and v_2 are the _____ and _____ components of **v**, respectively.

Skills and Applications

Determining Whether Two Vectors Are Equivalent In Exercises 9–14, determine whether **u** and **v** are equivalent. Explain.

9.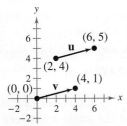

10.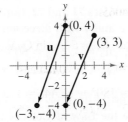

Vector	Initial Point	Terminal Point
11. **u**	(2, 2)	(−1, 4)
v	(−3, −1)	(−5, 2)
12. **u**	(2, 0)	(7, 4)
v	(−8, 1)	(2, 9)
13. **u**	(2, −1)	(5, −10)
v	(6, 1)	(9, −8)
14. **u**	(8, 1)	(13, −1)
v	(−2, 4)	(−7, 6)

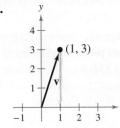

Finding the Component Form of a Vector In Exercises 15–24, find the component form and magnitude of the vector **v**.

15.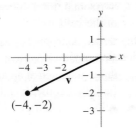

16.

17.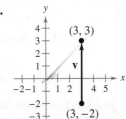

18.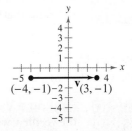

Initial Point	Terminal Point
19. (−3, −5)	(−11, 1)
20. (−2, 7)	(5, −17)
21. (1, 3)	(−8, −9)
22. (17, −5)	(9, 3)
23. (−1, 5)	(15, −21)
24. (−3, 11)	(9, 40)

Sketching the Graph of a Vector In Exercises 25–30, use the figure to sketch a graph of the specified vector. To print an enlarged copy of the graph, go to *MathGraphs.com*.

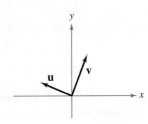

25. −**v**

26. 5**v**

27. **u** + **v**

28. **u** + 2**v**

29. **u** − **v**

30. **v** − $\frac{1}{2}$**u**

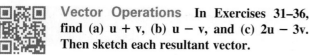

Vector Operations In Exercises 31–36, find (a) **u** + **v**, (b) **u** − **v**, and (c) 2**u** − 3**v**. Then sketch each resultant vector.

31. $\mathbf{u} = \langle 2, 1 \rangle$, $\mathbf{v} = \langle 1, 3 \rangle$
32. $\mathbf{u} = \langle 2, 3 \rangle$, $\mathbf{v} = \langle 4, 0 \rangle$
33. $\mathbf{u} = \langle -5, 3 \rangle$, $\mathbf{v} = \langle 0, 0 \rangle$
34. $\mathbf{u} = \langle 0, 0 \rangle$, $\mathbf{v} = \langle 2, 1 \rangle$
35. $\mathbf{u} = \langle 0, -7 \rangle$, $\mathbf{v} = \langle 1, -2 \rangle$
36. $\mathbf{u} = \langle -3, 1 \rangle$, $\mathbf{v} = \langle 2, -5 \rangle$

Finding the Magnitude of a Scalar Multiple In Exercises 37–40, find the magnitude of the scalar multiple, where $\mathbf{u} = \langle 2, 0 \rangle$ and $\mathbf{v} = \langle -3, 6 \rangle$.

37. $\|5\mathbf{u}\|$
38. $\|4\mathbf{v}\|$
39. $\|-3\mathbf{v}\|$
40. $\left\|-\frac{3}{4}\mathbf{u}\right\|$

Finding a Unit Vector In Exercises 41–46, find a unit vector **u** in the direction of **v**. Verify that $\|\mathbf{u}\| = 1$.

41. $\mathbf{v} = \langle 3, 0 \rangle$
42. $\mathbf{v} = \langle 0, -2 \rangle$
43. $\mathbf{v} = \langle -2, 2 \rangle$
44. $\mathbf{v} = \langle -5, 12 \rangle$
45. $\mathbf{v} = \langle 1, -6 \rangle$
46. $\mathbf{v} = \langle -8, -4 \rangle$

Finding a Vector In Exercises 47–50, find the vector **v** with the given magnitude and the same direction as **u**.

$\sqrt{9+16} = \sqrt{25}$
5

47. $\|\mathbf{v}\| = 10$, $\mathbf{u} = \langle -3, 4 \rangle$
48. $\|\mathbf{v}\| = 3$, $\mathbf{u} = \langle -12, -5 \rangle$
49. $\|\mathbf{v}\| = 9$, $\mathbf{u} = \langle 2, 5 \rangle$
50. $\|\mathbf{v}\| = 8$, $\mathbf{u} = \langle 3, 3 \rangle$

Writing a Linear Combination of Unit Vectors In Exercises 51–54, the initial and terminal points of a vector are given. Write the vector as a linear combination of the standard unit vectors **i** and **j**.

Initial Point	Terminal Point
51. $(-2, 1)$	$(3, -2)$
52. $(0, -2)$	$(3, 6)$
53. $(0, 1)$	$(-6, 4)$
54. $(2, 3)$	$(-1, -5)$

Vector Operations In Exercises 55–60, find the component form of **v** and sketch the specified vector operations geometrically, where $\mathbf{u} = 2\mathbf{i} - \mathbf{j}$ and $\mathbf{w} = \mathbf{i} + 2\mathbf{j}$.

55. $\mathbf{v} = \frac{3}{2}\mathbf{u}$
56. $\mathbf{v} = \frac{3}{4}\mathbf{w}$
57. $\mathbf{v} = \mathbf{u} + 2\mathbf{w}$
58. $\mathbf{v} = -\mathbf{u} + \mathbf{w}$
59. $\mathbf{v} = \mathbf{u} - 2\mathbf{w}$
60. $\mathbf{v} = \frac{1}{2}(3\mathbf{u} + \mathbf{w})$

Finding the Direction Angle of a Vector In Exercises 61–64, find the magnitude and direction angle of the vector **v**.

61. $\mathbf{v} = 6\mathbf{i} - 6\mathbf{j}$
62. $\mathbf{v} = -5\mathbf{i} + 4\mathbf{j}$
63. $\mathbf{v} = 3(\cos 60°\mathbf{i} + \sin 60°\mathbf{j})$
64. $\mathbf{v} = 8(\cos 135°\mathbf{i} + \sin 135°\mathbf{j})$

Finding the Component Form of a Vector In Exercises 65–70, find the component form of **v** given its magnitude and the angle it makes with the positive *x*-axis. Then sketch **v**.

Magnitude	Angle
65. $\|\mathbf{v}\| = 3$	$\theta = 0°$
66. $\|\mathbf{v}\| = 4\sqrt{3}$	$\theta = 90°$
67. $\|\mathbf{v}\| = \frac{7}{2}$	$\theta = 150°$
68. $\|\mathbf{v}\| = 2\sqrt{3}$	$\theta = 45°$
69. $\|\mathbf{v}\| = 3$	**v** in the direction $3\mathbf{i} + 4\mathbf{j}$
70. $\|\mathbf{v}\| = 2$	**v** in the direction $\mathbf{i} + 3\mathbf{j}$

Finding the Component Form of a Vector In Exercises 71 and 72, find the component form of the sum of **u** and **v** with direction angles $\theta_{\mathbf{u}}$ and $\theta_{\mathbf{v}}$.

71. $\|\mathbf{u}\| = 4$, $\theta_{\mathbf{u}} = 60°$
    $\|\mathbf{v}\| = 4$, $\theta_{\mathbf{v}} = 90°$

72. $\|\mathbf{u}\| = 20$, $\theta_{\mathbf{u}} = 45°$
    $\|\mathbf{v}\| = 50$, $\theta_{\mathbf{v}} = 180°$

Using the Law of Cosines In Exercises 73 and 74, use the Law of Cosines to find the angle α between the vectors. (Assume $0° \le \alpha \le 180°$.)

73. $\mathbf{v} = \mathbf{i} + \mathbf{j}$, $\mathbf{w} = 2\mathbf{i} - 2\mathbf{j}$
74. $\mathbf{v} = \mathbf{i} + 2\mathbf{j}$, $\mathbf{w} = 2\mathbf{i} - \mathbf{j}$

Resultant Force In Exercises 75 and 76, find the angle between the forces given the magnitude of their resultant. (*Hint:* Write force 1 as a vector in the direction of the positive *x*-axis and force 2 as a vector at an angle θ with the positive *x*-axis.)

	Force 1	Force 2	Resultant Force
75.	45 pounds	60 pounds	90 pounds
76.	3000 pounds	1000 pounds	3750 pounds

77. **Velocity** A gun with a muzzle velocity of 1200 feet per second is fired at an angle of 6° above the horizontal. Find the vertical and horizontal components of the velocity.

78. **Velocity** Pitcher Aroldis Chapman threw a pitch with a recorded velocity of 105 miles per hour. Assuming he threw the pitch at an angle of 3.5° below the horizontal, find the vertical and horizontal components of the velocity. (*Source: Guinness World Records*)

79. Resultant Force Forces with magnitudes of 125 newtons and 300 newtons act on a hook (see figure). The angle between the two forces is 45°. Find the direction and magnitude of the resultant of these forces. (*Hint:* Write the vector representing each force in component form, then add the vectors.)

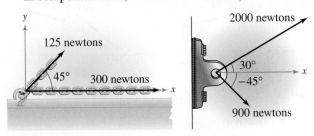

Figure for 79   Figure for 80

80. Resultant Force Forces with magnitudes of 2000 newtons and 900 newtons act on a machine part at angles of 30° and −45°, respectively, with the positive *x*-axis (see figure). Find the direction and magnitude of the resultant of these forces.

81. Resultant Force Three forces with magnitudes of 75 pounds, 100 pounds, and 125 pounds act on an object at angles of 30°, 45°, and 120°, respectively, with the positive *x*-axis. Find the direction and magnitude of the resultant of these forces.

82. Resultant Force Three forces with magnitudes of 70 pounds, 40 pounds, and 60 pounds act on an object at angles of −30°, 45°, and 135°, respectively, with the positive *x*-axis. Find the direction and magnitude of the resultant of these forces.

83. Cable Tension The cranes shown in the figure are lifting an object that weighs 20,240 pounds. Find the tension (in pounds) in the cable of each crane.

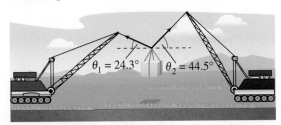

84. Cable Tension Repeat Exercise 83 for $\theta_1 = 35.6°$ and $\theta_2 = 40.4°$.

85. Rope Tension A tetherball weighing 1 pound is pulled outward from the pole by a horizontal force **u** until the rope makes a 45° angle with the pole (see figure). Determine the resulting tension (in pounds) in the rope and the magnitude of **u**.

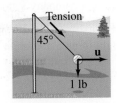

86. Physics Use the figure to determine the tension (in pounds) in each cable supporting the load.

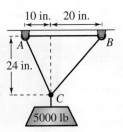

87. Tow Line Tension Two tugboats are towing a loaded barge and the magnitude of the resultant is 6000 pounds directed along the axis of the barge (see figure). Find the tension (in pounds) in the tow lines when they each make an 18° angle with the axis of the barge.

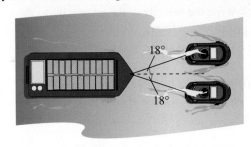

88. Rope Tension To carry a 100-pound cylindrical weight, two people lift on the ends of short ropes that are tied to an eyelet on the top center of the cylinder. Each rope makes a 20° angle with the vertical. Draw a diagram that gives a visual representation of the problem. Then find the tension (in pounds) in the ropes.

Inclined Ramp In Exercises 89–92, a force of *F* pounds is required to pull an object weighing *W* pounds up a ramp inclined at θ degrees from the horizontal.

89. Find *F* when *W* = 100 pounds and $\theta = 12°$.

90. Find *W* when *F* = 600 pounds and $\theta = 14°$.

91. Find θ when *F* = 5000 pounds and *W* = 15,000 pounds.

92. Find *F* when *W* = 5000 pounds and $\theta = 26°$.

93. Air Navigation An airplane travels in the direction of 148° with an airspeed of 875 kilometers per hour. Due to the wind, its groundspeed and direction are 800 kilometers per hour and 140°, respectively (see figure). Find the direction and speed of the wind.

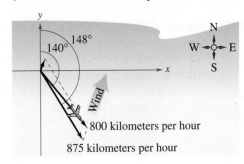

94. Air Navigation

A commercial jet travels from Miami to Seattle. The jet's velocity with respect to the air is 580 miles per hour, and its bearing is 332°. The jet encounters a wind with a velocity of 60 miles per hour from the southwest.

(a) Draw a diagram that gives a visual representation of the problem.

(b) Write the velocity of the wind as a vector in component form.

(c) Write the velocity of the jet relative to the air in component form.

(d) What is the speed of the jet with respect to the ground?

(e) What is the true direction of the jet?

Exploration

True or False? In Exercises 95–98, determine whether the statement is true or false. Justify your answer.

95. If **u** and **v** have the same magnitude and direction, then **u** and **v** are equivalent.

96. If **u** is a unit vector in the direction of **v**, then $\mathbf{v} = \|\mathbf{v}\|\mathbf{u}$.

97. If $\mathbf{v} = a\mathbf{i} + b\mathbf{j} = \mathbf{0}$, then $a = -b$.

98. If $\mathbf{u} = a\mathbf{i} + b\mathbf{j}$ is a unit vector, then $a^2 + b^2 = 1$.

99. **Error Analysis** Describe the error in finding the component form of the vector **u** that has initial point $(-3, 4)$ and terminal point $(6, -1)$.

The components are $u_1 = -3 - 6 = -9$ and $u_2 = 4 - (-1) = 5$. So, $\mathbf{u} = \langle -9, 5 \rangle$. ✗

100. **Error Analysis** Describe the error in finding the direction angle θ of the vector $\mathbf{v} = -5\mathbf{i} + 8\mathbf{j}$.

Because $\tan \theta = \dfrac{b}{a} = \dfrac{8}{-5}$, the reference angle is $\theta' = \left| \arctan\left(-\dfrac{8}{5}\right) \right| \approx |-57.99°| = 57.99°$ and $\theta \approx 360° - 57.99° = 302.01°$. ✗

101. **Proof** Prove that

$$(\cos \theta)\mathbf{i} + (\sin \theta)\mathbf{j}$$

is a unit vector for any value of θ.

102. **Technology** Write a program for your graphing utility that graphs two vectors and their difference given the vectors in component form.

Finding the Difference of Two Vectors In Exercises 103 and 104, use the program in Exercise 102 to find the difference of the vectors shown in the figure.

103.

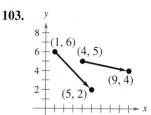

104.

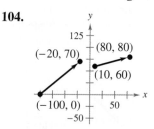

105. **Graphical Reasoning** Consider two forces

$$\mathbf{F}_1 = \langle 10, 0 \rangle \quad \text{and} \quad \mathbf{F}_2 = 5\langle \cos \theta, \sin \theta \rangle.$$

(a) Find $\|\mathbf{F}_1 + \mathbf{F}_2\|$ as a function of θ.

(b) Use a graphing utility to graph the function in part (a) for $0 \le \theta < 2\pi$.

(c) Use the graph in part (b) to determine the range of the function. What is its maximum, and for what value of θ does it occur? What is its minimum, and for what value of θ does it occur?

(d) Explain why the magnitude of the resultant is never 0.

106. **HOW DO YOU SEE IT?** Use the figure to determine whether each statement is true or false. Justify your answer.

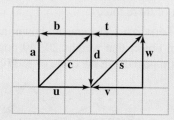

(a) $\mathbf{a} = -\mathbf{d}$ (b) $\mathbf{c} = \mathbf{s}$

(c) $\mathbf{a} + \mathbf{u} = \mathbf{c}$ (d) $\mathbf{v} + \mathbf{w} = -\mathbf{s}$

(e) $\mathbf{a} + \mathbf{w} = -2\mathbf{d}$ (f) $\mathbf{a} + \mathbf{d} = \mathbf{0}$

(g) $\mathbf{u} - \mathbf{v} = -2(\mathbf{b} + \mathbf{t})$ (h) $\mathbf{t} - \mathbf{w} = \mathbf{b} - \mathbf{a}$

107. **Writing** Give geometric descriptions of (a) vector addition and (b) scalar multiplication.

108. **Writing** Identify the quantity as a scalar or as a vector. Explain.

(a) The muzzle velocity of a bullet

(b) The price of a company's stock

(c) The air temperature in a room

(d) The weight of an automobile

6.4 Vectors and Dot Products

The dot product of two vectors has many real-life applications. For example, in Exercise 74 on page 436, you will use the dot product to find the force necessary to keep a sport utility vehicle from rolling down a hill.

- ■ Find the dot product of two vectors and use the properties of the dot product.
- ■ Find the angle between two vectors and determine whether two vectors are orthogonal.
- ■ Write a vector as the sum of two vector components.
- ■ Use vectors to determine the work done by a force.

The Dot Product of Two Vectors

So far, you have studied two vector operations—vector addition and multiplication by a scalar—each of which yields another vector. In this section, you will study a third vector operation, the **dot product.** This operation yields a scalar, rather than a vector.

Definition of the Dot Product

The **dot product** of $\mathbf{u} = \langle u_1, u_2 \rangle$ and $\mathbf{v} = \langle v_1, v_2 \rangle$ is $\mathbf{u} \cdot \mathbf{v} = u_1 v_1 + u_2 v_2$.

Properties of the Dot Product

Let $\mathbf{u}$, $\mathbf{v}$, and $\mathbf{w}$ be vectors in the plane or in space and let c be a scalar.

1. $\mathbf{u} \cdot \mathbf{v} = \mathbf{v} \cdot \mathbf{u}$

2. $\mathbf{0} \cdot \mathbf{v} = 0$

3. $\mathbf{u} \cdot (\mathbf{v} + \mathbf{w}) = \mathbf{u} \cdot \mathbf{v} + \mathbf{u} \cdot \mathbf{w}$

4. $\mathbf{v} \cdot \mathbf{v} = \|\mathbf{v}\|^2$

5. $c(\mathbf{u} \cdot \mathbf{v}) = c\mathbf{u} \cdot \mathbf{v} = \mathbf{u} \cdot c\mathbf{v}$

For proofs of the properties of the dot product, see Proofs in Mathematics on page 464.

EXAMPLE 1   **Finding Dot Products**

REMARK In Example 1, be sure you see that the dot product of two vectors is a scalar (a real number), not a vector. Moreover, notice that the dot product can be positive, zero, or negative.

a. $\langle 4, 5 \rangle \cdot \langle 2, 3 \rangle = 4(2) + 5(3)$
$$= 8 + 15$$
$$= 23$$

b. $\langle 2, -1 \rangle \cdot \langle 1, 2 \rangle = 2(1) + (-1)(2)$
$$= 2 - 2$$
$$= 0$$

c. $\langle 0, 3 \rangle \cdot \langle 4, -2 \rangle = 0(4) + 3(-2)$
$$= 0 - 6$$
$$= -6$$

✓ *Checkpoint* ◀))) *Audio-video solution in English & Spanish at LarsonPrecalculus.com*

Find each dot product.

a. $\langle 3, 4 \rangle \cdot \langle 2, -3 \rangle$ **b.** $\langle -3, -5 \rangle \cdot \langle 1, -8 \rangle$ **c.** $\langle -6, 5 \rangle \cdot \langle 5, 6 \rangle$

EXAMPLE 2   **Using Properties of the Dot Product**

Let $\mathbf{u} = \langle -1, 3 \rangle$, $\mathbf{v} = \langle 2, -4 \rangle$, and $\mathbf{w} = \langle 1, -2 \rangle$. Find each quantity.

a. $(\mathbf{u} \cdot \mathbf{v})\mathbf{w}$ **b.** $\mathbf{u} \cdot 2\mathbf{v}$ **c.** $\|\mathbf{u}\|$

Solution   Begin by finding the dot product of $\mathbf{u}$ and $\mathbf{v}$ and the dot product of $\mathbf{u}$ and $\mathbf{u}$.

$$\mathbf{u} \cdot \mathbf{v} = \langle -1, 3 \rangle \cdot \langle 2, -4 \rangle = -1(2) + 3(-4) = -14$$

$$\mathbf{u} \cdot \mathbf{u} = \langle -1, 3 \rangle \cdot \langle -1, 3 \rangle = -1(-1) + 3(3) = 10$$

a. $(\mathbf{u} \cdot \mathbf{v})\mathbf{w} = -14\langle 1, -2 \rangle = \langle -14, 28 \rangle$

b. $\mathbf{u} \cdot 2\mathbf{v} = 2(\mathbf{u} \cdot \mathbf{v}) = 2(-14) = -28$

c. Because $\|\mathbf{u}\|^2 = \mathbf{u} \cdot \mathbf{u} = 10$, it follows that $\|\mathbf{u}\| = \sqrt{\mathbf{u} \cdot \mathbf{u}} = \sqrt{10}$.

Notice that the product in part (a) is a vector, whereas the product in part (b) is a scalar. Can you see why?

✓ *Checkpoint*  Audio-video solution in English & Spanish at LarsonPrecalculus.com

Let $\mathbf{u} = \langle 3, 4 \rangle$ and $\mathbf{v} = \langle -2, 6 \rangle$. Find each quantity.

a. $(\mathbf{u} \cdot \mathbf{v})\mathbf{v}$ **b.** $\mathbf{u} \cdot (\mathbf{u} + \mathbf{v})$ **c.** $\|\mathbf{v}\|$

The Angle Between Two Vectors

The **angle between two nonzero vectors** is the angle θ, $0 \le \theta \le \pi$, between their respective standard position vectors, as shown in Figure 6.25. This angle can be found using the dot product.

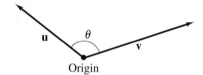

Figure 6.25

Angle Between Two Vectors

If θ is the angle between two nonzero vectors $\mathbf{u}$ and $\mathbf{v}$, then

$$\cos \theta = \frac{\mathbf{u} \cdot \mathbf{v}}{\|\mathbf{u}\| \|\mathbf{v}\|}.$$

For a proof of the angle between two vectors, see Proofs in Mathematics on page 464.

EXAMPLE 3   **Finding the Angle Between Two Vectors**

See LarsonPrecalculus.com for an interactive version of this type of example.

Find the angle θ between $\mathbf{u} = \langle 4, 3 \rangle$ and $\mathbf{v} = \langle 3, 5 \rangle$ (see Figure 6.26).

Solution

$$\cos \theta = \frac{\mathbf{u} \cdot \mathbf{v}}{\|\mathbf{u}\| \|\mathbf{v}\|} = \frac{\langle 4, 3 \rangle \cdot \langle 3, 5 \rangle}{\|\langle 4, 3 \rangle\| \|\langle 3, 5 \rangle\|} = \frac{4(3) + 3(5)}{\sqrt{4^2 + 3^2}\sqrt{3^2 + 5^2}} = \frac{27}{5\sqrt{34}}$$

This implies that the angle between the two vectors is

$$\theta = \cos^{-1} \frac{27}{5\sqrt{34}} \approx 0.3869 \text{ radian} \approx 22.17°. \qquad \text{Use a calculator.}$$

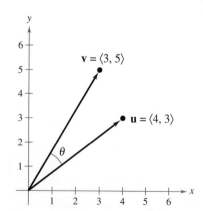

Figure 6.26

✓ *Checkpoint*  Audio-video solution in English & Spanish at LarsonPrecalculus.com

Find the angle θ between $\mathbf{u} = \langle 2, 1 \rangle$ and $\mathbf{v} = \langle 1, 3 \rangle$.

Rewriting the expression for the angle between two vectors in the form

$$\mathbf{u} \cdot \mathbf{v} = \|\mathbf{u}\| \, \|\mathbf{v}\| \cos \theta \qquad \text{Alternative form of dot product}$$

produces an alternative way to calculate the dot product. This form shows that $\mathbf{u} \cdot \mathbf{v}$ and $\cos \theta$ always have the same sign, because $\|\mathbf{u}\|$ and $\|\mathbf{v}\|$ are always positive. The figures below show the five possible orientations of two vectors.

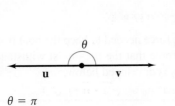

$\theta = \pi$

$\cos \theta = -1$

Opposite direction

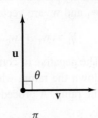

$\dfrac{\pi}{2} < \theta < \pi$

$-1 < \cos \theta < 0$

Obtuse angle

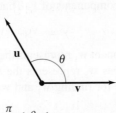

$\theta = \dfrac{\pi}{2}$

$\cos \theta = 0$

90° angle

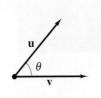

$0 < \theta < \dfrac{\pi}{2}$

$0 < \cos \theta < 1$

Acute angle

$\theta = 0$

$\cos \theta = 1$

Same direction

Definition of Orthogonal Vectors

The vectors $\mathbf{u}$ and $\mathbf{v}$ are **orthogonal** if and only if $\mathbf{u} \cdot \mathbf{v} = 0$.

The terms *orthogonal* and *perpendicular* have essentially the same meaning—meeting at right angles. Even though the angle between the zero vector and another vector is not defined, it is convenient to extend the definition of orthogonality to include the zero vector. In other words, the zero vector is orthogonal to every vector $\mathbf{u}$, because $\mathbf{0} \cdot \mathbf{u} = 0$.

EXAMPLE 4   **Determining Orthogonal Vectors**

Determine whether the vectors $\mathbf{u} = \langle 2, -3 \rangle$ and $\mathbf{v} = \langle 6, 4 \rangle$ are orthogonal.

Solution   Find the dot product of the two vectors.

$$\mathbf{u} \cdot \mathbf{v} = \langle 2, -3 \rangle \cdot \langle 6, 4 \rangle = 2(6) + (-3)(4) = 0$$

The dot product is 0, so the two vectors are orthogonal (see figure below).

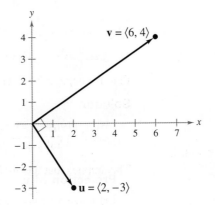

✓ *Checkpoint* ◀))) *Audio-video solution in English & Spanish at LarsonPrecalculus.com*

Determine whether the vectors $\mathbf{u} = \langle 6, 10 \rangle$ and $\mathbf{v} = \left\langle -\frac{1}{3}, \frac{1}{5} \right\rangle$ are orthogonal.

Figure 6.27

Finding Vector Components

You have seen applications in which you add two vectors to produce a resultant vector. Many applications in physics and engineering pose the reverse problem—decomposing a given vector into the sum of two **vector components.**

Consider a boat on an inclined ramp, as shown in Figure 6.27. The force $\mathbf{F}$ due to gravity pulls the boat *down* the ramp and *against* the ramp. These two orthogonal forces $\mathbf{w}_1$ and $\mathbf{w}_2$ are vector components of $\mathbf{F}$. That is,

$$\mathbf{F} = \mathbf{w}_1 + \mathbf{w}_2. \qquad \text{Vector components of } \mathbf{F}$$

The negative of component $\mathbf{w}_1$ represents the force needed to keep the boat from rolling down the ramp, whereas $\mathbf{w}_2$ represents the force that the tires must withstand against the ramp. A procedure for finding $\mathbf{w}_1$ and $\mathbf{w}_2$ is developed below.

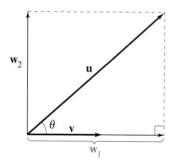

θ is acute.

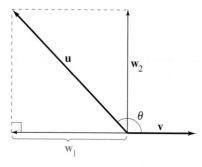

θ is obtuse.

Figure 6.28

Definition of Vector Components

Let $\mathbf{u}$ and $\mathbf{v}$ be nonzero vectors such that

$$\mathbf{u} = \mathbf{w}_1 + \mathbf{w}_2$$

where $\mathbf{w}_1$ and $\mathbf{w}_2$ are orthogonal and $\mathbf{w}_1$ is parallel to (or a scalar multiple of) $\mathbf{v}$, as shown in Figure 6.28. The vectors $\mathbf{w}_1$ and $\mathbf{w}_2$ are **vector components of $\mathbf{u}$.** The vector $\mathbf{w}_1$ is the **projection** of $\mathbf{u}$ onto $\mathbf{v}$ and is denoted by

$$\mathbf{w}_1 = \text{proj}_{\mathbf{v}}\mathbf{u}.$$

The vector $\mathbf{w}_2$ is given by

$$\mathbf{w}_2 = \mathbf{u} - \mathbf{w}_1.$$

To find the component $\mathbf{w}_2$, first find the projection of $\mathbf{u}$ onto $\mathbf{v}$. To find the projection, use the dot product.

$$\mathbf{u} = \mathbf{w}_1 + \mathbf{w}_2$$

$$\mathbf{u} = c\mathbf{v} + \mathbf{w}_2 \qquad \mathbf{w}_1 \text{ is a scalar multiple of } \mathbf{v}.$$

$$\mathbf{u} \cdot \mathbf{v} = (c\mathbf{v} + \mathbf{w}_2) \cdot \mathbf{v} \qquad \text{Dot product of each side with } \mathbf{v}$$

$$\mathbf{u} \cdot \mathbf{v} = c\mathbf{v} \cdot \mathbf{v} + \mathbf{w}_2 \cdot \mathbf{v} \qquad \text{Property 3 of the dot product}$$

$$\mathbf{u} \cdot \mathbf{v} = c\|\mathbf{v}\|^2 + 0 \qquad \mathbf{w}_2 \text{ and } \mathbf{v} \text{ are orthogonal.}$$

So,

$$c = \frac{\mathbf{u} \cdot \mathbf{v}}{\|\mathbf{v}\|^2}$$

and

$$\mathbf{w}_1 = \text{proj}_{\mathbf{v}}\mathbf{u} = c\mathbf{v} = \left(\frac{\mathbf{u} \cdot \mathbf{v}}{\|\mathbf{v}\|^2}\right)\mathbf{v}.$$

Projection of u onto v

Let $\mathbf{u}$ and $\mathbf{v}$ be nonzero vectors. The projection of $\mathbf{u}$ onto $\mathbf{v}$ is given by

$$\text{proj}_{\mathbf{v}}\mathbf{u} = \left(\frac{\mathbf{u} \cdot \mathbf{v}}{\|\mathbf{v}\|^2}\right)\mathbf{v}.$$

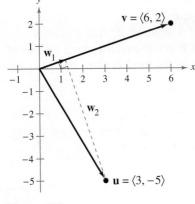

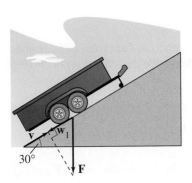

Figure 6.29

EXAMPLE 5 **Decomposing a Vector into Components**

Find the projection of $\mathbf{u} = \langle 3, -5 \rangle$ onto $\mathbf{v} = \langle 6, 2 \rangle$. Then write $\mathbf{u}$ as the sum of two orthogonal vectors, one of which is $\text{proj}_{\mathbf{v}}\mathbf{u}$.

Solution The projection of $\mathbf{u}$ onto $\mathbf{v}$ is

$$\mathbf{w}_1 = \text{proj}_{\mathbf{v}}\mathbf{u} = \left(\frac{\mathbf{u} \cdot \mathbf{v}}{\|\mathbf{v}\|^2}\right)\mathbf{v} = \left(\frac{8}{40}\right)\langle 6, 2 \rangle = \left\langle \frac{6}{5}, \frac{2}{5} \right\rangle$$

as shown in Figure 6.29. The component $\mathbf{w}_2$ is

$$\mathbf{w}_2 = \mathbf{u} - \mathbf{w}_1 = \langle 3, -5 \rangle - \left\langle \frac{6}{5}, \frac{2}{5} \right\rangle = \left\langle \frac{9}{5}, -\frac{27}{5} \right\rangle.$$

So,

$$\mathbf{u} = \mathbf{w}_1 + \mathbf{w}_2 = \left\langle \frac{6}{5}, \frac{2}{5} \right\rangle + \left\langle \frac{9}{5}, -\frac{27}{5} \right\rangle = \langle 3, -5 \rangle.$$

✓ *Checkpoint*  *Audio-video solution in English & Spanish at LarsonPrecalculus.com*

Find the projection of $\mathbf{u} = \langle 3, 4 \rangle$ onto $\mathbf{v} = \langle 8, 2 \rangle$. Then write $\mathbf{u}$ as the sum of two orthogonal vectors, one of which is $\text{proj}_{\mathbf{v}}\mathbf{u}$.

EXAMPLE 6 **Finding a Force**

A 200-pound cart is on a ramp inclined at $30°$, as shown in Figure 6.30. What force is required to keep the cart from rolling down the ramp?

Solution The force due to gravity is vertical and downward, so use the vector

$$\mathbf{F} = -200\mathbf{j} \qquad \text{Force due to gravity}$$

to represent the gravitational force. To find the force required to keep the cart from rolling down the ramp, project $\mathbf{F}$ onto a unit vector $\mathbf{v}$ in the direction of the ramp, where

$$\mathbf{v} = (\cos 30°)\mathbf{i} + (\sin 30°)\mathbf{j}$$
$$= \frac{\sqrt{3}}{2}\mathbf{i} + \frac{1}{2}\mathbf{j}. \qquad \text{Unit vector along ramp}$$

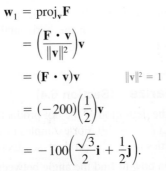

Figure 6.30

So, the projection of $\mathbf{F}$ onto $\mathbf{v}$ is

$$\mathbf{w}_1 = \text{proj}_{\mathbf{v}}\mathbf{F}$$
$$= \left(\frac{\mathbf{F} \cdot \mathbf{v}}{\|\mathbf{v}\|^2}\right)\mathbf{v}$$
$$= (\mathbf{F} \cdot \mathbf{v})\mathbf{v} \qquad \|\mathbf{v}\|^2 = 1$$
$$= (-200)\left(\frac{1}{2}\right)\mathbf{v}$$
$$= -100\left(\frac{\sqrt{3}}{2}\mathbf{i} + \frac{1}{2}\mathbf{j}\right).$$

The magnitude of this force is 100. So, a force of 100 pounds is required to keep the cart from rolling down the ramp.

✓ *Checkpoint* *Audio-video solution in English & Spanish at LarsonPrecalculus.com*

Rework Example 6 for a 150-pound cart that is on a ramp inclined at $15°$.

Force acts along the line of motion.
Figure 6.31

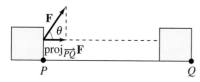

Force acts at angle θ with the line of motion.
Figure 6.32

Work

The work W done by a constant force $\mathbf{F}$ acting along the line of motion of an object is given by

$$W = \text{(magnitude of force)(distance)} = \|\mathbf{F}\| \|\overrightarrow{PQ}\|$$

as shown in Figure 6.31. When the constant force $\mathbf{F}$ is *not* directed along the line of motion, as shown in Figure 6.32, the work W done by the force is given by

$$W = \|\text{proj}_{\overrightarrow{PQ}} \mathbf{F}\| \|\overrightarrow{PQ}\| \qquad \text{Projection form for work}$$
$$= (\cos \theta)\|\mathbf{F}\| \|\overrightarrow{PQ}\| \qquad \|\text{proj}_{\overrightarrow{PQ}}\mathbf{F}\| = (\cos \theta)\|\mathbf{F}\|$$
$$= \mathbf{F} \cdot \overrightarrow{PQ}. \qquad \text{Alternate form of dot product}$$

The definition below summarizes the concept of work.

> ### Definition of Work
>
> The **work** W done by a constant force $\mathbf{F}$ as its point of application moves along the vector $\overrightarrow{PQ}$ is given by either formula below.
>
> 1. $W = \|\text{proj}_{\overrightarrow{PQ}}\mathbf{F}\| \|\overrightarrow{PQ}\|$ Projection form
> 2. $W = \mathbf{F} \cdot \overrightarrow{PQ}$ Dot product form

Figure 6.33

EXAMPLE 7   Determining Work

To close a sliding barn door, a person pulls on a rope with a constant force of 50 pounds at a constant angle of 60°, as shown in Figure 6.33. Determine the work done in moving the barn door 12 feet to its closed position.

Solution   Use a projection to find the work.

$$W = \|\text{proj}_{\overrightarrow{PQ}}\mathbf{F}\| \|\overrightarrow{PQ}\| = (\cos 60°)\|\mathbf{F}\| \|\overrightarrow{PQ}\| = \frac{1}{2}(50)(12) = 300 \text{ foot-pounds}$$

So, the work done is 300 foot-pounds. Verify this result by finding the vectors $\mathbf{F}$ and $\overrightarrow{PQ}$ and calculating their dot product.

✓ *Checkpoint*  *Audio-video solution in English & Spanish at LarsonPrecalculus.com*

A person pulls a wagon by exerting a constant force of 35 pounds on a handle that makes a 30° angle with the horizontal. Determine the work done in pulling the wagon 40 feet. ∎

Work is done only when an object is moved. It does not matter how much force is applied—if an object does not move, then no work is done.

> ## Summarize   (Section 6.4)
>
> 1. State the definition of the dot product and list the properties of the dot product (*page 429*). For examples of finding dot products and using the properties of the dot product, see Examples 1 and 2.
> 2. Explain how to find the angle between two vectors and how to determine whether two vectors are orthogonal (*page 430*). For examples involving the angle between two vectors, see Examples 3 and 4.
> 3. Explain how to write a vector as the sum of two vector components (*page 432*). For examples involving vector components, see Examples 5 and 6.
> 4. State the definition of work (*page 434*). For an example of determining work, see Example 7.

6.4 Exercises

See **CalcChat.com** for tutorial help and worked-out solutions to odd-numbered exercises.

Vocabulary: Fill in the blanks.

1. The _____ _____ of two vectors yields a scalar, rather than a vector.
2. The dot product of $\mathbf{u} = \langle u_1, u_2 \rangle$ and $\mathbf{v} = \langle v_1, v_2 \rangle$ is $\mathbf{u} \cdot \mathbf{v} =$ _____.
3. If θ is the angle between two nonzero vectors $\mathbf{u}$ and $\mathbf{v}$, then $\cos \theta =$ _____.
4. The vectors $\mathbf{u}$ and $\mathbf{v}$ are _____ if and only if $\mathbf{u} \cdot \mathbf{v} = 0$.
5. The projection of $\mathbf{u}$ onto $\mathbf{v}$ is given by $\text{proj}_{\mathbf{v}}\mathbf{u} =$ _____.
6. The work W done by a constant force $\mathbf{F}$ as its point of application moves along the vector $\overrightarrow{PQ}$ is given by $W =$ _____ or $W =$ _____.

Skills and Applications

 Finding a Dot Product In Exercises 7–12, find $\mathbf{u} \cdot \mathbf{v}$.

7. $\mathbf{u} = \langle 7, 1 \rangle$
   $\mathbf{v} = \langle -3, 2 \rangle$
8. $\mathbf{u} = \langle 6, 10 \rangle$
   $\mathbf{v} = \langle -2, 3 \rangle$
9. $\mathbf{u} = \langle -6, 2 \rangle$
   $\mathbf{v} = \langle 1, 3 \rangle$
10. $\mathbf{u} = \langle -2, 5 \rangle$
    $\mathbf{v} = \langle -1, -8 \rangle$
11. $\mathbf{u} = 4\mathbf{i} - 2\mathbf{j}$
    $\mathbf{v} = \mathbf{i} - \mathbf{j}$
12. $\mathbf{u} = \mathbf{i} - 2\mathbf{j}$
    $\mathbf{v} = -2\mathbf{i} - \mathbf{j}$

 Using Properties of the Dot Product In Exercises 13–22, use the vectors $\mathbf{u} = \langle 3, 3 \rangle$, $\mathbf{v} = \langle -4, 2 \rangle$, and $\mathbf{w} = \langle 3, -1 \rangle$ to find the quantity. State whether the result is a vector or a scalar.

13. $\mathbf{u} \cdot \mathbf{u}$
14. $3\mathbf{u} \cdot \mathbf{v}$
15. $(\mathbf{u} \cdot \mathbf{v})\mathbf{v}$
16. $(\mathbf{u} \cdot 2\mathbf{v})\mathbf{w}$
17. $(\mathbf{v} \cdot \mathbf{0})\mathbf{w}$
18. $(\mathbf{u} + \mathbf{v}) \cdot \mathbf{0}$
19. $\|\mathbf{w}\| - 1$
20. $2 - \|\mathbf{u}\|$
21. $(\mathbf{u} \cdot \mathbf{v}) - (\mathbf{u} \cdot \mathbf{w})$
22. $(\mathbf{v} \cdot \mathbf{u}) - (\mathbf{w} \cdot \mathbf{v})$

Finding the Magnitude of a Vector In Exercises 23–28, use the dot product to find the magnitude of $\mathbf{u}$.

23. $\mathbf{u} = \langle -8, 15 \rangle$
24. $\mathbf{u} = \langle 4, -6 \rangle$
25. $\mathbf{u} = 20\mathbf{i} + 25\mathbf{j}$
26. $\mathbf{u} = 12\mathbf{i} - 16\mathbf{j}$
27. $\mathbf{u} = 6\mathbf{j}$
28. $\mathbf{u} = -21\mathbf{i}$

 Finding the Angle Between Two Vectors In Exercises 29–38, find the angle θ (in radians) between the vectors.

29. $\mathbf{u} = \langle 1, 0 \rangle$
    $\mathbf{v} = \langle 0, -2 \rangle$
30. $\mathbf{u} = \langle 3, 2 \rangle$
    $\mathbf{v} = \langle 4, 0 \rangle$
31. $\mathbf{u} = 3\mathbf{i} + 4\mathbf{j}$
    $\mathbf{v} = -2\mathbf{j}$
32. $\mathbf{u} = 2\mathbf{i} - 3\mathbf{j}$
    $\mathbf{v} = \mathbf{i} - 2\mathbf{j}$

33. $\mathbf{u} = 2\mathbf{i} - \mathbf{j}$
    $\mathbf{v} = 6\mathbf{i} - 3\mathbf{j}$
34. $\mathbf{u} = 5\mathbf{i} + 5\mathbf{j}$
    $\mathbf{v} = -6\mathbf{i} + 6\mathbf{j}$
35. $\mathbf{u} = -6\mathbf{i} - 3\mathbf{j}$
    $\mathbf{v} = -8\mathbf{i} + 4\mathbf{j}$
36. $\mathbf{u} = 2\mathbf{i} - 3\mathbf{j}$
    $\mathbf{v} = 4\mathbf{i} + 3\mathbf{j}$
37. $\mathbf{u} = \cos\left(\dfrac{\pi}{3}\right)\mathbf{i} + \sin\left(\dfrac{\pi}{3}\right)\mathbf{j}$
    $\mathbf{v} = \cos\left(\dfrac{3\pi}{4}\right)\mathbf{i} + \sin\left(\dfrac{3\pi}{4}\right)\mathbf{j}$
38. $\mathbf{u} = \cos\left(\dfrac{\pi}{4}\right)\mathbf{i} + \sin\left(\dfrac{\pi}{4}\right)\mathbf{j}$
    $\mathbf{v} = \cos\left(\dfrac{5\pi}{4}\right)\mathbf{i} + \sin\left(\dfrac{5\pi}{4}\right)\mathbf{j}$

Finding the Angle Between Two Vectors In Exercises 39–42, find the angle θ (in degrees) between the vectors.

39. $\mathbf{u} = 3\mathbf{i} + 4\mathbf{j}$
    $\mathbf{v} = -7\mathbf{i} + 5\mathbf{j}$
40. $\mathbf{u} = 6\mathbf{i} - 3\mathbf{j}$
    $\mathbf{v} = -4\mathbf{i} - 4\mathbf{j}$
41. $\mathbf{u} = -5\mathbf{i} - 5\mathbf{j}$
    $\mathbf{v} = -8\mathbf{i} + 8\mathbf{j}$
42. $\mathbf{u} = 2\mathbf{i} - 3\mathbf{j}$
    $\mathbf{v} = 8\mathbf{i} + 3\mathbf{j}$

 Finding the Angles in a Triangle In Exercises 43–46, use vectors to find the interior angles of the triangle with the given vertices.

43. $(1, 2), (3, 4), (2, 5)$
44. $(-3, -4), (1, 7), (8, 2)$
45. $(-3, 0), (2, 2), (0, 6)$
46. $(-3, 5), (-1, 9), (7, 9)$

 Using the Angle Between Two Vectors In Exercises 47–50, find $\mathbf{u} \cdot \mathbf{v}$, where θ is the angle between $\mathbf{u}$ and $\mathbf{v}$.

47. $\|\mathbf{u}\| = 4$, $\|\mathbf{v}\| = 10$, $\theta = 2\pi/3$
48. $\|\mathbf{u}\| = 4$, $\|\mathbf{v}\| = 12$, $\theta = \pi/3$
49. $\|\mathbf{u}\| = 100$, $\|\mathbf{v}\| = 250$, $\theta = \pi/6$
50. $\|\mathbf{u}\| = 9$, $\|\mathbf{v}\| = 36$, $\theta = 3\pi/4$

Determining Orthogonal Vectors In
Exercises 51–56, determine whether **u** and **v**
are orthogonal.

51. $\mathbf{u} = \langle 3, 15 \rangle$
$\mathbf{v} = \langle -1, 5 \rangle$

52. $\mathbf{u} = \langle 30, 12 \rangle$
$\mathbf{v} = \langle \frac{1}{2}, -\frac{5}{4} \rangle$

53. $\mathbf{u} = 2\mathbf{i} - 2\mathbf{j}$
$\mathbf{v} = -\mathbf{i} - \mathbf{j}$

54. $\mathbf{u} = \frac{1}{4}(3\mathbf{i} - \mathbf{j})$
$\mathbf{v} = 5\mathbf{i} + 6\mathbf{j}$

55. $\mathbf{u} = \mathbf{i}$
$\mathbf{v} = -2\mathbf{i} + 2\mathbf{j}$

56. $\mathbf{u} = \langle \cos\theta, \sin\theta \rangle$
$\mathbf{v} = \langle \sin\theta, -\cos\theta \rangle$

 Decomposing a Vector into Components
In Exercises 57–60, find the projection of
u onto **v**. Then write **u** as the sum of two
orthogonal vectors, one of which is proj$_\mathbf{v}$**u**.

57. $\mathbf{u} = \langle 2, 2 \rangle$
$\mathbf{v} = \langle 6, 1 \rangle$

58. $\mathbf{u} = \langle 0, 3 \rangle$
$\mathbf{v} = \langle 2, 15 \rangle$

59. $\mathbf{u} = \langle 4, 2 \rangle$
$\mathbf{v} = \langle 1, -2 \rangle$

60. $\mathbf{u} = \langle -3, -2 \rangle$
$\mathbf{v} = \langle -4, -1 \rangle$

Finding the Projection of u onto v In Exercises
61–64, use the graph to find the projection of **u** onto **v**.
(The terminal points of the vectors in standard position
are given.) Use the formula for the projection of **u** onto
v to verify your result.

61.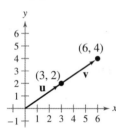

62.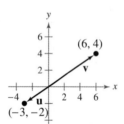

63.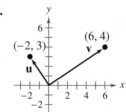

64.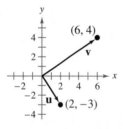

Finding Orthogonal Vectors In Exercises 65–68,
find two vectors in opposite directions that are orthogonal
to the vector **u**. (There are many correct answers.)

65. $\mathbf{u} = \langle 3, 5 \rangle$
66. $\mathbf{u} = \langle -8, 3 \rangle$
67. $\mathbf{u} = \frac{1}{2}\mathbf{i} - \frac{2}{3}\mathbf{j}$
68. $\mathbf{u} = -\frac{5}{2}\mathbf{i} - 3\mathbf{j}$

 Work In Exercises 69 and 70, determine
the work done in moving a particle from P to
Q when the magnitude and direction of the
force are given by **v**.

69. $P(0, 0)$, $Q(4, 7)$, $\mathbf{v} = \langle 1, 4 \rangle$
70. $P(1, 3)$, $Q(-3, 5)$, $\mathbf{v} = -2\mathbf{i} + 3\mathbf{j}$

Anthony Berenyi/Shutterstock.com

71. Business The vector $\mathbf{u} = \langle 1225, 2445 \rangle$ gives
the numbers of hours worked by employees of a
temporary work agency at two pay levels. The vector
$\mathbf{v} = \langle 12.20, 8.50 \rangle$ gives the hourly wage (in dollars)
paid at each level, respectively.

(a) Find the dot product $\mathbf{u} \cdot \mathbf{v}$ and interpret the result in
the context of the problem.

(b) Identify the vector operation used to increase wages
by 2%.

72. Revenue The vector $\mathbf{u} = \langle 3140, 2750 \rangle$ gives the
numbers of hamburgers and hot dogs, respectively,
sold at a fast-food stand in one month. The vector
$\mathbf{v} = \langle 2.25, 1.75 \rangle$ gives the prices (in dollars) of the food
items, respectively.

(a) Find the dot product $\mathbf{u} \cdot \mathbf{v}$ and interpret the result in
the context of the problem.

(b) Identify the vector operation used to increase the
prices by 2.5%.

73. Physics A truck with a gross weight of 30,000
pounds is parked on a slope of $d°$ (see figure). Assume
that the only force to overcome is the force of gravity.

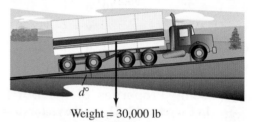

Weight = 30,000 lb

(a) Find the force required to keep the truck from
rolling down the hill in terms of d.

(b) Use a graphing utility to complete the table.

d	0°	1°	2°	3°	4°	5°
Force						

d	6°	7°	8°	9°	10°
Force					

(c) Find the force perpendicular to the hill when $d = 5°$.

74. Braking Load
A sport utility vehicle
with a gross weight
of 5400 pounds is
parked on a slope
of 10°. Assume
that the only force
to overcome is the
force of gravity.
Find the force
required to keep the vehicle from rolling down the
hill. Find the force perpendicular to the hill.

75. Work Determine the work done by a person lifting a 245-newton bag of sugar 3 meters.

76. Work Determine the work done by a crane lifting a 2400-pound car 5 feet.

77. Work A constant force of 45 pounds, exerted at an angle of 30° with the horizontal, is required to slide a table across a floor. Determine the work done in sliding the table 20 feet.

78. Work A constant force of 50 pounds, exerted at an angle of 25° with the horizontal, is required to slide a desk across a floor. Determine the work done in sliding the desk 15 feet.

79. Work A tractor pulls a log 800 meters, and the tension in the cable connecting the tractor and the log is approximately 15,691 newtons. The direction of the constant force is 35° above the horizontal. Determine the work done in pulling the log.

80. Work One of the events in a strength competition is to pull a cement block 100 feet. One competitor pulls the block by exerting a constant force of 250 pounds on a rope attached to the block at an angle of 30° with the horizontal (see figure). Determine the work done in pulling the block.

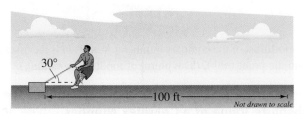

81. Work A child pulls a toy wagon by exerting a constant force of 25 pounds on a handle that makes a 20° angle with the horizontal (see figure). Determine the work done in pulling the wagon 50 feet.

82. Work A ski patroller pulls a rescue toboggan across a flat snow surface by exerting a constant force of 35 pounds on a handle that makes a 22° angle with the horizontal (see figure). Determine the work done in pulling the toboggan 200 feet.

Exploration

True or False? In Exercises 83 and 84, determine whether the statement is true or false. Justify your answer.

83. The work W done by a constant force $\mathbf{F}$ acting along the line of motion of an object is represented by a vector.

84. A sliding door moves along the line of vector $\overrightarrow{PQ}$. If a force is applied to the door along a vector that is orthogonal to $\overrightarrow{PQ}$, then no work is done.

Error Analysis In Exercises 85 and 86, describe the error in finding the quantity when $\mathbf{u} = \langle 2, -1 \rangle$ and $\mathbf{v} = \langle -3, 5 \rangle$.

85. $\mathbf{v} \cdot \mathbf{0} = \langle 0, 0 \rangle$ ✗

86. $\mathbf{u} \cdot 2\mathbf{v} = \langle 2, -1 \rangle \cdot \langle -6, 10 \rangle$
$\qquad = 2(-6) - (-1)(10)$
$\qquad = -12 + 10$
$\qquad = -2$

Finding an Unknown Vector Component In Exercises 87 and 88, find the value of k such that vectors $\mathbf{u}$ and $\mathbf{v}$ are orthogonal.

87. $\mathbf{u} = 8\mathbf{i} + 4\mathbf{j}$
$\quad \mathbf{v} = 2\mathbf{i} - k\mathbf{j}$

88. $\mathbf{u} = -3k\mathbf{i} + 5\mathbf{j}$
$\quad \mathbf{v} = 2\mathbf{i} - 4\mathbf{j}$

89. Think About It Let $\mathbf{u}$ be a unit vector. What is the value of $\mathbf{u} \cdot \mathbf{u}$? Explain.

90. **HOW DO YOU SEE IT?** What is known about θ, the angle between two nonzero vectors $\mathbf{u}$ and $\mathbf{v}$, under each condition (see figure)?

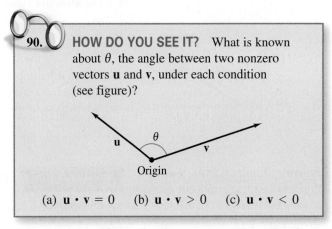

(a) $\mathbf{u} \cdot \mathbf{v} = 0$ (b) $\mathbf{u} \cdot \mathbf{v} > 0$ (c) $\mathbf{u} \cdot \mathbf{v} < 0$

91. Think About It What can be said about the vectors $\mathbf{u}$ and $\mathbf{v}$ under each condition?

(a) The projection of $\mathbf{u}$ onto $\mathbf{v}$ equals $\mathbf{u}$.

(b) The projection of $\mathbf{u}$ onto $\mathbf{v}$ equals $\mathbf{0}$.

92. Proof Use vectors to prove that the diagonals of a rhombus are perpendicular.

93. Proof Prove that

$$\|\mathbf{u} - \mathbf{v}\|^2 = \|\mathbf{u}\|^2 + \|\mathbf{v}\|^2 - 2\mathbf{u} \cdot \mathbf{v}.$$

6.5 The Complex Plane

■ Plot complex numbers in the complex plane and find absolute values of complex numbers.
■ Perform operations with complex numbers in the complex plane.
■ Use the Distance and Midpoint Formulas in the complex plane.

The Complex Plane

Just as a real number can be represented by a point on the real number line, a complex number $z = a + bi$ can be represented by the point (a, b) in a coordinate plane (the **complex plane**). In the complex plane, the horizontal axis is the **real axis** and the vertical axis is the **imaginary axis,** as shown in the figure below.

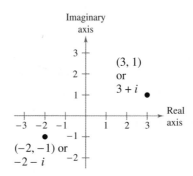

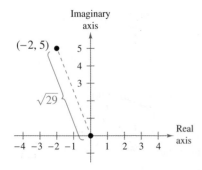

The complex plane has many practical applications. For example, in Exercise 49 on page 444, you will use the complex plane to write complex numbers that represent the positions of two ships.

The **absolute value,** or **modulus,** of the complex number $z = a + bi$ is the distance between the origin $(0, 0)$ and the point (a, b). (The plural of modulus is *moduli.*)

Definition of the Absolute Value of a Complex Number

The **absolute value** of the complex number $z = a + bi$ is

$$|a + bi| = \sqrt{a^2 + b^2}.$$

When the complex number $z = a + bi$ is a real number (that is, when $b = 0$), this definition agrees with that given for the absolute value of a real number

$$|a + 0i| = \sqrt{a^2 + 0^2}$$
$$= |a|.$$

EXAMPLE 1 **Finding the Absolute Value of a Complex Number**

See LarsonPrecalculus.com for an interactive version of this type of example.

Plot $z = -2 + 5i$ in the complex plane and find its absolute value.

Solution The number is plotted in Figure 6.34. It has an absolute value of

$$|z| = \sqrt{(-2)^2 + 5^2}$$
$$= \sqrt{29}.$$

✓ *Checkpoint* ◀))) *Audio-video solution in English & Spanish at LarsonPrecalculus.com*

Plot $z = 3 - 4i$ in the complex plane and find its absolute value.

Figure 6.34

Operations with Complex Numbers in the Complex Plane

In Section 6.3, you learned how to add and subtract vectors geometrically in the coordinate plane. In a similar way, you can add and subtract complex numbers geometrically in the complex plane.

The complex number $z = a + bi$ can be represented by the vector $\mathbf{u} = \langle a, b \rangle$. For example, the complex number $z = 1 + 2i$ can be represented by the vector $\mathbf{u} = \langle 1, 2 \rangle$. To add two complex numbers geometrically, first represent them as vectors $\mathbf{u}$ and $\mathbf{v}$. Then add the vectors, as shown in the next two figures. The sum of the vectors represents the sum of the complex numbers.

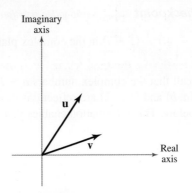

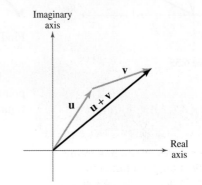

EXAMPLE 2   Adding in the Complex Plane

Find $(1 + 3i) + (2 + i)$ in the complex plane.

Solution

Let the vectors $\mathbf{u} = \langle 1, 3 \rangle$ and $\mathbf{v} = \langle 2, 1 \rangle$ represent the complex numbers $1 + 3i$ and $2 + i$, respectively. Graph the vectors $\mathbf{u}$, $\mathbf{v}$, and $\mathbf{u} + \mathbf{v}$, as shown at the right. From the graph, $\mathbf{u} + \mathbf{v} = \langle 3, 4 \rangle$, which implies that

$$(1 + 3i) + (2 + i) = 3 + 4i.$$

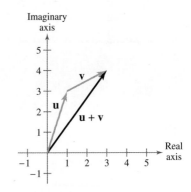

✓ *Checkpoint* 🔊)) *Audio-video solution in English & Spanish at LarsonPrecalculus.com*

Find $(3 + i) + (1 + 2i)$ in the complex plane.

To subtract two complex numbers geometrically, first represent them as vectors $\mathbf{u}$ and $\mathbf{v}$. Then subtract the vectors, as shown in the figure below. The difference of the vectors represents the difference of the complex numbers.

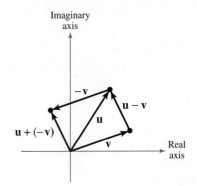

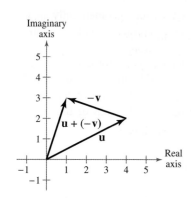

Figure 6.35

EXAMPLE 3 **Subtracting in the Complex Plane**

Find $(4 + 2i) - (3 - i)$ in the complex plane.

Solution

Let the vectors $\mathbf{u} = \langle 4, 2 \rangle$ and $\mathbf{v} = \langle 3, -1 \rangle$ represent the complex numbers $4 + 2i$ and $3 - i$, respectively. Graph the vectors $\mathbf{u}$, $-\mathbf{v}$, and $\mathbf{u} + (-\mathbf{v})$, as shown in Figure 6.35. From the graph, $\mathbf{u} - \mathbf{v} = \mathbf{u} + (-\mathbf{v}) = \langle 1, 3 \rangle$, which implies that

$$(4 + 2i) - (3 - i) = 1 + 3i.$$

✓ **Checkpoint** ◀))) *Audio-video solution in English & Spanish at LarsonPrecalculus.com*

Find $(2 - 4i) - (1 + i)$ in the complex plane.

Recall that the complex numbers $a + bi$ and $a - bi$ are *complex conjugates*. The points (a, b) and $(a, -b)$ are reflections of each other in the real axis, as shown in the figure below. This information enables you to find a complex conjugate geometrically.

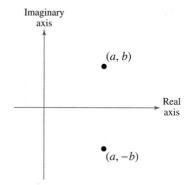

EXAMPLE 4 **Complex Conjugates in the Complex Plane**

Plot $z = -3 + i$ and its complex conjugate in the complex plane. Write the conjugate as a complex number.

Solution

The figure below shows the point $(-3, 1)$ and its reflection in the real axis, $(-3, -1)$. So, the complex conjugate of $-3 + i$ is $-3 - i$.

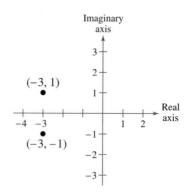

✓ **Checkpoint** ◀))) *Audio-video solution in English & Spanish at LarsonPrecalculus.com*

Plot $z = 2 - 3i$ and its complex conjugate in the complex plane. Write the conjugate as a complex number.

Distance and Midpoint Formulas in the Complex Plane

For two points in the complex plane, the distance between the points is the modulus (or absolute value) of the difference of the two corresponding complex numbers. Let (a, b) and (s, t) be points in the complex plane. One way to write the difference of the corresponding complex numbers is $(s + ti) - (a + bi) = (s - a) + (t - b)i$. The modulus of the difference is

$$|(s - a) + (t - b)i| = \sqrt{(s - a)^2 + (t - b)^2}.$$

So, $d = \sqrt{(s - a)^2 + (t - b)^2}$ is the distance between the points in the complex plane.

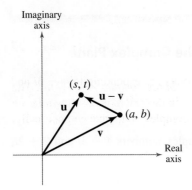

Imaginary axis

Real axis

Figure 6.36

> ### Distance Formula in the Complex Plane
> The distance d between the points (a, b) and (s, t) in the complex plane is
> $$d = \sqrt{(s - a)^2 + (t - b)^2}.$$

Figure 6.36 shows the points represented as vectors. The magnitude of the vector $\mathbf{u} - \mathbf{v}$ is the distance between (a, b) and (s, t).

$$\mathbf{u} - \mathbf{v} = \langle s - a, t - b \rangle$$
$$\|\mathbf{u} - \mathbf{v}\| = \sqrt{(s - a)^2 + (t - b)^2}$$

EXAMPLE 5   Finding Distance in the Complex Plane

Find the distance between $2 + 3i$ and $5 - 2i$ in the complex plane.

Solution

Let $a + bi = 2 + 3i$ and $s + ti = 5 - 2i$. The distance is

$$d = \sqrt{(s - a)^2 + (t - b)^2}$$
$$= \sqrt{(5 - 2)^2 + (-2 - 3)^2}$$
$$= \sqrt{3^2 + (-5)^2}$$
$$= \sqrt{34}$$
$$\approx 5.83 \text{ units}$$

as shown in the figure below.

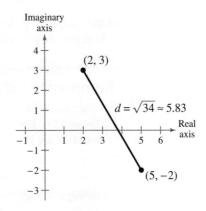

Imaginary axis

Real axis

$d = \sqrt{34} \approx 5.83$

(2, 3)

(5, −2)

✓ *Checkpoint* ◖))) Audio-video solution in English & Spanish at LarsonPrecalculus.com

Find the distance between $5 - 4i$ and $6 + 5i$ in the complex plane. ∎

To find the midpoint of the line segment joining two points in the complex plane, find the average values of the respective coordinates of the two endpoints.

Midpoint Formula in the Complex Plane

The midpoint of the line segment joining the points (a, b) and (s, t) in the complex plane is

$$\text{Midpoint} = \left(\frac{a + s}{2}, \frac{b + t}{2} \right).$$

EXAMPLE 6   Finding a Midpoint in the Complex Plane

Find the midpoint of the line segment joining the points corresponding to $4 - 3i$ and $2 + 2i$ in the complex plane.

Solution

Let the points $(4, -3)$ and $(2, 2)$ represent the complex numbers $4 - 3i$ and $2 + 2i$, respectively. Apply the Midpoint Formula.

$$\text{Midpoint} = \left(\frac{a + s}{2}, \frac{b + t}{2} \right) = \left(\frac{4 + 2}{2}, \frac{-3 + 2}{2} \right) = \left(3, -\frac{1}{2} \right)$$

The midpoint is $\left(3, -\frac{1}{2} \right)$, as shown in the figure below.

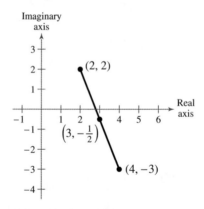

✓ **Checkpoint** 🔊))) *Audio-video solution in English & Spanish at LarsonPrecalculus.com*

Find the midpoint of the line segment joining the points corresponding to $2 + i$ and $5 - 5i$ in the complex plane. ∎

Summarize   (Section 6.5)

1. State the definition of the absolute value, or modulus, of a complex number *(page 438)*. For an example of finding the absolute value of a complex number, see Example 1.

2. Explain how to add, subtract, and find complex conjugates of complex numbers in the complex plane *(page 439)*. For examples of performing operations with complex numbers in the complex plane, see Examples 2–4.

3. Explain how to use the Distance and Midpoint Formulas in the complex plane *(page 441)*. For examples of using the Distance and Midpoint Formulas in the complex plane, see Examples 5 and 6.

6.5 Exercises

See **CalcChat.com** for tutorial help and worked-out solutions to odd-numbered exercises.

Vocabulary: Fill in the blanks.

1. In the complex plane, the horizontal axis is the _____ axis.
2. In the complex plane, the vertical axis is the _____ axis.
3. The _____ _____ of the complex number $a + bi$ is the distance between the origin and (a, b).
4. To subtract two complex numbers geometrically, first represent them as _____.
5. The points that represent a complex number and its complex conjugate are _____ of each other in the real axis.
6. The distance between two points in the complex plane is the _____ of the difference of the two corresponding complex numbers.

Skills and Applications

Matching In Exercises 7–14, match the complex number with its representation in the complex plane. [The representations are labeled (a)–(h).]

(a)
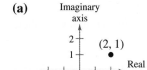

(b)

(c)
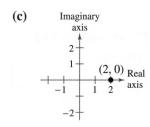

(d)
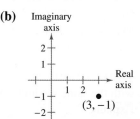

(e)
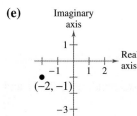

(f)
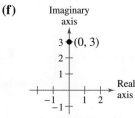

(g)
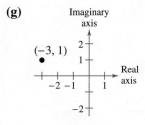

(h)
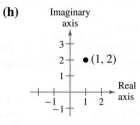

7. 2
8. $3i$
9. $1 + 2i$
10. $2 + i$
11. $3 - i$
12. $-3 + i$
13. $-2 - i$
14. $-1 - 3i$

 Finding the Absolute Value of a Complex Number In Exercises 15–20, plot the complex number and find its absolute value.

15. $-7i$
16. -7
17. $-6 + 8i$
18. $5 - 12i$
19. $4 - 6i$
20. $-8 + 3i$

 Adding in the Complex Plane In Exercises 21–28, find the sum of the complex numbers in the complex plane.

21. $(3 + i) + (2 + 5i)$
22. $(5 + 2i) + (3 + 4i)$
23. $(8 - 2i) + (2 + 6i)$
24. $(3 - i) + (-1 + 2i)$

25.
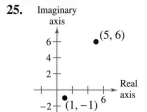

26.
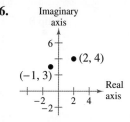

27.
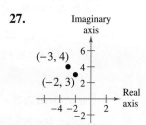

28.
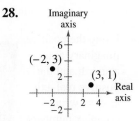

Subtracting in the Complex Plane In Exercises 29–36, find the difference of the complex numbers in the complex plane.

29. $(4 + 2i) - (6 + 4i)$
30. $(-3 + i) - (3 + i)$
31. $(5 - i) - (-5 + 2i)$
32. $(2 - 3i) - (3 + 2i)$
33. $2 - (2 + 6i)$
34. $-3 - (2 + 2i)$
35. $-2i - (3 - 5i)$
36. $3i - (-3 + 7i)$

Complex Conjugates in the Complex Plane In Exercises 37–40, plot the complex number and its complex conjugate. Write the conjugate as a complex number.

37. $2 + 3i$

38. $5 - 4i$

39. $-1 - 2i$

40. $-7 + 3i$

Finding Distance in the Complex Plane In Exercises 41–44, find the distance between the complex numbers in the complex plane.

41. $1 + 2i, -1 + 4i$

42. $-5 + i, -2 + 5i$

43. $6i, 3 - 4i$

44. $-7 - 3i, 3 + 5i$

Finding a Midpoint in the Complex Plane In Exercises 45–48, find the midpoint of the line segment joining the points corresponding to the complex numbers in the complex plane.

45. $2 + i, 6 + 5i$

46. $-3 + 4i, 1 - 2i$

47. $7i, 9 - 10i$

48. $-1 - \frac{3}{4}i, \frac{1}{2} + \frac{1}{4}i$

· · 49. Sailing · · · · · · · · · · · · · · · · · ·

Ship A is 3 miles east and 4 miles north of port. Ship B is 5 miles west and 2 miles north of port (see figure).

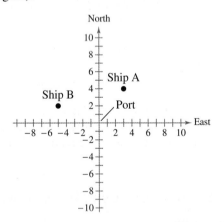

(a) Using the positive imaginary axis as north and the positive real axis as east, write complex numbers that represent the positions of Ship A and Ship B relative to port.

(b) How can you use the complex numbers in part (a) to find the distance between Ship A and Ship B?

· ·

50. Force Two forces are acting on a point. The first force has a horizontal component of 5 newtons and a vertical component of 3 newtons. The second force has a horizontal component of 4 newtons and a vertical component of 2 newtons.

(a) Plot the vectors that represent the two forces in the complex plane.

(b) Find the horizontal and vertical components of the resultant force acting on the point using the complex plane.

Exploration

True or False? In Exercises 51–54, determine whether the statement is true or false. Justify your answer.

51. The modulus of a complex number can be real or imaginary.

52. The distance between two points in the complex plane is always real.

53. The modulus of the sum of two complex numbers is equal to the sum of their moduli.

54. The modulus of the difference of two complex numbers is equal to the difference of their moduli.

55. Think About It What does the set of all points with the same modulus represent in the complex plane? Explain.

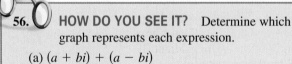

56. **HOW DO YOU SEE IT?** Determine which graph represents each expression.

(a) $(a + bi) + (a - bi)$

(b) $(a + bi) - (a - bi)$

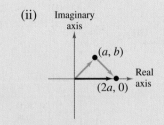

(i) Imaginary axis

(a, b)

Real axis

(a, −b)

(ii) Imaginary axis

(a, b)

Real axis

(2a, 0)

57. Think About It The points corresponding to a complex number and its complex conjugate are plotted in the complex plane. What type of triangle do these points form with the origin?

6.6 Trigonometric Form of a Complex Number

- Write trigonometric forms of complex numbers.
- Multiply and divide complex numbers written in trigonometric form.
- Use DeMoivre's Theorem to find powers of complex numbers.
- Find *n*th roots of complex numbers.

Trigonometric forms of complex numbers have applications in circuit analysis. For example, in Exercise 95 on page 453, you will use trigonometric forms of complex numbers to find the voltage of an alternating current circuit.

Trigonometric Form of a Complex Number

In Section 2.4, you learned how to add, subtract, multiply, and divide complex numbers. To work effectively with *powers* and *roots* of complex numbers, it is helpful to write complex numbers in trigonometric form. Consider the nonzero complex number $a + bi$, plotted at the right. By letting θ be the angle from the positive real axis (measured counterclockwise) to the line segment connecting the origin and the point (a, b), you can write $a = r \cos \theta$ and $b = r \sin \theta$, where $r = \sqrt{a^2 + b^2}$. Consequently, you have $a + bi = (r \cos \theta) + (r \sin \theta)i$, from which you can obtain the **trigonometric form of a complex number.**

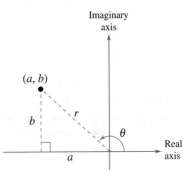

Trigonometric Form of a Complex Number

The **trigonometric form** of the complex number $z = a + bi$ is

$$z = r(\cos \theta + i \sin \theta)$$

where $a = r \cos \theta$, $b = r \sin \theta$, $r = \sqrt{a^2 + b^2}$, and $\tan \theta = b/a$. The number r is the **modulus** of z, and θ is an **argument** of z.

> •• **REMARK** For $0 \le \theta < 2\pi$, use the guidelines below. When z lies in Quadrant I, $\theta = \arctan(b/a)$. When z lies in Quadrant II or Quadrant III, $\theta = \pi + \arctan(b/a)$. When z lies in Quadrant IV, $\theta = 2\pi + \arctan(b/a)$.

The trigonometric form of a complex number is also called the *polar form*. There are infinitely many choices for θ, so the trigonometric form of a complex number is not unique. Normally, θ is restricted to the interval $0 \le \theta < 2\pi$, although on occasion it is convenient to use $\theta < 0$.

EXAMPLE 1 **Trigonometric Form of a Complex Number**

Write the complex number $z = -2 - 2\sqrt{3}i$ in trigonometric form.

Solution The modulus of z is $r = \sqrt{(-2)^2 + (-2\sqrt{3})^2} = \sqrt{16} = 4$, and the argument θ is determined from

$$\tan \theta = \frac{b}{a} = \frac{-2\sqrt{3}}{-2} = \sqrt{3}.$$

Because $z = -2 - 2\sqrt{3}i$ lies in Quadrant III, as shown in Figure 6.37, you have $\theta = \pi + \arctan\sqrt{3} = \pi + (\pi/3) = 4\pi/3$. So, the trigonometric form of z is

$$z = r(\cos \theta + i \sin \theta) = 4\left(\cos \frac{4\pi}{3} + i \sin \frac{4\pi}{3}\right).$$

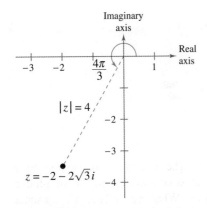

Figure 6.37

✓ **Checkpoint**))) Audio-video solution in English & Spanish at LarsonPrecalculus.com

Write the complex number $z = 6 - 6i$ in trigonometric form.

EXAMPLE 2 **Writing a Complex Number in Standard Form**

Write $z = \sqrt{8}[\cos(-\pi/3) + i\sin(-\pi/3)]$ in standard form $a + bi$.

Solution Because $\cos(-\pi/3) = 1/2$ and $\sin(-\pi/3) = -\sqrt{3}/2$, you can write

$$z = \sqrt{8}\left[\cos\left(-\frac{\pi}{3}\right) + i\sin\left(-\frac{\pi}{3}\right)\right] = 2\sqrt{2}\left(\frac{1}{2} - \frac{\sqrt{3}}{2}i\right) = \sqrt{2} - \sqrt{6}i.$$

✓ **Checkpoint** ◀)) Audio-video solution in English & Spanish at LarsonPrecalculus.com

Write $z = 8[\cos(2\pi/3) + i\sin(2\pi/3)]$ in standard form $a + bi$. ∎

Multiplication and Division of Complex Numbers

The trigonometric form adapts nicely to multiplication and division of complex numbers. Consider two complex numbers $z_1 = r_1(\cos\theta_1 + i\sin\theta_1)$ and $z_2 = r_2(\cos\theta_2 + i\sin\theta_2)$. The product of z_1 and z_2 is

$$z_1z_2 = r_1r_2(\cos\theta_1 + i\sin\theta_1)(\cos\theta_2 + i\sin\theta_2)$$
$$= r_1r_2[(\cos\theta_1\cos\theta_2 - \sin\theta_1\sin\theta_2) + i(\sin\theta_1\cos\theta_2 + \cos\theta_1\sin\theta_2)].$$

Using the sum and difference formulas for cosine and sine, this equation is equivalent to

$$z_1z_2 = r_1r_2[\cos(\theta_1 + \theta_2) + i\sin(\theta_1 + \theta_2)].$$

This establishes the first part of the rule below. The second part is left for you to verify (see Exercise 99).

> **Product and Quotient of Two Complex Numbers**
>
> Let $z_1 = r_1(\cos\theta_1 + i\sin\theta_1)$ and $z_2 = r_2(\cos\theta_2 + i\sin\theta_2)$ be complex numbers.
>
> $$z_1z_2 = r_1r_2[\cos(\theta_1 + \theta_2) + i\sin(\theta_1 + \theta_2)] \qquad \text{Product}$$
>
> $$\frac{z_1}{z_2} = \frac{r_1}{r_2}[\cos(\theta_1 - \theta_2) + i\sin(\theta_1 - \theta_2)], \quad z_2 \neq 0 \qquad \text{Quotient}$$

EXAMPLE 3 **Multiplying Complex Numbers**

Find the product z_1z_2 of $z_1 = 3\left(\cos\frac{\pi}{4} + i\sin\frac{\pi}{4}\right)$ and $z_2 = 2\left(\cos\frac{3\pi}{4} + i\sin\frac{3\pi}{4}\right)$.

Solution

$$z_1z_2 = 3\left(\cos\frac{\pi}{4} + i\sin\frac{\pi}{4}\right) \cdot 2\left(\cos\frac{3\pi}{4} + i\sin\frac{3\pi}{4}\right)$$
$$= 6\left[\cos\left(\frac{\pi}{4} + \frac{3\pi}{4}\right) + i\sin\left(\frac{\pi}{4} + \frac{3\pi}{4}\right)\right] \qquad \text{Multiply moduli and add arguments.}$$
$$= 6(\cos\pi + i\sin\pi)$$
$$= 6[-1 + i(0)]$$
$$= -6$$

✓ **Checkpoint** ◀)) Audio-video solution in English & Spanish at LarsonPrecalculus.com

Find the product z_1z_2 of $z_1 = 2\left(\cos\frac{5\pi}{6} + i\sin\frac{5\pi}{6}\right)$ and $z_2 = 5\left(\cos\frac{7\pi}{6} + i\sin\frac{7\pi}{6}\right)$. ∎

EXAMPLE 4 **Multiplying Complex Numbers**

$$2\left(\cos\frac{2\pi}{3} + i\sin\frac{2\pi}{3}\right) \cdot 8\left(\cos\frac{11\pi}{6} + i\sin\frac{11\pi}{6}\right)$$

$$= 16\left[\cos\left(\frac{2\pi}{3} + \frac{11\pi}{6}\right) + i\sin\left(\frac{2\pi}{3} + \frac{11\pi}{6}\right)\right] \qquad \text{Multiply moduli and add arguments.}$$

$$= 16\left(\cos\frac{5\pi}{2} + i\sin\frac{5\pi}{2}\right)$$

$$= 16\left(\cos\frac{\pi}{2} + i\sin\frac{\pi}{2}\right) \qquad \frac{5\pi}{2} \text{ and } \frac{\pi}{2} \text{ are coterminal.}$$

$$= 16i$$

> **REMARK** Check the solution to Example 4 by first converting the complex numbers to the standard forms $-1 + \sqrt{3}i$ and $4\sqrt{3} - 4i$ and then multiplying algebraically, as in Section 2.4.

✓ **Checkpoint** ◀))) *Audio-video solution in English & Spanish at LarsonPrecalculus.com*

Find the product z_1z_2 of $z_1 = 3\left(\cos\frac{\pi}{3} + i\sin\frac{\pi}{3}\right)$ and $z_2 = 4\left(\cos\frac{\pi}{6} + i\sin\frac{\pi}{6}\right)$.

EXAMPLE 5 **Dividing Complex Numbers**

$$\frac{24(\cos 300° + i\sin 300°)}{8(\cos 75° + i\sin 75°)} = 3[\cos(300° - 75°) + i\sin(300° - 75°)]$$

$$= 3(\cos 225° + i\sin 225°)$$

$$= -\frac{3\sqrt{2}}{2} - \frac{3\sqrt{2}}{2}i$$

> ▷ **TECHNOLOGY** Some graphing utilities can multiply and divide complex numbers in trigonometric form. If you have access to such a graphing utility, use it to check the solutions to Examples 3–5.

✓ **Checkpoint** ◀))) *Audio-video solution in English & Spanish at LarsonPrecalculus.com*

Find the quotient z_1/z_2 of $z_1 = \cos 40° + i\sin 40°$ and $z_2 = \cos 10° + i\sin 10°$. ∎

In Section 6.5, you added, subtracted, and found complex conjugates of complex numbers geometrically in the complex plane. In a similar way, you can multiply complex numbers geometrically in the complex plane.

EXAMPLE 6 **Multiplying in the Complex Plane**

Find the product z_1z_2 of $z_1 = 2\left(\cos\frac{\pi}{6} + i\sin\frac{\pi}{6}\right)$ and $z_2 = 2\left(\cos\frac{\pi}{3} + i\sin\frac{\pi}{3}\right)$ in the complex plane.

Solution

Let $\mathbf{u} = 2\langle\cos(\pi/6), \sin(\pi/6)\rangle = \langle\sqrt{3}, 1\rangle$ and $\mathbf{v} = 2\langle\cos(\pi/3), \sin(\pi/3)\rangle = \langle 1, \sqrt{3}\rangle$. Then $\|\mathbf{u}\| = \sqrt{(\sqrt{3})^2 + 1^2} = \sqrt{4} = 2$ and $\|\mathbf{v}\| = \sqrt{1^2 + (\sqrt{3})^2} = \sqrt{4} = 2$. So, the magnitude of the product vector is $2(2) = 4$. The sum of the direction angles is $(\pi/6) + (\pi/3) = \pi/2$. So, the product vector lies on the imaginary axis and is represented in vector form as $\langle 0, 4\rangle$, as shown in Figure 6.38. This implies that $z_1z_2 = 4i$.

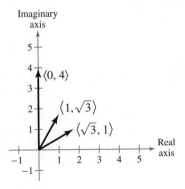

Figure 6.38

✓ **Checkpoint** ◀))) *Audio-video solution in English & Spanish at LarsonPrecalculus.com*

Find the product z_1z_2 of $z_1 = 2\left(\cos\frac{\pi}{4} + i\sin\frac{\pi}{4}\right)$ and $z_2 = 4\left(\cos\frac{3\pi}{4} + i\sin\frac{3\pi}{4}\right)$ in the complex plane. ∎

Powers of Complex Numbers

The trigonometric form of a complex number is used to raise a complex number to a power. To accomplish this, consider repeated use of the multiplication rule.

$$z = r(\cos\theta + i\sin\theta)$$

$$z^2 = r(\cos\theta + i\sin\theta)r(\cos\theta + i\sin\theta) = r^2(\cos 2\theta + i\sin 2\theta)$$

$$z^3 = r^2(\cos 2\theta + i\sin 2\theta)r(\cos\theta + i\sin\theta) = r^3(\cos 3\theta + i\sin 3\theta)$$

$$z^4 = r^4(\cos 4\theta + i\sin 4\theta)$$

$$z^5 = r^5(\cos 5\theta + i\sin 5\theta)$$

$$\vdots$$

This pattern leads to **DeMoivre's Theorem,** which is named after the French mathematician Abraham DeMoivre (1667–1754).

DeMoivre's Theorem

If $z = r(\cos\theta + i\sin\theta)$ is a complex number and n is a positive integer, then

$$z^n = [r(\cos\theta + i\sin\theta)]^n$$

$$= r^n(\cos n\theta + i\sin n\theta).$$

Abraham DeMoivre (1667–1754) is remembered for his work in probability theory and DeMoivre's Theorem. His book *The Doctrine of Chances* (published in 1718) includes the theory of recurring series and the theory of partial fractions.

EXAMPLE 7 **Finding a Power of a Complex Number**

Use DeMoivre's Theorem to find $\left(-1 + \sqrt{3}i\right)^{12}$.

Solution The modulus of $z = -1 + \sqrt{3}i$ is

$$r = \sqrt{(-1)^2 + \left(\sqrt{3}\right)^2} = 2$$

and the argument θ is determined from $\tan\theta = \sqrt{3}/(-1)$. Because $z = -1 + \sqrt{3}i$ lies in Quadrant II,

$$\theta = \pi + \arctan\frac{\sqrt{3}}{-1} = \pi + \left(-\frac{\pi}{3}\right) = \frac{2\pi}{3}.$$

So, the trigonometric form of z is

$$z = -1 + \sqrt{3}i = 2\left(\cos\frac{2\pi}{3} + i\sin\frac{2\pi}{3}\right).$$

Then, by DeMoivre's Theorem, you have

$$\left(-1 + \sqrt{3}i\right)^{12} = \left[2\left(\cos\frac{2\pi}{3} + i\sin\frac{2\pi}{3}\right)\right]^{12}$$

$$= 2^{12}\left[\cos\frac{12(2\pi)}{3} + i\sin\frac{12(2\pi)}{3}\right]$$

$$= 4096(\cos 8\pi + i\sin 8\pi)$$

$$= 4096(1 + 0)$$

$$= 4096.$$

✓ *Checkpoint*  *Audio-video solution in English & Spanish at LarsonPrecalculus.com*

Use DeMoivre's Theorem to find $(-1 - i)^4$.

Roots of Complex Numbers

Recall that a consequence of the Fundamental Theorem of Algebra is that a polynomial equation of degree n has n solutions in the complex number system. For example, the equation $x^6 = 1$ has six solutions. To find these solutions, use factoring and the Quadratic Formula.

$$x^6 - 1 = 0$$

$$(x^3 - 1)(x^3 + 1) = 0$$

$$(x - 1)(x^2 + x + 1)(x + 1)(x^2 - x + 1) = 0$$

Consequently, the solutions are

$$x = \pm 1, \quad x = \frac{-1 \pm \sqrt{3}i}{2}, \quad \text{and} \quad x = \frac{1 \pm \sqrt{3}i}{2}.$$

Each of these numbers is a sixth root of 1. In general, an **nth root of a complex number** is defined as follows.

Definition of an nth Root of a Complex Number

The complex number $u = a + bi$ is an **nth root** of the complex number z when

$$z = u^n$$

$$= (a + bi)^n.$$

To find a formula for an nth root of a complex number, let u be an nth root of z, where

$$u = s(\cos \beta + i \sin \beta)$$

and

$$z = r(\cos \theta + i \sin \theta).$$

By DeMoivre's Theorem and the fact that $u^n = z$, you have

$$s^n(\cos n\beta + i \sin n\beta) = r(\cos \theta + i \sin \theta).$$

Taking the absolute value of each side of this equation, it follows that $s^n = r$. Substituting back into the previous equation and dividing by r gives

$$\cos n\beta + i \sin n\beta = \cos \theta + i \sin \theta.$$

So, it follows that

$$\cos n\beta = \cos \theta$$

and

$$\sin n\beta = \sin \theta.$$

Both sine and cosine have a period of 2π, so these last two equations have solutions if and only if the angles differ by a multiple of 2π. Consequently, there must exist an integer k such that

$$n\beta = \theta + 2\pi k$$

$$\beta = \frac{\theta + 2\pi k}{n}.$$

Substituting this value of β and $s = \sqrt[n]{r}$ into the trigonometric form of u gives the result stated on the next page.

Finding *n*th Roots of a Complex Number

For a positive integer n, the complex number $z = r(\cos \theta + i \sin \theta)$ has exactly n distinct nth roots given by

$$z_k = \sqrt[n]{r}\left(\cos \frac{\theta + 2\pi k}{n} + i \sin \frac{\theta + 2\pi k}{n}\right)$$

where $k = 0, 1, 2, \ldots, n - 1$.

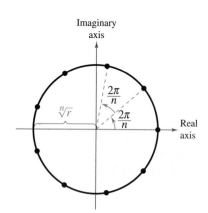

Imaginary axis

Real axis

Figure 6.39

When $k > n - 1$, the roots begin to repeat. For example, when $k = n$, the angle

$$\frac{\theta + 2\pi n}{n} = \frac{\theta}{n} + 2\pi$$

is coterminal with θ/n, which is also obtained when $k = 0$.

The formula for the nth roots of a complex number z has a geometrical interpretation, as shown in Figure 6.39. Note that the nth roots of z all have the same magnitude $\sqrt[n]{r}$, so they all lie on a circle of radius $\sqrt[n]{r}$ with center at the origin. Furthermore, successive nth roots have arguments that differ by $2\pi/n$, so the n roots are equally spaced around the circle.

You have already found the sixth roots of 1 by factoring and using the Quadratic Formula. Example 8 shows how to solve the same problem with the formula for nth roots.

EXAMPLE 8 **Finding the *n*th Roots of a Real Number**

Find all sixth roots of 1.

Solution First, write 1 in the trigonometric form $z = 1(\cos 0 + i \sin 0)$. Then, by the nth root formula with $n = 6$, $r = 1$, and $\theta = 0$, the roots have the form

$$z_k = \sqrt[6]{1}\left(\cos \frac{0 + 2\pi k}{6} + i \sin \frac{0 + 2\pi k}{6}\right) = \cos \frac{\pi k}{3} + i \sin \frac{\pi k}{3}.$$

So, for $k = 0, 1, 2, 3, 4,$ and 5, the roots are as listed below. (See Figure 6.40.)

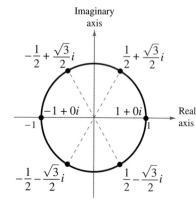

Imaginary axis

Real axis

Figure 6.40

$$z_0 = \cos 0 + i \sin 0 = 1$$

$$z_1 = \cos \frac{\pi}{3} + i \sin \frac{\pi}{3} = \frac{1}{2} + \frac{\sqrt{3}}{2}i \qquad \text{Increment by } \frac{2\pi}{n} = \frac{2\pi}{6} = \frac{\pi}{3}$$

$$z_2 = \cos \frac{2\pi}{3} + i \sin \frac{2\pi}{3} = -\frac{1}{2} + \frac{\sqrt{3}}{2}i$$

$$z_3 = \cos \pi + i \sin \pi = -1$$

$$z_4 = \cos \frac{4\pi}{3} + i \sin \frac{4\pi}{3} = -\frac{1}{2} - \frac{\sqrt{3}}{2}i$$

$$z_5 = \cos \frac{5\pi}{3} + i \sin \frac{5\pi}{3} = \frac{1}{2} - \frac{\sqrt{3}}{2}i$$

✓ **Checkpoint** ◄))) *Audio-video solution in English & Spanish at LarsonPrecalculus.com*

Find all fourth roots of 1. ∎

In Figure 6.40, notice that the roots obtained in Example 8 all have a magnitude of 1 and are equally spaced around the unit circle. Also notice that the complex roots occur in conjugate pairs, as discussed in Section 2.5. The n distinct nth roots of 1 are called the **nth roots of unity.**

EXAMPLE 9 Finding the *n*th Roots of a Complex Number

See LarsonPrecalculus.com for an interactive version of this type of example.

Find the three cube roots of $z = -2 + 2i$.

Solution The modulus of z is

$$r = \sqrt{(-2)^2 + 2^2} = \sqrt{8}$$

and the argument θ is determined from

$$\tan \theta = \frac{b}{a} = \frac{2}{-2} = -1.$$

Because z lies in Quadrant II, the trigonometric form of z is

$$z = -2 + 2i = \sqrt{8}(\cos 135° + i \sin 135°). \qquad \theta = \pi + \arctan(-1) = 3\pi/4 = 135°$$

By the nth root formula, the roots have the form

$$z_k = \sqrt[6]{8}\left(\cos \frac{135° + 360°k}{3} + i \sin \frac{135° + 360°k}{3}\right).$$

So, for $k = 0$, 1, and 2, the roots are as listed below. (See Figure 6.41.)

$$z_0 = \sqrt[6]{8}\left(\cos \frac{135° + 360°(0)}{3} + i \sin \frac{135° + 360°(0)}{3}\right)$$

$$= \sqrt{2}(\cos 45° + i \sin 45°)$$

$$= 1 + i$$

$$z_1 = \sqrt[6]{8}\left(\cos \frac{135° + 360°(1)}{3} + i \sin \frac{135° + 360°(1)}{3}\right)$$

$$= \sqrt{2}(\cos 165° + i \sin 165°)$$

$$\approx -1.3660 + 0.3660i$$

$$z_2 = \sqrt[6]{8}\left(\cos \frac{135° + 360°(2)}{3} + i \sin \frac{135° + 360°(2)}{3}\right)$$

$$= \sqrt{2}(\cos 285° + i \sin 285°)$$

$$\approx 0.3660 - 1.3660i.$$

✓ **Checkpoint** 🔊))) *Audio-video solution in English & Spanish at LarsonPrecalculus.com*

Find the three cube roots of $z = -6 + 6i$. ■

• • **REMARK** In Example 9, $r = \sqrt{8}$, so it follows that

$$\sqrt[n]{r} = \sqrt[3]{\sqrt{8}}$$

$$= \sqrt[3 \cdot 2]{8}$$

$$= \sqrt[6]{8}.$$

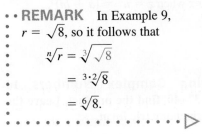

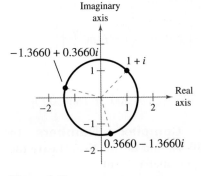

Figure 6.41

Summarize (Section 6.6)

1. State the trigonometric form of a complex number (*page 445*). For examples of writing complex numbers in trigonometric form and standard form, see Examples 1 and 2.

2. Explain how to multiply and divide complex numbers written in trigonometric form (*page 446*). For examples of multiplying and dividing complex numbers written in trigonometric form, see Examples 3–6.

3. Explain how to use DeMoivre's Theorem to find a power of a complex number (*page 448*). For an example of using DeMoivre's Theorem, see Example 7.

4. Explain how to find the *n*th roots of a complex number (*page 449*). For examples of finding *n*th roots of complex numbers, see Examples 8 and 9.

6.6 Exercises
See **CalcChat.com** for tutorial help and worked-out solutions to odd-numbered exercises.

Vocabulary: Fill in the blanks.

1. The _____ _____ of the complex number $z = a + bi$ is $z = r(\cos \theta + i \sin \theta)$, where r is the _____ of z and θ is an _____ of z.

2. _____ Theorem states that if $z = r(\cos \theta + i \sin \theta)$ is a complex number and n is a positive integer, then $z^n = r^n(\cos n\theta + i \sin n\theta)$.

3. The complex number $u = a + bi$ is an _____ _____ of the complex number z when $z = u^n = (a + bi)^n$.

4. Successive nth roots of a complex number have arguments that differ by _____.

Skills and Applications

 Trigonometric Form of a Complex Number In Exercises 5–24, plot the complex number. Then write the trigonometric form of the complex number.

5. $1 + i$
6. $5 - 5i$
7. $1 - \sqrt{3}i$
8. $4 - 4\sqrt{3}i$
9. $-2(1 + \sqrt{3}i)$
10. $\frac{5}{2}(\sqrt{3} - i)$
11. $-5i$
12. $12i$
13. 2
14. 4
15. $-7 + 4i$
16. $3 - i$
17. $2\sqrt{2} - i$
18. $-3 - i$
19. $5 + 2i$
20. $8 + 3i$
21. $3 + \sqrt{3}i$
22. $3\sqrt{2} - 7i$
23. $-8 - 5\sqrt{3}i$
24. $-9 - 2\sqrt{10}i$

 Writing a Complex Number in Standard Form In Exercises 25–32, write the standard form of the complex number. Then plot the complex number.

25. $2(\cos 60° + i \sin 60°)$
26. $5(\cos 135° + i \sin 135°)$
27. $\sqrt{48}[\cos(-30°) + i \sin(-30°)]$
28. $\sqrt{8}(\cos 225° + i \sin 225°)$
29. $\frac{9}{4}\left(\cos \frac{3\pi}{4} + i \sin \frac{3\pi}{4}\right)$
30. $6\left(\cos \frac{5\pi}{12} + i \sin \frac{5\pi}{12}\right)$
31. $5[\cos(198° \, 45') + i \sin(198° \, 45')]$
32. $9.75[\cos(280° \, 30') + i \sin(280° \, 30')]$

Writing a Complex Number in Standard Form In Exercises 33–36, use a graphing utility to write the complex number in standard form.

33. $5\left(\cos \frac{\pi}{9} + i \sin \frac{\pi}{9}\right)$
34. $10\left(\cos \frac{2\pi}{5} + i \sin \frac{2\pi}{5}\right)$
35. $2(\cos 155° + i \sin 155°)$
36. $9(\cos 58° + i \sin 58°)$

 Multiplying Complex Numbers In Exercises 37–40, find the product. Leave the result in trigonometric form.

37. $\left[2\left(\cos \frac{\pi}{4} + i \sin \frac{\pi}{4}\right)\right]\left[6\left(\cos \frac{\pi}{12} + i \sin \frac{\pi}{12}\right)\right]$

38. $\left[\frac{3}{4}\left(\cos \frac{\pi}{3} + i \sin \frac{\pi}{3}\right)\right]\left[4\left(\cos \frac{3\pi}{4} + i \sin \frac{3\pi}{4}\right)\right]$

39. $\left[\frac{5}{3}(\cos 120° + i \sin 120°)\right]\left[\frac{2}{3}(\cos 30° + i \sin 30°)\right]$

40. $\left[\frac{1}{2}(\cos 100° + i \sin 100°)\right]\left[\frac{4}{5}(\cos 300° + i \sin 300°)\right]$

 Dividing Complex Numbers In Exercises 41–44, find the quotient. Leave the result in trigonometric form.

41. $\dfrac{3(\cos 50° + i \sin 50°)}{9(\cos 20° + i \sin 20°)}$
42. $\dfrac{\cos 120° + i \sin 120°}{2(\cos 40° + i \sin 40°)}$

43. $\dfrac{\cos \pi + i \sin \pi}{\cos(\pi/3) + i \sin(\pi/3)}$
44. $\dfrac{5(\cos 4.3 + i \sin 4.3)}{4(\cos 2.1 + i \sin 2.1)}$

Multiplying or Dividing Complex Numbers In Exercises 45–50, (a) write the trigonometric forms of the complex numbers, (b) perform the operation using the trigonometric forms, and (c) perform the operation using the standard forms, and check your result with that of part (b).

45. $(2 + 2i)(1 - i)$
46. $(\sqrt{3} + i)(1 + i)$
47. $-2i(1 + i)$
48. $3i(1 - \sqrt{2}i)$
49. $\dfrac{3 + 4i}{1 - \sqrt{3}i}$
50. $\dfrac{1 + \sqrt{3}i}{6 - 3i}$

 Multiplying in the Complex Plane In Exercises 51 and 52, find the product in the complex plane.

51. $\left[2\left(\cos \frac{2\pi}{3} + i \sin \frac{2\pi}{3}\right)\right]\left[\frac{1}{2}\left(\cos \frac{\pi}{3} + i \sin \frac{\pi}{3}\right)\right]$

52. $\left[2\left(\cos \frac{\pi}{4} + i \sin \frac{\pi}{4}\right)\right]\left[3\left(\cos \frac{\pi}{4} + i \sin \frac{\pi}{4}\right)\right]$

Finding a Power of a Complex Number
In Exercises 53–68, use DeMoivre's Theorem to find the power of the complex number. Write the result in standard form.

53. $[5(\cos 20° + i \sin 20°)]^3$ **54.** $[3(\cos 60° + i \sin 60°)]^4$

55. $\left(\cos \dfrac{\pi}{4} + i \sin \dfrac{\pi}{4}\right)^{12}$ **56.** $\left[2\left(\cos \dfrac{\pi}{2} + i \sin \dfrac{\pi}{2}\right)\right]^8$

57. $[5(\cos 3.2 + i \sin 3.2)]^4$ **58.** $(\cos 0 + i \sin 0)^{20}$

59. $[3(\cos 15° + i \sin 15°)]^4$ **60.** $\left[2\left(\cos \dfrac{\pi}{8} + i \sin \dfrac{\pi}{8}\right)\right]^6$

61. $(1 + i)^5$ **62.** $(2 + 2i)^6$

63. $(-1 + i)^6$ **64.** $(3 - 2i)^8$

65. $2(\sqrt{3} + i)^{10}$ **66.** $4(1 - \sqrt{3}i)^3$

67. $(3 - 2i)^5$ **68.** $(\sqrt{5} - 4i)^3$

Graphing Powers of a Complex Number In Exercises 69 and 70, represent the powers z, z^2, z^3, and z^4 graphically. Describe the pattern.

69. $z = \dfrac{\sqrt{2}}{2}(1 + i)$ **70.** $z = \dfrac{1}{2}(1 + \sqrt{3}i)$

Finding the nth Roots of a Complex Number In Exercises 71–86, (a) use the formula on page 450 to find the roots of the complex number, (b) write each of the roots in standard form, and (c) represent each of the roots graphically.

71. Square roots of $5(\cos 120° + i \sin 120°)$

72. Square roots of $16(\cos 60° + i \sin 60°)$

73. Cube roots of $8\left(\cos \dfrac{2\pi}{3} + i \sin \dfrac{2\pi}{3}\right)$

74. Fifth roots of $32\left(\cos \dfrac{5\pi}{6} + i \sin \dfrac{5\pi}{6}\right)$

75. Cube roots of $-\dfrac{125}{2}(1 + \sqrt{3}i)$

76. Cube roots of $-4\sqrt{2}(-1 + i)$

77. Square roots of $-25i$

78. Fourth roots of $625i$

79. Fourth roots of 16 **80.** Fourth roots of i

81. Fifth roots of 1 **82.** Cube roots of 1000

83. Cube roots of -125 **84.** Fourth roots of -4

85. Fifth roots of $4(1 - i)$ **86.** Sixth roots of $64i$

Solving an Equation In Exercises 87–94, use the formula on page 450 to find all solutions of the equation and represent the solutions graphically.

87. $x^4 + i = 0$ **88.** $x^3 + 1 = 0$

89. $x^5 + 243 = 0$ **90.** $x^3 - 27 = 0$

91. $x^4 + 16i = 0$ **92.** $x^6 + 64i = 0$

93. $x^3 - (1 - i) = 0$ **94.** $x^4 + (1 + i) = 0$

95. Ohm's Law

Ohm's law for alternating current circuits is $E = IZ$, where E is the voltage in volts, I is the current in amperes, and Z is the impedance in ohms. Each variable is a complex number.

(a) Write E in trigonometric form when $I = 6(\cos 41° + i \sin 41°)$ amperes and $Z = 4[\cos(-11°) + i \sin(-11°)]$ ohms.

(b) Write the voltage from part (a) in standard form.

(c) A voltmeter measures the magnitude of the voltage in a circuit. What would be the reading on a voltmeter for the circuit described in part (a)?

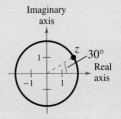

96. HOW DO YOU SEE IT?
The figure shows one of the fourth roots of a complex number z.

(a) How many roots are not shown?

(b) Describe the other roots.

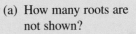
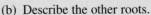

Exploration

True or False? In Exercises 97 and 98, determine whether the statement is true or false. Justify your answer.

97. Geometrically, the nth roots of any complex number z are all equally spaced around the unit circle.

98. The product of two complex numbers is zero only when the modulus of one (or both) of the complex numbers is zero.

99. Quotient of Two Complex Numbers Given two complex numbers $z_1 = r_1(\cos \theta_1 + i \sin \theta_1)$ and $z_2 = r_2(\cos \theta_2 + i \sin \theta_2)$, $z_2 \ne 0$, show that

$$\dfrac{z_1}{z_2} = \dfrac{r_1}{r_2}[\cos(\theta_1 - \theta_2) + i \sin(\theta_1 - \theta_2)].$$

100. Negative of a Complex Number Show that the negative of $z = r(\cos \theta + i \sin \theta)$ is $-z = r[\cos(\theta + \pi) + i \sin(\theta + \pi)]$.

101. Complex Conjugates Show that

$$\bar{z} = r[\cos(-\theta) + i \sin(-\theta)]$$

is the complex conjugate of $z = r(\cos \theta + i \sin \theta)$. Then find (a) $z\bar{z}$ and (b) $z/\bar{z}$, $\bar{z} \ne 0$.

Chapter Summary

	What Did You Learn?	Explanation/Examples	Review Exercises
Section 6.1	Use the Law of Sines to solve oblique triangles (AAS or ASA) (p. 400).	**Law of Sines** If ABC is a triangle with sides a, b, and c, then $$\frac{a}{\sin A} = \frac{b}{\sin B} = \frac{c}{\sin C}.$$	1–12
	Use the Law of Sines to solve oblique triangles (SSA) (p. 402).	If two sides and one opposite angle are given, then three possible situations can occur: (1) no such triangle exists, (2) one such triangle exists, or (3) two distinct triangles exist that satisfy the conditions.	1–12
	Find the areas of oblique triangles (p. 404).	Area $= \frac{1}{2}bc \sin A = \frac{1}{2}ab \sin C = \frac{1}{2}ac \sin B$	13–16
	Use the Law of Sines to model and solve real-life problems (p. 405).	The Law of Sines can be used to approximate the total distance of a boat race course. (See Example 7.)	17, 18
Section 6.2	Use the Law of Cosines to solve oblique triangles (SSS or SAS) (p. 409).	**Law of Cosines** **Standard Form** **Alternative Form** $a^2 = b^2 + c^2 - 2bc \cos A$ $\cos A = \dfrac{b^2 + c^2 - a^2}{2bc}$ $b^2 = a^2 + c^2 - 2ac \cos B$ $\cos B = \dfrac{a^2 + c^2 - b^2}{2ac}$ $c^2 = a^2 + b^2 - 2ab \cos C$ $\cos C = \dfrac{a^2 + b^2 - c^2}{2ab}$	19–30
	Use the Law of Cosines to model and solve real-life problems (p. 411).	The Law of Cosines can be used to find the distance between the pitcher's mound and first base on a women's softball field. (See Example 3.)	31, 32
	Use Heron's Area Formula to find areas of triangles (p. 412).	**Heron's Area Formula:** Given any triangle with sides of lengths a, b, and c, the area of the triangle is Area $= \sqrt{s(s - a)(s - b)(s - c)}$, where $s = (a + b + c)/2$.	33–36
Section 6.3	Represent vectors as directed line segments (p. 416).	Initial point P $\xrightarrow{\overrightarrow{PQ}}$ Q Terminal point	37, 38
	Write component forms of vectors (p. 417).	The component form of the vector with initial point $P(p_1, p_2)$ and terminal point $Q(q_1, q_2)$ is given by $\overrightarrow{PQ} = \langle q_1 - p_1, q_2 - p_2 \rangle = \langle v_1, v_2 \rangle = \mathbf{v}.$	39, 40
	Perform basic vector operations and represent vector operations graphically (p. 418).	Let $\mathbf{u} = \langle u_1, u_2 \rangle$ and $\mathbf{v} = \langle v_1, v_2 \rangle$ be vectors and let k be a scalar (a real number). $\mathbf{u} + \mathbf{v} = \langle u_1 + v_1, u_2 + v_2 \rangle$ $k\mathbf{u} = \langle ku_1, ku_2 \rangle$ $-\mathbf{v} = \langle -v_1, -v_2 \rangle$ $\mathbf{u} - \mathbf{v} = \langle u_1 - v_1, u_2 - v_2 \rangle$	41–48, 53–58
	Write vectors as linear combinations of unit vectors (p. 420).	The vector sum $\mathbf{v} = \langle v_1, v_2 \rangle = v_1\langle 1, 0 \rangle + v_2\langle 0, 1 \rangle = v_1\mathbf{i} + v_2\mathbf{j}$ is a linear combination of the vectors $\mathbf{i}$ and $\mathbf{j}$.	49–52

	What Did You Learn?	**Explanation/Examples**	**Review Exercises**
Section 6.3	Find direction angles of vectors (p. 422).	If $\mathbf{u} = a\mathbf{i} + b\mathbf{j}$, then the direction angle is determined from $\tan \theta = b/a$.	59–66
	Use vectors to model and solve real-life problems (p. 423).	Vectors can be used to find the resultant speed and true direction of an airplane. (See Example 11.)	67, 68
Section 6.4	Find the dot product of two vectors and use the properties of the dot product (p. 429).	The dot product of $\mathbf{u} = \langle u_1, u_2 \rangle$ and $\mathbf{v} = \langle v_1, v_2 \rangle$ is $\mathbf{u} \cdot \mathbf{v} = u_1 v_1 + u_2 v_2$.	69–80
	Find the angle between two vectors and determine whether two vectors are orthogonal (p. 430).	If θ is the angle between two nonzero vectors $\mathbf{u}$ and $\mathbf{v}$, then $\cos \theta = \dfrac{\mathbf{u} \cdot \mathbf{v}}{\|\mathbf{u}\|\|\mathbf{v}\|}$. The vectors $\mathbf{u}$ and $\mathbf{v}$ are orthogonal if and only if $\mathbf{u} \cdot \mathbf{v} = 0$.	81–88
	Write a vector as the sum of two vector components (p. 432).	Many applications in physics and engineering require the decomposition of a given vector into the sum of two vector components. (See Example 6.)	89–92
	Use vectors to determine the work done by a force (p. 434).	The work W done by a constant force $\mathbf{F}$ as its point of application moves along the vector $\overrightarrow{PQ}$ is given by **1.** $W = \|\text{proj}_{\overrightarrow{PQ}} \mathbf{F}\|\|\overrightarrow{PQ}\|$ or **2.** $W = \mathbf{F} \cdot \overrightarrow{PQ}$.	93–96
Section 6.5	Plot complex numbers in the complex plane and find absolute values of complex numbers (p. 438).	A complex number $z = a + bi$ can be represented by the point (a, b) in the complex plane. The horizontal axis is the real axis and the vertical axis is the imaginary axis. The absolute value, or modulus, of $z = a + bi$ is $|a + bi| = \sqrt{a^2 + b^2}$.	97–100
	Perform operations with complex numbers in the complex plane (p. 439).	Complex numbers can be added and subtracted geometrically in the complex plane. The points representing the complex conjugates $a + bi$ and $a - bi$ are reflections of each other in the real axis.	101–106
	Use the Distance and Midpoint Formulas in the complex plane (p. 441).	Let (a, b) and (s, t) be points in the complex plane. **Distance Formula** **Midpoint Formula** $d = \sqrt{(s - a)^2 + (t - b)^2}$ $\text{Midpoint} = \left(\dfrac{a + s}{2}, \dfrac{b + t}{2} \right)$	107–110
Section 6.6	Write trigonometric forms of complex numbers (p. 445).	The trigonometric form of the complex number $z = a + bi$ is $z = r(\cos \theta + i \sin \theta)$, where $a = r \cos \theta$, $b = r \sin \theta$, $r = \sqrt{a^2 + b^2}$, and $\tan \theta = b/a$.	111–116
	Multiply and divide complex numbers written in trigonometric form (p. 446).	Let $z_1 = r_1(\cos \theta_1 + i \sin \theta_1)$ and $z_2 = r_2(\cos \theta_2 + i \sin \theta_2)$ be complex numbers. $z_1 z_2 = r_1 r_2 [\cos(\theta_1 + \theta_2) + i \sin(\theta_1 + \theta_2)]$ $z_1/z_2 = (r_1/r_2)[\cos(\theta_1 - \theta_2) + i \sin(\theta_1 - \theta_2)], \quad z_2 \neq 0$	117–120
	Use DeMoivre's Theorem to find powers of complex numbers (p. 448).	**DeMoivre's Theorem:** If $z = r(\cos \theta + i \sin \theta)$ is a complex number and n is a positive integer, then $z^n = [r(\cos \theta + i \sin \theta)]^n = r^n(\cos n\theta + i \sin n\theta)$.	121–124
	Find nth roots of complex numbers (p. 449).	The complex number $u = a + bi$ is an nth root of the complex number z when $z = u^n = (a + bi)^n$.	125–132

Review Exercises See CalcChat.com for tutorial help and worked-out solutions to odd-numbered exercises.

6.1 **Using the Law of Sines** **In Exercises 1–12, use the Law of Sines to solve (if possible) the triangle. If two solutions exist, find both. Round your answers to two decimal places.**

1. $A = 38°$, $B = 70°$, $a = 8$
2. $A = 22°$, $B = 121°$, $a = 19$
3. $B = 72°$, $C = 82°$, $b = 54$
4. $B = 10°$, $C = 20°$, $c = 33$
5. $A = 16°$, $B = 98°$, $c = 8.4$
6. $A = 95°$, $B = 45°$, $c = 104.8$
7. $A = 24°$, $C = 48°$, $b = 27.5$
8. $B = 64°$, $C = 36°$, $a = 367$
9. $B = 150°$, $b = 30$, $c = 10$
10. $B = 150°$, $a = 10$, $b = 3$
11. $A = 75°$, $a = 51.2$, $b = 33.7$
12. $B = 25°$, $a = 6.2$, $b = 4$

Finding the Area of a Triangle **In Exercises 13–16, find the area of the triangle. Round your answers to one decimal place.**

13. $A = 33°$, $b = 7$, $c = 10$
14. $B = 80°$, $a = 4$, $c = 8$
15. $C = 119°$, $a = 18$, $b = 6$
16. $A = 11°$, $b = 22$, $c = 21$

17. **Height** From a certain distance, the angle of elevation to the top of a building is 17°. At a point 50 meters closer to the building, the angle of elevation is 31°. Find the height of the building.

18. **River Width** A surveyor finds that a tree on the opposite bank of a river flowing due east has a bearing of N 22° 30′ E from a certain point and a bearing of N 15° W from a point 400 feet downstream. Find the width of the river.

6.2 **Using the Law of Cosines** **In Exercises 19–26, use the Law of Cosines to solve the triangle. Round your answers to two decimal places.**

19. $a = 6$, $b = 9$, $c = 14$
20. $a = 75$, $b = 50$, $c = 110$
21. $a = 2.5$, $b = 5.0$, $c = 4.5$
22. $a = 16.4$, $b = 8.8$, $c = 12.2$
23. $B = 108°$, $a = 11$, $c = 11$
24. $B = 150°$, $a = 10$, $c = 20$
25. $C = 43°$, $a = 22.5$, $b = 31.4$
26. $A = 62°$, $b = 11.34$, $c = 19.52$

Solving a Triangle **In Exercises 27–30, determine whether the Law of Cosines is needed to solve the triangle. Then solve (if possible) the triangle. If two solutions exist, find both. Round your answers to two decimal places.**

27. $C = 64°$, $b = 9$, $c = 13$
28. $B = 52°$, $a = 4$, $c = 5$
29. $a = 13$, $b = 15$, $c = 24$
30. $A = 44°$, $B = 31°$, $c = 2.8$

31. **Geometry** The lengths of the diagonals of a parallelogram are 10 feet and 16 feet. Find the lengths of the sides of the parallelogram when the diagonals intersect at an angle of 28°.

32. **Air Navigation** Two planes leave an airport at approximately the same time. One flies 425 miles per hour at a bearing of 355°, and the other flies 530 miles per hour at a bearing of 67°. Draw a diagram that gives a visual representation of the problem and determine the distance between the planes after they fly for 2 hours.

Using Heron's Area Formula **In Exercises 33–36, use Heron's Area Formula to find the area of the triangle.**

33. $a = 3$, $b = 6$, $c = 8$
34. $a = 15$, $b = 8$, $c = 10$
35. $a = 12.3$, $b = 15.8$, $c = 3.7$
36. $a = \dfrac{4}{5}$, $b = \dfrac{3}{4}$, $c = \dfrac{5}{8}$

6.3 **Determining Whether Two Vectors Are Equivalent** **In Exercises 37 and 38, determine whether u and v are equivalent. Explain.**

37.
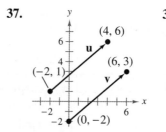

38.
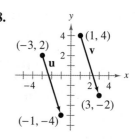

Finding the Component Form of a Vector **In Exercises 39 and 40, find the component form and magnitude of the vector v.**

39. Initial point: $(0, 10)$
    Terminal point: $(7, 3)$

40. Initial point: $(1, 5)$
    Terminal point: $(15, 9)$

Vector Operations In Exercises 41–48, find (a) **u** + **v**, (b) **u** − **v**, (c) 4**u**, and (d) 3**v** + 5**u**. Then sketch each resultant vector.

41. $\mathbf{u} = \langle -1, -3 \rangle, \quad \mathbf{v} = \langle -3, 6 \rangle$
42. $\mathbf{u} = \langle 4, 5 \rangle, \quad \mathbf{v} = \langle 0, -1 \rangle$
43. $\mathbf{u} = \langle -5, 2 \rangle, \quad \mathbf{v} = \langle 4, 4 \rangle$
44. $\mathbf{u} = \langle 1, -8 \rangle, \quad \mathbf{v} = \langle 3, -2 \rangle$
45. $\mathbf{u} = 2\mathbf{i} - \mathbf{j}, \quad \mathbf{v} = 5\mathbf{i} + 3\mathbf{j}$
46. $\mathbf{u} = -7\mathbf{i} - 3\mathbf{j}, \quad \mathbf{v} = 4\mathbf{i} - \mathbf{j}$
47. $\mathbf{u} = 4\mathbf{i}, \quad \mathbf{v} = -\mathbf{i} + 6\mathbf{j}$
48. $\mathbf{u} = -6\mathbf{j}, \quad \mathbf{v} = \mathbf{i} + \mathbf{j}$

Writing a Linear Combination of Unit Vectors In Exercises 49–52, the initial and terminal points of a vector are given. Write the vector as a linear combination of the standard unit vectors **i** and **j**.

Initial Point	Terminal Point
49. (2, 3)	(1, 8)
50. (4, −2)	(−2, −10)
51. (3, 4)	(9, 8)
52. (−2, 7)	(5, −9)

Vector Operations In Exercises 53–58, find the component form of **w** and sketch the specified vector operations geometrically, where $\mathbf{u} = 6\mathbf{i} - 5\mathbf{j}$ and $\mathbf{v} = 10\mathbf{i} + 3\mathbf{j}$.

53. $\mathbf{w} = 3\mathbf{v}$ 54. $\mathbf{w} = \frac{1}{2}\mathbf{v}$
55. $\mathbf{w} = 2\mathbf{u} + \mathbf{v}$ 56. $\mathbf{w} = 4\mathbf{u} - 5\mathbf{v}$
57. $\mathbf{w} = 5\mathbf{u} - 4\mathbf{v}$ 58. $\mathbf{w} = -3\mathbf{u} + 2\mathbf{v}$

Finding the Direction Angle of a Vector In Exercises 59–64, find the magnitude and direction angle of the vector **v**.

59. $\mathbf{v} = 5\mathbf{i} + 4\mathbf{j}$ 60. $\mathbf{v} = -4\mathbf{i} + 7\mathbf{j}$
61. $\mathbf{v} = -3\mathbf{i} - 3\mathbf{j}$ 62. $\mathbf{v} = 8\mathbf{i} - \mathbf{j}$
63. $\mathbf{v} = 7(\cos 60°\mathbf{i} + \sin 60°\mathbf{j})$
64. $\mathbf{v} = 3(\cos 150°\mathbf{i} + \sin 150°\mathbf{j})$

Finding the Component Form of a Vector In Exercises 65 and 66, find the component form of **v** given its magnitude and the angle it makes with the positive *x*-axis. Then sketch **v**.

Magnitude	Angle
65. $\|\mathbf{v}\| = 8$	$\theta = 120°$
66. $\|\mathbf{v}\| = \frac{1}{2}$	$\theta = 225°$

67. **Resultant Force** Forces with magnitudes of 85 pounds and 50 pounds act on a single point at angles of 45° and 60°, respectively, with the positive *x*-axis. Find the direction and magnitude of the resultant of these forces.

68. **Rope Tension** Two ropes support a 180-pound weight, as shown in the figure. Find the tension in each rope.

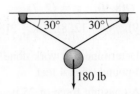

180 lb

6.4 Finding a Dot Product In Exercises 69–72, find **u** · **v**.

69. $\mathbf{u} = \langle 6, 7 \rangle$ 70. $\mathbf{u} = \langle -7, 12 \rangle$
    $\mathbf{v} = \langle -3, 9 \rangle$ $\mathbf{v} = \langle -4, -14 \rangle$
71. $\mathbf{u} = 3\mathbf{i} + 7\mathbf{j}$ 72. $\mathbf{u} = -7\mathbf{i} + 2\mathbf{j}$
    $\mathbf{v} = 11\mathbf{i} - 5\mathbf{j}$ $\mathbf{v} = 16\mathbf{i} - 12\mathbf{j}$

Using Properties of the Dot Product In Exercises 73–80, use the vectors $\mathbf{u} = \langle -4, 2 \rangle$ and $\mathbf{v} = \langle 5, 1 \rangle$ to find the quantity. State whether the result is a vector or a scalar.

73. $2\mathbf{u} \cdot \mathbf{u}$ 74. $3\mathbf{u} \cdot \mathbf{v}$
75. $4 - \|\mathbf{u}\|$ 76. $\|\mathbf{v}\|^2$
77. $\mathbf{u}(\mathbf{u} \cdot \mathbf{v})$ 78. $(\mathbf{u} \cdot \mathbf{v})\mathbf{v}$
79. $(\mathbf{u} \cdot \mathbf{u}) - (\mathbf{u} \cdot \mathbf{v})$ 80. $(\mathbf{v} \cdot \mathbf{v}) - (\mathbf{v} \cdot \mathbf{u})$

Finding the Angle Between Two Vectors In Exercises 81–84, find the angle θ (in degrees) between the vectors.

81. $\mathbf{u} = \langle 2\sqrt{2}, -4 \rangle, \quad \mathbf{v} = \langle -\sqrt{2}, 1 \rangle$
82. $\mathbf{u} = \langle 3, \sqrt{3} \rangle, \quad \mathbf{v} = \langle 4, 3\sqrt{3} \rangle$
83. $\mathbf{u} = \cos \frac{7\pi}{4}\mathbf{i} + \sin \frac{7\pi}{4}\mathbf{j}, \quad \mathbf{v} = \cos \frac{5\pi}{6}\mathbf{i} + \sin \frac{5\pi}{6}\mathbf{j}$
84. $\mathbf{u} = \cos 45°\mathbf{i} + \sin 45°\mathbf{j}, \quad \mathbf{v} = \cos 300°\mathbf{i} + \sin 300°\mathbf{j}$

Determining Orthogonal Vectors In Exercises 85–88, determine whether **u** and **v** are orthogonal.

85. $\mathbf{u} = \langle -3, 8 \rangle$ 86. $\mathbf{u} = \langle \frac{1}{4}, -\frac{1}{2} \rangle$
    $\mathbf{v} = \langle 8, 3 \rangle$ $\mathbf{v} = \langle -2, 4 \rangle$
87. $\mathbf{u} = -\mathbf{i}$ 88. $\mathbf{u} = -2\mathbf{i} + \mathbf{j}$
    $\mathbf{v} = \mathbf{i} + 2\mathbf{j}$ $\mathbf{v} = 3\mathbf{i} + 6\mathbf{j}$

Decomposing a Vector into Components In Exercises 89–92, find the projection of **u** onto **v**. Then write **u** as the sum of two orthogonal vectors, one of which is $\text{proj}_\mathbf{v} \mathbf{u}$.

89. $\mathbf{u} = \langle -4, 3 \rangle, \quad \mathbf{v} = \langle -8, -2 \rangle$
90. $\mathbf{u} = \langle 5, 6 \rangle, \quad \mathbf{v} = \langle 10, 0 \rangle$
91. $\mathbf{u} = \langle 2, 7 \rangle, \quad \mathbf{v} = \langle 1, -1 \rangle$
92. $\mathbf{u} = \langle -3, 5 \rangle, \quad \mathbf{v} = \langle -5, 2 \rangle$

Work In Exercises 93 and 94, determine the work done in moving a particle from P to Q when the magnitude and direction of the force are given by $\mathbf{v}$.

93. $P(5, 3)$, $Q(8, 9)$, $\mathbf{v} = \langle 2, 7 \rangle$

94. $P(-2, -9)$, $Q(-12, 8)$, $\mathbf{v} = 3\mathbf{i} - 6\mathbf{j}$

95. Work Determine the work done by a crane lifting an 18,000-pound truck 4 feet.

96. Work A constant force of 25 pounds, exerted at an angle of 20° with the horizontal, is required to slide a crate across a floor. Determine the work done in sliding the crate 12 feet.

6.5 **Finding the Absolute Value of a Complex Number** In Exercises 97–100, plot the complex number and find its absolute value.

97. $7i$

98. $-6i$

99. $5 + 3i$

100. $-10 - 4i$

Adding in the Complex Plane In Exercises 101 and 102, find the sum of the complex numbers in the complex plane.

101. $(2 + 3i) + (1 - 2i)$ **102.** $(-4 + 2i) + (2 + i)$

Subtracting in the Complex Plane In Exercises 103 and 104, find the difference of the complex numbers in the complex plane.

103. $(1 + 2i) - (3 + i)$ **104.** $(-2 + i) - (1 + 4i)$

Complex Conjugates in the Complex Plane In Exercises 105 and 106, plot the complex number and its complex conjugate. Write the conjugate as a complex number.

105. $3 + i$

106. $2 - 5i$

Finding Distance in the Complex Plane In Exercises 107 and 108, find the distance between the complex numbers in the complex plane.

107. $3 + 2i, 2 - i$ **108.** $1 + 5i, -1 + 3i$

Finding a Midpoint in the Complex Plane In Exercises 109 and 110, find the midpoint of the line segment joining the points corresponding to the complex numbers in the complex plane.

109. $1 + i, 4 + 3i$ **110.** $2 - i, 1 + 4i$

6.6 **Trigonometric Form of a Complex Number** In Exercises 111–116, plot the complex number. Then write the trigonometric form of the complex number.

111. $4i$

112. -7

113. $7 - 7i$

114. $5 + 12i$

115. $-5 - 12i$

116. $-3\sqrt{3} + 3i$

Multiplying Complex Numbers In Exercises 117 and 118, find the product. Leave the result in trigonometric form.

117. $\left[2\left(\cos \dfrac{\pi}{4} + i \sin \dfrac{\pi}{4} \right) \right] \left[2\left(\cos \dfrac{\pi}{3} + i \sin \dfrac{\pi}{3} \right) \right]$

118. $\left[4\left(\cos \dfrac{\pi}{3} + i \sin \dfrac{\pi}{3} \right) \right] \left[3\left(\cos \dfrac{5\pi}{6} + i \sin \dfrac{5\pi}{6} \right) \right]$

Dividing Complex Numbers In Exercises 119 and 120, find the quotient. Leave the result in trigonometric form.

119. $\dfrac{2(\cos 60° + i \sin 60°)}{3(\cos 15° + i \sin 15°)}$ **120.** $\dfrac{\cos 150° + i \sin 150°}{2(\cos 50° + i \sin 50°)}$

Finding a Power of a Complex Number In Exercises 121–124, use DeMoivre's Theorem to find the power of the complex number. Write the result in standard form.

121. $\left[5\left(\cos \dfrac{\pi}{12} + i \sin \dfrac{\pi}{12} \right) \right]^4$

122. $\left[2\left(\cos \dfrac{4\pi}{15} + i \sin \dfrac{4\pi}{15} \right) \right]^5$

123. $(2 + 3i)^6$

124. $(1 - i)^8$

Finding the nth Roots of a Complex Number In Exercises 125–128, (a) use the formula on page 450 to find the roots of the complex number, (b) write each of the roots in standard form, and (c) represent each of the roots graphically.

125. Sixth roots of $-729i$ **126.** Fourth roots of $256i$

127. Cube roots of 8 **128.** Fifth roots of -1024

Solving an Equation In Exercises 129–132, use the formula on page 450 to find all solutions of the equation and represent the solutions graphically.

129. $x^4 + 81 = 0$

130. $x^5 - 32 = 0$

131. $x^3 + 8i = 0$

132. $x^4 - 64i = 0$

Exploration

True or False? In Exercises 133 and 134, determine whether the statement is true or false. Justify your answer.

133. The Law of Sines is true when one of the angles in the triangle is a right angle.

134. When the Law of Sines is used, the solution is always unique.

135. Writing What characterizes a vector in the plane?

Take this test as you would take a test in class. When you are finished, check your work against the answers given in the back of the book.

240 mi ● C

37°

B

370 mi

24°

A ●

Figure for 8

In Exercises 1–6, determine whether the Law of Cosines is needed to solve the triangle. Then solve (if possible) the triangle. If two solutions exist, find both. Round your answers to two decimal places.

1. $A = 24°$, $B = 68°$, $a = 12.2$ 2. $B = 110°$, $C = 28°$, $a = 15.6$
3. $A = 24°$, $a = 11.2$, $b = 13.4$ 4. $a = 6.0$, $b = 7.3$, $c = 12.4$
5. $B = 100°$, $a = 23$, $b = 15$ 6. $C = 121°$, $a = 34$, $b = 55$

7. A triangular parcel of land has sides of lengths 60 meters, 70 meters, and 82 meters. Find the area of the parcel of land.

8. An airplane flies 370 miles from point A to point B with a bearing of 24°. Then it flies 240 miles from point B to point C with a bearing of 37° (see figure). Find the straight-line distance and bearing from point A to point C.

In Exercises 9 and 10, find the component form of the vector v.

9. Initial point of **v**: $(-3, 7)$; terminal point of **v**: $(11, -16)$
10. Magnitude of **v**: $\|\mathbf{v}\| = 12$; direction of **v**: $\mathbf{u} = \langle 3, -5 \rangle$

In Exercises 11–14, $\mathbf{u} = \langle 2, 7 \rangle$ and $\mathbf{v} = \langle -6, 5 \rangle$. Find the resultant vector and sketch its graph.

11. $\mathbf{u} + \mathbf{v}$ 12. $\mathbf{u} - \mathbf{v}$
13. $5\mathbf{u} - 3\mathbf{v}$ 14. $4\mathbf{u} + 2\mathbf{v}$

15. Find the distance between $4 + 3i$ and $1 - i$ in the complex plane.

16. Forces with magnitudes of 250 pounds and 130 pounds act on an object at angles of 45° and $-60°$, respectively, with the positive x-axis. Find the direction and magnitude of the resultant of these forces.

17. Find the angle θ (in degrees) between the vectors $\mathbf{u} = \langle -1, 5 \rangle$ and $\mathbf{v} = \langle 3, -2 \rangle$.

18. Determine whether the vectors $\mathbf{u} = \langle 6, -10 \rangle$ and $\mathbf{v} = \langle 5, 3 \rangle$ are orthogonal.

19. Find the projection of $\mathbf{u} = \langle 6, 7 \rangle$ onto $\mathbf{v} = \langle -5, -1 \rangle$. Then write $\mathbf{u}$ as the sum of two orthogonal vectors, one of which is $\text{proj}_{\mathbf{v}}\mathbf{u}$.

20. A 500-pound motorcycle is stopped at a red light on a hill inclined at 12°. Find the force required to keep the motorcycle from rolling down the hill.

21. Write the complex number $z = 4 - 4i$ in trigonometric form.

22. Write the complex number $z = 6(\cos 120° + i \sin 120°)$ in standard form.

In Exercises 23 and 24, use DeMoivre's Theorem to find the power of the complex number. Write the result in standard form.

23. $\left[3\left(\cos \dfrac{7\pi}{6} + i \sin \dfrac{7\pi}{6} \right) \right]^8$ 24. $(3 - 3i)^6$

25. Find the fourth roots of 256.

26. Find all solutions of the equation $x^3 - 27i = 0$ and represent the solutions graphically.

10. Skydiving   A skydiver falls at a constant downward velocity of 120 miles per hour. In the figure, vector **u** represents the skydiver's velocity. A steady breeze pushes the skydiver to the east at 40 miles per hour. Vector **v** represents the wind velocity.

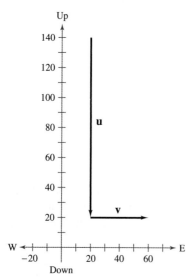

(a) Write the vectors **u** and **v** in component form.

(b) Let

    s = u + v.

Use the figure to sketch **s**. To print an enlarged copy of the graph, go to *MathGraphs.com*.

(c) Find the magnitude of **s**. What information does the magnitude give you about the skydiver's fall?

(d) Without wind, the skydiver would fall in a path perpendicular to the ground. At what angle to the ground is the path of the skydiver when affected by the 40-mile-per-hour wind from due west?

(e) The next day, the skydiver falls at a constant downward velocity of 120 miles per hour and a steady breeze pushes the skydiver to the west at 30 miles per hour. Draw a new figure that gives a visual representation of the problem and find the skydiver's new velocity.

11. Think About It   The vectors **u** and **v** have the same magnitudes in the two figures. In which figure is the magnitude of the sum greater? Explain.

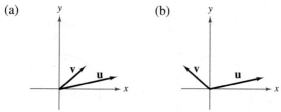

(a)   (b)

12. Speed and Velocity of an Airplane   Four basic forces are in action during flight: weight, lift, thrust, and drag. To fly through the air, an object must overcome its own *weight*. To do this, it must create an upward force called *lift*. To generate lift, a forward motion called *thrust* is needed. The thrust must be great enough to overcome air resistance, which is called *drag*.

For a commercial jet aircraft, a quick climb is important to maximize efficiency because the performance of an aircraft is enhanced at high altitudes. In addition, it is necessary to clear obstacles such as buildings and mountains and to reduce noise in residential areas. In the diagram, the angle θ is called the climb angle. The velocity of the plane can be represented by a vector **v** with a vertical component $\|\mathbf{v}\| \sin \theta$ (called climb speed) and a horizontal component $\|\mathbf{v}\| \cos \theta$, where $\|\mathbf{v}\|$ is the speed of the plane.

When taking off, a pilot must decide how much of the thrust to apply to each component. The more the thrust is applied to the horizontal component, the faster the airplane gains speed. The more the thrust is applied to the vertical component, the quicker the airplane climbs.

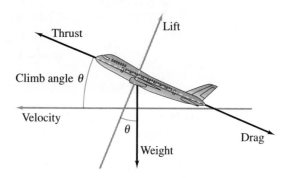

(a) Complete the table for an airplane that has a speed of $\|\mathbf{v}\| = 100$ miles per hour.

θ	0.5°	1.0°	1.5°	2.0°	2.5°	3.0°
$\|\mathbf{v}\| \sin \theta$						
$\|\mathbf{v}\| \cos \theta$						

(b) Does an airplane's speed equal the sum of the vertical and horizontal components of its velocity? If not, how could you find the speed of an airplane whose velocity components were known?

(c) Use the result of part (b) to find the speed of an airplane with the given velocity components.

    (i) $\|\mathbf{v}\| \sin \theta = 5.235$ miles per hour

       $\|\mathbf{v}\| \cos \theta = 149.909$ miles per hour

    (ii) $\|\mathbf{v}\| \sin \theta = 10.463$ miles per hour

       $\|\mathbf{v}\| \cos \theta = 149.634$ miles per hour

7 Systems of Equations and Inequalities

Thermodynamics *(Exercise 60, page 509)*

Target Heart Rate
(Exercise 68, page 518)

Global Positioning System *(page 495)*

Fuel Mixture
(Exercise 50, page 487)

Environmental Science *(Exercise 62, page 477)*

7.1 Linear and Nonlinear Systems of Equations

Graphs of systems of equations can help you solve real-life problems. For example, in Exercise 62 on page 477, you will use the graph of a system of equations to compare the consumption of wind energy and the consumption of geothermal energy.

- Use the method of substitution to solve systems of linear equations in two variables.
- Use the method of substitution to solve systems of nonlinear equations in two variables.
- Use a graphical method to solve systems of equations in two variables.
- Use systems of equations to model and solve real-life problems.

The Method of Substitution

Up to this point in the text, most problems have involved either a function of one variable or a single equation in two variables. However, many problems in science, business, and engineering involve two or more equations in two or more variables. To solve such a problem, you need to find the solutions of a **system of equations.** Here is an example of a system of two equations in two variables, x and y.

$$\begin{cases} 2x + y = 5 & \text{Equation 1} \\ 3x - 2y = 4 & \text{Equation 2} \end{cases}$$

A **solution** of this system is an ordered pair that satisfies each equation in the system. Finding the set of all solutions is called **solving the system of equations.** For example, the ordered pair $(2, 1)$ is a solution of this system. To check this, substitute 2 for x and 1 for y in *each* equation.

Check (2, 1) in Equation 1 and Equation 2:

$2x + y = 5$	Write Equation 1.
$2(2) + 1 \stackrel{?}{=} 5$	Substitute 2 for x and 1 for y.
$4 + 1 = 5$	Solution checks in Equation 1. ✓
$3x - 2y = 4$	Write Equation 2.
$3(2) - 2(1) \stackrel{?}{=} 4$	Substitute 2 for x and 1 for y.
$6 - 2 = 4$	Solution checks in Equation 2. ✓

In this chapter, you will study four ways to solve systems of equations, beginning with the **method of substitution.**

Method	Section	Type of System
1. Substitution	7.1	Linear or nonlinear, two variables
2. Graphical method	7.1	Linear or nonlinear, two variables
3. Elimination	7.2	Linear, two variables
4. Gaussian elimination	7.3	Linear, three or more variables

Method of Substitution

1. *Solve* one of the equations for one variable in terms of the other.

2. *Substitute* the expression found in Step 1 into the other equation to obtain an equation in one variable.

3. *Solve* the equation obtained in Step 2.

4. *Back-substitute* the value obtained in Step 3 into the expression obtained in Step 1 to find the value of the other variable.

5. *Check* that the solution satisfies *each* of the original equations.

EXAMPLE 1   **Solving a System of Equations by Substitution**

Solve the system of equations.

$$\begin{cases} x + y = 4 & \text{Equation 1} \\ x - y = 2 & \text{Equation 2} \end{cases}$$

Solution   Begin by solving for y in Equation 1.

$$y = 4 - x \qquad \text{Solve for } y \text{ in Equation 1.}$$

> **ALGEBRA HELP**   To review the techniques for solving different types of equations, see Appendix A.5.

Next, substitute this expression for y into Equation 2 and solve the resulting single-variable equation for x.

$x - y = 2$	Write Equation 2.
$x - (4 - x) = 2$	Substitute $4 - x$ for y.
$x - 4 + x = 2$	Distributive Property
$2x = 6$	Combine like terms.
$x = 3$	Divide each side by 2.

Finally, solve for y by *back-substituting* $x = 3$ into the equation $y = 4 - x$.

$y = 4 - x$	Write revised Equation 1.
$y = 4 - 3$	Substitute 3 for x.
$y = 1$	Solve for y.

So, the solution of the system is the ordered pair

$$(3, 1).$$

Check this solution as follows.

> **REMARK**   Many steps are required to solve a system of equations, so there are many possible ways to make errors in arithmetic. You should always check your solution by substituting it into *each* equation in the original system.

▷ **Check**

Substitute $(3, 1)$ into Equation 1:

$x + y = 4$	Write Equation 1.
$3 + 1 \overset{?}{=} 4$	Substitute for x and y.
$4 = 4$	Solution checks in Equation 1. ✓

Substitute $(3, 1)$ into Equation 2:

$x - y = 2$	Write Equation 2.
$3 - 1 \overset{?}{=} 2$	Substitute for x and y.
$2 = 2$	Solution checks in Equation 2. ✓

The point $(3, 1)$ satisfies both equations in the system. This confirms that $(3, 1)$ is a solution of the system of equations.

✓ *Checkpoint*   Audio-video solution in English & Spanish at LarsonPrecalculus.com

Solve the system of equations.

$$\begin{cases} x - y = 0 \\ 5x - 3y = 6 \end{cases}$$

The term *back-substitution* implies that you work *backwards*. First you solve for one of the variables, and then you substitute that value *back* into one of the equations in the system to find the value of the other variable.

| EXAMPLE 7 | **Movie Ticket Sales** |

Two new movies, a comedy and a drama, are released in the same week. In the first six weeks, the weekly ticket sales S (in millions of dollars) decrease for the comedy and increase for the drama according to the models

$$\begin{cases} S = 60 - 8x & \text{Comedy (Equation 1)} \\ S = 10 + 4.5x & \text{Drama (Equation 2)} \end{cases}$$

where x represents the time (in weeks), with $x = 1$ corresponding to the first week of release. According to the models, in what week are the ticket sales of the two movies equal?

Algebraic Solution

Both equations are already solved for S in terms of x, so substitute the expression for S from Equation 2 into Equation 1 and solve for x.

$10 + 4.5x = 60 - 8x$	Substitute for S in Equation 1.
$4.5x + 8x = 60 - 10$	Add $8x$ and -10 to each side.
$12.5x = 50$	Combine like terms.
$x = 4$	Divide each side by 12.5.

According to the models, the weekly ticket sales for the two movies are equal in the fourth week.

Numerical Solution

Create a table of values for each model.

Number of Weeks, x	1	2	3	4	5	6
Sales, S (comedy)	52	44	36	28	20	12
Sales, S (drama)	14.5	19	23.5	28	32.5	37

According to the table, the weekly ticket sales for the two movies are equal in the fourth week.

✓ **Checkpoint**  *Audio-video solution in English & Spanish at LarsonPrecalculus.com*

Two new movies, an animated movie and a horror movie, are released in the same week. In the first eight weeks, the weekly ticket sales S (in millions of dollars) decrease for the animated movie and increase for the horror movie according to the models

$$\begin{cases} S = 108 - 9.4x & \text{Animated} \\ S = 16 + 9x & \text{Horror} \end{cases}$$

where x represents the time (in weeks), with $x = 1$ corresponding to the first week of release. According to the models, in what week are the ticket sales of the two movies equal?

Summarize (Section 7.1)

1. Explain how to use the method of substitution to solve a system of linear equations in two variables *(page 468)*. For examples of using the method of substitution to solve systems of linear equations in two variables, see Examples 1 and 2.

2. Explain how to use the method of substitution to solve a system of nonlinear equations in two variables *(page 471)*. For examples of using the method of substitution to solve systems of nonlinear equations in two variables, see Examples 3 and 4.

3. Explain how to use a graphical method to solve a system of equations in two variables *(page 472)*. For an example of using a graphical approach to solve a system of equations in two variables, see Example 5.

4. Describe examples of how to use systems of equations to model and solve real-life problems *(pages 473 and 474, Examples 6 and 7)*.

62. Environmental S

The table shows the consu
Btus) of geothermal energ
United States from 2004 t
U.S. Energy Information

DATA Year	Geoth
2004	
2005	
2006	
2007	
2008	
2009	
2010	
2011	
2012	
2013	
2014	

Spreadsheet at LarsonPrecalculus.com

(a) Use a graphing utility
the geothermal energ
cubic model for the v
data. Let *t* represent t
corresponding to 200

(b) Use the
graphing utility
to graph the
data and the
two models
in the same
viewing window.

(c) Use the graph
from part (b) to appr
intersection of the gr
your answer in the c

(d) Describe the behavic
think the models can
consumption of geot
energy in the United
Explain.

(e) Use your school's lit
other reference sourc
and disadvantages of

Geometry In Exercise
equations to find the dime
the specified conditions.

63. The perimeter is 56 m
greater than the width.

64. The perimeter is 42 inc
the length.

ssuaphotos/Shutterstock.com

7.1 Exercises

See **CalcChat.com** for tutorial help and worked-out solutions to odd-numbered exercises.

Vocabulary: Fill in the blanks.

1. A _____ of a system of equations is an ordered pair that satisfies each equation in the system.

2. The first step in solving a system of equations by the method of _____ is to solve one of the equations for one variable in terms of the other.

3. Graphically, solutions of a system of two equations correspond to the _____ of _____ of the graphs of the two equations.

4. In business applications, the total revenue equals the total cost at the _____ point.

Skills and Applications

Checking Solutions In Exercises 5 and 6, determine whether each ordered pair is a solution of the system.

5. $\begin{cases} 2x - y = 4 \\ 8x + y = -9 \end{cases}$
   (a) $(0, -4)$
   (b) $(3, -1)$
   (c) $\left(\frac{3}{2}, -1\right)$
   (d) $\left(-\frac{1}{2}, -5\right)$

6. $\begin{cases} 4x^2 + y = 3 \\ -x - y = 11 \end{cases}$
   (a) $(2, -13)$
   (b) $(1, -2)$
   (c) $\left(-\frac{3}{2}, -\frac{31}{3}\right)$
   (d) $\left(-\frac{7}{4}, -\frac{37}{4}\right)$

Solving a System by Substitution In Exercises 7–14, solve the system by the method of substitution. Check your solution(s) graphically.

7. $\begin{cases} 2x + y = 6 \\ -x + y = 0 \end{cases}$

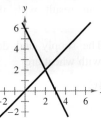

8. $\begin{cases} x - 4y = -11 \\ x + 3y = 3 \end{cases}$

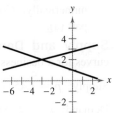

9. $\begin{cases} x - y = -4 \\ x^2 - y = -2 \end{cases}$

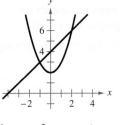

10. $\begin{cases} 3x + y = 2 \\ x^3 - 2 + y = 0 \end{cases}$

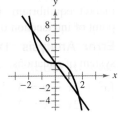

11. $\begin{cases} x^2 + y = 0 \\ x^2 - 4x - y = 0 \end{cases}$

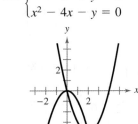

12. $\begin{cases} x + y = 0 \\ x^3 - 5x - y = 0 \end{cases}$

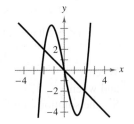

13. $\begin{cases} y = x^3 - 3x^2 + 1 \\ y = x^2 - 3x + 1 \end{cases}$

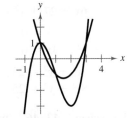

14. $\begin{cases} -\frac{1}{2}x + y = -\frac{5}{2} \\ x^2 + y^2 = 25 \end{cases}$

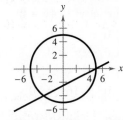

Solving a System by Substitution In Exercises 15–24, solve the system by the method of substitution.

15. $\begin{cases} x - y = 2 \\ 6x - 5y = 16 \end{cases}$

16. $\begin{cases} 2x + y = 9 \\ 3x - 5y = 20 \end{cases}$

17. $\begin{cases} 2x - y + 2 = 0 \\ 4x + y - 5 = 0 \end{cases}$

18. $\begin{cases} 6x - 3y - 4 = 0 \\ x + 2y - 4 = 0 \end{cases}$

19. $\begin{cases} 1.5x + 0.8y = 2.3 \\ 0.3x - 0.2y = 0.1 \end{cases}$

20. $\begin{cases} 0.5x + y = -3.5 \\ x - 3.2y = 3.4 \end{cases}$

21. $\begin{cases} \frac{1}{5}x + \frac{1}{2}y = 8 \\ x + y = 20 \end{cases}$

22. $\begin{cases} \frac{1}{2}x + \frac{3}{4}y = 10 \\ \frac{3}{4}x - y = 4 \end{cases}$

23. $\begin{cases} 6x + 5y = -3 \\ -x - \frac{5}{6}y = -7 \end{cases}$

24. $\begin{cases} -\frac{2}{3}x + y = 2 \\ 2x - 3y = 6 \end{cases}$

Solving a System by Substitution In Exercises 25–28, the given amount of annual interest is earned from a total of **$12,000** invested in two funds paying simple interest. Write and solve a system of equations to find the amount invested at each given rate.

Annual Interest	Rate 1	Rate 2
25. $500	2%	6%
26. $630	4%	7%
27. $396	2.8%	3.8%
28. $254	1.75%	2.25%

.0175x + .0225y = 254   X   Y

X + Y = 12,000

Solving a Syste
Equation In Ex
system by the meth

29. $\begin{cases} x^2 - y = 0 \\ 2x + y = 0 \end{cases}$ 30

31. $\begin{cases} x - y = -1 \\ x^2 - y = -4 \end{cases}$ 32

Solving a Sy
Graphically In l
system graphically.

33. $\begin{cases} -x + 2y = -2 \\ 3x + y = 20 \end{cases}$ 34

35. $\begin{cases} x - 3y = -3 \\ 5x + 3y = -6 \end{cases}$ 36

37. $\begin{cases} x + y = 4 \\ x^2 + y^2 - 4x = 0 \end{cases}$ 38

39. $\begin{cases} 3x - 2y = 0 \\ x^2 - y^2 = 4 \end{cases}$ 40

41. $\begin{cases} x^2 + y^2 = 25 \\ 3x^2 - 16y = 0 \end{cases}$ 42

Using Technology In
a graphing utility to solve
graphically. Round your so
places, if necessary.

43. $\begin{cases} y = e^x \\ x - y + 1 = 0 \end{cases}$ 44

45. $\begin{cases} y + 2 = \ln(x - 1) \\ 3y + 2x = 9 \end{cases}$ 46

Choosing a Solution Met
solve the system graphically
your choice of method.

47. $\begin{cases} y = 2x \\ y = x^2 + 1 \end{cases}$ 48

49. $\begin{cases} x - 2y = 4 \\ x^2 - y = 0 \end{cases}$ 50

51. $\begin{cases} y - e^{-x} = 1 \\ y - \ln x = 3 \end{cases}$ 5

53. $\begin{cases} xy - 1 = 0 \\ 2x - 4y + 7 = 0 \end{cases}$ 5

Break-Even An
and 56, use the eq
and total revenue
of units a compan
(Round to the nea

55. $C = 8650x + 250,000$, *F*

56. $C = 5.5\sqrt{x} + 10,000$, *R*

---

51. Investment Portfolio A total of $24,000 is invested in two corporate bonds that pay 3.5% and 5% simple interest. The investor wants an annual interest income of $930 from the investments. What amount should be invested in the 3.5% bond?

52. Investment Portfolio A total of $32,000 is invested in two municipal bonds that pay 5.75% and 6.25% simple interest. The investor wants an annual interest income of $1900 from the investments. What amount should be invested in the 5.75% bond?

53. Pharmacology The numbers of prescriptions P (in thousands) filled at two pharmacies from 2012 through 2016 are shown in the table.

Year	Pharmacy A	Pharmacy B
2012	19.2	20.4
2013	19.6	20.8
2014	20.0	21.1
2015	20.6	21.5
2016	21.3	22.0

(a) Use a graphing utility to create a scatter plot of the data for pharmacy A and find a linear model. Let t represent the year, with $t = 12$ corresponding to 2012. Repeat the procedure for pharmacy B.

(b) Assume that the models in part (a) can be used to represent future years. Will the number of prescriptions filled at pharmacy A ever exceed the number of prescriptions filled at pharmacy B? If so, when?

54. Daily Sales A store manager wants to know the demand for a product as a function of the price. The table shows the daily sales y for different prices x of the product.

Price, x	Demand, y
$1.00	45
$1.20	37
$1.50	23

(a) Find the least squares regression line $y = ax + b$ for the data by solving the system

$$\begin{cases} 3.00b + 3.70a = 105.00 \\ 3.70b + 4.69a = 123.90 \end{cases}$$

for a and b. Use a graphing utility to confirm the result.

(b) Use the linear model from part (a) to predict the demand when the price is $1.75.

Fitting a Line to Data One way to find the least squares regression line $y = ax + b$ for a set of points

$$(x_1, y_1), (x_2, y_2), \ldots, (x_n, y_n)$$

is by solving the system below for a and b.

$$\begin{cases} nb + \left(\sum_{i=1}^{n} x_i\right)a = \left(\sum_{i=1}^{n} y_i\right) \\ \left(\sum_{i=1}^{n} x_i\right)b + \left(\sum_{i=1}^{n} x_i^2\right)a = \left(\sum_{i=1}^{n} x_i y_i\right) \end{cases}$$

In Exercises 55 and 56, the sums have been evaluated. Solve the simplified system for a and b to find the least squares regression line for the points. Use a graphing utility to confirm the result. (*Note:* The symbol Σ is used to denote a sum of the terms of a sequence. You will learn how to use this notation in Section 9.1.)

55. $\begin{cases} 5b + 10a = 20.2 \\ 10b + 30a = 50.1 \end{cases}$

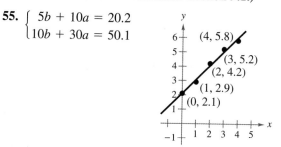

56. $\begin{cases} 6b + 15a = 23.6 \\ 15b + 55a = 48.8 \end{cases}$

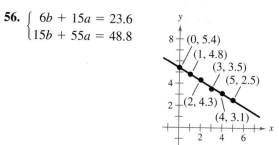

57. Agriculture An agricultural scientist used four test plots to determine the relationship between wheat yield y (in bushels per acre) and the amount of fertilizer x (in hundreds of pounds per acre). The table shows the results.

Fertilizer, x	1.0	1.5	2.0	2.5
Yield, y	32	41	48	53

(a) Find the least squares regression line $y = ax + b$ for the data by solving the system for a and b.

$$\begin{cases} 4b + 7a = 174 \\ 7b + 13.5a = 322 \end{cases}$$

(b) Use the linear model from part (a) to estimate the yield for a fertilizer application of 160 pounds per acre.

58. Gross Domestic Product The table shows the total gross domestic products y (in billions of dollars) of the United States for the years 2009 through 2015. (*Source: U.S. Office of Management and Budget*)

DATA	Year	GDP, y
	2009	14,414.6
	2010	14,798.5
	2011	15,379.2
	2012	16,027.2
	2013	16,498.1
	2014	17,183.5
	2015	17,803.4

Spreadsheet at LarsonPrecalculus.com

(a) Find the least squares regression line $y = at + b$ for the data, where t represents the year with $t = 9$ corresponding to 2009, by solving the system

$$\begin{cases} 7b + 84a = 112{,}104.5 \\ 84b + 1036a = 1{,}361{,}309.3 \end{cases}$$

for a and b. Use the *regression* feature of a graphing utility to confirm the result.

(b) Use the linear model to create a table of estimated values of y. Compare the estimated values with the actual data.

(c) Use the linear model to estimate the gross domestic product for 2016.

(d) Use the Internet, your school's library, or some other reference source to find the total national outlay for 2016. How does this value compare with your answer in part (c)?

(e) Is the linear model valid for long-term predictions of gross domestic products? Explain.

Exploration

True or False? In Exercises 59 and 60, determine whether the statement is true or false. Justify your answer.

59. If two lines do not have exactly one point of intersection, then they must be parallel.

60. Solving a system of equations graphically will always give an exact solution.

Finding the Value of a Constant In Exercises 61 and 62, find the value of k such that the system of linear equations is inconsistent.

61. $\begin{cases} 4x - 8y = -3 \\ 2x + ky = 16 \end{cases}$ **62.** $\begin{cases} 15x + 3y = 6 \\ -10x + ky = 9 \end{cases}$

63. Writing Briefly explain whether it is possible for a consistent system of linear equations to have exactly two solutions.

64. Think About It Give examples of systems of linear equations that have (a) no solution and (b) infinitely many solutions.

65. Comparing Methods Use the method of substitution to solve the system in Example 1. Do you prefer the method of substitution or the method of elimination? Explain.

66. HOW DO YOU SEE IT? Use the graphs of the two equations shown below.

(a) Describe the graphs of the two equations.

(b) Can you conclude that the system of equations whose graphs are shown is inconsistent? Explain.

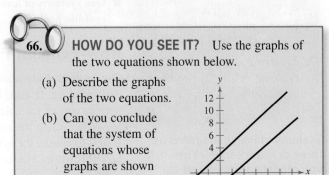

Think About It In Exercises 67 and 68, the graphs of the two equations appear to be parallel. Yet, when you solve the system algebraically, you find that the system does have a solution. Find the solution and explain why it does not appear on the portion of the graph shown.

67. $\begin{cases} 100y - x = 200 \\ 99y - x = -198 \end{cases}$ **68.** $\begin{cases} 21x - 20y = 0 \\ 13x - 12y = 120 \end{cases}$

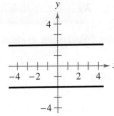   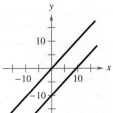

Advanced Applications In Exercises 69 and 70, solve the system of equations for u and v. While solving for these variables, consider the trigonometric functions as constants. (Systems of this type appear in a course in differential equations.)

69. $\begin{cases} u \sin x + v \cos x = 0 \\ u \cos x - v \sin x = \sec x \end{cases}$

70. $\begin{cases} u \cos 2x + v \sin 2x = 0 \\ u(-2 \sin 2x) + v(2 \cos 2x) = \csc 2x \end{cases}$

Project: College Expenses To work an extended application analyzing the average undergraduate tuition, room, and board charges at private degree-granting institutions in the United States from 1993 through 2013, visit this text's website at *LarsonPrecalculus.com*. (*Source: U.S. Department of Education*)

7.3 Multivariable Linear Systems

■ Use back-substitution to solve linear systems in row-echelon form.
■ Use Gaussian elimination to solve systems of linear equations.
■ Solve nonsquare systems of linear equations.
■ Use systems of linear equations in three or more variables to model and solve real-life problems.

Row-Echelon Form and Back-Substitution

Systems of linear equations in three or more variables can help you model and solve real-life problems. For example, in Exercise 70 on page 501, you will use a system of linear equations in three variables to analyze the reproductive rates of deer in a wildlife preserve.

The method of elimination can be applied to a system of linear equations in more than two variables. In fact, this method adapts to computer use for solving linear systems with dozens of variables.

When using the method of elimination to solve a system of linear equations, the goal is to rewrite the system in a form to which back-substitution can be applied. To see how this works, consider the following two systems of linear equations.

$$\begin{cases} x - 2y + 3z = 9 \\ -x + 3y \quad\;\; = -4 \\ 2x - 5y + 5z = 17 \end{cases}$$ System of three linear equations in three variables (See Example 3.)

$$\begin{cases} x - 2y + 3z = 9 \\ \quad\;\; y + 3z = 5 \\ \quad\quad\quad\;\; z = 2 \end{cases}$$ Equivalent system in row-echelon form (See Example 1.)

The second system is in **row-echelon form,** which means that it has a "stair-step" pattern with leading coefficients of 1. In comparing the two systems, notice that the row-echelon form can readily be solved using back-substitution.

EXAMPLE 1 Using Back-Substitution in Row-Echelon Form

Solve the system of linear equations.

$$\begin{cases} x - 2y + 3z = 9 & \text{Equation 1} \\ \quad\;\; y + 3z = 5 & \text{Equation 2} \\ \quad\quad\quad\;\; z = 2 & \text{Equation 3} \end{cases}$$

Solution From Equation 3, you know the value of z. To solve for y, back-substitute $z = 2$ into Equation 2.

$$y + 3(2) = 5 \qquad \text{Substitute 2 for } z.$$
$$y = -1 \qquad \text{Solve for } y.$$

To solve for x, back-substitute $y = -1$ and $z = 2$ into Equation 1.

$$x - 2(-1) + 3(2) = 9 \qquad \text{Substitute } -1 \text{ for } y \text{ and 2 for } z.$$
$$x = 1 \qquad \text{Solve for } x.$$

The solution is $x = 1$, $y = -1$, and $z = 2$, which can be written as the **ordered triple** $(1, -1, 2)$. Check this in the original system of equations.

✓ *Checkpoint* 🔊))) *Audio-video solution in English & Spanish at LarsonPrecalculus.com*

Solve the system of linear equations.

$$\begin{cases} x - y + 5z = 22 \\ \quad\;\; y + 3z = 6 \\ \quad\quad\quad\;\; z = 3 \end{cases}$$

Historically, one of the most influential Chinese mathematics books was the *Chui-chang suan-shu* or *Nine Chapters on the Mathematical Art*, a compilation of ancient Chinese problems published in 263 A.D. Chapter Eight of the *Nine Chapters* contained solutions of systems of linear equations such as the system below.

$$\begin{cases} 3x + 2y + \ z = 39 \\ 2x + 3y + \ z = 34 \\ \ x + 2y + 3z = 26 \end{cases}$$

This system was solved by performing column operations on a matrix, using the same strategies as Gaussian elimination. Matrices (plural for matrix) are discussed in the next chapter.

Gaussian Elimination

Recall from Section 7.2 that two systems of equations are *equivalent* when they have the same solution set. To solve a system that is not in row-echelon form, first convert it to an *equivalent* system that is in row-echelon form by using one or more of the row operations given below.

Operations That Produce Equivalent Systems

Each of the following **row operations** on a system of linear equations produces an *equivalent* system of linear equations.

1. Interchange two equations.

2. Multiply one of the equations by a nonzero constant.

3. Add a multiple of one equation to another equation to replace the latter equation.

To see how to use row operations, take another look at the method of elimination, as applied to a system of two linear equations.

EXAMPLE 2   Using Row Operations to Solve a System

Solve the system of linear equations.

$$\begin{cases} 3x - 2y = -1 \\ \ x - \ y = \ 0 \end{cases}$$

Solution Two strategies seem reasonable: eliminate the variable x or eliminate the variable y. The following steps show one way to eliminate x in the second equation. Start by interchanging the equations to obtain a leading coefficient of 1 in the first equation.

$$\begin{cases} \ x - \ y = \ 0 \\ 3x - 2y = -1 \end{cases}$$ Interchange the two equations in the system.

$$-3x + 3y = 0$$ Multiply the first equation by -3.

$$\begin{aligned} -3x + 3y &= \ 0 \\ \underline{3x - 2y} &= \underline{-1} \\ y &= -1 \end{aligned}$$ Add the multiple of the first equation to the second equation to obtain a new second equation.

$$\begin{cases} x - y = \ 0 \\ \ \ \ \ \ y = -1 \end{cases}$$ New system in row-echelon form

Now back-substitute $y = -1$ into the first equation of the new system in row-echelon form and solve for x.

$$x - (-1) = 0$$ Substitute -1 for y.

$$x = -1$$ Solve for x.

The solution is $(-1, -1)$. Check this in the original system.

✓ *Checkpoint* 🔊))) *Audio-video solution in English & Spanish at LarsonPrecalculus.com*

Solve the system of linear equations.

$$\begin{cases} 2x + \ y = 3 \\ \ x + 2y = 3 \end{cases}$$

EXAMPLE 5 **A System with Infinitely Many Solutions**

Solve the system of linear equations.

$$\begin{cases} x + y - 3z = -1 & \text{Equation 1} \\ \quad\quad y - z = 0 & \text{Equation 2} \\ -x + 2y \quad\quad = 1 & \text{Equation 3} \end{cases}$$

Solution

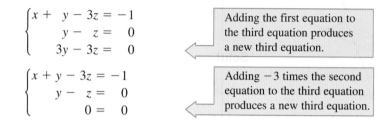

$$\begin{cases} x + y - 3z = -1 \\ \quad\quad y - z = 0 \\ \quad 3y - 3z = 0 \end{cases}$$

Adding the first equation to the third equation produces a new third equation.

$$\begin{cases} x + y - 3z = -1 \\ \quad\quad y - z = 0 \\ \quad\quad\quad 0 = 0 \end{cases}$$

Adding -3 times the second equation to the third equation produces a new third equation.

You have $0 = 0$, so Equation 3 depends on Equations 1 and 2 in the sense that it gives no additional information about the variables. So, the original system is equivalent to

$$\begin{cases} x + y - 3z = -1 \\ \quad\quad y - z = 0 \end{cases}.$$

In the second equation, solve for y in terms of z to obtain $y = z$. Back-substituting for y in the first equation yields $x = 2z - 1$. So, the system has infinitely many solutions consisting of all real values of x, y, and z for which

$$y = z \quad \text{and} \quad x = 2z - 1.$$

Letting $z = a$, where a is any real number, the solutions of the original system are all ordered triples of the form

$$(2a - 1, a, a).$$

✓ **Checkpoint** ◀))) *Audio-video solution in English & Spanish at LarsonPrecalculus.com*

Solve the system of linear equations.

$$\begin{cases} x + 2y - 7z = -4 \\ 2x + 3y + z = 5 \\ 3x + 7y - 36z = -25 \end{cases}$$

In Example 5, there are other ways to write the same infinite set of solutions. For instance, letting $x = b$, the solutions could have been written as

$$\left(b, \tfrac{1}{2}(b + 1), \tfrac{1}{2}(b + 1)\right). \qquad b \text{ is a real number.}$$

To convince yourself that this form produces the same set of solutions, consider the following.

Substitution	Solution	
$a = 0$	$(2(0) - 1, 0, 0) = (-1, 0, 0)$	Same solution
$b = -1$	$\left(-1, \tfrac{1}{2}(-1 + 1), \tfrac{1}{2}(-1 + 1)\right) = (-1, 0, 0)$	
$a = 1$	$(2(1) - 1, 1, 1) = (1, 1, 1)$	Same solution
$b = 1$	$\left(1, \tfrac{1}{2}(1 + 1), \tfrac{1}{2}(1 + 1)\right) = (1, 1, 1)$	
$a = 2$	$(2(2) - 1, 2, 2) = (3, 2, 2)$	Same solution
$b = 3$	$\left(3, \tfrac{1}{2}(3 + 1), \tfrac{1}{2}(3 + 1)\right) = (3, 2, 2)$	

Nonsquare Systems

So far, each system of linear equations you have looked at has been *square*, which means that the number of equations is equal to the number of variables. In a **nonsquare** system, the number of equations differs from the number of variables. A system of linear equations cannot have a unique solution unless there are at least as many equations as there are variables in the system.

| EXAMPLE 6 | **A System with Fewer Equations than Variables** |

See LarsonPrecalculus.com for an interactive version of this type of example.

Solve the system of linear equations.

$$\begin{cases} x - 2y + z = 2 & \text{Equation 1} \\ 2x - y - z = 1 & \text{Equation 2} \end{cases}$$

Solution   The system has three variables and only two equations, so the system does not have a unique solution. Begin by rewriting the system in row-echelon form.

$$\begin{cases} x - 2y + z = 2 \\ 3y - 3z = -3 \end{cases}$$

> Adding -2 times the first equation to the second equation produces a new second equation.

$$\begin{cases} x - 2y + z = 2 \\ y - z = -1 \end{cases}$$

> Multiplying the second equation by $\frac{1}{3}$ produces a new second equation.

Solve the new second equation for y in terms of z to obtain

$$y = z - 1.$$

Solve for x by back-substituting $y = z - 1$ into Equation 1.

$x - 2y + z = 2$	Write Equation 1.
$x - 2(z - 1) + z = 2$	Substitute $z - 1$ for y.
$x - 2z + 2 + z = 2$	Distributive Property
$x = z$	Solve for x.

Finally, by letting $z = a$, where a is any real number, you have the solution

$$x = a, \quad y = a - 1, \quad \text{and} \quad z = a.$$

So, the solution set of the system consists of all ordered triples of the form $(a, a - 1, a)$, where a is a real number.

✓ *Checkpoint*))) *Audio-video solution in English & Spanish at LarsonPrecalculus.com*

Solve the system of linear equations.

$$\begin{cases} x - y + 4z = 3 \\ 4x - z = 0 \end{cases}$$

In Example 6, choose several values of a to obtain different solutions of the system, such as

$$(1, 0, 1), \quad (2, 1, 2), \quad \text{and} \quad (3, 2, 3).$$

Then check each ordered triple in the original system to verify that it is a solution of the system.

The Global Positioning System (GPS) is a network of 24 satellites originally developed by the U.S. military as a navigational tool. Civilian applications of GPS receivers include determining directions, locating vessels lost at sea, and monitoring earthquakes. A GPS receiver works by using satellite readings to calculate its location. In a simplified mathematical model, nonsquare systems of three linear equations in four variables (three dimensions and time) determine the coordinates of the receiver as a function of time.

Applications

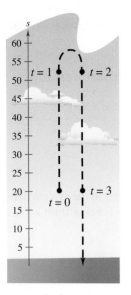

Figure 7.15

EXAMPLE 7   **Modeling Vertical Motion**

The height at time t of an object that is moving in a (vertical) line with constant acceleration a is given by the **position equation**

$$s = \tfrac{1}{2}at^2 + v_0 t + s_0$$

where s is the height in feet, a is the acceleration in feet per second squared, t is the time in seconds, v_0 is the initial velocity (at $t = 0$), and s_0 is the initial height. Find the values of a, v_0, and s_0 when $s = 52$ at $t = 1$, $s = 52$ at $t = 2$, and $s = 20$ at $t = 3$, and interpret the result. (See Figure 7.15.)

Solution   Substitute the three sets of values for t and s into the position equation to obtain three linear equations in a, v_0, and s_0.

When $t = 1$: $\tfrac{1}{2}a(1)^2 + v_0(1) + s_0 = 52$ ⟹ $a + 2v_0 + 2s_0 = 104$

When $t = 2$: $\tfrac{1}{2}a(2)^2 + v_0(2) + s_0 = 52$ ⟹ $2a + 2v_0 + s_0 = 52$

When $t = 3$: $\tfrac{1}{2}a(3)^2 + v_0(3) + s_0 = 20$ ⟹ $9a + 6v_0 + 2s_0 = 40$

Solve the resulting system of linear equations using Gaussian elimination.

$$\begin{cases} a + 2v_0 + 2s_0 = 104 \\ 2a + 2v_0 + s_0 = 52 \\ 9a + 6v_0 + 2s_0 = 40 \end{cases}$$

$$\begin{cases} a + 2v_0 + 2s_0 = 104 \\ - 2v_0 - 3s_0 = -156 \\ 9a + 6v_0 + 2s_0 = 40 \end{cases}$$

Adding -2 times the first equation to the second equation produces a new second equation.

$$\begin{cases} a + 2v_0 + 2s_0 = 104 \\ - 2v_0 - 3s_0 = -156 \\ - 12v_0 - 16s_0 = -896 \end{cases}$$

Adding -9 times the first equation to the third equation produces a new third equation.

$$\begin{cases} a + 2v_0 + 2s_0 = 104 \\ - 2v_0 - 3s_0 = -156 \\   2s_0 = 40 \end{cases}$$

Adding -6 times the second equation to the third equation produces a new third equation.

$$\begin{cases} a + 2v_0 + 2s_0 = 104 \\ v_0 + \tfrac{3}{2}s_0 = 78 \\   s_0 = 20 \end{cases}$$

Multiplying the second equation by $-\tfrac{1}{2}$ produces a new second equation and multiplying the third equation by $\tfrac{1}{2}$ produces a new third equation.

Using back-substitution, the solution of this system is

$$a = -32, \quad v_0 = 48, \quad \text{and} \quad s_0 = 20.$$

So, the position equation for the object is $s = -16t^2 + 48t + 20$, which implies that the object was thrown upward at a velocity of 48 feet per second from a height of 20 feet.

✓ **Checkpoint** 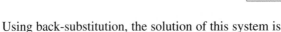 *Audio-video solution in English & Spanish at LarsonPrecalculus.com*

Use the position equation

$$s = \tfrac{1}{2}at^2 + v_0 t + s_0$$

from Example 7 to find the values of a, v_0, and s_0 when $s = 104$ at $t = 1$, $s = 76$ at $t = 2$, and $s = 16$ at $t = 3$, and interpret the result.   ▪

EXAMPLE 8   **Data Analysis: Curve-Fitting**

Find a quadratic equation $y = ax^2 + bx + c$ whose graph passes through the points $(-1, 3)$, $(1, 1)$, and $(2, 6)$.

Solution   Because the graph of $y = ax^2 + bx + c$ passes through the points $(-1, 3)$, $(1, 1)$, and $(2, 6)$, you can write the following.

When $x = -1$, $y = 3$: $a(-1)^2 + b(-1) + c = 3$
When $x = 1$, $y = 1$: $a(1)^2 + b(1) + c = 1$
When $x = 2$, $y = 6$: $a(2)^2 + b(2) + c = 6$

This yields the following system of linear equations.

$$\begin{cases} a - b + c = 3 \\ a + b + c = 1 \\ 4a + 2b + c = 6 \end{cases}$$

The solution of this system is $a = 2$, $b = -1$, and $c = 0$. So, the equation of the parabola is

$$y = 2x^2 - x$$

as shown below.

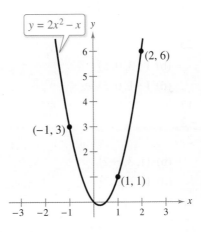

✓ *Checkpoint* ◀))) *Audio-video solution in English & Spanish at LarsonPrecalculus.com*

Find a quadratic equation $y = ax^2 + bx + c$ whose graph passes through the points $(0, 0)$, $(3, -3)$, and $(6, 0)$.

Summarize   (Section 7.3)

1. Explain what row-echelon form is *(page 490)*. For an example of solving a linear system in row-echelon form, see Example 1.

2. Describe the process of Gaussian elimination *(pages 491 and 492)*. For examples of using Gaussian elimination to solve systems of linear equations, see Examples 2–5.

3. Explain the difference between a square system of linear equations and a nonsquare system of linear equations *(page 495)*. For an example of solving a nonsquare system of linear equations, see Example 6.

4. Describe examples of how to use systems of linear equations in three or more variables to model and solve real-life problems *(pages 496 and 497, Examples 7 and 8)*.

7.3 Exercises

See CalcChat.com for tutorial help and worked-out solutions to odd-numbered exercises.

Vocabulary: Fill in the blanks.

1. A system of equations in _____ form has a "stair-step" pattern with leading coefficients of 1.
2. A solution of a system of three linear equations in three variables can be written as an _____ _____, which has the form (x, y, z).
3. The process used to write a system of linear equations in row-echelon form is called _____ elimination.
4. Interchanging two equations of a system of linear equations is a _____ _____ that produces an equivalent system.
5. In a _____ system, the number of equations differs from the number of variables in the system.
6. The equation $s = \frac{1}{2}at^2 + v_0 t + s_0$ is called the _____ equation, and it models the height s of an object at time t that is moving in a vertical line with a constant acceleration a.

Skills and Applications

Checking Solutions In Exercises 7–10, determine whether each ordered triple is a solution of the system of equations.

7. $\begin{cases} 6x - y + z = -1 \\ 4x \quad\;\; - 3z = -19 \\ \quad\;\; 2y + 5z = 25 \end{cases}$

   (a) $(0, 3, 1)$ (b) $(-3, 0, 5)$
   (c) $(0, -1, 4)$ (d) $(-1, 0, 5)$

8. $\begin{cases} 3x + 4y - z = 17 \\ 5x - y + 2z = -2 \\ 2x - 3y + 7z = -21 \end{cases}$

   (a) $(3, -1, 2)$ (b) $(1, 3, -2)$
   (c) $(1, 5, 6)$ (d) $(1, -2, 2)$

9. $\begin{cases} 4x + y - z = 0 \\ -8x - 6y + z = -\frac{7}{4} \\ 3x - y \quad\;\; = -\frac{9}{4} \end{cases}$

   (a) $\left(\frac{1}{2}, -\frac{3}{4}, -\frac{7}{4}\right)$ (b) $\left(\frac{3}{2}, -\frac{2}{5}, \frac{3}{5}\right)$
   (c) $\left(-\frac{1}{2}, \frac{3}{4}, -\frac{5}{4}\right)$ (d) $\left(-\frac{1}{2}, \frac{1}{6}, -\frac{3}{4}\right)$

10. $\begin{cases} -4x - y - 8z = -6 \\ \quad\;\; y + z = 0 \\ 4x - 7y \quad\;\; = 6 \end{cases}$

    (a) $(-2, -2, 2)$ (b) $\left(-\frac{33}{2}, -10, 10\right)$
    (c) $\left(\frac{1}{8}, -\frac{1}{2}, \frac{1}{2}\right)$ (d) $\left(-\frac{1}{2}, -2, 1\right)$

 Using Back-Substitution in Row-Echelon Form In Exercises 11–16, use back-substitution to solve the system of linear equations.

11. $\begin{cases} x - y + 5z = 37 \\ y + 2z = 6 \\ z = 8 \end{cases}$

12. $\begin{cases} x - 2y + 2z = 20 \\ y - z = 8 \\ z = -1 \end{cases}$

13. $\begin{cases} x + y - 3z = 7 \\ y + z = 12 \\ z = 2 \end{cases}$

14. $\begin{cases} x - y + 2z = 22 \\ y - 8z = 13 \\ z = -3 \end{cases}$

15. $\begin{cases} x - 2y + z = -\frac{1}{4} \\ y - z = -4 \\ z = 11 \end{cases}$

16. $\begin{cases} x \quad\;\; - 8z = \frac{1}{2} \\ y - 5z = 22 \\ z = -4 \end{cases}$

Performing Row Operations In Exercises 17 and 18, perform the row operation and write the equivalent system.

17. Add Equation 1 to Equation 2.

$$\begin{cases} x - 2y + 3z = 5 & \text{Equation 1} \\ -x + 3y - 5z = 4 & \text{Equation 2} \\ 2x \quad\;\; - 3z = 0 & \text{Equation 3} \end{cases}$$

What did this operation accomplish?

18. Add -2 times Equation 1 to Equation 3.

$$\begin{cases} x - 2y + 3z = 5 & \text{Equation 1} \\ -x + 3y - 5z = 4 & \text{Equation 2} \\ 2x \quad\;\; - 3z = 0 & \text{Equation 3} \end{cases}$$

What did this operation accomplish?

 Solving a System of Linear Equations In Exercises 19–22, solve the system of linear equations and check any solutions algebraically.

19. $\begin{cases} -2x + 3y = 10 \\ x + y = 0 \end{cases}$

20. $\begin{cases} 2x - y = 0 \\ x - y = 7 \end{cases}$

21. $\begin{cases} 3x - y = 9 \\ x - 2y = -2 \end{cases}$

22. $\begin{cases} x + 2y = 1 \\ 5x - 4y = -23 \end{cases}$

Solving a System of Linear Equations In Exercises 23–40, solve the system of linear equations and check any solutions algebraically.

23. $\begin{cases} x + y + z = 7 \\ 2x - y + z = 9 \\ 3x \quad\ - z = 10 \end{cases}$ 24. $\begin{cases} x + y + z = 5 \\ x - 2y + 4z = 13 \\ 3y + 4z = 13 \end{cases}$

25. $\begin{cases} 2x + 4y - z = 7 \\ 2x - 4y + 2z = -6 \\ x + 4y + z = 0 \end{cases}$ 26. $\begin{cases} 2x + 4y + z = 1 \\ x - 2y - 3z = 2 \\ x + y - z = -1 \end{cases}$

27. $\begin{cases} 2x + y - z = 7 \\ x - 2y + 2z = -9 \\ 3x - y + z = 5 \end{cases}$ 28. $\begin{cases} 5x - 3y + 2z = 3 \\ 2x + 4y - z = 7 \\ x - 11y + 4z = 3 \end{cases}$

29. $\begin{cases} 3x - 5y + 5z = 1 \\ 2x - 2y + 3z = 0 \\ 7x - y + 3z = 0 \end{cases}$ 30. $\begin{cases} 2x + y + 3z = 1 \\ 2x + 6y + 8z = 3 \\ 6x + 8y + 18z = 5 \end{cases}$

31. $\begin{cases} 2x + 3y = 0 \\ 4x + 3y - z = 0 \\ 8x + 3y + 3z = 0 \end{cases}$ 32. $\begin{cases} 4x + 3y + 17z = 0 \\ 5x + 4y + 22z = 0 \\ 4x + 2y + 19z = 0 \end{cases}$

33. $\begin{cases} x \quad\ + 4z = 1 \\ x + y + 10z = 10 \\ 2x - y + 2z = -5 \end{cases}$ 34. $\begin{cases} 2x - 2y - 6z = -4 \\ -3x + 2y + 6z = 1 \\ x - y - 5z = -3 \end{cases}$

35. $\begin{cases} 3x - 3y + 6z = 6 \\ x + 2y - z = 5 \\ 5x - 8y + 13z = 7 \end{cases}$ 36. $\begin{cases} x \quad\ + 2z = 5 \\ 3x - y - z = 1 \\ 6x - y + 5z = 16 \end{cases}$

37. $\begin{cases} x + 2y - 7z = -4 \\ 2x + y + z = 13 \\ 3x + 9y - 36z = -33 \end{cases}$

38. $\begin{cases} 2x + y - 3z = 4 \\ 4x \quad\ + 2z = 10 \\ -2x + 3y - 13z = -8 \end{cases}$

39. $\begin{cases} x \quad\quad\ + 3w = 4 \\ 2y - z - w = 0 \\ 3y \quad\ - 2w = 1 \\ 2x - y + 4z = 5 \end{cases}$

40. $\begin{cases} x + y + z + w = 6 \\ 2x + 3y \quad\ - w = 0 \\ -3x + 4y + z + 2w = 4 \\ x + 2y - z + w = 0 \end{cases}$

 Solving a Nonsquare System In Exercises 41–44, solve the system of linear equations and check any solutions algebraically.

41. $\begin{cases} x - 2y + 5z = 2 \\ 4x \quad\ - z = 0 \end{cases}$ 42. $\begin{cases} x - 3y + 2z = 18 \\ 5x - 13y + 12z = 80 \end{cases}$

43. $\begin{cases} 2x - 3y + z = -2 \\ -4x + 9y = 7 \end{cases}$ 44. $\begin{cases} 2x + 3y + 3z = 7 \\ 4x + 18y + 15z = 44 \end{cases}$

 Modeling Vertical Motion In Exercises 45 and 46, an object moving vertically is at the given heights at the specified times. Find the position equation $s = \frac{1}{2}at^2 + v_0t + s_0$ for the object.

45. At $t = 1$ second, $s = 128$ feet
    At $t = 2$ seconds, $s = 80$ feet
    At $t = 3$ seconds, $s = 0$ feet

46. At $t = 1$ second, $s = 132$ feet
    At $t = 2$ seconds, $s = 100$ feet
    At $t = 3$ seconds, $s = 36$ feet

 Finding the Equation of a Parabola In Exercises 47–52, find the equation

$$y = ax^2 + bx + c$$

of the parabola that passes through the points. To verify your result, use a graphing utility to plot the points and graph the parabola.

47. $(0, 0), (2, -2), (4, 0)$ 48. $(0, 3), (1, 4), (2, 3)$
49. $(2, 0), (3, -1), (4, 0)$ 50. $(1, 3), (2, 2), (3, -3)$
51. $\left(\frac{1}{2}, 1\right), (1, 3), (2, 13)$
52. $(-2, -3), (-1, 0), \left(\frac{1}{2}, -3\right)$

Finding the Equation of a Circle In Exercises 53–56, find the equation

$$x^2 + y^2 + Dx + Ey + F = 0$$

of the circle that passes through the points. To verify your result, use a graphing utility to plot the points and graph the circle.

53. $(0, 0), (5, 5), (10, 0)$ 54. $(0, 0), (0, 6), (3, 3)$
55. $(-3, -1), (2, 4), (-6, 8)$
56. $(0, 0), (0, -2), (3, 0)$

57. **Error Analysis** Describe the error.

The system
$$\begin{cases} x - 2y + 3x = 12 \\ y + 3z = 5 \\ 2z = 4 \end{cases}$$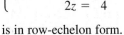

is in row-echelon form.

58. **Agriculture** A mixture of 5 pounds of fertilizer A, 13 pounds of fertilizer B, and 4 pounds of fertilizer C provides the optimal nutrients for a plant. Commercial brand X contains equal parts of fertilizer B and fertilizer C. Commercial brand Y contains one part of fertilizer A and two parts of fertilizer B. Commercial brand Z contains two parts of fertilizer A, five parts of fertilizer B, and two parts of fertilizer C. How much of each fertilizer brand is needed to obtain the desired mixture?

59. Finance To expand its clothing line, a small corporation borrowed $775,000 from three different lenders. The money was borrowed at 8%, 9%, and 10% simple interest. How much was borrowed at each rate when the annual interest owed was $67,500 and the amount borrowed at 8% was four times the amount borrowed at 10%?

60. Advertising A health insurance company advertises on television, on radio, and in the local newspaper. The marketing department has an advertising budget of $42,000 per month. A television ad costs $1000, a radio ad costs $200, and a newspaper ad costs $500. The department wants to run 60 ads per month and have as many television ads as radio and newspaper ads combined. How many of each type of ad can the department run each month?

Geometry In Exercises 61 and 62, find the values of *x*, *y*, and *z* in the figure.

61.

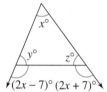

62.

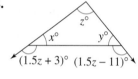

63. Geometry The perimeter of a triangle is 180 feet. The longest side of the triangle is 9 feet shorter than twice the shortest side. The sum of the lengths of the two shorter sides is 30 feet more than the length of the longest side. Find the lengths of the sides of the triangle.

64. Chemistry A chemist needs 10 liters of a 25% acid solution. The solution is to be mixed from three solutions whose concentrations are 10%, 20%, and 50%. How many liters of each solution will satisfy each condition?

(a) Use 2 liters of the 50% solution.

(b) Use as little as possible of the 50% solution.

(c) Use as much as possible of the 50% solution.

65. Electrical Network Applying Kirchhoff's Laws to the electrical network in the figure, the currents I_1, I_2, and I_3, are the solution of the system

$$\begin{cases} I_1 - I_2 + I_3 = 0 \\ 3I_1 + 2I_2 = 7. \\ 2I_2 + 4I_3 = 8 \end{cases}$$

Find the currents.

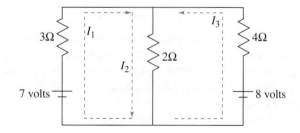

66. Pulley System A system of pulleys is loaded with 128-pound and 32-pound weights (see figure). The tensions t_1 and t_2 in the ropes and the acceleration a of the 32-pound weight are found by solving the system

$$\begin{cases} t_1 - 2t_2 = 0 \\ t_1 - 2a = 128 \\ t_2 + a = 32 \end{cases}$$

where t_1 and t_2 are in pounds and a is in feet per second squared. Solve this system.

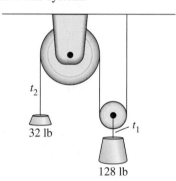

Fitting a Parabola One way to find the least squares regression parabola $y = ax^2 + bx + c$ for a set of points

$$(x_1, y_1), (x_2, y_2), \ldots, (x_n, y_n)$$

is by solving the system below for a, b, and c.

$$\begin{cases} nc + \left(\sum_{i=1}^{n} x_i\right)b + \left(\sum_{i=1}^{n} x_i^2\right)a = \sum_{i=1}^{n} y_i \\ \left(\sum_{i=1}^{n} x_i\right)c + \left(\sum_{i=1}^{n} x_i^2\right)b + \left(\sum_{i=1}^{n} x_i^3\right)a = \sum_{i=1}^{n} x_i y_i \\ \left(\sum_{i=1}^{n} x_i^2\right)c + \left(\sum_{i=1}^{n} x_i^3\right)b + \left(\sum_{i=1}^{n} x_i^4\right)a = \sum_{i=1}^{n} x_i^2 y_i \end{cases}$$

In Exercises 67 and 68, the sums have been evaluated. Solve the simplified system for *a*, *b*, and *c* to find the least squares regression parabola for the points. Use a graphing utility to confirm the result. (*Note:* The symbol Σ is used to denote a sum of the terms of a sequence. You will learn how to use this notation in Section 9.1.)

67. $\begin{cases} 4c + 9b + 29a = 20 \\ 9c + 29b + 99a = 70 \\ 29c + 99b + 353a = 254 \end{cases}$

68. $\begin{cases} 4c + 40a = 19 \\ 40b = -12 \\ 40c + 544a = 160 \end{cases}$

69. Stopping Distance In testing a new automobile braking system, engineers recorded the speed x (in miles per hour) and the stopping distance y (in feet). The table shows the results.

Speed, x	30	40	50	60	70
Stopping Distance, y	75	118	175	240	315

(a) Find the least squares regression parabola $y = ax^2 + bc + c$ for the data by solving the system.

$$\begin{cases} 5c + 250b + 13{,}500a = 923 \\ 250c + 13{,}500b + 775{,}000a = 52{,}170 \\ 13{,}500c + 775{,}000b + 46{,}590{,}000a = 3{,}101{,}300 \end{cases}$$

(b) Use a graphing utility to graph the model you found in part (a) and the data in the same viewing window. How well does the model fit the data? Explain.

(c) Use the model to estimate the stopping distance when the speed is 75 miles per hour.

70. Wildlife

A wildlife management team studied the reproductive rates of deer in four tracts of a wildlife preserve. In each tract, the number of females x and the percent of females y that had offspring the following year were recorded. The table shows the results.

Number, x	100	120	140	160
Percent, y	75	68	55	30

(a) Find the least squares regression parabola $y = ax^2 + bx + c$ for the data by solving the system.

$$\begin{cases} 4c + 520b + 69{,}600a = 228 \\ 520c + 69{,}600b + 9{,}568{,}000a = 28{,}160 \\ 69{,}600c + 9{,}568{,}000b + 1{,}346{,}880{,}000a = 3{,}575{,}200 \end{cases}$$

(b) Use a graphing utility to graph the model you found in part (a) and the data in the same viewing window. How well does the model fit the data? Explain.

(c) Use the model to estimate the percent of females that had offspring when there were 170 females.

(d) Use the model to estimate the number of females when 40% of the females had offspring.

Advanced Applications In Exercises 71 and 72, find values of x, y, and λ that satisfy the system. These systems arise in certain optimization problems in calculus, and λ is called a Lagrange multiplier.

71. $\begin{cases} 2x - 2x\lambda = 0 \\ -2y + \lambda = 0 \\ y - x^2 = 0 \end{cases}$ **72.** $\begin{cases} 2 + 2y + 2\lambda = 0 \\ 2x + 1 + \lambda = 0 \\ 2x + y - 100 = 0 \end{cases}$

Exploration

True or False? In Exercises 73 and 74, determine whether the statement is true or false. Justify your answer.

73. Every nonsquare system of linear equations has a unique solution.

74. If a system of three linear equations is inconsistent, then there are no points common to the graphs of all three equations of the system.

75. Think About It Are the following two systems of equations equivalent? Give reasons for your answer.

$\begin{cases} x + 3y - z = 6 \\ 2x - y + 2z = 1 \\ 3x + 2y - z = 2 \end{cases}$ $\begin{cases} x + 3y - z = 6 \\ -7y + 4z = 1 \\ -7y - 4z = -16 \end{cases}$

76. HOW DO YOU SEE IT? The number of sides x and the combined number of sides and diagonals y for each of three regular polygons are shown below. Write a system of linear equations to find an equation of the form $y = ax^2 + bx + c$ that represents the relationship between x and y for the three polygons.

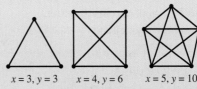

$x = 3, y = 3$ $x = 4, y = 6$ $x = 5, y = 10$

Finding Systems of Linear Equations In Exercises 77–80, find two systems of linear equations that have the ordered triple as a solution. (There are many correct answers.)

77. $(2, 0, -1)$ **78.** $(-5, 3, -2)$

79. $\left(\frac{1}{2}, -3, 0\right)$ **80.** $\left(4, \frac{2}{5}, \frac{1}{2}\right)$

Project: Earnings per Share To work an extended application analyzing the earnings per share for Wal-Mart Stores, Inc., from 2001 through 2015, visit this text's website at *LarsonPrecalculus.com*. (*Source: Wal-Mart Stores, Inc.*)

7.4 Partial Fractions

Partial fractions can help you analyze the behavior of a rational function. For example, in Exercise 60 on page 509, you will use partial fractions to analyze the exhaust temperatures of a diesel engine.

- Recognize partial fraction decompositions of rational expressions.
- Find partial fraction decompositions of rational expressions.

Introduction

In this section, you will learn to write a rational expression as the sum of two or more simpler rational expressions. For example, the rational expression

$$\frac{x + 7}{x^2 - x - 6}$$

can be written as the sum of two fractions with first-degree denominators. That is,

Partial fraction decomposition

of $\dfrac{x + 7}{x^2 - x - 6}$

$$\frac{x + 7}{x^2 - x - 6} = \overbrace{\frac{2}{x - 3} + \frac{-1}{x + 2}}.$$

Partial fraction   Partial fraction

Each fraction on the right side of the equation is a **partial fraction,** and together they make up the **partial fraction decomposition** of the left side.

▷ **ALGEBRA HELP** To review how to find the degree of a polynomial (such as $x - 3$ and $x + 2$), see Appendix A.3.

··**REMARK** Appendix A.4 shows you how to combine expressions such as

$$\frac{1}{x - 2} + \frac{-1}{x + 3} = \frac{5}{(x - 2)(x + 3)}.$$

The method of partial fraction decomposition shows you how to reverse this process and write

$$\frac{5}{(x - 2)(x + 3)} = \frac{1}{x - 2} + \frac{-1}{x + 3}.$$

····················▷

Decomposition of $N(x)/D(x)$ into Partial Fractions

1. *Divide when improper:* When $N(x)/D(x)$ is an improper fraction [degree of $N(x) \geq$ degree of $D(x)$], divide the denominator into the numerator to obtain

$$\frac{N(x)}{D(x)} = (\text{polynomial}) + \frac{N_1(x)}{D(x)}$$

and apply Steps 2, 3, and 4 to the proper rational expression

$$\frac{N_1(x)}{D(x)}.$$

Note that $N_1(x)$ is the remainder from the division of $N(x)$ by $D(x)$.

2. *Factor the denominator:* Completely factor the denominator into factors of the form

$$(px + q)^m \quad \text{and} \quad (ax^2 + bx + c)^n$$

where $(ax^2 + bx + c)$ is irreducible.

3. *Linear factors:* For *each* factor of the form $(px + q)^m$, the partial fraction decomposition must include the following sum of m fractions.

$$\frac{A_1}{(px + q)} + \frac{A_2}{(px + q)^2} + \cdots + \frac{A_m}{(px + q)^m}$$

4. *Quadratic factors:* For *each* factor of the form $(ax^2 + bx + c)^n$, the partial fraction decomposition must include the following sum of n fractions.

$$\frac{B_1x + C_1}{ax^2 + bx + c} + \frac{B_2x + C_2}{(ax^2 + bx + c)^2} + \cdots + \frac{B_nx + C_n}{(ax^2 + bx + c)^n}$$

Partial Fraction Decomposition

The examples in this section demonstrate algebraic techniques for determining the constants in the numerators of partial fractions. Note that the techniques vary slightly, depending on the type of factors of the denominator: linear or quadratic, distinct or repeated.

EXAMPLE 1   **Distinct Linear Factors**

Write the partial fraction decomposition of

$$\frac{x + 7}{x^2 - x - 6}.$$

Solution   The expression is proper, so begin by factoring the denominator.

$$x^2 - x - 6 = (x - 3)(x + 2)$$

Include one partial fraction with a constant numerator for each linear factor of the denominator. Write the form of the decomposition with A and B as the unknown constants.

$$\frac{x + 7}{x^2 - x - 6} = \frac{A}{x - 3} + \frac{B}{x + 2} \qquad \text{Write form of decomposition.}$$

Multiply each side of this equation by the least common denominator, $(x - 3)(x + 2)$, to obtain the **basic equation.**

$$x + 7 = A(x + 2) + B(x - 3) \qquad \text{Basic equation}$$

This equation is true for all x, so substitute any *convenient* values of x that will help determine the constants A and B. Values of x that are especially convenient are those that make the factors $(x + 2)$ and $(x - 3)$ equal to zero. For example, to solve for B, let $x = -2$.

$$-2 + 7 = A(-2 + 2) + B(-2 - 3) \qquad \text{Substitute } -2 \text{ for } x.$$
$$5 = A(0) + B(-5)$$
$$5 = -5B$$
$$-1 = B$$

To solve for A, let $x = 3$.

$$3 + 7 = A(3 + 2) + B(3 - 3) \qquad \text{Substitute 3 for } x.$$
$$10 = A(5) + B(0)$$
$$10 = 5A$$
$$2 = A$$

So, the partial fraction decomposition is

$$\frac{x + 7}{x^2 - x - 6} = \frac{2}{x - 3} + \frac{-1}{x + 2}.$$

Check this result by combining the two partial fractions on the right side of the equation, or by using your graphing utility.

✓ **Checkpoint** 🔊)) *Audio-video solution in English & Spanish at LarsonPrecalculus.com*

Write the partial fraction decomposition of

$$\frac{x + 5}{2x^2 - x - 1}.$$

▷ **TECHNOLOGY** To use a graphing utility to check the decomposition found in Example 1, graph

$$y_1 = \frac{x + 7}{x^2 - x - 6}$$

and

$$y_2 = \frac{2}{x - 3} + \frac{-1}{x + 2}$$

in the same viewing window. The graphs should be identical, as shown below.

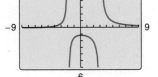

EXAMPLE 2 **Repeated Linear Factors**

Write the partial fraction decomposition of $\dfrac{x^4 + 2x^3 + 6x^2 + 20x + 6}{x^3 + 2x^2 + x}$.

▷ **ALGEBRA HELP** To review long division of polynomials, see Section 2.3. To review factoring of polynomials, see Appendix A.3.

Solution This rational expression is improper, so begin by dividing the numerator by the denominator.

$$
\begin{array}{r}
x \\
x^3 + 2x^2 + x \overline{)\,x^4 + 2x^3 + 6x^2 + 20x + 6} \\
\underline{x^4 + 2x^3 + x^2 } \\
5x^2 + 20x + 6
\end{array}
$$

The result is

$$x + \dfrac{5x^2 + 20x + 6}{x^3 + 2x^2 + x}.$$

The denominator of the remainder factors as

$$x^3 + 2x^2 + x = x(x^2 + 2x + 1) = x(x + 1)^2$$

so include a partial fraction with a constant numerator for each power of x and $(x + 1)$.

$$\dfrac{5x^2 + 20x + 6}{x(x + 1)^2} = \dfrac{A}{x} + \dfrac{B}{x + 1} + \dfrac{C}{(x + 1)^2} \qquad \text{Write form of decomposition.}$$

Multiply each side by the LCD, $x(x + 1)^2$, to obtain the basic equation.

$$5x^2 + 20x + 6 = A(x + 1)^2 + Bx(x + 1) + Cx \qquad \text{Basic equation}$$

▷

•• **REMARK** To obtain the basic equation, be sure to multiply *each* fraction by the LCD.

Let $x = -1$ to eliminate the A- and B-terms.

$$5(-1)^2 + 20(-1) + 6 = A(-1 + 1)^2 + B(-1)(-1 + 1) + C(-1)$$

$$5 - 20 + 6 = 0 + 0 - C$$

$$C = 9$$

Let $x = 0$ to eliminate the B- and C-terms.

$$5(0)^2 + 20(0) + 6 = A(0 + 1)^2 + B(0)(0 + 1) + C(0)$$

$$6 = A(1) + 0 + 0$$

$$6 = A$$

You have exhausted the most convenient values of x, but you can now use the known values of A and C to find the value of B. So, let $x = 1$, $A = 6$, and $C = 9$.

$$5(1)^2 + 20(1) + 6 = 6(1 + 1)^2 + B(1)(1 + 1) + 9(1)$$

$$31 = 6(4) + 2B + 9$$

$$-2 = 2B$$

$$-1 = B$$

So, the partial fraction decomposition is

$$\dfrac{x^4 + 2x^3 + 6x^2 + 20x + 6}{x^3 + 2x^2 + x} = x + \dfrac{6}{x} + \dfrac{-1}{x + 1} + \dfrac{9}{(x + 1)^2}.$$

✓ **Checkpoint** 🔊))) *Audio-video solution in English & Spanish at LarsonPrecalculus.com*

Write the partial fraction decomposition of $\dfrac{x^4 + x^3 + x + 4}{x^3 + x^2}$.

The procedure used to solve for the constants $A, B, \ldots$ in Examples 1 and 2 works well when the factors of the denominator are linear. When the denominator contains irreducible quadratic factors, a better process is to write the right side of the basic equation in polynomial form, *equate the coefficients* of like terms to form a system of equations, and solve the resulting system for the constants.

EXAMPLE 3   Distinct Linear and Quadratic Factors

Write the partial fraction decomposition of

$$\frac{3x^2 + 4x + 4}{x^3 + 4x}.$$

Solution This expression is proper, so begin by factoring the denominator. The denominator factors as

$$x^3 + 4x = x(x^2 + 4)$$

so when writing the form of the decomposition, include one partial fraction with a constant numerator and one partial fraction with a linear numerator.

$$\frac{3x^2 + 4x + 4}{x^3 + 4x} = \frac{A}{x} + \frac{Bx + C}{x^2 + 4} \qquad \text{Write form of decomposition.}$$

Multiply each side by the LCD, $x(x^2 + 4)$, to obtain the basic equation.

$$3x^2 + 4x + 4 = A(x^2 + 4) + (Bx + C)x \qquad \text{Basic equation}$$

Expand this basic equation and collect like terms.

$$3x^2 + 4x + 4 = Ax^2 + 4A + Bx^2 + Cx$$
$$= (A + B)x^2 + Cx + 4A \qquad \text{Polynomial form}$$

Use the fact that two polynomials are equal if and only if the coefficients of like terms are equal to write a system of linear equations.

$$3x^2 + 4x + 4 = (A + B)x^2 + Cx + 4A \qquad \text{Equate coefficients of like terms.}$$

$$\begin{cases} A + B &= 3 \qquad \text{Equation 1} \\ C &= 4 \qquad \text{Equation 2} \\ 4A &= 4 \qquad \text{Equation 3} \end{cases}$$

From Equation 3 and Equation 2, you have

$$A = 1 \quad \text{and} \quad C = 4.$$

Back-substituting $A = 1$ into Equation 1 yields

$$1 + B = 3 \quad \Longrightarrow \quad B = 2.$$

So, the partial fraction decomposition is

$$\frac{3x^2 + 4x + 4}{x^3 + 4x} = \frac{1}{x} + \frac{2x + 4}{x^2 + 4}.$$

✓ *Checkpoint* ◀))) *Audio-video solution in English & Spanish at LarsonPrecalculus.com*

Write the partial fraction decomposition of

$$\frac{2x^2 - 5}{x^3 + x}.$$

Johann Bernoulli (1667–1748), a Swiss mathematician, introduced the method of partial fractions and was instrumental in the early development of calculus. Bernoulli was a professor at the University of Basel and taught many outstanding students, including the renowned Leonhard Euler.

The next example shows how to find the partial fraction decomposition of a rational expression whose denominator has a *repeated* quadratic factor.

EXAMPLE 4 **Repeated Quadratic Factors**

See LarsonPrecalculus.com for an interactive version of this type of example.

Write the partial fraction decomposition of

$$\frac{8x^3 + 13x}{(x^2 + 2)^2}.$$

Solution Include one partial fraction with a linear numerator for each power of $(x^2 + 2)$.

$$\frac{8x^3 + 13x}{(x^2 + 2)^2} = \frac{Ax + B}{x^2 + 2} + \frac{Cx + D}{(x^2 + 2)^2} \qquad \text{Write form of decomposition.}$$

Multiply each side by the LCD, $(x^2 + 2)^2$, to obtain the basic equation.

$$8x^3 + 13x = (Ax + B)(x^2 + 2) + Cx + D \qquad \text{Basic equation}$$

$$= Ax^3 + 2Ax + Bx^2 + 2B + Cx + D$$

$$= Ax^3 + Bx^2 + (2A + C)x + (2B + D) \qquad \text{Polynomial form}$$

Equate coefficients of like terms on opposite sides of the equation to write a system of linear equations.

$$8x^3 + 0x^2 + 13x + 0 = Ax^3 + Bx^2 + (2A + C)x + (2B + D)$$

$$\begin{cases} A && = 8 & \qquad \text{Equation 1} \\ & B && = 0 & \qquad \text{Equation 2} \\ 2A + & & C & = 13 & \qquad \text{Equation 3} \\ & 2B + & D & = 0 & \qquad \text{Equation 4} \end{cases}$$

Use the values $A = 8$ and $B = 0$ to obtain the values of C and D.

$$2(8) + C = 13 \qquad \text{Substitute 8 for } A \text{ in Equation 3.}$$

$$C = -3$$

$$2(0) + D = 0 \qquad \text{Substitute 0 for } B \text{ in Equation 4.}$$

$$D = 0$$

So, using

$$A = 8, \quad B = 0, \quad C = -3, \quad \text{and} \quad D = 0$$

the partial fraction decomposition is

$$\frac{8x^3 + 13x}{(x^2 + 2)^2} = \frac{8x}{x^2 + 2} + \frac{-3x}{(x^2 + 2)^2}.$$

Check this result by combining the two partial fractions on the right side of the equation, or by using your graphing utility.

✓ **Checkpoint** ◗)) *Audio-video solution in English & Spanish at LarsonPrecalculus.com*

Write the partial fraction decomposition of $\dfrac{x^3 + 3x^2 - 2x + 7}{(x^2 + 4)^2}$.

EXAMPLE 5 Repeated Linear and Quadratic Factors

Write the partial fraction decomposition of $\dfrac{x+5}{x^2(x^2+1)^2}$.

Solution Include one partial fraction with a constant numerator for each power of x and one partial fraction with a linear numerator for each power of (x^2+1).

$$\frac{x+5}{x^2(x^2+1)^2} = \frac{A}{x} + \frac{B}{x^2} + \frac{Cx+D}{x^2+1} + \frac{Ex+F}{(x^2+1)^2} \qquad \text{Write form of decomposition.}$$

Multiply each side by the LCD, $x^2(x^2+1)^2$, to obtain the basic equation.

$$x+5 = Ax(x^2+1)^2 + B(x^2+1)^2 + (Cx+D)x^2(x^2+1) + (Ex+F)x^2 \qquad \substack{\text{Basic}\\\text{equation}}$$

$$= (A+C)x^5 + (B+D)x^4 + (2A+C+E)x^3 + (2B+D+F)x^2 + Ax + B$$

Write and solve the system of equations formed by equating coefficients on opposite sides of the equation to show that $A=1$, $B=5$, $C=-1$, $D=-5$, $E=-1$, and $F=-5$, and that the partial fraction decomposition is

$$\frac{x+5}{x^2(x^2+1)^2} = \frac{1}{x} + \frac{5}{x^2} - \frac{x+5}{x^2+1} - \frac{x+5}{(x^2+1)^2}.$$

✓ **Checkpoint**  Audio-video solution in English & Spanish at LarsonPrecalculus.com

Write the partial fraction decomposition of $\dfrac{4x-8}{x^2(x^2+2)^2}$.

Guidelines for Solving the Basic Equation

Linear Factors

1. Substitute the *zeros* of the distinct linear factors into the basic equation.
2. For repeated linear factors, use the coefficients determined in Step 1 to rewrite the basic equation. Then substitute *other* convenient values of x and solve for the remaining coefficients.

Quadratic Factors

1. Expand the basic equation.
2. Collect terms according to powers of x.
3. Equate the coefficients of like terms to obtain a system of equations involving the constants, A, B, C, ...
4. Use the system of linear equations to solve for A, B, C, ...

Keep in mind that for *improper* rational expressions, you must first divide before applying partial fraction decomposition.

Summarize (Section 7.4)

1. Explain what is meant by the partial fraction decomposition of a rational expression *(page 502)*.
2. Explain how to find the partial fraction decomposition of a rational expression *(pages 502–507)*. For examples of finding partial fraction decompositions of rational expressions, see Examples 1–5.

7.4 Exercises

See **CalcChat.com** for tutorial help and worked-out solutions to odd-numbered exercises.

Vocabulary: Fill in the blanks.

1. The result of writing a rational expression as the sum of two or more simpler rational expressions is called the _____ _____ _____.

2. If the degree of the numerator of a rational expression is greater than or equal to the degree of the denominator, then the fraction is _____.

3. Each fraction on the right side of the equation $\dfrac{x-1}{x^2-8x+15} = \dfrac{-1}{x-3} + \dfrac{2}{x-5}$ is a _____ _____.

4. You obtain the _____ _____ by multiplying each side of the partial fraction decomposition form by the least common denominator.

Skills and Applications

Matching In Exercises 5–8, match the rational expression with the form of its decomposition. [The decompositions are labeled (a), (b), (c), and (d).]

(a) $\dfrac{A}{x} + \dfrac{B}{x+2} + \dfrac{C}{x-2}$ (b) $\dfrac{A}{x} + \dfrac{B}{x-4}$

(c) $\dfrac{A}{x} + \dfrac{B}{x^2} + \dfrac{C}{x-4}$ (d) $\dfrac{A}{x} + \dfrac{B}{x-4} + \dfrac{C}{(x-4)^2}$

5. $\dfrac{3x-1}{x(x-4)}$

6. $\dfrac{3x-1}{x^2(x-4)}$

7. $\dfrac{3x-1}{x(x-4)^2}$

8. $\dfrac{3x-1}{x(x^2-4)}$

 Writing the Form of the Decomposition In Exercises 9–16, write the form of the partial fraction decomposition of the rational expression. Do not solve for the constants.

9. $\dfrac{3}{x^2-2x}$

10. $\dfrac{x-2}{x^2+4x+3}$

11. $\dfrac{6x+5}{(x+2)^4}$

12. $\dfrac{5x^2+3}{x^2(x-4)^2}$

13. $\dfrac{2x-3}{x^3+10x}$

14. $\dfrac{x-1}{x(x^2+1)^2}$

15. $\dfrac{8x}{x^2(x^2+3)^2}$

16. $\dfrac{x^2-9}{x^3(x^2+2)^2}$

 Writing the Partial Fraction Decomposition In Exercises 17–42, write the partial fraction decomposition of the rational expression. Check your result algebraically.

17. $\dfrac{1}{x^2+x}$

18. $\dfrac{3}{x^2-3x}$

19. $\dfrac{3}{x^2+x-2}$

20. $\dfrac{x+1}{x^2-x-6}$

21. $\dfrac{1}{x^2-1}$

22. $\dfrac{1}{4x^2-9}$

23. $\dfrac{x^2+12x+12}{x^3-4x}$

24. $\dfrac{x+2}{x(x^2-9)}$

25. $\dfrac{3x}{(x-3)^2}$

26. $\dfrac{2x-3}{(x-1)^2}$

27. $\dfrac{4x^2+2x-1}{x^2(x+1)}$

28. $\dfrac{6x^2+1}{x^2(x-1)^2}$

29. $\dfrac{x^2+2x+3}{x^3+x}$

30. $\dfrac{2x}{x^3-1}$

31. $\dfrac{x}{x^3-x^2-2x+2}$

32. $\dfrac{x+6}{x^3-3x^2-4x+12}$

33. $\dfrac{x}{16x^4-1}$

34. $\dfrac{3}{x^4+x}$

35. $\dfrac{x^2+5}{(x+1)(x^2-2x+3)}$

36. $\dfrac{x^2-4x+7}{(x+1)(x^2-2x+3)}$

37. $\dfrac{2x^2+x+8}{(x^2+4)^2}$

38. $\dfrac{3x^2+1}{(x^2+2)^2}$

39. $\dfrac{5x^2-2}{(x^2+3)^3}$

40. $\dfrac{x^2-4x+6}{(x^2+4)^3}$

41. $\dfrac{8x-12}{x^2(x^2+2)^2}$

42. $\dfrac{x+1}{x^3(x^2+1)^2}$

 Improper Rational Expression Decomposition In Exercises 43–50, write the partial fraction decomposition of the improper rational expression.

43. $\dfrac{x^2-x}{x^2+x+1}$

44. $\dfrac{x^2-4x}{x^2+x+6}$

45. $\dfrac{2x^3-x^2+x+5}{x^2+3x+2}$

46. $\dfrac{x^3+2x^2-x+1}{x^2+3x-4}$

47. $\dfrac{x^4}{(x-1)^3}$

48. $\dfrac{16x^4}{(2x-1)^3}$

49. $\dfrac{x^4+2x^3+4x^2+8x+2}{x^3+2x^2+x}$

50. $\dfrac{2x^4+8x^3+7x^2-7x-12}{x^3+4x^2+4x}$

Writing the Partial Fraction Decomposition In Exercises 51–58, write the partial fraction decomposition of the rational expression. Use a graphing utility to check your result.

51. $\dfrac{5 - x}{2x^2 + x - 1}$

52. $\dfrac{4x^2 - 1}{2x(x + 1)^2}$

53. $\dfrac{3x^2 - 7x - 2}{x^3 - x}$

54. $\dfrac{3x + 6}{x^3 + 2x}$

55. $\dfrac{x^2 + x + 2}{(x^2 + 2)^2}$

56. $\dfrac{x^3}{(x + 2)^2(x - 2)^2}$

57. $\dfrac{2x^3 - 4x^2 - 15x + 5}{x^2 - 2x - 8}$

58. $\dfrac{x^3 - x + 3}{x^2 + x - 2}$

59. Environmental Science The predicted cost C (in thousands of dollars) for a company to remove $p\%$ of a chemical from its waste water is given by the model

$$C = \frac{120p}{10{,}000 - p^2}, \quad 0 \le p < 100.$$

Write the partial fraction decomposition for the rational function. Verify your result by using a graphing utility to create a table comparing the original function with the partial fractions.

60. Thermodynamics

The magnitude of the range R of exhaust temperatures (in degrees Fahrenheit) in an experimental diesel engine is approximated by the model

$$R = \frac{5000(4 - 3x)}{(11 - 7x)(7 - 4x)}, \quad 0 < x \le 1$$

where x is the relative load (in foot-pounds).

(a) Write the partial fraction decomposition of the equation.

(b) The decomposition in part (a) is the difference of two fractions. The absolute values of the terms give the expected maximum and minimum temperatures of the exhaust gases for different loads.

$$\text{Ymax} = |\text{1st term}| \qquad \text{Ymin} = |\text{2nd term}|$$

Write the equations for Ymax and Ymin.

(c) Use a graphing utility to graph each equation from part (b) in the same viewing window.

(d) Determine the expected maximum and minimum temperatures for a relative load of 0.5.

PHOTO: Bosch

Exploration

True or False? In Exercises 61–63, determine whether the statement is true or false. Justify your answer.

61. For the rational expression $\dfrac{x}{(x + 10)(x - 10)^2}$, the partial fraction decomposition is of the form

$$\frac{A}{x + 10} + \frac{B}{(x - 10)^2}.$$

62. When writing the partial fraction decomposition of the expression $\dfrac{x^3 + x - 2}{x^2 - 5x - 14}$, the first step is to divide the numerator by the denominator.

63. In the partial fraction decomposition of a rational expression, the denominators of each partial fraction always have a lower degree than the denominator of the original expression.

64. **HOW DO YOU SEE IT?** Identify the graph of the rational function and the graph representing each partial fraction of its partial fraction decomposition. Then state any relationship between the vertical asymptotes of the graph of the rational function and the vertical asymptotes of the graphs representing the partial fractions of the decomposition. To print an enlarged copy of the graph, go to *MathGraphs.com*.

(a) $y = \dfrac{x - 12}{x(x - 4)}$

$\quad = \dfrac{3}{x} - \dfrac{2}{x - 4}$

(b) $y = \dfrac{2(4x - 3)}{x^2 - 9}$

$\quad = \dfrac{3}{x - 3} + \dfrac{5}{x + 3}$

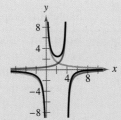
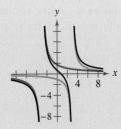

65. Error Analysis Describe the error in writing the basic equation for the partial fraction decomposition of the rational expression.

$$\frac{x^2 + 1}{x(x - 1)} = \frac{A}{x} + \frac{B}{x - 1}$$

$$x^2 + 1 = A(x - 1) + Bx$$

66. Writing Describe two ways of solving for the constants in a partial fraction decomposition.

7.5 Systems of Inequalities

Systems of inequalities in two variables can help you model and solve real-life problems. For example, in Exercise 68 on page 518, you will use a system of inequalities to analyze a person's recommended target heart rate during exercise.

- ◼ Sketch the graphs of inequalities in two variables.
- ◼ Solve systems of inequalities.
- ◼ Use systems of inequalities in two variables to model and solve real-life problems.

The Graph of an Inequality

The statements

$$3x - 2y < 6 \quad \text{and} \quad 2x^2 + 3y^2 \geq 6$$

are inequalities in two variables. An ordered pair (a, b) is a **solution of an inequality** in x and y when the inequality is true after a and b are substituted for x and y, respectively. The **graph of an inequality** is the collection of all solutions of the inequality. To sketch the graph of an inequality, begin by sketching the graph of the *corresponding equation*. The graph of the equation will usually separate the plane into two or more regions. In each such region, one of the following must be true.

1. *All* points in the region are solutions of the inequality.
2. *No* point in the region is a solution of the inequality.

So, you can determine whether the points in an entire region satisfy the inequality by testing *one* point in the region.

Sketching the Graph of an Inequality in Two Variables

1. Replace the inequality sign by an equal sign and sketch the graph of the equation. (Use a dashed line for $<$ or $>$ and a solid line for $\leq$ or $\geq$.)
2. Test one point in each of the regions formed by the graph in Step 1. If the point satisfies the inequality, then shade the entire region to denote that every point in the region satisfies the inequality.

• • **REMARK** Be careful when you are sketching the graph of an inequality in two variables. A dashed line means that the points on the line or curve *are not* solutions of the inequality. A solid line means that the points on the line or curve *are* solutions of the inequality.

▷

| **EXAMPLE 1** | **Sketching the Graph of an Inequality** |

See LarsonPrecalculus.com for an interactive version of this type of example.

Sketch the graph of $y \geq x^2 - 1$.

Solution Begin by graphing the corresponding equation $y = x^2 - 1$, as shown at the right. Test a point *above* the parabola, such as $(0, 0)$, and a point *below* the parabola, such as $(0, -2)$.

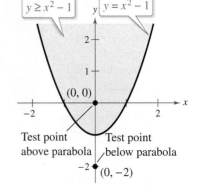

$$(0, 0): \quad 0 \overset{?}{\geq} 0^2 - 1$$

$$0 \geq -1 \qquad \text{(0, 0) is a solution.}$$

$$(0, -2): \quad -2 \overset{?}{\geq} 0^2 - 1$$

$$-2 \not\geq -1 \qquad \text{(0, -2) is not a solution.}$$

The points that satisfy the inequality $y \geq x^2 - 1$ are those lying above (or on) the parabola, as shown by the shaded region in the figure.

✓ *Checkpoint* ◀))) *Audio-video solution in English & Spanish at LarsonPrecalculus.com*

Sketch the graph of $(x + 2)^2 + (y - 2)^2 < 16$.

The inequality in Example 1 is a nonlinear inequality in two variables. Many of the examples in this section involve **linear inequalities** such as $ax + by < c$ (where a and b are not both zero). The graph of a linear inequality is a half-plane lying on one side of the line $ax + by = c$.

EXAMPLE 2 **Sketching the Graph of a Linear Inequality**

Sketch the graph of each linear inequality.

a. $x > -2$ **b.** $y \le 3$

Solution

▷ TECHNOLOGY A graphing utility can be used to graph an inequality or a system of inequalities. For example, to graph $y \ge x - 2$, enter $y = x - 2$ and use the *shade* feature of the graphing utility to shade the solution region as shown below. Consult the user's guide for your graphing utility for specific keystrokes.

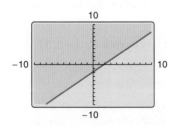

a. The graph of the corresponding equation $x = -2$ is a vertical line. The points that satisfy the inequality $x > -2$ are those lying to the right of this line, as shown in Figure 7.16.

b. The graph of the corresponding equation $y = 3$ is a horizontal line. The points that satisfy the inequality $y \le 3$ are those lying below (or on) this line, as shown in Figure 7.17.

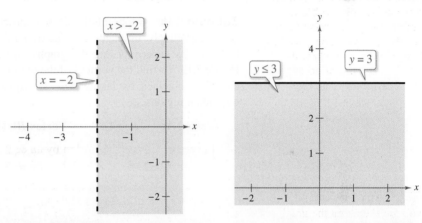

Figure 7.16

Figure 7.17

✓ **Checkpoint** ◀))) *Audio-video solution in English & Spanish at LarsonPrecalculus.com*

Sketch the graph of $x \ge 3$.

EXAMPLE 3 **Sketching the Graph of a Linear Inequality**

Sketch the graph of $x - y < 2$.

Solution The graph of the corresponding equation $x - y = 2$ is a line, as shown in Figure 7.18. The origin $(0, 0)$ satisfies the inequality, so the graph consists of the half-plane lying above the line. (Check a point below the line. Regardless of which point you choose, you will find that it does not satisfy the inequality.)

✓ **Checkpoint** ◀))) *Audio-video solution in English & Spanish at LarsonPrecalculus.com*

Sketch the graph of $x + y > -2$.

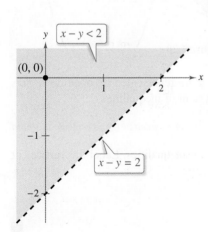

Figure 7.18

To graph a linear inequality, it sometimes helps to write the inequality in slope-intercept form. For example, writing $x - y < 2$ as

$$y > x - 2$$

helps you to see that the solution points lie *above* the line $x - y = 2$ (or $y = x - 2$), as shown in Figure 7.18.

Systems of Inequalities

Many practical problems in business, science, and engineering involve systems of linear inequalities. A **solution** of a system of inequalities in x and y is a point (x, y) that satisfies each inequality in the system.

To sketch the graph of a system of inequalities in two variables, first sketch the graph of each individual inequality (on the same coordinate system) and then find the region that is *common* to every graph in the system. This region represents the **solution set** of the system. For a system of *linear inequalities,* it is helpful to find the vertices of the solution region.

EXAMPLE 4 Solving a System of Inequalities

Sketch the graph of the solution set of the system of inequalities. Label the vertices of the region.

$$\begin{cases} x - y < 2 \\ x > -2 \\ y \le 3 \end{cases}$$

Solution The graphs of these inequalities are shown in Figures 7.18, 7.16, and 7.17, respectively, on page 511. The triangular region common to all three graphs can be found by superimposing the graphs on the same coordinate system, as shown in Figure 7.19. To find the vertices of the region, solve the three systems of corresponding equations obtained by taking *pairs* of equations representing the boundaries of the individual regions.

Vertex A: $(-2, -4)$

$$\begin{cases} x - y = 2 \\ x = -2 \end{cases}$$

Vertex B: $(5, 3)$

$$\begin{cases} x - y = 2 \\ y = 3 \end{cases}$$

Vertex C: $(-2, 3)$

$$\begin{cases} x = -2 \\ y = 3 \end{cases}$$

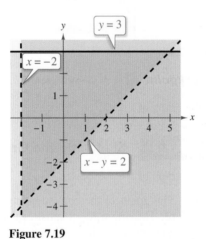

Figure 7.19

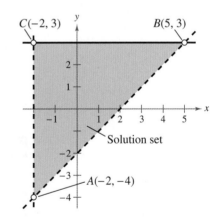

Figure 7.20

• **REMARK** Using a different colored pencil to shade the solution of each inequality in a system will make identifying the solution of the system of inequalities easier.

Note in Figure 7.20 that the vertices of the region are represented by open dots. This means that the vertices *are not* solutions of the system of inequalities.

✓ **Checkpoint** *Audio-video solution in English & Spanish at LarsonPrecalculus.com*

Sketch the graph of the solution set of the system of inequalities. Label the vertices of the region.

$$\begin{cases} x + y \ge 1 \\ -x + y \ge 1 \\ y \le 2 \end{cases}$$

For the triangular region shown in Example 4, each pair of boundary lines intersects at a vertex of the region. With more complicated regions, two boundary lines can sometimes intersect at a point that is not a vertex of the region, as shown below. As you sketch the graph of a solution set, use your sketch along with the inequalities of the system to determine which points of intersection are actually vertices of the region.

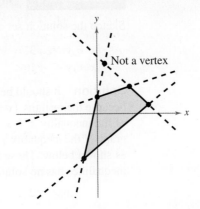

Not a vertex

EXAMPLE 5 Solving a System of Inequalities

Sketch the graph of the solution set of the system of inequalities.

$$\begin{cases} x^2 - y \le 1 & \text{Inequality 1} \\ -x + y \le 1 & \text{Inequality 2} \end{cases}$$

Solution The points that satisfy the inequality

$$x^2 - y \le 1 \qquad \text{Inequality 1}$$

are the points lying above (or on) the parabola

$$y = x^2 - 1. \qquad \text{Parabola}$$

The points satisfying the inequality

$$-x + y \le 1 \qquad \text{Inequality 2}$$

are the points lying below (or on) the line

$$y = x + 1. \qquad \text{Line}$$

To find the points of intersection of the parabola and the line, solve the system of corresponding equations.

$$\begin{cases} x^2 - y = 1 \\ -x + y = 1 \end{cases}$$

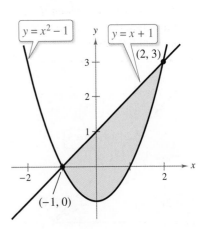

$y = x^2 - 1$ $y = x + 1$

$(2, 3)$

$(-1, 0)$

Using the method of substitution, you find that the solutions are $(-1, 0)$ and $(2, 3)$. These points are both solutions of the original system, so they are represented by closed dots in the graph of the solution region shown at the right.

✓ **Checkpoint** 🔊))) *Audio-video solution in English & Spanish at LarsonPrecalculus.com*

Sketch the graph of the solution set of the system of inequalities.

$$\begin{cases} x - y^2 > 0 \\ x + y < 2 \end{cases}$$

When solving a system of inequalities, be aware that the system might have no solution *or* its graph might be an unbounded region in the plane. Examples 6 and 7 show these two possibilities.

EXAMPLE 6 A System with No Solution

Sketch the solution set of the system of inequalities.

$$\begin{cases} x + y > & 3 \\ x + y < & -1 \end{cases}$$

Solution It should be clear from the way it is written that the system has no solution, because the quantity $(x + y)$ cannot be both less than -1 and greater than 3. The graph of the inequality $x + y > 3$ is the half-plane lying above the line $x + y = 3$, and the graph of the inequality $x + y < -1$ is the half-plane lying below the line $x + y = -1$, as shown below. These two half-planes have no points in common. So, the system of inequalities has no solution.

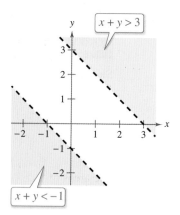

✓ *Checkpoint* ◀))) *Audio-video solution in English & Spanish at LarsonPrecalculus.com*

Sketch the solution set of the system of inequalities.

$$\begin{cases} 2x - y < & -3 \\ 2x - y > & 1 \end{cases}$$

EXAMPLE 7 An Unbounded Solution Set

Sketch the solution set of the system of inequalities.

$$\begin{cases} x + & y < 3 \\ x + 2y > 3 \end{cases}$$

Solution The graph of the inequality $x + y < 3$ is the half-plane that lies below the line $x + y = 3$. The graph of the inequality $x + 2y > 3$ is the half-plane that lies above the line $x + 2y = 3$. The intersection of these two half-planes is an *infinite wedge* that has a vertex at $(3, 0)$, as shown in Figure 7.21. So, the solution set of the system of inequalities is unbounded.

✓ *Checkpoint* ◀))) *Audio-video solution in English & Spanish at LarsonPrecalculus.com*

Sketch the solution set of the system of inequalities.

$$\begin{cases} x^2 - y < & 0 \\ x - y < & -2 \end{cases}$$

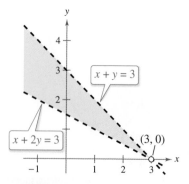

Figure 7.21

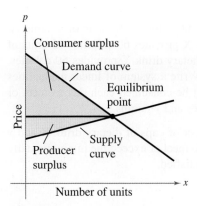

Figure 7.22

Applications

Example 9 in Section 7.2 discussed the equilibrium point for a system of demand and supply equations. The next example discusses two related concepts that economists call *consumer surplus* and *producer surplus*. As shown in Figure 7.22, the **consumer surplus** is the area of the region formed by the demand curve, the horizontal line passing through the equilibrium point, and the *p*-axis. Similarly, the **producer surplus** is the area of the region formed by the supply curve, the horizontal line passing through the equilibrium point, and the *p*-axis. The consumer surplus is a measure of the amount that consumers would have been willing to pay *above* what they actually paid, whereas the producer surplus is a measure of the amount that producers would have been willing to receive *below* what they actually received.

EXAMPLE 8   Consumer Surplus and Producer Surplus

The demand and supply equations for a new type of video game console are

$$\begin{cases} p = 180 - 0.00001x & \text{Demand equation} \\ p = 90 + 0.00002x & \text{Supply equation} \end{cases}$$

where *p* is the price per unit (in dollars) and *x* is the number of units. Find the consumer surplus and producer surplus for these two equations.

Solution   Begin by finding the equilibrium point (when supply and demand are equal) by solving the equation

$$90 + 0.00002x = 180 - 0.00001x.$$

In Example 9 in Section 7.2, you saw that the solution is $x = 3{,}000{,}000$ units, which corresponds to a price of $p = \$150$. So, the consumer surplus and producer surplus are the areas of the solution sets of the following systems of inequalities.

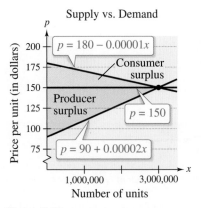

Figure 7.23

Consumer Surplus

$$\begin{cases} p \le 180 - 0.00001x \\ p \ge 150 \\ x \ge 0 \end{cases}$$

Producer Surplus

$$\begin{cases} p \ge 90 + 0.00002x \\ p \le 150 \\ x \ge 0 \end{cases}$$

In other words, the consumer and producer surpluses are the areas of the shaded triangles shown in Figure 7.23.

$$\begin{aligned} \text{Consumer surplus} &= \tfrac{1}{2}(\text{base})(\text{height}) \\ &= \tfrac{1}{2}(3{,}000{,}000)(30) \\ &= \$45{,}000{,}000 \end{aligned}$$

$$\begin{aligned} \text{Producer surplus} &= \tfrac{1}{2}(\text{base})(\text{height}) \\ &= \tfrac{1}{2}(3{,}000{,}000)(60) \\ &= \$90{,}000{,}000 \end{aligned}$$

✓ *Checkpoint* ◀))) *Audio-video solution in English & Spanish at LarsonPrecalculus.com*

The demand and supply equations for a flat-screen television are

$$\begin{cases} p = 567 - 0.00002x & \text{Demand equation} \\ p = 492 + 0.00003x & \text{Supply equation} \end{cases}$$

where *p* is the price per unit (in dollars) and *x* is the number of units. Find the consumer surplus and producer surplus for these two equations.

EXAMPLE 9 **Nutrition**

The liquid portion of a diet is to provide at least 300 calories, 36 units of vitamin A, and 90 units of vitamin C. A cup of dietary drink X provides 60 calories, 12 units of vitamin A, and 10 units of vitamin C. A cup of dietary drink Y provides 60 calories, 6 units of vitamin A, and 30 units of vitamin C. Write a system of linear inequalities that describes how many cups of each drink must be consumed each day to meet or exceed the minimum daily requirements for calories and vitamins.

Solution Begin by letting x represent the number of cups of dietary drink X and y represent the number of cups of dietary drink Y. To meet or exceed the minimum daily requirements, the following inequalities must be satisfied.

$$\begin{cases} 60x + 60y \geq 300 & \text{Calories} \\ 12x + 6y \geq 36 & \text{Vitamin A} \\ 10x + 30y \geq 90 & \text{Vitamin C} \\ x \geq 0 \\ y \geq 0 \end{cases}$$

The last two inequalities are included because x and y cannot be negative. The graph of this system of inequalities is shown below. (More is said about this application in Example 6 in Section 7.6.)

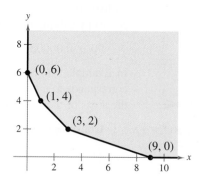

✓ *Checkpoint* 🔊))) *Audio-video solution in English & Spanish at LarsonPrecalculus.com*

A public aquarium is adding coral nutrients to a large reef tank. A bottle of brand X nutrients contains 8 units of nutrient A, 1 unit of nutrient B, and 2 units of nutrient C. A bottle of brand Y nutrients contains 2 units of nutrient A, 1 unit of nutrient B, and 7 units of nutrient C. The minimum amounts of nutrients A, B, and C that need to be added to the tank are 16 units, 5 units, and 20 units, respectively. Set up a system of linear inequalities that describes how many bottles of each brand must be added to meet or exceed the needs. ■

Summarize **(Section 7.5)**

1. Explain how to sketch the graph of an inequality in two variables *(page 510)*. For examples of sketching the graphs of inequalities in two variables, see Examples 1–3.

2. Explain how to solve a system of inequalities *(page 512)*. For examples of solving systems of inequalities, see Examples 4–7.

3. Describe examples of how to use systems of inequalities in two variables to model and solve real-life problems *(pages 515 and 516, Examples 8 and 9)*.

7.5 Exercises

See **CalcChat.com** for tutorial help and worked-out solutions to odd-numbered exercises.

Vocabulary: Fill in the blanks.

1. An ordered pair (a, b) is a _____ of an inequality in x and y when the inequality is true after a and b are substituted for x and y, respectively.

2. The _____ of an inequality is the collection of all solutions of the inequality.

3. A _____ of a system of inequalities in x and y is a point (x, y) that satisfies each inequality in the system.

4. The _____ _____ of a system of inequalities in two variables is represented by the region that is common to every graph in the system.

Skills and Applications

 Graphing an Inequality In Exercises 5–18, sketch the graph of the inequality.

5. $y < 5 - x^2$
6. $y^2 - x < 0$
7. $x \geq 6$
8. $x < -4$
9. $y > -7$
10. $10 \geq y$
11. $y < 2 - x$
12. $y > 4x - 3$
13. $2y - x \geq 4$
14. $5x + 3y \geq -15$
15. $x^2 + (y - 3)^2 < 4$
16. $(x + 2)^2 + y^2 > 9$
17. $y > -\dfrac{2}{x^2 + 1}$
18. $y \leq \dfrac{3}{x^2 + x + 1}$

 Graphing an Inequality In Exercises 19–26, use a graphing utility to graph the inequality.

19. $y \geq -\ln(x - 1)$
20. $y < \ln(x + 3) - 1$
21. $y < 2^x$
22. $y \geq 3^{-x} - 2$
23. $y \leq 2 - \frac{1}{5}x$
24. $y > -2.4x + 3.3$
25. $\frac{2}{3}y + 2x^2 - 5 \geq 0$
26. $-\frac{1}{6}x^2 - \frac{2}{7}y < -\frac{1}{3}$

 Writing an Inequality In Exercises 27–30, write an inequality for the shaded region shown in the figure.

27.
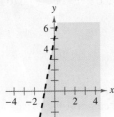

28.
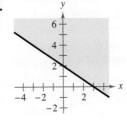

29.
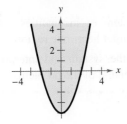

30.
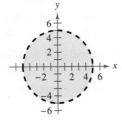

 Solving a System of Inequalities In Exercises 31–38, sketch the graph of the solution set of the system of inequalities. Label the vertices of the region.

31. $\begin{cases} x + y \leq 1 \\ -x + y \leq 1 \\ y \geq 0 \end{cases}$

32. $\begin{cases} 3x + 4y < 12 \\ x > 0 \\ y > 0 \end{cases}$

33. $\begin{cases} -3x + 2y < 6 \\ x - 4y > -2 \\ 2x + y < 3 \end{cases}$

34. $\begin{cases} x - 7y > -36 \\ 5x + 2y > 5 \\ 6x - 5y > 6 \end{cases}$

35. $\begin{cases} 2x + y > 2 \\ 6x + 3y < 2 \end{cases}$

36. $\begin{cases} x - 2y < -6 \\ 5x - 3y > -9 \end{cases}$

37. $\begin{cases} 2x - 3y > 7 \\ 5x + y < 9 \end{cases}$

38. $\begin{cases} 4x - 6y > 2 \\ -2x + 3y \geq 5 \end{cases}$

 Solving a System of Inequalities In Exercises 39–44, sketch the graph of the solution set of the system.

39. $\begin{cases} x^2 + y \leq 7 \\ x \geq -2 \\ y \geq 0 \end{cases}$

40. $\begin{cases} 4x^2 + y \geq 2 \\ x \leq 1 \\ y \leq 1 \end{cases}$

41. $\begin{cases} x - y^2 > 0 \\ x - y > 2 \end{cases}$

42. $\begin{cases} x^2 + y^2 \leq 25 \\ 4x - 3y \leq 0 \end{cases}$

43. $\begin{cases} 3x + 4 \geq y^2 \\ x - y < 0 \end{cases}$

44. $\begin{cases} x < 2y - y^2 \\ 0 < x + y \end{cases}$

 Solving a System of Inequalities In Exercises 45–50, use a graphing utility to graph the solution set of the system of inequalities.

45. $\begin{cases} y \leq \sqrt{3x} + 1 \\ y \geq x^2 + 1 \end{cases}$

46. $\begin{cases} y < 2\sqrt{x} - 1 \\ y \geq x^2 - 1 \end{cases}$

47. $\begin{cases} y < -x^2 + 2x + 3 \\ y > x^2 - 4x + 3 \end{cases}$

48. $\begin{cases} y \geq x^4 - 2x^2 + 1 \\ y \leq 1 - x^2 \end{cases}$

49. $\begin{cases} x^2 y \geq 1 \\ 0 < x \leq 4 \\ y \leq 4 \end{cases}$

50. $\begin{cases} y \leq e^{-x^2/2} \\ y \geq 0 \\ -2 \leq x \leq 2 \end{cases}$

Writing a System of Inequalities In Exercises 51–58, write a system of inequalities that describes the region.

51.

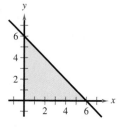

52.

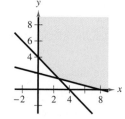

53.

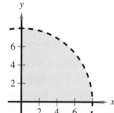

54.

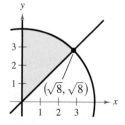

$(\sqrt{8}, \sqrt{8})$

55. Rectangle: vertices at $(4, 3), (9, 3), (9, 9), (4, 9)$

56. Parallelogram: vertices at $(0, 0), (4, 0), (1, 4), (5, 4)$

57. Triangle: vertices at $(0, 0), (6, 0), (1, 5)$

58. Triangle: vertices at $(-1, 0), (1, 0), (0, 1)$

Consumer Surplus and Producer Surplus In Exercises 59–62, (a) graph the systems of inequalities representing the consumer surplus and producer surplus for the supply and demand equations and (b) find the consumer surplus and producer surplus.

Demand	Supply
59. $p = 50 - 0.5x$	$p = 0.125x$
60. $p = 100 - 0.05x$	$p = 25 + 0.1x$
61. $p = 140 - 0.00002x$	$p = 80 + 0.00001x$
62. $p = 400 - 0.0002x$	$p = 225 + 0.0005x$

63. Investment Analysis A person plans to invest up to $20,000 in two different interest-bearing accounts. Each account must contain at least $5000. The amount in one account is to be at least twice the amount in the other account. Write and graph a system of inequalities that describes the various amounts that can be deposited in each account.

64. Ticket Sales For a concert event, there are $30 reserved seat tickets and $20 general admission tickets. There are 2000 reserved seats available, and fire regulations limit the number of paid ticket holders to 3000. The promoter must take in at least $75,000 in ticket sales. Write and graph a system of inequalities that describes the different numbers of tickets that can be sold.

65. Production A furniture company produces tables and chairs. Each table requires 1 hour in the assembly center and $1\frac{1}{3}$ hours in the finishing center. Each chair requires $1\frac{1}{2}$ hours in the assembly center and $1\frac{1}{2}$ hours in the finishing center. The assembly center is available 12 hours per day, and the finishing center is available 15 hours per day. Write and graph a system of inequalities that describes all possible production levels.

66. Inventory A store sells two models of laptop computers. The store stocks at least twice as many units of model A as of model B. The costs to the store for the two models are $800 and $1200, respectively. The management does not want more than $20,000 in computer inventory at any one time, and it wants at least four model A laptop computers and two model B laptop computers in inventory at all times. Write and graph a system of inequalities that describes all possible inventory levels.

67. Nutrition A dietician prescribes a special dietary plan using two different foods. Each ounce of food X contains 180 milligrams of calcium, 6 milligrams of iron, and 220 milligrams of magnesium. Each ounce of food Y contains 100 milligrams of calcium, 1 milligram of iron, and 40 milligrams of magnesium. The minimum daily requirements of the diet are 1000 milligrams of calcium, 18 milligrams of iron, and 400 milligrams of magnesium.

(a) Write and graph a system of inequalities that describes the different amounts of food X and food Y that can be prescribed.

(b) Find two solutions of the system and interpret their meanings in the context of the problem.

68. Target Heart Rate

One formula for a person's maximum heart rate is $220 - x$, where x is the person's age in years for $20 \le x \le 70$. The American Heart Association recommends that when a person exercises, the person should strive for a heart rate that is at least 50% of the maximum and at most 85% of the maximum. *(Source: American Heart Association)*

(a) Write and graph a system of inequalities that describes the exercise target heart rate region.

(b) Find two solutions of the system and interpret their meanings in the context of the problem.

69. Shipping A warehouse supervisor has instructions to ship at least 50 bags of gravel that weigh 55 pounds each and at least 40 bags of stone that weigh 70 pounds each. The maximum weight capacity of the truck being used is 7500 pounds.

(a) Write and graph a system that describes the numbers of bags of stone and gravel that can be shipped.

(b) Find two solutions of the system and interpret their meanings in the context of the problem.

70. Physical Fitness Facility A physical fitness facility is constructing an indoor running track with space for exercise equipment inside the track (see figure). The track must be at least 125 meters long, and the exercise space must have an area of at least 500 square meters.

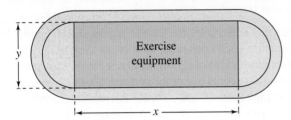

(a) Write and graph a system of inequalities that describes the requirements of the facility.

(b) Find two solutions of the system and interpret their meanings in the context of the problem.

Exploration

True or False? **In Exercises 71 and 72, determine whether the statement is true or false. Justify your answer.**

71. The area of the figure described by the system

$$\begin{cases} x \geq -3 \\ x \leq 6 \\ y \leq 5 \\ y \geq -6 \end{cases}$$

is 99 square units.

72. The graph shows the solution of the system

$$\begin{cases} y \leq 6 \\ -4x - 9y > 6. \\ 3x + y^2 \geq 2 \end{cases}$$

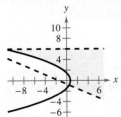

73. Think About It After graphing the boundary line of the inequality $x + y < 3$, explain how to determine the region that you need to shade.

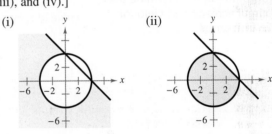

74. **HOW DO YOU SEE IT?** The graph of the solution of the inequality $x + 2y < 6$ is shown in the figure. Describe how the solution set would change for each inequality.

(a) $x + 2y \leq 6$

(b) $x + 2y > 6$

75. Matching Match the system of inequalities with the graph of its solution. [The graphs are labeled (i), (ii), (iii), and (iv).]

(i)

(ii)

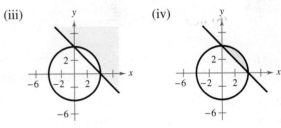

(iii)

(iv)

(a) $\begin{cases} x^2 + y^2 \leq 16 \\ x + y \geq 4 \end{cases}$

(b) $\begin{cases} x^2 + y^2 \leq 16 \\ x + y \leq 4 \end{cases}$

(c) $\begin{cases} x^2 + y^2 \geq 16 \\ x + y \geq 4 \end{cases}$

(d) $\begin{cases} x^2 + y^2 \geq 16 \\ x + y \leq 4 \end{cases}$

76. Graphical Reasoning Two concentric circles have radii x and y, where $y > x$. The area between the circles is at least 10 square units.

(a) Write a system of inequalities that describes the constraints on the circles.

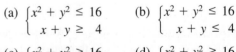

(b) Use a graphing utility to graph the system of inequalities in part (a). Graph the line $y = x$ in the same viewing window.

(c) Identify the graph of the line in relation to the boundary of the inequality. Explain its meaning in the context of the problem.

7.6 Linear Programming

Linear programming is often used to make real-life decisions. For example, in Exercise 43 on page 528, you will use linear programming to determine the optimal acreage and yield for two fruit crops.

■ Solve linear programming problems.
■ Use linear programming to model and solve real-life problems.

Linear Programming: A Graphical Approach

Many applications in business and economics involve a process called **optimization,** in which you find the minimum or maximum value of a quantity. In this section, you will study an optimization strategy called **linear programming.**

A two-dimensional linear programming problem consists of a linear **objective function** and a system of linear inequalities called **constraints.** The objective function gives the quantity to be maximized (or minimized), and the constraints determine the set of **feasible solutions.** For example, one such problem is to maximize the value of

$$z = ax + by \qquad \text{Objective function}$$

subject to a set of constraints that determines the shaded region shown below. Every point in the shaded region satisfies each constraint, so it is not clear how you should find the point that yields a maximum value of z. Fortunately, it can be shown that when there is an optimal solution, it must occur at one of the vertices. So, *to find the maximum value of z, evaluate z at each of the vertices* and compare the resulting z-values.

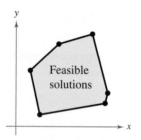

Optimal Solution of a Linear Programming Problem

If a linear programming problem has an optimal solution, then it must occur at a vertex of the set of feasible solutions.

A linear programming problem can include hundreds, and sometimes even thousands, of variables. However, in this section, you will solve linear programming problems that involve only two variables. The guidelines for solving a linear programming problem in two variables are listed below.

Solving a Linear Programming Problem

1. Sketch the region corresponding to the system of constraints. (The points inside or on the boundary of the region are *feasible solutions.*)

2. Find the vertices of the region.

3. Evaluate the objective function at each of the vertices and select the values of the variables that optimize the objective function. For a bounded region, both a minimum and a maximum value will exist. (For an unbounded region, *if* an optimal solution exists, then it will occur at a vertex.)

| EXAMPLE 1 | **Solving a Linear Programming Problem** |

Find the maximum value of

$$z = 3x + 2y \qquad \text{Objective function}$$

subject to the following constraints.

$$\left. \begin{array}{r} x \geq 0 \\ y \geq 0 \\ x + 2y \leq 4 \\ x - y \leq 1 \end{array} \right\} \text{Constraints}$$

Solution   The constraints form the region shown in Figure 7.24. At the four vertices of this region, the objective function has the following values.

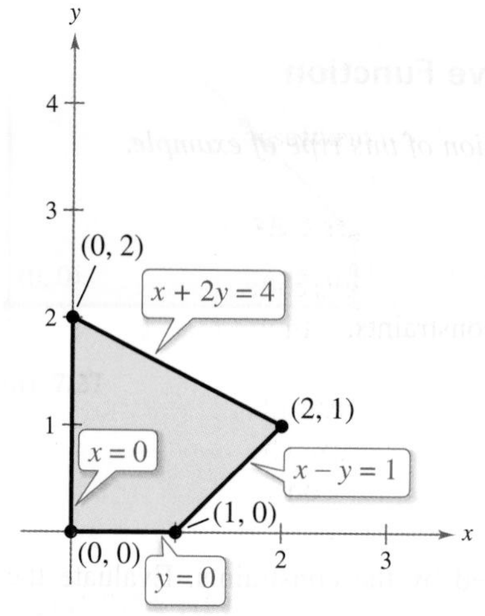

At $(0, 0)$: $z = 3(0) + 2(0) = 0$

At $(0, 2)$: $z = 3(0) + 2(2) = 4$

At $(2, 1)$: $z = 3(2) + 2(1) = 8$ Maximum value of z

At $(1, 0)$: $z = 3(1) + 2(0) = 3$

So, the maximum value of z is 8, and this occurs when $x = 2$ and $y = 1$.

✔ **Checkpoint**  Audio-video solution in English & Spanish at LarsonPrecalculus.com

Find the maximum value of

$$z = 4x + 5y$$

subject to the following constraints.

$$\begin{array}{r} x \geq 0 \\ y \geq 0 \\ x + y \leq 6 \end{array}$$

In Example 1, consider some of the *interior* points in the region. You will see that the corresponding values of z are less than 8. Here are some examples.

At $(1, 1)$: $z = 3(1) + 2(1) = 5$

At $\left(\frac{1}{2}, \frac{3}{2}\right)$: $z = 3\left(\frac{1}{2}\right) + 2\left(\frac{3}{2}\right) = \frac{9}{2}$

At $\left(\frac{3}{2}, 1\right)$: $z = 3\left(\frac{3}{2}\right) + 2(1) = \frac{13}{2}$

To see why the maximum value of the objective function in Example 1 must occur at a vertex, consider writing the objective function in slope-intercept form.

$$y = -\frac{3}{2}x + \frac{z}{2} \qquad \text{Family of lines}$$

Notice that the y-intercept

$$b = \frac{z}{2}$$

varies according to the value of z. This equation represents a family of lines, each of slope $-\frac{3}{2}$. Of these infinitely many lines, you want the one that has the largest z-value while still intersecting the region determined by the constraints. In other words, of all the lines whose slope is $-\frac{3}{2}$, you want the one that has the largest y-intercept *and* intersects the region, as shown in Figure 7.25. Notice from the graph that this line will pass through one point of the region, the vertex $(2, 1)$.

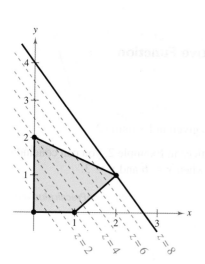

Figure 7.24

Figure 7.25

40. Optimal Labor A manufacturer has two different factories that produce three grades of steel: structural steel, rail steel, and pipe steel. They must produce 32 tons of structural, 26 tons of rail, and 30 tons of pipe steel to fill an order. The table shows the number of employees at each factory and the amounts of steel they produce hourly. How many hours should each factory operate to fill the orders at the minimum labor (in employee-hours)? What is the minimum labor?

	Factory X	Factory Y
Employees	120	80
Structural steel	2	5
Rail steel	8	2
Pipe steel	3	3

41. Optimal Revenue An accounting firm has 780 hours of staff time and 272 hours of reviewing time available each week. The firm charges $1600 for an audit and $250 for a tax return. Each audit requires 60 hours of staff time and 16 hours of review time. Each tax return requires 10 hours of staff time and 4 hours of review time. What numbers of audits and tax returns will yield an optimal revenue? What is the optimal revenue?

42. Optimal Revenue The accounting firm in Exercise 41 lowers its charge for an audit to $1400. What numbers of audits and tax returns will yield an optimal revenue? What is the optimal revenue?

· · **43. Agriculture** · · · · · · · · · · · · · · · · · ·

A fruit grower raises crops A and B. The yield is 300 bushels per acre for crop A and 500 bushels per acre for crop B. Research and available resources indicate the following constraints.

- The fruit grower has 150 acres of land available.

- It takes 1 day to trim the trees on an acre of crop A and 2 days to trim an acre of crop B, and there are 240 days per year available for trimming.

- It takes 0.3 day to pick an acre of crop A and 0.1 day to pick an acre of crop B, and there are 30 days per year available for picking.

What is the optimal acreage for each fruit? What is the optimal yield?

44. Optimal Profit In Exercise 43, the profit is $185 per acre for crop A and $245 per acre for crop B. What is the optimal profit?

45. Media Selection A company budgets a maximum of $1,000,000 for national advertising of an allergy medication. Each TV ad costs $100,000 and each one-page newspaper ad costs $20,000. Each TV ad is expected to be viewed by 20 million viewers, and each newspaper ad is expected to be seen by 5 million readers. The company's marketing department recommends that at most 80% of the budget be spent on TV ads. What is the optimal amount that should be spent on each type of ad? What is the optimal total audience?

46. Investment Portfolio An investor has up to $450,000 to invest in two types of investments. Type A pays 6% annually and type B pays 10% annually. To have a well-balanced portfolio, the investor imposes the following conditions. At least one-half of the total portfolio is to be allocated to type A investments and at least one-fourth of the portfolio is to be allocated to type B investments. What is the optimal amount that should be invested in each type of investment? What is the optimal return?

Exploration

True or False? In Exercises 47–49, determine whether the statement is true or false. Justify your answer.

47. If an objective function has a maximum value at the vertices $(4, 7)$ and $(8, 3)$, then it also has a maximum value at the points $(4.5, 6.5)$ and $(7.8, 3.2)$.

48. If an objective function has a minimum value at the vertex $(20, 0)$, then it also has a minimum value at $(0, 0)$.

49. If the constraint region of a linear programming problem lies in Quadrant I and is unbounded, the objective function cannot have a maximum value.

50. HOW DO YOU SEE IT? Using the constraint region shown below, determine which of the following objective functions has (a) a maximum at vertex A, (b) a maximum at vertex B, (c) a maximum at vertex C, and (d) a minimum at vertex C.

(i) $z = 2x + y$

(ii) $z = 2x - y$

(iii) $z = -x + 2y$

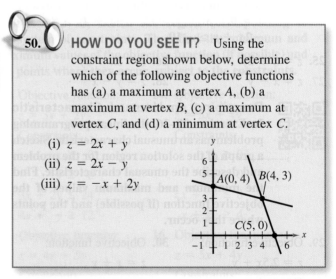

51. Think About It A linear programming problem has an objective function $z = 3x + 5y$ and an infinite number of optimal solutions that lie on the line segment connecting two points. What is the slope between the points?

Chapter Summary

What Did You Learn?	Explanation/Examples	Review Exercises
Section 7.1 Use the method of substitution to solve systems of linear equations in two variables *(p. 468)*.	**Method of Substitution** 1. *Solve* one of the equations for one variable in terms of the other. 2. *Substitute* the expression found in Step 1 into the other equation to obtain an equation in one variable. 3. *Solve* the equation obtained in Step 2. 4. *Back-substitute* the value obtained in Step 3 into the expression obtained in Step 1 to find the value of the other variable. 5. *Check* that the solution satisfies *each* of the original equations.	1–6
Use the method of substitution to solve systems of nonlinear equations in two variables *(p. 471)*.	The method of substitution (see steps above) can be used to solve systems in which one or both of the equations are nonlinear. (See Examples 3 and 4.)	7–10
Use a graphical method to solve systems of equations in two variables *(p. 472)*.	One intersection point Two intersection points No intersection points	11–18
Use systems of equations to model and solve real-life problems *(p. 473)*.	A system of equations can help you find the break-even point for a company. (See Example 6.)	19–22
Section 7.2 Use the method of elimination to solve systems of linear equations in two variables *(p. 478)*.	**Method of Elimination** 1. *Obtain coefficients* for x (or y) that differ only in sign. 2. *Add* the equations to eliminate one variable. 3. *Solve* the equation obtained in Step 2. 4. *Back-substitute* the value obtained in Step 3 into either of the original equations and solve for the other variable. 5. *Check* that the solution satisfies *each* of the original equations.	23–28
Interpret graphically the numbers of solutions of systems of linear equations in two variables *(p. 482)*.	Exactly one solution Infinitely many solutions No solution	29–32
Use systems of linear equations in two variables to model and solve real-life problems *(p. 484)*.	A system of linear equations in two variables can help you find the equilibrium point for a market. (See Example 9.)	33, 34

What Did You Learn?	Explanation/Examples	Review Exercises
Section 7.3 Use back-substitution to solve linear systems in row-echelon form *(p. 490)*.	**Row-Echelon Form** $$\begin{cases} x - 2y + 3z = 9 \\ -x + 3y \quad\quad = -4 \\ 2x - 5y + 5z = 17 \end{cases} \implies \begin{cases} x - 2y + 3z = 9 \\ y + 3z = 5 \\ z = 2 \end{cases}$$	35, 36
Use Gaussian elimination to solve systems of linear equations *(p. 491)*.	To produce an equivalent system of linear equations, use one or more of the following row operations. (1) Interchange two equations. (2) Multiply one equation by a nonzero constant. (3) Add a multiple of one of the equations to another equation to replace the latter equation.	37–42
Solve nonsquare systems of linear equations *(p. 495)*.	In a nonsquare system, the number of equations differs from the number of variables. A system of linear equations cannot have a unique solution unless there are at least as many equations as there are variables in the system.	43, 44
Use systems of linear equations in three or more variables to model and solve real-life problems *(p. 496)*.	A system of linear equations in three variables can help you find the position equation of an object that is moving in a (vertical) line with constant acceleration. (See Example 7.)	45–54
Section 7.4 Recognize partial fraction decompositions of rational expressions *(p. 502)*.	$$\frac{9}{x^3 - 6x^2} = \frac{9}{x^2(x - 6)} = \frac{A}{x} + \frac{B}{x^2} + \frac{C}{x - 6}$$	55–58
Find partial fraction decompositions of rational expressions *(p. 503)*.	The techniques used for determining the constants in the numerators of partial fractions vary slightly, depending on the type of factors of the denominator: linear or quadratic, distinct or repeated.	59–66
Section 7.5 Sketch the graphs of inequalities in two variables *(p. 510)*, and solve systems of inequalities *(p. 512)*.	A solution of a system of inequalities in x and y is a point (x, y) that satisfies each inequality in the system. $$\begin{cases} x^2 + y \le 5 \\ x \ge -1 \\ y \ge 0 \end{cases}$$	67–80
Use systems of inequalities in two variables to model and solve real-life problems *(p. 515)*.	A system of inequalities in two variables can help you find the consumer surplus and producer surplus for given demand and supply equations. (See Example 8.)	81–86
Section 7.6 Solve linear programming problems *(p. 520)*.	To solve a linear programming problem, (1) sketch the region corresponding to the system of constraints, (2) find the vertices of the region, and (3) evaluate the objective function at each of the vertices and select the values of the variables that optimize the objective function.	87–90
Use linear programming to model and solve real-life problems *(p. 524)*.	Linear programming can help you find the maximum profit in business applications. (See Example 5.)	91, 92

Review Exercises   <small>See CalcChat.com for tutorial help and worked-out solutions to odd-numbered exercises.</small>

7.1 **Solving a System by Substitution** **In Exercises 1–10, solve the system by the method of substitution.**

1. $\begin{cases} x + y = 2 \\ x - y = 0 \end{cases}$

2. $\begin{cases} 2x - 3y = 3 \\ x - y = 0 \end{cases}$

3. $\begin{cases} 4x - y - 1 = 0 \\ 8x + y - 17 = 0 \end{cases}$

4. $\begin{cases} 10x + 6y + 14 = 0 \\ x + 9y + 7 = 0 \end{cases}$

5. $\begin{cases} 0.5x + y = 0.75 \\ 1.25x - 4.5y = -2.5 \end{cases}$

6. $\begin{cases} -x + \frac{2}{5}y = \frac{3}{5} \\ -x + \frac{1}{5}y = -\frac{4}{5} \end{cases}$

7. $\begin{cases} x^2 - y^2 = 9 \\ x - y = 1 \end{cases}$

8. $\begin{cases} x^2 + y^2 = 169 \\ 3x + 2y = 39 \end{cases}$

9. $\begin{cases} y = 2x^2 \\ y = x^4 - 2x^2 \end{cases}$

10. $\begin{cases} x = y + 3 \\ x = y^2 + 1 \end{cases}$

Solving a System of Equations Graphically **In Exercises 11–14, solve the system graphically.**

11. $\begin{cases} 2x - y = 10 \\ x + 5y = -6 \end{cases}$

12. $\begin{cases} 8x - 3y = -3 \\ 2x + 5y = 28 \end{cases}$

13. $\begin{cases} y = 2x^2 - 4x + 1 \\ y = x^2 - 4x + 3 \end{cases}$

14. $\begin{cases} y^2 - 2y + x = 0 \\ x + y = 0 \end{cases}$

Using Technology **In Exercises 15–18, use a graphing utility to solve the systems of equations. Round your solution(s) to two decimal places.**

15. $\begin{cases} y = -2e^{-x} \\ 2e^x + y = 0 \end{cases}$

16. $\begin{cases} x^2 + y^2 = 100 \\ 2x - 3y = -12 \end{cases}$

17. $\begin{cases} y = 2 + \log x \\ y = \frac{3}{4}x + 5 \end{cases}$

18. $\begin{cases} y = \ln(x - 1) - 3 \\ y = 4 - \frac{1}{2}x \end{cases}$

19. Body Mass Index Body Mass Index (BMI) is a measure of body fat based on height and weight. The 85th percentile BMI for females, ages 9 to 20, increases more slowly than that for males of the same age range. Models that represent the 85th percentile BMI for males and females, ages 9 to 20, are

$\begin{cases} B = 0.78a + 11.7 & \text{Males} \\ B = 0.68a + 13.5 & \text{Females} \end{cases}$

where B is the BMI (kg/m^2) and a represents the age, with $a = 9$ corresponding to 9 years old. Use a graphing utility to determine when the BMI for males exceeds the BMI for females. *(Source: Centers for Disease Control and Prevention)*

20. Choice of Two Jobs You receive two sales job offers. One company offers an annual salary of $55,000 plus a year-end bonus of 1.5% of your total sales. The other company offers an annual salary of $52,000 plus a year-end bonus of 2% of your total sales. How much would you have to sell to make the second job offer better?

21. Geometry The perimeter of a rectangle is 68 feet and its width is $\frac{8}{9}$ times its length. Use a system of equations to find the dimensions of the rectangle.

22. Geometry The perimeter of a rectangle is 40 inches. The area of the rectangle is 96 square inches. Use a system of equations to find the dimensions of the rectangle.

7.2 **Solving a System by Elimination** **In Exercises 23–28, solve the system by the method of elimination and check any solutions algebraically.**

23. $\begin{cases} 2x - y = 2 \\ 6x + 8y = 39 \end{cases}$

24. $\begin{cases} 12x + 42y = -17 \\ 30x - 18y = 19 \end{cases}$

25. $\begin{cases} 3x - 2y = 0 \\ 3x + 2y = 0 \end{cases}$

26. $\begin{cases} 7x + 12y = 63 \\ 2x + 3y = 15 \end{cases}$

27. $\begin{cases} 1.25x - 2y = 3.5 \\ 5x - 8y = 14 \end{cases}$

28. $\begin{cases} 1.5x + 2.5y = 8.5 \\ 6x + 10y = 24 \end{cases}$

Matching a System with Its Graph **In Exercises 29–32, match the system of linear equations with its graph. Describe the number of solutions and state whether the system is consistent or inconsistent. [The graphs are labeled (a), (b), (c), and (d).]**

(a)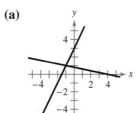

(b)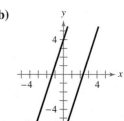

(c)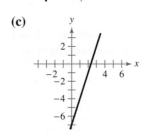

(d)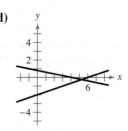

29. $\begin{cases} x + 5y = 4 \\ x - 3y = 6 \end{cases}$

30. $\begin{cases} -3x + y = -7 \\ 9x - 3y = 21 \end{cases}$

31. $\begin{cases} 3x - y = 7 \\ -6x + 2y = 8 \end{cases}$

32. $\begin{cases} 2x - y = -3 \\ x + 5y = 4 \end{cases}$

Finding the Equilibrium Point **In Exercises 33 and 34**, find the equilibrium point of the demand and supply equations.

Demand	Supply
33. $p = 43 - 0.0002x$	$p = 22 + 0.00001x$
34. $p = 120 - 0.0001x$	$p = 45 + 0.0002x$

7.3 **Using Back-Substitution in Row-Echelon Form** **In Exercises 35 and 36**, use back-substitution to solve the system of linear equations.

35. $\begin{cases} x - 4y + 3z = 3 \\ \quad\quad y - z = 1 \\ \quad\quad\quad\quad z = -5 \end{cases}$

36. $\begin{cases} x - 7y + 8z = 85 \\ \quad\quad y - 9z = -35 \\ \quad\quad\quad\quad z = 3 \end{cases}$

Solving a System of Linear Equations **In Exercises 37–42**, solve the system of linear equations and check any solutions algebraically.

37. $\begin{cases} 4x - 3y - 2z = -65 \\ \quad\quad 8y - 7z = -14 \\ 4x \quad\quad - 2z = -44 \end{cases}$

38. $\begin{cases} 5x \quad\quad - 7z = 9 \\ \quad\quad 3y - 8z = -4 \\ 5x - 3y \quad\quad = 20 \end{cases}$

39. $\begin{cases} x + 2y + 6z = 4 \\ -3x + 2y - z = -4 \\ 4x \quad\quad + 2z = 16 \end{cases}$

40. $\begin{cases} x - 2y + z = -6 \\ 2x - 3y \quad\quad = -7 \\ -x + 3y - 3z = 11 \end{cases}$

41. $\begin{cases} 2x \quad\quad + 6z = -9 \\ 3x - 2y + 11z = -16 \\ 3x - y + 7z = -11 \end{cases}$

42. $\begin{cases} x \quad\quad\quad + 4w = 1 \\ 3y + z - w = 4 \\ 2y \quad\quad - 3w = 2 \\ 4x - y + 2z \quad\quad = 5 \end{cases}$

Solving a Nonsquare System **In Exercises 43 and 44**, solve the system of linear equations and check any solutions algebraically.

43. $\begin{cases} 5x - 12y + 7z = 16 \\ 3x - 7y + 4z = 9 \end{cases}$

44. $\begin{cases} 2x + 5y - 19z = 34 \\ 3x + 8y - 31z = 54 \end{cases}$

Finding the Equation of a Parabola **In Exercises 45 and 46**, find the equation of the parabola

$$y = ax^2 + bx + c$$

that passes through the points. To verify your result, use a graphing utility to plot the points and graph the parabola.

45.
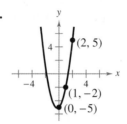

46.
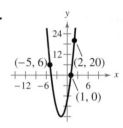

Finding the Equation of a Circle **In Exercises 47 and 48**, find the equation of the circle

$$x^2 + y^2 + Dx + Ey + F = 0$$

that passes through the points. To verify your result, use a graphing utility to plot the points and graph the circle.

47.
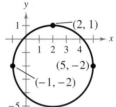

48.
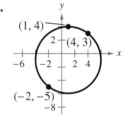

49. Agriculture A mixture of 6 gallons of chemical A, 8 gallons of chemical B, and 13 gallons of chemical C is required to kill a destructive crop insect. Commercial spray X contains one, two, and two parts, respectively, of these chemicals. Commercial spray Y contains only chemical C. Commercial spray Z contains chemicals A, B, and C in equal amounts. How much of each type of commercial spray gives the desired mixture?

50. Sports The Old Course at St Andrews Links in St Andrews, Scotland, is one of the oldest golf courses in the world. It is an 18-hole course that consists of par-3 holes, par-4 holes, and par-5 holes. There are seven times as many par-4 holes as par-5 holes, and the sum of the numbers of par-3 and par-5 holes is four. Find the numbers of par-3, par-4, and par-5 holes on the course. *(Source: St Andrews Links Trust)*

51. Investment An inheritance of $40,000 is divided among three investments yielding $3500 in interest per year. The interest rates for the three investments are 7%, 9%, and 11% simple interest. Find the amount placed in each investment when the second and third amounts are $3000 and $5000 less than the first, respectively.

52. Investment An amount of $46,000 is divided among three investments yielding $3020 in interest per year. The interest rates for the three investments are 5%, 7%, and 8% simple interest. Find the amount placed in each investment when the second and third amounts are $2000 and $3000 less than the first, respectively.

Modeling Vertical Motion In Exercises 53 and 54, an object moving vertically is at the given heights at the specified times. Find the position equation

$$s = \frac{1}{2}at^2 + v_0 t + s_0$$

for the object.

53. At $t = 1$ second, $s = 134$ feet

At $t = 2$ seconds, $s = 86$ feet

At $t = 3$ seconds, $s = 6$ feet

54. At $t = 1$ second, $s = 184$ feet

At $t = 2$ seconds, $s = 116$ feet

At $t = 3$ seconds, $s = 16$ feet

7.4 Writing the Form of the Decomposition In Exercises 55–58, write the form of the partial fraction decomposition of the rational expression. Do not solve for the constants.

55. $\dfrac{3}{x^2 + 20x}$

56. $\dfrac{x - 8}{x^2 - 3x - 28}$

57. $\dfrac{3x - 4}{x^3 - 5x^2}$

58. $\dfrac{x - 2}{x(x^2 + 2)^2}$

Writing the Partial Fraction Decomposition In Exercises 59–66, write the partial fraction decomposition of the rational expression. Check your result algebraically.

59. $\dfrac{4 - x}{x^2 + 6x + 8}$

60. $\dfrac{-x}{x^2 + 3x + 2}$

61. $\dfrac{x^2}{x^2 + 2x - 15}$

62. $\dfrac{9}{x^2 - 9}$

63. $\dfrac{x^2 + 2x}{x^3 - x^2 + x - 1}$

64. $\dfrac{4x}{3(x - 1)^2}$

65. $\dfrac{3x^2 + 4x}{(x^2 + 1)^2}$

66. $\dfrac{4x^2}{(x - 1)(x^2 + 1)}$

7.5 Graphing an Inequality In Exercises 67–72, sketch the graph of the inequality.

67. $y \geq 5$ **68.** $x < -3$

69. $y \leq 5 - 2x$ **70.** $3y - x \geq 7$

71. $(x - 1)^2 + (y - 3)^2 < 16$

72. $x^2 + (y + 5)^2 > 1$

Solving a System of Inequalities In Exercises 73–76, sketch the graph of the solution set of the system of inequalities. Label the vertices of the region.

73. $\begin{cases} x + 2y \leq 2 \\ -x + 2y \leq 2 \\ \qquad y \geq 0 \end{cases}$

74. $\begin{cases} 2x + 3y < 6 \\ x \qquad > 0 \\ \qquad y > 0 \end{cases}$

75. $\begin{cases} 2x - y < -1 \\ -3x + 2y > \quad 4 \\ \qquad y > \quad 0 \end{cases}$

76. $\begin{cases} 3x - 2y > -4 \\ 6x - y < \quad 5 \\ \qquad y < \quad 1 \end{cases}$

Solving a System of Inequalities In Exercises 77–80, sketch the graph of the solution set of the system of inequalities.

77. $\begin{cases} y < x + 1 \\ y > x^2 - 1 \end{cases}$ **78.** $\begin{cases} y \leq 6 - 2x - x^2 \\ y \geq x + 6 \end{cases}$

79. $\begin{cases} x^2 + y^2 > 4 \\ x^2 + y^2 \leq 9 \end{cases}$

80. $\begin{cases} x^2 + y^2 \leq 169 \\ x + y \leq \quad 7 \end{cases}$

81. Geometry Write a system of inequalities to describe the region of a rectangle with vertices at $(3, 1)$, $(7, 1)$, $(7, 10)$, and $(3, 10)$.

82. Geometry Write a system of inequalities that describes the triangular region with vertices $(0, 5)$, $(5, 0)$, and $(0, 0)$.

Consumer Surplus and Producer Surplus In Exercises 83 and 84, (a) graph the systems of inequalities representing the consumer surplus and producer surplus for the supply and demand equations and (b) find the consumer surplus and producer surplus.

Demand	Supply
83. $p = 160 - 0.0001x$	$p = 70 + 0.0002x$
84. $p = 130 - 0.0002x$	$p = 30 + 0.0003x$

85. Inventory Costs A warehouse operator has 24,000 square feet of floor space in which to store two products. Each unit of product I requires 20 square feet of floor space and costs $12 per day to store. Each unit of product II requires 30 square feet of floor space and costs $8 per day to store. The total storage cost per day cannot exceed $12,400. Write and graph a system that describes all possible inventory levels.

86. Nutrition A dietician prescribes a special dietary plan using two different foods. Each ounce of food X contains 200 milligrams of calcium, 3 milligrams of iron, and 100 milligrams of magnesium. Each ounce of food Y contains 150 milligrams of calcium, 2 milligrams of iron, and 80 milligrams of magnesium. The minimum daily requirements of the diet are 800 milligrams of calcium, 10 milligrams of iron, and 200 milligrams of magnesium.

(a) Write and graph a system of inequalities that describes the different amounts of food X and food Y that can be prescribed.

(b) Find two solutions to the system and interpret their meanings in the context of the problem.

7.6 Solving a Linear Programming Problem In Exercises 87–90, sketch the region corresponding to the system of constraints. Then find the minimum and maximum values of the objective function (if possible) and the points where they occur, subject to the constraints.

87. Objective function:

$z = 3x + 4y$

Constraints:

$x \geq 0$
$y \geq 0$
$2x + 5y \leq 50$
$4x + y \leq 28$

88. Objective function:

$z = 10x + 7y$

Constraints:

$x \geq 0$
$y \geq 0$
$2x + y \geq 100$
$x + y \geq 75$

89. Objective function:

$z = 1.75x + 2.25y$

Constraints:

$x \geq 0$
$y \geq 0$
$2x + y \geq 25$
$3x + 2y \geq 45$

90. Objective function:

$z = 50x + 70y$

Constraints:

$x \geq 0$
$y \geq 0$
$x + 2y \leq 1500$
$5x + 2y \leq 3500$

91. Optimal Revenue A student is working part time as a hairdresser to pay college expenses. The student may work no more than 24 hours per week. Haircuts cost $25 and require an average of 20 minutes, and permanents cost $70 and require an average of 1 hour and 10 minutes. How many haircuts and/or permanents will yield an optimal revenue? What is the optimal revenue?

92. Optimal Profit A manufacturer produces two models of bicycles. The table shows the times (in hours) required for assembling, painting, and packaging each model.

Process	Hours, Model A	Hours, Model B
Assembling	2	2.5
Painting	4	1
Packaging	1	0.75

The total times available for assembling, painting, and packaging are 4000 hours, 4800 hours, and 1500 hours, respectively. The profits per unit are $45 for model A and $50 for model B. What is the optimal production level for each model? What is the optimal profit?

Exploration

True or False? In Exercises 93 and 94, determine whether the statement is true or false. Justify your answer.

93. The system

$$\begin{cases} y \leq 2 \\ y \leq -2 \\ y \leq 4x - 10 \\ y \leq -4x + 26 \end{cases}$$

represents a region in the shape of an isosceles trapezoid.

94. For the rational expression $\dfrac{2x + 3}{x^2(x + 2)^2}$, the partial fraction decomposition is of the form

$$\frac{Ax + B}{x^2} + \frac{Cx + D}{(x + 2)^2}.$$

Writing a System of Linear Equations In Exercises 95–98, write a system of linear equations that has the ordered pair as a solution. (There are many correct answers.)

95. $(-8, 10)$ **96.** $(5, -4)$

97. $\left(\frac{4}{3}, 3\right)$ **98.** $\left(-2, \frac{11}{5}\right)$

Writing a System of Linear Equations In Exercises 99–102, write a system of linear equations that has the ordered triple as a solution. (There are many correct answers.)

99. $(4, -1, 3)$ **100.** $(-3, 5, 6)$

101. $\left(5, \frac{3}{2}, 2\right)$ **102.** $\left(-\frac{1}{2}, -2, -\frac{3}{4}\right)$

103. Writing Explain what is meant by an inconsistent system of linear equations.

104. Graphical Reasoning How can you tell graphically that a system of linear equations in two variables has no solution? Give an example.

Chapter Test

See CalcChat.com for tutorial help and worked-out solutions to odd-numbered exercises.

Take this test as you would take a test in class. When you are finished, check your work against the answers given in the back of the book.

In Exercises 1–3, solve the system of equations by the method of substitution.

1. $\begin{cases} x + y = -9 \\ 5x - 8y = 20 \end{cases}$
2. $\begin{cases} y = x + 1 \\ y = (x - 1)^3 \end{cases}$
3. $\begin{cases} 2x - y^2 = 0 \\ x - y = 4 \end{cases}$

In Exercises 4–6, solve the system of equations graphically.

4. $\begin{cases} 3x - 6y = 0 \\ 2x + 5y = 18 \end{cases}$
5. $\begin{cases} y = 9 - x^2 \\ y = x + 3 \end{cases}$
6. $\begin{cases} y - \ln x = 4 \\ 7x - 2y - 5 = -6 \end{cases}$

In Exercises 7 and 8, solve the system of equations by the method of elimination.

7. $\begin{cases} 3x + 4y = -26 \\ 7x - 5y = 11 \end{cases}$
8. $\begin{cases} 1.4x - y = 17 \\ 0.8x + 6y = -10 \end{cases}$

In Exercises 9 and 10, solve the system of linear equations and check any solutions algebraically.

9. $\begin{cases} x - 2y + 3z = 11 \\ 2x \quad - z = 3 \\ 3y + z = -8 \end{cases}$
10. $\begin{cases} 3x + 2y + z = 17 \\ -x + y + z = 4 \\ x - y - z = 3 \end{cases}$

In Exercises 11–14, write the partial fraction decomposition of the rational expression. Check your result algebraically.

11. $\dfrac{2x + 5}{x^2 - x - 2}$
12. $\dfrac{3x^2 - 2x + 4}{x^2(2 - x)}$
13. $\dfrac{x^4 + 5}{x^3 - x}$
14. $\dfrac{x^2 - 4}{x^3 + 2x}$

In Exercises 15–17, sketch the graph of the solution set of the system of inequalities.

15. $\begin{cases} 2x + y \le 4 \\ 2x - y \ge 0 \\ x \ge 0 \end{cases}$
16. $\begin{cases} y < -x^2 + x + 4 \\ y > 4x \end{cases}$
17. $\begin{cases} x^2 + y^2 \le 36 \\ x \ge 2 \\ y \ge -4 \end{cases}$

18. Find the minimum and maximum values of the objective function $z = 20x + 12y$ and the points where they occur, subject to the following constraints.

$$\left. \begin{array}{r} x \ge 0 \\ y \ge 0 \\ x + 4y \le 32 \\ 3x + 2y \le 36 \end{array} \right\} \text{Constraints}$$

19. A total of $50,000 is invested in two funds that pay 4% and 5.5% simple interest. The yearly interest is $2390. How much is invested at each rate?

20. Find the equation of the parabola $y = ax^2 + bx + c$ that passes through the points $(0, 6)$, $(-2, 2)$, and $\left(3, \frac{9}{2}\right)$.

21. A manufacturer produces two models of television stands. The table at the left shows the times (in hours) required for assembling, staining, and packaging the two models. The total times available for assembling, staining, and packaging are 3750 hours, 8950 hours, and 2650 hours, respectively. The profits per unit are $30 for model I and $40 for model II. What is the optimal inventory level for each model? What is the optimal profit?

	Model I	Model II
Assembling	0.5	0.75
Staining	2.0	1.5
Packaging	0.5	0.5

Table for 21

Proofs in Mathematics ▪ ▪ ▪ ▪ ▪ ▪ ▪ ▪ ▪ ▪ ▪ ▪ ▪ ▪

An **indirect proof** can be useful in proving statements of the form "p implies q." Recall that the conditional statement $p \rightarrow q$ is false only when p is true and q is false. To prove a conditional statement indirectly, assume that p is true and q is false. If this assumption leads to an impossibility, then you have proved that the conditional statement is true. An indirect proof is also called a **proof by contradiction.**

An indirect proof can be used to prove the conditional statement

"If a is a positive integer and a^2 is divisible by 2, then a is divisible by 2."

The proof is as follows.

Proof

First, assume that p, "a is a positive integer and a^2 is divisible by 2," is true and q, "a is divisible by 2," is false. This means that a is not divisible by 2. If so, then a is odd and can be written as $a = 2n + 1$, where n is an integer.

$a = 2n + 1$	Definition of an odd integer
$a^2 = 4n^2 + 4n + 1$	Square each side.
$a^2 = 2(2n^2 + 2n) + 1$	Distributive Property

So, by the definition of an odd integer, a^2 is odd. This contradicts the assumption, and you can conclude that a is divisible by 2. ▪

EXAMPLE Using an Indirect Proof

Use an indirect proof to prove that $\sqrt{2}$ is an irrational number.

Solution Begin by assuming that $\sqrt{2}$ is *not* an irrational number. Then $\sqrt{2}$ can be written as the quotient of two integers a and b ($b \neq 0$) that have no common factors.

$\sqrt{2} = \dfrac{a}{b}$	Assume that $\sqrt{2}$ is a rational number.
$2 = \dfrac{a^2}{b^2}$	Square each side.
$2b^2 = a^2$	Multiply each side by b^2.

This implies that 2 is a factor of a^2. So, 2 is also a factor of a, and a can be written as $2c$, where c is an integer.

$2b^2 = (2c)^2$	Substitute $2c$ for a.
$2b^2 = 4c^2$	Simplify.
$b^2 = 2c^2$	Divide each side by 2.

This implies that 2 is a factor of b^2 and also a factor of b. So, 2 is a factor of both a and b. This contradicts the assumption that a and b have no common factors. So, you can conclude that $\sqrt{2}$ is an irrational number. ▪

P.S. Problem Solving ▪ ▪ ▪ ▪ ▪ ▪ ▪ ▪ ▪ ▪ ▪ ▪ ▪ ▪ ▪ ▪

1. **Geometry** A theorem from geometry states that if a triangle is inscribed in a circle such that one side of the triangle is a diameter of the circle, then the triangle is a right triangle. Show that this theorem is true for the circle

$$x^2 + y^2 = 100$$

and the triangle formed by the lines

$$y = 0$$
$$y = \tfrac{1}{2}x + 5$$

and

$$y = -2x + 20.$$

2. **Finding Values of Constants** Find values of k_1 and k_2 such that the system of equations has an infinite number of solutions.

$$\begin{cases} 3x - 5y = 8 \\ 2x + k_1 y = k_2 \end{cases}$$

3. **Finding Conditions on Constants** Under what condition(s) will the system of equations in x and y have exactly one solution?

$$\begin{cases} ax + by = e \\ cx + dy = f \end{cases}$$

4. **Finding Values of Constants** Find values of a, b, and c (if possible) such that the system of linear equations has (a) a unique solution, (b) no solution, and (c) an infinite number of solutions.

$$\begin{cases} x + y \quad\;\; = 2 \\ \quad\; y + z = 2 \\ x \quad\;\; + z = 2 \\ ax + by + cz = 0 \end{cases}$$

5. **Graphical Analysis** Graph the lines determined by each system of linear equations. Then use Gaussian elimination to solve each system. At each step of the elimination process, graph the corresponding lines. How do the graphs at the different steps compare?

(a) $\begin{cases} x - 4y = -3 \\ 5x - 6y = 13 \end{cases}$

(b) $\begin{cases} 2x - 3y = 7 \\ -4x + 6y = -14 \end{cases}$

6. **Maximum Numbers of Solutions** A system of two equations in two variables has a finite number of solutions. Determine the maximum number of solutions of the system satisfying each condition.

(a) Both equations are linear.

(b) One equation is linear and the other is quadratic.

(c) Both equations are quadratic.

7. **Vietnam Veterans Memorial** The Vietnam Veterans Memorial (or "The Wall") in Washington, D.C., was designed by Maya Ying Lin when she was a student at Yale University. This monument has two vertical, triangular sections of black granite with a common side (see figure). The bottom of each section is level with the ground. The tops of the two sections can be approximately modeled by the equations

$$-2x + 50y = 505$$

and

$$2x + 50y = 505$$

when the x-axis is superimposed at the base of the wall. Each unit in the coordinate system represents 1 foot. How high is the memorial at the point where the two sections meet? How long is each section?

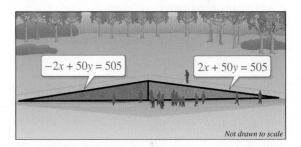

Not drawn to scale

8. **Finding Atomic Weights** Weights of atoms and molecules are measured in atomic mass units (u). A molecule of C_2H_6 (ethane) is made up of two carbon atoms and six hydrogen atoms and weighs 30.069 u. A molecule of C_3H_8 (propane) is made up of three carbon atoms and eight hydrogen atoms and weighs 44.096 u. Find the weights of a carbon atom and a hydrogen atom.

9. **DVD Connector Cables** Connecting a DVD player to a television set requires a cable with special connectors at both ends. You buy a six-foot cable for $15.50 and a three-foot cable for $10.25. Assuming that the cost of a cable is the sum of the cost of the two connectors and the cost of the cable itself, what is the cost of a four-foot cable?

10. **Distance** A hotel 35 miles from an airport runs a shuttle service to and from the airport. The 9:00 A.M. bus leaves for the airport traveling at 30 miles per hour. The 9:15 A.M. bus leaves for the airport traveling at 40 miles per hour.

(a) Write a system of linear equations that represents distance as a function of time for the buses.

(b) Graph and solve the system.

(c) How far from the airport will the 9:15 A.M. bus catch up to the 9:00 A.M. bus?

11. Systems with Rational Expressions Solve each system of equations by letting $X = 1/x$, $Y = 1/y$, and $Z = 1/z$.

(a) $\begin{cases} \dfrac{12}{x} - \dfrac{12}{y} = 7 \\[2mm] \dfrac{3}{x} + \dfrac{4}{y} = 0 \end{cases}$

(b) $\begin{cases} \dfrac{2}{x} + \dfrac{1}{y} - \dfrac{3}{z} = 4 \\[2mm] \dfrac{4}{x} + \dfrac{2}{z} = 10 \\[2mm] -\dfrac{2}{x} + \dfrac{3}{y} - \dfrac{13}{z} = -8 \end{cases}$

12. Finding Values of Constants For what values of a, b, and c does the linear system have $(-1, 2, -3)$ as its only solution?

$\begin{cases} x + 2y - 3z = a & \text{Equation 1} \\ -x - y + z = b & \text{Equation 2} \\ 2x + 3y - 2z = c & \text{Equation 3} \end{cases}$

13. System of Linear Equations The following system has one solution: $x = 1$, $y = -1$, and $z = 2$.

$\begin{cases} 4x - 2y + 5z = 16 & \text{Equation 1} \\ x + y = 0 & \text{Equation 2} \\ -x - 3y + 2z = 6 & \text{Equation 3} \end{cases}$

Solve each system of two equations that consists of (a) Equation 1 and Equation 2, (b) Equation 1 and Equation 3, and (c) Equation 2 and Equation 3. (d) How many solutions does each of these systems have?

14. System of Linear Equations Solve the system of linear equations algebraically.

$\begin{cases} x_1 - x_2 + 2x_3 + 2x_4 + 6x_5 = 6 \\ 3x_1 - 2x_2 + 4x_3 + 4x_4 + 12x_5 = 14 \\ - x_2 - x_3 - x_4 - 3x_5 = -3 \\ 2x_1 - 2x_2 + 4x_3 + 5x_4 + 15x_5 = 10 \\ 2x_1 - 2x_2 + 4x_3 + 4x_4 + 13x_5 = 13 \end{cases}$

15. Biology Each day, an average adult moose can process about 32 kilograms of terrestrial vegetation (twigs and leaves) and aquatic vegetation. From this food, it needs to obtain about 1.9 grams of sodium and 11,000 calories of energy. Aquatic vegetation has about 0.15 gram of sodium per kilogram and about 193 calories of energy per kilogram, whereas terrestrial vegetation has minimal sodium and about four times as much energy as aquatic vegetation. Write and graph a system of inequalities that describes the amounts t and a of terrestrial and aquatic vegetation, respectively, for the daily diet of an average adult moose. *(Source: Biology by Numbers)*

16. Height and Weight For a healthy person who is 4 feet 10 inches tall, the recommended minimum weight is about 91 pounds and increases by about 3.6 pounds for each additional inch of height. The recommended maximum weight is about 115 pounds and increases by about 4.5 pounds for each additional inch of height. *(Source: National Institutes of Health)*

(a) Let x be the number of inches by which a person's height exceeds 4 feet 10 inches and let y be the person's weight (in pounds). Write a system of inequalities that describes the possible values of x and y for a healthy person.

(b) Use a graphing utility to graph the system of inequalities from part (a).

(c) What is the recommended weight range for a healthy person who is 6 feet tall?

17. Cholesterol Cholesterol in human blood is necessary, but too much can lead to health problems. There are three main types of cholesterol: HDL (high-density lipoproteins), LDL (low-density lipoproteins), and VLDL (very low-density lipoproteins). HDL is considered "good" cholesterol; LDL and VLDL are considered "bad" cholesterol.

A standard fasting cholesterol blood test measures total cholesterol, HDL cholesterol, and triglycerides. These numbers are used to estimate LDL and VLDL, which are difficult to measure directly. Your doctor recommends that your combined LDL/VLDL cholesterol level be less than 130 milligrams per deciliter, your HDL cholesterol level be at least 60 milligrams per deciliter, and your total cholesterol level be no more than 200 milligrams per deciliter.

(a) Write a system of linear inequalities for the recommended cholesterol levels. Let x represent the HDL cholesterol level, and let y represent the combined LDL/VLDL cholesterol level.

(b) Graph the system of inequalities from part (a). Label any vertices of the solution region.

(c) Is the following set of cholesterol levels within the recommendations? Explain.

LDL/VLDL: 120 milligrams per deciliter

HDL: 90 milligrams per deciliter

Total: 210 milligrams per deciliter

(d) Give an example of cholesterol levels in which the LDL/VLDL cholesterol level is too high but the HDL cholesterol level is acceptable.

(e) Another recommendation is that the ratio of total cholesterol to HDL cholesterol be less than 4 (that is, less than 4 to 1). Identify a point in the solution region from part (b) that meets this recommendation, and explain why it meets the recommendation.

8 Matrices and Determinants

Sudoku *(page 581)*

Data Encryption *(page 592)*

Beam Deflection *(page 569)*

Flight Crew Scheduling
(page 562)

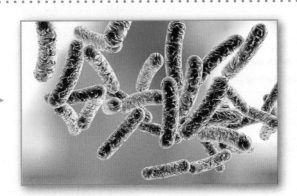

Waterborne Disease *(Exercise 93, page 552)*

8.1 Matrices and Systems of Equations

- ■ **Write matrices and determine their dimensions.**
- ■ **Perform elementary row operations on matrices.**
- ■ **Use matrices and Gaussian elimination to solve systems of linear equations.**
- ■ **Use matrices and Gauss-Jordan elimination to solve systems of linear equations.**

Matrices

In this section, you will study a streamlined technique for solving systems of linear equations. This technique involves the use of a rectangular array of numbers called a **matrix.** The plural of matrix is *matrices.*

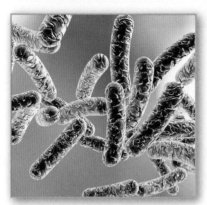

Matrices can help you solve real-life problems that are represented by systems of equations. For example, in Exercise 93 on page 552, you will use a matrix to find a model for the numbers of new cases of a waterborne disease in a small city.

Definition of Matrix

If m and n are positive integers, then an $m \times n$ (read "m by n") matrix is a rectangular array

$$
\begin{array}{c}
 \\
\begin{array}{ccccc}
\text{Column 1} & \text{Column 2} & \text{Column 3} & \cdots & \text{Column } n
\end{array} \\
\begin{array}{c}
\text{Row 1} \\
\text{Row 2} \\
\text{Row 3} \\
\vdots \\
\text{Row } m
\end{array}
\begin{bmatrix}
a_{11} & a_{12} & a_{13} & \cdots & a_{1n} \\
a_{21} & a_{22} & a_{23} & \cdots & a_{2n} \\
a_{31} & a_{32} & a_{33} & \cdots & a_{3n} \\
\vdots & \vdots & \vdots & & \vdots \\
a_{m1} & a_{m2} & a_{m3} & \cdots & a_{mn}
\end{bmatrix}
\end{array}
$$

in which each **entry** a_{ij} of the matrix is a number. An $m \times n$ matrix has m rows and n columns.

The entry in the ith row and jth column of a matrix is denoted by the *double subscript* notation a_{ij}. For example, a_{23} refers to the entry in the second row, third column. A matrix having m rows and n columns is said to be of **dimension** $m \times n$. If $m = n$, then the matrix is **square** of dimension $m \times m$ (or $n \times n$). For a square matrix, the entries $a_{11}, a_{22}, a_{33}, \ldots$ are the **main diagonal** entries. A matrix with only one row is called a **row matrix,** and a matrix with only one column is called a **column matrix.**

EXAMPLE 1 **Dimensions of Matrices**

Determine the dimension of each matrix.

a. $\begin{bmatrix} 2 \end{bmatrix}$ **b.** $\begin{bmatrix} 1 & -3 & 0 & \frac{1}{2} \end{bmatrix}$ **c.** $\begin{bmatrix} 0 & 0 \\ 0 & 0 \end{bmatrix}$ **d.** $\begin{bmatrix} 5 & 0 \\ 2 & -2 \\ -7 & 4 \end{bmatrix}$

Solution

a. This matrix has *one* row and *one* column. The dimension of the matrix is 1×1.

b. This matrix has *one* row and *four* columns. The dimension of the matrix is 1×4.

c. This matrix has *two* rows and *two* columns. The dimension of the matrix is 2×2.

d. This matrix has *three* rows and *two* columns. The dimension of the matrix is 3×2.

✓ **Checkpoint** ◀))) *Audio-video solution in English & Spanish at LarsonPrecalculus.com*

Determine the dimension of the matrix $\begin{bmatrix} 14 & 7 & 10 \\ -2 & -3 & -8 \end{bmatrix}$. ■

A matrix derived from a system of linear equations (each written in standard form with the constant term on the right) is the **augmented matrix** of the system. Moreover, the matrix derived from the coefficients of the system (but not including the constant terms) is the **coefficient matrix** of the system.

$$System: \begin{cases} x - 4y + 3z = 5 \\ -x + 3y - z = -3 \\ 2x - 4z = 6 \end{cases}$$

Augmented
matrix:
$$\begin{bmatrix} 1 & -4 & 3 & \vdots & 5 \\ -1 & 3 & -1 & \vdots & -3 \\ 2 & 0 & -4 & \vdots & 6 \end{bmatrix}$$

Coefficient
matrix:
$$\begin{bmatrix} 1 & -4 & 3 \\ -1 & 3 & -1 \\ 2 & 0 & -4 \end{bmatrix}$$

REMARK The vertical dots in an augmented matrix separate the coefficients of the linear system from the constant terms.

Note the use of 0 for the coefficient of the missing y-variable in the third equation, and also note the fourth column of constant terms in the augmented matrix.

When forming either the coefficient matrix or the augmented matrix of a system, you should begin by vertically aligning the variables in the equations and using zeros for the coefficients of the missing variables.

EXAMPLE 2 **Writing an Augmented Matrix**

Write the augmented matrix for the system of linear equations.

$$\begin{cases} x + 3y - w = 9 \\ -y + 4z + 2w = -2 \\ x - 5z - 6w = 0 \\ 2x + 4y - 3z = 4 \end{cases}$$

What is the dimension of the augmented matrix?

Solution

Begin by rewriting the linear system and aligning the variables.

$$\begin{cases} x + 3y - w = 9 \\ -y + 4z + 2w = -2 \\ x - 5z - 6w = 0 \\ 2x + 4y - 3z = 4 \end{cases}$$

Next, use the coefficients and constant terms as the matrix entries. Include zeros for the coefficients of the missing variables.

$$\begin{matrix} R_1 \\ R_2 \\ R_3 \\ R_4 \end{matrix} \begin{bmatrix} 1 & 3 & 0 & -1 & \vdots & 9 \\ 0 & -1 & 4 & 2 & \vdots & -2 \\ 1 & 0 & -5 & -6 & \vdots & 0 \\ 2 & 4 & -3 & 0 & \vdots & 4 \end{bmatrix}$$

The augmented matrix has four rows and five columns, so it is a 4×5 matrix. The notation R_n is used to designate each row in the matrix. For example, Row 1 is represented by R_1.

✓ *Checkpoint* ◀))) *Audio-video solution in English & Spanish at LarsonPrecalculus.com*

Write the augmented matrix for the system of linear equations. What is the dimension of the augmented matrix?

$$\begin{cases} x + y + z = 2 \\ 2x - y + 3z = -1 \\ -x + 2y - z = 4 \end{cases}$$

Elementary Row Operations

In Section 7.3, you studied three operations that can be used on a system of linear equations to produce an equivalent system.

1. Interchange two equations.

2. Multiply an equation by a nonzero constant.

3. Add a multiple of an equation to another equation.

In matrix terminology, these three operations correspond to **elementary row operations.** An elementary row operation on an augmented matrix of a given system of linear equations produces a new augmented matrix corresponding to a new (but equivalent) system of linear equations. Two matrices are **row-equivalent** when one can be obtained from the other by a sequence of elementary row operations.

▷

· · REMARK Although elementary row operations are simple to perform, they involve many arithmetic calculations, with many ways to make a mistake. So, get in the habit of noting the elementary row operations performed in each step to make it more convenient to go back and check your work.

> **Elementary Row Operations**
>
> | Operation | Notation |
> |---|---|
> | **1.** Interchange two rows. | $R_a \leftrightarrow R_b$ |
> | **2.** Multiply a row by a nonzero constant. | $cR_a \quad (c \neq 0)$ |
> | **3.** Add a multiple of a row to another row. | $cR_a + R_b$ |

EXAMPLE 3 **Elementary Row Operations**

a. Interchange the first and second rows of the original matrix.

Original Matrix

$$\begin{bmatrix} 0 & 1 & 3 & 4 \\ -1 & 2 & 0 & 3 \\ 2 & -3 & 4 & 1 \end{bmatrix}$$

New Row-Equivalent Matrix

$$\begin{matrix} R_2 \\ R_1 \end{matrix} \begin{bmatrix} -1 & 2 & 0 & 3 \\ 0 & 1 & 3 & 4 \\ 2 & -3 & 4 & 1 \end{bmatrix}$$

b. Multiply the first row of the original matrix by $\frac{1}{2}$.

Original Matrix

$$\begin{bmatrix} 2 & -4 & 6 & -2 \\ 1 & 3 & -3 & 0 \\ 5 & -2 & 1 & 2 \end{bmatrix}$$

New Row-Equivalent Matrix

$$\frac{1}{2}R_1 \rightarrow \begin{bmatrix} 1 & -2 & 3 & -1 \\ 1 & 3 & -3 & 0 \\ 5 & -2 & 1 & 2 \end{bmatrix}$$

▷ **TECHNOLOGY** Most graphing utilities can perform elementary row operations on matrices. Consult the user's guide for your graphing utility for specific keystrokes.
  After performing a row operation, the new row-equivalent matrix that is displayed on your graphing utility is stored in the *answer* variable. So, use the *answer* variable and not the original matrix for subsequent row operations.

c. Add -2 times the first row of the original matrix to the third row.

Original Matrix

$$\begin{bmatrix} 1 & 2 & -4 & 3 \\ 0 & 3 & -2 & -1 \\ 2 & 1 & 5 & -2 \end{bmatrix}$$

New Row-Equivalent Matrix

$$\begin{bmatrix} 1 & 2 & -4 & 3 \\ 0 & 3 & -2 & -1 \\ -2R_1 + R_3 \rightarrow & 0 & -3 & 13 & -8 \end{bmatrix}$$

Note that the elementary row operation is written beside the row that is *changed*.

✓ **Checkpoint** ◀))) *Audio-video solution in English & Spanish at LarsonPrecalculus.com*

Identify the elementary row operation performed to obtain the new row-equivalent matrix.

Original Matrix

$$\begin{bmatrix} 1 & 0 & 2 \\ 3 & 1 & 7 \\ 2 & -6 & 14 \end{bmatrix}$$

New Row-Equivalent Matrix

$$\begin{bmatrix} 1 & 0 & 2 \\ 0 & 1 & 1 \\ 2 & -6 & 14 \end{bmatrix}$$

Gaussian Elimination with Back-Substitution

In Example 3 in Section 7.3, you used Gaussian elimination with back-substitution to solve a system of linear equations. The next example demonstrates the matrix version of Gaussian elimination. The two methods are essentially the same. The basic difference is that with matrices you do not need to keep writing the variables.

EXAMPLE 4 Comparing Linear Systems and Matrix Operations

Linear System

$$\begin{cases} x - 2y + 3z = 9 \\ -x + 3y \quad\quad = -4 \\ 2x - 5y + 5z = 17 \end{cases}$$

Add the first equation to the second equation.

$$\begin{cases} x - 2y + 3z = 9 \\ y + 3z = 5 \\ 2x - 5y + 5z = 17 \end{cases}$$

Add -2 times the first equation to the third equation.

$$\begin{cases} x - 2y + 3z = 9 \\ y + 3z = 5 \\ -y - z = -1 \end{cases}$$

Add the second equation to the third equation.

$$\begin{cases} x - 2y + 3z = 9 \\ y + 3z = 5 \\ 2z = 4 \end{cases}$$

Multiply the third equation by $\frac{1}{2}$.

$$\begin{cases} x - 2y + 3z = 9 \\ y + 3z = 5 \\ z = 2 \end{cases}$$

Associated Augmented Matrix

$$\begin{bmatrix} 1 & -2 & 3 & \vdots & 9 \\ -1 & 3 & 0 & \vdots & -4 \\ 2 & -5 & 5 & \vdots & 17 \end{bmatrix}$$

Add the first row to the second row: $R_1 + R_2$.

$$R_1 + R_2 \rightarrow \begin{bmatrix} 1 & -2 & 3 & \vdots & 9 \\ 0 & 1 & 3 & \vdots & 5 \\ 2 & -5 & 5 & \vdots & 17 \end{bmatrix}$$

Add -2 times the first row to the third row: $-2R_1 + R_3$.

$$-2R_1 + R_3 \rightarrow \begin{bmatrix} 1 & -2 & 3 & \vdots & 9 \\ 0 & 1 & 3 & \vdots & 5 \\ 0 & -1 & -1 & \vdots & -1 \end{bmatrix}$$

Add the second row to the third row: $R_2 + R_3$.

$$R_2 + R_3 \rightarrow \begin{bmatrix} 1 & -2 & 3 & \vdots & 9 \\ 0 & 1 & 3 & \vdots & 5 \\ 0 & 0 & 2 & \vdots & 4 \end{bmatrix}$$

Multiply the third row by $\frac{1}{2}$: $\frac{1}{2}R_3$.

$$\tfrac{1}{2}R_3 \rightarrow \begin{bmatrix} 1 & -2 & 3 & \vdots & 9 \\ 0 & 1 & 3 & \vdots & 5 \\ 0 & 0 & 1 & \vdots & 2 \end{bmatrix}$$

At this point, use back-substitution to find x and y.

$$y + 3(2) = 5 \qquad \text{Substitute 2 for } z.$$
$$y = -1 \qquad \text{Solve for } y.$$
$$x - 2(-1) + 3(2) = 9 \qquad \text{Substitute } -1 \text{ for } y \text{ and 2 for } z.$$
$$x = 1 \qquad \text{Solve for } x.$$

The solution is $(1, -1, 2)$.

• **REMARK** Remember that you should check a solution by substituting the values of x, y, and z into each equation of the original system. For example, check the solution to Example 4 as shown below.

Equation 1:
$1 - 2(-1) + 3(2) = 9$ ✓

Equation 2:
$-1 + 3(-1) = -4$ ✓

Equation 3:
$2(1) - 5(-1) + 5(2) = 17$ ✓

✓ **Checkpoint** ◀))) *Audio-video solution in English & Spanish at LarsonPrecalculus.com*

Compare solving the linear system below to solving it using its associated augmented matrix.

$$\begin{cases} 2x + y - z = -3 \\ 4x - 2y + 2z = -2 \\ -6x + 5y + 4z = 10 \end{cases}$$

The last matrix in Example 4 is in *row-echelon form.* The term *echelon* refers to the stair-step pattern formed by the nonzero entries of the matrix. The row-echelon form and *reduced row-echelon form* of matrices are described below.

Row-Echelon Form and Reduced Row-Echelon Form

A matrix in **row-echelon form** has the following properties.

1. Any rows consisting entirely of zeros occur at the bottom of the matrix.

2. For each row that does not consist entirely of zeros, the first nonzero entry is 1 (called a **leading 1**).

3. For two successive (nonzero) rows, the leading 1 in the higher row is farther to the left than the leading 1 in the lower row.

A matrix in *row-echelon form* is in **reduced row-echelon form** when every column that has a leading 1 has zeros in every position above and below its leading 1.

It is worth noting that the row-echelon form of a matrix is not unique. That is, two different sequences of elementary row operations may yield different row-echelon forms. The *reduced* row-echelon form of a matrix, however, is unique.

EXAMPLE 5   **Row-Echelon Form**

Determine whether each matrix is in row-echelon form. If it is, determine whether it is in reduced row-echelon form.

a. $\begin{bmatrix} 1 & 2 & -1 & 4 \\ 0 & 1 & 0 & 3 \\ 0 & 0 & 1 & -2 \end{bmatrix}$
b. $\begin{bmatrix} 1 & 2 & -1 & 2 \\ 0 & 0 & 0 & 0 \\ 0 & 1 & 2 & -4 \end{bmatrix}$

c. $\begin{bmatrix} 1 & -5 & 2 & -1 & 3 \\ 0 & 0 & 1 & 3 & -2 \\ 0 & 0 & 0 & 1 & 4 \\ 0 & 0 & 0 & 0 & 1 \end{bmatrix}$
d. $\begin{bmatrix} 1 & 0 & 0 & -1 \\ 0 & 1 & 0 & 2 \\ 0 & 0 & 1 & 3 \\ 0 & 0 & 0 & 0 \end{bmatrix}$

e. $\begin{bmatrix} 1 & 2 & -3 & 4 \\ 0 & 2 & 1 & -1 \\ 0 & 0 & 1 & -3 \end{bmatrix}$
f. $\begin{bmatrix} 0 & 1 & 0 & 5 \\ 0 & 0 & 1 & 3 \\ 0 & 0 & 0 & 0 \end{bmatrix}$

Solution   The matrices in (a), (c), (d), and (f) are in row-echelon form. The matrices in (d) and (f) are in *reduced* row-echelon form because every column that has a leading 1 has zeros in every position above and below its leading 1. The matrix in (b) is not in row-echelon form because a row of all zeros occurs above a row that is not all zeros. The matrix in (e) is not in row-echelon form because the first nonzero entry in Row 2 is not a leading 1.

✓ *Checkpoint* ◄))) *Audio-video solution in English & Spanish at LarsonPrecalculus.com*

Determine whether the matrix is in row-echelon form. If it is, determine whether it is in reduced row-echelon form.

$$\begin{bmatrix} 1 & 0 & -2 & 4 \\ 0 & 1 & 11 & 3 \\ 0 & 0 & 0 & 0 \end{bmatrix}$$

Every matrix is row-equivalent to a matrix in row-echelon form. For instance, in Example 5, you can change the matrix in part (e) to row-echelon form by multiplying its second row by $\frac{1}{2}$.

Gaussian elimination with back-substitution works well for solving systems of linear equations by hand or with a computer. For this algorithm, the order in which the elementary row operations are performed is important. You should operate from left to right by columns, using elementary row operations to obtain zeros in all entries directly below the leading 1's.

EXAMPLE 6   **Gaussian Elimination with Back-Substitution**

Solve the system

$$\begin{cases} y + z - 2w = -3 \\ x + 2y - z \quad\quad = 2 \\ 2x + 4y + z - 3w = -2 \\ x - 4y - 7z - w = -19 \end{cases}$$

Solution

$$\begin{bmatrix} 0 & 1 & 1 & -2 & \vdots & -3 \\ 1 & 2 & -1 & 0 & \vdots & 2 \\ 2 & 4 & 1 & -3 & \vdots & -2 \\ 1 & -4 & -7 & -1 & \vdots & -19 \end{bmatrix}$$
Write augmented matrix.

$$\begin{matrix} {\scriptstyle \curvearrowright R_2} \\ {\scriptstyle \curvearrowright R_1} \end{matrix}\begin{bmatrix} 1 & 2 & -1 & 0 & \vdots & 2 \\ 0 & 1 & 1 & -2 & \vdots & -3 \\ 2 & 4 & 1 & -3 & \vdots & -2 \\ 1 & -4 & -7 & -1 & \vdots & -19 \end{bmatrix}$$
Interchange R_1 and R_2 so first column has leading 1 in upper left corner.

$$\begin{matrix} \\ \\ {\scriptstyle -2R_1 + R_3 \rightarrow} \\ {\scriptstyle -R_1 + R_4 \rightarrow} \end{matrix}\begin{bmatrix} 1 & 2 & -1 & 0 & \vdots & 2 \\ 0 & 1 & 1 & -2 & \vdots & -3 \\ 0 & 0 & 3 & -3 & \vdots & -6 \\ 0 & -6 & -6 & -1 & \vdots & -21 \end{bmatrix}$$
Perform operations on R_3 and R_4 so first column has zeros below its leading 1.

$$\begin{matrix} \\ \\ \\ {\scriptstyle 6R_2 + R_4 \rightarrow} \end{matrix}\begin{bmatrix} 1 & 2 & -1 & 0 & \vdots & 2 \\ 0 & 1 & 1 & -2 & \vdots & -3 \\ 0 & 0 & 3 & -3 & \vdots & -6 \\ 0 & 0 & 0 & -13 & \vdots & -39 \end{bmatrix}$$
Perform operations on R_4 so second column has zeros below its leading 1.

$$\begin{matrix} \\ \\ {\scriptstyle \frac{1}{3}R_3 \rightarrow} \\ {\scriptstyle -\frac{1}{13}R_4 \rightarrow} \end{matrix}\begin{bmatrix} 1 & 2 & -1 & 0 & \vdots & 2 \\ 0 & 1 & 1 & -2 & \vdots & -3 \\ 0 & 0 & 1 & -1 & \vdots & -2 \\ 0 & 0 & 0 & 1 & \vdots & 3 \end{bmatrix}$$
Perform operations on R_3 and R_4 so third and fourth columns have leading 1's.

The matrix is now in row-echelon form, and the corresponding system is

$$\begin{cases} x + 2y - z \quad\quad = 2 \\ y + z - 2w = -3 \\ z - w = -2 \\ w = 3 \end{cases}$$

Using back-substitution, the solution is $(-1, 2, 1, 3)$.

REMARK   Note that the order of the variables in the system of equations is x, y, z, and w. The coordinates of the solution are given in this order.

✓ Checkpoint ◀))) *Audio-video solution in English & Spanish at LarsonPrecalculus.com*

Solve the system

$$\begin{cases} -3x + 5y + 3z = -19 \\ 3x + 4y + 4z = 8. \\ 4x - 8y - 6z = 26 \end{cases}$$

The steps below summarize the procedure used in Example 6.

> ### Gaussian Elimination with Back-Substitution
> 1. Write the augmented matrix of the system of linear equations.
> 2. Use elementary row operations to rewrite the augmented matrix in row-echelon form.
> 3. Write the system of linear equations corresponding to the matrix in row-echelon form and use back-substitution to find the solution.

When solving a system of linear equations, remember that it is possible for the system to have no solution. If, in the elimination process, you obtain a row of all zeros except for the last entry, then the system has no solution, or is *inconsistent*.

EXAMPLE 7   A System with No Solution

Solve the system $\begin{cases} x - y + 2z = 4 \\ x \quad\;\; + z = 6 \\ 2x - 3y + 5z = 4 \\ 3x + 2y - z = 1 \end{cases}$.

Solution

$$\begin{bmatrix} 1 & -1 & 2 & \vdots & 4 \\ 1 & 0 & 1 & \vdots & 6 \\ 2 & -3 & 5 & \vdots & 4 \\ 3 & 2 & -1 & \vdots & 1 \end{bmatrix}$$

Write augmented matrix.

$$\begin{matrix} \\ -R_1 + R_2 \rightarrow \\ -2R_1 + R_3 \rightarrow \\ -3R_1 + R_4 \rightarrow \end{matrix} \begin{bmatrix} 1 & -1 & 2 & \vdots & 4 \\ 0 & 1 & -1 & \vdots & 2 \\ 0 & -1 & 1 & \vdots & -4 \\ 0 & 5 & -7 & \vdots & -11 \end{bmatrix}$$

Perform row operations.

$$\begin{matrix} \\ \\ R_2 + R_3 \rightarrow \\ \\ \end{matrix} \begin{bmatrix} 1 & -1 & 2 & \vdots & 4 \\ 0 & 1 & -1 & \vdots & 2 \\ 0 & 0 & 0 & \vdots & -2 \\ 0 & 5 & -7 & \vdots & -11 \end{bmatrix}$$

Perform row operations.

Note that the third row of this matrix consists entirely of zeros except for the last entry. This means that the original system of linear equations is inconsistent. You can see why this is true by converting back to a system of linear equations.

$$\begin{cases} x - y + 2z = 4 \\ y - z = 2 \\ 0 = -2 \\ 5y - 7z = -11 \end{cases}$$

The third equation is not possible, so the system has no solution.

✓ *Checkpoint* ◀))) *Audio-video solution in English & Spanish at LarsonPrecalculus.com*

Solve the system $\begin{cases} x + y + z = 1 \\ x + 2y + 2z = 2 \\ x - y - z = 1 \end{cases}$.

Gauss-Jordan Elimination

With Gaussian elimination, elementary row operations are applied to a matrix to obtain a (row-equivalent) row-echelon form of the matrix. A second method of elimination, called **Gauss-Jordan elimination,** after Carl Friedrich Gauss and Wilhelm Jordan (1842–1899), continues the reduction process until the *reduced* row-echelon form is obtained. This procedure is demonstrated in Example 8.

EXAMPLE 8   Gauss-Jordan Elimination

See LarsonPrecalculus.com for an interactive version of this type of example.

▷ **TECHNOLOGY** For a demonstration of a graphical approach to Gauss-Jordan elimination on a 2 × 3 matrix, see the program called "Visualizing Row Operations," available at *CengageBrain.com*.

Use Gauss-Jordan elimination to solve the system $\begin{cases} x - 2y + 3z = 9 \\ -x + 3y = -4. \\ 2x - 5y + 5z = 17 \end{cases}$

Solution   In Example 4, Gaussian elimination was used to obtain the row-echelon form of the linear system above.

$$\begin{bmatrix} 1 & -2 & 3 & \vdots & 9 \\ 0 & 1 & 3 & \vdots & 5 \\ 0 & 0 & 1 & \vdots & 2 \end{bmatrix}$$

Now, rather than using back-substitution, apply elementary row operations until you obtain zeros above each of the leading 1's.

$$\begin{array}{c} 2R_2 + R_1 \rightarrow \\ \\ \end{array} \begin{bmatrix} 1 & 0 & 9 & \vdots & 19 \\ 0 & 1 & 3 & \vdots & 5 \\ 0 & 0 & 1 & \vdots & 2 \end{bmatrix}$$

Perform operations on R_1 so second column has a zero above its leading 1.

$$\begin{array}{c} -9R_3 + R_1 \rightarrow \\ -3R_3 + R_2 \rightarrow \\ \end{array} \begin{bmatrix} 1 & 0 & 0 & \vdots & 1 \\ 0 & 1 & 0 & \vdots & -1 \\ 0 & 0 & 1 & \vdots & 2 \end{bmatrix}$$

Perform operations on R_1 and R_2 so third column has zeros above its leading 1.

•• **REMARK**   The advantage of using Gauss-Jordan elimination to solve a system of linear equations is that the solution of the system is easily found without using back-substitution, as illustrated in Example 8. ▷

The matrix is now in reduced row-echelon form. Converting back to a system of linear equations, you have

$$\begin{cases} x = 1 \\ y = -1. \\ z = 2 \end{cases}$$

So, the solution is $(1, -1, 2)$.

✓ *Checkpoint* ◀))) *Audio-video solution in English & Spanish at LarsonPrecalculus.com*

Use Gauss-Jordan elimination to solve the system $\begin{cases} -3x + 7y + 2z = 1 \\ -5x + 3y - 5z = -8. \\ 2x - 2y - 3z = 15 \end{cases}$ ∎

The elimination procedures described in this section sometimes result in fractional coefficients. For example, consider the system

$$\begin{cases} 2x - 5y + 5z = 17 \\ 3x - 2y + 3z = 11. \\ -3x + 3y = -6 \end{cases}$$

Multiplying the first row by $\frac{1}{2}$ to produce a leading 1 results in fractional coefficients. You can sometimes avoid fractions by judiciously choosing the order in which you apply elementary row operations.

EXAMPLE 9 **A System with an Infinite Number of Solutions**

Solve the system $\begin{cases} 2x + 4y - 2z = 0 \\ 3x + 5y \quad\quad = 1 \end{cases}$.

Solution

$$\begin{bmatrix} 2 & 4 & -2 & \vdots & 0 \\ 3 & 5 & 0 & \vdots & 1 \end{bmatrix}$$

$$\frac{1}{2}R_1 \rightarrow \begin{bmatrix} 1 & 2 & -1 & \vdots & 0 \\ 3 & 5 & 0 & \vdots & 1 \end{bmatrix}$$

$$-3R_1 + R_2 \rightarrow \begin{bmatrix} 1 & 2 & -1 & \vdots & 0 \\ 0 & -1 & 3 & \vdots & 1 \end{bmatrix}$$

$$-R_2 \rightarrow \begin{bmatrix} 1 & 2 & -1 & \vdots & 0 \\ 0 & 1 & -3 & \vdots & -1 \end{bmatrix}$$

$$-2R_2 + R_1 \rightarrow \begin{bmatrix} 1 & 0 & 5 & \vdots & 2 \\ 0 & 1 & -3 & \vdots & -1 \end{bmatrix}$$

The corresponding system of equations is

$$\begin{cases} x + 5z = \quad 2 \\ y - 3z = -1 \end{cases}.$$

Solving for x and y in terms of z, you have

$$x = -5z + 2 \quad \text{and} \quad y = 3z - 1.$$

To write a solution of the system that does not use any of the three variables of the system, let a represent any real number and let $z = a$. Substitute a for z in the equations for x and y.

$$x = -5z + 2 = -5a + 2 \quad \text{and} \quad y = 3z - 1 = 3a - 1$$

So, the solution set can be written as an ordered triple of the form

$$(-5a + 2, 3a - 1, a)$$

where a is any real number. Remember that a solution set of this form represents an infinite number of solutions. Substitute values for a to obtain a few solutions. Then check each solution in the original system of equations.

✓ **Checkpoint** *Audio-video solution in English & Spanish at LarsonPrecalculus.com*

Solve the system $\begin{cases} 2x - 6y + 6z = 46 \\ 2x - 3y \quad\quad = 31 \end{cases}$.

Summarize (Section 8.1)

1. State the definition of a matrix *(page 540)*. For examples of writing matrices and determining their dimensions, see Examples 1 and 2.

2. List the elementary row operations *(page 542)*. For an example of performing elementary row operations, see Example 3.

3. Explain how to use matrices and Gaussian elimination to solve systems of linear equations *(page 543)*. For examples of using Gaussian elimination, see Examples 4, 6, and 7.

4. Explain how to use matrices and Gauss-Jordan elimination to solve systems of linear equations *(page 547)*. For examples of using Gauss-Jordan elimination, see Examples 8 and 9.

8.1 Exercises

See **CalcChat.com** for tutorial help and worked-out solutions to odd-numbered exercises.

Vocabulary: Fill in the blanks.

1. A matrix is _____ when the number of rows equals the number of columns.

2. For a square matrix, the entries a_{11}, a_{22}, a_{33}, . . . are the _____ _____ entries.

3. A matrix derived from a system of linear equations (each written in standard form with the constant term on the right) is the _____ matrix of the system.

4. A matrix derived from the coefficients of a system of linear equations (but not including the constant terms) is the _____ matrix of the system.

5. Two matrices are _____ when one can be obtained from the other by a sequence of elementary row operations.

6. A matrix in row-echelon form is in _____ _____ _____ when every column that has a leading 1 has zeros in every position above and below its leading 1.

Skills and Applications

 Dimension of a Matrix **In Exercises 7–14, determine the dimension of the matrix.**

7. $\begin{bmatrix} 7 & 0 \end{bmatrix}$

8. $\begin{bmatrix} 5 & -3 & 8 & 7 \end{bmatrix}$

9. $\begin{bmatrix} 2 \\ 36 \\ 3 \end{bmatrix}$

10. $\begin{bmatrix} -3 & 7 & 15 & 0 \\ 0 & 0 & 3 & 3 \\ 1 & 1 & 6 & 7 \end{bmatrix}$

11. $\begin{bmatrix} 33 & 45 \\ -9 & 20 \end{bmatrix}$

12. $\begin{bmatrix} -7 & 6 & 4 \\ 0 & -5 & 1 \end{bmatrix}$

13. $\begin{bmatrix} 1 & 6 & -1 \\ 8 & 0 & 3 \\ 3 & -9 & 9 \end{bmatrix}$

14. $\begin{bmatrix} 3 & -1 \\ 4 & 1 \\ -5 & 9 \end{bmatrix}$

 Writing an Augmented Matrix **In Exercises 15–20, write the augmented matrix for the system of linear equations.**

15. $\begin{cases} 2x - y = 7 \\ x + y = 2 \end{cases}$

16. $\begin{cases} 5x + 2y = 13 \\ -3x + 4y = -24 \end{cases}$

17. $\begin{cases} x - y + 2z = 2 \\ 4x - 3y + z = -1 \\ 2x + y = 0 \end{cases}$

18. $\begin{cases} -2x - 4y + z = 13 \\ 6x - 7z = 22 \\ 3x - y + z = 9 \end{cases}$

19. $\begin{cases} 3x - 5y + 2z = 12 \\ 12x - 7z = 10 \end{cases}$

20. $\begin{cases} 9x + y - 3z = 21 \\ -15y + 13z = -8 \end{cases}$

Writing a System of Equations **In Exercises 21–26, write the system of linear equations represented by the augmented matrix. (Use variables x, y, z, and w, if applicable.)**

21. $\begin{bmatrix} 1 & 1 & \vdots & 3 \\ 5 & -3 & \vdots & -1 \end{bmatrix}$

22. $\begin{bmatrix} 5 & 2 & \vdots & 9 \\ 3 & -8 & \vdots & 0 \end{bmatrix}$

23. $\begin{bmatrix} 2 & 0 & 5 & \vdots & -12 \\ 0 & 1 & -2 & \vdots & 7 \\ 6 & 3 & 0 & \vdots & 2 \end{bmatrix}$

24. $\begin{bmatrix} 4 & -5 & -1 & \vdots & 18 \\ -11 & 0 & 6 & \vdots & 25 \\ 3 & 8 & 0 & \vdots & -29 \end{bmatrix}$

25. $\begin{bmatrix} 9 & 12 & 3 & 0 & \vdots & 0 \\ -2 & 18 & 5 & 2 & \vdots & 10 \\ 1 & 7 & -8 & 0 & \vdots & -4 \\ 3 & 0 & 2 & 0 & \vdots & -10 \end{bmatrix}$

26. $\begin{bmatrix} 6 & 2 & -1 & -5 & \vdots & -25 \\ -1 & 0 & 7 & 3 & \vdots & 7 \\ 4 & -1 & -10 & 6 & \vdots & 23 \\ 0 & 8 & 1 & -11 & \vdots & -21 \end{bmatrix}$

 Identifying an Elementary Row Operation **In Exercises 27–30, identify the elementary row operation(s) performed to obtain the new row-equivalent matrix.**

| Original Matrix | New Row-Equivalent Matrix |

27. $\begin{bmatrix} -2 & 5 & 1 \\ 3 & -1 & -8 \end{bmatrix}$ $\begin{bmatrix} 13 & 0 & -39 \\ 3 & -1 & -8 \end{bmatrix}$

28. $\begin{bmatrix} 3 & -1 & -4 \\ -4 & 3 & 7 \end{bmatrix}$ $\begin{bmatrix} 3 & -1 & -4 \\ 5 & 0 & -5 \end{bmatrix}$

29. $\begin{bmatrix} 0 & -1 & -5 & 5 \\ -1 & 3 & -7 & 6 \\ 4 & -5 & 1 & 3 \end{bmatrix}$ $\begin{bmatrix} -1 & 3 & -7 & 6 \\ 0 & -1 & -5 & 5 \\ 0 & 7 & -27 & 27 \end{bmatrix}$

30. $\begin{bmatrix} -1 & -2 & 3 & -2 \\ 2 & -5 & 1 & -7 \\ 5 & 4 & -7 & 6 \end{bmatrix}$ $\begin{bmatrix} -1 & -2 & 3 & -2 \\ 0 & -9 & 7 & -11 \\ 0 & -6 & 8 & -4 \end{bmatrix}$

Elementary Row Operations In Exercises 31–38, fill in the blank(s) using elementary row operations to form a row-equivalent matrix.

31. $\begin{bmatrix} 3 & 6 & 8 \\ 4 & -3 & 6 \end{bmatrix}$

$\begin{bmatrix} 1 & \blacksquare & \frac{8}{3} \\ 4 & -3 & 6 \end{bmatrix}$

32. $\begin{bmatrix} 1 & 4 & 3 \\ 2 & 10 & 5 \end{bmatrix}$

$\begin{bmatrix} 1 & 4 & 3 \\ 0 & \blacksquare & -1 \end{bmatrix}$

33. $\begin{bmatrix} 1 & 1 & 1 \\ 5 & -2 & 4 \end{bmatrix}$

$\begin{bmatrix} 1 & 1 & 1 \\ 0 & \blacksquare & -1 \end{bmatrix}$

34. $\begin{bmatrix} -3 & 3 & 12 \\ 18 & -8 & 4 \end{bmatrix}$

$\begin{bmatrix} 1 & -1 & \blacksquare \\ 18 & -8 & 4 \end{bmatrix}$

35. $\begin{bmatrix} 1 & 5 & 4 & -1 \\ 0 & 1 & -2 & 2 \\ 0 & 0 & 1 & -7 \end{bmatrix}$

$\begin{bmatrix} 1 & 0 & \blacksquare & \blacksquare \\ 0 & 1 & -2 & 2 \\ 0 & 0 & 1 & -7 \end{bmatrix}$

36. $\begin{bmatrix} 1 & 0 & 6 & 1 \\ 0 & -1 & 0 & 7 \\ 0 & 0 & -1 & 3 \end{bmatrix}$

$\begin{bmatrix} 1 & 0 & 6 & 1 \\ 0 & 1 & 0 & \blacksquare \\ 0 & 0 & 1 & \blacksquare \end{bmatrix}$

37. $\begin{bmatrix} 1 & 1 & 4 & -1 \\ 3 & 8 & 10 & 3 \\ -2 & 1 & 12 & 6 \end{bmatrix}$

$\begin{bmatrix} 1 & 1 & 4 & -1 \\ 0 & 5 & \blacksquare & \blacksquare \\ 0 & 3 & \blacksquare & \blacksquare \end{bmatrix}$

$\begin{bmatrix} 1 & 1 & 4 & -1 \\ 0 & 1 & -\frac{2}{5} & \frac{6}{5} \\ 0 & 3 & \blacksquare & \blacksquare \end{bmatrix}$

38. $\begin{bmatrix} 2 & 4 & 8 & 3 \\ 1 & -1 & -3 & 2 \\ 2 & 6 & 4 & 9 \end{bmatrix}$

$\begin{bmatrix} 1 & \blacksquare & \blacksquare & \blacksquare \\ 1 & -1 & -3 & 2 \\ 2 & 6 & 4 & 9 \end{bmatrix}$

$\begin{bmatrix} 1 & 2 & 4 & \frac{3}{2} \\ 0 & \blacksquare & -7 & \frac{1}{2} \\ 0 & 2 & \blacksquare & \blacksquare \end{bmatrix}$

Comparing Linear Systems and Matrix Operations In Exercises 39 and 40, (a) perform the row operations to solve the augmented matrix, (b) write and solve the system of linear equations (in variables x, y, and z, if applicable) represented by the augmented matrix, and (c) compare the two solution methods. Which do you prefer?

39. $\begin{bmatrix} -3 & 4 & \vdots & 22 \\ 6 & -4 & \vdots & -28 \end{bmatrix}$

(i) Add R_2 to R_1.

(ii) Add -2 times R_1 to R_2.

(iii) Multiply R_2 by $-\frac{1}{4}$.

(iv) Multiply R_1 by $\frac{1}{3}$.

40. $\begin{bmatrix} 7 & 13 & 1 & \vdots & -4 \\ -3 & -5 & -1 & \vdots & -4 \\ 3 & 6 & 1 & \vdots & -2 \end{bmatrix}$

(i) Add R_2 to R_1.

(ii) Multiply R_1 by $\frac{1}{4}$.

(iii) Add R_3 to R_2.

(iv) Add -3 times R_1 to R_3.

(v) Add -2 times R_2 to R_1.

Row-Echelon Form In Exercises 41–44, determine whether the matrix is in row-echelon form. If it is, determine whether it is in reduced row-echelon form.

41. $\begin{bmatrix} 1 & 0 & 0 & 0 \\ 0 & 1 & 1 & 5 \\ 0 & 0 & 0 & 0 \end{bmatrix}$

42. $\begin{bmatrix} 1 & 0 & 0 & 5 \\ 0 & 1 & 0 & 3 \\ 0 & 1 & 0 & 0 \end{bmatrix}$

43. $\begin{bmatrix} 1 & 0 & 0 & 1 \\ 0 & 1 & 0 & -1 \\ 0 & 0 & 0 & 2 \end{bmatrix}$

44. $\begin{bmatrix} 1 & 0 & 0 & 5 \\ 0 & 1 & -2 & 3 \\ 0 & 0 & 1 & 0 \end{bmatrix}$

Writing a Matrix in Row-Echelon Form In Exercises 45–48, write the matrix in row-echelon form. (Remember that the row-echelon form of a matrix is not unique.)

45. $\begin{bmatrix} 1 & 1 & 0 & 5 \\ -2 & -1 & 2 & -10 \\ 3 & 6 & 7 & 14 \end{bmatrix}$

46. $\begin{bmatrix} 1 & 2 & -1 & 3 \\ 3 & 7 & -5 & 14 \\ -2 & -1 & -3 & 8 \end{bmatrix}$

47. $\begin{bmatrix} 1 & -1 & -1 & 1 \\ 5 & -4 & 1 & 8 \\ -6 & 8 & 18 & 0 \end{bmatrix}$

48. $\begin{bmatrix} 1 & -3 & 0 & -7 \\ -3 & 10 & 1 & 23 \\ 4 & -10 & 2 & -24 \end{bmatrix}$

Using a Graphing Utility In Exercises 49–54, use the matrix capabilities of a graphing utility to write the matrix in reduced row-echelon form.

49. $\begin{bmatrix} -1 & 2 & 1 \\ 3 & 4 & 9 \\ 2 & 1 & -2 \end{bmatrix}$

50. $\begin{bmatrix} 1 & 3 & 2 \\ 5 & 15 & 9 \\ 2 & 6 & 10 \end{bmatrix}$

51. $\begin{bmatrix} 1 & 2 & 3 & -5 \\ 1 & 2 & 4 & -9 \\ -2 & -4 & -4 & 3 \\ 4 & 8 & 11 & -14 \end{bmatrix}$

52. $\begin{bmatrix} -2 & 3 & -1 & -2 \\ 4 & -2 & 5 & 8 \\ 1 & 5 & -2 & 0 \\ 3 & 8 & -10 & -30 \end{bmatrix}$

53. $\begin{bmatrix} -3 & 5 & 1 & 12 \\ 1 & -1 & 1 & 4 \end{bmatrix}$

54. $\begin{bmatrix} 5 & 1 & 2 & 4 \\ -1 & 5 & 10 & -32 \end{bmatrix}$

Using Back-Substitution In Exercises 55–58, write the system of linear equations represented by the augmented matrix. Then use back-substitution to solve the system. (Use variables x, y, and z, if applicable.)

55. $\begin{bmatrix} 1 & -2 & \vdots & 4 \\ 0 & 1 & \vdots & -1 \end{bmatrix}$

56. $\begin{bmatrix} 1 & 5 & \vdots & 0 \\ 0 & 1 & \vdots & 6 \end{bmatrix}$

57. $\begin{bmatrix} 1 & -1 & 2 & \vdots & 4 \\ 0 & 1 & -1 & \vdots & 2 \\ 0 & 0 & 1 & \vdots & -2 \end{bmatrix}$

58. $\begin{bmatrix} 1 & 2 & -2 & \vdots & -1 \\ 0 & 1 & 1 & \vdots & 9 \\ 0 & 0 & 1 & \vdots & -3 \end{bmatrix}$

Gaussian Elimination with Back-Substitution In Exercises 59–68, use matrices to solve the system of linear equations, if possible. Use Gaussian elimination with back-substitution.

59. $\begin{cases} x + 2y = 7 \\ -x + y = 8 \end{cases}$ **60.** $\begin{cases} 2x + 6y = 16 \\ 2x + 3y = 7 \end{cases}$

61. $\begin{cases} 3x - 2y = -27 \\ x + 3y = 13 \end{cases}$ **62.** $\begin{cases} -x + y = 4 \\ 2x - 4y = -34 \end{cases}$

63. $\begin{cases} x + 2y - 3z = -28 \\ 4y + 2z = 0 \\ -x + y - z = -5 \end{cases}$

64. $\begin{cases} 3x - 2y + z = 15 \\ -x + y + 2z = -10 \\ x - y - 4z = 14 \end{cases}$

65. $\begin{cases} -3x + 2y = -22 \\ 3x + 4y = 4 \\ 4x - 8y = 32 \end{cases}$ **66.** $\begin{cases} x + 2y = 0 \\ x + y = 6 \\ 3x - 2y = 8 \end{cases}$

67. $\begin{cases} 3x + 2y - z + w = 0 \\ x - y + 4z + 2w = 25 \\ -2x + y + 2z - w = 2 \\ x + y + z + w = 6 \end{cases}$

68. $\begin{cases} x - 4y + 3z - 2w = 9 \\ 3x - 2y + z - 4w = -13 \\ -4x + 3y - 2z + w = -4 \\ -2x + y - 4z + 3w = -10 \end{cases}$

Interpreting Reduced Row-Echelon Form In Exercises 69 and 70, an augmented matrix that represents a system of linear equations (in variables x, y, and z, if applicable) has been reduced using Gauss-Jordan elimination. Write the solution represented by the augmented matrix.

69. $\begin{bmatrix} 1 & 0 & \vdots & 3 \\ 0 & 1 & \vdots & -4 \end{bmatrix}$

70. $\begin{bmatrix} 1 & 0 & 0 & \vdots & 5 \\ 0 & 1 & 0 & \vdots & -3 \\ 0 & 0 & 1 & \vdots & 0 \end{bmatrix}$

Gauss-Jordan Elimination In Exercises 71–78, use matrices to solve the system of linear equations, if possible. Use Gauss-Jordan elimination.

71. $\begin{cases} -2x + 6y = -22 \\ x + 2y = -9 \end{cases}$ **72.** $\begin{cases} 5x - 5y = -5 \\ -2x - 3y = 7 \end{cases}$

73. $\begin{cases} x + 2y + z = 8 \\ 3x + 7y + 6z = 26 \end{cases}$ **74.** $\begin{cases} x + y + 4z = 5 \\ 2x + y - z = 9 \end{cases}$

75. $\begin{cases} x - 3z = -2 \\ 3x + y - 2z = 5 \\ 2x + 2y + z = 4 \end{cases}$ **76.** $\begin{cases} 2x - y + 3z = 24 \\ 2y - z = 14 \\ 7x - 5y = 6 \end{cases}$

77. $\begin{cases} -x + y - z = -14 \\ 2x - y + z = 21 \\ 3x + 2y + z = 19 \end{cases}$

78. $\begin{cases} 2x + 2y - z = 2 \\ x - 3y + z = -28 \\ -x + y = 14 \end{cases}$

Using a Graphing Utility In Exercises 79–84, use the matrix capabilities of a graphing utility to write the augmented matrix corresponding to the system of linear equations in reduced row-echelon form. Then solve the system.

79. $\begin{cases} 3x + 3y + 12z = 6 \\ x + y + 4z = 2 \\ 2x + 5y + 20z = 10 \\ -x + 2y + 8z = 4 \end{cases}$

80. $\begin{cases} 2x + 10y + 2z = 6 \\ x + 5y + 2z = 6 \\ x + 5y + z = 3 \\ -3x - 15y - 3z = -9 \end{cases}$

81. $\begin{cases} 2x + y - z + 2w = -6 \\ 3x + 4y + w = 1 \\ x + 5y + 2z + 6w = -3 \\ 5x + 2y - z - w = 3 \end{cases}$

82. $\begin{cases} x + 2y + 2z + 4w = 11 \\ 3x + 6y + 5z + 12w = 30 \\ x + 3y - 3z + 2w = -5 \\ 6x - y - z + w = -9 \end{cases}$

83. $\begin{cases} x + y + z + w = 0 \\ 2x + 3y + z - 2w = 0 \\ 3x + 5y + z = 0 \end{cases}$

84. $\begin{cases} x + 2y + z + 3w = 0 \\ x - y + w = 0 \\ y - z + 2w = 0 \end{cases}$

85. Error Analysis Describe the error.

The matrix

$\begin{bmatrix} 3 \\ 0 \\ 8 \\ -1 \end{bmatrix}$

has four rows and one column, so the dimension of the matrix is 1×4.

86. Error Analysis Describe the error.

The matrix

$\begin{bmatrix} 1 & 2 & 7 & 2 \\ 0 & 1 & 4 & 0 \\ 0 & 0 & 1 & \frac{1}{3} \end{bmatrix}$

is in reduced row-echelon form.

Curve Fitting In Exercises 87–92, use a system of linear equations to find the quadratic function $f(x) = ax^2 + bx + c$ that satisfies the given conditions. Solve the system using matrices.

87. $f(1) = 1$, $f(2) = -1$, $f(3) = -5$

88. $f(1) = 2$, $f(2) = 9$, $f(3) = 20$

89. $f(-2) = -15$, $f(-1) = 7$, $f(1) = -3$

90. $f(-2) = -3$, $f(1) = -3$, $f(2) = -11$

91. $f(1) = 8$, $f(2) = 13$, $f(3) = 20$

92. $f(1) = 9$, $f(2) = 8$, $f(3) = 5$

93. Waterborne Disease

From 2005 through 2016, the numbers of new cases of a waterborne disease in a small city increased in a pattern that was approximately linear (see figure). Find the least squares regression line

$y = at + b$

for the data shown in the figure by solving the system below using matrices. Let t represent the year, with $t = 0$ corresponding to 2005.

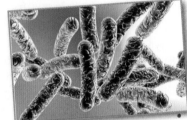

$$\begin{cases} 12b + 66a = 831 \\ 66b + 506a = 5643 \end{cases}$$

Use the result to predict the number of new cases of the waterborne disease in 2020. Is the estimate reasonable? Explain.

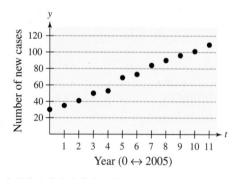

Year (0 ↔ 2005)

94. Museum A natural history museum borrows $2,000,000 at simple annual interest to purchase new exhibits. Some of the money is borrowed at 7%, some at 8.5%, and some at 9.5%. Use a system of linear equations to determine how much is borrowed at each rate given that the total annual interest is $169,750 and the amount borrowed at 8.5% is four times the amount borrowed at 9.5%. Solve the system of linear equations using matrices.

95. Breeding Facility A city zoo borrows $2,000,000 at simple annual interest to construct a breeding facility. Some of the money is borrowed at 8%, some at 9%, and some at 12%. Use a system of linear equations to determine how much is borrowed at each rate given that the total annual interest is $186,000 and the amount borrowed at 8% is twice the amount borrowed at 12%. Solve the system of linear equations using matrices.

96. Mathematical Modeling A video of the path of a ball thrown by a baseball player was analyzed with a grid covering the TV screen. The video was paused three times, and the position of the ball was measured each time. The coordinates obtained are shown in the table. (x and y are measured in feet.)

Horizontal Distance, x	0	15	30
Height, y	5.0	9.6	12.4

(a) Use a system of equations to find the equation of the parabola $y = ax^2 + bx + c$ that passes through the three points. Solve the system using matrices.

(b) Use a graphing utility to graph the parabola.

(c) Graphically approximate the maximum height of the ball and the point at which the ball struck the ground.

(d) Analytically find the maximum height of the ball and the point at which the ball struck the ground.

(e) Compare your results from parts (c) and (d).

Exploration

True or False? In Exercises 97 and 98, determine whether the statement is true or false. Justify your answer.

97. $\begin{bmatrix} 5 & 0 & -2 & 7 \\ -1 & 3 & -6 & 0 \end{bmatrix}$ is a 4×2 matrix.

98. The method of Gaussian elimination reduces a matrix until a reduced row-echelon form is obtained.

99. Think About It What is the relationship between the three elementary row operations performed on an augmented matrix and the operations that lead to equivalent systems of equations?

100. **HOW DO YOU SEE IT?** Determine whether the matrix below is in row-echelon form, reduced row-echelon form, or neither when it satisfies the given conditions.

$$\begin{bmatrix} 1 & b \\ c & 1 \end{bmatrix}$$

(a) $b = 0$, $c = 0$ (b) $b \neq 0$, $c = 0$

(c) $b = 0$, $c \neq 0$ (d) $b \neq 0$, $c \neq 0$

8.2 Operations with Matrices

■ Determine whether two matrices are equal.
■ Add and subtract matrices, and multiply matrices by scalars.
■ Multiply two matrices.
■ Use matrices to transform vectors.
■ Use matrix operations to model and solve real-life problems.

Matrix operations have many practical applications. For example, in Exercise 80 on page 567, you will use matrix multiplication to analyze the calories burned by individuals of different body weights while performing different types of exercises.

Equality of Matrices

In Section 8.1, you used matrices to solve systems of linear equations. There is a rich mathematical theory of matrices, and its applications are numerous. This section and the next two sections introduce some fundamental concepts of matrix theory. It is standard mathematical convention to represent matrices in any of the three ways listed below.

Representation of Matrices

1. A matrix can be denoted by an uppercase letter such as A, B, or C.

2. A matrix can be denoted by a representative element enclosed in brackets, such as $[a_{ij}]$, $[b_{ij}]$, or $[c_{ij}]$.

3. A matrix can be denoted by a rectangular array of numbers such as

$$A = [a_{ij}] = \begin{bmatrix} a_{11} & a_{12} & a_{13} & \cdots & a_{1n} \\ a_{21} & a_{22} & a_{23} & \cdots & a_{2n} \\ a_{31} & a_{32} & a_{33} & \cdots & a_{3n} \\ \vdots & \vdots & \vdots & & \vdots \\ a_{m1} & a_{m2} & a_{m3} & \cdots & a_{mn} \end{bmatrix}.$$

Two matrices $A = [a_{ij}]$ and $B = [b_{ij}]$ are **equal** when they have the same dimension $(m \times n)$ and $a_{ij} = b_{ij}$ for $1 \le i \le m$ and $1 \le j \le n$. In other words, two matrices are equal when their corresponding entries are equal.

EXAMPLE 1 Equality of Matrices

Solve for a_{11}, a_{12}, a_{21}, and a_{22} in the matrix equation $\begin{bmatrix} a_{11} & a_{12} \\ a_{21} & a_{22} \end{bmatrix} = \begin{bmatrix} 2 & -1 \\ -3 & 0 \end{bmatrix}.$

Solution Two matrices are equal when their corresponding entries are equal, so $a_{11} = 2$, $a_{12} = -1$, $a_{21} = -3$, and $a_{22} = 0$.

✓ **Checkpoint** ◀))) *Audio-video solution in English & Spanish at LarsonPrecalculus.com*

Solve for a_{11}, a_{12}, a_{21}, and a_{22} in the matrix equation $\begin{bmatrix} a_{11} & a_{12} \\ a_{21} & a_{22} \end{bmatrix} = \begin{bmatrix} 6 & 3 \\ -2 & 4 \end{bmatrix}.$ ■

Be sure you see that for two matrices to be equal, they must have the same dimension *and* their corresponding entries must be equal. For example,

$$\begin{bmatrix} 2 & -1 \\ \sqrt{4} & \frac{1}{2} \end{bmatrix} = \begin{bmatrix} 2 & -1 \\ 2 & 0.5 \end{bmatrix} \quad \text{but} \quad \begin{bmatrix} 2 & -1 & 0 \\ 3 & 4 & 0 \end{bmatrix} \ne \begin{bmatrix} 2 & -1 \\ 3 & 4 \end{bmatrix}.$$

Matrix Addition and Scalar Multiplication

Two basic matrix operations are matrix addition and scalar multiplication. With matrix addition, you add two matrices (of the same dimension) by adding their corresponding entries.

> **Definition of Matrix Addition**
>
> If $A = [a_{ij}]$ and $B = [b_{ij}]$ are matrices of dimension $m \times n$, then their sum is the $m \times n$ matrix
>
> $$A + B = [a_{ij} + b_{ij}].$$
>
> The sum of two matrices of different dimensions is undefined.

EXAMPLE 2 **Addition of Matrices**

a. $\begin{bmatrix} -1 & 2 \\ 0 & 1 \end{bmatrix} + \begin{bmatrix} 1 & 3 \\ -1 & 2 \end{bmatrix} = \begin{bmatrix} -1+1 & 2+3 \\ 0+(-1) & 1+2 \end{bmatrix} = \begin{bmatrix} 0 & 5 \\ -1 & 3 \end{bmatrix}$

b. $\begin{bmatrix} 0 & 1 & -2 \\ 1 & 2 & 3 \end{bmatrix} + \begin{bmatrix} 0 & 0 & 0 \\ 0 & 0 & 0 \end{bmatrix} = \begin{bmatrix} 0 & 1 & -2 \\ 1 & 2 & 3 \end{bmatrix}$

c. $\begin{bmatrix} 1 \\ -3 \\ -2 \end{bmatrix} + \begin{bmatrix} -1 \\ 3 \\ 2 \end{bmatrix} = \begin{bmatrix} 0 \\ 0 \\ 0 \end{bmatrix}$

d. The sum of

$$A = \begin{bmatrix} 2 & 1 & 0 \\ 4 & 0 & -1 \\ 3 & -2 & 2 \end{bmatrix} \quad \text{and} \quad B = \begin{bmatrix} 0 & 1 \\ -1 & 3 \\ 2 & 4 \end{bmatrix}$$

is undefined because A is of dimension 3×3 and B is of dimension 3×2.

✓ **Checkpoint** 🔊))) *Audio-video solution in English & Spanish at LarsonPrecalculus.com*

Find each sum, if possible.

a. $\begin{bmatrix} 4 & -1 \\ 2 & -3 \end{bmatrix} + \begin{bmatrix} 2 & -1 \\ 0 & 6 \end{bmatrix}$ **b.** $\begin{bmatrix} 2 & -1 \\ 3 & 4 \\ 0 & -2 \end{bmatrix} + \begin{bmatrix} -2 & 1 \\ -3 & -4 \\ 0 & 2 \end{bmatrix}$

c. $\begin{bmatrix} 3 & 9 & 6 \\ 0 & 4 & -2 \\ 1 & -1 & 0 \end{bmatrix} + \begin{bmatrix} 3 & 9 & 6 \\ 0 & 2 & -4 \end{bmatrix}$ **d.** $\begin{bmatrix} 1 \\ -1 \\ 1 \end{bmatrix} + \begin{bmatrix} -1 \\ 1 \\ 1 \end{bmatrix}$

In operations with matrices, numbers are usually referred to as **scalars.** In this text, scalars will always be real numbers. To multiply a matrix A by a scalar c, multiply each entry in A by c.

> **Definition of Scalar Multiplication**
>
> If $A = [a_{ij}]$ is an $m \times n$ matrix and c is a scalar, then the **scalar multiple** of A by c is the $m \times n$ matrix
>
> $$cA = [ca_{ij}].$$

The symbol $-A$ represents the **negation** of A, which is the scalar product $(-1)A$. Moreover, if A and B are of the same dimension, then $A - B$ represents the sum of A and $(-1)B$. That is,

$$A - B = A + (-1)B. \qquad \text{Subtraction of matrices}$$

EXAMPLE 3 Operations with Matrices

For the matrices below, find (a) $3A$, (b) $-B$, and (c) $3A - B$.

$$A = \begin{bmatrix} 2 & 2 & 4 \\ -3 & 0 & -1 \\ 2 & 1 & 2 \end{bmatrix} \quad \text{and} \quad B = \begin{bmatrix} 2 & 0 & 0 \\ 1 & -4 & 3 \\ -1 & 3 & 2 \end{bmatrix}$$

Solution

a. $3A = 3 \begin{bmatrix} 2 & 2 & 4 \\ -3 & 0 & -1 \\ 2 & 1 & 2 \end{bmatrix}$ Scalar multiplication

$$= \begin{bmatrix} 3(2) & 3(2) & 3(4) \\ 3(-3) & 3(0) & 3(-1) \\ 3(2) & 3(1) & 3(2) \end{bmatrix}$$ Multiply each entry by 3.

$$= \begin{bmatrix} 6 & 6 & 12 \\ -9 & 0 & -3 \\ 6 & 3 & 6 \end{bmatrix}$$ Simplify.

b. $-B = (-1) \begin{bmatrix} 2 & 0 & 0 \\ 1 & -4 & 3 \\ -1 & 3 & 2 \end{bmatrix}$ Definition of negation

$$= \begin{bmatrix} -2 & 0 & 0 \\ -1 & 4 & -3 \\ 1 & -3 & -2 \end{bmatrix}$$ Multiply each entry by -1.

c. $3A - B = \begin{bmatrix} 6 & 6 & 12 \\ -9 & 0 & -3 \\ 6 & 3 & 6 \end{bmatrix} + \begin{bmatrix} -2 & 0 & 0 \\ -1 & 4 & -3 \\ 1 & -3 & -2 \end{bmatrix}$ $3A - B = 3A + (-1)B$

$$= \begin{bmatrix} 4 & 6 & 12 \\ -10 & 4 & -6 \\ 7 & 0 & 4 \end{bmatrix}$$ Add corresponding entries.

>•**REMARK** The order of operations for matrix expressions is similar to that for real numbers. As shown in Example 3(c), you perform scalar multiplication before matrix addition and subtraction.

✓ *Checkpoint* ◀))) Audio-video solution in English & Spanish at LarsonPrecalculus.com

For the matrices below, find (a) $A - B$, (b) $3A$, and (c) $3A - 2B$.

$$A = \begin{bmatrix} 4 & -1 \\ 0 & 4 \\ -3 & 8 \end{bmatrix} \quad \text{and} \quad B = \begin{bmatrix} 0 & 4 \\ -1 & 3 \\ 1 & 7 \end{bmatrix}$$

It is often convenient to rewrite the scalar multiple cA by factoring c out of every entry in the matrix. The example below shows factoring the scalar $\frac{1}{2}$ out of a matrix.

$$\begin{bmatrix} \frac{1}{2} & -\frac{3}{2} \\ \frac{5}{2} & \frac{1}{2} \end{bmatrix} = \begin{bmatrix} \frac{1}{2}(1) & \frac{1}{2}(-3) \\ \frac{1}{2}(5) & \frac{1}{2}(1) \end{bmatrix} = \frac{1}{2} \begin{bmatrix} 1 & -3 \\ 5 & 1 \end{bmatrix}$$

▷ **ALGEBRA HELP** To review the properties of addition and multiplication of real numbers (and other properties of real numbers), see Appendix A.1.

The properties of matrix addition and scalar multiplication are similar to those of addition and multiplication of real numbers.

Properties of Matrix Addition and Scalar Multiplication

Let A, B, and C be $m \times n$ matrices and let c and d be scalars.

1. $A + B = B + A$ Commutative Property of Matrix Addition

2. $A + (B + C) = (A + B) + C$ Associative Property of Matrix Addition

3. $(cd)A = c(dA)$ Associative Property of Scalar Multiplication

4. $1A = A$ Scalar Identity Property

5. $c(A + B) = cA + cB$ Distributive Property

6. $(c + d)A = cA + dA$ Distributive Property

Note that the Associative Property of Matrix Addition allows you to write expressions such as $A + B + C$ without ambiguity because the same sum occurs no matter how the matrices are grouped. This same reasoning applies to sums of four or more matrices.

▷ **TECHNOLOGY** Most graphing utilities can perform matrix operations. Consult the user's guide for your graphing utility for specific keystrokes. Use a graphing utility to find the sum of the matrices

$$A = \begin{bmatrix} 2 & -3 \\ -1 & 0 \end{bmatrix}$$

and

$$B = \begin{bmatrix} -1 & 4 \\ 2 & -5 \end{bmatrix}.$$

EXAMPLE 4   **Addition of More than Two Matrices**

$$\begin{bmatrix} 1 \\ 2 \\ -3 \end{bmatrix} + \begin{bmatrix} -1 \\ -1 \\ 2 \end{bmatrix} + \begin{bmatrix} 0 \\ 1 \\ 4 \end{bmatrix} + \begin{bmatrix} 2 \\ -3 \\ -2 \end{bmatrix} = \begin{bmatrix} 2 \\ -1 \\ 1 \end{bmatrix} \qquad \text{Add corresponding entries.}$$

✓ **Checkpoint** ◀))) *Audio-video solution in English & Spanish at LarsonPrecalculus.com*

Evaluate the expression.

$$\begin{bmatrix} 3 & -8 \\ 0 & 2 \end{bmatrix} + \begin{bmatrix} -2 & 3 \\ 6 & -5 \end{bmatrix} + \begin{bmatrix} 0 & 7 \\ 4 & -1 \end{bmatrix}$$

EXAMPLE 5   **Evaluating an Expression**

$$3\left(\begin{bmatrix} -2 & 0 \\ 4 & 1 \end{bmatrix} + \begin{bmatrix} 4 & -2 \\ 3 & 7 \end{bmatrix} \right) = 3\begin{bmatrix} -2 & 0 \\ 4 & 1 \end{bmatrix} + 3\begin{bmatrix} 4 & -2 \\ 3 & 7 \end{bmatrix}$$

$$= \begin{bmatrix} -6 & 0 \\ 12 & 3 \end{bmatrix} + \begin{bmatrix} 12 & -6 \\ 9 & 21 \end{bmatrix}$$

$$= \begin{bmatrix} 6 & -6 \\ 21 & 24 \end{bmatrix}$$

✓ **Checkpoint** ◀))) *Audio-video solution in English & Spanish at LarsonPrecalculus.com*

Evaluate the expression.

$$2\left(\begin{bmatrix} 1 & 3 \\ -2 & 2 \end{bmatrix} + \begin{bmatrix} -4 & 0 \\ -3 & 1 \end{bmatrix} \right)$$

In Example 5, you could add the two matrices first and then multiply the resulting matrix by 3. The result would be the same.

One important property of addition of real numbers is that the number 0 is the additive identity. That is, $c + 0 = c$ for any real number c. For matrices, a similar property holds. That is, if A is an $m \times n$ matrix and O is the $m \times n$ **zero matrix** consisting entirely of zeros, then

$A + O = A$.

In other words, O is the **additive identity** for the set of all $m \times n$ matrices. For example, the matrices below are the additive identities for the sets of all 2×3 and 2×2 matrices.

$$O = \begin{bmatrix} 0 & 0 & 0 \\ 0 & 0 & 0 \end{bmatrix} \quad \text{and} \quad O = \begin{bmatrix} 0 & 0 \\ 0 & 0 \end{bmatrix}$$

$\underbrace{}_{2 \times 3 \text{ zero matrix}}$ $\underbrace{}_{2 \times 2 \text{ zero matrix}}$

The algebra of real numbers and the algebra of matrices have many similarities. For example, compare the solutions below.

Real Numbers **(Solve for x.)**	**$m \times n$ Matrices** **(Solve for X.)**
$x + a = b$	$X + A = B$
$x + a + (-a) = b + (-a)$	$X + A + (-A) = B + (-A)$
$x + 0 = b - a$	$X + O = B - A$
$x = b - a$	$X = B - A$

The algebra of real numbers and the algebra of matrices also have important differences (see Example 9 and Exercises 83–88).

> • **REMARK** When you solve for X in a matrix equation, you are solving for a *matrix X* that makes the equation true.

EXAMPLE 6 **Solving a Matrix Equation**

Solve for X in the equation $3X + A = B$, where

$$A = \begin{bmatrix} 1 & -2 \\ 0 & 3 \end{bmatrix} \quad \text{and} \quad B = \begin{bmatrix} -3 & 4 \\ 2 & 1 \end{bmatrix}.$$

Solution Begin by solving the matrix equation for X.

$3X + A = B$

$3X = B - A$

$X = \frac{1}{3}(B - A)$

Now, substituting the matrices A and B, you have

$$X = \frac{1}{3}\left(\begin{bmatrix} -3 & 4 \\ 2 & 1 \end{bmatrix} - \begin{bmatrix} 1 & -2 \\ 0 & 3 \end{bmatrix} \right) \qquad \text{Substitute the matrices.}$$

$$= \frac{1}{3}\begin{bmatrix} -4 & 6 \\ 2 & -2 \end{bmatrix} \qquad \text{Subtract matrix } A \text{ from matrix } B.$$

$$= \begin{bmatrix} -\frac{4}{3} & 2 \\ \frac{2}{3} & -\frac{2}{3} \end{bmatrix} \qquad \text{Multiply the resulting matrix by } \frac{1}{3}.$$

✓ *Checkpoint* ◀))) *Audio-video solution in English & Spanish at LarsonPrecalculus.com*

Solve for X in the equation $2X - A = B$, where

$$A = \begin{bmatrix} 6 & 1 \\ 0 & 3 \end{bmatrix} \quad \text{and} \quad B = \begin{bmatrix} 4 & -1 \\ -2 & 5 \end{bmatrix}.$$

Matrix Multiplication

Another basic matrix operation is **matrix multiplication.** At first glance, the definition may seem unusual. You will see later, however, that this definition of the product of two matrices has many practical applications.

Definition of Matrix Multiplication

If $A = [a_{ij}]$ is an $m \times n$ matrix and $B = [b_{ij}]$ is an $n \times p$ matrix, then the product AB is an $m \times p$ matrix given by $AB = [c_{ij}]$, where

$$c_{ij} = a_{i1}b_{1j} + a_{i2}b_{2j} + a_{i3}b_{3j} + \cdots + a_{in}b_{nj}.$$

$$
\begin{array}{ccccc}
A & \times & B & = & AB \\
m \times n & & n \times p & & m \times p
\end{array}
$$

Equal

Dimension of AB

The definition of matrix multiplication uses a *row-by-column* multiplication, where the entry in the ith row and jth column of the product AB is obtained by multiplying the entries in the ith row of A by the corresponding entries in the jth column of B and then adding the results. So, for the product of two matrices to be defined, the number of columns of the first matrix must equal the number of rows of the second matrix. That is, the middle two indices must be the same. The outside two indices give the dimension of the product, as shown at the left. The general pattern for matrix multiplication is shown below.

$$
\begin{bmatrix}
a_{11} & a_{12} & a_{13} & \cdots & a_{1n} \\
a_{21} & a_{22} & a_{23} & \cdots & a_{2n} \\
a_{31} & a_{32} & a_{33} & \cdots & a_{3n} \\
\vdots & \vdots & \vdots & & \vdots \\
a_{i1} & a_{i2} & a_{i3} & \cdots & a_{in} \\
\vdots & \vdots & \vdots & & \vdots \\
a_{m1} & a_{m2} & a_{m3} & \cdots & a_{mn}
\end{bmatrix}
\begin{bmatrix}
b_{11} & b_{12} & \cdots & b_{1j} & \cdots & b_{1p} \\
b_{21} & b_{22} & \cdots & b_{2j} & \cdots & b_{2p} \\
b_{31} & b_{32} & \cdots & b_{3j} & \cdots & b_{3p} \\
\vdots & \vdots & & \vdots & & \vdots \\
b_{n1} & b_{n2} & \cdots & b_{nj} & \cdots & b_{np}
\end{bmatrix}
=
\begin{bmatrix}
c_{11} & c_{12} & \cdots & c_{1j} & \cdots & c_{1p} \\
c_{21} & c_{22} & \cdots & c_{2j} & \cdots & c_{2p} \\
\vdots & \vdots & & \vdots & & \vdots \\
c_{i1} & c_{i2} & \cdots & c_{ij} & \cdots & c_{ip} \\
\vdots & \vdots & & \vdots & & \vdots \\
c_{m1} & c_{m2} & \cdots & c_{mj} & \cdots & c_{mp}
\end{bmatrix}
$$

$$a_{i1}b_{1j} + a_{i2}b_{2j} + a_{i3}b_{3j} + \cdots + a_{in}b_{nj} = c_{ij}$$

EXAMPLE 7   **Finding the Product of Two Matrices**

Find the product AB, where $A = \begin{bmatrix} -1 & 3 \\ 4 & -2 \\ 5 & 0 \end{bmatrix}$ and $B = \begin{bmatrix} -3 & 2 \\ -4 & 1 \end{bmatrix}$.

Solution   To find the entries of the product, multiply each row of A by each column of B.

$$
AB = \begin{bmatrix} -1 & 3 \\ 4 & -2 \\ 5 & 0 \end{bmatrix} \begin{bmatrix} -3 & 2 \\ -4 & 1 \end{bmatrix}
$$

$$
= \begin{bmatrix} (-1)(-3) + & 3(-4) & (-1)(2) + & 3(1) \\ 4(-3) + (-2)(-4) & & 4(2) + (-2)(1) \\ 5(-3) + & 0(-4) & 5(2) + & 0(1) \end{bmatrix}
$$

$$
= \begin{bmatrix} -9 & 1 \\ -4 & 6 \\ -15 & 10 \end{bmatrix}
$$

> **REMARK** In Example 7, the product AB is defined because the number of columns of A is equal to the number of rows of B. Also, note that the product AB has dimension 3×2.

✓ Checkpoint ◄))) *Audio-video solution in English & Spanish at LarsonPrecalculus.com*

Find the product AB, where $A = \begin{bmatrix} -1 & 4 \\ 2 & 0 \\ 1 & 2 \end{bmatrix}$ and $B = \begin{bmatrix} 1 & -2 \\ 0 & 7 \end{bmatrix}$. ∎

EXAMPLE 8   **Finding the Product of Two Matrices**

Find the product AB, where $A = \begin{bmatrix} 1 & 0 & 3 \\ 2 & -1 & -2 \end{bmatrix}$ and $B = \begin{bmatrix} -2 & 4 \\ 1 & 0 \\ -1 & 1 \end{bmatrix}$.

Solution   Note that the dimension of A is 2×3 and the dimension of B is 3×2. So, the product AB has dimension 2×2.

$$AB = \begin{bmatrix} 1 & 0 & 3 \\ 2 & -1 & -2 \end{bmatrix} \begin{bmatrix} -2 & 4 \\ 1 & 0 \\ -1 & 1 \end{bmatrix}$$

$$= \begin{bmatrix} 1(-2) + 0(1) + 3(-1) & 1(4) + 0(0) + 3(1) \\ 2(-2) + (-1)(1) + (-2)(-1) & 2(4) + (-1)(0) + (-2)(1) \end{bmatrix}$$

$$= \begin{bmatrix} -5 & 7 \\ -3 & 6 \end{bmatrix}$$

✓ *Checkpoint* ◀))) *Audio-video solution in English & Spanish at LarsonPrecalculus.com*

Find the product AB, where $A = \begin{bmatrix} 0 & 4 & -3 \\ 2 & 1 & 7 \\ 3 & -2 & 1 \end{bmatrix}$ and $B = \begin{bmatrix} -2 & 0 \\ 0 & -4 \\ 1 & 2 \end{bmatrix}$.

EXAMPLE 9   **Matrix Multiplication**

See LarsonPrecalculus.com for an interactive version of this type of example.

a. $\underset{2 \times 2}{\begin{bmatrix} 3 & 4 \\ -2 & 5 \end{bmatrix}} \underset{2 \times 2}{\begin{bmatrix} 1 & 0 \\ 0 & 1 \end{bmatrix}} = \underset{2 \times 2}{\begin{bmatrix} 3 & 4 \\ -2 & 5 \end{bmatrix}}$

b. $\underset{3 \times 3}{\begin{bmatrix} 6 & 2 & 0 \\ 3 & -1 & 2 \\ 1 & 4 & 6 \end{bmatrix}} \underset{3 \times 1}{\begin{bmatrix} 1 \\ 2 \\ -3 \end{bmatrix}} = \underset{3 \times 1}{\begin{bmatrix} 10 \\ -5 \\ -9 \end{bmatrix}}$

c. $\underset{1 \times 3}{\begin{bmatrix} 1 & -2 & -3 \end{bmatrix}} \underset{3 \times 1}{\begin{bmatrix} 2 \\ -1 \\ 1 \end{bmatrix}} = \underset{1 \times 1}{\begin{bmatrix} 1 \end{bmatrix}}$

> **REMARK** In Examples 9(c) and 9(d), note that the two products are different. Even when both AB and BA are defined, matrix multiplication is not, in general, commutative. That is, for most matrices, $AB \neq BA$. This is one way in which the algebra of real numbers and the algebra of matrices differ.

d. $\underset{3 \times 1}{\begin{bmatrix} 2 \\ -1 \\ 1 \end{bmatrix}} \underset{1 \times 3}{\begin{bmatrix} 1 & -2 & -3 \end{bmatrix}} = \underset{3 \times 3}{\begin{bmatrix} 2 & -4 & -6 \\ -1 & 2 & 3 \\ 1 & -2 & -3 \end{bmatrix}}$

e. The product $\underset{3 \times 2}{\begin{bmatrix} -2 & 1 \\ 1 & -3 \\ 1 & 4 \end{bmatrix}} \underset{3 \times 4}{\begin{bmatrix} -2 & 3 & 1 & 4 \\ 0 & 1 & -1 & 2 \\ 2 & -1 & 0 & 1 \end{bmatrix}}$ is not defined.

✓ *Checkpoint* ◀))) *Audio-video solution in English & Spanish at LarsonPrecalculus.com*

Find each product, if possible.

a. $\begin{bmatrix} 1 \\ -3 \end{bmatrix} \begin{bmatrix} 3 & -1 \end{bmatrix}$ b. $\begin{bmatrix} 3 & -1 \end{bmatrix} \begin{bmatrix} 1 \\ -3 \end{bmatrix}$ c. $\begin{bmatrix} 3 & 1 & 2 \\ 7 & 0 & -2 \end{bmatrix} \begin{bmatrix} 6 & 4 \\ 2 & -1 \end{bmatrix}$

EXAMPLE 10 **Squaring a Matrix**

Find A^2, where $A = \begin{bmatrix} 3 & 1 \\ -1 & 2 \end{bmatrix}$. (*Note:* $A^2 = AA$.)

Solution

$$A^2 = \begin{bmatrix} 3 & 1 \\ -1 & 2 \end{bmatrix}\begin{bmatrix} 3 & 1 \\ -1 & 2 \end{bmatrix}$$

$$= \begin{bmatrix} 8 & 5 \\ -5 & 3 \end{bmatrix}$$

✓ *Checkpoint* ◀)) *Audio-video solution in English & Spanish at LarsonPrecalculus.com*

Find A^2, where $A = \begin{bmatrix} 2 & 1 \\ 3 & -2 \end{bmatrix}$.

Properties of Matrix Multiplication

Let A, B, and C be matrices and let c be a scalar.

1. $A(BC) = (AB)C$ Associative Property of Matrix Multiplication

2. $A(B + C) = AB + AC$ Left Distributive Property

3. $(A + B)C = AC + BC$ Right Distributive Property

4. $c(AB) = (cA)B = A(cB)$ Associative Property of Scalar Multiplication

---

Definition of the Identity Matrix

The $n \times n$ matrix that consists of 1's on its main diagonal and 0's elsewhere is called the **identity matrix of dimension $n \times n$** and is denoted by

$$I_n = \begin{bmatrix} 1 & 0 & 0 & \cdots & 0 \\ 0 & 1 & 0 & \cdots & 0 \\ 0 & 0 & 1 & \cdots & 0 \\ \vdots & \vdots & \vdots & & \vdots \\ 0 & 0 & 0 & \cdots & 1 \end{bmatrix}. \qquad \text{Identity matrix}$$

Note that an identity matrix must be *square*. When the dimension is understood to be $n \times n$, you can denote I_n simply by I.

If A is an $n \times n$ matrix, then the identity matrix has the property that $AI_n = A$ and $I_nA = A$. For example,

$$\begin{bmatrix} 3 & -2 & 5 \\ 1 & 0 & 4 \\ -1 & 2 & -3 \end{bmatrix}\begin{bmatrix} 1 & 0 & 0 \\ 0 & 1 & 0 \\ 0 & 0 & 1 \end{bmatrix} = \begin{bmatrix} 3 & -2 & 5 \\ 1 & 0 & 4 \\ -1 & 2 & -3 \end{bmatrix} \qquad AI = A$$

and

$$\begin{bmatrix} 1 & 0 & 0 \\ 0 & 1 & 0 \\ 0 & 0 & 1 \end{bmatrix}\begin{bmatrix} 3 & -2 & 5 \\ 1 & 0 & 4 \\ -1 & 2 & -3 \end{bmatrix} = \begin{bmatrix} 3 & -2 & 5 \\ 1 & 0 & 4 \\ -1 & 2 & -3 \end{bmatrix}. \qquad IA = A$$

Using Matrices to Transform Vectors

In Section 6.3, you performed vector operations with vectors written in component form and with vectors written as linear combinations of the standard unit vectors **i** and **j**. Another way to perform vector operations is with the vectors written as column matrices.

EXAMPLE 11   Vector Operations

Let $\mathbf{v} = \langle 2, 4 \rangle$ and $\mathbf{w} = \langle 6, 2 \rangle$. Use matrices to find each vector.

a. $\mathbf{v} + \mathbf{w}$ **b.** $\mathbf{w} - 2\mathbf{v}$

Solution   Begin by writing **v** and **w** as column matrices.

$$\mathbf{v} = \begin{bmatrix} 2 \\ 4 \end{bmatrix}, \qquad \mathbf{w} = \begin{bmatrix} 6 \\ 2 \end{bmatrix}$$

a. $\mathbf{v} + \mathbf{w} = \begin{bmatrix} 2 \\ 4 \end{bmatrix} + \begin{bmatrix} 6 \\ 2 \end{bmatrix} = \begin{bmatrix} 8 \\ 6 \end{bmatrix} = \langle 8, 6 \rangle$

Figure 8.1 shows a sketch of $\mathbf{v} + \mathbf{w}$.

b. $\mathbf{w} - 2\mathbf{v} = \begin{bmatrix} 6 \\ 2 \end{bmatrix} - 2\begin{bmatrix} 2 \\ 4 \end{bmatrix} = \begin{bmatrix} 6 \\ 2 \end{bmatrix} - \begin{bmatrix} 4 \\ 8 \end{bmatrix} = \begin{bmatrix} 2 \\ -6 \end{bmatrix} = \langle 2, -6 \rangle$

Figure 8.2 shows a sketch of $\mathbf{w} - 2\mathbf{v} = \mathbf{w} + (-2\mathbf{v})$.

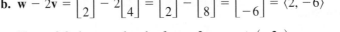 ✓ **Checkpoint** 🔊))) *Audio-video solution in English & Spanish at LarsonPrecalculus.com*

Let $\mathbf{v} = \langle 3, 6 \rangle$ and $\mathbf{w} = \langle 8, 5 \rangle$. Use matrices to find each vector.

a. $\mathbf{v} - \mathbf{w}$ **b.** $3\mathbf{v} + \mathbf{w}$

One way to transform a vector **v** is to multiply **v** by a square **transformation matrix** A to produce another vector $A\mathbf{v}$. A column matrix with two rows can represent a vector **v**, so the transformation matrix must have two columns (and also two rows) for $A\mathbf{v}$ to be defined.

EXAMPLE 12   Describing a Vector Transformation

Find the product $A\mathbf{v}$, where $A = \begin{bmatrix} 1 & 0 \\ 0 & -1 \end{bmatrix}$ and $\mathbf{v} = \langle 1, 3 \rangle$, and describe the transformation.

Solution   First note that A has two columns and **v**, written as the column matrix $\mathbf{v} = \begin{bmatrix} 1 \\ 3 \end{bmatrix}$, has two rows, so $A\mathbf{v}$ is defined.

$$A\mathbf{v} = \begin{bmatrix} 1 & 0 \\ 0 & -1 \end{bmatrix}\begin{bmatrix} 1 \\ 3 \end{bmatrix} = \begin{bmatrix} 1 \\ -3 \end{bmatrix} = \langle 1, -3 \rangle$$

Figure 8.3 shows a sketch of the vectors **v** and $A\mathbf{v}$. The matrix A transforms **v** by reflecting **v** in the x-axis.

 ✓ **Checkpoint** 🔊))) *Audio-video solution in English & Spanish at LarsonPrecalculus.com*

Find the product $A\mathbf{v}$, where $A = \begin{bmatrix} -1 & 0 \\ 0 & 1 \end{bmatrix}$ and $\mathbf{v} = \langle 3, 1 \rangle$, and describe the transformation.

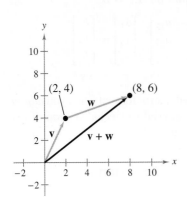

Figure 8.1

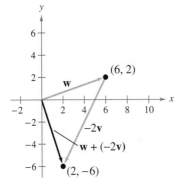

Figure 8.2

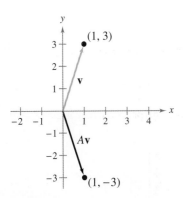

Figure 8.3

Many real-life applications of linear systems involve enormous numbers of equations and variables. For example, a flight crew scheduling problem for American Airlines required the manipulation of matrices with 837 rows and 12,753,313 columns. *(Source: Very Large-Scale Linear Programming. A Case Study in Combining Interior Point and Simplex Methods, Bixby, Robert E., et al., Operations Research, 40, no. 5)*

Applications

Matrix multiplication can be used to represent a system of linear equations. Note how the system below can be written as the matrix equation $AX = B$, where A is the *coefficient matrix* of the system and X and B are column matrices. The column matrix B is also called a *constant matrix*. Its entries are the constant terms in the system of equations.

System

$$\begin{cases} a_{11}x_1 + a_{12}x_2 + a_{13}x_3 = b_1 \\ a_{21}x_1 + a_{22}x_2 + a_{23}x_3 = b_2 \\ a_{31}x_1 + a_{32}x_2 + a_{33}x_3 = b_3 \end{cases}$$

Matrix Equation $AX = B$

$$\underset{A}{\begin{bmatrix} a_{11} & a_{12} & a_{13} \\ a_{21} & a_{22} & a_{23} \\ a_{31} & a_{32} & a_{33} \end{bmatrix}} \underset{X}{\begin{bmatrix} x_1 \\ x_2 \\ x_3 \end{bmatrix}} = \underset{B}{\begin{bmatrix} b_1 \\ b_2 \\ b_3 \end{bmatrix}}$$

In Example 13, $[A \vdots B]$ represents the augmented matrix formed when you *adjoin* matrix B to matrix A. Also, $[I \vdots X]$ represents the reduced row-echelon form of the augmented matrix that yields the solution of the system.

EXAMPLE 13 Solving a System of Linear Equations

For the system of linear equations, (a) write the system as a matrix equation, $AX = B$, and (b) use Gauss-Jordan elimination on $[A \vdots B]$ to solve for the matrix X.

$$\begin{cases} x_1 - 2x_2 + x_3 = -4 \\ x_2 + 2x_3 = 4 \\ 2x_1 + 3x_2 - 2x_3 = 2 \end{cases}$$

Solution

a. In matrix form, $AX = B$, the system is

$$\begin{bmatrix} 1 & -2 & 1 \\ 0 & 1 & 2 \\ 2 & 3 & -2 \end{bmatrix} \begin{bmatrix} x_1 \\ x_2 \\ x_3 \end{bmatrix} = \begin{bmatrix} -4 \\ 4 \\ 2 \end{bmatrix}.$$

b. Form the augmented matrix by adjoining matrix B to matrix A.

$$[A \vdots B] = \begin{bmatrix} 1 & -2 & 1 & \vdots & -4 \\ 0 & 1 & 2 & \vdots & 4 \\ 2 & 3 & -2 & \vdots & 2 \end{bmatrix}$$

Using Gauss-Jordan elimination, rewrite this matrix as

$$[I \vdots X] = \begin{bmatrix} 1 & 0 & 0 & \vdots & -1 \\ 0 & 1 & 0 & \vdots & 2 \\ 0 & 0 & 1 & \vdots & 1 \end{bmatrix}.$$

So, the solution of the matrix equation is

$$X = \begin{bmatrix} x_1 \\ x_2 \\ x_3 \end{bmatrix} = \begin{bmatrix} -1 \\ 2 \\ 1 \end{bmatrix}.$$

✓ *Checkpoint* ◀))) Audio-video solution in English & Spanish at LarsonPrecalculus.com

For the system of linear equations, (a) write the system as a matrix equation, $AX = B$, and (b) use Gauss-Jordan elimination on $[A \vdots B]$ to solve for the matrix X.

$$\begin{cases} -2x_1 - 3x_2 = -4 \\ 6x_1 + x_2 = -36 \end{cases}$$

EXAMPLE 14 Softball Team Expenses

Two softball teams submit equipment lists to their sponsors.

Equipment	Women's Team	Men's Team
Bats	12	15
Balls	45	38
Gloves	15	17

Each bat costs $80, each ball costs $4, and each glove costs $90. Use matrices to find the total cost of equipment for each team.

Solution   Write the equipment lists E and the costs per item C in matrix form as

$$E = \begin{bmatrix} 12 & 15 \\ 45 & 38 \\ 15 & 17 \end{bmatrix}$$

and

$$C = \begin{bmatrix} 80 & 4 & 90 \end{bmatrix}.$$

The total cost of equipment for each team is the product

$$CE = \begin{bmatrix} 80 & 4 & 90 \end{bmatrix} \begin{bmatrix} 12 & 15 \\ 45 & 38 \\ 15 & 17 \end{bmatrix}$$

$$= \begin{bmatrix} 80(12) + 4(45) + 90(15) & 80(15) + 4(38) + 90(17) \end{bmatrix}$$

$$= \begin{bmatrix} 2490 & 2882 \end{bmatrix}.$$

So, the total cost of equipment for the women's team is $2490, and the total cost of equipment for the men's team is $2882.

• • **REMARK** Notice in Example 14 that it is not possible to find the total cost using the product EC because EC is not defined. That is, the number of columns of E (2 columns) does not equal the number of rows of C (1 row).

✓ *Checkpoint* ◀))) *Audio-video solution in English & Spanish at LarsonPrecalculus.com*

Repeat Example 14 when each bat costs $100, each ball costs $3, and each glove costs $65.

Summarize   (Section 8.2)

1. State the conditions under which two matrices are equal *(page 553)*. For an example involving matrix equality, see Example 1.

2. Explain how to add matrices *(page 554)*. For an example of matrix addition, see Example 2.

3. Explain how to multiply a matrix by a scalar *(page 554)*. For an example of scalar multiplication, see Example 3.

4. List the properties of matrix addition and scalar multiplication *(page 556)*. For examples of using these properties, see Examples 4–6.

5. Explain how to multiply two matrices *(page 558)*. For examples of matrix multiplication, see Examples 7–10.

6. Explain how to use matrices to transform vectors *(page 561)*. For examples involving matrices and vectors, see Examples 11 and 12.

7. Describe real-life applications of matrix operations *(pages 562 and 563, Examples 13 and 14)*.

8.2 Exercises

See **CalcChat.com** for tutorial help and worked-out solutions to odd-numbered exercises.

Vocabulary: Fill in the blanks.

1. Two matrices are _____ when their corresponding entries are equal.
2. When performing matrix operations, real numbers are usually referred to as _____.
3. A matrix consisting entirely of zeros is called a _____ matrix and is denoted by _____.
4. The $n \times n$ matrix that consists of 1's on its main diagonal and 0's elsewhere is called the _____ matrix of dimension $n \times n$.

Skills and Applications

 Equality of Matrices In Exercises 5–8, solve for x and y.

5. $\begin{bmatrix} x & -2 \\ 7 & 23 \end{bmatrix} = \begin{bmatrix} -4 & -2 \\ 7 & y \end{bmatrix}$ 6. $\begin{bmatrix} -5 & x \\ 3y & 8 \end{bmatrix} = \begin{bmatrix} -5 & 13 \\ 12 & 8 \end{bmatrix}$

7. $\begin{bmatrix} 16 & 4 & x & 4 \\ 0 & 2 & 4 & 0 \end{bmatrix} = \begin{bmatrix} 16 & 4 & 2x+1 & 4 \\ 0 & 2 & 3y-5 & 0 \end{bmatrix}$

8. $\begin{bmatrix} x+2 & 8 & -3 \\ 1 & 18 & -8 \\ 7 & -2 & y+2 \end{bmatrix} = \begin{bmatrix} 2x+6 & 8 & -3 \\ 1 & 18 & -8 \\ 7 & -2 & x \end{bmatrix}$

 Operations with Matrices In Exercises 9–16, if possible, find (a) $A + B$, (b) $A - B$, (c) $3A$, and (d) $3A - 2B$.

9. $A = \begin{bmatrix} 1 & -1 \\ 2 & -1 \end{bmatrix}$, $B = \begin{bmatrix} 2 & -1 \\ -1 & 8 \end{bmatrix}$

10. $A = \begin{bmatrix} 1 & 2 \\ 2 & 1 \end{bmatrix}$, $B = \begin{bmatrix} -3 & -2 \\ 4 & 2 \end{bmatrix}$

11. $A = \begin{bmatrix} 6 & 0 & 3 \\ -1 & -4 & 0 \end{bmatrix}$, $B = \begin{bmatrix} 8 & -1 \\ 4 & -3 \end{bmatrix}$

12. $A = \begin{bmatrix} 3 \\ 2 \\ -1 \end{bmatrix}$, $B = \begin{bmatrix} -4 & 6 & 2 \end{bmatrix}$

13. $A = \begin{bmatrix} 8 & -1 \\ 2 & 3 \\ -4 & 5 \end{bmatrix}$, $B = \begin{bmatrix} 1 & 6 \\ -1 & -5 \\ 1 & 10 \end{bmatrix}$

14. $A = \begin{bmatrix} 1 & -1 & 3 \\ 0 & 6 & 9 \end{bmatrix}$, $B = \begin{bmatrix} -2 & 0 & -5 \\ -3 & 4 & -7 \end{bmatrix}$

15. $A = \begin{bmatrix} 4 & 5 & -1 & 3 & 4 \\ 1 & 2 & -2 & -1 & 0 \end{bmatrix}$,
$B = \begin{bmatrix} 1 & 0 & -1 & 1 & 0 \\ -6 & 8 & 2 & -3 & -7 \end{bmatrix}$

16. $A = \begin{bmatrix} -1 & 4 & 0 \\ 3 & -2 & 2 \\ 5 & 4 & -1 \\ 0 & 8 & -6 \\ -4 & -1 & 0 \end{bmatrix}$, $B = \begin{bmatrix} -3 & 5 & 1 \\ 2 & -4 & -7 \\ 10 & -9 & -1 \\ 3 & 2 & -4 \\ 0 & 1 & -2 \end{bmatrix}$

 Evaluating an Expression In Exercises 17–22, evaluate the expression.

17. $\begin{bmatrix} -5 & 0 \\ 3 & -6 \end{bmatrix} + \begin{bmatrix} 7 & 1 \\ -2 & -1 \end{bmatrix} + \begin{bmatrix} -10 & -8 \\ 14 & 6 \end{bmatrix}$

18. $\begin{bmatrix} 6 & 8 \\ -1 & 0 \end{bmatrix} + \begin{bmatrix} 0 & 5 \\ -3 & -1 \end{bmatrix} + \begin{bmatrix} -11 & -7 \\ 2 & -1 \end{bmatrix}$

19. $4\left(\begin{bmatrix} -4 & 0 & 1 \\ 0 & 2 & 3 \end{bmatrix} - \begin{bmatrix} 2 & 1 & -2 \\ 3 & -6 & 0 \end{bmatrix} \right)$

20. $\frac{1}{2}\left(\begin{bmatrix} 5 & -2 & 4 & 0 \end{bmatrix} + \begin{bmatrix} 14 & 6 & -18 & 9 \end{bmatrix} \right)$

21. $-3\left(\begin{bmatrix} 0 & -3 \\ 7 & 2 \end{bmatrix} + \begin{bmatrix} -6 & 3 \\ 8 & 1 \end{bmatrix} \right) - 2\begin{bmatrix} 4 & -4 \\ 7 & -9 \end{bmatrix}$

22. $-1\begin{bmatrix} 4 & 11 \\ -2 & -1 \\ 9 & 3 \end{bmatrix} + \frac{1}{6}\left(\begin{bmatrix} -5 & -1 \\ 3 & 4 \\ 0 & 13 \end{bmatrix} + \begin{bmatrix} 7 & 5 \\ -9 & -1 \\ 6 & -1 \end{bmatrix} \right)$

Operations with Matrices In Exercises 23–26, use the matrix capabilities of a graphing utility to evaluate the expression.

23. $\frac{11}{25}\begin{bmatrix} 2 & 5 \\ -1 & -4 \end{bmatrix} + 6\begin{bmatrix} -3 & 0 \\ 2 & 2 \end{bmatrix}$

24. $55\left(\begin{bmatrix} 14 & -11 \\ -22 & 19 \end{bmatrix} - \begin{bmatrix} -8 & 20 \\ 13 & 6 \end{bmatrix} \right)$

25. $-2\begin{bmatrix} 1.23 & 4.19 & -3.85 \\ 7.21 & -2.60 & 6.54 \end{bmatrix} - \begin{bmatrix} 8.35 & -3.02 & 7.30 \\ -0.38 & -5.49 & 1.68 \end{bmatrix}$

26. $-1\begin{bmatrix} 10 & 15 \\ -20 & 10 \\ 12 & 4 \end{bmatrix} + \frac{1}{8}\left(\begin{bmatrix} -13 & 11 \\ 7 & 0 \\ 6 & 9 \end{bmatrix} + \begin{bmatrix} -3 & 13 \\ -3 & 8 \\ -14 & 15 \end{bmatrix} \right)$

 Solving a Matrix Equation In Exercises 27–34, solve for X in the equation, where

$$A = \begin{bmatrix} -2 & 1 & 3 \\ -1 & 0 & 4 \end{bmatrix} \quad \text{and} \quad B = \begin{bmatrix} 0 & 2 & -4 \\ 3 & 0 & 1 \end{bmatrix}.$$

27. $X = 2A + 2B$ 28. $X = 3A - 2B$
29. $2X = 2A - B$ 30. $2X = A + B$
31. $2X + 3A = B$ 32. $3X - 4A = 2B$
33. $4B = -2X - 2A$ 34. $5A = 6B - 3X$

Finding the Product of Two Matrices In Exercises 35–40, if possible, find AB and state the dimension of the result.

35. $A = \begin{bmatrix} -1 & 6 \\ -4 & 5 \\ 0 & 3 \end{bmatrix}$, $B = \begin{bmatrix} 2 & 3 \\ 0 & 9 \end{bmatrix}$

36. $A = \begin{bmatrix} 0 & -1 & 2 \\ 6 & 0 & 3 \\ 7 & -1 & 8 \end{bmatrix}$, $B = \begin{bmatrix} 2 & -1 \\ 4 & -5 \\ 1 & 6 \end{bmatrix}$

37. $A = \begin{bmatrix} 2 & 1 \\ -3 & 4 \\ 1 & 6 \end{bmatrix}$, $B = \begin{bmatrix} 0 & -1 & 0 \\ 4 & 0 & 2 \\ 8 & -1 & 7 \end{bmatrix}$

38. $A = \begin{bmatrix} 1 & 0 & 3 & -2 \\ 6 & 13 & 8 & -17 \end{bmatrix}$, $B = \begin{bmatrix} 1 & 6 \\ 4 & 2 \end{bmatrix}$

39. $A = \begin{bmatrix} 5 & 0 & 0 \\ 0 & -8 & 0 \\ 0 & 0 & 7 \end{bmatrix}$, $B = \begin{bmatrix} \frac{1}{5} & 0 & 0 \\ 0 & -\frac{1}{8} & 0 \\ 0 & 0 & \frac{1}{2} \end{bmatrix}$

40. $A = \begin{bmatrix} 0 & 0 & 5 \\ 0 & 0 & -3 \\ 0 & 0 & 4 \end{bmatrix}$, $B = \begin{bmatrix} 6 & -11 & 4 \\ 8 & 16 & 4 \\ 0 & 0 & 0 \end{bmatrix}$

Finding the Product of Two Matrices In Exercises 41–44, use the matrix capabilities of a graphing utility to find AB, if possible.

41. $A = \begin{bmatrix} 7 & 5 & -4 \\ -2 & 5 & 1 \\ 10 & -4 & -7 \end{bmatrix}$, $B = \begin{bmatrix} 2 & -2 & 3 \\ 8 & 1 & 4 \\ -4 & 2 & -8 \end{bmatrix}$

42. $A = \begin{bmatrix} 11 & -12 & 4 \\ 14 & 10 & 12 \\ 6 & -2 & 9 \end{bmatrix}$, $B = \begin{bmatrix} 12 & 10 \\ -5 & 12 \\ 15 & 16 \end{bmatrix}$

43. $A = \begin{bmatrix} -3 & 8 & -6 & 8 \\ -12 & 15 & 9 & 6 \\ 5 & -1 & 1 & 5 \end{bmatrix}$, $B = \begin{bmatrix} 3 & 1 & 6 \\ 24 & 15 & 14 \\ 16 & 10 & 21 \\ 8 & -4 & 10 \end{bmatrix}$

44. $A = \begin{bmatrix} -2 & 4 & 8 \\ 21 & 5 & 6 \\ 13 & 2 & 6 \end{bmatrix}$, $B = \begin{bmatrix} 2 & 0 \\ -7 & 15 \\ 32 & 14 \\ 0.5 & 1.6 \end{bmatrix}$

Operations with Matrices In Exercises 45–52, if possible, find (a) AB, (b) BA, and (c) A^2.

45. $A = \begin{bmatrix} 1 & 2 \\ 4 & 2 \end{bmatrix}$, $B = \begin{bmatrix} 2 & -1 \\ -1 & 8 \end{bmatrix}$

46. $A = \begin{bmatrix} 6 & 3 \\ -2 & -4 \end{bmatrix}$, $B = \begin{bmatrix} -2 & 0 \\ 2 & 4 \end{bmatrix}$

47. $A = \begin{bmatrix} 5 & -9 & 0 \\ 3 & 0 & -8 \\ -1 & 4 & 11 \end{bmatrix}$, $B = \begin{bmatrix} 1 & 0 & 0 \\ 0 & 1 & 0 \\ 0 & 0 & 1 \end{bmatrix}$

48. $A = \begin{bmatrix} 2 & -2 \\ -3 & 0 \\ 7 & 6 \end{bmatrix}$, $B = \begin{bmatrix} 1 & 0 \\ 0 & 1 \end{bmatrix}$

49. $A = \begin{bmatrix} -4 & -1 \\ 2 & 12 \end{bmatrix}$, $B = \begin{bmatrix} -6 \\ 5 \end{bmatrix}$

50. $A = \begin{bmatrix} 1 & 3 & -2 \\ -5 & 10 & 1 \end{bmatrix}$, $B = \begin{bmatrix} 3 \\ 3 \\ 3 \end{bmatrix}$

51. $A = \begin{bmatrix} 7 \\ 8 \\ -1 \end{bmatrix}$, $B = \begin{bmatrix} 1 & 1 & 2 \end{bmatrix}$

52. $A = \begin{bmatrix} 3 & 2 & 1 & 4 \end{bmatrix}$, $B = \begin{bmatrix} 2 \\ 3 \\ 0 \\ 1 \end{bmatrix}$

Operations with Matrices In Exercises 53–56, evaluate the expression. Use the matrix capabilities of a graphing utility to verify your answer.

53. $\begin{bmatrix} 3 & 1 \\ 0 & -2 \end{bmatrix}\begin{bmatrix} 1 & 0 \\ -2 & 2 \end{bmatrix}\begin{bmatrix} 1 & 0 \\ 2 & 4 \end{bmatrix}$

54. $-3\left(\begin{bmatrix} 6 & 5 & -1 \\ 1 & -2 & 0 \end{bmatrix}\begin{bmatrix} 0 & 3 \\ -1 & -3 \\ 4 & 1 \end{bmatrix}\right)$

55. $\begin{bmatrix} 0 & 2 & -2 \\ 4 & 1 & 2 \end{bmatrix}\left(\begin{bmatrix} 4 & 0 \\ 0 & -1 \\ -1 & 2 \end{bmatrix} + \begin{bmatrix} -2 & 3 \\ -3 & 5 \\ 0 & -3 \end{bmatrix}\right)$

56. $\begin{bmatrix} 3 \\ -1 \\ 5 \\ 7 \end{bmatrix}\left(\begin{bmatrix} 5 & -6 \end{bmatrix} + \begin{bmatrix} 7 & -1 \end{bmatrix} + \begin{bmatrix} -8 & 9 \end{bmatrix}\right)$

Vector Operations In Exercises 57–60, use matrices to find (a) $\mathbf{u} + \mathbf{v}$, (b) $\mathbf{u} - \mathbf{v}$, and (c) $3\mathbf{v} - \mathbf{u}$.

57. $\mathbf{u} = \langle 1, 5 \rangle$, $\mathbf{v} = \langle 3, 2 \rangle$

58. $\mathbf{u} = \langle 4, 2 \rangle$, $\mathbf{v} = \langle 6, -3 \rangle$

59. $\mathbf{u} = \langle -2, 2 \rangle$, $\mathbf{v} = \langle 5, 4 \rangle$

60. $\mathbf{u} = \langle 7, -4 \rangle$, $\mathbf{v} = \langle 2, 1 \rangle$

Describing a Vector Transformation In Exercises 61–66, find $A\mathbf{v}$, where $\mathbf{v} = \langle 4, 2 \rangle$, and describe the transformation.

61. $A = \begin{bmatrix} 1 & 0 \\ 0 & -1 \end{bmatrix}$

62. $A = \begin{bmatrix} -1 & 0 \\ 0 & 1 \end{bmatrix}$

63. $A = \begin{bmatrix} 0 & 1 \\ 1 & 0 \end{bmatrix}$

64. $A = \begin{bmatrix} 0 & -1 \\ -1 & 0 \end{bmatrix}$

65. $A = \begin{bmatrix} 2 & 0 \\ 0 & 1 \end{bmatrix}$

66. $A = \begin{bmatrix} 1 & 0 \\ 0 & 3 \end{bmatrix}$

In Example 2, note that the two systems of linear equations have the *same coefficient matrix A*. Rather than solve the two systems represented by

$$\begin{bmatrix} 1 & 4 & \vdots & 1 \\ -1 & -3 & \vdots & 0 \end{bmatrix}$$

and

$$\begin{bmatrix} 1 & 4 & \vdots & 0 \\ -1 & -3 & \vdots & 1 \end{bmatrix}$$

separately, you can solve them *simultaneously* by *adjoining* the identity matrix to the coefficient matrix to obtain

$$\begin{bmatrix} \overset{A}{} & & & \overset{I}{} & \\ 1 & 4 & \vdots & 1 & 0 \\ -1 & -3 & \vdots & 0 & 1 \end{bmatrix}.$$

This "doubly augmented" matrix can be represented as

$$[A \; \vdots \; I].$$

By applying Gauss-Jordan elimination to this matrix, you can solve *both* systems with a single elimination process.

$$\begin{bmatrix} 1 & 4 & \vdots & 1 & 0 \\ -1 & -3 & \vdots & 0 & 1 \end{bmatrix}$$

$$\begin{matrix} \\ R_1 + R_2 \rightarrow \end{matrix} \begin{bmatrix} 1 & 4 & \vdots & 1 & 0 \\ 0 & 1 & \vdots & 1 & 1 \end{bmatrix}$$

$$\begin{matrix} -4R_2 + R_1 \rightarrow \\ \\ \end{matrix} \begin{bmatrix} 1 & 0 & \vdots & -3 & -4 \\ 0 & 1 & \vdots & 1 & 1 \end{bmatrix}$$

So, from the "doubly augmented" matrix $[A \; \vdots \; I]$, you obtain the matrix $[I \; \vdots \; A^{-1}]$.

$$\begin{bmatrix} \overset{A}{} & & & \overset{I}{} & \\ 1 & 4 & \vdots & 1 & 0 \\ -1 & -3 & \vdots & 0 & 1 \end{bmatrix} \implies \begin{bmatrix} \overset{I}{} & & & \overset{A^{-1}}{} & \\ 1 & 0 & \vdots & -3 & -4 \\ 0 & 1 & \vdots & 1 & 1 \end{bmatrix}$$

This procedure (or algorithm) works for any square matrix that has an inverse.

▷ **TECHNOLOGY** Most graphing utilities can find the inverse of a square matrix. To do so, you may have to use the inverse key ⎡x⁻¹⎤. Consult the user's guide for your graphing utility for specific keystrokes.

Finding an Inverse Matrix

Let A be a square matrix of dimension $n \times n$.

1. Write the $n \times 2n$ matrix that consists of the given matrix A on the left and the $n \times n$ identity matrix I on the right to obtain

   $$[A \; \vdots \; I].$$

2. If possible, row reduce A to I using elementary row operations on the *entire* matrix

   $$[A \; \vdots \; I].$$

   The result will be the matrix

   $$[I \; \vdots \; A^{-1}].$$

   If this is not possible, then A is not invertible.

3. Check your work by multiplying to see that

   $$AA^{-1} = I = A^{-1}A.$$

EXAMPLE 3 **Finding the Inverse of a Matrix**

Find the inverse of

$$A = \begin{bmatrix} 1 & -1 & 0 \\ 1 & 0 & -1 \\ 6 & -2 & -3 \end{bmatrix}.$$

Solution Begin by adjoining the identity matrix to A to form the matrix

$$[A \;\vdots\; I] = \begin{bmatrix} 1 & -1 & 0 & \vdots & 1 & 0 & 0 \\ 1 & 0 & -1 & \vdots & 0 & 1 & 0 \\ 6 & -2 & -3 & \vdots & 0 & 0 & 1 \end{bmatrix}.$$

Use elementary row operations to obtain the form $[I \;\vdots\; A^{-1}]$.

$$\begin{matrix} \\ -R_1 + R_2 \rightarrow \\ -6R_1 + R_3 \rightarrow \end{matrix} \begin{bmatrix} 1 & -1 & 0 & \vdots & 1 & 0 & 0 \\ 0 & 1 & -1 & \vdots & -1 & 1 & 0 \\ 0 & 4 & -3 & \vdots & -6 & 0 & 1 \end{bmatrix}$$

$$\begin{matrix} R_2 + R_1 \rightarrow \\ \\ -4R_2 + R_3 \rightarrow \end{matrix} \begin{bmatrix} 1 & 0 & -1 & \vdots & 0 & 1 & 0 \\ 0 & 1 & -1 & \vdots & -1 & 1 & 0 \\ 0 & 0 & 1 & \vdots & -2 & -4 & 1 \end{bmatrix}$$

$$\begin{matrix} R_3 + R_1 \rightarrow \\ R_3 + R_2 \rightarrow \\ \end{matrix} \begin{bmatrix} 1 & 0 & 0 & \vdots & -2 & -3 & 1 \\ 0 & 1 & 0 & \vdots & -3 & -3 & 1 \\ 0 & 0 & 1 & \vdots & -2 & -4 & 1 \end{bmatrix} = [I \;\vdots\; A^{-1}]$$

So, the matrix A is invertible and its inverse is

$$A^{-1} = \begin{bmatrix} -2 & -3 & 1 \\ -3 & -3 & 1 \\ -2 & -4 & 1 \end{bmatrix}.$$

Check

$$AA^{-1} = \begin{bmatrix} 1 & -1 & 0 \\ 1 & 0 & -1 \\ 6 & -2 & -3 \end{bmatrix} \begin{bmatrix} -2 & -3 & 1 \\ -3 & -3 & 1 \\ -2 & -4 & 1 \end{bmatrix} = \begin{bmatrix} 1 & 0 & 0 \\ 0 & 1 & 0 \\ 0 & 0 & 1 \end{bmatrix} = I$$

REMARK Be sure to check your solution because it is not uncommon to make arithmetic errors when using elementary row operations.

✓ **Checkpoint** ◀))) *Audio-video solution in English & Spanish at LarsonPrecalculus.com*

Find the inverse of

$$A = \begin{bmatrix} 1 & -2 & -1 \\ 0 & -1 & 2 \\ 1 & -2 & 0 \end{bmatrix}.$$

The process shown in Example 3 applies to any $n \times n$ matrix A. When using this algorithm, if the matrix A does not reduce to the identity matrix, then A does not have an inverse. For example, the matrix below has no inverse.

$$A = \begin{bmatrix} 1 & 2 & 0 \\ 3 & -1 & 2 \\ -2 & 3 & -2 \end{bmatrix}$$

To confirm that this matrix has no inverse, adjoin the identity matrix to A to form $[A \;\vdots\; I]$ and try to apply Gauss-Jordan elimination to the matrix. You will find that it is impossible to obtain the identity matrix I on the left. So, A is not invertible.

The Inverse of a 2 × 2 Matrix

Using Gauss-Jordan elimination to find the inverse of a matrix works well (even as a computer technique) for matrices of dimension 3 × 3 or greater. For 2 × 2 matrices, however, many people prefer to use a formula for the inverse rather than Gauss-Jordan elimination. This simple formula, which works *only* for 2 × 2 matrices, is explained as follows. A 2 × 2 matrix A given by

$$A = \begin{bmatrix} a & b \\ c & d \end{bmatrix}$$

is invertible if and only if

$$ad - bc \neq 0.$$

Moreover, if $ad - bc \neq 0$, then the inverse is given by

$$A^{-1} = \frac{1}{ad - bc} \begin{bmatrix} d & -b \\ -c & a \end{bmatrix}. \qquad \text{Formula for the inverse of a 2 × 2 matrix}$$

The denominator

$$ad - bc$$

is the **determinant** of the 2 × 2 matrix A. You will study determinants in the next section.

EXAMPLE 4   Finding the Inverse of a 2 × 2 Matrix

See LarsonPrecalculus.com for an interactive version of this type of example.

If possible, find the inverse of each matrix.

a. $A = \begin{bmatrix} 3 & -1 \\ -2 & 2 \end{bmatrix}$ **b.** $B = \begin{bmatrix} 3 & -1 \\ -6 & 2 \end{bmatrix}$

Solution

a. The determinant of a matrix A is

$$ad - bc = 3(2) - (-1)(-2) = 4.$$

This quantity is not zero, so the matrix is invertible. The inverse is formed by interchanging the entries on the main diagonal, changing the signs of the other two entries, and multiplying by the scalar $\frac{1}{4}$.

$$A^{-1} = \frac{1}{ad - bc} \begin{bmatrix} d & -b \\ -c & a \end{bmatrix} \qquad \text{Formula for the inverse of a 2 × 2 matrix}$$

$$= \frac{1}{4} \begin{bmatrix} 2 & 1 \\ 2 & 3 \end{bmatrix} \qquad \text{Substitute for } a, b, c, d, \text{ and the determinant.}$$

$$= \begin{bmatrix} \dfrac{1}{2} & \dfrac{1}{4} \\ \dfrac{1}{2} & \dfrac{3}{4} \end{bmatrix} \qquad \text{Multiply by the scalar } \tfrac{1}{4}.$$

b. The determinant of matrix B is

$$ad - bc = 3(2) - (-1)(-6) = 0.$$

Because $ad - bc = 0$, B is not invertible.

✓ **Checkpoint** ◀))) *Audio-video solution in English & Spanish at LarsonPrecalculus.com*

If possible, find the inverse of $A = \begin{bmatrix} 5 & -1 \\ 3 & 4 \end{bmatrix}$. ∎

Systems of Linear Equations

You know that a system of linear equations can have exactly one solution, infinitely many solutions, or no solution. If the coefficient matrix A of a *square* system (a system that has the same number of equations as variables) is invertible, then the system has a unique solution, which can be found using an inverse matrix as follows.

A System of Equations with a Unique Solution

If A is an invertible matrix, then the system of linear equations represented by $AX = B$ has a unique solution given by $X = A^{-1}B$.

EXAMPLE 5   Solving a System Using an Inverse Matrix

▷ **TECHNOLOGY** On most graphing utilities, to solve a linear system that has an invertible coefficient matrix, you can use the formula $X = A^{-1}B$. That is, enter the $n \times n$ coefficient matrix $[A]$ and the $n \times 1$ column matrix $[B]$. The solution matrix X is given by

$$[A]^{-1}[B].$$

Use an inverse matrix to solve the system

$$\begin{cases} x + \quad y + \quad z = 10{,}000 \\ 0.06x + 0.075y + 0.095z = \quad 730. \\ x \quad\quad - \quad 2z = \quad 0 \end{cases}$$

Solution   Begin by writing the system in the matrix form $AX = B$.

$$\begin{bmatrix} 1 & 1 & 1 \\ 0.06 & 0.075 & 0.095 \\ 1 & 0 & -2 \end{bmatrix} \begin{bmatrix} x \\ y \\ z \end{bmatrix} = \begin{bmatrix} 10{,}000 \\ 730 \\ 0 \end{bmatrix}$$

Then, use Gauss-Jordan elimination to find A^{-1}.

$$A^{-1} = \begin{bmatrix} 15 & -200 & -2 \\ -21.5 & 300 & 3.5 \\ 7.5 & -100 & -1.5 \end{bmatrix}$$

Finally, multiply B by A^{-1} on the left to obtain the solution.

$$X = A^{-1}B = \begin{bmatrix} 15 & -200 & -2 \\ -21.5 & 300 & 3.5 \\ 7.5 & -100 & -1.5 \end{bmatrix} \begin{bmatrix} 10{,}000 \\ 730 \\ 0 \end{bmatrix} = \begin{bmatrix} 4000 \\ 4000 \\ 2000 \end{bmatrix}$$

The solution of the system is $x = 4000$, $y = 4000$, and $z = 2000$, or $(4000, 4000, 2000)$.

✓ *Checkpoint* ◀))) *Audio-video solution in English & Spanish at LarsonPrecalculus.com*

Use an inverse matrix to solve the system $\begin{cases} 2x + 3y + z = -1 \\ 3x + 3y + z = \quad 1. \\ 2x + 4y + z = -2 \end{cases}$

Summarize   (Section 8.3)

1. State the definition of the inverse of a square matrix *(page 568)*. For an example of how to show that a matrix is the inverse of another matrix, see Example 1.

2. Explain how to find an inverse matrix *(pages 569 and 570)*. For examples of finding inverse matrices, see Examples 2 and 3.

3. State the formula for the inverse of a 2×2 matrix *(page 572)*. For an example of using this formula to find an inverse matrix, see Example 4.

4. Explain how to use an inverse matrix to solve a system of linear equations *(page 573)*. For an example of using an inverse matrix to solve a system of linear equations, see Example 5.

8.3 Exercises

See **CalcChat.com** for tutorial help and worked-out solutions to odd-numbered exercises.

Vocabulary: Fill in the blanks.

1. If there exists an $n \times n$ matrix A^{-1} such that $AA^{-1} = I_n = A^{-1}A$, then A^{-1} is the _____ of A.
2. A matrix that has an inverse is invertible or _____. A matrix that does not have an inverse is _____.
3. A 2×2 matrix is invertible if and only if its _____ is not zero.
4. If A is an invertible matrix, then the system of linear equations represented by $AX = B$ has a unique solution given by $X =$ _____.

Skills and Applications

 The Inverse of a Matrix In Exercises 5–12, show that B is the inverse of A.

5. $A = \begin{bmatrix} 2 & 1 \\ 5 & 3 \end{bmatrix}$, $B = \begin{bmatrix} 3 & -1 \\ -5 & 2 \end{bmatrix}$

6. $A = \begin{bmatrix} 1 & -1 \\ -1 & 2 \end{bmatrix}$, $B = \begin{bmatrix} 2 & 1 \\ 1 & 1 \end{bmatrix}$

7. $A = \begin{bmatrix} 3 & 2 \\ 1 & 4 \end{bmatrix}$, $B = \frac{1}{10}\begin{bmatrix} 4 & -2 \\ -1 & 3 \end{bmatrix}$

8. $A = \begin{bmatrix} 1 & -1 \\ 2 & 3 \end{bmatrix}$, $B = \frac{1}{5}\begin{bmatrix} 3 & 1 \\ -2 & 1 \end{bmatrix}$

9. $A = \begin{bmatrix} 2 & -17 & 11 \\ -1 & 11 & -7 \\ 0 & 3 & -2 \end{bmatrix}$, $B = \begin{bmatrix} 1 & 1 & 2 \\ 2 & 4 & -3 \\ 3 & 6 & -5 \end{bmatrix}$

10. $A = \begin{bmatrix} -4 & 1 & 5 \\ -1 & 2 & 4 \\ 0 & -1 & -1 \end{bmatrix}$,

$B = \frac{1}{4}\begin{bmatrix} -2 & 4 & 6 \\ 1 & -4 & -11 \\ -1 & 4 & 7 \end{bmatrix}$

11. $A = \begin{bmatrix} 2 & 0 & 2 & 1 \\ 3 & 0 & 0 & 1 \\ -1 & 1 & -2 & 1 \\ 3 & -1 & 1 & 0 \end{bmatrix}$,

$B = \frac{1}{3}\begin{bmatrix} -1 & 3 & -2 & -2 \\ -2 & 9 & -7 & -10 \\ 1 & 0 & -1 & -1 \\ 3 & -6 & 6 & 6 \end{bmatrix}$

12. $A = \begin{bmatrix} -1 & 1 & 0 & -1 \\ 1 & -1 & 1 & 0 \\ -1 & 1 & 2 & 0 \\ 0 & -1 & 1 & 1 \end{bmatrix}$,

$B = \frac{1}{3}\begin{bmatrix} -3 & 1 & 1 & -3 \\ -3 & -1 & 2 & -3 \\ 0 & 1 & 1 & 0 \\ -3 & -2 & 1 & 0 \end{bmatrix}$

Finding the Inverse of a Matrix In Exercises 13–24, find the inverse of the matrix, if possible.

13. $\begin{bmatrix} 2 & 1 \\ 5 & 3 \end{bmatrix}$

14. $\begin{bmatrix} 1 & 2 \\ 3 & 7 \end{bmatrix}$

15. $\begin{bmatrix} 1 & -2 \\ 2 & -3 \end{bmatrix}$

16. $\begin{bmatrix} -7 & 33 \\ 4 & -19 \end{bmatrix}$

17. $\begin{bmatrix} 3 & 1 \\ 4 & 2 \end{bmatrix}$

18. $\begin{bmatrix} 4 & -1 \\ -3 & 1 \end{bmatrix}$

19. $\begin{bmatrix} 1 & 1 & 1 \\ 3 & 5 & 4 \\ 3 & 6 & 5 \end{bmatrix}$

20. $\begin{bmatrix} 1 & 2 & 2 \\ 3 & 7 & 9 \\ -1 & -4 & -7 \end{bmatrix}$

21. $\begin{bmatrix} -5 & 0 & 0 \\ 2 & 0 & 0 \\ -1 & 5 & 7 \end{bmatrix}$

22. $\begin{bmatrix} 1 & 0 & 0 \\ 3 & 0 & 0 \\ 2 & 5 & 5 \end{bmatrix}$

23. $\begin{bmatrix} -8 & 0 & 0 & 0 \\ 0 & 1 & 0 & 0 \\ 0 & 0 & 4 & 0 \\ 0 & 0 & 0 & -5 \end{bmatrix}$

24. $\begin{bmatrix} 1 & 3 & -2 & 0 \\ 0 & 2 & 4 & 6 \\ 0 & 0 & -2 & 1 \\ 0 & 0 & 0 & 5 \end{bmatrix}$

Finding the Inverse of a Matrix In Exercises 25–32, use the matrix capabilities of a graphing utility to find the inverse of the matrix, if possible.

25. $\begin{bmatrix} 1 & 2 & -1 \\ 3 & 7 & -10 \\ -5 & -7 & -15 \end{bmatrix}$

26. $\begin{bmatrix} 10 & 5 & -7 \\ -5 & 1 & 4 \\ 3 & 2 & -2 \end{bmatrix}$

27. $\begin{bmatrix} -\frac{1}{2} & \frac{3}{4} & \frac{1}{4} \\ 1 & 0 & -\frac{3}{2} \\ 0 & -1 & \frac{1}{2} \end{bmatrix}$

28. $\begin{bmatrix} -\frac{5}{6} & \frac{1}{3} & \frac{11}{6} \\ 0 & \frac{2}{3} & 2 \\ 1 & -\frac{1}{2} & -\frac{5}{2} \end{bmatrix}$

29. $\begin{bmatrix} 0.1 & 0.2 & 0.3 \\ -0.3 & 0.2 & 0.2 \\ 0.5 & 0.4 & 0.4 \end{bmatrix}$

30. $\begin{bmatrix} 0.6 & 0 & -0.3 \\ 0.7 & -1 & 0.2 \\ 1 & 0 & -0.9 \end{bmatrix}$

31. $\begin{bmatrix} -1 & 0 & 1 & 0 \\ 0 & 2 & 0 & -1 \\ 2 & 0 & -1 & 0 \\ 0 & -1 & 0 & 1 \end{bmatrix}$

32. $\begin{bmatrix} 1 & -2 & -1 & -2 \\ 3 & -5 & -2 & -3 \\ 2 & -5 & -2 & -5 \\ -1 & 4 & 4 & 11 \end{bmatrix}$

Finding the Inverse of a 2 × 2 Matrix In Exercises 33–38, use the formula on page 572 to find the inverse of the 2 × 2 matrix, if possible.

33. $\begin{bmatrix} 2 & 3 \\ -1 & 5 \end{bmatrix}$

34. $\begin{bmatrix} 1 & -2 \\ -3 & 2 \end{bmatrix}$

35. $\begin{bmatrix} -4 & -6 \\ 2 & 3 \end{bmatrix}$

36. $\begin{bmatrix} -12 & 3 \\ 5 & -2 \end{bmatrix}$

37. $\begin{bmatrix} 0.5 & 0.3 \\ 1.5 & 0.6 \end{bmatrix}$

38. $\begin{bmatrix} -1.25 & 0.625 \\ 0.16 & 0.32 \end{bmatrix}$

Solving a System Using an Inverse Matrix In Exercises 39–42, use the inverse matrix found in Exercise 15 to solve the system of linear equations.

39. $\begin{cases} x - 2y = 5 \\ 2x - 3y = 10 \end{cases}$

40. $\begin{cases} x - 2y = 0 \\ 2x - 3y = 3 \end{cases}$

41. $\begin{cases} x - 2y = 4 \\ 2x - 3y = 2 \end{cases}$

42. $\begin{cases} x - 2y = 1 \\ 2x - 3y = -2 \end{cases}$

Solving a System Using an Inverse Matrix In Exercises 43 and 44, use the inverse matrix found in Exercise 19 to solve the system of linear equations.

43. $\begin{cases} x + y + z = 0 \\ 3x + 5y + 4z = 5 \\ 3x + 6y + 5z = 2 \end{cases}$

44. $\begin{cases} x + y + z = -1 \\ 3x + 5y + 4z = 2 \\ 3x + 6y + 5z = 0 \end{cases}$

Solving a System Using an Inverse Matrix In Exercises 45 and 46, use the inverse matrix found in Exercise 32 to solve the system of linear equations.

45. $\begin{cases} x_1 - 2x_2 - x_3 - 2x_4 = 0 \\ 3x_1 - 5x_2 - 2x_3 - 3x_4 = 1 \\ 2x_1 - 5x_2 - 2x_3 - 5x_4 = -1 \\ -x_1 + 4x_2 + 4x_3 + 11x_4 = 2 \end{cases}$

46. $\begin{cases} x_1 - 2x_2 - x_3 - 2x_4 = 1 \\ 3x_1 - 5x_2 - 2x_3 - 3x_4 = -2 \\ 2x_1 - 5x_2 - 2x_3 - 5x_4 = 0 \\ -x_1 + 4x_2 + 4x_3 + 11x_4 = -3 \end{cases}$

Solving a System Using an Inverse Matrix In Exercises 47–54, use an inverse matrix to solve the system of linear equations, if possible.

47. $\begin{cases} 5x + 4y = -1 \\ 2x + 5y = 3 \end{cases}$

48. $\begin{cases} 18x + 12y = 13 \\ 30x + 24y = 23 \end{cases}$

49. $\begin{cases} -0.4x + 0.8y = 1.6 \\ 2x - 4y = 5 \end{cases}$

50. $\begin{cases} 0.2x - 0.6y = 2.4 \\ -x + 1.4y = -8.8 \end{cases}$

51. $\begin{cases} 2.3x - 1.9y = 6 \\ 1.5x + 0.75y = -12 \end{cases}$

52. $\begin{cases} 5.1x - 3.4y = -20 \\ 0.9x - 0.6y = -51 \end{cases}$

53. $\begin{cases} 4x - y + z = -5 \\ 2x + 2y + 3z = 10 \\ 5x - 2y + 6z = 1 \end{cases}$

54. $\begin{cases} 4x - 2y + 3z = -2 \\ 2x + 2y + 5z = 16 \\ 8x - 5y - 2z = 4 \end{cases}$

Using a Graphing Utility In Exercises 55 and 56, use the matrix capabilities of a graphing utility to solve the system of linear equations, if possible.

55. $\begin{cases} 5x - 3y + 2z = 2 \\ 2x + 2y - 3z = 3 \\ x - 7y + 7z = -4 \end{cases}$

56. $\begin{cases} 2x + 3y + 5z = 4 \\ 3x + 5y + 9z = 7 \\ 5x + 9y + 16z = 13 \end{cases}$

Investment Portfolio In Exercises 57 and 58, you invest in AAA-rated bonds, A-rated bonds, and B-rated bonds. The average yields are 4.5% on AAA bonds, 5% on A bonds, and 9% on B bonds. You invest twice as much in B bonds as in A bonds. Let x, y, and z represent the amounts invested in AAA, A, and B bonds, respectively.

$$\begin{cases} x + y + z = \text{(total investment)} \\ 0.045x + 0.05y + 0.09z = \text{(annual return)} \\ 2y - z = 0 \end{cases}$$

Use the inverse of the coefficient matrix of this system to find the amount invested in each type of bond for the given total investment and annual return.

	Total Investment	Annual Return
57.	$10,000	$650
58.	$12,000	$835

Circuit Analysis

In Exercises 59–62, consider the circuit shown in the figure. The currents I_1, I_2, and I_3 (in amperes) are the solution of the system

$$\begin{cases} 2I_1 + 4I_3 = E_1 \\ I_2 + 4I_3 = E_2 \\ I_1 + I_2 - I_3 = 0 \end{cases}$$

where E_1 and E_2 are voltages. Use the inverse of the coefficient matrix of this system to find the unknown currents for the given voltages.

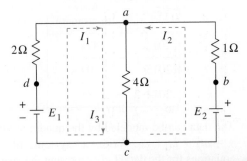

59. $E_1 = 15$ volts, $E_2 = 17$ volts

60. $E_1 = 10$ volts, $E_2 = 10$ volts

61. $E_1 = 28$ volts, $E_2 = 21$ volts

62. $E_1 = 24$ volts, $E_2 = 23$ volts

Raw Materials In Exercises 63 and 64, find the numbers of bags of potting soil that a company can produce for seedlings, general potting, and hardwood plants with the given amounts of raw materials. The raw materials used in one bag of each type of potting soil are shown below.

	Sand	Loam	Peat Moss
Seedlings	2 units	1 unit	1 unit
General	1 unit	2 units	1 unit
Hardwoods	2 units	2 units	2 units

63. 500 units of sand
500 units of loam
400 units of peat moss

64. 500 units of sand
750 units of loam
450 units of peat moss

65. Floral Design A florist is creating 10 centerpieces. Roses cost $2.50 each, lilies cost $4 each, and irises cost $2 each. The customer has a budget of $300 allocated for the centerpieces and wants each centerpiece to contain 12 flowers, with twice as many roses as the number of irises and lilies combined.

(a) Write a system of linear equations that represents the situation. Then write a matrix equation that corresponds to your system.

(b) Solve your system of linear equations using an inverse matrix. Find the number of flowers of each type that the florist can use to create the 10 centerpieces.

66. International Travel The table shows the numbers of visitors y (in thousands) to the United States from China from 2012 through 2014. (*Source: U.S. Department of Commerce*)

Year	Visitors, y (in thousands)
2012	1474
2013	1807
2014	2188

(a) The data can be modeled by the quadratic function $y = at^2 + bt + c$. Write a system of linear equations for the data. Let t represent the year, with $t = 12$ corresponding to 2012.

(b) Use the matrix capabilities of a graphing utility to find the inverse of the coefficient matrix of the system from part (a).

(c) Use the result of part (b) to solve the system and write the model $y = at^2 + bt + c$.

(d) Use the graphing utility to graph the model with the data.

Exploration

True or False? In Exercises 67 and 68, determine whether the statement is true or false. Justify your answer.

67. Multiplication of an invertible matrix and its inverse is commutative.

68. When the product of two square matrices is the identity matrix, the matrices are inverses of one another.

69. Writing Explain how to determine whether the inverse of a 2×2 matrix exists, as well as how to find the inverse when it exists.

70. Writing Explain how to write a system of three linear equations in three variables as a matrix equation $AX = B$, as well as how to solve the system using an inverse matrix.

Think About It In Exercises 71 and 72, find the value of k that makes the matrix singular.

71. $\begin{bmatrix} 4 & 3 \\ -2 & k \end{bmatrix}$

72. $\begin{bmatrix} 2k+1 & 3 \\ -7 & 1 \end{bmatrix}$

73. Conjecture Consider matrices of the form

$$A = \begin{bmatrix} a_{11} & 0 & 0 & 0 & \cdots & 0 \\ 0 & a_{22} & 0 & 0 & \cdots & 0 \\ 0 & 0 & a_{33} & 0 & \cdots & 0 \\ \vdots & \vdots & \vdots & \vdots & & \vdots \\ 0 & 0 & 0 & 0 & \cdots & a_{nn} \end{bmatrix}.$$

(a) Write a 2×2 matrix and a 3×3 matrix in the form of A. Find the inverse of each.

(b) Use the result of part (a) to make a conjecture about the inverses of matrices in the form of A.

74. HOW DO YOU SEE IT?   Consider the matrix

$$A = \begin{bmatrix} x & y \\ 0 & z \end{bmatrix}.$$

Use the determinant of A to state the conditions for which (a) A^{-1} exists and (b) $A^{-1} = A$.

75. Verifying a Formula Verify that the inverse of an invertible 2×2 matrix

$$A = \begin{bmatrix} a & b \\ c & d \end{bmatrix}$$

is given by $A^{-1} = \dfrac{1}{ad - bc} \begin{bmatrix} d & -b \\ -c & a \end{bmatrix}$.

Project: Consumer Credit To work an extended application analyzing the outstanding consumer credit in the United States, visit this text's website at *LarsonPrecalculus.com*. (*Source: Board of Governors of the Federal Reserve System*)

8.4 The Determinant of a Square Matrix

■ Find the determinants of 2 × 2 matrices.
■ Find minors and cofactors of square matrices.
■ Find the determinants of square matrices.

Determinants are often used in other branches of mathematics. For example, the types of determinants in Exercises 87–92 on page 584 occur when changes of variables are made in calculus.

The Determinant of a 2 × 2 Matrix

Every *square* matrix can be associated with a real number called its **determinant.** Determinants have many uses, and several will be discussed in this section and the next section. Historically, the use of determinants arose from special number patterns that occur when systems of linear equations are solved. For example, the system

$$\begin{cases} a_1x + b_1y = c_1 \\ a_2x + b_2y = c_2 \end{cases}$$

has a solution

$$x = \frac{c_1b_2 - c_2b_1}{a_1b_2 - a_2b_1} \quad \text{and} \quad y = \frac{a_1c_2 - a_2c_1}{a_1b_2 - a_2b_1}$$

provided that $a_1b_2 - a_2b_1 \neq 0$. Note that the denominators of the two fractions are the same. This denominator is called the *determinant* of the coefficient matrix of the system.

Coefficient Matrix   **Determinant**

$$A = \begin{bmatrix} a_1 & b_1 \\ a_2 & b_2 \end{bmatrix} \qquad \det(A) = a_1b_2 - a_2b_1$$

The determinant of matrix A can also be denoted by vertical bars on both sides of the matrix, as shown in the definition below.

Definition of the Determinant of a 2 × 2 Matrix

The **determinant** of the matrix

$$A = \begin{bmatrix} a_1 & b_1 \\ a_2 & b_2 \end{bmatrix}$$

is given by

$$\det(A) = |A| = \begin{vmatrix} a_1 & b_1 \\ a_2 & b_2 \end{vmatrix} = a_1b_2 - a_2b_1.$$

In this text, $\det(A)$ and $|A|$ are used interchangeably to represent the determinant of A. Although vertical bars are also used to denote the absolute value of a real number, the context will show which use is intended.

A convenient method for remembering the formula for the determinant of a 2 × 2 matrix is shown below.

$$\det(A) = \begin{vmatrix} a_1 & b_1 \\ a_2 & b_2 \end{vmatrix} = a_1b_2 - a_2b_1$$

Note that the determinant is the difference of the products of the two diagonals of the matrix.

In Example 1, you will see that the determinant of a matrix can be positive, zero, or negative.

EXAMPLE 1 **The Determinant of a 2 × 2 Matrix**

Find the determinant of each matrix.

a. $A = \begin{bmatrix} 2 & -3 \\ 1 & 2 \end{bmatrix}$

b. $B = \begin{bmatrix} 2 & 1 \\ 4 & 2 \end{bmatrix}$

c. $C = \begin{bmatrix} 0 & \frac{3}{2} \\ 2 & 4 \end{bmatrix}$

Solution

a. $\det(A) = \begin{vmatrix} 2 & -3 \\ 1 & 2 \end{vmatrix} = 2(2) - 1(-3) = 4 + 3 = 7$

b. $\det(B) = \begin{vmatrix} 2 & 1 \\ 4 & 2 \end{vmatrix} = 2(2) - 4(1) = 4 - 4 = 0$

c. $\det(C) = \begin{vmatrix} 0 & \frac{3}{2} \\ 2 & 4 \end{vmatrix} = 0(4) - 2\left(\frac{3}{2}\right) = 0 - 3 = -3$

✓ **Checkpoint** ◀))) *Audio-video solution in English & Spanish at LarsonPrecalculus.com*

Find the determinant of each matrix.

a. $A = \begin{bmatrix} 1 & 2 \\ 3 & -1 \end{bmatrix}$

b. $B = \begin{bmatrix} 5 & 0 \\ -4 & 2 \end{bmatrix}$

c. $C = \begin{bmatrix} 3 & 6 \\ 2 & 4 \end{bmatrix}$

The determinant of a matrix of dimension 1×1 is defined simply as the entry of the matrix. For example, if $A = [-2]$, then $\det(A) = -2$.

▷ **TECHNOLOGY** Most graphing utilities can find the determinant of a matrix. For example, to find the determinant of

$$A = \begin{bmatrix} 2.4 & 0.8 \\ -0.6 & -3.2 \end{bmatrix}$$

use the *matrix editor* to enter the matrix as $[A]$ and then choose the *determinant* feature. The result is -7.2, as shown below.

```
[A]
            [2.4    .8]
            [-.6  -3.2]
det([A])
                  -7.2
```

Consult the user's guide for your graphing utility for specific keystrokes.

Minors and Cofactors

To define the determinant of a square matrix of dimension 3×3 or greater, it is helpful to introduce the concepts of **minors** and **cofactors.**

Sign Pattern for Cofactors

$$\begin{bmatrix} + & - & + \\ - & + & - \\ + & - & + \end{bmatrix}$$

3×3 matrix

$$\begin{bmatrix} + & - & + & - \\ - & + & - & + \\ + & - & + & - \\ - & + & - & + \end{bmatrix}$$

4×4 matrix

$$\begin{bmatrix} + & - & + & - & + & \cdots \\ - & + & - & + & - & \cdots \\ + & - & + & - & + & \cdots \\ - & + & - & + & - & \cdots \\ + & - & + & - & + & \cdots \\ \vdots & \vdots & \vdots & \vdots & \vdots & \end{bmatrix}$$

$n \times n$ matrix

> **Minors and Cofactors of a Square Matrix**
>
> If A is a square matrix, then the **minor** M_{ij} of the entry a_{ij} is the determinant of the matrix obtained by deleting the ith row and jth column of A. The **cofactor** C_{ij} of the entry a_{ij} is
>
> $$C_{ij} = (-1)^{i+j} M_{ij}.$$

In the sign pattern for cofactors at the left, notice that *odd* positions (where $i + j$ is odd) have negative signs and *even* positions (where $i + j$ is even) have positive signs.

EXAMPLE 2   **Finding the Minors and Cofactors of a Matrix**

Find all the minors and cofactors of

$$A = \begin{bmatrix} 0 & 2 & 1 \\ 3 & -1 & 2 \\ 4 & 0 & 1 \end{bmatrix}.$$

Solution   To find the minor M_{11}, delete the first row and first column of A and find the determinant of the resulting matrix.

$$\begin{bmatrix} 0 & 2 & 1 \\ 3 & -1 & 2 \\ 4 & 0 & 1 \end{bmatrix}, \quad M_{11} = \begin{vmatrix} -1 & 2 \\ 0 & 1 \end{vmatrix} = -1(1) - 0(2) = -1$$

Similarly, to find M_{12}, delete the first row and second column.

$$\begin{bmatrix} 0 & 2 & 1 \\ 3 & -1 & 2 \\ 4 & 0 & 1 \end{bmatrix}, \quad M_{12} = \begin{vmatrix} 3 & 2 \\ 4 & 1 \end{vmatrix} = 3(1) - 4(2) = -5$$

Continuing this pattern, you obtain the minors.

$$\begin{array}{lll} M_{11} = -1 & M_{12} = -5 & M_{13} = 4 \\ M_{21} = 2 & M_{22} = -4 & M_{23} = -8 \\ M_{31} = 5 & M_{32} = -3 & M_{33} = -6 \end{array}$$

Now, to find the cofactors, combine these minors with the checkerboard pattern of signs for a 3×3 matrix shown at the upper left.

$$\begin{array}{lll} C_{11} = -1 & C_{12} = 5 & C_{13} = 4 \\ C_{21} = -2 & C_{22} = -4 & C_{23} = 8 \\ C_{31} = 5 & C_{32} = 3 & C_{33} = -6 \end{array}$$

✓ *Checkpoint* ◄))) Audio-video solution in English & Spanish at LarsonPrecalculus.com

Find all the minors and cofactors of

$$A = \begin{bmatrix} 1 & 2 & 3 \\ 0 & -1 & 5 \\ 2 & 1 & 4 \end{bmatrix}.$$

The Determinant of a Square Matrix

The definition below is *inductive* because it uses determinants of matrices of dimension $(n - 1) \times (n - 1)$ to define determinants of matrices of dimension $n \times n$.

Determinant of a Square Matrix

If A is a square matrix (of dimension 2×2 or greater), then the determinant of A is the sum of the entries in any row (or column) of A multiplied by their respective cofactors. For example, expanding along the first row yields

$$|A| = a_{11}C_{11} + a_{12}C_{12} + \cdots + a_{1n}C_{1n}.$$

Applying this definition to find a determinant is called **expanding by cofactors.**

Verify that for a 2×2 matrix

$$A = \begin{bmatrix} a_1 & b_1 \\ a_2 & b_2 \end{bmatrix}$$

this definition of the determinant yields

$$|A| = a_1 b_2 - a_2 b_1$$

as previously defined.

EXAMPLE 3   **The Determinant of a 3 × 3 Matrix**

See LarsonPrecalculus.com for an interactive version of this type of example.

Find the determinant of $A = \begin{bmatrix} 0 & 2 & 1 \\ 3 & -1 & 2 \\ 4 & 0 & 1 \end{bmatrix}$.

Solution   Note that this is the same matrix used in Example 2. There you found that the cofactors of the entries in the first row are

$$C_{11} = -1, \quad C_{12} = 5, \quad \text{and} \quad C_{13} = 4.$$

Use the definition of the determinant of a square matrix to expand along the first row.

$$|A| = a_{11}C_{11} + a_{12}C_{12} + a_{13}C_{13} \qquad \text{First-row expansion}$$

$$= 0(-1) + 2(5) + 1(4)$$

$$= 14$$

✓ *Checkpoint* ◀))) *Audio-video solution in English & Spanish at LarsonPrecalculus.com*

Find the determinant of $A = \begin{bmatrix} 3 & 4 & -2 \\ 3 & 5 & 0 \\ -1 & 4 & 1 \end{bmatrix}$.

In Example 3, it was efficient to expand by cofactors along the first row, but any row or column can be used. For example, expanding along the second row gives the same result.

$$|A| = a_{21}C_{21} + a_{22}C_{22} + a_{23}C_{23} \qquad \text{Second-row expansion}$$

$$= 3(-2) + (-1)(-4) + 2(8)$$

$$= 14$$

The goal of Sudoku is to fill in a 9×9 grid so that each column, row, and 3×3 sub-grid contains all the numbers 1 through 9 without repetition. When solved correctly, no two rows or two columns are the same. Note that when a matrix has two rows or two columns that are the same, the determinant is zero.

When expanding by cofactors, you do not need to find cofactors of zero entries, because zero times its cofactor is zero. So, the row (or column) containing the most zeros is usually the best choice for expansion by cofactors. This is demonstrated in the next example.

EXAMPLE 4 The Determinant of a 4×4 Matrix

Find the determinant of $A = \begin{bmatrix} 1 & -2 & 3 & 0 \\ -1 & 1 & 0 & 2 \\ 0 & 2 & 0 & 3 \\ 3 & 4 & 0 & 2 \end{bmatrix}$.

Solution Notice that three of the entries in the third column are zeros. So, to eliminate some of the work in the expansion, expand along the third column.

$$|A| = 3(C_{13}) + 0(C_{23}) + 0(C_{33}) + 0(C_{43})$$

The cofactors C_{23}, C_{33}, and C_{43} have zero coefficients, so the only cofactor you need to find is C_{13}. Start by deleting the first row and third column of A to form the determinant that gives the minor M_{13}.

$$C_{13} = (-1)^{1+3} \begin{vmatrix} -1 & 1 & 2 \\ 0 & 2 & 3 \\ 3 & 4 & 2 \end{vmatrix} \qquad \text{Delete 1st row and 3rd column.}$$

$$= \begin{vmatrix} -1 & 1 & 2 \\ 0 & 2 & 3 \\ 3 & 4 & 2 \end{vmatrix} \qquad \text{Simplify.}$$

Now, expand by cofactors along the second row.

$$C_{13} = 0(-1)^3 \begin{vmatrix} 1 & 2 \\ 4 & 2 \end{vmatrix} + 2(-1)^4 \begin{vmatrix} -1 & 2 \\ 3 & 2 \end{vmatrix} + 3(-1)^5 \begin{vmatrix} -1 & 1 \\ 3 & 4 \end{vmatrix}$$

$$= 0 + 2(1)(-8) + 3(-1)(-7)$$

$$= 5$$

So, $|A| = 3C_{13} = 3(5) = 15$.

✓ **Checkpoint** ◀))) *Audio-video solution in English & Spanish at LarsonPrecalculus.com*

Find the determinant of $A = \begin{bmatrix} 2 & 6 & -4 & 2 \\ 2 & -2 & 3 & 6 \\ 1 & 5 & 0 & 1 \\ 3 & 1 & 0 & -5 \end{bmatrix}$.

Summarize *(Section 8.4)*

1. State the definition of the determinant of a 2×2 matrix *(page 577)*. For an example of finding the determinants of 2×2 matrices, see Example 1.

2. State the definitions of minors and cofactors of a square matrix *(page 579)*. For an example of finding the minors and cofactors of a square matrix, see Example 2.

3. State the definition of the determinant of a square matrix using expanding by cofactors *(page 580)*. For examples of finding determinants using expanding by cofactors, see Examples 3 and 4.

• • Entries Involving Expressions • • • • • • • • •

In Exercises 87–92, find the determinant in which the entries are functions. Determinants of this type occur when changes of variables are made in calculus.

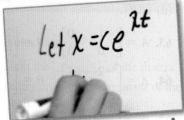

Let $x = Ce^{\lambda t}$

87. $\begin{vmatrix} 4u & -1 \\ -1 & 2v \end{vmatrix}$

88. $\begin{vmatrix} 3x^2 & -3y^2 \\ 1 & 1 \end{vmatrix}$

89. $\begin{vmatrix} e^{2x} & e^{3x} \\ 2e^{2x} & 3e^{3x} \end{vmatrix}$

90. $\begin{vmatrix} e^{-x} & xe^{-x} \\ -e^{-x} & (1-x)e^{-x} \end{vmatrix}$

91. $\begin{vmatrix} x & \ln x \\ 1 & 1/x \end{vmatrix}$

92. $\begin{vmatrix} x & x\ln x \\ 1 & 1+\ln x \end{vmatrix}$

Exploration

True or False? In Exercises 93 and 94, determine whether the statement is true or false. Justify your answer.

93. If a square matrix has an entire row of zeros, then the determinant of the matrix is zero.

94. If the rows of a 2 × 2 matrix are the same, then the determinant of the matrix is zero.

95. **Think About It** Find square matrices A and B such that $|A + B| \neq |A| + |B|$.

96. **Conjecture** Consider square matrices in which the entries are consecutive integers. An example of such a matrix is

$$\begin{bmatrix} 4 & 5 & 6 \\ 7 & 8 & 9 \\ 10 & 11 & 12 \end{bmatrix}.$$

(a) Use the matrix capabilities of a graphing utility to find the determinants of four matrices of this type. Make a conjecture based on the results.

(b) Verify your conjecture.

97. **Error Analysis** Describe the error.

$$\begin{vmatrix} 1 & 1 & 4 \\ 3 & 2 & 0 \\ 2 & 1 & 3 \end{vmatrix} = 3(1)\begin{vmatrix} 1 & 4 \\ 1 & 3 \end{vmatrix} + 2(-1)\begin{vmatrix} 1 & 4 \\ 2 & 3 \end{vmatrix}$$

$$+ 0(1)\begin{vmatrix} 1 & 1 \\ 2 & 1 \end{vmatrix}$$

$$= 3(-1) - 2(-5) + 0$$

$$= 7$$

98. **Think About It** Let A be a 3 × 3 matrix such that $|A| = 5$. Is it possible to find $|2A|$? Explain.

Properties of Determinants In Exercises 99–101, explain why each equation is an example of the given property of determinants (A and B are square matrices). Use a graphing utility to verify the results.

99. If B is obtained from A by interchanging two rows of A or interchanging two columns of A, then $|B| = -|A|$.

(a) $\begin{vmatrix} 1 & 3 & 4 \\ -7 & 2 & -5 \\ 6 & 1 & 2 \end{vmatrix} = -\begin{vmatrix} 1 & 4 & 3 \\ -7 & -5 & 2 \\ 6 & 2 & 1 \end{vmatrix}$

(b) $\begin{vmatrix} 1 & 3 & 4 \\ -2 & 2 & 0 \\ 1 & 6 & 2 \end{vmatrix} = -\begin{vmatrix} 1 & 6 & 2 \\ -2 & 2 & 0 \\ 1 & 3 & 4 \end{vmatrix}$

100. If B is obtained from A by adding a multiple of a row of A to another row of A or by adding a multiple of a column of A to another column of A, then $|B| = |A|$.

(a) $\begin{vmatrix} 1 & -3 \\ 5 & 2 \end{vmatrix} = \begin{vmatrix} 1 & -3 \\ 0 & 17 \end{vmatrix}$

(b) $\begin{vmatrix} 5 & 4 & 2 \\ 2 & -3 & 4 \\ 7 & 6 & 3 \end{vmatrix} = \begin{vmatrix} 1 & 10 & -6 \\ 2 & -3 & 4 \\ 7 & 6 & 3 \end{vmatrix}$

101. If B is obtained from A by multiplying a row by a nonzero constant c or by multiplying a column by a nonzero constant c, then $|B| = c|A|$.

(a) $\begin{vmatrix} 5 & 10 \\ 2 & -3 \end{vmatrix} = 5\begin{vmatrix} 1 & 2 \\ 2 & -3 \end{vmatrix}$

(b) $\begin{vmatrix} 1 & 8 & -3 \\ 3 & -12 & 6 \\ 7 & 4 & 9 \end{vmatrix} = 12\begin{vmatrix} 1 & 2 & -1 \\ 3 & -3 & 2 \\ 7 & 1 & 3 \end{vmatrix}$

102. **HOW DO YOU SEE IT?** Explain why the determinant of each matrix is equal to zero.

(a) $\begin{bmatrix} 2 & -4 & 5 \\ 1 & -2 & 3 \\ 0 & 0 & 0 \end{bmatrix}$

(b) $\begin{bmatrix} 4 & -4 & 5 & 7 \\ 2 & -2 & 3 & 1 \\ 4 & -4 & 5 & 7 \\ 6 & 1 & -3 & -3 \end{bmatrix}$

103. **Conjecture** A **diagonal matrix** is a square matrix in which each entry not on the main diagonal is zero. Find the determinant of each diagonal matrix. Make a conjecture based on your results.

(a) $\begin{bmatrix} 7 & 0 \\ 0 & 4 \end{bmatrix}$ (b) $\begin{bmatrix} -1 & 0 & 0 \\ 0 & 5 & 0 \\ 0 & 0 & 2 \end{bmatrix}$ (c) $\begin{bmatrix} 2 & 0 & 0 & 0 \\ 0 & -2 & 0 & 0 \\ 0 & 0 & 1 & 0 \\ 0 & 0 & 0 & 3 \end{bmatrix}$

8.5 Applications of Matrices and Determinants

- Use Cramer's Rule to solve systems of linear equations.
- Use determinants to find areas of triangles.
- Use determinants to test for collinear points and find equations of lines passing through two points.
- Use 2 × 2 matrices to perform transformations in the plane and find areas of parallelograms.
- Use matrices to encode and decode messages.

Determinants have many applications in real life. For example, in Exercise 21 on page 595, you will use a determinant to find the area of a region of forest infested with gypsy moths.

Cramer's Rule

So far, you have studied four methods for solving a system of linear equations: substitution, graphing, elimination with equations, and elimination with matrices. In this section, you will study one more method, **Cramer's Rule,** named after the Swiss mathematician Gabriel Cramer (1704–1752). This rule uses determinants to write the solution of a system of linear equations. To see how Cramer's Rule works, consider the system described at the beginning of Section 8.4, which is shown below.

$$\begin{cases} a_1x + b_1y = c_1 \\ a_2x + b_2y = c_2 \end{cases}$$

This system has a solution

$$x = \frac{c_1b_2 - c_2b_1}{a_1b_2 - a_2b_1} \quad \text{and} \quad y = \frac{a_1c_2 - a_2c_1}{a_1b_2 - a_2b_1}$$

provided that

$$a_1b_2 - a_2b_1 \neq 0.$$

Each numerator and denominator in this solution can be expressed as a determinant.

$$x = \frac{c_1b_2 - c_2b_1}{a_1b_2 - a_2b_1} = \frac{\begin{vmatrix} c_1 & b_1 \\ c_2 & b_2 \end{vmatrix}}{\begin{vmatrix} a_1 & b_1 \\ a_2 & b_2 \end{vmatrix}} \qquad y = \frac{a_1c_2 - a_2c_1}{a_1b_2 - a_2b_1} = \frac{\begin{vmatrix} a_1 & c_1 \\ a_2 & c_2 \end{vmatrix}}{\begin{vmatrix} a_1 & b_1 \\ a_2 & b_2 \end{vmatrix}}$$

Relative to the original system, the denominators for x and y are the determinant of the *coefficient* matrix of the system. This determinant is denoted by D. The numerators for x and y are denoted by D_x and D_y, respectively, and are formed by using the column of constants as replacements for the coefficients of x and y.

Coefficient Matrix	D	D_x	D_y
$\begin{bmatrix} a_1 & b_1 \\ a_2 & b_2 \end{bmatrix}$	$\begin{vmatrix} a_1 & b_1 \\ a_2 & b_2 \end{vmatrix}$	$\begin{vmatrix} c_1 & b_1 \\ c_2 & b_2 \end{vmatrix}$	$\begin{vmatrix} a_1 & c_1 \\ a_2 & c_2 \end{vmatrix}$

For example, given the system

$$\begin{cases} 2x - 5y = 3 \\ -4x + 3y = 8 \end{cases}$$

the coefficient matrix, D, D_x, and D_y are as follows.

Coefficient Matrix	D	D_x	D_y
$\begin{bmatrix} 2 & -5 \\ -4 & 3 \end{bmatrix}$	$\begin{vmatrix} 2 & -5 \\ -4 & 3 \end{vmatrix}$	$\begin{vmatrix} 3 & -5 \\ 8 & 3 \end{vmatrix}$	$\begin{vmatrix} 2 & 3 \\ -4 & 8 \end{vmatrix}$

Cramer's Rule generalizes to systems of n equations in n variables. The value of each variable is given as the quotient of two determinants. The denominator is the determinant of the coefficient matrix, and the numerator is the determinant of the matrix formed by replacing the column in the coefficient matrix corresponding to the variable being solved for with the column representing the constants. For example, the solution for x_3 in the system below is shown.

$$\begin{cases} a_{11}x_1 + a_{12}x_2 + a_{13}x_3 = b_1 \\ a_{21}x_1 + a_{22}x_2 + a_{23}x_3 = b_2 \\ a_{31}x_1 + a_{32}x_2 + a_{33}x_3 = b_3 \end{cases} \qquad x_3 = \frac{|A_3|}{|A|} = \frac{\begin{vmatrix} a_{11} & a_{12} & b_1 \\ a_{21} & a_{22} & b_2 \\ a_{31} & a_{32} & b_3 \end{vmatrix}}{\begin{vmatrix} a_{11} & a_{12} & a_{13} \\ a_{21} & a_{22} & a_{23} \\ a_{31} & a_{32} & a_{33} \end{vmatrix}}$$

Cramer's Rule

If a system of n linear equations in n variables has a coefficient matrix A with a nonzero determinant $|A|$, then the solution of the system is

$$x_1 = \frac{|A_1|}{|A|}, \quad x_2 = \frac{|A_2|}{|A|}, \quad \ldots, \quad x_n = \frac{|A_n|}{|A|}$$

where the ith column of A_i is the column of constants in the system of equations. If the determinant of the coefficient matrix is zero, then the system has either no solution or infinitely many solutions.

EXAMPLE 1 **Using Cramer's Rule for a 2 × 2 System**

Use Cramer's Rule (if possible) to solve the system

$$\begin{cases} 4x - 2y = 10 \\ 3x - 5y = 11 \end{cases}.$$

Solution To begin, find the determinant of the coefficient matrix.

$$D = \begin{vmatrix} 4 & -2 \\ 3 & -5 \end{vmatrix} = -20 - (-6) = -14$$

This determinant is not zero, so you can apply Cramer's Rule.

$$x = \frac{D_x}{D} = \frac{\begin{vmatrix} 10 & -2 \\ 11 & -5 \end{vmatrix}}{-14} = \frac{-50 - (-22)}{-14} = \frac{-28}{-14} = 2$$

$$y = \frac{D_y}{D} = \frac{\begin{vmatrix} 4 & 10 \\ 3 & 11 \end{vmatrix}}{-14} = \frac{44 - 30}{-14} = \frac{14}{-14} = -1$$

The solution is $(2, -1)$. Check this in the original system.

✓ **Checkpoint** ◀))) *Audio-video solution in English & Spanish at LarsonPrecalculus.com*

Use Cramer's Rule (if possible) to solve the system

$$\begin{cases} 3x + 4y = 1 \\ 5x + 3y = 9 \end{cases}.$$

EXAMPLE 2   **Using Cramer's Rule for a 3 × 3 System**

Use Cramer's Rule (if possible) to solve the system $\begin{cases} -x + 2y - 3z = 1 \\ 2x \qquad\ + z = 0. \\ 3x - 4y + 4z = 2 \end{cases}$

Solution   To find the determinant of the coefficient matrix

$$\begin{bmatrix} -1 & 2 & -3 \\ 2 & 0 & 1 \\ 3 & -4 & 4 \end{bmatrix}$$

expand along the second row.

$$D = 2(-1)^3 \begin{vmatrix} 2 & -3 \\ -4 & 4 \end{vmatrix} + 0(-1)^4 \begin{vmatrix} -1 & -3 \\ 3 & 4 \end{vmatrix} + 1(-1)^5 \begin{vmatrix} -1 & 2 \\ 3 & -4 \end{vmatrix}$$

$$= -2(-4) + 0 - 1(-2)$$

$$= 10$$

This determinant is not zero, so you can apply Cramer's Rule.

$$x = \frac{D_x}{D} = \frac{\begin{vmatrix} 1 & 2 & -3 \\ 0 & 0 & 1 \\ 2 & -4 & 4 \end{vmatrix}}{10} = \frac{8}{10} = \frac{4}{5}$$

$$y = \frac{D_y}{D} = \frac{\begin{vmatrix} -1 & 1 & -3 \\ 2 & 0 & 1 \\ 3 & 2 & 4 \end{vmatrix}}{10} = \frac{-15}{10} = -\frac{3}{2}$$

$$z = \frac{D_z}{D} = \frac{\begin{vmatrix} -1 & 2 & 1 \\ 2 & 0 & 0 \\ 3 & -4 & 2 \end{vmatrix}}{10} = \frac{-16}{10} = -\frac{8}{5}$$

The solution is

$\left(\frac{4}{5}, -\frac{3}{2}, -\frac{8}{5}\right)$.

Check this in the original system.

✓ **Checkpoint** 🔊))) *Audio-video solution in English & Spanish at LarsonPrecalculus.com*

Use Cramer's Rule (if possible) to solve the system $\begin{cases} 4x - y + z = 12 \\ 2x + 2y + 3z = 1. \\ 5x - 2y + 6z = 22 \end{cases}$ ▪

Remember that Cramer's Rule does not apply when the determinant of the coefficient matrix is zero. This would create division by zero, which is undefined. For example, consider the system of linear equations below.

$$\begin{cases} -x \qquad\ + z = 4 \\ 2x - y + z = -3 \\ \qquad y - 3z = 1 \end{cases}$$

The determinant of the coefficient matrix is zero, so you cannot apply Cramer's Rule.

Area of a Triangle

Another application of matrices and determinants is finding the area of a triangle whose vertices are given as three points in a coordinate plane.

Area of a Triangle

The area of a triangle with vertices (x_1, y_1), (x_2, y_2), and (x_3, y_3) is

$$\text{Area} = \pm\frac{1}{2}\begin{vmatrix} x_1 & y_1 & 1 \\ x_2 & y_2 & 1 \\ x_3 & y_3 & 1 \end{vmatrix}$$

where you choose the sign $(\pm)$ so that the area is positive.

For a proof of this formula for the area of a triangle, see Proofs in Mathematics on page 605.

EXAMPLE 3 **Finding the Area of a Triangle**

See LarsonPrecalculus.com for an interactive version of this type of example.

Find the area of the triangle whose vertices are $(1, 0)$, $(2, 2)$, and $(4, 3)$, as shown at the right.

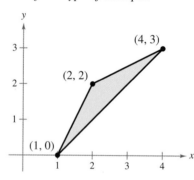

Solution Letting $(x_1, y_1) = (1, 0)$, $(x_2, y_2) = (2, 2)$, and $(x_3, y_3) = (4, 3)$, you have

$$\begin{vmatrix} x_1 & y_1 & 1 \\ x_2 & y_2 & 1 \\ x_3 & y_3 & 1 \end{vmatrix} = \begin{vmatrix} 1 & 0 & 1 \\ 2 & 2 & 1 \\ 4 & 3 & 1 \end{vmatrix}$$

$$= 1(-1)^2\begin{vmatrix} 2 & 1 \\ 3 & 1 \end{vmatrix} + 0(-1)^3\begin{vmatrix} 2 & 1 \\ 4 & 1 \end{vmatrix} + 1(-1)^4\begin{vmatrix} 2 & 2 \\ 4 & 3 \end{vmatrix}$$

$$= -3.$$

Using this value, the area of the triangle is

$$\text{Area} = -\frac{1}{2}\begin{vmatrix} 1 & 0 & 1 \\ 2 & 2 & 1 \\ 4 & 3 & 1 \end{vmatrix} \qquad \text{Choose } (-) \text{ so that the area is positive.}$$

$$= -\frac{1}{2}(-3)$$

$$= \frac{3}{2} \text{ square units.}$$

• • **REMARK** Recall from
Section 6.2 that another way
to find the area of a triangle is
to use Heron's Area Formula.
Verify the result of Example 3
using Heron's Area Formula.
Which method do you prefer?

✓ **Checkpoint** ◄))) *Audio-video solution in English & Spanish at LarsonPrecalculus.com*

Find the area of the triangle whose vertices are $(0, 0)$, $(4, 1)$, and $(2, 5)$.

Lines in a Plane

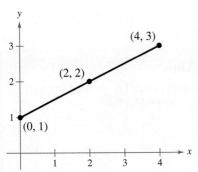

Figure 8.4

In Example 3, what would have happened if the three points were collinear (lying on the same line)? The answer is that the determinant would have been zero. Consider, for example, the three collinear points $(0, 1)$, $(2, 2)$, and $(4, 3)$, as shown in Figure 8.4. The area of the "triangle" that has these three points as vertices is

$$\frac{1}{2}\begin{vmatrix} 0 & 1 & 1 \\ 2 & 2 & 1 \\ 4 & 3 & 1 \end{vmatrix} = \frac{1}{2}\left[0(-1)^2\begin{vmatrix} 2 & 1 \\ 3 & 1 \end{vmatrix} + 1(-1)^3\begin{vmatrix} 2 & 1 \\ 4 & 1 \end{vmatrix} + 1(-1)^4\begin{vmatrix} 2 & 2 \\ 4 & 3 \end{vmatrix} \right]$$

$$= \frac{1}{2}[0 - 1(-2) + 1(-2)]$$

$$= 0.$$

A generalization of this result is below.

Test for Collinear Points

Three points

$$(x_1, y_1), \quad (x_2, y_2), \quad \text{and} \quad (x_3, y_3)$$

are collinear (lie on the same line) if and only if

$$\begin{vmatrix} x_1 & y_1 & 1 \\ x_2 & y_2 & 1 \\ x_3 & y_3 & 1 \end{vmatrix} = 0.$$

EXAMPLE 4   **Testing for Collinear Points**

Determine whether the points

$$(-2, -2), \quad (1, 1), \quad \text{and} \quad (7, 5)$$

are collinear. (See Figure 8.5.)

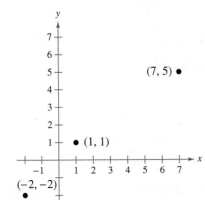

Figure 8.5

Solution   Letting $(x_1, y_1) = (-2, -2)$, $(x_2, y_2) = (1, 1)$, and $(x_3, y_3) = (7, 5)$, you have

$$\begin{vmatrix} x_1 & y_1 & 1 \\ x_2 & y_2 & 1 \\ x_3 & y_3 & 1 \end{vmatrix} = \begin{vmatrix} -2 & -2 & 1 \\ 1 & 1 & 1 \\ 7 & 5 & 1 \end{vmatrix}$$

$$= -2(-1)^2\begin{vmatrix} 1 & 1 \\ 5 & 1 \end{vmatrix} + (-2)(-1)^3\begin{vmatrix} 1 & 1 \\ 7 & 1 \end{vmatrix} + 1(-1)^4\begin{vmatrix} 1 & 1 \\ 7 & 5 \end{vmatrix}$$

$$= -2(-4) + 2(-6) + 1(-2)$$

$$= -6.$$

The value of this determinant is *not* zero, so the three points are not collinear. Note that the area of the triangle with vertices at these points is $\left(-\frac{1}{2}\right)(-6) = 3$ square units.

✓ *Checkpoint* *Audio-video solution in English & Spanish at LarsonPrecalculus.com*

Determine whether the points

$$(-2, 4), \quad (3, -1), \quad \text{and} \quad (6, -4)$$

are collinear.

The test for collinear points can be adapted for another use. Given two points on a rectangular coordinate system, you can find an equation of the line passing through the two points.

Two-Point Form of the Equation of a Line

An equation of the line passing through the distinct points (x_1, y_1) and (x_2, y_2) is given by

$$\begin{vmatrix} x & y & 1 \\ x_1 & y_1 & 1 \\ x_2 & y_2 & 1 \end{vmatrix} = 0.$$

EXAMPLE 5 **Finding an Equation of a Line**

Find an equation of the line passing through the points $(2, 4)$ and $(-1, 3)$, as shown in the figure.

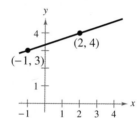

Solution Let $(x_1, y_1) = (2, 4)$ and $(x_2, y_2) = (-1, 3)$. Applying the determinant formula for the equation of a line produces

$$\begin{vmatrix} x & y & 1 \\ 2 & 4 & 1 \\ -1 & 3 & 1 \end{vmatrix} = 0.$$

Evaluate this determinant to find an equation of the line.

$$x(-1)^2 \begin{vmatrix} 4 & 1 \\ 3 & 1 \end{vmatrix} + y(-1)^3 \begin{vmatrix} 2 & 1 \\ -1 & 1 \end{vmatrix} + 1(-1)^4 \begin{vmatrix} 2 & 4 \\ -1 & 3 \end{vmatrix} = 0$$

$$x(1) - y(3) + (1)(10) = 0$$

$$x - 3y + 10 = 0$$

✓ *Checkpoint* 🔊))) *Audio-video solution in English & Spanish at LarsonPrecalculus.com*

Find an equation of the line passing through the points $(-3, -1)$ and $(3, 5)$. ∎

Note that this method of finding an equation of a line works for all lines, including horizontal and vertical lines. For example, an equation of the vertical line passing through $(2, 0)$ and $(2, 2)$ is

$$\begin{vmatrix} x & y & 1 \\ 2 & 0 & 1 \\ 2 & 2 & 1 \end{vmatrix} = 0$$

$$-2x + 4 = 0$$

$$x = 2.$$

Further Applications of 2 × 2 Matrices

In addition to transforming vectors (discussed in Section 8.2), you can use transformation matrices to transform figures in the coordinate plane. Several transformations and their corresponding transformation matrices are listed below.

Transformation Matrices

Reflection in the *y*-axis
$$\begin{bmatrix} -1 & 0 \\ 0 & 1 \end{bmatrix}$$

Reflection in the *x*-axis
$$\begin{bmatrix} 1 & 0 \\ 0 & -1 \end{bmatrix}$$

Horizontal stretch ($k > 1$) or shrink ($0 < k < 1$)
$$\begin{bmatrix} k & 0 \\ 0 & 1 \end{bmatrix}$$

Vertical stretch ($k > 1$) or shrink ($0 < k < 1$)
$$\begin{bmatrix} 1 & 0 \\ 0 & k \end{bmatrix}$$

EXAMPLE 6 **Transforming a Square**

To find the image of the square whose vertices are $(0, 0)$, $(2, 0)$, $(0, 2)$, and $(2, 2)$ after a reflection in the *y*-axis, first write the vertices as column matrices. Then multiply each column matrix by the appropriate transformation matrix on the left.

$$\begin{bmatrix} -1 & 0 \\ 0 & 1 \end{bmatrix}\begin{bmatrix} 0 \\ 0 \end{bmatrix} = \begin{bmatrix} 0 \\ 0 \end{bmatrix} \qquad \begin{bmatrix} -1 & 0 \\ 0 & 1 \end{bmatrix}\begin{bmatrix} 2 \\ 0 \end{bmatrix} = \begin{bmatrix} -2 \\ 0 \end{bmatrix}$$

$$\begin{bmatrix} -1 & 0 \\ 0 & 1 \end{bmatrix}\begin{bmatrix} 0 \\ 2 \end{bmatrix} = \begin{bmatrix} 0 \\ 2 \end{bmatrix} \qquad \begin{bmatrix} -1 & 0 \\ 0 & 1 \end{bmatrix}\begin{bmatrix} 2 \\ 2 \end{bmatrix} = \begin{bmatrix} -2 \\ 2 \end{bmatrix}$$

So, the vertices of the image are $(0, 0)$, $(-2, 0)$, $(0, 2)$, and $(-2, 2)$. Figure 8.6 shows a sketch of the square and its image.

✓ **Checkpoint** ◀))) *Audio-video solution in English & Spanish at LarsonPrecalculus.com*

Find the image of the square in Example 6 after a vertical stretch by a factor of $k = 2$. ■

You can find the area of a parallelogram using the determinant of a 2 × 2 matrix.

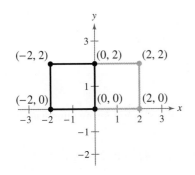

Figure 8.6

Area of a Parallelogram

The area of a parallelogram with vertices $(0, 0)$, (a, b), (c, d), and $(a + c, b + d)$ is

$$\text{Area} = \left|\det(A)\right| \qquad \left|\det(A)\right| \text{ is the absolute value of the determinant.}$$

where $A = \begin{bmatrix} a & b \\ c & d \end{bmatrix}$.

REMARK For an informal *proof without words* of this formula, see Proofs in Mathematics on page 606.

EXAMPLE 7 **Finding the Area of a Parallelogram**

To find the area of the parallelogram shown in Figure 8.7 using the formula above, let $(a, b) = (2, 0)$ and $(c, d) = (1, 3)$. Then

$$A = \begin{bmatrix} 2 & 0 \\ 1 & 3 \end{bmatrix}$$

and the area of the parallelogram is

$$\text{Area} = \left|\det(A)\right| = |6| = 6 \text{ square units.}$$

✓ **Checkpoint** ◀))) *Audio-video solution in English & Spanish at LarsonPrecalculus.com*

Find the area of the parallelogram with vertices $(0, 0)$, $(5, 5)$, $(2, 4)$, and $(7, 9)$. ■

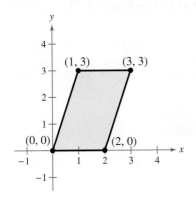

Figure 8.7

Information security is of the utmost importance when conducting business online, and can include the use of data *encryption.* This is the process of encoding information so that the only way to decode it, apart from an "exhaustion attack," is to use a *key.* Data encryption technology uses algorithms based on the material presented here, but on a much more sophisticated level.

Cryptography

A **cryptogram** is a message written according to a secret code. (The Greek word *kryptos* means "hidden.") Matrix multiplication can be used to encode and decode messages. To begin, assign a number to each letter in the alphabet (with 0 assigned to a blank space), as listed below.

0 = _	9 = I	18 = R
1 = A	10 = J	19 = S
2 = B	11 = K	20 = T
3 = C	12 = L	21 = U
4 = D	13 = M	22 = V
5 = E	14 = N	23 = W
6 = F	15 = O	24 = X
7 = G	16 = P	25 = Y
8 = H	17 = Q	26 = Z

Then convert the message to numbers and partition the numbers into **uncoded row matrices,** each having n entries, as demonstrated in Example 8.

EXAMPLE 8   **Forming Uncoded Row Matrices**

Write the uncoded 1×3 row matrices for the message

MEET ME MONDAY.

Solution   Partitioning the message (including blank spaces, but ignoring punctuation) into groups of three produces the uncoded row matrices below.

$$\begin{bmatrix} 13 & 5 & 5 \end{bmatrix} \begin{bmatrix} 20 & 0 & 13 \end{bmatrix} \begin{bmatrix} 5 & 0 & 13 \end{bmatrix} \begin{bmatrix} 15 & 14 & 4 \end{bmatrix} \begin{bmatrix} 1 & 25 & 0 \end{bmatrix}$$
$$\ \ M\ \ \ E\ \ \ E\ \ \ \ \ T\ \ \ \ M\ \ \ E\ \ \ \ \ M\ \ \ O\ \ \ N\ \ \ \ D\ \ \ A\ \ \ Y$$

Note the use of a blank space to fill out the last uncoded row matrix.

✓ *Checkpoint* ◀)))   *Audio-video solution in English & Spanish at LarsonPrecalculus.com*

Write the uncoded 1×3 row matrices for the message

OWLS ARE NOCTURNAL.

To encode a message, create an $n \times n$ invertible matrix A, called an **encoding matrix,** such as

$$A = \begin{bmatrix} 1 & -2 & 2 \\ -1 & 1 & 3 \\ 1 & -1 & -4 \end{bmatrix}.$$

Multiply the uncoded row matrices by A (on the right) to obtain the **coded row matrices.** Here is an example.

Uncoded Matrix	Encoding Matrix A	Coded Matrix

$$\begin{bmatrix} 13 & 5 & 5 \end{bmatrix} \begin{bmatrix} 1 & -2 & 2 \\ -1 & 1 & 3 \\ 1 & -1 & -4 \end{bmatrix} = \begin{bmatrix} 13 & -26 & 21 \end{bmatrix}$$

EXAMPLE 9   **Encoding a Message**

Use the invertible matrix below to encode the message MEET ME MONDAY.

$$A = \begin{bmatrix} 1 & -2 & 2 \\ -1 & 1 & 3 \\ 1 & -1 & -4 \end{bmatrix}$$

Solution   Obtain the coded row matrices by multiplying each of the uncoded row matrices found in Example 8 by the matrix A.

Uncoded Matrix	Encoding Matrix A	Coded Matrix

$$\begin{bmatrix} 13 & 5 & 5 \end{bmatrix} \begin{bmatrix} 1 & -2 & 2 \\ -1 & 1 & 3 \\ 1 & -1 & -4 \end{bmatrix} = \begin{bmatrix} 13 & -26 & 21 \end{bmatrix}$$

$$\begin{bmatrix} 20 & 0 & 13 \end{bmatrix} \begin{bmatrix} 1 & -2 & 2 \\ -1 & 1 & 3 \\ 1 & -1 & -4 \end{bmatrix} = \begin{bmatrix} 33 & -53 & -12 \end{bmatrix}$$

$$\begin{bmatrix} 5 & 0 & 13 \end{bmatrix} \begin{bmatrix} 1 & -2 & 2 \\ -1 & 1 & 3 \\ 1 & -1 & -4 \end{bmatrix} = \begin{bmatrix} 18 & -23 & -42 \end{bmatrix}$$

$$\begin{bmatrix} 15 & 14 & 4 \end{bmatrix} \begin{bmatrix} 1 & -2 & 2 \\ -1 & 1 & 3 \\ 1 & -1 & -4 \end{bmatrix} = \begin{bmatrix} 5 & -20 & 56 \end{bmatrix}$$

$$\begin{bmatrix} 1 & 25 & 0 \end{bmatrix} \begin{bmatrix} 1 & -2 & 2 \\ -1 & 1 & 3 \\ 1 & -1 & -4 \end{bmatrix} = \begin{bmatrix} -24 & 23 & 77 \end{bmatrix}$$

So, the sequence of coded row matrices is

$$\begin{bmatrix} 13 & -26 & 21 \end{bmatrix} \begin{bmatrix} 33 & -53 & -12 \end{bmatrix} \begin{bmatrix} 18 & -23 & -42 \end{bmatrix} \begin{bmatrix} 5 & -20 & 56 \end{bmatrix} \begin{bmatrix} -24 & 23 & 77 \end{bmatrix}.$$

Finally, removing the matrix notation produces the cryptogram

$$13 \quad -26 \quad 21 \quad 33 \quad -53 \quad -12 \quad 18 \quad -23 \quad -42 \quad 5 \quad -20 \quad 56 \quad -24 \quad 23 \quad 77.$$

✓ **Checkpoint**  *Audio-video solution in English & Spanish at LarsonPrecalculus.com*

Use the invertible matrix below to encode the message OWLS ARE NOCTURNAL.

$$A = \begin{bmatrix} 1 & -1 & 0 \\ 1 & 0 & -1 \\ 6 & -2 & -3 \end{bmatrix}$$

If you do not know the encoding matrix A, decoding a cryptogram such as the one found in Example 9 can be difficult. But if you know the encoding matrix A, decoding is straightforward. You just multiply the coded row matrices by A^{-1} (on the right) to obtain the uncoded row matrices. Here is an example.

$$\underbrace{\begin{bmatrix} 13 & -26 & 21 \end{bmatrix}}_{\text{Coded}} \underbrace{\begin{bmatrix} -1 & -10 & -8 \\ -1 & -6 & -5 \\ 0 & -1 & -1 \end{bmatrix}}_{A^{-1}} = \underbrace{\begin{bmatrix} 13 & 5 & 5 \end{bmatrix}}_{\text{Uncoded}}$$

HISTORICAL NOTE

During World War II, Navajo soldiers created a code using their native language to send messages between battalions. The soldiers assigned native words to represent characters in the English alphabet, and they created a number of expressions for important military terms, such as *iron-fish* to mean *submarine*. Without the Navajo Code Talkers, the Second World War might have had a very different outcome.

EXAMPLE 10   **Decoding a Message**

Use the inverse of A in Example 9 to decode the cryptogram

13 −26 21 33 −53 −12 18 −23 −42 5 −20 56 −24 23 77.

Solution   Find the decoding matrix A^{-1}, partition the message into groups of three to form the coded row matrices and multiply each coded row matrix by A^{-1} (on the right).

Coded Matrix	Decoding Matrix A^{-1}	Decoded Matrix

$$[13 \; -26 \;\; 21]\begin{bmatrix} -1 & -10 & -8 \\ -1 & -6 & -5 \\ 0 & -1 & -1 \end{bmatrix} = [13 \;\; 5 \;\; 5]$$

$$[33 \; -53 \; -12]\begin{bmatrix} -1 & -10 & -8 \\ -1 & -6 & -5 \\ 0 & -1 & -1 \end{bmatrix} = [20 \;\; 0 \;\; 13]$$

$$[18 \; -23 \; -42]\begin{bmatrix} -1 & -10 & -8 \\ -1 & -6 & -5 \\ 0 & -1 & -1 \end{bmatrix} = [5 \;\; 0 \;\; 13]$$

$$[5 \; -20 \;\; 56]\begin{bmatrix} -1 & -10 & -8 \\ -1 & -6 & -5 \\ 0 & -1 & -1 \end{bmatrix} = [15 \;\; 14 \;\; 4]$$

$$[-24 \;\; 23 \;\; 77]\begin{bmatrix} -1 & -10 & -8 \\ -1 & -6 & -5 \\ 0 & -1 & -1 \end{bmatrix} = [1 \;\; 25 \;\; 0]$$

So, the message is

[13   5   5] [20   0   13] [5   0   13] [15   14   4] [1   25   0].

M   E   E   T   M   E   M   O   N   D   A   Y

✓ **Checkpoint** 🔊))) *Audio-video solution in English & Spanish at LarsonPrecalculus.com*

Use the inverse of A in the Checkpoint with Example 9 to decode the cryptogram

110 −39 −59 25 −21 −3 23 −18 −5 47 −20 −24
149 −56 −75 87 −38 −37.

Summarize   (Section 8.5)

1. Explain how to use Cramer's Rule to solve systems of linear equations *(page 586)*. For examples of using Cramer's Rule, see Examples 1 and 2.

2. State the formula for finding the area of a triangle using a determinant *(page 588)*. For an example of using this formula to find the area of a triangle, see Example 3.

3. Explain how to use determinants to test for collinear points *(page 589)* and find equations of lines passing through two points *(page 590)*. For examples of these applications, see Examples 4 and 5.

4. Explain how to use 2 × 2 matrices to perform transformations in the plane and find areas of parallelograms *(page 591)*. For examples of these applications, see Examples 6 and 7.

5. Explain how to use matrices to encode and decode messages *(pages 592–594)*. For examples involving encoding and decoding messages, see Examples 8–10.

8.5 Exercises

See **CalcChat.com** for tutorial help and worked-out solutions to odd-numbered exercises.

Vocabulary: Fill in the blanks.

1. The method of using determinants to solve a system of linear equations is called _____ _____.

2. Three points are _____ when they lie on the same line.

3. The area A of a triangle with vertices (x_1, y_1), (x_2, y_2), and (x_3, y_3) is given by _____.

4. A message written according to a secret code is a _____.

5. To encode a message, create an invertible matrix A and multiply the _____ row matrices by A (on the right) to obtain the _____ row matrices.

6. A message encoded using an invertible matrix A can be decoded by multiplying the coded row matrices by _____ (on the right).

Skills and Applications

Using Cramer's Rule **In Exercises 7–14, use Cramer's Rule (if possible) to solve the system of equations.**

7. $\begin{cases} -5x + 9y = -14 \\ 3x - 7y = 10 \end{cases}$ 8. $\begin{cases} 4x - 3y = -10 \\ 6x + 9y = 12 \end{cases}$

9. $\begin{cases} 3x + 2y = -2 \\ 6x + 4y = 4 \end{cases}$ 10. $\begin{cases} 12x - 7y = -4 \\ -11x + 8y = 10 \end{cases}$

11. $\begin{cases} 4x - y + z = -5 \\ 2x + 2y + 3z = 10 \\ 5x - 2y + 6z = 1 \end{cases}$ 12. $\begin{cases} 4x - 2y + 3z = -2 \\ 2x + 2y + 5z = 16 \\ 8x - 5y - 2z = 4 \end{cases}$

13. $\begin{cases} x + 2y + 3z = -3 \\ -2x + y - z = 6 \\ 3x - 3y + 2z = -11 \end{cases}$ 14. $\begin{cases} 5x - 4y + z = -14 \\ -x + 2y - 2z = 10 \\ 3x + y + z = 1 \end{cases}$

Finding the Area of a Triangle **In Exercises 15–18, use a determinant to find the area of the triangle with the given vertices.**

15.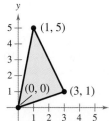

16.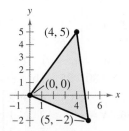

17. $(0, 4), (-2 -3), (2, -3)$

18. $(-2, 1), (1, 6), (3, -1)$

Finding a Coordinate **In Exercises 19 and 20, find a value of y such that the triangle with the given vertices has an area of 4 square units.**

19. $(-5, 1), (0, 2), (-2, y)$

20. $(-4, 2), (-3, 5), (-1, y)$

• **21. Area of Infestation**
A large region of forest is infested with gypsy moths. The region is triangular, as shown in the figure. From vertex A, the distances to the other vertices are 25 miles south

and 10 miles east (for vertex B), and 20 miles south and 28 miles east (for vertex C). Use a graphing utility to find the area (in square miles) of the region.

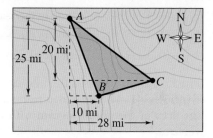

22. Botany A botanist is studying the plants growing in the triangular region shown in the figure. Starting at vertex A, the botanist walks 65 feet east and 50 feet north to vertex B, and then walks 85 feet west and 30 feet north to vertex C. Use a graphing utility to find the area (in square feet) of the region.

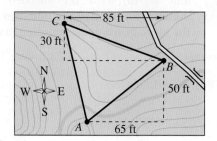

Testing for Collinear Points In Exercises 23–28, use a determinant to determine whether the points are collinear.

23. $(2, -6), (0, -2), (3, -8)$

24. $(3, -5), (6, 1), (4, 2)$

25. $\left(2, -\frac{1}{2}\right), (-4, 4), (6, -3)$

26. $(0, 1), \left(-2, \frac{7}{2}\right), \left(1, -\frac{1}{4}\right)$

27. $(0, 2), (1, 2.4), (-1, 1.6)$

28. $(3, 7), (4, 9.5), (-1, -5)$

Finding a Coordinate In Exercises 29 and 30, find the value of y such that the points are collinear.

29. $(2, -5), (4, y), (5, -2)$ **30.** $(-6, 2), (-5, y), (-3, 5)$

Finding an Equation of a Line In Exercises 31–36, use a determinant to find an equation of the line passing through the points.

31. $(0, 0), (5, 3)$

32. $(0, 0), (-2, 2)$

33. $(-4, 3), (2, 1)$

34. $(10, 7), (-2, -7)$

35. $\left(-\frac{1}{2}, 3\right), \left(\frac{5}{2}, 1\right)$

36. $\left(\frac{2}{3}, 4\right), (6, 12)$

Transforming a Square In Exercises 37–40, use matrices to find the vertices of the image of the square with the given vertices after the given transformation. Then sketch the square and its image.

37. $(0, 0), (0, 3), (3, 0), (3, 3)$; horizontal stretch, $k = 2$

38. $(1, 2), (3, 2), (1, 4), (3, 4)$; reflection in the x-axis

39. $(4, 3), (5, 3), (4, 4), (5, 4)$; reflection in the y-axis

40. $(1, 1), (3, 2), (0, 3), (2, 4)$; vertical shrink, $k = \frac{1}{2}$

Finding the Area of a Parallelogram In Exercises 41–44, use a determinant to find the area of the parallelogram with the given vertices.

41. $(0, 0), (1, 0), (2, 2), (3, 2)$

42. $(0, 0), (3, 0), (4, 1), (7, 1)$

43. $(0, 0), (-2, 0), (3, 5), (1, 5)$

44. $(0, 0), (0, 8), (8, -6), (8, 2)$

Encoding a Message In Exercises 45 and 46, (a) write the uncoded 1×2 row matrices for the message, and then (b) encode the message using the encoding matrix.

Message	Encoding Matrix
45. COME HOME SOON	$\begin{bmatrix} 1 & 2 \\ 3 & 5 \end{bmatrix}$
46. HELP IS ON THE WAY	$\begin{bmatrix} -2 & 3 \\ -1 & 1 \end{bmatrix}$

Encoding a Message In Exercises 47 and 48, (a) write the uncoded 1×3 row matrices for the message, and then (b) encode the message using the encoding matrix.

Message	Encoding Matrix
47. CALL ME TOMORROW	$\begin{bmatrix} 1 & -1 & 0 \\ 1 & 0 & -1 \\ -6 & 2 & 3 \end{bmatrix}$
48. PLEASE SEND MONEY	$\begin{bmatrix} 4 & 2 & 1 \\ -3 & -3 & -1 \\ 3 & 2 & 1 \end{bmatrix}$

Encoding a Message In Exercises 49–52, write a cryptogram for the message using the matrix

$$A = \begin{bmatrix} 1 & 2 & 2 \\ 3 & 7 & 9 \\ -1 & -4 & -7 \end{bmatrix}.$$

49. LANDING SUCCESSFUL

50. ICEBERG DEAD AHEAD

51. HAPPY BIRTHDAY

52. OPERATION OVERLOAD

Decoding a Message In Exercises 53–56, use A^{-1} to decode the cryptogram.

53. $A = \begin{bmatrix} 1 & 2 \\ 3 & 5 \end{bmatrix}$

11   21   64   112   25   50   29   53   23   46   40
75   55   92

54. $A = \begin{bmatrix} 2 & 3 \\ 3 & 4 \end{bmatrix}$

85   120   6   8   10   15   84   117   42   56   90
125   60   80   30   45   19   26

55. $A = \begin{bmatrix} 1 & -1 & 0 \\ 1 & 0 & -1 \\ -6 & 2 & 3 \end{bmatrix}$

9   -1   -9   38   -19   -19   28   -9   -19
-80   25   41   -64   21   31   9   -5   -4

56. $A = \begin{bmatrix} 3 & -4 & 2 \\ 0 & 2 & 1 \\ 4 & -5 & 3 \end{bmatrix}$

112   -140   83   19   -25   13   72   -76   61   95
-118   71   20   21   38   35   -23   36   42   -48   32

Decoding a Message In Exercises 57 and 58, decode the cryptogram by using the inverse of A in Exercises 49–52.

57. 20   17   -15   -12   -56   -104   1   -25   -65
62   143   181

58. 13   -9   -59   61   112   106   -17   -73
-131   11   24   29   65   144   172

59. Decoding a Message The cryptogram below was encoded with a 2×2 matrix.

8 21 −15 −10 −13 −13 5 10 5 25
5 19 −1 6 20 40 −18 −18 1 16

The last word of the message is _RON. What is the message?

60. Decoding a Message The cryptogram below was encoded with a 2×2 matrix.

5 2 25 11 −2 −7 −15 −15 32 14
−8 −13 38 19 −19 −19 37 16

The last word of the message is _SUE. What is the message?

61. Circuit Analysis Consider the circuit shown in the figure. The currents I_1, I_2, and I_3 (in amperes) are the solution of the system

$$\begin{cases} 4I_1 \quad\;\; + 8I_3 = 2 \\ \quad\;\; 2I_2 + 8I_3 = 6. \\ I_1 + I_2 - I_3 = 0 \end{cases}$$

Use Cramer's Rule to find the three currents.

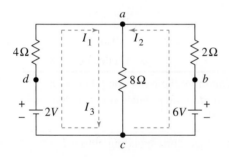

62. Pulley System A system of pulleys is loaded with 192-pound and 64-pound weights (see figure). The tensions t_1 and t_2 in the ropes and the acceleration a of the 64-pound weight are found by solving the system of equations

$$\begin{cases} t_1 - 2t_2 \qquad = \quad 0 \\ t_1 \qquad - 3a = 192 \\ \quad\; t_2 + 2a = \quad 64 \end{cases}$$

where t_1 and t_2 are measured in pounds and a is in feet per second squared. Use Cramer's Rule to find t_1, t_2, and a.

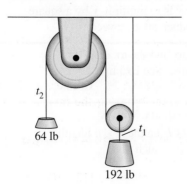

64 lb

192 lb

Exploration

True or False? In Exercises 63 and 64, determine whether the statement is true or false. Justify your answer.

63. In Cramer's Rule, the numerator is the determinant of the coefficient matrix.

64. Cramer's Rule cannot be used to solve a system of linear equations when the determinant of the coefficient matrix is zero.

65. Error Analysis Describe the error.

Consider the system

$$\begin{cases} 2x - 3y = 0 \\ 4x - 6y = 0 \end{cases}.$$

The determinant of the coefficient matrix is

$$D = \begin{vmatrix} 2 & -3 \\ 4 & -6 \end{vmatrix}$$

$$= -12 - (-12)$$

$$= 0$$

so the system has no solution.

66. 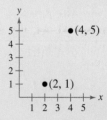 **HOW DO YOU SEE IT?** At this point in the text, you know several methods for finding an equation of a line that passes through two given points. Briefly describe the methods that can be used to find an equation of the line that passes through the two points shown. Discuss the advantages and disadvantages of each method.

67. Finding the Area of a Triangle Use a determinant to find the area of the triangle whose vertices are $(3, -1)$, $(7, -1)$, and $(7, 5)$. Confirm your answer by plotting the points in a coordinate plane and using the formula

$$\text{Area} = \tfrac{1}{2}(\text{base})(\text{height}).$$

68. Writing Use your school's library, the Internet, or some other reference source to research a few current real-life uses of cryptography. Write a short summary of these uses. Include a description of how messages are encoded and decoded in each case.

Chapter Summary

	What Did You Learn?	**Explanation/Examples**	**Review Exercises**
Section 8.1	Write matrices and determine their dimensions (*p. 540*).	$\begin{bmatrix} -1 & 1 \\ 4 & 7 \end{bmatrix}$ $\begin{bmatrix} -2 & 3 & 0 \end{bmatrix}$ $\begin{bmatrix} 4 & -3 \\ 5 & 0 \\ -2 & 1 \end{bmatrix}$ $\begin{bmatrix} 8 \\ -8 \end{bmatrix}$ 2×2 $\qquad$ 1×3 $\qquad$ 3×2 $\quad$ 2×1	1–8
	Perform elementary row operations on matrices (*p. 542*).	**Elementary Row Operations** 1. Interchange two rows. 2. Multiply a row by a nonzero constant. 3. Add a multiple of a row to another row.	9, 10
	Use matrices and Gaussian elimination to solve systems of linear equations (*p. 543*).	**Gaussian Elimination with Back-Substitution** 1. Write the augmented matrix of the system of linear equations. 2. Use elementary row operations to rewrite the augmented matrix in row-echelon form. 3. Write the system of linear equations corresponding to the matrix in row-echelon form and use back-substitution to find the solution.	11–26
	Use matrices and Gauss-Jordan elimination to solve systems of linear equations (*p. 547*).	Gauss-Jordan elimination continues the reduction process on a matrix in row-echelon form until the *reduced* row-echelon form is obtained. (See Example 8.)	27–32
Section 8.2	Determine whether two matrices are equal (*p. 553*).	Two matrices are equal when their corresponding entries are equal.	33–36
	Add and subtract matrices and multiply matrices by scalars (*p. 554*).	If $A = [a_{ij}]$ and $B = [b_{ij}]$ are matrices of dimension $m \times n$, then their sum is the $m \times n$ matrix $A + B = [a_{ij} + b_{ij}]$. If $A = [a_{ij}]$ is an $m \times n$ matrix and c is a scalar, then the scalar multiple of A by c is the $m \times n$ matrix $cA = [ca_{ij}]$.	37–48
	Multiply two matrices (*p. 558*).	If $A = [a_{ij}]$ is an $m \times n$ matrix and $B = [b_{ij}]$ is an $n \times p$ matrix, then the product AB is an $m \times p$ matrix given by $$AB = [c_{ij}]$$ where $c_{ij} = a_{i1}b_{1j} + a_{i2}b_{2j} + a_{i3}b_{3j} + \cdots + a_{in}b_{nj}$.	49–58
	Use matrices to transform vectors (*p. 561*).	One way to transform a vector $\mathbf{v}$ is to multiply $\mathbf{v}$ by a square transformation matrix A to produce another vector $A\mathbf{v}$.	59–62
	Use matrix operations to model and solve real-life problems (*p. 562*).	Matrix operations can be used to find the total cost of equipment for two softball teams. (See Example 14.)	63, 64
Section 8.3	Verify that two matrices are inverses of each other (*p. 568*).	**Definition of the Inverse of a Square Matrix** Let A be an $n \times n$ matrix and let I_n be the $n \times n$ identity matrix. If there exists a matrix A^{-1} such that $$AA^{-1} = I_n = A^{-1}A$$ then A^{-1} is the inverse of A.	65–68

	What Did You Learn?	**Explanation/Examples**	**Review Exercises**
Section 8.3	Use Gauss-Jordan elimination to find the inverses of matrices *(p. 570)*.	**Finding an Inverse Matrix** Let A be a square matrix of dimension $n \times n$. **1.** Write the $n \times 2n$ matrix that consists of the given matrix A on the left and the $n \times n$ identity matrix I on the right to obtain $[A \;\vdots\; I]$. **2.** If possible, row reduce A to I using elementary row operations on the *entire* matrix $[A \;\vdots\; I]$. The result will be the matrix $[I \;\vdots\; A^{-1}]$. If this is not possible, then A is not invertible. **3.** Check your work by multiplying to see that $AA^{-1} = I = A^{-1}A$.	69–74
	Use a formula to find the inverses of 2×2 matrices *(p. 572)*.	If $A = \begin{bmatrix} a & b \\ c & d \end{bmatrix}$ and $ad - bc \neq 0$, then $$A^{-1} = \frac{1}{ad - bc}\begin{bmatrix} d & -b \\ -c & a \end{bmatrix}.$$	75–78
	Use inverse matrices to solve systems of linear equations *(p. 573)*.	If A is an invertible matrix, then the system of linear equations represented by $AX = B$ has a unique solution given by $X = A^{-1}B$.	79–92
Section 8.4	Find the determinants of 2×2 matrices *(p. 577)*.	The determinant of the matrix $A = \begin{bmatrix} a_1 & b_1 \\ a_2 & b_2 \end{bmatrix}$ is given by $$\det(A) = |A| = \begin{vmatrix} a_1 & b_1 \\ a_2 & b_2 \end{vmatrix} = a_1b_2 - a_2b_1.$$	93–96
	Find minors and cofactors of square matrices *(p. 579)*.	If A is a square matrix, then the minor M_{ij} of the entry a_{ij} is the determinant of the matrix obtained by deleting the ith row and jth column of A. The cofactor C_{ij} of the entry a_{ij} is $C_{ij} = (-1)^{i+j}M_{ij}$.	97–100
	Find the determinants of square matrices *(p. 580)*.	If A is a square matrix (of dimension 2×2 or greater), then the determinant of A is the sum of the entries in any row (or column) of A multiplied by their respective cofactors.	101–106
Section 8.5	Use Cramer's Rule to solve systems of linear equations *(p. 586)*.	Cramer's Rule uses determinants to write the solution of a system of linear equations.	107–110
	Use determinants to find areas of triangles *(p. 588)*, test for collinear points *(p. 589)*, and find equations of lines passing through two points *(p. 590)*.	The area of a triangle with vertices (x_1, y_1), (x_2, y_2), and (x_3, y_3) is $$\text{Area} = \pm\frac{1}{2}\begin{vmatrix} x_1 & y_1 & 1 \\ x_2 & y_2 & 1 \\ x_3 & y_3 & 1 \end{vmatrix}$$ where you choose the sign ($\pm$) so that the area is positive.	111–118
	Use 2×2 matrices to perform transformations in the plane and find areas of parallelograms *(p. 591)*.	The area of a parallelogram with vertices $(0, 0)$, (a, b), (c, d), and $(a + c, b + d)$ is $$\text{Area} = |\det(A)|, \text{ where } A = \begin{bmatrix} a & b \\ c & d \end{bmatrix}.$$	119, 120
	Use matrices to encode and decode messages *(p. 592)*.	The inverse of a matrix can be used to decode a cryptogram. (See Example 10.)	121, 122

Review Exercises

See **CalcChat.com** for tutorial help and worked-out solutions to odd-numbered exercises.

8.1 **Dimension of a Matrix** In Exercises 1–4, determine the dimension of the matrix.

1. $\begin{bmatrix} -1 & 3 \end{bmatrix}$

2. $\begin{bmatrix} 3 & 1 \\ 5 & -2 \end{bmatrix}$

3. $\begin{bmatrix} 2 & 1 & 0 & 4 & -1 \\ 6 & 2 & 1 & 8 & 0 \end{bmatrix}$

4. $\begin{bmatrix} 5 \end{bmatrix}$

Writing an Augmented Matrix In Exercises 5 and 6, write the augmented matrix for the system of linear equations.

5. $\begin{cases} 3x - 10y = 15 \\ 5x + 4y = 22 \end{cases}$

6. $\begin{cases} 8x - 7y + 4z = 12 \\ 3x - 5y + 2z = 20 \end{cases}$

Writing a System of Equations In Exercises 7 and 8, write the system of linear equations represented by the augmented matrix. (Use variables x, y, z, and w, if applicable.)

7. $\begin{bmatrix} 1 & 0 & 2 & \vdots & -8 \\ 2 & -2 & 3 & \vdots & 12 \\ 4 & 7 & 1 & \vdots & 3 \end{bmatrix}$

8. $\begin{bmatrix} 2 & 10 & 8 & 5 & \vdots & -1 \\ -3 & 4 & 0 & 9 & \vdots & 2 \end{bmatrix}$

Writing a Matrix in Row-Echelon Form In Exercises 9 and 10, write the matrix in row-echelon form. (Remember that the row-echelon form of a matrix is not unique.)

9. $\begin{bmatrix} 0 & 1 & 1 \\ 1 & 2 & 3 \\ 2 & 2 & 2 \end{bmatrix}$

10. $\begin{bmatrix} 4 & 8 & 16 \\ 3 & -1 & 2 \\ -2 & 10 & 12 \end{bmatrix}$

Using Back-Substitution In Exercises 11–14, write the system of linear equations represented by the augmented matrix. Then use back-substitution to solve the system. (Use variables x, y, and z, if applicable.)

11. $\begin{bmatrix} 1 & 2 & 3 & \vdots & 9 \\ 0 & 1 & -2 & \vdots & 2 \\ 0 & 0 & 1 & \vdots & -1 \end{bmatrix}$

12. $\begin{bmatrix} 1 & 3 & -9 & \vdots & 4 \\ 0 & 1 & -1 & \vdots & 10 \\ 0 & 0 & 1 & \vdots & -2 \end{bmatrix}$

13. $\begin{bmatrix} 1 & 3 & 4 & \vdots & 1 \\ 0 & 1 & 2 & \vdots & 3 \\ 0 & 0 & 1 & \vdots & 4 \end{bmatrix}$

14. $\begin{bmatrix} 1 & -8 & 0 & \vdots & -2 \\ 0 & 1 & -1 & \vdots & -7 \\ 0 & 0 & 1 & \vdots & 1 \end{bmatrix}$

Gaussian Elimination with Back-Substitution In Exercises 15–26, use matrices to solve the system of linear equations, if possible. Use Gaussian elimination with back-substitution.

15. $\begin{cases} 5x + 4y = 2 \\ -x + y = -22 \end{cases}$

16. $\begin{cases} 2x - 5y = 2 \\ 3x - 7y = 1 \end{cases}$

17. $\begin{cases} 0.3x - 0.1y = -0.13 \\ 0.2x - 0.3y = -0.25 \end{cases}$

18. $\begin{cases} 0.2x - 0.1y = 0.07 \\ 0.4x - 0.5y = -0.01 \end{cases}$

19. $\begin{cases} -x + 2y = 3 \\ 2x - 4y = 6 \end{cases}$

20. $\begin{cases} -x + 2y = 3 \\ 2x - 4y = -6 \end{cases}$

21. $\begin{cases} x - 2y + z = 7 \\ 2x + y - 2z = -4 \\ -x + 3y + 2z = -3 \end{cases}$

22. $\begin{cases} x - 2y + z = 4 \\ 2x + y - 2z = -24 \\ -x + 3y + 2z = 20 \end{cases}$

23. $\begin{cases} 2x + y + 2z = 4 \\ 2x + 2y = 5 \\ 2x - y + 6z = 2 \end{cases}$

24. $\begin{cases} x + 2y + 6z = 1 \\ 2x + 5y + 15z = 4 \\ 3x + y + 3z = -6 \end{cases}$

25. $\begin{cases} 2x + 3y + z = 10 \\ 2x - 3y - 3z = 22 \\ 4x - 2y + 3z = -2 \end{cases}$

26. $\begin{cases} 2x + 3y + 3z = 3 \\ 6x + 6y + 12z = 13 \\ 12x + 9y - z = 2 \end{cases}$

Gauss-Jordan Elimination In Exercises 27–30, use matrices to solve the system of linear equations, if possible. Use Gauss-Jordan elimination.

27. $\begin{cases} x + 2y - z = 3 \\ x - y - z = -3 \\ 2x + y + 3z = 10 \end{cases}$

28. $\begin{cases} x - 3y + z = 2 \\ 3x - y - z = -6 \\ -x + y - 3z = -2 \end{cases}$

29. $\begin{cases} -x + y + 2z = 1 \\ 2x + 3y + z = -2 \\ 5x + 4y + 2z = 4 \end{cases}$

30. $\begin{cases} 4x + 4y + 4z = 5 \\ 4x - 2y - 8z = 1 \\ 5x + 3y + 8z = 6 \end{cases}$

Using a Graphing Utility In Exercises 31 and 32, use the matrix capabilities of a graphing utility to write the augmented matrix corresponding to the system of linear equations in reduced row-echelon form. Then solve the system, if possible.

31. $\begin{cases} 3x - y + 5z - 2w = -44 \\ x + 6y + 4z - w = 1 \\ 5x - y + z + 3w = -15 \\ 4y - z - 8w = 58 \end{cases}$

32. $\begin{cases} 4x + 12y + 2z = 20 \\ x + 6y + 4z = 12 \\ x + 6y + z = 8 \\ -2x - 10y - 2z = -10 \end{cases}$

8.2 **Equality of Matrices**   In Exercises 33–36, solve for x and y.

33. $\begin{bmatrix} -1 & x \\ y & 9 \end{bmatrix} = \begin{bmatrix} -1 & 12 \\ 11 & 9 \end{bmatrix}$

34. $\begin{bmatrix} -1 & 0 \\ x & 5 \\ -4 & -3 \end{bmatrix} = \begin{bmatrix} -1 & 0 \\ 8 & 5 \\ -4 & y \end{bmatrix}$

35. $\begin{bmatrix} x + 3 & -4 & 44 \\ 0 & -3 & 2 \\ -2 & y + 5 & 6 \end{bmatrix} = \begin{bmatrix} 5x - 1 & -4 & 44 \\ 0 & -3 & 2 \\ -2 & 16 & 6 \end{bmatrix}$

36. $\begin{bmatrix} -9 & 4 & 2 & -5 \\ 0 & -3 & 7 & 2y \\ 6 & -1 & 1 & 0 \end{bmatrix} = \begin{bmatrix} -9 & 4 & x - 10 & -5 \\ 0 & -3 & 7 & -6 \\ 6 & -1 & 1 & 0 \end{bmatrix}$

Operations with Matrices   In Exercises 37–40, if possible, find (a) $A + B$, (b) $A - B$, (c) $4A$, and (d) $2A + 2B$.

37. $A = \begin{bmatrix} 2 & -2 \\ 3 & 5 \end{bmatrix}$, $B = \begin{bmatrix} -3 & 10 \\ 12 & 8 \end{bmatrix}$

38. $A = \begin{bmatrix} 4 & 3 \\ -6 & 1 \\ 10 & 1 \end{bmatrix}$, $B = \begin{bmatrix} 3 & 11 \\ 15 & 25 \\ 20 & 29 \end{bmatrix}$

39. $A = \begin{bmatrix} 5 & 4 \\ -7 & 2 \\ 11 & 2 \end{bmatrix}$, $B = \begin{bmatrix} 0 & 3 \\ 4 & 12 \\ 20 & 40 \end{bmatrix}$

40. $A = \begin{bmatrix} 6 & -5 & 7 \end{bmatrix}$, $B = \begin{bmatrix} -1 \\ 4 \\ 8 \end{bmatrix}$

Evaluating an Expression   In Exercises 41–44, evaluate the expression.

41. $\begin{bmatrix} 7 & 3 \\ -1 & 5 \end{bmatrix} + \begin{bmatrix} 10 & -20 \\ 14 & -3 \end{bmatrix} + \begin{bmatrix} 5 & 0 \\ 1 & 9 \end{bmatrix}$

42. $\begin{bmatrix} -11 & -7 \\ 16 & -2 \\ 19 & 1 \end{bmatrix} - \begin{bmatrix} 6 & 0 \\ 8 & -4 \\ -2 & 10 \end{bmatrix} + \begin{bmatrix} -3 & 1 \\ 2 & 28 \\ 12 & -2 \end{bmatrix}$

43. $-2 \left(\begin{bmatrix} 1 & 2 \\ 5 & -4 \\ 6 & 0 \end{bmatrix} + \begin{bmatrix} 7 & 1 \\ 1 & 2 \\ 1 & 4 \end{bmatrix} \right)$

44. $5 \left(\begin{bmatrix} 8 & -1 & 8 \\ -2 & 4 & 12 \\ 0 & -6 & 0 \end{bmatrix} - \begin{bmatrix} -2 & 0 & -4 \\ 3 & -1 & 1 \\ 6 & 12 & -8 \end{bmatrix} \right)$

Solving a Matrix Equation   In Exercises 45–48, solve for X in the equation, where

$A = \begin{bmatrix} -4 & 0 \\ 1 & -5 \\ -3 & 2 \end{bmatrix}$ and $B = \begin{bmatrix} 1 & 2 \\ -2 & 1 \\ 4 & 4 \end{bmatrix}$.

45. $X = 2A - 3B$

46. $6X = 4A + 3B$

47. $3X + 2A = B$

48. $2A - 5B = 3X$

Finding the Product of Two Matrices   In Exercises 49–52, if possible, find AB and state the dimension of the result.

49. $A = \begin{bmatrix} 2 & -2 \\ 3 & 5 \end{bmatrix}$, $B = \begin{bmatrix} -3 & 10 \\ 12 & 8 \end{bmatrix}$

50. $A = \begin{bmatrix} 5 & 4 \\ -7 & 2 \\ 11 & 2 \end{bmatrix}$, $B = \begin{bmatrix} 4 & 12 \\ 20 & 40 \\ 15 & 30 \end{bmatrix}$

51. $A = \begin{bmatrix} 5 & 4 \\ -7 & 2 \\ 11 & 2 \end{bmatrix}$, $B = \begin{bmatrix} 4 & 12 \\ 20 & 40 \end{bmatrix}$

52. $A = \begin{bmatrix} 6 & -5 & 7 \end{bmatrix}$, $B = \begin{bmatrix} -1 \\ 4 \\ 8 \end{bmatrix}$

Finding the Product of Two Matrices   In Exercises 53–56, use the matrix capabilities of a graphing utility to find AB, if possible.

53. $A = \begin{bmatrix} 4 & 1 \\ 11 & -7 \\ 12 & 3 \end{bmatrix}$, $B = \begin{bmatrix} 3 & -5 & 6 \\ 2 & -2 & -2 \end{bmatrix}$

54. $A = \begin{bmatrix} -2 & 3 & 10 \\ 4 & -2 & 2 \end{bmatrix}$, $B = \begin{bmatrix} 1 & 1 \\ -5 & 2 \\ 3 & 2 \end{bmatrix}$

55. $A = \begin{bmatrix} 1 & 2 & -1 \\ 0 & 4 & -2 \\ 1 & 1 & 3 \end{bmatrix}$, $B = \begin{bmatrix} 1 & -1 & 2 \end{bmatrix}$

56. $A = \begin{bmatrix} 4 & -2 & 6 \end{bmatrix}$, $B = \begin{bmatrix} -2 & 1 \\ 0 & -3 \\ 2 & 0 \end{bmatrix}$

Operations with Matrices   In Exercises 57 and 58, if possible, find (a) AB, (b) BA, and (c) A^2.

57. $A = \begin{bmatrix} 1 & 3 \\ 4 & 1 \end{bmatrix}$, $B = \begin{bmatrix} 5 & -1 \\ -2 & 0 \end{bmatrix}$

58. $A = \begin{bmatrix} 2 & 3 \\ 8 & -1 \\ 0 & 2 \end{bmatrix}$, $B = \begin{bmatrix} 4 \\ 1 \end{bmatrix}$

Describing a Vector Transformation   In Exercises 59–62, find $A\mathbf{v}$, where $\mathbf{v} = \langle 2, 5 \rangle$, and describe the transformation.

59. $A = \begin{bmatrix} 1 & 0 \\ 0 & -1 \end{bmatrix}$

60. $A = \begin{bmatrix} 0 & -1 \\ -1 & 0 \end{bmatrix}$

61. $A = \begin{bmatrix} \frac{1}{2} & 0 \\ 0 & 1 \end{bmatrix}$

62. $A = \begin{bmatrix} 1 & 0 \\ 0 & 6 \end{bmatrix}$

63. Manufacturing A tire corporation has three factories that manufacture two models of tires. The production levels are represented by A.

$$A = \begin{bmatrix} 80 & 120 & 140 \\ 40 & 100 & 80 \end{bmatrix} \begin{matrix} A \\ B \end{matrix} \text{Model}$$

Factory $\overbrace{\quad 1 \quad 2 \quad 3 \quad}$

Find the production levels when production decreases by 5%.

64. Cell Phone Charges The pay-as-you-go charges (per minute) of two cell phone companies for calls inside the coverage area, regional roaming calls, and calls outside the coverage area are represented by C.

Company $\overbrace{\quad A \qquad B \quad}$

$$C = \begin{bmatrix} \$0.07 & \$0.095 \\ \$0.10 & \$0.08 \\ \$0.28 & \$0.25 \end{bmatrix} \begin{matrix} \text{Inside} \\ \text{Regional Roaming} \\ \text{Outside} \end{matrix} \text{Coverage area}$$

The numbers of minutes you plan to use in the coverage areas per month are represented by the matrix

$$T = \begin{bmatrix} 120 & 80 & 20 \end{bmatrix}.$$

Compute TC and interpret the result.

8.3 **The Inverse of a Matrix** In Exercises 65–68, show that B is the inverse of A.

65. $A = \begin{bmatrix} -4 & -1 \\ 7 & 2 \end{bmatrix}, \quad B = \begin{bmatrix} -2 & -1 \\ 7 & 4 \end{bmatrix}$

66. $A = \begin{bmatrix} 5 & -1 \\ 11 & -2 \end{bmatrix}, \quad B = \begin{bmatrix} -2 & 1 \\ -11 & 5 \end{bmatrix}$

67. $A = \begin{bmatrix} 1 & 1 & 0 \\ 1 & 0 & 1 \\ 6 & 2 & 3 \end{bmatrix}, \quad B = \begin{bmatrix} -2 & -3 & 1 \\ 3 & 3 & -1 \\ 2 & 4 & -1 \end{bmatrix}$

68. $A = \begin{bmatrix} 1 & -1 & 0 \\ -1 & 0 & -1 \\ 8 & -4 & 2 \end{bmatrix},$

$B = \begin{bmatrix} -2 & 1 & \frac{1}{2} \\ -3 & 1 & \frac{1}{2} \\ 2 & -2 & -\frac{1}{2} \end{bmatrix}$

Finding the Inverse of a Matrix In Exercises 69–72, find the inverse of the matrix, if possible.

69. $\begin{bmatrix} -6 & 5 \\ -5 & 4 \end{bmatrix}$ **70.** $\begin{bmatrix} 3 & 4 \\ 6 & 8 \end{bmatrix}$

71. $\begin{bmatrix} 2 & 0 & 3 \\ -1 & 1 & 1 \\ 2 & -2 & 1 \end{bmatrix}$ **72.** $\begin{bmatrix} 0 & -2 & 1 \\ -5 & -2 & -3 \\ 7 & 3 & 4 \end{bmatrix}$

Finding the Inverse of a Matrix In Exercises 73 and 74, use the matrix capabilities of a graphing utility to find the inverse of the matrix, if possible.

73. $\begin{bmatrix} -1 & -2 & -2 \\ 3 & 7 & 9 \\ 1 & 4 & 7 \end{bmatrix}$

74. $\begin{bmatrix} 8 & 0 & 2 & 8 \\ 4 & -2 & 0 & -2 \\ 1 & 2 & 1 & 4 \\ -1 & 4 & 1 & 1 \end{bmatrix}$

Finding the Inverse of a 2 × 2 Matrix In Exercises 75–78, use the formula on page 572 to find the inverse of the 2 × 2 matrix, if possible.

75. $\begin{bmatrix} -7 & 2 \\ -8 & 2 \end{bmatrix}$ **76.** $\begin{bmatrix} 10 & 4 \\ 7 & 3 \end{bmatrix}$

77. $\begin{bmatrix} -12 & 6 \\ 10 & -5 \end{bmatrix}$ **78.** $\begin{bmatrix} -18 & -15 \\ -6 & -5 \end{bmatrix}$

Solving a System Using an Inverse Matrix In Exercises 79–88, use an inverse matrix to solve the system of linear equations, if possible.

79. $\begin{cases} -x + 4y = 8 \\ 2x - 7y = -5 \end{cases}$ **80.** $\begin{cases} 5x - y = 13 \\ -9x + 2y = -24 \end{cases}$

81. $\begin{cases} -3x + 10y = 8 \\ 5x - 17y = -13 \end{cases}$ **82.** $\begin{cases} 4x - 2y = -10 \\ -19x + 9y = 47 \end{cases}$

83. $\begin{cases} \frac{1}{2}x + \frac{1}{3}y = 2 \\ -3x + 2y = 0 \end{cases}$ **84.** $\begin{cases} -\frac{5}{6}x + \frac{3}{8}y = -2 \\ 4x - 3y = 0 \end{cases}$

85. $\begin{cases} 0.3x + 0.7y = 10.2 \\ 0.4x + 0.6y = 7.6 \end{cases}$ **86.** $\begin{cases} 3.5x - 4.5y = 8 \\ 2.5x - 7.5y = 25 \end{cases}$

87. $\begin{cases} 3x + 2y - z = 6 \\ x - y + 2z = -1 \\ 5x + y + z = 7 \end{cases}$ **88.** $\begin{cases} 4x + 5y - 6z = -6 \\ 3x + 2y + 2z = 8 \\ 2x + y + z = 3 \end{cases}$

Using a Graphing Utility In Exercises 89–92, use the matrix capabilities of a graphing utility to solve the system of linear equations, if possible.

89. $\begin{cases} x + 2y = -1 \\ 3x + 4y = -5 \end{cases}$ **90.** $\begin{cases} x + 3y = 23 \\ -6x + 2y = -18 \end{cases}$

91. $\begin{cases} \frac{6}{5}x - \frac{4}{7}y = \frac{6}{5} \\ -\frac{12}{5}x + \frac{12}{7}y = -\frac{17}{15} \end{cases}$ **92.** $\begin{cases} 5x + 10y = 7 \\ 2x + y = -98 \end{cases}$

8.4 **Finding the Determinant of a Matrix** In Exercises 93–96, find the determinant of the matrix.

93. $\begin{bmatrix} 2 & 5 \\ -4 & 3 \end{bmatrix}$ **94.** $\begin{bmatrix} -3 & 1 \\ 5 & -2 \end{bmatrix}$

95. $\begin{bmatrix} 10 & -2 \\ 18 & 8 \end{bmatrix}$ **96.** $\begin{bmatrix} -30 & 10 \\ 5 & 2 \end{bmatrix}$

Finding the Minors and Cofactors of a Matrix In Exercises 97–100, find all the (a) minors and (b) cofactors of the matrix.

97. $\begin{bmatrix} 2 & -1 \\ 7 & 4 \end{bmatrix}$ **98.** $\begin{bmatrix} 3 & 6 \\ 5 & -4 \end{bmatrix}$

99. $\begin{bmatrix} 3 & 2 & -1 \\ -2 & 5 & 0 \\ 1 & 8 & 6 \end{bmatrix}$ **100.** $\begin{bmatrix} 8 & 3 & 4 \\ 6 & 5 & -9 \\ -4 & 1 & 2 \end{bmatrix}$

Finding the Determinant of a Matrix In Exercises 101–106, find the determinant of the matrix. Expand by cofactors using the row or column that appears to make the computations easiest.

101. $\begin{bmatrix} -2 & 0 & 0 \\ 2 & -1 & 0 \\ -1 & 1 & -3 \end{bmatrix}$ **102.** $\begin{bmatrix} 0 & 1 & -2 \\ 0 & 1 & 2 \\ -1 & -1 & 3 \end{bmatrix}$

103. $\begin{bmatrix} 4 & 1 & -1 \\ 2 & 3 & 2 \\ 1 & -1 & 0 \end{bmatrix}$ **104.** $\begin{bmatrix} -1 & -2 & 1 \\ 2 & 3 & 0 \\ -5 & -1 & 3 \end{bmatrix}$

105. $\begin{bmatrix} -2 & 4 & 1 \\ -6 & 0 & 2 \\ 5 & 3 & 4 \end{bmatrix}$ **106.** $\begin{bmatrix} 1 & 1 & 4 \\ -4 & 1 & 2 \\ 0 & 1 & -1 \end{bmatrix}$

8.5 **Using Cramer's Rule** In Exercises 107–110, use Cramer's Rule (if possible) to solve the system of equations.

107. $\begin{cases} 5x - 2y = 6 \\ -11x + 3y = -23 \end{cases}$ **108.** $\begin{cases} 3x + 8y = -7 \\ 9x - 5y = 37 \end{cases}$

109. $\begin{cases} -2x + 3y - 5z = -11 \\ 4x - y + z = -3 \\ -x - 4y + 6z = 15 \end{cases}$

110. $\begin{cases} 5x - 2y + z = 15 \\ 3x - 3y - z = -7 \\ 2x - y - 7z = -3 \end{cases}$

Finding the Area of a Triangle In Exercises 111 and 112, use a determinant to find the area of the triangle with the given vertices.

111. 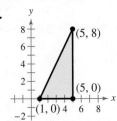   **112.**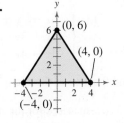

Testing for Collinear Points In Exercises 113 and 114, use a determinant to determine whether the points are collinear.

113. $(-1, 7), (3, -9), (-3, 15)$
114. $(0, -5), (-2, -6), (8, -1)$

Finding an Equation of a Line In Exercises 115–118, use a determinant to find an equation of the line passing through the points.

115. $(-4, 0), (4, 4)$ **116.** $(2, 5), (6, -1)$
117. $\left(-\frac{5}{2}, 3\right), \left(\frac{7}{2}, 1\right)$ **118.** $(-0.8, 0.2), (0.7, 3.2)$

Finding the Area of a Parallelogram In Exercises 119 and 120, use a determinant to find the area of the parallelogram with the given vertices.

119. $(0, 0), (2, 0), (1, 4), (3, 4)$
120. $(0, 0), (-3, 0), (1, 3), (-2, 3)$

Decoding a Message In Exercises 121 and 122, decode the cryptogram using the inverse of the matrix

$$A = \begin{bmatrix} -5 & 4 & -3 \\ 10 & -7 & 6 \\ 8 & -6 & 5 \end{bmatrix}.$$

121. $-5 \quad 11 \quad -2 \quad 370 \quad -265 \quad 225 \quad -57 \quad 48 \quad -33 \quad 32$
$-15 \quad 20 \quad 245 \quad -171 \quad 147$

122. $145 \quad -105 \quad 92 \quad 264 \quad -188 \quad 160 \quad 23 \quad -16 \quad 15$
$129 \quad -84 \quad 78 \quad -9 \quad 8 \quad -5 \quad 159 \quad -118 \quad 100 \quad 219$
$-152 \quad 133 \quad 370 \quad -265 \quad 225 \quad -105 \quad 84 \quad -63$

Exploration

True or False? In Exercises 123 and 124, determine whether the statement is true or false. Justify your answer.

123. It is possible to find the determinant of a 4×5 matrix.

124. $\begin{vmatrix} a_{11} & a_{12} & a_{13} \\ a_{21} & a_{22} & a_{23} \\ a_{31} + c_1 & a_{32} + c_2 & a_{33} + c_3 \end{vmatrix}$

$= \begin{vmatrix} a_{11} & a_{12} & a_{13} \\ a_{21} & a_{22} & a_{23} \\ a_{31} & a_{32} & a_{33} \end{vmatrix} + \begin{vmatrix} a_{11} & a_{12} & a_{13} \\ a_{21} & a_{22} & a_{23} \\ c_1 & c_2 & c_3 \end{vmatrix}$

125. Writing What is the cofactor of an entry of a matrix? How are cofactors used to find the determinant of the matrix?

126. Think About It Three people are solving a system of equations using an augmented matrix. Each person writes the matrix in row-echelon form. Their reduced matrices are shown below.

$\begin{bmatrix} 1 & 2 & \vdots & 3 \\ 0 & 1 & \vdots & 1 \end{bmatrix}$

$\begin{bmatrix} 1 & 0 & \vdots & 1 \\ 0 & 1 & \vdots & 1 \end{bmatrix}$

$\begin{bmatrix} 1 & 2 & \vdots & 3 \\ 0 & 0 & \vdots & 0 \end{bmatrix}$

Can all three be right? Explain.

Chapter Test

See CalcChat.com for tutorial help and worked-out solutions to odd-numbered exercises.

Take this test as you would take a test in class. When you are finished, check your work against the answers given in the back of the book.

In Exercises 1 and 2, write the matrix in reduced row-echelon form.

1. $\begin{bmatrix} 1 & -1 & 5 \\ 6 & 2 & 3 \\ 5 & 3 & -3 \end{bmatrix}$

2. $\begin{bmatrix} 1 & 0 & -1 & 2 \\ -1 & 1 & 1 & -3 \\ 1 & 1 & -1 & 1 \\ 3 & 2 & -3 & 4 \end{bmatrix}$

3. Write the augmented matrix for the system of equations and solve the system.

$$\begin{cases} 4x + 3y - 2z = 14 \\ -x - y + 2z = -5 \\ 3x + y - 4z = 8 \end{cases}$$

4. If possible, find (a) $A - B$, (b) $3C$, (c) $3A - 2B$, (d) BC, and (e) C^2.

$$A = \begin{bmatrix} 6 & 5 \\ -5 & -5 \end{bmatrix}, \quad B = \begin{bmatrix} 5 & 0 \\ -5 & -1 \end{bmatrix}, \quad C = \begin{bmatrix} 2 & -1 & 4 \\ 0 & 6 & -3 \end{bmatrix}$$

5. Find the product $A\mathbf{v}$, where $A = \begin{bmatrix} 0 & -1 \\ -1 & 0 \end{bmatrix}$ and $\mathbf{v} = \langle 2, 3 \rangle$, and describe the transformation.

In Exercises 6 and 7, find the inverse of the matrix, if possible.

6. $\begin{bmatrix} -4 & 3 \\ 5 & -2 \end{bmatrix}$

7. $\begin{bmatrix} -2 & 4 & -6 \\ 2 & 1 & 0 \\ 4 & -2 & 5 \end{bmatrix}$

8. Use the result of Exercise 6 to solve the system.

$$\begin{cases} -4x + 3y = 6 \\ 5x - 2y = 24 \end{cases}$$

In Exercises 9–11, find the determinant of the matrix.

9. $\begin{bmatrix} -6 & 4 \\ 10 & 12 \end{bmatrix}$

10. $\begin{bmatrix} \frac{5}{2} & -\frac{3}{8} \\ -8 & \frac{6}{5} \end{bmatrix}$

11. $\begin{bmatrix} 6 & -7 & 2 \\ 3 & -2 & 0 \\ 1 & 5 & 1 \end{bmatrix}$

In Exercises 12 and 13, use Cramer's Rule (if possible) to solve the system of equations.

12. $\begin{cases} 7x + 6y = 9 \\ -2x - 11y = -49 \end{cases}$

13. $\begin{cases} 6x - y + 2z = -4 \\ -2x + 3y - z = 10 \\ 4x - 4y + z = -18 \end{cases}$

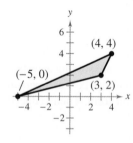

Figure for 14

14. Use a determinant to find the area of the triangle at the left.

15. Write the uncoded 1×3 row matrices for the message KNOCK ON WOOD. Then encode the message using the encoding matrix A at the right.

$$A = \begin{bmatrix} 1 & -1 & 0 \\ 1 & 0 & -1 \\ 6 & -2 & -3 \end{bmatrix}$$

16. One hundred liters of a 50% solution is obtained by mixing a 60% solution with a 20% solution. Use a system of linear equations to determine how many liters of each solution are required to obtain the desired mixture. Solve the system using matrices.

Proofs in Mathematics ▪ ▪ ▪ ▪ ▪ ▪ ▪ ▪ ▪ ▪ ▪ ▪ ▪ ▪

Area of a Triangle *(p. 588)*

The area of a triangle with vertices (x_1, y_1), (x_2, y_2), and (x_3, y_3) is

$$\text{Area} = \pm\frac{1}{2}\begin{vmatrix} x_1 & y_1 & 1 \\ x_2 & y_2 & 1 \\ x_3 & y_3 & 1 \end{vmatrix}$$

where you choose the sign $(\pm)$ so that the area is positive.

Proof

Prove the case for $y_i > 0$. Assume that

$$x_1 \le x_3 \le x_2$$

and that (x_3, y_3) lies above the line segment connecting (x_1, y_1) and (x_2, y_2), as shown in the figure below.

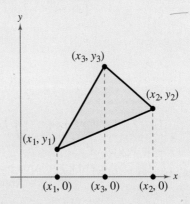

Consider the three trapezoids whose vertices are

Trapezoid 1: $(x_1, 0)$, (x_1, y_1), (x_3, y_3), $(x_3, 0)$
Trapezoid 2: $(x_3, 0)$, (x_3, y_3), (x_2, y_2), $(x_2, 0)$
Trapezoid 3: $(x_1, 0)$, (x_1, y_1), (x_2, y_2), $(x_2, 0)$.

The area of the triangle is the sum of the areas of the first two trapezoids minus the area of the third trapezoid. So,

$$\text{Area} = \frac{1}{2}(y_1 + y_3)(x_3 - x_1) + \frac{1}{2}(y_3 + y_2)(x_2 - x_3) - \frac{1}{2}(y_1 + y_2)(x_2 - x_1)$$

$$= \frac{1}{2}(x_1 y_2 + x_2 y_3 + x_3 y_1 - x_1 y_3 - x_2 y_1 - x_3 y_2)$$

$$= \frac{1}{2}\begin{vmatrix} x_1 & y_1 & 1 \\ x_2 & y_2 & 1 \\ x_3 & y_3 & 1 \end{vmatrix}.$$

If the vertices do not occur in the order

$$x_1 \le x_3 \le x_2$$

or if the vertex (x_3, y_3) does not lie above the line segment connecting the other two vertices, then the formula above may yield the negative of the area. So, use $\pm$ and choose the correct sign so that the area is positive. ▪

A proof without words is a picture or diagram that gives a visual understanding of why a theorem or statement is true. It can also provide a starting point for writing a formal proof.

In Section 8.5 (page 591), you learned that the area of a parallelogram with vertices $(0, 0)$, (a, b), (c, d), and $(a + c, b + d)$ is the absolute value of the determinant of the matrix A, where

$$A = \begin{bmatrix} a & b \\ c & d \end{bmatrix}.$$

The color-coded visual proof below shows this for a case in which the determinant is positive. Also shown is a brief explanation of why this proof works.

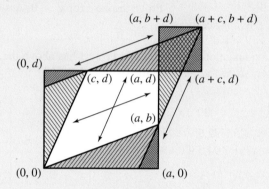

$$\begin{vmatrix} a & b \\ c & d \end{vmatrix} = ad - bc = \|\square\| - \|\square\| = \|\square\|$$

Area of $\square$ = Area of orange $\triangle$ + Area of yellow $\triangle$ + Area of blue $\triangle$
+ Area of pink $\triangle$ + Area of white quadrilateral

Area of $\square$ = Area of orange $\triangle$ + Area of pink $\triangle$ + Area of green quadrilateral

Area of $\square$ = Area of white quadrilateral + Area of blue $\triangle$ + Area of yellow $\triangle$
− Area of green quadrilateral

= Area of $\square$ − Area of $\square$

The formula in Section 8.5 is a generalization, taking into consideration the possibility that the coordinates could yield a negative determinant. Area is always positive, which is the reason the formula uses absolute value. Verify the formula using values of a, b, c, and d that produce a negative determinant. ■

From "Proof Without Words: A 2 × 2 Determinant Is the Area of a Parallelogram" by Solomon W. Golomb, *Mathematics Magazine,* Vol. 58, No. 2, pg. 107.

P.S. Problem Solving ■ ■ ■ ■ ■ ■ ■ ■ ■ ■ ■ ■ ■ ■ ■

1. Multiplying by a Transformation Matrix The columns of matrix T show the coordinates of the vertices of a triangle. Matrix A is a transformation matrix.

$$A = \begin{bmatrix} 0 & -1 \\ 1 & 0 \end{bmatrix} \quad T = \begin{bmatrix} 1 & 2 & 3 \\ 1 & 4 & 2 \end{bmatrix}$$

(a) Find AT and AAT. Then sketch the original triangle and the two images of the triangle. What transformation does A represent?

(b) Given the triangle determined by AAT, describe the transformation that produces the triangle determined by AT and then the triangle determined by T.

2. Population The matrices show the male and female populations in the United States in 2011 and 2014. The male and female populations are separated into three age groups. *(Source: U.S. Census Bureau)*

2011

	0–19	20–64	65+
Male	42,376,825	92,983,543	17,934,267
Female	40,463,751	94,530,885	23,432,361

2014

	0–19	20–64	65+
Male	41,969,399	94,615,796	20,351,292
Female	40,166,203	95,862,447	25,891,919

(a) The total population in 2011 was 311,721,632 and the total population in 2014 was 318,857,056. Rewrite the matrices to give the information as percents of the total population.

(b) Write a matrix that gives the change in the percent of the population for each gender and age group from 2011 to 2014.

(c) Based on the result of part (b), which gender(s) and age group(s) had percents that decreased from 2011 to 2014?

3. Determining Whether Matrices are Idempotent A square matrix is **idempotent** when $A^2 = A$. Determine whether each matrix is idempotent.

(a) $\begin{bmatrix} 1 & 0 \\ 0 & 0 \end{bmatrix}$ (b) $\begin{bmatrix} 0 & 1 \\ 1 & 0 \end{bmatrix}$

(c) $\begin{bmatrix} 2 & 3 \\ -1 & -2 \end{bmatrix}$ (d) $\begin{bmatrix} 2 & 3 \\ 1 & 2 \end{bmatrix}$

(e) $\begin{bmatrix} 0 & 0 & 1 \\ 0 & 1 & 0 \\ 1 & 0 & 0 \end{bmatrix}$ (f) $\begin{bmatrix} 0 & 1 & 0 \\ 1 & 0 & 0 \\ 0 & 0 & 1 \end{bmatrix}$

4. Finding a Matrix Find a singular 2×2 matrix satisfying $A^2 = A$.

5. Quadratic Matrix Equation Let

$$A = \begin{bmatrix} 1 & 2 \\ -2 & 1 \end{bmatrix}.$$

(a) Show that $A^2 - 2A + 5I = O$, where I is the identity matrix of dimension 2×2.

(b) Show that $A^{-1} = \frac{1}{5}(2I - A)$.

(c) Show that for any square matrix satisfying

$$A^2 - 2A + 5I = O$$

the inverse of A is given by

$$A^{-1} = \frac{1}{5}(2I - A).$$

6. Satellite Television Two competing companies offer satellite television to a city with 100,000 households. Gold Satellite System has 25,000 subscribers and Galaxy Satellite Network has 30,000 subscribers. (The other 45,000 households do not subscribe.) The matrix shows the percent changes in satellite subscriptions each year.

		Percent Changes		
		From Gold	From Galaxy	From Non-subscriber
Percent Changes	To Gold	0.70	0.15	0.15
	To Galaxy	0.20	0.80	0.15
	To Nonsubscriber	0.10	0.05	0.70

(a) Find the number of subscribers each company will have in 1 year using matrix multiplication. Explain how you obtained your answer.

(b) Find the number of subscribers each company will have in 2 years using matrix multiplication. Explain how you obtained your answer.

(c) Find the number of subscribers each company will have in 3 years using matrix multiplication. Explain how you obtained your answer.

(d) What is happening to the number of subscribers to each company? What is happening to the number of nonsubscribers?

7. The Transpose of a Matrix The **transpose** of a matrix, denoted A^T, is formed by writing its rows as columns. Find the transpose of each matrix and verify that $(AB)^T = B^T A^T$.

$$A = \begin{bmatrix} -1 & 1 & -2 \\ 2 & 0 & 1 \end{bmatrix}, \quad B = \begin{bmatrix} -3 & 0 \\ 1 & 2 \\ 1 & -1 \end{bmatrix}$$

8. Finding a Value Find x such that the matrix is equal to its own inverse.

$$A = \begin{bmatrix} 3 & x \\ -2 & -3 \end{bmatrix}$$

607

9. Finding a Value Find x such that the matrix is singular.

$$A = \begin{bmatrix} 4 & x \\ -2 & -3 \end{bmatrix}$$

10. Verifying an Equation Verify the following equation.

$$\begin{vmatrix} 1 & 1 & 1 \\ a & b & c \\ a^2 & b^2 & c^2 \end{vmatrix} = (a - b)(b - c)(c - a)$$

11. Verifying an Equation Verify the following equation.

$$\begin{vmatrix} 1 & 1 & 1 \\ a & b & c \\ a^3 & b^3 & c^3 \end{vmatrix} = (a - b)(b - c)(c - a)(a + b + c)$$

12. Verifying an Equation Verify the following equation.

$$\begin{vmatrix} x & 0 & c \\ -1 & x & b \\ 0 & -1 & a \end{vmatrix} = ax^2 + bx + c$$

13. Finding a Matrix Find a 4×4 matrix whose determinant is equal to $ax^3 + bx^2 + cx + d$. (*Hint:* Use the equation in Exercise 12 as a model.)

14. Finding the Determinant of a Matrix Let A be an $n \times n$ matrix each of whose rows sum to zero. Find $|A|$.

15. Finding Atomic Masses The table shows the masses (in atomic mass units) of three compounds. Use a linear system and Cramer's Rule to find the atomic masses of sulfur (S), nitrogen (N), and fluorine (F).

Compound	Formula	Mass
Tetrasulfur tetranitride	S_4N_4	184
Sulfur hexafluoride	SF_6	146
Dinitrogen tetrafluoride	N_2F_4	104

16. Finding the Costs of Items A walkway lighting package includes a transformer, a certain length of wire, and a certain number of lights on the wire. The price of each lighting package depends on the length of wire and the number of lights on the wire. Use the information below to find the cost of a transformer, the cost per foot of wire, and the cost of a light. Assume that the cost of each item is the same in each lighting package.

• A package that contains a transformer, 25 feet of wire, and 5 lights costs $20.

• A package that contains a transformer, 50 feet of wire, and 15 lights costs $35.

• A package that contains a transformer, 100 feet of wire, and 20 lights costs $50.

17. Decoding a Message Use the inverse of A to decode the cryptogram.

$$A = \begin{bmatrix} 1 & -2 & 2 \\ 1 & 1 & -3 \\ 1 & -1 & 4 \end{bmatrix}$$

23 13 −34 31 −34 63 25 −17 61
24 14 −37 41 −17 −8 20 −29 40 38
−56 116 13 −11 1 22 −3 −6 41
−53 85 28 −32 16

18. Decoding a Message A code breaker intercepts the encoded message below.

45 −35 38 −30 18 −18 35 −30 81 −60
42 −28 75 −55 2 −2 22 −21 15 −10

Let $A^{-1} = \begin{bmatrix} w & x \\ y & z \end{bmatrix}$.

(a) You know that

$$[45 \quad -35]A^{-1} = [10 \quad 15]$$

$$[38 \quad -30]A^{-1} = [8 \quad 14]$$

where A^{-1} is the inverse of the encoding matrix A. Write and solve two systems of equations to find w, x, y, and z.

(b) Decode the message.

19. Conjecture Let

$$A = \begin{bmatrix} 6 & 4 & 1 \\ 0 & 2 & 3 \\ 1 & 1 & 2 \end{bmatrix}.$$

Use a graphing utility to find A^{-1}. Compare $|A^{-1}|$ with $|A|$. Make a conjecture about the determinant of the inverse of a matrix.

20. Conjecture Consider matrices of the form

$$A = \begin{bmatrix} 0 & a_{12} & a_{13} & a_{14} & \cdots & a_{1n} \\ 0 & 0 & a_{23} & a_{24} & \cdots & a_{2n} \\ 0 & 0 & 0 & a_{34} & \cdots & a_{3n} \\ \vdots & \vdots & \vdots & \vdots & & \vdots \\ 0 & 0 & 0 & 0 & \cdots & a_{(n-1)n} \\ 0 & 0 & 0 & 0 & \cdots & 0 \end{bmatrix}.$$

(a) Write a 2×2 matrix and a 3×3 matrix in the form of A.

(b) Use a graphing utility to raise each of the matrices to higher powers. Describe the result.

(c) Use the result of part (b) to make a conjecture about powers of A when A is a 4×4 matrix. Use the graphing utility to test your conjecture.

(d) Use the results of parts (b) and (c) to make a conjecture about powers of A when A is an $n \times n$ matrix.

9 Sequences, Series, and Probability

Tossing Dice
(Example 3, page 668)

Horse Racing *(Example 6, page 659)*

Electricity *(Exercise 86, page 655)*

Dominoes *(page 639)*

Physical Activity *(Exercise 98, page 619)*

9.1 Sequences and Series

- Use sequence notation to write the terms of sequences.
- Use factorial notation.
- Use summation notation to write sums.
- Find the sums of series.
- Use sequences and series to model and solve real-life problems.

Sequences

In mathematics, the word *sequence* is used in much the same way as in ordinary English. Saying that a collection is listed in *sequence* means that it is ordered so that it has a first member, a second member, a third member, and so on. Two examples are 1, 2, 3, 4, . . . and 1, 3, 5, 7, ...

Mathematically, you can think of a sequence as a *function* whose domain is the set of positive integers. Rather than using function notation, however, sequences are usually written using subscript notation, as shown in the following definition.

Sequences and series model many real-life situations over time. For example, in Exercise 98 on page 619, a sequence models the percent of United States adults who met federal physical activity guidelines from 2007 through 2014.

Definition of Sequence

An **infinite sequence** is a function whose domain is the set of positive integers. The function values

$$a_1, a_2, a_3, a_4, \ldots, a_n, \ldots$$

are the **terms** of the sequence. When the domain of the function consists of the first n positive integers only, the sequence is a **finite sequence.**

On occasion, it is convenient to begin subscripting a sequence with 0 instead of 1 so that the terms of the sequence become

$$a_0, a_1, a_2, a_3, \ldots.$$

When this is the case, the domain includes 0.

REMARK The subscripts of a sequence make up the domain of the sequence and serve to identify the positions of terms within the sequence. For example, a_4 is the fourth term of the sequence, and a_n is the nth term of the sequence. Any variable can be a subscript. The most commonly used variable subscripts in sequence and series notation are $i, j, k,$ and n.

▷ **EXAMPLE 1** Writing the Terms of a Sequence

a. The first four terms of the sequence given by $a_n = 3n - 2$ are

$a_1 = 3(1) - 2 = 1$ 1st term

$a_2 = 3(2) - 2 = 4$ 2nd term

$a_3 = 3(3) - 2 = 7$ 3rd term

$a_4 = 3(4) - 2 = 10.$ 4th term

b. The first four terms of the sequence given by $a_n = 3 + (-1)^n$ are

$a_1 = 3 + (-1)^1 = 3 - 1 = 2$ 1st term

$a_2 = 3 + (-1)^2 = 3 + 1 = 4$ 2nd term

$a_3 = 3 + (-1)^3 = 3 - 1 = 2$ 3rd term

$a_4 = 3 + (-1)^4 = 3 + 1 = 4.$ 4th term

✓ **Checkpoint** ◀))) *Audio-video solution in English & Spanish at LarsonPrecalculus.com*

Write the first four terms of the sequence given by $a_n = 2n + 1$.

•• **REMARK** Write the first four terms of the sequence given by

$$a_n = \frac{(-1)^{n+1}}{2n + 1}.$$

Are they the same as the first four terms of the sequence in Example 2? If not, then how do they differ?
•••••••••••••••••••▷

EXAMPLE 2 **A Sequence Whose Terms Alternate in Sign**

Write the first four terms of the sequence given by $a_n = \dfrac{(-1)^n}{2n + 1}$.

Solution   The first four terms of the sequence are as follows.

$$a_1 = \frac{(-1)^1}{2(1) + 1} = \frac{-1}{2 + 1} = -\frac{1}{3} \qquad \text{1st term}$$

$$a_2 = \frac{(-1)^2}{2(2) + 1} = \frac{1}{4 + 1} = \frac{1}{5} \qquad \text{2nd term}$$

$$a_3 = \frac{(-1)^3}{2(3) + 1} = \frac{-1}{6 + 1} = -\frac{1}{7} \qquad \text{3rd term}$$

$$a_4 = \frac{(-1)^4}{2(4) + 1} = \frac{1}{8 + 1} = \frac{1}{9} \qquad \text{4th term}$$

✓ **Checkpoint** ◀))) *Audio-video solution in English & Spanish at LarsonPrecalculus.com*

Write the first four terms of the sequence given by $a_n = \dfrac{2 + (-1)^n}{n}$.

Simply listing the first few terms is not sufficient to define a unique sequence—the *n*th term *must be given*. To see this, consider the following sequences, both of which have the same first three terms.

$$\frac{1}{2}, \frac{1}{4}, \frac{1}{8}, \frac{1}{16}, \cdots, \frac{1}{2^n}, \cdots$$

$$\frac{1}{2}, \frac{1}{4}, \frac{1}{8}, \frac{1}{15}, \cdots, \frac{6}{(n + 1)(n^2 - n + 6)}, \cdots$$

▷ **TECHNOLOGY**
• To graph a sequence using a
• graphing utility, set the mode to
• *sequence* and *dot* and enter the
• expression for a_n. The graph of
• the sequence in Example 3(a)
• is shown below. To identify the
• terms, use the *trace* feature or
• *value* feature.

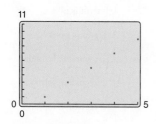

EXAMPLE 3   **Finding the *n*th Term of a Sequence**

Write an expression for the apparent *n*th term (a_n) of each sequence.

a. $1, 3, 5, 7, \ldots$ **b.** $2, -5, 10, -17, \ldots$

Solution

a.   *n*: 1   2   3   4 . . . *n*
    Terms: 1   3   5   7 . . . a_n

Apparent pattern: Each term is 1 less than twice *n*. So, the apparent *n*th term is

$$a_n = 2n - 1.$$

b.   *n*: 1   2   3   4 . . . *n*
    Terms: 2   −5   10   −17 . . . a_n

Apparent pattern: The absolute value of each term is 1 more than the square of *n*, and the terms have alternating signs, with those in the even positions being negative. So, the apparent *n*th term is

$$a_n = (-1)^{n+1}(n^2 + 1).$$

✓ **Checkpoint** ◀))) *Audio-video solution in English & Spanish at LarsonPrecalculus.com*

Write an expression for the apparent *n*th term (a_n) of each sequence.

a. $1, 5, 9, 13, \ldots$ **b.** $2, -4, 6, -8, \ldots$

Some sequences are defined **recursively.** To define a sequence recursively, you need to be given one or more of the first few terms. All other terms of the sequence are then defined using previous terms.

EXAMPLE 4   **A Recursive Sequence**

Write the first five terms of the sequence defined recursively as

$$a_1 = 3$$
$$a_k = 2a_{k-1} + 1, \quad \text{where } k \geq 2.$$

Solution

$a_1 = 3$	1st term is given.
$a_2 = 2a_{2-1} + 1 = 2a_1 + 1 = 2(3) + 1 = 7$	Use recursion formula.
$a_3 = 2a_{3-1} + 1 = 2a_2 + 1 = 2(7) + 1 = 15$	Use recursion formula.
$a_4 = 2a_{4-1} + 1 = 2a_3 + 1 = 2(15) + 1 = 31$	Use recursion formula.
$a_5 = 2a_{5-1} + 1 = 2a_4 + 1 = 2(31) + 1 = 63$	Use recursion formula.

✓ *Checkpoint* ◀))) *Audio-video solution in English & Spanish at LarsonPrecalculus.com*

Write the first five terms of the sequence defined recursively as

$$a_1 = 6$$
$$a_{k+1} = a_k + 1, \quad \text{where } k \geq 1.$$

In the next example, you will study a well-known recursive sequence, the Fibonacci sequence.

EXAMPLE 5   **The Fibonacci Sequence: A Recursive Sequence**

The Fibonacci sequence is defined recursively, as follows.

$$a_0 = 1$$
$$a_1 = 1$$
$$a_k = a_{k-2} + a_{k-1}, \quad \text{where } k \geq 2$$

Write the first six terms of this sequence.

Solution

$a_0 = 1$	0th term is given.
$a_1 = 1$	1st term is given.
$a_2 = a_{2-2} + a_{2-1} = a_0 + a_1 = 1 + 1 = 2$	Use recursion formula.
$a_3 = a_{3-2} + a_{3-1} = a_1 + a_2 = 1 + 2 = 3$	Use recursion formula.
$a_4 = a_{4-2} + a_{4-1} = a_2 + a_3 = 2 + 3 = 5$	Use recursion formula.
$a_5 = a_{5-2} + a_{5-1} = a_3 + a_4 = 3 + 5 = 8$	Use recursion formula.

✓ *Checkpoint* ◀))) *Audio-video solution in English & Spanish at LarsonPrecalculus.com*

Write the first five terms of the sequence defined recursively as

$$a_0 = 1, \quad a_1 = 3, \quad a_k = a_{k-2} + a_{k-1}, \quad \text{where } k \geq 2.$$

Factorial Notation

Many sequences involve terms defined using special products called **factorials**.

> ### Definition of Factorial
> If n is a positive integer, then n **factorial** is defined as
> $$n! = 1 \cdot 2 \cdot 3 \cdot 4 \cdots (n-1) \cdot n.$$
> As a special case, zero factorial is defined as $0! = 1$.

•• **REMARK** The value of n does not have to be very large before the value of $n!$ becomes extremely large. For example, $10! = 3,628,800$.

Notice that $1! = 1$, $2! = 1 \cdot 2 = 2$, $3! = 1 \cdot 2 \cdot 3 = 6$, and $4! = 1 \cdot 2 \cdot 3 \cdot 4 = 24$. Factorials follow the same conventions for order of operations as exponents. So,

$$2n! = 2(n!) = 2(1 \cdot 2 \cdot 3 \cdot 4 \cdots n), \quad \text{whereas} \quad (2n)! = 1 \cdot 2 \cdot 3 \cdot 4 \cdots 2n.$$

EXAMPLE 6 **Writing the Terms of a Sequence Involving Factorials**

Write the first five terms of the sequence given by $a_n = \dfrac{2^n}{n!}$. Begin with $n = 0$.

Algebraic Solution

$$a_0 = \frac{2^0}{0!} = \frac{1}{1} = 1 \qquad \text{0th term}$$

$$a_1 = \frac{2^1}{1!} = \frac{2}{1} = 2 \qquad \text{1st term}$$

$$a_2 = \frac{2^2}{2!} = \frac{4}{2} = 2 \qquad \text{2nd term}$$

$$a_3 = \frac{2^3}{3!} = \frac{8}{6} = \frac{4}{3} \qquad \text{3rd term}$$

$$a_4 = \frac{2^4}{4!} = \frac{16}{24} = \frac{2}{3} \qquad \text{4th term}$$

Graphical Solution

Using a graphing utility set to *dot* and *sequence* modes, enter the expression for a_n. Next, graph the sequence. Use the graph to estimate the first five terms.

$$u_0 = 1$$
$$u_1 = 2$$
$$u_2 = 2$$
$$u_3 \approx 1.333 \approx \tfrac{4}{3}$$
$$u_4 \approx 0.667 \approx \tfrac{2}{3}$$

Use the *trace* feature to approximate the first five terms.

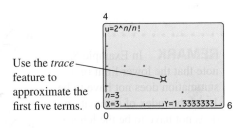

✓ **Checkpoint**))) *Audio-video solution in English & Spanish at LarsonPrecalculus.com*

Write the first five terms of the sequence given by $a_n = \dfrac{3^n + 1}{n!}$. Begin with $n = 0$. ■

▷ **ALGEBRA HELP** Here is another way to simplify the expression in Example 7(a).

$$\frac{8!}{2! \cdot 6!} = \frac{8 \cdot 7 \cdot \cancel{6!}}{2 \cdot 1 \cdot \cancel{6!}} = 28$$

When fractions involve factorials, you can often divide out common factors.

EXAMPLE 7 **Simplifying Factorial Expressions**

a. $\dfrac{8!}{2! \cdot 6!} = \dfrac{1 \cdot 2 \cdot 3 \cdot 4 \cdot 5 \cdot 6 \cdot 7 \cdot 8}{1 \cdot 2 \cdot 1 \cdot 2 \cdot 3 \cdot 4 \cdot 5 \cdot 6} = \dfrac{7 \cdot 8}{2} = 28$

b. $\dfrac{n!}{(n-1)!} = \dfrac{1 \cdot 2 \cdot 3 \cdots (n-1) \cdot n}{1 \cdot 2 \cdot 3 \cdots (n-1)} = n$

✓ **Checkpoint**))) *Audio-video solution in English & Spanish at LarsonPrecalculus.com*

Simplify the factorial expression $\dfrac{4!(n+1)!}{3!n!}$. ■

Writing the Terms of a Recursive Sequence In Exercises 51–56, write the first five terms of the sequence defined recursively.

51. $a_1 = 28$, $a_{k+1} = a_k - 4$
52. $a_1 = 3$, $a_{k+1} = 2(a_k - 1)$
53. $a_1 = 81$, $a_{k+1} = \frac{1}{3}a_k$
54. $a_1 = 14$, $a_{k+1} = (-2)a_k$
55. $a_0 = 1$, $a_1 = 2$, $a_k = a_{k-2} + \frac{1}{2}a_{k-1}$
56. $a_0 = -1$, $a_1 = 1$, $a_k = a_{k-2} + a_{k-1}$

Fibonacci Sequence In Exercises 57 and 58, use the Fibonacci sequence. (See Example 5.)

57. Write the first 12 terms of the Fibonacci sequence whose nth term is a_n and the first 10 terms of the sequence given by
$$b_n = \frac{a_{n+1}}{a_n}, \quad n \geq 1.$$

58. Using the definition for b_n in Exercise 57, show that b_n can be defined recursively by
$$b_n = 1 + \frac{1}{b_{n-1}}.$$

Writing the Terms of a Sequence Involving Factorials In Exercises 59–62, write the first five terms of the sequence. (Assume that n begins with 0.)

59. $a_n = \dfrac{5}{n!}$
60. $a_n = \dfrac{1}{(n+1)!}$
61. $a_n = \dfrac{(-1)^n(n+3)!}{n!}$
62. $a_n = \dfrac{(-1)^{2n+1}}{(2n+1)!}$

Simplifying a Factorial Expression In Exercises 63–66, simplify the factorial expression.

63. $\dfrac{4!}{6!}$
64. $\dfrac{12!}{4! \cdot 8!}$
65. $\dfrac{(n+1)!}{n!}$
66. $\dfrac{(2n-1)!}{(2n+1)!}$

Finding a Sum In Exercises 67–74, find the sum.

67. $\displaystyle\sum_{i=0}^{4} 3i^2$
68. $\displaystyle\sum_{k=1}^{4} 10$
69. $\displaystyle\sum_{j=3}^{5} \frac{1}{j^2 - 3}$
70. $\displaystyle\sum_{i=1}^{5} (2i - 1)$
71. $\displaystyle\sum_{k=2}^{5} (k+1)^2(k-3)$
72. $\displaystyle\sum_{i=1}^{4} [(i-1)^2 + (i+1)^3]$

73. $\displaystyle\sum_{i=1}^{4} \frac{i!}{2^i}$
74. $\displaystyle\sum_{j=0}^{5} \frac{(-1)^j}{j!}$

Finding a Sum In Exercises 75–78, use a graphing utility to find the sum.

75. $\displaystyle\sum_{k=0}^{4} \frac{(-1)^k}{k!}$
76. $\displaystyle\sum_{k=0}^{4} \frac{(-1)^k}{k+1}$
77. $\displaystyle\sum_{n=0}^{25} \frac{1}{4^n}$
78. $\displaystyle\sum_{n=0}^{10} \frac{n!}{2^n}$

Using Sigma Notation to Write a Sum In Exercises 79–88, use sigma notation to write the sum.

79. $\dfrac{1}{3(1)} + \dfrac{1}{3(2)} + \dfrac{1}{3(3)} + \cdots + \dfrac{1}{3(9)}$
80. $\dfrac{5}{1+1} + \dfrac{5}{1+2} + \dfrac{5}{1+3} + \cdots + \dfrac{5}{1+15}$
81. $\left[2\left(\frac{1}{8}\right) + 3\right] + \left[2\left(\frac{2}{8}\right) + 3\right] + \cdots + \left[2\left(\frac{8}{8}\right) + 3\right]$
82. $\left[1 - \left(\frac{1}{6}\right)^2\right] + \left[1 - \left(\frac{2}{6}\right)^2\right] + \cdots + \left[1 - \left(\frac{6}{6}\right)^2\right]$
83. $3 - 9 + 27 - 81 + 243 - 729$
84. $1 - \frac{1}{2} + \frac{1}{4} - \frac{1}{8} + \cdots - \frac{1}{128}$
85. $\dfrac{1^2}{2} + \dfrac{2^2}{6} + \dfrac{3^2}{24} + \dfrac{4^2}{120} + \cdots + \dfrac{7^2}{40{,}320}$
86. $\dfrac{1}{1 \cdot 3} + \dfrac{1}{2 \cdot 4} + \dfrac{1}{3 \cdot 5} + \cdots + \dfrac{1}{10 \cdot 12}$
87. $\frac{1}{4} + \frac{3}{8} + \frac{7}{16} + \frac{15}{32} + \frac{31}{64}$
88. $\frac{1}{2} + \frac{2}{4} + \frac{6}{8} + \frac{24}{16} + \frac{120}{32} + \frac{720}{64}$

Finding a Partial Sum of a Series In Exercises 89–92, find the (a) third, (b) fourth, and (c) fifth partial sums of the series.

89. $\displaystyle\sum_{i=1}^{\infty} \left(\frac{1}{2}\right)^i$
90. $\displaystyle\sum_{i=1}^{\infty} 2\left(\frac{1}{3}\right)^i$
91. $\displaystyle\sum_{n=1}^{\infty} 4\left(-\frac{1}{2}\right)^n$
92. $\displaystyle\sum_{n=1}^{\infty} 5\left(-\frac{1}{4}\right)^n$

Finding the Sum of an Infinite Series In Exercises 93–96, find the sum of the infinite series.

93. $\displaystyle\sum_{i=1}^{\infty} \frac{6}{10^i}$
94. $\displaystyle\sum_{k=1}^{\infty} \left(\frac{1}{10}\right)^k$
95. $\displaystyle\sum_{k=1}^{\infty} 7\left(\frac{1}{10}\right)^k$
96. $\displaystyle\sum_{i=1}^{\infty} \frac{2}{10^i}$

97. Compound Interest An investor deposits $10,000 in an account that earns 3.5% interest compounded quarterly. The balance in the account after n quarters is given by

$$A_n = 10{,}000\left(1 + \frac{0.035}{4}\right)^n, \quad n = 1, 2, 3, \ldots .$$

(a) Write the first eight terms of the sequence.

(b) Find the balance in the account after 10 years by computing the 40th term of the sequence.

(c) Is the balance after 20 years twice the balance after 10 years? Explain.

· · 98. Physical Activity · · · · · · · · · · · · · · ·

The percent p_n of United States adults who met federal physical activity guidelines from 2007 through 2014 can be approximated by

$p_n = 0.0061n^3 - 0.419n^2 + 7.85n + 4.9,$

$n = 7, 8, \ldots, 14$

where n is the year, with $n = 7$ corresponding to 2007. *(Source: National Center for Health Statistics)*

(a) Write the terms of this finite sequence. Use a graphing utility to construct a bar graph that represents the sequence.

(b) What can you conclude from the bar graph in part (a)?

· ·

Exploration

True or False? In Exercises 99 and 100, determine whether the statement is true or false. Justify your answer.

99. $\displaystyle\sum_{i=1}^{4}(i^2 + 2i) = \sum_{i=1}^{4}i^2 + 2\sum_{i=1}^{4}i$

100. $\displaystyle\sum_{j=1}^{4}2^j = \sum_{j=3}^{6}2^{j-2}$

Arithmetic Mean In Exercises 101–103, use the following definition of the arithmetic mean $\bar{x}$ of a set of n measurements $x_1, x_2, x_3, \ldots, x_n$.

$$\bar{x} = \frac{1}{n}\sum_{i=1}^{n}x_i$$

101. Find the arithmetic mean of the six checking account balances $327.15, $785.69, $433.04, $265.38, $604.12, and $590.30. Use the statistical capabilities of a graphing utility to verify your result.

102. Proof Prove that $\displaystyle\sum_{i=1}^{n}(x_i - \bar{x}) = 0.$

103. Proof Prove that

$$\sum_{i=1}^{n}(x_i - \bar{x})^2 = \sum_{i=1}^{n}x_i^2 - \frac{1}{n}\left(\sum_{i=1}^{n}x_i\right)^2.$$

104. HOW DO YOU SEE IT? The graph represents the first 10 terms of a sequence. Complete each expression for the apparent nth term (a_n) of the sequence. Which expressions are appropriate to represent the cost a_n to buy n MP3 songs at a cost of $1 per song? Explain.

(a) $a_n = 1 \cdot \boxed{}$

(b) $a_n = \dfrac{\boxed{}!}{(n-1)!}$

(c) $a_n = \displaystyle\sum_{k=1}^{n}\boxed{}$

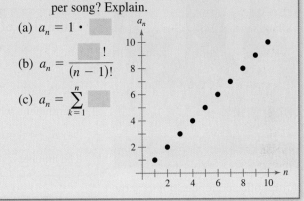

Error Analysis In Exercises 105 and 106, describe the error in finding the sum.

105. $\displaystyle\sum_{k=1}^{4}(3 + 2k^2) = \sum_{k=1}^{4}3 + \sum_{k=1}^{4}2k^2$

$= 3 + (2 + 8 + 18 + 32)$

$= 63$

106. $\displaystyle\sum_{n=0}^{3}(-1)^n n! = (-1)(1) + (1)(2) + (-1)(6)$

$= -5$

107. Cube A $3 \times 3 \times 3$ cube is made up of 27 unit cubes (a unit cube has a length, width, and height of 1 unit), and only the faces of each cube that are visible are painted blue, as shown in the figure.

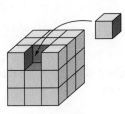

(a) Determine how many unit cubes of the $3 \times 3 \times 3$ cube have 0 blue faces, 1 blue face, 2 blue faces, and 3 blue faces.

(b) Repeat part (a) for a $4 \times 4 \times 4$ cube, a $5 \times 5 \times 5$ cube, and a $6 \times 6 \times 6$ cube.

(c) Write formulas you could use to repeat part (a) for an $n \times n \times n$ cube.

9.2 Arithmetic Sequences and Partial Sums

Arithmetic sequences have many real-life applications. For example, in Exercise 73 on page 627, you will use an arithmetic sequence to determine how far an object falls in 7 seconds when dropped from the top of the Willis Tower in Chicago.

- Recognize, write, and find the *n*th terms of arithmetic sequences.
- Find *n*th partial sums of arithmetic sequences.
- Use arithmetic sequences to model and solve real-life problems.

Arithmetic Sequences

A sequence whose consecutive terms have a common difference is an **arithmetic sequence.**

Definition of Arithmetic Sequence

A sequence is **arithmetic** when the differences between consecutive terms are the same. So, the sequence

$$a_1, a_2, a_3, a_4, \ldots, a_n, \ldots$$

is arithmetic when there is a number d such that

$$a_2 - a_1 = a_3 - a_2 = a_4 - a_3 = \cdots = d.$$

The number d is the **common difference** of the arithmetic sequence.

EXAMPLE 1 Examples of Arithmetic Sequences

a. The sequence whose *n*th term is $4n + 3$ is arithmetic. The common difference between consecutive terms is 4.

$$7, 11, 15, 19, \ldots, 4n + 3, \ldots \qquad \text{Begin with } n = 1.$$
$$\underbrace{\qquad}_{11 - 7 = 4}$$

b. The sequence whose *n*th term is $7 - 5n$ is arithmetic. The common difference between consecutive terms is -5.

$$2, -3, -8, -13, \ldots, 7 - 5n, \ldots \qquad \text{Begin with } n = 1.$$
$$\underbrace{\qquad}_{-3 - 2 = -5}$$

c. The sequence whose *n*th term is $\frac{1}{4}(n + 3)$ is arithmetic. The common difference between consecutive terms is $\frac{1}{4}$.

$$1, \frac{5}{4}, \frac{3}{2}, \frac{7}{4}, \ldots, \frac{n + 3}{4}, \ldots \qquad \text{Begin with } n = 1.$$
$$\underbrace{\qquad}_{\frac{5}{4} - 1 = \frac{1}{4}}$$

✓ *Checkpoint*))) *Audio-video solution in English & Spanish at LarsonPrecalculus.com*

Write the first four terms of the arithmetic sequence whose *n*th term is $3n - 1$. Then find the common difference between consecutive terms. ∎

The sequence 1, 4, 9, 16, . . . , whose *n*th term is n^2, is *not* arithmetic. The difference between the first two terms is

$$a_2 - a_1 = 4 - 1 = 3$$

but the difference between the second and third terms is

$$a_3 - a_2 = 9 - 4 = 5.$$

The nth term of an arithmetic sequence can be derived from the pattern below.

$a_1 = a_1$ 1st term

$a_2 = a_1 + d$ 2nd term

$a_3 = a_1 + 2d$ 3rd term

$a_4 = a_1 + 3d$ 4th term

$a_5 = a_1 + 4d$ 5th term

1 less

$\vdots$

$a_n = a_1 + (n - 1)d$ nth term

1 less

The following definition summarizes this result.

The nth Term of an Arithmetic Sequence

The nth term of an arithmetic sequence has the form

$$a_n = a_1 + (n - 1)d$$

where d is the common difference between consecutive terms of the sequence and a_1 is the first term.

EXAMPLE 2   **Finding the nth Term**

Find a formula for the nth term of the arithmetic sequence whose common difference is 3 and whose first term is 2.

Solution You know that the formula for the nth term is of the form $a_n = a_1 + (n - 1)d$. Moreover, the common difference is $d = 3$ and the first term is $a_1 = 2$, so the formula must have the form

$$a_n = 2 + 3(n - 1). \qquad \text{Substitute 2 for } a_1 \text{ and 3 for } d.$$

So, the formula for the nth term is $a_n = 3n - 1$.

✓ **Checkpoint** 🔊))) *Audio-video solution in English & Spanish at LarsonPrecalculus.com*

Find a formula for the nth term of the arithmetic sequence whose common difference is 5 and whose first term is -1.

The sequence in Example 2 is as follows.

$$2, 5, 8, 11, 14, \ldots, 3n - 1, \ldots$$

The figure below shows a graph of the first 15 terms of this sequence. Notice that the points lie on a line. This makes sense because a_n is a linear function of n. In other words, the terms "arithmetic" and "linear" are closely connected.

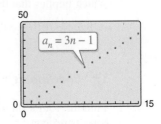

•• REMARK   Another way to find a_1 in Example 3 is to use the definition of the *n*th term of an arithmetic sequence, as shown below.

$a_n = a_1 + (n - 1)d$

$a_4 = a_1 + (4 - 1)d$

$20 = a_1 + (4 - 1)5$

$20 = a_1 + 15$

$5 = a_1$

EXAMPLE 3   **Writing the Terms of an Arithmetic Sequence**

The 4th term of an arithmetic sequence is 20, and the 13th term is 65. Write the first 11 terms of this sequence.

Solution   You know that $a_4 = 20$ and $a_{13} = 65$. So, you must add the common difference d nine times to the 4th term to obtain the 13th term. Therefore, the 4th and 13th terms of the sequence are related by

$a_{13} = a_4 + 9d.$ a_4 and a_{13} are nine terms apart.

Using $a_4 = 20$ and $a_{13} = 65$, you have $65 = 20 + 9d$. Solve for d to find that the common difference is $d = 5$. Use the common difference with the known term a_4 to write the other terms of the sequence.

a_1	a_2	a_3	a_4	a_5	a_6	a_7	a_8	a_9	a_{10}	a_{11} . . .
5	10	15	20	25	30	35	40	45	50	55 . . .

✓ *Checkpoint*  *Audio-video solution in English & Spanish at LarsonPrecalculus.com*

The 8th term of an arithmetic sequence is 25, and the 12th term is 41. Write the first 11 terms of this sequence.

When you know the *n*th term of an arithmetic sequence *and* you know the common difference of the sequence, you can find the $(n + 1)$th term by using the *recursion formula*

$a_{n+1} = a_n + d.$ Recursion formula

With this formula, you can find any term of an arithmetic sequence, *provided* that you know the preceding term. For example, when you know the first term, you can find the second term. Then, knowing the second term, you can find the third term, and so on.

EXAMPLE 4   **Using a Recursion Formula**

Find the ninth term of the arithmetic sequence whose first two terms are 2 and 9.

Solution   The common difference between consecutive terms of this sequence is

$d = 9 - 2 = 7.$

There are two ways to find the ninth term. One way is to write the first nine terms (by repeatedly adding 7).

2, 9, 16, 23, 30, 37, 44, 51, 58

Another way to find the ninth term is to first find a formula for the *n*th term. The common difference is $d = 7$ and the first term is $a_1 = 2$, so the formula must have the form

$a_n = 2 + 7(n - 1).$ Substitute 2 for a_1 and 7 for d.

Therefore, a formula for the *n*th term is

$a_n = 7n - 5$

which implies that the ninth term is

$a_9 = 7(9) - 5$

$= 58.$

✓ *Checkpoint*  *Audio-video solution in English & Spanish at LarsonPrecalculus.com*

Find the 10th term of the arithmetic sequence that begins with 7 and 15.

The Sum of a Finite Arithmetic Sequence

There is a formula for the *sum* of a finite arithmetic sequence.

•• **REMARK** Note that
this formula works only for
arithmetic sequences.

> ### The Sum of a Finite Arithmetic Sequence
>
> The sum of a finite arithmetic sequence with n terms is given by $S_n = \dfrac{n}{2}(a_1 + a_n)$.

For a proof of this formula, see Proofs in Mathematics on page 687.

EXAMPLE 5   Sum of a Finite Arithmetic Sequence

Find the sum: $1 + 3 + 5 + 7 + 9 + 11 + 13 + 15 + 17 + 19$.

Solution   To begin, notice that the sequence is arithmetic (with a common difference of 2). Moreover, the sequence has 10 terms. So, the sum of the sequence is

$$S_n = \frac{n}{2}(a_1 + a_n) \qquad \text{Sum of a finite arithmetic sequence}$$

$$= \frac{10}{2}(1 + 19) \qquad \text{Substitute 10 for } n, \text{ 1 for } a_1, \text{ and 19 for } a_n.$$

$$= 5(20) = 100. \qquad \text{Simplify.}$$

✓ *Checkpoint*  *Audio-video solution in English & Spanish at LarsonPrecalculus.com*

Find the sum: $40 + 37 + 34 + 31 + 28 + 25 + 22$.

EXAMPLE 6   Sum of a Finite Arithmetic Sequence

Find the sum of the integers (a) from 1 to 100 and (b) from 1 to N.

Solution

a. The integers from 1 to 100 form an arithmetic sequence that has 100 terms. So, use the formula for the sum of a finite arithmetic sequence.

$$S_n = 1 + 2 + 3 + 4 + 5 + 6 + \cdots + 99 + 100$$

$$= \frac{n}{2}(a_1 + a_n) \qquad \text{Sum of a finite arithmetic sequence}$$

$$= \frac{100}{2}(1 + 100) \qquad \text{Substitute 100 for } n, \text{ 1 for } a_1, \text{ and 100 for } a_n.$$

$$= 50(101) = 5050 \qquad \text{Simplify.}$$

b. $S_n = 1 + 2 + 3 + 4 + \cdots + N$

$$= \frac{n}{2}(a_1 + a_n) \qquad \text{Sum of a finite arithmetic sequence}$$

$$= \frac{N}{2}(1 + N) \qquad \text{Substitute } N \text{ for } n, \text{ 1 for } a_1, \text{ and } N \text{ for } a_n.$$

✓ *Checkpoint*  *Audio-video solution in English & Spanish at LarsonPrecalculus.com*

Find the sum of the integers (a) from 1 to 35 and (b) from 1 to $2N$.

A teacher of Carl Friedrich Gauss (1777–1855) asked him to add all the integers from 1 to 100. When Gauss returned with the correct answer after only a few moments, the teacher could only look at him in astounded silence. This is what Gauss did:

$$\begin{array}{r} S_n = 1 + 2 + 3 + \cdots + 100 \\ S_n = 100 + 99 + 98 + \cdots + 1 \\ \hline 2S_n = 101 + 101 + 101 + \cdots + 101 \end{array}$$

$$S_n = \frac{100 \times 101}{2} = 5050.$$

Recall that the sum of the first n terms of an infinite sequence is the *nth partial sum*. The nth partial sum of an arithmetic sequence can be found by using the formula for the sum of a finite arithmetic sequence.

EXAMPLE 7   Partial Sum of an Arithmetic Sequence

Find the 150th partial sum of the arithmetic sequence

5, 16, 27, 38, 49, ...

Solution   For this arithmetic sequence, $a_1 = 5$ and $d = 16 - 5 = 11$. So,

$$a_n = 5 + 11(n - 1)$$

and the nth term is

$$a_n = 11n - 6.$$

Therefore, $a_{150} = 11(150) - 6 = 1644$, and the sum of the first 150 terms is

$$S_{150} = \frac{n}{2}(a_1 + a_{150}) \qquad \textit{nth partial sum formula}$$

$$= \frac{150}{2}(5 + 1644) \qquad \textit{Substitute 150 for } n, \text{ 5 for } a_1, \text{ and 1644 for } a_{150}.$$

$$= 75(1649) \qquad \textit{Simplify.}$$

$$= 123{,}675. \qquad \textit{nth partial sum}$$

✓ *Checkpoint*  *Audio-video solution in English & Spanish at LarsonPrecalculus.com*

Find the 120th partial sum of the arithmetic sequence

6, 12, 18, 24, 30, ...

EXAMPLE 8   Partial Sum of an Arithmetic Sequence

Find the 16th partial sum of the arithmetic sequence

100, 95, 90, 85, 80, ...

Solution   For this arithmetic sequence, $a_1 = 100$ and $d = 95 - 100 = -5$. So,

$$a_n = 100 + (-5)(n - 1)$$

and the nth term is

$$a_n = -5n + 105.$$

Therefore, $a_{16} = -5(16) + 105 = 25$, and the sum of the first 16 terms is

$$S_{16} = \frac{n}{2}(a_1 + a_{16}) \qquad \textit{nth partial sum formula}$$

$$= \frac{16}{2}(100 + 25) \qquad \textit{Substitute 16 for } n, \text{ 100 for } a_1, \text{ and 25 for } a_{16}.$$

$$= 8(125) \qquad \textit{Simplify.}$$

$$= 1000. \qquad \textit{nth partial sum}$$

✓ *Checkpoint*  *Audio-video solution in English & Spanish at LarsonPrecalculus.com*

Find the 30th partial sum of the arithmetic sequence

78, 76, 74, 72, 70, ...

Application

EXAMPLE 9 **Total Sales**

See LarsonPrecalculus.com for an interactive version of this type of example.

A small business sells $10,000 worth of skin care products during its first year. The owner of the business has set a goal of increasing annual sales by $7500 each year for 9 years. Assuming that this goal is met, find the total sales during the first 10 years this business is in operation.

Solution When the goal is met the annual sales form an arithmetic sequence with

$$a_1 = 10,000 \quad \text{and} \quad d = 7500.$$

So,

$$a_n = 10,000 + 7500(n - 1)$$

and the nth term of the sequence is

$$a_n = 7500n + 2500.$$

Therefore, the 10th term of the sequence is

$$a_{10} = 7500(10) + 2500$$

$$= 77,500. \qquad \text{See figure.}$$

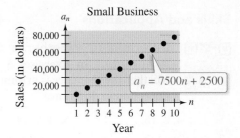

The sum of the first 10 terms of the sequence is

$$S_{10} = \frac{n}{2}(a_1 + a_{10}) \qquad n\text{th partial sum formula}$$

$$= \frac{10}{2}(10,000 + 77,500) \qquad \text{Substitute 10 for } n, \text{ 10,000 for } a_1, \text{ and 77,500 for } a_{10}.$$

$$= 5(87,500) \qquad \text{Simplify.}$$

$$= 437,500. \qquad \text{Multiply.}$$

So, the total sales for the first 10 years will be $437,500.

✓ *Checkpoint*  Audio-video solution in English & Spanish at LarsonPrecalculus.com

A company sells $160,000 worth of printing paper during its first year. The sales manager has set a goal of increasing annual sales of printing paper by $20,000 each year for 9 years. Assuming that this goal is met, find the total sales of printing paper during the first 10 years this company is in operation.

Summarize   (Section 9.2)

1. State the definition of an arithmetic sequence *(page 620)*, and state the formula for the nth term of an arithmetic sequence *(page 621)*. For examples of recognizing, writing, and finding the nth terms of arithmetic sequences, see Examples 1–4.

2. State the formula for the sum of a finite arithmetic sequence and explain how to use it to find the nth partial sum of an arithmetic sequence *(pages 623 and 624)*. For examples of finding sums of arithmetic sequences, see Examples 5–8.

3. Describe an example of how to use an arithmetic sequence to model and solve a real-life problem *(page 625, Example 9)*.

9.2 Exercises

See CalcChat.com for tutorial help and worked-out solutions to odd-numbered exercises.

Vocabulary: Fill in the blanks.

1. A sequence is _____ when the differences between consecutive terms are the same. This difference is the _____ difference.

2. The nth term of an arithmetic sequence has the form $a_n =$ _____.

3. When you know the nth term of an arithmetic sequence *and* you know the common difference of the sequence, you can find the $(n+1)$th term by using the _____ formula $a_{n+1} = a_n + d$.

4. The formula $S_n = \dfrac{n}{2}(a_1 + a_n)$ gives the sum of a _____ _____ _____ with n terms.

Skills and Applications

 Determining Whether a Sequence Is Arithmetic In Exercises 5–12, determine whether the sequence is arithmetic. If so, find the common difference.

5. $1, 2, 4, 8, 16, \ldots$
6. $4, 9, 14, 19, 24, \ldots$
7. $10, 8, 6, 4, 2, \ldots$
8. $80, 40, 20, 10, 5, \ldots$
9. $\frac{5}{4}, \frac{3}{2}, \frac{7}{4}, 2, \frac{9}{4}, \ldots$
10. $6.6, 5.9, 5.2, 4.5, 3.8, \ldots$
11. $1^2, 2^2, 3^2, 4^2, 5^2, \ldots$
12. $\ln 1, \ln 2, \ln 4, \ln 8, \ln 16, \ldots$

Writing the Terms of a Sequence In Exercises 13–20, write the first five terms of the sequence. Determine whether the sequence is arithmetic. If so, find the common difference. (Assume that n begins with 1.)

13. $a_n = 5 + 3n$
14. $a_n = 100 - 3n$
15. $a_n = 3 - 4(n - 2)$
16. $a_n = 1 + (n - 1)n$
17. $a_n = (-1)^n$
18. $a_n = n - (-1)^n$
19. $a_n = (2^n)n$
20. $a_n = \dfrac{3(-1)^n}{n}$

 Finding the nth Term In Exercises 21–30, find a formula for a_n for the arithmetic sequence.

21. $a_1 = 1, d = 3$
22. $a_1 = 15, d = 4$
23. $a_1 = 100, d = -8$
24. $a_1 = 0, d = -\frac{2}{3}$
25. $4, \frac{3}{2}, -1, -\frac{7}{2}, \ldots$
26. $10, 5, 0, -5, -10, \ldots$
27. $a_1 = 5, a_4 = 15$
28. $a_1 = -4, a_5 = 16$
29. $a_3 = 94, a_6 = 103$
30. $a_5 = 190, a_{10} = 115$

 Writing the Terms of an Arithmetic Sequence In Exercises 31–36, write the first five terms of the arithmetic sequence.

31. $a_1 = 5, d = 6$
32. $a_1 = 5, d = -\frac{3}{4}$
33. $a_1 = 2, a_{12} = -64$
34. $a_4 = 16, a_{10} = 46$
35. $a_8 = 26, a_{12} = 42$
36. $a_3 = 19, a_{15} = -1.7$

Writing the Terms of an Arithmetic Sequence In Exercises 37–40, write the first five terms of the arithmetic sequence defined recursively.

37. $a_1 = 15, a_{n+1} = a_n + 4$
38. $a_1 = 200, a_{n+1} = a_n - 10$
39. $a_5 = 7, a_{n+1} = a_n - 2$
40. $a_3 = 0.5, a_{n+1} = a_n + 0.75$

 Using a Recursion Formula In Exercises 41–44, the first two terms of the arithmetic sequence are given. Find the missing term.

41. $a_1 = 5, a_2 = -1, a_{10} =$ ▢
42. $a_1 = 3, a_2 = 13, a_9 =$ ▢
43. $a_1 = \dfrac{1}{8}, a_2 = \dfrac{3}{4}, a_7 =$ ▢
44. $a_1 = -0.7, a_2 = -13.8, a_8 =$ ▢

 Sum of a Finite Arithmetic Sequence In Exercises 45–50, find the sum of the finite arithmetic sequence.

45. $2 + 4 + 6 + 8 + 10 + 12 + 14 + 16 + 18 + 20$
46. $1 + 4 + 7 + 10 + 13 + 16 + 19$
47. $-1 + (-3) + (-5) + (-7) + (-9)$
48. $-5 + (-3) + (-1) + 1 + 3 + 5$
49. Sum of the first 100 positive odd integers
50. Sum of the integers from -100 to 30

 Partial Sum of an Arithmetic Sequence In Exercises 51–54, find the nth partial sum of the arithmetic sequence for the given value of n.

51. $8, 20, 32, 44, \ldots, \quad n = 50$
52. $-6, -2, 2, 6, \ldots, \quad n = 100$
53. $0, -9, -18, -27, \ldots, \quad n = 40$
54. $75, 70, 65, 60, \ldots, \quad n = 25$

Finding a Sum In Exercises 55–60, find the partial sum.

55. $\sum_{n=1}^{50} n$

56. $\sum_{n=51}^{100} 7n$

57. $\sum_{n=1}^{500} (n + 8)$

58. $\sum_{n=1}^{250} (1000 - n)$

59. $\sum_{n=1}^{100} (-6n + 20)$

60. $\sum_{n=1}^{75} (12n - 9)$

Matching an Arithmetic Sequence with Its Graph In Exercises 61–64, match the arithmetic sequence with its graph. [The graphs are labeled (a)–(d).]

(a)

(b)
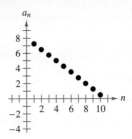

(c)
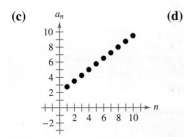

(d)
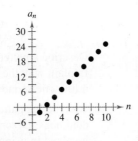

61. $a_n = -\frac{3}{4}n + 8$

62. $a_n = 3n - 5$

63. $a_n = 2 + \frac{3}{4}n$

64. $a_n = 25 - 3n$

Graphing the Terms of a Sequence In Exercises 65–68, use a graphing utility to graph the first 10 terms of the sequence. (Assume that n begins with 1.)

65. $a_n = 15 - \frac{3}{2}n$

66. $a_n = -5 + 2n$

67. $a_n = 0.2n + 3$

68. $a_n = -0.3n + 8$

Job Offer In Exercises 69 and 70, consider a job offer with the given starting salary and annual raise. (a) Determine the salary during the sixth year of employment. (b) Determine the total compensation from the company through six full years of employment.

	Starting Salary	Annual Raise
69.	$32,500	$1500
70.	$36,800	$1750

71. Seating Capacity Determine the seating capacity of an auditorium with 36 rows of seats when there are 15 seats in the first row, 18 seats in the second row, 21 seats in the third row, and so on.

72. Brick Pattern A triangular brick wall is made by cutting some bricks in half to use in the first column of every other row (see figure). The wall has 28 rows. The top row is one-half brick wide and the bottom row is 14 bricks wide. How many bricks are in the finished wall?

73. Falling Object

An object with negligible air resistance is dropped from the top of the Willis Tower in Chicago at a height of 1451 feet. During the first second of fall, the object falls 16 feet; during the second second, it  falls 48 feet; during the third second, it falls 80 feet; during the fourth second, it falls 112 feet. Assuming this pattern continues, how many feet does the object fall in the first 7 seconds after it is dropped?

74. Prize Money A county fair is holding a baked goods competition in which the top eight bakers receive cash prizes. First place receives $200, second place receives $175, third place receives $150, and so on.

(a) Write the nth term (a_n) of a sequence that represents the cash prize received in terms of the place n the baked good is awarded.

(b) Find the total amount of prize money awarded at the competition.

75. Total Sales An entrepreneur sells $15,000 worth of sports memorabilia during one year and sets a goal of increasing annual sales by $5000 each year for the next 9 years. Assuming that the entrepreneur meets this goal, find the total sales during the first 10 years of this business. What kinds of economic factors could prevent the business from meeting its goals?

76. Borrowing Money You borrow $5000 from your parents to purchase a used car. The arrangements of the loan are such that you make payments of $250 per month toward the balance plus 1% interest on the unpaid balance from the previous month.

(a) Find the first year's monthly payments and the unpaid balance after each month.

(b) Find the total amount of interest paid over the term of the loan.

77. Business The table shows the net numbers of new stores opened by H&M from 2011 through 2015. (*Source: H&M Hennes & Mauritz AB*)

Year	New Stores
2011	266
2012	304
2013	356
2014	379
2015	413

Spreadsheet at LarsonPrecalculus.com

(a) Construct a bar graph showing the annual net numbers of new stores opened by H&M from 2011 through 2015.

(b) Find the nth term (a_n) of an arithmetic sequence that approximates the data. Let n represent the year, with $n = 1$ corresponding to 2011. (*Hint:* Use the average change per year for d.)

(c) Use a graphing utility to graph the terms of the finite sequence you found in part (b).

(d) Use summation notation to represent the *total* number of new stores opened from 2011 through 2015. Use this sum to approximate the total number of new stores opened during these years.

78. Business In Exercise 77, there are a total number of 2206 stores at the end of 2010. Write the terms of a sequence that represents the total number of stores at the end of each year from 2011 through 2015. Is the sequence approximately arithmetic? Explain.

Exploration

True or False? **In Exercises 79 and 80, determine whether the statement is true or false. Justify your answer.**

79. Given an arithmetic sequence for which only the first two terms are known, it is possible to find the nth term.

80. When the first term, the nth term, and n are known for an arithmetic sequence, you have enough information to find the nth partial sum of the sequence.

81. Comparing Graphs of a Sequence and a Line

(a) Graph the first 10 terms of the arithmetic sequence $a_n = 2 + 3n$.

(b) Graph the equation of the line $y = 3x + 2$.

(c) Discuss any differences between the graph of $a_n = 2 + 3n$ and the graph of $y = 3x + 2$.

(d) Compare the slope of the line in part (b) with the common difference of the sequence in part (a). What can you conclude about the slope of a line and the common difference of an arithmetic sequence?

82. Writing Describe two ways to use the first two terms of an arithmetic sequence to find the 13th term.

Finding the Terms of a Sequence **In Exercises 83 and 84, find the first 10 terms of the sequence.**

83. $a_1 = x, d = 2x$

84. $a_1 = -y, d = 5y$

85. Error Analysis Describe the error in finding the sum of the first 50 odd integers.

$$S_n = \frac{n}{2}(a_1 + a_n) = \frac{50}{2}(1 + 101) = 2550 \quad \times$$

86. 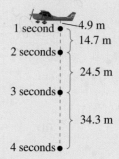 **HOW DO YOU SEE IT?** A steel ball with negligible air resistance is dropped from an airplane. The figure shows the distance that the ball falls during each of the first four seconds after it is dropped.

1 second — 4.9 m
2 seconds — 14.7 m
3 seconds — 24.5 m
4 seconds — 34.3 m

(a) Describe a pattern in the distances shown. Explain why the distances form a finite arithmetic sequence.

(b) Assume the pattern described in part (a) continues. Describe the steps and formulas involved in using the sum of a finite sequence to find the total distance the ball falls in n seconds, where n is a whole number.

87. Pattern Recognition

(a) Compute the following sums of consecutive positive odd integers.

$$1 + 3 = \quad\rule{1cm}{0.4cm}$$
$$1 + 3 + 5 = \quad\rule{1cm}{0.4cm}$$
$$1 + 3 + 5 + 7 = \quad\rule{1cm}{0.4cm}$$
$$1 + 3 + 5 + 7 + 9 = \quad\rule{1cm}{0.4cm}$$
$$1 + 3 + 5 + 7 + 9 + 11 = \quad\rule{1cm}{0.4cm}$$

(b) Use the sums in part (a) to make a conjecture about the sums of consecutive positive odd integers. Check your conjecture for the sum

$$1 + 3 + 5 + 7 + 9 + 11 + 13 = \quad\rule{1cm}{0.4cm}.$$

(c) Verify your conjecture algebraically.

Project: Net Sales To work an extended application analyzing the net sales for Dollar Tree from 2001 through 2014, visit the textbook's website at *LarsonPrecalculus.com*. (*Source: Dollar Tree, Inc.*)

9.3 Geometric Sequences and Series

■ Recognize, write, and find the *n*th terms of geometric sequences.
■ Find the sum of a finite geometric sequence.
■ Find the sum of an infinite geometric series.
■ Use geometric sequences to model and solve real-life problems.

Geometric Sequences

In Section 9.2, you learned that a sequence whose consecutive terms have a common *difference* is an arithmetic sequence. In this section, you will study another important type of sequence called a **geometric sequence.** Consecutive terms of a geometric sequence have a common *ratio.*

Geometric sequences can help you model and solve real-life problems. For example, in Exercise 84 on page 636, you will use a geometric sequence to model the population of Argentina from 2009 through 2015.

· · · · · · · · · · · · · · · · ▷

· · **REMARK** Be sure you
understand that a sequence such
as 1, 4, 9, 16, . . . , whose *n*th
term is n^2, is *not* geometric. The
ratio of the second term to the
first term is

$$\frac{a_2}{a_1} = \frac{4}{1} = 4$$

but the ratio of the third term to
the second term is

$$\frac{a_3}{a_2} = \frac{9}{4}.$$

> **Definition of Geometric Sequence**
>
> A sequence is **geometric** when the ratios of consecutive terms are the same.
> So, the sequence $a_1, a_2, a_3, a_4, \ldots, a_n, \ldots$ is geometric when there is a
> number r such that
>
> $$\frac{a_2}{a_1} = \frac{a_3}{a_2} = \frac{a_4}{a_3} = \cdots = r, \quad r \neq 0.$$
>
> The number r is the **common ratio** of the geometric sequence.

| **EXAMPLE 1** | **Examples of Geometric Sequences** |

a. The sequence whose *n*th term is 2^n is geometric. The common ratio of consecutive
terms is 2.

$$2, 4, 8, 16, \ldots, 2^n, \ldots \qquad \text{Begin with } n = 1.$$

$$\underbrace{}_{\frac{4}{2} = 2}$$

b. The sequence whose *n*th term is $4(3^n)$ is geometric. The common ratio of consecutive
terms is 3.

$$12, 36, 108, 324, \ldots, 4(3^n), \ldots \qquad \text{Begin with } n = 1.$$

$$\underbrace{}_{\frac{36}{12} = 3}$$

c. The sequence whose *n*th term is $\left(-\frac{1}{3}\right)^n$ is geometric. The common ratio of consecutive
terms is $-\frac{1}{3}$.

$$-\frac{1}{3}, \frac{1}{9}, -\frac{1}{27}, \frac{1}{81}, \ldots, \left(-\frac{1}{3}\right)^n, \ldots \qquad \text{Begin with } n = 1.$$

$$\underbrace{\phantom{-\frac{1}{3}, \frac{1}{9}}}_{\frac{1/9}{-1/3} = -\frac{1}{3}}$$

✓ *Checkpoint* ◀))) *Audio-video solution in English & Spanish at LarsonPrecalculus.com*

Write the first four terms of the geometric sequence whose *n*th term is $6(-2)^n$.
Then find the common ratio of the consecutive terms. ■

In Example 1, notice that each of the geometric sequences has an *n*th term that is
of the form ar^n, where the common ratio of the sequence is r. A geometric sequence
may be thought of as an exponential function whose domain is the set of natural numbers.

The *n*th Term of a Geometric Sequence

The *n*th term of a geometric sequence has the form

$$a_n = a_1 r^{n-1}$$

where r is the common ratio of consecutive terms of the sequence. So, every geometric sequence can be written in the form below.

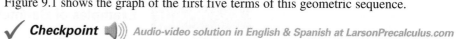

$$a_1, a_1r, a_1r^2, a_1r^3, a_1r^4, \ldots, a_1r^{n-1}, \ldots$$

When you know the *n*th term of a geometric sequence, multiply by r to find the $(n + 1)$th term. That is, $a_{n+1} = a_n r$.

EXAMPLE 2 Writing the Terms of a Geometric Sequence

Write the first five terms of the geometric sequence whose first term is $a_1 = 3$ and whose common ratio is $r = 2$. Then graph the terms on a set of coordinate axes.

Solution Starting with 3, repeatedly multiply by 2 to obtain the terms below.

$a_1 = 3$ 1st term $a_4 = 3(2^3) = 24$ 4th term

$a_2 = 3(2^1) = 6$ 2nd term $a_5 = 3(2^4) = 48$ 5th term

$a_3 = 3(2^2) = 12$ 3rd term

Figure 9.1 shows the graph of the first five terms of this geometric sequence.

✓ Checkpoint  Audio-video solution in English & Spanish at LarsonPrecalculus.com

Write the first five terms of the geometric sequence whose first term is $a_1 = 2$ and whose common ratio is $r = 4$. Then graph the terms on a set of coordinate axes.

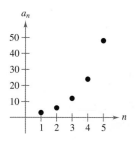

Figure 9.1

EXAMPLE 3 Finding a Term of a Geometric Sequence

Find the 15th term of the geometric sequence whose first term is 20 and whose common ratio is 1.05.

Algebraic Solution

$a_n = a_1 r^{n-1}$ Formula for *n*th term of a geometric sequence

$a_{15} = 20(1.05)^{15-1}$ Substitute 20 for a_1, 1.05 for r, and 15 for n.

≈ 39.60 Use a calculator.

Numerical Solution

For this sequence, $r = 1.05$ and $a_1 = 20$. So, $a_n = 20(1.05)^{n-1}$. Use a graphing utility to create a table that shows the terms of the sequence.

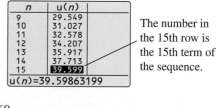

The number in the 15th row is the 15th term of the sequence.

So, $a_{15} \approx 39.60$.

✓ Checkpoint  Audio-video solution in English & Spanish at LarsonPrecalculus.com

Find the 12th term of the geometric sequence whose first term is 14 and whose common ratio is 1.2.

EXAMPLE 4 **Writing the *n*th Term of a Geometric Sequence**

Find a formula for the *n*th term of the geometric sequence

$$5, 15, 45, \ldots .$$

What is the 12th term of the sequence?

Solution The common ratio of this sequence is $r = 15/5 = 3$. The first term is $a_1 = 5$, so the formula for the *n*th term is

$$a_n = a_1 r^{n-1}$$
$$= 5(3)^{n-1}.$$

Use the formula for a_n to find the 12th term of the sequence.

$$a_{12} = 5(3)^{12-1} \qquad \text{Substitute 12 for } n.$$
$$= 5(177{,}147) \qquad \text{Use a calculator.}$$
$$= 885{,}735. \qquad \text{Multiply.}$$

✓ *Checkpoint* ◀))) *Audio-video solution in English & Spanish at LarsonPrecalculus.com*

Find a formula for the *n*th term of the geometric sequence

$$4, 20, 100, \ldots .$$

What is the 12th term of the sequence? ∎

When you know *any* two terms of a geometric sequence, you can use that information to find *any other* term of the sequence.

EXAMPLE 5 **Finding a Term of a Geometric Sequence**

The 4th term of a geometric sequence is 125, and the 10th term is 125/64. Find the 14th term. (Assume that the terms of the sequence are positive.)

Solution The 10th term is related to the 4th term by the equation

$$a_{10} = a_4 r^6. \qquad \text{Multiply fourth term by } r^{10-4}.$$

Use $a_{10} = 125/64$ and $a_4 = 125$ to solve for r.

$$\frac{125}{64} = 125 r^6 \qquad \text{Substitute } \tfrac{125}{64} \text{ for } a_{10} \text{ and 125 for } a_4.$$

$$\frac{1}{64} = r^6 \qquad \text{Divide each side by 125.}$$

$$\frac{1}{2} = r \qquad \text{Take the sixth root of each side.}$$

Multiply the 10th term by $r^{14-10} = r^4$ to obtain the 14th term.

$$a_{14} = a_{10} r^4 = \frac{125}{64}\left(\frac{1}{2}\right)^4 = \frac{125}{64}\left(\frac{1}{16}\right) = \frac{125}{1024}$$

✓ *Checkpoint* ◀))) *Audio-video solution in English & Spanish at LarsonPrecalculus.com*

The second term of a geometric sequence is 6, and the fifth term is 81/4. Find the eighth term. (Assume that the terms of the sequence are positive.) ∎

▷ **ALGEBRA HELP**

Remember that r is the common ratio of consecutive terms of a geometric sequence. So, in Example 5

$$a_{10} = a_1 r^9$$
$$= a_1 \cdot r \cdot r \cdot r \cdot r^6$$
$$= a_1 \cdot \frac{a_2}{a_1} \cdot \frac{a_3}{a_2} \cdot \frac{a_4}{a_3} \cdot r^6$$
$$= a_4 r^6.$$

The Sum of a Finite Geometric Sequence

The formula for the sum of a *finite* geometric sequence is as follows.

The Sum of a Finite Geometric Sequence

The sum of the finite geometric sequence

$$a_1, a_1r, a_1r^2, a_1r^3, a_1r^4, \ldots, a_1r^{n-1}$$

with common ratio $r \neq 1$ is given by $S_n = \displaystyle\sum_{i=1}^{n} a_1 r^{i-1} = a_1\left(\dfrac{1 - r^n}{1 - r}\right)$.

For a proof of this formula for the sum of a finite geometric sequence, see Proofs in Mathematics on page 687.

EXAMPLE 6 **Sum of a Finite Geometric Sequence**

Find the sum $\displaystyle\sum_{i=1}^{12} 4(0.3)^{i-1}$.

Solution You have

$$\sum_{i=1}^{12} 4(0.3)^{i-1} = 4(0.3)^0 + 4(0.3)^1 + 4(0.3)^2 + \cdots + 4(0.3)^{11}.$$

Using $a_1 = 4$, $r = 0.3$, and $n = 12$, apply the formula for the sum of a finite geometric sequence.

$$S_n = a_1\left(\frac{1 - r^n}{1 - r}\right) \qquad \text{Sum of a finite geometric sequence}$$

$$\sum_{i=1}^{12} 4(0.3)^{i-1} = 4\left[\frac{1 - (0.3)^{12}}{1 - 0.3}\right] \qquad \text{Substitute 4 for } a_1, \text{ 0.3 for } r, \text{ and 12 for } n.$$

$$\approx 5.714 \qquad \text{Use a calculator.}$$

✓ **Checkpoint** ◖))) *Audio-video solution in English & Spanish at LarsonPrecalculus.com*

Find the sum $\displaystyle\sum_{i=1}^{10} 2(0.25)^{i-1}$.

When using the formula for the sum of a finite geometric sequence, make sure that the sum is of the form

$$\sum_{i=1}^{n} a_1 r^{i-1}. \qquad \text{Exponent for } r \text{ is } i - 1.$$

For a sum that is not of this form, you must rewrite the sum before applying the formula. For example, the sum $\displaystyle\sum_{i=1}^{12} 4(0.3)^i$ is evaluated as follows.

$$\sum_{i=1}^{12} 4(0.3)^i = \sum_{i=1}^{12} 4[(0.3)(0.3)^{i-1}] \qquad \text{Property of exponents}$$

$$= \sum_{i=1}^{12} 4(0.3)(0.3)^{i-1} \qquad \text{Associative Property}$$

$$= 4(0.3)\left[\frac{1 - (0.3)^{12}}{1 - 0.3}\right] \qquad a_1 = 4(0.3), r = 0.3, n = 12$$

$$\approx 1.714$$

Geometric Series

The sum of the terms of an infinite geometric *sequence* is called an **infinite geometric series** or simply a **geometric series.**

The formula for the sum of a *finite geometric sequence* can, depending on the value of r, be extended to produce a formula for the sum of an *infinite geometric series*. Specifically, if the common ratio r has the property that $|r| < 1$, then it can be shown that r^n approaches zero as n increases without bound. Consequently,

$$a_1 \left(\frac{1 - r^n}{1 - r} \right) \rightarrow a_1 \left(\frac{1 - 0}{1 - r} \right) \quad \text{as} \quad n \rightarrow \infty.$$

The following summarizes this result.

The Sum of an Infinite Geometric Series

If $|r| < 1$, then the infinite geometric series

$$a_1 + a_1 r + a_1 r^2 + a_1 r^3 + \cdots + a_1 r^{n-1} + \cdots$$

has the sum

$$S = \sum_{i=0}^{\infty} a_1 r^i = \frac{a_1}{1 - r}.$$

Note that when $|r| \geq 1$, the series does not have a sum.

EXAMPLE 7   **Finding the Sum of an Infinite Geometric Series**

Find each sum.

a. $\displaystyle \sum_{n=0}^{\infty} 4(0.6)^n$

b. $3 + 0.3 + 0.03 + 0.003 + \cdots$

Solution

a. $\displaystyle \sum_{n=0}^{\infty} 4(0.6)^n = 4 + 4(0.6) + 4(0.6)^2 + 4(0.6)^3 + \cdots + 4.(0.6)^n + \cdots$

$$= \frac{4}{1 - 0.6} \qquad \frac{a_1}{1 - r}$$

$$= 10$$

b. $3 + 0.3 + 0.03 + 0.003 + \cdots = 3 + 3(0.1) + 3(0.1)^2 + 3(0.1)^3 + \cdots$

$$= \frac{3}{1 - 0.1} \qquad \frac{a_1}{1 - r}$$

$$= \frac{10}{3}$$

$$\approx 3.33$$

✓ **Checkpoint** 🔊)) *Audio-video solution in English & Spanish at LarsonPrecalculus.com*

Find each sum.

a. $\displaystyle \sum_{n=0}^{\infty} 5(0.5)^n$

b. $5 + 1 + 0.2 + 0.04 + \cdots$

Application

EXAMPLE 8   **Increasing Annuity**

See LarsonPrecalculus.com for an interactive version of this type of example.

An investor deposits $50 on the first day of each month in an account that pays 3% interest, compounded monthly. What is the balance at the end of 2 years? (This type of investment plan is called an **increasing annuity.**)

Solution   To find the balance in the account after 24 months, consider each of the 24 deposits separately. The first deposit will gain interest for 24 months, and its balance will be

$$A_{24} = 50\left(1 + \frac{0.03}{12}\right)^{24}$$

$$= 50(1.0025)^{24}.$$

The second deposit will gain interest for 23 months, and its balance will be

$$A_{23} = 50\left(1 + \frac{0.03}{12}\right)^{23}$$

$$= 50(1.0025)^{23}.$$

The last deposit will gain interest for only 1 month, and its balance will be

$$A_1 = 50\left(1 + \frac{0.03}{12}\right)^{1}$$

$$= 50(1.0025).$$

The total balance in the annuity will be the sum of the balances of the 24 deposits. Using the formula for the sum of a finite geometric sequence, with $A_1 = 50(1.0025)$, $r = 1.0025$, and $n = 24$, you have

$$S_n = A_1\left(\frac{1 - r^n}{1 - r}\right) \qquad \text{Sum of a finite geometric sequence}$$

$$S_{24} = 50(1.0025)\left[\frac{1 - (1.0025)^{24}}{1 - 1.0025}\right] \qquad \text{Substitute } 50(1.0025) \text{ for } A_1, \\ 1.0025 \text{ for } r, \text{ and } 24 \text{ for } n.$$

$$\approx \$1238.23. \qquad \text{Use a calculator.}$$

✓ **Checkpoint**   Audio-video solution in English & Spanish at LarsonPrecalculus.com

An investor deposits $70 on the first day of each month in an account that pays 2% interest, compounded monthly. What is the balance at the end of 4 years?   ◼

·· REMARK Recall from Section 3.1 that the formula for compound interest (for n compoundings per year) is

$$A = P\left(1 + \frac{r}{n}\right)^{nt}.$$

So, in Example 8, $50 is the principal P, 0.03 is the annual interest rate r, 12 is the number n of compoundings per year, and 2 is the time t in years. When you substitute these values into the formula, you obtain

$$A = 50\left(1 + \frac{0.03}{12}\right)^{12(2)}$$

$$= 50\left(1 + \frac{0.03}{12}\right)^{24}.$$

Summarize   (Section 9.3)

1. State the definition of a geometric sequence *(page 629)* and state the formula for the *n*th term of a geometric sequence *(page 630)*. For examples of recognizing, writing, and finding the *n*th terms of geometric sequences, see Examples 1–5.

2. State the formula for the sum of a finite geometric sequence *(page 632)*. For an example of finding the sum of a finite geometric sequence, see Example 6.

3. State the formula for the sum of an infinite geometric series *(page 633)*. For an example of finding the sums of infinite geometric series, see Example 7.

4. Describe an example of how to use a geometric sequence to model and solve a real-life problem *(page 634, Example 8)*.

9.3 Exercises

See CalcChat.com for tutorial help and worked-out solutions to odd-numbered exercises.

Vocabulary: Fill in the blanks.

1. A sequence is _____ when the ratios of consecutive terms are the same. This ratio is the _____ ratio.
2. The term of a geometric sequence has the form $a_n =$ _____.
3. The sum of a finite geometric sequence with common ratio $r \neq 1$ is given by $S_n =$ _____.
4. The sum of the terms of an infinite geometric sequence is called a _____ _____.

Skills and Applications

 Determining Whether a Sequence Is Geometric In Exercises 5–12, determine whether the sequence is geometric. If so, find the common ratio.

5. 3, 6, 12, 24, . . .
6. 5, 10, 15, 20, . . .
7. $\frac{1}{27}, \frac{1}{9}, \frac{1}{3}, 1, . . .$
8. 27, −9, 3, −1, . . .
9. $1, \frac{1}{2}, \frac{1}{3}, \frac{1}{4}, . . .$
10. 5, 1, 0.2, 0.04, . . .
11. $1, -\sqrt{7}, 7, -7\sqrt{7}, . . .$
12. $2, \frac{4}{\sqrt{3}}, \frac{8}{3}, \frac{16}{3\sqrt{3}}, . . .$

 Writing the Terms of a Geometric Sequence In Exercises 13–22, write the first five terms of the geometric sequence.

13. $a_1 = 4, r = 3$
14. $a_1 = 7, r = 4$
15. $a_1 = 1, r = \frac{1}{2}$
16. $a_1 = 6, r = -\frac{1}{4}$
17. $a_1 = 1, r = e$
18. $a_1 = 2, r = \pi$
19. $a_1 = 3, r = \sqrt{5}$
20. $a_1 = 4, r = -1/\sqrt{2}$
21. $a_1 = 2, r = 3x$
22. $a_1 = 4, r = x/5$

 Finding a Term of a Geometric Sequence In Exercises 23–32, write an expression for the nth term of the geometric sequence. Then find the missing term.

23. $a_1 = 4, r = \frac{1}{2}, a_{10} =$
24. $a_1 = 5, r = \frac{7}{2}, a_8 =$
25. $a_1 = 6, r = -\frac{1}{3}, a_{12} =$
26. $a_1 = 64, r = -\frac{1}{4}, a_{10} =$
27. $a_1 = 100, r = e^x, a_9 =$
28. $a_1 = 1, r = e^{-x}, a_4 =$
29. $a_1 = 1, r = \sqrt{2}, a_{12} =$
30. $a_1 = 1, r = \sqrt{3}, a_8 =$
31. $a_1 = 500, r = 1.02, a_{40} =$
32. $a_1 = 1000, r = 1.005, a_{60} =$

 Writing the nth Term of a Geometric Sequence In Exercises 33–38, find a formula for the nth term of the sequence.

33. 64, 32, 16, . . .
34. 81, 27, 9, . . .
35. 9, 18, 36, . . .
36. 5, −10, 20, . . .
37. $6, -9, \frac{27}{2}, . . .$
38. 80, −40, 20, . . .

 Finding a Term of a Geometric Sequence In Exercises 39–46, find the specified term of the geometric sequence.

39. 8th term: 6, 18, 54, . . .
40. 7th term: 5, 20, 80, . . .
41. 9th term: $\frac{1}{3}, -\frac{1}{6}, \frac{1}{12}, . . .$
42. 8th term: $\frac{3}{2}, -1, \frac{2}{3}, . . .$
43. a_3: $a_1 = 16, a_4 = \frac{27}{4}$
44. a_1: $a_2 = 3, a_5 = \frac{3}{64}$
45. a_6: $a_4 = -18, a_7 = \frac{2}{3}$
46. a_5: $a_2 = 2, a_3 = -\sqrt{2}$

Matching a Geometric Sequence with Its Graph In Exercises 47–50, match the geometric sequence with its graph. [The graphs are labeled (a), (b), (c), and (d).]

(a)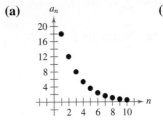

(b)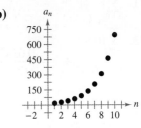

(c)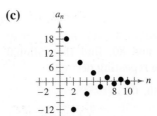

(d)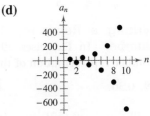

47. $a_n = 18\left(\frac{2}{3}\right)^{n-1}$
48. $a_n = 18\left(-\frac{2}{3}\right)^{n-1}$
49. $a_n = 18\left(\frac{3}{2}\right)^{n-1}$
50. $a_n = 18\left(-\frac{3}{2}\right)^{n-1}$

Graphing the Terms of a Sequence In Exercises 51–54, use a graphing utility to graph the first 10 terms of the sequence.

51. $a_n = 14(1.4)^{n-1}$
52. $a_n = 18(0.7)^{n-1}$
53. $a_n = 8(-0.3)^{n-1}$
54. $a_n = 11(-1.9)^{n-1}$

Sum of a Finite Geometric Sequence
In Exercises 55–64, find the sum of the finite geometric sequence.

55. $\displaystyle\sum_{n=1}^{7} 4^{n-1}$

56. $\displaystyle\sum_{n=1}^{10} \left(\frac{3}{2}\right)^{n-1}$

57. $\displaystyle\sum_{n=1}^{6} (-7)^{n-1}$

58. $\displaystyle\sum_{n=1}^{8} 5\left(-\frac{5}{2}\right)^{n-1}$

59. $\displaystyle\sum_{n=0}^{20} 3\left(\frac{3}{2}\right)^{n}$

60. $\displaystyle\sum_{n=0}^{40} 5\left(\frac{3}{5}\right)^{n}$

61. $\displaystyle\sum_{n=0}^{5} 200(1.05)^{n}$

62. $\displaystyle\sum_{n=0}^{6} 500(1.04)^{n}$

63. $\displaystyle\sum_{n=0}^{40} 2\left(-\frac{1}{4}\right)^{n}$

64. $\displaystyle\sum_{n=0}^{50} 10\left(\frac{2}{3}\right)^{n-1}$

Using Summation Notation In Exercises 65–68, use summation notation to write the sum.

65. $10 + 30 + 90 + \cdots + 7290$

66. $15 - 3 + \frac{3}{5} - \cdots - \frac{3}{625}$

67. $0.1 + 0.4 + 1.6 + \cdots + 102.4$

68. $32 + 24 + 18 + 13.5 + 10.125$

Sum of an Infinite Geometric Series
In Exercises 69–78, find the sum of the infinite geometric series.

69. $\displaystyle\sum_{n=0}^{\infty} \left(\frac{1}{2}\right)^{n}$

70. $\displaystyle\sum_{n=0}^{\infty} 2\left(\frac{3}{4}\right)^{n}$

71. $\displaystyle\sum_{n=0}^{\infty} \left(-\frac{1}{2}\right)^{n}$

72. $\displaystyle\sum_{n=0}^{\infty} 2\left(-\frac{2}{3}\right)^{n}$

73. $\displaystyle\sum_{n=0}^{\infty} (0.8)^{n}$

74. $\displaystyle\sum_{n=0}^{\infty} 4(0.2)^{n}$

75. $8 + 6 + \frac{9}{2} + \frac{27}{8} + \cdots$

76. $9 + 6 + 4 + \frac{8}{3} + \cdots$

77. $\frac{1}{9} - \frac{1}{3} + 1 - 3 + \cdots$

78. $-\frac{125}{36} + \frac{25}{6} - 5 + 6 - \cdots$

Writing a Repeating Decimal as a Rational Number In Exercises 79 and 80, find the rational number representation of the repeating decimal.

79. $0.\overline{36}$

80. $0.3\overline{18}$

Graphical Reasoning In Exercises 81 and 82, use a graphing utility to graph the function. Identify the horizontal asymptote of the graph and determine its relationship to the sum.

81. $f(x) = 6\left[\dfrac{1 - (0.5)^{x}}{1 - (0.5)}\right]$, $\quad\displaystyle\sum_{n=0}^{\infty} 6\left(\frac{1}{2}\right)^{n}$

82. $f(x) = 2\left[\dfrac{1 - (0.8)^{x}}{1 - (0.8)}\right]$, $\quad\displaystyle\sum_{n=0}^{\infty} 2\left(\frac{4}{5}\right)^{n}$

83. Depreciation A tool and die company buys a machine for $175,000 and it depreciates at a rate of 30% per year. (In other words, at the end of each year the depreciated value is 70% of what it was at the beginning of the year.) Find the depreciated value of the machine after 5 full years.

84. Population

The table shows the mid-year populations of Argentina (in millions) from 2009 through 2015. (Source: U.S. Census Bureau)

DATA	Year	Population
	2009	40.9
	2010	41.3
	2011	41.8
	2012	42.2
	2013	42.6
	2014	43.0
	2015	43.4

Spreadsheet at LarsonPrecalculus.com

(a) Use the *exponential regression* feature of a graphing utility to find the nth term (a_n) of a geometric sequence that models the data. Let n represent the year, with $n = 9$ corresponding to 2009.

(b) Use the sequence from part (a) to describe the rate at which the population of Argentina is growing.

(c) Use the sequence from part (a) to predict the population of Argentina in 2025. The U.S. Census Bureau predicts the population of Argentina will be 47.2 million in 2025. How does this value compare with your prediction?

(d) Use the sequence from part (a) to predict when the population of Argentina will reach 50.0 million.

85. Annuity An investor deposits P dollars on the first day of each month in an account with an annual interest rate r, compounded monthly. The balance A after t years is

$$A = P\left(1 + \frac{r}{12}\right) + \cdots + P\left(1 + \frac{r}{12}\right)^{12t}.$$

Show that the balance is

$$A = P\left[\left(1 + \frac{r}{12}\right)^{12t} - 1\right]\left(1 + \frac{12}{r}\right).$$

86. Annuity An investor deposits $100 on the first day of each month in an account that pays 2% interest, compounded monthly. The balance A in the account at the end of 5 years is

$$A = 100\left(1 + \frac{0.02}{12}\right)^1 + \cdots + 100\left(1 + \frac{0.02}{12}\right)^{60}.$$

Use the result of Exercise 85 to find A.

Multiplier Effect In Exercises 87 and 88, use the following information. A state government gives property owners a tax rebate with the anticipation that each property owner will spend approximately $p\%$ of the rebate, and in turn each recipient of this amount will spend $p\%$ of what he or she receives, and so on. Economists refer to this exchange of money and its circulation within the economy as the "multiplier effect." The multiplier effect operates on the idea that the expenditures of one individual become the income of another individual. For the given tax rebate, find the total amount of spending that results, assuming that this effect continues without end.

	Tax rebate	$p\%$
87.	$400	75%
88.	$600	72.5%

89. Geometry The sides of a square are 27 inches in length. New squares are formed by dividing the original square into nine squares. The center square is then shaded (see figure). This process is repeated three more times. Determine the total area of the shaded region.

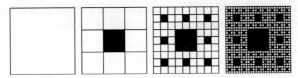

90. Distance A ball is dropped from a height of 6 feet and begins bouncing as shown in the figure. The height of each bounce is three-fourths the height of the previous bounce. Find the total vertical distance the ball travels before coming to rest.

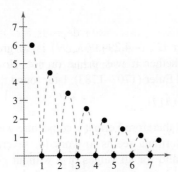

91. Salary An investment firm has a job opening with a salary of $45,000 for the first year. During the next 39 years, there is a 5% raise each year. Find the total compensation over the 40-year period.

92. HOW DO YOU SEE IT? Use the figures shown below.

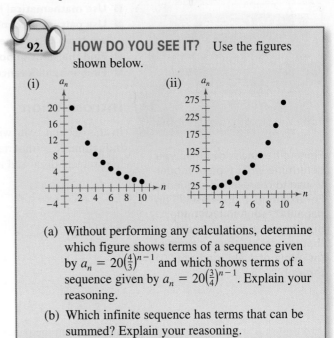

(a) Without performing any calculations, determine which figure shows terms of a sequence given by $a_n = 20\left(\frac{4}{3}\right)^{n-1}$ and which shows terms of a sequence given by $a_n = 20\left(\frac{3}{4}\right)^{n-1}$. Explain your reasoning.

(b) Which infinite sequence has terms that can be summed? Explain your reasoning.

Exploration

True or False? In Exercises 93 and 94, determine whether the statement is true or false. Justify your answer.

93. A sequence is geometric when the ratios of consecutive differences of consecutive terms are the same.

94. To find the nth term of a geometric sequence, multiply its common ratio by the first term of the sequence raised to the $(n - 1)$th power.

95. Graphical Reasoning Consider the graph of

$$y = \frac{1 - r^x}{1 - r}.$$

(a) Use a graphing utility to graph y for $r = \frac{1}{2}$, $\frac{2}{3}$, and $\frac{4}{5}$. What happens as $x \to \infty$?

(b) Use the graphing utility to graph y for $r = 1.5, 2,$ and 3. What happens as $x \to \infty$?

96. Writing Write a brief paragraph explaining why the terms of a geometric sequence decrease in magnitude when $-1 < r < 1$.

Project: Population To work an extended application analyzing the population of Delaware, visit this text's website at *LarsonPrecalculus.com*. (*Source: U.S. Census Bureau*)

9.4 Mathematical Induction

Skip

Finite differences can help you determine what type of model to use to represent a sequence. For example, in Exercise 75 on page 647, you will use finite differences to find a model that represents the numbers of residents of Alabama from 2010 through 2015.

■ Use mathematical induction to prove statements involving a positive integer n.
■ Use pattern recognition and mathematical induction to write a formula for the nth term of a sequence.
■ Find the sums of powers of integers.
■ Find finite differences of sequences.

Introduction

In this section, you will study a form of mathematical proof called **mathematical induction.** It is important that you see the logical need for it, so take a closer look at the problem discussed in Example 5 in Section 9.2.

$$S_1 = 1 = 1^2$$

$$S_2 = 1 + 3 = 2^2$$

$$S_3 = 1 + 3 + 5 = 3^2$$

$$S_4 = 1 + 3 + 5 + 7 = 4^2$$

$$S_5 = 1 + 3 + 5 + 7 + 9 = 5^2$$

$$S_6 = 1 + 3 + 5 + 7 + 9 + 11 = 6^2$$

Judging from the pattern formed by these first six sums, it appears that the sum of the first n odd integers is

$$S_n = 1 + 3 + 5 + 7 + 9 + 11 + \cdots + (2n - 1) = n^2.$$

Although this particular formula *is* valid, it is important for you to see that recognizing a pattern and then simply *jumping to the conclusion* that the pattern must be true for all values of n is *not* a logically valid method of proof. There are many examples in which a pattern appears to be developing for small values of n, but then at some point the pattern fails. One of the most famous cases of this was the conjecture by the French mathematician Pierre de Fermat (1601–1665), who speculated that all numbers of the form

$$F_n = 2^{2^n} + 1, \quad n = 0, 1, 2, \ldots$$

are prime. For $n = 0, 1, 2, 3,$ and 4, the conjecture is true.

$$F_0 = 3$$

$$F_1 = 5$$

$$F_2 = 17$$

$$F_3 = 257$$

$$F_4 = 65,537$$

The size of the next Fermat number ($F_5 = 4{,}294{,}967{,}297$) is so great that it was difficult for Fermat to determine whether it was prime or not. However, another well-known mathematician, Leonhard Euler (1707–1783), later found the factorization

$$F_5 = 4{,}294{,}967{,}297 = 641(6{,}700{,}417)$$

which proved that F_5 is not prime and therefore Fermat's conjecture was false.

Just because a rule, pattern, or formula seems to work for several values of n, you cannot simply decide that it is valid for all values of n without going through a *legitimate proof.* Mathematical induction is one method of proof.

REMARK It is important to recognize that in order to prove a statement by induction, both parts of the Principle of Mathematical Induction are necessary.

The Principle of Mathematical Induction

Let P_n be a statement involving the positive integer n. If

1. P_1 is true, and
2. for every positive integer k, the truth of P_k implies the truth of P_{k+1}

then the statement P_n must be true for all positive integers n.

To apply the Principle of Mathematical Induction, you need to be able to determine the statement P_{k+1} for a given statement P_k. To determine P_{k+1}, substitute the quantity $k + 1$ for k in the statement P_k.

EXAMPLE 1　A Preliminary Example

Find the statement P_{k+1} for each given statement P_k.

a. P_k: $S_k = \dfrac{k^2(k + 1)^2}{4}$

b. P_k: $S_k = 1 + 5 + 9 + \cdots + [4(k - 1) - 3] + (4k - 3)$

c. P_k: $k + 3 < 5k^2$

d. P_k: $3^k \geq 2k + 1$

Solution

a. P_{k+1}: $S_{k+1} = \dfrac{(k + 1)^2(k + 1 + 1)^2}{4}$ 　　Replace k with $k + 1$.

$= \dfrac{(k + 1)^2(k + 2)^2}{4}$ 　　Simplify.

b. P_{k+1}: $S_{k+1} = 1 + 5 + 9 + \cdots + \{4[(k + 1) - 1] - 3\} + [4(k + 1) - 3]$

$= 1 + 5 + 9 + \cdots + (4k - 3) + (4k + 1)$

c. P_{k+1}: $(k + 1) + 3 < 5(k + 1)^2$

$k + 4 < 5(k^2 + 2k + 1)$

d. P_{k+1}: $3^{k+1} \geq 2(k + 1) + 1$

$3^{k+1} \geq 2k + 3$

✓ **Checkpoint** ◀))) *Audio-video solution in English & Spanish at LarsonPrecalculus.com*

Find the statement P_{k+1} for each given statement P_k.

a. P_k: $S_k = \dfrac{6}{k(k + 3)}$ 　　**b.** P_k: $k + 2 \leq 3(k - 1)^2$ 　　**c.** P_k: $2^{4k-2} + 1 > 5k$ ■

An unending line of dominoes can illustrate how the Principle of Mathematical Induction works.

A well-known illustration of how the Principle of Mathematical Induction works is an unending line of dominoes. It is clear that you could not knock down an infinite number of dominoes *one domino* at a time. However, if it were true that each domino would knock down the next one as it fell, then you could knock them all down by pushing the first one and starting a chain reaction. Mathematical induction works in the same way. If the truth of P_k implies the truth of P_{k+1} and if P_1 is true, then the chain reaction proceeds as follows: P_1 implies P_2, P_2 implies P_3, P_3 implies P_4, and so on.

When using mathematical induction to prove a *summation* formula (such as the one in Example 2), it is helpful to think of S_{k+1} as $S_{k+1} = S_k + a_{k+1}$, where a_{k+1} is the $(k + 1)$th term of the original sum.

EXAMPLE 2 **Using Mathematical Induction**

Use mathematical induction to prove the formula

$$S_n = 1 + 3 + 5 + 7 + \cdots + (2n - 1)$$
$$= n^2$$

for all integers $n \geq 1$.

Solution Mathematical induction consists of two distinct parts.

1. First, you must show that the formula is true when $n = 1$. When $n = 1$, the formula is valid, because

    $$S_1 = 1$$
    $$= 1^2.$$

2. The second part of mathematical induction has two steps. The first step is to *assume* that the formula is valid for some integer k. The second step is to use this assumption to prove that the formula is valid for the *next* integer, $k + 1$. Assuming that the formula

    $$S_k = 1 + 3 + 5 + 7 + \cdots + (2k - 1) \qquad a_k = 2k - 1$$
    $$= k^2$$

    is true, you must show that the formula $S_{k+1} = (k + 1)^2$ is true.

    $$S_{k+1} = 1 + 3 + 5 + 7 + \cdots + (2k - 1)$$
    $$\qquad\qquad\qquad\qquad + [2(k + 1) - 1] \qquad S_{k+1} = S_k + a_{k+1}$$
    $$= [1 + 3 + 5 + 7 + \cdots + (2k - 1)] + (2k + 2 - 1)$$
    $$= S_k + (2k + 1) \qquad\qquad\qquad \text{Group terms to form } S_k.$$
    $$= k^2 + 2k + 1 \qquad\qquad\qquad \text{By assumption}$$
    $$= (k + 1)^2 \qquad\qquad\qquad\qquad S_k \text{ implies } S_{k+1}.$$

Combining the results of parts (1) and (2), you can conclude by mathematical induction that the formula is valid for all integers $n \geq 1$.

✓ **Checkpoint**))) *Audio-video solution in English & Spanish at LarsonPrecalculus.com*

Use mathematical induction to prove the formula

$$S_n = 5 + 7 + 9 + 11 + \cdots + (2n + 3) = n(n + 4)$$

for all integers $n \geq 1$. ◼

It occasionally happens that a statement involving natural numbers is not true for the first $k - 1$ positive integers but is true for all values of $n \geq k$. In these instances, you use a slight variation of the Principle of Mathematical Induction in which you verify P_k rather than P_1. This variation is called the *Extended Principle of Mathematical Induction*. To see the validity of this, note in the unending line of dominoes discussed on page 639 that all but the first $k - 1$ dominoes can be knocked down by knocking over the kth domino. This suggests that you can prove a statement P_n to be true for $n \geq k$ by showing that P_k is true and that P_k implies P_{k+1}. In Exercises 25–28 of this section, you will apply the Extended Principle of Mathematical Induction.

EXAMPLE 3 Using Mathematical Induction

Use mathematical induction to prove the formula

$$S_n = 1^2 + 2^2 + 3^2 + 4^2 + \cdots + n^2$$

$$= \frac{n(n + 1)(2n + 1)}{6}$$

for all integers $n \geq 1$.

Solution

1. When $n = 1$, the formula is valid, because

$$S_1 = 1^2$$

$$= \frac{1(2)(3)}{6}.$$

2. Assuming that the formula

$$S_k = 1^2 + 2^2 + 3^2 + 4^2 + \cdots + k^2 \qquad\qquad a_k = k^2$$

$$= \frac{k(k + 1)(2k + 1)}{6}$$

is true, you must show that

$$S_{k+1} = \frac{(k + 1)(k + 1 + 1)[2(k + 1) + 1]}{6}$$

$$= \frac{(k + 1)(k + 2)(2k + 3)}{6}$$

is true.

$$S_{k+1} = 1^2 + 2^2 + 3^2 + 4^2 + \cdots + k^2 + (k + 1)^2 \qquad S_{k+1} = S_k + a_{k+1}$$

$$= S_k + (k + 1)^2 \qquad\qquad \text{Group terms to form } S_k.$$

$$= \frac{k(k + 1)(2k + 1)}{6} + (k + 1)^2 \qquad\qquad \text{By assumption}$$

$$= \frac{k(k + 1)(2k + 1) + 6(k + 1)^2}{6} \qquad\qquad \text{Combine fractions.}$$

$$= \frac{(k + 1)[k(2k + 1) + 6(k + 1)]}{6} \qquad\qquad \text{Factor.}$$

$$= \frac{(k + 1)(2k^2 + 7k + 6)}{6} \qquad\qquad \text{Simplify.}$$

$$= \frac{(k + 1)(k + 2)(2k + 3)}{6} \qquad\qquad S_k \text{ implies } S_{k+1}.$$

•• **REMARK** Remember that when adding rational expressions, you must first find a common denominator. Example 3 uses the *least* common denominator of 6.

Combining the results of parts (1) and (2), you can conclude by mathematical induction that the formula is valid for all integers $n \geq 1$.

✓ **Checkpoint** 🔊)) *Audio-video solution in English & Spanish at LarsonPrecalculus.com*

Use mathematical induction to prove the formula

$$S_n = 1(1 - 1) + 2(2 - 1) + 3(3 - 1) + \cdots + n(n - 1) = \frac{n(n - 1)(n + 1)}{3}$$

for all integers $n \geq 1$.

When proving a formula using mathematical induction, the only statement that you *need* to verify is P_1. As a check, however, it is a good idea to try verifying some of the other statements. For instance, in Example 3, try verifying P_2 and P_3.

EXAMPLE 4 **Proving an Inequality**

Prove that

$$n < 2^n$$

for all integers $n \geq 1$.

Solution

1. For $n = 1$, the statement is true because

$$1 < 2^1.$$

2. Assuming that $k < 2^k$, you need to show that $k + 1 < 2^{k+1}$. To do this, use the fact that $2^{n+1} = 2(2^n) = 2^n + 2^n$. Then for $n = k$, you have

$$2^k + 2^k > 2^k + 1 > k + 1. \qquad \text{By assumption}$$

It follows that

$$2^{k+1} > 2^k + 1 > k + 1 \quad \text{or} \quad k + 1 < 2^{k+1}.$$

Combining the results of parts (1) and (2), you can conclude by mathematical induction that $n < 2^n$ for all integers $n \geq 1$.

✓ **Checkpoint**  Audio-video solution in English & Spanish at LarsonPrecalculus.com

Prove that

$$n! \geq n$$

for all integers $n \geq 1$.

EXAMPLE 5 **Proving a Property**

Prove that 3 is a factor of $4^n - 1$ for all integers $n \geq 1$.

Solution

1. For $n = 1$, the statement is true because

$$4^1 - 1 = 3.$$

So, 3 is a factor.

2. Assuming that 3 is a factor of $4^k - 1$, you must show that 3 is a factor of $4^{k+1} - 1$. To do this, write the following.

$$4^{k+1} - 1 = 4^{k+1} - 4^k + 4^k - 1 \qquad \text{Subtract and add } 4^k.$$
$$= 4^k(4 - 1) + (4^k - 1) \qquad \text{Regroup terms.}$$
$$= 4^k \cdot 3 + (4^k - 1) \qquad \text{Simplify.}$$

Because 3 is a factor of $4^k \cdot 3$ and 3 is also a factor of $4^k - 1$, it follows that 3 is a factor of $4^{k+1} - 1$. Combining the results of parts (1) and (2), you can conclude by mathematical induction that 3 is a factor of $4^n - 1$ for all integers $n \geq 1$.

✓ **Checkpoint**  Audio-video solution in English & Spanish at LarsonPrecalculus.com

Prove that 2 is a factor of $3^n + 1$ for all integers $n \geq 1$.

Pattern Recognition

Although choosing a formula on the basis of a few observations does *not* guarantee the validity of the formula, pattern recognition *is* important. Once you have a pattern or formula that you think works, try using mathematical induction to prove your formula.

Finding a Formula for the *n*th Term of a Sequence

To find a formula for the *n*th term of a sequence, consider these guidelines.

1. Calculate the first several terms of the sequence. It is often a good idea to write the terms in both simplified and factored forms.

2. Try to find a recognizable pattern for the terms and write a formula for the *n*th term of the sequence. This is your *hypothesis* or *conjecture*. You might compute one or two more terms in the sequence to test your hypothesis.

3. Use mathematical induction to prove your hypothesis.

EXAMPLE 6 Finding a Formula for a Finite Sum

Find a formula for the finite sum and prove its validity.

$$\frac{1}{1 \cdot 2} + \frac{1}{2 \cdot 3} + \frac{1}{3 \cdot 4} + \frac{1}{4 \cdot 5} + \cdots + \frac{1}{n(n + 1)}$$

Solution Begin by writing the first few sums.

$$S_1 = \frac{1}{1 \cdot 2} = \frac{1}{2} = \frac{1}{1 + 1}$$

$$S_2 = \frac{1}{1 \cdot 2} + \frac{1}{2 \cdot 3} = \frac{4}{6} = \frac{2}{3} = \frac{2}{2 + 1}$$

$$S_3 = \frac{1}{1 \cdot 2} + \frac{1}{2 \cdot 3} + \frac{1}{3 \cdot 4} = \frac{9}{12} = \frac{3}{4} = \frac{3}{3 + 1}$$

From this sequence, it appears that the formula for the *k*th sum is

$$S_k = \frac{1}{1 \cdot 2} + \frac{1}{2 \cdot 3} + \frac{1}{3 \cdot 4} + \frac{1}{4 \cdot 5} + \cdots + \frac{1}{k(k + 1)} = \frac{k}{k + 1}.$$

To prove the validity of this hypothesis, use mathematical induction. Note that you have already verified the formula for $n = 1$, so begin by assuming that the formula is valid for $n = k$ and trying to show that it is valid for $n = k + 1$.

$$S_{k+1} = \left[\frac{1}{1 \cdot 2} + \frac{1}{2 \cdot 3} + \frac{1}{3 \cdot 4} + \frac{1}{4 \cdot 5} + \cdots + \frac{1}{k(k + 1)} \right] + \frac{1}{(k + 1)(k + 2)}$$

$$= \frac{k}{k + 1} + \frac{1}{(k + 1)(k + 2)} \qquad \text{By assumption}$$

$$= \frac{k(k + 2) + 1}{(k + 1)(k + 2)} = \frac{k^2 + 2k + 1}{(k + 1)(k + 2)} = \frac{(k + 1)^2}{(k + 1)(k + 2)} = \frac{k + 1}{k + 2}$$

So, by mathematical induction the hypothesis is valid.

✓ *Checkpoint* 🔊))) *Audio-video solution in English & Spanish at LarsonPrecalculus.com*

Find a formula for the finite sum and prove its validity.

$$3 + 7 + 11 + 15 + \cdots + 4n - 1$$

9.4 Exercises   See CalcChat.com for tutorial help and worked-out solutions to odd-numbered exercises.

Vocabulary: Fill in the blanks.

1. The first step in proving a formula by _____ _____ is to show that the formula is true when $n = 1$.
2. To find the _____ differences of a sequence, subtract consecutive terms.
3. A sequence is an _____ sequence when the first differences are all the same nonzero number.
4. If the _____ differences of a sequence are all the same nonzero number, then the sequence has a perfect quadratic model.

Skills and Applications

 Finding P_{k+1} Given P_k In Exercises 5–10, find the statement P_{k+1} for the given statement P_k.

5. $P_k = \dfrac{5}{k(k+1)}$

6. $P_k = \dfrac{1}{2(k+2)}$

7. $P_k = k^2(k+3)^2$

8. $P_k = \frac{1}{3}k(2k+1)$

9. $P_k = \dfrac{3}{(k+2)(k+3)}$

10. $P_k = \dfrac{k^2}{2(k+1)^2}$

 Using Mathematical Induction In Exercises 11–24, use mathematical induction to prove the formula for all integers $n \geq 1$.

11. $2 + 4 + 6 + 8 + \cdots + 2n = n(n+1)$

12. $6 + 12 + 18 + 24 + \cdots + 6n = 3n(n+1)$

13. $2 + 7 + 12 + 17 + \cdots + (5n - 3) = \dfrac{n}{2}(5n - 1)$

14. $1 + 4 + 7 + 10 + \cdots + (3n - 2) = \dfrac{n}{2}(3n - 1)$

15. $1 + 2 + 2^2 + 2^3 + \cdots + 2^{n-1} = 2^n - 1$

16. $2(1 + 3 + 3^2 + 3^3 + \cdots + 3^{n-1}) = 3^n - 1$

17. $1 + 2 + 3 + 4 + \cdots + n = \dfrac{n(n+1)}{2}$

18. $1^3 + 2^3 + 3^3 + 4^3 + \cdots + n^3 = \dfrac{n^2(n+1)^2}{4}$

19. $1^2 + 3^2 + 5^2 + \cdots + (2n - 1)^2 = \dfrac{n(2n-1)(2n+1)}{3}$

20. $\left(1 + \dfrac{1}{1}\right)\left(1 + \dfrac{1}{2}\right)\left(1 + \dfrac{1}{3}\right)\cdots\left(1 + \dfrac{1}{n}\right) = n + 1$

21. $\displaystyle\sum_{i=1}^{n} i^5 = \dfrac{n^2(n+1)^2(2n^2 + 2n - 1)}{12}$

22. $\displaystyle\sum_{i=1}^{n} i^4 = \dfrac{n(n+1)(2n+1)(3n^2 + 3n - 1)}{30}$

23. $\displaystyle\sum_{i=1}^{n} i(i+1) = \dfrac{n(n+1)(n+2)}{3}$

24. $\displaystyle\sum_{i=1}^{n} \dfrac{1}{(2i-1)(2i+1)} = \dfrac{n}{2n+1}$

 Proving an Inequality In Exercises 25–30, use mathematical induction to prove the inequality for the specified integer values of n.

25. $n! > 2^n$, $n \geq 4$

26. $\left(\frac{4}{3}\right)^n > n$, $n \geq 7$

27. $\dfrac{1}{\sqrt{1}} + \dfrac{1}{\sqrt{2}} + \dfrac{1}{\sqrt{3}} + \cdots + \dfrac{1}{\sqrt{n}} > \sqrt{n}$, $n \geq 2$

28. $2n^2 > (n+1)^2$, $n \geq 3$

29. $\left(\dfrac{x}{y}\right)^{n+1} < \left(\dfrac{x}{y}\right)^n$, $n \geq 1$ and $0 < x < y$

30. $(1 + a)^n \geq na$, $n \geq 1$ and $a > 0$

 Proving a Property In Exercises 31–40, use mathematical induction to prove the property for all integers $n \geq 1$.

31. A factor of $n^3 + 3n^2 + 2n$ is 3.

32. A factor of $n^4 - n + 4$ is 2.

33. A factor of $2^{2n+1} + 1$ is 3.

34. A factor of $2^{2n-1} + 3^{2n-1}$ is 5.

35. $(ab)^n = a^n b^n$

36. $\left(\dfrac{a}{b}\right)^n = \dfrac{a^n}{b^n}$

37. If $x_1 \neq 0, x_2 \neq 0, \ldots, x_n \neq 0$, then
$(x_1 x_2 x_3 \cdots x_n)^{-1} = x_1^{-1} x_2^{-1} x_3^{-1} \cdots x_n^{-1}$.

38. If $x_1 > 0, x_2 > 0, \ldots, x_n > 0$, then
$\ln(x_1 x_2 \cdots x_n) = \ln x_1 + \ln x_2 + \cdots + \ln x_n$.

39. $x(y_1 + y_2 + \cdots + y_n) = xy_1 + xy_2 + \cdots + xy_n$

40. $(a + bi)^n$ and $(a - bi)^n$ are complex conjugates.

Finding a Formula for a Finite Sum In Exercises 41–44, find a formula for the sum of the first n terms of the sequence. Prove the validity of your formula.

41. $1, 5, 9, 13, \ldots$

42. $3, -\frac{9}{2}, \frac{27}{4}, -\frac{81}{8}, \ldots$

43. $\dfrac{1}{4}, \dfrac{1}{12}, \dfrac{1}{24}, \dfrac{1}{40}, \ldots, \dfrac{1}{2n(n+1)}, \ldots$

44. $\dfrac{1}{2 \cdot 3}, \dfrac{1}{3 \cdot 4}, \dfrac{1}{4 \cdot 5}, \dfrac{1}{5 \cdot 6}, \ldots, \dfrac{1}{(n+1)(n+2)}, \ldots$

Finding a Sum In Exercises 45–54, find the sum using the formulas for the sums of powers of integers.

45. $\displaystyle\sum_{n=1}^{15} n$

46. $\displaystyle\sum_{n=1}^{30} n$

47. $\displaystyle\sum_{n=1}^{6} n^2$

48. $\displaystyle\sum_{n=1}^{10} n^3$

49. $\displaystyle\sum_{n=1}^{5} n^4$

50. $\displaystyle\sum_{n=1}^{8} n^5$

51. $\displaystyle\sum_{n=1}^{6} (n^2 - n)$

52. $\displaystyle\sum_{n=1}^{20} (n^3 - n)$

53. $\displaystyle\sum_{i=1}^{6} (6i - 8i^3)$

54. $\displaystyle\sum_{j=1}^{10} \left(3 - \tfrac{1}{2}j + \tfrac{1}{2}j^2\right)$

Finding a Linear or Quadratic Model In Exercises 55–60, decide whether the sequence can be represented perfectly by a linear or a quadratic model. Then find the model.

55. 5, 14, 23, 32, 41, 50, . . .

56. 3, 9, 15, 21, 27, 33, . . .

57. 4, 10, 20, 34, 52, 74, . . .

58. 0, 9, 24, 45, 72, 105, . . .

59. −1, 11, 31, 59, 95, 139, . . .

60. −2, 13, 38, 73, 118, 173, . . .

Linear Model, Quadratic Model, or Neither? In Exercises 61–68, write the first six terms of the sequence beginning with the term a_1. Then calculate the first and second differences of the sequence. State whether the sequence has a perfect linear model, a perfect quadratic model, or neither.

61. $a_1 = 0$
$a_n = a_{n-1} + 3$

62. $a_1 = 2$
$a_n = a_{n-1} + 2$

63. $a_1 = 4$
$a_n = a_{n-1} + 3n$

64. $a_1 = 3$
$a_n = 2a_{n-1}$

65. $a_1 = 3$
$a_n = a_{n-1} + n^2$

66. $a_1 = 0$
$a_n = a_{n-1} - 2n$

67. $a_1 = 5$
$a_n = 4n - a_{n-1}$

68. $a_1 = -2$
$a_n = a_{n-1} + 4n$

Finding a Quadratic Model In Exercises 69–74, find the quadratic model for the sequence with the given terms.

69. $a_0 = 3, a_1 = 3, a_4 = 15$

70. $a_0 = 7, a_1 = 6, a_3 = 10$

71. $a_0 = -1, a_2 = 5, a_4 = 15$

72. $a_0 = 3, a_2 = -3, a_6 = 21$

73. $a_1 = 0, a_2 = 7, a_4 = 27$

74. $a_0 = -7, a_2 = -3, a_6 = -43$

• 75. Residents •

The table shows the numbers a_n (in thousands) of residents of Alabama from 2010 through 2015. *(Source: U.S. Census Bureau)*

Year	Number of Residents, a_n
2010	4785
2011	4801
2012	4816
2013	4831
2014	4846
2015	4859

DATA — Spreadsheet at LarsonPrecalculus.com

(a) Find the first differences of the data shown in the table. Then find a linear model that approximates the data. Let n represent the year, with $n = 10$ corresponding to 2010.

(b) Use a graphing utility to find a linear model for the data. Compare this model with the model from part (a).

(c) Use the models found in parts (a) and (b) to predict the number of residents in 2021. How do these values compare?

76. HOW DO YOU SEE IT? Find a formula for the sum of the angles (in degrees) of a regular polygon. Then use mathematical induction to prove this formula for a general n-sided polygon.

Equilateral triangle (180°)   Square (360°)   Regular pentagon (540°)

Exploration

True or False? In Exercises 77 and 78, determine whether the statement is true or false. Justify your answer.

77. If the statement P_k is true and P_k implies P_{k+1}, then P_1 is also true.

78. A sequence with n terms has $n - 1$ second differences.

9.5 The Binomial Theorem

Binomial coefficients have many applications in real life. For example, in Exercise 86 on page 655, you will use binomial coefficients to write the expansion of a model that represents the average prices of residential electricity in the United States.

■ Use the Binomial Theorem to find binomial coefficients.
■ Use Pascal's Triangle to find binomial coefficients.
■ Use binomial coefficients to write binomial expansions.

Binomial Coefficients

Recall that a *binomial* is a polynomial that has two terms. In this section, you will study a formula that provides a quick method of finding the terms that result from raising a binomial to a power, or **expanding a binomial.** To begin, look at the expansion of

$$(x + y)^n$$

for several values of n.

$$(x + y)^0 = 1$$
$$(x + y)^1 = x + y$$
$$(x + y)^2 = x^2 + 2xy + y^2$$
$$(x + y)^3 = x^3 + 3x^2y + 3xy^2 + y^3$$
$$(x + y)^4 = x^4 + 4x^3y + 6x^2y^2 + 4xy^3 + y^4$$
$$(x + y)^5 = x^5 + 5x^4y + 10x^3y^2 + 10x^2y^3 + 5xy^4 + y^5$$

There are several observations you can make about these expansions.

1. In each expansion, there are $n + 1$ terms.

2. In each expansion, x and y have symmetric roles. The powers of x decrease by 1 in successive terms, whereas the powers of y increase by 1.

3. The sum of the powers of each term is n. For example, in the expansion of $(x + y)^5$, the sum of the powers of each term is 5.

$$4 + 1 = 5 \quad 3 + 2 = 5$$

$$(x + y)^5 = x^5 + 5x^4y^1 + 10x^3y^2 + 10x^2y^3 + 5x^1y^4 + y^5$$

4. The coefficients increase and then decrease in a symmetric pattern.

The coefficients of a binomial expansion are called **binomial coefficients.** To find them, you can use the **Binomial Theorem.**

The Binomial Theorem

In the expansion of $(x + y)^n$

$$(x + y)^n = x^n + nx^{n-1}y + \cdots + {}_nC_r x^{n-r}y^r + \cdots + nxy^{n-1} + y^n$$

the coefficient of $x^{n-r} y^r$ is

$${}_nC_r = \frac{n!}{(n - r)!r!}.$$

The symbol $\binom{n}{r}$ is often used in place of ${}_nC_r$ to denote binomial coefficients.

For a proof of the Binomial Theorem, see Proofs in Mathematics on page 688.

▷ TECHNOLOGY
Most graphing utilities can evaluate $_nC_r$. If yours can, use it to check Example 1.

EXAMPLE 1 Finding Binomial Coefficients

Find each binomial coefficient.

a. $_8C_2$ **b.** $\begin{pmatrix} 10 \\ 3 \end{pmatrix}$ **c.** $_7C_0$ **d.** $\begin{pmatrix} 8 \\ 8 \end{pmatrix}$

Solution

a. $_8C_2 = \dfrac{8!}{6! \cdot 2!} = \dfrac{(8 \cdot 7) \cdot \cancel{6!}}{\cancel{6!} \cdot 2!} = \dfrac{8 \cdot 7}{2 \cdot 1} = 28$

b. $\begin{pmatrix} 10 \\ 3 \end{pmatrix} = \dfrac{10!}{7! \cdot 3!} = \dfrac{(10 \cdot 9 \cdot 8) \cdot \cancel{7!}}{\cancel{7!} \cdot 3!} = \dfrac{10 \cdot 9 \cdot 8}{3 \cdot 2 \cdot 1} = 120$

c. $_7C_0 = \dfrac{\cancel{7!}}{\cancel{7!} \cdot 0!} = 1$ **d.** $\begin{pmatrix} 8 \\ 8 \end{pmatrix} = \dfrac{\cancel{8!}}{0! \cdot \cancel{8!}} = 1$

✓ **Checkpoint**  Audio-video solution in English & Spanish at LarsonPrecalculus.com

Find each binomial coefficient.

a. $\begin{pmatrix} 11 \\ 5 \end{pmatrix}$ **b.** $_9C_2$ **c.** $\begin{pmatrix} 5 \\ 0 \end{pmatrix}$ **d.** $_{15}C_{15}$

When $r \neq 0$ and $r \neq n$, as in parts (a) and (b) above, there is a pattern for evaluating binomial coefficients that works because there will always be factorial terms that divide out of the expression.

$$\underset{\text{2 factors}}{\underbrace{_8C_2 = \overset{\text{2 factors}}{\overbrace{\dfrac{8 \cdot 7}{2 \cdot 1}}}}} \quad \text{and} \quad \underset{\text{3 factors}}{\underbrace{\begin{pmatrix} 10 \\ 3 \end{pmatrix} = \overset{\text{3 factors}}{\overbrace{\dfrac{10 \cdot 9 \cdot 8}{3 \cdot 2 \cdot 1}}}}}$$

EXAMPLE 2 Finding Binomial Coefficients

a. $_7C_3 = \dfrac{7 \cdot \cancel{6} \cdot 5}{\cancel{3} \cdot \cancel{2} \cdot 1} = 35$

b. $\begin{pmatrix} 7 \\ 4 \end{pmatrix} = \dfrac{7 \cdot \cancel{6} \cdot 5 \cdot \cancel{4}}{\cancel{4} \cdot \cancel{3} \cdot \cancel{2} \cdot 1} = 35$

c. $_{12}C_1 = \dfrac{12}{1} = 12$

d. $\begin{pmatrix} 12 \\ 11 \end{pmatrix} = \dfrac{12 \cdot \cancel{11} \cdot \cancel{10} \cdot \cancel{9} \cdot \cancel{8} \cdot \cancel{7} \cdot \cancel{6} \cdot \cancel{5} \cdot \cancel{4} \cdot \cancel{3} \cdot \cancel{2}}{\cancel{11} \cdot \cancel{10} \cdot \cancel{9} \cdot \cancel{8} \cdot \cancel{7} \cdot \cancel{6} \cdot \cancel{5} \cdot \cancel{4} \cdot \cancel{3} \cdot \cancel{2} \cdot 1} = \dfrac{12}{1} = 12$

✓ **Checkpoint**  Audio-video solution in English & Spanish at LarsonPrecalculus.com

Find each binomial coefficient.

a. $_7C_5$ **b.** $\begin{pmatrix} 7 \\ 2 \end{pmatrix}$ **c.** $_{14}C_{13}$ **d.** $\begin{pmatrix} 14 \\ 1 \end{pmatrix}$

•• REMARK   The property $_nC_r = {}_nC_{n-r}$ produces the symmetric pattern of binomial coefficients identified earlier.

In Example 2, it is not a coincidence that the results in parts (a) and (b) are the same and that the results in parts (c) and (d) are the same. In general, it is true that $_nC_r = {}_nC_{n-r}$, for all integers r and n, where $0 \leq r \leq n$.

Pascal's Triangle

There is a convenient way to remember the pattern for binomial coefficients. By arranging the coefficients in a triangular pattern, you obtain the array shown below, which is called **Pascal's Triangle.** This triangle is named after the French mathematician Blaise Pascal (1623–1662).

$$
\begin{array}{ccccccccccccccc}
 & & & & & & & 1 & & & & & & & \\
 & & & & & & 1 & & 1 & & & & & & \\
 & & & & & 1 & & 2 & & 1 & & & & & \\
 & & & & 1 & & 3 & & 3 & & 1 & & & & \\
 & & & 1 & & 4 & & 6 & & 4 & & 1 & & & \\
 & & 1 & & 5 & & 10 & & 10 & & 5 & & 1 & & \\
 & 1 & & 6 & & 15 & & 20 & & 15 & & 6 & & 1 & \\
1 & & 7 & & 21 & & 35 & & 35 & & 21 & & 7 & & 1
\end{array}
$$

$4 + 6 = 10$

$15 + 6 = 21$

Note the pattern in Pascal's Triangle. The first and last numbers in each row are 1. Each other number in a row is the sum of the two numbers immediately above it. Pascal noticed that numbers in this triangle are precisely the same numbers as the coefficients of binomial expansions.

$$(x + y)^0 = 1 \qquad \text{0th row}$$
$$(x + y)^1 = 1x + 1y \qquad \text{1st row}$$
$$(x + y)^2 = 1x^2 + 2xy + 1y^2 \qquad \text{2nd row}$$
$$(x + y)^3 = 1x^3 + 3x^2y + 3xy^2 + 1y^3 \qquad \text{3rd row}$$
$$(x + y)^4 = 1x^4 + 4x^3y + 6x^2y^2 + 4xy^3 + 1y^4 \qquad \vdots$$
$$(x + y)^5 = 1x^5 + 5x^4y + 10x^3y^2 + 10x^2y^3 + 5xy^4 + 1y^5$$
$$(x + y)^6 = 1x^6 + 6x^5y + 15x^4y^2 + 20x^3y^3 + 15x^2y^4 + 6xy^5 + 1y^6$$
$$(x + y)^7 = 1x^7 + 7x^6y + 21x^5y^2 + 35x^4y^3 + 35x^3y^4 + 21x^2y^5 + 7xy^6 + 1y^7$$

The top row in Pascal's Triangle is called the *zeroth row* because it corresponds to the binomial expansion $(x + y)^0 = 1$. Similarly, the next row is called the *first row* because it corresponds to the binomial expansion

$$(x + y)^1 = 1x + 1y.$$

In general, the *nth row* in Pascal's Triangle gives the coefficients of $(x + y)^n$.

EXAMPLE 3 **Using Pascal's Triangle**

Use the seventh row of Pascal's Triangle to find the binomial coefficients.

$$_8C_0,\ _8C_1,\ _8C_2,\ _8C_3,\ _8C_4,\ _8C_5,\ _8C_6,\ _8C_7,\ _8C_8$$

Solution

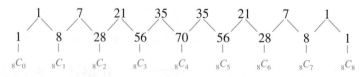

$$
\begin{array}{ccccccccc}
1 & & 7 & & 21 & & 35 & & 35 & & 21 & & 7 & & 1 \\
1 & & 8 & & 28 & & 56 & & 70 & & 56 & & 28 & & 8 & & 1 \\
_8C_0 & & _8C_1 & & _8C_2 & & _8C_3 & & _8C_4 & & _8C_5 & & _8C_6 & & _8C_7 & & _8C_8
\end{array}
$$

✓ **Checkpoint** ◀))) *Audio-video solution in English & Spanish at LarsonPrecalculus.com*

Use the eighth row of Pascal's Triangle to find the binomial coefficients.

$$_9C_0,\ _9C_1,\ _9C_2,\ _9C_3,\ _9C_4,\ _9C_5,\ _9C_6,\ _9C_7,\ _9C_8,\ _9C_9$$

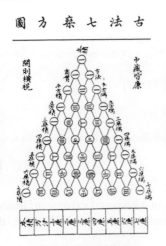

PRECIOUS MIRROR OF THE FOUR ELEMENTS

Eastern cultures were familiar with "Pascal's" Triangle and forms of the Binomial Theorem prior to the Western "discovery" of the theorem. A Chinese text entitled *Precious Mirror of the Four Elements* contains a triangle of binomial expansions through the eighth power.

Binomial Expansions

The formula for binomial coefficients and Pascal's Triangle give you a systematic way to write the coefficients of a binomial expansion, as demonstrated in the next four examples.

EXAMPLE 4 Expanding a Binomial

Write the expansion of the expression

$(x + 1)^3$.

Solution The binomial coefficients from the third row of Pascal's Triangle are

$1, 3, 3, 1$.

So, the expansion is

$$(x + 1)^3 = (1)x^3 + (3)x^2(1) + (3)x(1^2) + (1)(1^3)$$
$$= x^3 + 3x^2 + 3x + 1.$$

✓ *Checkpoint* 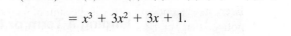 *Audio-video solution in English & Spanish at LarsonPrecalculus.com*

Write the expansion of the expression

$(x + 2)^4$.

To expand binomials representing *differences* rather than sums, you alternate signs. Here are two examples.

$$(x - 1)^2 = x^2 - 2x + 1$$
$$(x - 1)^3 = x^3 - 3x^2 + 3x - 1$$

▷ **ALGEBRA HELP** The solutions to Example 5 use the property of exponents

$(ab)^m = a^m b^m$.

For instance, in Example 5(a),

$(2x)^4 = 2^4 x^4 = 16x^4$.

To review properties of exponents, see Appendix A.2.

EXAMPLE 5 Expanding a Binomial

See LarsonPrecalculus.com for an interactive version of this type of example.

Write the expansion of each expression.

a. $(2x - 3)^4$

b. $(x - 2y)^4$

Solution The binomial coefficients from the fourth row of Pascal's Triangle are

$1, 4, 6, 4, 1$.

The expansions are given below.

a. $(2x - 3)^4 = (1)(2x)^4 - (4)(2x)^3(3) + (6)(2x)^2(3^2) - (4)(2x)(3^3) + (1)(3^4)$

$\qquad = 16x^4 - 96x^3 + 216x^2 - 216x + 81$

b. $(x - 2y)^4 = (1)x^4 - (4)x^3(2y) + (6)x^2(2y)^2 - (4)x(2y)^3 + (1)(2y)^4$

$\qquad = x^4 - 8x^3y + 24x^2y^2 - 32xy^3 + 16y^4$

✓ *Checkpoint* 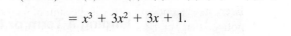 *Audio-video solution in English & Spanish at LarsonPrecalculus.com*

Write the expansion of each expression.

a. $(y - 2)^4$

b. $(2x - y)^5$

▷ **TECHNOLOGY** Use a graphing utility to check the expansion in Example 6. Graph the original binomial expression and the expansion in the same viewing window. The graphs should coincide, as shown in the figure below.

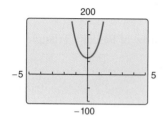

| **EXAMPLE 6** | **Expanding a Binomial** |

Write the expansion of $(x^2 + 4)^3$.

Solution   Use the third row of Pascal's Triangle.

$$(x^2 + 4)^3 = (1)(x^2)^3 + (3)(x^2)^2(4) + (3)x^2(4^2) + (1)(4^3)$$

$$= x^6 + 12x^4 + 48x^2 + 64$$

✓ *Checkpoint*  *Audio-video solution in English & Spanish at LarsonPrecalculus.com*

Write the expansion of $(5 + y^2)^3$.

Sometimes you will need to find a specific term in a binomial expansion. Instead of writing the entire expansion, use the fact that, from the Binomial Theorem, the $(r + 1)$th term is $_nC_r x^{n-r} y^r$.

| **EXAMPLE 7** | **Finding a Term or Coefficient** |

a. Find the sixth term of $(a + 2b)^8$.

b. Find the coefficient of the term a^6b^5 in the expansion of $(3a - 2b)^{11}$.

Solution

a. Remember that the formula is for the $(r + 1)$th term, so r is one less than the number of the term you need. So, to find the sixth term in this binomial expansion, use $r = 5$, $n = 8$, $x = a$, and $y = 2b$.

$$_nC_r x^{n-r} y^r = {_8C_5} a^3 (2b)^5$$

$$= 56a^3 (32b^5)$$

$$= 1792a^3 b^5$$

b. In this case, $n = 11$, $r = 5$, $x = 3a$, and $y = -2b$. Substitute these values to obtain

$$_nC_r x^{n-r} y^r = {_{11}C_5} (3a)^6 (-2b)^5$$

$$= (462)(729a^6)(-32b^5)$$

$$= -10,777,536a^6 b^5.$$

So, the coefficient is $-10,777,536$.

✓ *Checkpoint*  *Audio-video solution in English & Spanish at LarsonPrecalculus.com*

a. Find the fifth term of $(a + 2b)^8$.

b. Find the coefficient of the term a^4b^7 in the expansion of $(3a - 2b)^{11}$.

Summarize   (Section 9.5)

1. State the Binomial Theorem *(page 648)*. For examples of using the Binomial Theorem to find binomial coefficients, see Examples 1 and 2.

2. Explain how to use Pascal's Triangle to find binomial coefficients *(page 650)*. For an example of using Pascal's Triangle to find binomial coefficients, see Example 3.

3. Explain how to use binomial coefficients to write a binomial expansion *(page 651)*. For examples of using binomial coefficients to write binomial expansions, see Examples 4–6.

9.5 Exercises

Vocabulary: Fill in the blanks.

1. When you find the terms that result from raising a binomial to a power, you are _____ the binomial.
2. The coefficients of a binomial expansion are called _____ _____.
3. To find binomial coefficients, you can use the _____ _____ or _____ _____.
4. The symbol used to denote a binomial coefficient is _____ or _____.

Skills and Applications

 Finding a Binomial Coefficient In Exercises 5–12, find the binomial coefficient.

5. $_5C_3$

6. $_7C_6$

7. $_{12}C_0$

8. $_{20}C_{20}$

9. $\begin{pmatrix} 10 \\ 4 \end{pmatrix}$

10. $\begin{pmatrix} 10 \\ 6 \end{pmatrix}$

11. $\begin{pmatrix} 100 \\ 98 \end{pmatrix}$

12. $\begin{pmatrix} 100 \\ 2 \end{pmatrix}$

 Using Pascal's Triangle In Exercises 13–16, evaluate using Pascal's Triangle.

13. $_6C_3$

14. $_4C_2$

15. $\begin{pmatrix} 5 \\ 1 \end{pmatrix}$

16. $\begin{pmatrix} 7 \\ 4 \end{pmatrix}$

 Expanding a Binomial In Exercises 17–24, use the Binomial Theorem to write the expansion of the expression.

17. $(x + 1)^6$

18. $(x + 1)^4$

19. $(y - 3)^3$

20. $(y - 2)^5$

21. $(r + 3s)^3$

22. $(x + 2y)^4$

23. $(3a - 4b)^5$

24. $(2x - 5y)^5$

 Expanding an Expression In Exercises 25–38, expand the expression by using Pascal's Triangle to determine the coefficients.

25. $(a + 6)^4$

26. $(a + 5)^5$

27. $(y - 1)^6$

28. $(y - 4)^4$

29. $(3 - 2z)^4$

30. $(3v + 2)^6$

31. $(x + 2y)^5$

32. $(2t - s)^5$

33. $(x^2 + y^2)^4$

34. $(x^2 + y^2)^6$

35. $\left(\dfrac{1}{x} + y \right)^5$

36. $\left(\dfrac{1}{x} + 2y \right)^6$

37. $2(x - 3)^4 + 5(x - 3)^2$

38. $(4x - 1)^3 - 2(4x - 1)^4$

 Finding a Term In Exercises 39–46, find the specified nth term in the expansion of the binomial.

39. $(x + y)^{10}$, $n = 4$

40. $(x - y)^6$, $n = 2$

41. $(x - 6y)^5$, $n = 3$

42. $(x + 2z)^7$, $n = 4$

43. $(4x + 3y)^9$, $n = 8$

44. $(5a + 6b)^5$, $n = 5$

45. $(10x - 3y)^{12}$, $n = 10$

46. $(7x + 2y)^{15}$, $n = 7$

 Finding a Coefficient In Exercises 47–54, find the coefficient a of the term in the expansion of the binomial.

Binomial	Term
47. $(x + 2)^6$	ax^3
48. $(x - 2)^6$	ax^3
49. $(4x - y)^{10}$	ax^2y^8
50. $(x - 2y)^{10}$	ax^8y^2
51. $(2x - 5y)^9$	ax^4y^5
52. $(3x + 4y)^8$	ax^6y^2
53. $(x^2 + y)^{10}$	ax^8y^6
54. $(z^2 - t)^{10}$	az^4t^8

Expanding an Expression In Exercises 55–60, use the Binomial Theorem to write the expansion of the expression.

55. $\left(\sqrt{x} + 5 \right)^3$

56. $\left(2\sqrt{t} - 1 \right)^3$

57. $(x^{2/3} - y^{1/3})^3$

58. $(u^{3/5} + 2)^5$

59. $\left(3\sqrt{t} + \sqrt[4]{t} \right)^4$

60. $(x^{3/4} - 2x^{5/4})^4$

∫ **Simplifying a Difference Quotient** In Exercises 61–66, simplify the difference quotient, using the Binomial Theorem if necessary.

$$\dfrac{f(x + h) - f(x)}{h} \qquad \text{Difference quotient}$$

61. $f(x) = x^3$

62. $f(x) = x^4$

63. $f(x) = x^6$

64. $f(x) = x^7$

65. $f(x) = \sqrt{x}$

66. $f(x) = \dfrac{1}{x}$

Expanding a Complex Number In Exercises 67–72, use the Binomial Theorem to expand the complex number. Simplify your result.

67. $(1 + i)^4$

68. $(2 - i)^5$

69. $(2 - 3i)^6$

70. $\left(5 + \sqrt{-9}\right)^3$

71. $\left(-\dfrac{1}{2} + \dfrac{\sqrt{3}}{2}i\right)^3$

72. $\left(5 - \sqrt{3}i\right)^4$

Approximation In Exercises 73–76, use the Binomial Theorem to approximate the quantity accurate to three decimal places. For example, in Exercise 73, use the expansion

$$(1.02)^8 = (1 + 0.02)^8$$

$$= 1 + 8(0.02) + 28(0.02)^2 + \cdots + (0.02)^8.$$

73. $(1.02)^8$

74. $(2.005)^{10}$

75. $(2.99)^{12}$

76. $(1.98)^9$

Probability In Exercises 77–80, consider n independent trials of an experiment in which each trial has two possible outcomes: "success" or "failure." The probability of a success on each trial is p, and the probability of a failure is $q = 1 - p$. In this context, the term $_nC_k\,p^k q^{n-k}$ in the expansion of $(p + q)^n$ gives the probability of k successes in the n trials of the experiment.

77. You toss a fair coin seven times. To find the probability of obtaining four heads, evaluate the term

$$_7C_4\left(\tfrac{1}{2}\right)^4\left(\tfrac{1}{2}\right)^3$$

in the expansion of $\left(\tfrac{1}{2} + \tfrac{1}{2}\right)^7$.

78. The probability of a baseball player getting a hit during any given time at bat is $\tfrac{1}{4}$. To find the probability that the player gets three hits during the next 10 times at bat, evaluate the term

$$_{10}C_3\left(\tfrac{1}{4}\right)^3\left(\tfrac{3}{4}\right)^7$$

in the expansion of $\left(\tfrac{1}{4} + \tfrac{3}{4}\right)^{10}$.

79. The probability of a sales representative making a sale with any one customer is $\tfrac{1}{3}$. The sales representative makes eight contacts a day. To find the probability of making four sales, evaluate the term

$$_8C_4\left(\tfrac{1}{3}\right)^4\left(\tfrac{2}{3}\right)^4$$

in the expansion of $\left(\tfrac{1}{3} + \tfrac{2}{3}\right)^8$.

80. To find the probability that the sales representative in Exercise 79 makes four sales when the probability of a sale with any one customer is $\tfrac{1}{2}$, evaluate the term

$$_8C_4\left(\tfrac{1}{2}\right)^4\left(\tfrac{1}{2}\right)^4$$

in the expansion of $\left(\tfrac{1}{2} + \tfrac{1}{2}\right)^8$.

Graphical Reasoning In Exercises 81 and 82, use a graphing utility to graph f and g in the same viewing window. What is the relationship between the two graphs? Use the Binomial Theorem to write the polynomial function g in standard form.

81. $f(x) = x^3 - 4x$

$g(x) = f(x + 4)$

82. $f(x) = -x^4 + 4x^2 - 1$

$g(x) = f(x - 3)$

83. Finding a Pattern Describe the pattern formed by the sums of the numbers along the diagonal line segments shown in Pascal's Triangle (see figure).

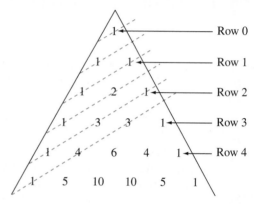

84. Error Analysis Describe the error.

$$(x - 3)^3 = {_3C_0}x^3 + {_3C_1}x^2(3) + {_3C_2}x(3)^2$$
$$+ {_3C_3}(3)^3$$
$$= 1x^3 + 3x^2(3) + 3x(3)^2 + 1(3)^3$$
$$= x^3 + 9x^2 + 27x + 27$$

85. Child Support The amounts $f(t)$ (in billions of dollars) of child support collected in the United States from 2005 through 2014 can be approximated by the model

$$f(t) = -0.056t^2 + 1.62t + 16.4, \quad 5 \le t \le 14$$

where t represents the year, with $t = 5$ corresponding to 2005. *(Source: U.S. Department of Health and Human Services)*

(a) You want to adjust the model so that $t = 5$ corresponds to 2010 rather than 2005. To do this, you shift the graph of f five units *to the left* to obtain $g(t) = f(t + 5)$. Use binomial coefficients to write $g(t)$ in standard form.

(b) Use a graphing utility to graph f and g in the same viewing window.

(c) Use the graphs to estimate when the child support collections exceeded $27 billion.

86. Electricity

The table shows the average prices $f(t)$ (in cents per kilowatt-hour) of residential electricity in the United States from 2007 through 2014.   (*Source: U.S. Energy Information Administration*)

DATA	Year	Average Price, $f(t)$
	2007	10.65
	2008	11.26
	2009	11.51
	2010	11.54
	2011	11.72
	2012	11.88
	2013	12.13
	2014	12.52

Spreadsheet at LarsonPrecalculus.com

(a) Use the *regression* feature of a graphing utility to find a cubic model for the data. Let t represent the year, with $t = 7$ corresponding to 2007.

(b) Use the graphing utility to plot the data and the model in the same viewing window.

(c) You want to adjust the model so that $t = 7$ corresponds to 2012 rather than 2007. To do this, you shift the graph of f five units *to the left* to obtain $g(t) = f(t + 5)$. Use binomial coefficients to write $g(t)$ in standard form.

(d) Use the graphing utility to graph g in the same viewing window as f.

(e) Use both models to predict the average price in 2015. Do you obtain the same answer?

(f) Do your answers to part (e) seem reasonable? Explain.

(g) What factors do you think contributed to the change in the average price?

Exploration

True or False?   **In Exercises 87 and 88, determine whether the statement is true or false. Justify your answer.**

87. The Binomial Theorem could be used to produce each row of Pascal's Triangle.

88. A binomial that represents a difference cannot always be accurately expanded using the Binomial Theorem.

89. Writing   Explain how to form the rows of Pascal's Triangle.

90. Forming Rows of Pascal's Triangle   Form rows 8–10 of Pascal's Triangle.

91. Graphical Reasoning   Use a graphing utility to graph the functions in the same viewing window. Which two functions have identical graphs, and why?

$$f(x) = (1 - x)^3$$
$$g(x) = 1 - x^3$$
$$h(x) = 1 + 3x + 3x^2 + x^3$$
$$k(x) = 1 - 3x + 3x^2 - x^3$$
$$p(x) = 1 + 3x - 3x^2 + x^3$$

92.   **HOW DO YOU SEE IT?**   The expansions of $(x + y)^4$, $(x + y)^5$, and $(x + y)^6$ are shown below.

$$(x + y)^4 = 1x^4 + 4x^3y + 6x^2y^2 + 4xy^3 + 1y^4$$
$$(x + y)^5 = 1x^5 + 5x^4y + 10x^3y^2 + 10x^2y^3 + 5xy^4 + 1y^5$$
$$(x + y)^6 = 1x^6 + 6x^5y + 15x^4y^2 + 20x^3y^3 + 15x^2y^4 + 6xy^5 + 1y^6$$

(a) Explain how the exponent of a binomial is related to the number of terms in its expansion.

(b) How many terms are in the expansion of $(x + y)^n$?

Proof   **In Exercises 93–96, prove the property for all integers r and n, where $0 \le r \le n$.**

93. $_nC_r = {}_nC_{n-r}$

94. $_nC_0 - {}_nC_1 + {}_nC_2 - \cdots \pm {}_nC_n = 0$

95. $_{n+1}C_r = {}_nC_r + {}_nC_{r-1}$

96. The sum of the numbers in the nth row of Pascal's Triangle is 2^n.

97. Binomial Coefficients and Pascal's Triangle   Complete the table. What characteristic of Pascal's Triangle does this table illustrate?

n	r	$_nC_r$	$_nC_{n-r}$
9	5		
7	1		
12	4		
6	0		
10	7		

9.6 Counting Principles

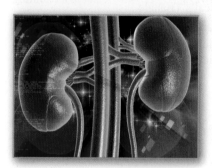

Counting principles are useful for helping you solve counting problems that occur in real life. For example, in Exercise 39 on page 664, you will use counting principles to determine the number of possible orders there are for best match, second-best match, and third-best match kidney donors.

- ◼ Solve simple counting problems.
- ◼ Use the Fundamental Counting Principle to solve counting problems.
- ◼ Use permutations to solve counting problems.
- ◼ Use combinations to solve counting problems.

Simple Counting Problems

This section and Section 9.7 present a brief introduction to some of the basic counting principles and their applications to probability. In Section 9.7, you will see that much of probability has to do with counting the number of ways an event can occur. The two examples below describe simple counting problems.

EXAMPLE 1   Selecting Pairs of Numbers at Random

You place eight pieces of paper, numbered from 1 to 8, in a box. You draw one piece of paper at random from the box, record its number, and *replace* the paper in the box. Then, you draw a second piece of paper at random from the box and record its number. Finally, you add the two numbers. How many different ways can you obtain a sum of 12?

Solution   To solve this problem, count the different ways to obtain a sum of 12 using two numbers from 1 to 8.

First number	4	5	6	7	8
Second number	8	7	6	5	4

So, a sum of 12 can occur in five different ways.

✓ **Checkpoint** 🔊))) *Audio-video solution in English & Spanish at LarsonPrecalculus.com*

In Example 1, how many different ways can you obtain a sum of 14?

EXAMPLE 2   Selecting Pairs of Numbers at Random

You place eight pieces of paper, numbered from 1 to 8, in a box. You draw one piece of paper at random from the box, record its number, and *do not* replace the paper in the box. Then, you draw a second piece of paper at random from the box and record its number. Finally, you add the two numbers. How many different ways can you obtain a sum of 12?

Solution   To solve this problem, count the different ways to obtain a sum of 12 using two *different* numbers from 1 to 8.

First number	4	5	7	8
Second number	8	7	5	4

So, a sum of 12 can occur in four different ways.

✓ **Checkpoint** 🔊))) *Audio-video solution in English & Spanish at LarsonPrecalculus.com*

In Example 2, how many different ways can you obtain a sum of 14? ◼

Notice the difference between the counting problems in Examples 1 and 2. The random selection in Example 1 occurs **with replacement,** whereas the random selection in Example 2 occurs **without replacement,** which eliminates the possibility of choosing two 6's.

Kannanimages/Shutterstock.com

The Fundamental Counting Principle

Examples 1 and 2 describe simple counting problems and *list* each possible way that an event can occur. When it is possible, this is always the best way to solve a counting problem. However, some events can occur in so many different ways that it is not feasible to write the entire list. In such cases, you must rely on formulas and counting principles. The most important of these is the **Fundamental Counting Principle**.

Fundamental Counting Principle

Let E_1 and E_2 be two events. The first event E_1 can occur in m_1 different ways. After E_1 has occurred, E_2 can occur in m_2 different ways. The number of ways the two events can occur is $m_1 \cdot m_2$.

The Fundamental Counting Principle can be extended to three or more events. For example, the number of ways that three events E_1, E_2, and E_3 can occur is

$$m_1 \cdot m_2 \cdot m_3.$$

EXAMPLE 3 **Using the Fundamental Counting Principle**

How many different pairs of letters from the English alphabet are possible?

Solution There are two events in this situation. The first event is the choice of the first letter, and the second event is the choice of the second letter. The English alphabet contains 26 letters, so it follows that the number of two-letter pairs is

$$26 \cdot 26 = 676.$$

✓ *Checkpoint* 🔊)) *Audio-video solution in English & Spanish at LarsonPrecalculus.com*

A combination lock will open when you select the right choice of three numbers (from 1 to 30, inclusive). How many different lock combinations are possible?

EXAMPLE 4 **Using the Fundamental Counting Principle**

Telephone numbers in the United States have 10 digits. The first three digits are the *area code* and the next seven digits are the *local telephone number*. How many different telephone numbers are possible within each area code? (Note that a local telephone number cannot begin with 0 or 1.)

Solution The first digit of a local telephone number cannot be 0 or 1, so there are only eight choices for the first digit. For each of the other six digits, there are 10 choices.

So, the number of telephone numbers that are possible within each area code is

$$8 \cdot 10 \cdot 10 \cdot 10 \cdot 10 \cdot 10 \cdot 10 = 8{,}000{,}000.$$

✓ *Checkpoint* 🔊)) *Audio-video solution in English & Spanish at LarsonPrecalculus.com*

A product's catalog number is made up of one letter from the English alphabet followed by a five-digit number. How many different catalog numbers are possible? ■

Permutations

One important application of the Fundamental Counting Principle is in determining the number of ways that n elements can be arranged (in order). An ordering of n elements is called a **permutation** of the elements.

Definition of a Permutation

A **permutation** of n different elements is an ordering of the elements such that one element is first, one is second, one is third, and so on.

EXAMPLE 5   **Finding the Number of Permutations**

How many permutations of the letters

  A, B, C, D, E, and F

are possible?

Solution   Consider the reasoning below.

  First position: Any of the *six* letters

  Second position: Any of the remaining *five* letters

  Third position: Any of the remaining *four* letters

  Fourth position: Any of the remaining *three* letters

  Fifth position: Either of the remaining *two* letters

  Sixth position: The *one* remaining letter

So, the numbers of choices for the six positions are as shown in the figure.

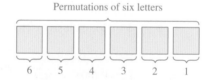

Permutations of six letters

The total number of permutations of the six letters is

  $6! = 6 \cdot 5 \cdot 4 \cdot 3 \cdot 2 \cdot 1 = 720.$

✓ *Checkpoint* 🔊))) *Audio-video solution in English & Spanish at LarsonPrecalculus.com*

How many permutations of the letters

  W, X, Y, and Z

are possible?

Generalizing the result in Example 5, the number of permutations of n different elements is $n!$.

Number of Permutations of *n* Elements

The number of permutations of n elements is

  $n \cdot (n - 1) \cdots 4 \cdot 3 \cdot 2 \cdot 1 = n!.$

In other words, there are $n!$ different ways of ordering n elements.

As of 2015, twelve thoroughbred racehorses hold the title of Triple Crown winner for winning the Kentucky Derby, the Preakness Stakes, and the Belmont Stakes in the same year. Fifty-two horses have won two out of the three races.

It is useful, on occasion, to order a *subset* of a collection of elements rather than the entire collection. For example, you may want to order r elements out of a collection of n elements. Such an ordering is called a **permutation of n elements taken r at a time.** The next example demonstrates this ordering.

EXAMPLE 6 **Counting Horse Race Finishes**

Eight horses are running in a race. In how many different ways can these horses come in first, second, and third? (Assume that there are no ties.)

Solution Here are the different possibilities.

Win (first position): *Eight* choices

Place (second position): *Seven* choices

Show (third position): *Six* choices

The numbers of choices for the three positions are as shown in the figure.

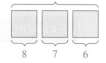

Different orders of horses

8   7   6

So, using the Fundamental Counting Principle, there are

$$8 \cdot 7 \cdot 6 = 336$$

different ways in which the eight horses can come in first, second, and third.

 ✔ *Checkpoint* ◀)))) *Audio-video solution in English & Spanish at LarsonPrecalculus.com*

A coin club has five members. In how many different ways can there be a president and a vice-president?

Generalizing the result in Example 6 gives the formula below.

> **Permutations of n Elements Taken r at a Time**
>
> The number of permutations of n elements taken r at a time is
>
> $$_nP_r = \frac{n!}{(n-r)!} = n(n-1)(n-2)\cdots(n-r+1).$$

▷ **TECHNOLOGY**
Most graphing utilities can evaluate $_nP_r$. If yours can, use it to evaluate several permutations. Check your results algebraically by hand.

Using this formula, rework Example 6 to find that the number of permutations of eight horses taken three at a time is

$$_8P_3 = \frac{8!}{(8-3)!}$$

$$= \frac{8!}{5!}$$

$$= \frac{8 \cdot 7 \cdot 6 \cdot 5!}{5!}$$

$$= 336$$

which is the same answer obtained in the example.

A ♥	A ♦	A ♣	A ♠
2 ♥	2 ♦	2 ♣	2 ♠
3 ♥	3 ♦	3 ♣	3 ♠
4 ♥	4 ♦	4 ♣	4 ♠
5 ♥	5 ♦	5 ♣	5 ♠
6 ♥	6 ♦	6 ♣	6 ♠
7 ♥	7 ♦	7 ♣	7 ♠
8 ♥	8 ♦	8 ♣	8 ♠
9 ♥	9 ♦	9 ♣	9 ♠
10 ♥	10 ♦	10 ♣	10 ♠
J ♥	J ♦	J ♣	J ♠
Q ♥	Q ♦	Q ♣	Q ♠
K ♥	K ♦	K ♣	K ♠

Ranks and suits in a standard deck of playing cards
Figure 9.2

EXAMPLE 9   **Counting Card Hands**

A standard poker hand consists of five cards dealt from a deck of 52 (see Figure 9.2). How many different poker hands are possible? (Order is not important.)

Solution   To determine the number of different poker hands, find the number of combinations of 52 elements taken five at a time.

$$_{52}C_5 = \frac{52!}{(52 - 5)!5!}$$

$$= \frac{52!}{47!5!}$$

$$= \frac{52 \cdot 51 \cdot 50 \cdot 49 \cdot 48 \cdot \cancel{47!}}{\cancel{47!} \cdot 5 \cdot 4 \cdot 3 \cdot 2 \cdot 1}$$

$$= 2{,}598{,}960$$

✓ *Checkpoint* *Audio-video solution in English & Spanish at LarsonPrecalculus.com*

In three-card poker, a hand consists of three cards dealt from a deck of 52. How many different three-card poker hands are possible? (Order is not important.)

EXAMPLE 10   **Forming a Team**

You are forming a 12-member swim team from 10 girls and 15 boys. The team must consist of five girls and seven boys. How many different 12-member teams are possible?

Solution   There are $_{10}C_5$ ways of choosing five girls. There are $_{15}C_7$ ways of choosing seven boys. By the Fundamental Counting Principle, there are $_{10}C_5 \cdot {}_{15}C_7$ ways of choosing five girls and seven boys.

$$_{10}C_5 \cdot {}_{15}C_7 = \frac{10!}{5! \cdot 5!} \cdot \frac{15!}{8! \cdot 7!} = 252 \cdot 6435 = 1{,}621{,}620$$

So, there are 1,621,620 12-member swim teams possible.

✓ *Checkpoint* 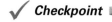 *Audio-video solution in English & Spanish at LarsonPrecalculus.com*

In Example 10, the team must consist of six boys and six girls. How many different 12-member teams are possible?

• • REMARK   When solving problems involving counting principles, you need to distinguish among the various counting principles to determine which is necessary to solve the problem. To do this, ask yourself the questions below.

1. Is the order of the elements important? *Permutation*

2. Is the order of the elements not important? *Combination*

3. Does the problem involve two or more separate events? *Fundamental Counting Principle*

Summarize   (Section 9.6)

1. Explain how to solve a simple counting problem *(page 656)*. For examples of solving simple counting problems, see Examples 1 and 2.

2. State the Fundamental Counting Principle *(page 657)*. For examples of using the Fundamental Counting Principle to solve counting problems, see Examples 3 and 4.

3. Explain how to find the number of permutations of n elements *(page 658)*, the number of permutations of n elements taken r at a time *(page 659)*, and the number of distinguishable permutations *(page 660)*. For examples of using permutations to solve counting problems, see Examples 5–7.

4. Explain how to find the number of combinations of n elements taken r at a time *(page 661)*. For examples of using combinations to solve counting problems, see Examples 8–10.

9.6 Exercises

See **CalcChat.com** for tutorial help and worked-out solutions to odd-numbered exercises.

Vocabulary: Fill in the blanks.

1. The _____ _____ _____ states that when there are m_1 different ways for one event to occur and m_2 different ways for a second event to occur, there are $m_1 \cdot m_2$ ways for both events to occur.

2. An ordering of n elements is a _____ of the elements.

3. The number of permutations of n elements taken r at a time is given by _____.

4. The number of _____ _____ of n objects is given by $\dfrac{n!}{n_1! \cdot n_2! \cdot n_3! \cdot \cdots \cdot n_k!}$.

5. When selecting subsets of a larger set in which order is not important, you are finding the number of _____ of n elements taken r at a time.

6. The number of combinations of n elements taken r at a time is given by _____.

Skills and Applications

Random Selection **In Exercises 7–14, determine the number of ways a computer can randomly generate one or more such integers from 1 through 12.**

7. An odd integer

8. An even integer

9. A prime integer

10. An integer that is greater than 9

11. An integer that is divisible by 4

12. An integer that is divisible by 3

13. Two *distinct* integers whose sum is 9

14. Two *distinct* integers whose sum is 8

15. **Entertainment Systems** A customer can choose one of three amplifiers, one of two compact disc players, and one of five speaker models for an entertainment system. Determine the number of possible system configurations. *Amp 3 disc 2 Speakers 5*

16. **Job Applicants** A small college needs two additional faculty members: a chemist and a statistician. There are five applicants for the chemistry position and three applicants for the statistics position. In how many ways can the college fill these positions?

17. **Course Schedule** A college student is preparing a course schedule for the next semester. The student may select one of two mathematics courses, one of three science courses, and one of five courses from the social sciences. How many schedules are possible?

18. **Physiology** In a physiology class, a student must dissect three different specimens. The student can select one of nine earthworms, one of four frogs, and one of seven fetal pigs. In how many ways can the student select the specimens?

19. **True-False Exam** In how many ways can you answer a six-question true-false exam? (Assume that you do not omit any questions.)

20. **True-False Exam** In how many ways can you answer a 12-question true-false exam? (Assume that you do not omit any questions.)

21. **License Plate Numbers** In the state of Pennsylvania, each standard automobile license plate number consists of three letters followed by a four-digit number. How many distinct license plate numbers are possible in Pennsylvania?

22. **License Plate Numbers** In a certain state, each automobile license plate number consists of two letters followed by a four-digit number. To avoid confusion between "O" and "zero" and between "I" and "one," the letters "O" and "I" are not used. How many distinct license plate numbers are possible in this state?

23. **Three-Digit Numbers** How many three-digit numbers are possible under each condition?
    (a) The leading digit cannot be zero.
    (b) The leading digit cannot be zero and no repetition of digits is allowed.
    (c) The leading digit cannot be zero and the number must be a multiple of 5.
    (d) The number is at least 400.

24. **Four-Digit Numbers** How many four-digit numbers are possible under each condition?
    (a) The leading digit cannot be zero.
    (b) The leading digit cannot be zero and no repetition of digits is allowed.
    (c) The leading digit cannot be zero and the number must be less than 5000.
    (d) The leading digit cannot be zero and the number must be even.

25. **Combination Lock** A combination lock will open when you select the right choice of three numbers (from 1 to 40, inclusive). How many different lock combinations are possible?

26. Combination Lock A combination lock will open when you select the right choice of three numbers (from 1 to 50, inclusive). How many different lock combinations are possible?

27. Concert Seats Four couples reserve seats in one row for a concert. In how many different ways can they sit when

(a) there are no seating restrictions?

(b) the two members of each couple wish to sit together?

28. Single File In how many orders can four girls and four boys walk through a doorway single file when

(a) there are no restrictions?

(b) the girls walk through before the boys?

29. Posing for a Photograph In how many ways can five children posing for a photograph line up in a row?

30. Riding in a Car In how many ways can six people sit in a six-passenger car?

 Evaluating $_nP_r$ In Exercises 31–34, evaluate $_nP_r$.

31. $_5P_2$ **32.** $_6P_6$ **33.** $_{12}P_2$ **34.** $_6P_5$

Evaluating $_nP_r$ In Exercises 35–38, use a graphing utility to evaluate $_nP_r$.

35. $_{15}P_3$ **36.** $_{100}P_4$ **37.** $_{50}P_4$ **38.** $_{10}P_5$

• • 39. Kidney Donors • • • • • • • • • • • • • • • • • •

A patient with end-stage kidney disease has nine family members who are potential kidney donors. How many possible orders are there for a best match, a second-best match, and a third-best match?

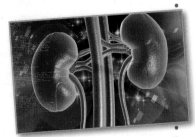

40. Choosing Officers From a pool of 12 candidates, the offices of president, vice-president, secretary, and treasurer need to be filled. In how many different ways can the offices be filled?

41. Batting Order A baseball coach is creating a nine-player batting order by selecting from a team of 15 players. How many different batting orders are possible?

42. Athletics Eight sprinters qualify for the finals in the 100-meter dash at the NCAA national track meet. In how many ways can the sprinters come in first, second, and third? (Assume there are no ties.)

 Number of Distinguishable Permutations In Exercises 43–46, find the number of distinguishable permutations of the group of letters.

43. A, A, G, E, E, E, M   **44.** B, B, B, T, T, T, T, T

45. A, L, G, E, B, R, A   **46.** M, I, S, S, I, S, S, I, P, P, I

47. Writing Permutations Write all permutations of the letters A, B, C, and D.

48. Writing Permutations Write all permutations of the letters A, B, C, and D when letters B and C must remain between A and D.

 Evaluating $_nC_r$ In Exercises 49–52, evaluate $_nC_r$ using the formula from this section.

49. $_6C_4$ **50.** $_5C_4$ **51.** $_9C_9$ **52.** $_{12}C_0$

Evaluating $_nC_r$ In Exercises 53–56, use a graphing utility to evaluate $_nC_r$.

53. $_{16}C_2$ **54.** $_{17}C_5$ **55.** $_{20}C_6$ **56.** $_{50}C_8$

57. Writing Combinations Write all combinations of two letters that can be formed from the letters A, B, C, D, E, and F. (Order is not important.)

58. Forming an Experimental Group To conduct an experiment, researchers randomly select five students from a class of 20. How many different groups of five students are possible?

59. Jury Selection In how many different ways can a jury of 12 people be randomly selected from a group of 40 people?

60. Committee Members A U.S. Senate Committee has 14 members. Assuming party affiliation is not a factor in selection, how many different committees are possible from the 100 U.S. senators?

61. Lottery Choices In the Massachusetts Mass Cash game, a player randomly chooses five distinct numbers from 1 to 35. In how many ways can a player select the five numbers?

62. Lottery Choices In the Louisiana Lotto game, a player randomly chooses six distinct numbers from 1 to 40. In how many ways can a player select the six numbers?

63. Defective Units A shipment of 25 television sets contains three defective units. In how many ways can a vending company purchase four of these units and receive (a) all good units, (b) two good units, and (c) at least two good units?

64. Interpersonal Relationships The complexity of interpersonal relationships increases dramatically as the size of a group increases. Determine the numbers of different two-person relationships in groups of people of sizes (a) 3, (b) 8, (c) 12, and (d) 20.

65. **Poker Hand** You are dealt five cards from a standard deck of 52 playing cards. In how many ways can you get (a) a full house and (b) a five-card combination containing two jacks and three aces? (A full house consists of three of one kind and two of another. For example, A-A-A-5-5 and K-K-K-10-10 are full houses.)

66. **Job Applicants** An employer interviews 12 people for four openings at a company. Five of the 12 people are women. All 12 applicants are qualified. In how many ways can the employer fill the four positions when (a) the selection is random and (b) exactly two selections are women?

67. **Forming a Committee** A local college is forming a six-member research committee with one administrator, three faculty members, and two students. There are seven administrators, 12 faculty members, and 20 students in contention for the committee. How many six-member committees are possible?

68. **Law Enforcement** A police department uses computer imaging to create digital photographs of alleged perpetrators from eyewitness accounts. One software package contains 195 hairlines, 99 sets of eyes and eyebrows, 89 noses, 105 mouths, and 74 chin and cheek structures.

    (a) Find the possible number of different faces that the software could create.

    (b) An eyewitness can clearly recall the hairline and eyes and eyebrows of a suspect. How many different faces are possible with this information?

Geometry In Exercises 69–72, find the number of diagonals of the polygon. (A *diagonal* is a line segment connecting any two nonadjacent vertices of a polygon.)

69. Pentagon

70. Hexagon

71. Octagon

72. Decagon (10 sides)

73. **Geometry** Three points that are not collinear determine three lines. How many lines are determined by nine points, no three of which are collinear?

74. **Lottery** Powerball is a lottery game that is operated by the Multi-State Lottery Association and is played in 44 states, Washington D.C., Puerto Rico, and the U.S. Virgin Islands. The game is played by drawing five white balls out of a drum of 69 white balls (numbered 1–69) and one red powerball out of a drum of 26 red balls (numbered 1–26). The jackpot is won by matching all five white balls in any order and the red powerball.

    (a) Find the possible number of winning Powerball numbers.

    (b) Find the possible number of winning Powerball numbers when you win the jackpot by matching all five white balls in order and the red powerball.

Solving an Equation In Exercises 75–82, solve for n.

75. $4 \cdot {}_{n+1}P_2 = {}_{n+2}P_3$

76. $5 \cdot {}_{n-1}P_1 = {}_nP_2$

77. ${}_{n+1}P_3 = 4 \cdot {}_nP_2$

78. ${}_{n+2}P_3 = 6 \cdot {}_{n+2}P_1$

79. $14 \cdot {}_nP_3 = {}_{n+2}P_4$

80. ${}_nP_5 = 18 \cdot {}_{n-2}P_4$

81. ${}_nP_4 = 10 \cdot {}_{n-1}P_3$

82. ${}_nP_6 = 12 \cdot {}_{n-1}P_5$

Exploration

True or False? In Exercises 83 and 84, determine whether the statement is true or false. Justify your answer.

83. The number of letter pairs that can be formed in any order from any two of the first 13 letters in the alphabet (A–M) is an example of a permutation.

84. The number of permutations of n elements can be determined by using the Fundamental Counting Principle.

85. **Think About It** Without calculating, determine which of the following is greater. Explain.

    (a) The number of combinations of 10 elements taken six at a time

    (b) The number of permutations of 10 elements taken six at a time

86. **HOW DO YOU SEE IT?** Without calculating, determine whether the value of ${}_nP_r$ is greater than the value of ${}_nC_r$ for the values of n and r given in the table. Complete the table using yes (Y) or no (N). Is the value of ${}_nP_r$ always greater than the value of ${}_nC_r$? Explain.

n \ r	0	1	2	3	4	5	6	7
1								
2								
3								
4								
5								
6								
7								

Proof In Exercises 87–90, prove the identity.

87. ${}_nP_{n-1} = {}_nP_n$

88. ${}_nC_n = {}_nC_0$

89. ${}_nC_{n-1} = {}_nC_1$

90. ${}_nC_r = \dfrac{{}_nP_r}{r!}$

91. **Think About It** Can your graphing utility evaluate ${}_{100}P_{80}$? If not, explain why.

9.7 Probability

Probability applies to many real-life applications. For example, in Exercise 59 on page 676, you will find probabilities that relate to a communication network and an independent backup system for a space vehicle.

■ Find probabilities of events.
■ Find probabilities of mutually exclusive events.
■ Find probabilities of independent events.
■ Find the probability of the complement of an event.

The Probability of an Event

Any happening for which the result is uncertain is an **experiment.** The possible results of the experiment are **outcomes,** the set of all possible outcomes of the experiment is the **sample space** of the experiment, and any subcollection of a sample space is an **event.**

For example, when you toss a six-sided die, the numbers 1 through 6 can represent the sample space. For this experiment, each of the outcomes is *equally likely.*

To describe sample spaces in such a way that each outcome is equally likely, you must sometimes distinguish between or among various outcomes in ways that appear artificial. Example 1 illustrates such a situation.

EXAMPLE 1 **Finding a Sample Space**

Find the sample space for each experiment.

a. You toss one coin.

b. You toss two coins.

c. You toss three coins.

Solution

a. The coin will land either heads up (denoted by H) or tails up (denoted by T), so the sample space is

$$S = \{H, T\}.$$

b. Either coin can land heads up or tails up, so the possible outcomes are as follows.

HH = heads up on both coins

HT = heads up on the first coin and tails up on the second coin

TH = tails up on the first coin and heads up on the second coin

TT = tails up on both coins

So, the sample space is

$$S = \{HH, HT, TH, TT\}.$$

Note that this list distinguishes between the two cases HT and TH, even though these two outcomes appear to be similar.

c. Using notation similar to that used in part (b), the sample space is

$$S = \{HHH, HHT, HTH, HTT, THH, THT, TTH, TTT\}.$$

Note that this list distinguishes among the cases HHT, HTH, and THH, and among the cases HTT, THT, and TTH.

✓ **Checkpoint**   *Audio-video solution in English & Spanish at LarsonPrecalculus.com*

Find the sample space for the experiment.

You toss a coin twice and a six-sided die once.

To find the probability of an event, count the number of outcomes in the event and in the sample space. The *number of equally likely outcomes* in event E is denoted by $n(E)$, and the number of equally likely outcomes in the sample space S is denoted by $n(S)$. The probability that event E will occur is given by $n(E)/n(S)$.

The Probability of an Event

If an event E has $n(E)$ equally likely outcomes and its sample space S has $n(S)$ equally likely outcomes, then the **probability** of event E is

$$P(E) = \frac{n(E)}{n(S)}.$$

The number of outcomes in an event must be less than or equal to the number of outcomes in the sample space, so the probability of an event must be a number between 0 and 1, inclusive. That is,

$$0 \le P(E) \le 1$$

as shown in the figure. If $P(E) = 0$, then event E *cannot occur,* and E is an **impossible event.** If $P(E) = 1$, then event E *must occur,* and E is a **certain event.**

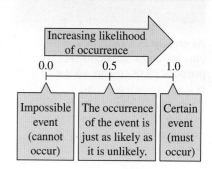

EXAMPLE 2   Finding the Probability of an Event

See LarsonPrecalculus.com for an interactive version of this type of example.

a. You toss two coins. What is the probability that both land heads up?

b. You draw one card at random from a standard deck of 52 playing cards. What is the probability that it is an ace?

Solution

a. Using the results of Example 1(b), let

$$E = \{HH\} \quad \text{and} \quad S = \{HH, HT, TH, TT\}.$$

The probability of getting two heads is

$$P(E) = \frac{n(E)}{n(S)} = \frac{1}{4}.$$

b. The deck has four aces (one in each suit), so the probability of drawing an ace is

$$P(E) = \frac{n(E)}{n(S)} = \frac{4}{52} = \frac{1}{13}.$$

✓ **Checkpoint**  *Audio-video solution in English & Spanish at LarsonPrecalculus.com*

a. You toss three coins. What is the probability that all three land tails up?

b. You draw one card at random from a standard deck of 52 playing cards. What is the probability that it is a diamond?

In some cases, the number of outcomes in the sample space may not be given. In these cases, either write out the sample space or use the counting principles discussed in Section 9.6. Example 3 on the next page uses the Fundamental Counting Principle.

> **REMARK**
> You can write a probability as a fraction, a decimal, or a percent. For instance, in Example 2(a), the probability of getting two heads can be written as $\frac{1}{4}$, 0.25, or 25%.

As shown in Example 3, when you toss two six-sided dice, the probability of rolling a total of 7 is $\frac{1}{6}$.

Felix Furo/Shutterstock.com

EXAMPLE 3 **Finding the Probability of an Event**

You toss two six-sided dice. What is the probability that the total of the two dice is 7?

Solution There are six possible outcomes on each die, so by the Fundamental Counting Principle, there are 6 · 6 or 36 different outcomes when you toss two dice. To find the probability of rolling a total of 7, you must first count the number of ways in which this can occur.

First Die	Second Die
1	6
2	5
3	4
4	3
5	2
6	1

So, a total of 7 can be rolled in six ways, which means that the probability of rolling a total of 7 is

$$P(E) = \frac{n(E)}{n(S)} = \frac{6}{36} = \frac{1}{6}.$$

 **Checkpoint** Audio-video solution in English & Spanish at LarsonPrecalculus.com

You toss two six-sided dice. What is the probability that the total of the two dice is 5?

EXAMPLE 4 **Finding the Probability of an Event**

Twelve-sided dice, as shown in Figure 9.3, can be constructed (in the shape of regular dodecahedrons) such that each of the numbers from 1 to 6 occurs twice on each die. Show that these dice can be used in any game requiring ordinary six-sided dice without changing the probabilities of the various events.

Solution For an ordinary six-sided die, each of the numbers

1, 2, 3, 4, 5, and 6

occurs once, so the probability of rolling any one of these numbers is

$$P(E) = \frac{n(E)}{n(S)} = \frac{1}{6}.$$

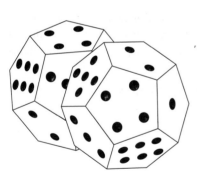

Figure 9.3

For one of the 12-sided dice, each number occurs twice, so the probability of rolling each number is

$$P(E) = \frac{n(E)}{n(S)} = \frac{2}{12} = \frac{1}{6}.$$

 **Checkpoint** Audio-video solution in English & Spanish at LarsonPrecalculus.com

Show that the probability of drawing a club at random from a standard deck of 52 playing cards is the same as the probability of drawing the ace of hearts at random from a set of four cards consisting of the aces of hearts, diamonds, clubs, and spades.

EXAMPLE 5   **Random Selection**

The figure shows the numbers of degree-granting postsecondary institutions in various regions of the United States in 2015. What is the probability that an institution selected at random is in one of the three southern regions?   *(Source: National Center for Education Statistics)*

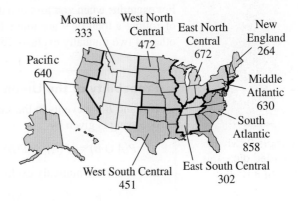

Solution   From the figure, the total number of institutions is 4622. There are $858 + 302 + 451 = 1611$ institutions in the three southern regions, so the probability that the institution is in one of these regions is

$$P(E) = \frac{n(E)}{n(S)} = \frac{1611}{4622} \approx 0.349.$$

✓ *Checkpoint* ◀))) Audio-video solution in English & Spanish at LarsonPrecalculus.com

In Example 5, what is the probability that an institution selected at random is in the Pacific region?

EXAMPLE 6   **Finding the Probability of Winning a Lottery**

In Arizona's The Pick game, a player chooses six different numbers from 1 to 44. If these six numbers match the six numbers drawn (in any order), the player wins (or shares) the top prize. What is the probability of winning the top prize when the player buys one ticket?

Solution   To find the number of outcomes in the sample space, use the formula for the number of combinations of 44 numbers taken six at a time.

$$n(S) = {}_{44}C_6$$

$$= \frac{44 \cdot 43 \cdot 42 \cdot 41 \cdot 40 \cdot 39}{6 \cdot 5 \cdot 4 \cdot 3 \cdot 2 \cdot 1}$$

$$= 7{,}059{,}052$$

When a player buys one ticket, the probability of winning is

$$P(E) = \frac{1}{7{,}059{,}052}.$$

✓ *Checkpoint* ◀))) Audio-video solution in English & Spanish at LarsonPrecalculus.com

In Pennsylvania's Cash 5 game, a player chooses five different numbers from 1 to 43. If these five numbers match the five numbers drawn (in any order), the player wins (or shares) the top prize. What is the probability of winning the top prize when the player buys one ticket?

Mutually Exclusive Events

Two events A and B (from the same sample space) are **mutually exclusive** when A and B have no outcomes in common. In the terminology of sets, the intersection of A and B is the empty set, which implies that

$$P(A \cap B) = 0.$$

For example, when you toss two dice, the event A of rolling a total of 6 and the event B of rolling a total of 9 are mutually exclusive. To find the probability that one or the other of two mutually exclusive events will occur, *add* their individual probabilities.

Probability of the Union of Two Events

If A and B are events in the same sample space, then the probability of A or B occurring is given by

$$P(A \cup B) = P(A) + P(B) - P(A \cap B).$$

If A and B are mutually exclusive, then

$$P(A \cup B) = P(A) + P(B).$$

EXAMPLE 7   **Probability of a Union of Events**

You draw one card at random from a standard deck of 52 playing cards. What is the probability that the card is either a heart or a face card?

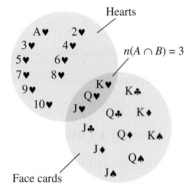

Hearts

$n(A \cap B) = 3$

Face cards

Figure 9.4

Solution   The deck has 13 hearts, so the probability of drawing a heart (event A) is

$$P(A) = \frac{13}{52}.$$

Similarly, the deck has 12 face cards, so the probability of drawing a face card (event B) is

$$P(B) = \frac{12}{52}.$$

Three of the cards are hearts *and* face cards (see Figure 9.4), so it follows that

$$P(A \cap B) = \frac{3}{52}.$$

Finally, applying the formula for the probability of the union of two events, the probability of drawing either a heart or a face card is

$$P(A \cup B) = P(A) + P(B) - P(A \cap B)$$

$$= \frac{13}{52} + \frac{12}{52} - \frac{3}{52}$$

$$= \frac{22}{52}$$

$$\approx 0.423.$$

✓ *Checkpoint* 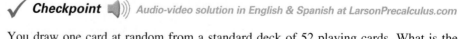 *Audio-video solution in English & Spanish at LarsonPrecalculus.com*

You draw one card at random from a standard deck of 52 playing cards. What is the probability that the card is either an ace or a spade?

EXAMPLE 8 Probability of Mutually Exclusive Events

The human resources department of a company has compiled data showing the number of years of service for each employee. The table shows the results.

DATA	Years of Service	Number of Employees
	0–4	157
	5–9	89
	10–14	74
	15–19	63
	20–24	42
	25–29	38
	30–34	35
	35–39	21
	40–44	8
	45 or more	2

Spreadsheet at LarsonPrecalculus.com

a. What is the probability that an employee chosen at random has 4 or fewer years of service?

b. What is the probability that an employee chosen at random has 9 or fewer years of service?

Solution

a. To begin, add the number of employees to find that the total is 529. Next, let event A represent choosing an employee with 0 to 4 years of service. Then the probability of choosing an employee who has 4 or fewer years of service is

$$P(A) = \frac{157}{529}$$

$$\approx 0.297.$$

b. Let event B represent choosing an employee with 5 to 9 years of service. Then

$$P(B) = \frac{89}{529}.$$

Event A from part (a) and event B have no outcomes in common, so these two events are mutually exclusive and

$$P(A \cup B) = P(A) + P(B)$$

$$= \frac{157}{529} + \frac{89}{529}$$

$$= \frac{246}{529}$$

$$\approx 0.465.$$

So, the probability of choosing an employee who has 9 or fewer years of service is about 0.465.

✓ **Checkpoint**  *Audio-video solution in English & Spanish at LarsonPrecalculus.com*

In Example 8, what is the probability that an employee chosen at random has 30 or more years of service?

Independent Events

Two events are **independent** when the occurrence of one has no effect on the occurrence of the other. For example, rolling a total of 12 with two six-sided dice has no effect on the outcome of future rolls of the dice. To find the probability that two independent events will occur, *multiply* the probabilities of each.

Probability of Independent Events

If A and B are independent events, then the probability that both A and B will occur is

$$P(A \text{ and } B) = P(A) \cdot P(B).$$

This rule can be extended to any number of independent events.

EXAMPLE 9 Probability of Independent Events

A random number generator selects three integers from 1 to 20. What is the probability that all three numbers are less than or equal to 5?

Solution Let event A represent selecting a number from 1 to 5. Then the probability of selecting a number from 1 to 5 is

$$P(A) = \frac{5}{20} = \frac{1}{4}.$$

So, the probability that all three numbers are less than or equal to 5 is

$$P(A) \cdot P(A) \cdot P(A) = \left(\frac{1}{4}\right)\left(\frac{1}{4}\right)\left(\frac{1}{4}\right)$$

$$= \frac{1}{64}.$$

✓ ***Checkpoint*** *Audio-video solution in English & Spanish at LarsonPrecalculus.com*

A random number generator selects two integers from 1 to 30. What is the probability that both numbers are less than 12?

EXAMPLE 10 Probability of Independent Events

In 2015, approximately 65% of Americans expected much of the workforce to be automated within 50 years. In a survey, researchers selected 10 people at random from the population. What is the probability that all 10 people expected much of the workforce to be automated within 50 years? *(Source: Pew Research Center)*

Solution Let event A represent selecting a person who expected much of the workforce to be automated within 50 years. The probability of event A is 0.65. Each of the 10 occurrences of event A is an independent event, so the probability that all 10 people expected much of the workforce to be automated within 50 years is

$$[P(A)]^{10} = (0.65)^{10}$$

$$\approx 0.013.$$

✓ ***Checkpoint*** *Audio-video solution in English & Spanish at LarsonPrecalculus.com*

In Example 10, researchers selected five people at random from the population. What is the probability that all five people expected much of the workforce to be automated within 50 years? ■

The Complement of an Event

The **complement of an event** A is the collection of all outcomes in the sample space that are *not* in A. The complement of event A is denoted by A'. Because $P(A \text{ or } A') = 1$ and A and A' are mutually exclusive, it follows that $P(A) + P(A') = 1$. So, the probability of A' is

$$P(A') = 1 - P(A).$$

Probability of a Complement

Let A be an event and let A' be its complement. If the probability of A is $P(A)$, then the probability of the complement is

$$P(A') = 1 - P(A).$$

For example, if the probability of *winning* a game is $P(A) = \frac{1}{4}$, then the probability of *losing* the game is $P(A') = 1 - \frac{1}{4} = \frac{3}{4}$.

EXAMPLE 11 **Probability of a Complement**

A manufacturer has determined that a machine averages one faulty unit for every 1000 it produces. What is the probability that an order of 200 units will have one or more faulty units?

Solution To solve this problem as stated, you would need to find the probabilities of having exactly one faulty unit, exactly two faulty units, exactly three faulty units, and so on. However, using complements, it is much less tedious to find the probability that all units are perfect and then subtract this value from 1. The probability that any given unit is perfect is $999/1000$, so the probability that all 200 units are perfect is

$$P(A) = \left(\frac{999}{1000} \right)^{200} \approx 0.819$$

and the probability that at least one unit is faulty is

$$P(A') = 1 - P(A) \approx 1 - 0.819 = 0.181.$$

✓ *Checkpoint* 🔊))) *Audio-video solution in English & Spanish at LarsonPrecalculus.com*

A manufacturer has determined that a machine averages one faulty unit for every 500 it produces. What is the probability that an order of 300 units will have one or more faulty units? ◼

Summarize (Section 9.7)

1. State the definition of the probability of an event *(page 667)*. For examples of finding the probabilities of events, see Examples 2–6.

2. State the definition of mutually exclusive events and explain how to find the probability of the union of two events *(page 670)*. For examples of finding the probabilities of the unions of two events, see Examples 7 and 8.

3. State the definition of, and explain how to find the probability of, independent events *(page 672)*. For examples of finding the probabilities of independent events, see Examples 9 and 10.

4. State the definition of, and explain how to find the probability of, the complement of an event *(page 673)*. For an example of finding the probability of the complement of an event, see Example 11.

9.7 Exercises

See CalcChat.com for tutorial help and worked-out solutions to odd-numbered exercises.

Vocabulary

In Exercises 1–7, fill in the blanks.

1. An _____ is any happening for which the result is uncertain, and the possible results are called _____.

2. The set of all possible outcomes of an experiment is the _____ _____.

3. The formula for the _____ of an event is $P(E) = \dfrac{n(E)}{n(S)}$, where $n(E)$ is the number of equally likely outcomes in the event and $n(S)$ is the number of equally likely outcomes in the sample space.

4. If $P(E) = 0$, then E is an _____ event, and if $P(E) = 1$, then E is a _____ event.

5. Two events A and B (from the same sample space) are _____ _____ when A and B have no outcomes in common.

6. Two events are _____ when the occurrence of one has no effect on the occurrence of the other.

7. The _____ of an event A is the collection of all outcomes in the sample space that are not in A.

8. Match the probability formula with the correct probability name.
   (a) Probability of the union of two events
   (b) Probability of mutually exclusive events
   (c) Probability of independent events
   (d) Probability of a complement

   (i) $P(A \cup B) = P(A) + P(B)$
   (ii) $P(A') = 1 - P(A)$
   (iii) $P(A \cup B) = P(A) + P(B) - P(A \cap B)$
   (iv) $P(A \text{ and } B) = P(A) \cdot P(B)$

Skills and Applications

 Finding a Sample Space **In Exercises 9–14, find the sample space for the experiment.**

9. You toss a coin and a six-sided die.

10. You toss a six-sided die twice and record the sum.

11. A taste tester ranks three varieties of yogurt, A, B, and C, according to preference.

12. You select two marbles (without replacement) from a bag containing two red marbles, two blue marbles, and one yellow marble. You record the color of each marble.

13. Two county supervisors are selected from five supervisors, A, B, C, D, and E, to study a recycling plan.

14. A sales representative visits three homes per day. In each home, there may be a sale (denote by S) or there may be no sale (denote by F).

 Tossing a Coin **In Exercises 15–20, find the probability for the experiment of tossing a coin three times.**

15. The probability of getting exactly one tail

16. The probability of getting exactly two tails

17. The probability of getting a head on the first toss

18. The probability of getting a tail on the last toss

19. The probability of getting at least one head

20. The probability of getting at least two heads

Drawing a Card **In Exercises 21–24, find the probability for the experiment of drawing a card at random from a standard deck of 52 playing cards.**

21. The card is a face card.

22. The card is not a face card.

23. The card is a red face card.

24. The card is a 9 or lower. (Aces are low.)

 Tossing a Die **In Exercises 25–30, find the probability for the experiment of tossing a six-sided die twice.**

25. The sum is 6.

26. The sum is at least 8.

27. The sum is less than 11.

28. The sum is 2, 3, or 12.

29. The sum is odd and no more than 7.

30. The sum is odd or prime.

Drawing Marbles **In Exercises 31–34, find the probability for the experiment of drawing two marbles at random (without replacement) from a bag containing one green, two yellow, and three red marbles.**

31. Both marbles are red.   32. Both marbles are yellow.

33. Neither marble is yellow.

34. The marbles are different colors.

35. Unemployment In 2015, there were approximately 8.3 million unemployed workers in the United States. The circle graph shows the age profile of these unemployed workers. *(Source: U.S. Bureau of Labor Statistics)*

Ages of Unemployed Workers

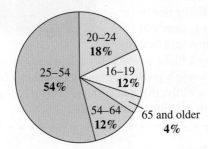

(a) Estimate the number of unemployed workers in the 16–19 age group.

(b) What is the probability that a person selected at random from the population of unemployed workers is in the 20–24 age group?

(c) What is the probability that a person selected at random from the population of unemployed workers is in the 25–54 age group?

(d) What is the probability that a person selected at random from the population of unemployed workers is 55 or older?

36. Political Poll An independent polling organization interviewed 100 college students to determine their political party affiliations and whether they favor a balanced-budget amendment to the Constitution. The table lists the results of the study. In the table, *D* represents Democrat and *R* represents Republican.

	Favor	Not Favor	Unsure	Total
D	23	25	7	55
R	32	9	4	45
Total	55	34	11	100

Find the probability that a person selected at random from the sample is as described.

(a) A person who does not favor the amendment

(b) A Republican

(c) A Democrat who favors the amendment

37. Education In a high school graduating class of 128 students, 52 are on the honor roll. Of these, 48 are going on to college. Of the 76 students not on the honor roll, 56 are going on to college. What is the probability that a student selected at random from the class is (a) going to college, (b) not going to college, and (c) not going to college and on the honor roll?

38. Alumni Association A college sends a survey to members of the class of 2016. Of the 1254 people who graduated that year, 672 are women, of whom 124 went on to graduate school. Of the 582 male graduates, 198 went on to graduate school. Find the probability that a class of 2016 alumnus selected at random is as described.

(a) Female

(b) Male

(c) Female and did not attend graduate school

39. Winning an Election Three people are running for president of a class. The results of a poll show that the first candidate has an estimated 37% chance of winning and the second candidate has an estimated 44% chance of winning. What is the probability that the third candidate will win?

40. Payroll Error The employees of a company work in six departments: 31 are in sales, 54 are in research, 42 are in marketing, 20 are in engineering, 47 are in finance, and 58 are in production. The payroll clerk loses one employee's paycheck. What is the probability that the employee works in the research department?

41. Exam Questions A class receives a list of 20 study problems, from which 10 will be part of an upcoming exam. A student knows how to solve 15 of the problems. Find the probability that the student will be able to answer (a) all 10 questions on the exam, (b) exactly eight questions on the exam, and (c) at least nine questions on the exam.

42. Payroll Error A payroll clerk addresses five paychecks and envelopes to five different people and randomly inserts the paychecks into the envelopes. Find the probability of each event.

(a) Exactly one paycheck is inserted in the correct envelope.

(b) At least one paycheck is inserted in the correct envelope.

43. Game Show On a game show, you are given five digits to arrange in the proper order to form the price of a car. If you are correct, you win the car. What is the probability of winning, given the following conditions?

(a) You guess the position of each digit.

(b) You know the first digit and guess the positions of the other digits.

44. Card Game The deck for a card game contains 108 cards. Twenty-five each are red, yellow, blue, and green, and eight are wild cards. Each player is randomly dealt a seven-card hand.

(a) What is the probability that a hand will contain exactly two wild cards?

(b) What is the probability that a hand will contain two wild cards, two red cards, and three blue cards?

45. Drawing a Card You draw one card at random from a standard deck of 52 playing cards. Find the probability that (a) the card is an even-numbered card, (b) the card is a heart or a diamond, and (c) the card is a nine or a face card.

46. Drawing Cards You draw five cards at random from a standard deck of 52 playing cards. What is the probability that the hand drawn is a full house? (A full house consists of three of one kind and two of another.)

47. Shipment A shipment of 12 microwave ovens contains three defective units. A vending company purchases four units at random. What is the probability that (a) all four units are good, (b) exactly two units are good, and (c) at least two units are good?

48. PIN Code ATM personal identification number (PIN) codes typically consist of four-digit sequences of numbers. Find the probability that if you forget your PIN, you can guess the correct sequence (a) at random and (b) when you recall the first two digits.

49. Random Number Generator A random number generator selects two integers from 1 through 40. What is the probability that (a) both numbers are even, (b) one number is even and one number is odd, (c) both numbers are less than 30, and (d) the same number is selected twice?

50. Flexible Work Hours In a recent survey, people were asked whether they would prefer to work flexible hours—even when it meant slower career advancement—so they could spend more time with their families. The figure shows the results of the survey. What is the probability that three people chosen at random would prefer flexible work hours?

Flexible Work Hours

Flexible hours **78%**

Don't know **9%**

Rigid hours **13%**

 Probability of a Complement In Exercises 51–54, you are given the probability that an event *will* happen. Find the probability that the event *will not* happen.

51. $P(E) = 0.73$ **52.** $P(E) = 0.28$

53. $P(E) = \dfrac{1}{5}$ **54.** $P(E) = \dfrac{2}{7}$

Probability of a Complement In Exercises 55–58, you are given the probability that an event *will not* happen. Find the probability that the event *will* happen.

55. $P(E') = 0.29$ **56.** $P(E') = 0.89$

57. $P(E') = \dfrac{14}{25}$ **58.** $P(E') = \dfrac{79}{100}$

59. Backup System

A space vehicle has an independent backup system for one of its communication networks. The probability that either system will function satisfactorily during a flight is 0.985. What is the probability that during a given flight (a) both systems function satisfactorily, (b) both systems fail, and (c) at least one system functions satisfactorily?

60. Backup Vehicle A fire department keeps two rescue vehicles. Due to the demand on the vehicles and the chance of mechanical failure, the probability that a specific vehicle is available when needed is 90%. The availability of one vehicle is independent of the availability of the other. Find the probability that (a) both vehicles are available at a given time, (b) neither vehicle is available at a given time, and (c) at least one vehicle is available at a given time.

61. Roulette American roulette is a game in which a wheel turns on a spindle and is divided into 38 pockets. Thirty-six of the pockets are numbered 1–36, of which half are red and half are black. Two of the pockets are green and are numbered 0 and 00 (see figure). The dealer spins the wheel and a small ball in opposite directions. As the ball slows to a stop, it has an equal probability of landing in any of the numbered pockets.

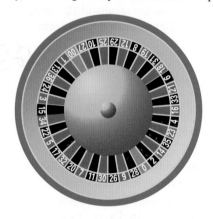

(a) Find the probability of landing in the number 00 pocket.

(b) Find the probability of landing in a red pocket.

(c) Find the probability of landing in a green pocket or a black pocket.

(d) Find the probability of landing in the number 14 pocket on two consecutive spins.

(e) Find the probability of landing in a red pocket on three consecutive spins.

62. A Boy or a Girl? Assume that the probability of the birth of a child of a particular sex is 50%. In a family with four children, find the probability of each event.

 (a) All the children are boys.

 (b) All the children are the same sex.

 (c) There is at least one boy.

63. Geometry You and a friend agree to meet at your favorite restaurant between 5:00 P.M. and 6:00 P.M. The one who arrives first will wait 15 minutes for the other, and then will leave (see figure). What is the probability that the two of you will actually meet, assuming that your arrival times are random within the hour?

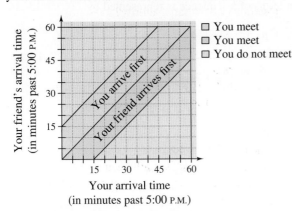

Your arrival time
(in minutes past 5:00 P.M.)

64. Estimating π You drop a coin of diameter d onto a paper that contains a grid of squares d units on a side (see figure).

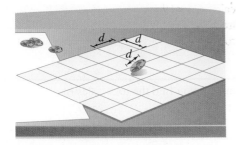

 (a) Find the probability that the coin covers a vertex of one of the squares on the grid.

 (b) Perform the experiment 100 times and use the results to approximate π.

Exploration

True or False? **In Exercises 65 and 66, determine whether the statement is true or false. Justify your answer.**

65. If A and B are independent events with nonzero probabilities, then A can occur when B occurs.

66. Rolling a number less than 3 on a normal six-sided die has a probability of $\frac{1}{3}$. The complement of this event is rolling a number greater than 3, which has a probability of $\frac{1}{2}$.

67. Pattern Recognition Consider a group of n people.

 (a) Explain why the pattern below gives the probabilities that the n people have distinct birthdays.

$$n = 2: \frac{365}{365} \cdot \frac{364}{365} = \frac{365 \cdot 364}{365^2}$$

$$n = 3: \frac{365}{365} \cdot \frac{364}{365} \cdot \frac{363}{365} = \frac{365 \cdot 364 \cdot 363}{365^3}$$

 (b) Use the pattern in part (a) to write an expression for the probability that $n = 4$ people have distinct birthdays.

 (c) Let P_n be the probability that the n people have distinct birthdays. Verify that this probability can be obtained recursively by

$$P_1 = 1 \text{ and } P_n = \frac{365 - (n - 1)}{365} P_{n-1}.$$

 (d) Explain why $Q_n = 1 - P_n$ gives the probability that at least two people in a group of n people have the same birthday.

 (e) Use the results of parts (c) and (d) to complete the table.

n	10	15	20	23	30	40	50
P_n							
Q_n							

 (f) How many people must be in a group so that the probability of at least two of them having the same birthday is greater than $\frac{1}{2}$? Explain.

68. **HOW DO YOU SEE IT?** The circle graphs show the percents of undergraduate students by class level at two colleges. A student is chosen at random from the combined undergraduate population of the two colleges. The probability that the student is a freshman, sophomore, or junior is 81%. Which college has a greater number of undergraduate students? Explain.

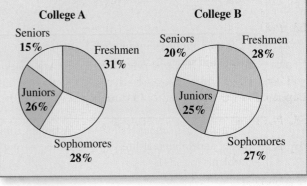

College A

Seniors 15%
Freshmen 31%
Juniors 26%
Sophomores 28%

College B

Seniors 20%
Freshmen 28%
Juniors 25%
Sophomores 27%

Chapter Summary

	What Did You Learn?	Explanation/Examples	Review Exercises
Section 9.1	Use sequence notation to write the terms of sequences (p. 610).	$a_n = 7n - 4;\quad a_1 = 7(1) - 4 = 3,\ a_2 = 7(2) - 4 = 10,$ $a_3 = 7(3) - 4 = 17,\ a_4 = 7(4) - 4 = 24$	1–8
	Use factorial notation (p. 613).	If n is a positive integer, then $n! = 1 \cdot 2 \cdot 3 \cdot 4 \cdots (n - 1) \cdot n.$	9–12
	Use summation notation to write sums (p. 614).	The sum of the first n terms of a sequence is represented by $\displaystyle\sum_{i=1}^{n} a_i = a_1 + a_2 + a_3 + a_4 + \cdots + a_n.$	13–16
	Find the sums of series (p. 615).	$\displaystyle\sum_{i=1}^{\infty} \frac{8}{10^i} = \frac{8}{10^1} + \frac{8}{10^2} + \frac{8}{10^3} + \frac{8}{10^4} + \frac{8}{10^5} + \cdots$ $= 0.8 + 0.08 + 0.008 + 0.0008 + 0.00008 + \cdots$ $= 0.88888\ldots = \dfrac{8}{9}$	17, 18
	Use sequences and series to model and solve real-life problems (p. 616).	A sequence can help you model the balance after n compoundings in an account that earns compound interest. (See Example 10.)	19, 20
Section 9.2	Recognize, write, and find the nth terms of arithmetic sequences (p. 620).	$a_n = 9n + 5;\quad a_1 = 9(1) + 5 = 14,\ a_2 = 9(2) + 5 = 23,$ $a_3 = 9(3) + 5 = 32,\ a_4 = 9(4) + 5 = 41;$ common difference: $d = 9$	21–30
	Find nth partial sums of arithmetic sequences (p. 623).	The sum of a finite arithmetic sequence with n terms is given by $S_n = (n/2)(a_1 + a_n).$	31–36
	Use arithmetic sequences to model and solve real-life problems (p. 625).	An arithmetic sequence can help you find the total sales of a small business. (See Example 9.)	37, 38
Section 9.3	Recognize, write, and find the nth terms of geometric sequences (p. 629).	$a_n = 3(4^n);\quad a_1 = 3(4^1) = 12,\ a_2 = 3(4^2) = 48,$ $a_3 = 3(4^3) = 192,\ a_4 = 3(4^4) = 768;$ common ratio: $r = 4$	39–50
	Find the sum of a finite geometric sequence (p. 632).	The sum of the finite geometric sequence $a_1, a_1r, a_1r^2, \ldots, a_1r^{n-1}$ with common ratio $r \neq 1$ is given by $S_n = \displaystyle\sum_{i=1}^{n} a_1 r^{i-1} = a_1 \left(\dfrac{1 - r^n}{1 - r} \right).$	51–58
	Find the sum of an infinite geometric series (p. 633).	If $|r| < 1$, then the infinite geometric series $a_1 + a_1r + a_1r^2 + \cdots + a_1r^{n-1} + \cdots$ has the sum $S = \displaystyle\sum_{i=0}^{\infty} a_1 r^i = \dfrac{a_1}{1 - r}.$	59–62
	Use geometric sequences to model and solve real-life problems (p. 634).	A finite geometric sequence can help you find the balance of an increasing annuity at the end of two years. (See Example 8.)	63, 64

	What Did You Learn?	**Explanation/Examples**	**Review Exercises**
Section 9.4	Use mathematical induction to prove statements involving a positive integer n (p. 638).	Let P_n be a statement involving the positive integer n. If (1) P_1 is true, and (2) for every positive integer k, the truth of P_k implies the truth of P_{k+1}, then the statement P_n must be true for all positive integers n.	65–68
	Use pattern recognition and mathematical induction to write a formula for the nth term of a sequence (p. 643).	To find a formula for the nth term of a sequence, (1) calculate the first several terms of the sequence, (2) try to find a pattern for the terms and write a formula for the nth term of the sequence (hypothesis), and (3) use mathematical induction to prove your hypothesis.	69–72
	Find the sums of powers of integers (p. 644).	$\sum\limits_{i=1}^{8} i^2 = \dfrac{n(n+1)(2n+1)}{6} = \dfrac{8(8+1)(16+1)}{6} = 204$	73, 74
	Find finite differences of sequences (p. 645).	The first differences of a sequence are found by subtracting consecutive terms. The second differences are found by subtracting consecutive first differences.	75, 76
Section 9.5	Use the Binomial Theorem to find binomial coefficients (p. 648).	**The Binomial Theorem:** In the expansion of $(x+y)^n =$ $x^n + nx^{n-1}y + \cdots + {}_nC_r x^{n-r}y^r + \cdots + nxy^{n-1} + y^n$, the coefficient of $x^{n-r}y^r$ is ${}_nC_r = \dfrac{n!}{(n-r)!r!}$.	77, 78
	Use Pascal's Triangle to find binomial coefficients (p. 650).	First several rows of Pascal's Triangle: $\begin{matrix} & & & 1 \\ & & 1 & & 1 \\ & 1 & & 2 & & 1 \\ 1 & & 3 & & 3 & & 1 \\ 1 & 4 & & 6 & & 4 & & 1 \end{matrix}$	79, 80
	Use binomial coefficients to write binomial expansions (p. 651).	$(x+1)^3 = x^3 + 3x^2 + 3x + 1$ $(x-1)^4 = x^4 - 4x^3 + 6x^2 - 4x + 1$	81–84
Section 9.6	Solve simple counting problems (p. 656).	A computer randomly generates an integer from 1 through 15. The computer can generate an integer that is divisible by 3 in 5 ways (3, 6, 9, 12, and 15).	85, 86
	Use the Fundamental Counting Principle to solve counting problems (p. 657).	**Fundamental Counting Principle:** Let E_1 and E_2 be two events. The first event E_1 can occur in m_1 different ways. After E_1 has occurred, E_2 can occur in m_2 different ways. The number of ways the two events can occur is $m_1 \cdot m_2$.	87, 88
	Use permutations (p. 658) and combinations (p. 661) to solve counting problems.	The number of permutations of n elements taken r at a time is ${}_nP_r = n!/(n-r)!$. The number of combinations of n elements taken r at a time is ${}_nC_r = n!/[(n-r)!r!]$, or ${}_nC_r = {}_nP_r/r!$.	89–92
Section 9.7	Find probabilities of events (p. 667).	If an event E has $n(E)$ equally likely outcomes and its sample space S has $n(S)$ equally likely outcomes, then the probability of event E is $P(E) = n(E)/n(S)$.	93, 94
	Find probabilities of mutually exclusive events (p. 670) and independent events (p. 672), and find the probability of the complement of an event (p. 673).	If A and B are events in the same sample space, then the probability of A or B occurring is $P(A \cup B) = P(A) + P(B) - P(A \cap B)$. If A and B are mutually exclusive, then $P(A \cup B) = P(A) + P(B)$. If A and B are independent, then $P(A \text{ and } B) = P(A) \cdot P(B)$. The probability of the complement of A is $P(A') = 1 - P(A)$.	95–100

Review Exercises
See CalcChat.com for tutorial help and worked-out solutions to odd-numbered exercises.

9.1 Writing the Terms of a Sequence In Exercises 1–4, write the first five terms of the sequence. (Assume that n begins with 1.)

1. $a_n = 3 + \dfrac{12}{n}$

2. $a_n = \dfrac{(-1)^n 5n}{2n - 1}$

3. $a_n = \dfrac{120}{n!}$

4. $a_n = (n + 1)(n + 2)$

Finding the nth Term of a Sequence In Exercises 5–8, write an expression for the apparent nth term (a_n) of the sequence. (Assume that n begins with 1.)

5. $-2, 2, -2, 2, -2, \ldots$

6. $-1, 2, 7, 14, 23, \ldots$

7. $4, 2, \frac{4}{3}, 1, \frac{4}{5}, \ldots$

8. $1, -\frac{1}{2}, \frac{1}{3}, -\frac{1}{4}, \frac{1}{5}, \ldots$

Simplifying a Factorial Expression In Exercises 9–12, simplify the factorial expression.

9. $\dfrac{3!}{5!}$

10. $\dfrac{7!}{3! \cdot 4!}$

11. $\dfrac{(n - 1)!}{(n + 1)!}$

12. $\dfrac{n!}{(n + 2)!}$

Finding a Sum In Exercises 13 and 14, find the sum.

13. $\displaystyle\sum_{j=1}^{4} \dfrac{6}{j^2}$

14. $\displaystyle\sum_{k=1}^{10} 2k^3$

Using Sigma Notation to Write a Sum In Exercises 15 and 16, use sigma notation to write the sum.

15. $\dfrac{1}{2(1)} + \dfrac{1}{2(2)} + \dfrac{1}{2(3)} + \cdots + \dfrac{1}{2(20)}$

16. $\dfrac{1}{2} + \dfrac{2}{3} + \dfrac{3}{4} + \cdots + \dfrac{9}{10}$

Finding the Sum of an Infinite Series In Exercises 17 and 18, find the sum of the infinite series.

17. $\displaystyle\sum_{i=1}^{\infty} \dfrac{4}{10^i}$

18. $\displaystyle\sum_{k=1}^{\infty} 8\left(\dfrac{1}{10}\right)^k$

19. **Compound Interest** An investor deposits $10,000 in an account that earns 2.25% interest compounded monthly. The balance in the account after n months is given by

$$A_n = 10{,}000\left(1 + \dfrac{0.0225}{12}\right)^n, \quad n = 1, 2, 3, \ldots.$$

(a) Write the first 10 terms of the sequence.

(b) Find the balance in the account after 10 years by computing the 120th term of the sequence.

20. **Population** The population a_n (in thousands) of Miami, Florida, from 2010 through 2014 can be approximated by

$$a_n = -0.34n^2 + 14.8n + 288, \quad n = 10, 11, \ldots, 14$$

where n is the year with $n = 10$ corresponding to 2010. Write the terms of this finite sequence. Use a graphing utility to construct a bar graph that represents the sequence. *(Source: U.S. Census Bureau)*

9.2 Determining Whether a Sequence Is Arithmetic In Exercises 21–24, determine whether the sequence is arithmetic. If so, find the common difference.

21. $5, -1, -7, -13, -19, \ldots$

22. $0, 1, 3, 6, 10, \ldots$

23. $\frac{1}{8}, \frac{1}{4}, \frac{1}{2}, 1, 2, \ldots$

24. $1, \frac{15}{16}, \frac{7}{8}, \frac{13}{16}, \frac{3}{4}, \ldots$

Finding the nth Term In Exercises 25–28, find a formula for a_n for the arithmetic sequence.

25. $a_1 = 7, d = 12$

26. $a_1 = 34, d = -4$

27. $a_3 = 96, a_7 = 24$

28. $a_7 = 8, a_{13} = 6$

Writing the Terms of an Arithmetic Sequence In Exercises 29 and 30, write the first five terms of the arithmetic sequence.

29. $a_1 = 4, d = 17$

30. $a_1 = 25, a_{n+1} = a_n + 3$

31. **Sum of a Finite Arithmetic Sequence** Find the sum of the first 100 positive multiples of 9.

32. **Sum of a Finite Arithmetic Sequence** Find the sum of the integers from 30 to 80.

Finding a Sum In Exercises 33–36, find the sum.

33. $\displaystyle\sum_{j=1}^{10} (2j - 3)$

34. $\displaystyle\sum_{j=1}^{8} (20 - 3j)$

35. $\displaystyle\sum_{k=1}^{11} \left(\dfrac{2}{3}k + 4\right)$

36. $\displaystyle\sum_{k=1}^{25} \left(\dfrac{3k + 1}{4}\right)$

37. **Job Offer** The starting salary for a job is $43,800 with a guaranteed increase of $1950 per year. Determine (a) the salary during the fifth year and (b) the total compensation through five full years of employment.

38. **Baling Hay** In the first two trips baling hay around a large field, a farmer obtains 123 bales and 112 bales, respectively. Each round gets shorter, so the farmer estimates that the same pattern will continue. Estimate the total number of bales made after the farmer takes another six trips around the field.

9.3 Determining Whether a Sequence Is Geometric In Exercises 39–42, determine whether the sequence is geometric. If so, find the common ratio.

39. 2, 6, 18, 54, 162, . . .   **40.** 48, −24, 12, −6, . . .

41. $\frac{1}{5}, -\frac{3}{5}, \frac{9}{5}, -\frac{27}{5}, \ldots$ **42.** $\frac{1}{4}, \frac{2}{5}, \frac{3}{6}, \frac{4}{7}, \ldots$

Writing the Terms of a Geometric Sequence In Exercises 43–46, write the first five terms of the geometric sequence.

43. $a_1 = 2, \quad r = 15$

44. $a_1 = 6, \quad r = -\frac{1}{3}$

45. $a_1 = 9, \quad a_3 = 4$

46. $a_1 = 2, \quad a_3 = 12$

Finding a Term of a Geometric Sequence In Exercises 47–50, write an expression for the nth term of the geometric sequence. Then find the 10th term of the sequence.

47. $a_1 = 100, \quad r = 1.05$

48. $a_1 = 5, \quad r = 0.2$

49. $a_1 = 18, \quad a_2 = -9$

50. $a_3 = 6, \quad a_4 = 1$

Sum of a Finite Geometric Sequence In Exercises 51–58, find the sum of the finite geometric sequence.

51. $\displaystyle\sum_{i=1}^{7} 2^{i-1}$ **52.** $\displaystyle\sum_{i=1}^{5} 3^{i-1}$

53. $\displaystyle\sum_{i=1}^{4} \left(\frac{1}{2}\right)^{i}$ **54.** $\displaystyle\sum_{i=1}^{6} \left(\frac{1}{3}\right)^{i-1}$

55. $\displaystyle\sum_{i=1}^{5} (2)^{i-1}$ **56.** $\displaystyle\sum_{i=1}^{4} 6(3)^{i}$

57. $\displaystyle\sum_{i=1}^{5} 10(0.6)^{i-1}$ **58.** $\displaystyle\sum_{i=1}^{4} 20(0.2)^{i-1}$

Sum of an Infinite Geometric Sequence In Exercises 59–62, find the sum of the infinite geometric series.

59. $\displaystyle\sum_{i=0}^{\infty} \left(\frac{7}{8}\right)^{i}$ **60.** $\displaystyle\sum_{i=0}^{\infty} (0.5)^{i}$

61. $\displaystyle\sum_{k=1}^{\infty} 4\left(\frac{2}{3}\right)^{k-1}$ **62.** $\displaystyle\sum_{k=1}^{\infty} 1.3\left(\frac{1}{10}\right)^{k-1}$

63. Depreciation A paper manufacturer buys a machine for $120,000. It depreciates at a rate of 30% per year. (In other words, at the end of each year the depreciated value is 70% of what it was at the beginning of the year.)

(a) Find the formula for the nth term of a geometric sequence that gives the value of the machine t full years after it is purchased.

(b) Find the depreciated value of the machine after 5 full years.

64. Annuity An investor deposits $800 in an account on the first day of each month for 10 years. The account pays 3%, compounded monthly. What is the balance at the end of 10 years?

9.4 Using Mathematical Induction In Exercises 65–68, use mathematical induction to prove the formula for all integers $n \geq 1$.

65. $3 + 5 + 7 + \cdots + (2n + 1) = n(n + 2)$

66. $1 + \frac{3}{2} + 2 + \frac{5}{2} + \cdots + \frac{1}{2}(n + 1) = \frac{n}{4}(n + 3)$

67. $\displaystyle\sum_{i=0}^{n-1} ar^i = \frac{a(1 - r^n)}{1 - r}$

68. $\displaystyle\sum_{k=0}^{n-1} (a + kd) = \frac{n}{2}[2a + (n - 1)d]$

Finding a Formula for a Finite Sum In Exercises 69–72, find a formula for the sum of the first n terms of the sequence. Prove the validity of your formula.

69. 9, 13, 17, 21, . . .

70. 68, 60, 52, 44, . . .

71. $1, \frac{3}{5}, \frac{9}{25}, \frac{27}{125}, \ldots$

72. $12, -1, \frac{1}{12}, -\frac{1}{144}, \ldots$

Finding a Sum In Exercises 73 and 74, find the sum using the formulas for the sums of powers of integers.

73. $\displaystyle\sum_{n=1}^{75} n$ **74.** $\displaystyle\sum_{n=1}^{6} (n^5 - n^2)$

Linear Model, Quadratic Model, or Neither? In Exercises 75 and 76, write the first five terms of the sequence beginning with the term a_1. Then calculate the first and second differences of the sequence. State whether the sequence has a perfect linear model, a perfect quadratic model, or neither.

75. $a_1 = 5$
$a_n = a_{n-1} + 5$

76. $a_1 = -3$
$a_n = a_{n-1} - 2n$

9.5 Finding a Binomial Coefficient In Exercises 77 and 78, find the binomial coefficient.

77. $_6C_4$ **78.** $_{12}C_3$

Using Pascal's Triangle In Exercises 79 and 80, evaluate using Pascal's Triangle.

79. $\dbinom{7}{2}$ **80.** $\dbinom{10}{4}$

Expanding a Bionomial In Exercises 81–84, use the Binomial Theorem to write the expansion of the expression.

81. $(x + 4)^4$ **82.** $(5 + 2z)^4$

83. $(4 - 5x)^3$ **84.** $(a - 3b)^5$

9.6 **Random Selection** In Exercises 85 and 86, determine the number of ways a computer can generate the sum using randomly selected integers from 1 through 14.

85. Two *distinct* integers whose sum is 7

86. Two *distinct* integers whose sum is 12

87. **Telephone Numbers** All of the landline telephone numbers in a small town use the same three-digit prefix. How many different telephone numbers are possible by changing only the last four digits?

88. **Course Schedule** A college student is preparing a course schedule for the next semester. The student may select one of three mathematics courses, one of four science courses, and one of six history courses. How many schedules are possible?

89. **Genetics** A geneticist is using gel electrophoresis to analyze five DNA samples. The geneticist treats each sample with a different restriction enzyme and then injects it into one of five wells formed in a bed of gel. In how many orders can the geneticist inject the five samples into the wells?

90. **Race** There are 10 bicyclists entered in a race. In how many different ways can the top three places be decided?

91. **Jury Selection** In how many different ways can a jury of 12 people be randomly selected from a group of 32 people?

92. **Menu Choices** A local sandwich shop offers five different breads, four different meats, three different cheeses, and six different vegetables. A customer can choose one bread, one or no meat, one or no cheese, and up to three vegetables. Find the total number of combinations of sandwiches possible.

9.7

93. **Apparel** A drawer contains six white socks, two blue socks, and two gray socks.

   (a) What is the probability of randomly selecting one blue sock?

   (b) What is the probability of randomly selecting one white sock?

94. **Bookshelf Order** A child returns a five-volume set of books to a bookshelf. The child is not able to read, and so cannot distinguish one volume from another. What is the probability that the child shelves the books in the correct order?

95. **Students by Class** At a university, 31% of the students are freshmen, 26% are sophomores, 25% are juniors, and 18% are seniors. One student receives a cash scholarship randomly by lottery. Find the probability that the scholarship winner is as described.

   (a) A junior or senior

   (b) A freshman, sophomore, or junior

96. **Opinion Poll** In a survey, a sample of college students, faculty members, and administrators were asked whether they favor a proposed increase in the annual activity fee to enhance student life on campus. The table lists the results of the survey.

	Students	Faculty	Admin.	Total
Favor	237	37	18	292
Oppose	163	38	7	208
Total	400	75	25	500

Find the probability that a person selected at random from the sample is as described.

   (a) A person who opposes the proposal

   (b) A student

   (c) A faculty member who favors the proposal

97. **Tossing a Die** You toss a six-sided die four times. What is the probability of getting four 5's?

98. **Tossing a Die** You toss a six-sided die six times. What is the probability of getting each number exactly once?

99. **Drawing a Card** You draw one card at random from a standard deck of 52 playing cards. What is the probability that the card is not a club?

100. **Tossing a Coin** You toss a coin five times. What is the probability of getting at least one tail?

Exploration

True or False? In Exercises 101–104, determine whether the statement is true or false. Justify your answer.

101. $\dfrac{(n+2)!}{n!} = \dfrac{n+2}{n}$

102. $\sum_{i=1}^{5} (i^3 + 2i) = \sum_{i=1}^{5} i^3 + \sum_{i=1}^{5} 2i$

103. $\sum_{k=1}^{8} 3k = 3 \sum_{k=1}^{8} k$

104. $\sum_{j=1}^{6} 2^j = \sum_{j=3}^{8} 2^{j-2}$

105. **Think About It** An infinite sequence beginning with a_1 is a function. What is the domain of the function?

106. **Think About It** How do the two sequences differ?

   (a) $a_n = \dfrac{(-1)^n}{n}$ (b) $a_n = \dfrac{(-1)^{n+1}}{n}$

107. **Writing** Explain what is meant by a recursion formula.

108. **Writing** Write a brief paragraph explaining how to identify the graph of an arithmetic sequence and the graph of a geometric sequence.

Chapter Test

Take this test as you would take a test in class. When you are finished, check your work against the answers given in the back of the book.

1. Write the first five terms of the sequence $a_n = \dfrac{(-1)^n}{3n + 2}$. (Assume that n begins with 1.)

2. Write an expression for the apparent nth term (a_n) of the sequence. (Assume that n begins with 1.)

$$\frac{3}{1!}, \frac{4}{2!}, \frac{5}{3!}, \frac{6}{4!}, \frac{7}{5!}, \cdots$$

3. Write the next three terms of the series. Then find the seventh partial sum of the series.

$$8 + 21 + 34 + 47 + \cdots$$

4. The 5th term of an arithmetic sequence is 45, and the 12th term is 24. Find the nth term.

5. The second term of a geometric sequence is 14, and the sixth term is 224. Find the nth term. (Assume that the terms of the sequence are positive.)

In Exercises 6–9, find the sum.

6. $\displaystyle\sum_{i=1}^{50} (2i^2 + 5)$

7. $\displaystyle\sum_{n=1}^{9} (12n - 7)$

8. $\displaystyle\sum_{i=1}^{\infty} 4\left(\frac{1}{2}\right)^i$

9. $\displaystyle\sum_{n=1}^{\infty} \left(-\frac{1}{3}\right)^n$

10. Use mathematical induction to prove the formula for all integers $n \geq 1$.

$$5 + 10 + 15 + \cdots + 5n = \frac{5n(n + 1)}{2}$$

11. Use the Binomial Theorem to write the expansion of $(x + 6y)^4$.

12. Expand $3(x - 2)^5 + 4(x - 2)^3$ by using Pascal's Triangle to determine the coefficients.

13. Find the coefficient of the term a^4b^3 in the expansion of $(3a - 2b)^7$.

In Exercises 14 and 15, evaluate each expression.

14. (a) $_9P_2$ (b) $_{70}P_3$

15. (a) $_{11}C_4$ (b) $_{66}C_4$

16. How many distinct license plate numbers consisting of one letter followed by a three-digit number are possible?

17. Eight people are going for a ride in a boat that seats eight people. One person will drive, and only three of the remaining people are willing to ride in the two bow seats. How many seating arrangements are possible?

18. You attend a karaoke night and hope to hear your favorite song. The karaoke song book has 300 different songs (your favorite song is among them). Assuming that the singers are equally likely to pick any song and no song repeats, what is the probability that your favorite song is one of the 20 that you hear that night?

19. You and three of your friends are at a party. Names of all of the 30 guests are placed in a hat and drawn randomly to award four door prizes. Each guest can win only one prize. What is the probability that you and your friends win all four prizes?

20. The weather report calls for a 90% chance of snow. According to this report, what is the probability that it will *not* snow?

Cumulative Test for Chapters 7–9

See CalcChat.com for tutorial help and worked-out solutions to odd-numbered exercises.

Take this test as you would take a test in class. When you are finished, check your work against the answers given in the back of the book.

In Exercises 1–4, solve the system by the specified method.

1. Substitution

$$\begin{cases} y = 3 - x^2 \\ 2(y - 2) = x - 1 \end{cases}$$

2. Elimination

$$\begin{cases} x + 3y = -6 \\ 2x + 4y = -10 \end{cases}$$

3. Gaussian Elimination

$$\begin{cases} -2x + 4y - z = -16 \\ x - 2y + 2z = 5 \\ x - 3y - z = 13 \end{cases}$$

4. Gauss-Jordan Elimination

$$\begin{cases} x + 3y - 2z = -7 \\ -2x + y - z = -5 \\ 4x + y + z = 3 \end{cases}$$

5. A custom-blend bird seed is made by mixing two types of bird seeds costing $0.75 per pound and $1.25 per pound. How many pounds of each type of seed mixture are used to make 200 pounds of custom-blend bird seed costing $0.95 per pound?

6. Find a quadratic equation $y = ax^2 + bx + c$ whose graph passes through the points $(0, 6)$, $(2, 3)$, and $(4, 2)$.

7. Write the partial fraction decomposition of the rational expression $\dfrac{2x^2 - x - 6}{x^3 + 2x}$.

In Exercises 8 and 9, sketch the graph of the solution set of the system of inequalities. Label the vertices of the region.

8. $\begin{cases} 2x + y \geq -3 \\ x - 3y \leq 2 \end{cases}$

9. $\begin{cases} x - y > 6 \\ 5x + 2y < 10 \end{cases}$

10. Sketch the region corresponding to the system of constraints. Then find the minimum and maximum values of the objective function $z = 3x + 2y$ and the points where they occur, subject to the constraints.

$$\begin{aligned} x + 4y &\leq 20 \\ 2x + y &\leq 12 \\ x &\geq 0 \\ y &\geq 0 \end{aligned}$$

$$\begin{cases} -x + 2y - z = 9 \\ 2x - y + 2z = -9 \\ 3x + 3y - 4z = 7 \end{cases}$$

System for 11 and 12

In Exercises 11 and 12, use the system of linear equations shown at the left.

11. Write the augmented matrix for the system.

12. Solve the system using the matrix found in Exercise 11 and Gauss-Jordan elimination.

In Exercises 13–18, perform the operation(s) using the matrices below, if possible.

$$A = \begin{bmatrix} -1 & 3 \\ 6 & 2 \end{bmatrix}, \quad B = \begin{bmatrix} -2 & 5 \\ 0 & -1 \end{bmatrix}, \quad C = \begin{bmatrix} 4 & 0 & 1 \\ -3 & 2 & -1 \end{bmatrix}$$

13. $A + B$

14. $2A - 5B$

15. AC

16. CB

17. A^2

18. $BA - B^2$

19. Find the inverse of the matrix, if possible: $\begin{bmatrix} 1 & 2 & -1 \\ 3 & 7 & -10 \\ -5 & -7 & -15 \end{bmatrix}$.

$$\begin{bmatrix} 7 & 1 & 0 \\ -2 & 4 & -1 \\ 3 & 8 & 5 \end{bmatrix}$$

Matrix for 20

	Gym shoes	Jogging shoes	Walking shoes
14–17	0.079	0.064	0.029
Age group 18–24	0.050	0.060	0.022
25–34	0.103	0.259	0.085

Matrix for 22

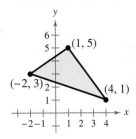

Figure for 25

20. Find the determinant of the matrix shown at the left.

21. Use matrices to find the vertices of the image of the square with vertices $(0, 2)$, $(0, 5)$, $(3, 2)$, and $(3, 5)$ after a reflection in the x-axis.

22. The matrix at the left shows the percents (in decimal form) of the total amounts spent on three types of footwear in a recent year. The total amounts (in millions of dollars) spent by the age groups on the three types of footwear were \$479.88 (14–17 age group), \$365.88 (18–24 age group), and \$1248.89 (25–34 age group). How many dollars worth of gym shoes, jogging shoes, and walking shoes were sold that year? *(Source: National Sporting Goods Association)*

In Exercises 23 and 24, use Cramer's Rule to solve the system of equations.

23. $\begin{cases} 8x - 3y = -52 \\ 3x + 5y = 5 \end{cases}$

24. $\begin{cases} 5x + 4y + 3z = 7 \\ -3x - 8y + 7z = -9 \\ 7x - 5y - 6z = -53 \end{cases}$

25. Use a determinant to find the area of the triangle shown at the left.

26. Write the first five terms of the sequence $a_n = \dfrac{(-1)^{n+1}}{2n + 3}$. (Assume that n begins with 1.)

27. Write an expression for the apparent nth term (a_n) of the sequence.

$$\frac{2!}{4}, \frac{3!}{5}, \frac{4!}{6}, \frac{5!}{7}, \frac{6!}{8}, \dots$$

28. Find the 16th partial sum of the arithmetic sequence $6, 18, 30, 42, \dots$.

29. The sixth term of an arithmetic sequence is 20.6, and the ninth term is 30.2.

    (a) Find the 20th term.

    (b) Find the nth term.

30. Write the first five terms of the sequence $a_n = 3(2)^{n-1}$. (Assume that n begins with 1.)

31. Find the sum: $\displaystyle\sum_{i=0}^{\infty} 1.9\left(\tfrac{1}{10}\right)^{i-1}$.

32. Use mathematical induction to prove the inequality

$$(n + 1)! > 2^n, \quad n \ge 2.$$

33. Use the Binomial Theorem to write the expansion of $(w - 9)^4$.

In Exercises 34–37, evaluate the expression.

34. $_{14}P_3$

35. $_{25}P_2$

36. $\begin{pmatrix} 8 \\ 4 \end{pmatrix}$

37. $_{11}C_6$

In Exercises 38 and 39, find the number of distinguishable permutations of the group of letters.

38. B, A, S, K, E, T, B, A, L, L

39. A, N, T, A, R, C, T, I, C, A

40. There are 10 applicants for three sales positions at a department store. All of the applicants are qualified. In how many ways can the department store fill the three positions?

41. On a game show, a contestant is given the digits 3, 4, and 5 to arrange in the proper order to form the price of an appliance. If the contestant is correct, he or she wins the appliance. What is the probability of winning when the contestant knows that the price is at least \$400?

Proofs in Mathematics ▪ ▪ ▪ ▪ ▪ ▪ ▪ ▪ ▪ ▪ ▪ ▪ ▪ ▪

Properties of Sums *(p. 614)*

1. $\displaystyle\sum_{i=1}^{n} c = cn$, $\quad c$ is a constant.

2. $\displaystyle\sum_{i=1}^{n} ca_i = c \sum_{i=1}^{n} a_i$, $\quad c$ is a constant.

3. $\displaystyle\sum_{i=1}^{n} (a_i + b_i) = \sum_{i=1}^{n} a_i + \sum_{i=1}^{n} b_i$

4. $\displaystyle\sum_{i=1}^{n} (a_i - b_i) = \sum_{i=1}^{n} a_i - \sum_{i=1}^{n} b_i$

INFINITE SERIES

People considered the study of infinite series a novelty in the fourteenth century. Logician Richard Suiseth, whose nickname was Calculator, solved this problem.

If throughout the first half of a given time interval a variation continues at a certain intensity; throughout the next quarter of the interval at double the intensity; throughout the following eighth at triple the intensity and so ad infinitum; The average intensity for the whole interval will be the intensity of the variation during the second subinterval (or double the intensity).

This is the same as saying that the sum of the infinite series

$$\frac{1}{2} + \frac{2}{4} + \frac{3}{8} + \cdots$$
$$+ \frac{n}{2^n} + \cdots$$

is 2.

Proof

Each of these properties follows directly from the properties of real numbers.

1. $\displaystyle\sum_{i=1}^{n} c = c + c + c + \cdots + c = cn \qquad n \text{ terms}$

The proof of Property 2 uses the Distributive Property.

2. $\displaystyle\sum_{i=1}^{n} ca_i = ca_1 + ca_2 + ca_3 + \cdots + ca_n$

$\qquad = c(a_1 + a_2 + a_3 + \cdots + a_n)$

$\qquad = c \displaystyle\sum_{i=1}^{n} a_i$

The proof of Property 3 uses the Commutative and Associative Properties of Addition.

3. $\displaystyle\sum_{i=1}^{n} (a_i + b_i) = (a_1 + b_1) + (a_2 + b_2) + (a_3 + b_3) + \cdots + (a_n + b_n)$

$\qquad = (a_1 + a_2 + a_3 + \cdots + a_n) + (b_1 + b_2 + b_3 + \cdots + b_n)$

$\qquad = \displaystyle\sum_{i=1}^{n} a_i + \sum_{i=1}^{n} b_i$

The proof of Property 4 uses the Commutative and Associative Properties of Addition and the Distributive Property.

4. $\displaystyle\sum_{i=1}^{n} (a_i - b_i) = (a_1 - b_1) + (a_2 - b_2) + (a_3 - b_3) + \cdots + (a_n - b_n)$

$\qquad = (a_1 + a_2 + a_3 + \cdots + a_n) + (-b_1 - b_2 - b_3 - \cdots - b_n)$

$\qquad = (a_1 + a_2 + a_3 + \cdots + a_n) - (b_1 + b_2 + b_3 + \cdots + b_n)$

$\qquad = \displaystyle\sum_{i=1}^{n} a_i - \sum_{i=1}^{n} b_i$

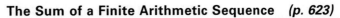

The Sum of a Finite Arithmetic Sequence   *(p. 623)*

The sum of a finite arithmetic sequence with n terms is given by

$$S_n = \frac{n}{2}(a_1 + a_n).$$

Proof

Begin by generating the terms of the arithmetic sequence in two ways. In the first way, repeatedly add d to the first term.

$$S_n = a_1 + a_2 + a_3 + \cdots + a_{n-2} + a_{n-1} + a_n$$
$$= a_1 + [a_1 + d] + [a_1 + 2d] + \cdots + [a_1 + (n-1)d]$$

In the second way, repeatedly subtract d from the nth term.

$$S_n = a_n + a_{n-1} + a_{n-2} + \cdots + a_3 + a_2 + a_1$$
$$= a_n + [a_n - d] + [a_n - 2d] + \cdots + [a_n - (n-1)d]$$

Add these two versions of S_n. The multiples of d sum to zero and you obtain the formula.

$$2S_n = (a_1 + a_n) + (a_1 + a_n) + (a_1 + a_n) + \cdots + (a_1 + a_n) \qquad n \text{ terms}$$
$$2S_n = n(a_1 + a_n)$$
$$S_n = \frac{n}{2}(a_1 + a_n)$$

The Sum of a Finite Geometric Sequence   *(p. 632)*

The sum of the finite geometric sequence

$$a_1, a_1 r, a_1 r^2, a_1 r^3, a_1 r^4, \ldots, a_1 r^{n-1}$$

with common ratio $r \neq 1$ is given by $S_n = \displaystyle\sum_{i=1}^{n} a_1 r^{i-1} = a_1\left(\frac{1 - r^n}{1 - r}\right).$

Proof

$$S_n = a_1 + a_1 r + a_1 r^2 + \cdots + a_1 r^{n-2} + a_1 r^{n-1}$$
$$rS_n = a_1 r + a_1 r^2 + a_1 r^3 + \cdots + a_1 r^{n-1} + a_1 r^n \qquad \text{Multiply by } r.$$

Subtracting the second equation from the first yields

$$S_n - rS_n = a_1 - a_1 r^n.$$

So, $S_n(1 - r) = a_1(1 - r^n)$, and, because $r \neq 1$, you have $S_n = a_1\left(\dfrac{1 - r^n}{1 - r}\right).$

The Binomial Theorem *(p. 648)*

In the expansion of $(x + y)^n$

$$(x + y)^n = x^n + nx^{n-1}y + \cdots + {_nC_r}x^{n-r}y^r + \cdots + nxy^{n-1} + y^n$$

the coefficient of $x^{n-r}y^r$ is

$${_nC_r} = \frac{n!}{(n - r)!r!}.$$

Proof

Use mathematical induction. The steps are straightforward but look a little messy, so only an outline of the proof is given below.

1. For $n = 1$, you have $(x + y)^1 = x^1 + y^1 = {_1C_0}x + {_1C_1}y$, and the formula is valid.

2. Assuming that the formula is true for $n = k$, the coefficient of $x^{k-r}y^r$ is

$${_kC_r} = \frac{k!}{(k - r)!r!} = \frac{k(k - 1)(k - 2)\cdots(k - r + 1)}{r!}.$$

To show that the formula is true for $n = k + 1$, look at the coefficient of $x^{k+1-r}y^r$ in the expansion of

$$(x + y)^{k+1} = (x + y)^k(x + y).$$

On the right-hand side, the term involving $x^{k+1-r}y^r$ is the sum of two products.

$$\left({_kC_r}x^{k-r}y^r\right)(x) + \left({_kC_{r-1}}x^{k+1-r}y^{r-1}\right)(y)$$

$$= \left[\frac{k!}{(k - r)!r!} + \frac{k!}{(k + 1 - r)!(r - 1)!}\right]x^{k+1-r}y^r$$

$$= \left[\frac{(k + 1 - r)k!}{(k + 1 - r)!r!} + \frac{k!r}{(k + 1 - r)!r!}\right]x^{k+1-r}y^r$$

$$= \left[\frac{k!(k + 1 - r + r)}{(k + 1 - r)!r!}\right]x^{k+1-r}y^r$$

$$= \left[\frac{(k + 1)!}{(k + 1 - r)!r!}\right]x^{k+1-r}y^r$$

$$= {_{k+1}C_r}x^{k+1-r}y^r$$

So, by mathematical induction, the Binomial Theorem is valid for all positive integers n.

■

P.S. Problem Solving

 1. Decreasing Sequence Consider the sequence

$$a_n = \frac{n+1}{n^2+1}.$$

(a) Use a graphing utility to graph the first 10 terms of the sequence.

(b) Use the graph from part (a) to estimate the value of a_n as n approaches infinity.

(c) Complete the table.

n	1	10	100	1000	10,000
a_n					

(d) Use the table from part (c) to determine (if possible) the value of a_n as n approaches infinity.

 2. Alternating Sequence Consider the sequence

$$a_n = 3 + (-1)^n.$$

(a) Use a graphing utility to graph the first 10 terms of the sequence.

(b) Use the graph from part (a) to describe the behavior of the graph of the sequence.

(c) Complete the table.

n	1	10	101	1000	10,001
a_n					

(d) Use the table from part (c) to determine (if possible) the value of a_n as n approaches infinity.

3. Greek Mythology Can the Greek hero Achilles, running at 20 feet per second, ever catch a tortoise, starting 20 feet ahead of Achilles and running at 10 feet per second? The Greek mathematician Zeno said no. When Achilles runs 20 feet, the tortoise will be 10 feet ahead. Then, when Achilles runs 10 feet, the tortoise will be 5 feet ahead. Achilles will keep cutting the distance in half but will never catch the tortoise. The table shows Zeno's reasoning. In the table, both the distances and the times required to achieve them form infinite geometric series. Using the table, show that both series have finite sums. What do these sums represent?

DATA	Distance (in feet)	Time (in seconds)
	20	1
	10	0.5
	5	0.25
	2.5	0.125
	1.25	0.0625
	0.625	0.03125

Spreadsheet at LarsonPrecalculus.com

 4. Conjecture Let $x_0 = 1$ and consider the sequence x_n given by

$$x_n = \frac{1}{2}x_{n-1} + \frac{1}{x_{n-1}}, \quad n = 1, 2, \ldots$$

Use a graphing utility to compute the first 10 terms of the sequence and make a conjecture about the value of x_n as n approaches infinity.

5. Operations on an Arithmetic Sequence Determine whether each operation results in an arithmetic sequence when performed on an arithmetic sequence. If so, state the common difference.

(a) A constant C is added to each term.

(b) Each term is multiplied by a nonzero constant C.

(c) Each term is squared.

6. Sequences of Powers The following sequence of perfect squares is not arithmetic.

1, 4, 9, 16, 25, 36, 49, 64, 81, . . .

The related sequence formed from the first differences of this sequence, however, is arithmetic.

(a) Write the first eight terms of the related arithmetic sequence described above. What is the nth term of this sequence?

(b) Explain how to find an arithmetic sequence that is related to the following sequence of perfect cubes.

1, 8, 27, 64, 125, 216, 343, 512, 729, . . .

(c) Write the first seven terms of the related arithmetic sequence in part (b) and find the nth term of the sequence.

(d) Explain how to find an arithmetic sequence that is related to the following sequence of perfect fourth powers.

1, 16, 81, 256, 625, 1296, 2401, 4096, 6561, . . .

(e) Write the first six terms of the related arithmetic sequence in part (d) and find the nth term of the sequence.

7. Piecewise-Defined Sequence A sequence can be defined using a piecewise formula. An example of a piecewise-defined sequence is given below.

$$a_1 = 7, \, a_n = \begin{cases} \frac{1}{2}a_{n-1}, & \text{when } a_{n-1} \text{ is even.} \\ 3a_{n-1} + 1, & \text{when } a_{n-1} \text{ is odd.} \end{cases}$$

(a) Write the first 20 terms of the sequence.

(b) Write the first 10 terms of the sequences for which $a_1 = 4$, $a_1 = 5$, and $a_1 = 12$ (using a_n as defined above). What conclusion can you make about the behavior of each sequence?

8. Fibonacci Sequence Let $f_1, f_2, \ldots, f_n, \ldots$ be the Fibonacci sequence.

(a) Use mathematical induction to prove that

$$f_1 + f_2 + \cdots + f_n = f_{n+2} - 1.$$

(b) Find the sum of the first 20 terms of the Fibonacci sequence.

9. Pentagonal Numbers The numbers 1, 5, 12, 22, 35, 51, . . . are called pentagonal numbers because they represent the numbers of dots in the sequence of figures shown below. Use mathematical induction to prove that the nth pentagonal number P_n is given by

$$P_n = \frac{n(3n - 1)}{2}.$$

10. Think About It What conclusion can be drawn about the sequence of statements P_n for each situation?

(a) P_3 is true and P_k implies P_{k+1}.

(b) $P_1, P_2, P_3, \ldots, P_{50}$ are all true.

(c) P_1, P_2, and P_3 are all true, but the truth of P_k does not imply that P_{k+1} is true.

(d) P_2 is true and P_{2k} implies P_{2k+2}.

11. Sierpinski Triangle Recall that a *fractal* is a geometric figure that consists of a pattern that is repeated infinitely on a smaller and smaller scale. One well-known fractal is the *Sierpinski Triangle*. In the first stage, the midpoints of the three sides are used to create the vertices of a new triangle, which is then removed, leaving three triangles. The figure below shows the first two stages. Note that each remaining triangle is similar to the original triangle. Assume that the length of each side of the original triangle is one unit. Write a formula that describes the side length of the triangles generated in the nth stage. Write a formula for the area of the triangles generated in the nth stage.

12. Job Offer You work for a company that pays $0.01 the first day, $0.02 the second day, $0.04 the third day, and so on. If the daily wage keeps doubling, what will your total income be for working 30 days?

13. Multiple Choice A multiple choice question has five possible answers. You know that the answer is not B or D, but you are not sure about answers A, C, and E. What is the probability that you will get the right answer when you take a guess?

14. Throwing a Dart You throw a dart at the circular target shown below. The dart is equally likely to hit any point inside the target. What is the probability that it hits the region outside the triangle?

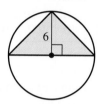

15. Odds The odds in favor of an event occurring is the ratio of the probability that the event will occur to the probability that the event will not occur. The reciprocal of this ratio represents the odds against the event occurring.

(a) A bag contains three blue marbles and seven yellow marbles. What are the odds in favor of choosing a blue marble? What are the odds against choosing a blue marble?

(b) Six of the marbles in a bag are red. The odds against choosing a red marble are 4 to 1. How many marbles are in the bag?

(c) Write a formula for converting the odds in favor of an event to the probability of the event.

(d) Write a formula for converting the probability of an event to the odds in favor of the event.

16. Expected Value An event A has n possible outcomes, which have the values $x_1, x_2, \ldots, x_n$. The probabilities of the n outcomes occurring are $p_1, p_2, \ldots, p_n$. The **expected value** V of an event A is the sum of the products of the outcomes' probabilities and their values,

$$V = p_1 x_1 + p_2 x_2 + \cdots + p_n x_n.$$

(a) To win California's Super Lotto Plus game, you must match five different numbers chosen from the numbers 1 to 47, plus one MEGA number chosen from the numbers 1 to 27. You purchase a ticket for $1. If the jackpot for the next drawing is $12,000,000, what is the expected value of the ticket?

(b) You are playing a dice game in which you need to score 60 points to win. On each turn, you toss two six-sided dice. Your score for the turn is 0 when the dice do not show the same number. Your score for the turn is the product of the numbers on the dice when they do show the same number. What is the expected value of each turn? How many turns will it take on average to score 60 points?

10 Topics in Analytic Geometry

Satellite Orbit
(Exercise 62, page 764)

Microphone Pickup Pattern *(Exercise 69, page 758)*

Nuclear Cooling Towers *(page 718)*

Halley's Comet *(page 712)*

Suspension Bridge *(Exercise 72, page 706)*

10.1 Lines

One practical application of the inclination of a line is in measuring heights indirectly. For example, in Exercise 86 on page 698, you will use the inclination of a line to determine the change in elevation from the base to the top of an incline railway.

- ■ Find the inclination of a line.
- ■ Find the angle between two lines.
- ■ Find the distance between a point and a line.

Inclination of a Line

In Section 1.3, you learned that the slope of a line is the ratio of the change in y to the change in x. In this section, you will look at the slope of a line in terms of the angle of inclination of the line.

Every nonhorizontal line must intersect the x-axis. The angle formed by such an intersection determines the **inclination** of the line, as specified in the definition below.

> ### Definition of Inclination
>
> The **inclination** of a nonhorizontal line is the positive angle θ (less than π) measured counterclockwise from the x-axis to the line. (See figures below.)

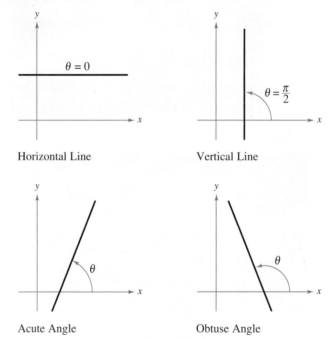

The inclination of a line is related to its slope in the manner described below.

> ### Inclination and Slope
>
> If a nonvertical line has inclination θ and slope m, then
>
> $$m = \tan \theta.$$

For a proof of this relation between inclination and slope, see Proofs in Mathematics on page 772.

Note that if $m \geq 0$, then $\theta = \arctan m$ because $0 \leq \theta < \pi/2$. On the other hand, if $m < 0$, then $\theta = \pi + \arctan m$ because $\pi/2 < \theta < \pi$.

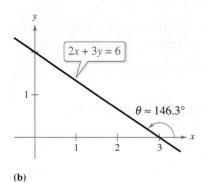

(a)

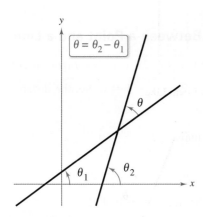

(b)

Figure 10.1

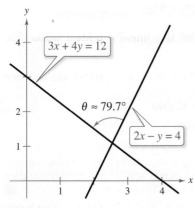

Figure 10.2

EXAMPLE 1 **Finding Inclinations of Lines**

Find the inclination of (a) $x - y = 2$ and (b) $2x + 3y = 6$.

Solution

a. The slope of this line is $m = 1$. So, use $\tan \theta = 1$ to determine its inclination. Note that $m \geq 0$. This means that

$$\theta = \arctan 1 = \pi/4 \text{ radian} = 45°$$

as shown in Figure 10.1(a).

b. The slope of this line is $m = -\frac{2}{3}$. So, use $\tan \theta = -\frac{2}{3}$ to determine its inclination. Note that $m < 0$. This means that

$$\theta = \pi + \arctan\left(-\frac{2}{3}\right) \approx \pi + (-0.5880) \approx 2.5536 \text{ radians} \approx 146.3°$$

as shown in Figure 10.1(b).

✓ **Checkpoint** ◀))) *Audio-video solution in English & Spanish at LarsonPrecalculus.com*

Find the inclination of (a) $4x - 5y = 7$ and (b) $x + y = -1$. ■

The Angle Between Two Lines

When two distinct lines intersect and are nonperpendicular, their intersection forms two pairs of opposite angles. One pair is acute and the other pair is obtuse. The smaller of these angles is the **angle between the two lines.** If two lines have inclinations θ_1 and θ_2, where $\theta_1 < \theta_2$ and $\theta_2 - \theta_1 < \pi/2$, then the angle between the two lines is $\theta = \theta_2 - \theta_1$, as shown in Figure 10.2. Use the formula for the tangent of the difference of two angles

$$\tan \theta = \tan(\theta_2 - \theta_1) = \frac{\tan \theta_2 - \tan \theta_1}{1 + \tan \theta_1 \tan \theta_2}$$

to obtain a formula for the tangent of the angle between two lines.

Angle Between Two Lines

If two nonperpendicular lines have slopes m_1 and m_2, then the tangent of the angle between the two lines is

$$\tan \theta = \left| \frac{m_2 - m_1}{1 + m_1 m_2} \right|.$$

EXAMPLE 2 **Finding the Angle Between Two Lines**

Find the angle between $2x - y = 4$ and $3x + 4y = 12$.

Solution The two lines have slopes of $m_1 = 2$ and $m_2 = -\frac{3}{4}$, respectively. So, the tangent of the angle between the two lines is

$$\tan \theta = \left| \frac{m_2 - m_1}{1 + m_1 m_2} \right| = \left| \frac{(-3/4) - 2}{1 + (2)(-3/4)} \right| = \left| \frac{-11/4}{-2/4} \right| = \frac{11}{2}$$

and the angle is $\theta = \arctan \frac{11}{2} \approx 1.3909$ radians $\approx 79.7°$, as shown in Figure 10.3.

✓ **Checkpoint** ◀))) *Audio-video solution in English & Spanish at LarsonPrecalculus.com*

Find the angle between $-4x + 5y = 10$ and $3x + 2y = -5$. ■

Figure 10.3

The Distance Between a Point and a Line

Finding the distance between a line and a point not on the line is an application of perpendicular lines. This distance is the length of the perpendicular line segment joining the point and the line, shown as d at the right.

Recall from Section 1.1 that d is given by

$$d = \sqrt{(x_2 - x_1)^2 + (y_2 - y_1)^2}.$$

This formula can be written in terms of the coordinates x_1 and y_1 and the coefficients A, B, and C in the general form of the equation of a line, $Ax + By + C = 0$.

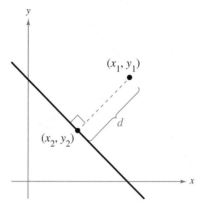

Distance Between a Point and a Line

The distance between the point (x_1, y_1) and the line $Ax + By + C = 0$ is

$$d = \frac{|Ax_1 + By_1 + C|}{\sqrt{A^2 + B^2}}.$$

For a proof of this formula for the distance between a point and a line, see Proofs in Mathematics on page 772.

EXAMPLE 3   **Finding the Distance Between a Point and a Line**

Find the distance between the point $(4, 1)$ and the line $y = 2x + 1$. (See Figure 10.4.)

Solution   The general form of the equation is $-2x + y - 1 = 0$. So, the distance between the point and the line is

$$d = \frac{|-2(4) + 1(1) + (-1)|}{\sqrt{(-2)^2 + 1^2}} = \frac{8}{\sqrt{5}} \approx 3.58 \text{ units.}$$

✓ *Checkpoint* 🔊)) *Audio-video solution in English & Spanish at LarsonPrecalculus.com*

Find the distance between the point $(5, -1)$ and the line $y = -3x + 2$.

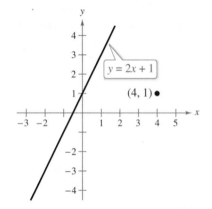

Figure 10.4

EXAMPLE 4   **Finding the Distance Between a Point and a Line**

See LarsonPrecalculus.com for an interactive version of this type of example.

Find the distance between the point $(2, -1)$ and the line $7x + 5y = -13$. (See Figure 10.5.)

Solution   The general form of the equation is $7x + 5y + 13 = 0$. So, the distance between the point and the line is

$$d = \frac{|7(2) + 5(-1) + 13|}{\sqrt{7^2 + 5^2}} = \frac{22}{\sqrt{74}} \approx 2.56 \text{ units.}$$

✓ *Checkpoint* 🔊)) *Audio-video solution in English & Spanish at LarsonPrecalculus.com*

Find the distance between the point $(3, 2)$ and the line $-3x + 5y = -2$.

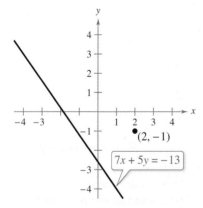

Figure 10.5

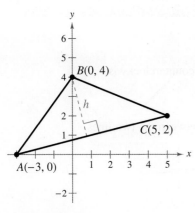

Figure 10.6

EXAMPLE 5 An Application of Two Distance Formulas

Figure 10.6 shows a triangle with vertices $A(-3, 0)$, $B(0, 4)$, and $C(5, 2)$.

a. Find the altitude h from vertex B to side AC.

b. Find the area of the triangle.

Solution

a. First, find the equation of line AC.

$$\text{Slope: } m = \frac{2 - 0}{5 - (-3)} = \frac{2}{8} = \frac{1}{4}$$

$$\text{Equation: } \quad y - 0 = \frac{1}{4}(x + 3) \qquad \text{Point-slope form}$$

$$4y = x + 3 \qquad \text{Multiply each side by 4.}$$

$$x - 4y + 3 = 0 \qquad \text{General form}$$

Then, use the formula for the distance between line AC and the point $(0, 4)$ to find the altitude.

$$\text{Altitude} = h = \frac{|1(0) + (-4)(4) + 3|}{\sqrt{1^2 + (-4)^2}} = \frac{13}{\sqrt{17}} \text{ units.}$$

b. Use the formula for the distance between two points to find the length of the base AC.

$$b = \sqrt{[5 - (-3)]^2 + (2 - 0)^2} \qquad \text{Distance Formula}$$

$$= \sqrt{8^2 + 2^2} \qquad \text{Simplify.}$$

$$= 2\sqrt{17} \text{ units} \qquad \text{Simplify.}$$

So, the area of the triangle is

$$A = \frac{1}{2}bh \qquad \text{Formula for the area of a triangle}$$

$$= \frac{1}{2}(2\sqrt{17})\left(\frac{13}{\sqrt{17}}\right) \qquad \text{Substitute for } b \text{ and } h.$$

$$= 13 \text{ square units.} \qquad \text{Simplify.}$$

✓ **Checkpoint** 🔊))) *Audio-video solution in English & Spanish at LarsonPrecalculus.com*

A triangle has vertices $A(-2, 0)$, $B(0, 5)$, and $C(4, 3)$.

a. Find the altitude from vertex B to side AC.

b. Find the area of the triangle.

Summarize (Section 10.1)

1. Explain how to find the inclination of a line *(page 692)*. For an example of finding the inclinations of lines, see Example 1.

2. Explain how to find the angle between two lines *(page 693)*. For an example of finding the angle between two lines, see Example 2.

3. Explain how to find the distance between a point and a line *(page 694)*. For examples of finding the distances between points and lines, see Examples 3–5.

10.1 Exercises

See **CalcChat.com** for tutorial help and worked-out solutions to odd-numbered exercises.

Vocabulary: Fill in the blanks.

1. The _____ of a nonhorizontal line is the positive angle θ (less than π) measured counterclockwise from the x-axis to the line.

2. If a nonvertical line has inclination θ and slope m, then $m =$ _____.

3. If two nonperpendicular lines have slopes m_1 and m_2, then the tangent of the angle between the two lines is $\tan \theta =$ _____.

4. The distance between the point (x_1, y_1) and the line $Ax + By + C = 0$ is $d =$ _____.

Skills and Applications

Finding the Slope of a Line In Exercises 5–16, find the slope of the line with inclination θ.

5. $\theta = \dfrac{\pi}{6}$ radian

6. $\theta = \dfrac{\pi}{4}$ radian

7. $\theta = \dfrac{3\pi}{4}$ radians

8. $\theta = \dfrac{2\pi}{3}$ radians

9. $\theta = \dfrac{\pi}{3}$ radians

10. $\theta = \dfrac{5\pi}{6}$ radians

11. $\theta = 0.39$ radian

12. $\theta = 0.63$ radian

13. $\theta = 1.27$ radians

14. $\theta = 1.35$ radians

15. $\theta = 1.81$ radians

16. $\theta = 2.88$ radians

Finding the Inclination of a Line In Exercises 17–24, find the inclination θ (in radians and degrees) of the line with slope m.

17. $m = 1$

18. $m = \sqrt{3}$

19. $m = \dfrac{2}{3}$

20. $m = \dfrac{1}{4}$

21. $m = -1$

22. $m = -\sqrt{3}$

23. $m = -\dfrac{3}{2}$

24. $m = -\dfrac{5}{9}$

 Finding the Inclination of a Line In Exercises 25–34, find the inclination θ (in radians and degrees) of the line passing through the points.

25. $\left(\sqrt{3}, 2\right), (0, 1)$

26. $\left(1, 2\sqrt{3}\right), \left(0, \sqrt{3}\right)$

27. $\left(-\sqrt{3}, -1\right), (0, -2)$

28. $\left(3, \sqrt{3}\right), \left(6, -2\sqrt{3}\right)$

29. $(6, 1), (10, 8)$

30. $(12, 8), (-4, -3)$

31. $(-2, 20), (10, 0)$

32. $(0, 100), (50, 0)$

33. $\left(\frac{1}{4}, \frac{3}{2}\right), \left(\frac{1}{3}, \frac{1}{2}\right)$

34. $\left(\frac{2}{5}, -\frac{3}{4}\right), \left(-\frac{11}{10}, -\frac{1}{4}\right)$

 Finding the Inclination of a Line In Exercises 35–44, find the inclination θ (in radians and degrees) of the line.

35. $2x + 2y - 5 = 0$

36. $x - \sqrt{3}y + 1 = 0$

37. $3x - 3y + 1 = 0$

38. $\sqrt{3}x - y + 2 = 0$

39. $x + \sqrt{3}y + 2 = 0$

40. $-2\sqrt{3}x - 2y = 0$

41. $6x - 2y + 8 = 0$

42. $2x - 6y - 12 = 0$

43. $4x + 5y - 9 = 0$

44. $5x + 3y = 0$

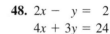 **Finding the Angle Between Two Lines** In Exercises 45–54, find the angle θ (in radians and degrees) between the lines.

45. $3x + y = 3$
    $x - y = 2$

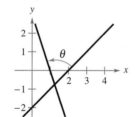

46. $x + 3y = 2$
    $x - 2y = -3$

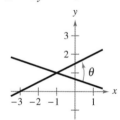

47. $x - y = 0$
    $3x - 2y = -1$

48. $2x - y = 2$
    $4x + 3y = 24$

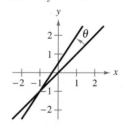

49. $x - 2y = 7$
    $6x + 2y = 5$

50. $5x + 2y = 16$
    $3x - 5y = -1$

51. $x + 2y = 8$
    $x - 2y = 2$

52. $3x - 5y = 3$
    $3x + 5y = 12$

53. $0.05x - 0.03y = 0.21$
    $0.07x + 0.02y = 0.16$

54. $0.02x - 0.05y = -0.19$
    $0.03x + 0.04y = 0.52$

Angle Measurement   In Exercises 55–58, find the slope of each side of the triangle, and use the slopes to find the measures of the interior angles.

55.

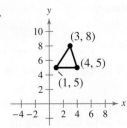

56.

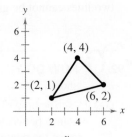

57.

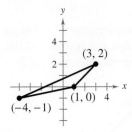

58.

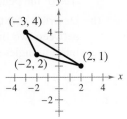

 Finding the Distance Between a Point and a Line   In Exercises 59–72, find the distance between the point and the line.

Point	Line
59. $(1, 2)$	$y = x + 2$
60. $(3, 1)$	$y = x + 3$
61. $(2, 3)$	$y = 2x - 3$
62. $(1, 5)$	$y = 4x + 5$
63. $(-2, 4)$	$y = -x + 6$
64. $(3, -3)$	$y = -3x - 4$
65. $(1, -2)$	$y = 3x - 6$
66. $(-3, 7)$	$y = -4x + 3$
67. $(2, 3)$	$3x + y = 1$
68. $(2, 1)$	$-2x + y = 2$
69. $(6, 2)$	$-3x + 4y = -5$
70. $(1, -4)$	$2x - 3y = -5$
71. $(-2, 4)$	$4x + 3y = 5$
72. $(-3, -5)$	$-3x - 4y = 4$

 An Application of Two Distance Formulas   In Exercises 73–78, the points represent the vertices of a triangle. (a) Draw triangle ABC in the coordinate plane, (b) find the altitude from vertex B of the triangle to side AC, and (c) find the area of the triangle.

73. $A(-1, 0), B(0, 3), C(3, 1)$

74. $A(-4, 0), B(0, 5), C(3, 3)$

75. $A(-3, 0), B(0, -2), C(2, 3)$

76. $A(-2, 0), B(0, -3), C(5, 1)$

77. $A(1, 1), B(2, 4), C(3, 5)$

78. $A(-3, -2), B(-1, -4), C(3, -1)$

Finding the Distance Between Parallel Lines   In Exercises 79 and 80, find the distance between the parallel lines.

79. $x + y = 1$
$x + y = 5$

80. $3x - 4y = 1$
$3x - 4y = 10$

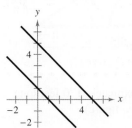

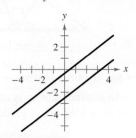

81. Road Grade   A straight road rises with an inclination of 0.1 radian from the horizontal (see figure). Find the slope of the road and the change in elevation over a two-mile section of the road.

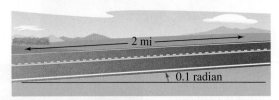

82. Road Grade   A straight road rises with an inclination of 0.2 radian from the horizontal. Find the slope of the road and the change in elevation over a one-mile section of the road.

83. Pitch of a Roof   A roof has a rise of 3 feet for every horizontal change of 5 feet (see figure). Find the inclination θ of the roof.

84. Conveyor Design   A moving conveyor rises 1 meter for every 3 meters of horizontal travel (see figure).

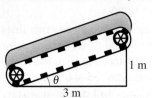

(a) Find the inclination θ of the conveyor.

(b) The conveyor runs between two floors in a factory. The distance between the floors is 5 meters. Find the length of the conveyor.

85. Truss Find the angles α and β shown in the drawing of the roof truss.

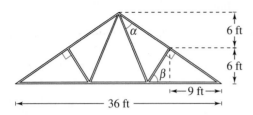

86. Incline Railway

An incline railway is approximately 170 feet long with a 36% uphill grade (see figure).

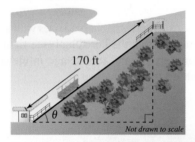

170 ft

θ

Not drawn to scale

(a) Find the inclination θ of the railway.

(b) Find the change in elevation from the base to the top of the railway.

(c) Using the origin of a rectangular coordinate system as the base of the inclined plane, find an equation of the line that models the railway track.

(d) Sketch a graph of the equation you found in part (c).

Exploration

True or False? In Exercises 87–90, determine whether the statement is true or false. Justify your answer.

87. A line that has an inclination of 0 radians has a slope of 0.

88. A line that has an inclination greater than $\dfrac{\pi}{2}$ radians has a negative slope.

89. To find the angle between two lines whose angles of inclination θ_1 and θ_2 are known, substitute θ_1 and θ_2 for m_1 and m_2, respectively, in the formula for the tangent of the angle between two lines.

90. The inclination of a line is the angle between the line and the x-axis.

91. Writing Explain why the inclination of a line can be an angle that is greater than $\dfrac{\pi}{2}$, but the angle between two lines cannot be greater than $\dfrac{\pi}{2}$.

92. **HOW DO YOU SEE IT?** Use the pentagon shown below.

(a) Describe how to use the formula for the distance between a point and a line to find the area of the pentagon.

(b) Describe how to use the formula for the angle between two lines to find the measures of the interior angles of the pentagon.

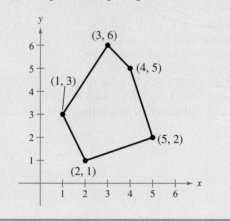

93. Think About It Consider a line with slope m and x-intercept $(0, 4)$.

(a) Write the distance d between the origin and the line as a function of m.

(b) Graph the function in part (a).

(c) Find the slope that yields the maximum distance between the origin and the line.

(d) Find the asymptote of the graph in part (b) and interpret its meaning in the context of the problem.

94. Think About It Consider a line with slope m and y-intercept $(0, 4)$.

(a) Write the distance d between the point $(3, 1)$ and the line as a function of m.

(b) Graph the function in part (a).

(c) Find the slope that yields the maximum distance between the point and the line.

(d) Is it possible for the distance to be 0? If so, what is the slope of the line that yields a distance of 0?

(e) Find the asymptote of the graph in part (b) and interpret its meaning in the context of the problem.

10.2 Introduction to Conics: Parabolas

■ Recognize a conic as the intersection of a plane and a double-napped cone.
■ Write equations of parabolas in standard form.
■ Use the reflective property of parabolas to write equations of tangent lines.

Parabolas have many real-life applications and are often used to model and solve engineering problems. For example, in Exercise 72 on page 706, you will use a parabola to model the cables of the Golden Gate Bridge.

Conics

The earliest basic descriptions of conic sections took place during the classical Greek period, 500 to 336 B.C. This early Greek study was largely concerned with the geometric properties of conics. It was not until the early 17th century that the broad applicability of conics became apparent and played a prominent role in the early development of calculus.

A **conic section** (or simply **conic**) is the intersection of a plane and a double-napped cone. Notice in Figure 10.7 that in the formation of the four basic conics, the intersecting plane does not pass through the vertex of the cone. When the plane does pass through the vertex, the resulting figure is a **degenerate conic,** as shown in Figure 10.8.

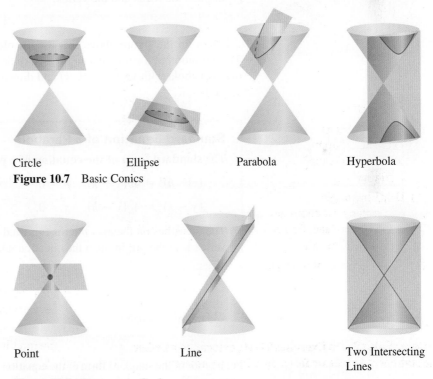

Circle   Ellipse   Parabola   Hyperbola
Figure 10.7   Basic Conics

Point   Line   Two Intersecting Lines

Figure 10.8   Degenerate Conics

There are several ways to approach the study of conics. You could begin by defining conics in terms of the intersections of planes and cones, as the Greeks did, or you could define them algebraically, in terms of the general second-degree equation

$$Ax^2 + Bxy + Cy^2 + Dx + Ey + F = 0.$$

However, you will study a third approach, in which each of the conics is defined as a *locus* (collection) of points satisfying a given geometric property. For example, in Section 1.2, you saw how the definition of a circle as *the collection of all points* (x, y) *that are equidistant from a fixed point* (h, k) led to the standard form of the equation of a circle

$$(x - h)^2 + (y - k)^2 = r^2. \qquad \text{Equation of a circle}$$

Recall that the center of a circle is at (h, k) and that the radius of the circle is r.

Parabolas

In Section 2.1, you learned that the graph of the quadratic function

$$f(x) = ax^2 + bx + c$$

is a parabola that opens upward or downward. The definition of a parabola below is more general in the sense that it is independent of the orientation of the parabola.

Definition of a Parabola

A **parabola** is the set of all points (x, y) in a plane that are equidistant from a fixed line, the **directrix,** and a fixed point, the **focus,** not on the line. (See figure.) The **vertex** is the midpoint between the focus and the directrix. The **axis** of the parabola is the line passing through the focus and the vertex.

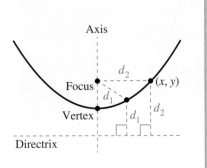

Note in the figure above that a parabola is symmetric with respect to its axis. The definition of a parabola can be used to derive the **standard form of the equation of a parabola** with vertex at (h, k) and directrix parallel to the x-axis or to the y-axis, stated below.

Standard Equation of a Parabola

The **standard form of the equation of a parabola** with vertex at (h, k) is

$$(x - h)^2 = 4p(y - k), \quad p \neq 0 \qquad \text{Vertical axis; directrix: } y = k - p$$

$$(y - k)^2 = 4p(x - h), \quad p \neq 0. \qquad \text{Horizontal axis; directrix: } x = h - p$$

The focus lies on the axis p units (directed distance) from the vertex. If the vertex is at the origin, then the equation takes one of two forms.

$$x^2 = 4py \qquad \text{Vertical axis}$$

$$y^2 = 4px \qquad \text{Horizontal axis}$$

See the figures below.

For a proof of the standard form of the equation of a parabola, see Proofs in Mathematics on page 773.

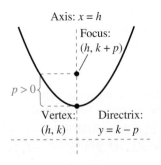

$(x - h)^2 = 4p(y - k)$
Vertical axis: $p > 0$

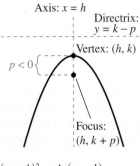

$(x - h)^2 = 4p(y - k)$
Vertical axis: $p < 0$

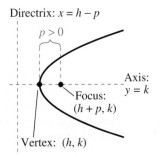

$(y - k)^2 = 4p(x - h)$
Horizontal axis: $p > 0$

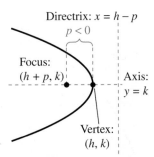

$(y - k)^2 = 4p(x - h)$
Horizontal axis: $p < 0$

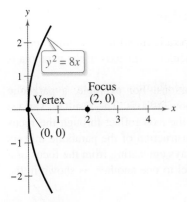

Figure 10.9

EXAMPLE 1　Finding the Standard Equation of a Parabola

Find the standard form of the equation of the parabola with vertex at the origin and focus $(2, 0)$.

Solution　The axis of the parabola is horizontal, passing through $(0, 0)$ and $(2, 0)$, as shown in Figure 10.9. The equation is of the form $y^2 = 4px$, where $p = 2$. So, the standard form of the equation is $y^2 = 8x$. You can use a graphing utility to confirm this equation. Let $y_1 = \sqrt{8x}$ to graph the upper portion of the parabola and let $y_2 = -\sqrt{8x}$ to graph the lower portion of the parabola.

✔ **Checkpoint** ◀))) *Audio-video solution in English & Spanish at LarsonPrecalculus.com*

Find the standard form of the equation of the parabola with vertex at the origin and focus $\left(0, \frac{3}{8}\right)$.

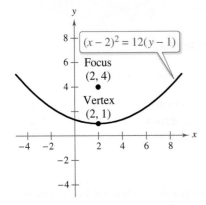

Figure 10.10

EXAMPLE 2　Finding the Standard Equation of a Parabola

See LarsonPrecalculus.com for an interactive version of this type of example.

Find the standard form of the equation of the parabola with vertex $(2, 1)$ and focus $(2, 4)$.

Solution　The axis of the parabola is vertical, passing through $(2, 1)$ and $(2, 4)$. The equation is of the form

$$(x - h)^2 = 4p(y - k)$$

where $h = 2$, $k = 1$, and $p = 4 - 1 = 3$. So, the standard form of the equation is

$$(x - 2)^2 = 12(y - 1).$$

Figure 10.10 shows the graph of this parabola.

✔ **Checkpoint** ◀))) *Audio-video solution in English & Spanish at LarsonPrecalculus.com*

Find the standard form of the equation of the parabola with vertex $(2, -3)$ and focus $(4, -3)$.

▷ **ALGEBRA HELP**　The technique of completing the square is used to write the equation in Example 3 in standard form. To review completing the square, see Appendix A.5.

EXAMPLE 3　Finding the Focus of a Parabola

Find the focus of the parabola $y = -\frac{1}{2}x^2 - x + \frac{1}{2}$.

Solution　Convert to standard form by completing the square.

$y = -\frac{1}{2}x^2 - x + \frac{1}{2}$	Write original equation.
$-2y = x^2 + 2x - 1$	Multiply each side by -2.
$1 - 2y = x^2 + 2x$	Add 1 to each side.
$1 + 1 - 2y = x^2 + 2x + 1$	Complete the square.
$2 - 2y = x^2 + 2x + 1$	Combine like terms.
$-2(y - 1) = (x + 1)^2$	Write in standard form.

Comparing this equation with

$$(x - h)^2 = 4p(y - k)$$

shows that $h = -1$, $k = 1$, and $p = -\frac{1}{2}$. The parabola opens downward, as shown in Figure 10.11, because p is negative. So, the focus of the parabola is $(h, k + p) = \left(-1, \frac{1}{2}\right)$.

✔ **Checkpoint** ◀))) *Audio-video solution in English & Spanish at LarsonPrecalculus.com*

Find the focus of the parabola $x = \frac{1}{4}y^2 + \frac{3}{2}y + \frac{13}{4}$.

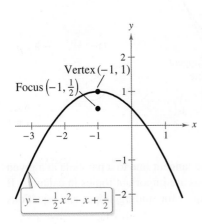

Figure 10.11

One important application of parabolas is in astronomy. Radio telescopes use parabolic dishes to collect radio waves from space.

The Reflective Property of Parabolas

A line segment that passes through the focus of a parabola and has endpoints on the parabola is a **focal chord.** The focal chord perpendicular to the axis of the parabola is called the **latus rectum.**

Parabolas occur in a wide variety of applications. For example, a parabolic reflector can be formed by revolving a parabola about its axis. The resulting surface has the property that all incoming rays parallel to the axis reflect through the focus of the parabola. This is the principle behind the construction of the parabolic mirrors used in reflecting telescopes. Conversely, the light rays emanating from the focus of a parabolic reflector used in a flashlight are all parallel to one another, as shown in the figure below.

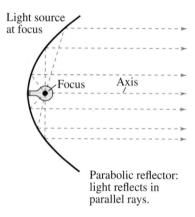

Light source at focus

Focus   Axis

Parabolic reflector: light reflects in parallel rays.

A line is **tangent** to a parabola at a point on the parabola when the line intersects, but does not cross, the parabola at the point. Tangent lines to parabolas have special properties related to the use of parabolas in constructing reflective surfaces.

Reflective Property of a Parabola

The tangent line to a parabola at a point P makes equal angles with the following two lines (see figure below).

1. The line passing through P and the focus
2. The axis of the parabola

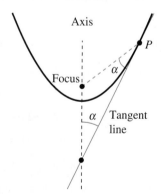

Axis

Focus α P

α Tangent line

Example 4 shows how to find an equation of a tangent line to a parabola at a given point. Finding slopes and equations of tangent lines are important topics in calculus. If you take a calculus course, you will study techniques for finding slopes and equations of tangent lines to parabolas and other curves.

| EXAMPLE 4 | **Finding the Tangent Line at a Point on a Parabola** |

Find an equation of the tangent line to the parabola $y = x^2$ at the point $(1, 1)$.

Solution   For this parabola, the vertex is at the origin, the axis is vertical, and $p = \frac{1}{4}$, so the focus is $\left(0, \frac{1}{4}\right)$, as shown in the figure below.

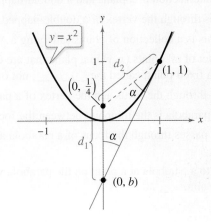

To find the y-intercept $(0, b)$ of the tangent line, equate the lengths of the two sides of the isosceles triangle shown in the figure:

$$d_1 = \tfrac{1}{4} - b$$

and

$$d_2 = \sqrt{(1 - 0)^2 + \left(1 - \tfrac{1}{4}\right)^2} = \tfrac{5}{4}.$$

▷ **TECHNOLOGY**   Use a graphing utility to confirm the result of Example 4. Graph

$$y_1 = x^2 \quad \text{and} \quad y_2 = 2x - 1$$

in the same viewing window and verify that the line touches the parabola at the point $(1, 1)$.

Note that $d_1 = \frac{1}{4} - b$ rather than $b - \frac{1}{4}$. The order of subtraction for the distance is important because the distance must be positive. Setting $d_1 = d_2$ produces

$$\tfrac{1}{4} - b = \tfrac{5}{4}$$

$$b = -1.$$

So, the slope of the tangent line is

$$m = \frac{1 - (-1)}{1 - 0} = 2$$

and the equation of the tangent line in slope-intercept form is

$$y = 2x - 1.$$

▷ **ALGEBRA HELP**   To review techniques for writing linear equations, see Section 1.3.

✓ **Checkpoint** 🔊))) *Audio-video solution in English & Spanish at LarsonPrecalculus.com*

Find an equation of the tangent line to the parabola $y = 3x^2$ at the point $(1, 3)$.   ◼

Summarize   (Section 10.2)

1. List the four basic conic sections and the degenerate conics. Use sketches to show how to form each basic conic section and degenerate conic from the intersection of a plane and a double-napped cone *(page 699)*.

2. State the definition of a parabola and the standard form of the equation of a parabola *(page 700)*. For examples involving writing equations of parabolas in standard form, see Examples 1–3.

3. State the reflective property of a parabola *(page 702)*. For an example of using this property to write an equation of a tangent line, see Example 4.

10.2 Exercises

Vocabulary: **Fill in the blanks.**

1. A _____ is the intersection of a plane and a double-napped cone.
2. When a plane passes through the vertex of a double-napped cone, the intersection is a _____ _____.
3. A _____ of points is a collection of points satisfying a given geometric property.
4. A _____ is the set of all points (x, y) in a plane that are equidistant from a fixed line, called the _____, and a fixed point, called the _____, not on the line.
5. The line that passes through the focus and the vertex of a parabola is the _____ of the parabola.
6. The _____ of a parabola is the midpoint between the focus and the directrix.
7. A line segment that passes through the focus of a parabola and has endpoints on the parabola is a _____ _____.
8. A line is _____ to a parabola at a point on the parabola when the line intersects, but does not cross, the parabola at the point.

Skills and Applications

Matching   In Exercises 9–12, match the equation with its graph. [The graphs are labeled (a), (b), (c), and (d).]

(a)

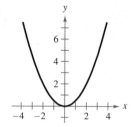

(b)

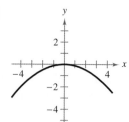

(c)

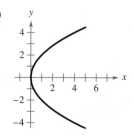

(d)

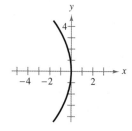

9. $y^2 = 4x$
10. $x^2 = 2y$
11. $x^2 = -8y$
12. $y^2 = -12x$

 Finding the Standard Equation of a Parabola   In Exercises 13–26, find the standard form of the equation of the parabola with the given characteristic(s) and vertex at the origin.

13.

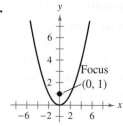

14.

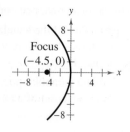

15. Focus: $\left(0, \frac{1}{2}\right)$
16. Focus: $\left(\frac{3}{2}, 0\right)$
17. Focus: $(-2, 0)$
18. Focus: $(0, -1)$
19. Directrix: $y = 2$
20. Directrix: $y = -4$
21. Directrix: $x = -1$
22. Directrix: $x = 3$
23. Vertical axis; passes through the point $(4, 6)$
24. Vertical axis; passes through the point $(-3, -3)$
25. Horizontal axis; passes through the point $(-2, 5)$
26. Horizontal axis; passes through the point $(3, -2)$

 Finding the Standard Equation of a Parabola   In Exercises 27–36, find the standard form of the equation of the parabola with the given characteristics.

27.

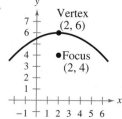

28.

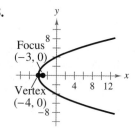

29. Vertex: $(6, 3)$; focus: $(4, 3)$
30. Vertex: $(1, -8)$; focus: $(3, -8)$
31. Vertex: $(0, 2)$; directrix: $y = 4$
32. Vertex: $(1, 2)$; directrix: $y = -1$
33. Focus: $(2, 2)$; directrix: $x = -2$
34. Focus: $(0, 0)$; directrix: $y = 8$
35. Vertex: $(3, -3)$; vertical axis; passes through the point $(0, 0)$
36. Vertex: $(-1, 6)$; horizontal axis; passes through the point $(-9, 2)$

Finding the Vertex, Focus, and Directrix of a Parabola In Exercises 37–50, find the vertex, focus, and directrix of the parabola. Then sketch the parabola.

37. $y = \frac{1}{2}x^2$

38. $y = -4x^2$

39. $y^2 = -6x$

40. $y^2 = 3x$

41. $x^2 + 12y = 0$

42. $x + y^2 = 0$

43. $(x - 1)^2 + 8(y + 2) = 0$

44. $(x + 5) + (y - 1)^2 = 0$

45. $(y + 7)^2 = 4\left(x - \frac{3}{2}\right)$

46. $\left(x + \frac{1}{2}\right)^2 = 4(y - 1)$

47. $y = \frac{1}{4}(x^2 - 2x + 5)$

48. $x = \frac{1}{4}(y^2 + 2y + 33)$

49. $y^2 + 6y + 8x + 25 = 0$

50. $x^2 - 4x - 4y = 0$

 Finding the Vertex, Focus, and Directrix of a Parabola In Exercises 51–54, find the vertex, focus, and directrix of the parabola. Use a graphing utility to graph the parabola.

51. $x^2 + 4x - 6y = -10$

52. $x^2 - 2x + 8y = -9$

53. $y^2 + x + y = 0$

54. $y^2 - 4x - 4 = 0$

 Finding the Tangent Line at a Point on a Parabola In Exercises 55–60, find an equation of the tangent line to the parabola at the given point.

55.

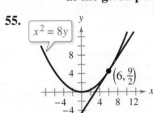

$x^2 = 8y$

$\left(6, \frac{9}{2}\right)$

56.

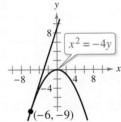

$x^2 = -4y$

$(-6, -9)$

57. $x^2 = 2y$, $(4, 8)$

58. $x^2 = 2y$, $\left(-3, \frac{9}{2}\right)$

59. $y = -2x^2$, $(-1, -2)$

60. $y = -2x^2$, $(2, -8)$

61. Flashlight The light bulb in a flashlight is at the focus of the parabolic reflector, 1.5 centimeters from the vertex of the reflector (see figure). Write an equation for a cross section of the flashlight's reflector with its focus on the positive x-axis and its vertex at the origin.

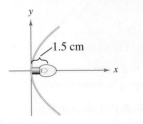

Figure for 61   Figure for 62

62. Satellite Dish The receiver of a parabolic satellite dish is at the focus of the parabola (see figure). Write an equation for a cross section of the satellite dish.

63. Highway Design Highway engineers use a parabolic curve to design an entrance ramp from a straight street to an interstate highway (see figure). Write an equation of the parabola.

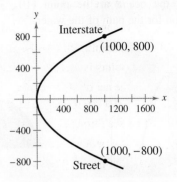

64. Road Design Roads are often designed with parabolic surfaces to allow rain to drain off. A particular road is 32 feet wide and 0.4 foot higher in the center than it is on the sides (see figure).

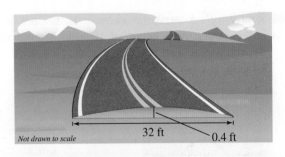

Not drawn to scale   32 ft   0.4 ft

(a) Write an equation of the parabola with its vertex at the origin that models the road surface.

(b) How far from the center of the road is the road surface 0.1 foot lower than the center?

65. Beam Deflection A simply supported beam is 64 feet long and has a load at the center (see figure). The deflection of the beam at its center is 1 inch. The shape of the deflected beam is parabolic.

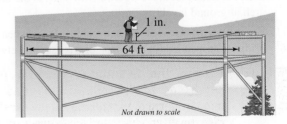

Not drawn to scale

(a) Write an equation of the parabola with its vertex at the origin that models the shape of the beam.

(b) How far from the center of the beam is the deflection $\frac{1}{2}$ inch?

66. Beam Deflection Repeat Exercise 65 when the length of the beam is 36 feet and the deflection of the beam at its center is 2 inches.

67. Fluid Flow   Water is flowing from a horizontal pipe 48 feet above the ground. The falling stream of water has the shape of a parabola whose vertex $(0, 48)$ is at the end of the pipe (see figure). The stream of water strikes the ocean at the point $\left(10\sqrt{3}, 0\right)$. Write an equation for the path of the water.

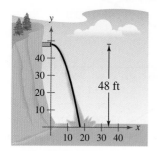
Figure for 67

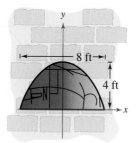
Figure for 68

68. Window Design   A church window is bounded above by a parabola (see figure). Write an equation of the parabola.

69. Archway   A parabolic archway is 12 meters high at the vertex. At a height of 10 meters, the width of the archway is 8 meters (see figure). How wide is the archway at ground level?

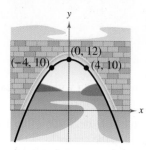
Figure for 69

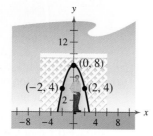
Figure for 70

70. Lattice Arch   A parabolic lattice arch is 8 feet high at the vertex. At a height of 4 feet, the width of the lattice arch is 4 feet (see figure). How wide is the lattice arch at ground level?

71. Suspension Bridge   Each cable of a suspension bridge is suspended (in the shape of a parabola) between two towers (see figure).

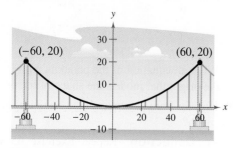

(a) Find the coordinates of the focus.

(b) Write an equation that models the cables.

72. Suspension Bridge

Each cable of the Golden Gate Bridge is suspended (in the shape of a parabola) between two towers that are 1280 meters apart. The top of each tower is 152 meters above the roadway. The cables touch the roadway at the midpoint between the towers.

(a) Sketch the bridge on a rectangular coordinate system with the cables touching the roadway at the origin. Label the coordinates of the known points.

(b) Write an equation that models the cables.

(c) Complete the table by finding the height y of the cables over the roadway at a distance of x meters from the point where the cables touch the roadway.

Distance, x	Height, y
0	
100	
250	
400	
500	

73. Satellite Orbit   A satellite in a 100-mile-high circular orbit around Earth has a velocity of approximately 17,500 miles per hour. When this velocity is multiplied by $\sqrt{2}$, the satellite has the minimum velocity necessary to escape Earth's gravity and follow a parabolic path with the center of Earth as the focus (see figure).

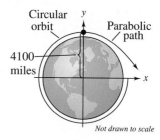
Not drawn to scale

(a) Find the escape velocity of the satellite.

(b) Write an equation for the parabolic path of the satellite. (Assume that the radius of Earth is 4000 miles.)

74. Path of a Softball The path of a softball is modeled by

$$-12.5(y - 7.125) = (x - 6.25)^2$$

where x and y are measured in feet, with $x = 0$ corresponding to the position from which the ball was thrown.

(a) Use a graphing utility to graph the trajectory of the softball.

(b) Use the *trace* feature of the graphing utility to approximate the highest point and the range of the trajectory.

Projectile Motion **In Exercises 75 and 76, consider the path of an object projected horizontally with a velocity of v feet per second at a height of s feet, where the model for the path is**

$$x^2 = -\frac{v^2}{16}(y - s).$$

In this model (in which air resistance is disregarded), y is the height (in feet) of the projectile and x is the horizontal distance (in feet) the projectile travels.

75. A ball is thrown from the top of a 100-foot tower with a velocity of 28 feet per second.

(a) Write an equation for the parabolic path.

(b) How far does the ball travel horizontally before it strikes the ground?

76. A cargo plane is flying at an altitude of 500 feet and a speed of 255 miles per hour. A supply crate is dropped from the plane. How many *feet* will the crate travel horizontally before it hits the ground?

Exploration

True or False? **In Exercises 77–79, determine whether the statement is true or false. Justify your answer.**

77. It is possible for a parabola to intersect its directrix.

78. A tangent line to a parabola always intersects the directrix.

79. When the vertex and focus of a parabola are on a horizontal line, the directrix of the parabola is vertical.

80. Slope of a Tangent Line Let (x_1, y_1) be the coordinates of a point on the parabola $x^2 = 4py$. The equation of the line tangent to the parabola at the point is

$$y - y_1 = \frac{x_1}{2p}(x - x_1).$$

What is the slope of the tangent line?

81. Think About It Explain what each equation represents, and how equations (a) and (b) are equivalent.

(a) $y = a(x - h)^2 + k,\quad a \neq 0$

(b) $(x - h)^2 = 4p(y - k),\quad p \neq 0$

(c) $(y - k)^2 = 4p(x - h),\quad p \neq 0$

82. **HOW DO YOU SEE IT?**
In parts (a)–(d), describe how a plane could intersect the double-napped cone to form each conic section (see figure).

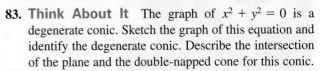

(a) Circle   (b) Ellipse

(c) Parabola   (d) Hyperbola

83. Think About It The graph of $x^2 + y^2 = 0$ is a degenerate conic. Sketch the graph of this equation and identify the degenerate conic. Describe the intersection of the plane and the double-napped cone for this conic.

84. Graphical Reasoning Consider the parabola $x^2 = 4py$.

(a) Use a graphing utility to graph the parabola for $p = 1$, $p = 2$, $p = 3$, and $p = 4$. Describe the effect on the graph when p increases.

(b) Find the focus for each parabola in part (a).

(c) For each parabola in part (a), find the length of the latus rectum (see figure). How can the length of the latus rectum be determined directly from the standard form of the equation of the parabola?

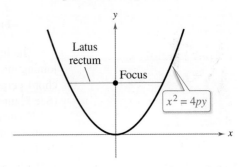

(d) How can you use the result of part (c) as a sketching aid when graphing parabolas?

85. Geometry The area of the shaded region in the figure is $A = \dfrac{8}{3}p^{1/2}b^{3/2}$.

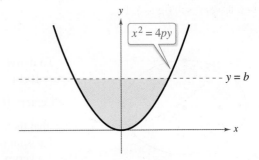

(a) Find the area when $p = 2$ and $b = 4$.

(b) Give a geometric explanation of why the area approaches 0 as p approaches 0.

10.3 Ellipses

Ellipses have many real-life applications. For example, Exercise 55 on page 715 shows how a lithotripter machine uses the focal properties of an ellipse to break up kidney stones.

- ■ Write equations of ellipses in standard form and sketch ellipses.
- ■ Use properties of ellipses to model and solve real-life problems.
- ■ Find eccentricities of ellipses.

Introduction

Another type of conic is an **ellipse.** It is defined below.

> ### Definition of an Ellipse
> An **ellipse** is the set of all points (x, y) in a plane, the sum of whose distances from two distinct fixed points **(foci)** is constant. See Figure 10.12.

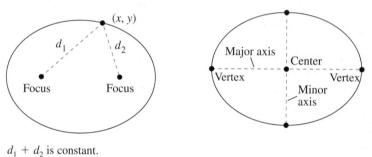

$d_1 + d_2$ is constant.

Figure 10.12

Figure 10.13

The line through the foci intersects the ellipse at two points **(vertices).** The chord joining the vertices is the **major axis,** and its midpoint is the **center** of the ellipse. The chord perpendicular to the major axis at the center is the **minor axis** of the ellipse. (See Figure 10.13.)

To visualize the definition of an ellipse, imagine two thumbtacks placed at the foci, as shown in the figure below. When the ends of a fixed length of string are fastened to the thumbtacks and the string is drawn taut with a pencil, the path traced by the pencil is an ellipse.

To derive the standard form of the equation of an ellipse, consider the ellipse in Figure 10.14 with the points listed below.

Center: (h, k) Vertices: $(h \pm a, k)$ Foci: $(h \pm c, k)$

Note that the center is also the midpoint of the segment joining the foci.

The sum of the distances from any point on the ellipse to the two foci is constant. Using a vertex point, this constant sum is

$$(a + c) + (a - c) = 2a \qquad \text{Length of major axis}$$

which is the length of the major axis.

$2\sqrt{b^2 + c^2} = 2a$
$b^2 + c^2 = a^2$

Figure 10.14

Now, if you let (x, y) be *any* point on the ellipse, then the sum of the distances between (x, y) and the two foci must also be $2a$. That is,

$$\sqrt{[x - (h - c)]^2 + (y - k)^2} + \sqrt{[x - (h + c)]^2 + (y - k)^2} = 2a$$

which, after expanding and regrouping, reduces to

$$(a^2 - c^2)(x - h)^2 + a^2(y - k)^2 = a^2(a^2 - c^2).$$

From Figure 10.14,

$$b^2 + c^2 = a^2$$

$$b^2 = a^2 - c^2$$

which implies that the equation of the ellipse is

$$b^2(x - h)^2 + a^2(y - k)^2 = a^2b^2$$

$$\frac{(x - h)^2}{a^2} + \frac{(y - k)^2}{b^2} = 1.$$

You would obtain a similar equation in the derivation by starting with a vertical major axis. A summary of these results is given below.

> ·· **REMARK** Consider the equation of the ellipse
>
> $$\frac{(x - h)^2}{a^2} + \frac{(y - k)^2}{b^2} = 1.$$
>
> If you let $a = b = r$, then the equation can be rewritten as
>
> $$(x - h)^2 + (y - k)^2 = r^2$$
>
> which is the standard form of the equation of a circle with radius r. Geometrically, when $a = b$ for an ellipse, the major and minor axes are of equal length, and so the graph is a circle.

Standard Equation of an Ellipse

The **standard form of the equation of an ellipse** with center (h, k) and major and minor axes of lengths $2a$ and $2b$, respectively, where $0 < b < a$, is

$$\frac{(x - h)^2}{a^2} + \frac{(y - k)^2}{b^2} = 1 \qquad \text{Major axis is horizontal.}$$

$$\frac{(x - h)^2}{b^2} + \frac{(y - k)^2}{a^2} = 1. \qquad \text{Major axis is vertical.}$$

The foci lie on the major axis, c units from the center, with

$$c^2 = a^2 - b^2.$$

If the center is at the origin, then the equation takes one of the forms below.

$$\frac{x^2}{a^2} + \frac{y^2}{b^2} = 1 \qquad \text{Major axis is horizontal.}$$

$$\frac{x^2}{b^2} + \frac{y^2}{a^2} = 1 \qquad \text{Major axis is vertical.}$$

The figures below show generalized horizontal and vertical orientations for ellipses.

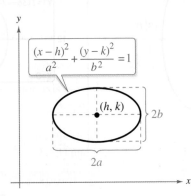

$$\frac{(x - h)^2}{a^2} + \frac{(y - k)^2}{b^2} = 1$$

Major axis is horizontal.

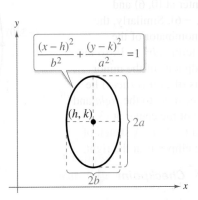

$$\frac{(x - h)^2}{b^2} + \frac{(y - k)^2}{a^2} = 1$$

Major axis is vertical.

Application

Ellipses have many practical and aesthetic uses. For example, machine gears, supporting arches, and acoustic designs often involve elliptical shapes. The orbits of satellites and planets are also ellipses. Example 5 investigates the elliptical orbit of the moon about Earth.

| EXAMPLE 5 | **An Application Involving an Elliptical Orbit** |

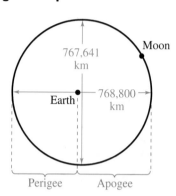

................▷

REMARK Note in Example 5 that Earth is *not* the center of the moon's orbit.

The moon travels about Earth in an elliptical orbit with the center of Earth at one focus, as shown in the figure at the right. The major and minor axes of the orbit have lengths of 768,800 kilometers and 767,641 kilometers, respectively. Find the greatest and least distances (the *apogee* and *perigee,* respectively) from Earth's center to the moon's center. Then use a graphing utility to graph the orbit of the moon.

Solution Because $2a = 768,800$ and $2b = 767,641$, you have $a = 384,400$ and $b = 383,820.5$, which implies that

$$c = \sqrt{a^2 - b^2} = \sqrt{384,400^2 - 383,820.5^2} \approx 21,099.$$

So, the greatest distance between the center of Earth and the center of the moon is

$$a + c \approx 384,400 + 21,099 = 405,499 \text{ kilometers}$$

and the least distance is

$$a - c \approx 384,400 - 21,099 = 363,301 \text{ kilometers}.$$

To use a graphing utility to graph the orbit of the moon, first let $a = 384,400$ and $b = 383,820.5$ in the standard form of an equation of an ellipse centered at the origin, and then solve for y.

$$\frac{x^2}{384,400^2} + \frac{y^2}{383,820.5^2} = 1 \implies y = \pm 383,820.5\sqrt{1 - \frac{x^2}{384,400^2}}$$

Graph the upper and lower portions in the same viewing window, as shown below.

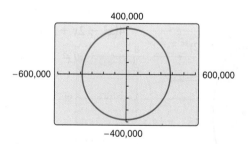

In Exercise 56, you will investigate the elliptical orbit of Halley's comet about the sun. Halley's comet is visible from Earth approximately every 76.1 years. The comet's latest appearance was in 1986.

✓ **Checkpoint** ◀))) *Audio-video solution in English & Spanish at LarsonPrecalculus.com*

Encke's comet travels about the sun in an elliptical orbit with the center of the sun at one focus. The major and minor axes of the orbit have lengths of approximately 4.420 astronomical units and 2.356 astronomical units, respectively. (An astronomical unit is about 93 million miles.) Find the greatest and least distances (the *aphelion* and *perihelion,* respectively). from the sun's center to the comet's center. Then use a graphing utility to graph the orbit of the comet.

Eccentricity

It was difficult for early astronomers to detect that the orbits of the planets are ellipses because the foci of the planetary orbits are relatively close to their centers, and so the orbits are nearly circular. You can measure the "ovalness" of an ellipse by using the concept of **eccentricity.**

Definition of Eccentricity

The **eccentricity** e of an ellipse is the ratio $e = \dfrac{c}{a}$.

Note that $0 < e < 1$ for *every* ellipse.

To see how this ratio describes the shape of an ellipse, note that because the foci of an ellipse are located along the major axis between the vertices and the center, it follows that $0 < c < a$. For an ellipse that is nearly circular, the foci are close to the center and the ratio c/a is close to 0, as shown in Figure 10.18. On the other hand, for an elongated ellipse, the foci are close to the vertices and the ratio c/a is close to 1, as shown in Figure 10.19.

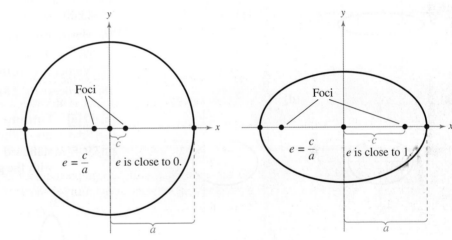

Figure 10.18   **Figure 10.19**

The orbit of the moon has an eccentricity of $e \approx 0.0549$. The eccentricities of the eight planetary orbits are listed below.

Mercury: $e \approx 0.2056$	Jupiter: $e \approx 0.0489$
Venus: $e \approx 0.0067$	Saturn: $e \approx 0.0565$
Earth: $e \approx 0.0167$	Uranus: $e \approx 0.0457$
Mars: $e \approx 0.0935$	Neptune: $e \approx 0.0113$

Summarize (Section 10.3)

1. State the definition of an ellipse and the standard form of the equation of an ellipse *(page 708)*. For examples involving the equations and graphs of ellipses, see Examples 1–4.

2. Describe a real-life application of an ellipse *(page 712, Example 5)*.

3. State the definition of the eccentricity of an ellipse and explain how eccentricity describes the shape of an ellipse *(page 713)*.

The time it takes Saturn to orbit the sun is about 29.5 Earth years.

10.3 Exercises

See CalcChat.com for tutorial help and worked-out solutions to odd-numbered exercises.

Vocabulary: Fill in the blanks.

1. An _____ is the set of all points (x, y) in a plane, the sum of whose distances from two distinct fixed points, called _____, is constant.

2. The chord joining the vertices of an ellipse is the _____ _____, and its midpoint is the _____ of the ellipse.

3. The chord perpendicular to the major axis at the center of an ellipse is the _____ _____ of the ellipse.

4. You can measure the "ovalness" of an ellipse by using the concept of _____.

Skills and Applications

Matching In Exercises 5–8, match the equation with its graph. [The graphs are labeled (a), (b), (c), and (d).]

(a)
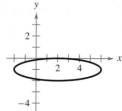

(b)
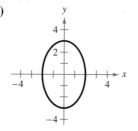

(c)
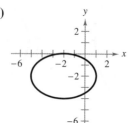

(d)

5. $\dfrac{x^2}{4} + \dfrac{y^2}{9} = 1$

6. $\dfrac{x^2}{9} + \dfrac{y^2}{4} = 1$

7. $\dfrac{(x-2)^2}{16} + (y+1)^2 = 1$

8. $\dfrac{(x+2)^2}{9} + \dfrac{(y+2)^2}{4} = 1$

 An Ellipse Centered at the Origin In Exercises 9–18, find the standard form of the equation of the ellipse with the given characteristics and center at the origin.

9.
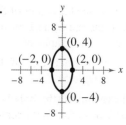

10.
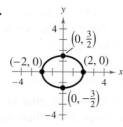

11. Vertices: $(\pm 7, 0)$; foci: $(\pm 2, 0)$

12. Vertices: $(0, \pm 8)$; foci: $(0, \pm 4)$

13. Foci: $(\pm 4, 0)$; major axis of length 10

14. Foci: $(0, \pm 3)$; major axis of length 8

15. Vertical major axis; passes through the points $(0, 6)$ and $(3, 0)$

16. Horizontal major axis; passes through the points $(5, 0)$ and $(0, 2)$

17. Vertices: $(\pm 6, 0)$; passes through the point $(4, 1)$

18. Vertices: $(0, \pm 8)$; passes through the point $(3, 4)$

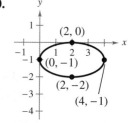 **Finding the Standard Equation of an Ellipse** In Exercises 19–30, find the standard form of the equation of the ellipse with the given characteristics.

19.
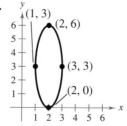

20.

21. Vertices: $(2, 0)$, $(10, 0)$; minor axis of length 4

22. Vertices: $(3, 1)$, $(3, 11)$; minor axis of length 2

23. Foci: $(0, 0)$, $(4, 0)$; major axis of length 6

24. Foci: $(0, 0)$, $(0, 8)$; major axis of length 16

25. Center: $(1, 3)$; vertex: $(-2, 3)$; minor axis of length 4

26. Center: $(2, -1)$; vertex: $\left(2, \tfrac{1}{2}\right)$; minor axis of length 2

27. Center: $(1, 4)$; $a = 2c$; vertices: $(1, 0)$, $(1, 8)$

28. Center: $(3, 2)$; $a = 3c$; foci: $(1, 2)$, $(5, 2)$

29. Vertices: $(0, 2)$, $(4, 2)$; endpoints of the minor axis: $(2, 3)$, $(2, 1)$

30. Vertices: $(5, 0)$, $(5, 12)$; endpoints of the minor axis: $(1, 6)$, $(9, 6)$

Sketching an Ellipse In Exercises 31–46, find the center, vertices, foci, and eccentricity of the ellipse. Then sketch the ellipse.

31. $\dfrac{x^2}{25} + \dfrac{y^2}{16} = 1$

32. $\dfrac{x^2}{16} + \dfrac{y^2}{81} = 1$

33. $9x^2 + y^2 = 36$

34. $x^2 + 16y^2 = 64$

35. $\dfrac{(x-4)^2}{16} + \dfrac{(y+1)^2}{25} = 1$

36. $\dfrac{(x+3)^2}{12} + \dfrac{(y-2)^2}{16} = 1$

37. $\dfrac{(x+5)^2}{9/4} + (y-1)^2 = 1$

38. $(x+2)^2 + \dfrac{(y+4)^2}{1/4} = 1$

39. $9x^2 + 4y^2 + 36x - 24y + 36 = 0$

40. $9x^2 + 4y^2 - 54x + 40y + 37 = 0$

41. $x^2 + 5y^2 - 8x - 30y - 39 = 0$

42. $3x^2 + y^2 + 18x - 2y - 8 = 0$

43. $6x^2 + 2y^2 + 18x - 10y + 2 = 0$

44. $x^2 + 4y^2 - 6x + 20y - 2 = 0$

45. $12x^2 + 20y^2 - 12x + 40y - 37 = 0$

46. $36x^2 + 9y^2 + 48x - 36y + 43 = 0$

Graphing an Ellipse In Exercises 47–50, use a graphing utility to graph the ellipse. Find the center, foci, and vertices.

47. $5x^2 + 3y^2 = 15$

48. $3x^2 + 4y^2 = 12$

49. $x^2 + 9y^2 - 10x + 36y + 52 = 0$

50. $4x^2 + 3y^2 - 8x + 18y + 19 = 0$

51. **Using Eccentricity** Find an equation of the ellipse with vertices $(\pm 5, 0)$ and eccentricity $e = \dfrac{4}{5}$.

52. **Using Eccentricity** Find an equation of the ellipse with vertices $(0, \pm 8)$ and eccentricity $e = \dfrac{1}{2}$.

53. **Architecture** Statuary Hall is an elliptical room in the United States Capitol in Washington, D.C. The room is also called the Whispering Gallery because a person standing at one focus of the room can hear even a whisper spoken by a person standing at the other focus. The dimensions of Statuary Hall are 46 feet wide by 97 feet long.

    (a) Find an equation of the shape of the room.

    (b) Determine the distance between the foci.

54. **Architecture** A mason is building a semielliptical fireplace arch that has a height of 2 feet at the center and a width of 6 feet along the base (see figure). The mason draws the semiellipse on the wall by the method shown on page 708. Find the positions of the thumbtacks and the length of the string.

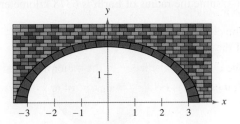

55. **Lithotripter**

A lithotripter machine uses an elliptical reflector to break up kidney stones nonsurgically. A spark plug in the reflector generates energy waves at one focus of an ellipse. The reflector directs these waves toward the kidney stone, positioned at the other focus of the ellipse, with enough energy to break up the stone (see figure). The lengths of the major and minor axes of the ellipse are 280 millimeters and 160 millimeters, respectively. How far is the spark plug from the kidney stone?

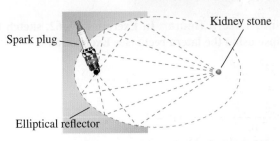

56. **Astronomy** Halley's comet has an elliptical orbit with the center of the sun at one focus. The eccentricity of the orbit is approximately 0.967. The length of the major axis of the orbit is approximately 35.88 astronomical units. (An astronomical unit is about 93 million miles.)

    (a) Find an equation of the orbit. Place the center of the orbit at the origin and place the major axis on the x-axis.

    (b) Use a graphing utility to graph the equation of the orbit.

    (c) Find the greatest and least distances (the aphelion and perihelion, respectively) from the sun's center to the comet's center.

57. Astronomy   The first artificial satellite to orbit Earth was Sputnik I (launched by the former Soviet Union in 1957). Its highest point above Earth's surface was 939 kilometers, and its lowest point was 215 kilometers (see figure). The center of Earth was at one focus of the elliptical orbit. Find the eccentricity of the orbit. (Assume the radius of Earth is 6378 kilometers.)

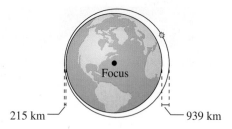

215 km —   — 939 km

58. Geometry   A line segment through a focus of an ellipse with endpoints on the ellipse and perpendicular to the major axis is called a **latus rectum** of the ellipse. An ellipse has two latera recta. Knowing the length of the latera recta is helpful in sketching an ellipse because it yields other points on the curve (see figure). Show that the length of each latus rectum is $2b^2/a$.

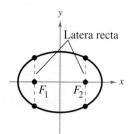

Using Latera Recta   In Exercises 59–62, sketch the ellipse using the latera recta (see Exercise 58).

59. $\dfrac{x^2}{9} + \dfrac{y^2}{16} = 1$

60. $\dfrac{x^2}{4} + \dfrac{y^2}{1} = 1$

61. $5x^2 + 3y^2 = 15$

62. $9x^2 + 4y^2 = 36$

Exploration

True or False?   In Exercises 63 and 64, determine whether the statement is true or false. Justify your answer.

63. The graph of $x^2 + 4y^4 - 4 = 0$ is an ellipse.

64. It is easier to distinguish the graph of an ellipse from the graph of a circle when the eccentricity of the ellipse is close to 1.

65. Think About It   Find an equation of an ellipse such that for any point on the ellipse, the sum of the distances from the point to the points $(2, 2)$ and $(10, 2)$ is 36.

66. Think About It   At the beginning of this section, you learned that an ellipse can be drawn using two thumbtacks, a string of fixed length (greater than the distance between the two thumbtacks), and a pencil. When the ends of the string are fastened to the thumbtacks and the string is drawn taut with the pencil, the path traced by the pencil is an ellipse.

(a) What is the length of the string in terms of a?

(b) Explain why the path is an ellipse.

67. Conjecture   Consider the ellipse

$$\frac{x^2}{a^2} + \frac{y^2}{b^2} = 1, \quad a + b = 20.$$

(a) The area of the ellipse is given by $A = \pi ab$. Write the area of the ellipse as a function of a.

(b) Find the equation of an ellipse with an area of 264 square centimeters.

(c) Complete the table using your equation from part (a). Then make a conjecture about the shape of the ellipse with maximum area.

a	8	9	10	11	12	13
A						

(d) Use a graphing utility to graph the area function and use the graph to support your conjecture in part (c).

68. HOW DO YOU SEE IT?   Without performing any calculations, order the eccentricities of the ellipses from least to greatest.

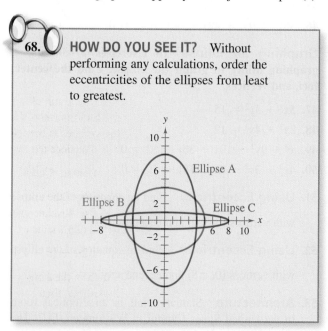

69. Proof   Show that $a^2 = b^2 + c^2$ for the ellipse

$$\frac{x^2}{a^2} + \frac{y^2}{b^2} = 1$$

where $a > 0, b > 0$, and the distance from the center of the ellipse $(0, 0)$ to a focus is c.

10.4 Hyperbolas

Hyperbolas have many types of real-life applications. For example, in Exercise 53 on page 725, you will investigate the use of hyperbolas in long distance radio navigation for aircraft and ships.

- ■ Write equations of hyperbolas in standard form.
- ■ Find asymptotes of and sketch hyperbolas.
- ■ Use properties of hyperbolas to solve real-life problems.
- ■ Classify conics from their general equations.

Introduction

The definition of a **hyperbola** is similar to that of an ellipse. For an ellipse, the *sum* of the distances between the foci and a point on the ellipse is constant. For a hyperbola, the absolute value of the *difference* of the distances between the foci and a point on the hyperbola is constant.

> ### Definition of a Hyperbola
>
> A **hyperbola** is the set of all points (x, y) in a plane for which the absolute value of the difference of the distances from two distinct fixed points (**foci**) is constant. See Figure 10.20.

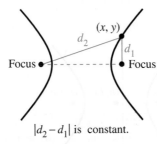

$|d_2 - d_1|$ is constant.

Figure 10.20

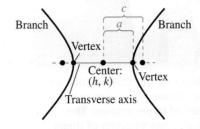

Figure 10.21

The graph of a hyperbola has two disconnected parts (**branches**). The line through the foci intersects the hyperbola at two points (**vertices**). The line segment connecting the vertices is the **transverse axis,** and its midpoint is the **center** of the hyperbola.

Consider the hyperbola in Figure 10.21 with the points listed below.

Center: (h, k) Vertices: $(h \pm a, k)$ Foci: $(h \pm c, k)$

Note that the center is also the midpoint of the segment joining the foci.

The absolute value of the difference of the distances from *any* point on the hyperbola to the two foci is constant. Using a vertex point, this constant value is

$$|[2a + (c - a)] - (c - a)| = |2a| = 2a \qquad \text{Length of transverse axis}$$

which is the length of the transverse axis. Now, if you let (x, y) be *any* point on the hyperbola, then

$$|d_2 - d_1| = 2a$$

(see Figure 10.20). You would obtain the same result for a hyperbola with a vertical transverse axis.

The development of the standard form of the equation of a hyperbola is similar to that of an ellipse. Note in the definition on the next page that a, b, and c are related differently for hyperbolas than for ellipses. For a hyperbola, the distance between the foci and the center is greater than the distance between the vertices and the center.

EXAMPLE 3   **Sketching a Hyperbola**

Sketch the hyperbola $4x^2 - 3y^2 + 8x + 16 = 0$.

Solution

$4x^2 - 3y^2 + 8x + 16 = 0$	Write original equation.
$(4x^2 + 8x) - 3y^2 = -16$	Group terms.
$4(x^2 + 2x) - 3y^2 = -16$	Factor 4 out of x-terms.
$4(x^2 + 2x + 1) - 3y^2 = -16 + 4(1)$	Complete the square.
$4(x + 1)^2 - 3y^2 = -12$	Write in completed square form.
$-\dfrac{(x + 1)^2}{3} + \dfrac{y^2}{4} = 1$	Divide each side by -12.
$\dfrac{y^2}{2^2} - \dfrac{(x + 1)^2}{(\sqrt{3})^2} = 1$	Write in standard form.

The center of the hyperbola is $(-1, 0)$. The y^2-term is positive, so the transverse axis is vertical. The vertices occur at $(-1, 2)$ and $(-1, -2)$, and the endpoints of the conjugate axis occur at $(-1 - \sqrt{3}, 0)$ and $(-1 + \sqrt{3}, 0)$. Draw a rectangle through the vertices and the endpoints of the conjugate axes. Sketch the asymptotes by drawing lines through the opposite corners of the rectangle. Using $a = 2$ and $b = \sqrt{3}$, the equations of the asymptotes are

$$y = \frac{2}{\sqrt{3}}(x + 1) \quad \text{and} \quad y = -\frac{2}{\sqrt{3}}(x + 1).$$

Finally, from $c^2 = a^2 + b^2$, you have $c = \sqrt{2^2 + (\sqrt{3})^2} = \sqrt{7}$. So, the foci of the hyperbola are $(-1, \sqrt{7})$ and $(-1, -\sqrt{7})$. Figure 10.26 shows a sketch of the hyperbola.

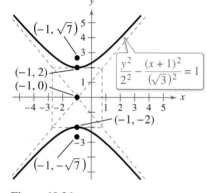

Figure 10.26

✓ **Checkpoint** ◀))) Audio-video solution in English & Spanish at LarsonPrecalculus.com

Sketch the hyperbola $9x^2 - 4y^2 + 8y - 40 = 0$. ◼

▷ **TECHNOLOGY** To use a graphing utility to graph a hyperbola, graph the upper and lower portions in the same viewing window. For instance, to graph the hyperbola in Example 3, first solve for y to get

$$y_1 = 2\sqrt{1 + \frac{(x + 1)^2}{3}} \quad \text{and} \quad y_2 = -2\sqrt{1 + \frac{(x + 1)^2}{3}}.$$

Use a viewing window in which $-9 \le x \le 9$ and $-6 \le y \le 6$. You should obtain the graph shown below. Notice that the graphing utility does not draw the asymptotes. However, by graphing the asymptotes in the same viewing window, you can see that the values of the hyperbola approach the asymptotes.

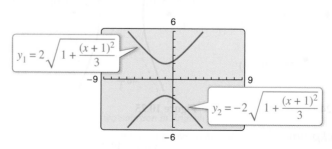

EXAMPLE 4 **Using Asymptotes to Find the Standard Equation**

See LarsonPrecalculus.com for an interactive version of this type of example.

Find the standard form of the equation of the hyperbola with vertices $(3, -5)$ and $(3, 1)$ and asymptotes

$$y = 2x - 8 \quad \text{and} \quad y = -2x + 4$$

as shown in Figure 10.27.

Solution The center of the hyperbola is $(3, -2)$. Furthermore, the hyperbola has a vertical transverse axis with $a = 3$. The slopes of the asymptotes are

$$m_1 = 2 = \frac{a}{b} \quad \text{and} \quad m_2 = -2 = -\frac{a}{b}$$

and $a = 3$, so

$$2 = \frac{a}{b} \implies 2 = \frac{3}{b} \implies b = \frac{3}{2}.$$

The standard form of the equation of the hyperbola is

$$\frac{(y + 2)^2}{3^2} - \frac{(x - 3)^2}{\left(\frac{3}{2}\right)^2} = 1.$$

 Checkpoint ◀))) Audio-video solution in English & Spanish at LarsonPrecalculus.com

Find the standard form of the equation of the hyperbola with vertices $(3, 2)$ and $(9, 2)$ and asymptotes

$$y = -2 + \frac{2}{3}x \quad \text{and} \quad y = 6 - \frac{2}{3}x.$$

As with ellipses, the *eccentricity* of a hyperbola is

$$e = \frac{c}{a}. \qquad \text{Eccentricity}$$

You know that $c > a$ for a hyperbola, so it follows that $e > 1$. When the eccentricity is large, the branches of the hyperbola are nearly flat, as shown in Figure 10.28. When the eccentricity is close to 1, the branches of the hyperbola are more narrow, as shown in Figure 10.29.

Figure 10.27

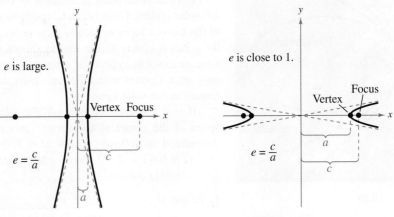

Figure 10.28   Figure 10.29

Applications

The next example shows how the properties of hyperbolas are used in radar and other detection systems. The United States and Great Britain developed this application during World War II.

EXAMPLE 5 **An Application Involving Hyperbolas**

Two microphones, 1 mile apart, record an explosion. Microphone A receives the sound 2 seconds before microphone B. Where did the explosion occur?

Solution Assuming sound travels at 1100 feet per second, you know that the explosion took place 2200 feet farther from B than from A, as shown in the figure. The locus of all points that are 2200 feet closer to A than to B is one branch of the hyperbola of the form

$$\frac{x^2}{a^2} - \frac{y^2}{b^2} = 1$$

where

$$a = \frac{2200}{2} = 1100.$$

Because

$$c = \frac{5280}{2} = 2640$$

it follows that

$$b^2 = c^2 - a^2$$
$$= 2640^2 - 1100^2$$
$$= 5{,}759{,}600.$$

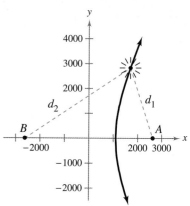

$2c = 1 \text{ mi} = 5280 \text{ ft}$
$|d_2 - d_1| = 2a = 2200 \text{ ft}$

So, the explosion occurred somewhere on the right branch of the hyperbola

$$\frac{x^2}{1{,}210{,}000} - \frac{y^2}{5{,}759{,}600} = 1.$$

✓ **Checkpoint**))) *Audio-video solution in English & Spanish at LarsonPrecalculus.com*

Repeat Example 5 when microphone A receives the sound 4 seconds before microphone B.

Another interesting application of conic sections involves the orbits of comets in our solar system. Comets can have elliptical, parabolic, or hyperbolic orbits. The center of the sun is a focus of each of these orbits, and each orbit has a vertex at the point where the comet is closest to the sun, as shown in Figure 10.30. Undoubtedly, many comets with parabolic or hyperbolic orbits have not been identified. You get to see such comets only *once*. Comets with elliptical orbits, such as Halley's comet, are the only ones that remain in our solar system.

If p is the distance between the vertex and the focus (in meters), and v is the speed of the comet at the vertex (in meters per second), then the type of orbit is determined as follows, where $M = 1.989 \times 10^{30}$ kilograms (the mass of the sun) and $G \approx 6.67 \times 10^{-11}$ cubic meter per kilogram-second squared (the universal gravitational constant).

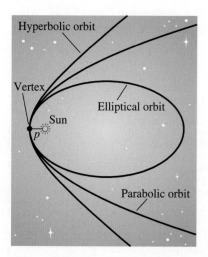

Figure 10.30

1. Elliptical: $v < \sqrt{2GM/p}$

2. Parabolic: $v = \sqrt{2GM/p}$

3. Hyperbolic: $v > \sqrt{2GM/p}$

General Equations of Conics

Classifying a Conic from Its General Equation

The graph of $Ax^2 + Cy^2 + Dx + Ey + F = 0$ is one of the following.

1. *Circle:* $A = C$ $A \neq 0$
2. *Parabola:* $AC = 0$ $A = 0$ or $C = 0$, but not both.
3. *Ellipse:* $AC > 0$ $A \neq C$ and A and C have like signs.
4. *Hyperbola:* $AC < 0$ A and C have unlike signs.

The test above is valid when the graph is a conic. The test does not apply to equations such as $x^2 + y^2 = -1$, whose graph is not a conic.

EXAMPLE 6   Classifying Conics from General Equations

a. For the equation $4x^2 - 9x + y - 5 = 0$, you have

$$AC = 4(0) = 0. \qquad \text{Parabola}$$

So, the graph is a parabola.

b. For the equation $4x^2 - y^2 + 8x - 6y + 4 = 0$, you have

$$AC = 4(-1) < 0. \qquad \text{Hyperbola}$$

So, the graph is a hyperbola.

c. For the equation $2x^2 + 4y^2 - 4x + 12y = 0$, you have

$$AC = 2(4) > 0. \qquad \text{Ellipse}$$

So, the graph is an ellipse.

d. For the equation $2x^2 + 2y^2 - 8x + 12y + 2 = 0$, you have

$$A = C = 2. \qquad \text{Circle}$$

So, the graph is a circle.

✓ *Checkpoint* ))) *Audio-video solution in English & Spanish at LarsonPrecalculus.com*

Classify the graph of each equation.

a. $3x^2 + 3y^2 - 6x + 6y + 5 = 0$ **b.** $2x^2 - 4y^2 + 4x + 8y - 3 = 0$

c. $3x^2 + y^2 + 6x - 2y + 3 = 0$ **d.** $2x^2 + 4x + y - 2 = 0$

Caroline Herschel (1750–1848) was the first woman to be credited with discovering a comet. During her long life, this German astronomer discovered a total of eight comets.

Summarize   (Section 10.4)

1. State the definition of a hyperbola and the standard form of the equation of a hyperbola *(page 717)*. For an example of finding the standard form of the equation of a hyperbola, see Example 1.

2. Explain how to find asymptotes of and sketch a hyperbola *(page 719)*. For examples involving asymptotes and graphs of hyperbolas, see Examples 2–4.

3. Describe a real-life application of a hyperbola *(page 722, Example 5)*.

4. Explain how to classify a conic from its general equation *(page 723)*. For an example of classifying conics from their general equations, see Example 6.

10.4 Exercises

See CalcChat.com for tutorial help and worked-out solutions to odd-numbered exercises.

Vocabulary: Fill in the blanks.

1. A _____ is the set of all points (x, y) in a plane for which the absolute value of the difference of the distances from two distinct fixed points, called _____, is constant.

2. The graph of a hyperbola has two disconnected parts called _____.

3. The line segment connecting the vertices of a hyperbola is the _____ _____, and its midpoint is the _____ of the hyperbola.

4. Every hyperbola has two _____ that intersect at the center of the hyperbola.

Skills and Applications

Matching In Exercises 5–8, match the equation with its graph. [The graphs are labeled (a), (b), (c), and (d).]

(a)
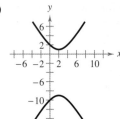

(b)
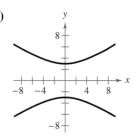

(c)
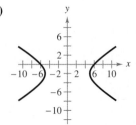

(d)
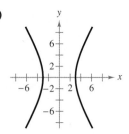

5. $\dfrac{y^2}{9} - \dfrac{x^2}{25} = 1$

6. $\dfrac{x^2}{9} - \dfrac{y^2}{25} = 1$

7. $\dfrac{x^2}{25} - \dfrac{(y + 2)^2}{9} = 1$

8. $\dfrac{(y + 4)^2}{25} - \dfrac{(x - 2)^2}{9} = 1$

Finding the Standard Equation of a Hyperbola In Exercises 9–18, find the standard form of the equation of the hyperbola with the given characteristics.

9. Vertices: $(0, \pm 2)$; foci: $(0, \pm 4)$

10. Vertices: $(\pm 4, 0)$; foci: $(\pm 6, 0)$

11. Vertices: $(2, 0), (6, 0)$; foci: $(0, 0), (8, 0)$

12. Vertices: $(2, 3), (2, -3)$; foci: $(2, 6), (2, -6)$

13. Vertices: $(4, 1), (4, 9)$; foci: $(4, 0), (4, 10)$

14. Vertices: $(-1, 1), (3, 1)$; foci: $(-2, 1), (4, 1)$

15. Vertices: $(2, 3), (2, -3)$; passes through the point $(0, 5)$

16. Vertices: $(-2, 1), (2, 1)$; passes through the point $(5, 4)$

17. Vertices: $(0, -3), (4, -3)$; passes through the point $(-4, 5)$

18. Vertices: $(1, -3), (1, -7)$; passes through the point $(5, -11)$

 Sketching a Hyperbola In Exercises 19–32, find the center, vertices, foci, and the equations of the asymptotes of the hyperbola. Then sketch the hyperbola using the asymptotes as an aid.

19. $x^2 - y^2 = 1$

20. $\dfrac{x^2}{9} - \dfrac{y^2}{16} = 1$

21. $\dfrac{1}{36}y^2 - \dfrac{1}{100}x^2 = 1$

22. $\dfrac{1}{144}x^2 - \dfrac{1}{169}y^2 = 1$

23. $2y^2 - \dfrac{x^2}{2} = 2$

24. $\dfrac{y^2}{3} - 3x^2 = 3$

25. $\dfrac{(x - 1)^2}{4} - \dfrac{(y + 2)^2}{1} = 1$

26. $\dfrac{(x + 3)^2}{144} - \dfrac{(y - 2)^2}{25} = 1$

27. $\dfrac{(y + 6)^2}{1/9} - \dfrac{(x - 2)^2}{1/4} = 1$

28. $\dfrac{(y - 1)^2}{1/4} - \dfrac{(x + 3)^2}{1/16} = 1$

29. $9x^2 - y^2 - 36x - 6y + 18 = 0$

30. $x^2 - 9y^2 + 36y - 72 = 0$

31. $4x^2 - y^2 + 8x + 2y - 1 = 0$

32. $16y^2 - x^2 + 2x + 64y + 64 = 0$

Graphing a Hyperbola In Exercises 33–38, use a graphing utility to graph the hyperbola and its asymptotes. Find the center, vertices, and foci.

33. $2x^2 - 3y^2 = 6$

34. $6y^2 - 3x^2 = 18$

35. $25y^2 - 9x^2 = 225$

36. $25x^2 - 4y^2 = 100$

37. $9y^2 - x^2 + 2x + 54y + 62 = 0$

38. $9x^2 - y^2 + 54x + 10y + 55 = 0$

Finding the Standard Equation of a Hyperbola In Exercises 39–48, find the standard form of the equation of the hyperbola with the given characteristics.

39. Vertices: $(\pm 1, 0)$; asymptotes: $y = \pm 5x$

40. Vertices: $(0, \pm 3)$; asymptotes: $y = \pm 3x$

41. Foci: $(0, \pm 8)$; asymptotes: $y = \pm 4x$

42. Foci: $(\pm 10, 0)$; asymptotes: $y = \pm \frac{3}{4}x$

43. Vertices: $(1, 2), (3, 2)$;
asymptotes: $y = x, y = 4 - x$

44. Vertices: $(3, 0), (3, 6)$;
asymptotes: $y = 6 - x, y = x$

45. Vertices: $(3, 0), (3, 4)$;
asymptotes: $y = \frac{2}{3}x, y = 4 - \frac{2}{3}x$

46. Vertices: $(-4, 1), (0, 1)$;
asymptotes: $y = x + 3, y = -x - 1$

47. Foci: $(-1, -1), (9, -1)$;
asymptotes: $y = \frac{3}{4}x - 4, y = -\frac{3}{4}x + 2$

48. Foci: $\left(9, \pm 2\sqrt{10}\right)$;
asymptotes: $y = 3x - 27, y = -3x + 27$

49. Art A cross section of a sculpture can be modeled by a hyperbola (see figure).

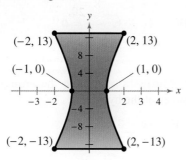

(a) Write an equation that models the curved sides of the sculpture.

(b) Each unit in the coordinate plane represents 1 foot. Find the width of the sculpture at a height of 18 feet.

50. Clock The base of a clock has the shape of a hyperbola (see figure).

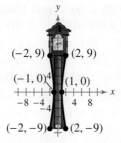

(a) Write an equation of the cross section of the base.

(b) Each unit in the coordinate plane represents $\frac{1}{2}$ foot. Find the width of the base 4 inches from the bottom.

51. Sound Location You and a friend live 4 miles apart. You hear a clap of thunder from lightning 18 seconds before your friend hears it. Where did the lightning occur? (Assume sound travels at 1100 feet per second.)

52. Sound Location Listening station A and listening station B are located at $(3300, 0)$ and $(-3300, 0)$, respectively. Station A detects an explosion 4 seconds before station B. (Assume the coordinate system is measured in feet and sound travels at 1100 feet per second.)

(a) Where did the explosion occur?

(b) Station C is located at $(3300, 1100)$ and detects the explosion 1 second after station A. Find the coordinates of the explosion.

53. Navigation

Long-distance radio navigation for aircraft and ships uses synchronized pulses transmitted by widely separated transmitting stations. These pulses travel at the speed of light (186,000 miles per second). The difference in the times of arrival of these pulses at an aircraft or ship is constant on a hyperbola having the transmitting stations as foci.

Assume that two stations 300 miles apart are positioned on a rectangular coordinate system with coordinates $(-150, 0)$ and $(150, 0)$ and that a ship is traveling on a hyperbolic path with coordinates $(x, 75)$ (see figure).

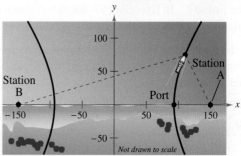

(a) Find the x-coordinate of the position of the ship when the time difference between the pulses from the transmitting stations is 1000 microseconds (0.001 second).

(b) Determine the distance between the port and station A.

(c) Find a linear equation that approximates the ship's path as it travels far away from the shore.

54. Hyperbolic Mirror A hyperbolic mirror (used in some telescopes) has the property that a light ray directed at focus A is reflected to focus B (see figure). Find the vertex of the mirror when its mount at the top edge of the mirror has coordinates (24, 24).

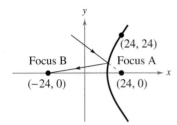

 Classifying a Conic from a General Equation In Exercises 55–66, classify the graph of the equation as a circle, a parabola, an ellipse, or a hyperbola.

55. $9x^2 + 4y^2 - 18x + 16y - 119 = 0$

56. $x^2 + y^2 - 4x - 6y - 23 = 0$

57. $4x^2 - y^2 - 4x - 3 = 0$

58. $y^2 - 6y - 4x + 21 = 0$

59. $y^2 - 4x^2 + 4x - 2y - 4 = 0$

60. $y^2 + 12x + 4y + 28 = 0$

61. $4x^2 + 25y^2 + 16x + 250y + 541 = 0$

62. $4y^2 - 2x^2 - 4y - 8x - 15 = 0$

63. $25x^2 - 10x - 200y - 119 = 0$

64. $4y^2 + 4x^2 - 24x + 35 = 0$

65. $100x^2 + 100y^2 - 100x + 400y + 409 = 0$

66. $9x^2 + 4y^2 - 90x + 8y + 228 = 0$

Exploration

True or False? In Exercises 67–69, determine whether the statement is true or false. Justify your answer.

67. In the standard form of the equation of a hyperbola, the larger the ratio of b to a, the larger the eccentricity of the hyperbola.

68. If the asymptotes of the hyperbola

$$\frac{x^2}{a^2} - \frac{y^2}{b^2} = 1$$

where $a, b > 0$, intersect at right angles, then $a = b$.

69. The graph of

$$x^2 - y^2 + 4x - 4y = 0$$

is a hyperbola.

70. Think About It Write an equation whose graph is the bottom half of the hyperbola

$$9x^2 - 54x - 4y^2 + 8y + 41 = 0.$$

71. Writing Explain how to use a rectangle to sketch the asymptotes of a hyperbola.

 72. **HOW DO YOU SEE IT?** Match each equation with its graph.

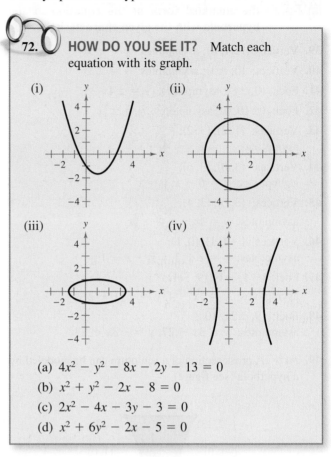

(i)

(ii)

(iii)

(iv)

(a) $4x^2 - y^2 - 8x - 2y - 13 = 0$

(b) $x^2 + y^2 - 2x - 8 = 0$

(c) $2x^2 - 4x - 3y - 3 = 0$

(d) $x^2 + 6y^2 - 2x - 5 = 0$

73. Error Analysis Describe the error in finding the asymptotes of the hyperbola

$$\frac{(y + 5)^2}{9} - \frac{(x - 3)^2}{4} = 1.$$

$$y = k \pm \frac{b}{a}(x - h)$$

$$= -5 \pm \tfrac{2}{3}(x - 3)$$

The asymptotes are
$y = \tfrac{2}{3}x - 7$ and $y = -\tfrac{2}{3}x - 3$.

74. Think About It Consider a hyperbola centered at the origin with a horizontal transverse axis. Use the definition of a hyperbola to derive its standard form.

75. Points of Intersection Sketch the circle $x^2 + y^2 = 4$. Then find the values of C so that the parabola $y = x^2 + C$ intersects the circle at the given number of points.

(a) 0 points

(b) 1 point

(c) 2 points

(d) 3 points

(e) 4 points

10.5 Rotation of Conics

- Rotate the coordinate axes to eliminate the *xy*-term in equations of conics.
- Use the discriminant to classify conics.

Rotation

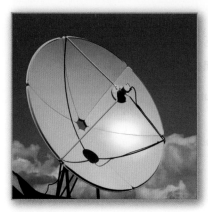

Rotated conics can model objects in real life. For example, in Exercise 63 on page 734, you will use a rotated parabola to model the cross section of a satellite dish.

In the preceding section, you classified conics whose equations were written in the general form

$$Ax^2 + Cy^2 + Dx + Ey + F = 0.$$

The graphs of such conics have axes that are parallel to one of the coordinate axes. Conics whose axes are rotated so that they are not parallel to either the *x*-axis or the *y*-axis have general equations that contain an *xy*-term.

$$Ax^2 + Bxy + Cy^2 + Dx + Ey + F = 0 \qquad \text{Equation in } xy\text{-plane}$$

To eliminate this *xy*-term, use a procedure called **rotation of axes.** The objective is to rotate the *x*- and *y*-axes until they are parallel to the axes of the conic. The rotated axes are denoted as the *x'*-axis and the *y'*-axis, as shown in Figure 10.31. After the rotation, the equation of the conic in the *x'y'*-plane will have the form

$$A'(x')^2 + C'(y')^2 + D'x' + E'y' + F' = 0. \qquad \text{Equation in } x'y'\text{-plane}$$

This equation has no *x'y'*-term, so you can obtain a standard form by completing the square. The theorem below identifies how much to rotate the axes to eliminate the *xy*-term and also the equations for determining the new coefficients *A'*, *C'*, *D'*, *E'*, and *F'*.

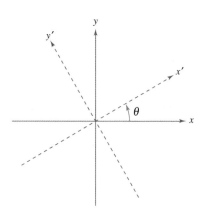

Figure 10.31

Rotation of Axes to Eliminate an *xy*-Term

The general second-degree equation

$$Ax^2 + Bxy + Cy^2 + Dx + Ey + F = 0$$

where $B \neq 0$, can be rewritten as

$$A'(x')^2 + C'(y')^2 + D'x' + E'y' + F' = 0$$

by rotating the coordinate axes through an angle θ, where

$$\cot 2\theta = \frac{A - C}{B}.$$

The coefficients of the new equation are obtained by making the substitutions

$$x = x' \cos \theta - y' \sin \theta$$

and

$$y = x' \sin \theta + y' \cos \theta.$$

Remember that the substitutions

$$x = x' \cos \theta - y' \sin \theta$$

and

$$y = x' \sin \theta + y' \cos \theta$$

should eliminate the *x'y'*-term in the rotated system. Use this as a check of your work. If you obtain an equation of a conic in the *x'y'*-plane that contains an *x'y'*-term, you know that you have made a mistake.

Figure 10.34

| EXAMPLE 3 | **Rotation of Axes for a Parabola** |

See LarsonPrecalculus.com for an interactive version of this type of example.

Rotate the axes to eliminate the xy-term in the equation

$$x^2 - 4xy + 4y^2 + 5\sqrt{5}y + 1 = 0.$$

Then write the equation in standard form and sketch its graph.

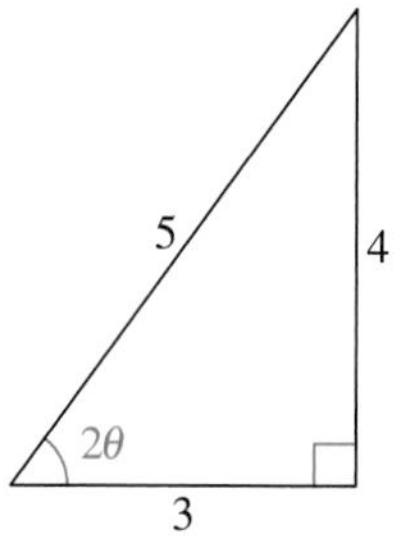

Solution   Because $A = 1$, $B = -4$, and $C = 4$, you have

$$\cot 2\theta = \frac{A - C}{B} = \frac{1 - 4}{-4} = \frac{3}{4}.$$

Use this information to draw a right triangle, as shown in Figure 10.34. From the figure, $\cos 2\theta = \frac{3}{5}$. To find the values of $\sin \theta$ and $\cos \theta$, use the half-angle formulas

$$\sin \theta = \sqrt{\frac{1 - \cos 2\theta}{2}} \quad \text{and} \quad \cos \theta = \sqrt{\frac{1 + \cos 2\theta}{2}}.$$

So,

$$\sin \theta = \sqrt{\frac{1 - \frac{3}{5}}{2}} = \sqrt{\frac{1}{5}} = \frac{1}{\sqrt{5}} \quad \text{and} \quad \cos \theta = \sqrt{\frac{1 + \frac{3}{5}}{2}} = \sqrt{\frac{4}{5}} = \frac{2}{\sqrt{5}}.$$

Consequently, use the substitutions

$$x = x' \cos \theta - y' \sin \theta = x'\left(\frac{2}{\sqrt{5}}\right) - y'\left(\frac{1}{\sqrt{5}}\right) = \frac{2x' - y'}{\sqrt{5}}$$

and

$$y = x' \sin \theta + y' \cos \theta = x'\left(\frac{1}{\sqrt{5}}\right) + y'\left(\frac{2}{\sqrt{5}}\right) = \frac{x' + 2y'}{\sqrt{5}}.$$

Substituting these expressions into the original equation, you have

$$x^2 - 4xy + 4y^2 + 5\sqrt{5}y + 1 = 0$$

$$\left(\frac{2x' - y'}{\sqrt{5}}\right)^2 - 4\left(\frac{2x' - y'}{\sqrt{5}}\right)\left(\frac{x' + 2y'}{\sqrt{5}}\right) + 4\left(\frac{x' + 2y'}{\sqrt{5}}\right)^2 + 5\sqrt{5}\left(\frac{x' + 2y'}{\sqrt{5}}\right) + 1 = 0$$

which simplifies to

$$5(y')^2 + 5x' + 10y' + 1 = 0$$

$$5[(y')^2 + 2y'] = -5x' - 1 \qquad \text{Group terms.}$$

$$5[(y')^2 + 2y' + 1] = -5x' - 1 + 5(1) \qquad \text{Complete the square.}$$

$$5(y' + 1)^2 = -5x' + 4 \qquad \text{Write in completed square form.}$$

$$(y' + 1)^2 = (-1)\left(x' - \frac{4}{5}\right). \qquad \text{Write in standard form.}$$

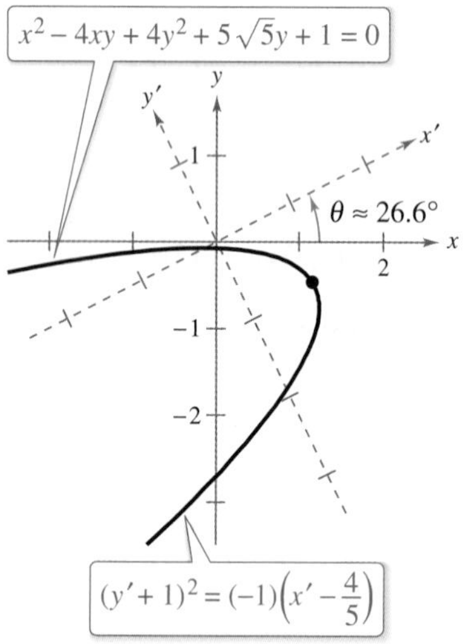

In the $x'y'$-system, this is the equation of a parabola with vertex $\left(\frac{4}{5}, -1\right)$. Its axis is parallel to the x'-axis in the $x'y'$-system, and $\theta = \sin^{-1}(1/\sqrt{5}) \approx 0.4636$ radian $\approx 26.6°$, as shown in Figure 10.35.

Vertex:

In $x'y'$-system: $\left(\frac{4}{5}, -1\right)$

In xy-system: $\left(\frac{13}{5\sqrt{5}}, -\frac{6}{5\sqrt{5}}\right)$

Figure 10.35

✓ *Checkpoint* ◀))) *Audio-video solution in English & Spanish at LarsonPrecalculus.com*

Rotate the axes to eliminate the xy-term in the equation

$$4x^2 + 4xy + y^2 - 2\sqrt{5}x + 4\sqrt{5}y - 30 = 0.$$

Then write the equation in standard form and sketch its graph.

Invariants Under Rotation

In the rotation of axes theorem stated at the beginning of this section, the constant term is the same in both equations, that is, $F' = F$. Such quantities are **invariant under rotation.** The next theorem lists this and other rotation invariants.

Rotation Invariants

The rotation of the coordinate axes through an angle θ that transforms the equation $Ax^2 + Bxy + Cy^2 + Dx + Ey + F = 0$ into the form

$$A'(x')^2 + C'(y')^2 + D'x' + E'y' + F' = 0$$

has the rotation invariants listed below.

1. $F = F'$

2. $A + C = A' + C'$

3. $B^2 - 4AC = (B')^2 - 4A'C'$

You can use the results of this theorem to classify the graph of a second-degree equation *with* an xy-term in much the same way you do for a second-degree equation *without* an xy-term. Note that $B' = 0$, so the invariant $B^2 - 4AC$ reduces to

$$B^2 - 4AC = -4A'C'. \qquad \text{Discriminant}$$

This quantity is the **discriminant** of the equation

$$Ax^2 + Bxy + Cy^2 + Dx + Ey + F = 0.$$

Now, from the classification procedure given in Section 10.4, you know that the value of $A'C'$ determines the type of graph for the equation

$$A'(x')^2 + C'(y')^2 + D'x' + E'y' + F' = 0.$$

Consequently, the value of $B^2 - 4AC$ will determine the type of graph for the original equation, as given in the classification below.

• • **REMARK** When there is an xy-term in the equation of a conic, you should realize that the conic is rotated. Before rotating the axes, you should use the discriminant to classify the conic. Use this classification to verify the types of graphs in Examples 1–3.

Classification of Conics by the Discriminant

The graph of the equation $Ax^2 + Bxy + Cy^2 + Dx + Ey + F = 0$ is, except in degenerate cases, determined by its discriminant as follows.

1. *Ellipse or circle:* $B^2 - 4AC < 0$

2. *Parabola:* $\quad B^2 - 4AC = 0$

3. *Hyperbola:* $\quad B^2 - 4AC > 0$

For example, in the general equation

$$3x^2 + 7xy + 5y^2 - 6x - 7y + 15 = 0$$

you have $A = 3$, $B = 7$, and $C = 5$. So, the discriminant is

$$B^2 - 4AC = 7^2 - 4(3)(5)$$

$$= 49 - 60$$

$$= -11.$$

The graph of the equation is an ellipse or a circle because $-11 < 0$.

Rotation and Graphing Utilities In Exercises 37–44, (a) use the discriminant to classify the graph of the equation, (b) use the Quadratic Formula to solve for y, and (c) use a graphing utility to graph the equation.

37. $16x^2 - 8xy + y^2 - 10x + 5y = 0$

38. $x^2 - 4xy - 2y^2 - 6 = 0$

39. $12x^2 - 6xy + 7y^2 - 45 = 0$

40. $2x^2 + 4xy + 5y^2 + 3x - 4y - 20 = 0$

41. $x^2 - 6xy - 5y^2 + 4x - 22 = 0$

42. $36x^2 - 60xy + 25y^2 + 9y = 0$

43. $x^2 + 4xy + 4y^2 - 5x - y - 3 = 0$

44. $x^2 + xy + 4y^2 + x + y - 4 = 0$

Sketching the Graph of a Degenerate Conic In Exercises 45–54, sketch the graph of the degenerate conic.

45. $y^2 - 16x^2 = 0$ 46. $y^2 - 25x^2 = 0$

47. $15x^2 - 2xy - y^2 = 0$

48. $32x^2 - 4xy - y^2 = 0$

49. $x^2 - 2xy + y^2 = 0$

50. $x^2 + 4xy + 4y^2 = 0$

51. $x^2 + y^2 + 2x - 4y + 5 = 0$

52. $x^2 + y^2 - 2x + 6y + 10 = 0$

53. $x^2 + 2xy + y^2 - 1 = 0$

54. $4x^2 + 4xy + y^2 - 1 = 0$

Finding Points of Intersection In Exercises 55–62, find any points of intersection of the graphs of the equations algebraically and then verify using a graphing utility.

55. $x^2 - 4y^2 - 20x - 64y - 172 = 0$
         $16x^2 + 4y^2 - 320x + 64y + 1600 = 0$

56. $x^2 - y^2 - 12x + 16y - 64 = 0$
    $x^2 + y^2 - 12x - 16y + 64 = 0$

57. $x^2 + 4y^2 - 2x - 8y + 1 = 0$
    $-x^2 + 2x - 4y - 1 = 0$

58. $-16x^2 - y^2 + 24y - 80 = 0$
    $16x^2 + 25y^2 - 400 = 0$

59. $x^2 + y^2 - 4 = 0$
    $3x - y^2 = 0$

60. $4x^2 + 9y^2 - 36y = 0$
    $x^2 + 9y - 27 = 0$

61. $-x^2 - y^2 - 8x + 20y - 7 = 0$
    $x^2 + 9y^2 + 8x + 4y + 7 = 0$

62. $x^2 + 2y^2 - 4x + 6y - 5 = 0$
    $x^2 - 4x - y + 4 = 0$

63. Satellite Dish

The parabolic cross section of a satellite dish is modeled by a portion of the graph of the equation

$$x^2 - 2xy - 27\sqrt{2}x + y^2 + 9\sqrt{2}y + 378 = 0$$

where all measurements are in feet.

(a) Rotate the axes to eliminate the xy-term in the equation. Then write the equation in standard form.

(b) A receiver is located at the focus of the cross section. Find the distance from the vertex of the cross section to the receiver.

64. HOW DO YOU SEE IT? Match each graph with the discriminant of its corresponding equation.

(a) -7

(b) 0

(c) 1

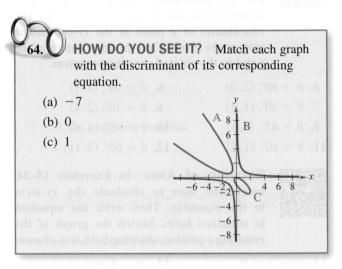

Exploration

True or False? In Exercises 65 and 66, determine whether the statement is true or false. Justify your answer.

65. The graph of the equation

    $x^2 + xy + ky^2 + 6x + 10 = 0$

    where k is any constant less than $\frac{1}{4}$, is a hyperbola.

66. After a rotation of axes is used to eliminate the xy-term from an equation of the form

    $Ax^2 + Bxy + Cy^2 + Dx + Ey + F = 0$

    the coefficients of the x^2- and y^2-terms remain A and C, respectively.

67. **Rotating a Circle** Show that the equation

    $x^2 + y^2 = r^2$

    is invariant under rotation of axes.

68. **Finding Lengths of Axes** Find the lengths of the major and minor axes of the ellipse in Exercise 19.

10.6 Parametric Equations

- Evaluate sets of parametric equations for given values of the parameter.
- Sketch curves represented by sets of parametric equations.
- Rewrite sets of parametric equations as single rectangular equations by eliminating the parameter.
- Find sets of parametric equations for graphs.

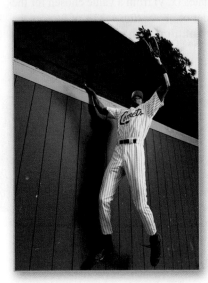

One application of parametric equations is modeling the path of an object. For example, in Exercise 93 on page 743, you will write a set of parametric equations that models the path of a baseball.

Plane Curves

Up to this point, you have been representing a graph by a single equation involving *two* variables such as x and y. In this section, you will study situations in which it is useful to introduce a *third* variable to represent a curve in the plane.

To see the usefulness of this procedure, consider the path of an object propelled into the air at an angle of 45°. When the initial velocity of the object is 48 feet per second, it can be shown that the object follows the parabolic path

$$y = -\frac{x^2}{72} + x. \qquad \text{Rectangular equation}$$

However, this equation does not tell the whole story. Although it does tell you *where* the object has been, it does not tell you *when* the object was at a given point (x, y) on the path. To determine this time, you can introduce a third variable t, called a **parameter**. It is possible to write both x and y as functions of t to obtain the **parametric equations**

$$x = 24\sqrt{2}t \qquad \text{Parametric equation for } x$$

$$y = -16t^2 + 24\sqrt{2}t. \qquad \text{Parametric equation for } y$$

This set of equations shows that at time $t = 0$, the object is at the point $(0, 0)$. Similarly, at time $t = 1$, the object is at the point $\left(24\sqrt{2}, 24\sqrt{2} - 16\right)$, and so on, as shown in the figure below.

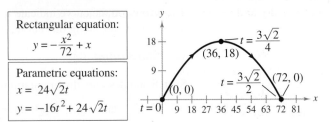

Curvilinear Motion: Two Variables for Position, One Variable for Time

For this motion problem, x and y are continuous functions of t, and the resulting path is a **plane curve**. (Recall that a *continuous function* is one whose graph has no breaks, holes, or gaps.)

> ### Definition of Plane Curve
>
> If f and g are continuous functions of t on an interval I, then the set of ordered pairs $(f(t), g(t))$ is a **plane curve** C. The equations
>
> $$x = f(t) \quad \text{and} \quad y = g(t)$$
>
> are **parametric equations** for C, and t is the **parameter**.

Sketching a Plane Curve

One way to sketch a curve represented by a pair of parametric equations is to plot points in the *xy*-plane. You determine each set of coordinates (x, y) from a value chosen for the parameter *t*. Plotting the resulting points in the order of *increasing* values of *t* traces the curve in a specific direction. This is called the **orientation** of the curve.

EXAMPLE 1   Sketching a Curve

See LarsonPrecalculus.com for an interactive version of this type of example.

Sketch and describe the orientation of the curve given by the parametric equations

$$x = t^2 - 4 \quad \text{and} \quad y = \frac{t}{2}, \quad -2 \le t \le 3.$$

Solution   Using values of *t* in the specified interval, the parametric equations yield the values of *x* and *y* shown in the table. By plotting the points (x, y) in the order of increasing values of *t*, you obtain the curve shown in the figure.

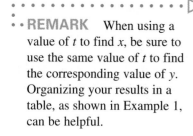

.. **REMARK**   When using a value of *t* to find *x*, be sure to use the same value of *t* to find the corresponding value of *y*. Organizing your results in a table, as shown in Example 1, can be helpful.

t	x	y
−2	0	−1
−1	−3	$-\frac{1}{2}$
0	−4	0
1	−3	$\frac{1}{2}$
2	0	1
3	5	$\frac{3}{2}$

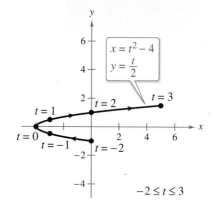

The arrows on the curve indicate its orientation as *t* increases from −2 to 3. So, when a particle moves along this curve, it starts at $(0, -1)$ and ends at $\left(5, \frac{3}{2}\right)$.

✓ *Checkpoint* 🔊))) *Audio-video solution in English & Spanish at LarsonPrecalculus.com*

Sketch and describe the orientation of the curve given by the parametric equations

$$x = 2t \quad \text{and} \quad y = 4t^2 + 2, \quad -2 \le t \le 2.$$

Note that the graph in Example 1 does not define *y* as a function of *x*. This points out one benefit of parametric equations—they can represent graphs that are not necessarily graphs of functions.

Two different sets of parametric equations can have the same graph. For example, the set of parametric equations

$$x = 4t^2 - 4 \quad \text{and} \quad y = t, \quad -1 \le t \le \frac{3}{2}$$

has the same graph as the set of parametric equations given in Example 1 (see Figure 10.39). However, comparing the values of *t* in the two graphs shows that the second graph is traced out more *rapidly* (considering *t* as time) than the first graph. So, in applications, different parametric representations can represent various *speeds* at which objects travel along a given path.

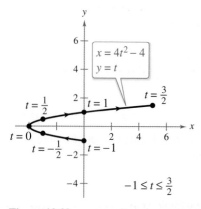

Figure 10.39

Eliminating the Parameter

Sketching a curve represented by a pair of parametric equations can sometimes be simplified by finding a rectangular equation (in x and y) that has the same graph. This process is called **eliminating the parameter,** and is illustrated below using the parametric equations from Example 1.

| Parametric equations | ⇨ | Solve for t in one equation. | ⇨ | Substitute into other equation. | ⇨ | Rectangular equation |

$$x = t^2 - 4 \qquad\qquad t = 2y \qquad\qquad x = (2y)^2 - 4 \qquad x = 4y^2 - 4$$
$$y = \frac{t}{2}$$

The equation $x = 4y^2 - 4$ represents a parabola with a horizontal axis and vertex at $(-4, 0)$. You graphed a portion of this parabola in Example 1.

When converting equations from parametric to rectangular form, you may need to alter the domain of the rectangular equation so that its graph matches the graph of the parametric equations. Example 2 demonstrates such a situation.

EXAMPLE 2   **Eliminating the Parameter**

Sketch the curve represented by the equations

$$x = \frac{1}{\sqrt{t+1}} \quad \text{and} \quad y = \frac{t}{t+1}$$

by eliminating the parameter and adjusting the domain of the resulting rectangular equation.

Solution   Solve for t in the equation for x.

$$x^2 = \frac{1}{t+1} \quad\Longrightarrow\quad t + 1 = \frac{1}{x^2} \quad\Longrightarrow\quad t = \frac{1}{x^2} - 1 = \frac{1 - x^2}{x^2}$$

Then substitute for t in the equation for y to obtain the rectangular equation

$$y = \frac{t}{t+1} = \frac{\dfrac{1 - x^2}{x^2}}{\dfrac{1 - x^2}{x^2} + 1} = \frac{\dfrac{1 - x^2}{x^2}}{\dfrac{1 - x^2}{x^2} + 1} \cdot \frac{x^2}{x^2} = \frac{1 - x^2}{1 - x^2 + x^2} = \frac{1 - x^2}{1} = 1 - x^2.$$

This rectangular equation shows that the curve is a parabola that opens downward and has its vertex at $(0, 1)$. Also, this rectangular equation is defined for all values of x. The parametric equation for x, however, is defined only when

$$\sqrt{t+1} > 0 \quad\Longrightarrow\quad t + 1 > 0 \quad\Longrightarrow\quad t > -1.$$

This implies that you should restrict the domain of x to positive values. Figure 10.40 shows a sketch of the curve.

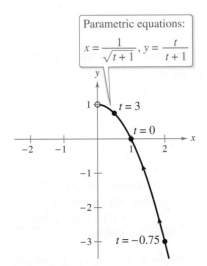

Parametric equations:
$$x = \frac{1}{\sqrt{t+1}}, y = \frac{t}{t+1}$$

$t = 3$

$t = 0$

$t = -0.75$

Figure 10.40

✓ *Checkpoint* ◀))) *Audio-video solution in English & Spanish at LarsonPrecalculus.com*

Sketch the curve represented by the equations

$$x = \frac{1}{\sqrt{t-1}} \quad \text{and} \quad y = \frac{t+1}{t-1}$$

by eliminating the parameter and adjusting the domain of the resulting rectangular equation.

It is not necessary for the parameter in a set of parametric equations to represent time. The next example uses an *angle* as the parameter.

•• **REMARK** To eliminate
the parameter in equations
involving trigonometric
functions, you may need to
use fundamental trigonometric
identities, as shown in
Example 3.
........................▷

EXAMPLE 3　Eliminating Angle Parameters

Sketch the curve represented by each set of equations by eliminating the parameter.

a. $x = 3\cos\theta$ and $y = 4\sin\theta$, $0 \le \theta < 2\pi$

b. $x = 1 + 3\sec\theta$ and $y = -3 + \tan\theta$, $\pi/2 < \theta < 3\pi/2$

Solution

a. Solve for $\cos\theta$ and $\sin\theta$ in the equations.

$$\cos\theta = \frac{x}{3} \quad \text{and} \quad \sin\theta = \frac{y}{4} \qquad \text{Solve for } \cos\theta \text{ and } \sin\theta.$$

Then use the identity $\sin^2\theta + \cos^2\theta = 1$ to form an equation involving only x and y.

$$\cos^2\theta + \sin^2\theta = 1 \qquad \text{Pythagorean identity}$$

$$\left(\frac{x}{3}\right)^2 + \left(\frac{y}{4}\right)^2 = 1 \qquad \text{Substitute } \frac{x}{3} \text{ for } \cos\theta \text{ and } \frac{y}{4} \text{ for } \sin\theta.$$

$$\frac{x^2}{9} + \frac{y^2}{16} = 1 \qquad \text{Rectangular equation}$$

The graph of this rectangular equation is an ellipse centered at $(0, 0)$, with vertices $(0, 4)$ and $(0, -4)$, and minor axis of length $2b = 6$, as shown in Figure 10.41. Note that the elliptic curve is traced out *counterclockwise* as θ increases on the interval $[0, 2\pi)$.

b. Solve for $\sec\theta$ and $\tan\theta$ in the equations.

$$\sec\theta = \frac{x-1}{3} \quad \text{and} \quad \tan\theta = y + 3 \qquad \text{Solve for } \sec\theta \text{ and } \tan\theta.$$

Then use the identity $\sec^2\theta - \tan^2\theta = 1$ to form an equation involving only x and y.

$$\sec^2\theta - \tan^2\theta = 1 \qquad \text{Pythagorean identity}$$

$$\left(\frac{x-1}{3}\right)^2 - (y+3)^2 = 1 \qquad \text{Substitute } \frac{x-1}{3} \text{ for } \sec\theta \text{ and } y + 3 \text{ for } \tan\theta.$$

$$\frac{(x-1)^2}{9} - \frac{(y+3)^2}{1} = 1 \qquad \text{Rectangular equation}$$

The graph of this rectangular equation is a hyperbola centered at $(1, -3)$ with a horizontal transverse axis of length $2a = 6$. However, the restriction on θ corresponds to a restriction on the domain of x to $x \le -2$, which corresponds to the *left* branch of the hyperbola only. Figure 10.42 shows the graph.

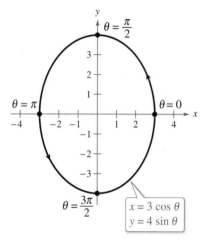

Figure 10.41

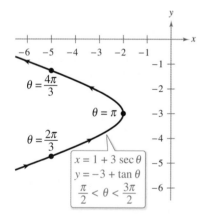

Figure 10.42

✓ *Checkpoint*  *Audio-video solution in English & Spanish at LarsonPrecalculus.com*

Sketch the curve represented by each set of equations by eliminating the parameter.

a. $x = 5\cos\theta$ and $y = 3\sin\theta$, $0 \le \theta < 2\pi$

b. $x = -1 + \tan\theta$ and $y = 2 + 2\sec\theta$, $\pi/2 < \theta < 3\pi/2$ ∎

•• **REMARK** It is important
to realize that eliminating the
parameter is primarily an aid to
curve sketching, as demonstrated
in Examples 2 and 3.
........................▷

When parametric equations represent the path of a moving object, the graph of the corresponding rectangular equation is not sufficient to describe the object's motion. You still need the parametric equations to tell you the *position, direction,* and *speed* at a given time.

Finding Parametric Equations for a Graph

You have been studying techniques for sketching the graph represented by a set of parametric equations. Now consider the *reverse* problem—that is, how can you find a set of parametric equations for a given graph or a given physical description? From the discussion after Example 1, you know that such a representation is not unique. That is, the equations

$$x = 4t^2 - 4 \quad \text{and} \quad y = t, \quad -1 \le t \le \frac{3}{2}$$

produced the same graph as the equations

$$x = t^2 - 4 \quad \text{and} \quad y = \frac{t}{2}, \quad -2 \le t \le 3.$$

Example 4 further demonstrates this.

EXAMPLE 4 Finding Parametric Equations for a Graph

Find a set of parametric equations to represent the graph of $y = 1 - x^2$, using each parameter.

a. $t = x$

b. $t = 1 - x$

Solution

a. Letting $t = x$, you obtain the parametric equations

$$x = t \quad \text{and} \quad y = 1 - x^2 = 1 - t^2.$$

Figure 10.43 shows the curve represented by the parametric equations.

b. Letting $t = 1 - x$, you obtain the parametric equations

$$x = 1 - t \quad \text{and} \quad y = 1 - x^2 = 1 - (1 - t)^2 = 2t - t^2.$$

Figure 10.44 shows the curve represented by the parametric equations. Note that the graphs in Figures 10.43 and 10.44 have opposite orientations.

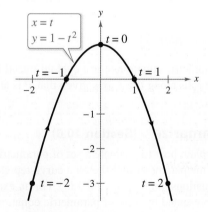

Figure 10.43

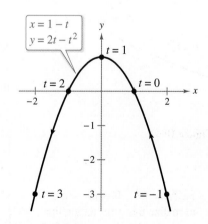

Figure 10.44

✓ **Checkpoint** 🔊))) *Audio-video solution in English & Spanish at LarsonPrecalculus.com*

Find a set of parametric equations to represent the graph of $y = x^2 + 2$, using each parameter.

a. $t = x$ **b.** $t = 2 - x$

A **cycloid** is a curve traced by a point P on a circle as the circle rolls along a straight line in a plane.

EXAMPLE 5   Parametric Equations for a Cycloid

Write parametric equations for a cycloid traced by a point P on a circle of radius a units as the circle rolls along the x-axis given that P is at a minimum when $x = 0$.

Solution   Let the parameter θ be the measure of the circle's rotation, and let the point $P(x, y)$ begin at the origin. When $\theta = 0$, P is at the origin; when $\theta = \pi$, P is at a maximum point $(\pi a, 2a)$; and when $\theta = 2\pi$, P is back on the x-axis at $(2\pi a, 0)$. From the figure below, $\angle APC = \pi - \theta$. So, you have

$$\sin \theta = \sin(\pi - \theta) = \sin(\angle APC) = \frac{AC}{a} = \frac{BD}{a}$$

$$\cos \theta = -\cos(\pi - \theta) = -\cos(\angle APC) = -\frac{AP}{a}$$

•• **REMARK**   In Example 5,
$\overset{\frown}{PD}$ represents the arc of the
circle between points P and D.
.... ▷

which implies that $BD = a \sin \theta$ and $AP = -a \cos \theta$. The circle rolls along the x-axis, so you know that $OD = \overset{\frown}{PD} = a\theta$. Furthermore, $BA = DC = a$, so you have

$$x = OD - BD = a\theta - a \sin \theta$$

and

$$y = BA + AP = a - a \cos \theta.$$

The parametric equations are $x = a(\theta - \sin \theta)$ and $y = a(1 - \cos \theta)$.

▷ **TECHNOLOGY**   Use a
graphing utility in *parametric*
mode to obtain a graph similar
to the one in Example 5 by
graphing

$$X_{1T} = T - \sin T$$

and

$$Y_{1T} = 1 - \cos T.$$

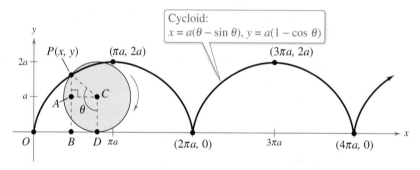

Cycloid:
$x = a(\theta - \sin \theta), y = a(1 - \cos \theta)$

✓ **Checkpoint** 🔊 *Audio-video solution in English & Spanish at LarsonPrecalculus.com*

Write parametric equations for a cycloid traced by a point P on a circle of radius a as the circle rolls along the x-axis given that P is at a maximum when $x = 0$. ◼

Summarize   (Section 10.6)

1. Explain how to evaluate a set of parametric equations for given values of the parameter and sketch a curve represented by a set of parametric equations (*pages 735 and 736*). For an example of sketching a curve represented by a set of parametric equations, see Example 1.

2. Explain how to rewrite a set of parametric equations as a single rectangular equation by eliminating the parameter (*page 737*). For examples of sketching curves by eliminating the parameter, see Examples 2 and 3.

3. Explain how to find a set of parametric equations for a graph (*page 739*). For examples of finding sets of parametric equations for graphs, see Examples 4 and 5.

10.6 Exercises

See CalcChat.com for tutorial help and worked-out solutions to odd-numbered exercises.

Vocabulary: Fill in the blanks.

1. If f and g are continuous functions of t on an interval I, then the set of ordered pairs $(f(t), g(t))$ is a _____ _____ C.

2. The _____ of a curve is the direction in which the curve is traced for increasing values of the parameter.

3. The process of converting a set of parametric equations to a corresponding rectangular equation is called _____ the _____.

4. A curve traced by a point on the circumference of a circle as the circle rolls along a straight line in a plane is a _____.

Skills and Applications

5. **Sketching a Curve** Consider the parametric equations $x = \sqrt{t}$ and $y = 3 - t$.
   (a) Create a table of x- and y-values using $t = 0, 1, 2, 3,$ and 4.
   (b) Plot the points (x, y) generated in part (a), and sketch a graph of the parametric equations.
   (c) Sketch the graph of $y = 3 - x^2$. How do the graphs differ?

6. **Sketching a Curve** Consider the parametric equations $x = 4 \cos^2 \theta$ and $y = 2 \sin \theta$.
   (a) Create a table of x- and y-values using $\theta = -\pi/2, -\pi/4, 0, \pi/4,$ and $\pi/2$.
   (b) Plot the points (x, y) generated in part (a), and sketch a graph of the parametric equations.
   (c) Sketch the graph of $x = -y^2 + 4$. How do the graphs differ?

 Sketching a Curve In Exercises 7–12, sketch and describe the orientation of the curve given by the parametric equations.

7. $x = t, \quad y = -5t$
8. $x = 2t - 1, \quad y = t + 4$
9. $x = t^2, \quad y = 3t$
10. $x = \sqrt{t}, \quad y = 2t - 1$
11. $x = 3 \cos \theta, \quad y = 2 \sin^2 \theta, \quad 0 \le \theta \le \pi$
12. $x = \cos \theta, \quad y = 2 \sin \theta, \quad 0 \le \theta \le 2\pi$

 Sketching a Curve In Exercises 13–38, (a) sketch the curve represented by the parametric equations (indicate the orientation of the curve) and (b) eliminate the parameter and write the resulting rectangular equation whose graph represents the curve. Adjust the domain of the rectangular equation, if necessary.

13. $x = t, \quad y = 4t$
14. $x = t, \quad y = -\frac{1}{2}t$
15. $x = -t + 1, \quad y = -3t$
16. $x = 3 - 2t, \quad y = 2 + 3t$
17. $x = \frac{1}{4}t, \quad y = t^2$
18. $x = t, \quad y = t^3$
19. $x = t^2, \quad y = -2t$
20. $x = -t^2, \quad y = \dfrac{t}{3}$
21. $x = \sqrt{t}, \quad y = 1 - t$
22. $x = \sqrt{t + 2}, \quad y = t - 1$
23. $x = \sqrt{t} - 3, \quad y = t^3$
24. $x = \sqrt{t - 1}, \quad y = \sqrt[3]{t - 1}$
25. $x = t + 1$
    $y = \dfrac{t}{t + 1}$
26. $x = t - 1$
    $y = \dfrac{t}{t - 1}$
27. $x = 4 \cos \theta$
    $y = 2 \sin \theta$
28. $x = 2 \cos \theta$
    $y = 3 \sin \theta$
29. $x = 1 + \cos \theta$
    $y = 1 + 2 \sin \theta$
30. $x = 2 + 5 \cos \theta$
    $y = -6 + 4 \sin \theta$
31. $x = 2 \sec \theta, \quad y = \tan \theta, \quad \pi/2 \le \theta \le 3\pi/2$
32. $x = 3 \cot \theta, \quad y = 4 \csc \theta, \quad 0 \le \theta \le \pi$
33. $x = 3 \cos \theta$
    $y = 3 \sin \theta$
34. $x = 6 \sin 2\theta$
    $y = 6 \cos 2\theta$
35. $x = e^t, \quad y = e^{3t}$
36. $x = e^{-t}, \quad y = e^{3t}$
37. $x = t^3, \quad y = 3 \ln t$
38. $x = \ln 2t, \quad y = 2t^2$

Graphing a Curve In Exercises 39–48, use a graphing utility to graph the curve represented by the parametric equations.

39. $x = t$
    $y = \sqrt{t}$
40. $x = t + 1$
    $y = \sqrt{2 - t}$
41. $x = 2t$
    $y = |t + 1|$
42. $x = |t + 2|$
    $y = 3 - t$
43. $x = 4 + 3 \cos \theta$
    $y = -2 + \sin \theta$
44. $x = 4 + 3 \cos \theta$
    $y = -2 + 2 \sin \theta$
45. $x = 2 \csc \theta$
    $y = 4 \cot \theta$
46. $x = \sec \theta$
    $y = \tan \theta$
47. $x = \frac{1}{2}t$
    $t = \ln(t^2 + 1)$
48. $x = 10 - 0.01e^t$
    $y = 0.4t^2$

Comparing Plane Curves In Exercises 49 and 50, determine how the plane curves differ from each other.

49. (a) $x = t$
        $y = 2t + 1$

    (b) $x = \cos \theta$
        $y = 2 \cos \theta + 1$

    (c) $x = e^{-t}$
        $y = 2e^{-t} + 1$

    (d) $x = e^t$
        $y = 2e^t + 1$

50. (a) $x = t$
        $y = t^2 - 1$

    (b) $x = t^2$
        $y = t^4 - 1$

    (c) $x = \sin t$
        $y = \sin^2 t - 1$

    (d) $x = e^t$
        $y = e^{2t} - 1$

 Eliminating the Parameter In Exercises 51–54, eliminate the parameter and obtain the standard form of the rectangular equation.

51. Line passing through (x_1, y_1) and (x_2, y_2):
    $x = x_1 + t(x_2 - x_1), \quad y = y_1 + t(y_2 - y_1)$

52. Circle: $x = h + r \cos \theta, \quad y = k + r \sin \theta$

53. Ellipse with horizontal major axis:
    $x = h + a \cos \theta, \quad y = k + b \sin \theta$

54. Hyperbola with horizontal transverse axis:
    $x = h + a \sec \theta, \quad y = k + b \tan \theta$

 Finding Parametric Equations for a Graph In Exercises 55–62, use the results of Exercises 51–54 to find a set of parametric equations to represent the graph of the line or conic.

55. Line: passes through $(0, 0)$ and $(3, 6)$

56. Line: passes through $(3, 2)$ and $(-6, 3)$

57. Circle: center: $(3, 2)$; radius: 4

58. Circle: center: $(-2, -5)$; radius: 7

59. Ellipse: vertices: $(\pm 5, 0)$; foci: $(\pm 4, 0)$

60. Ellipse: vertices: $(7, 3), (-1, 3)$; foci: $(5, 3), (1, 3)$

61. Hyperbola: vertices: $(1, 0), (9, 0)$; foci: $(0, 0), (10, 0)$

62. Hyperbola: vertices: $(4, 1), (8, 1)$; foci: $(2, 1), (10, 1)$

 Finding Parametric Equations for a Graph In Exercises 63–66, use the results of Exercises 51 and 54 to find a set of parametric equations to represent the section of the graph of the line or conic. (*Hint:* Adjust the domain of the standard form of the rectangular equation to determine the appropriate interval for the parameter.)

63. Line segment between $(0, 0)$ and $(-5, 2)$

64. Line segment between $(1, -4)$ and $(9, 0)$

65. Left branch of the hyperbola with vertices $(\pm 3, 0)$ and foci $(\pm 5, 0)$

66. Right branch of the hyperbola with vertices $(-4, 3)$ and $(6, 3)$ and foci $(-12, 3)$ and $(14, 3)$

 Finding Parametric Equations for a Graph In Exercises 67–78, find a set of parametric equations to represent the graph of the rectangular equation using (a) $t = x$ and (b) $t = 2 - x$.

67. $y = 3x - 2$

68. $y = 2 - x$

69. $x = 2y + 1$

70. $x = 3y - 2$

71. $y = x^2 + 1$

72. $y = 6x^2 - 5$

73. $y = 1 - 2x^2$

74. $y = 2 - 5x^2$

75. $y = \dfrac{1}{x}$

76. $y = \dfrac{1}{2x}$

77. $y = e^x$

78. $y = e^{2x}$

Graphing a Curve In Exercises 79–86, use a graphing utility to graph the curve represented by the parametric equations.

79. Cycloid: $x = 4(\theta - \sin \theta), \quad y = 4(1 - \cos \theta)$

80. Cycloid: $x = \theta + \sin \theta, \quad y = 1 - \cos \theta$

81. Prolate cycloid: $x = 2\theta - 4 \sin \theta, \quad y = 2 - 4 \cos \theta$

82. Epicycloid: $x = 8 \cos \theta - 2 \cos 4\theta$
    $y = 8 \sin \theta - 2 \sin 4\theta$

83. Hypocycloid: $x = 3 \cos^3 \theta, \quad y = 3 \sin^3 \theta$

84. Curtate cycloid: $x = 8\theta - 4 \sin \theta, \quad y = 8 - 4 \cos \theta$

85. Witch of Agnesi: $x = 2 \cot \theta, \quad y = 2 \sin^2 \theta$

86. Folium of Descartes: $x = \dfrac{3t}{1 + t^3}, \quad y = \dfrac{3t^2}{1 + t^3}$

Matching In Exercises 87–90, match the parametric equations with the correct graph and describe the domain and range. [The graphs are labeled (a)–(d).]

(a)
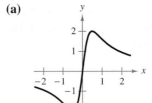

(b)
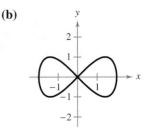

(c)
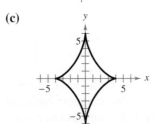

(d)
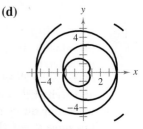

87. Lissajous curve: $x = 2 \cos \theta, \quad y = \sin 2\theta$

88. Evolute of ellipse: $x = 4 \cos^3 \theta, \quad y = 6 \sin^3 \theta$

89. Involute of circle: $x = \frac{1}{2}(\cos \theta + \theta \sin \theta)$
    $y = \frac{1}{2}(\sin \theta - \theta \cos \theta)$

90. Serpentine curve: $x = \frac{1}{2} \cot \theta, \quad y = 4 \sin \theta \cos \theta$

Projectile Motion Consider a projectile launched at a height of h feet above the ground at an angle of θ with the horizontal. The initial velocity is v_0 feet per second, and the path of the projectile is modeled by the parametric equations

$x = (v_0 \cos \theta)t$

and

$y = h + (v_0 \sin \theta)t - 16t^2.$

In Exercises 91 and 92, use a graphing utility to graph the paths of a projectile launched from ground level at each value of θ and v_0. For each case, use the graph to approximate the maximum height and the range of the projectile.

91. (a) $\theta = 60°$, $v_0 = 88$ feet per second
    (b) $\theta = 60°$, $v_0 = 132$ feet per second
    (c) $\theta = 45°$, $v_0 = 88$ feet per second
    (d) $\theta = 45°$, $v_0 = 132$ feet per second

92. (a) $\theta = 15°$, $v_0 = 50$ feet per second
    (b) $\theta = 15°$, $v_0 = 120$ feet per second
    (c) $\theta = 10°$, $v_0 = 50$ feet per second
    (d) $\theta = 10°$, $v_0 = 120$ feet per second

93. **Path of a Baseball**

The center field fence in a baseball stadium is 7 feet high and 408 feet from home plate. A baseball player hits a baseball at a point 3 feet above the ground. The ball leaves the bat at an angle of θ degrees with the horizontal at a speed of 100 miles per hour (see figure).

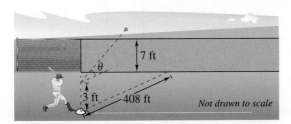

7 ft

θ

3 ft   408 ft   *Not drawn to scale*

(a) Write a set of parametric equations that model the path of the baseball. (See Exercises 91 and 92.)

(b) Use a graphing utility to graph the path of the baseball when $\theta = 15°$. Is the hit a home run?

(c) Use the graphing utility to graph the path of the baseball when $\theta = 23°$. Is the hit a home run?

(d) Find the minimum angle required for the hit to be a home run.

94. **Path of an Arrow** An archer releases an arrow from a bow at a point 5 feet above the ground. The arrow leaves the bow at an angle of 15° with the horizontal and at an initial speed of 225 feet per second.

(a) Write a set of parametric equations that model the path of the arrow. (See Exercises 91 and 92.)

(b) Assuming the ground is level, find the distance the arrow travels before it hits the ground. (Ignore air resistance.)

(c) Use a graphing utility to graph the path of the arrow and approximate its maximum height.

(d) Find the total time the arrow is in the air.

95. **Path of a Football** A quarterback releases a pass at a height of 7 feet above the playing field, and a receiver catches the football at a height of 4 feet, 30 yards directly downfield. The pass is released at an angle of 35° with the horizontal.

(a) Write a set of parametric equations for the path of the football. (See Exercises 91 and 92.)

(b) Find the speed of the football when it is released.

(c) Use a graphing utility to graph the path of the football and approximate its maximum height.

(d) Find the time the receiver has to position himself after the quarterback releases the football.

96. **Projectile Motion** Eliminate the parameter t in the parametric equations

$x = (v_0 \cos \theta)t$

and

$y = h + (v_0 \sin \theta)t - 16t^2$

for the motion of a projectile to show that the rectangular equation is

$$y = -\frac{16 \sec^2 \theta}{v_0^2}x^2 + (\tan \theta)x + h.$$

97. **Path of a Projectile** The path of a projectile is given by the rectangular equation

$y = 7 + x - 0.02x^2.$

(a) Find the values of h, v_0, and θ. Then write a set of parametric equations that model the path. (See Exercise 96.)

(b) Use a graphing utility to graph the rectangular equation for the path of the projectile. Confirm your answer in part (a) by sketching the curve represented by the parametric equations.

(c) Use the graphing utility to approximate the maximum height of the projectile and its range.

98. **Path of a Projectile** Repeat Exercise 97 for a projectile with a path given by the rectangular equation

$y = 6 + x - 0.08x^2.$

99. Curtate Cycloid A wheel of radius a units rolls along a straight line without slipping. The curve traced by a point P that is b units from the center $(b < a)$ is called a **curtate cycloid** (see figure). Use the angle θ shown in the figure to find a set of parametric equations for the curve.

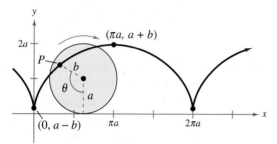

100. Epicycloid A circle of radius one unit rolls around the outside of a circle of radius two units without slipping. The curve traced by a point on the circumference of the smaller circle is called an **epicycloid** (see figure). Use the angle θ shown in the figure to find a set of parametric equations for the curve.

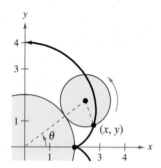

Exploration

True or False? In Exercises 101–104, determine whether the statement is true or false. Justify your answer.

101. The two sets of parametric equations

$$x = t, \quad y = t^2 + 1 \quad \text{and} \quad x = 3t, \quad y = 9t^2 + 1$$

correspond to the same rectangular equation.

102. The graphs of the parametric equations

$$x = t^2, \quad y = t^2 \quad \text{and} \quad x = t, \quad y = t$$

both represent the line $y = x$, so they are the same plane curve.

103. If y is a function of t and x is a function of t, then y must be a function of x.

104. The parametric equations

$$x = at + h \quad \text{and} \quad y = bt + k$$

where $a \neq 0$ and $b \neq 0$, represent a circle centered at (h, k) when $a = b$.

105. Writing Write a short paragraph explaining why parametric equations are useful.

106. Writing Explain what is meant by the orientation of a plane curve.

107. Error Analysis Describe the error in finding the rectangular equation for the parametric equations

$$x = \sqrt{t - 1} \quad \text{and} \quad y = 2t.$$

$$x = \sqrt{t - 1} \quad \Longrightarrow \quad t = x^2 + 1$$

$$y = 2(x^2 + 1) = 2x^2 + 2$$

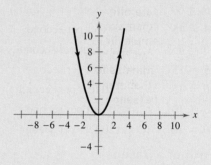

108. **HOW DO YOU SEE IT?** The graph of the parametric equations $x = t$ and $y = t^2$ is shown below. Determine whether the graph would change for each set of parametric equations. If so, how would it change?

(a) $x = -t, \; y = t^2$

(b) $x = t + 1, \; y = t^2$

(c) $x = t, \; y = t^2 + 1$

109. Think About It The graph of the parametric equations $x = t^3$ and $y = t - 1$ is shown below. Would the graph change for the parametric equations $x = (-t)^3$ and $y = -t - 1$? If so, how would it change?

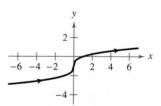

110. Think About It The graph of the parametric equations $x = t^2$ and $y = t + 1$ is shown below. Would the graph change for the parametric equations $x = (t + 1)^2$ and $y = t + 2$? If so, how would it change?

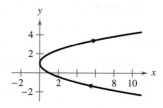

10.7   Polar Coordinates

Polar coordinates are often useful tools in mathematical modeling. For example, in Exercise 109 on page 750, you will use polar coordinates to write an equation that models the position of a passenger car on a Ferris wheel.

- Plot points in the polar coordinate system.
- Convert points from rectangular to polar form and vice versa.
- Convert equations from rectangular to polar form and vice versa.

Introduction

So far, you have been representing graphs of equations as collections of points (x, y) in the rectangular coordinate system, where x and y represent the directed distances from the coordinate axes to the point (x, y). In this section, you will study a different system called the **polar coordinate system.**

To form the polar coordinate system in the plane, fix a point O, called the **pole** (or **origin**), and construct from O an initial ray called the **polar axis,** as shown in the figure at the right. Then each point P in the plane can be assigned **polar coordinates** (r, θ), where r and θ are defined below.

1. $r = $ *directed distance* from O to P

2. $\theta = $ *directed angle,* counterclockwise from the polar axis to segment $\overline{OP}$

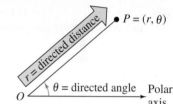

EXAMPLE 1   Plotting Points in the Polar Coordinate System

Plot each point given in polar coordinates.

a. $(2, \pi/3)$ **b.** $(3, -\pi/6)$ **c.** $(3, 11\pi/6)$

Solution

a. The point $(r, \theta) = (2, \pi/3)$ lies two units from the pole on the terminal side of the angle $\theta = \pi/3$, as shown in Figure 10.45.

b. The point $(r, \theta) = (3, -\pi/6)$ lies three units from the pole on the terminal side of the angle $\theta = -\pi/6$, as shown in Figure 10.46.

c. The point $(r, \theta) = (3, 11\pi/6)$ coincides with the point $(3, -\pi/6)$, as shown in Figure 10.47.

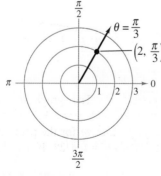

Figure 10.45

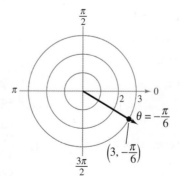

Figure 10.46

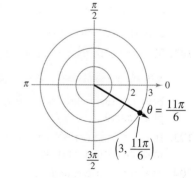

Figure 10.47

✓ **Checkpoint** *Audio-video solution in English & Spanish at LarsonPrecalculus.com*

Plot each point given in polar coordinates.

a. $(3, \pi/4)$ **b.** $(2, -\pi/3)$ **c.** $(2, 5\pi/3)$

In rectangular coordinates, each point (x, y) has a unique representation. This is not true for polar coordinates. For example, the coordinates

$$(r, \theta) \quad \text{and} \quad (r, \theta + 2\pi)$$

represent the same point, as illustrated in Example 1. Another way to obtain multiple representations of a point is to use negative values for r. Because r is a *directed distance,* the coordinates

$$(r, \theta) \quad \text{and} \quad (-r, \theta + \pi)$$

represent the same point. In general, the point (r, θ) can be represented by

$$(r, \theta) = (r, \theta \pm 2n\pi) \quad \text{or} \quad (r, \theta) = (-r, \theta \pm (2n + 1)\pi)$$

where n is any integer. Moreover, the pole is represented by $(0, \theta)$, where θ is any angle.

EXAMPLE 2   **Multiple Representations of Points**

Plot the point

$$\left(3, -\frac{3\pi}{4}\right)$$

and find three additional polar representations of this point, using

$$-2\pi < \theta < 2\pi.$$

Solution   The point is shown below. Three other representations are

$$\left(3, -\frac{3\pi}{4} + 2\pi\right) = \left(3, \frac{5\pi}{4}\right), \qquad \text{Add } 2\pi \text{ to } \theta.$$

$$\left(-3, -\frac{3\pi}{4} - \pi\right) = \left(-3, -\frac{7\pi}{4}\right), \qquad \text{Replace } r \text{ with } -r \text{ and subtract } \pi \text{ from } \theta.$$

and

$$\left(-3, -\frac{3\pi}{4} + \pi\right) = \left(-3, \frac{\pi}{4}\right). \qquad \text{Replace } r \text{ with } -r \text{ and add } \pi \text{ to } \theta.$$

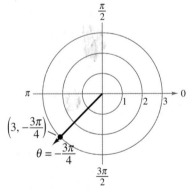

$$\left(3, -\tfrac{3\pi}{4}\right) = \left(3, \tfrac{5\pi}{4}\right) = \left(-3, -\tfrac{7\pi}{4}\right) = \left(-3, \tfrac{\pi}{4}\right)$$

✓ *Checkpoint* 🔊 *Audio-video solution in English & Spanish at LarsonPrecalculus.com*

Plot the point

$$\left(-1, \frac{3\pi}{4}\right)$$

and find three additional polar representations of this point, using $-2\pi < \theta < 2\pi$. ∎

Coordinate Conversion

To establish the relationship between polar and rectangular coordinates, let the polar axis coincide with the positive x-axis and the pole with the origin, as shown in Figure 10.48. Because (x, y) lies on a circle of radius r, it follows that $r^2 = x^2 + y^2$. Moreover, for $r > 0$, the definitions of the trigonometric functions imply that

$$\tan \theta = \frac{y}{x}, \quad \cos \theta = \frac{x}{r}, \quad \text{and} \quad \sin \theta = \frac{y}{r}.$$

Show that the same relationships hold for $r < 0$.

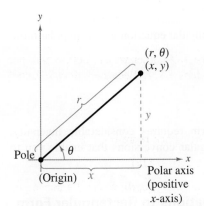

Figure 10.48

Coordinate Conversion

The polar coordinates (r, θ) and the rectangular coordinates (x, y) are related as follows.

Polar-to-Rectangular	**Rectangular-to-Polar**
$x = r \cos \theta$	$\tan \theta = \dfrac{y}{x}$
$y = r \sin \theta$	$r^2 = x^2 + y^2$

EXAMPLE 3 **Polar-to-Rectangular Conversion**

Convert $\left(\sqrt{3}, \dfrac{\pi}{6}\right)$ to rectangular coordinates.

Solution Substitute $r = \sqrt{3}$ and $\theta = \pi/6$ to find the x- and y-coordinates.

$$x = r \cos \theta = \sqrt{3} \cos \frac{\pi}{6} = \sqrt{3}\left(\frac{\sqrt{3}}{2}\right) = \frac{3}{2}$$

$$y = r \sin \theta = \sqrt{3} \sin \frac{\pi}{6} = \sqrt{3}\left(\frac{1}{2}\right) = \frac{\sqrt{3}}{2}$$

The rectangular coordinates are $(x, y) = \left(\dfrac{3}{2}, \dfrac{\sqrt{3}}{2}\right)$. (See Figure 10.49.)

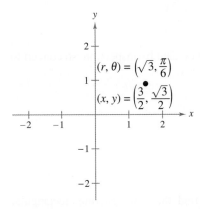

Figure 10.49

✓ **Checkpoint** ◀))) *Audio-video solution in English & Spanish at LarsonPrecalculus.com*

Convert $(2, \pi)$ to rectangular coordinates.

EXAMPLE 4 **Rectangular-to-Polar Conversion**

Convert $(-1, 1)$ to polar coordinates.

Solution The point $(x, y) = (-1, 1)$ lies in the second quadrant.

$$\tan \theta = \frac{y}{x} = \frac{1}{-1} = -1 \implies \theta = \pi + \arctan(-1) = \frac{3\pi}{4}$$

The angle θ lies in the same quadrant as (x, y), so use the positive value for r.

$$r = \sqrt{x^2 + y^2} = \sqrt{(-1)^2 + (1)^2} = \sqrt{2}$$

So, *one* set of polar coordinates is $(r, \theta) = \left(\sqrt{2}, 3\pi/4\right)$, as shown in Figure 10.50.

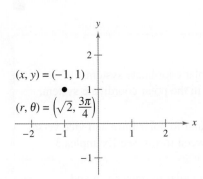

Figure 10.50

✓ **Checkpoint** ◀))) *Audio-video solution in English & Spanish at LarsonPrecalculus.com*

Convert $(0, 2)$ to polar coordinates.

Equation Conversion

To convert a rectangular equation to polar form, replace x with $r \cos \theta$ and y with $r \sin \theta$. For example, here is how to write the rectangular equation $y = x^2$ in polar form.

$$y = x^2 \qquad \text{Rectangular equation}$$

$$r \sin \theta = (r \cos \theta)^2 \qquad \text{Polar equation}$$

$$r = \sec \theta \tan \theta \qquad \text{Solve for } r.$$

Converting a polar equation to rectangular form requires considerable ingenuity. Example 5 demonstrates several polar-to-rectangular conversions that enable you to sketch the graphs of some polar equations.

EXAMPLE 5   Converting Polar Equations to Rectangular Form

See LarsonPrecalculus.com for an interactive version of this type of example.

a. The graph of the polar equation $r = 2$ consists of all points that are two units from the pole. In other words, this graph is a circle centered at the origin with a radius of 2, as shown in Figure 10.51. Confirm this by converting to rectangular form, using the relationship $r^2 = x^2 + y^2$.

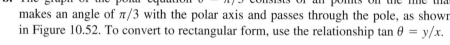

$$\underbrace{r = 2}_{\text{Polar equation}} \implies r^2 = 2^2 \implies \underbrace{x^2 + y^2 = 2^2}_{\text{Rectangular equation}}$$

b. The graph of the polar equation $\theta = \pi/3$ consists of all points on the line that makes an angle of $\pi/3$ with the polar axis and passes through the pole, as shown in Figure 10.52. To convert to rectangular form, use the relationship $\tan \theta = y/x$.

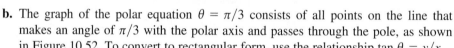

$$\underbrace{\theta = \pi/3}_{\text{Polar equation}} \implies \tan \theta = \sqrt{3} \implies \underbrace{y = \sqrt{3}x}_{\text{Rectangular equation}}$$

c. The graph of the polar equation $r = \sec \theta$ is not evident by inspection, so convert to rectangular form using the relationship $r \cos \theta = x$.

$$\underbrace{r = \sec \theta}_{\text{Polar equation}} \implies r \cos \theta = 1 \implies \underbrace{x = 1}_{\text{Rectangular equation}}$$

The graph is a vertical line, as shown in Figure 10.53.

✓ **Checkpoint** ◀))) Audio-video solution in English & Spanish at LarsonPrecalculus.com

Describe the graph of each polar equation and find the corresponding rectangular equation.

a. $r = 7$ **b.** $\theta = \pi/4$ **c.** $r = 6 \sin \theta$ ■

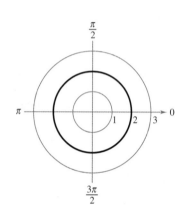

Figure 10.51

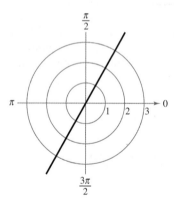

Figure 10.52

Figure 10.53

Summarize   (Section 10.7)

1. Explain how to plot the point (r, θ) in the polar coordinate system *(page 745)*. For examples of plotting points in the polar coordinate system, see Examples 1 and 2.

2. Explain how to convert points from rectangular to polar form and vice versa *(page 747)*. For examples of converting between forms, see Examples 3 and 4.

3. Explain how to convert equations from rectangular to polar form and vice versa *(page 748)*. For an example of converting polar equations to rectangular form, see Example 5.

10.7 Exercises

See **CalcChat.com** for tutorial help and worked-out solutions to odd-numbered exercises.

Vocabulary: Fill in the blanks.

1. The origin of the polar coordinate system is called the _____.

2. For the point (r, θ), r is the _____ _____ from O to P and θ is the _____ _____, counterclockwise from the polar axis to the line segment $\overline{OP}$.

3. To plot the point (r, θ), use the _____ coordinate system.

4. The polar coordinates (r, θ) and the rectangular coordinates (x, y) are related as follows:

$x =$ _____ $y =$ _____ $\tan \theta =$ _____ $r^2 =$ _____

Skills and Applications

 Plotting a Point in the Polar Coordinate System In Exercises 5–18, plot the point given in polar coordinates and find three additional polar representations of the point, using $-2\pi < \theta < 2\pi$.

5. $(2, \pi/6)$ **6.** $(3, 5\pi/4)$

7. $(4, -\pi/3)$ **8.** $(1, -3\pi/4)$

9. $(2, 3\pi)$ **10.** $(4, 5\pi/2)$

11. $(-2, 2\pi/3)$ **12.** $(-3, 11\pi/6)$

13. $(0, 7\pi/6)$

14. $(0, -7\pi/2)$

15. $\left(\sqrt{2}, 2.36\right)$

16. $\left(2\sqrt{2}, 4.71\right)$

17. $(-3, -1.57)$

18. $(-5, -2.36)$

 Polar-to-Rectangular Conversion In Exercises 19–28, a point is given in polar coordinates. Convert the point to rectangular coordinates.

19. $(0, \pi)$ **20.** $(0, -\pi)$

21. $(3, \pi/2)$ **22.** $(3, 3\pi/2)$

23. $(2, 3\pi/4)$ **24.** $(1, 5\pi/4)$

25. $(-2, 7\pi/6)$ **26.** $(-3, 5\pi/6)$

27. $(-3, -\pi/3)$ **28.** $(-2, -4\pi/3)$

Using a Graphing Utility to Find Rectangular Coordinates In Exercises 29–38, use a graphing utility to find the rectangular coordinates of the point given in polar coordinates. Round your results to two decimal places.

29. $(2, 7\pi/8)$ **30.** $(3/2, 6\pi/5)$

31. $(1, 5\pi/12)$ **32.** $(4, 7\pi/9)$

33. $(-2.5, 1.1)$ **34.** $(-2, 5.76)$

35. $(2.5, -2.9)$ **36.** $(8.75, -6.5)$

37. $(-3.1, 7.92)$ **38.** $(-2.04, -5.3)$

 Rectangular-to-Polar Conversion In Exercises 39–50, a point is given in rectangular coordinates. Convert the point to polar coordinates. (There are many correct answers.)

39. $(1, 1)$ **40.** $(2, 2)$

41. $(-3, -3)$ **42.** $(-4, -4)$

43. $(3, 0)$ **44.** $(-6, 0)$

45. $(0, -5)$ **46.** $(0, 8)$

47. $\left(-\sqrt{3}, -\sqrt{3}\right)$ **48.** $\left(-\sqrt{3}, \sqrt{3}\right)$

49. $\left(\sqrt{3}, -1\right)$ **50.** $\left(-1, \sqrt{3}\right)$

 Using a Graphing Utility to Find Polar Coordinates In Exercises 51–58, use a graphing utility to find one set of polar coordinates of the point given in rectangular coordinates. Round your results to two decimal places.

51. $(3, -2)$ **52.** $(6, 3)$

53. $(-5, 2)$ **54.** $(7, -2)$

55. $\left(-\sqrt{3}, -4\right)$ **56.** $\left(5, -\sqrt{2}\right)$

57. $\left(\frac{5}{2}, \frac{4}{3}\right)$ **58.** $\left(-\frac{7}{9}, -\frac{3}{4}\right)$

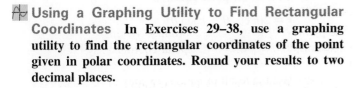 **Converting a Rectangular Equation to Polar Form** In Exercises 59–78, convert the rectangular equation to polar form. Assume $a > 0$.

59. $x^2 + y^2 = 9$ **60.** $x^2 + y^2 = 16$

61. $y = x$ **62.** $y = -x$

63. $x = 10$ **64.** $y = -2$

65. $3x - y + 2 = 0$ **66.** $3x + 5y - 2 = 0$

67. $xy = 16$ **68.** $2xy = 1$

69. $x = a$ **70.** $y = a$

71. $x^2 + y^2 = a^2$ **72.** $x^2 + y^2 = 9a^2$

73. $x^2 + y^2 - 2ax = 0$ **74.** $x^2 + y^2 - 2ay = 0$

75. $(x^2 + y^2)^2 = x^2 - y^2$ **76.** $(x^2 + y^2)^2 = 9(x^2 - y^2)$

77. $y^3 = x^2$ **78.** $y^2 = x^3$

Converting a Polar Equation to Rectangular Form **In Exercises 79–100, convert the polar equation to rectangular form.**

79. $r = 5$

80. $r = -7$

81. $\theta = 2\pi/3$

82. $\theta = -5\pi/3$

83. $\theta = \pi/2$

84. $\theta = 3\pi/2$

85. $r = 4 \csc \theta$

86. $r = 2 \csc \theta$

87. $r = -3 \sec \theta$

88. $r = -\sec \theta$

89. $r = -2 \cos \theta$

90. $r = 4 \sin \theta$

91. $r^2 = \cos \theta$

92. $r^2 = 2 \sin \theta$

93. $r^2 = \sin 2\theta$

94. $r^2 = \cos 2\theta$

95. $r = 2 \sin 3\theta$

96. $r = 3 \cos 2\theta$

97. $r = \dfrac{2}{1 + \sin \theta}$

98. $r = \dfrac{1}{1 - \cos \theta}$

99. $r = \dfrac{6}{2 - 3 \sin \theta}$

100. $r = \dfrac{5}{\sin \theta - 4 \cos \theta}$

Converting a Polar Equation to Rectangular Form **In Exercises 101–108, describe the graph of the polar equation and find the corresponding rectangular equation.**

101. $r = 6$

102. $r = 8$

103. $\theta = \pi/6$

104. $\theta = 3\pi/4$

105. $r = 3 \sec \theta$

106. $r = 2 \csc \theta$

107. $r = 2 \sin \theta$

108. $r = -6 \cos \theta$

109. Ferris Wheel

The center of a Ferris wheel lies at the pole of the polar coordinate system, where the distances are in feet. Passengers enter a car at $(30, -\pi/2)$. It takes 45 seconds for the wheel to complete one clockwise revolution.

(a) Write a polar equation that models the possible positions of a passenger car.

(b) Passengers enter a car. Find and interpret their coordinates after 15 seconds of rotation.

(c) Convert the point in part (b) to rectangular coordinates. Interpret the coordinates.

110. Ferris Wheel Repeat Exercise 109 when the distance from a passenger car to the center is 35 feet and it takes 60 seconds to complete one clockwise revolution.

Exploration

True or False? **In Exercises 111 and 112, determine whether the statement is true or false. Justify your answer.**

111. If $\theta_1 = \theta_2 + 2\pi n$ for some integer n, then (r, θ_1) and (r, θ_2) represent the same point in the polar coordinate system.

112. If $|r_1| = |r_2|$, then (r_1, θ) and (r_2, θ) represent the same point in the polar coordinate system.

113. **Error Analysis** Describe the error in converting the rectangular coordinates $\left(1, -\sqrt{3}\right)$ to polar form.

$$\tan \theta = -\sqrt{3}/1 \implies \theta = \frac{2\pi}{3}$$

$$r = \sqrt{1^2 + \left(-\sqrt{3}\right)^2} = \sqrt{4} = 2$$

$$(r, \theta) = \left(2, \frac{2\pi}{3}\right)$$

114. **HOW DO YOU SEE IT?** Use the polar coordinate system shown below.

(a) Identify the polar coordinates of points A–E.

(b) Which points lie on the graph of $r = 3$?

(c) Which points lie on the graph of $\theta = \pi/4$?

115. **Think About It**

(a) Convert the polar equation

$$r = 2(h \cos \theta + k \sin \theta)$$

to rectangular form and verify that it represents a circle.

(b) Use the result of part (a) to convert

$$r = \cos \theta + 3 \sin \theta$$

to rectangular form and find the center and radius of the circle it represents.

10.8 Graphs of Polar Equations

Graphs of polar equations are often useful visual tools in mathematical modeling. For example, in Exercise 69 on page 758, you will use the graph of a polar equation to analyze the pickup pattern of a microphone.

- Graph polar equations by point plotting.
- Use symmetry, zeros, and maximum *t*-values to sketch graphs of polar equations.
- Recognize special polar graphs.

Introduction

In previous chapters, you sketched graphs in the rectangular coordinate system. You began with the basic point-plotting method. Then you used sketching aids such as symmetry, intercepts, asymptotes, periods, and shifts to further investigate the natures of graphs. This section approaches curve sketching in the polar coordinate system similarly, beginning with a demonstration of point plotting.

EXAMPLE 1 **Graphing a Polar Equation by Point Plotting**

Sketch the graph of the polar equation $r = 4 \sin \theta$.

Solution The sine function is periodic, so to obtain a full range of *r*-values, consider values of θ in the interval $0 \le \theta \le 2\pi$, as shown in the table below.

θ	0	$\dfrac{\pi}{6}$	$\dfrac{\pi}{3}$	$\dfrac{\pi}{2}$	$\dfrac{2\pi}{3}$	$\dfrac{5\pi}{6}$	π	$\dfrac{7\pi}{6}$	$\dfrac{3\pi}{2}$	$\dfrac{11\pi}{6}$	2π
r	0	2	$2\sqrt{3}$	4	$2\sqrt{3}$	2	0	-2	-4	-2	0

By plotting these points, it appears that the graph is a circle of radius 2 whose center is at the point $(x, y) = (0, 2)$, as shown in the figure below.

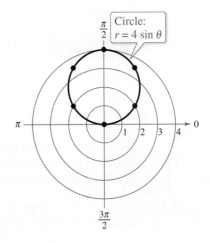

✓ **Checkpoint** ◀))) *Audio-video solution in English & Spanish at LarsonPrecalculus.com*

Sketch the graph of the polar equation $r = 6 \cos \theta$.

One way to confirm the graph in Example 1 is to convert the polar equation to rectangular form and then sketch the graph of the rectangular equation. You can also use a graphing utility set to *polar* mode and graph the polar equation, or use a graphing utility set to *parametric* mode and graph a parametric representation.

Symmetry, Zeros, and Maximum *r*-Values

Note in Example 1 that as θ increases from 0 to 2π, the graph is traced twice. Moreover, note that the graph is *symmetric with respect to the line* $\theta = \pi/2$. Had you known about this symmetry and retracing ahead of time, you could have used fewer points. The figures below show the three important types of symmetry to consider in polar curve sketching.

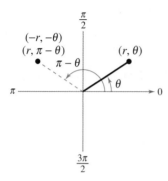

Symmetry with Respect to the
Line $\theta = \dfrac{\pi}{2}$

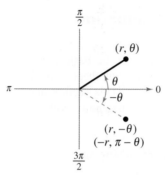

Symmetry with Respect to the
Polar Axis

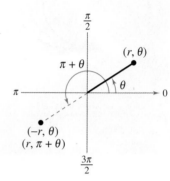

Symmetry with Respect to the
Pole

Tests for Symmetry in Polar Coordinates

The graph of a polar equation is symmetric with respect to the following when the given substitution yields an equivalent equation.

1. The line $\theta = \pi/2$: Replace (r, θ) with $(r, \pi - \theta)$ or $(-r, -\theta)$.

2. The polar axis: Replace (r, θ) with $(r, -\theta)$ or $(-r, \pi - \theta)$.

3. The pole: Replace (r, θ) with $(r, \pi + \theta)$ or $(-r, \theta)$.

EXAMPLE 2   **Using Symmetry to Sketch a Polar Graph**

Use symmetry to sketch the graph of $r = 3 + 2 \cos \theta$.

Solution   Replacing (r, θ) with $(r, -\theta)$ produces

$$r = 3 + 2 \cos(-\theta) = 3 + 2 \cos \theta. \qquad \cos(-\theta) = \cos \theta$$

So, the curve is symmetric with respect to the polar axis. Plotting the points in the table below and using polar axis symmetry, you obtain the graph shown in Figure 10.54. This graph is called a **limaçon.**

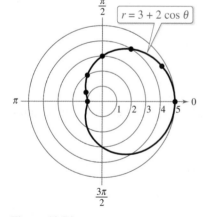

Figure 10.54

θ	0	$\dfrac{\pi}{6}$	$\dfrac{\pi}{3}$	$\dfrac{\pi}{2}$	$\dfrac{2\pi}{3}$	$\dfrac{5\pi}{6}$	π
r	5	$3 + \sqrt{3}$	4	3	2	$3 - \sqrt{3}$	1

✓ **Checkpoint** ◄))) *Audio-video solution in English & Spanish at LarsonPrecalculus.com*

Use symmetry to sketch the graph of $r = 3 + 2 \sin \theta$.

Example 2 uses the property that the cosine function is *even*. Recall from Section 4.2 that the cosine function is even because $\cos(-\theta) = \cos \theta$, and the sine function is odd because $\sin(-\theta) = -\sin \theta$.

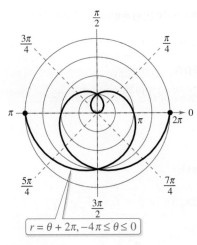

$r = \theta + 2\pi,\ -4\pi \le \theta \le 0$

Figure 10.55

The tests for symmetry in polar coordinates listed on the preceding page are sufficient to guarantee symmetry, but a graph may have symmetry even though its equation does not satisfy the tests. For example, Figure 10.55 shows the graph of

$$r = \theta + 2\pi$$

to be symmetric with respect to the line $\theta = \pi/2$, and yet the corresponding test fails to reveal this. That is, neither of the replacements below yields an equivalent equation.

Original Equation	Replacement	New Equation
$r = \theta + 2\pi$	(r, θ) with $(r, \pi - \theta)$	$r = -\theta + 3\pi$
$r = \theta + 2\pi$	(r, θ) with $(-r, -\theta)$	$-r = -\theta + 2\pi$

The equations $r = 4 \sin \theta$ and $r = 3 + 2 \cos \theta$, discussed in Examples 1 and 2, are of the form

$$r = f(\sin \theta) \quad \text{and} \quad r = g(\cos \theta)$$

respectively. Graphs of equations of these forms have symmetry in polar coordinates as listed below.

Quick Tests for Symmetry in Polar Coordinates

1. The graph of $r = f(\sin \theta)$ is symmetric with respect to the line $\theta = \dfrac{\pi}{2}$.

2. The graph of $r = g(\cos \theta)$ is symmetric with respect to the polar axis.

Two additional aids to sketching graphs of polar equations involve knowing the θ-values for which $|r|$ is maximum and knowing the θ-values for which $r = 0$. For instance, in Example 1, the maximum value of $|r|$ for $r = 4 \sin \theta$ is $|r| = 4$, and this occurs when $\theta = \pi/2$. Moreover, $r = 0$ when $\theta = 0$.

EXAMPLE 3   **Sketching a Polar Graph**

Sketch the graph of $r = 1 - 2 \cos \theta$.

Solution   From the equation $r = 1 - 2 \cos \theta$, you obtain the following features of the graph.

Symmetry: With respect to the polar axis

Maximum value of $|r|$: $r = 3$ when $\theta = \pi$

Zero of r: $r = 0$ when $\theta = \pi/3$

The table shows several θ-values in the interval $[0, \pi]$. Plot the corresponding points and sketch the graph, as shown in Figure 10.56.

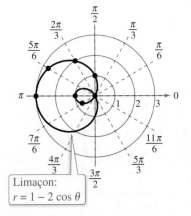

Limaçon:
$r = 1 - 2 \cos \theta$

Figure 10.56

θ	0	$\dfrac{\pi}{6}$	$\dfrac{\pi}{3}$	$\dfrac{\pi}{2}$	$\dfrac{2\pi}{3}$	$\dfrac{5\pi}{6}$	π
r	-1	$1 - \sqrt{3}$	0	1	2	$1 + \sqrt{3}$	3

Note that the negative r-values determine the *inner loop* of the graph in Figure 10.56. This graph, like the graph in Example 2, is a limaçon.

✓ *Checkpoint*  *Audio-video solution in English & Spanish at LarsonPrecalculus.com*

Sketch the graph of $r = 1 + 2 \sin \theta$.

Some curves reach their zeros and maximum r-values at more than one point, as shown in Example 4.

EXAMPLE 4 Sketching a Polar Graph

See LarsonPrecalculus.com for an interactive version of this type of example.

Sketch the graph of $r = 2 \cos 3\theta$.

Solution

Symmetry: With respect to the polar axis

Maximum value of $|r|$: $|r| = 2$ when $3\theta = 0, \pi, 2\pi, 3\pi$ or $\theta = 0, \dfrac{\pi}{3}, \dfrac{2\pi}{3}, \pi$

Zeros of r: $r = 0$ when $3\theta = \dfrac{\pi}{2}, \dfrac{3\pi}{2}, \dfrac{5\pi}{2}$ or $\theta = \dfrac{\pi}{6}, \dfrac{\pi}{2}, \dfrac{5\pi}{6}$

θ	0	$\dfrac{\pi}{12}$	$\dfrac{\pi}{6}$	$\dfrac{\pi}{4}$	$\dfrac{\pi}{3}$	$\dfrac{5\pi}{12}$	$\dfrac{\pi}{2}$
r	2	$\sqrt{2}$	0	$-\sqrt{2}$	-2	$-\sqrt{2}$	0

Plot these points and use the specified symmetry, zeros, and maximum values to obtain the graph, as shown in the figures below. This graph is called a **rose curve,** and each loop on the graph is called a *petal*. Note how the entire curve is traced as θ increases from 0 to π.

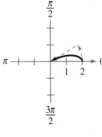

$0 \leq \theta \leq \dfrac{\pi}{6}$

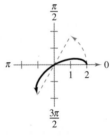

$0 \leq \theta \leq \dfrac{\pi}{3}$

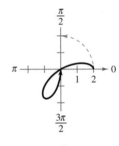

$0 \leq \theta \leq \dfrac{\pi}{2}$

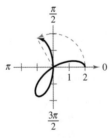

$0 \leq \theta \leq \dfrac{2\pi}{3}$

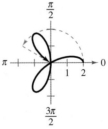

$0 \leq \theta \leq \dfrac{5\pi}{6}$

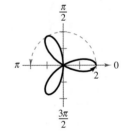

$0 \leq \theta \leq \pi$

▷ **TECHNOLOGY** Use a graphing utility in *polar* mode to verify the graph of $r = 2 \cos 3\theta$ shown in Example 4.

✓ *Checkpoint* 🔊))) *Audio-video solution in English & Spanish at LarsonPrecalculus.com*

Sketch the graph of $r = 2 \sin 3\theta$.

Special Polar Graphs

Several important types of graphs have equations that are simpler in polar form than in rectangular form. For example, the circle with the polar equation $r = 4 \sin \theta$ in Example 1 has the more complicated rectangular equation $x^2 + (y - 2)^2 = 4$. Several types of graphs that have simpler polar equations are shown below.

Limaçons

$r = a \pm b \cos \theta, r = a \pm b \sin \theta \quad (a > 0, b > 0)$

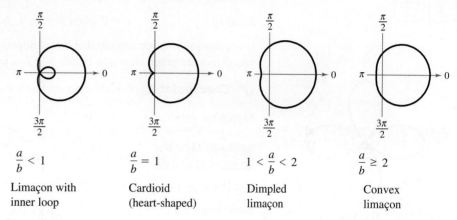

$\dfrac{a}{b} < 1$	$\dfrac{a}{b} = 1$	$1 < \dfrac{a}{b} < 2$	$\dfrac{a}{b} \geq 2$
Limaçon with inner loop	Cardioid (heart-shaped)	Dimpled limaçon	Convex limaçon

Rose Curves

n petals when n is odd, $2n$ petals when n is even $\quad (n \geq 2)$

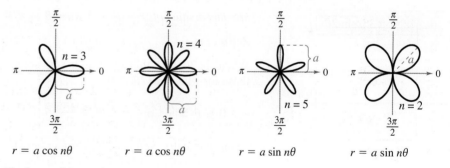

$r = a \cos n\theta$	$r = a \cos n\theta$	$r = a \sin n\theta$	$r = a \sin n\theta$

Circles and Lemniscates

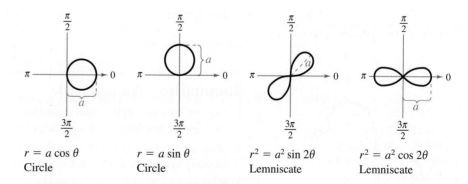

$r = a \cos \theta$	$r = a \sin \theta$	$r^2 = a^2 \sin 2\theta$	$r^2 = a^2 \cos 2\theta$
Circle	Circle	Lemniscate	Lemniscate

The quick tests for symmetry presented on page 753 can be especially useful when graphing many of the curves shown above. For example, limaçons have the form $r = f(\sin \theta)$ or the form $r = g(\cos \theta)$, so you know that a limaçon will be either symmetric with respect to the line $\theta = \pi/2$ or symmetric with respect to the polar axis.

θ	r
0	3
$\dfrac{\pi}{6}$	$\dfrac{3}{2}$
$\dfrac{\pi}{4}$	0
$\dfrac{\pi}{3}$	$-\dfrac{3}{2}$

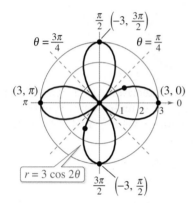

Figure 10.57

θ	$r = \pm 3\sqrt{\sin 2\theta}$
0	0
$\dfrac{\pi}{12}$	$\pm\dfrac{3}{\sqrt{2}}$
$\dfrac{\pi}{4}$	± 3
$\dfrac{5\pi}{12}$	$\pm\dfrac{3}{\sqrt{2}}$
$\dfrac{\pi}{2}$	0

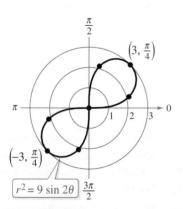

Figure 10.58

EXAMPLE 5 **Sketching a Rose Curve**

Sketch the graph of $r = 3 \cos 2\theta$.

Solution

Type of curve: Rose curve with $2n = 4$ petals

Symmetry: With respect to the line $\theta = \pi/2$, the polar axis, and the pole

Maximum value of $|r|$: $|r| = 3$ when $\theta = 0,\ \pi/2,\ \pi,\ 3\pi/2$

Zeros of r: $r = 0$ when $\theta = \pi/4,\ 3\pi/4$

Using this information and plotting the additional points included in the table at the left, you obtain the graph shown in Figure 10.57.

✓ *Checkpoint* ◀))) *Audio-video solution in English & Spanish at LarsonPrecalculus.com*

Sketch the graph of $r = 3 \cos 3\theta$.

EXAMPLE 6 **Sketching a Lemniscate**

Sketch the graph of $r^2 = 9 \sin 2\theta$.

Solution

Type of curve: Lemniscate

Symmetry: With respect to the pole

Maximum value of $|r|$: $|r| = 3$ when $\theta = \pi/4$

Zeros of r: $r = 0$ when $\theta = 0,\ \pi/2$

When $\sin 2\theta < 0$, this equation has no solution points. So, restrict the values of θ to those for which $\sin 2\theta \geq 0$.

$$0 \leq \theta \leq \frac{\pi}{2} \quad \text{or} \quad \pi \leq \theta \leq \frac{3\pi}{2}$$

Using symmetry, you need to consider only the first of these two intervals. By finding a few additional points (included in the table at the left), you obtain the graph shown in Figure 10.58.

✓ *Checkpoint* ◀))) *Audio-video solution in English & Spanish at LarsonPrecalculus.com*

Sketch the graph of $r^2 = 4 \cos 2\theta$.

Summarize **(Section 10.8)**

1. Explain how to graph a polar equation by point plotting *(page 751)*. For an example of graphing a polar equation by point plotting, see Example 1.

2. State the tests for symmetry in polar coordinates *(page 752)*. For an example of using symmetry to sketch the graph of a polar equation, see Example 2.

3. Explain how to use zeros and maximum r-values to sketch the graph of a polar equation *(page 753)*. For examples of using zeros and maximum r-values to sketch graphs of polar equations, see Examples 3 and 4.

4. State and give examples of the special polar graphs discussed in this lesson *(page 755)*. For examples of sketching special polar graphs, see Examples 5 and 6.

10.8 Exercises

See **CalcChat.com** for tutorial help and worked-out solutions to odd-numbered exercises.

Vocabulary: Fill in the blanks.

1. The graph of $r = f(\sin \theta)$ is symmetric with respect to the line _____.
2. The graph of $r = g(\cos \theta)$ is symmetric with respect to the _____ _____.
3. The equation $r = 2 + \cos \theta$ represents a _____ _____.
4. The equation $r = 2 \cos \theta$ represents a _____.
5. The equation $r^2 = 4 \sin 2\theta$ represents a _____.
6. The equation $r = 1 + \sin \theta$ represents a _____.

Skills and Applications

Identifying Types of Polar Graphs In Exercises 7–12, identify the type of polar graph.

7.

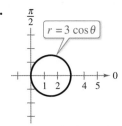

8.

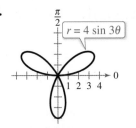

9.

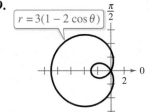

10.

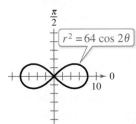

11.

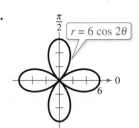

12.

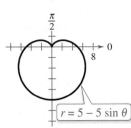

 Testing for Symmetry In Exercises 13–18, test for symmetry with respect to the line $\theta = \pi/2$, the polar axis, and the pole.

13. $r = 6 + 3 \cos \theta$
14. $r = 9 \cos 3\theta$
15. $r = \dfrac{2}{1 + \sin \theta}$
16. $r = \dfrac{3}{2 + \cos \theta}$
17. $r^2 = 36 \cos 2\theta$
18. $r^2 = 25 \sin 2\theta$

 Finding the Maximum Value of $|r|$ and Zeros of r In Exercises 19–22, find the maximum value of $|r|$ and any zeros of r.

19. $r = 10 - 10 \sin \theta$
20. $r = 6 + 12 \cos \theta$
21. $r = 4 \cos 3\theta$
22. $r = 3 \sin 2\theta$

 Sketching the Graph of a Polar Equation In Exercises 23–48, sketch the graph of the polar equation using symmetry, zeros, maximum r-values, and any other additional points.

23. $r = 5$
24. $r = -8$
25. $r = \pi/4$
26. $r = -2\pi/3$
27. $r = 3 \sin \theta$
28. $r = 4 \cos \theta$
29. $r = 3(1 - \cos \theta)$
30. $r = 4(1 - \sin \theta)$
31. $r = 4(1 + \sin \theta)$
32. $r = 6(1 + \cos \theta)$
33. $r = 5 + 2 \cos \theta$
34. $r = 5 - 2 \sin \theta$
35. $r = 1 - 3 \sin \theta$
36. $r = 2 - 5 \cos \theta$
37. $r = 3 - 6 \cos \theta$
38. $r = 4 + 6 \sin \theta$
39. $r = 5 \sin 2\theta$
40. $r = 2 \cos 2\theta$
41. $r = 6 \cos 3\theta$
42. $r = 3 \sin 3\theta$
43. $r = 2 \sec \theta$
44. $r = 5 \csc \theta$
45. $r = \dfrac{3}{\sin \theta - 2 \cos \theta}$
46. $r = \dfrac{6}{2 \sin \theta - 3 \cos \theta}$
47. $r^2 = 9 \cos 2\theta$
48. $r^2 = 16 \sin \theta$

 Graphing a Polar Equation In Exercises 49–58, use a graphing utility to graph the polar equation.

49. $r = 9/4$
50. $r = -5/2$
51. $r = 5\pi/8$
52. $r = -\pi/10$
53. $r = 8 \cos \theta$
54. $r = \cos 2\theta$
55. $r = 3(2 - \sin \theta)$
56. $r = 2 \cos(3\theta - 2)$
57. $r = 8 \sin \theta \cos^2 \theta$
58. $r = 2 \csc \theta + 5$

Finding an Interval In Exercises 59–64, use a graphing utility to graph the polar equation. Find an interval for θ for which the graph is traced only once.

59. $r = 3 - 8 \cos \theta$
60. $r = 5 + 4 \cos \theta$
61. $r = 2 \cos(3\theta/2)$
62. $r = 3 \sin(5\theta/2)$
63. $r^2 = 16 \sin 2\theta$
64. $r^2 = 1/\theta$

Asymptote of a Graph of a Polar Equation In Exercises 65–68, use a graphing utility to graph the polar equation and show that the given line is an asymptote of the graph.

Name of Graph	Polar Equation	Asymptote
65. Conchoid	$r = 2 - \sec\theta$	$x = -1$
66. Conchoid	$r = 2 + \csc\theta$	$y = 1$
67. Hyperbolic spiral	$r = \dfrac{3}{\theta}$	$y = 3$
68. Strophoid	$r = 2\cos 2\theta \sec\theta$	$x = -2$

• • 69. Microphone • • • • • • • • • • • • • • •

The pickup pattern of a microphone is modeled by the polar equation

$r = 5 + 5\cos\theta$

where $|r|$ measures how sensitive the microphone is to sounds coming from the angle θ.

(a) Sketch the graph of the model and identify the type of polar graph.

(b) At what angle is the microphone most sensitive to sound?

70. Area The total area of the region bounded by the lemniscate $r^2 = a^2 \cos 2\theta$ is a^2.

(a) Sketch the graph of $r^2 = 16 \cos 2\theta$.

(b) Find the area of one loop of the graph from part (a).

Exploration

True or False? In Exercises 71 and 72, determine whether the statement is true or false. Justify your answer.

71. The graph of $r = 10 \sin 5\theta$ is a rose curve with five petals.

72. A rose curve is always symmetric with respect to the line $\theta = \pi/2$.

73. Graphing a Polar Equation Consider the equation $r = 3 \sin k\theta$.

(a) Use a graphing utility to graph the equation for $k = 1.5$. Find the interval for θ over which the graph is traced only once.

(b) Use the graphing utility to graph the equation for $k = 2.5$. Find the interval for θ over which the graph is traced only once.

(c) Is it possible to find an interval for θ over which the graph is traced only once for any rational number k? Explain.

74. **HOW DO YOU SEE IT?** Match each polar equation with its graph.

(i) 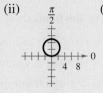   (ii) 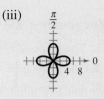   (iii)

(a) $r = 5 \sin\theta$

(b) $r = 2 + 5 \sin\theta$

(c) $r = 5 \cos 2\theta$

75. Sketching the Graph of a Polar Equation Sketch the graph of $r = 10 \cos\theta$ over each interval. Describe the part of the graph obtained in each case.

(a) $0 \le \theta \le \dfrac{\pi}{2}$ (b) $\dfrac{\pi}{2} \le \theta \le \pi$

(c) $-\dfrac{\pi}{2} \le \theta \le \dfrac{\pi}{2}$ (d) $\dfrac{\pi}{4} \le \theta \le \dfrac{3\pi}{4}$

76. Graphical Reasoning Use a graphing utility to graph the polar equation $r = 6[1 + \cos(\theta - \phi)]$ for each value of ϕ. Use the graphs to describe the effect of the angle ϕ. Write the equation as a function of $\sin\theta$ for part (c).

(a) $\phi = 0$ (b) $\phi = \pi/4$ (c) $\phi = \pi/2$

77. Rotating Polar Graphs The graph of $r = f(\theta)$ is rotated about the pole through an angle ϕ. Show that the equation of the rotated graph is $r = f(\theta - \phi)$.

78. Rotating Polar Graphs Consider the graph of $r = f(\sin\theta)$.

(a) Show that when the graph is rotated counterclockwise $\pi/2$ radians about the pole, the equation of the rotated graph is $r = f(-\cos\theta)$.

(b) Show that when the graph is rotated counterclockwise π radians about the pole, the equation of the rotated graph is $r = f(-\sin\theta)$.

(c) Show that when the graph is rotated counterclockwise $3\pi/2$ radians about the pole, the equation of the rotated graph is $r = f(\cos\theta)$.

Rotating Polar Graphs In Exercises 79 and 80, use the results of Exercises 77 and 78.

79. Write an equation for the limaçon $r = 2 - \sin\theta$ after it rotated through each angle.

(a) $\dfrac{\pi}{4}$ (b) $\dfrac{\pi}{2}$ (c) π (d) $\dfrac{3\pi}{2}$

80. Write an equation for the rose curve $r = 2 \sin 2\theta$ after it is rotated through each angle.

(a) $\dfrac{\pi}{6}$ (b) $\dfrac{\pi}{2}$ (c) $\dfrac{2\pi}{3}$ (d) π

10.9 Polar Equations of Conics

- Define conics in terms of eccentricity, and write and graph polar equations of conics.
- Use equations of conics in polar form to model real-life problems.

Alternative Definition and Polar Equations of Conics

In Sections 10.3 and 10.4, you learned that the rectangular equations of ellipses and hyperbolas take simpler forms when the origin lies at their *centers*. There are many important applications of conics in which it is more convenient to use a *focus* as the origin. In these cases, it is convenient to use polar coordinates.

To begin, consider an alternative definition of a conic that uses the concept of *eccentricity*.

Polar equation of conics can model the orbits of planets and satellites. For example, in Exercise 62 on page 764, you will use a polar equation to model the parabolic path of a satellite.

> ### Alternative Definition of a Conic
> The locus of a point in the plane that moves such that its distance from a fixed point (focus) is in a constant ratio to its distance from a fixed line (directrix) is a **conic.** The constant ratio is the *eccentricity* of the conic and is denoted by e. Moreover, the conic is an **ellipse** when $0 < e < 1$, a **parabola** when $e = 1$, and a **hyperbola** when $e > 1$. (See the figures below.)

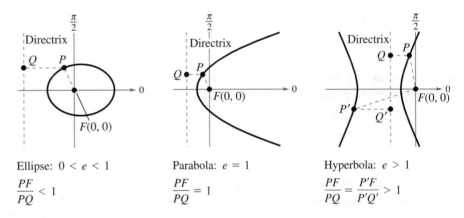

Ellipse: $0 < e < 1$
$$\frac{PF}{PQ} < 1$$

Parabola: $e = 1$
$$\frac{PF}{PQ} = 1$$

Hyperbola: $e > 1$
$$\frac{PF}{PQ} = \frac{P'F}{P'Q'} > 1$$

In the figures, note that for each type of conic, a focus is at the pole. The benefit of locating a focus of a conic at the pole is that the equation of the conic takes on a simpler form.

> ### Polar Equations of Conics
> The graph of a polar equation of the form
>
> **1.** $r = \dfrac{ep}{1 \pm e \cos \theta}$ or **2.** $r = \dfrac{ep}{1 \pm e \sin \theta}$
>
> is a conic, where $e > 0$ is the eccentricity and $|p|$ is the distance between the focus (pole) and the directrix.

For a proof of the polar equations of conics, see Proofs in Mathematics on page 774.

An equation of the form

$$r = \frac{ep}{1 \pm e \cos \theta} \qquad \text{Vertical directrix}$$

corresponds to a conic with a vertical directrix and symmetry with respect to the polar axis. An equation of the form

$$r = \frac{ep}{1 \pm e \sin \theta} \qquad \text{Horizontal directrix}$$

corresponds to a conic with a horizontal directrix and symmetry with respect to the line $\theta = \pi/2$. Moreover, the converse is also true—that is, any conic with a focus at the pole and having a horizontal or vertical directrix can be represented by one of these equations.

EXAMPLE 1   Identifying a Conic from Its Equation

See LarsonPrecalculus.com for an interactive version of this type of example.

Identify the type of conic represented by the equation

$$r = \frac{15}{3 - 2 \cos \theta}.$$

Algebraic Solution

To identify the type of conic, rewrite the equation in the form

$$r = \frac{ep}{1 \pm e \cos \theta}.$$

$$r = \frac{15}{3 - 2 \cos \theta} \qquad \text{Write original equation.}$$

$$= \frac{5}{1 - (2/3) \cos \theta} \qquad \begin{array}{l}\text{Divide numerator and} \\ \text{denominator by 3.}\end{array}$$

Because $e = \frac{2}{3} < 1$, the graph is an ellipse.

Graphical Solution

Use a graphing utility in *polar* mode and be sure to use a square setting, as shown in the figure below.

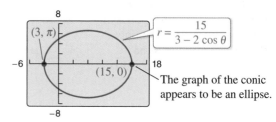

The graph of the conic appears to be an ellipse.

✓ **Checkpoint** ◀))) *Audio-video solution in English & Spanish at LarsonPrecalculus.com*

Identify the type of conic represented by the equation

$$r = \frac{8}{2 - 3 \sin \theta}.$$

For the ellipse in Example 1, the major axis is horizontal and the vertices lie at $(r, \theta) = (15, 0)$ and $(r, \theta) = (3, \pi)$. So, the length of the major axis is $2a = 18$. To find the length of the *minor* axis, use the definition of eccentricity $e = c/a$ and the relation $a^2 = b^2 + c^2$ for ellipses to conclude that

$$b^2 = a^2 - c^2 = a^2 - (ea)^2 = a^2(1 - e^2). \qquad \text{Ellipse}$$

Because $a = 18/2 = 9$ and $e = 2/3$, you have

$$b^2 = 9^2\left[1 - \left(\tfrac{2}{3}\right)^2\right] = 45$$

which implies that $b = \sqrt{45} = 3\sqrt{5}$. So, the length of the minor axis is $2b = 6\sqrt{5}$.

A similar analysis holds for hyperbolas. Using $e = c/a$ and the relation $c^2 = a^2 + b^2$ for hyperbolas yields

$$b^2 = c^2 - a^2 = (ea)^2 - a^2 = a^2(e^2 - 1). \qquad \text{Hyperbola}$$

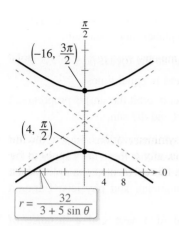

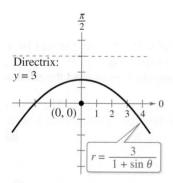

Figure 10.59

| EXAMPLE 2 | **Sketching a Conic from Its Polar Equation** |

Identify the type of conic represented by $r = \dfrac{32}{3 + 5 \sin \theta}$ and sketch its graph.

Solution   Dividing the numerator and denominator by 3, you have

$$r = \frac{32/3}{1 + (5/3) \sin \theta}.$$

Because $e = \frac{5}{3} > 1$, the graph is a hyperbola. The transverse axis of the hyperbola lies on the line $\theta = \pi/2$, and the vertices occur at $(r, \theta) = (4, \pi/2)$ and $(r, \theta) = (-16, 3\pi/2)$. The length of the transverse axis is 12, so $a = 6$. To find b, write

$$b^2 = a^2(e^2 - 1) = 6^2\left[\left(\tfrac{5}{3}\right)^2 - 1\right] = 64$$

which implies that $b = 8$. Use a and b to determine that the asymptotes of the hyperbola are $y = 10 \pm \frac{3}{4}x$. Figure 10.59 shows the graph.

✓ **Checkpoint** ◄))) *Audio-video solution in English & Spanish at LarsonPrecalculus.com*

Identify the conic $r = \dfrac{3}{2 - 4 \sin \theta}$ and sketch its graph.

In the next example, you will find a polar equation of a specified conic. To do this, let p be the distance between the pole and the directrix.

▷ **TECHNOLOGY** Use a graphing utility set in *polar* mode to verify the four orientations listed at the right. Remember that e must be positive, but p can be positive or negative.

1. Horizontal directrix above the pole: $r = \dfrac{ep}{1 + e \sin \theta}$

2. Horizontal directrix below the pole: $r = \dfrac{ep}{1 - e \sin \theta}$

3. Vertical directrix to the right of the pole: $r = \dfrac{ep}{1 + e \cos \theta}$

4. Vertical directrix to the left of the pole: $r = \dfrac{ep}{1 - e \cos \theta}$

| EXAMPLE 3 | **Finding the Polar Equation of a Conic** |

Find a polar equation of the parabola whose focus is the pole and whose directrix is the line $y = 3$.

Solution   The directrix is horizontal and above the pole, so use an equation of the form

$$r = \frac{ep}{1 + e \sin \theta}.$$

Moreover, the eccentricity of a parabola is $e = 1$ and the distance between the pole and the directrix is $p = 3$, so you have the equation

$$r = \frac{3}{1 + \sin \theta}.$$

Figure 10.60 shows the parabola.

Figure 10.60

✓ **Checkpoint** ◄))) *Audio-video solution in English & Spanish at LarsonPrecalculus.com*

Find a polar equation of the parabola whose focus is the pole and whose directrix is the line $x = -2$.

Application

Kepler's Laws (listed below), named after the German astronomer Johannes Kepler (1571–1630), can be used to describe the orbits of the planets about the sun.

1. Each planet moves in an elliptical orbit with the sun at one focus.

2. A ray from the sun to a planet sweeps out equal areas in equal times.

3. The square of the period (the time it takes for a planet to orbit the sun) is proportional to the cube of the mean distance between the planet and the sun.

Although Kepler stated these laws on the basis of observation, Isaac Newton (1642–1727) later validated them. In fact, Newton showed that these laws apply to the orbits of all heavenly bodies, including comets and satellites. The next example, which involves the comet named after the English mathematician and physicist Edmund Halley (1656–1742), illustrates this.

If you use Earth as a reference with a period of 1 year and a distance of 1 astronomical unit (about 93 million miles), then the proportionality constant in Kepler's third law is 1. For example, Mars has a mean distance to the sun of $d \approx 1.524$ astronomical units. Solve for its period P in $d^3 = P^2$ to find that the period of Mars is $P \approx 1.88$ years.

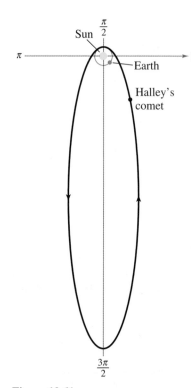

Figure 10.61

EXAMPLE 4 **Halley's Comet**

Halley's comet has an elliptical orbit with an eccentricity of $e \approx 0.967$. The length of the major axis of the orbit is approximately 35.88 astronomical units. Find a polar equation for the orbit. How close does Halley's comet come to the sun?

Solution Using a vertical major axis, as shown in Figure 10.61, choose an equation of the form $r = ep/(1 + e \sin \theta)$. The vertices of the ellipse occur when $\theta = \pi/2$ and $\theta = 3\pi/2$, and the length of the major axis is the sum of the r-values of the vertices. That is,

$$2a = \frac{0.967p}{1 + 0.967} + \frac{0.967p}{1 - 0.967} \approx 29.79p \approx 35.88.$$

So, $p \approx 1.204$ and $ep \approx (0.967)(1.204) \approx 1.164$. Substituting this value for ep in the equation, you have

$$r = \frac{1.164}{1 + 0.967 \sin \theta}$$

where r is measured in astronomical units. To find the closest point to the sun (a focus), substitute $\theta = \pi/2$ into this equation to obtain

$$r = \frac{1.164}{1 + 0.967 \sin(\pi/2)} \approx 0.59 \text{ astronomical unit} \approx 55,000,000 \text{ miles.}$$

✓ **Checkpoint**  *Audio-video solution in English & Spanish at LarsonPrecalculus.com*

Encke's comet has an elliptical orbit with an eccentricity of $e \approx 0.847$. The length of the major axis of the orbit is approximately 4.420 astronomical units. Find a polar equation for the orbit. How close does Encke's comet come to the sun? ◾

Summarize **(Section 10.9)**

1. State the definition of a conic in terms of eccentricity *(page 759)*. For examples of writing and graphing polar equations of conics, see Examples 2 and 3.

2. Describe a real-life application of an equation of a conic in polar form *(page 762, Example 4)*.

10.9 Exercises

See **CalcChat.com** for tutorial help and worked-out solutions to odd-numbered exercises.

Vocabulary

In Exercises 1–3, fill in the blanks.

1. The locus of a point in the plane that moves such that its distance from a fixed point (focus) is in a constant ratio to its distance from a fixed line (directrix) is a _____.

2. The constant ratio is the _____ of the conic and is denoted by _____.

3. An equation of the form $r = \dfrac{ep}{1 - e\cos\theta}$ has a _____ directrix to the _____ of the pole.

4. Match the conic with its eccentricity.

   (a) $0 < e < 1$ (b) $e = 1$ (c) $e > 1$

   (i) Parabola   (ii) Hyperbola   (iii) Ellipse

Skills and Applications

 Identifying a Conic In Exercises 5–8, write the polar equation of the conic for each value of e. Identify the type of conic represented by each equation. Verify your answers with a graphing utility.

   (a) $e = 1$ (b) $e = 0.5$ (c) $e = 1.5$

5. $r = \dfrac{2e}{1 + e\cos\theta}$

6. $r = \dfrac{2e}{1 - e\cos\theta}$

7. $r = \dfrac{2e}{1 - e\sin\theta}$

8. $r = \dfrac{2e}{1 + e\sin\theta}$

Matching   In Exercises 9–12, match the polar equation with its graph. [The graphs are labeled (a), (b), (c), and (d).]

(a)
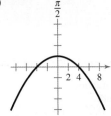

(b)
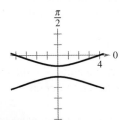

(c)
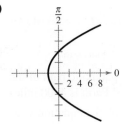

(d)
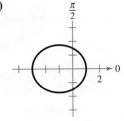

9. $r = \dfrac{4}{1 - \cos\theta}$

10. $r = \dfrac{3}{2 + \cos\theta}$

11. $r = \dfrac{4}{1 + \sin\theta}$

12. $r = \dfrac{4}{1 - 3\sin\theta}$

Sketching a Conic In Exercises 13–24, identify the conic represented by the equation and sketch its graph.

13. $r = \dfrac{3}{1 - \cos\theta}$

14. $r = \dfrac{7}{1 + \sin\theta}$

15. $r = \dfrac{5}{1 - \sin\theta}$

16. $r = \dfrac{6}{1 + \cos\theta}$

17. $r = \dfrac{2}{2 - \cos\theta}$

18. $r = \dfrac{4}{4 + \sin\theta}$

19. $r = \dfrac{6}{2 + \sin\theta}$

20. $r = \dfrac{6}{3 - 2\sin\theta}$

21. $r = \dfrac{3}{2 + 4\sin\theta}$

22. $r = \dfrac{5}{-1 + 2\cos\theta}$

23. $r = \dfrac{3}{2 - 6\cos\theta}$

24. $r = \dfrac{3}{2 + 6\sin\theta}$

Graphing a Polar Equation   In Exercises 25–32, use a graphing utility to graph the polar equation. Identify the conic.

25. $r = \dfrac{-1}{1 - \sin\theta}$

26. $r = \dfrac{-5}{2 + 4\sin\theta}$

27. $r = \dfrac{3}{-4 + 2\cos\theta}$

28. $r = \dfrac{4}{1 - 2\cos\theta}$

29. $r = \dfrac{4}{3 - \cos\theta}$

30. $r = \dfrac{10}{1 + \cos\theta}$

31. $r = \dfrac{14}{14 + 17\sin\theta}$

32. $r = \dfrac{12}{2 - \cos\theta}$

Graphing a Rotated Conic **In Exercises 33–36, use a graphing utility to graph the rotated conic.**

33. $r = \dfrac{3}{1 - \cos[\theta - (\pi/4)]}$ (See Exercise 13.)

34. $r = \dfrac{4}{4 + \sin[\theta - (\pi/3)]}$ (See Exercise 18.)

35. $r = \dfrac{6}{2 + \sin[\theta + (\pi/6)]}$ (See Exercise 19.)

36. $r = \dfrac{3}{2 + 6\sin[\theta + (2\pi/3)]}$ (See Exercise 24.)

Finding the Polar Equation of a Conic **In Exercises 37–52, find a polar equation of the indicated conic with the given characteristics and focus at the pole.**

	Conic	Eccentricity	Directrix
37.	Parabola	$e = 1$	$x = -1$
38.	Parabola	$e = 1$	$y = -4$
39.	Ellipse	$e = \frac{1}{2}$	$x = 3$
40.	Ellipse	$e = \frac{3}{4}$	$y = -2$
41.	Hyperbola	$e = 2$	$x = 1$
42.	Hyperbola	$e = \frac{3}{2}$	$y = -2$

	Conic	Vertex or Vertices
43.	Parabola	$(2, 0)$
44.	Parabola	$(10, \pi/2)$
45.	Parabola	$(5, \pi)$
46.	Parabola	$(1, -\pi/2)$
47.	Ellipse	$(2, 0), (10, \pi)$
48.	Ellipse	$(2, \pi/2), (4, 3\pi/2)$
49.	Ellipse	$(20, 0), (4, \pi)$
50.	Hyperbola	$(2, 0), (8, 0)$
51.	Hyperbola	$(1, 3\pi/2), (9, 3\pi/2)$
52.	Hyperbola	$(4, \pi/2), (1, \pi/2)$

53. **Astronomy** The planets travel in elliptical orbits with the sun at one focus. Assume that the focus is at the pole, the major axis lies on the polar axis, and the length of the major axis is $2a$ (see figure). Show that the polar equation of the orbit is $r = a(1 - e^2)/(1 - e\cos\theta)$, where e is the eccentricity.

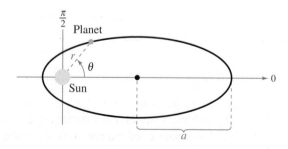

54. **Astronomy** Use the result of Exercise 53 to show that the minimum distance (*perihelion*) from the sun to the planet is

$r = a(1 - e)$

and the maximum distance (*aphelion*) is

$r = a(1 + e)$.

Planetary Motion **In Exercises 55–60, use the results of Exercises 53 and 54 to find (a) the polar equation of the planet's orbit and (b) the perihelion and aphelion.**

55. Earth $a \approx 9.2957 \times 10^7$ miles, $e \approx 0.0167$
56. Saturn $a \approx 1.4335 \times 10^9$ kilometers, $e \approx 0.0565$
57. Venus $a \approx 1.0821 \times 10^8$ kilometers, $e \approx 0.0067$
58. Mercury $a \approx 3.5984 \times 10^7$ miles, $e \approx 0.2056$
59. Mars $a \approx 1.4162 \times 10^8$ miles, $e \approx 0.0935$
60. Jupiter $a \approx 7.7857 \times 10^8$ kilometers, $e \approx 0.0489$

61. **Error Analysis** Describe the error.

For the polar equation $r = \dfrac{3}{2 + \sin\theta}$, $e = 1$.

So, the equation represents a parabola.

62. Satellite Orbit

A satellite in a 100-mile-high circular orbit around Earth has a velocity of approximately 17,500 miles per hour. If this velocity is multiplied by $\sqrt{2}$, then the satellite will have the minimum velocity necessary to escape Earth's gravity and will follow a parabolic path with the center of Earth as the focus (see figure).

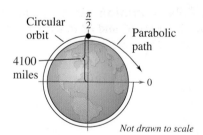

Not drawn to scale

(a) Find a polar equation of the parabolic path of the satellite. Assume the radius of Earth is 4000 miles.

(b) Use a graphing utility to graph the equation you found in part (a).

(c) Find the distance between the surface of the Earth and the satellite when $\theta = 30°$.

(d) Find the distance between the surface of Earth and the satellite when $\theta = 60°$.

Exploration

True or False?   **In Exercises 63–66, determine whether the statement is true or false. Justify your answer.**

63. For values of $e > 1$ and $0 \le \theta \le 2\pi$, the graphs of

$$r = \frac{ex}{1 - e\cos\theta} \quad \text{and} \quad r = \frac{e(-x)}{1 + e\cos\theta}$$

are the same.

64. The graph of

$$r = \frac{4}{-3 - 3\sin\theta}$$

has a horizontal directrix above the pole.

65. The conic represented by

$$r^2 = \frac{16}{9 - 4\cos\left(\theta + \dfrac{\pi}{4}\right)}$$

is an ellipse.

66. The conic represented by

$$r = \frac{6}{3 - 2\cos\theta}$$

is a parabola.

67. Verifying a Polar Equation   Show that the polar equation of the ellipse represented by

$$\frac{x^2}{a^2} + \frac{y^2}{b^2} = 1 \quad \text{is} \quad r^2 = \frac{b^2}{1 - e^2\cos^2\theta}.$$

68. Verifying a Polar Equation   Show that the polar equation of the hyperbola represented by

$$\frac{x^2}{a^2} - \frac{y^2}{b^2} = 1 \quad \text{is} \quad r^2 = \frac{-b^2}{1 - e^2\cos^2\theta}.$$

Writing a Polar Equation   In Exercises 69–74, use the results of Exercises 67 and 68 to write the polar form of the equation of the conic.

69. $\dfrac{x^2}{169} + \dfrac{y^2}{144} = 1$ **70.** $\dfrac{x^2}{25} + \dfrac{y^2}{16} = 1$

71. $\dfrac{x^2}{9} - \dfrac{y^2}{16} = 1$

72. $\dfrac{x^2}{36} - \dfrac{y^2}{4} = 1$

73. Hyperbola
   One focus: $(5, 0)$
   Vertices: $(4, 0), (4, \pi)$

74. Ellipse
   One focus: $(4, 0)$
   Vertices: $(5, 0), (5, \pi)$

75. Writing Explain how the graph of each equation differs from the conic represented by $r = \dfrac{5}{1 - \sin\theta}$. (See Exercise 15.)

(a) $r = \dfrac{5}{1 - \cos\theta}$ (b) $r = \dfrac{5}{1 + \sin\theta}$

(c) $r = \dfrac{5}{1 + \cos\theta}$ (d) $r = \dfrac{5}{1 - \sin[\theta - (\pi/4)]}$

76. **HOW DO YOU SEE IT?**   The graph of

$$r = \frac{e}{1 - e\sin\theta}$$

is shown for different values of e. Determine which graph matches each value of e.

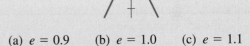

(a) $e = 0.9$ (b) $e = 1.0$ (c) $e = 1.1$

77. Reasoning
   (a) Identify the type of conic represented by

$$r = \frac{4}{1 - 0.4\cos\theta}$$

   without graphing the equation.

   (b) Without graphing the equations, describe how the graph of each equation below differs from the polar equation given in part (a).

$$r_1 = \frac{4}{1 + 0.4\cos\theta} \qquad r_2 = \frac{4}{1 - 0.4\sin\theta}$$

   (c) Use a graphing utility to verify your results in part (b).

78. Reasoning   The equation

$$r = \frac{ep}{1 \pm e\sin\theta}$$

represents an ellipse with $e < 1$. What happens to the lengths of both the major axis and the minor axis when the value of e remains fixed and the value of p changes? Use an example to explain your reasoning.

Chapter Summary

	What Did You Learn?	**Explanation/Examples**	**Review Exercises**
Section 10.1	Find the inclination of a line *(p. 692)*.	If a nonvertical line has inclination θ and slope m, then $m = \tan\theta$.	1–4
	Find the angle between two lines *(p. 693)*.	If two nonperpendicular lines have slopes m_1 and m_2, then the tangent of the angle between the two lines is $\tan\theta = \left|(m_2 - m_1)/(1 + m_1 m_2)\right|$.	5–8
	Find the distance between a point and a line *(p. 694)*.	The distance between the point (x_1, y_1) and the line $Ax + By + C = 0$ is $d = \left|Ax_1 + By_1 + C\right|/\sqrt{A^2 + B^2}$.	9, 10
Section 10.2	Recognize a conic as the intersection of a plane and a double-napped cone *(p. 699)*.	In the formation of the four basic conics, the intersecting plane does not pass through the vertex of the cone. (See Figure 10.7.)	11, 12
	Write equations of parabolas in standard form *(p. 700)*.	**Horizontal Axis** $(y - k)^2 = 4p(x - h), p \neq 0$ **Vertical Axis** $(x - h)^2 = 4p(y - k), p \neq 0$	13–16, 19, 20
	Use the reflective property of parabolas to write equations of tangent lines *(p. 702)*.	The tangent line to a parabola at a point P makes equal angles with (1) the line passing through P and the focus and (2) the axis of the parabola.	17, 18
Section 10.3	Write equations of ellipses in standard form and sketch ellipses *(p. 709)*.	**Horizontal Major Axis** $\dfrac{(x - h)^2}{a^2} + \dfrac{(y - k)^2}{b^2} = 1$ **Vertical Major Axis** $\dfrac{(x - h)^2}{b^2} + \dfrac{(y - k)^2}{a^2} = 1$	21–24, 27–30
	Use properties of ellipses to model and solve real-life problems *(p. 712)*.	Properties of ellipses can be used to find distances from Earth's center to the moon's center in the moon's orbit. (See Example 5.)	25, 26
	Find eccentricities *(p. 713)*.	The eccentricity e of an ellipse is the ratio $e = c/a$.	27–30
Section 10.4	Write equations of hyperbolas in standard form *(p. 718)*, and find asymptotes of and sketch hyperbolas *(p. 719)*.	**Horizontal Transverse Axis** $\dfrac{(x - h)^2}{a^2} - \dfrac{(y - k)^2}{b^2} = 1$ **Vertical Transverse Axis** $\dfrac{(y - k)^2}{a^2} - \dfrac{(x - h)^2}{b^2} = 1$	31–38
	Use properties of hyperbolas to solve real-life problems *(p. 722)*.	Properties of hyperbolas can be used in radar and other detection systems. (See Example 5.)	39, 40
	Classify conics from their general equations *(p. 723)*.	The graph of $Ax^2 + Cy^2 + Dx + Ey + F = 0$ is, except in degenerate cases, a circle ($A = C$), a parabola ($AC = 0$), an ellipse ($A \neq C$ and $AC > 0$), or a hyperbola ($AC < 0$).	41–44
Section 10.5	Rotate the coordinate axes to eliminate the xy-term in equations of conics *(p. 727)*.	The equation $Ax^2 + Bxy + Cy^2 + Dx + Ey + F = 0$, where $B \neq 0$, can be rewritten as $A'(x')^2 + C'(y')^2 + D'x' + E'y' + F' = 0$ by rotating the coordinate axes through an angle θ, where $\cot 2\theta = (A - C)/B$.	45–48
	Use the discriminant to classify conics *(p. 731)*.	The graph of $Ax^2 + Bxy + Cy^2 + Dx + Ey + F = 0$ is, except in degenerate cases, an ellipse or a circle ($B^2 - 4AC < 0$), a parabola ($B^2 - 4AC = 0$), or a hyperbola ($B^2 - 4AC > 0$).	49–52

What Did You Learn?	Explanation/Examples	Review Exercises
Section 10.6 Evaluate sets of parametric equations for given values of the parameter *(p. 735)*.	If f and g are continuous functions of t on an interval I, then the set of ordered pairs $(f(t), g(t))$ is a plane curve C. The equations $x = f(t)$ and $y = g(t)$ are parametric equations for C, and t is the parameter.	53, 54
Sketch curves represented by sets of parametric equations *(p. 736)*.	One way to sketch a curve represented by a pair of parametric equations is to plot points in the xy-plane. You determine each set of coordinates (x, y) from a value chosen for the parameter t.	55–60
Rewrite sets of parametric equations as single rectangular equations by eliminating the parameter *(p. 737)*.	To eliminate the parameter in a pair of parametric equations, solve for t in one equation and substitute the expression for t into the other equation. The result is the corresponding rectangular equation.	55–60
Find sets of parametric equations for graphs *(p. 739)*.	A set of parametric equations that represent a graph is not unique. (See Example 4.)	61–66
Section 10.7 Plot points in the polar coordinate system *(p. 745)*.		67–70
Convert points *(p. 747)* and equations *(p. 748)* from rectangular to polar form and vice versa.	**Polar Coordinates (r, θ) and Rectangular Coordinates (x, y)** Polar-to-Rectangular: $x = r \cos \theta$, $y = r \sin \theta$ Rectangular-to-Polar: $\tan \theta = \dfrac{y}{x}$, $r^2 = x^2 + y^2$	71–90
Section 10.8 Graph polar equations by point plotting *(p. 751)*.	Graphing a polar equation by point plotting is similar to graphing a rectangular equation. (See Example 1.)	91–100
Use symmetry, zeros, and maximum r-values to sketch graphs of polar equations *(p. 752)*.	The graph of a polar equation is symmetric with respect to the following when the given substitution yields an equivalent equation. **1.** The line $\theta = \dfrac{\pi}{2}$: Replace (r, θ) with $(r, \pi - \theta)$ or $(-r, -\theta)$. **2.** The polar axis: Replace (r, θ) with $(r, -\theta)$ or $(-r, \pi - \theta)$. **3.** The pole: Replace (r, θ) with $(r, \pi + \theta)$ or $(-r, \theta)$. Additional aids to graphing polar equations are the θ-values for which $|r|$ is maximum and the θ-values for which $r = 0$.	91–100
Recognize special polar graphs *(p. 755)*.	Several important types of graphs, such as limaçons, rose curves, circles, and lemniscates, have equations that are simpler in polar form than in rectangular form.	101–104
Section 10.9 Define conics in terms of eccentricity, and write and graph polar equations of conics *(p. 759)*.	**Ellipse:** $0 < e < 1$ **Parabola:** $e = 1$ **Hyperbola:** $e > 1$ The graph of a polar equation of the form **(1)** $r = (ep)/(1 \pm e \cos \theta)$ or **(2)** $r = (ep)/(1 \pm e \sin \theta)$ is a conic, where $e > 0$ is the eccentricity and $|p|$ is the distance between the focus (pole) and the directrix.	105–112
Use equations of conics in polar form to model real-life problems *(p. 762)*.	The equation of a conic in polar form can be used to model the orbit of Halley's comet. (See Example 4.)	113, 114

Review Exercises See CalcChat.com for tutorial help and worked-out solutions to odd-numbered exercises.

10.1 **Finding the Inclination of a Line** In Exercises 1–4, find the inclination θ (in radians and degrees) of the line with the given characteristics.

1. Passes through the points $(-1, 2)$ and $(2, 5)$
2. Passes through the points $(3, 4)$ and $(-2, 7)$
3. Equation: $5x + 2y + 4 = 0$
4. Equation: $2x - 5y - 7 = 0$

Finding the Angle Between Two Lines In Exercises 5–8, find the angle θ (in radians and degrees) between the lines.

5. $4x + y = 2$
     $-5x + y = -1$

6. $-5x + 3y = 3$
    $-2x + 3y = 1$

7. $2x - 7y = 8$
    $\dfrac{2}{5}x + y = 0$

8. $0.03x + 0.05y = 0.16$
    $0.07x - 0.02y = 0.15$

Finding the Distance Between a Point and a Line In Exercises 9 and 10, find the distance between the point and the line.

Point	Line
9. $(4, 3)$	$y = 2x - 1$
10. $(-2, 1)$	$y = -4x + 2$

10.2 **Forming a Conic Section** In Exercises 11 and 12, state the type of conic formed by the intersection of the plane and the double-napped cone.

11.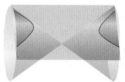

12.

Finding the Standard Equation of a Parabola In Exercises 13–16, find the standard form of the equation of the parabola with the given characteristics. Then sketch the parabola.

13. Vertex: $(0, 0)$
    Focus: $(0, 3)$

14. Vertex: $(4, 0)$
    Focus: $(0, 0)$

15. Vertex: $(0, 2)$
    Directrix: $x = -3$

16. Vertex: $(-3, -3)$
    Directrix: $y = 0$

Finding the Tangent Line at a Point on a Parabola In Exercises 17 and 18, find an equation of the tangent line to the parabola at the given point.

17. $y = 2x^2$, $(-1, 2)$
18. $x^2 = -2y$, $(-4, -8)$

19. **Architecture** A parabolic archway is 10 meters high at the vertex. At a height of 8 meters, the width of the archway is 6 meters (see figure). How wide is the archway at ground level?

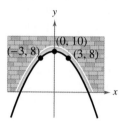
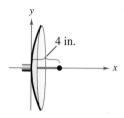

Figure for 19   Figure for 20

20. **Parabolic Microphone** The receiver of a parabolic microphone is at the focus of the parabolic reflector, 4 inches from the vertex (see figure). Write an equation for a cross section of the reflector with its focus on the positive x-axis and its vertex at the origin.

10.3 **Finding the Standard Equation of an Ellipse** In Exercises 21–24, find the standard form of the equation of the ellipse with the given characteristics.

21. Vertices: $(2, 0)$, $(2, 16)$; minor axis of length 6
22. Vertices: $(0, 3)$, $(10, 3)$; minor axis of length 4
23. Vertices: $(0, 1)$, $(4, 1)$; endpoints of the minor axis: $(2, 0)$, $(2, 2)$
24. Vertices: $(-4, -1)$, $(-4, 11)$; endpoints of the minor axis: $(-6, 5)$, $(-2, 5)$

25. **Architecture** A mason is building a semielliptical arch that has a height of 4 feet and a width of 10 feet. Where should the foci be placed in order to sketch the arch?

26. **Wading Pool** You are building a wading pool that is in the shape of an ellipse. An equation for the elliptical shape of the pool is $(x^2/324) + (y^2/196) = 1$, where x and y are measured in feet. Find the longest distance across the pool, the shortest distance, and the distance between the foci.

Sketching an Ellipse In Exercises 27–30, find the center, vertices, foci, and eccentricity of the ellipse. Then sketch the ellipse.

27. $\dfrac{(x + 2)^2}{64} + \dfrac{(y - 5)^2}{36} = 1$

28. $\dfrac{(x - 4)^2}{25} + \dfrac{(y + 3)^2}{49} = 1$

29. $16x^2 + 9y^2 - 32x + 72y + 16 = 0$
30. $4x^2 + 25y^2 + 16x - 150y + 141 = 0$

10.4 Finding the Standard Equation of a Hyperbola   In Exercises 31–34, find the standard form of the equation of the hyperbola with the given characteristics.

31. Vertices: $(0, \pm 6)$

Foci: $(0, \pm 8)$

32. Vertices: $(5, 2), (-5, 2)$

Foci: $(6, 2), (-6, 2)$

33. Foci: $(\pm 5, 0)$

Asymptotes: $y = \pm \frac{3}{4}x$

34. Foci: $(0, \pm 13)$

Asymptotes: $y = \pm \frac{5}{12}x$

Sketching a Hyperbola   In Exercises 35–38, find the center, vertices, foci, and the equations of the asymptotes of the hyperbola. Then sketch the hyperbola using the asymptotes as an aid.

35. $\dfrac{(x-4)^2}{49} - \dfrac{(y+2)^2}{25} = 1$

36. $\dfrac{(y-3)^2}{9} - x^2 = 1$

37. $9x^2 - 16y^2 - 18x - 32y - 151 = 0$

38. $-4x^2 + 25y^2 - 8x + 150y + 121 = 0$

39. Sound Location   Two microphones, 2 miles apart, record an explosion. Microphone A receives the sound 6 seconds before microphone B. Where did the explosion occur? (Assume sound travels at 1100 feet per second.)

40. Navigation   Radio transmitting station A is located 200 miles east of transmitting station B. A ship is in an area to the north and 40 miles west of station A. Synchronized radio pulses transmitted at 186,000 miles per second by the two stations are received 0.0005 second sooner from station A than from station B. How far north is the ship?

Classifying a Conic from a General Equation   In Exercises 41–44, classify the graph of the equation as a circle, a parabola, an ellipse, or a hyperbola.

41. $5x^2 - 2y^2 + 10x - 4y + 17 = 0$

42. $-4y^2 + 5x + 3y + 7 = 0$

43. $3x^2 + 2y^2 - 12x + 12y + 29 = 0$

44. $4x^2 + 4y^2 - 4x + 8y - 11 = 0$

10.5 Rotation of Axes   In Exercises 45–48, rotate the axes to eliminate the xy-term in the equation. Then write the equation in standard form. Sketch the graph of the resulting equation, showing both sets of axes.

45. $xy + 5 = 0$ 　　**46.** $x^2 - 4xy + y^2 + 9 = 0$

47. $5x^2 - 2xy + 5y^2 - 12 = 0$

48. $4x^2 + 8xy + 4y^2 + 7\sqrt{2}x + 9\sqrt{2}y = 0$

Rotation and Graphing Utilities   In Exercises 49–52, (a) use the discriminant to classify the graph of the equation, (b) use the Quadratic Formula to solve for y, and (c) use a graphing utility to graph the equation.

49. $16x^2 - 24xy + 9y^2 - 30x - 40y = 0$

50. $13x^2 - 8xy + 7y^2 - 45 = 0$

51. $x^2 - 10xy + y^2 + 1 = 0$

52. $x^2 + y^2 + 2xy + 2\sqrt{2}x - 2\sqrt{2}y + 2 = 0$

10.6 Sketching a Curve   In Exercises 53 and 54, (a) create a table of x- and y-values for the parametric equations using $t = -2, -1, 0, 1,$ and 2, and (b) plot the points (x, y) generated in part (a) and sketch the graph of the parametric equations.

53. $x = 3t - 2$ and $y = 7 - 4t$

54. $x = \dfrac{1}{4}t$ and $y = \dfrac{6}{t+3}$

Sketching a Curve   In Exercises 55–60, (a) sketch the curve represented by the parametric equations (indicate the orientation of the curve) and (b) eliminate the parameter and write the resulting rectangular equation whose graph represents the curve. Adjust the domain of the rectangular equation, if necessary. Verify your result with a graphing utility.

55. $x = 2t$ 　　**56.** $x = 1 + 4t$

$y = 4t$ 　　　　$y = 2 - 3t$

57. $x = t^2$ 　　**58.** $x = t + 4$

$y = \sqrt{t}$ 　　　$y = t^2$

59. $x = 3\cos\theta$ 　　**60.** $x = 3 + 3\cos\theta$

$y = 3\sin\theta$ 　　　$y = 2 + 5\sin\theta$

Finding Parametric Equations for a Graph   In Exercises 61–66, find a set of parametric equations to represent the graph of the rectangular equation using (a) $t = x$, (b) $t = x + 1$, and (c) $t = 3 - x$.

61. $y = 2x + 3$

62. $y = 4 - 3x$

63. $y = x^2 + 3$

64. $y = 2 - x^2$

65. $y = 1 - 4x^2$

66. $y = 2x^2 + 2$

10.7 Plotting a Point in the Polar Coordinate System   In Exercises 67–70, plot the point given in polar coordinates and find three additional polar representations of the point, using $-2\pi < \theta < 2\pi$.

67. $\left(4, \dfrac{5\pi}{6}\right)$ 　　**68.** $\left(-3, -\dfrac{\pi}{4}\right)$

69. $(-7, 4.19)$ 　　**70.** $\left(\sqrt{3}, 2.62\right)$

Polar-to-Rectangular Conversion In Exercises 71–74, a point is given in polar coordinates. Convert the point to rectangular coordinates.

71. $\left(0, \dfrac{\pi}{2}\right)$ **72.** $\left(2, \dfrac{5\pi}{4}\right)$

73. $\left(-1, \dfrac{\pi}{3}\right)$ **74.** $\left(3, -\dfrac{3\pi}{4}\right)$

Rectangular-to-Polar Conversion In Exercises 75–78, a point is given in rectangular coordinates. Convert the point to polar coordinates. (There are many correct answers.)

75. $(3, 3)$ **76.** $(3, -4)$

77. $\left(-\sqrt{5}, \sqrt{5}\right)$ **78.** $\left(-\sqrt{2}, -\sqrt{2}\right)$

Converting a Rectangular Equation to Polar Form In Exercises 79–84, convert the rectangular equation to polar form.

79. $x^2 + y^2 = 81$ **80.** $x^2 + y^2 = 48$

81. $x = 5$ **82.** $y = 4$

83. $xy = 5$ **84.** $xy = -2$

Converting a Polar Equation to Rectangular Form In Exercises 85–90, convert the polar equation to rectangular form.

85. $r = 4$ **86.** $r = 12$

87. $r = 3 \cos \theta$ **88.** $r = 8 \sin \theta$

89. $r^2 = \sin \theta$ **90.** $r^2 = 4 \cos 2\theta$

10.8 **Sketching the Graph of a Polar Equation** In Exercises 91–100, sketch the graph of the polar equation using symmetry, zeros, maximum r-values, and any other additional points.

91. $r = 6$ **92.** $r = 11$

93. $r = -2(1 + \cos \theta)$ **94.** $r = 1 - 4 \cos \theta$

95. $r = 4 \sin 2\theta$ **96.** $r = \cos 5\theta$

97. $r = 2 + 6 \sin \theta$ **98.** $r = 5 - 5 \cos \theta$

99. $r^2 = 9 \sin \theta$ **100.** $r^2 = \cos 2\theta$

Identifying Types of Polar Graphs In Exercises 101–104, identify the type of polar graph and use a graphing utility to graph the equation.

101. $r = 3(2 - \cos \theta)$ **102.** $r = 5(1 - 2 \cos \theta)$

103. $r = 8 \cos 3\theta$ **104.** $r^2 = 2 \sin 2\theta$

10.9 **Sketching a Conic** In Exercises 105–108, identify the conic represented by the equation and sketch its graph.

105. $r = \dfrac{1}{1 + 2 \sin \theta}$ **106.** $r = \dfrac{6}{1 + \sin \theta}$

107. $r = \dfrac{4}{5 - 3 \cos \theta}$ **108.** $r = \dfrac{16}{4 + 5 \cos \theta}$

Finding the Polar Equation of a Conic In Exercises 109–112, find a polar equation of the indicated conic with the given characteristics and focus at the pole.

	Conic	Vertex or Vertices
109.	Parabola	$(2, \pi)$
110.	Parabola	$(2, \pi/2)$
111.	Ellipse	$(5, 0), (1, \pi)$
112.	Hyperbola	$(1, 0), (7, 0)$

113. **Explorer 18** In 1963, the United States launched Explorer 18. Its low and high points above the surface of Earth were 110 miles and 122,800 miles, respectively. The center of Earth was at one focus of the orbit (see figure). Find the polar equation of the orbit and the distance between the surface of Earth and the satellite when $\theta = \pi/3$. Assume Earth has a radius of 4000 miles.

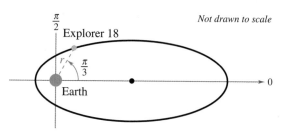

114. **Asteroid** An asteroid takes a parabolic path with Earth as its focus. It is about 6,000,000 miles from Earth at its closest approach. Write the polar equation of the path of the asteroid with its vertex at $\theta = \pi/2$. Find the distance between the asteroid and Earth when $\theta = -\pi/3$.

Exploration

True or False? In Exercises 115–117, determine whether the statement is true or false. Justify your answer.

115. The graph of $\frac{1}{4}x^2 - y^4 = 1$ is a hyperbola.

116. Only one set of parametric equations can represent the line $y = 3 - 2x$.

117. There is a unique polar coordinate representation of each point in the plane.

118. **Think About It** Consider an ellipse with the major axis horizontal and 10 units in length. The number b in the standard form of the equation of the ellipse must be less than what real number? Explain the change in the shape of the ellipse as b approaches this number.

119. **Think About It** What is the relationship between the graphs of the rectangular and polar equations?

(a) $x^2 + y^2 = 25$, $r = 5$

(b) $x - y = 0$, $\theta = \dfrac{\pi}{4}$

Chapter Test

See CalcChat.com for tutorial help and worked-out solutions to odd-numbered exercises.

Take this test as you would take a test in class. When you are finished, check your work against the answers given in the back of the book.

1. Find the inclination θ (in radians and degrees) of $4x - 7y + 6 = 0$.
2. Find the angle θ (in radians and degrees) between $3x + y = 6$ and $5x - 2y = -4$.
3. Find the distance between the point $(2, 9)$ and the line $y = 3x + 4$.

In Exercises 4–7, identify the conic and write the equation in standard form. Find the center, vertices, foci, and the equations of the asymptotes, if applicable. Then sketch the conic.

4. $y^2 - 2x + 2 = 0$
5. $x^2 - 4y^2 - 4x = 0$
6. $9x^2 + 16y^2 + 54x - 32y - 47 = 0$
7. $2x^2 + 2y^2 - 8x - 4y + 9 = 0$

8. Find the standard form of the equation of the parabola with vertex $(3, -4)$ and focus $(6, -4)$.
9. Find the standard form of the equation of the hyperbola with foci $(0, \pm 2)$ and asymptotes $y = \pm\frac{1}{9}x$.
10. Rotate the axes to eliminate the xy-term in the equation $xy + 1 = 0$. Then write the equation in standard form. Sketch the graph of the resulting equation, showing both sets of axes.
11. Sketch the curve represented by the parametric equations $x = 2 + 3\cos\theta$ and $y = 2\sin\theta$. Eliminate the parameter and write the resulting rectangular equation.
12. Find a set of parametric equations to represent the graph of the rectangular equation $y = 3 - x^2$ using (a) $t = x$ and (b) $t = x + 2$.
13. Convert the polar coordinates $\left(-2, \dfrac{5\pi}{6}\right)$ to rectangular coordinates.
14. Convert the rectangular coordinates $(2, -2)$ to polar coordinates and find three additional polar representations of the point, using $-2\pi < \theta < 2\pi$.
15. Convert the rectangular equation $x^2 + y^2 = 64$ to polar form.

In Exercises 16–19, identify the type of graph represented by the polar equation. Then sketch the graph.

16. $r = \dfrac{4}{1 + \cos\theta}$
17. $r = \dfrac{4}{2 + \sin\theta}$
18. $r = 2 + 3\sin\theta$
19. $r = 2\sin 4\theta$

20. Find a polar equation of the ellipse with focus at the pole, eccentricity $e = \frac{1}{4}$, and directrix $y = 4$.
21. A straight road rises with an inclination of 0.15 radian from the horizontal. Find the slope of the road and the change in elevation over a one-mile section of the road.
22. A baseball is hit at a point 3 feet above the ground toward the left field fence. The fence is 10 feet high and 375 feet from home plate. The path of the baseball can be modeled by the parametric equations $x = (115\cos\theta)t$ and $y = 3 + (115\sin\theta)t - 16t^2$. Does the baseball go over the fence when it is hit at an angle of $\theta = 30°$? Does the baseball go over the fence when $\theta = 35°$?

Proofs in Mathematics ■ ■ ■ ■ ■ ■ ■ ■ ■ ■ ■ ■ ■ ■

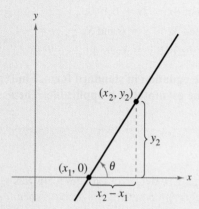

Inclination and Slope   *(p. 692)*

If a nonvertical line has inclination θ and slope m, then $m = \tan \theta$.

Proof

If $m = 0$, then the line is horizontal and $\theta = 0$. So, the result is true for horizontal lines because $m = 0 = \tan 0$.

If the line has a positive slope, then it will intersect the x-axis. Label this point $(x_1, 0)$, as shown in the figure. If (x_2, y_2) is a second point on the line, then the slope is

$$m = \frac{y_2 - 0}{x_2 - x_1} = \frac{y_2}{x_2 - x_1} = \tan \theta.$$

The case in which the line has a negative slope can be proved in a similar manner. ■

Distance Between a Point and a Line   *(p. 694)*

The distance between the point (x_1, y_1) and the line $Ax + By + C = 0$ is

$$d = \frac{|Ax_1 + By_1 + C|}{\sqrt{A^2 + B^2}}.$$

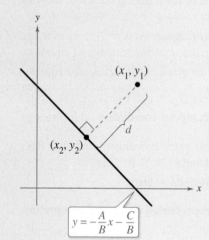

Proof

For simplicity, assume that the given line is neither horizontal nor vertical (see figure). Writing the equation $Ax + By + C = 0$ in slope-intercept form

$$y = -\frac{A}{B}x - \frac{C}{B}$$

shows that the line has a slope of $m = -A/B$. So, the slope of the line passing through (x_1, y_1) and perpendicular to the given line is B/A, and its equation is $y - y_1 = (B/A)(x - x_1)$. These two lines intersect at the point (x_2, y_2), where

$$x_2 = \frac{B^2x_1 - ABy_1 - AC}{A^2 + B^2} \quad \text{and} \quad y_2 = \frac{-ABx_1 + A^2y_1 - BC}{A^2 + B^2}.$$

Finally, the distance between (x_1, y_1) and (x_2, y_2) is

$$d = \sqrt{(x_2 - x_1)^2 + (y_2 - y_1)^2}$$

$$= \sqrt{\left(\frac{B^2x_1 - ABy_1 - AC}{A^2 + B^2} - x_1\right)^2 + \left(\frac{-ABx_1 + A^2y_1 - BC}{A^2 + B^2} - y_1\right)^2}$$

$$= \sqrt{\frac{A^2(Ax_1 + By_1 + C)^2 + B^2(Ax_1 + By_1 + C)^2}{(A^2 + B^2)^2}}$$

$$= \sqrt{\frac{(Ax_1 + By_1 + C)^2(A^2 + B^2)}{(A^2 + B^2)^2}}$$

$$= \sqrt{\frac{(Ax_1 + By_1 + C)^2}{A^2 + B^2}}$$

$$= \frac{|Ax_1 + By_1 + C|}{\sqrt{A^2 + B^2}}. ■$$

PARABOLIC PATHS

There are many natural occurrences of parabolas in real life. For example, Italian astronomer and mathematician Galileo Galilei discovered in the 17th century that an object projected upward and obliquely to the pull of gravity travels in a parabolic path. Examples of this include the path of a jumping dolphin and the path of water molecules from a drinking water fountain.

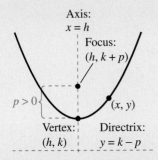

Parabola with vertical axis

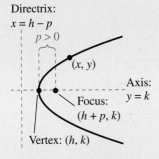

Parabola with horizontal axis

Standard Equation of a Parabola   *(p. 700)*

The **standard form of the equation of a parabola** with vertex at (h, k) is

$$(x - h)^2 = 4p(y - k), \quad p \neq 0 \qquad \text{Vertical axis; directrix: } y = k - p$$

$$(y - k)^2 = 4p(x - h), \quad p \neq 0. \qquad \text{Horizontal axis; directrix: } x = h - p$$

The focus lies on the axis p units (directed distance) from the vertex. If the vertex is at the origin, then the equation takes one of two forms.

$$x^2 = 4py \qquad \qquad \text{Vertical axis}$$

$$y^2 = 4px \qquad \qquad \text{Horizontal axis}$$

Proof

First, examine the case in which the directrix is parallel to the x-axis and the focus lies above the vertex, as shown in the top figure. If (x, y) is any point on the parabola, then, by definition, it is equidistant from the focus $(h, k + p)$ and the directrix $y = k - p$. Apply the Distance Formula to obtain

$$\sqrt{(x - h)^2 + [y - (k + p)]^2} = y - (k - p)$$

$$(x - h)^2 + [y - (k + p)]^2 = [y - (k - p)]^2$$

$$(x - h)^2 + y^2 - 2y(k + p) + (k + p)^2 = y^2 - 2y(k - p) + (k - p)^2$$

$$(x - h)^2 + y^2 - 2ky - 2py + k^2 + 2pk + p^2 = y^2 - 2ky + 2py + k^2 - 2pk + p^2$$

$$(x - h)^2 - 2py + 2pk = 2py - 2pk$$

$$(x - h)^2 = 4p(y - k).$$

Next, examine the case in which the directrix is parallel to the y-axis and the focus lies to the right of the vertex, as shown in the bottom figure. If (x, y) is any point on the parabola, then, by definition, it is equidistant from the focus $(h + p, k)$ and the directrix $x = h - p$. Apply the Distance Formula to obtain

$$\sqrt{[x - (h + p)]^2 + (y - k)^2} = x - (h - p)$$

$$[x - (h + p)]^2 + (y - k)^2 = [x - (h - p)]^2$$

$$x^2 - 2x(h + p) + (h + p)^2 + (y - k)^2 = x^2 - 2x(h - p) + (h - p)^2$$

$$x^2 - 2hx - 2px + h^2 + 2ph + p^2 + (y - k)^2 = x^2 - 2hx + 2px + h^2 - 2ph + p^2$$

$$-2px + 2ph + (y - k)^2 = 2px - 2ph$$

$$(y - k)^2 = 4p(x - h).$$

Note that if a parabola is centered at the origin, then the two equations above would simplify to $x^2 = 4py$ and $y^2 = 4px$, respectively. The cases in which the focus lies (1) below the vertex and (2) to the left of the vertex can be proved in manners similar to the above. ■

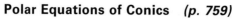

Polar Equations of Conics   *(p. 759)*

The graph of a polar equation of the form

1. $r = \dfrac{ep}{1 \pm e \cos \theta}$

or

2. $r = \dfrac{ep}{1 \pm e \sin \theta}$

is a conic, where $e > 0$ is the eccentricity and $|p|$ is the distance between the focus (pole) and the directrix.

Proof

A proof for

$$r = \frac{ep}{1 + e \cos \theta}$$

with $p > 0$ is shown here. The proofs of the other cases are similar. In the figure at the left, consider a vertical directrix, p units to the right of the focus $F(0, 0)$. If $P(r, \theta)$ is a point on the graph of

$$r = \frac{ep}{1 + e \cos \theta}$$

then the distance between P and the directrix is

$$
\begin{aligned}
PQ &= |p - x| \\
&= |p - r \cos \theta| \\
&= \left| p - \left(\frac{ep}{1 + e \cos \theta} \right) \cos \theta \right| \\
&= \left| p \left(1 - \frac{e \cos \theta}{1 + e \cos \theta} \right) \right| \\
&= \left| \frac{p}{1 + e \cos \theta} \right| \\
&= \left| \frac{r}{e} \right|.
\end{aligned}
$$

Moreover, the distance between P and the pole is $PF = |r|$, so the ratio of PF to PQ is

$$
\begin{aligned}
\frac{PF}{PQ} &= \frac{|r|}{\left| \dfrac{r}{e} \right|} \\
&= |e| \\
&= e
\end{aligned}
$$

and, by definition, the graph of the equation must be a conic.

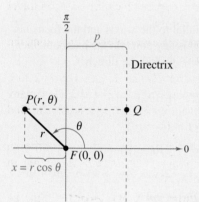

P.S. Problem Solving

1. **Mountain Climbing** Several mountain climbers are located in a mountain pass between two peaks. The angles of elevation to the two peaks are 0.84 radian and 1.10 radians. A range finder shows that the distances to the peaks are 3250 feet and 6700 feet, respectively (see figure).

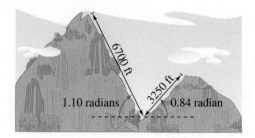

(a) Find the angle between the two lines.

(b) Approximate the amount of vertical climb that is necessary to reach the summit of each peak.

2. **Finding the Equation of a Parabola** Find the general equation of a parabola that has the x-axis as the axis of symmetry and the focus at the origin.

3. **Area** Find the area of the square whose vertices lie on the graph of the ellipse, as shown below.

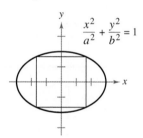

$$\frac{x^2}{a^2} + \frac{y^2}{b^2} = 1$$

4. **Involute** The *involute* of a circle can be described by the endpoint P of a string that is held taut as it is unwound from a spool (see figure below). The spool does not rotate. Show that the parametric equations

$x = r(\cos \theta + \theta \sin \theta)$

and

$y = r(\sin \theta - \theta \cos \theta)$

represent the involute of a circle.

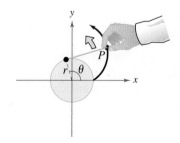

5. **Tour Boat** A tour boat travels between two islands that are 12 miles apart (see figure). There is enough fuel for a 20-mile trip.

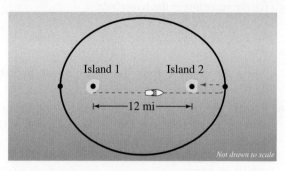

Not drawn to scale

(a) Explain why the region in which the boat can travel is bounded by an ellipse.

(b) Let $(0, 0)$ represent the center of the ellipse. Find the coordinates of each island.

(c) The boat travels from Island 1, past Island 2 to a vertex of the ellipse, and then to Island 2. How many miles does the boat travel? Use your answer to find the coordinates of the vertex.

(d) Use the results from parts (b) and (c) to write an equation of the ellipse that bounds the region in which the boat can travel.

6. **Finding an Equation of a Hyperbola** Find an equation of the hyperbola such that for any point on the hyperbola, the absolute value of the difference of its distances from the points $(2, 2)$ and $(10, 2)$ is 6.

7. **Proof** Prove that the graph of the equation

$Ax^2 + Cy^2 + Dx + Ey + F = 0$

is one of the following (except in degenerate cases).

Conic	Condition
(a) Circle	$A = C, A \neq 0$
(b) Parabola	$A = 0$ or $C = 0$, but not both.
(c) Ellipse	$AC > 0, A \neq C$
(d) Hyperbola	$AC < 0$

8. **Projectile Motion** The two sets of parametric equations below model projectile motion.

$x_1 = (v_0 \cos \theta)t, \quad y_1 = (v_0 \sin \theta)t$

$x_2 = (v_0 \cos \theta)t, \quad y_2 = h + (v_0 \sin \theta)t - 16t^2$

(a) Under what circumstances is it appropriate to use each model?

(b) Eliminate the parameter for each set of equations.

(c) In which case is the path of the moving object not affected by a change in the velocity v? Explain.

9. Proof Prove that

$$c^2 = a^2 + b^2$$

for the equation of the hyperbola

$$\frac{x^2}{a^2} - \frac{y^2}{b^2} = 1$$

where the distance from the center of the hyperbola $(0, 0)$ to a focus is c.

10. Proof Prove that the angle θ used to eliminate the xy-term in $Ax^2 + Bxy + Cy^2 + Dx + Ey + F = 0$ by a rotation of axes is given by

$$\cot 2\theta = \frac{A - C}{B}.$$

11. Orientation of an Ellipse As t increases, the ellipse given by the parametric equations

$$x = \cos t$$

and

$$y = 2 \sin t$$

is traced *counterclockwise*. Find a set of parametric equations that represent the same ellipse traced *clockwise*.

12. Writing Use a graphing utility to graph the polar equation

$$r = \cos 5\theta + n \cos \theta$$

for the integers $n = -5$ to $n = 5$ using $0 \le \theta \le \pi$. As you graph these equations, you should see the graph's shape change from a heart to a bell. Write a short paragraph explaining what values of n produce the heart portion of the curve and what values of n produce the bell portion.

13. Strophoid The curve given by the polar equation

$$r = 2 \cos 2\theta \sec \theta$$

is called a **strophoid.**

(a) Find a rectangular equation of the strophoid.

(b) Find a pair of parametric equations that represent the strophoid.

(c) Use a graphing utility to graph the strophoid.

14. Think About It

(a) Show that the distance between the points (r_1, θ_1) and (r_2, θ_2) is $\sqrt{r_1^2 + r_2^2 - 2r_1 r_2 \cos(\theta_1 - \theta_2)}$.

(b) Simplify the Distance Formula for $\theta_1 = \theta_2$. Is the simplification what you expected? Explain.

(c) Simplify the Distance Formula for $\theta_1 - \theta_2 = 90°$. Is the simplification what you expected? Explain.

15. Hypocycloid A **hypocycloid** has the parametric equations

$$x = (a - b) \cos t + b \cos\left(\frac{a - b}{b}t\right)$$

and

$$y = (a - b) \sin t - b \sin\left(\frac{a - b}{b}t\right).$$

Use a graphing utility to graph the hypocycloid for each pair of values. Describe each graph.

(a) $a = 2, b = 1$

(b) $a = 3, b = 1$

(c) $a = 4, b = 1$

(d) $a = 10, b = 1$

(e) $a = 3, b = 2$

(f) $a = 4, b = 3$

16. Butterfly Curve The graph of the polar equation

$$r = e^{\cos \theta} - 2 \cos 4\theta + \sin^5\left(\frac{\theta}{12}\right)$$

is called the *butterfly curve,* as shown in the figure.

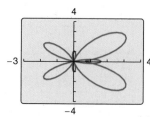

$$r = e^{\cos \theta} - 2 \cos 4\theta + \sin^5\left(\frac{\theta}{12}\right)$$

(a) The graph shown was produced using $0 \le \theta \le 2\pi$. Does this show the entire graph? Explain.

(b) Approximate the maximum r-value of the graph. Does this value change when you use $0 \le \theta \le 4\pi$ instead of $0 \le \theta \le 2\pi$? Explain.

17. Rose Curves The rose curves described in this chapter are of the form

$$r = a \cos n\theta$$

or

$$r = a \sin n\theta$$

where n is a positive integer that is greater than or equal to 2. Use a graphing utility to graph $r = a \cos n\theta$ and $r = a \sin n\theta$ for some noninteger values of n. Describe the graphs.

11 Analytic Geometry in Three Dimensions

Product Design
(Exercise 60, page 807)

Torque *(Exercise 59, page 798)*

Equilibrium *(Example 8, page 789)*

Neutron Star *(page 781)*

Geography *(Exercise 72, page 784)*

11.1 The Three-Dimensional Coordinate System

One practical application of the three-dimensional coordinate system is modeling solids in space. For example, in Exercise 72 on page 784, you will model the spherical shape of Earth in a three-dimensional coordinate system.

■ Plot points in the three-dimensional coordinate system.
■ Find distances between points in space and find midpoints of line segments joining points in space.
■ Write equations of spheres in standard form and sketch traces of surfaces in space.

The Three-Dimensional Coordinate System

Recall that the Cartesian plane is formed by two perpendicular number lines, the x-axis and the y-axis. These axes determine a two-dimensional coordinate system for identifying points in a plane. To identify a point in space, you must introduce a third dimension to the model. The geometry of this three-dimensional model is **solid analytic geometry.**

You can construct a **three-dimensional coordinate system** by passing a z-axis perpendicular to both the x- and y-axes at the origin, as shown in the figure at the right. Taken as pairs, the axes determine three **coordinate planes: the *xy*-plane, the *xz*-plane,** and the ***yz*-plane.** These planes separate the three-dimensional coordinate system into eight **octants.** The first octant is the one in which all three coordinates are positive. A point P in space is determined by an ordered triple (x, y, z), where x, y, and z are as described below.

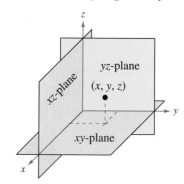

x = directed distance from yz-plane to P

y = directed distance from xz-plane to P

z = directed distance from xy-plane to P

A three-dimensional coordinate system can have either a **left-handed** or a **right-handed** orientation. In this text, you will work exclusively with right-handed systems, as illustrated at the right. In a right-handed system, Octants II, III, and IV are found by rotating counterclockwise around the positive z-axis. Octant V is vertically below Octant I. Octants VI, VII, and VIII are then found by rotating counterclockwise around the negative z-axis. See below.

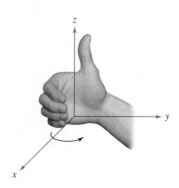

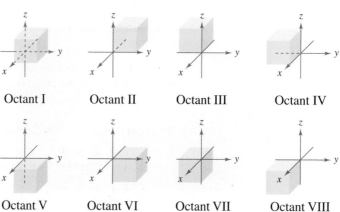

| Octant I | Octant II | Octant III | Octant IV |
| Octant V | Octant VI | Octant VII | Octant VIII |

EXAMPLE 1 Plotting Points in Space

Plot each point in space.

a. $(2, -3, 3)$ **b.** $(-2, 6, 2)$ **c.** $(1, 4, 0)$ **d.** $(2, 2, -3)$

Solution To plot the point $(2, -3, 3)$, notice that $x = 2$, $y = -3$, and $z = 3$. To help visualize the point, locate the point $(2, -3)$ in the xy-plane (denoted by a cross in Figure 11.1). The point $(2, -3, 3)$ lies three units above the cross. The other three points are also shown in Figure 11.1.

✓ **Checkpoint** ◄))) *Audio-video solution in English & Spanish at LarsonPrecalculus.com*

Plot each point in space.

a. $(-1, 2, -2)$ **b.** $(3, -1, -1)$ **c.** $(-1, -2, -1)$ **d.** $(-3, -2, 1)$ ■

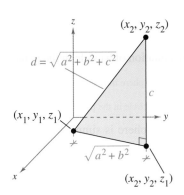

Figure 11.1

The Distance and Midpoint Formulas

Many of the formulas established for the two-dimensional coordinate system can be extended to three dimensions. For example, to find the distance between two points in space, you can use the Pythagorean Theorem twice, as shown in Figure 11.2. Note that

$$a = |x_2 - x_1|, \quad b = |y_2 - y_1|, \quad \text{and} \quad c = |z_2 - z_1|.$$

Distance Formula in Space

The distance between the points (x_1, y_1, z_1) and (x_2, y_2, z_2) given by the **Distance Formula in Space** is

$$d = \sqrt{(x_2 - x_1)^2 + (y_2 - y_1)^2 + (z_2 - z_1)^2}.$$

EXAMPLE 2   Finding the Distance Between Two Points in Space

Find the distance between $(1, 0, 2)$ and $(2, 4, -3)$.

Solution

$d = \sqrt{(x_2 - x_1)^2 + (y_2 - y_1)^2 + (z_2 - z_1)^2}$	Distance Formula in Space
$= \sqrt{(2 - 1)^2 + (4 - 0)^2 + (-3 - 2)^2}$	Substitute.
$= \sqrt{1 + 16 + 25}$	Simplify.
$= \sqrt{42}$	Simplify.

✓ **Checkpoint** ◄))) *Audio-video solution in English & Spanish at LarsonPrecalculus.com*

Find the distance between $(0, 1, 3)$ and $(1, 4, -2)$.   ■

Figure 11.2

Notice the similarity between the Distance Formulas in the plane and in space. The Midpoint Formulas in the plane and in space are also similar.

Midpoint Formula in Space

The midpoint of the line segment joining the points (x_1, y_1, z_1) and (x_2, y_2, z_2) given by the **Midpoint Formula in Space** is

$$\left(\frac{x_1 + x_2}{2}, \frac{y_1 + y_2}{2}, \frac{z_1 + z_2}{2} \right).$$

Using the Midpoint Formula in Space
In Exercises 41–48, find the midpoint of the line segment joining the points.

41. $(3, 0, 1), (2, -6, 3)$ **42.** $(1, 5, -1), (2, 2, 2)$

43. $(3, -6, 10), (-3, 4, 4)$ **44.** $(-1, 5, -3), (3, 7, -1)$

45. $(-5, -2, 5), (6, 3, -7)$ **46.** $(0, -2, 5), (4, 2, 7)$

47. $(-2, 8, 10), (7, -4, 2)$ **48.** $(9, -5, 1), (9, -2, -4)$

Finding the Equation of a Sphere
In Exercises 49–54, find the standard equation of the sphere with the given characteristics.

49. Center: $(3, 2, 4)$; radius: 4

50. Center: $(2, 5, 2)$; radius: 5

51. Center: $(-3, 7, 5)$; diameter: 10

52. Center: $(0, 5, -9)$; diameter: 8

53. Endpoints of a diameter: $(3, 0, 0), (0, 0, 6)$

54. Endpoints of a diameter: $(1, 0, 0), (0, 5, 0)$

Finding the Center and Radius of a Sphere
In Exercises 55–62, find the center and radius of the sphere.

55. $x^2 + y^2 + z^2 - 6x = 0$

56. $x^2 + y^2 + z^2 - 9x = 0$

57. $x^2 + y^2 + z^2 - 4x + 2y - 6z + 10 = 0$

58. $x^2 + y^2 + z^2 + 4x - 8z + 19 = 0$

59. $9x^2 + 9y^2 + 9z^2 - 18x - 6y - 72z + 73 = 0$

60. $2x^2 + 2y^2 + 2z^2 - 2x - 6y - 4z + 5 = 0$

61. $9x^2 + 9y^2 + 9z^2 - 6x + 18y + 1 = 0$

62. $4x^2 + 4y^2 + 4z^2 - 4x - 32y + 8z + 33 = 0$

Sketching a Trace of a Surface
In Exercises 63–66, sketch the graph of the equation and the specified trace.

63. $(x - 1)^2 + y^2 + z^2 = 36$; xz-trace

64. $x^2 + (y + 3)^2 + z^2 = 25$; yz-trace

65. $(x + 2)^2 + (y - 3)^2 + z^2 = 9$; yz-trace

66. $x^2 + (y - 1)^2 + (z + 1)^2 = 4$; xy-trace

Graphing a Sphere
In Exercises 67–70, use a three-dimensional graphing utility to graph the sphere.

67. $x^2 + y^2 + z^2 - 6x - 8y - 10z + 46 = 0$

68. $x^2 + y^2 + z^2 + 6y - 8z + 21 = 0$

69. $4x^2 + 4y^2 + 4z^2 - 8x - 16y + 8z - 25 = 0$

70. $9x^2 + 9y^2 + 9z^2 + 18x - 18y + 36z + 35 = 0$

71. Architecture A spherical building has a diameter of 205 feet. The center of the building is placed at the origin of a three-dimensional coordinate system. What is the standard equation of the sphere?

72. Geography

Assume that Earth is a sphere with a radius of 4000 miles. The center of Earth is placed at the origin of a three-dimensional coordinate system.

(a) What is the standard equation of the sphere?

(b) What trace(s) could represent lines of longitude that run north-south? What shape would the trace(s) form?

(c) What trace(s) could represent lines of latitude that run east-west? What shape would the trace(s) form?

Exploration

True or False? In Exercises 73 and 74, determine whether the statement is true or false. Justify your answer.

73. In the ordered triple (x, y, z) that represents point P in space, x is the directed distance from the xy-plane to P.

74. A sphere always has a circle as its xy-trace.

75. Think About It What is the z-coordinate of any point in the xy-plane? What is the y-coordinate of any point in the xz-plane? What is the x-coordinate of any point in the yz-plane?

76. **HOW DO YOU SEE IT?** Approximate the coordinates of each point, and state the octant in which each point lies.

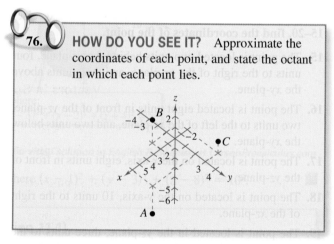

77. Finding an Endpoint A line segment has (x_1, y_1, z_1) as one endpoint and (x_m, y_m, z_m) as its midpoint. Find the other endpoint (x_2, y_2, z_2) of the line segment in terms of $x_1, y_1, z_1, x_m, y_m,$ and z_m.

78. Finding an Endpoint Use the result of Exercise 77 to find the coordinates of the endpoint of a line segment when the coordinates of the other endpoint and the midpoint are $(3, 0, 2)$ and $(5, 8, 7)$, respectively.

11.2 Vectors in Space

- ■ Find the component forms of the unit vectors in the same direction of, the magnitudes of, the dot products of, and the angles between vectors in space.
- ■ Determine whether vectors in space are orthogonal or parallel.
- ■ Use vectors in space to solve real-life problems.

Vectors in space can represent physical forces. For example, in Exercise 66 on page 791, you will find the tension in each of the cables used to support auditorium lights.

Vectors in Space

Physical forces and velocities are not confined to the plane, so it is natural to extend the concept of vectors from a two-dimensional plane to three-dimensional space. In space, vectors are denoted by ordered triples

$$\mathbf{v} = \langle v_1, v_2, v_3 \rangle. \qquad \text{Component form}$$

The **zero vector** is denoted by $\mathbf{0} = \langle 0, 0, 0 \rangle$. Using the unit vectors $\mathbf{i} = \langle 1, 0, 0 \rangle$, $\mathbf{j} = \langle 0, 1, 0 \rangle$, and $\mathbf{k} = \langle 0, 0, 1 \rangle$, the **standard unit vector notation** for $\mathbf{v}$ is

$$\mathbf{v} = v_1\mathbf{i} + v_2\mathbf{j} + v_3\mathbf{k} \qquad \text{Unit vector form}$$

as shown in Figure 11.8. When $\mathbf{v}$ is represented by the directed line segment from $P(p_1, p_2, p_3)$ to $Q(q_1, q_2, q_3)$, as shown in Figure 11.9, the **component form** of $\mathbf{v}$ is produced by subtracting the coordinates of the initial point from the corresponding coordinates of the terminal point, as follows.

$$\mathbf{v} = \langle v_1, v_2, v_3 \rangle = \langle q_1 - p_1, q_2 - p_2, q_3 - p_3 \rangle$$

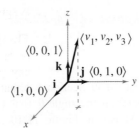

Figure 11.8

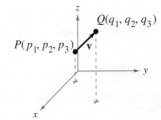

Figure 11.9

Vectors in Space

1. Two vectors are **equal** if and only if their corresponding components are equal.

2. The **magnitude** (or **length**) of $\mathbf{u} = \langle u_1, u_2, u_3 \rangle$ is $\|\mathbf{u}\| = \sqrt{u_1^2 + u_2^2 + u_3^2}$.

3. A **unit vector u** in the direction of $\mathbf{v}$ is $\mathbf{u} = \dfrac{\mathbf{v}}{\|\mathbf{v}\|}, \quad \mathbf{v} \neq \mathbf{0}$.

4. The **sum** of $\mathbf{u} = \langle u_1, u_2, u_3 \rangle$ and $\mathbf{v} = \langle v_1, v_2, v_3 \rangle$ is

    $$\mathbf{u} + \mathbf{v} = \langle u_1 + v_1, u_2 + v_2, u_3 + v_3 \rangle. \qquad \text{Vector addition}$$

5. The **scalar multiple** of the real number c and $\mathbf{u} = \langle u_1, u_2, u_3 \rangle$ is

    $$c\mathbf{u} = \langle cu_1, cu_2, cu_3 \rangle. \qquad \text{Scalar multiplication}$$

6. The **dot product** of $\mathbf{u} = \langle u_1, u_2, u_3 \rangle$ and $\mathbf{v} = \langle v_1, v_2, v_3 \rangle$ is

    $$\mathbf{u} \cdot \mathbf{v} = u_1v_1 + u_2v_2 + u_3v_3. \qquad \text{Dot product}$$

Find the component form and magnitude of the vector **v** that has initial point $(3, 4, 2)$ and terminal point $(3, 6, 4)$. Then find a unit vector in the direction of **v**.

Solution   The component form of **v** is $\mathbf{v} = \langle 3 - 3, 6 - 4, 4 - 2 \rangle = \langle 0, 2, 2 \rangle$. The magnitude of **v** is

$$\|\mathbf{v}\| = \sqrt{0^2 + 2^2 + 2^2} = \sqrt{8} = 2\sqrt{2}.$$

A unit vector in the direction of **v** is

$$\mathbf{u} = \frac{\mathbf{v}}{\|\mathbf{v}\|} = \frac{1}{2\sqrt{2}} \langle 0, 2, 2 \rangle = \left\langle 0, \frac{1}{\sqrt{2}}, \frac{1}{\sqrt{2}} \right\rangle = \left\langle 0, \frac{\sqrt{2}}{2}, \frac{\sqrt{2}}{2} \right\rangle.$$

Find the component form and magnitude of the vector **v** that has initial point $(1, -4, 3)$ and terminal point $(2, 2, -1)$. Then find a unit vector in the direction of **v**.

EXAMPLE 2   **Finding the Dot Product of Two Vectors**

The dot product of $\langle 0, 3, -2 \rangle$ and $\langle 4, -2, 3 \rangle$ is

$$\langle 0, 3, -2 \rangle \cdot \langle 4, -2, 3 \rangle = 0(4) + 3(-2) + (-2)(3) = 0 - 6 - 6 = -12.$$

Note that the dot product of two vectors is a real number, not a vector.

Find the dot product of $\langle 4, 0, 1 \rangle$ and $\langle -1, 3, 2 \rangle$.

You learned in Section 6.4 that in the coodinate plane, the angle between two nonzero vectors is the angle θ, $0 \le \theta \le \pi$, between their respective standard position vectors, as shown in Figure 11.10. This angle can be found using the dot product. This is also true for vectors in space. (Note that the angle between the zero vector and another vector is not defined.)

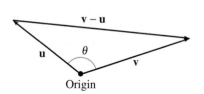

Origin

Figure 11.10

Angle Between Two Vectors

If θ is the angle between two nonzero vectors **u** and **v**, then

$$\cos \theta = \frac{\mathbf{u} \cdot \mathbf{v}}{\|\mathbf{u}\|\|\mathbf{v}\|}.$$

EXAMPLE 3   **Finding the Angle Between Two Vectors**

Find the angle θ between $\mathbf{u} = \langle 1, 0, 2 \rangle$ and $\mathbf{v} = \langle 3, 1, 0 \rangle$ shown in Figure 11.11.

Solution

$$\cos \theta = \frac{\mathbf{u} \cdot \mathbf{v}}{\|\mathbf{u}\|\|\mathbf{v}\|} = \frac{\langle 1, 0, 2 \rangle \cdot \langle 3, 1, 0 \rangle}{\|\langle 1, 0, 2 \rangle\|\|\langle 3, 1, 0 \rangle\|} = \frac{3}{\sqrt{50}} \implies \theta \approx 64.9°$$

Find the angle θ between $\mathbf{u} = \langle 1, -2, 4 \rangle$ and $\mathbf{v} = \langle 2, -3, -1 \rangle$.

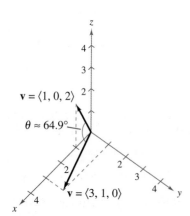

$\mathbf{v} = \langle 1, 0, 2 \rangle$

$\theta \approx 64.9°$

$\mathbf{v} = \langle 3, 1, 0 \rangle$

Figure 11.11

Orthogonal and Parallel Vectors

Recall that if the dot product of two nonzero vectors is zero, then the angle between the vectors is 90° and the vectors are orthogonal. Note that the standard unit vectors **i**, **j**, and **k** are orthogonal to each other.

> **EXAMPLE 4**   **Determining Orthogonal Vectors**

Determine whether each vector is orthogonal to the vector $\mathbf{w} = \langle -2, 1, 3 \rangle$.

a. $\mathbf{u} = \langle -3, -1, 2 \rangle$ **b.** $\mathbf{v} = \langle 3, 0, 2 \rangle$

Solution

a. The dot product of **w** and **u** is

$$\mathbf{w} \cdot \mathbf{u} = \langle -2, 1, 3 \rangle \cdot \langle -3, -1, 2 \rangle = -2(-3) + 1(-1) + 3(2) = 11.$$

The dot product is not 0, so the vectors **w** and **u** are *not* orthogonal.

b. The dot product of **w** and **v** is

$$\mathbf{w} \cdot \mathbf{v} = \langle -2, 1, 3 \rangle \cdot \langle 3, 0, 2 \rangle = -2(3) + 1(0) + 3(2) = 0.$$

The dot product is 0, so the vectors **w** and **v** *are* orthogonal.

✓ **Checkpoint**  Audio-video solution in English & Spanish at LarsonPrecalculus.com

Determine whether each vector is orthogonal to the vector $\mathbf{w} = \langle 4, 2, -1 \rangle$.

a. $\mathbf{u} = \langle -1, 2, -4 \rangle$ **b.** $\mathbf{v} = \langle 1, -1, 2 \rangle$

Recall from the definition of scalar multiplication that positive scalar multiples of a nonzero vector **v** have the same direction as **v**, whereas negative multiples have the direction opposite of **v**. In general, two nonzero vectors **u** and **v** are **parallel** when there is some scalar c such that $\mathbf{u} = c\mathbf{v}$. For example, in Figure 11.12, the vectors **u**, **v**, and **w** are parallel because $\mathbf{u} = 2\mathbf{v}$ and $\mathbf{w} = -\mathbf{v}$.

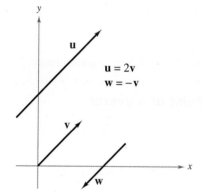

$\mathbf{u} = 2\mathbf{v}$
$\mathbf{w} = -\mathbf{v}$

Figure 11.12

> **EXAMPLE 5**   **Determining Parallel Vectors**

Vector **w** has initial point $(1, -2, 0)$ and terminal point $(3, 2, 1)$. Determine whether each vector is parallel to **w**.

a. $\mathbf{u} = \langle 4, 8, 2 \rangle$ **b.** $\mathbf{v} = \langle 4, 8, 4 \rangle$

Solution   Begin by writing **w** in component form.

$$\mathbf{w} = \langle 3 - 1, 2 - (-2), 1 - 0 \rangle = \langle 2, 4, 1 \rangle$$

a. Because $\mathbf{u} = \langle 4, 8, 2 \rangle = 2\langle 2, 4, 1 \rangle = 2\mathbf{w}$, you can conclude that **u** *is* parallel to **w**.

b. You need to find a scalar c such that

$$\langle 4, 8, 4 \rangle = c\langle 2, 4, 1 \rangle.$$

However, equating corresponding components produces $c = 2$ for the first two components and $c = 4$ for the third. So, the equation has no solution, and the vectors **v** and **w** are *not* parallel.

✓ **Checkpoint**  Audio-video solution in English & Spanish at LarsonPrecalculus.com

Vector **w** has initial point $(2, -2, 1)$ and terminal point $(0, 4, -3)$. Determine whether each vector is parallel to **w**.

a. $\mathbf{u} = \langle -1, 3, -2 \rangle$ **b.** $\mathbf{v} = \langle -2, 6, -5 \rangle$

You can use vectors to determine whether three points are collinear (lie on the same line). The points P, Q, and R are **collinear** if and only if the vectors $\overrightarrow{PQ}$ and $\overrightarrow{PR}$ are parallel.

EXAMPLE 6 Using Vectors to Determine Collinear Points

Determine whether the points $P(2, -1, 4)$, $Q(5, 4, 6)$, and $R(-4, -11, 0)$ are collinear.

Solution The component forms of $\overrightarrow{PQ}$ and $\overrightarrow{PR}$ are

$$\overrightarrow{PQ} = \langle 5 - 2, 4 - (-1), 6 - 4 \rangle = \langle 3, 5, 2 \rangle$$

and

$$\overrightarrow{PR} = \langle -4 - 2, -11 - (-1), 0 - 4 \rangle = \langle -6, -10, -4 \rangle.$$

Because $\overrightarrow{PR} = -2\overrightarrow{PQ}$, you can conclude that $\overrightarrow{PQ}$ and $\overrightarrow{PR}$ are parallel. So, the points P, Q, and R are collinear, as shown in the figure below.

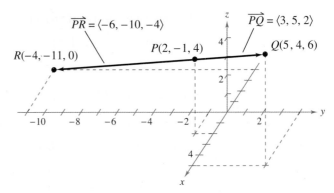

✓ **Checkpoint** ◀))) Audio-video solution in English & Spanish at LarsonPrecalculus.com

Determine whether the points $P(-2, 7, -3)$, $Q(1, 4, 3)$, and $R(3, 2, 7)$ are collinear.

EXAMPLE 7 Finding the Terminal Point of a Vector

The initial point of the vector $\mathbf{v} = \langle 4, 2, -1 \rangle$ is $P(3, -1, 6)$. What is the terminal point of this vector?

Solution Using the component form of the vector whose initial point is $P(3, -1, 6)$ and whose terminal point is $Q(q_1, q_2, q_3)$, you have

$$\overrightarrow{PQ} = \langle q_1 - p_1, q_2 - p_2, q_3 - p_3 \rangle$$
$$= \langle q_1 - 3, q_2 - (-1), q_3 - 6 \rangle$$
$$= \langle 4, 2, -1 \rangle.$$

This implies that

$$q_1 - 3 = 4, \quad q_2 + 1 = 2, \quad \text{and} \quad q_3 - 6 = -1.$$

The solutions of these three equations are

$$q_1 = 7, \quad q_2 = 1, \quad \text{and} \quad q_3 = 5.$$

So, the terminal point is $Q(7, 1, 5)$.

✓ **Checkpoint** ◀))) Audio-video solution in English & Spanish at LarsonPrecalculus.com

The initial point of the vector $\mathbf{v} = \langle 2, -3, 6 \rangle$ is $P(1, -2, -5)$. What is the terminal point of this vector?

Application

The next example shows how to use vectors to solve an equilibrium problem in space.

EXAMPLE 8 **Solving an Equilibrium Problem**

A weight of 480 pounds is supported by three ropes. As shown in Figure 11.13, the weight is located at $S(0, 2, -1)$. The ropes are attached at the points $P(2, 0, 0)$, $Q(0, 4, 0)$, and $R(-2, 0, 0)$. Find the force (or tension) on each rope.

Solution The (downward) force of the weight is represented by the vector

$$\mathbf{w} = \langle 0, 0, -480 \rangle.$$

Find the force vectors corresponding to the ropes.

$$\mathbf{u} = \|\mathbf{u}\| \frac{\overrightarrow{SP}}{\|\overrightarrow{SP}\|} = \|\mathbf{u}\| \frac{\langle 2 - 0, 0 - 2, 0 - (-1) \rangle}{3} = \|\mathbf{u}\| \left\langle \frac{2}{3}, -\frac{2}{3}, \frac{1}{3} \right\rangle$$

$$\mathbf{v} = \|\mathbf{v}\| \frac{\overrightarrow{SQ}}{\|\overrightarrow{SQ}\|} = \|\mathbf{v}\| \frac{\langle 0 - 0, 4 - 2, 0 - (-1) \rangle}{\sqrt{5}} = \|\mathbf{v}\| \left\langle 0, \frac{2}{\sqrt{5}}, \frac{1}{\sqrt{5}} \right\rangle$$

$$\mathbf{z} = \|\mathbf{z}\| \frac{\overrightarrow{SR}}{\|\overrightarrow{SR}\|} = \|\mathbf{z}\| \frac{\langle -2 - 0, 0 - 2, 0 - (-1) \rangle}{3} = \|\mathbf{z}\| \left\langle -\frac{2}{3}, -\frac{2}{3}, \frac{1}{3} \right\rangle$$

For the system to be in equilibrium, it must be true that

$$\mathbf{u} + \mathbf{v} + \mathbf{z} + \mathbf{w} = \mathbf{0} \quad \text{or} \quad \mathbf{u} + \mathbf{v} + \mathbf{z} = -\mathbf{w}.$$

This yields the system of linear equations below.

$$\frac{2}{3}\|\mathbf{u}\| \qquad\qquad -\frac{2}{3}\|\mathbf{z}\| = \quad 0$$

$$-\frac{2}{3}\|\mathbf{u}\| + \frac{2}{\sqrt{5}}\|\mathbf{v}\| - \frac{2}{3}\|\mathbf{z}\| = \quad 0$$

$$\frac{1}{3}\|\mathbf{u}\| + \frac{1}{\sqrt{5}}\|\mathbf{v}\| + \frac{1}{3}\|\mathbf{z}\| = 480$$

Using the techniques demonstrated in Chapter 7, the solution of the system is

$$\|\mathbf{u}\| = 360.0, \quad \|\mathbf{v}\| \approx 536.7, \quad \text{and} \quad \|\mathbf{z}\| = 360.0.$$

So, the rope attached at point P has 360 pounds of tension, the rope attached at point Q has about 536.7 pounds of tension, and the rope attached at point R has 360 pounds of tension.

✓ *Checkpoint* 🔊))) *Audio-video solution in English & Spanish at LarsonPrecalculus.com*

A weight of 240 pounds, located at $S(0, 3, -2)$, is supported by ropes attached at the points $P(3, 0, 0)$, $Q(0, 6, 0)$, and $R(-3, 0, 0)$. Find the force (or tension) on each rope.

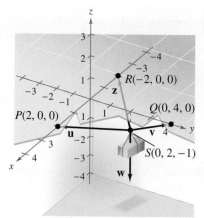

Figure 11.13

For a system in equilibrium, the sum of all the forces in the system must be zero.

Summarize (Section 11.2)

1. Explain how to extend the concept of a vector to three-dimensional space (*pages 785 and 786, Examples 1–3*).

2. Explain how to determine whether vectors in space are orthogonal or parallel (*page 787, Examples 4 and 5*).

3. Describe a real-life application of vectors in space (*page 789, Example 8*).

11.2   Exercises   See CalcChat.com for tutorial help and worked-out solutions to odd-numbered exercises.

Vocabulary: Fill in the blanks.

1. In space, the _____ vector is denoted by $\mathbf{0} = \langle 0, 0, 0 \rangle$.

2. In space, the standard unit vector notation for a vector $\mathbf{v}$ is _____.

3. The _____ _____ of a vector $\mathbf{v}$ is produced by subtracting the coordinates of the initial point from the corresponding coordinates of the terminal point.

4. When the dot product of two nonzero vectors is zero, the angle between the vectors is 90° and the vectors are _____.

5. Two nonzero vectors $\mathbf{u}$ and $\mathbf{v}$ are _____ when there is some scalar c such that $\mathbf{u} = c\mathbf{v}$.

6. The points P, Q, and R are _____ if and only if the vectors $\overrightarrow{PQ}$ and $\overrightarrow{PR}$ are parallel.

Skills and Applications

 Finding the Component Form of a Vector In Exercises 7–14, find the component form of the vector $\mathbf{v}$.

7.

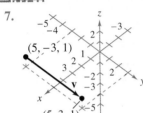

8.

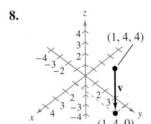

9.

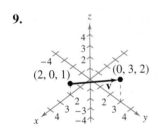

10.

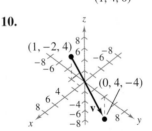

Initial point	Terminal point
11. $(-5, 4, -3)$	$(2, -1, 4)$
12. $(-8, 2, 6)$	$(0, 3, 0)$
13. $(-2, 3, -5)$	$(1, 4, -5)$
14. $(-1, 0, 2)$	$(-4, 3, -1)$

Sketching Vectors In Exercises 15–18, sketch (a) $2\mathbf{v}$, (b) $-\mathbf{v}$, (c) $\frac{3}{2}\mathbf{v}$, and (d) $-\frac{1}{2}\mathbf{v}$ with the initial point at the origin.

15. $\mathbf{v} = \langle 1, 1, 3 \rangle$ **16.** $\mathbf{v} = \langle -1, 2, 2 \rangle$

17. $\mathbf{v} = 2\mathbf{i} + 2\mathbf{j} - \mathbf{k}$

18. $\mathbf{v} = \mathbf{i} - 2\mathbf{j} + \mathbf{k}$

Using Vector Operations In Exercises 19–24, find the vector $\mathbf{z}$, given $\mathbf{u} = \langle -1, 3, 2 \rangle$, $\mathbf{v} = \langle 5, -2, 1 \rangle$, and $\mathbf{w} = \langle 4, -1, -4 \rangle$.

19. $\mathbf{z} = \mathbf{u} - 2\mathbf{v}$ **20.** $\mathbf{z} = 3\mathbf{w} - 2\mathbf{v} + \mathbf{u}$

21. $\mathbf{z} = 2\mathbf{u} - 3\mathbf{v} + \frac{1}{2}\mathbf{w}$ **22.** $\mathbf{z} = 7\mathbf{u} + \mathbf{v} - \frac{1}{4}\mathbf{w}$

23. $2\mathbf{z} - 4\mathbf{u} = \mathbf{w}$ **24.** $\mathbf{u} + \mathbf{v} + \mathbf{z} = \mathbf{0}$

 Finding the Magnitude of a Vector In Exercises 25–34, find the magnitude of $\mathbf{v}$.

25. $\mathbf{v} = \langle 2, 5, 4 \rangle$ **26.** $\mathbf{v} = \langle 1, 2, 3 \rangle$

27. $\mathbf{v} = \langle 4, 3, -5 \rangle$ **28.** $\mathbf{v} = \langle -1, 0, 3 \rangle$

29. $\mathbf{v} = \mathbf{i} + 3\mathbf{j} - \mathbf{k}$ **30.** $\mathbf{v} = -\mathbf{i} - 4\mathbf{j} + 3\mathbf{k}$

31. $\mathbf{v} = 4\mathbf{i} - 3\mathbf{j} - 7\mathbf{k}$ **32.** $\mathbf{v} = 2\mathbf{i} - \mathbf{j} + 6\mathbf{k}$

33. Initial point: $(1, -3, 4)$; terminal point: $(1, 0, -1)$

34. Initial point: $(0, -1, 0)$; terminal point: $(1, 2, -2)$

 Finding a Unit Vector In Exercises 35–38, find a unit vector (a) in the direction of $\mathbf{v}$ and (b) in the direction opposite of $\mathbf{v}$.

35. $\mathbf{v} = \langle 2, 0, 5 \rangle$ **36.** $\mathbf{v} = \langle 3, 7, 0 \rangle$

37. $\mathbf{v} = 8\mathbf{i} + 3\mathbf{j} - \mathbf{k}$ **38.** $\mathbf{v} = -3\mathbf{i} + 5\mathbf{j} + 10\mathbf{k}$

 Finding the Dot Product of Two Vectors In Exercises 39–42, find the dot product of $\mathbf{u}$ and $\mathbf{v}$.

39. $\mathbf{u} = \langle 4, 4, -1 \rangle$ **40.** $\mathbf{u} = \langle 3, -1, 6 \rangle$
    $\mathbf{v} = \langle 2, -5, -8 \rangle$ $\mathbf{v} = \langle 4, -10, 1 \rangle$

41. $\mathbf{u} = 2\mathbf{i} - 5\mathbf{j} + 3\mathbf{k}$ **42.** $\mathbf{u} = 3\mathbf{j} - 6\mathbf{k}$
    $\mathbf{v} = 9\mathbf{i} + 3\mathbf{j} - \mathbf{k}$ $\mathbf{v} = 6\mathbf{i} - 4\mathbf{j} - 2\mathbf{k}$

 Finding the Angle Between Two Vectors In Exercises 43–46, find the angle θ between the vectors.

43. $\mathbf{u} = \langle 0, 2, 2 \rangle$ **44.** $\mathbf{u} = \langle -1, 3, 0 \rangle$
    $\mathbf{v} = \langle 3, 0, -4 \rangle$ $\mathbf{v} = \langle 1, 2, -1 \rangle$

45. $\mathbf{u} = 10\mathbf{i} + 40\mathbf{j}$ **46.** $\mathbf{u} = 8\mathbf{j} - 20\mathbf{k}$
    $\mathbf{v} = -3\mathbf{j} + 8\mathbf{k}$ $\mathbf{v} = 10\mathbf{i} - 5\mathbf{k}$

Determining Orthogonal and Parallel Vectors In Exercises 47–54, determine whether **u** and **v** are orthogonal, parallel, or neither.

47. $\mathbf{u} = \langle -12, 6, 15 \rangle$
$\mathbf{v} = \langle 8, -4, -10 \rangle$

48. $\mathbf{u} = \langle -1, 3, -1 \rangle$
$\mathbf{v} = \langle 2, -1, 5 \rangle$

49. $\mathbf{u} = \langle 0, 1, 6 \rangle$
$\mathbf{v} = \langle 1, -2, -1 \rangle$

50. $\mathbf{u} = \langle 0, 4, -1 \rangle$
$\mathbf{v} = \langle 1, 0, 0 \rangle$

51. $\mathbf{u} = \frac{3}{4}\mathbf{i} - \frac{1}{2}\mathbf{j} + 2\mathbf{k}$
$\mathbf{v} = 4\mathbf{i} + 10\mathbf{j} + \mathbf{k}$

52. $\mathbf{u} = -\mathbf{i} + \frac{1}{2}\mathbf{j} - \mathbf{k}$
$\mathbf{v} = 8\mathbf{i} - 4\mathbf{j} + 8\mathbf{k}$

53. $\mathbf{u} = -2\mathbf{i} + 3\mathbf{j} - \mathbf{k}$
$\mathbf{v} = 2\mathbf{i} + \mathbf{j} - \mathbf{k}$

54. $\mathbf{u} = 2\mathbf{i} - 3\mathbf{j} + \mathbf{k}$
$\mathbf{v} = -\mathbf{i} - \mathbf{j} - \mathbf{k}$

Using Vectors to Determine Collinear Points In Exercises 55–58, use vectors to determine whether the points are collinear.

55. $(5, 4, 1), (7, 3, -1), (4, 5, 3)$

56. $(-2, 7, 4), (-4, 8, 1), (0, 6, 7)$

57. $(1, 3, 2), (-1, 2, 5), (3, 4, -1)$

58. $(0, 4, 4), (-1, 5, 6), (-2, 6, 7)$

Finding the Terminal Point of a Vector In Exercises 59 and 60, the vector **v** and its initial point are given. Find the terminal point.

59. $\mathbf{v} = \langle 2, -4, 7 \rangle$
Initial point: $(1, 5, 0)$

60. $\mathbf{v} = \langle 4, -1, -1 \rangle$
Initial point: $(6, -4, 3)$

Finding a Scalar In Exercises 61 and 62, determine the values of c that satisfy the given conditions.

61. $\|c\mathbf{u}\| = 3, \quad \mathbf{u} = \mathbf{i} + 2\mathbf{j} + 3\mathbf{k}$

62. $\|c\mathbf{u}\| = 12, \quad \mathbf{u} = -2\mathbf{i} + 2\mathbf{j} - 4\mathbf{k}$

Finding a Vector In Exercises 63 and 64, find the component form of **v**.

63. Vector **v** lies in the yz-plane, has magnitude 4, and makes an angle of $45°$ with the positive y-axis.

64. Vector **v** lies in the xz-plane, has magnitude 10, and makes an angle of $60°$ with the positive z-axis.

65. Tension The weight of a crate is 500 newtons. Find the tension in each of the supporting cables shown in the figure.

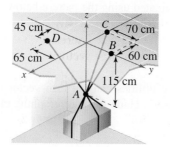

66. Tension

The lights in an auditorium are 24-pound disks of radius 18 inches. Each disk is supported by three equally spaced cables that are L inches long (see figure).

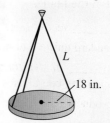

(a) Write the tension T in each cable as a function of L. Determine the domain of the function.

(b) Use a graphing utility to graph the function in part (a). What are the asymptotes of the graph? Interpret their meaning in the context of the problem.

(c) Determine the minimum length of each cable when a cable can carry a maximum load of 10 pounds.

Exploration

True or False? In Exercises 67 and 68, determine whether the statement is true or false. Justify your answer.

67. If the dot product of two nonzero vectors is zero, then the angle between the vectors is a right angle.

68. If $\overrightarrow{AB}$ and $\overrightarrow{AC}$ are parallel vectors, then points A, B, and C are collinear.

69. Think About It What is known about the nonzero vectors **u** and **v** when $\mathbf{u} \cdot \mathbf{v} < 0$? Explain.

70. HOW DO YOU SEE IT? Use the figure below.

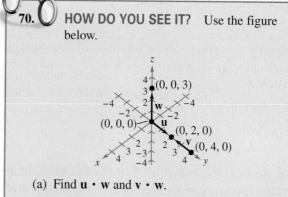

(a) Find $\mathbf{u} \cdot \mathbf{w}$ and $\mathbf{v} \cdot \mathbf{w}$.

(b) A vector **z** is parallel to **u** and **v**. What is the dot product of **z** and **w**? Explain.

11.3 The Cross Product of Two Vectors

- ◼ Find cross products of vectors in space.
- ◼ Use geometric properties of cross products of vectors in space.
- ◼ Use triple scalar products to find volumes of parallelepipeds.

The Cross Product

Many applications in physics, engineering, and geometry involve finding a vector in space that is orthogonal to two given vectors. In this section, you will study a product that yields such a vector. It is called the **cross product,** and it is conveniently defined and calculated using the standard unit vector notation.

The cross product of two vectors in space has many applications in physics and engineering. For example, in Exercise 59 on page 798, you will use a cross product to find the torque on the crank of a bicycle pedal.

Definition of the Cross Product of Two Vectors in Space

Let

$$\mathbf{u} = u_1\mathbf{i} + u_2\mathbf{j} + u_3\mathbf{k}$$

and

$$\mathbf{v} = v_1\mathbf{i} + v_2\mathbf{j} + v_3\mathbf{k}$$

be vectors in space. The **cross product** of $\mathbf{u}$ and $\mathbf{v}$ is the vector

$$\mathbf{u} \times \mathbf{v} = (u_2v_3 - u_3v_2)\mathbf{i} - (u_1v_3 - u_3v_1)\mathbf{j} + (u_1v_2 - u_2v_1)\mathbf{k}.$$

Note that this definition applies only to three-dimensional vectors. The cross product is not defined for two-dimensional vectors.

A convenient way to calculate $\mathbf{u} \times \mathbf{v}$ is to use the *determinant form* with cofactor expansion shown below. (This 3×3 determinant form is used simply to help remember the formula for the cross product—the corresponding array is technically not a matrix because its entries are not all numbers.)

$$\mathbf{u} \times \mathbf{v} = \begin{vmatrix} \mathbf{i} & \mathbf{j} & \mathbf{k} \\ u_1 & u_2 & u_3 \\ v_1 & v_2 & v_3 \end{vmatrix} \quad \Longleftarrow \text{ Components of } \mathbf{u}$$
$$\Longleftarrow \text{ Components of } \mathbf{v}$$

$$= \begin{vmatrix} \mathbf{i} & \mathbf{j} & \mathbf{k} \\ u_1 & u_2 & u_3 \\ v_1 & v_2 & v_3 \end{vmatrix}\mathbf{i} - \begin{vmatrix} \mathbf{i} & \mathbf{j} & \mathbf{k} \\ u_1 & u_2 & u_3 \\ v_1 & v_2 & v_3 \end{vmatrix}\mathbf{j} + \begin{vmatrix} \mathbf{i} & \mathbf{j} & \mathbf{k} \\ u_1 & u_2 & u_3 \\ v_1 & v_2 & v_3 \end{vmatrix}\mathbf{k}$$

$$= \begin{vmatrix} u_2 & u_3 \\ v_2 & v_3 \end{vmatrix}\mathbf{i} - \begin{vmatrix} u_1 & u_3 \\ v_1 & v_3 \end{vmatrix}\mathbf{j} + \begin{vmatrix} u_1 & u_2 \\ v_1 & v_2 \end{vmatrix}\mathbf{k}$$

$$= (u_2v_3 - u_3v_2)\mathbf{i} - (u_1v_3 - u_3v_1)\mathbf{j} + (u_1v_2 - u_2v_1)\mathbf{k}$$

Note the minus sign in front of the **j**-component. Recall from Section 8.4 that each of the three 2×2 determinants can be evaluated using the pattern below.

$$\begin{vmatrix} a_1 & b_1 \\ a_2 & b_2 \end{vmatrix} = a_1b_2 - a_2b_1$$

▷ **TECHNOLOGY** Some graphing utilities can perform vector operations, such as the cross product. Consult the user's guide for your graphing utility for specific instructions.

EXAMPLE 1   **Finding Cross Products**

Given $\mathbf{u} = \mathbf{i} + 2\mathbf{j} + \mathbf{k}$ and $\mathbf{v} = 3\mathbf{i} + \mathbf{j} + 2\mathbf{k}$, find each cross product.

a. $\mathbf{u} \times \mathbf{v}$ **b.** $\mathbf{v} \times \mathbf{u}$ **c.** $\mathbf{v} \times \mathbf{v}$

Solution

a. $\mathbf{u} \times \mathbf{v} = \begin{vmatrix} \mathbf{i} & \mathbf{j} & \mathbf{k} \\ 1 & 2 & 1 \\ 3 & 1 & 2 \end{vmatrix}$

$= \begin{vmatrix} 2 & 1 \\ 1 & 2 \end{vmatrix}\mathbf{i} - \begin{vmatrix} 1 & 1 \\ 3 & 2 \end{vmatrix}\mathbf{j} + \begin{vmatrix} 1 & 2 \\ 3 & 1 \end{vmatrix}\mathbf{k}$

$= (4 - 1)\mathbf{i} - (2 - 3)\mathbf{j} + (1 - 6)\mathbf{k}$

$= 3\mathbf{i} + \mathbf{j} - 5\mathbf{k}$

b. $\mathbf{v} \times \mathbf{u} = \begin{vmatrix} \mathbf{i} & \mathbf{j} & \mathbf{k} \\ 3 & 1 & 2 \\ 1 & 2 & 1 \end{vmatrix}$

$= \begin{vmatrix} 1 & 2 \\ 2 & 1 \end{vmatrix}\mathbf{i} - \begin{vmatrix} 3 & 2 \\ 1 & 1 \end{vmatrix}\mathbf{j} + \begin{vmatrix} 3 & 1 \\ 1 & 2 \end{vmatrix}\mathbf{k}$

$= (1 - 4)\mathbf{i} - (3 - 2)\mathbf{j} + (6 - 1)\mathbf{k}$

$= -3\mathbf{i} - \mathbf{j} + 5\mathbf{k}$

Note that this result is the negative of that in part (a).

c. $\mathbf{v} \times \mathbf{v} = \begin{vmatrix} \mathbf{i} & \mathbf{j} & \mathbf{k} \\ 3 & 1 & 2 \\ 3 & 1 & 2 \end{vmatrix} = \mathbf{0}$

✓ **Checkpoint** 🔊))) *Audio-video solution in English & Spanish at LarsonPrecalculus.com*

Given $\mathbf{u} = \mathbf{i} - 3\mathbf{j} + 2\mathbf{k}$ and $\mathbf{v} = 2\mathbf{i} - \mathbf{j} + 2\mathbf{k}$, find each cross product.

a. $\mathbf{u} \times \mathbf{v}$ **b.** $\mathbf{v} \times \mathbf{u}$ **c.** $\mathbf{v} \times \mathbf{v}$ ◼

The results obtained in Example 1 suggest some interesting algebraic properties of the cross product. For example,

$$\mathbf{u} \times \mathbf{v} = -(\mathbf{v} \times \mathbf{u}) \quad \text{and} \quad \mathbf{v} \times \mathbf{v} = \mathbf{0}.$$

These properties, and several others, are summarized below.

Algebraic Properties of the Cross Product

Let $\mathbf{u}$, $\mathbf{v}$, and $\mathbf{w}$ be vectors in space and let c be a scalar.

1. $\mathbf{u} \times \mathbf{v} = -(\mathbf{v} \times \mathbf{u})$
2. $\mathbf{u} \times (\mathbf{v} + \mathbf{w}) = (\mathbf{u} \times \mathbf{v}) + (\mathbf{u} \times \mathbf{w})$
3. $c(\mathbf{u} \times \mathbf{v}) = (c\mathbf{u}) \times \mathbf{v} = \mathbf{u} \times (c\mathbf{v})$
4. $\mathbf{u} \times \mathbf{0} = \mathbf{0} \times \mathbf{u} = \mathbf{0}$
5. $\mathbf{u} \times \mathbf{u} = \mathbf{0}$
6. $\mathbf{u} \cdot (\mathbf{v} \times \mathbf{w}) = (\mathbf{u} \times \mathbf{v}) \cdot \mathbf{w}$

For proofs of the Algebraic Properties of the Cross Product, see Proofs in Mathematics on page 813.

Geometric Properties of the Cross Product

The first property listed on the preceding page indicates that the cross product is *not commutative*. In particular, this property indicates that the vectors **u** × **v** and **v** × **u** have equal lengths but opposite directions. The list below gives some *geometric* properties of the cross product of two vectors.

> ### Geometric Properties of the Cross Product
>
> Let **u** and **v** be nonzero vectors in space, and let θ be the angle between **u** and **v**.
>
> **1.** **u** × **v** is orthogonal to both **u** and **v**.
>
> **2.** $\|\mathbf{u} \times \mathbf{v}\| = \|\mathbf{u}\|\|\mathbf{v}\| \sin \theta$
>
> **3.** **u** × **v** = **0** if and only if **u** and **v** are scalar multiples of each other.
>
> **4.** $\|\mathbf{u} \times \mathbf{v}\|$ = area of parallelogram that has **u** and **v** as adjacent sides.

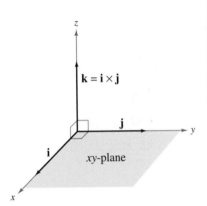

$\mathbf{k} = \mathbf{i} \times \mathbf{j}$

xy-plane

Figure 11.14

For proofs of the Geometric Properties of the Cross Product, see Proofs in Mathematics on page 814.

Both **u** × **v** and **v** × **u** are perpendicular to the plane determined by **u** and **v**. One way to remember the orientations of the vectors **u**, **v**, and **u** × **v** is to compare them with the unit vectors **i**, **j**, and **k** = **i** × **j**, respectively, as shown in Figure 11.14. The three vectors **u**, **v**, and **u** × **v** form a *right-handed system*.

EXAMPLE 2   Using the Cross Product

See LarsonPrecalculus.com for an interactive version of this type of example.

Find a unit vector that is orthogonal to both

$$\mathbf{u} = 3\mathbf{i} - 4\mathbf{j} + \mathbf{k}$$

and

$$\mathbf{v} = -3\mathbf{i} + 6\mathbf{j}.$$

Solution   Find the cross product **u** × **v**.

$$\mathbf{u} \times \mathbf{v} = \begin{vmatrix} \mathbf{i} & \mathbf{j} & \mathbf{k} \\ 3 & -4 & 1 \\ -3 & 6 & 0 \end{vmatrix}$$

$$= -6\mathbf{i} - 3\mathbf{j} + 6\mathbf{k}$$

The result is orthogonal to both **u** and **v**, as shown in Figure 11.15. The length of **u** × **v** is

$$\|\mathbf{u} \times \mathbf{v}\| = \sqrt{(-6)^2 + (-3)^2 + 6^2}$$

$$= \sqrt{81}$$

$$= 9.$$

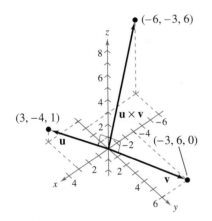

Figure 11.15

So, a unit vector orthogonal to both **u** and **v** is

$$\frac{\mathbf{u} \times \mathbf{v}}{\|\mathbf{u} \times \mathbf{v}\|} = -\frac{2}{3}\mathbf{i} - \frac{1}{3}\mathbf{j} + \frac{2}{3}\mathbf{k}.$$

✓ *Checkpoint* 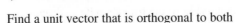 *Audio-video solution in English & Spanish at LarsonPrecalculus.com*

Find a unit vector that is orthogonal to both

$$\mathbf{u} = 2\mathbf{i} + 4\mathbf{j} - 3\mathbf{k} \quad \text{and} \quad \mathbf{v} = -2\mathbf{i} - 3\mathbf{j} + 2\mathbf{k}.$$

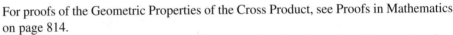

In Example 2, note that you could have used the cross product $\mathbf{v} \times \mathbf{u}$ to form a unit vector that is orthogonal to both $\mathbf{u}$ and $\mathbf{v}$. With that choice, you would have obtained the *negative* of the unit vector found in the example.

The fourth geometric property of the cross product states that $\|\mathbf{u} \times \mathbf{v}\|$ is the area of the parallelogram that has $\mathbf{u}$ and $\mathbf{v}$ as adjacent sides. It follows that the area of a triangle that has vectors $\mathbf{u}$ and $\mathbf{v}$ as adjacent sides is $\frac{1}{2}\|\mathbf{u} \times \mathbf{v}\|$.

EXAMPLE 3   Geometric Application of the Cross Product

Show that the points A, B, C, and D are the vertices of a parallelogram. Then find the area of the parallelogram.

$$A(5, 2, 0), \quad B(2, 6, 1), \quad C(2, 4, 7), \quad D(5, 0, 6)$$

Solution   From Figure 11.16, you can see that the sides of quadrilateral $ABCD$ correspond to the four vectors below.

$$\overrightarrow{AB} = -3\mathbf{i} + 4\mathbf{j} + \mathbf{k} \qquad \overrightarrow{CD} = 3\mathbf{i} - 4\mathbf{j} - \mathbf{k} = -\overrightarrow{AB}$$
$$\overrightarrow{AD} = 0\mathbf{i} - 2\mathbf{j} + 6\mathbf{k} \qquad \overrightarrow{CB} = 0\mathbf{i} + 2\mathbf{j} - 6\mathbf{k} = -\overrightarrow{AD}$$

Because $\overrightarrow{CD} = -\overrightarrow{AB}$ and $\overrightarrow{CB} = -\overrightarrow{AD}$, you can conclude that $\overrightarrow{AB}$ is parallel to $\overrightarrow{CD}$ and $\overrightarrow{AD}$ is parallel to $\overrightarrow{CB}$. It follows that the quadrilateral is a parallelogram with $\overrightarrow{AB}$ and $\overrightarrow{AD}$ as adjacent sides. Moreover,

$$\overrightarrow{AB} \times \overrightarrow{AD} = \begin{vmatrix} \mathbf{i} & \mathbf{j} & \mathbf{k} \\ -3 & 4 & 1 \\ 0 & -2 & 6 \end{vmatrix} = 26\mathbf{i} + 18\mathbf{j} + 6\mathbf{k}$$

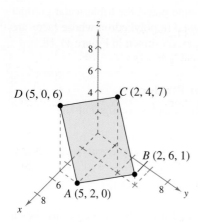

Figure 11.16

so the area of the parallelogram is

$$\|\overrightarrow{AB} \times \overrightarrow{AD}\| = \sqrt{26^2 + 18^2 + 6^2} = \sqrt{1036} \approx 32.19.$$

✓ *Checkpoint* ◀))) *Audio-video solution in English & Spanish at LarsonPrecalculus.com*

Show that the points A, B, C, and D are the vertices of a parallelogram. Then find the area of the parallelogram.

$$A(4, 3, 0), \quad B(3, 4, 2), \quad C(3, 2, 6), \quad D(4, 1, 4)$$

EXAMPLE 4   Finding the Area of a Triangle

Find the area of the triangle with vertices

$$A(4, 3, 0), \quad B(5, 4, 6), \quad \text{and} \quad C(4, 1, 5).$$

Solution   Figure 11.17 shows the triangle. The cross product of

$$\overrightarrow{AB} = \mathbf{i} + \mathbf{j} + 6\mathbf{k} \quad \text{and} \quad \overrightarrow{AC} = 0\mathbf{i} - 2\mathbf{j} + 5\mathbf{k}$$

is

$$\overrightarrow{AB} \times \overrightarrow{AC} = \begin{vmatrix} \mathbf{i} & \mathbf{j} & \mathbf{k} \\ 1 & 1 & 6 \\ 0 & -2 & 5 \end{vmatrix} = 17\mathbf{i} - 5\mathbf{j} - 2\mathbf{k}.$$

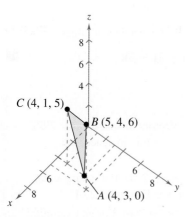

Figure 11.17

So, the area of the triangle is

$$A = \tfrac{1}{2}\|\overrightarrow{AB} \times \overrightarrow{AC}\| = \tfrac{1}{2}\sqrt{17^2 + (-5)^2 + (-2)^2} = \tfrac{1}{2}\sqrt{318} \approx 8.9.$$

✓ *Checkpoint* ◀))) *Audio-video solution in English & Spanish at LarsonPrecalculus.com*

Find the area of the triangle with vertices

$$A(3, 2, 1), \quad B(3, 1, 7), \quad \text{and} \quad C(5, 2, 6).$$

The Triple Scalar Product

> ### The Triple Scalar Product
>
> For $\mathbf{u} = u_1\mathbf{i} + u_2\mathbf{j} + u_3\mathbf{k}$, $\mathbf{v} = v_1\mathbf{i} + v_2\mathbf{j} + v_3\mathbf{k}$, and $\mathbf{w} = w_1\mathbf{i} + w_2\mathbf{j} + w_3\mathbf{k}$, the **triple scalar product** is
>
> $$\mathbf{u} \cdot (\mathbf{v} \times \mathbf{w}) = \begin{vmatrix} u_1 & u_2 & u_3 \\ v_1 & v_2 & v_3 \\ w_1 & w_2 & w_3 \end{vmatrix}.$$

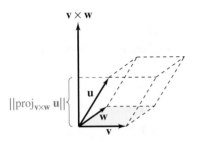

Area of base $= \|\mathbf{v} \times \mathbf{w}\|$
Volume of parallelepiped $= |\mathbf{u} \cdot (\mathbf{v} \times \mathbf{w})|$
Figure 11.18

When the vectors $\mathbf{u}$, $\mathbf{v}$, and $\mathbf{w}$ do not lie in the same plane, the triple scalar product $\mathbf{u} \cdot (\mathbf{v} \times \mathbf{w})$ determines the volume of a *parallelepiped* (a polyhedron whose faces are all parallelograms) with $\mathbf{u}$, $\mathbf{v}$, and $\mathbf{w}$ as adjacent edges, as shown in Figure 11.18.

> ### Geometric Property of the Triple Scalar Product
>
> The volume V of a parallelepiped with vectors $\mathbf{u}$, $\mathbf{v}$, and $\mathbf{w}$ as adjacent edges is
>
> $$V = |\mathbf{u} \cdot (\mathbf{v} \times \mathbf{w})|.$$

EXAMPLE 5   **Volume by the Triple Scalar Product**

Find the volume of the parallelepiped that has $\mathbf{u}$, $\mathbf{v}$, and $\mathbf{w}$ as adjacent edges, as shown in Figure 11.19.

Solution   Find the value of the triple scalar product.

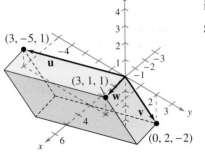

Figure 11.19

$$\mathbf{u} \cdot (\mathbf{v} \times \mathbf{w}) = \begin{vmatrix} 3 & -5 & 1 \\ 0 & 2 & -2 \\ 3 & 1 & 1 \end{vmatrix}$$

$$= 3\begin{vmatrix} 2 & -2 \\ 1 & 1 \end{vmatrix} - (-5)\begin{vmatrix} 0 & -2 \\ 3 & 1 \end{vmatrix} + 1\begin{vmatrix} 0 & 2 \\ 3 & 1 \end{vmatrix}$$

$$= 3(4) + 5(6) + 1(-6)$$

$$= 36$$

So, the volume of the parallelepiped is $|\mathbf{u} \cdot (\mathbf{v} \times \mathbf{w})| = |36| = 36$.

✓ *Checkpoint* 🔊))) *Audio-video solution in English & Spanish at LarsonPrecalculus.com*

Find the volume of the parallelepiped that has $\mathbf{u}$, $\mathbf{v}$, and $\mathbf{w}$ as adjacent edges.

$$\mathbf{u} = 2\mathbf{i} - \mathbf{j} + 4\mathbf{k}, \quad \mathbf{v} = \mathbf{i} + 3\mathbf{j} - 2\mathbf{k}, \quad \mathbf{w} = 2\mathbf{i} - 2\mathbf{j} + 3\mathbf{k}$$

> ### Summarize   (Section 11.3)
>
> 1. State the definition of the cross product of two vectors in space *(page 792)*. For an example of finding cross products of two vectors in space, see Example 1.
> 2. List geometric properties of the cross product *(page 794)*. For examples of using geometric properties of the cross product, see Examples 2–4.
> 3. State the definition of the triple scalar product *(page 796)*. For an example of using a triple scalar product to find the volume of a parallelepiped, see Example 5.

11.3   Exercises

See **CalcChat.com** for tutorial help and worked-out solutions to odd-numbered exercises.

Vocabulary: Fill in the blanks.

1. To find a vector in space that is orthogonal to two given vectors, find the _____ _____ of the two vectors.

2. The dot product of **u** and **v** × **w** is called the _____ _____ _____ of **u**, **v**, and **w**.

Skills and Applications

 Performing Vector Operations In Exercises 3–12, given u = 3i − j + 2k and v = 2i + j − k, evaluate the expression.

3. **u** × **v**   4. **v** × **u**

5. **v** × **v**   6. **v** × (**u** × **u**)

7. (2**u**) × **v**   8. **u** × (3**v**)

9. **u** × (−**v**)   10. (−2**u**) × **v**

11. **u** · (**u** × **v**)   12. **v** · (**v** × **u**)

 Finding the Cross Product In Exercises 13–24, find u × v and show that it is orthogonal to both u and v.

13. **u** = ⟨1, −1, 0⟩   14. **u** = ⟨−1, 1, 0⟩
    v = ⟨0, 1, −1⟩   **v** = ⟨1, 0, −1⟩

15. **u** = ⟨3, −2, 4⟩   16. **u** = ⟨4, 5, 2⟩
    v = ⟨0, 1, −1⟩   **v** = ⟨5, −2, −3⟩

17. **u** = 2**i** + 4**j** + 3**k**   18. **u** = 3**i** − 2**j** + **k**
    v = −**i** + 3**j** − 2**k**   **v** = −2**i** + **j** + 2**k**

19. **u** = 6**i** + 2**j** + **k**   20. **u** = **i** + $\frac{3}{2}$**j** − $\frac{5}{2}$**k**
    v = **i** + 3**j** − 2**k**   **v** = $\frac{1}{2}$**i** − $\frac{3}{4}$**j** + $\frac{1}{4}$**k**

21. **u** = 6**k**   22. **u** = $\frac{1}{3}$**i**
    v = −**i** + 3**j** + **k**   **v** = $\frac{2}{3}$**j** − 9**k**

23. **u** = −**i** + **k**   24. **u** = **i** − 2**k**
    v = **j** − 2**k**   **v** = −**j** + **k**

 Using the Cross Product In Exercises 25–32, find a unit vector that is orthogonal to both u and v.

25. **u** = ⟨2, −3, 4⟩   26. **u** = ⟨2, −1, 3⟩
    v = ⟨0, −1, 1⟩   **v** = ⟨1, 0, −2⟩

27. **u** = 3**i** + **j**   28. **u** = **i** + 2**j**
    v = **j** + **k**   **v** = **i** − 3**k**

29. **u** = **i** + **j** − **k**   30. **u** = **i** − 2**j** + 2**k**
    v = **i** + **j** + **k**   **v** = 2**i** − **j** − 2**k**

31. **u** = −8**i** + 4**j** + 6**k**   32. **u** = 9**i** + 3**j** − 12**k**
    v = $\frac{1}{2}$**i** − $\frac{1}{2}$**j** + $\frac{1}{4}$**k**   **v** = −$\frac{2}{3}$**i** − $\frac{1}{3}$**j** + $\frac{1}{3}$**k**

 Finding the Area of a Parallelogram In Exercises 33–40, find the area of the parallelogram that has u and v as adjacent sides.

33. **u** = ⟨0, 3, 0⟩   34. **u** = ⟨0, 2, 1⟩
    v = ⟨2, 1, 0⟩   **v** = ⟨2, 4, 0⟩

35. **u** = ⟨4, 4, −6⟩   36. **u** = ⟨4, −3, 2⟩
    v = ⟨0, 4, 6⟩   **v** = ⟨5, 0, 1⟩

37. **u** = **k**   38. **u** = **i** + 2**j** + 2**k**
    v = **i** + **k**   **v** = **i** + **k**

39. **u** = 4**i** + 3**j** + 7**k**   40. **u** = −2**i** + 3**j** + 2**k**
    v = 2**i** − 4**j** + 6**k**   **v** = **i** + 2**j** + 4**k**

Geometric Application of the Cross Product In Exercises 41–44, (a) show that the points are the vertices of a parallelogram and (b) find its area.

41. $A(2, −1, 4)$, $B(3, 1, 2)$, $C(0, 5, 6)$, $D(−1, 3, 8)$
42. $A(1, 1, 1)$, $B(6, 5, 2)$, $C(7, 7, 5)$, $D(2, 3, 4)$
43. $A(3, 2, −1)$, $B(−2, 2, −3)$, $C(3, 5, −2)$, $D(−2, 5, −4)$
44. $A(2, 1, 1)$, $B(2, 3, 1)$, $C(−2, 4, 1)$, $D(−2, 6, 1)$

Finding the Area of a Triangle In Exercises 45–48, find the area of the triangle with the given vertices.

45. $(0, 0, 0)$, $(1, 2, 3)$, $(3, 0, 0)$
46. $(1, 4, 3)$, $(2, 0, 2)$, $(2, 2, 0)$
47. $(2, 3, −5)$, $(−2, −2, 0)$, $(3, 0, 6)$
48. $(2, 4, 0)$, $(−2, −4, 0)$, $(0, 0, 4)$

Finding the Triple Scalar Product In Exercises 49–52, find the triple scalar product of the vectors.

49. **u** = ⟨3, 4, 4⟩, **v** = ⟨2, 3, 0⟩, **w** = ⟨0, 0, 6⟩
50. **u** = ⟨4, 0, 1⟩, **v** = ⟨0, 5, 0⟩, **w** = ⟨0, 0, 1⟩
51. **u** = 2**i** + 3**j** + **k**, **v** = **i** − **j**, **w** = 4**i** + 3**j** + **k**
52. **u** = **i** + 4**j** − 7**k**, **v** = 2**i** + 4**k**, **w** = −3**j** + 6**k**

Finding Volume **In Exercises 53–58, find the volume of the parallelepiped that has u, v, and w as adjacent edges.**

53.

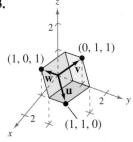

54.

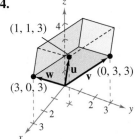

55.

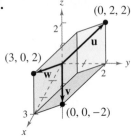

56.

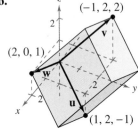

57. $\mathbf{u} = \langle 4, 0, 0 \rangle$, $\mathbf{v} = \langle 0, -2, 3 \rangle$, $\mathbf{w} = \langle 0, 5, 3 \rangle$

58. $\mathbf{u} = \mathbf{i} + \mathbf{j}$, $\mathbf{v} = \mathbf{i} + 2\mathbf{k}$, $\mathbf{w} = \mathbf{j} + \mathbf{k}$

• •59. Torque • • • • • • • • • • • • • • • • • • •

The brakes on a
bicycle are applied
by using a downward
force of p pounds on
the pedal when the
six-inch crank makes
a 40° angle with the
horizontal (see figure).
Vectors representing
the position of the crank
and the force are $\mathbf{V} = \frac{1}{2}(\cos 40°\mathbf{j} + \sin 40°\mathbf{k})$ and
$\mathbf{F} = -p\mathbf{k}$, respectively.

(a) The magnitude of the torque T on the crank is
given by $T = \|\mathbf{V} \times \mathbf{F}\|$. Write T as a function of p.

(b) Use the function from part (a) to complete the table.

p	15	20	25	30	35	40	45
T							

60. Torque Both the magnitude and direction of the
force on a crankshaft change as the crankshaft rotates.
Use the technique given in Exercise 59 to find the
magnitude of the torque on the crankshaft for a
downward force of 200 pounds using the position and
data shown in the figure.

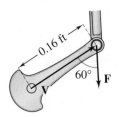

Exploration

True or False? **In Exercises 61 and 62, determine
whether the statement is true or false. Justify your
answer.**

61. The cross product is not defined for vectors in the plane.

62. If $\mathbf{u}$ and $\mathbf{v}$ are vectors in space that are nonzero and
nonparallel, then

$$\mathbf{u} \times \mathbf{v} = \mathbf{v} \times \mathbf{u}.$$

63. Error Analysis Describe the error.

$$2\mathbf{u} \times 2\mathbf{v} = 2(\mathbf{u} \times \mathbf{v}) \quad \text{X}$$

64. **HOW DO YOU SEE IT?** Explain how you
can use the cross product to find the surface
area of the parallelepiped shown below.

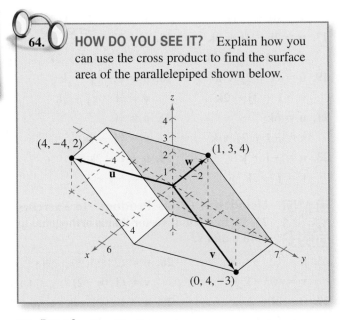

65. Proof Consider the vectors

$$\mathbf{u} = \langle \cos \alpha, \sin \alpha, 0 \rangle \quad \text{and} \quad \mathbf{v} = \langle \cos \beta, \sin \beta, 0 \rangle$$

where $\alpha > \beta$. Find the cross product of the vectors and
use the result to prove the identity

$$\sin(\alpha - \beta) = \sin \alpha \cos \beta - \cos \alpha \sin \beta.$$

11.4 Lines and Planes in Space

■ **Find parametric and symmetric equations of lines in space.**
■ **Find equations of planes in space.**
■ **Sketch planes in space.**
■ **Find distances between points and planes in space.**

Lines in Space

In the plane, *slope* is used to determine an equation of a line. In space, it is more convenient to use *vectors* to determine equations of lines.

In the figure below, consider the line L through the point $P(x_1, y_1, z_1)$ and parallel to the vector

$$\mathbf{v} = \langle a, b, c \rangle.$$ Direction vector for L

Normal vectors to two planes are useful in finding the angle between the planes. For example, in Exercise 60 on page 807, you will use normal vectors to find the angle between two adjacent sides of a tapered bread pan.

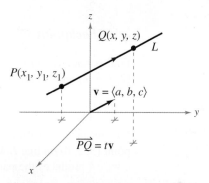

The vector $\mathbf{v}$ is the **direction vector** for the line L, and a, b, and c are the **direction numbers.** One way of describing the line L is to say that it consists of all points $Q(x, y, z)$ for which the vector $\overrightarrow{PQ}$ is parallel to $\mathbf{v}$. This means that $\overrightarrow{PQ}$ is a scalar multiple of $\mathbf{v}$, and you can write

$$\overrightarrow{PQ} = t\mathbf{v}$$

where t is a scalar. So,

$$\overrightarrow{PQ} = \langle x - x_1, y - y_1, z - z_1 \rangle$$
$$= \langle at, bt, ct \rangle$$
$$= t\mathbf{v}.$$

By equating corresponding components, you obtain **parametric equations of a line in space.**

Parametric Equations of a Line in Space

A line L parallel to the vector $\mathbf{v} = \langle a, b, c \rangle$ and passing through the point $P(x_1, y_1, z_1)$ is represented by the parametric equations

$$x = x_1 + at, \quad y = y_1 + bt, \quad \text{and} \quad z = z_1 + ct.$$

When the direction numbers a, b, and c are all nonzero, solve each parametric equation for t and set the resulting expressions equal to each other to find a set of **symmetric equations** of a line.

$$\frac{x - x_1}{a} = \frac{y - y_1}{b} = \frac{z - z_1}{c} \qquad \text{Symmetric equations}$$

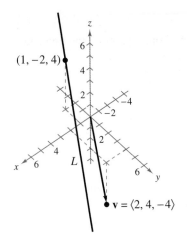

(1, −2, 4)

$\mathbf{v} = \langle 2, 4, -4 \rangle$

L

Figure 11.20

EXAMPLE 1 Finding Parametric and Symmetric Equations

Find parametric and symmetric equations of the line L that passes through the point $(1, -2, 4)$ and is parallel to $\mathbf{v} = \langle 2, 4, -4 \rangle$.

Solution To find a set of parametric equations of the line, use the coordinates

$$x_1 = 1, \quad y_1 = -2, \quad \text{and} \quad z_1 = 4$$

and direction numbers

$$a = 2, \quad b = 4, \quad \text{and} \quad c = -4$$

as shown in Figure 11.20.

$$x = 1 + 2t, \quad y = -2 + 4t, \quad z = 4 + (-4)t \qquad \text{Parametric equations}$$

Because a, b, and c are all nonzero, a set of symmetric equations is

$$\frac{x - 1}{2} = \frac{y + 2}{4} = \frac{z - 4}{-4}. \qquad \text{Symmetric equations}$$

✓ **Checkpoint** Audio-video solution in English & Spanish at LarsonPrecalculus.com

Find parametric and symmetric equations of the line L that passes through the point $(2, 1, -3)$ and is parallel to $\mathbf{v} = \langle 4, -2, 7 \rangle$.

Neither the parametric equations nor the symmetric equations of a line are unique. For instance, in Example 1, by letting $t = 1$, the parametric equations yield another point $(3, 2, 0)$ on line L. Using this point with the direction numbers $a = 2$, $b = 4$, and $c = -4$ produces the parametric equations

$$x = 3 + 2t, \quad y = 2 + 4t, \quad \text{and} \quad z = -4t.$$

EXAMPLE 2 Finding Equations of a Line Through Two Points

Find parametric and symmetric equations of the line that passes through the points $(-2, 1, 0)$ and $(1, 3, 5)$.

Solution Begin by letting $P = (-2, 1, 0)$ and $Q = (1, 3, 5)$. Then a direction vector for the line passing through P and Q is

$$\mathbf{v} = \overrightarrow{PQ}$$
$$= \langle 1 - (-2), 3 - 1, 5 - 0 \rangle$$
$$= \langle 3, 2, 5 \rangle$$
$$= \langle a, b, c \rangle.$$

Use the direction numbers $a = 3$, $b = 2$, and $c = 5$ with the initial point $P(-2, 1, 0)$ to find that

$$x = -2 + 3t, \quad y = 1 + 2t, \quad \text{and} \quad z = 5t. \qquad \text{Parametric equations}$$

Because a, b, and c are all nonzero, a set of symmetric equations is

$$\frac{x + 2}{3} = \frac{y - 1}{2} = \frac{z}{5}. \qquad \text{Symmetric equations}$$

✓ **Checkpoint** Audio-video solution in English & Spanish at LarsonPrecalculus.com

Find parametric and symmetric equations of the line that passes through the points $(1, -6, 3)$ and $(3, 5, 8)$.

• • **REMARK** To check the answer to Example 2, verify that the two original points lie on the line. One way to verify that a point lies on a given line in space is to show that its coordinates satisfy the symmetric equations of the line. Another way is to show that the coordinates of the point yield the same value of t in each of the parametric equations.

Planes in Space

You have seen how to find equations of a line in space from a point on the line and a vector *parallel* to the line. An equation of a plane in space can be found using a point in the plane and a vector *normal* (perpendicular) to the plane.

Consider the plane containing the point

$$P(x_1, y_1, z_1)$$

and having a nonzero normal vector

$$\mathbf{n} = \langle a, b, c \rangle$$

as shown at the right. This plane consists of all points

$$Q(x, y, z)$$

for which the vector $\overrightarrow{PQ}$ is orthogonal to $\mathbf{n}$.
So, you can use the dot product to write the following.

$$\mathbf{n} \cdot \overrightarrow{PQ} = 0 \qquad \text{$\overrightarrow{PQ}$ is orthogonal to $\mathbf{n}$.}$$

$$\langle a, b, c \rangle \cdot \langle x - x_1, y - y_1, z - z_1 \rangle = 0$$

$$a(x - x_1) + b(y - y_1) + c(z - z_1) = 0$$

The third equation of the plane is in standard form.

Standard Equation of a Plane in Space

The plane containing the point (x_1, y_1, z_1) and having normal vector $\mathbf{n} = \langle a, b, c \rangle$ can be represented by the **standard form of the equation of a plane**

$$a(x - x_1) + b(y - y_1) + c(z - z_1) = 0.$$

Applying the Distributive Property and letting $-ax_1 - by_1 - cz_1 = d$ yields the **general form of the equation of a plane** in space

$$ax + by + cz + d = 0. \qquad \text{General form of equation of plane}$$

The general form of the equation of a plane can be used to find a normal vector to the plane. Simply use the coefficients of x, y, and z to write $\mathbf{n} = \langle a, b, c \rangle$.

EXAMPLE 3   Finding an Equation of a Plane in Three-Space

Find the general form of the equation of the plane passing through the point $(1, 3, 6)$ and perpendicular to the vector $\mathbf{n} = -3\mathbf{i} + 2\mathbf{j} + \mathbf{k}$.

Solution   Use the direction numbers for $\mathbf{n}$ and the initial point $(x_1, y_1, z_1) = (1, 3, 6)$ to determine an equation of the plane.

$$a(x - x_1) + b(y - y_1) + c(z - z_1) = 0$$

$$-3(x - 1) + 2(y - 3) + 1(z - 6) = 0 \qquad \text{Standard form}$$

$$-3x + 2y + z - 9 = 0 \qquad \text{General form}$$

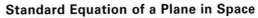 ✓ **Checkpoint** ◀))) *Audio-video solution in English & Spanish at LarsonPrecalculus.com*

Find the general form of the equation of the plane passing through the point $(-2, 1, 3)$ and perpendicular to the vector $\mathbf{n} = 2\mathbf{i} - \mathbf{j} + 6\mathbf{k}$. ◼

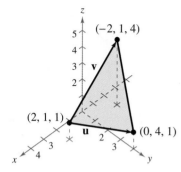

Figure 11.21

> **EXAMPLE 4** **Finding an Equation of a Plane in Three-Space**

See LarsonPrecalculus.com for an interactive version of this type of example.

Find the general form of the equation of the plane passing through the points

$$(2, 1, 1), \quad (0, 4, 1), \quad \text{and} \quad (-2, 1, 4).$$

Solution To find the equation of the plane, you need a point in the plane and a vector that is normal to the plane. To obtain a normal vector, use the cross product of vectors **u** and **v**, as shown in Figure 11.21. Start by finding the component forms of **u** and **v**.

$$\mathbf{u} = \langle 0 - 2, 4 - 1, 1 - 1 \rangle = \langle -2, 3, 0 \rangle$$

$$\mathbf{v} = \langle -2 - 2, 1 - 1, 4 - 1 \rangle = \langle -4, 0, 3 \rangle$$

Use **u** and **v** to find a normal vector.

$$\mathbf{n} = \mathbf{u} \times \mathbf{v}$$

$$= \begin{vmatrix} \mathbf{i} & \mathbf{j} & \mathbf{k} \\ -2 & 3 & 0 \\ -4 & 0 & 3 \end{vmatrix}$$

$$= 9\mathbf{i} + 6\mathbf{j} + 12\mathbf{k}$$

Use the direction numbers for **n** and the initial point $(x_1, y_1, z_1) = (2, 1, 1)$ to determine an equation of the plane.

$$a(x - x_1) + b(y - y_1) + c(z - z_1) = 0$$

$$9(x - 2) + 6(y - 1) + 12(z - 1) = 0 \qquad \text{Standard form}$$

$$9x + 6y + 12z - 36 = 0$$

$$3x + 2y + 4z - 12 = 0 \qquad \text{General form}$$

Check that each of the three points satisfies the equation $3x + 2y + 4z - 12 = 0$.

✓ **Checkpoint** 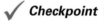 *Audio-video solution in English & Spanish at LarsonPrecalculus.com*

Find the general form of the equation of the plane passing through the points $(1, 0, 5)$, $(-1, -2, 9)$, and $(2, 3, 2)$. ◼

Two distinct planes in three-space either are parallel or intersect in a line. When they intersect, you can determine the angle θ ($0° \leq \theta \leq 90°$) between them from the angle between their normal vectors, as shown at the right. Specifically, if vectors $\mathbf{n}_1$ and $\mathbf{n}_2$ are normal to two intersecting planes, then the angle θ between the normal vectors is equal to the **angle between the two planes** and

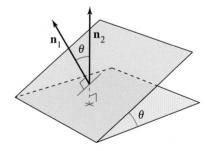

$$\cos \theta = \frac{|\mathbf{n}_1 \cdot \mathbf{n}_2|}{\|\mathbf{n}_1\| \|\mathbf{n}_2\|}.$$

Consequently, two planes with normal vectors $\mathbf{n}_1$ and $\mathbf{n}_2$ are

1. *perpendicular* when $\mathbf{n}_1 \cdot \mathbf{n}_2 = 0$.

2. *parallel* when $\mathbf{n}_1$ is a scalar multiple of $\mathbf{n}_2$.

EXAMPLE 5   **Finding the Line of Intersection of Two Planes**

Find the angle θ between the two planes given by

$$x - 2y + z = 0 \qquad \text{Equation for plane 1}$$

$$2x + 3y - 2z = 0 \qquad \text{Equation for plane 2}$$

and find parametric equations of their line of intersection.

Solution   The normal vectors to the planes are $\mathbf{n}_1 = \langle 1, -2, 1 \rangle$ and $\mathbf{n}_2 = \langle 2, 3, -2 \rangle$. Consequently,

$$\cos \theta = \frac{|\mathbf{n}_1 \cdot \mathbf{n}_2|}{\|\mathbf{n}_1\|\|\mathbf{n}_2\|} = \frac{|-6|}{\sqrt{6}\sqrt{17}} = \frac{6}{\sqrt{102}} \approx 0.59409.$$

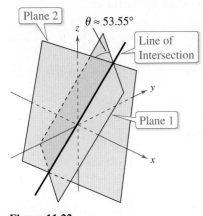

Plane 2 · $\theta \approx 53.55°$ · Line of Intersection · Plane 1

Figure 11.22

This implies that the angle between the two planes is $\theta \approx 53.55°$, as shown in Figure 11.22. To find the line of intersection of the two planes, simultaneously solve the two linear equations representing the planes. One way to do this is to multiply the first equation by -2, then add the new first equation to the second equation.

$$\begin{array}{ll} x - 2y + z = 0 \\ 2x + 3y - 2z = 0 \end{array} \implies \begin{array}{l} -2x + 4y - 2z = 0 \\ \underline{2x + 3y - 2z = 0} \end{array}$$

$$7y - 4z = 0 \implies y = \frac{4z}{7}$$

Substitute $y = 4z/7$ back into one of the original equations to determine that $x = z/7$. Finally, let $t = z/7$ to obtain the parametric equations

$$x = t = x_1 + at \qquad \text{Parametric equation for } x$$

$$y = 4t = y_1 + bt \qquad \text{Parametric equation for } y$$

$$z = 7t = z_1 + ct. \qquad \text{Parametric equation for } z$$

The point $(x_1, y_1, z_1) = (0, 0, 0)$ lies in both planes, so substitute for x_1, y_1, and z_1 in these parametric equations to find that $a = 1$, $b = 4$, and $c = 7$ are direction numbers for the line of intersection.

✓ **Checkpoint** 🔊 *Audio-video solution in English & Spanish at LarsonPrecalculus.com*

Find the angle θ between the two planes given by

$$2x + y + 3z = 0 \qquad \text{Equation for plane 1}$$

$$4x - 2y - 2z = 0 \qquad \text{Equation for plane 2}$$

and find parametric equations of their line of intersection.   ■

Notice that the direction numbers in Example 5 can also be obtained from the cross product of the two normal vectors.

$$\mathbf{n}_1 \times \mathbf{n}_2 = \begin{vmatrix} \mathbf{i} & \mathbf{j} & \mathbf{k} \\ 1 & -2 & 1 \\ 2 & 3 & -2 \end{vmatrix}$$

$$= \begin{vmatrix} -2 & 1 \\ 3 & -2 \end{vmatrix} \mathbf{i} - \begin{vmatrix} 1 & 1 \\ 2 & -2 \end{vmatrix} \mathbf{j} + \begin{vmatrix} 1 & -2 \\ 2 & 3 \end{vmatrix} \mathbf{k}$$

$$= \mathbf{i} + 4\mathbf{j} + 7\mathbf{k}$$

This means that the *line of intersection of the two planes is parallel to the cross product of their normal vectors.*

Sketching Planes in Space

▷ TECHNOLOGY Most three-dimensional graphing utilities and computer algebra systems can graph a plane in space. Consult the user's guide for your graphing utility for specific instructions.

As discussed in Section 11.1, when a plane in space intersects one of the coordinate planes, the line of intersection is called the *trace* of the given plane in the coordinate plane. To sketch a plane in space, it is helpful to find its points of intersection with the coordinate axes and its traces in the coordinate planes. For example, consider the plane

$3x + 2y + 4z = 12.$ Equation of plane

To find the *xy*-trace, let $z = 0$ and sketch the line

$3x + 2y = 12$ *xy*-trace

in the *xy*-plane. This line intersects the *x*-axis at $(4, 0, 0)$ and the *y*-axis at $(0, 6, 0)$. Continue this process as shown in Figure 11.23 to find the *yz*-trace and the *xz*-trace and then shade the triangular region formed in the first octant by the traces.

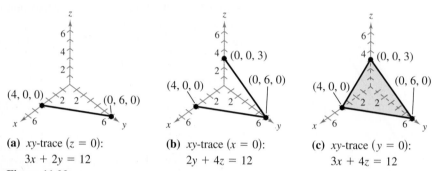

(a) *xy*-trace $(z = 0)$:
$3x + 2y = 12$

(b) *xy*-trace $(x = 0)$:
$2y + 4z = 12$

(c) *xy*-trace $(y = 0)$:
$3x + 4z = 12$

Figure 11.23

If the equation of a plane has a missing variable, such as

$2x + z = 1$

then the plane must be *parallel to the axis* represented by the missing variable, as shown in Figure 11.24.

If two variables are missing from the equation of a plane, then it is *parallel to the coordinate plane* represented by the missing variables, as shown in Figure 11.25. For instance, the graph of

$z + 2 = 0$

is parallel to the *xy*-plane. Confirm this using a three-dimensional graphing utility.

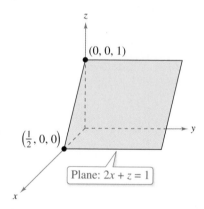

Plane is parallel to *y*-axis.
Figure 11.24

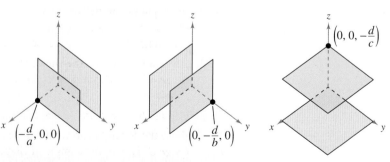

(a) Plane $ax + d = 0$ is parallel to *yz*-plane.

(b) Plane $by + d = 0$ is parallel to *xz*-plane.

(c) Plane $cz + d = 0$ is parallel to *xy*-plane.

Figure 11.25

Distance Between a Point and a Plane

The distance D between a point Q and a plane is the length of the shortest line segment connecting Q to the plane, as shown in Figure 11.26. When P is *any* point in the plane, find this distance by projecting the vector $\overrightarrow{PQ}$ onto the normal vector $\mathbf{n}$. The length of this projection is the desired distance.

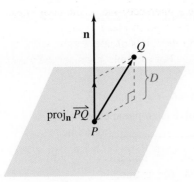

$D = \|\text{proj}_{\mathbf{n}}\, \overrightarrow{PQ}\|$
Figure 11.26

Distance Between a Point and a Plane

The **distance between a plane and a point** Q (not in the plane) is

$$D = \|\text{proj}_{\mathbf{n}}\overrightarrow{PQ}\| = \frac{|\overrightarrow{PQ} \cdot \mathbf{n}|}{\|\mathbf{n}\|}$$

where P is a point in the plane and $\mathbf{n}$ is normal to the plane.

To find a point in the plane given by $ax + by + cz + d = 0$, where $a \neq 0$, let $y = 0$ and $z = 0$. Then, solving the equation $ax + d = 0$, you find that the point $(-d/a, 0, 0)$ lies in the plane.

EXAMPLE 6   **Finding the Distance Between a Point and a Plane**

Find the distance between the point $Q(1, 5, -4)$ and the plane given by $3x - y + 2z = 6$.

Solution   You know that $\mathbf{n} = \langle 3, -1, 2 \rangle$ is normal to the given plane. Let $y = 0$ and $z = 0$ to find the point $P(2, 0, 0)$ in the plane. The vector from P to Q is

$$\overrightarrow{PQ} = \langle 1 - 2, 5 - 0, -4 - 0 \rangle = \langle -1, 5, -4 \rangle.$$

The formula for the distance between a point and a plane produces

$$D = \frac{|\overrightarrow{PQ} \cdot \mathbf{n}|}{\|\mathbf{n}\|} = \frac{|\langle -1, 5, -4 \rangle \cdot \langle 3, -1, 2 \rangle|}{\sqrt{9 + 1 + 4}} = \frac{|-3 - 5 - 8|}{\sqrt{14}} = \frac{16}{\sqrt{14}}.$$

✓ **Checkpoint** 🔊))) *Audio-video solution in English & Spanish at LarsonPrecalculus.com*

Find the distance between the point $Q(-6, 2, 1)$ and the plane given by $2x + 5y - z = 3$.

The choice of the point P in Example 6 is arbitrary. Choose a different point in the plane to verify that you find the same distance.

Summarize   (Section 11.4)

1. Explain how to find the parametric and symmetric equations of a line in space parallel to the vector $\mathbf{v} = \langle a, b, c \rangle$ and passing through the point $P(x_1, y_1, z_1)$ *(page 799)*. For examples of finding parametric and symmetric equations of a line in space, see Examples 1 and 2.

2. Explain how to find the standard form of the equation of a plane in space *(page 801)*. For examples of finding equations of planes, see Examples 3 and 4.

3. Explain how to sketch a plane in space *(page 804)*.

4. State the formula for the distance between a point and a plane *(page 805)*. For an example of finding the distance between a point and a plane, see Example 6.

11.4 Exercises

See CalcChat.com for tutorial help and worked-out solutions to odd-numbered exercises.

Vocabulary: Fill in the blanks.

1. The _____ vector for a line L is parallel to L.
2. The _____ _____ of a line in space are $x = x_1 + at$, $y = y_1 + bt$, and $z = z_1 + ct$.
3. When the direction numbers a, b, and c of the vector $\mathbf{v} = \langle a, b, c \rangle$ are all nonzero, solve each parametric equation for t and set the resulting expressions equal to each other to find a set of _____ _____ of a line.
4. A vector that is perpendicular to a plane is called a _____ vector.

Skills and Applications

 Finding Equations In Exercises 5–10, find (a) a set of parametric equations and (b) a set of symmetric equations of the line passing through the point and parallel to the given vector or line.

Point	Parallel to
5. $(4, 1, 2)$	$\mathbf{v} = \langle 5, 3, 6 \rangle$
6. $(-3, 6, 1)$	$\mathbf{v} = \langle -2, -8, 7 \rangle$
7. $(2, -5, 0)$	$\mathbf{v} = 4\mathbf{i} + 3\mathbf{j} - \mathbf{k}$
8. $(-2, 0, 3)$	$\mathbf{v} = 2\mathbf{i} + 4\mathbf{j} - 2\mathbf{k}$
9. $(2, -3, 5)$	$x = 5 + 2t, y = 7 - 3t, z = -2 + t$
10. $(1, 0, 1)$	$x = 3 + 3t, y = 5 - 2t, z = -7 + t$

 Finding Equations In Exercises 11–16, find (a) a set of parametric equations and (b) if possible, a set of symmetric equations of the line that passes through the points.

11. $(2, 0, 2), (1, 4, -3)$ **12.** $(2, 3, 0), (10, 8, 12)$
13. $(-3, 8, 15), (1, -2, 16)$ **14.** $(2, 3, -1), (1, -5, 3)$
15. $(3, 1, 2), (-1, 1, 5)$ **16.** $(2, -1, 5), (2, 1, -3)$

Using Parametric Equations In Exercises 17 and 18, sketch a graph of the line.

17. $x = 3t, y = 1 + t, z = 2 + 2t$
18. $x = 3 - 2t, y = 2 + t, z = -1 - t$

 Finding an Equation of a Plane in Three-Space In Exercises 19–22, find the general form of the equation of the plane passing through the point and perpendicular to the given vector or line.

Point	Perpendicular to
19. $(5, 6, 3)$	$\mathbf{n} = -2\mathbf{i} + \mathbf{j} - 2\mathbf{k}$
20. $(3, 5, 2)$	$\mathbf{n} = 6\mathbf{i} - 4\mathbf{j} + 7\mathbf{k}$
21. $(2, 1, 3)$	$x = 1 - 2t, y = 3 + t, z = 5 + 2t$
22. $(4, 7, 6)$	$x = 3 - t, y = 4 - 2t, z = -3 + 4t$

 Finding an Equation of a Plane in Three-Space In Exercises 23–26, find the general form of the equation of the plane passing through the three points.

23. $(3, 1, 2), (2, 3, 4), (-3, 4, 2)$
24. $(4, -1, 3), (2, 5, 1), (-1, 2, 1)$
25. $(2, 3, -2), (3, 4, 2), (1, -1, 0)$
26. $(5, -1, 4), (1, -1, 2), (2, 1, -3)$

 Finding an Equation of a Plane in Three-Space In Exercises 27–32, find the general form of the equation of the plane with the given characteristics.

27. Passes through $(2, 5, 3)$ and is parallel to the xz-plane
28. Passes through $(1, 2, 3)$ and is parallel to the yz-plane
29. Passes through $(0, 2, 4)$ and $(-1, -2, 0)$ and is perpendicular to the yz-plane
30. Passes through $(1, -2, 4)$ and $(4, 0, -1)$ and is perpendicular to the xz-plane
31. Passes through $(2, 2, 1)$ and $(-1, 1, -1)$ and is perpendicular to $2x - 3y + z = 3$
32. Passes through $(1, 2, 0)$ and $(-1, -1, 2)$ and is perpendicular to $2x - 3y + z = 6$

 Parallel or Perpendicular Planes In Exercises 33–38, determine whether the planes are parallel, perpendicular, or neither. If they are neither parallel nor perpendicular, find the angle of intersection.

33. $5x - 3y + z = 4$ **34.** $x + 2y - 5z = 2$
   $x + 4y + 7z = 1$ $3x - 4y - 2z = 4$

35. $x - 5y - z = 1$ **36.** $2x - z = 1$
   $5x - 25y - 5z = -3$ $4x + y + 8z = 10$

37. $3x + 2z = 4$
   $6x - 2y - 3z = 8$

38. $3x + y - 4z = 3$
   $-9x - 3y + 12z = 4$

Finding Parametric Equations of a Line In Exercises 39–42, find a set of parametric equations of the line. (There are many correct answers.)

39. Passes through $(2, 3, 4)$ and is parallel to the xz-plane and the yz-plane

40. Passes through $(-4, 5, 2)$ and is parallel to the xy-plane and the yz-plane

41. Passes through $(2, 3, 4)$ and is perpendicular to $3x + 2y - z = 6$

42. Passes through $(-4, 5, 2)$ and is perpendicular to $-x + 2y + z = 5$

 Finding the Line of Intersection of Two Planes In Exercises 43–46, (a) find the angle θ between the two planes and (b) find parametric equations of their line of intersection.

43. $x + y - 2z = 0$
$2x - y + 3z = 0$

44. $x - 3y + 2z = 0$
$3x + 2y - 5z = 0$

45. $x + y - z = 0$
$2x - 5y - z = 1$

46. $x - 3y + z = -2$
$2x + 5z + 3 = 0$

Sketching a Plane in Space In Exercises 47–52, plot the intercepts and sketch a graph of the plane.

47. $x + 2y + 3z = 6$

48. $2x - y + 4z = 4$

49. $3x + 2y - z = 6$

50. $2x - 3y + 4z = 12$

51. $x + 2y = 4$

52. $y + z = 5$

 Finding the Distance Between a Point and a Plane In Exercises 53–58, find the distance between the point and the plane.

53. $(1, 0, 4)$
$x - 2y + z = 4$

54. $(3, 2, 1)$
$x - y + 2z = 4$

55. $(1, 3, 4)$
$4x - 5y + 2z = 6$

56. $(5, 4, 2)$
$2x + 6y + 3z = 8$

57. $(-2, 4, 3)$
$2x + 3y + 2z = 4$

58. $(-1, 3, -6)$
$x - 2y + 2z = 3$

59. Mechanical Design A chute at the top of a grain elevator of a combine (see figure) funnels the grain into a bin. Find the angle θ between two adjacent sides.

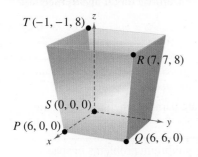

60. Product Design

A bread pan is tapered so that a loaf of bread can be removed easily. The figure shows the shape and dimensions of the pan. Find the angle θ between two adjacent sides of the pan.

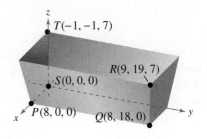

Exploration

True or False? In Exercises 61 and 62, determine whether the statement is true or false. Justify your answer.

61. Two lines in space are either parallel or they intersect.

62. Two nonparallel planes in space will always intersect.

63. Error Analysis Describe the error.

Parametric equations of the line that passes through the point $(-2, 5, 3)$ and is parallel to $\mathbf{v} = \langle -3, -4, 2 \rangle$ are $x = -3 - 2t$, $y = -4 + 5t$, and $z = 2 + 3t$.

64. HOW DO YOU SEE IT? The figure shows two planes and their normal vectors.

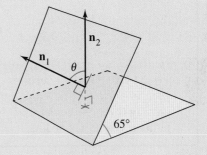

(a) What is the angle between the normal vectors? Explain.

(b) A third plane is parallel to $\mathbf{n}_1$. What is the angle between this plane and the blue plane? Explain.

Chapter Summary

	What Did You Learn?	Explanation/Examples	Review Exercises
Section 11.1	Plot points in the three-dimensional coordinate system *(p. 778).*		1–4
	Find distances between points in space and find midpoints of line segments joining points in space *(p. 779).*	The distance between the points (x_1, y_1, z_1) and (x_2, y_2, z_2) given by the Distance Formula in Space is $$d = \sqrt{(x_2 - x_1)^2 + (y_2 - y_1)^2 + (z_2 - z_1)^2}.$$ The midpoint of the line segment joining the points (x_1, y_1, z_1) and (x_2, y_2, z_2) given by the Midpoint Formula in Space is $$\left(\frac{x_1 + x_2}{2}, \frac{y_1 + y_2}{2}, \frac{z_1 + z_2}{2}\right).$$	5–14
	Write equations of spheres in standard form and sketch traces of surfaces in space *(p. 780).*	The standard equation of a sphere with center (h, k, j) and radius r is $(x - h)^2 + (y - k)^2 + (z - j)^2 = r^2$.	15–26
Section 11.2	Find the component forms of the unit vectors in the same direction of, the magnitudes of, the dot products of, and the angles between vectors in space *(p. 785).*	**Vectors in Space** 1. Two vectors are equal if and only if their corresponding components are equal. 2. The magnitude (or length) of $\mathbf{u} = \langle u_1, u_2, u_3 \rangle$ is $$\|\mathbf{u}\| = \sqrt{u_1^2 + u_2^2 + u_3^2}.$$ 3. A unit vector $\mathbf{u}$ in the direction of $\mathbf{v}$ is $\mathbf{u} = \dfrac{\mathbf{v}}{\|\mathbf{v}\|}, \mathbf{v} \neq \mathbf{0}$. 4. The sum of $\mathbf{u} = \langle u_1, u_2, u_3 \rangle$ and $\mathbf{v} = \langle v_1, v_2, v_3 \rangle$ is $$\mathbf{u} + \mathbf{v} = \langle u_1 + v_1, u_2 + v_2, u_3 + v_3 \rangle.$$ 5. The scalar multiple of the real number c and $\mathbf{u} = \langle u_1, u_2, u_3 \rangle$ is $c\mathbf{u} = \langle cu_1, cu_2, cu_3 \rangle$. 6. The dot product of $\mathbf{u} = \langle u_1, u_2, u_3 \rangle$ and $\mathbf{v} = \langle v_1, v_2, v_3 \rangle$ is $\mathbf{u} \cdot \mathbf{v} = u_1v_1 + u_2v_2 + u_3v_3$. **Angle Between Two Vectors** If θ is the angle between two nonzero vectors $\mathbf{u}$ and $\mathbf{v}$, then $$\cos\theta = \frac{\mathbf{u} \cdot \mathbf{v}}{\|\mathbf{u}\|\|\mathbf{v}\|}.$$	27–36
	Determine whether vectors in space are orthogonal or parallel *(p. 787).*	If the dot product of two nonzero vectors is zero, then the angle between the vectors is 90° and the vectors are orthogonal. Two nonzero vectors $\mathbf{u}$ and $\mathbf{v}$ are parallel when there is some scalar c such that $\mathbf{u} = c\mathbf{v}$.	37–44
	Use vectors in space to solve real-life problems *(p. 789).*	Vectors can be used to solve equilibrium problems in space. (See Example 8.)	45, 46

What Did You Learn?	Explanation/Examples	Review Exercises

<table>
<tr><td rowspan="3">Section 11.3</td><td>Find cross products of vectors in space (p. 792).</td><td>

Definition of the Cross Product of Two Vectors in Space

Let $\mathbf{u} = u_1\mathbf{i} + u_2\mathbf{j} + u_3\mathbf{k}$ and $\mathbf{v} = v_1\mathbf{i} + v_2\mathbf{j} + v_3\mathbf{k}$ be vectors in space. The cross product of $\mathbf{u}$ and $\mathbf{v}$ is the vector

$$\mathbf{u} \times \mathbf{v} = (u_2v_3 - u_3v_2)\mathbf{i} - (u_1v_3 - u_3v_1)\mathbf{j} + (u_1v_2 - u_2v_1)\mathbf{k}.$$

</td><td>47–50</td></tr>
<tr><td>Use geometric properties of cross products of vectors in space (p. 794).</td><td>

Geometric Properties of the Cross Product

Let $\mathbf{u}$ and $\mathbf{v}$ be nonzero vectors in space, and let θ be the angle between $\mathbf{u}$ and $\mathbf{v}$.

1. $\mathbf{u} \times \mathbf{v}$ is orthogonal to both $\mathbf{u}$ and $\mathbf{v}$.

2. $\|\mathbf{u} \times \mathbf{v}\| = \|\mathbf{u}\|\|\mathbf{v}\| \sin \theta$

3. $\mathbf{u} \times \mathbf{v} = \mathbf{0}$ if and only if $\mathbf{u}$ and $\mathbf{v}$ are scalar multiples of each other.

4. $\|\mathbf{u} \times \mathbf{v}\| = $ area of parallelogram that has $\mathbf{u}$ and $\mathbf{v}$ as adjacent sides.

</td><td>51–56</td></tr>
<tr><td>Use triple scalar products to find volumes of parallelepipeds (p. 796).</td><td>

The Triple Scalar Product

For $\mathbf{u} = u_1\mathbf{i} + u_2\mathbf{j} + u_3\mathbf{k}$, $\mathbf{v} = v_1\mathbf{i} + v_2\mathbf{j} + v_3\mathbf{k}$, and $\mathbf{w} = w_1\mathbf{i} + w_2\mathbf{j} + w_3\mathbf{k}$, the triple scalar product is

$$\mathbf{u} \cdot (\mathbf{v} \times \mathbf{w}) = \begin{vmatrix} u_1 & u_2 & u_3 \\ v_1 & v_2 & v_3 \\ w_1 & w_2 & w_3 \end{vmatrix}.$$

Geometric Property of the Triple Scalar Product

The volume V of a parallelepiped with vectors $\mathbf{u}$, $\mathbf{v}$, and $\mathbf{w}$ as adjacent edges is $V = |\mathbf{u} \cdot (\mathbf{v} \times \mathbf{w})|$.

</td><td>57, 58</td></tr>
<tr><td rowspan="4">Section 11.4</td><td>Find parametric and symmetric equations of lines in space (p. 799).</td><td>

Parametric Equations of a Line in Space

A line L parallel to the vector $\mathbf{v} = \langle a, b, c \rangle$ and passing through the point $P(x_1, y_1, z_1)$ is represented by the parametric equations $x = x_1 + at$, $y = y_1 + bt$, and $z = z_1 + ct$.

</td><td>59–62</td></tr>
<tr><td>Find equations of planes in space (p. 801).</td><td>

Standard Equation of a Plane in Space

The plane containing the point (x_1, y_1, z_1) and having normal vector $\mathbf{n} = \langle a, b, c \rangle$ can be represented by the standard form of the equation of a plane

$$a(x - x_1) + b(y - y_1) + c(z - z_1) = 0.$$

</td><td>63–66</td></tr>
<tr><td>Sketch planes in space (p. 804).</td><td>

See Figure 11.23, which shows how to sketch the plane

$3x + 2y + 4z = 12$.

</td><td>67, 68</td></tr>
<tr><td>Find distances between points and planes in space (p. 805).</td><td>

Distance Between a Point and a Plane

The distance between a plane and a point Q (not in the plane) is

$$D = \|\text{proj}_\mathbf{n} \overrightarrow{PQ}\|$$

$$= \frac{|\overrightarrow{PQ} \cdot \mathbf{n}|}{\|\mathbf{n}\|}$$

where P is a point in the plane and $\mathbf{n}$ is normal to the plane.

</td><td>69–72</td></tr>
</table>

Review Exercises   See CalcChat.com for tutorial help and worked-out solutions to odd-numbered exercises.

11.1   **Plotting Points in Space**   In Exercises 1 and 2, plot the points in the same three-dimensional coordinate system.

1. $(2, 1, 4)$, $(1, -3, -2)$ **2.** $(0, 0, 2)$, $(-3, -2, -5)$

Finding the Coordinates of a Point   In Exercises 3 and 4, find the coordinates of the point.

3. The point is located in the xy-plane, four units to the right of the xz-plane, and five units behind the yz-plane.

4. The point is located on the y-axis, seven units to the left of the xz-plane.

Finding the Distance Between Two Points in Space   In Exercises 5–8, find the distance between the points.

5. $(4, 0, 7)$, $(5, 2, 1)$ **6.** $(6, 2, 3)$, $(0, 1, 4)$

7. $(2, 3, -4)$, $(-1, -3, 0)$ **8.** $(-7, -5, 6)$, $(1, 1, 6)$

Using the Pythagorean Theorem   In Exercises 9 and 10, find the lengths of the sides of the right triangle with the given vertices. Show that these lengths satisfy the Pythagorean Theorem.

9.
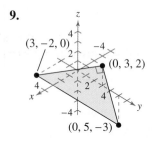

10.
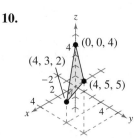

Using the Midpoint Formula in Space   In Exercises 11–14, find the midpoint of the line segment joining the points.

11. $(3, 6, 5)$, $(7, 2, 5)$ **12.** $(7, 2, 3)$, $(9, 4, 9)$

13. $(-8, 5, 2)$, $(4, -1, 2)$

14. $(-7, 2, -8)$, $(5, -4, -8)$

Finding the Equation of a Sphere   In Exercises 15–20, find the standard equation of the sphere with the given characteristics.

15. Center: $(1, 5, 4)$; radius: 2

16. Center: $(2, 6, 3)$; radius: 6

17. Center: $(1, -4, 2)$; diameter: 14

18. Center: $(-3, 4, -1)$; diameter: 18

19. Endpoints of a diameter: $(-2, -2, -2)$, $(2, 2, 2)$

20. Endpoints of a diameter: $(4, -1, -3)$, $(-2, 5, 3)$

Finding the Center and Radius of a Sphere   In Exercises 21–24, find the center and radius of the sphere.

21. $x^2 + y^2 + z^2 - 8z = 0$

22. $x^2 + y^2 + z^2 - 4x - 6y + 4 = 0$

23. $x^2 + y^2 + z^2 - 10x + 6y - 4z + 34 = 0$

24. $2x^2 + 2y^2 + 2z^2 + 2x + 2y + 2z + 1 = 0$

Sketching a Trace of a Surface   In Exercises 25 and 26, sketch the graph of the equation and each of the specified traces.

25. $x^2 + (y - 3)^2 + z^2 = 16$

(a) xz-trace   (b) yz-trace

26. $(x + 2)^2 + (y - 1)^2 + z^2 = 9$

(a) xy-trace   (b) yz-trace

11.2   **Finding the Component Form of a Vector**   In Exercises 27–30, find (a) the component form of the vector **v**, (b) the magnitude of **v**, and (c) a unit vector in the direction of **v**.

	Initial point	Terminal point
27.	$(3, -2, 1)$	$(4, 4, 0)$
28.	$(2, -1, 2)$	$(-3, 2, 3)$
29.	$(7, -4, 3)$	$(-3, 2, 10)$
30.	$(0, 3, -1)$	$(5, -8, 6)$

Finding the Dot Product of Two Vectors   In Exercises 31–34, find the dot product of **u** and **v**.

31. $\mathbf{u} = \langle -1, 4, 3 \rangle$ **32.** $\mathbf{u} = \langle 8, -4, 2 \rangle$
$\mathbf{v} = \langle 0, -6, 5 \rangle$ $\mathbf{v} = \langle 2, 5, 2 \rangle$

33. $\mathbf{u} = 2\mathbf{i} - \mathbf{j} + \mathbf{k}$ **34.** $\mathbf{u} = 2\mathbf{i} + \mathbf{j} - 2\mathbf{k}$
$\mathbf{v} = \mathbf{i} - \mathbf{k}$ $\mathbf{v} = \mathbf{i} - 3\mathbf{j} + 2\mathbf{k}$

Finding the Angle Between Two Vectors   In Exercises 35 and 36, find the angle θ between the vectors.

35. $\mathbf{u} = \langle 2, -1, 0 \rangle$ **36.** $\mathbf{u} = \langle 3, 1, -1 \rangle$
$\mathbf{v} = \langle 1, 4, 1 \rangle$ $\mathbf{v} = \langle 4, 5, 2 \rangle$

Determining Orthogonal and Parallel Vectors   In Exercises 37–40, determine whether **u** and **v** are orthogonal, parallel, or neither.

37. $\mathbf{u} = \langle 39, -12, 21 \rangle$ **38.** $\mathbf{u} = \langle 8, 5, -8 \rangle$

$\mathbf{v} = \langle -26, 8, -14 \rangle$ $\mathbf{v} = \left\langle -2, 4, \dfrac{1}{2} \right\rangle$

39. $\mathbf{u} = 6\mathbf{i} + 5\mathbf{j} + 9\mathbf{k}$ **40.** $\mathbf{u} = 2\mathbf{i} - 3\mathbf{j} + \mathbf{k}$
$\mathbf{v} = 5\mathbf{i} + 3\mathbf{j} - 5\mathbf{k}$ $\mathbf{v} = -2\mathbf{i} + 3\mathbf{j} + \mathbf{k}$

Using Vectors to Determine Collinear Points In Exercises 41–44, use vectors to determine whether the points are collinear.

41. $(6, 3, -1), (5, 8, 3), (7, -2, -5)$

42. $(5, 2, 0), (2, 6, 1), (2, 4, 7)$

43. $(5, -4, 7), (8, -5, 5), (11, 6, 3)$

44. $(3, 4, -1), (-1, 6, 9), (5, 3, -6)$

45. Tension A load of 300 pounds is supported by three cables, as shown in the figure. Find the tension in each of the supporting cables.

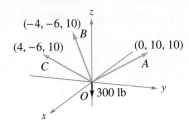

46. Tension Determine the tension in each of the supporting cables in Exercise 45 when the load is 200 pounds.

11.3 **Finding the Cross Product** In Exercises 47–50, find u × v and show that it is orthogonal to both u and v.

47. $\mathbf{u} = \langle -2, 8, 2 \rangle$ **48.** $\mathbf{u} = \langle 10, 15, 5 \rangle$
     $\mathbf{v} = \langle 1, 1, -1 \rangle$ $\mathbf{v} = \langle 5, -3, 0 \rangle$

49. $\mathbf{u} = 2\mathbf{i} + 3\mathbf{j} + 2\mathbf{k}$ **50.** $\mathbf{u} = -\mathbf{i} + 2\mathbf{j} - 2\mathbf{k}$
     $\mathbf{v} = 3\mathbf{i} + \mathbf{j} + 2\mathbf{k}$ $\mathbf{v} = \mathbf{i}$

Using the Cross Product In Exercises 51–54, find a unit vector that is orthogonal to both u and v.

51. $\mathbf{u} = 2\mathbf{i} - \mathbf{j} + \mathbf{k}$ **52.** $\mathbf{u} = \mathbf{j} + 4\mathbf{k}$
     $\mathbf{v} = -\mathbf{i} + \mathbf{j} - 2\mathbf{k}$ $\mathbf{v} = -\mathbf{i} + 3\mathbf{j}$

53. $\mathbf{u} = -3\mathbf{i} + 2\mathbf{j} - 5\mathbf{k}$ **54.** $\mathbf{u} = 4\mathbf{k}$
     $\mathbf{v} = 10\mathbf{i} - 15\mathbf{j} + 2\mathbf{k}$ $\mathbf{v} = \mathbf{i} + 12\mathbf{k}$

Geometric Application of the Cross Product In Exercises 55 and 56, (a) show that the points are the vertices of a parallelogram and (b) find its area.

55. $A(0, 1, 1), B(2, -1, 1), C(5, 1, 4), D(3, 3, 4)$

56. $A(0, 4, 0), B(1, 4, 1), C(1, 6, 1), D(0, 6, 0)$

Finding the Volume of a Parallelepiped In Exercises 57 and 58, find the volume of the parallelepiped that has u, v, and w as adjacent edges.

57. $\mathbf{u} = \langle 3, 0, 0 \rangle$ **58.** $\mathbf{u} = 2\mathbf{i}$
     $\mathbf{v} = \langle 2, 0, 5 \rangle$ $\mathbf{v} = 4\mathbf{j}$
     $\mathbf{w} = \langle 0, 5, 1 \rangle$ $\mathbf{w} = 6\mathbf{k}$

11.4 **Finding Equations** In Exercises 59–62, find (a) a set of parametric equations and (b) a set of symmetric equations of the line with the given characteristics.

59. Passes through $(0, 0, 0)$ and is parallel to

     $\mathbf{v} = \langle -2, 5, 1 \rangle$

60. Passes through $(1, 3, 5)$ and is parallel to

     $\mathbf{v} = \langle 3, -4, -2 \rangle$

61. Passes through $(-1, 3, 5)$ and $(3, 6, -1)$

62. Passes through $(0, -10, 3)$ and $(5, 10, 0)$

Finding an Equation of a Plane in Three-Space In Exercises 63–66, find the general form of the equation of the plane with the given characteristics.

63. Passes through $(0, 0, 0), (5, 0, 2),$ and $(2, 3, 8)$

64. Passes through $(-1, 3, 4), (4, -2, 2),$ and $(2, 8, 6)$

65. Passes through $(5, 3, 2)$ and is parallel to the xy-plane

66. Passes through $(0, 0, 6)$ and $(2, 1, 3)$ and is perpendicular to $2x + 3y - z = 5$

Sketching a Plane in Space In Exercises 67 and 68, plot the intercepts and sketch a graph of the plane.

67. $3x - 2y + 3z = 6$

68. $5x - y - 5z = 5$

Finding the Distance Between a Point and a Plane In Exercises 69–72, find the distance between the point and the plane.

69. $(1, 2, 3)$ **70.** $(2, 3, 10)$
     $2x - y + z = 4$ $x - 10y + 3z = 3$

71. $(-1, 3, -5)$ **72.** $(2, -4, 3)$
     $3x - 2y + 2z = 8$ $4x - 3y - z = 6$

Exploration

True or False? In Exercises 73 and 74, determine whether the statement is true or false. Justify your answer.

73. If **u** and **v** are vectors in space that are nonzero and nonparallel, and c is a scalar, then

     $c(\mathbf{u} \times \mathbf{v}) = c\mathbf{u} \times c\mathbf{v}.$

74. The triple scalar product of three vectors in space is a scalar.

Verifying Properties In Exercises 75 and 76, let $\mathbf{u} = \langle u_1, u_2, u_3 \rangle$, $\mathbf{v} = \langle v_1, v_2, v_3 \rangle$, and $\mathbf{w} = \langle w_1, w_2, w_3 \rangle$.

75. Show that $\mathbf{u} \cdot (\mathbf{v} + \mathbf{w}) = \mathbf{u} \cdot \mathbf{v} + \mathbf{u} \cdot \mathbf{w}$.

76. Show that $\mathbf{u} \times (\mathbf{v} + \mathbf{w}) = (\mathbf{u} \times \mathbf{v}) + (\mathbf{u} \times \mathbf{w})$.

Chapter Test

See **CalcChat.com** for tutorial help and worked-out solutions to odd-numbered exercises.

Take this test as you would take a test in class. When you are finished, check your work against the answers given in the back of the book.

1. Plot $(4, 1, 2)$ and $(-2, 4, -3)$ in the same three-dimensional coordinate system.

In Exercises 2–4, use the points $A(8, -2, 5)$, $B(6, 4, -1)$, and $C(-4, 3, 0)$ to solve the problem.

2. Consider the triangle with vertices A, B, and C. Is it a right triangle? Explain.

3. Find the midpoint of the line segment joining points A and B.

4. Find the standard equation of the sphere for which A and B are the endpoints of a diameter. Sketch the sphere and its xz-trace.

In Exercises 5–9, let $\mathbf{u}$ and $\mathbf{v}$ be the vectors from $A(8, -2, 5)$ to $B(6, 4, -1)$ and from A to $C(-4, 3, 0)$, respectively.

5. Write $\mathbf{u}$ and $\mathbf{v}$ in component form.

6. Find (a) $\mathbf{u} \cdot \mathbf{v}$ and (b) $\mathbf{u} \times \mathbf{v}$.

7. Find a unit vector in the direction of (a) $\mathbf{u}$ and (b) $\mathbf{v}$.

8. Find the angle θ between $\mathbf{u}$ and $\mathbf{v}$.

9. Find (a) a set of parametric equations and (b) a set of symmetric equations of the line passing through points A and B.

In Exercises 10–12, determine whether u and v are orthogonal, parallel, or neither.

10. $\mathbf{u} = \langle 4, -1, 6 \rangle$
    $\mathbf{v} = \left\langle -2, \frac{1}{2}, -3 \right\rangle$

11. $\mathbf{u} = -3\mathbf{i} + 2\mathbf{j} - \mathbf{k}$
    $\mathbf{v} = \mathbf{i} + \mathbf{j} - \mathbf{k}$

12. $\mathbf{u} = \mathbf{i} - 2\mathbf{j} - \mathbf{k}$
    $\mathbf{v} = \mathbf{j} + 6\mathbf{k}$

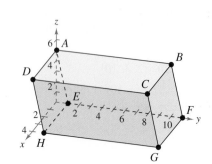

13. Consider the points $A(2, -3, 1)$, $B(6, 5, -1)$, $C(3, -6, 4)$, and $D(7, 2, 2)$.
    (a) Show that the points are the vertices of a parallelogram.
    (b) Find the area of the parallelogram.

14. Find the volume of the parallelepiped with the given vertices, shown at the left.

Figure for 14

   $A(0, 0, 5)$, $B(0, 10, 5)$, $C(4, 10, 5)$, $D(4, 0, 5)$,
   $E(0, 1, 0)$, $F(0, 11, 0)$, $G(4, 11, 0)$, $H(4, 1, 0)$

In Exercises 15 and 16, plot the intercepts and sketch a graph of the plane.

15. $7x + 2y - 2z = 14$

16. $5x - 3y + 5z = 15$

17. Find the general form of the equation of the plane passing through the points $(2, -3, -1)$, $(2, 5, -3)$, and $(-2, 0, 1)$.

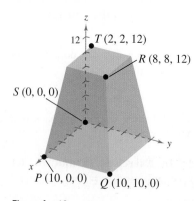

18. Find the distance between the point $(2, -1, 6)$ and the plane $3x - 2y + z = 6$.

19. A tractor fuel tank has the shape and dimensions shown in the figure. Find the angle θ between two adjacent sides.

Figure for 19

Proofs in Mathematics ■ ■ ■ ■ ■ ■ ■ ■ ■ ■ ■ ■ ■ ■ ■

Algebraic Properties of the Cross Product *(p. 793)*

Let $\mathbf{u}$, $\mathbf{v}$, and $\mathbf{w}$ be vectors in space and let c be a scalar.

1. $\mathbf{u} \times \mathbf{v} = -(\mathbf{v} \times \mathbf{u})$

2. $\mathbf{u} \times (\mathbf{v} + \mathbf{w}) = (\mathbf{u} \times \mathbf{v}) + (\mathbf{u} \times \mathbf{w})$

3. $c(\mathbf{u} \times \mathbf{v}) = (c\mathbf{u}) \times \mathbf{v} = \mathbf{u} \times (c\mathbf{v})$

4. $\mathbf{u} \times \mathbf{0} = \mathbf{0} \times \mathbf{u} = \mathbf{0}$

5. $\mathbf{u} \times \mathbf{u} = \mathbf{0}$

6. $\mathbf{u} \cdot (\mathbf{v} \times \mathbf{w}) = (\mathbf{u} \times \mathbf{v}) \cdot \mathbf{w}$

Proof

Let $\mathbf{u} = u_1\mathbf{i} + u_2\mathbf{j} + u_3\mathbf{k}$, $\mathbf{v} = v_1\mathbf{i} + v_2\mathbf{j} + v_3\mathbf{k}$, $\mathbf{w} = w_1\mathbf{i} + w_2\mathbf{j} + w_3\mathbf{k}$, $\mathbf{0} = 0\mathbf{i} + 0\mathbf{j} + 0\mathbf{k}$, and c be a scalar.

1. $\mathbf{u} \times \mathbf{v} = (u_2v_3 - u_3v_2)\mathbf{i} - (u_1v_3 - u_3v_1)\mathbf{j} + (u_1v_2 - u_2v_1)\mathbf{k}$

   $\mathbf{v} \times \mathbf{u} = (v_2u_3 - v_3u_2)\mathbf{i} - (v_1u_3 - v_3u_1)\mathbf{j} + (v_1u_2 - v_2u_1)\mathbf{k}$

   This implies $\mathbf{u} \times \mathbf{v} = -(\mathbf{v} \times \mathbf{u})$.

2. $\mathbf{u} \times (\mathbf{v} + \mathbf{w}) = [u_2(v_3 + w_3) - u_3(v_2 + w_2)]\mathbf{i} - [u_1(v_3 + w_3)$
   $\qquad\qquad\qquad - u_3(v_1 + w_1)]\mathbf{j} + [u_1(v_2 + w_2) - u_2(v_1 + w_1)]\mathbf{k}$

   $= (u_2v_3 - u_3v_2)\mathbf{i} - (u_1v_3 - u_3v_1)\mathbf{j} + (u_1v_2 - u_2v_1)\mathbf{k}$
   $\qquad + (u_2w_3 - u_3w_2)\mathbf{i} - (u_1w_3 - u_3w_1)\mathbf{j} + (u_1w_2 - u_2w_1)\mathbf{k}$

   $= (\mathbf{u} \times \mathbf{v}) + (\mathbf{u} \times \mathbf{w})$

3. $(c\mathbf{u}) \times \mathbf{v} = (cu_2v_3 - cu_3v_2)\mathbf{i} - (cu_1v_3 - cu_3v_1)\mathbf{j} + (cu_1v_2 - cu_2v_1)\mathbf{k}$

   $= c[(u_2v_3 - u_3v_2)\mathbf{i} - (u_1v_3 - u_3v_1)\mathbf{j} + (u_1v_2 - u_2v_1)\mathbf{k}]$

   $= c(\mathbf{u} \times \mathbf{v})$

4. $\mathbf{u} \times \mathbf{0} = (u_2 \cdot 0 - u_3 \cdot 0)\mathbf{i} - (u_1 \cdot 0 - u_3 \cdot 0)\mathbf{j} + (u_1 \cdot 0 - u_2 \cdot 0)\mathbf{k}$

   $= 0\mathbf{i} + 0\mathbf{j} + 0\mathbf{k} = \mathbf{0}$

   $\mathbf{0} \times \mathbf{u} = (0 \cdot u_3 - 0 \cdot u_2)\mathbf{i} - (0 \cdot u_3 - 0 \cdot u_1)\mathbf{j} + (0 \cdot u_2 - 0 \cdot u_1)\mathbf{k}$

   $= 0\mathbf{i} + 0\mathbf{j} + 0\mathbf{k} = \mathbf{0}$

   So, $\mathbf{u} \times \mathbf{0} = \mathbf{0} \times \mathbf{u} = \mathbf{0}$.

5. $\mathbf{u} \times \mathbf{u} = (u_2u_3 - u_3u_2)\mathbf{i} - (u_1u_3 - u_3u_1)\mathbf{j} + (u_1u_2 - u_2u_1)\mathbf{k} = \mathbf{0}$

6. $\mathbf{u} \cdot (\mathbf{v} \times \mathbf{w}) = \begin{vmatrix} u_1 & u_2 & u_3 \\ v_1 & v_2 & v_3 \\ w_1 & w_2 & w_3 \end{vmatrix}$ and

   $(\mathbf{u} \times \mathbf{v}) \cdot \mathbf{w} = \mathbf{w} \cdot (\mathbf{u} \times \mathbf{v}) = \begin{vmatrix} w_1 & w_2 & w_3 \\ u_1 & u_2 & u_3 \\ v_1 & v_2 & v_3 \end{vmatrix}$

   $\mathbf{u} \cdot (\mathbf{v} \times \mathbf{w}) = u_1(v_2w_3 - v_3w_2) - u_2(v_1w_3 - v_3w_1) + u_3(v_1w_2 - v_2w_1)$

   $= u_1v_2w_3 - u_1v_3w_2 - u_2v_1w_3 + u_2v_3w_1 + u_3v_1w_2 - u_3v_2w_1$

   $= u_2v_3w_1 - u_3v_2w_1 - u_1v_3w_2 + u_3v_1w_2 + u_1v_2w_3 - u_2v_1w_3$

   $= (u_2v_3 - u_3v_2)w_1 - (u_1v_3 - u_3v_1)w_2 + (u_1v_2 - u_2v_1)w_3$

   $= (\mathbf{u} \times \mathbf{v}) \cdot \mathbf{w}$

Geometric Properties of the Cross Product *(p. 794)*

Let $\mathbf{u}$ and $\mathbf{v}$ be nonzero vectors in space, and let θ be the angle between $\mathbf{u}$ and $\mathbf{v}$.

1. $\mathbf{u} \times \mathbf{v}$ is orthogonal to both $\mathbf{u}$ and $\mathbf{v}$.

2. $\|\mathbf{u} \times \mathbf{v}\| = \|\mathbf{u}\|\|\mathbf{v}\| \sin \theta$

3. $\mathbf{u} \times \mathbf{v} = \mathbf{0}$ if and only if $\mathbf{u}$ and $\mathbf{v}$ are scalar multiples of each other.

4. $\|\mathbf{u} \times \mathbf{v}\| = $ area of parallelogram that has $\mathbf{u}$ and $\mathbf{v}$ as adjacent sides.

Proof

Let $\mathbf{u} = u_1\mathbf{i} + u_2\mathbf{j} + u_3\mathbf{k}$, $\mathbf{v} = v_1\mathbf{i} + v_2\mathbf{j} + v_3\mathbf{k}$, and $\mathbf{0} = 0\mathbf{i} + 0\mathbf{j} + 0\mathbf{k}$.

1.
$$\mathbf{u} \times \mathbf{v} = (u_2v_3 - u_3v_2)\mathbf{i} - (u_1v_3 - u_3v_1)\mathbf{j} + (u_1v_2 - u_2v_1)\mathbf{k}$$

$$(\mathbf{u} \times \mathbf{v}) \cdot \mathbf{u} = (u_2v_3 - u_3v_2)u_1 - (u_1v_3 - u_3v_1)u_2 + (u_1v_2 - u_2v_1)u_3$$

$$= u_1u_2v_3 - u_1u_3v_2 - u_1u_2v_3 + u_2u_3v_1 + u_1u_3v_2 - u_2u_3v_1$$

$$= 0$$

$$(\mathbf{u} \times \mathbf{v}) \cdot \mathbf{v} = (u_2v_3 - u_3v_2)v_1 - (u_1v_3 - u_3v_1)v_2 + (u_1v_2 - u_2v_1)v_3$$

$$= u_2v_1v_3 - u_3v_1v_2 - u_1v_2v_3 + u_3v_1v_2 + u_1v_2v_3 - u_2v_1v_3$$

$$= 0$$

Two vectors are orthogonal when their dot product is zero, so it follows that $\mathbf{u} \times \mathbf{v}$ is orthogonal to both $\mathbf{u}$ and $\mathbf{v}$.

2. Note that $\cos \theta = \dfrac{\mathbf{u} \cdot \mathbf{v}}{\|\mathbf{u}\|\|\mathbf{v}\|}$. So,

$$\|\mathbf{u}\|\|\mathbf{v}\| \sin \theta = \|\mathbf{u}\|\|\mathbf{v}\|\sqrt{1 - \cos^2 \theta}$$

$$= \|\mathbf{u}\|\|\mathbf{v}\|\sqrt{1 - \frac{(\mathbf{u} \cdot \mathbf{v})^2}{\|\mathbf{u}\|^2\|\mathbf{v}\|^2}}$$

$$= \sqrt{\|\mathbf{u}\|^2\|\mathbf{v}\|^2 - (\mathbf{u} \cdot \mathbf{v})^2}$$

$$= \sqrt{(u_1^2 + u_2^2 + u_3^2)(v_1^2 + v_2^2 + v_3^2) - (u_1v_1 + u_2v_2 + u_3v_3)^2}$$

$$= \sqrt{(u_2v_3 - u_3v_2)^2 + (u_1v_3 - u_3v_1)^2 + (u_1v_2 - u_2v_1)^2}$$

$$= \|\mathbf{u} \times \mathbf{v}\|.$$

3. If $\mathbf{u}$ and $\mathbf{v}$ are scalar multiples of each other, then $\mathbf{u} = c\mathbf{v}$ for some scalar c.

$$\mathbf{u} \times \mathbf{v} = (c\mathbf{v}) \times \mathbf{v} = c(\mathbf{v} \times \mathbf{v}) = c(\mathbf{0}) = \mathbf{0}$$

If $\mathbf{u} \times \mathbf{v} = \mathbf{0}$, then $\|\mathbf{u}\|\|\mathbf{v}\| \sin \theta = 0$. (Assume $\mathbf{u} \neq \mathbf{0}$ and $\mathbf{v} \neq \mathbf{0}$.) So, $\sin \theta = 0$, and $\theta = 0$ or $\theta = \pi$. In either case, θ is the angle between the vectors, so $\mathbf{u}$ and $\mathbf{v}$ are parallel. So, $\mathbf{u} = c\mathbf{v}$ for some scalar c.

4. The figure at the left is a parallelogram that has $\mathbf{u}$ and $\mathbf{v}$ as adjacent sides. The height of the parallelogram is $\|\mathbf{v}\| \sin \theta$, so the area is

$$\text{Area} = (\text{base})(\text{height})$$

$$= \|\mathbf{u}\|\|\mathbf{v}\| \sin \theta$$

$$= \|\mathbf{u} \times \mathbf{v}\|.$$

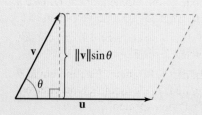

P.S. Problem Solving

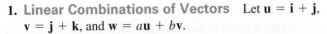

1. **Linear Combinations of Vectors** Let $\mathbf{u} = \mathbf{i} + \mathbf{j}$, $\mathbf{v} = \mathbf{j} + \mathbf{k}$, and $\mathbf{w} = a\mathbf{u} + b\mathbf{v}$.

 (a) Sketch $\mathbf{u}$ and $\mathbf{v}$.

 (b) When $\mathbf{w} = \mathbf{0}$, show that a and b must both be zero.

 (c) Find a and b such that $\mathbf{w} = \mathbf{i} + 2\mathbf{j} + \mathbf{k}$.

 (d) Show that no choice of a and b yields $\mathbf{w} = \mathbf{i} + 2\mathbf{j} + 3\mathbf{k}$.

2. **Describing a Set of Points** The initial and terminal points of $\mathbf{v}$ are (x_1, y_1, z_1) and (x, y, z), respectively. Describe the set of all points (x, y, z) such that $\|\mathbf{v}\| = 4$.

3. **Classifying Triangles** The vertices of a triangle are given. Classify the triangle as acute, obtuse, or right. Explain.

 (a) $(1, 2, 0), (0, 0, 0), (-2, 1, 0)$

 (b) $(-3, 0, 0), (0, 0, 0), (1, 2, 3)$

 (c) $(2, -3, 4), (0, 1, 2), (-1, 2, 0)$

 (d) $(2, -7, 3), (-1, 5, 8), (4, 6, -1)$

4. **Representing Force as a Vector** A television camera weighing 120 pounds is supported by a tripod (see figure). Represent the force exerted on each leg of the tripod as a vector.

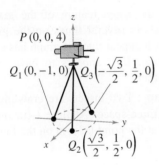

5. **Finding Total Force** A precast concrete wall is temporarily kept in its vertical position by ropes (see figure). The tensions in AB and AC are 420 pounds and 650 pounds, respectively. Find the total force exerted on the pin at position A.

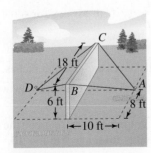

6. **Writing a Program** You are given the component forms of the vectors $\mathbf{u}$ and $\mathbf{v}$. Write a program for a graphing utility in which the output is (a) the component form of $\mathbf{u} + \mathbf{v}$, (b) $\|\mathbf{u} + \mathbf{v}\|$, (c) $\|\mathbf{u}\|$, and (d) $\|\mathbf{v}\|$.

7. **Running a Program** Run the program you wrote in Exercise 6 for the vectors $\mathbf{u} = \langle -1, 3, 4 \rangle$ and $\mathbf{v} = \langle 5, 4.5, -6 \rangle$.

8. **Proof** Prove

$$\|\mathbf{u} \times \mathbf{v}\| = \|\mathbf{u}\|\|\mathbf{v}\|$$

 when $\mathbf{u}$ and $\mathbf{v}$ are orthogonal.

9. **Proof** Prove $\mathbf{u} \times (\mathbf{v} \times \mathbf{w}) = (\mathbf{u} \cdot \mathbf{w})\mathbf{v} - (\mathbf{u} \cdot \mathbf{v})\mathbf{w}$.

10. **Proof** Prove that the triple scalar product of $\mathbf{u}$, $\mathbf{v}$, and $\mathbf{w}$ is

$$\mathbf{u} \cdot (\mathbf{v} \times \mathbf{w}) = \begin{vmatrix} u_1 & u_2 & u_3 \\ v_1 & v_2 & v_3 \\ w_1 & w_2 & w_3 \end{vmatrix}.$$

11. **Proof** Prove that the volume V of a parallelepiped with vectors $\mathbf{u}$, $\mathbf{v}$, and $\mathbf{w}$ as adjacent edges is given by

$$V = |\mathbf{u} \cdot (\mathbf{v} \times \mathbf{w})|.$$

12. **Think About It** Find a vector that is normal to each plane. What can you say about the positions of these planes in space relative to each other?

 (a) $2x + 3y - z = 2$

 (b) $4x + 6y - 2z = 5$

 (c) $-2x - 3y + z = -2$

 (d) $-6x - 9y + 3z = 11$

13. **Torque** In physics, the cross product can be used to measure *torque*, or the *moment* $\mathbf{M}$ of a force $\mathbf{F}$ about a point P. If the point of application of the force is Q, then the moment $\mathbf{M}$ of $\mathbf{F}$ about P is given by

$$\mathbf{M} = \overrightarrow{PQ} \times \mathbf{F}.$$

 A force of 60 pounds acts on the pipe wrench shown in the figure.

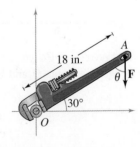

 (a) Find the magnitude of the moment about O. Use a graphing utility to graph the resulting function of θ.

 (b) Use the result of part (a) to determine the magnitude of the moment when $\theta = 45°$.

 (c) Use the graph in part (a) to determine the angle θ such that the magnitude of the moment is maximized. Is the answer what you expected? Explain.

14. Magnitudes of Moments A force of 200 pounds acts on the bracket shown in the figure.

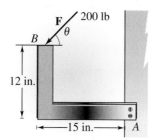

(a) Determine the vector $\overrightarrow{AB}$ and the vector $\mathbf{F}$ representing the force. ($\mathbf{F}$ will be in terms of θ.)

(b) Find the magnitude of the moment (torque) about A by evaluating $\|\overrightarrow{AB} \times \mathbf{F}\|$. Use a graphing utility to graph the resulting function of θ for $0° \le \theta \le 180°$.

(c) Use the result of part (b) to determine the magnitude of the moment when $\theta = 30°$.

(d) Use the graph in part (b) to determine the angle θ such that the magnitude of the moment is maximized.

(e) Use the graph in part (b) to approximate the zero of the function. Interpret the meaning of the zero in the context of the problem.

15. Proof Using vectors, prove the Law of Sines: If $\mathbf{a}$, $\mathbf{b}$, and $\mathbf{c}$ are three sides of a triangle (see figure), then

$$\frac{\sin A}{\|\mathbf{a}\|} = \frac{\sin B}{\|\mathbf{b}\|} = \frac{\sin C}{\|\mathbf{c}\|}.$$

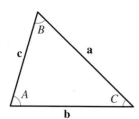

16. Distance Between a Point and a Line The distance D between a point Q and a line in space is given by

$$D = \frac{\|\overrightarrow{PQ} \times \mathbf{u}\|}{\|\mathbf{u}\|}$$

where $\mathbf{u}$ is a direction vector for the line and P is a point on the line. Find the distance between each point and the line given by each set of parametric equations.

(a) $(1, 5, -2)$

$x = -2 + 4t$

$y = 3$

$z = 1 - t$

(b) $(1, -2, 4)$

$x = 2t$

$y = -3 + t$

$z = 2 + 2t$

17. Distance Between a Point and a Line Use the formula given in Exercise 16.

(a) Find the shortest distance between the point $Q(2, 0, 0)$ and the line passing through the points $P_1(0, 0, 1)$ and $P_2(0, 1, 2)$.

(b) Find the shortest distance between the point $Q(2, 0, 0)$ and the line segment from $P_1(0, 0, 1)$ to $P_2(0, 1, 2)$.

18. Distance Between a Point and a Line Consider the line given by the parametric equations

$x = -t + 3$

$y = \frac{1}{2}t + 1$

$z = 2t - 1$

and the point $(4, 3, s)$ for any real number s.

(a) Write the distance between the point and the line as a function of s. (*Hint:* Use the formula given in Exercise 16.)

(b) Use a graphing utility to graph the function from part (a). Use the graph to find the value of s such that the distance between the point and the line is a minimum.

(c) Use the *zoom* feature of the graphing utility to zoom out several times on the graph in part (b). Does it appear that the graph has slant asymptotes? Explain. If it appears to have slant asymptotes, find them.

19. Distance Two insects are crawling along different lines in three-space. At time t (in minutes), the first insect is at the point (x, y, z) on the line given by

$x = 6 + t$

$y = 8 - t$

$z = 3 + t.$

Also, at time t, the second insect is at the point (x, y, z) on the line given by

$x = 1 + t$

$y = 2 + t$

$z = 2t.$

Assume distances are given in inches.

(a) Find the distance between the two insects at time $t = 0$.

(b) Use a graphing utility to graph the distance between the insects from $t = 0$ to $t = 10$.

(c) Using the graph in part (b), what can you conclude about the distance between the insects?

(d) Using the graph in part (b), determine how close the insects get to each other.

12 Limits and an Introduction to Calculus

Land Surveying *(Exercise 45, page 865)*

Mobile Phone Protective Cases
(Example 3, page 852)

Temperature in Dallas, Texas *(Example 2, page 840)*

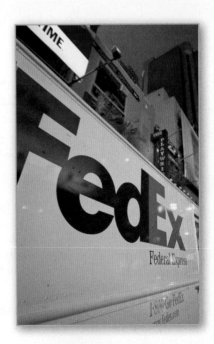

Delivery Charges
(Example 8, page 834)

Maximum Volume *(Exercise 3, page 826)*

12.1 Introduction to Limits

The concept of a limit is useful in applications involving maximization. For example, in Exercise 3 on page 826, you will use the concept of a limit to verify the maximum volume of an open box.

- Understand the limit concept.
- Use the definition of a limit to estimate limits.
- Determine whether limits of functions exist.
- Use properties of limits and direct substitution to evaluate limits.

The Limit Concept

The notion of a limit is a *fundamental* concept of calculus. In this chapter, you will learn how to evaluate limits and how to use them in the two basic problems of calculus: the tangent line problem and the area problem.

EXAMPLE 1   **Finding a Rectangle of Maximum Area**

Find the dimensions of a rectangle with perimeter 24 inches that yield the maximum area.

Solution   Let w represent the width of the rectangle and let l represent the length of the rectangle. From

$$2w + 2l = 24 \qquad \text{Perimeter is 24.}$$

it follows that $l = 12 - w$, as shown in the figure below. So, the area of the rectangle is

$$A = lw \qquad\qquad \text{Formula for area}$$
$$= (12 - w)w \qquad \text{Substitute } 12 - w \text{ for } l.$$
$$= 12w - w^2. \qquad \text{Simplify.}$$

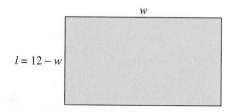

Using this model for area, experiment with different values of w to see how to obtain the maximum area. After checking several values, it appears that the maximum area occurs when $w = 6$, as shown in the table.

Width, w	5.0	5.5	5.9	6.0	6.1	6.5	7.0
Area, A	35.00	35.75	35.99	36.00	35.99	35.75	35.00

In limit terminology, you say "the limit of A as w approaches 6 is 36" and write

$$\lim_{w \to 6} A = \lim_{w \to 6} (12w - w^2) = 36.$$

So, the dimensions of a rectangle with perimeter 24 inches that yield the maximum area are $w = 6$ inches and $l = 12 - 6 = 6$ inches.

✓ **Checkpoint** ◀))) *Audio-video solution in English & Spanish at LarsonPrecalculus.com*

Find the dimensions of a rectangle with perimeter 52 inches that yield the maximum area.

Definition of a Limit

• • **REMARK** An alternative
notation for $\lim\limits_{x \to c} f(x) = L$ is

$$f(x) \to L \text{ as } x \to c$$

which is read as "$f(x)$ approaches
L as x approaches c." ▷

> ### Definition of a Limit
>
> If $f(x)$ becomes arbitrarily close to a unique number L as x approaches c from either side, then the **limit** of $f(x)$ as x approaches c is L. This is written as
>
> $$\lim_{x \to c} f(x) = L.$$

EXAMPLE 2 Estimating a Limit Numerically

Use a table to estimate the limit numerically: $\lim\limits_{x \to 2} (3x - 2)$.

Solution Let $f(x) = 3x - 2$. Then construct a table that shows values of $f(x)$ for two sets of x-values—one that approaches 2 from the left and one that approaches 2 from the right.

x	1.9	1.99	1.999	2.0	2.001	2.01	2.1
$f(x)$	3.700	3.970	3.997	?	4.003	4.030	4.300

From the table, it appears that the closer x gets to 2, the closer $f(x)$ gets to 4. So, estimate the limit to be 4. Figure 12.1 verifies this conclusion.

✓ *Checkpoint* ◀))) *Audio-video solution in English & Spanish at LarsonPrecalculus.com*

Use a table to estimate the limit numerically: $\lim\limits_{x \to 3} (3 - 2x)$. ◼

In Figure 12.1, note that the graph of $f(x) = 3x - 2$ is continuous. For graphs that are not continuous, finding a limit can be more challenging.

EXAMPLE 3 Estimating a Limit Numerically

Use a table to estimate the limit numerically.

$$\lim_{x \to 0} \frac{x}{\sqrt{x + 1} - 1}$$

Solution Let $f(x) = x/\left(\sqrt{x + 1} - 1\right)$. Then construct a table that shows values of $f(x)$ for two sets of x-values—one that approaches 0 from the left and one that approaches 0 from the right.

x	-0.01	-0.001	-0.0001	0	0.0001	0.001	0.01
$f(x)$	1.99499	1.99950	1.99995	?	2.00005	2.00050	2.00499

From the table, it appears that the limit is 2. Figure 12.2 verifies this conclusion.

✓ *Checkpoint* ◀))) *Audio-video solution in English & Spanish at LarsonPrecalculus.com*

Use a table to estimate the limit numerically.

$$\lim_{x \to 1} \frac{x - 1}{x^2 + 3x - 4}$$ ◼

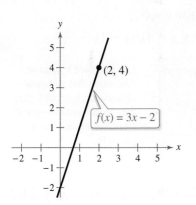

Figure 12.1

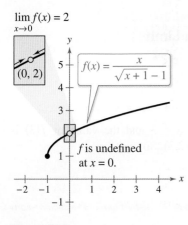

Figure 12.2

• • • • • • • • • • • • • • • • • ▷
• • **REMARK** In Example 3,
note that $f(0)$ is undefined, so
it is not possible to *reach* the
limit. In Example 2, note that
$f(2) = 4$, so it *is* possible to
reach the limit.

In Example 3, note that $f(x)$ has a limit when $x \to 0$ even though the function is not defined when $x = 0$. This often happens, and it is important to realize that *the existence or nonexistence of $f(x)$ at $x = c$ has no bearing on the existence of the limit of $f(x)$ as x approaches c.*

EXAMPLE 4 Estimating a Limit

Estimate the limit: $\lim\limits_{x \to 1} \dfrac{x^3 - x^2 + x - 1}{x - 1}$.

Numerical Solution

Let $f(x) = (x^3 - x^2 + x - 1)/(x - 1)$. Then construct a table that shows values of $f(x)$ for two sets of x-values—one that approaches 1 from the left and one that approaches 1 from the right.

x	0.9	0.99	0.999	1.0
$f(x)$	1.8100	1.9801	1.9980	?

x	1.001	1.01	1.1
$f(x)$	2.0020	2.0201	2.2100

From the table, it appears that the limit is 2.

Graphical Solution

Use a graphing utility to graph

$$f(x) = (x^3 - x^2 + x - 1)/(x - 1).$$

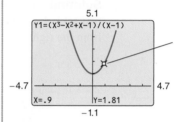

Use the *trace* feature to determine that as x gets closer and closer to 1, $f(x)$ gets closer and closer to 2 from the left and from the right.

From the graph, estimate the limit to be 2. As you use the *trace* feature, notice that there is no value given for y when $x = 1$, and that there is a hole or break in the graph at $x = 1$.

✓ **Checkpoint** ◀))) *Audio-video solution in English & Spanish at LarsonPrecalculus.com*

Estimate the limit: $\lim\limits_{x \to 2} \dfrac{x^3 - 2x^2 + 3x - 6}{x - 2}$.

EXAMPLE 5 Using a Graph to Find a Limit

Find the limit of $f(x)$ as x approaches 3.

$$f(x) = \begin{cases} 2, & x \neq 3 \\ 0, & x = 3 \end{cases}$$

Solution Because $f(x) = 2$ for all x other than $x = 3$ and the value of $f(3)$ is immaterial, it follows that the limit is 2 (see Figure 12.3). So, write

$$\lim_{x \to 3} f(x) = 2.$$

✓ **Checkpoint** ◀))) *Audio-video solution in English & Spanish at LarsonPrecalculus.com*

Find the limit of $f(x)$ as x approaches 2.

$$f(x) = \begin{cases} -3, & x \neq 2 \\ 0, & x = 2 \end{cases}$$

Figure 12.3

In Example 5, the fact that $f(3) = 0$ has no bearing on the existence or value of the limit as x approaches 3. For example, if the function were defined as

$$f(x) = \begin{cases} 2, & x \neq 3 \\ 4, & x = 3 \end{cases}$$

then the limit as x approaches 3 would still equal 2.

Limits That Fail to Exist

Next, you will examine some limits that fail to exist.

EXAMPLE 6   **Comparing Left and Right Behavior**

Show that the limit does not exist.

$$\lim_{x \to 0} \frac{|x|}{x}$$

Solution   Consider the graph of $f(x) = |x|/x$, shown in Figure 12.4. Notice that for positive x-values

$$\frac{|x|}{x} = 1, \quad x > 0$$

and for negative x-values

$$\frac{|x|}{x} = -1, \quad x < 0.$$

This means that no matter how close x gets to 0, there are both positive and negative x-values that yield $f(x) = 1$ and $f(x) = -1$, respectively. This implies that the limit does not exist.

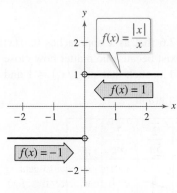

Figure 12.4

✓ *Checkpoint*  *Audio-video solution in English & Spanish at LarsonPrecalculus.com*

Show that the limit does not exist.

$$\lim_{x \to 1} \frac{2|x - 1|}{x - 1}$$

EXAMPLE 7   **Unbounded Behavior**

Discuss the existence of the limit.

$$\lim_{x \to 0} \frac{1}{x^2}$$

Solution   Let $f(x) = 1/x^2$. In Figure 12.5, note that as x approaches 0 from either the right or the left, $f(x)$ increases without bound. This means that choosing x close enough to 0 enables you to force $f(x)$ to be as large as you want. For example, $f(x)$ is larger than 100 when you choose x that is within $\frac{1}{10}$ of 0. That is,

$$0 < |x| < \frac{1}{10} \quad \Longrightarrow \quad f(x) = \frac{1}{x^2} > 100.$$

Similarly, you can force $f(x)$ to be larger than 1,000,000 by choosing x that is within $\frac{1}{1000}$ of 0, as shown below.

$$0 < |x| < \frac{1}{1000} \quad \Longrightarrow \quad f(x) = \frac{1}{x^2} > 1,000,000$$

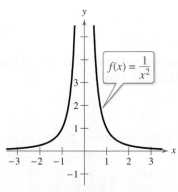

Figure 12.5

Because $f(x)$ is not approaching a unique real number L as x approaches 0, the limit does not exist.

✓ *Checkpoint*  *Audio-video solution in English & Spanish at LarsonPrecalculus.com*

Discuss the existence of the limit.

$$\lim_{x \to 0} \left(-\frac{1}{x^2} \right)$$

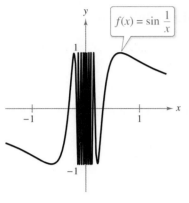

Figure 12.6

EXAMPLE 8 **Oscillating Behavior**

See LarsonPrecalculus.com for an interactive version of this type of example.

Discuss the existence of the limit.

$$\lim_{x \to 0} \sin \frac{1}{x}$$

Solution Let $f(x) = \sin(1/x)$. Notice in Figure 12.6 that as x approaches 0, $f(x)$ oscillates between -1 and 1. So, the limit does not exist because no matter how close you are to 0, it is possible to choose values of x_1 and x_2 such that $\sin(1/x_1) = 1$ and $\sin(1/x_2) = -1$, as shown in the table.

x	$-\dfrac{2}{\pi}$	$-\dfrac{2}{3\pi}$	$-\dfrac{2}{5\pi}$	0	$\dfrac{2}{5\pi}$	$\dfrac{2}{3\pi}$	$\dfrac{2}{\pi}$
$\sin \dfrac{1}{x}$	-1	1	-1	?	1	-1	1

✔ *Checkpoint* 🔊)) Audio-video solution in English & Spanish at LarsonPrecalculus.com

Discuss the existence of the limit.

$$\lim_{x \to 1} \cos \frac{1}{x-1}$$

Examples 6, 7, and 8 show three of the most common types of behavior associated with the *nonexistence* of a limit.

Conditions Under Which Limits Do Not Exist

The limit of $f(x)$ as $x \to c$ does not exist when any of the conditions listed below are true.

1. $f(x)$ approaches a different number from the right side of c than it approaches from the left side of c. Example 6

2. $f(x)$ increases or decreases without bound as x approaches c. Example 7

3. $f(x)$ oscillates between two fixed values as x approaches c. Example 8

▷ **TECHNOLOGY** A graphing utility can help you discover the behavior of a function near the x-value at which you are evaluating a limit. When you do this, however, realize that you should not always trust the graphs that graphing utilities display. For instance, when you use a graphing utility to graph the function in Example 8 over an interval containing 0, you will most likely obtain an incorrect graph, as shown at the right. The reason that a graphing utility cannot show the correct graph is that the graph has infinitely many oscillations over any interval that contains 0.

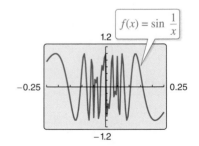

Properties of Limits and Direct Substitution

Sometimes, as in Example 2, the limit of $f(x)$ as $x \to c$ is $f(c)$. In such cases, the limit can be evaluated by **direct substitution.** That is,

$$\lim_{x \to c} f(x) = f(c). \qquad \text{Substitute } c \text{ for } x.$$

There are many "well-behaved" functions, such as polynomial functions and rational functions with nonzero denominators, that have this property. The list below includes some basic limits.

Basic Limits

Let b and c be real numbers and let n be a positive integer.

1. $\lim_{x \to c} b = b$ Limit of a constant function

2. $\lim_{x \to c} x = c$ Limit of the identity function

3. $\lim_{x \to c} x^n = c^n$ Limit of a power function

4. $\lim_{x \to c} \sqrt[n]{x} = \sqrt[n]{c}$,   valid for all c when n is odd   Limit of a radical function
   and valid for $c > 0$ when n is even

For a proof of the limit of a power function, see Proofs in Mathematics on page 874. This list can also include trigonometric functions. For example,

$$\lim_{x \to \pi} \sin x = \sin \pi = 0$$

and

$$\lim_{x \to 0} \cos x = \cos 0 = 1.$$

By combining the basic limits listed above with the properties of limits listed below, you can find limits for a wide variety of functions.

Properties of Limits

Let b and c be real numbers, let n be a positive integer, and let f and g be functions with the limits

$$\lim_{x \to c} f(x) = L \quad \text{and} \quad \lim_{x \to c} g(x) = K.$$

1. Scalar multiple: $\lim_{x \to c} [bf(x)] = bL$

2. Sum or difference: $\lim_{x \to c} [f(x) \pm g(x)] = L \pm K$

3. Product: $\lim_{x \to c} [f(x)g(x)] = LK$

4. Quotient: $\lim_{x \to c} \dfrac{f(x)}{g(x)} = \dfrac{L}{K}, \quad K \ne 0$

5. Power: $\lim_{x \to c} [f(x)]^n = L^n$

▷ **TECHNOLOGY** When evaluating limits, remember that there are several ways to solve most problems. Often, a problem can be solved *numerically, graphically,* or *algebraically.* A graphing utility can be used to confirm limits numerically with the *table* feature or graphically with the *zoom* and *trace* features.

EXAMPLE 9 **Direct Substitution and Properties of Limits**

Find each limit.

a. $\lim\limits_{x\to 4} x^2$ **b.** $\lim\limits_{x\to 4} 5x$ **c.** $\lim\limits_{x\to \pi} \dfrac{\tan x}{x}$

d. $\lim\limits_{x\to 9} \sqrt{x}$ **e.** $\lim\limits_{x\to \pi} (x \cos x)$ **f.** $\lim\limits_{x\to 3} (x + 4)^2$

Solution Use the properties of limits and direct substitution to evaluate each limit.

a. $\lim\limits_{x\to 4} x^2 = (4)^2 = 16$ Use direct substitution.

b. $\lim\limits_{x\to 4} 5x = 5 \lim\limits_{x\to 4} x = 5(4) = 20$ Use the Scalar Multiple Property and direct substitution.

c. $\lim\limits_{x\to \pi} \dfrac{\tan x}{x} = \dfrac{\lim\limits_{x\to \pi} \tan x}{\lim\limits_{x\to \pi} x} = \dfrac{\tan \pi}{\pi} = \dfrac{0}{\pi} = 0$ Use the Quotient Property and direct substitution.

d. $\lim\limits_{x\to 9} \sqrt{x} = \sqrt{9} = 3$ Use direct substitution.

e. $\lim\limits_{x\to \pi} (x \cos x) = \left(\lim\limits_{x\to \pi} x\right)\left(\lim\limits_{x\to \pi} \cos x\right) = \pi(\cos \pi) = -\pi$ Use the Product Property and direct substitution.

f. $\lim\limits_{x\to 3} (x + 4)^2 = \left[\left(\lim\limits_{x\to 3} x\right) + \left(\lim\limits_{x\to 3} 4\right)\right]^2 = (3 + 4)^2 = 49$ Use the Power and Sum Properties and direct substitution.

✓ **Checkpoint** 🔊))) *Audio-video solution in English & Spanish at LarsonPrecalculus.com*

Find each limit.

a. $\lim\limits_{x\to 3} \dfrac{1}{4}$ **b.** $\lim\limits_{x\to 3} x^3$ **c.** $\lim\limits_{x\to \pi} \dfrac{\cos x}{x}$

d. $\lim\limits_{x\to 12} \sqrt{x}$ **e.** $\lim\limits_{x\to \pi} (x \tan x)$ **f.** $\lim\limits_{x\to 3} (1 - x)^2$

Example 9 shows algebraic solutions. To verify the limit in Example 9(a) numerically, create a table that shows values of x^2 for two sets of x-values—one set that approaches 4 from the left and one that approaches 4 from the right.

x	3.9	3.99	3.999	4.0	4.001	4.01	4.1
x^2	15.2100	15.9201	15.9920	?	16.0080	16.0801	16.8100

The table shows that the limit as x approaches 4 is 16. To verify the limit in Example 9(a) graphically, sketch the graph of $y = x^2$, as shown below. The graph also shows that the limit as x approaches 4 is 16.

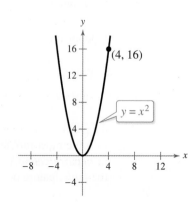

The results of using direct substitution to evaluate limits of polynomial and rational functions are summarized below.

Limits of Polynomial and Rational Functions

1. If p is a polynomial function and c is a real number, then

$$\lim_{x \to c} p(x) = p(c).$$

2. If r is a rational function $r(x) = p(x)/q(x)$, and c is a real number such that $q(c) \neq 0$, then

$$\lim_{x \to c} r(x) = r(c) = \frac{p(c)}{q(c)}.$$

For a proof of the limit of a polynomial function, see Proofs in Mathematics on page 874.

EXAMPLE 10 **Evaluating Limits by Direct Substitution**

Find each limit.

a. $\displaystyle \lim_{x \to -1} (x^2 + x - 6)$ **b.** $\displaystyle \lim_{x \to -1} \frac{x^3 - 5x}{x}$ **c.** $\displaystyle \lim_{x \to -1} \frac{x^2 + x - 6}{x + 3}$

Solution The first function is a polynomial function. The second and third functions are rational functions (with nonzero denominators at $x = -1$). So, you can evaluate the limits by direct substitution.

a. $\displaystyle \lim_{x \to -1} (x^2 + x - 6) = (-1)^2 + (-1) - 6 = -6$

b. $\displaystyle \lim_{x \to -1} \frac{x^3 - 5x}{x} = \frac{(-1)^3 - 5(-1)}{-1} = -\frac{4}{1} = -4$

c. $\displaystyle \lim_{x \to -1} \frac{x^2 + x - 6}{x + 3} = \frac{(-1)^2 + (-1) - 6}{-1 + 3} = -\frac{6}{2} = -3$

✓ **Checkpoint** ◀))) *Audio-video solution in English & Spanish at LarsonPrecalculus.com*

Find each limit.

a. $\displaystyle \lim_{x \to 3} (x^2 - 3x + 7)$ **b.** $\displaystyle \lim_{x \to 3} \frac{x^2 - 3x + 7}{x}$ **c.** $\displaystyle \lim_{x \to 3} \frac{x + 3}{x^2 + 3x}$

Summarize (Section 12.1)

1. State two uses of the limit concept in calculus *(page 818)*. For an example that uses limit terminology to state a maximum area, see Example 1.

2. State the definition of a limit *(page 819)*. For examples of estimating limits and using graphs to find limits, see Examples 2–5.

3. List the three most common types of behavior associated with the nonexistence of a limit *(pages 821 and 822)*. For examples of functions that do not have limits, see Examples 6–8.

4. Explain how to use properties of limits and direct substitution to evaluate a limit *(page 823)*. For examples of using properties of limits and direct substitution to evaluate limits, see Examples 9 and 10.

12.1 Exercises

See **CalcChat.com** for tutorial help and worked-out solutions to odd-numbered exercises.

Vocabulary: Fill in the blanks.

1. If $f(x)$ becomes arbitrarily close to a unique number L as x approaches c from either side, then the _____ of $f(x)$ as x approaches c is L.

2. To evaluate the limit of a polynomial function, use _____ _____.

Skills and Applications

3. Geometry

You create an open box from a square piece of material 24 centimeters on a side. You cut equal squares with sides of length x from the corners and turn up the sides.

(a) Draw and label a diagram that represents the box.

(b) Write a function V that represents the volume of the box.

(c) The box has a maximum volume when $x = 4$. Use a graphing utility to complete the table and observe the behavior of the function as x approaches 4. Use the table to find $\lim\limits_{x \to 4} V(x)$.

x	3	3.5	3.9	4	4.1	4.5	5
$V(x)$							

(d) Use the graphing utility to graph the volume function. Verify that the volume is maximum when $x = 4$.

4. Geometry A right triangle has a hypotenuse of $\sqrt{18}$ inches.

(a) Draw and label a diagram that shows the base x and height y of the triangle.

(b) Write a function A in terms of x that represents the area of the triangle.

(c) The triangle has a maximum area when $x = 3$ inches. Use a graphing utility to complete the table and observe the behavior of the function as x approaches 3. Use the table to find $\lim\limits_{x \to 3} A(x)$.

x	2	2.5	2.9	3	3.1	3.5	4
$A(x)$							

(d) Use the graphing utility to graph the area function. Verify that the area is maximum when $x = 3$ inches.

Estimating a Limit Numerically In Exercises 5–10, complete the table and use the result to estimate the limit numerically. Determine whether it is possible to reach the limit.

5. $\lim\limits_{x \to 1} (7x + 3)$

x	0.9	0.99	0.999	1	1.001	1.01	1.1
$f(x)$				?			

6. $\lim\limits_{x \to -1} (3x^2 + 2x - 6)$

x	-1.1	-1.01	-1.001	-1
$f(x)$				?

x	-0.999	-0.99	-0.9
$f(x)$			

7. $\lim\limits_{x \to -2} \dfrac{x + 2}{x^2 - 4}$

x	-2.1	-2.01	-2.001	-2
$f(x)$				?

x	-1.999	-1.99	-1.9
$f(x)$			

8. $\lim\limits_{x \to 3} \dfrac{x - 3}{x^2 - 2x - 3}$

x	2.9	2.99	2.999	3	3.001	3.01	3.1
$f(x)$				?			

9. $\lim\limits_{x \to 0} \dfrac{\sin 2x}{x}$

x	-0.1	-0.01	-0.001	0
$f(x)$				?

x	0.001	0.01	0.1
$f(x)$			

10. $\lim\limits_{x \to 0} \dfrac{\tan x}{2x}$

x	-0.1	-0.01	-0.001	0
$f(x)$				?

x	0.001	0.01	0.1
$f(x)$			

 Using a Graphing Utility to Estimate a Limit In Exercises 11–22, use a graphing utility to create a table of values for the function and estimate the limit numerically. Confirm your result graphically.

11. $\lim\limits_{x \to 1} \dfrac{x - 1}{x^2 + 2x - 3}$

12. $\lim\limits_{x \to -2} \dfrac{x + 2}{x^2 + 5x + 6}$

13. $\lim\limits_{x \to 0} \dfrac{\sqrt{x + 5} - \sqrt{5}}{x}$

14. $\lim\limits_{x \to -3} \dfrac{\sqrt{1 - x} - 2}{x + 3}$

15. $\lim\limits_{x \to -4} \dfrac{\dfrac{x}{x + 2} - 2}{x + 4}$

16. $\lim\limits_{x \to 2} \dfrac{\dfrac{1}{x + 2} - \dfrac{1}{4}}{x - 2}$

17. $\lim\limits_{x \to 0} \dfrac{\sin^2 x}{x}$

18. $\lim\limits_{x \to 0} \dfrac{2x}{\tan 4x}$

19. $\lim\limits_{x \to 0} \dfrac{e^{2x} - 1}{2x}$

20. $\lim\limits_{x \to 0} \dfrac{1 - e^{-4x}}{x}$

21. $\lim\limits_{x \to 1} \dfrac{\ln(2x - 1)}{x - 1}$

22. $\lim\limits_{x \to 1} \dfrac{\ln x^2}{x - 1}$

 Using a Graph to Find a Limit In Exercises 23 and 24, graph the function and find the limit (if it exists) as x approaches 2.

23. $f(x) = \begin{cases} 2x + 1, & x < 2 \\ x + 3, & x \geq 2 \end{cases}$

24. $f(x) = \begin{cases} -2x, & x \leq 2 \\ x^2 - 4x + 1, & x > 2 \end{cases}$

Using a Graph to Find a Limit In Exercises 25–32, use the graph to find the limit, if it exists. If the limit does not exist, explain why.

25. $\lim\limits_{x \to -2} (x^2 - 2)$

26. $\lim\limits_{x \to 2} \dfrac{3x^2 - 12}{x - 2}$

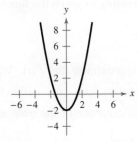 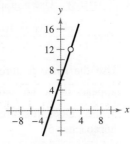

27. $\lim\limits_{x \to -2} -\dfrac{|x + 2|}{x + 2}$

28. $\lim\limits_{x \to 1} \dfrac{|x - 1|}{x - 1}$

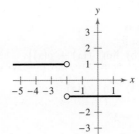

29. $\lim\limits_{x \to -3} -\dfrac{2}{(x + 3)^2}$

30. $\lim\limits_{x \to 1} \dfrac{1}{x - 1}$

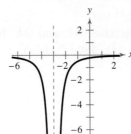 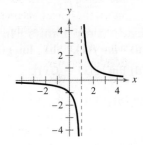

31. $\lim\limits_{x \to 0} 2 \cos \dfrac{\pi}{x}$

32. $\lim\limits_{x \to -1} \sin \dfrac{\pi x}{2}$

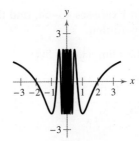 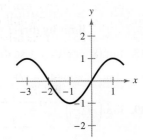

 Determining Whether a Limit Exists In Exercises 33–40, use a graphing utility to graph the function and use the graph to determine whether the limit exists. If the limit does not exist, explain why.

33. $f(x) = \dfrac{5}{2 + e^{1/x}}, \quad \lim\limits_{x \to 0} f(x)$

34. $f(x) = \ln(7 - x), \quad \lim\limits_{x \to -1} f(x)$

35. $f(x) = \cos \dfrac{1}{x}, \quad \lim\limits_{x \to 0} f(x)$

36. $f(x) = \sin \pi x, \quad \lim\limits_{x \to 1} f(x)$

37. $f(x) = \dfrac{\sqrt{x + 3} - 1}{x - 4}, \quad \lim\limits_{x \to 4} f(x)$

38. $f(x) = \dfrac{\sqrt{x + 5} - 4}{x - 2}, \quad \lim\limits_{x \to 2} f(x)$

39. $f(x) = \dfrac{x - 1}{x^2 - 4x + 3}, \quad \lim\limits_{x \to 1} f(x)$

40. $f(x) = \dfrac{7}{x - 3}, \quad \lim\limits_{x \to 3} f(x)$

Evaluating Limits In Exercises 41 and 42, use the given information to evaluate each limit.

41. $\lim\limits_{x \to c} f(x) = 3, \quad \lim\limits_{x \to c} g(x) = 6$

(a) $\lim\limits_{x \to c} [-2g(x)]$

(b) $\lim\limits_{x \to c} [f(x) + g(x)]$

(c) $\lim\limits_{x \to c} \dfrac{f(x)}{g(x)}$

(d) $\lim\limits_{x \to c} \sqrt{f(x)}$

42. $\lim\limits_{x \to c} f(x) = 5, \quad \lim\limits_{x \to c} g(x) = -2$

(a) $\lim\limits_{x \to c} [f(x) + g(x)]^2$

(b) $\lim\limits_{x \to c} [6f(x)g(x)]$

(c) $\lim\limits_{x \to c} \dfrac{5g(x)}{4f(x)}$

(d) $\lim\limits_{x \to c} \dfrac{1}{\sqrt{f(x)}}$

Evaluating Limits In Exercises 43 and 44, find (a) $\lim\limits_{x \to 2} f(x)$, (b) $\lim\limits_{x \to 2} g(x)$, (c) $\lim\limits_{x \to 2} [f(x)g(x)]$, and (d) $\lim\limits_{x \to 2} [g(x) - f(x)]$.

43. $f(x) = x^3, \quad g(x) = \dfrac{\sqrt{x^2 + 5}}{2x^2}$

44. $f(x) = \dfrac{x}{3 - x}, \quad g(x) = \sin \pi x$

Evaluating a Limit by Direct Substitution In Exercises 45–64, find the limit by direct substitution.

45. $\lim\limits_{x \to 4} (8 - x^2)$

46. $\lim\limits_{x \to -2} \left(\tfrac{1}{2}x^3 - 5x\right)$

47. $\lim\limits_{x \to -3} (2x^2 + 4x + 1)$

48. $\lim\limits_{x \to -3} (x^3 - 3x + 8)$

49. $\lim\limits_{x \to 2} \left(-\dfrac{12}{x}\right)$

50. $\lim\limits_{x \to -4} \dfrac{8}{x - 4}$

51. $\lim\limits_{x \to -2} \dfrac{2x}{2x^2 - 3}$

52. $\lim\limits_{x \to 5} \dfrac{x - 2}{x^2 - 3x + 2}$

53. $\lim\limits_{x \to -1} \dfrac{6x + 5}{3x - 7}$

54. $\lim\limits_{x \to 3} \dfrac{x^2 + 1}{x}$

55. $\lim\limits_{x \to -3} \sqrt{6 - x}$

56. $\lim\limits_{x \to 3} \sqrt[3]{x^2 - 1}$

57. $\lim\limits_{x \to 7} \dfrac{5x}{\sqrt{x + 2}}$

58. $\lim\limits_{x \to 8} \dfrac{\sqrt{x + 1}}{x - 4}$

59. $\lim\limits_{x \to 3} e^x$

60. $\lim\limits_{x \to e} \ln x$

61. $\lim\limits_{x \to \pi} \cos x$

62. $\lim\limits_{x \to \pi/2} \tan 2x$

63. $\lim\limits_{x \to 1/2} \arcsin x$

64. $\lim\limits_{x \to 1} \arccos \dfrac{x}{2}$

Exploration

True or False? In Exercises 65 and 66, determine whether the statement is true or false. Justify your answer.

65. The limit of a function as x approaches c does not exist when the function approaches -3 from the left of c and 3 from the right of c.

66. The limit of the product of two functions is equal to the product of the limits of the two functions.

67. Think About It From Exercises 5–10, select a limit that is possible to reach and one that is not possible to reach.

(a) Use a graphing utility to graph the corresponding functions using a standard viewing window. Do the graphs reveal whether it is possible to reach the limit? Explain.

(b) Use the graphing utility to graph the corresponding functions using a *decimal* setting. Do the graphs reveal whether it is possible to reach the limit? Explain.

68. Think About It Use the results of Exercise 67 to draw a conclusion as to whether you can use the graph generated by a graphing utility to determine reliably whether it is possible to reach a limit.

69. Writing Write a brief description of the meaning of the notation $\lim\limits_{x \to 5} f(x) = 12$.

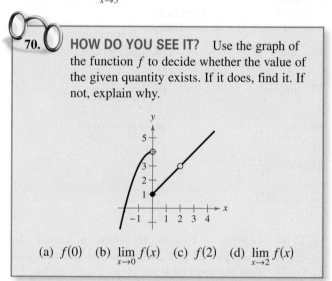

70. **HOW DO YOU SEE IT?** Use the graph of the function f to decide whether the value of the given quantity exists. If it does, find it. If not, explain why.

(a) $f(0)$ (b) $\lim\limits_{x \to 0} f(x)$ (c) $f(2)$ (d) $\lim\limits_{x \to 2} f(x)$

Error Analysis In Exercises 71 and 72, describe the error.

71. When $f(2) = 4$, $\lim\limits_{x \to 2} f(x) = 4$. ✗

72. When $\lim\limits_{x \to 2} f(x) = 4$, $f(2) = 4$. ✗

73. Think About It Use a graphing utility to graph the tangent function. What are $\lim\limits_{x \to 0} \tan x$ and $\lim\limits_{x \to \pi/4} \tan x$? What can you conclude about the existence of the limit $\lim\limits_{x \to \pi/2} \tan x$?

74. Writing Use a graphing utility to graph the function

$$f(x) = \dfrac{x^2 - 3x - 10}{x - 5}.$$

Use the *trace* feature to approximate $\lim\limits_{x \to 4} f(x)$. What appears to be the value of $\lim\limits_{x \to 5} f(x)$? Is f defined at $x = 5$? Does this affect the existence of the limit as x approaches 5?

12.2 Techniques for Evaluating Limits

- Use the dividing out technique to evaluate limits of functions.
- Use the rationalizing technique to evaluate limits of functions.
- Use technology to approximate limits of functions numerically and graphically.
- Evaluate one-sided limits of functions.
- Evaluate limits from calculus.

Limits have applications in real-life situations. For instance, in Exercises 41 and 42 on page 838, you will use limits to find the velocity of a free-falling object at different times.

Dividing Out Technique

In Section 12.1, you studied several types of functions whose limits can be evaluated by direct substitution. In this section, you will study several techniques for evaluating limits of functions for which direct substitution fails. For example, consider the limit

$$\lim_{x \to -3} \frac{x^2 + x - 6}{x + 3}.$$

Direct substitution produces 0 in both the numerator and denominator.

$$(-3)^2 + (-3) - 6 = 0 \qquad \text{Numerator is 0 when } x = -3.$$

$$-3 + 3 = 0 \qquad \text{Denominator is 0 when } x = -3.$$

The resulting fraction, $\frac{0}{0}$, has no meaning as a real number. It is called an **indeterminate form** because you cannot, from the form alone, determine the limit. By using a table, however, it appears that the limit of the function as x approaches -3 is -5.

x	-3.01	-3.001	-3.0001	-3	-2.9999	-2.999	-2.99
$\dfrac{x^2 + x - 6}{x + 3}$	-5.01	-5.001	-5.0001	?	-4.9999	-4.999	-4.99

When you attempt to evaluate a limit of a rational function by direct substitution and encounter the indeterminate form $\frac{0}{0}$, the numerator and denominator must have a common factor. After factoring and dividing out, use direct substitution again. Examples 1 and 2 show this **dividing out technique.**

EXAMPLE 1 **Dividing Out Technique**

See LarsonPrecalculus.com for an interactive version of this type of example.

Find the limit: $\lim_{x \to -3} \dfrac{x^2 + x - 6}{x + 3}$.

Solution From the discussion above, you know that direct substitution fails. So, begin by factoring the numerator and dividing out any common factors.

$$\lim_{x \to -3} \frac{x^2 + x - 6}{x + 3} = \lim_{x \to -3} \frac{(x - 2)(x + 3)}{x + 3} \qquad \text{Factor numerator.}$$

$$= \lim_{x \to -3} (x - 2) \qquad \text{Divide out common factor and simplify.}$$

$$= -5 \qquad \text{Direct substitution}$$

✓ *Checkpoint* ◀))) Audio-video solution in English & Spanish at LarsonPrecalculus.com

Find the limit: $\lim_{x \to 4} \dfrac{x^2 - 7x + 12}{x - 4}$.

The validity of the dividing out technique stems from the fact that when two functions agree at all but a single number c, they must have identical limit behavior at $x = c$. In Example 1, the functions

$$f(x) = \frac{x^2 + x - 6}{x + 3} \quad \text{and} \quad g(x) = x - 2$$

agree at all values of x other than $x = -3$. So, you can use $g(x)$ to find the limit of $f(x)$.

EXAMPLE 2 Dividing Out Technique

Find the limit.

$$\lim_{x \to 1} \frac{x - 1}{x^3 - x^2 + x - 1}$$

Solution Begin by substituting $x = 1$ into the numerator and denominator.

$$1 - 1 = 0 \qquad \text{Numerator is 0 when } x = 1.$$

$$1^3 - 1^2 + 1 - 1 = 0 \qquad \text{Denominator is 0 when } x = 1.$$

Both the numerator and denominator are zero when $x = 1$, so direct substitution will not yield the limit. To find the limit, factor the numerator and denominator, divide out any common factors, and then use direct substitution again.

$$\lim_{x \to 1} \frac{x - 1}{x^3 - x^2 + x - 1} = \lim_{x \to 1} \frac{x - 1}{(x - 1)(x^2 + 1)} \qquad \text{Factor denominator.}$$

$$= \lim_{x \to 1} \frac{\cancel{x - 1}}{\cancel{(x - 1)}(x^2 + 1)} \qquad \text{Divide out common factor.}$$

$$= \lim_{x \to 1} \frac{1}{x^2 + 1} \qquad \text{Simplify.}$$

$$= \frac{1}{1^2 + 1} \qquad \text{Direct substitution}$$

$$= \frac{1}{2} \qquad \text{Simplify.}$$

• • **REMARK** To obtain the factorization of the denominator, divide by $(x - 1)$ or factor by grouping.

$$x^3 - x^2 + x - 1$$
$$= x^2(x - 1) + (x - 1)$$
$$= (x - 1)(x^2 + 1)$$

The graph below verifies this result.

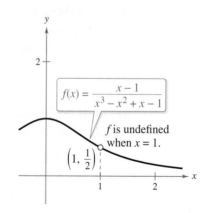

$$f(x) = \frac{x - 1}{x^3 - x^2 + x - 1}$$

f is undefined when $x = 1$.

$\left(1, \frac{1}{2}\right)$

✓ **Checkpoint**)))) Audio-video solution in English & Spanish at LarsonPrecalculus.com

Find the limit: $\displaystyle\lim_{x \to 7} \frac{x - 7}{x^3 - 7x^2 + 7x - 49}$.

Rationalizing Technique

▷ **ALGEBRA HELP** To review techniques for rationalizing numerators and denominators, see Appendix A.2.

A way to find the limits of some functions is to first rationalize the numerator of the function. This is the **rationalizing technique.** Recall that to rationalize a numerator of the form $a \pm b\sqrt{m}$ or $b\sqrt{m} \pm a$, multiply the numerator and denominator by the *conjugate* of the numerator. For example, the conjugate of $\sqrt{x} + 4$ is $\sqrt{x} - 4$.

EXAMPLE 3   Rationalizing Technique

Find the limit.

$$\lim_{x \to 0} \frac{\sqrt{x+1} - 1}{x}$$

Solution By direct substitution, you obtain the indeterminate form $\frac{0}{0}$.

$$\lim_{x \to 0} \frac{\sqrt{x+1} - 1}{x} = \frac{\sqrt{0+1} - 1}{0} = \frac{0}{0} \qquad \text{Indeterminate form}$$

In this case, rewrite the fraction by rationalizing the numerator.

$$\frac{\sqrt{x+1} - 1}{x} = \left(\frac{\sqrt{x+1} - 1}{x} \right)\left(\frac{\sqrt{x+1} + 1}{\sqrt{x+1} + 1} \right)$$

$$= \frac{(x+1) - 1}{x\left(\sqrt{x+1} + 1 \right)} \qquad \text{Multiply.}$$

$$= \frac{x}{x\left(\sqrt{x+1} + 1 \right)} \qquad \text{Simplify.}$$

$$= \frac{\cancel{x}}{\cancel{x}\left(\sqrt{x+1} + 1 \right)} \qquad \text{Divide out common factor.}$$

$$= \frac{1}{\sqrt{x+1} + 1}, \quad x \neq 0 \qquad \text{Simplify.}$$

Now, evaluate the limit by direct substitution.

$$\lim_{x \to 0} \frac{\sqrt{x+1} - 1}{x} = \lim_{x \to 0} \frac{1}{\sqrt{x+1} + 1} = \frac{1}{\sqrt{0+1} + 1} = \frac{1}{1+1} = \frac{1}{2}$$

To verify your conclusion that the limit is $\frac{1}{2}$, construct a table, such as the one shown below, or sketch a graph, as shown in Figure 12.7.

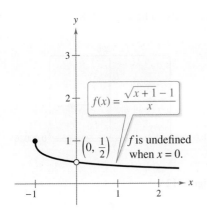

$f(x) = \dfrac{\sqrt{x+1} - 1}{x}$

$\left(0, \dfrac{1}{2} \right)$

f is undefined when $x = 0$.

Figure 12.7

x	-0.1	-0.01	-0.001	0	0.001	0.01	0.1
$f(x)$	0.5132	0.5013	0.5001	?	0.4999	0.4988	0.4881

✓ **Checkpoint** 🔊))) *Audio-video solution in English & Spanish at LarsonPrecalculus.com*

Find the limit.

$$\lim_{x \to 0} \frac{1 - \sqrt{1-x}}{x}$$

The rationalizing technique for evaluating limits is based on multiplication by a convenient form of 1. In Example 3, the convenient form is

$$1 = \frac{\sqrt{x+1} + 1}{\sqrt{x+1} + 1}.$$

Using Technology

The dividing out and rationalizing techniques may not work when finding limits of nonalgebraic functions. You often need to use more sophisticated analytic techniques to find limits of nonalgebraic functions, as shown in Examples 4 and 5.

EXAMPLE 4 **Approximating a Limit**

Approximate the limit.

$$\lim_{x \to 0} (1 + x)^{1/x}$$

Numerical Solution

Let $f(x) = (1 + x)^{1/x}$. Use the *table* feature of a graphing utility to create a table that shows the values of f for x starting at $x = -0.003$ with a step of 0.001, as shown in the figure below. Because 0 is halfway between -0.001 and 0.001, use the average of the values of f at these two x-values to estimate the limit.

$$\lim_{x \to 0} (1 + x)^{1/x} \approx \frac{2.7196 + 2.7169}{2} = 2.71825$$

The actual limit can be found algebraically to be $e \approx 2.71828$.

X	Y1
-.003	2.7224
-.002	2.721
-.001	2.7196
0	ERROR
.001	2.7169
.002	2.7156
.003	2.7142
X=0	

Graphical Solution

To approximate the limit graphically, graph the function $f(x) = (1 + x)^{1/x}$, as shown in Figure 12.8. Using the *zoom* and *trace* features of the graphing utility, choose two points on the graph of f, such as

$$(-0.00017, 2.7185) \quad \text{and} \quad (0.00017, 2.7181)$$

as shown in Figure 12.9. The x-coordinates of these two points are equidistant from 0, so approximate the limit to be the average of the y-coordinates. That is,

$$\lim_{x \to 0} (1 + x)^{1/x} \approx \frac{2.7185 + 2.7181}{2} = 2.7183.$$

The actual limit can be found algebraically to be $e \approx 2.71828$.

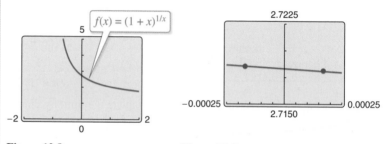

Figure 12.8   **Figure 12.9**

✓ *Checkpoint* 🔊))) *Audio-video solution in English & Spanish at LarsonPrecalculus.com*

Approximate the limit: $\lim_{x \to 0} \dfrac{e^x - 1}{x}$.

EXAMPLE 5 **Approximating a Limit Graphically**

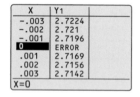

Approximate the limit: $\lim_{x \to 0} \dfrac{\sin x}{x}$.

Solution Direct substitution produces the indeterminate form $\frac{0}{0}$. To approximate the limit, begin by using a graphing utility to graph $f(x) = (\sin x)/x$, as shown in Figure 12.10. Then use the *zoom* and *trace* features of the graphing utility to choose a point on each side of 0, such as $(-0.0012467, 0.9999997)$ and $(0.0012467, 0.9999997)$. Finally, approximate the limit as the average of the y-coordinates of these two points, $\lim_{x \to 0} (\sin x)/x \approx 0.9999997$. Example 10 shows that this limit is exactly 1.

Figure 12.10

✓ *Checkpoint* 🔊))) *Audio-video solution in English & Spanish at LarsonPrecalculus.com*

Approximate the limit: $\lim_{x \to 0} \dfrac{1 - \cos x}{x}$.

▷ TECHNOLOGY The graphs shown in Figures 12.8 and 12.10 appear to be continuous at $x = 0$. However, when you use the *trace* or the *value* feature of a graphing utility to determine $f(0)$, no value is given. Some graphing utilities show breaks or holes in a graph when you use an appropriate viewing window. The holes in the graphs in Figures 12.8 and 12.10 occur on the y-axis, so the holes are not visible.

One-Sided Limits

In Section 12.1, you saw that one way in which a limit can fail to exist is when a function approaches a different value from the left side of c than it approaches from the right side of c. This type of behavior can be described more concisely with the concept of a **one-sided limit.**

$$\lim_{x \to c^-} f(x) = L_1 \quad \text{or} \quad f(x) \to L_1 \text{ as } x \to c^- \qquad \text{Limit from the left}$$

$$\lim_{x \to c^+} f(x) = L_2 \quad \text{or} \quad f(x) \to L_2 \text{ as } x \to c^+ \qquad \text{Limit from the right}$$

EXAMPLE 6   **Evaluating One-Sided Limits**

Find the limit of $f(x)$ as x approaches 0 from the left and from the right.

$$f(x) = \frac{|2x|}{x}$$

Solution   From the graph of f, shown below, notice that $f(x) = -2$ for all $x < 0$.

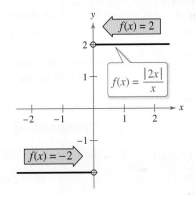

So, the limit from the left is

$$\lim_{x \to 0^-} \frac{|2x|}{x} = -2. \qquad \text{Limit from the left: } f(x) \to -2 \text{ as } x \to 0^-$$

Also from the graph, notice that $f(x) = 2$ for all $x > 0$, so the limit from the right is

$$\lim_{x \to 0^+} \frac{|2x|}{x} = 2. \qquad \text{Limit from the right: } f(x) \to 2 \text{ as } x \to 0^+$$

✓ *Checkpoint*  *Audio-video solution in English & Spanish at LarsonPrecalculus.com*

Find the limit of $f(x)$ as x approaches 3 from the left and from the right.

$$f(x) = \frac{|x - 3|}{x - 3}$$

In Example 6, note that the function approaches different limits from the left and from the right. In such cases, the limit of $f(x)$ as $x \to c$ does not exist. For the limit of a function to exist as $x \to c$, it must be true that both one-sided limits exist and are equal.

Existence of a Limit

If f is a function and c and L are real numbers, then

$$\lim_{x \to c} f(x) = L$$

if and only if both the left and right limits *exist* and are *equal* to L.

EXAMPLE 7 **Evaluating One-Sided Limits**

Find the limit of $f(x)$ as x approaches 1.

$$f(x) = \begin{cases} 4 - x, & x < 1 \\ 4x - x^2, & x > 1 \end{cases}$$

Solution Remember that you are concerned about the value of f *near* $x = 1$ rather than *at* $x = 1$. So, for $x < 1$, $f(x)$ is given by $4 - x$. Use direct substitution to obtain

$$\lim_{x \to 1^-} f(x) = \lim_{x \to 1^-} (4 - x)$$
$$= 4 - 1$$
$$= 3.$$

For $x > 1$, $f(x)$ is given by $4x - x^2$.
Use direct substitution to obtain

$$\lim_{x \to 1^+} f(x) = \lim_{x \to 1^+} (4x - x^2)$$
$$= 4(1) - 1^2$$
$$= 3.$$

Both one-sided limits exist and are equal
to 3, so it follows that $\lim_{x \to 1} f(x) = 3$. The
graph at the right confirms this conclusion.

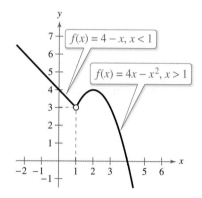

✓ **Checkpoint** ◀))) *Audio-video solution in English & Spanish at LarsonPrecalculus.com*

Find the limit of $f(x)$ as x approaches -1.

$$f(x) = \begin{cases} -x^2 - 3x, & x < -1 \\ x + 3, & x > -1 \end{cases}$$

EXAMPLE 8 **Comparing Limits from the Left and Right**

As illustrated in Example 8, a
step function can model delivery
charges. The limit of such a
function does not exist at a
"jump" because the one-sided
limits are not equal.

For 2-day shipping, a delivery service charges $24 for the first pound and $4 for each additional pound or portion of a pound. Let x represent the weight (in pounds) of a package and let $f(x)$ represent the shipping cost. Show that the limit of $f(x)$ as $x \to 2$ does not exist.

$$f(x) = \begin{cases} \$24, & 0 < x \le 1 \\ \$28, & 1 < x \le 2 \\ \$32, & 2 < x \le 3 \end{cases}$$

Solution The graph of f is at the right. The
limit of $f(x)$ as x approaches 2 from the left is

$$\lim_{x \to 2^-} f(x) = 28$$

whereas the limit of $f(x)$ as x approaches 2
from the right is

$$\lim_{x \to 2^+} f(x) = 32.$$

These one-sided limits are not equal, so the
limit of $f(x)$ as $x \to 2$ does not exist.

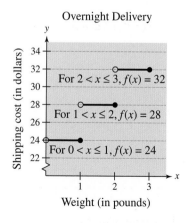

✓ **Checkpoint** ◀))) *Audio-video solution in English & Spanish at LarsonPrecalculus.com*

In Example 8, show that the limit of $f(x)$ as $x \to 1$ does not exist.

Limits from Calculus

▷ **ALGEBRA HELP** To review evaluating difference quotients, see Section 1.4.

In the next example, you will study an important type of limit from calculus—the limit of a *difference quotient*.

EXAMPLE 9 **Evaluating a Limit from Calculus**

For the function

$$f(x) = x^2 - 1$$

find

$$\lim_{h \to 0} \frac{f(3 + h) - f(3)}{h}.$$

Solution Begin by substituting for $f(3 + h)$ and $f(3)$ and simplifying.

$$\lim_{h \to 0} \frac{f(3 + h) - f(3)}{h} = \lim_{h \to 0} \frac{[(3 + h)^2 - 1] - (3^2 - 1)}{h}$$

$$= \lim_{h \to 0} \frac{9 + 6h + h^2 - 1 - 9 + 1}{h}$$

$$= \lim_{h \to 0} \frac{6h + h^2}{h}$$

By factoring and dividing out, you obtain

$$\lim_{h \to 0} \frac{f(3 + h) - f(3)}{h} = \lim_{h \to 0} \frac{\cancel{h}(6 + h)}{\cancel{h}}$$

$$= \lim_{h \to 0} (6 + h)$$

$$= 6 + 0$$

$$= 6.$$

So, the limit is 6.

✓ **Checkpoint** ◀))) *Audio-video solution in English & Spanish at LarsonPrecalculus.com*

For the function $f(x) = 4x - x^2$, find

$$\lim_{h \to 0} \frac{f(2 + h) - f(2)}{h}.$$

•• **REMARK** Note that for any x-value, the limit of a difference quotient is an expression of the form

$$\lim_{h \to 0} \frac{f(x + h) - f(x)}{h}.$$

Direct substitution into the difference quotient always produces the indeterminate form $\frac{0}{0}$.

$$\lim_{h \to 0} \frac{f(x + h) - f(x)}{h}$$

$$= \frac{f(x + 0) - f(x)}{0}$$

$$= \frac{f(x) - f(x)}{0}$$

$$= \frac{0}{0}$$

The theorem below concerns the limit of a function that is squeezed between two other functions, each of which has the same limit at a given x-value, as shown in Figure 12.11.

$h(x) \leq f(x) \leq g(x)$

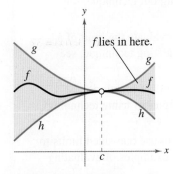

The Squeeze Theorem
Figure 12.11

The Squeeze Theorem

If $h(x) \leq f(x) \leq g(x)$ for all x in an open interval containing c, except possibly at c itself, and if

$$\lim_{x \to c} h(x) = L = \lim_{x \to c} g(x)$$

then $\lim_{x \to c} f(x)$ exists and is equal to L.

A proof of the Squeeze Theorem is beyond the scope of this text, but you may be able to see, intuitively, that this theorem makes sense. Example 10 shows an application of the Squeeze Theorem.

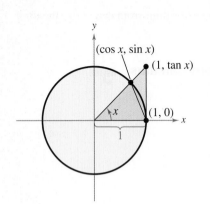

Figure 12.12

• • • • • • • • • • • • • • • ▷
•
•• **REMARK** Recall from
Section 4.1 that the area of a
sector of a circle of radius r is
given by

$$A = \frac{1}{2}r^2\theta$$

where θ is the measure of the
central angle in radians.

EXAMPLE 10 **A Special Trigonometric Limit**

Find the limit: $\lim\limits_{x \to 0} \dfrac{\sin x}{x}$.

Solution One way to find this limit is to consider a sector of a circle of radius 1
with central angle x, "squeezed" between two triangles (see Figure 12.12), where x is
an acute positive angle measured in radians.

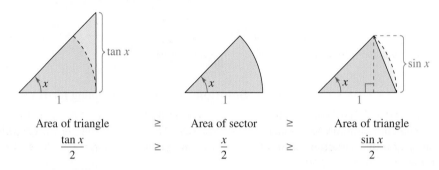

Area of triangle	$\geq$	Area of sector	$\geq$	Area of triangle
$\dfrac{\tan x}{2}$	$\geq$	$\dfrac{x}{2}$	$\geq$	$\dfrac{\sin x}{2}$

Multiplying each expression by $2/\sin x$ produces

$$\frac{1}{\cos x} \geq \frac{x}{\sin x} \geq 1$$

and taking reciprocals and reversing the inequality symbols yields

$$\cos x \leq \frac{\sin x}{x} \leq 1.$$

Because $\cos x = \cos(-x)$ and $(\sin x)/x = [\sin(-x)]/(-x)$, this inequality is valid
for *all* nonzero x in the open interval $(-\pi/2, \pi/2)$. Finally, $\lim\limits_{x \to 0} \cos x = 1$ and
$\lim\limits_{x \to 0} 1 = 1$, so apply the Squeeze Theorem to conclude that $\lim\limits_{x \to 0} (\sin x)/x = 1$.

✓ *Checkpoint*  *Audio-video solution in English & Spanish at LarsonPrecalculus.com*

Find the limit: $\lim\limits_{x \to 0} \dfrac{1 - \cos x}{x}$.

Summarize (Section 12.2)

1. Explain how to use the dividing out technique to evaluate the limit of
   a function *(page 829)*. For examples of the dividing out technique, see
   Examples 1 and 2.

2. Explain how to use the rationalizing technique to evaluate the limit of a
   function *(page 831)*. For an example of the rationalizing technique, see
   Example 3.

3. Explain how to use technology to approximate limits of functions
   numerically and graphically *(page 832)*. For examples of using technology
   to approximate limits, see Examples 4 and 5.

4. Explain the concept of a one-sided limit and how to use one-sided limits to
   determine the existence of a limit *(page 833)*. For examples of evaluating
   one-sided limits, see Examples 6–8.

5. Explain how to evaluate the limit of a difference quotient and how to
   evaluate a limit using the Squeeze Theorem *(page 835)*. For examples of
   evaluating such limits from calculus, see Examples 9 and 10.

12.2 Exercises

See **CalcChat.com** for tutorial help and worked-out solutions to odd-numbered exercises.

Vocabulary: Fill in the blanks.

1. The fraction $\frac{0}{0}$ has no meaning as a real number and is called an _____ _____.

2. To evaluate the limit of a rational function that has common factors in its numerator and denominator when direct substitution fails, use the _____ _____ _____.

3. The limit $\lim\limits_{x \to c^-} f(x) = L_1$ is an example of a _____ _____.

4. The _____ _____ enables you to find the limit of a function that is squeezed between two functions, each of which has the same limit at a given x-value.

Skills and Applications

Using a Graph to Determine Limits In Exercises 5 and 6, use the graph to find each limit visually. Then identify another function that agrees with the given function at all but one point.

5. $g(x) = \dfrac{x^3 - x}{x - 1}$

6. $f(x) = \dfrac{x^2 - 1}{x + 1}$

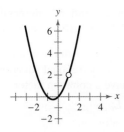

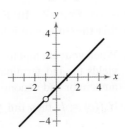

(a) $\lim\limits_{x \to 1} g(x)$

(b) $\lim\limits_{x \to -1} g(x)$

(c) $\lim\limits_{x \to 0} g(x)$

(a) $\lim\limits_{x \to 1} f(x)$

(b) $\lim\limits_{x \to 2} f(x)$

(c) $\lim\limits_{x \to -1} f(x)$

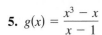 **Finding a Limit** In Exercises 7–20, find the limit algebraically, if it exists. Use a graphing utility to verify your result graphically.

7. $\lim\limits_{x \to 6} \dfrac{x^2 - 36}{x - 6}$

8. $\lim\limits_{x \to 1} \dfrac{x^2 - 1}{x - 1}$

9. $\lim\limits_{x \to 3} \dfrac{x - 3}{x^2 - x - 6}$

10. $\lim\limits_{x \to -4} \dfrac{x + 4}{2x^2 + 9x + 4}$

11. $\lim\limits_{x \to 0} \dfrac{\sqrt{x + 25} - 5}{x}$

12. $\lim\limits_{x \to 0} \dfrac{\sqrt{x + 4} - 2}{x}$

13. $\lim\limits_{x \to -3} \dfrac{\sqrt{x + 7} - 2}{x + 3}$

14. $\lim\limits_{x \to 2} \dfrac{4 - \sqrt{18 - x}}{x - 2}$

15. $\lim\limits_{x \to 0} \dfrac{\dfrac{1}{x + 1} - 1}{x}$

16. $\lim\limits_{x \to 0} \dfrac{\dfrac{1}{x - 8} + \dfrac{1}{8}}{x}$

17. $\lim\limits_{x \to 0} \dfrac{\sec x}{\tan x}$

18. $\lim\limits_{x \to \pi} \dfrac{\csc x}{\cot x}$

19. $\lim\limits_{x \to 0} \dfrac{\cos x - 1}{\sin x}$

20. $\lim\limits_{x \to \pi/2} \dfrac{\cos x}{1 - \sin x}$

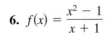 **Approximating a Limit Graphically** In Exercises 21–30, use a graphing utility to graph the function and approximate the limit accurate to three decimal places.

21. $\lim\limits_{x \to 0} \dfrac{e^{3x} - 1}{x}$

22. $\lim\limits_{x \to 0} \dfrac{1 - e^{-x}}{x}$

23. $\lim\limits_{x \to 0^+} (x \ln x)$

24. $\lim\limits_{x \to 0^+} (x^2 \ln x)$

25. $\lim\limits_{x \to 0} (1 - x)^{2/x}$

26. $\lim\limits_{x \to 0} (1 + 2x)^{1/x}$

27. $\lim\limits_{x \to 0} \dfrac{\sin 2x}{x}$

28. $\lim\limits_{x \to 0} \dfrac{1 - \cos 2x}{x}$

29. $\lim\limits_{x \to 1} \dfrac{1 - \sqrt[3]{x}}{1 - x}$

30. $\lim\limits_{x \to 1} \dfrac{\sqrt[3]{x} - x}{x - 1}$

 Using Different Methods In Exercises 31–34, (a) graphically approximate the limit (if it exists) by using a graphing utility to graph the function, (b) numerically approximate the limit (if it exists) by using the *table* feature of the graphing utility to create a table, and (c) algebraically evaluate the limit (if it exists) by using the appropriate technique(s).

31. $\lim\limits_{x \to 2} \dfrac{x^4 - 2x^2 - 8}{x^4 - 6x^2 + 8}$

32. $\lim\limits_{x \to 2} \dfrac{x^4 - 1}{x^4 - 3x^2 - 4}$

33. $\lim\limits_{x \to 16^+} \dfrac{4 - \sqrt{x}}{x - 16}$

34. $\lim\limits_{x \to 0^-} \dfrac{\sqrt{x + 2} - \sqrt{2}}{x}$

 Evaluating One-Sided Limits In Exercises 35–38, graph the function. Find the limit (if it exists) by evaluating the corresponding one-sided limits.

35. $\lim\limits_{x \to 1} \dfrac{|x - 1|}{x - 1}$

36. $\lim\limits_{x \to -3} \dfrac{|x + 3|}{x}$

37. $\lim\limits_{x \to 2} f(x)$, where $f(x) = \begin{cases} 3x - 2, & x < 2 \\ 8 - x^2, & x \ge 2 \end{cases}$

38. $\lim\limits_{x \to 3} f(x)$, where $f(x) = \begin{cases} x^2 - 4, & x \le 3 \\ x + 3, & x > 3 \end{cases}$

Finding Limits   **In Exercises 39 and 40, state whether each limit can be evaluated by using direct substitution. Then evaluate or approximate each limit.**

39. (a) $\lim\limits_{x \to 0} x^2 \sin x^2$ **40.** (a) $\lim\limits_{x \to 0} \dfrac{x}{\cos x}$

   (b) $\lim\limits_{x \to 0} \dfrac{\sin x^2}{x^2}$ (b) $\lim\limits_{x \to 0} \dfrac{1 - \cos x}{x^2}$

•••• **Free-Falling Object** ••••••••••••••••

In Exercises 41 and 42, use the position function

$$s(t) = -16t^2 + 256$$

which gives the height (in feet) of a free-falling object. The velocity at time $t = a$ seconds is given by

$$\lim\limits_{t \to a} \dfrac{s(a) - s(t)}{a - t}.$$

41. Find the velocity when $t = 1$ second.

42. Find the velocity when $t = 2$ seconds.

•••••••••••••••••••••••••••••••

43. Salary Contract   A union contract guarantees an 8% salary increase each year for 3 years. For a current salary of \$30,000, the salaries S (in thousands of dollars) for the next 3 years are given by

$$S(t) = \begin{cases} 30.000, & 0 < t \le 1 \\ 32.400, & 1 < t \le 2 \\ 34.992, & 2 < t \le 3 \end{cases}$$

where t represents the time in years. Show that the limit of f as $t \to 2$ does not exist.

44. Consumer Awareness   The cost C (in dollars) of making x photocopies at a copy shop is given by

$$C(x) = \begin{cases} 0.15x, & 0 < x \le 25 \\ 0.10x, & 25 < x \le 100 \\ 0.07x, & 100 < x \le 500 \\ 0.05x, & x > 500 \end{cases}$$

(a) Sketch a graph of the function.

(b) Find each limit and interpret your result in the context of the situation.

   (i) $\lim\limits_{x \to 15} C(x)$ (ii) $\lim\limits_{x \to 99} C(x)$ (iii) $\lim\limits_{x \to 305} C(x)$

(c) Create a table of values to show numerically that each limit does not exist.

   (i) $\lim\limits_{x \to 25} C(x)$ (ii) $\lim\limits_{x \to 100} C(x)$ (iii) $\lim\limits_{x \to 500} C(x)$

(d) Explain how to use the graph in part (a) to verify that the limits in part (c) do not exist.

 Evaluating a Limit from Calculus   **In Exercises 45–48, find**

$$\lim\limits_{h \to 0} \dfrac{f(x + h) - f(x)}{h}.$$

45. $f(x) = x^2 + 2x - 4$ **46.** $f(x) = 3x^2 - 5x + 1$

47. $f(x) = \sqrt{x - 2}$ **48.** $f(x) = \dfrac{1}{x - 1}$

 Using the Squeeze Theorem   **In Exercises 49–52, use a graphing utility to graph the given function and the equations $y = |x|$ and $y = -|x|$ in the same viewing window. Using the graphs to observe the Squeeze Theorem visually, find $\lim\limits_{x \to 0} f(x)$.**

49. $f(x) = |x| \sin x$ **50.** $f(x) = |x| \cos x$

51. $h(x) = x \cos \dfrac{1}{x}$ **52.** $f(x) = x \sin \dfrac{1}{x}$

Exploration

True or False?   **In Exercises 53 and 54, determine whether the statement is true or false. Justify your answer.**

53. When your attempt to find the limit of a rational function yields the indeterminate form $\frac{0}{0}$, the rational function's numerator and denominator have a common factor.

54. If $f(c) = L$, then $\lim\limits_{x \to c} f(x) = L$.

55. Error Analysis   Describe the error evaluating

$$\lim\limits_{x \to -3} \dfrac{x^2 - 2x - 15}{x + 3}.$$

The limit does not exist because division by 0 is undefined.

56.   **HOW DO YOU SEE IT?**   The graph of

$$f(x) = \begin{cases} 2x, & x \le 0 \\ x^2 + 1, & x > 0 \end{cases}$$

is shown at the right. Find each limit. If the limit does not exist, explain why.

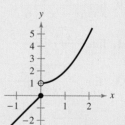

(a) $\lim\limits_{x \to 0^-} f(x)$

(b) $\lim\limits_{x \to 0^+} f(x)$

(c) $\lim\limits_{x \to 0} f(x)$

57. Think About It   Sketch the graph of a function for which the limit of $f(x)$ as x approaches 1 is 4 but for which $f(1) \ne 4$.

12.3 The Tangent Line Problem

The slope of a tangent line to the graph of a function is useful for analyzing rates of change in real-life situations. For example, in Exercise 71 on page 847, you will use the slope of a tangent line to a graph to analyze the path of a ball.

- Understand the tangent line problem.
- Use a tangent line to approximate the slope of a graph at a point.
- Use the limit definition of slope to find exact slopes of graphs.
- Find derivatives of functions and use derivatives to find slopes of graphs.

Tangent Line to a Graph

Calculus is a branch of mathematics that studies rates of change of functions. When you take a course in calculus, you will learn that rates of change have many applications.

Earlier in the text, you learned that the slope of a line is a measure of the rate at which a line rises or falls. On a line, this rate (or slope) is the same at every point. For graphs other than lines, the rate at which the graph rises or falls changes from point to point. For example, the parabola shown below rises more quickly at the point (x_1, y_1) than at the point (x_2, y_2). At the vertex (x_3, y_3), the graph does not rise or fall (the slope is 0), and at the point (x_4, y_4), the graph falls.

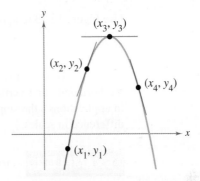

To determine the rate at which a graph rises or falls at a *single point*, you can find the slope of the tangent line at that point. The **tangent line** to the graph of a function f at a point $P(x_1, y_1)$ is the line whose slope best approximates the slope of the graph at the point. The graphs below show other examples of tangent lines.

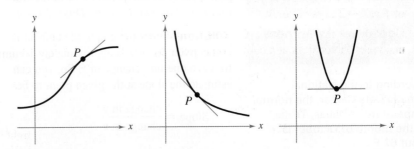

From geometry, you know that a line is tangent to a circle when the line intersects the circle at only one point, as shown at the right. Tangent lines to noncircular graphs, however, can intersect the graph at more than one point. For example, in the leftmost graph above, if the tangent line were extended, it would intersect the graph at a point other than the point of tangency.

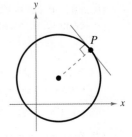

Slope of a Graph

A tangent line approximates the slope of the graph at a point, so the problem of finding the slope of a graph at a point is the same as finding the slope of the tangent line at the point.

EXAMPLE 1 **Visually Approximating the Slope of a Graph**

Use Figure 12.13 to approximate the slope of the graph of $f(x) = x^2$ at the point $(1, 1)$.

Solution From the graph of $f(x) = x^2$, notice that the tangent line at $(1, 1)$ rises approximately two units for each unit change in x. So, you can estimate the slope of the tangent line at $(1, 1)$ to be

$$\text{Slope} = \frac{\text{change in } y}{\text{change in } x} \approx \frac{2}{1} = 2.$$

The tangent line at the point $(1, 1)$ has a slope of about 2, so the graph of f has a slope of about 2 at the point $(1, 1)$.

✓ *Checkpoint*  Audio-video solution in English & Spanish at LarsonPrecalculus.com

Use Figure 12.13 to approximate the slope of the graph of $f(x) = x^2$ at the point $(2, 4)$.

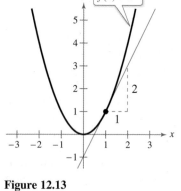

Figure 12.13

When you visually approximate the slope of a graph, remember that the scales on the horizontal and vertical axes may differ. When this happens (as it frequently does in applications), the slope of the tangent line is distorted, and you must account for the difference in scales.

EXAMPLE 2 **Visually Approximating the Slope of a Graph**

The figure at the right graphically depicts the monthly normal temperatures (in degrees Fahrenheit) for Dallas, Texas. Approximate the slope of this graph at the point shown and give a physical interpretation of the result. (*Source: National Climatic Data Center*)

Solution From the graph, the tangent line at the given point falls approximately 10 units for each one-unit change in x. So, you can estimate the slope at the given point to be

$$\text{Slope} = \frac{\text{change in } y}{\text{change in } x}$$

$$\approx \frac{-10}{1}$$

$$= -10 \text{ degrees per month.}$$

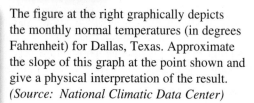

This means that the monthly normal temperature in November is about 10 degrees lower than the monthly normal temperature in October.

✓ *Checkpoint*  Audio-video solution in English & Spanish at LarsonPrecalculus.com

In Example 2, approximate the slope of the graph at the point $(4, 66)$ and give a physical interpretation of the result.

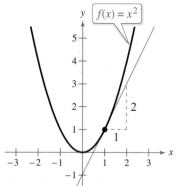

According to the National Climatic Data Center, the normal temperature in Dallas, Texas, for the month of October is about 67°F.

Slope and the Limit Process

In Examples 1 and 2, you approximated the slope of a graph at a point by creating a graph and then "eyeballing" the tangent line at the point of tangency. A more precise method of approximating tangent lines uses a **secant line** through the point of tangency and a second point on the graph, as shown at the right. If

$$(x, f(x)) \quad \text{and} \quad (x + h, f(x + h))$$

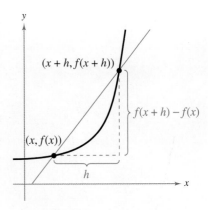

are two points on the graph of f, and $(x, f(x))$ is the point of tangency, then the slope of the secant line through the two points is

$$m_{\text{sec}} = \frac{\text{change in } y}{\text{change in } x} = \frac{f(x + h) - f(x)}{h}. \qquad \text{Slope of secant line}$$

Notice that the right side of this equation is a difference quotient. The denominator h is the *change in x*, and the numerator is the *change in y*. Using this method, you obtain more and more accurate approximations of the slope of the tangent line by choosing points closer and closer to the point of tangency, as shown in the figures below.

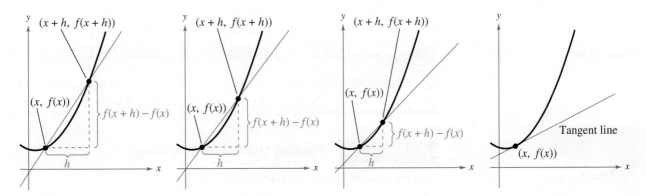

As h approaches 0, the secant line approaches the tangent line.

Using the limit process, you can find the *exact* slope of the tangent line at $(x, f(x))$.

Definition of the Slope of a Graph

The **slope** m of the graph of f at the point $(x, f(x))$ is equal to the slope of its tangent line at $(x, f(x))$, and is given by

$$m = \lim_{h \to 0} m_{\text{sec}}$$

$$= \lim_{h \to 0} \frac{f(x + h) - f(x)}{h}$$

provided this limit exists.

As suggested in Section 12.2, the difference quotient is used frequently in calculus. Using the difference quotient to find the slope of a tangent line to a graph is a major concept of calculus.

EXAMPLE 3   **Finding the Slope of a Graph**

Find the slope of the graph of $f(x) = x^2$ at the point $(-2, 4)$.

Solution   Find an expression that represents the slope of a secant line at $(-2, 4)$.

$$m_{sec} = \frac{f(-2 + h) - f(-2)}{h} \qquad \text{Set up difference quotient.}$$

$$= \frac{(-2 + h)^2 - (-2)^2}{h} \qquad \text{Substitute into } f(x) = x^2.$$

$$= \frac{4 - 4h + h^2 - 4}{h} \qquad \text{Expand terms.}$$

$$= \frac{-4h + h^2}{h} \qquad \text{Simplify.}$$

$$= \frac{\cancel{h}(-4 + h)}{\cancel{h}} \qquad \text{Factor and divide out.}$$

$$= -4 + h, \quad h \neq 0 \qquad \text{Simplify.}$$

Next, find the limit of m_{sec} as h approaches 0.

$$m = \lim_{h \to 0} m_{sec}$$

$$= \lim_{h \to 0} (-4 + h)$$

$$= -4$$

The graph has a slope of -4 at the point $(-2, 4)$, as shown in Figure 12.14.

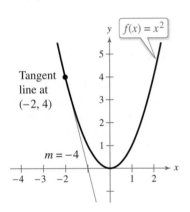

Figure 12.14

✓ *Checkpoint* ◀))) *Audio-video solution in English & Spanish at LarsonPrecalculus.com*

Use the limit process to find the slope of the graph of $g(x) = x^2 - 2x$ at the point $(3, 3)$.

EXAMPLE 4   **Finding the Slope of a Graph**

Find the slope of $f(x) = -2x + 4$.

Solution

$$m = \lim_{h \to 0} \frac{f(x + h) - f(x)}{h} \qquad \text{Set up difference quotient.}$$

$$= \lim_{h \to 0} \frac{[-2(x + h) + 4] - (-2x + 4)}{h} \qquad \text{Substitute into } f(x) = -2x + 4.$$

$$= \lim_{h \to 0} \frac{-2x - 2h + 4 + 2x - 4}{h} \qquad \text{Expand terms.}$$

$$= \lim_{h \to 0} \frac{-2\cancel{h}}{\cancel{h}} \qquad \text{Divide out.}$$

$$= -2 \qquad \text{Simplify.}$$

You know from your study of linear functions that the line $f(x) = -2x + 4$ has a slope of -2, as shown in Figure 12.15. This conclusion is consistent with that obtained by the limit definition of slope, as shown above.

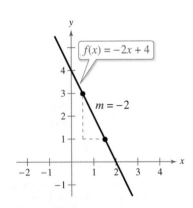

Figure 12.15

✓ *Checkpoint* ◀))) *Audio-video solution in English & Spanish at LarsonPrecalculus.com*

Use the limit process to find the slope of $f(x) = -3x + 4$.

It is important that you see the difference between the ways the difference quotients were set up in Examples 3 and 4. In Example 3, you found the slope of a graph at a specific point $(c, f(c))$. To find the slope in such a case, use the form of the difference quotient below.

$$m = \lim_{h \to 0} \frac{f(c + h) - f(c)}{h} \qquad \text{Slope at specific point}$$

In Example 4, however, you found a *formula* for the slope at *any* point on the graph. In such cases, you should use x, rather than c, in the difference quotient.

$$m = \lim_{h \to 0} \frac{f(x + h) - f(x)}{h} \qquad \text{Formula for slope}$$

This form will always produce a function of x, which can then be evaluated to find the slope at any desired point on the graph.

▷ TECHNOLOGY Use the slopes found in Example 5 and the point-slope form of the equation of a line to find that the equations of the tangent lines at $(-1, 2)$ and $(2, 5)$ are $y = -2x$ and $y = 4x - 3$, respectively. Then use a graphing utility to graph the function and the tangent lines

$y_1 = x^2 + 1$

$y_2 = -2x$

$y_3 = 4x - 3$

in the same viewing window. You can verify the equations of the tangent lines using the *tangent* feature of the graphing utility.

EXAMPLE 5   **Finding a Formula for the Slope of a Graph**

Find a formula for the slope of the graph of $f(x) = x^2 + 1$. What are the slopes at the points $(-1, 2)$ and $(2, 5)$?

Solution

$$m_{\text{sec}} = \frac{f(x + h) - f(x)}{h} \qquad \text{Set up difference quotient.}$$

$$= \frac{[(x + h)^2 + 1] - (x^2 + 1)}{h} \qquad \text{Substitute into } f(x) = x^2 + 1.$$

$$= \frac{x^2 + 2xh + h^2 + 1 - x^2 - 1}{h} \qquad \text{Expand terms.}$$

$$= \frac{2xh + h^2}{h} \qquad \text{Simplify.}$$

$$= \frac{\cancel{h}(2x + h)}{\cancel{h}} \qquad \text{Factor and divide out.}$$

$$= 2x + h, \quad h \neq 0 \qquad \text{Simplify.}$$

Next, find the limit of m_{sec} as h approaches 0.

$$m = \lim_{h \to 0} m_{\text{sec}} \qquad \text{Formula for slope}$$

$$= \lim_{h \to 0} (2x + h)$$

$$= 2x$$

Using the formula $m = 2x$ for the slope at $(x, f(x))$, find the slope at the specified points. At $(-1, 2)$, the slope is

$$m = 2(-1) = -2$$

and at $(2, 5)$, the slope is

$$m = 2(2) = 4.$$

The graph of f is at the right.

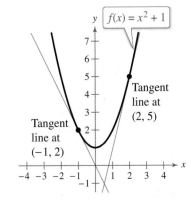

✓ **Checkpoint** 🔊))) *Audio-video solution in English & Spanish at LarsonPrecalculus.com*

Find a formula for the slope of the graph of $f(x) = 2x^2 - 3$. What is the slope at the points $(-3, 15)$ and $(2, 5)$?

12.3 Exercises

See CalcChat.com for tutorial help and worked-out solutions to odd-numbered exercises.

Vocabulary: Fill in the blanks.

1. _____ is a branch of mathematics that studies rates of change of functions.

2. The _____ _____ to the graph of a function at a point is the line whose slope best approximates the slope of the graph at the point.

3. The slope of the tangent line to a graph at $(x, f(x))$ is given by $m =$ _____.

4. The _____ $f'(x)$ is the formula for the slope of the tangent line to the graph of f at the point $(x, f(x))$.

Skills and Applications

 Approximating the Slope of a Graph
In Exercises 5–8, use the figure to approximate the slope of the graph at the point (x, y).

5.
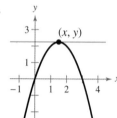

6.
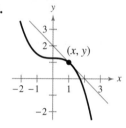

7.
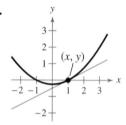

8.
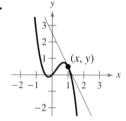

Approximating the Slope of a Tangent Line In Exercises 9–14, sketch a graph of the function and the tangent line at the point $(1, f(1))$. Use the graph to approximate the slope of the tangent line.

9. $f(x) = x^2 - 2$
10. $f(x) = x^2 - 2x + 1$
11. $f(x) = \sqrt{2 - x}$
12. $f(x) = \sqrt{x + 3}$
13. $f(x) = \dfrac{4}{x + 1}$
14. $f(x) = \dfrac{3}{2 - x}$

 Finding the Slope of a Graph In Exercises 15–22, use the limit process to find the slope of the graph of the function at the specified point.

15. $g(x) = x^2 - 6x$, $(1, -5)$
16. $f(x) = 10x - 2x^2$, $(3, 12)$
17. $g(x) = 9 - 3x$, $(2, 3)$
18. $h(x) = 3x + 4$, $(-1, 1)$
19. $g(x) = \dfrac{4}{x}$, $(2, 2)$
20. $g(x) = \dfrac{1}{x - 4}$, $(3, -1)$
21. $h(x) = \sqrt{x}$, $(9, 3)$
22. $h(x) = \sqrt{x + 8}$, $(-4, 2)$

 Finding a Formula for the Slope of a Graph In Exercises 23–28, find a formula for the slope of the graph of f at the point $(x, f(x))$. Then use it to find the slope at the two given points.

23. $f(x) = 4 - x^2$
(a) $(0, 4)$
(b) $(-1, 3)$

24. $f(x) = x^3$
(a) $(1, 1)$
(b) $(-2, -8)$

25. $f(x) = \dfrac{1}{x + 4}$
(a) $\left(0, \frac{1}{4}\right)$
(b) $\left(-2, \frac{1}{2}\right)$

26. $f(x) = \dfrac{1}{x + 2}$
(a) $\left(0, \frac{1}{2}\right)$
(b) $(-1, 1)$

27. $f(x) = \sqrt{x - 1}$
(a) $(5, 2)$
(b) $(10, 3)$

28. $f(x) = \sqrt{x - 4}$
(a) $(5, 1)$
(b) $(8, 2)$

 Finding a Derivative In Exercises 29–42, find the derivative of the function.

29. $f(x) = 6$
30. $f(x) = -8$
31. $g(x) = 2x - 7$
32. $f(x) = -5x + 1$
33. $f(x) = 2x^2 + 3x$
34. $f(x) = x^2 - 3x + 4$
35. $f(x) = x^{-2}$
36. $f(x) = x^{-3}$
37. $f(x) = \sqrt{x - 7}$
38. $g(x) = \sqrt{x + 9}$
39. $h(x) = \dfrac{1}{x + 1}$
40. $f(x) = \dfrac{1}{x - 8}$
41. $f(x) = \dfrac{1}{\sqrt{x - 4}}$
42. $h(x) = \dfrac{1}{\sqrt{x + 1}}$

 Using the Derivative In Exercises 43–50, (a) find the slope of the graph of f at the given point, (b) find an equation of the tangent line to the graph at the point, and (c) graph the function and the tangent line.

43. $f(x) = x^2 - 1$, $(2, 3)$
44. $f(x) = 6x - x^2$, $(1, 5)$
45. $f(x) = x^3 - 2x$, $(1, -1)$
46. $f(x) = x^3 - x^2$, $(2, 4)$

47. $f(x) = \sqrt{x+1}$, $(3, 2)$

48. $f(x) = \sqrt{x-2}$, $(3, 1)$

49. $f(x) = \dfrac{1}{x+5}$, $(-4, 1)$

50. $f(x) = \dfrac{1}{x-3}$, $(4, 1)$

Graphing a Function Over an Interval In Exercises 51–54, use a graphing utility to graph f over the interval $[-2, 2]$ and complete the table. Compare the value of the first derivative with a visual approximation of the slope of the graph.

x	-2	-1.5	-1	-0.5	0	0.5	1	1.5	2
$f(x)$									
$f'(x)$									

51. $f(x) = \frac{1}{2}x^2$

52. $f(x) = \frac{1}{4}x^3$

53. $f(x) = \sqrt{x+3}$

54. $f(x) = \dfrac{x^2 - 4}{x + 4}$

Using the Derivative In Exercises 55–58, find an equation of the line that is tangent to the graph of f and parallel to the given line.

Function	Line
55. $f(x) = -\frac{1}{4}x^2$	$x + y = 0$
56. $f(x) = x^2 + 1$	$2x + y = 0$
57. $f(x) = -\frac{1}{2}x^3$	$6x + y + 4 = 0$
58. $f(x) = x^2 - x$	$x + 2y - 6 = 0$

Using the Derivative In Exercises 59–62, find the derivative of f. Use the derivative to determine any points on the graph of f at which the tangent line is horizontal. Use a graphing utility to verify your results.

59. $f(x) = x^2 - 4x + 3$ **60.** $f(x) = x^2 - 6x + 4$

61. $f(x) = 3x^3 - 9x$ **62.** $f(x) = x^3 + 3x$

Using the Derivative In Exercises 63–70, use the function and its derivative to determine any points on the graph of f at which the tangent line is horizontal. Use a graphing utility to verify your results.

63. $f(x) = x^4 - 2x^2$, $f'(x) = 4x^3 - 4x$

64. $f(x) = 3x^4 + 4x^3$, $f'(x) = 12x^3 + 12x^2$

65. $f(x) = 2\cos x + x$, $f'(x) = -2\sin x + 1$, over the interval $(0, 2\pi)$

66. $f(x) = x - 2\sin x$, $f'(x) = 1 - 2\cos x$, over the interval $(0, 2\pi)$

67. $f(x) = x \ln x$, $f'(x) = \ln x + 1$

68. $f(x) = \dfrac{\ln x}{x}$, $f'(x) = \dfrac{1 - \ln x}{x^2}$

69. $f(x) = x^2 e^x$, $f'(x) = x^2 e^x + 2x e^x$

70. $f(x) = x e^{-x}$, $f'(x) = e^{-x} - x e^{-x}$

71. Path of a Ball

The path of a ball thrown by a child is modeled by

$y = -x^2 + 5x + 2$

where y is the height of the ball (in feet) and x is the horizontal distance (in feet) from the point from which the child threw the ball. Using your knowledge of the slopes of tangent lines, show that the height of the ball is increasing on the interval $[0, 2]$ and decreasing on the interval $[3, 5]$. Explain.

72. Profit The profit P (in hundreds of dollars) that a company makes depends on the amount x (in hundreds of dollars) the company spends on advertising. The profit function is

$P(x) = 200 + 30x - 0.5x^2$.

Using your knowledge of the slopes of tangent lines, show that the profit is increasing on the interval $[0, 20]$ and decreasing on the interval $[40, 60]$.

73. Revenue The table shows the revenues y (in millions of dollars) for Netflix, Inc. from 2006 through 2015. (*Source: Netflix, Inc.*)

DATA Year	Revenue, y
2006	996.7
2007	1205.3
2008	1364.7
2009	1670.3
2010	2162.6
2011	3204.6
2012	3609.3
2013	4374.6
2014	5504.7
2015	6779.5

Spreadsheet at LarsonPrecalculus.com

(a) Use the *regression* feature of a graphing utility to find a quadratic model for the data. Let x represent the year, with $x = 6$ corresponding to 2006.

(b) Use the graphing utility to graph the model found in part (a). Estimate the slope of the graph when $x = 10$ and give an interpretation of the result.

(c) Use the graphing utility to graph the tangent line to the model when $x = 10$. Compare the slope given by the graphing utility with the estimate in part (b).

74. Market Research The data in the table show the number N (in thousands) of books sold when the price per book is p (in dollars).

Price, p	Number of Books, N
$15	630
$20	396
$25	227
$30	102
$35	36

(a) Use the *regression* feature of a graphing utility to find a quadratic model for the data.

(b) Use the graphing utility to graph the model found in part (a). Estimate the slopes of the graph when $p = \$20$ and $p = \$30$.

(c) Use the graphing utility to graph the tangent lines to the model when $p = \$20$ and $p = \$30$. Compare the slopes given by the graphing utility with your estimates in part (b).

(d) The slopes of the tangent lines at $p = \$20$ and $p = \$30$ are not the same. Explain what this means to the company selling the books.

Exploration

True or False? **In Exercises 75 and 76, determine whether the statement is true or false. Justify your answer.**

75. The slope of the graph of $f(x) = x^2$ is different at every point on the graph of f.

76. A tangent line to a graph can intersect the graph only at the point of tangency.

77. Conjecture Consider the functions $f(x) = x^2$ and $g(x) = x^3$.

(a) Sketch the graphs of f and f' on the same set of coordinate axes.

(b) Sketch the graphs of g and g' on the same set of coordinate axes.

(c) Identify any pattern between the functions f and g and their respective derivatives. Use the pattern to make a conjecture about $h'(x)$ when $h(x) = x^n$, where n is an integer and $n \geq 2$.

78. Conjecture Consider the function $f(x) = 3x^2 - 2x$.

(a) Use a graphing utility to graph the function.

(b) Use the *trace* feature to approximate the coordinates of the vertex of this parabola.

(c) Use the derivative of $f(x) = 3x^2 - 2x$ to find the slope of the tangent line at the vertex.

(d) Make a conjecture about the slope of the tangent line at the vertex of an arbitrary parabola.

79. Think About It Sketch the graph of a function whose derivative is always positive.

80. Think About It Sketch the graph of a function whose derivative is always negative.

81. Think About It Sketch the graph of a function for which $f'(x) < 0$ for $x < 1$, $f'(x) \geq 0$ for $x > 1$, and $f'(1) = 0$.

82. HOW DO YOU SEE IT? Use the graphs to explain the definition of the derivative of a function f at a point $(x, f(x))$.

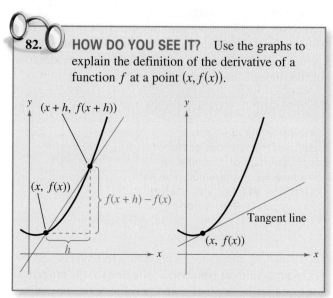

Matching a Function with the Graph of Its Derivative **In Exercises 83–86, match the function with the graph of its *derivative*. Do not find the derivative of the function. [The graphs are labeled (a), (b), (c), and (d).]**

(a)
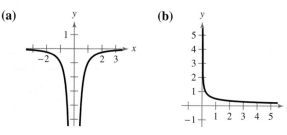

(b)

(c)
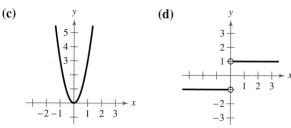

(d)

83. $f(x) = \sqrt{x}$

84. $f(x) = 1/x$

85. $f(x) = |x|$

86. $f(x) = x^3$

Project: Veterinarians To work an extended application analyzing the number of veterinarians employed in the United States, visit this text's website at *LarsonPrecalculus.com*. *(Source: U.S. Bureau of Labor Statistics)*

12.4 Limits at Infinity and Limits of Sequences

Finding limits at infinity is useful in many types of real-life applications. For example, in Exercise 39 on page 856, you will use a limit at infinity to analyze the oxygen level in a pond.

- ■ Evaluate limits of functions at infinity.
- ■ Find limits of sequences.

Limits at Infinity and Horizontal Asymptotes

As pointed out at the beginning of this chapter, there are two basic problems in calculus: finding tangent lines and finding the area of a region. In Section 12.3, you saw how limits can help you solve the tangent line problem. In this section and the next, you will see how a different type of limit, a *limit at infinity*, can help you solve the area problem. To get an idea of what is meant by a limit at infinity, consider the function

$$f(x) = \frac{x + 1}{2x}.$$

The graph of f is shown below.

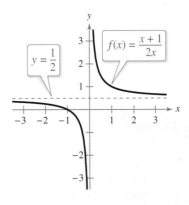

▷ **ALGEBRA HELP** The function

$$f(x) = \frac{x + 1}{2x}$$

is a rational function. To review rational functions, see Section 2.6.

From Section 2.6, you know that

$$y = \frac{1}{2}$$

is a horizontal asymptote of the graph of this function. Using limit notation, this is represented by the statements below.

$$\lim_{x \to -\infty} f(x) = \frac{1}{2} \qquad \text{Horizontal asymptote to the left}$$

$$\lim_{x \to \infty} f(x) = \frac{1}{2} \qquad \text{Horizontal asymptote to the right}$$

These limits mean that the value of $f(x)$ gets arbitrarily close to $\frac{1}{2}$ as x decreases or increases without bound.

Definition of Limits at Infinity

If f is a function and L_1 and L_2 are real numbers, then the statements

$$\lim_{x \to -\infty} f(x) = L_1 \qquad \text{Limit as } x \text{ approaches } -\infty$$

and

$$\lim_{x \to \infty} f(x) = L_2 \qquad \text{Limit as } x \text{ approaches } \infty$$

denote the **limits at infinity.** The first statement is read *"the limit of $f(x)$ as x approaches $-\infty$ is L_1,"* and the second is read *"the limit of $f(x)$ as x approaches ∞ is L_2."*

Many of the properties of limits listed in Section 12.1 apply to limits at infinity. You can use these properties with the limits at infinity below to find more complicated limits, as shown in Example 1.

Limits at Infinity

If r is a positive real number, then

$$\lim_{x \to \infty} \frac{1}{x^r} = 0.$$ Limit toward the right

Furthermore, if x^r is defined when $x < 0$, then

$$\lim_{x \to -\infty} \frac{1}{x^r} = 0.$$ Limit toward the left

---

EXAMPLE 1   Evaluating a Limit at Infinity

Find the limit.

$$\lim_{x \to \infty} \left(4 - \frac{3}{x^2} \right)$$

Algebraic Solution

Use the properties of limits listed in Section 12.1.

$$\lim_{x \to \infty} \left(4 - \frac{3}{x^2} \right) = \lim_{x \to \infty} 4 - \lim_{x \to \infty} \frac{3}{x^2}$$

$$= \lim_{x \to \infty} 4 - 3\left(\lim_{x \to \infty} \frac{1}{x^2} \right)$$

$$= 4 - 3(0)$$

$$= 4$$

So, the limit of $f(x) = 4 - \dfrac{3}{x^2}$ as x approaches ∞ is 4.

Graphical Solution

Use a graphing utility to graph

$$y = 4 - \frac{3}{x^2}.$$

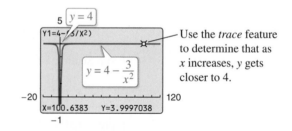

Use the *trace* feature to determine that as x increases, y gets closer to 4.

From the results shown above, estimate that the limit is 4. Also, note that the line $y = 4$ appears to be a horizontal asymptote to the right.

✓ **Checkpoint** ◀))) Audio-video solution in English & Spanish at LarsonPrecalculus.com

Find the limit: $\lim\limits_{x \to \infty} \dfrac{2}{x^2}.$

In the graphical solution to Example 1, it appears that the line $y = 4$ is also a horizontal asymptote *to the left*. Verify this by showing that

$$\lim_{x \to -\infty} \left(4 - \frac{3}{x^2} \right) = 4.$$

The graph of a rational function need not have a horizontal asymptote. When it does, however, its left and right horizontal asymptotes must be the same.

When evaluating limits at infinity for more complicated rational functions, divide the numerator and denominator by the highest power of the variable in the denominator. This enables you to evaluate each limit using the limits at infinity at the top of this page.

EXAMPLE 2   **Evaluating Limits at Infinity**

See LarsonPrecalculus.com for an interactive version of this type of example.

Find the limit as x approaches ∞ (if it exists) for each function.

a. $f(x) = \dfrac{-2x + 3}{3x^2 + 1}$

b. $f(x) = \dfrac{-2x^2 + 3}{3x^2 + 1}$

c. $f(x) = \dfrac{-2x^3 + 3}{3x^2 + 1}$

Solution   In each case, begin by dividing both the numerator and denominator by x^2, the highest power of x in the denominator.

a. $\displaystyle\lim_{x\to\infty} \frac{-2x + 3}{3x^2 + 1} = \lim_{x\to\infty} \frac{-\dfrac{2}{x} + \dfrac{3}{x^2}}{3 + \dfrac{1}{x^2}} = \frac{-0 + 0}{3 + 0} = 0$

b. $\displaystyle\lim_{x\to\infty} \frac{-2x^2 + 3}{3x^2 + 1} = \lim_{x\to\infty} \frac{-2 + \dfrac{3}{x^2}}{3 + \dfrac{1}{x^2}} = \frac{-2 + 0}{3 + 0} = -\frac{2}{3}$

c. $\displaystyle\lim_{x\to\infty} \frac{-2x^3 + 3}{3x^2 + 1} = \lim_{x\to\infty} \frac{-2x + \dfrac{3}{x^2}}{3 + \dfrac{1}{x^2}}$

In this case, the limit does not exist because the numerator decreases without bound as the denominator approaches 3.

✓ **Checkpoint** ◀))) *Audio-video solution in English & Spanish at LarsonPrecalculus.com*

Find the limit as x approaches ∞ (if it exists) for each function.

a. $f(x) = \dfrac{2x}{1 - x^2}$

b. $f(x) = \dfrac{2x}{1 - x}$

c. $f(x) = \dfrac{2x^2}{1 - x}$

In Example 2, observe that when the degree of the numerator is less than the degree of the denominator, as in part (a), the limit is 0. When the degrees of the numerator and denominator are equal, as in part (b), the limit is the ratio of the leading coefficients. When the degree of the numerator is greater than the degree of the denominator, as in part (c), the limit does not exist.

This result seems reasonable when you realize that for large values of x, the highest-powered term of a polynomial is the most "influential" term. That is, a polynomial tends to behave as its highest-powered term behaves as x approaches positive or negative infinity.

Limits at Infinity for Rational Functions

Consider the rational function

$$f(x) = \frac{N(x)}{D(x)}$$

where

$$N(x) = a_n x^n + \cdots + a_0 \quad \text{and} \quad D(x) = b_m x^m + \cdots + b_0.$$

The limit of $f(x)$ as x approaches positive or negative infinity is as follows.

$$\lim_{x \to \pm\infty} f(x) = \begin{cases} 0, & n < m \\ \dfrac{a_n}{b_m}, & n = m \end{cases}$$

If $n > m$, then the limit does not exist.

EXAMPLE 3 **Finding the Average Cost**

Mobile phone protective cases accounted for 20.5% of the global mobile phone accessories market in 2014. Industry analysts predict the global annual sales of mobile phone accessories to exceed $120 billion by 2025. *(Source: Future Market Insights)*

You are manufacturing mobile phone protective cases that cost \$4.50 per case to produce. Your initial investment is \$20,000, which implies that the total cost C of producing x cases is given by $C = 4.50x + 20,000$. The average cost $\overline{C}$ per case is given by

$$\overline{C} = \frac{C}{x} = \frac{4.50x + 20,000}{x}.$$

Find the average cost per case when (a) $x = 1000$, (b) $x = 10,000$, and (c) $x = 100,000$. (d) What is the limit of $\overline{C}$ as x approaches infinity?

Solution

a. When $x = 1000$,

$$\overline{C} = \frac{4.50(1000) + 20,000}{1000}$$

$$= \$24.50.$$

b. When $x = 10,000$,

$$\overline{C} = \frac{4.50(10,000) + 20,000}{10,000}$$

$$= \$6.50.$$

c. When $x = 100,000$,

$$\overline{C} = \frac{4.50(100,000) + 20,000}{100,000}$$

$$= \$4.70.$$

d. As x approaches infinity, the limit of $\overline{C}$ is

$$\lim_{x \to \infty} \frac{4.50x + 20,000}{x} = \$4.50.$$

The graph of $\overline{C}$ is at the right.

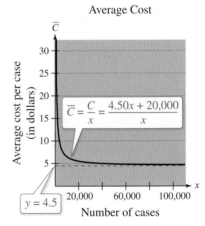

Average Cost

$$\overline{C} = \frac{C}{x} = \frac{4.50x + 20,000}{x}$$

$y = 4.5$

Average cost per case (in dollars)

Number of cases

As $x \to \infty$, the average cost per case approaches \$4.50.

✓ **Checkpoint** ◀))) *Audio-video solution in English & Spanish at LarsonPrecalculus.com*

Repeat Example 3 for the same initial investment, but a cost per case of \$4.00. ■

▷ **ALGEBRA HELP** To
review sequences, see
Sections 9.1–9.3.

▷ **TECHNOLOGY** There
are a number of ways to use
a graphing utility to generate
the terms of a sequence. For
example, to display the first
10 terms of the sequence

$$a_n = \frac{1}{2^n}$$

use the *sequence* feature or the
table feature.

Limits of Sequences

Limits of sequences have many of the same properties as limits of functions. For example, consider the sequence whose nth term is $a_n = 1/2^n$.

$$\frac{1}{2}, \frac{1}{4}, \frac{1}{8}, \frac{1}{16}, \frac{1}{32}, \ldots$$

As n increases without bound, the terms of this sequence get closer and closer to 0, and the sequence is said to **converge** to 0. Using limit notation, this is represented as

$$\lim_{n \to \infty} (1/2^n) = 0.$$

So, the limit of a function can be used to evaluate the limit of a sequence.

Limit of a Sequence

Let f be a function of a real variable such that

$$\lim_{x \to \infty} f(x) = L.$$

If $\{a_n\}$ is a sequence such that $f(n) = a_n$ for every positive integer n, then

$$\lim_{n \to \infty} a_n = L.$$

A sequence that does not converge is said to **diverge**. For example, the terms of the sequence $1, -1, 1, -1, 1, \ldots$ oscillate between 1 and -1. This sequence diverges because it does not converge to a unique number.

EXAMPLE 4   **Finding the Limit of a Sequence**

Write the first five terms of each sequence and find the limit of the sequence. (Assume that n begins with 1.)

a. $a_n = \dfrac{2n + 1}{n + 4}$ **b.** $a_n = \dfrac{2n + 1}{n^2 + 4}$ **c.** $a_n = \dfrac{2n^2 + 1}{4n^2}$

Solution

a. The first five terms are $a_1 = \frac{3}{5}$, $a_2 = \frac{5}{6}$, $a_3 = 1$, $a_4 = \frac{9}{8}$, and $a_5 = \frac{11}{9}$. The limit of the sequence is

$$\lim_{n \to \infty} \frac{2n + 1}{n + 4} = 2.$$

b. The first five terms are $a_1 = \frac{3}{5}$, $a_2 = \frac{5}{8}$, $a_3 = \frac{7}{13}$, $a_4 = \frac{9}{20}$, and $a_5 = \frac{11}{29}$. The limit of the sequence is

$$\lim_{n \to \infty} \frac{2n + 1}{n^2 + 4} = 0.$$

c. The first five terms are $a_1 = \frac{3}{4}$, $a_2 = \frac{9}{16}$, $a_3 = \frac{19}{36}$, $a_4 = \frac{33}{64}$, and $a_5 = \frac{51}{100}$. The limit of the sequence is

$$\lim_{n \to \infty} \frac{2n^2 + 1}{4n^2} = \frac{1}{2}.$$

✓ *Checkpoint* ◀))) *Audio-video solution in English & Spanish at LarsonPrecalculus.com*

Write the first five terms of each sequence and find the limit of the sequence. (Assume that n begins with 1.)

a. $a_n = \dfrac{n + 2}{2n - 1}$ **b.** $a_n = \dfrac{n^3 + 2}{3n^3}$ **c.** $a_n = \dfrac{n + 2}{3n^2}$

In the next section, you will encounter limits of sequences such as that shown in Example 5. A strategy for evaluating such limits is to begin by writing the nth term in standard rational function form. Then determine the limit by comparing the degrees of the numerator and denominator, as shown on page 852.

EXAMPLE 5 Finding the Limit of a Sequence

Find the limit of the sequence whose nth term is

$$a_n = \frac{8}{n^3}\left[\frac{n(n+1)(2n+1)}{6}\right].$$

Algebraic Solution

Begin by writing the nth term in standard rational function form—as the ratio of two polynomials.

$$a_n = \frac{8}{n^3}\left[\frac{n(n+1)(2n+1)}{6}\right] \qquad \text{Write original } n\text{th term.}$$

$$= \frac{8(n)(n+1)(2n+1)}{6n^3} \qquad \text{Multiply fractions.}$$

$$= \frac{8n^3 + 12n^2 + 4n}{3n^3} \qquad \text{Write in standard rational form.}$$

This form shows that the degree of the numerator is equal to the degree of the denominator. So, the limit of the sequence is the ratio of the leading coefficients.

$$\lim_{n\to\infty} \frac{8n^3 + 12n^2 + 4n}{3n^3} = \frac{8}{3}$$

Numerical Solution

Construct a table that shows the value of a_n as n becomes larger and larger.

n	a_n
1	8
10	3.08
100	2.7068
1000	2.6707
10,000	2.6671
100,000	2.6667
1,000,000	2.6667

Notice from the table that as n approaches ∞, a_n gets closer and closer to

$$2.6667 \approx \frac{8}{3}.$$

 **✓ Checkpoint** *Audio-video solution in English & Spanish at LarsonPrecalculus.com*

Find the limit of each sequence whose nth term is given.

a. $a_n = \dfrac{5}{n^3}\left[\dfrac{(n+1)(n-1)(2n)}{7}\right]$

b. $a_n = \dfrac{2}{n}\left(n - \dfrac{2}{n}\left[\dfrac{n(n-1)}{4}\right]\right)$

Summarize (Section 12.4)

1. Define and explain how to evaluate limits of a function at infinity *(pages 849 and 850)*. For examples of evaluating limits at infinity, see Examples 1 and 2.

2. Explain how to evaluate limits at infinity for rational functions *(pages 851 and 852)*. For a real-life example of evaluating the limit at infinity for a rational function, see Example 3.

3. Explain how to find the limit of a sequence *(page 853)*. For examples of finding limits of sequences, see Examples 4 and 5.

12.4 Exercises

See CalcChat.com for tutorial help and worked-out solutions to odd-numbered exercises.

Vocabulary: Fill in the blanks.

1. A _____ at _____ can help you solve the area problem in calculus.
2. For a rational function, when the degrees of the numerator and denominator are equal, the limit is the _____ of the leading coefficients.
3. A sequence that has a limit is said to _____.
4. A sequence that does not have a limit is said to _____.

Skills and Applications

Matching a Function with Its Graph In Exercises 5–8, match the function with its graph, using the asymptotes as aids. [The graphs are labeled (a), (b), (c), and (d).]

(a)
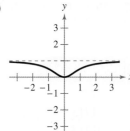

(b)
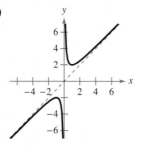

(c)
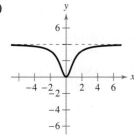

(d)
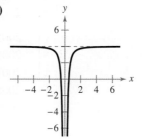

5. $f(x) = \dfrac{4x^2}{x^2 + 1}$

6. $f(x) = \dfrac{x^2}{x^2 + 1}$

7. $f(x) = 4 - \dfrac{1}{x^2}$

8. $f(x) = x + \dfrac{1}{x}$

 Evaluating a Limit at Infinity In Exercises 9–28, find the limit, if it exists. If the limit does not exist, explain why. Use a graphing utility to verify your result graphically.

9. $\lim\limits_{x \to \infty} \left(2 + \dfrac{3}{x^2}\right)$

10. $\lim\limits_{x \to \infty} \left(4 - \dfrac{1}{x^2}\right)$

11. $\lim\limits_{x \to \infty} \dfrac{1 - x}{1 + x}$

12. $\lim\limits_{x \to \infty} \dfrac{1 + 5x}{1 - 4x}$

13. $\lim\limits_{x \to -\infty} \dfrac{3x^2 - 4}{1 - x^2}$

14. $\lim\limits_{x \to \infty} \dfrac{1 - 2x}{x + 2}$

15. $\lim\limits_{x \to -\infty} \dfrac{5x - 1}{3x^2 + 2}$

16. $\lim\limits_{x \to -\infty} \dfrac{4 + x}{2x^2 + 1}$

17. $\lim\limits_{t \to \infty} \dfrac{t^2}{t + 3}$

18. $\lim\limits_{y \to \infty} \dfrac{4y^4}{y^2 + 3}$

19. $\lim\limits_{t \to \infty} \dfrac{4t^2 - 2t + 1}{-3t^2 + 2t + 2}$

20. $\lim\limits_{x \to -\infty} \dfrac{2x^2 - 5x - 12}{1 - 6x - 8x^2}$

21. $\lim\limits_{x \to -\infty} \dfrac{-(x^2 + 3)}{(2 - x)^2}$

22. $\lim\limits_{x \to \infty} \dfrac{2x^2 - 6}{(x - 1)^2}$

23. $\lim\limits_{x \to \infty} \dfrac{x^3 + 3}{x^2}$

24. $\lim\limits_{x \to -\infty} \left(\dfrac{1}{2}x - \dfrac{4}{x^2}\right)$

25. $\lim\limits_{x \to -\infty} \left[\dfrac{x}{(x + 1)^2} - 4\right]$

26. $\lim\limits_{x \to \infty} \left[7 + \dfrac{2x^2}{(x + 3)^2}\right]$

27. $\lim\limits_{t \to \infty} \left(\dfrac{1}{3t^2} - \dfrac{5t}{t + 2}\right)$

28. $\lim\limits_{x \to \infty} \left[\dfrac{1}{2} + \dfrac{x^2}{(x - 3)^2}\right]$

 Using Horizontal Asymptotes In Exercises 29–34, use a graphing utility to graph the function and estimate the horizontal asymptote and the limits at infinity for the function.

29. $y = \dfrac{3x}{1 - x}$

30. $y = \dfrac{x^2}{x^2 + 4}$

31. $y = \dfrac{5x}{1 - x^2}$

32. $y = \dfrac{2x + 1}{x^2 - 1}$

33. $y = 1 - \dfrac{3}{x^2}$

34. $y = 2 + \dfrac{1}{x}$

 Estimating a Limit In Exercises 35–38, (a) complete the table and numerically estimate the limit as x approaches infinity for the function, and (b) use a graphing utility to graph the function and estimate the limit graphically.

x	10^0	10^1	10^2	10^3	10^4	10^5	10^6
$f(x)$							

35. $f(x) = x - \sqrt{x^2 + 2}$
36. $f(x) = 3x - \sqrt{9x^2 + 1}$
37. $f(x) = 3\left(2x - \sqrt{4x^2 + x}\right)$
38. $f(x) = 4\left(4x - \sqrt{16x^2 - x}\right)$

39. Oxygen Level

The function $f(t)$ measures the level of oxygen in a pond at week t, where $f(t) = 1$ is the normal (unpolluted) level. When $t = 0$, polluters dump organic waste into the pond, and as the waste material oxidizes, the level of oxygen in the pond is given by

$$f(t) = \frac{t^2 - t + 1}{t^2 + 1}.$$

(a) What is the limit of f as t approaches infinity?

(b) Use a graphing utility to graph the function and verify the result of part (a).

(c) Explain the meaning of the limit in the context of the problem.

40. Typing Speed The average typing speed S (in words per minute) for a student after t weeks of lessons is given by

$$S = \frac{100t^2}{65 + t^2}, \quad t > 0.$$

(a) What is the limit of S as t approaches infinity?

(b) Use a graphing utility to graph the function and verify the result of part (a).

(c) Explain the meaning of the limit in the context of the problem.

41. Average Cost The cost function for a certain model of MP3 player is $C = 73x + 25{,}000$, where C is in dollars and x is the number of MP3 players produced.

(a) Write a model for the average cost per unit produced.

(b) Find the average costs per unit when $x = 1000$, $x = 5000$, and $x = 10{,}000$.

(c) Determine the limit of the average cost function as x approaches infinity. Explain the meaning of the limit in the context of the problem.

42. Average Cost The cost function for a company to recycle x tons of material is $C = 1.25x + 10{,}500$, where C is in dollars.

(a) Write a model for the average cost per ton of material recycled.

(b) Find the average costs of recycling 100, 1000, and 10,000 tons of material.

(c) Determine the limit of the average cost function as x approaches infinity. Explain the meaning of the limit in the context of the problem.

43. Social Security The average monthly Social Security benefit B (in dollars) for a retired worker age 62 or over from 2005 through 2014 can be modeled by

$$B = \frac{812.97 + 29.70t + 0.617t^2}{1.0 - 0.01t + 0.0008t^2}, \quad 5 \le t \le 14$$

where t represents the year, with $t = 5$ corresponding to 2005. *(Source: U.S. Social Security Administration)*

(a) Use the model to predict the average monthly benefit in 2020.

(b) What is the limit of the function as t approaches infinity? Explain the meaning of the limit in the context of the problem.

(c) Discuss why this model is not realistic for long-term predictions of average monthly Social Security benefits.

44. Military The table shows the numbers N (in thousands) of U.S. military retirees receiving retired pay for the years 2000 through 2012. *(Source: U.S. Department of Defense)*

Year	Number, N
2000	1701
2001	1713
2002	1713
2003	1722
2004	1792
2005	1812
2006	1833
2007	1860
2008	1881
2009	1904
2010	1917
2011	1933
2012	1944

Spreadsheet at LarsonPrecalculus.com

A model for the data is

$$N = \frac{1680.4 - 54.17t}{1.0 - 0.047t + 0.00067t^2}, \quad 0 \le t \le 12$$

where t represents the year, with $t = 0$ corresponding to 2000.

(a) Use a graphing utility to create a scatter plot of the data and graph the model in the same viewing window. How well does the model fit the data?

(b) Use the model to predict the number of military retirees in 2019.

(c) What is the limit of the function as t approaches infinity? Explain the meaning of the limit in the context of the problem. Do you think the limit is realistic? Explain.

Finding the Limit of a Sequence In Exercises 45–54, write the first five terms of the sequence and find the limit of the sequence, if it exists. If the limit does not exist, explain why. (Assume that n begins with 1.)

45. $a_n = \dfrac{n+1}{n^2+1}$

46. $a_n = \dfrac{3n}{n^2+2}$

47. $a_n = \dfrac{n}{2n+1}$

48. $a_n = \dfrac{4n-1}{n+3}$

49. $a_n = \dfrac{n^2}{2n+3}$

50. $a_n = \dfrac{4n^2+1}{2n}$

51. $a_n = \dfrac{(n+1)!}{n!}$

52. $a_n = \dfrac{(3n-1)!}{(3n+1)!}$

53. $a_n = \dfrac{(-1)^n}{n}$

54. $a_n = \dfrac{(-1)^{n+1}}{n^2}$

Finding the Limit of a Sequence In Exercises 55–58, find the limit of the sequence. Then verify the limit numerically by using a graphing utility to complete the table.

n	10^0	10^1	10^2	10^3	10^4	10^5	10^6
a_n							

55. $a_n = \dfrac{1}{n}\left(n + \dfrac{1}{n}\left[\dfrac{n(n+1)}{2}\right]\right)$

56. $a_n = \dfrac{4}{n}\left(n + \dfrac{4}{n}\left[\dfrac{n(n+1)}{2}\right]\right)$

57. $a_n = \dfrac{16}{n^3}\left[\dfrac{n(n+1)(2n+1)}{6}\right]$

58. $a_n = \dfrac{n(n+1)}{n^2} - \dfrac{1}{n^4}\left[\dfrac{n(n+1)}{2}\right]^2$

Exploration

True or False? In Exercises 59–62, determine whether the statement is true or false. Justify your answer.

59. Every rational function has a horizontal asymptote.

60. If $f(x)$ increases without bound as x approaches c, then the limit of $f(x)$ exists.

61. If a sequence converges, then it has a limit.

62. When the degree of the numerator of a rational function is less than the degree of the denominator, the limits at infinity do not exist.

63. Error Analysis Describe the error.

$$\lim_{x \to \infty}\left(5 - \dfrac{2}{x^2}\right) = 5\left(\lim_{x \to \infty}\dfrac{-2}{x^2}\right)$$
$$= -10\left(\lim_{x \to \infty}\dfrac{1}{x^2}\right)$$
$$= 0$$

64. Think About It Find two functions f and g, and a real number c such that both $f(x)$ and $g(x)$ increase without bound as x approaches c, but $\lim_{x \to c}[f(x) - g(x)]$ exists.

65. Think About It Use a graphing utility to graph the function

$$f(x) = \dfrac{x}{\sqrt{x^2+1}}.$$

How many horizontal asymptotes does the function appear to have? What can you conclude about the limits of the function at infinity?

66. **HOW DO YOU SEE IT?** Use each graph to estimate $\lim_{x \to \infty} f(x)$, $\lim_{x \to -\infty} f(x)$, and the horizontal asymptote of the graph of f.

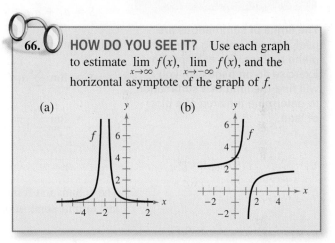

Determining Convergence or Divergence In Exercises 67–70, create a scatter plot of the terms of the sequence. Determine whether the sequence converges or diverges. Estimate the limit of the sequence, if it exists.

67. $a_n = 4\left(\dfrac{2}{3}\right)^n$

68. $a_n = 3\left(\dfrac{3}{2}\right)^n$

69. $a_n = \dfrac{3[1-(1.5)^n]}{1-1.5}$

70. $a_n = \dfrac{3[1-(0.5)^n]}{1-0.5}$

71. Think About It Use a graphing utility to graph the two functions

$$y_1 = \dfrac{1}{\sqrt{x}} \quad \text{and} \quad y_2 = \dfrac{1}{\sqrt[3]{x}}$$

in the same viewing window. Why is no part of the graph of y_1 to the left of the y-axis? Compare this to the graph of y_2. How are the graphs of y_1 and y_2 related to the statement at the top of page 850 about the infinite limit

$$\lim_{x \to -\infty}\dfrac{1}{x^r}?$$

72. Conjecture Create a table to verify that

$$\lim_{x \to \infty}\dfrac{1}{x} = 0.$$

Then make a conjecture about $\lim_{x \to 0}(1/x)$.

12.5 The Area Problem

The limits of summations are useful in determining areas of plane regions. For example, in Exercise 45 on page 865, you will find the limit of a summation to determine the area of a plot of land.

- Find limits of summations.
- Use rectangles to approximate areas of plane regions.
- Use limits of summations to find areas of plane regions.

Limits of Summations

In Section 9.3, you used the concept of a limit to obtain a formula for the sum S of an infinite geometric series

$$S = a_1 + a_1 r + a_1 r^2 + \cdots = \sum_{i=1}^{\infty} a_1 r^{i-1} = \frac{a_1}{1-r}, \quad |r| < 1.$$

Using limit notation, this sum can be written as

$$S = \lim_{n \to \infty} \sum_{i=1}^{n} a_1 r^{i-1}$$

$$= \lim_{n \to \infty} \frac{a_1(1 - r^n)}{1 - r} \qquad \sum_{i=1}^{n} a_1 r^{i-1} = a_1 \left(\frac{1 - r^n}{1 - r} \right)$$

$$= \frac{a_1}{1-r}. \qquad \lim_{n \to \infty} r^n = 0 \text{ for } |r| < 1$$

The summation formulas and properties listed below are useful for evaluating finite and infinite summations.

Summation Formulas and Properties

1. $\displaystyle\sum_{i=1}^{n} c = cn, \; c$ is a constant.   **2.** $\displaystyle\sum_{i=1}^{n} i = \frac{n(n+1)}{2}$

3. $\displaystyle\sum_{i=1}^{n} i^2 = \frac{n(n+1)(2n+1)}{6}$ **4.** $\displaystyle\sum_{i=1}^{n} i^3 = \frac{n^2(n+1)^2}{4}$

5. $\displaystyle\sum_{i=1}^{n} (a_i \pm b_i) = \sum_{i=1}^{n} a_i \pm \sum_{i=1}^{n} b_i$ **6.** $\displaystyle\sum_{i=1}^{n} k a_i = k \sum_{i=1}^{n} a_i, \; k$ is a constant.

> **EXAMPLE 1**   **Evaluating a Summation**

Evaluate the summation.

$$\sum_{i=1}^{200} i = 1 + 2 + 3 + 4 + \cdots + 200$$

Solution   Using the second summation formula with $n = 200$,

$$\sum_{i=1}^{n} i = \frac{n(n+1)}{2}$$

$$\sum_{i=1}^{200} i = \frac{200(200+1)}{2} = \frac{40{,}200}{2} = 20{,}100.$$

✓ **Checkpoint**))) Audio-video solution in English & Spanish at LarsonPrecalculus.com

Evaluate the summation $\displaystyle\sum_{i=1}^{10} i^2$.

EXAMPLE 2 **Evaluating a Summation**

Evaluate the summation

$$S = \sum_{i=1}^{n} \frac{i+2}{n^2} = \frac{3}{n^2} + \frac{4}{n^2} + \frac{5}{n^2} + \cdots + \frac{n+2}{n^2}$$

for $n = 10, 100, 1000,$ and $10,000.$

Solution Begin by applying summation formulas and properties to simplify S. In the second line of the solution, note that

$$\frac{1}{n^2}$$

factors out of the sum because n is considered to be constant. You cannot factor i out of the first summation in the third line of the solution because i is the index of summation.

$$S = \sum_{i=1}^{n} \frac{i+2}{n^2} \qquad \text{Write original form of summation.}$$

$$= \frac{1}{n^2} \sum_{i=1}^{n} (i+2) \qquad \text{Factor constant } 1/n^2 \text{ out of sum.}$$

$$= \frac{1}{n^2} \left(\sum_{i=1}^{n} i + \sum_{i=1}^{n} 2 \right) \qquad \text{Write as two sums.}$$

$$= \frac{1}{n^2} \left[\frac{n(n+1)}{2} + 2n \right] \qquad \text{Apply Formulas 1 and 2.}$$

$$= \frac{1}{n^2} \left(\frac{n^2 + 5n}{2} \right) \qquad \text{Add fractions.}$$

$$= \frac{n+5}{2n} \qquad \text{Simplify.}$$

Now, evaluate the sum by substituting the appropriate values of n. The table below shows the results.

n	10	100	1000	10,000
$\sum_{i=1}^{n} \dfrac{i+2}{n^2} = \dfrac{n+5}{2n}$	0.75	0.525	0.5025	0.50025

✓ *Checkpoint* ◀))) *Audio-video solution in English & Spanish at LarsonPrecalculus.com*

Evaluate the summation

$$S = \sum_{i=1}^{n} \frac{3i+2}{n^2} = \frac{5}{n^2} + \frac{8}{n^2} + \frac{11}{n^2} + \cdots + \frac{3n+2}{n^2}$$

for $n = 10, 100, 1000,$ and $10,000.$

In Example 2, note that the sum appears to approach a limit as n increases. To find the limit of

$$\frac{n+5}{2n}$$

as n approaches infinity, use the techniques from Section 12.4 to write

$$\lim_{n \to \infty} \frac{n+5}{2n} = \frac{1}{2}.$$

Be sure you notice the strategy used in Example 2. Rather than separately evaluating the sums

$$\sum_{i=1}^{10} \frac{i+2}{n^2}, \qquad \sum_{i=1}^{100} \frac{i+2}{n^2}, \qquad \sum_{i=1}^{1000} \frac{i+2}{n^2}, \qquad \sum_{i=1}^{10,000} \frac{i+2}{n^2}$$

it was more efficient to first convert to rational form using the summation formulas and properties listed on page 858.

$$S = \underbrace{\sum_{i=1}^{n} \frac{i+2}{n^2}}_{\substack{\text{Summation} \\ \text{form}}} = \underbrace{\frac{n+5}{2n}}_{\substack{\text{Rational} \\ \text{form}}}$$

Then, each sum was evaluated by substituting the appropriate values of n into the rational form and simplifying.

EXAMPLE 3   Finding the Limit of a Summation

Find the limit of $S(n)$ as $n \to \infty$.

$$S(n) = \sum_{i=1}^{n} \left(1 + \frac{i}{n}\right)^2 \left(\frac{1}{n}\right)$$

▷ **ALGEBRA HELP**   Notice the algebra involved in rewriting the summation in rational form in Example 3. To review operations with rational expressions and simplifying rational expressions, see Appendix A.4.

Solution   Begin by rewriting the summation in rational form.

$$S(n) = \sum_{i=1}^{n} \left(1 + \frac{i}{n}\right)^2 \left(\frac{1}{n}\right) \qquad \text{Write original form of summation.}$$

$$= \sum_{i=1}^{n} \left(\frac{n^2 + 2ni + i^2}{n^2}\right)\left(\frac{1}{n}\right) \qquad \text{Square } [1 + (i/n)] \text{ and write as a single fraction.}$$

$$= \frac{1}{n^3} \sum_{i=1}^{n} (n^2 + 2ni + i^2) \qquad \text{Factor constant } 1/n^3 \text{ out of the sum.}$$

$$= \frac{1}{n^3}\left(\sum_{i=1}^{n} n^2 + \sum_{i=1}^{n} 2ni + \sum_{i=1}^{n} i^2\right) \qquad \text{Write as three sums.}$$

$$= \frac{1}{n^3}\left\{n^3 + 2n\left[\frac{n(n+1)}{2}\right] + \frac{n(n+1)(2n+1)}{6}\right\} \qquad \text{Use summation formulas.}$$

$$= \frac{14n^3 + 9n^2 + n}{6n^3} \qquad \text{Simplify.}$$

Now, use this rational form to find the limit as $n \to \infty$.

$$\lim_{n \to \infty} S(n) = \lim_{n \to \infty} \frac{14n^3 + 9n^2 + n}{6n^3}$$

$$= \frac{14}{6}$$

$$= \frac{7}{3}$$

✓ *Checkpoint* ◀))) *Audio-video solution in English & Spanish at LarsonPrecalculus.com*

Find the limit of $S(n)$ as $n \to \infty$.

$$S(n) = \sum_{i=1}^{n} \left(\frac{2}{n} + \frac{i}{n^2}\right)\left(\frac{1}{n}\right)$$

∎

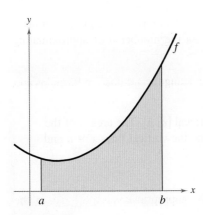

Figure 12.17

The Area Problem

You now have the tools needed to solve the second basic problem of calculus: the area problem. The problem is to find the *area* of the region R bounded by the graph of a nonnegative, continuous function f, the x-axis, and the vertical lines $x = a$ and $x = b$, as shown in Figure 12.17.

When the region R is a rectangle, triangle, trapezoid, or semicircle, you can find its area using a geometric formula. For many other types of regions, however, you must use a different approach—one that involves the limit of a summation. The basic strategy is to use a collection of rectangles of equal width that approximates the region R, as illustrated in Example 4.

EXAMPLE 4   Approximating the Area of a Region

See LarsonPrecalculus.com for an interactive version of this type of example.

Use the five rectangles in Figure 12.18 to approximate the area of the region bounded by the graph of $f(x) = 6 - x^2$, the x-axis, and the lines $x = 0$ and $x = 2$.

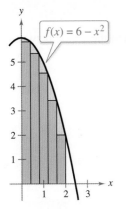

Figure 12.18

Solution   The length of the interval along the x-axis is 2 and there are five rectangles, so the width of each rectangle is $\frac{2}{5}$. To obtain the height of each rectangle, evaluate $f(x)$ at the right endpoint of each interval. The five intervals are

$$\left[0, \tfrac{2}{5}\right], \quad \left[\tfrac{2}{5}, \tfrac{4}{5}\right], \quad \left[\tfrac{4}{5}, \tfrac{6}{5}\right], \quad \left[\tfrac{6}{5}, \tfrac{8}{5}\right], \quad \text{and} \quad \left[\tfrac{8}{5}, \tfrac{10}{5}\right].$$

Notice that the right endpoint of each interval is $\frac{2}{5}i$ for $i = 1, 2, 3, 4, 5$. The sum of the areas of the five rectangles is

$$\sum_{i=1}^{5} f\overbrace{\left(\frac{2i}{5}\right)}^{\text{Height}}\overbrace{\left(\frac{2}{5}\right)}^{\text{Width}} = \sum_{i=1}^{5}\left[6 - \left(\frac{2i}{5}\right)^2\right]\left(\frac{2}{5}\right)$$

$$= \frac{2}{5}\left(\sum_{i=1}^{5}6 - \frac{4}{25}\sum_{i=1}^{5}i^2\right)$$

$$= \frac{2}{5}\left[6(5) - \frac{4}{25}\cdot\frac{5(5+1)(10+1)}{6}\right]$$

$$= \frac{2}{5}\left(30 - \frac{44}{5}\right)$$

$$= \frac{212}{25}.$$

So, the area of the region is approximately $212/25 = 8.48$ square units.

✓ **Checkpoint** ◀))) *Audio-video solution in English & Spanish at LarsonPrecalculus.com*

Use the four rectangles in Figure 12.19 to approximate the area of the region bounded by the graph of $f(x) = x + 1$, the x-axis, and the lines $x = 0$ and $x = 2$.   ◼

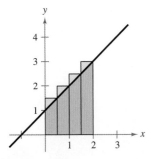

Figure 12.19

By increasing the number of rectangles used in Example 4, you obtain a more accurate approximation of the area of the region. For instance, using 25 rectangles each of width $\frac{2}{25}$, you can approximate the area to be $A \approx 9.17$ square units. The table below includes even better approximations.

Number of Rectangles	5	25	100	1000	5000
Approximate Area	8.48	9.17	9.29	9.33	9.33

The Exact Area of a Plane Region

Based on the procedure illustrated in Example 4, the *exact* **area of a plane region** R can be found by using the limit process to increase the number n of approximating rectangles without bound.

Area of a Plane Region

Let f be continuous and nonnegative on the interval $[a, b]$. The area A of the region bounded by the graph of f, the x-axis, and the vertical lines $x = a$ and $x = b$ is given by

$$A = \lim_{n \to \infty} \sum_{i=1}^{n} \underbrace{f\left(a + \frac{(b-a)i}{n}\right)}_{\text{Height}} \underbrace{\left(\frac{b-a}{n}\right)}_{\text{Width}}.$$

EXAMPLE 5 **Finding the Area of a Region**

Find the area of the region bounded by the graph of $f(x) = x^2$ and the x-axis between $x = 0$ and $x = 1$, as shown in Figure 12.20.

Solution Begin by finding the dimensions of the rectangles.

$$\text{Width:}\quad \frac{b-a}{n} = \frac{1-0}{n} = \frac{1}{n}$$

$$\text{Height:}\quad f\left(a + \frac{(b-a)i}{n}\right) = f\left(0 + \frac{(1-0)i}{n}\right) = f\left(\frac{i}{n}\right) = \frac{i^2}{n^2}$$

Next, approximate the area as the sum of the areas of n rectangles.

$$A \approx \sum_{i=1}^{n} f\left(a + \frac{(b-a)i}{n}\right)\left(\frac{b-a}{n}\right)$$

$$= \sum_{i=1}^{n} \left(\frac{i^2}{n^2}\right)\left(\frac{1}{n}\right) \qquad \text{Summation form}$$

$$= \sum_{i=1}^{n} \frac{i^2}{n^3}$$

$$= \frac{1}{n^3} \sum_{i=1}^{n} i^2$$

$$= \frac{1}{n^3}\left[\frac{n(n+1)(2n+1)}{6}\right]$$

$$= \frac{2n^3 + 3n^2 + n}{6n^3} \qquad \text{Rational form}$$

Finally, find the exact area by taking the limit as n approaches ∞.

$$A = \lim_{n \to \infty} \frac{2n^3 + 3n^2 + n}{6n^3} = \frac{1}{3}$$

So, the area of the region is $\frac{1}{3}$ square unit.

✓ **Checkpoint** ◀))) *Audio-video solution in English & Spanish at LarsonPrecalculus.com*

Find the area of the region bounded by the graph of $f(x) = 3x$ and the x-axis between $x = 0$ and $x = 1$. ◼

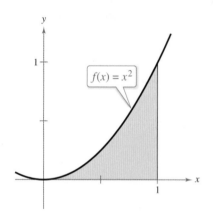

Figure 12.20

EXAMPLE 6 **Finding the Area of a Region**

Find the area of the region bounded by the graph of $f(x) = 3x - x^2$, the x-axis, and the lines $x = 1$ and $x = 2$, as shown in Figure 12.21.

Solution Begin by finding the dimensions of the rectangles.

$$\text{Width: } \frac{b - a}{n} = \frac{2 - 1}{n} = \frac{1}{n}$$

$$\text{Height: } f\left(a + \frac{(b - a)i}{n}\right) = f\left(1 + \frac{i}{n}\right)$$

$$= 3\left(1 + \frac{i}{n}\right) - \left(1 + \frac{i}{n}\right)^2$$

$$= 2 + \frac{i}{n} - \frac{i^2}{n^2}$$

Next, approximate the area as the sum of the areas of n rectangles.

$$A \approx \sum_{i=1}^{n} f\left(a + \frac{(b - a)i}{n}\right)\left(\frac{b - a}{n}\right)$$

$$= \sum_{i=1}^{n} \left(2 + \frac{i}{n} - \frac{i^2}{n^2}\right)\left(\frac{1}{n}\right)$$

$$= \frac{1}{n} \sum_{i=1}^{n} 2 + \frac{1}{n^2} \sum_{i=1}^{n} i - \frac{1}{n^3} \sum_{i=1}^{n} i^2$$

$$= \frac{1}{n}(2n) + \frac{1}{n^2}\left[\frac{n(n + 1)}{2}\right] - \frac{1}{n^3}\left[\frac{n(n + 1)(2n + 1)}{6}\right]$$

$$= \frac{12n^3}{6n^3} + \frac{3n^3 + 3n^2}{6n^3} - \frac{2n^3 + 3n^2 + n}{6n^3}$$

$$= \frac{13n^3 - n}{6n^3}$$

Finally, find the exact area by taking the limit as n approaches ∞.

$$A = \lim_{n \to \infty} \frac{13n^3 - n}{6n^3} = \frac{13}{6}$$

✓ **Checkpoint** 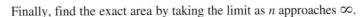 Audio-video solution in English & Spanish at LarsonPrecalculus.com

Find the area of the region bounded by the graph of $f(x) = x^2 + 2$, the x-axis, and the lines $x = 1$ and $x = 3$. ◾

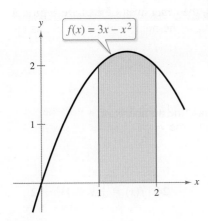

Figure 12.21

Summarize (Section 12.5)

1. Explain how to find the limit of a summation *(pages 858–860)*. For an example of finding the limit of a summation, see Example 3.

2. Explain how to use rectangles to approximate the area of a plane region *(page 861)*. For an example of using rectangles to approximate the area of a plane region, see Example 4.

3. Explain how to use the limit of a summation to find the area of a plane region *(page 862)*. For examples of using limits of summations to find areas of plane regions, see Examples 5 and 6.

12.5 Exercises

See **CalcChat.com** for tutorial help and worked-out solutions to odd-numbered exercises.

Vocabulary: Fill in the blanks.

1. $\sum_{i=1}^{n} c = $ _____, c is a constant.   **2.** $\sum_{i=1}^{n} i = $ _____   **3.** $\sum_{i=1}^{n} i^3 = $ _____

4. The exact _____ of a plane region R can be found using the limit process to increase the number n of approximating rectangles without bound.

Skills and Applications

 Evaluating a Summation In Exercises 5–12, evaluate the sum using the summation formulas and properties.

5. $\sum_{i=1}^{60} 4$ **6.** $\sum_{i=1}^{45} 9$

7. $\sum_{i=1}^{20} i^3$ **8.** $\sum_{i=1}^{40} i^2$

9. $\sum_{k=1}^{15} (k^3 + 2)$ **10.** $\sum_{k=1}^{50} (2k + 1)$

11. $\sum_{j=1}^{25} (j^2 + j)$ **12.** $\sum_{j=1}^{10} (j^3 - 3j^2)$

 Finding the Limit of a Summation In Exercises 13–20, (a) rewrite the sum as a rational function $S(n)$, (b) use $S(n)$ to complete the table, and (c) find $\lim_{n \to \infty} S(n)$.

n	10^0	10^1	10^2	10^3	10^4
$S(n)$					

13. $\sum_{i=1}^{n} \frac{i^3}{n^4}$ **14.** $\sum_{i=1}^{n} \frac{i}{n^2}$

15. $\sum_{i=1}^{n} \frac{3}{n^3}(1 + i^2)$ **16.** $\sum_{i=1}^{n} \frac{2i + 3}{n^2}$

17. $\sum_{i=1}^{n} \left(\frac{i^2}{n^3} + \frac{2}{n}\right)\left(\frac{1}{n}\right)$ **18.** $\sum_{i=1}^{n} \left[3 - 2\left(\frac{i}{n}\right)\right]\left(\frac{1}{n}\right)$

19. $\sum_{i=1}^{n} \left[1 - \left(\frac{i}{n}\right)^2\right]\left(\frac{1}{n}\right)$ **20.** $\sum_{i=1}^{n} \left(\frac{4}{n} + \frac{2i}{n^2}\right)\left(\frac{2i}{n}\right)$

Approximating the Area of a Region In Exercises 21–24, approximate the area of the region using the rectangles of equal width shown.

21. $f(x) = x + 4$ **22.** $f(x) = 2 - x^2$

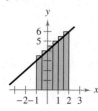   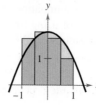

23. $f(x) = \frac{1}{8}x^3 + 1$ **24.** $f(x) = \frac{1}{4}(x - 1)^3 + 2$

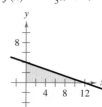

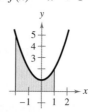 **Approximating the Area of a Region** In Exercises 25–30, complete the table to show the approximate area of the region using the indicated numbers n of rectangles of equal width.

n	4	8	20	50
Approximate Area				

25. $f(x) = -\frac{1}{3}x + 4$ **26.** $f(x) = \frac{3}{2}x + 1$

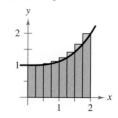

27. $f(x) = x^2 + 1$ **28.** $f(x) = 9 - x^2$

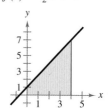

29. $f(x) = \frac{1}{9}x^3 + 3$ **30.** $f(x) = 3 - \frac{1}{4}x^3$

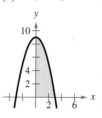 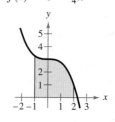

Finding the Area of a Region In Exercises 31–36, complete the table to show the approximate area of the region bounded by the graph of f and the x-axis over the given interval using the indicated numbers n of rectangles of equal width. Then find the exact area as $n \to \infty$.

n	4	8	20	50	100
Approximate Area					

Function	Interval
31. $f(x) = 3x + 4$	$[0, 4]$
32. $f(x) = 15 - x$	$[2, 6]$
33. $f(x) = 9 - x^2$	$[0, 2]$
34. $f(x) = x^2 + 1$	$[4, 6]$
35. $f(x) = \dfrac{1}{2}x + 4$	$[-1, 3]$
36. $f(x) = \dfrac{1}{2}x + 1$	$[-2, 2]$

 Finding the Area of a Region In Exercises 37–44, use the limit process to find the area of the region bounded by the graph of the function and the x-axis over the given interval.

Function	Interval
37. $f(x) = 5 - 2x$	$[0, 1]$
38. $f(x) = 2x + 7$	$[0, 2]$
39. $f(x) = -2x + 3$	$[0, 1]$
40. $f(x) = 3x - 4$	$[2, 4]$
41. $f(x) = 2 - x^2$	$[-1, 1]$
42. $f(x) = x^2 + 2$	$[0, 1]$
43. $f(x) = \dfrac{1}{4}(x^2 + 4x)$	$[1, 4]$
44. $f(x) = x^2 - x^3$	$[-1, 1]$

45. Land Surveying

A plot of land is modeled in the coordinate plane as the region bounded by the positive x- and y-axes, and the equation

$$y = (-3.0 \times 10^{-6})x^3 + 0.002x^2 - 1.05x + 400.$$

Use a graphing utility to graph the equation. Then, use the limit process to find the area of the plot of land. Assume all distances are in feet.

46. Land Surveying The table shows the measurements (in feet) of a plot of land bounded by a stream and two straight roads that meet at right angles (see figure).

x	0	50	100	150	200	250	300
y	450	362	305	268	245	156	0

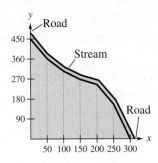

(a) Use the *regression* feature of a graphing utility to find a model of the form $y = ax^3 + bx^2 + cx + d$.

(b) Use the graphing utility to plot the data and graph the model in the same viewing window.

(c) Use the model in part (a) to estimate the area of the plot of land.

47. Writing Describe the process of finding the area of a region bounded by the graph of a nonnegative, continuous function f, the x-axis, and the vertical lines $x = a$ and $x = b$.

48. **HOW DO YOU SEE IT?** Determine which value best approximates the area of the region shown in the graph. (Make your selection on the basis of the sketch of the region and not by performing any calculations.)

(a) -2 (b) 1 (c) 4 (d) 6 (e) 9

Exploration

True or False? In Exercises 49 and 50, determine whether the statement is true or false. Justify your answer.

49. The sum of the first n positive integers is $[n(n + 1)]/2$.

50. The exact area of a plane region R can be found using the limit process to increase the number n of approximating rectangles without bound.

Chapter Summary

	What Did You Learn?	Explanation/Examples	Review Exercises
Section 12.1	Understand the limit concept (p. 818) and use the definition of a limit to estimate limits (p. 819).	If $f(x)$ becomes arbitrarily close to a unique number L as x approaches c from either side, then the limit of $f(x)$ as x approaches c is L. This is written as $\lim\limits_{x \to c} f(x) = L$.	1–4
	Determine whether limits of functions exist (p. 821).	**Conditions Under Which Limits Do Not Exist** The limit of $f(x)$ as $x \to c$ does not exist when any of the conditions listed below are true. **1.** $f(x)$ approaches a different number from the right side of c than it approaches from the left side of c. **2.** $f(x)$ increases or decreases without bound as x approaches c. **3.** $f(x)$ oscillates between two fixed values as x approaches c.	5–8
	Use properties of limits and direct substitution to evaluate limits (p. 823).	Let b and c be real numbers and let n be a positive integer. **1.** $\lim\limits_{x \to c} b = b$ **2.** $\lim\limits_{x \to c} x = c$ **3.** $\lim\limits_{x \to c} x^n = c^n$ **4.** $\lim\limits_{x \to c} \sqrt[n]{x} = \sqrt[n]{c}$, valid for all c when n is odd and valid for $c > 0$ when n is even **Properties of Limits** Let b and c be real numbers, let n be a positive integer, and let f and g be functions with the limits $\lim\limits_{x \to c} f(x) = L$ and $\lim\limits_{x \to c} g(x) = K$. **1.** $\lim\limits_{x \to c} [bf(x)] = bL$ **2.** $\lim\limits_{x \to c} [f(x) \pm g(x)] = L \pm K$ **3.** $\lim\limits_{x \to c} [f(x)g(x)] = LK$ **4.** $\lim\limits_{x \to c} \dfrac{f(x)}{g(x)} = \dfrac{L}{K},\ K \neq 0$ **5.** $\lim\limits_{x \to c} [f(x)]^n = L^n$	9–24
Section 12.2	Use the dividing out technique to evaluate limits of functions (p. 829).	When you attempt to evaluate a limit of a rational function by direct substitution, you may encounter the indeterminate form $\frac{0}{0}$. In this case, factor and divide out any common factors, then use direct substitution again. (See Examples 1 and 2.)	25–32
	Use the rationalizing technique to evaluate limits of functions (p. 831).	The rationalizing technique of finding the limit of a function involves rationalizing the numerator of the function. (See Example 3.)	33–36
	Use technology to approximate limits of functions numerically and graphically (p. 832).	The *table* feature or *zoom* and *trace* features of a graphing utility can be used to approximate limits. (See Examples 4 and 5.)	37–44
	Evaluate one-sided limits of functions (p. 833).	**Limit from the left:** $\lim\limits_{x \to c^-} f(x) = L_1$ or $f(x) \to L_1$ as $x \to c^-$ **Limit from the right:** $\lim\limits_{x \to c^+} f(x) = L_2$ or $f(x) \to L_2$ as $x \to c^+$	45–50
	Evaluate limits from calculus (p. 835).	For any x-value, the limit of a difference quotient is an expression of the form $\lim\limits_{h \to 0} \dfrac{f(x + h) - f(x)}{h}$.	51, 52

What Did You Learn?	Explanation/Examples	Review Exercises
Section 12.3 — Understand the tangent line problem *(p. 839)* and use a tangent line to approximate the slope of a graph at a point *(p. 840)*.	The tangent line to the graph of a function f at a point $P(x_1, y_1)$ is the line whose slope best approximates the slope of the graph at the point.	53–58
Use the limit definition of slope to find exact slopes of graphs *(p. 841)*.	**Definition of the Slope of a Graph** The slope m of the graph of f at the point $(x, f(x))$ is equal to the slope of its tangent line at $(x, f(x))$ and is given by $$m = \lim_{h \to 0} m_{\text{sec}} = \lim_{h \to 0} \frac{f(x + h) - f(x)}{h}$$ provided this limit exists.	59–62
Find derivatives of functions and use derivatives to find slopes of graphs *(p. 844)*.	The derivative of f at x is given by $$f'(x) = \lim_{h \to 0} \frac{f(x + h) - f(x)}{h}$$ provided this limit exists. The derivative $f'(x)$ is a formula for the slope of the tangent line to the graph of f at the point $(x, f(x))$.	63–76
Section 12.4 — Evaluate limits of functions at infinity *(p. 849)*.	If f is a function and L_1 and L_2 are real numbers, then the statements $\lim\limits_{x \to -\infty} f(x) = L_1$ and $\lim\limits_{x \to \infty} f(x) = L_2$ denote the limits at infinity.	77–86
Find limits of sequences *(p. 853)*.	**Limit of a Sequence** Let f be a function of a real variable such that $\lim\limits_{x \to \infty} f(x) = L$. If $\{a_n\}$ is a sequence such that $f(n) = a_n$ for every positive integer n, then $\lim\limits_{n \to \infty} a_n = L$.	87–92
Section 12.5 — Find limits of summations *(p. 858)*.	**Summation Formulas and Properties** **1.** $\sum_{i=1}^{n} c = cn$, c is a constant. **2.** $\sum_{i=1}^{n} i = \dfrac{n(n + 1)}{2}$ **3.** $\sum_{i=1}^{n} i^2 = \dfrac{n(n + 1)(2n + 1)}{6}$ **4.** $\sum_{i=1}^{n} i^3 = \dfrac{n^2(n + 1)^2}{4}$ **5.** $\sum_{i=1}^{n} (a_i \pm b_i) = \sum_{i=1}^{n} a_i \pm \sum_{i=1}^{n} b_i$ **6.** $\sum_{i=1}^{n} ka_i = k\sum_{i=1}^{n} a_i$, k is a constant.	93, 94
Use rectangles to approximate areas of plane regions *(p. 861)*.	A collection of rectangles of equal width can be used to approximate the area of a region. Increasing the number of rectangles gives a more accurate approximation. (See Example 4.)	95–98
Use limits of summations to find areas of plane regions *(p. 862)*.	**Area of a Plane Region** Let f be continuous and nonnegative on $[a, b]$. The area A of the region bounded by the graph of f, the x-axis, and the vertical lines $x = a$ and $x = b$ is given by $$A = \lim_{n \to \infty} \sum_{i=1}^{n} f\!\left(a + \frac{(b - a)i}{n}\right)\!\left(\frac{b - a}{n}\right).$$	99–105

Review Exercises See CalcChat.com for tutorial help and worked-out solutions to odd-numbered exercises.

12.1 **Estimating a Limit Numerically** In Exercises 1–4, complete the table and use the result to estimate the limit numerically. Determine whether it is possible to reach the limit.

1. $\lim\limits_{x \to 3} (6x - 1)$

x	2.9	2.99	2.999	3	3.001	3.01	3.1
$f(x)$				?			

2. $\lim\limits_{x \to 2} (x^2 - 3x + 1)$

x	1.9	1.99	1.999	2	2.001	2.01	2.1
$f(x)$				?			

3. $\lim\limits_{x \to 3} \dfrac{x - 3}{x^2 - x - 6}$

x	2.9	2.99	2.999	3	3.001	3.01	3.1
$f(x)$				?			

4. $\lim\limits_{x \to 0} \dfrac{\ln(1 - x)}{x}$

x	-0.1	-0.01	-0.001	0	0.001	0.01	0.1
$f(x)$				?			

Using a Graph to Find a Limit In Exercises 5–8, use the graph to find the limit, if it exists. If the limit does not exist, explain why.

5. $\lim\limits_{x \to 1} (3 - x)$

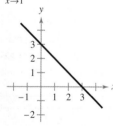

6. $\lim\limits_{x \to -1} (2x^2 + 1)$

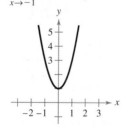

7. $\lim\limits_{x \to 1} \dfrac{x^2 - 1}{x - 1}$

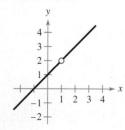

8. $\lim\limits_{x \to 2} \dfrac{1}{x - 2}$

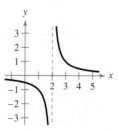

Evaluating Limits In Exercises 9 and 10, use the given information to evaluate each limit.

9. $\lim\limits_{x \to c} f(x) = 4, \ \lim\limits_{x \to c} g(x) = 5$

   (a) $\lim\limits_{x \to c} [f(x)]^3$ (b) $\lim\limits_{x \to c} [3f(x) - g(x)]$

   (c) $\lim\limits_{x \to c} [f(x)g(x)]$ (d) $\lim\limits_{x \to c} \dfrac{f(x)}{g(x)}$

10. $\lim\limits_{x \to c} f(x) = 27, \ \lim\limits_{x \to c} g(x) = 12$

   (a) $\lim\limits_{x \to c} \sqrt[3]{f(x)}$ (b) $\lim\limits_{x \to c} \dfrac{f(x)}{18}$

   (c) $\lim\limits_{x \to c} [f(x)g(x)]$ (d) $\lim\limits_{x \to c} [f(x) - 2g(x)]$

Evaluating a Limit by Direct Substitution In Exercises 11–24, find the limit by direct substitution.

11. $\lim\limits_{x \to 3} (3x - 1)$ **12.** $\lim\limits_{x \to 4} \left(\dfrac{1}{2}x + 3\right)$

13. $\lim\limits_{x \to -2} (x^2 - 4x + 12)$ **14.** $\lim\limits_{x \to 2} (5x - 3)(3x + 5)$

15. $\lim\limits_{x \to 9} \log_9 x$ **16.** $\lim\limits_{x \to 1} 2e^x$

17. $\lim\limits_{x \to -1} \sqrt{5 - x}$ **18.** $\lim\limits_{x \to -2} \sqrt[3]{4x}$

19. $\lim\limits_{x \to 2} \dfrac{3x + 5}{5x - 3}$ **20.** $\lim\limits_{x \to 2} \dfrac{x^2 - 1}{x^3 + 2}$

21. $\lim\limits_{x \to \pi} \sin 3x$ **22.** $\lim\limits_{x \to 0} \tan x$

23. $\lim\limits_{x \to 0} \arctan x$ **24.** $\lim\limits_{x \to 1/2} \arccos x$

12.2 **Finding a Limit** In Exercises 25–36, find the limit algebraically, if it exists. Use a graphing utility to verify your result graphically.

25. $\lim\limits_{x \to 4} \dfrac{x^2 - 16}{x - 4}$ **26.** $\lim\limits_{x \to 7} \dfrac{7 - x}{x^2 - 49}$

27. $\lim\limits_{x \to 5} \dfrac{x - 5}{x^2 + 5x - 50}$ **28.** $\lim\limits_{x \to -1} \dfrac{x + 1}{x^2 - 5x - 6}$

29. $\lim\limits_{x \to -2} \dfrac{x^2 - 4}{x^3 + 8}$ **30.** $\lim\limits_{t \to -3} \dfrac{t^3 + 27}{t + 3}$

31. $\lim\limits_{x \to -1} \dfrac{\dfrac{1}{x + 2} - 1}{x + 1}$ **32.** $\lim\limits_{x \to 0} \dfrac{\dfrac{2}{x + 2} - 1}{x}$

33. $\lim\limits_{u \to 0} \dfrac{\sqrt{16 - u} - 4}{u}$ **34.** $\lim\limits_{v \to 0} \dfrac{\sqrt{v + 9} - 3}{v}$

35. $\lim\limits_{x \to 5} \dfrac{\sqrt{x - 1} - 2}{x - 5}$

36. $\lim\limits_{x \to 1} \dfrac{\sqrt{3} - \sqrt{x + 2}}{1 - x}$

Using Different Methods **In Exercises 37–44,** (a) graphically approximate the limit (if it exists) by using a graphing utility to graph the function, and (b) numerically approximate the limit (if it exists) by using the *table* feature of the graphing utility to create a table.

37. $\lim\limits_{x\to3} \dfrac{x-3}{x^2-9}$ **38.** $\lim\limits_{x\to-4} \dfrac{x^4-3x^2-4}{x^4-15x^2-16}$

39. $\lim\limits_{x\to0} e^{-2/x}$ **40.** $\lim\limits_{x\to0} e^{-4/x^2}$

41. $\lim\limits_{x\to0} \dfrac{\sin 4x}{2x}$ **42.** $\lim\limits_{x\to0} \dfrac{\tan 2x}{x}$

43. $\lim\limits_{x\to1^+} \dfrac{\sqrt{2x+1}-\sqrt{3}}{x-1}$ **44.** $\lim\limits_{x\to1^+} \dfrac{1-\sqrt{x}}{x-1}$

Evaluating One-Sided Limits **In Exercises 45–50,** graph the function. Find the limit (if it exists) by evaluating the corresponding one-sided limits.

45. $\lim\limits_{x\to4} \dfrac{|x|}{x-4}$ **46.** $\lim\limits_{x\to-2} \dfrac{|x+2|}{x+2}$

47. $\lim\limits_{x\to2} \dfrac{2}{x^2-4}$ **48.** $\lim\limits_{x\to-3} \dfrac{1}{x^2+9}$

49. $\lim\limits_{x\to2} f(x)$ where $f(x) = \begin{cases} 2-x, & x \le 2 \\ x^2-3, & x > 2 \end{cases}$

50. $\lim\limits_{x\to1} f(x)$ where $f(x) = \begin{cases} -x^2-4, & x < 1 \\ x-6, & x \ge 1 \end{cases}$

Evaluating a Limit from Calculus **In Exercises 51 and 52, find**

$$\lim\limits_{h\to0} \dfrac{f(x+h)-f(x)}{h}.$$

51. $f(x) = -5x + 2$ **52.** $f(x) = x^2 + 3x - 2$

12.3 **Approximating the Slope of a Graph** **In Exercises 53 and 54, use the figure to approximate the slope of the graph at the point (x, y).**

53. 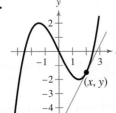   **54.**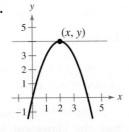

Approximating the Slope of a Tangent Line **In Exercises 55–58, sketch a graph of the function and the tangent line at the point $(2, f(2))$. Use the graph to approximate the slope of the tangent line.**

55. $f(x) = x^2 - 2x$ **56.** $f(x) = \sqrt{x+2}$

57. $f(x) = \dfrac{6}{x-4}$ **58.** $f(x) = \dfrac{1}{3-x}$

Finding a Formula for the Slope of a Graph **In Exercises 59–62, find a formula for the slope of the graph of f at the point $(x, f(x))$. Then use it to find the slope at the two given points.**

59. $f(x) = x^2 - 4x$
  (a) $(0, 0)$ (b) $(2, -4)$

60. $f(x) = \dfrac{1}{4}x^4$
  (a) $(-2, 4)$ (b) $\left(1, \dfrac{1}{4}\right)$

61. $f(x) = \dfrac{4}{x-6}$
  (a) $(7, 4)$ (b) $(8, 2)$

62. $f(x) = \sqrt{x+8}$
  (a) $(1, 3)$ (b) $(8, 4)$

Finding a Derivative **In Exercises 63–74, find the derivative of the function.**

63. $f(x) = 5$ **64.** $g(x) = -3$

65. $h(x) = 5 - \dfrac{1}{2}x$ **66.** $f(x) = 12x$

67. $g(x) = 4x^2 - x$ **68.** $f(x) = -x^3 + 4x$

69. $f(t) = \sqrt{t+5}$ **70.** $g(t) = \sqrt{t-3}$

71. $g(s) = \dfrac{4}{s+5}$ **72.** $g(t) = \dfrac{6}{5-t}$

73. $g(x) = \dfrac{1}{\sqrt{x+4}}$ **74.** $f(x) = \dfrac{1}{\sqrt{12-x}}$

Using the Derivative **In Exercises 75 and 76, (a) find the slope of the graph of f at the given point, (b) find an equation of the tangent line to the graph at the point, and (c) graph the function and the tangent line.**

75. $f(x) = 2x^2 - 1,\ (0, -1)$
76. $f(x) = x^2 + 10,\ (2, 14)$

12.4 **Evaluating a Limit at Infinity** **In Exercises 77–86, find the limit, if it exists. If the limit does not exist, explain why. Use a graphing utility to verify your result graphically.**

77. $\lim\limits_{x\to\infty} \dfrac{4x}{2x-3}$ **78.** $\lim\limits_{x\to\infty} \dfrac{7x}{14x+2}$

79. $\lim\limits_{x\to\infty} \left(3 - \dfrac{1}{x^2}\right)$ **80.** $\lim\limits_{x\to\infty} \left(7 + \dfrac{3}{2x^2}\right)$

81. $\lim\limits_{x\to-\infty} \dfrac{2x}{x^2-25}$ **82.** $\lim\limits_{x\to-\infty} \dfrac{3x}{(1-x)^3}$

83. $\lim\limits_{x\to\infty} \dfrac{x^2}{2x+3}$ **84.** $\lim\limits_{y\to\infty} \dfrac{3y^4}{y^2+1}$

85. $\lim\limits_{x\to\infty} \left[\dfrac{x}{(x-2)^2} + 3\right]$ **86.** $\lim\limits_{x\to\infty} \left[2 - \dfrac{2x^2}{(x+1)^2}\right]$

Finding the Limit of a Sequence In Exercises 87–92, write the first five terms of the sequence and find the limit of the sequence, if it exists. If the limit does not exist, then explain why. (Assume n begins with 1.)

87. $a_n = \dfrac{2n - 1}{5n + 2}$

88. $a_n = \dfrac{n}{n^2 + 1}$

89. $a_n = \dfrac{(-1)^n}{n^3}$

90. $a_n = \dfrac{(-1)^{n+1}}{n}$

91. $a_n = \dfrac{n^2}{3n + 2}$

92. $a_n = \dfrac{1}{2n^2}[3 - 2n(n + 1)]$

12.5 **Finding the Limit of a Summation** In Exercises 93 and 94, (a) rewrite the sum as a rational function $S(n)$, (b) use $S(n)$ to complete the table, and (c) find $\lim\limits_{n \to \infty} S(n)$.

n	10^0	10^1	10^2	10^3	10^4
$S(n)$					

93. $\displaystyle\sum_{i=1}^{n} \left(\dfrac{4i^2}{n^2} - \dfrac{i}{n} \right)\left(\dfrac{1}{n} \right)$

94. $\displaystyle\sum_{i=1}^{n} \left[4 - \left(\dfrac{3i}{n} \right)^2 \right]\left(\dfrac{3i}{n^2} \right)$

Approximating the Area of a Region In Exercises 95 and 96, approximate the area of the region using the rectangles of equal width shown.

95. $f(x) = 4 - x$

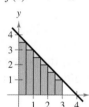

96. $f(x) = 4 - x^2$

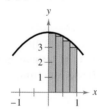

Approximating the Area of a Region In Exercises 97 and 98, complete the table to show the approximate area of the region using the indicated numbers n of rectangles of equal width.

n	4	8	20	50
Approximate Area				

97. $f(x) = \frac{1}{4}x^2 + 2$

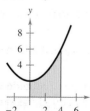

98. $f(x) = 4x - x^2$

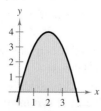

Finding the Area of a Region In Exercises 99–104, use the limit process to find the area of the region bounded by the graph of the function and the x-axis over the specified interval.

Function	Interval
99. $f(x) = 10 - x$	$[0, 4]$
100. $f(x) = 2x - 6$	$[3, 6]$
101. $f(x) = x^2 + 4$	$[-1, 2]$
102. $f(x) = 8(x - x^2)$	$[0, 1]$
103. $f(x) = x^3 + 1$	$[1, 2]$
104. $f(x) = 4 - (x - 2)^2$	$[0, 4]$

105. **Land Surveying** The table shows the measurements (in feet) of a plot of land bounded by a stream and two straight roads that meet at right angles (see figure).

x	0	100	200	300	400	500
y	125	125	120	112	90	90

x	600	700	800	900	1000
y	95	88	75	35	0

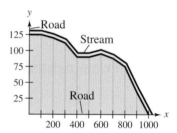

(a) Use the *regression* feature of a graphing utility to find a model of the form $y = ax^3 + bx^2 + cx + d$.

(b) Use the graphing utility to plot the data and graph the model in the same viewing window.

(c) Use the model in part (a) to estimate the area of the plot of land.

Exploration

106. **Writing** List several reasons why the limit of a function may not exist.

True or False? In Exercises 107 and 108, determine whether the statement is true or false. Justify your answer.

107. The limit of the sum of two functions is the sum of the limits of the two functions.

108. If the degree of the numerator $N(x)$ of a rational function $f(x) = N(x)/D(x)$ is greater than the degree of its denominator $D(x)$, then the limit of the rational function as x approaches ∞ is 0.

Chapter Test

See CalcChat.com for tutorial help and worked-out solutions to odd-numbered exercises.

Take this test as you would take a test in class. When you are finished, check your work against the answers given in the back of the book.

In Exercises 1–3, sketch a graph of the function and approximate the limit, if it exists. Then find the limit (if it exists) algebraically by using appropriate technique(s).

1. $\lim\limits_{x\to3} x^2 - 3x - 4$

2. $\lim\limits_{x\to1} \dfrac{-x^2 + 5x - 3}{1 - x}$

3. $\lim\limits_{x\to5} \dfrac{\sqrt{x} - 2}{x - 5}$

 In Exercises 4 and 5, use a graphing utility to graph the function and approximate the limit. Write an approximation that is accurate to four decimal places. Then create a table of values for the function and use the result to verify your approximation numerically.

4. $\lim\limits_{x\to0} \dfrac{\sin 3x}{x}$

5. $\lim\limits_{x\to0} \dfrac{e^{2x} - 1}{x}$

6. Find a formula for the slope of the graph of f at the point $(x, f(x))$. Then use the formula to find the slope at the given point.

   (a) $f(x) = 4x^2 - x + 12$, $(3, 2)$ (b) $f(x) = 2x^3 + 6x$, $(-1, -8)$

In Exercises 7–9, find the derivative of the function.

7. $f(x) = 9$

8. $f(x) = 2x^2 + 4x - 1$

9. $f(x) = \dfrac{1}{x + 3}$

In Exercises 10–12, find the limit, if it exists. If the limit does not exist, explain why. Use a graphing utility to verify your result graphically.

10. $\lim\limits_{x\to\infty} \dfrac{6}{5x - 1}$

11. $\lim\limits_{x\to\infty} \dfrac{1 - 3x^2}{x^2 - 5}$

12. $\lim\limits_{x\to-\infty} \dfrac{x^2}{3x + 2}$

In Exercises 13 and 14, write the first five terms of the sequence and find the limit of the sequence, if it exists. If the limit does not exist, explain why. (Assume n begins with 1.)

13. $a_n = \dfrac{n^2 + 3n - 4}{2n^2 + n - 2}$

14. $a_n = \dfrac{1 + (-1)^n}{n}$

15. Approximate the area of the region bounded by the graph of $f(x) = 8 - 2x^2$ shown at the left using the rectangles of equal width shown.

In Exercises 16 and 17, use the limit process to find the area of the region bounded by the graph of the function and the x-axis over the given interval.

16. $f(x) = x + 2$; interval: $[-2, 2]$

17. $f(x) = 5 - x^2$; interval: $[0, 2]$

18. The table shows the altitude y (in feet) of a space shuttle during its first 5 seconds of motion.

    (a) Use the *regression* feature of a graphing utility to find a quadratic model $y = ax^2 + bx + c$ for the data.

   (b) The value of the derivative of the model is the rate of change of altitude with respect to time, or the velocity, at that instant. Find the velocity of the shuttle at $x = 5$ seconds.

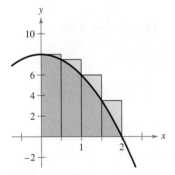

Figure for 15

DATA Time, x	Altitude, y
0	0
1	1
2	23
3	60
4	115
5	188

Table for 18

Spreadsheet at LarsonPrecalculus.com

Cumulative Test for Chapters 10–12

See CalcChat.com for tutorial help and worked-out solutions to odd-numbered exercises.

Take this test as you would take a test in class. When you are finished, check your work against the answers given in the back of the book.

1. Find the angle θ (in radians and degrees) between

$$x - 3y = -1 \quad \text{and} \quad -8x + 2y = 12.$$

In Exercises 2 and 3, identify the conic represented by the equation and sketch its graph.

2. $(x - 2)^2 + 8(y + 5) = 0$ 3. $y^2 - x^2 - 2x - 4y + 1 = 0$

4. Find the standard form of the equation of the ellipse with vertices $(0, 0)$ and $(0, 4)$ and endpoints of the minor axis $(1, 2)$ and $(-1, 2)$.

5. Rotate the axis to eliminate the xy-term in the equation $x^2 + 6xy + y^2 - 6 = 0$. Then write the equation in standard form. Sketch the graph of the resulting equation, showing both sets of axes.

6. Sketch the curve represented by the parametric equations $x = 4 \ln t$ and $y = \frac{1}{2}t^2$. Eliminate the parameter and write the resulting rectangular equation.

7. Plot the point $(-2, -3\pi/4)$ (given in polar coordinates) and find three additional polar representations of the point, using $-2\pi < \theta < 2\pi$.

8. Convert the rectangular equation $-8x - 3y + 5 = 0$ to polar form.

9. Convert the polar equation $r = \dfrac{2}{4 - 5 \cos \theta}$ to rectangular form.

In Exercises 10–12, identify the type of graph represented by the polar equation. Then sketch the graph.

10. $r = -\dfrac{\pi}{6}$ 11. $r = \dfrac{9}{3 - 2 \sin \theta}$ 12. $r = 2 + 5 \cos \theta$

In Exercises 13 and 14, find the coordinates of the point.

13. The point is located six units behind the yz-plane, one unit to the right of the xz-plane, and three units above the xy-plane.

14. The point is located on the y-axis, four units to the left of the xz-plane.

15. Find the distance between $(-2, 3, -6)$ and $(4, -5, 1)$.

16. Find the lengths of the sides of the right triangle shown at the left. Show that these lengths satisfy the Pythagorean Theorem.

17. Find the midpoint of the line segment joining $(-6, 1, 4)$ and $(2, -1, 9)$.

18. Find the standard form of the equation of the sphere for which the endpoints of a diameter are $(0, 0, 0)$ and $(4, 4, 8)$.

19. Sketch the graph of the equation $(x - 2)^2 + (y + 1)^2 + z^2 = 4$, the xy-trace, and the yz-trace.

20. For the vectors $\mathbf{u} = \langle 5, -1, -2 \rangle$ and $\mathbf{v} = \langle 0, -3, 4 \rangle$, find $\mathbf{u} \cdot \mathbf{v}$ and $\mathbf{u} \times \mathbf{v}$.

21. Find the area of a triangle with vertices $A(1, 0, 4)$, $B(-2, -3, 1)$, and $C(4, -1, 5)$.

In Exercises 22–24, determine whether u and v are orthogonal, parallel, or neither.

22. $\mathbf{u} = \langle 4, 4, 0 \rangle, \quad \mathbf{v} = \langle 0, -8, 6 \rangle$

23. $\mathbf{u} = \langle 4, -2, 10 \rangle, \quad \mathbf{v} = \langle -2, 6, 2 \rangle$

24. $\mathbf{u} = \langle -1, 6, -3 \rangle, \quad \mathbf{v} = \langle 3, -18, 9 \rangle$

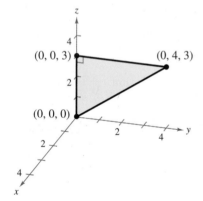

Figure for 16

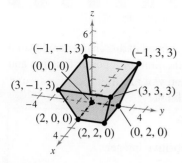

Figure for 29

25. Find sets of (a) parametric equations and (b) symmetric equations of the line passing through the point $(-1, 2, 0)$ and perpendicular to $2x - 4y + z = 8$.

26. Find the general form of the equation of the plane passing through the points $(0, 0, 0)$, $(-2, 3, 0)$, and $(5, 8, 25)$.

27. Plot the intercepts and sketch a graph of the plane $3x - 6y - 12z = 24$.

28. Find the distance between the point $(0, 0, 25)$ and the plane $2x - 5y + z = 10$.

29. The figure at the left shows the shape and dimensions of a plastic wastebasket. In fabricating a mold for making the wastebasket, it is necessary to know the angle between two adjacent sides. Find the angle.

In Exercises 30–35, find the limit, if it exists. If the limit does not exist, explain why. Use a graphing utility to verify your result graphically.

30. $\lim\limits_{x \to 0} \dfrac{\sqrt{x + 4} - 2}{x}$

31. $\lim\limits_{x \to 4^-} \dfrac{|x - 4|}{x - 4}$

32. $\lim\limits_{x \to 0} \sin \dfrac{\pi}{x}$

33. $\lim\limits_{x \to 0} \dfrac{\dfrac{1}{x - 3} + \dfrac{1}{3}}{x}$

34. $\lim\limits_{x \to 0} \dfrac{\sqrt{x + 16} - 4}{x}$

35. $\lim\limits_{x \to 2^-} \dfrac{x - 2}{x^2 - 4}$

In Exercises 36–38, find a formula for the slope of the graph of f at the point $(x, f(x))$. Then use the formula to find the slope at the given point.

36. $f(x) = \sqrt{x + 3}, \quad (-2, 1)$

37. $f(x) = \dfrac{1}{x + 3}, \quad \left(1, \dfrac{1}{4}\right)$

38. $f(x) = x^2 - 2x, \quad (2, 0)$

In Exercises 39–42, find the limit, if it exists. If the limit does not exist, explain why. Use a graphing utility to verify your result graphically.

39. $\lim\limits_{x \to \infty} \dfrac{2x^4 - x^3 + 4}{x^2 - 9}$

40. $\lim\limits_{x \to \infty} \dfrac{3 - 7x}{x + 4}$

41. $\lim\limits_{x \to \infty} \dfrac{3x^2 + 1}{x^2 + 4}$

42. $\lim\limits_{x \to \infty} \dfrac{3 - x}{x^2 + 1}$

In Exercises 43 and 44, evaluate the sum using the summation formulas and properties.

43. $\sum\limits_{k=1}^{20} (2k^2 - 7k)$

44. $\sum\limits_{i=1}^{40} (12 + i^3)$

In Exercises 45 and 46, approximate the area of the region using the rectangles of equal width shown.

45.

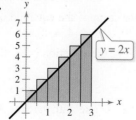

46.

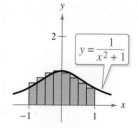

In Exercises 47 and 48, use the limit process to find the area of the region bounded by the graph of the function and the x-axis over the given interval.

47. $f(x) = x + 2$

Interval: $[0, 1]$

48. $f(x) = 4 - x^2$

Interval: $[0, 2]$

Proofs in Mathematics ■ ■ ■ ■ ■ ■ ■ ■ ■ ■ ■ ■ ■ ■ ■

Many of the proofs of the definitions and properties presented in this chapter are beyond the scope of this text. Included below are proofs for the limit of a power function and the limit of a polynomial function.

Limit of a Power Function   *(p. 823)*

$\lim\limits_{x \to c} x^n = c^n$, where c is a real number and n is a positive integer.

PROVING LIMITS

To prove most of the definitions and properties in this chapter, you must use the *formal* definition of a limit. This definition is called the *epsilon-delta definition* and was first introduced by German mathematician Karl Weierstrass (1815–1897). When you take a course in calculus, you will use this definition of a limit extensively.

Proof

$$\lim_{x \to c} x^n = \lim_{x \to c} (\underbrace{x \cdot x \cdot x \cdots \cdot x}_{n \text{ factors}})$$

$$= \underbrace{\lim_{x \to c} x \cdot \lim_{x \to c} x \cdot \lim_{x \to c} x \cdots \cdot \lim_{x \to c} x}_{n \text{ factors}} \qquad \text{Product Property}$$

$$= \underbrace{c \cdot c \cdot c \cdots \cdot c}_{n \text{ factors}} \qquad \text{Limit of the identity function}$$

$$= c^n \qquad \text{Exponential form} \qquad ■$$

Limit of a Polynomial Function   *(p. 825)*

If p is a polynomial function and c is a real number, then

$$\lim_{x \to c} p(x) = p(c).$$

Proof

Let p be a polynomial function such that

$$p(x) = a_n x^n + a_{n-1} x^{n-1} + \cdots + a_2 x^2 + a_1 x + a_0.$$

A polynomial function is the sum of monomial functions, so you can write

$$\lim_{x \to c} p(x) = \lim_{x \to c} (a_n x^n + a_{n-1} x^{n-1} + \cdots + a_2 x^2 + a_1 x + a_0)$$

$$= \lim_{x \to c} a_n x^n + \lim_{x \to c} a_{n-1} x^{n-1} + \cdots + \lim_{x \to c} a_2 x^2 + \lim_{x \to c} a_1 x + \lim_{x \to c} a_0$$

$$= a_n c^n + a_{n-1} c^{n-1} + \cdots + a_2 c^2 + a_1 c + a_0 \qquad \text{Scalar Multiple Property, and limits of power, identity, and constant functions}$$

$$= p(c). \qquad p \text{ evaluated at } c \qquad ■$$

P.S. Problem Solving

1. **Graphical Reasoning** For each condition of the function f, determine which graphs could be the graph of f. [The graphs are labeled (i), (ii), (iii), and (iv).]

   (a) $\lim\limits_{x \to 2} f(x) = 3$ (b) $\lim\limits_{x \to 2^-} f(x) = 3$

   (c) $\lim\limits_{x \to 2^+} f(x) = 3$

   (i)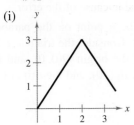
   (ii)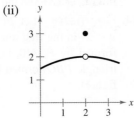

   (iii)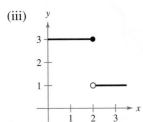
   (iv)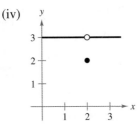

2. **Tangent Lines to a Circle** Let $P(3, 4)$ be a point on the circle $x^2 + y^2 = 25$. (See figure.)

   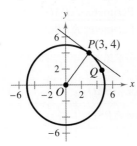

   (a) What is the slope of the line joining P and $O(0, 0)$?

   (b) Find an equation of the tangent line to the circle at P.

   (c) Let $Q(x, y)$ be another point on the circle in the first quadrant. (See figure.) Find the slope m_x of the line joining P and Q in terms of x.

   (d) Evaluate $\lim\limits_{x \to 3} m_x$. How does this number relate to your answer in part (b)?

3. **Finding a Function** Find a function of the form $f(x) = a + b\sqrt{x}$ that is tangent to the line $2y - 3x = 5$ at the point $(1, 4)$.

4. **Evaluating Limits**

   (a) Sketch the graph of the function $f(x) = [\![x]\!] + [\![-x]\!]$.

   (b) Evaluate $f(1), f(0), f\!\left(\frac{1}{2}\right)$, and $f(-2.7)$.

   (c) Evaluate each limit.

   $$\lim_{x \to 1^-} f(x), \quad \lim_{x \to 1^+} f(x), \quad \lim_{x \to 1/2} f(x)$$

5. **Evaluating Limits**

   (a) Sketch the graph of the function $f(x) = \left[\!\!\left[\dfrac{1}{x}\right]\!\!\right]$.

   (b) Evaluate $f\!\left(\frac{1}{4}\right), f(3)$, and $f(1)$.

   (c) Evaluate each limit.

   $$\lim_{x \to 1^-} f(x), \quad \lim_{x \to 1^+} f(x), \quad \lim_{x \to (1/2)^-} f(x), \quad \lim_{x \to (1/2)^+} f(x)$$

6. **Finding Values of Constants** Find the values of the constants a and b such that

   $$\lim_{x \to 0} \frac{\sqrt{a + bx} - \sqrt{3}}{x} = \sqrt{3}.$$

7. **Evaluating Limits** Consider the function

   $$f(x) = \frac{\sqrt{3 + x^{1/3}} - 2}{x - 1}.$$

   (a) Find the domain of f.

   (b) Use a graphing utility to graph the function.

   (c) Evaluate $\lim\limits_{x \to -27^+} f(x)$. Verify your result using the graph in part (b).

   (d) Evaluate $\lim\limits_{x \to 1} f(x)$. Verify your result using the graph in part (b).

8. **Limits of Piecewise-Defined Functions** Let

   $$f(x) = \begin{cases} 0, & \text{when } x \text{ is rational} \\ 1, & \text{when } x \text{ is irrational} \end{cases}$$

   and

   $$g(x) = \begin{cases} 0, & \text{when } x \text{ is rational} \\ x, & \text{when } x \text{ is irrational} \end{cases}.$$

   Find $\lim\limits_{x \to 0} f(x)$ and $\lim\limits_{x \to 0} g(x)$, if they exist. If the limits do not exist, explain why.

9. **Two Lines Tangent to Two Parabolas** Graph the two parabolas

   $$y = x^2 \quad \text{and} \quad y = -x^2 + 2x - 5$$

   in the same coordinate plane. Find equations of the two lines that are each simultaneously tangent to both parabolas.

10. **Tangent Lines and Normal Lines**

    (a) Find an equation of the tangent line to the parabola $y = x^2$ at the point $(2, 4)$.

    (b) Find an equation of the normal line to $y = x^2$ at the point $(2, 4)$. (The **normal line** is perpendicular to the tangent line.) Where does this line intersect the parabola other than at $(2, 4)$?

    (c) Find equations of the tangent line and normal line to $y = x^2$ at the point $(0, 0)$.

11. Limits of a Distance   A line with slope m passes through the point $(0, 4)$.

(a) Recall that the distance d between a point (x_1, y_1) and the line $Ax + By + C = 0$ is given by

$$d = \frac{|Ax_1 + By_1 + C|}{\sqrt{A^2 + B^2}}.$$

Write the distance d between the line and the point $(3, 1)$ as a function of m.

(b) Use a graphing utility to graph the function from part (a).

(c) Find $\lim\limits_{m \to \infty} d(m)$ and $\lim\limits_{m \to -\infty} d(m)$. Give a geometric interpretation of the results.

12. Temperature   The heat exchanger of a heating system has a probe attached to it. The probe records the temperature T (in degrees Celsius) t seconds after the furnace is started. The table shows the results for the first 2 minutes.

DATA	t	T
	0	25.2°
	15	36.9°
	30	45.5°
	45	51.4°
	60	56.0°
	75	59.6°
	90	62.0°
	105	64.0°
	120	65.2°

Spreadsheet at LarsonPrecalculus.com

(a) Use the *regression* feature of a graphing utility to find a model of the form

$$T_1 = at^2 + bt + c$$

for the data.

(b) Use the graphing utility to graph T_1 with the original data. How well does the model fit the data?

(c) A rational model for the data is given by

$$T_2 = \frac{86t + 1451}{t + 58}.$$

Use the graphing utility to graph T_2 with the original data. How well does the model fit the data?

(d) Evaluate $T_1(0)$ and $T_2(0)$.

(e) Find $\lim\limits_{t \to \infty} T_2$. Verify your result using the graph in part (c).

(f) Interpret the result of part (e) in the context of the problem. Is it possible to perform this type of analysis using T_1? Explain.

13. Error Analysis   When using a graphing utility to generate a table to approximate

$$\lim_{x \to 0} \frac{\tan 2x}{x}$$

a student concludes that the limit is 0.03491 rather than 2. Determine the probable cause of the error.

14. Geometry   Let $P(x, y)$ be a point on the parabola $y = x^2$ in the first quadrant. Consider the triangle PAO formed by P, $A(0, 1)$, and the origin $O(0, 0)$, and the triangle PBO formed by P, $B(1, 0)$, and the origin (see figure).

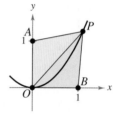

(a) Write the perimeter of each triangle in terms of x.

(b) Complete the table. Let $r(x)$ be the ratio of the perimeters of the two triangles.

$$r(x) = \frac{\text{Perimeter } \triangle PAO}{\text{Perimeter } \triangle PBO}$$

x	4	2	1	0.1	0.01
Perimeter $\triangle PAO$					
Perimeter $\triangle PBO$					
$r(x)$					

(c) Find $\lim\limits_{x \to 0^+} r(x)$.

15. Parabolic Arch   Archimedes showed that the area of a parabolic arch is equal to $\frac{2}{3}$ of the product of the base and the height (see figure).

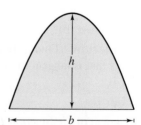

(a) Graph the parabolic arch bounded by $y = 9 - x^2$ and the x-axis.

(b) Use the limit process to find the area of the parabolic arch.

(c) Find the base and height of the arch and verify Archimedes' formula.

13 Concepts in Statistics

Hardware Retailer *(page 899)*

Product Lifetime
(Exercise 42, page 895)

Seal Sanctuary *(Example 11, page 893)*

Gasoline Prices
(Exercise 17, page 884)

Retirement Contributions *(page 882)*

Line Plots

A **line plot** uses a portion of a real number line to order numbers. Line plots are especially useful for ordering small sets of numbers (about 50 or less) by hand.

EXAMPLE 2 **Constructing a Line Plot**

Use a line plot to organize the test scores listed below. Which score occurs with the greatest frequency? *(Spreadsheet at LarsonPrecalculus.com)*

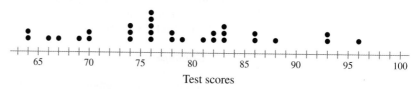

DATA 93, 70, 76, 67, 86, 93, 82, 78, 83, 86, 64, 78, 76, 66, 83, 83, 96, 74, 69, 76, 64, 74, 79, 76, 88, 76, 81, 82, 74, 70

Solution Begin by determining the least and greatest data values. For this data set, the least value is 64 and the greatest is 96. Next, draw a portion of a real number line that includes the interval [64, 96]. To create the line plot, start with the first number, 93, and enter a ● above 93 on the number line. Continue recording ●'s for the numbers in the list until you obtain the line plot below. The line plot shows that 76 occurs with the greatest frequency.

Test scores

✓ *Checkpoint* 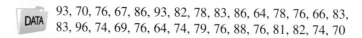 *Audio-video solution in English & Spanish at LarsonPrecalculus.com*

Use a line plot to organize the test scores listed below. Which score occurs with the greatest frequency? *(Spreadsheet at LarsonPrecalculus.com)*

DATA 68, 73, 67, 95, 71, 82, 85, 74, 82, 61, 87, 92, 78, 74, 64, 71, 74, 82, 71, 83, 92, 82, 78, 72, 82, 64, 85, 67, 71, 62

EXAMPLE 3 **Analyzing a Line Plot**

The line plot shows the daily high temperatures (in degrees Fahrenheit) in a city during the month of June.

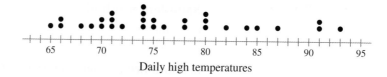

Daily high temperatures

a. What is the range of daily high temperatures?

b. On how many days was the high temperature in the 80s?

Solution

a. The line plot shows that the maximum daily high temperature was 93°F and the minimum daily high temperature was 65°F. So, the range is 93 − 65 = 28°F.

b. There are 7 ●'s in the interval [80, 90). So, the high temperature was in the 80s on 7 days.

✓ *Checkpoint*  *Audio-video solution in English & Spanish at LarsonPrecalculus.com*

In Example 3, on how many days was the high temperature less than 80°F? ◼

13.1 Exercises

See **CalcChat.com** for tutorial help and worked-out solutions to odd-numbered exercises.

Vocabulary: Fill in the blanks.

1. _____ is the branch of mathematics that studies techniques for collecting, organizing, and interpreting information.

2. _____ are collections of *all* of the outcomes, measurements, counts, or responses that are of interest.

3. _____ are subsets, or parts, of a population.

4. _____ statistics involves organizing, summarizing, and displaying data, whereas _____ statistics involves using a sample to draw conclusions about a population.

5. _____ data consist of numerical values, whereas _____ data consist of attributes, labels, or nonnumerical values.

6. In a _____ sample, every member of the population has an equal chance of being selected.

7. In a _____ _____ sample, researchers divide the population into distinct groups and select members at random from each group.

8. In a _____ sample, researchers only select members of the population who are easily accessible.

9. In a _____ sample, members of the population select themselves by volunteering.

10. A _____ has a portion of a real number line as its horizontal axis, and the bars are not separated by spaces.

Skills and Applications

 Venn Diagrams In Exercises 11 and 12, use the Venn diagram to determine the population and the sample.

11.
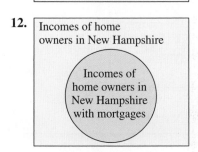

Students donating at a blood drive

Students donating who have type O blood

12.

Incomes of home owners in New Hampshire

Incomes of home owners in New Hampshire with mortgages

13. **Financial Support** In a survey of 750 parents, 31% plan to financially support their children through college graduation and 6% plan to financially support their children through the start of college. *(Source: Yahoo Financial)*

   (a) Which part of the study represents descriptive statistics?

   (b) Use inferential statistics to draw a conclusion from the study.

14. **Superstition** In a sample of 800 adults, 16% said that they are superstitious. *(Rasmussen Reports)*

   (a) Which part of the study represents descriptive statistics?

   (b) Use inferential statistics to draw a conclusion from the study.

 Quiz and Exam Scores In Exercises 15 and 16, use the given scores from a math class of 30 students. The scores are for two 25-point quizzes and two 100-point exams. *(Spreadsheet at LarsonPrecalculus.com)*

 Quiz #1 20, 15, 14, 20, 16, 19, 10, 21, 24, 15, 15, 14, 15, 21, 19, 15, 20, 18, 18, 22, 18, 16, 18, 19, 21, 19, 16, 20, 14, 12

Quiz #2 22, 22, 23, 22, 21, 24, 22, 19, 21, 23, 23, 25, 24, 22, 22, 23, 23, 23, 23, 22, 24, 23, 22, 24, 21, 24, 16, 21, 16, 14

Exam #1 77, 100, 77, 70, 83, 89, 87, 85, 81, 84, 81, 78, 89, 78, 88, 85, 90, 92, 75, 81, 85, 100, 98, 81, 78, 75, 85, 89, 82, 75

Exam #2 76, 78, 73, 59, 70, 81, 71, 66, 66, 73, 68, 67, 63, 67, 77, 84, 87, 71, 78, 78, 90, 80, 77, 70, 80, 64, 74, 68, 68, 68

15. Construct a line plot for each quiz. For each quiz, which score occurred with the greatest frequency?

16. Construct a line plot for each exam. For each exam, which score occurred with the greatest frequency?

17. Gasoline Prices

The line plot shows a sample of prices per gallon of unleaded regular gasoline from 25 different cities.

(a) In how many cities was the price of gasoline less than $2.80?

(b) What is the range of prices?

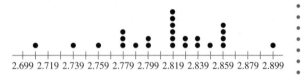

2.699 2.719 2.739 2.759 2.779 2.799 2.819 2.839 2.859 2.879 2.899

18. Cattle Weights The line plot shows the weights (to the nearest hundred pounds) of 30 cattle.

(a) How many cattle weighed about 800 to 1000 pounds?

(b) What is the range of weights?

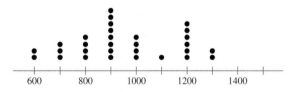

600   800   1000   1200   1400

 Exam Scores In Exercises 19 and 20, use the given scores from a math class of 30 students. The scores are for two 100-point exams. (*Spreadsheet at LarsonPrecalculus.com*)

 Exam #1 77, 100, 77, 70, 83, 89, 87, 85, 81, 84, 81, 78, 89, 78, 88, 85, 90, 92, 75, 81, 85, 100, 98, 81, 78, 75, 85, 89, 82, 75

Exam #2 76, 78, 73, 59, 70, 81, 71, 66, 66, 73, 68, 67, 63, 67, 77, 84, 87, 71, 78, 78, 90, 80, 77, 70, 80, 64, 74, 68, 68, 68

19. Construct a stem-and-leaf plot for Exam #1. Use the plot to identify the highest and lowest scores on the test.

20. Construct a double stem-and-leaf plot to compare the scores for Exam #1 and Exam #2. Which set of scores is higher as a group?

21. Retirement Contributions The amounts (in dollars) that 35 employees contribute to their personal retirement plans are listed. Use a frequency distribution and a histogram to organize the data. (*Spreadsheet at LarsonPrecalculus.com*)

 100, 200, 130, 136, 161, 156, 209, 126, 135, 98, 114, 117, 168, 133, 140, 124, 172, 127, 143, 157, 124, 152, 104, 126, 155, 92, 194, 115, 120, 136, 148, 112, 116, 146, 96

22. Agriculture The table shows the total number of farms (in thousands) in each of the 50 states in 2015. Use a frequency distribution and a histogram to organize the data. (*Source: U.S. Dept. of Agriculture*)

AK	1	AL	43	AR	44	AZ	20	CA	78
CO	34	CT	6	DE	3	FL	47	GA	41
HI	7	IA	88	ID	24	IL	74	IN	58
KS	60	KY	76	LA	27	MA	8	MD	12
ME	8	MI	52	MN	74	MO	97	MS	37
MT	28	NC	49	ND	30	NE	49	NH	4
NJ	9	NM	25	NV	4	NY	36	OH	74
OK	78	OR	35	PA	58	RI	1	SC	24
SD	31	TN	67	TX	242	UT	18	VA	45
VT	7	WA	36	WI	69	WV	21	WY	12

23. Meteorology The seasonal snowfall amounts (in inches) for Chicago, Illinois, for the years 1982 through 2014 are listed. (The amounts are listed in order by year.) How would you organize the data? Explain your reasoning. (*Source: National Weather Service*) (*Spreadsheet at LarsonPrecalculus.com*)

26.6, 49.0, 39.1, 29.0, 26.2, 42.6, 24.5, 33.8, 36.7, 28.4, 46.9, 41.8, 24.1, 23.9, 40.6, 29.6, 50.9, 30.3, 39.2, 31.1, 28.6, 24.8, 39.4, 26.0, 35.6, 60.3, 52.7, 54.2, 57.9, 19.8, 30.1, 82.0, 50.7

24. **HOW DO YOU SEE IT?** Describe and correct the error in creating the histogram below using the frequency distribution at the right.

Interval	Tally
[0, 10)	|||
[10, 20)	‖‖	
[20, 30)	‖‖
[30, 40)	||||

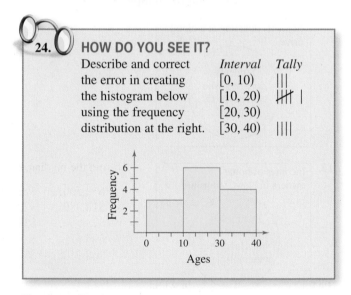

Exploration

True or False? In Exercises 25 and 26, determine whether the statement is true or false. Justify your answer.

25. A census surveys an entire population.

26. The ranges of data from two random samples from the same population must be equal.

27. Writing Describe how you could choose samples of students at your school using each of the five sampling methods listed on page 879.

13.2 Analyzing Data

- Find and interpret the mean, median, and mode of a data set.
- Determine the measure of central tendency that best represents a data set.
- Find the standard deviation of a data set.
- Create and use box-and-whisker plots.
- Interpret normally distributed data.

Measures of central tendency and dispersion provide a convenient way to describe and compare data sets. For example, in Exercise 42 on page 895, you will use box-and-whisker plots to analyze the lifetime of a machine part.

Mean, Median, and Mode

It is often helpful to describe a data set by a single number that is most representative of the entire collection of numbers. Such a number is a **measure of central tendency.** The most commonly used measures are listed below.

1. The **mean,** or **average,** of n numbers is the sum of the numbers divided by n.
2. The numerical **median** of n numbers is the middle number when the numbers are written in order. When n is even, the median is the average of the two middle numbers.
3. The **mode** of n numbers is the number that occurs most frequently. When two numbers tie for most frequent occurrence, the collection has two modes and is called **bimodal.** When no entry occurs more than once, the data set has no mode.

EXAMPLE 1 **Finding Measures of Central Tendency**

The annual incomes of 25 employees of a company are listed below. What are the mean, median, and mode of the incomes? *(Spreadsheet at LarsonPrecalculus.com)*

DATA
$17,305,	$478,320,	$45,678,	$18,980,	$17,408,
$25,676,	$28,906,	$12,500,	$24,540,	$33,450,
$12,500,	$33,855,	$37,450,	$20,432,	$28,956,
$34,983,	$36,540,	$250,921,	$36,853,	$16,430,
$32,654,	$98,213,	$48,980,	$94,024,	$35,671

Solution The mean of the incomes is

$$\text{Mean} = \frac{17,305 + 478,320 + 45,678 + 18,980 + \cdots + 35,671}{25} = \$60,849.$$

To find the median, order the incomes.

$12,500,	$12,500,	$16,430,	$17,305,	$17,408,
$18,980,	$20,432,	$24,540,	$25,676,	$28,906,
$28,956,	$32,654,	$33,450,	$33,855,	$34,983,
$35,671,	$36,540,	$36,853,	$37,450,	$45,678,
$48,980,	$94,024,	$98,213,	$250,921,	$478,320

In this list, the median income is the middle income, $33,450. The income $12,500 occurs twice and is the only income that occurs more than once. So, the mode is $12,500.

✓ Checkpoint ◀))) *Audio-video solution in English & Spanish at LarsonPrecalculus.com*

Find the mean, median, and mode of the data set below.

68, 73, 67, 95, 71, 82, 85, 74, 82, 61

In Example 1, note that the three measures of central tendency differ considerably. The mean is inflated by the two highest salaries and the mode is the least value. So, in this case, the median best represents a "typical" income.

• • REMARK The way in which the data is used can affect which measure best represents the data. For instance, in Example 1, the median best represents a "typical" income to a potential employee, but the mean is more useful to an accountant estimating a payroll budget.

Choosing a Measure of Central Tendency

There are different ways to determine which measure of central tendency best represents a data set. In general, the measure that is close to most of the data is the most representative of the data set. For example, in both data sets below, the mean is 6, the median is 6, and the modes are 1 and 11.

Data set A: 1, 1, 1, 1, 1, 11, 11, 11, 11, 11

Data set B: 1, 1, 5.7, 5.8, 5.9, 6.1, 6.2, 6.3, 11, 11

The modes are the most representative measure of the data in set A, and the mean or median is the most representative measure of the data in set B.

EXAMPLE 2 **Choosing Measures of Central Tendency**

Determine which measure of central tendency is the most representative of the data shown in each frequency distribution.

a.

Number	1	2	3	4	5	6	7	8	9
Frequency	0	0	2	5	6	8	2	1	1

b.

Number	1	2	3	4	5	6	7	8	9
Frequency	15	2	1	0	0	0	1	2	15

c.

Number	1	2	3	4	5	6	7	8	9
Frequency	1	6	7	5	1	0	0	1	1

Solution

a. For these data, the mean is 5.4, the median is 5, and the mode is 6. Most of the data are close to the mean and the median, so the mean or median is the most representative measure.

b. For these data, the mean and median are each 5, and the modes are 1 and 9 (the distribution is bimodal). Most of the data are close to the modes, so the modes are the most representative measure.

c. For these data, the mean is approximately 3.45, the median is 3, and the mode is 3. Most of the data are close to the median and the mode, so the median or mode is the most representative measure.

✓ **Checkpoint** ◀))) *Audio-video solution in English & Spanish at LarsonPrecalculus.com*

Determine which measure of central tendency is the most representative of the data shown in each frequency distribution.

a.

Number	1	2	3	4	5	6	7	8	9
Frequency	4	20	3	0	0	0	4	20	3

b.

Number	1	2	3	4	5	6	7	8	9
Frequency	1	0	0	0	1	3	5	7	8

c.

Number	1	2	3	4	5	6	7	8	9
Frequency	0	2	5	4	4	4	3	3	0

Variance and Standard Deviation

Very different sets of numbers can have the same mean. You will now study two **measures of dispersion,** which give you an idea of how much the numbers in a data set differ from the mean of the set. These two measures are the *variance* of the set and the *standard deviation* of the set.

> ### Definitions of Variance and Standard Deviation
>
> Consider a set of numbers $\{x_1, x_2, \ldots, x_n\}$ with a mean of $\bar{x}$. The **variance** of the set is
>
> $$v = \frac{(x_1 - \bar{x})^2 + (x_2 - \bar{x})^2 + \cdots + (x_n - \bar{x})^2}{n}$$
>
> and the **standard deviation** of the set is $\sigma = \sqrt{v}$ (σ is the lowercase Greek letter *sigma*).

The greater the standard deviation of a data set, the more the numbers in the set *vary* from the mean. For example, each of the data sets below has a mean of 5.

$$\{5, 5, 5, 5\}, \quad \{4, 4, 6, 6\}, \quad \text{and} \quad \{3, 3, 7, 7\}$$

The standard deviations of the data sets are 0, 1, and 2, respectively.

$$\sigma_1 = \sqrt{\frac{(5-5)^2 + (5-5)^2 + (5-5)^2 + (5-5)^2}{4}} = 0$$

$$\sigma_2 = \sqrt{\frac{(4-5)^2 + (4-5)^2 + (6-5)^2 + (6-5)^2}{4}} = 1$$

$$\sigma_3 = \sqrt{\frac{(3-5)^2 + (3-5)^2 + (7-5)^2 + (7-5)^2}{4}} = 2$$

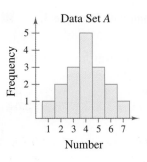

Data Set A

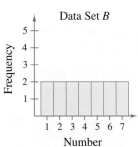

Data Set B

EXAMPLE 3 **Comparing Standard Deviations**

The three data sets represented by the histograms in Figure 13.2 all have a mean of 4. Which data set has the least standard deviation? Which has the greatest?

Solution Of the three data sets, the numbers in data set A are grouped most closely to the mean of 4 and the numbers in data set C are the most dispersed from the mean. So, data set A has the least standard deviation and data set C has the greatest standard deviation.

✓ *Checkpoint* ◀))) *Audio-video solution in English & Spanish at LarsonPrecalculus.com*

The three data sets represented by the histograms shown below all have a mean of 4. Which data set has the least standard deviation? Which has the greatest?

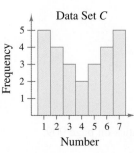

Data Set C

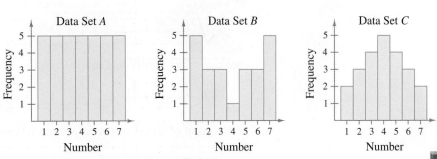

Figure 13.2

EXAMPLE 4 **Finding Standard Deviation**

Find the standard deviation of each data set in Example 3.

Solution Each data set has a mean of $\bar{x} = 4$. The standard deviation of data set A is

$$\sigma = \sqrt{\frac{(-3)^2 + 2(-2)^2 + 3(-1)^2 + 5(0)^2 + 3(1)^2 + 2(2)^2 + (3)^2}{17}}$$

$$\approx 1.53.$$

The standard deviation of data set B is

$$\sigma = \sqrt{\frac{2(-3)^2 + 2(-2)^2 + 2(-1)^2 + 2(0)^2 + 2(1)^2 + 2(2)^2 + 2(3)^2}{14}}$$

$$= 2.$$

The standard deviation of data set C is

$$\sigma = \sqrt{\frac{5(-3)^2 + 4(-2)^2 + 3(-1)^2 + 2(0)^2 + 3(1)^2 + 4(2)^2 + 5(3)^2}{26}}$$

$$\approx 2.22.$$

These values confirm the results of Example 3. That is, data set A has the least standard deviation and data set C has the greatest.

✓ *Checkpoint*  *Audio-video solution in English & Spanish at LarsonPrecalculus.com*

Find the standard deviation of each data set in the Checkpoint with Example 3. ◼

The alternative formula below provides a more efficient way to compute the standard deviation.

Alternative Formula for Standard Deviation

The standard deviation of $\{x_1, x_2, \ldots, x_n\}$ is

$$\sigma = \sqrt{\frac{x_1^2 + x_2^2 + \cdots + x_n^2}{n} - \bar{x}^2}.$$

Conceptually, the process of proving this formula is straightforward. It consists of showing that the expressions

$$\sqrt{\frac{(x_1 - \bar{x})^2 + (x_2 - \bar{x})^2 + \cdots + (x_n - \bar{x})^2}{n}}$$

and

$$\sqrt{\frac{x_1^2 + x_2^2 + \cdots + x_n^2}{n} - \bar{x}^2}$$

are equivalent. Verify this equivalence for the set $\{x_1, x_2, x_3\}$ with

$$\bar{x} = \frac{x_1 + x_2 + x_3}{3}.$$

For a proof of the more general case, see Proofs in Mathematics on page 909.

EXAMPLE 5   **Using the Alternative Formula**

Use the alternative formula for standard deviation to find the standard deviation of the set of numbers below.

5, 6, 6, 7, 7, 8, 8, 8, 9, 10

Solution   Begin by finding the mean of the data set, which is 7.4. So, the standard deviation is

$$\sigma = \sqrt{\frac{5^2 + 2(6^2) + 2(7^2) + 3(8^2) + 9^2 + 10^2}{10} - (7.4)^2}$$

$$= \sqrt{\frac{568}{10} - 54.76}$$

$$\approx 1.43.$$

Check this result using the *one-variable statistics* feature of a graphing utility.

✓ **Checkpoint** 🔊))) *Audio-video solution in English & Spanish at LarsonPrecalculus.com*

Use the alternative formula for standard deviation to find the standard deviation of the set of numbers below.

3, 3, 3, 4, 4, 5, 5, 6, 6, 7

A well-known theorem in statistics, called *Chebychev's Theorem*, states that the portion of any data set lying within k standard deviations of the mean is at least

$$1 - \frac{1}{k^2}$$

where $k > 1$. So, at least 75% of the numbers in a data set must lie within two standard deviations of the mean, and at least $88.\overline{8}\%$ of the numbers must lie within three standard deviations of the mean. For most distributions, the percent of the numbers within k standard deviations of the mean is greater than the percent given by Chebychev's Theorem. For instance, in all three distributions shown in Example 3, 100% of the numbers lie within two standard deviations of the mean.

EXAMPLE 6   **Describing a Distribution**

The table at the left shows the minimum wages of the 50 states and the District of Columbia in 2016. Find the mean and standard deviation of the data. What percent of the data values lie within one standard deviation of the mean? *(Source: U.S. Department of Labor)*

Solution   Begin by entering the numbers into a graphing utility. Then use the *one-variable statistics* feature to obtain $\bar{x} \approx 8.12$ and $\sigma \approx 0.94$. The interval that contains all numbers that lie within one standard deviation of the mean is

$$[8.12 - 0.94, 8.12 + 0.94] \quad \text{or} \quad [7.18, 9.06].$$

From the table, you can see that all but nine of the data values (about 82%) lie in this interval.

✓ **Checkpoint** 🔊))) *Audio-video solution in English & Spanish at LarsonPrecalculus.com*

In Example 6, what percent of the data values lie within two standard deviations of the mean?

DATA

Spreadsheet at LarsonPrecalculus.com

AK	9.75	MT	8.05
AL	7.25	NC	7.25
AR	8.00	ND	7.25
AZ	8.05	NE	9.00
CA	10.00	NH	7.25
CO	8.31	NJ	8.38
CT	9.60	NM	7.50
DC	10.50	NV	8.25
DE	8.25	NY	9.00
FL	8.05	OH	8.10
GA	7.25	OK	7.25
HI	8.50	OR	9.25
IA	7.25	PA	7.25
ID	7.25	RI	9.60
IL	8.25	SC	7.25
IN	7.25	SD	8.55
KS	7.25	TN	7.25
KY	7.25	TX	7.25
LA	7.25	UT	7.25
MA	10.00	VA	7.25
MD	8.25	VT	9.60
ME	7.50	WA	9.47
MI	8.50	WI	7.25
MN	9.00	WV	8.75
MO	7.65	WY	7.25
MS	7.25		

Box-and-Whisker Plots

Standard deviation is the measure of dispersion that is associated with the mean. **Quartiles** measure dispersion associated with the median.

Definition of Quartiles

Consider an ordered set of numbers whose median is m. The **lower quartile** is the median of the numbers that occur before m. The **upper quartile** is the median of the numbers that occur after m.

EXAMPLE 7 **Finding Quartiles of a Data Set**

Find the lower and upper quartiles for the data set below.

42, 14, 24, 16, 12, 18, 20, 24, 16, 26, 13, 27

Solution   Begin by ordering the data.

12, 13, 14,	16, 16, 18,	20, 24, 24,	26, 27, 42
1st 25%	2nd 25%	3rd 25%	4th 25%

The median of the entire data set is $(18 + 20)/2 = 19$. The median of the six numbers that are less than 19 is $(14 + 16)/2 = 15$. So, the lower quartile is 15. The median of the six numbers that are greater than 19 is 25. So, the upper quartile is $(24 + 26)/2 = 25$.

✓ **Checkpoint** 🔊))) *Audio-video solution in English & Spanish at LarsonPrecalculus.com*

Find the lower and upper quartiles for the data set below.

39, 47, 81, 43, 23, 23, 27, 86, 15, 3, 74, 55

The **interquartile range** of a data set is the difference of the upper quartile and the lower quartile. The interquartile range of the data set in Example 7 is

$$25 - 15 = 10.$$

A value that is widely separated from the rest of the data in a data set is called an **outlier.** Typically, a data value is considered to be an outlier when it is greater than the upper quartile by more than 1.5 times the interquartile range or when it is less than the lower quartile by more than 1.5 times the interquartile range. Verify that, in Example 7, the value 42 is an outlier.

Quartiles are represented graphically by a **box-and-whisker plot,** as shown in the figure below. In the plot, notice that five numbers are listed: the least number, the lower quartile, the median, the upper quartile, and the greatest number. These numbers are the **five-number summary** of the data set. Also notice that the numbers are spaced proportionally, as though they were on a real number line.

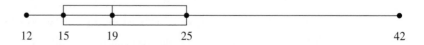

The next example shows how to find quartiles when the number of elements in a data set is not divisible by 4.

▷ TECHNOLOGY Some graphing utilities have the capability of creating box-and-whisker plots. If your graphing utility has this capability, use it to recreate the box-and-whisker plot shown above for the data set in Example 7.

EXAMPLE 8 **Sketching Box-and-Whisker Plots**

See LarsonPrecalculus.com for an interactive version of this type of example.

Sketch a box-and-whisker plot for each data set.

a. 27, 28, 30, 42, 45, 50, 50, 61, 62, 64, 66

b. 82, 82, 83, 85, 87, 89, 90, 94, 95, 95, 96, 98, 99

c. 11, 13, 13, 15, 17, 18, 20, 24, 24, 27

Solution

a. The median is 50. The lower quartile is 30 (the median of the first five numbers). The upper quartile is 62 (the median of the last five numbers). See the plot below.

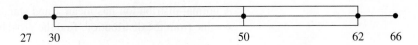

b. The median is 90. The lower quartile is 84 (the median of the first six numbers). The upper quartile is 95.5 (the median of the last six numbers). See the plot below.

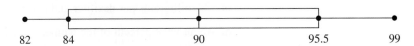

c. The median is 17.5. The lower quartile is 13 (the median of the first five numbers). The upper quartile is 24 (the median of the last five numbers). See the plot below.

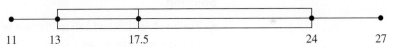

✓ *Checkpoint* *Audio-video solution in English & Spanish at LarsonPrecalculus.com*

Sketch a box-and-whisker plot for the data set: 2, 7, 9, 38, 44, 54, 56, 62, 79, 93. ◼

Normal Distributions

Recall that a **normal distribution** is modeled by a bell-shaped curve. This is called a **normal curve** and it is symmetric with respect to the mean.

Areas Under a Normal Curve

A normal distribution with mean $\bar{x}$ and standard deviation σ has the following properties:

- The total area under the normal curve is 1.
- About 68% of the area lies within 1 standard deviation of the mean.
- About 95% of the area lies within 2 standard deviations of the mean.
- About 99.7% of the area lies within 3 standard deviations of the mean.

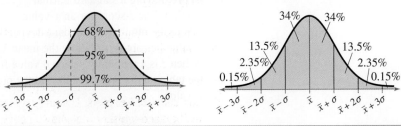

EXAMPLE 9 **Finding a Normal Probability**

A normal distribution has mean $\bar{x}$ and standard deviation σ. For a randomly selected x-value from the distribution, find the probability that $\bar{x} - 2\sigma \leq x \leq \bar{x}$.

Solution The probability that

$$\bar{x} - 2\sigma \leq x \leq \bar{x}$$

is the shaded area under the normal curve shown at the right.

$$P(\bar{x} - 2\sigma \leq x \leq \bar{x}) \approx 0.135 + 0.34 = 0.475$$

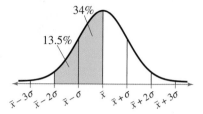

✓ **Checkpoint** ◀))) *Audio-video solution in English & Spanish at LarsonPrecalculus.com*

A normal distribution has mean $\bar{x}$ and standard deviation σ. For a randomly selected x-value from the distribution, find the probability that $\bar{x} - \sigma \leq x \leq \bar{x} + 2\sigma$.

EXAMPLE 10 **Interpreting Normally Distributed Data**

The blood cholesterol levels for a group of women are normally distributed with a mean of 172 milligrams per deciliter and a standard deviation of 14 milligrams per deciliter.

a. About what percent of the women have levels between 158 and 186 milligrams per deciliter?

b. Levels less than 158 milligrams per deciliter are considered desirable. About what percent of the levels are desirable?

Solution

a. The levels of 158 and 186 milligrams per deciliter represent one standard deviation on either side of the mean. So, about 68% of the women have levels between 158 and 186 milligrams per deciliter.

b. A level of 158 milligrams per deciliter is one standard deviation to the left of the mean. So, the percent of the levels that are desirable is about 0.15% + 2.35% + 13.5% = 16%.

✓ **Checkpoint** ◀))) *Audio-video solution in English & Spanish at LarsonPrecalculus.com*

In Example 10, about what percent of the women have levels between 172 and 200 milligrams per deciliter?

The **standard normal distribution** is a normal distribution with mean 0 and standard deviation 1. Using the formula

$$z = \frac{x - \bar{x}}{\sigma}$$

transforms an x-value from a normal distribution with mean $\bar{x}$ and standard deviation σ into a corresponding z-value, or **z-score,** having a standard normal distribution. The z-score for an x-value is equal to the number of standard deviations the x-value lies above or below the mean $\bar{x}$. (See figure.)

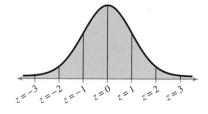

When z is a randomly selected value from a standard normal distribution, you can use the table on the next page to find the probability that z is less than or equal to some given value.

Jody/Shutterstock.com

Standard Normal Table

z	.0	.1	.2	.3	.4	.5	.6	.7	.8	.9
−3	.0013	.0010	.0007	.0005	.0003	.0002	.0002	.0001	.0001	.0000+
−2	.0228	.0179	.0139	.0107	.0082	.0062	.0047	.0035	.0026	.0019
−1	.1587	.1357	.1151	.0968	.0808	.0668	.0548	.0446	.0359	.0287
−0	.5000	.4602	.4207	.3821	.3446	.3085	.2743	.2420	.2119	.1841
0	.5000	.5398	.5793	.6179	.6554	.6915	.7257	.7580	.7881	.8159
1	.8413	.8643	.8849	.9032	.9192	.9332	.9452	.9554	.9641	.9713
2	.9772	.9821	.9861	.9893	.9918	.9938	.9953	.9965	.9974	.9981
3	.9987	.9990	.9993	.9995	.9997	.9998	.9998	.9999	.9999	1.0000−

> **REMARK** In the table, the value .0000+ means "slightly more than 0" and the value 1.0000− means "slightly less than 1."

For example, to find $P(z \le -0.4)$, find the value in the table where the row labeled "−0" and the column labeled ".4" intersect. The table shows that

$$P(z \le -0.4) = 0.3446.$$

You can also use the standard normal table to find probabilities for normal distributions by first converting values from the distribution to z-scores. Example 11 illustrates this.

EXAMPLE 11 **Using the Standard Normal Table**

Scientists conducted aerial surveys of a seal sanctuary and recorded the number x of seals they observed during each survey. The numbers recorded were normally distributed with a mean of 73 and a standard deviation of 14.1. Find the probability that the scientists observed at most 50 seals during a survey.

Solution Begin by finding the z-score that corresponds to an x-value of 50.

$$z = \frac{x - \bar{x}}{\sigma} = \frac{50 - 73}{14.1} \approx -1.6$$

Then use the table to find $P(x \le 50) \approx P(z \le -1.6) = 0.0548$. So, the probability that the scientists observed at most 50 seals during a survey is about 0.0548.

A seal sanctuary has protected beaches where harbor seals give birth to their young in their natural habitat.

✓ *Checkpoint* ◀))) *Audio-video solution in English & Spanish at LarsonPrecalculus.com*

In Example 11, find the probability that the scientists observed at most 90 seals during a survey. ◼

Summarize (Section 13.2)

1. State the definitions of the mean, median, and mode of a data set *(page 885)*. For an example of finding measures of central tendency, see Example 1.

2. Explain how to determine the measure of central tendency that best represents a data set *(page 886)*. For an example of determining the measures of central tendency that best represent data sets, see Example 2.

3. Explain how to find the standard deviation of a data set *(pages 887 and 888)*. For examples of finding the standard deviations of data sets, see Examples 4–6.

4. Explain how to sketch a box-and-whisker plot *(page 890)*. For an example of sketching box-and-whisker plots, see Example 8.

5. Describe what is meant by normally distributed data *(page 891)*. For examples of using normally distributed data, see Examples 9–11.

13.2   Exercises

Vocabulary: Fill in the blanks.

1. The _____ of *n* numbers is the sum of the numbers divided by *n*.

2. When there is an even number of data values in a data set, the _____ is the average of the two middle numbers.

3. When two numbers of a data set tie for most frequent occurrence, the collection has two _____ and is called _____.

4. Two measures of dispersion associated with the mean are the _____ and the _____ _____ of a data set.

5. _____ measure dispersion associated with the median.

6. The _____ _____ of a data set is the difference of the upper quartile and the lower quartile.

7. A value that is widely separated from the rest of the data in a data set is called an _____.

8. Quartiles are represented graphically by a _____ _____.

9. The _____ _____ distribution is a normal distribution with mean 0 and standard deviation 1.

10. The _____ for an *x*-value is equal to the number of standard deviations the *x*-value lies above or below the mean.

Skills and Applications

 Finding Measures of Central Tendency In Exercises 11–16, find the mean, median, and mode(s) of the data set.

11. 5, 12, 7, 14, 8, 9, 7, 12

12. 30, 37, 32, 39, 33, 34, 32

13. 5, 12, 7, 24, 8, 9, 7, 12

14. 20, 37, 32, 39, 33, 34, 32

15. 5, 6, 12, 7, 14, 9, 7, 7, 6

16. 30, 37, 32, 39, 30, 34, 32, 30, 32, 37

17. **Electric Bills**   A homeowner's monthly electric bills for a year are listed. What are the mean and median of the data?   *(Spreadsheet at LarsonPrecalculus.com)*

DATA			
January	$67.92	February	$59.84
March	$52.00	April	$52.50
May	$57.99	June	$65.35
July	$81.76	August	$74.98
September	$87.82	October	$83.18
November	$65.35	December	$57.00

18. **Car Rental**   A car rental company records the number of cars rented each day for a week. The results are listed. What are the mean, median, and mode of the data?   *(Spreadsheet at LarsonPrecalculus.com)*

DATA			
Monday	410	Tuesday	260
Wednesday	320	Thursday	320
Friday	460	Saturday	150
Sunday	180		

19. **Families**   Researchers conducted a study on families with six children. The table shows the numbers of daughters in each family. Determine the mean, median, and mode of this data set.

Number of Daughters	0	1	2	3	4	5	6
Frequency	1	24	45	54	50	19	7

Spreadsheet at LarsonPrecalculus.com

20. **Sports**   The table shows the numbers of games in which a baseball player had 0, 1, 2, 3, and 4 hits during the last 50 games.

Number of Hits	0	1	2	3	4
Frequency	14	26	7	2	1

(a) Determine the mean number of hits per game.

(b) The player had a total of 200 at-bats in the 50 games. Determine the player's batting average.

21. **Test Scores**   A professor records the students' scores for a 100-point exam. The results are listed. *(Spreadsheet at LarsonPrecalculus.com)*

99, 64, 80, 77, 59, 72, 87,
79, 92, 88, 90, 42, 20, 89,
42, 100, 98, 84, 78, 91

Which measure of central tendency best describes these test scores?

22. Test Scores The histograms represent the test scores of two classes of a college course in mathematics. Which histogram shows the lesser standard deviation? Explain.

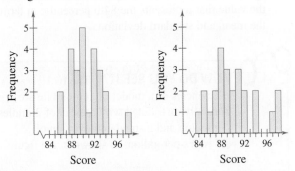

Finding Mean and Standard Deviation In Exercises 23 and 24, each line plot represents a data set. Find the mean and standard deviation of each data set.

23. (a)
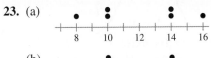

(b)
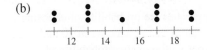

24. (a)
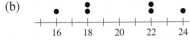

(b)

 Finding Mean, Variance, and Standard Deviation In Exercises 25–30, find the mean ($\bar{x}$), variance (v), and standard deviation (σ) of the data set.

25. 4, 10, 8, 2   **26.** 2, 12, 4, 7, 5

27. 0, 1, 1, 2, 2, 2, 3, 3, 4

28. 2, 2, 2, 2, 2, 2

29. 42, 50, 61, 47, 56, 68

30. 1.2, 0.6, 2.8, 1.7, 0.9

 Using the Alternative Formula for Standard Deviation In Exercises 31–36, use the alternative formula for standard deviation to find the standard deviation of the data set.

31. 3, 5, 7, 7, 18, 8

32. 24, 20, 38, 15, 52, 33, 46

33. 246, 336, 473, 167, 219

34. 5.1, 6.7, 4.5, 10.2, 9.1

35. 8.6, 6.4, 2.9, 5.0, 6.7

36. 9.2, 10.6, 7.2, 4.3, 7.0

 Quartiles and Box-and-Whisker Plots In Exercises 37–40, find the lower and upper quartiles and sketch a box-and-whisker plot for each data set.

37. 11, 10, 11, 14, 17, 16, 14, 11, 8, 14, 20

38. 46, 48, 48, 50, 52, 47, 51, 47, 49, 53

39. 19, 12, 14, 9, 14, 15, 17, 13, 19, 11, 10, 19

40. 20.1, 43.4, 34.9, 23.9, 33.5, 24.1, 22.5, 42.4, 25.7, 17.4, 23.8, 33.3, 17.3, 36.4, 21.8

41. Price of Gold The data represent the average prices of gold (in dollars per troy ounce) for the years 1996 through 2015. Use a graphing utility to find the mean, variance, and standard deviation of the data. What percent of the data lies within one standard deviations of the mean? *(Source: U.S. Bureau of Mines and U.S. Geological Survey) (Spreadsheet at LarsonPrecalculus.com)*

DATA
389, 332, 295, 280, 280,
272, 311, 365, 411, 446,
606, 699, 874, 975, 1227,
1572, 1673, 1415, 1269, 1170

42. Product Lifetime

A manufacturer redesigns a machine part to increase its average lifetime. The two data sets list the lifetimes (in months) of 20 randomly selected parts of each design.
(Spreadsheet at LarsonPrecalculus.com)

Original Design

15.1	78.3	56.3	68.9	30.6
27.2	12.5	42.7	72.7	20.2
53.0	13.5	11.0	18.4	85.2
10.8	38.3	85.1	10.0	12.6

New Design

55.8	71.5	25.6	19.0	23.1
37.2	60.0	35.3	18.9	80.5
46.7	31.1	67.9	23.5	99.5
54.0	23.2	45.5	24.8	87.8

(a) Construct a box-and-whisker plot for each data set.

(b) Explain the differences between the box-and-whisker plots you constructed in part (a).

Finding a Normal Probability In Exercises 43–46, a normal distribution has mean $\bar{x}$ and standard deviation σ. Find the indicated probability for a randomly selected x-value from the distribution.

43. $P(x \leq \bar{x} - \sigma)$ 44. $P(x \geq \bar{x} + 2\sigma)$

45. $P(\bar{x} - \sigma \leq x \leq \bar{x} + \sigma)$ 46. $P(\bar{x} - 3\sigma \leq x \leq \bar{x})$

Normal Distribution In Exercises 47–50, a normal distribution has a mean of 33 and a standard deviation of 4. Find the probability that a randomly selected x-value from the distribution is in the given interval.

47. Between 29 and 37   48. Between 33 and 45

49. At least 25   50. At most 37

Using the Standard Normal Table In Exercises 51–56, a normal distribution has a mean of 64 and a standard deviation of 7. Use the standard normal table to find the indicated probability for a randomly selected x-value from the distribution.

51. $P(x \leq 68)$ 52. $P(x \leq 71)$

53. $P(x \geq 45)$ 54. $P(x \geq 75)$

55. $P(60 \leq x \leq 75)$ 56. $P(45 \leq x \leq 65)$

57. **Biology** A study found that the wing lengths of houseflies are normally distributed with a mean of about 4.6 millimeters and a standard deviation of about 0.4 millimeter. What is the probability that a randomly selected housefly has a wing length of at least 5 millimeters?

58. **Boxes of Cereal** A machine fills boxes of cereal. The weights of cereal in the boxes are normally distributed with a mean of 20 ounces and a standard deviation of 0.25 ounce.

    (a) Find the z-scores for weights of 19.4 ounces and 20.4 ounces.

    (b) What is the probability that a randomly selected cereal box weighs between 19.4 and 20.4 ounces?

Exploration

True or False? In Exercises 59–62, determine whether the statement is true or false. Justify your answer.

59. Some quantitative data sets do not have medians.

60. About one quarter of the data in a set is less than the lower quartile.

61. To construct a box-and-whisker plot for a data set, you need to know the least number, the lower quartile, the mean, the upper quartile, and the greatest number of the data set.

62. It is not possible for a value from a normal distribution to have a z-score equal to 0.

63. **Writing** When $n\%$ of the values in a data set are less than or equal to a certain value, that value represents the **nth percentile.** For normally distributed data, describe the value that represents the 84th percentile in terms of the mean and standard deviation.

64. **HOW DO YOU SEE IT?** The fuel efficiencies of a model of automobile are normally distributed with a mean of 26 miles per gallon and a standard deviation of 2.4 miles per gallon, as shown in the figure.

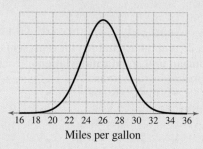

Miles per gallon

    (a) Which is greater, the probability of choosing a car at random that gets between 26 and 28 miles per gallon or the probability of choosing a car at random that gets between 22 and 24 miles per gallon?

    (b) Which is greater, the probability of choosing a car at random that gets between 20 and 22 miles per gallon or the probability of choosing a car at random that gets at least 30 miles per gallon?

65. **Reasoning** Compare your answers for Exercises 11 and 13 and those for Exercises 12 and 14. Which of the measures of central tendency is affected by outliers? Explain.

66. **Reasoning**

    (a) Add 6 to each measurement in Exercise 11 and calculate the mean, median, and modes of the revised measurements. How do the measures of central tendency change?

    (b) Make a conjecture about how the measures of central tendency of a data set change when you add a constant k to each data value.

67. **Reasoning** Without calculating the standard deviation, explain why the data set $\{4, 4, 20, 20\}$ has a standard deviation of 8.

68. **Think About It** Construct a collection of numbers that has the following measures of central tendency.

    Mean = 6

    Median = 4

    Mode = 4

13.3 Modeling Data

- ■ Use the correlation coefficient to measure how well a model fits a data set.
- ■ Use the sum of the squared differences to measure how well a model fits a data set.
- ■ Find the least squares regression line for a data set.
- ■ Find the least squares regression parabola for a data set.
- ■ Analyze misleading graphs.

The method of least squares provides a way of creating mathematical models for data sets. For example, in Exercise 21 on page 902, you will find the least squares regression line for the average annual costs of healthcare for a family of four in the United States from 2012 through 2015.

Throughout this text, you have been using or finding models for data sets. For example, in Section 1.10, you determined how closely a given model represents a data set, and you used the *regression* feature of a graphing utility to find a linear model for a data set. This section expands upon the concepts learned in Section 1.10 regarding the use of least squares regression to fit models to data.

Correlation

A **correlation** is a relationship between two variables. A scatter plot is helpful in determining whether a correlation exists between two variables. The scatter plots below illustrate several types of correlations.

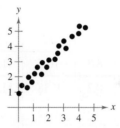

Positive linear correlation: *y* tends to increase as *x* increases.

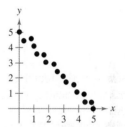

Negative linear correlation: *y* tends to decrease as *x* increases.

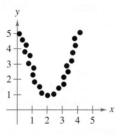

Nonlinear correlation

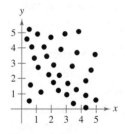

No correlation

• • **REMARK** Recall from Section 1.10 that the correlation coefficient can be used to determine whether a linear model is a good fit for a data set.

The **correlation coefficient** r gives a measure of the strength and direction of a linear correlation between the two variables. The range of the possible values of r is $[-1, 1]$. For a linear model, r close to 1 indicates a strong positive linear correlation between x and y, r close to -1 indicates a strong negative linear correlation, and r close to 0 indicates a weak or no *linear* correlation. A formula for r is

▷ **ALGEBRA HELP** Note that the formula for r uses summation notation. To review summation notation, see Section 9.1.

$$r = \frac{n\sum\limits_{i=1}^{n} x_i y_i - \left(\sum\limits_{i=1}^{n} x_i\right)\left(\sum\limits_{i=1}^{n} y_i\right)}{\sqrt{n\sum\limits_{i=1}^{n} x_i^2 - \left(\sum\limits_{i=1}^{n} x_i\right)^2}\sqrt{n\sum\limits_{i=1}^{n} y_i^2 - \left(\sum\limits_{i=1}^{n} y_i\right)^2}}$$

where n is the number of data points (x, y).

Sum of the Squared Differences

The *regression* feature of a graphing utility uses the **method of least squares** to find a mathematical model for a data set. As a measure of how well a model fits a set of data points

$$\{(x_1, y_1), (x_2, y_2), \ldots, (x_n, y_n)\}$$

add the squares of the differences between the actual y-values and the values given by the model to obtain the **sum of the squared differences.**

EXAMPLE 1 **Finding the Sum of the Squared Differences**

The table shows the heights x (in feet) and the diameters y (in inches) of eight trees. Find the sum of the squared differences for the linear model $y^* = 0.54x - 29.5$.

DATA x	70	72	75	76	77	78	80	85
y	8.3	10.5	11.0	11.4	16.3	14.0	18.0	12.9

Spreadsheet at LarsonPrecalculus.com

Solution Evaluate $y^* = 0.54x - 29.5$ for each given value of x. Then find the difference $y - y^*$ using each value of y and the corresponding value of y^*, and square the difference.

x	70	72	75	76	77	78	80	85
y	8.3	10.5	11.0	11.4	16.3	14.0	18.0	12.9
y^*	8.3	9.38	11.0	11.54	12.08	12.62	13.7	16.4
$y - y^*$	0	1.12	0	-0.14	4.22	1.38	4.3	-3.5
$(y - y^*)^2$	0	1.2544	0	0.0196	17.8084	1.9044	18.49	12.25

Finally, add the values in the last row to find the sum of the squared differences.

$$0 + 1.2544 + 0 + 0.0196 + 17.8084 + 1.9044 + 18.49 + 12.25 = 51.7268$$

•• **REMARK** In Example 1, note that the sum of the squared differences is relatively large, so the model is likely not a good fit for the data set.

✓ **Checkpoint** ◀))) *Audio-video solution in English & Spanish at LarsonPrecalculus.com*

The table shows the shoe sizes x and the heights y (in inches) of eight men. Find the sum of the squared differences for the linear model $y^* = 1.87x + 51.2$.

DATA x	8.5	9.0	9.0	9.5	10.0	10.5	11.0	12.0
y	66.0	68.5	67.5	70.0	72.0	69.5	71.5	73.5

Spreadsheet at LarsonPrecalculus.com

Least Squares Regression Lines

The linear model that has the least sum of the squared differences is the **least squares regression line** for the data and is the best-fitting linear model for the data. To find the least squares regression line $y = ax + b$ for the points $(x_1, y_1), (x_2, y_2), \ldots, (x_n, y_n)$ algebraically, solve the system below for a and b.

•• **REMARK** Deriving this system requires the use of *partial derivatives*, which you will study in a calculus course.

$$\begin{cases} nb + \left(\displaystyle\sum_{i=1}^{n} x_i \right)a = \displaystyle\sum_{i=1}^{n} y_i \\ \left(\displaystyle\sum_{i=1}^{n} x_i \right)b + \left(\displaystyle\sum_{i=1}^{n} x_i^2 \right)a = \displaystyle\sum_{i=1}^{n} x_i y_i \end{cases}$$

EXAMPLE 2 **Finding a Least Squares Regression Line**

Find the least squares regression line for the points $(-3, 0)$, $(-1, 1)$, $(0, 2)$, and $(2, 3)$.

Solution Construct a table to find the coefficients and constants of the system of equations for the least squares regression line.

x	y	xy	x^2
-3	0	0	9
-1	1	-1	1
0	2	0	0
2	3	6	4
$\displaystyle\sum_{i=1}^{n} x_i = -2$	$\displaystyle\sum_{i=1}^{n} y_i = 6$	$\displaystyle\sum_{i=1}^{n} x_i y_i = 5$	$\displaystyle\sum_{i=1}^{n} x_i^2 = 14$

Apply the system for the least squares regression line with $n = 4$.

$$\begin{cases} nb + \left(\displaystyle\sum_{i=1}^{n} x_i\right)a = \displaystyle\sum_{i=1}^{n} y_i \\ \left(\displaystyle\sum_{i=1}^{n} x_i\right)b + \left(\displaystyle\sum_{i=1}^{n} x_i^2\right)a = \displaystyle\sum_{i=1}^{n} x_i y_i \end{cases} \implies \begin{cases} 4b - 2a = 6 \\ -2b + 14a = 5 \end{cases}$$

Solving this system of equations produces

$$a = \frac{8}{13} \quad \text{and} \quad b = \frac{47}{26}.$$

So, the least squares regression line is $y = \frac{8}{13}x + \frac{47}{26}$, as shown in Figure 13.3.

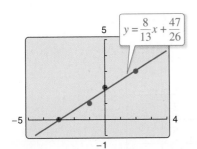

$y = \frac{8}{13}x + \frac{47}{26}$

Figure 13.3

✓ *Checkpoint*  Audio-video solution in English & Spanish at LarsonPrecalculus.com

Find the least squares regression line for the points $(-1, -1)$, $(0, 0)$, $(1, 2)$, and $(2, 4)$.

EXAMPLE 3 **Finding a Correlation Coefficient**

Find the correlation coefficient r for the data given in Example 2. How well does the linear model fit the data?

Solution Using the formula given on page 897 with $n = 4$,

$$r = \frac{n\displaystyle\sum_{i=1}^{n} x_i y_i - \left(\displaystyle\sum_{i=1}^{n} x_i\right)\left(\displaystyle\sum_{i=1}^{n} y_i\right)}{\sqrt{n\displaystyle\sum_{i=1}^{n} x_i^2 - \left(\displaystyle\sum_{i=1}^{n} x_i\right)^2}\sqrt{n\displaystyle\sum_{i=1}^{n} y_i^2 - \left(\displaystyle\sum_{i=1}^{n} y_i\right)^2}}$$

$$= \frac{4(5) - (-2)(6)}{\sqrt{4(14) - (-2)^2}\sqrt{4(14) - 6^2}}$$

$$\approx 0.992.$$

Least squares regression analysis has a wide variety of real-life applications. For example, in Exercise 23, you will find the least squares regression line that models the demand for a tool at a hardware retailer as a function of price.

Because r is close to 1, there is a strong positive linear correlation between x and y. So, the least squares regression line found in Example 2 fits the data very well.

✓ *Checkpoint* 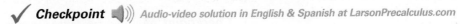 Audio-video solution in English & Spanish at LarsonPrecalculus.com

Find the correlation coefficient r for the data given in the Checkpoint with Example 2. How well does the linear model fit the data?

Least Squares Regression Parabolas

To find the **least squares regression parabola** $y = ax^2 + bx + c$ for the points

$$(x_1, y_1), (x_2, y_2), \ldots, (x_n, y_n)$$

algebraically, solve the system below for a, b, and c.

$$
\begin{cases}
nc + \left(\sum_{i=1}^{n} x_i\right)b + \left(\sum_{i=1}^{n} x_i^2\right)a = \sum_{i=1}^{n} y_i \\
\left(\sum_{i=1}^{n} x_i\right)c + \left(\sum_{i=1}^{n} x_i^2\right)b + \left(\sum_{i=1}^{n} x_i^3\right)a = \sum_{i=1}^{n} x_i y_i \\
\left(\sum_{i=1}^{n} x_i^2\right)c + \left(\sum_{i=1}^{n} x_i^3\right)b + \left(\sum_{i=1}^{n} x_i^4\right)a = \sum_{i=1}^{n} x_i^2 y_i
\end{cases}
$$

▷ **REMARK**   You may recall finding least square regression lines and parabolas in Section 7.2 (Exercises 54–58) and Section 7.3 (Exercises 67–70), respectively. In these exercises, the systems of equations were given because you had not yet studied summation notation.

EXAMPLE 4   **Finding a Least Squares Regression Parabola**

See LarsonPrecalculus.com for an interactive version of this type of example.

Find the least squares regression parabola for the points $(1, 2)$, $(2, 1)$, $(3, 2)$, and $(4, 4)$.

Solution   Begin by constructing a table, as shown below.

x	x^2	x^3	x^4	y	xy	x^2y
1	1	1	1	2	2	2
2	4	8	16	1	2	4
3	9	27	81	2	6	18
4	16	64	256	4	16	64

Use the table to find the sums needed to write the system of equations.

$$\sum_{i=1}^{n} x_i = 10, \quad \sum_{i=1}^{n} x_i^2 = 30, \quad \sum_{i=1}^{n} x_i^3 = 100, \quad \sum_{i=1}^{n} x_i^4 = 354,$$

$$\sum_{i=1}^{n} y_i = 9, \quad \sum_{i=1}^{n} x_i y_i = 26, \quad \sum_{i=1}^{n} x_i^2 y_i = 88$$

Apply the system for the least squares regression parabola with $n = 4$.

$$
\begin{cases}
nc + \left(\sum_{i=1}^{n} x_i\right)b + \left(\sum_{i=1}^{n} x_i^2\right)a = \sum_{i=1}^{n} y_i \\
\left(\sum_{i=1}^{n} x_i\right)c + \left(\sum_{i=1}^{n} x_i^2\right)b + \left(\sum_{i=1}^{n} x_i^3\right)a = \sum_{i=1}^{n} x_i y_i \\
\left(\sum_{i=1}^{n} x_i^2\right)c + \left(\sum_{i=1}^{n} x_i^3\right)b + \left(\sum_{i=1}^{n} x_i^4\right)a = \sum_{i=1}^{n} x_i^2 y_i
\end{cases}
\implies
\begin{cases}
4c + 10b + 30a = 9 \\
10c + 30b + 100a = 26 \\
30c + 100b + 354a = 88
\end{cases}
$$

Solving this system of equations produces $a = \frac{3}{4}$, $b = -\frac{61}{20}$, and $c = \frac{17}{4}$. So, the least squares regression parabola is

$$y = \frac{3}{4}x^2 - \frac{61}{20}x + \frac{17}{4}$$

as shown in Figure 13.4.

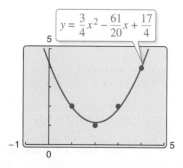

$$y = \frac{3}{4}x^2 - \frac{61}{20}x + \frac{17}{4}$$

Figure 13.4

✓ **Checkpoint** ◀))) *Audio-video solution in English & Spanish at LarsonPrecalculus.com*

Find the least squares regression parabola for the points $(-3, -2)$, $(-2, 0)$, $(-1, 1)$, and $(1, 0)$.

Analyze Misleading Graphs

A *misleading graph* is a graph that is not drawn appropriately. This type of graph can misrepresent data and lead to false conclusions. For example, the bar graph below has a break in the vertical axis, which gives the impression that first quarter sales were disproportionately more than sales in other quarters.

One way to redraw this graph so it is not misleading is to show it without the break, as shown below. Notice that in the redrawn graph, the differences in the quarterly sales do not appear to be as dramatic.

> **REMARK** To recognize when a graph is trying to deceive or mislead, ask yourself these questions:
> - Does the graph have a title?
> - Does the graph need a key?
> - Are the numbers of the scale evenly spaced?
> - Are all the axes or sections of the graph labeled?
> - Does the scale begin at zero? If not, is there a break?
> - Are all the components of the graph, such as the bars, the same size?

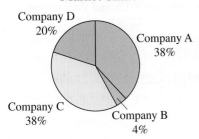

Figure 13.5

Another example of a misleading graph is the circle graph at the right. This graph is shown at an angle that makes the market share of Company C appear larger than it is. Redrawing the graph as shown in Figure 13.5 makes it clearer that Company A and Company C have equal market shares.

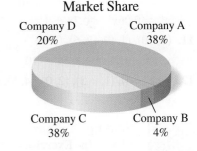

Summarize (Section 13.3)

1. Explain how to use the correlation coefficient to measure how well a model fits a data set *(page 897)*. For an example of finding a correlation coefficient, see Example 3.

2. Explain how to use the sum of the squared differences to measure how well a model fits a data set *(page 898)*. For an example of finding the sum of the squared differences for a linear model, see Example 1.

3. List the steps involved in finding the least squares regression line for a data set *(page 898)*. For an example of finding the least squares regression line for a data set, see Example 2.

4. List the steps involved in finding the least squares regression parabola for a data set *(page 900)*. For an example of finding the least squares regression parabola for a data set, see Example 4.

5. Give examples of misleading graphs *(page 901)*.

13.3   Exercises

See **CalcChat.com** for tutorial help and worked-out solutions to odd-numbered exercises.

Vocabulary: Fill in the blanks.

1. A _____ is a relationship between two variables.
2. A graphing utility uses the _____ of _____ _____ to find a mathematical model for a data set.
3. The sum of the _____ _____ measures how well a model fits a set of data points.
4. The _____ _____ _____ line for a data set is the linear model that has the least sum of the squared differences.

Skills and Applications

Finding the Sum of the Squared Differences In Exercises 5–12, find the sum of the squared differences for the given data points and the model.

5. $(0, 2), (1, 1), (2, 2), (3, 4), (5, 6), \quad y^* = 0.8x + 2$
6. $(0, 7), (2, 5), (3, 3), (4, 3), (6, 0), \quad y^* = -1.2x + 7$
7. $(-3, -1), (-1, 0), (0, 2), (2, 3), (4, 4)$
   $y^* = 0.5x + 0.5$
8. $(-2, 6), (-1, 4), (0, 2), (1, 1), (2, 1)$
   $y^* = -1.7x + 2.7$
9. $(-3, 2), (-2, 2), (-1, 4), (0, 6), (1, 8)$
   $y^* = 0.29x^2 + 2.2x + 6$
10. $(-3, 4), (-1, 2), (1, 1), (3, 0)$
    $y^* = 0.06x^2 - 0.7x + 1$
11. $(0, 10), (1, 9), (2, 6), (3, 0)$
    $y^* = -1.25x^2 + 0.5x + 10$
12. $(-1, -4), (1, -3), (2, 0), (4, 5), (6, 9)$
    $y^* = 0.14x^2 + 1.3x - 3$

Finding a Least Squares Regression Line In Exercises 13–16, find the least squares regression line for the points. Use a graphing utility to verify your answer.

13. $(-4, 1), (-3, 3), (-2, 4), (-1, 6)$
14. $(0, -1), (2, 0), (4, 3), (6, 5)$
15. $(-3, 18), (-1, 9), (2, 3), (4, -8)$
16. $(0, -1), (2, 1), (3, 2), (5, 3)$

17. **Finding a Correlation Coefficient** Find the correlation coefficient r for the data given in Exercise 13. How well does the linear model fit the data?
18. **Finding a Correlation Coefficient** Find the correlation coefficient r for the data given in Exercise 14. How well does the linear model fit the data?
19. **Finding a Correlation Coefficient** Find the correlation coefficient r for the data given in Exercise 15. How well does the linear model fit the data?

20. **Finding a Correlation Coefficient** Find the correlation coefficient r for the data given in Exercise 16. How well does the linear model fit the data?

21. **Healthcare Costs**

The average annual costs of healthcare for a family of four in the United States from 2012 through 2015 are represented by the ordered pairs (x, y), where x represents the year, with $x = 12$ corresponding to 2012, and y represents the average cost (in thousands of dollars). Find the least squares regression line for the data. What is the sum of the squared differences? *(Source: Milliman)*

$(12, 20.73), (13, 22.03), (14, 23.22), (15, 24.67)$

22. **College Enrollment** The projected enrollments in U.S. private colleges from 2017 through 2020 are represented by the ordered pairs (x, y), where x represents the year, with $x = 17$ corresponding to 2017, and y represents the projected enrollment (in millions of students). Find the least squares regression line for the data. What is the sum of the squared differences? *(Source: U.S. National Center for Education Statistics)*

$(17, 6.137), (18, 6.217), (19, 6.290), (20, 6.340)$

23. **Demand** A hardware retailer wants to know the demand y for a tool as a function of price x. The table lists the monthly sales for four different prices of the tool.

Price, x	$25	$30	$35	$40
Demand, y	82	75	67	55

(a) Find the least squares regression line for the data.
(b) Estimate the demand when the price is $32.95.
(c) What price will create a demand of 83 tools?

24. Agriculture An agronomist used four test plots to determine the relationship between the amount of fertilizer x (in pounds per acre) and the wheat yield y (in bushels per acre). The table shows the results.

Fertilizer, x	100	150	200	250
Yield, y	35	44	50	56

(a) Find the least squares regression line for the data.

(b) Estimate the yield for a fertilizer application of 160 pounds per acre.

(c) How much fertilizer should the agronomist apply to achieve a yield of 70 bushels per acre?

 Finding a Least Squares Regression Parabola In Exercises 25–32, find the least squares regression parabola for the points. Use a graphing utility to verify your answer.

25. $(0, 0), (2, 4), (4, 2)$ **26.** $(-2, 6), (-1, 2), (1, 3)$

27. $(-1, 4), (0, 2), (1, 0), (3, 4)$

28. $(-3, -1), (-1, 2), (1, 2), (3, 0)$

29. $(-2, 18), (-1, 9), (0, 4), (1, 3), (2, 6)$

30. $(-2, -17), (0, 0), (2, 4), (4, -3), (6, -25)$

31.
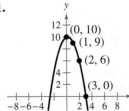

32.
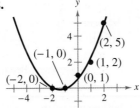

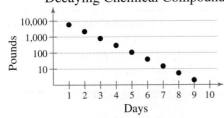 **Misleading Graph** In Exercises 33 and 34, explain why the graph could be misleading.

33.
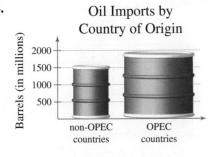

34.
Oil Imports by Country of Origin

True or False? In Exercises 35 and 36, determine whether the statement is true or false. Justify your answer.

35. A correlation coefficient of $r \approx 0.021$ implies that the data have weak or no linear correlation.

36. A linear regression model with a positive correlation coefficient has a graph with a slope that is greater than 0.

37. Error Analysis Describe the error in interpreting the graph.

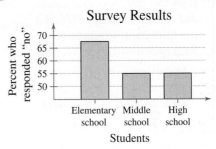

The percent of elementary school students who responded "no" is more than twice the percent of middle school students who responded "no."

38. HOW DO YOU SEE IT? Match each correlation coefficient r with the corresponding scatter plot. Explain your reasoning. [The scatter plots are labeled (i), (ii), (iii), and (iv).]

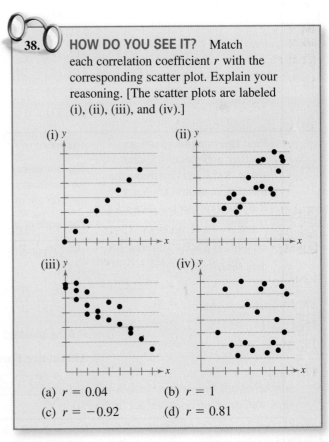

(a) $r = 0.04$ (b) $r = 1$
(c) $r = -0.92$ (d) $r = 0.81$

39. Writing A linear model for predicting prize winnings at a race is based on data for 3 years. Write a paragraph discussing the potential accuracy or inaccuracy of such a model.

Chapter Summary

What Did You Learn?	Explanation/Examples	Review Exercises

Section 13.1

Understand terminology associated with statistics *(p. 878)*, use line plots to order and analyze data *(p. 880)*, use stem-and-leaf plots to organize and compare data *(p. 881)*, and use histograms to represent frequency distributions *(p. 882).*

Data consist of information that comes from observations, responses, counts, or measurements. Samples are subsets, or parts, of a population. Types of sampling methods include random, stratified random, systematic, convenience, and self-selected. See pages 878 and 879 for more terminology.

Line Plot

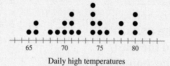

Daily high temperatures

Stem-and-Leaf Plot

Stems	Leaves
5	8
6	4 4 6 9
7	0 0 4 4 4 6 6 6 6 6 8 8 9
8	1 2 2 3 3 3 6 6 8
9	3 3 6

Key: $5|8 = 58$

Frequency Distribution

Interval	Tally
[9, 11)	‖
[11, 13)	‖‖‖
[13, 15)	‖‖‖ ‖‖‖ ‖‖‖ ‖‖‖‖
[15, 17)	‖‖‖ ‖‖‖ ‖‖‖ ‖‖‖ ‖‖
[17, 19)	‖‖
[19, 21)	‖

Histogram

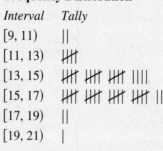

Percent of population 65 or older

1–4

Section 13.2

Find and interpret the mean, median, and mode of a data set *(p. 885),* and determine the measure of central tendency that best represents a data set *(p. 886).*

The mean of n numbers is the sum of the numbers divided by n.

The numerical median of n numbers is the middle number when the numbers are written in order. When n is even, the median is the average of the two middle numbers.

The mode of n numbers is the number that occurs most frequently.

5–10

Find the standard deviation of a data set *(p. 887).*

The variance of a set of numbers $\{x_1, x_2, \ldots, x_n\}$ with a mean of $\bar{x}$ is

$$v = \frac{(x_1 - \bar{x})^2 + (x_2 - \bar{x})^2 + \cdots + (x_n - \bar{x})^2}{n}$$

and the standard deviation is $\sigma = \sqrt{v}$.

Alternative Formula for Standard Deviation

$$\sigma = \sqrt{\frac{x_1^2 + x_2^2 + \cdots + x_n^2}{n} - \bar{x}^2}$$

11–16

Create and use box-and-whisker plots *(p. 890).*

Box-and-Whisker Plot

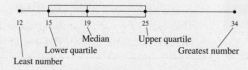

17, 18

What Did You Learn?	**Explanation/Examples**	**Review Exercises**

Section 13.2

| Interpret normally distributed data (p. 891). | Normal distribution with mean $\bar{x}$ and standard deviation σ: Using the formula $z = (x - \bar{x})/\sigma$ transforms an x-value from a normal distribution with mean $\bar{x}$ and standard deviation σ into a corresponding z-score having a standard normal distribution. | 19–30 |

Section 13.3

Use the correlation coefficient to measure how well a model fits a data set (p. 897).	The correlation coefficient r gives a measure of how well a model fits a data set. A formula for r is $$r = \frac{n\sum_{i=1}^{n} x_i y_i - \left(\sum_{i=1}^{n} x_i\right)\left(\sum_{i=1}^{n} y_i\right)}{\sqrt{n\sum_{i=1}^{n} x_i^2 - \left(\sum_{i=1}^{n} x_i\right)^2}\sqrt{n\sum_{i=1}^{n} y_i^2 - \left(\sum_{i=1}^{n} y_i\right)^2}}$$ where n is the number of data points (x, y). For a linear model, r close to 1 indicates a strong positive linear correlation between x and y, r close to -1 indicates a strong negative linear correlation, and r close to 0 indicates a weak or no *linear* correlation.	31, 32
Use the sum of the squared differences to measure how well a model fits a data set (p. 898).	As a measure of how well a model fits a set of data points, add the squares of the differences between the actual y-values and the values given by the model to obtain the sum of the squared differences.	33, 34
Find the least squares regression line (p. 898) and the least squares regression parabola (p. 900) for data sets.	To find the least squares regression line $y = ax + b$ for the points $(x_1, y_1), (x_2, y_2), \ldots, (x_n, y_n)$ algebraically, solve the system below for a and b. $$\begin{cases} nb + \left(\sum_{i=1}^{n} x_i\right)a = \sum_{i=1}^{n} y_i \\ \left(\sum_{i=1}^{n} x_i\right)b + \left(\sum_{i=1}^{n} x_i^2\right)a = \sum_{i=1}^{n} x_i y_i \end{cases}$$ To find the least squares regression parabola $y = ax^2 + bx + c$ for the points $(x_1, y_1), (x_2, y_2), \ldots, (x_n, y_n)$ algebraically, solve the system below for a, b, and c. $$\begin{cases} nc + \left(\sum_{i=1}^{n} x_i\right)b + \left(\sum_{i=1}^{n} x_i^2\right)a = \sum_{i=1}^{n} y_i \\ \left(\sum_{i=1}^{n} x_i\right)c + \left(\sum_{i=1}^{n} x_i^2\right)b + \left(\sum_{i=1}^{n} x_i^3\right)a = \sum_{i=1}^{n} x_i y_i \\ \left(\sum_{i=1}^{n} x_i^2\right)c + \left(\sum_{i=1}^{n} x_i^3\right)b + \left(\sum_{i=1}^{n} x_i^4\right)a = \sum_{i=1}^{n} x_i^2 y_i \end{cases}$$	35–38
Analyze misleading graphs (p. 901).	A misleading graph is a graph that is not drawn appropriately. This type of graph can misrepresent data and lead to false conclusions. (See page 901 for examples.)	39, 40

Review Exercises See CalcChat.com for tutorial help and worked-out solutions to odd-numbered exercises.

13.1 Constructing a Line Plot In Exercises 1 and 2, construct a line plot for the data.

1. 11, 14, 21, 17, 13, 13, 17, 17, 18, 14, 21, 11, 14, 15

2. 6, 8, 7, 9, 4, 4, 6, 8, 8, 8, 9, 5, 5, 9, 9, 4, 2, 2, 6, 8

3. Home Runs The numbers of home runs hit by the 20 baseball players with the best single-season batting averages in Major League Baseball since 1900 are shown below. Construct a stem-and-leaf plot for the data. (*Source: Major League Baseball*) (*Spreadsheet at LarsonPrecalculus.com*)

DATA 18, 4, 42, 33, 7, 19, 20, 9, 29, 15, 17, 15, 16, 2, 23, 6, 9, 44, 6, 7

4. Sandwich Prices The prices (in dollars) of sandwiches at a restaurant are shown below. Use a frequency distribution and a histogram to organize the data. (*Spreadsheet at LarsonPrecalculus.com*)

DATA 4.00, 4.00, 4.25, 4.50, 4.75, 4.25, 5.95, 5.50, 5.50, 5.75, 6.25

13.2 Finding Measures of Central Tendency In Exercises 5–8, find the mean, median, and mode(s) of the data set.

5. 6, 13, 8, 15, 9, 10, 8

6. 29, 36, 31, 38, 32, 31

7. 4, 11, 6, 13, 8, 6

8. 31, 38, 33, 40, 35, 33, 31

Choosing a Measure of Central Tendency In Exercises 9 and 10, determine which measure of central tendency is the most representative of the data shown in the frequency distribution.

9.

Number	1	2	3	4	5	6	7	8	9
Frequency	3	2	5	6	2	1	0	0	1

10.

Number	1	2	3	4	5	6	7	8
Frequency	0	5	12	1	1	12	3	1

Finding Mean and Standard Deviation In Exercises 11 and 12, each line plot represents a data set. Find the mean and standard deviation of each data set.

11. (a)

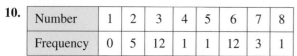

(b)

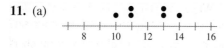

12. (a)

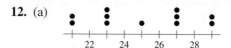

(b)

Finding Mean, Variance, and Standard Deviation In Exercises 13 and 14, find the mean ($\bar{x}$), variance (v), and standard deviation (σ) of the data set.

13. 1, 2, 3, 4, 5, 6, 7

14. 1, 1, 1, 5, 5, 5

Using the Alternative Formula for Standard Deviation In Exercises 15 and 16, use the alternative formula for standard deviation to find the standard deviation of the data set.

15. 2, 5, 5, 7, 7, 9, 11

16. 9.3, 8.5, 12.0, 11.6, 10.1

Quartiles and Box-and-Whisker Plots In Exercises 17 and 18, find the lower and upper quartiles and sketch a box-and-whisker plot for the data set.

17. 23, 15, 14, 23, 13, 14, 13, 20, 12

18. 78.4, 76.3, 107.5, 78.5, 93.2, 90.3, 77.8, 37.1, 97.1, 75.5, 58.8, 65.6

Using a Normal Curve In Exercises 19 and 20, determine the percent of the area under the normal curve represented by the shaded region.

19.

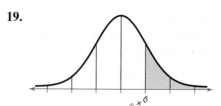

20.

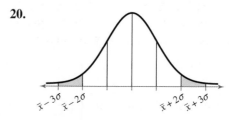

Finding a Normal Probability In Exercises 21–24, a normal distribution has mean $\bar{x}$ and standard deviation σ. Find the indicated probability for a randomly selected x-value from the distribution.

21. $P(x \leq \bar{x} + \sigma)$ **22.** $P(x \geq \bar{x} - \sigma)$

23. $P(21 \leq x \leq 41)$, when $\bar{x} = 33$ and $\sigma = 4$

24. $P(x \geq 25)$, when $\bar{x} = 33$ and $\sigma = 4$

Using the Standard Normal Table In Exercises 25–28, a normal distribution has a mean of 64 and a standard deviation of 7. Use the standard normal table to find the indicated probability for a randomly selected x-value from the distribution.

25. $P(x \leq 75)$ **26.** $P(x \geq 66)$

27. $P(52 \leq x \leq 59)$

28. $P(x \leq 58 \text{ or } x \geq 70)$

29. Fire Department The time a fire department takes to arrive at the scene of an emergency is normally distributed with a mean of 6 minutes and a standard deviation of 1 minute.

(a) What is the probability that the fire department takes at most 8 minutes to arrive at the scene of an emergency?

(b) What is the probability that the fire department takes between 4 and 7 minutes to arrive at the scene of an emergency?

30. College Entrance Tests Two students took different college entrance tests. The scores on the test that the first student took are normally distributed with a mean of 20 points and a standard deviation of 4.2 points. The scores on the test that the second student took are normally distributed with a mean of 500 points and a standard deviation of 90 points. The students scored 30 and 610 on their tests, respectively.

(a) Find the z-score for the first student's test score.

(b) Find the z-score for the second student's test score.

(c) Which student scored better on their college entrance test? Explain.

13.3 Finding the Correlation Coefficient In Exercises 31 and 32, find the correlation coefficient r of the data set and describe the correlation.

31. $(-3, 2), (-2, 3), (-1, 4), (0, 6)$

32. $(-3, 4), (-1, 2), (1, 1), (3, 0)$

Finding the Sum of the Squared Differences In Exercises 33 and 34, find the sum of the squared differences for the set of data points and the model.

33. $(0, 11), (1, 9), (2, 5), (3, 0)$

$y^* = -3.7x + 12$

34. $(-1, -4), (1, -3), (2, 0), (4, 5), (6, 9)$

$y^* = 2.0x - 3$

Finding a Least Squares Regression Line In Exercises 35 and 36, find the least squares regression line for the points. Use a graphing utility to verify your answer.

35. $(0, 1), (1, 7), (2, 12)$

36. $(-1, 5), (0, 3), (1, 1), (2, -1)$

Finding a Least Squares Regression Parabola In Exercises 37 and 38, find the least squares regression parabola for the points. Use a graphing utility to verify your answer.

37. $(0, 1), (1, 8), (2, 1)$

38. $(-5, 5), (-4, 1), (-3, 0), (-2, 1), (-1, 4)$

Misleading Graph In Exercises 39 and 40, explain why the graph could be misleading.

39.

Sales

40.

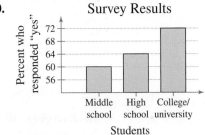

Survey Results

Exploration

True or False? In Exercises 41–46, determine whether the statement is true or false. Justify your answer.

41. Populations are collections of some of the outcomes, measurements, counts, or responses that are of interest.

42. Inferential statistics involves drawing conclusions about a sample using a population.

43. The measure of central tendency that is most likely to be affected by the presence of an outlier is the mean.

44. It is possible for the mean, median, and mode of a data set to be equal.

45. Data that are modeled by

$y = 3.29 - 4.17x$

have a negative linear correlation.

46. When the correlation coefficient for a linear regression model is close to -1, the regression line is not a good fit for describe the data.

47. Reasoning When the standard deviation of a data set of numbers is 0, what does this imply about the set?

48. Think About It Construct a collection of numbers that has the following measures of central tendency.

Mean $= 6$,   median $= 6$,   mode $= 4$

Chapter Test

See **CalcChat.com** for tutorial help and worked-out solutions to odd-numbered exercises.

Take this test as you would take a test in class. When you are finished, check your work against the answers given in the back of the book.

In Exercises 1–5, use the data set of the ages of all customers at an electronics store listed below.

14, 16, 45, 7, 32, 32, 14, 16, 26, 46, 27, 43, 33, 27, 37, 12, 20, 40, 34, 16

1. Construct a line plot for the data.

2. Construct a stem-and-leaf plot for the data.

3. Construct a frequency distribution for the data.

4. Construct a histogram for the data.

5. You arrange the ages in increasing order, and then select the 4th, 8th, 12th, 16th, and 20th data value. Identify the type of sample you selected.

In Exercises 6 and 7, find the mean, median, and mode of the data set.

6. 3, 10, 5, 12, 6, 7, 5

7. 25, 39, 36, 43, 37, 36

In Exercises 8 and 9, find the mean ($\bar{x}$), variance (v), and standard deviation (σ) of the data set.

8. 0, 1, 2, 3, 4, 5, 6

9. 2, 2, 4, 5, 7, 7

In Exercises 10 and 11, find the lower and upper quartiles and sketch a box-and-whisker plot for the data set.

10. 9, 5, 5, 5, 6, 5, 4, 12, 7, 10, 7, 11, 8, 9, 9

11. 25, 20, 22, 28, 24, 28, 25, 19, 27, 29, 28, 21

In Exercises 12 and 13, a normal distribution has a mean of 25 and a standard deviation of 3. Find the probability that a randomly selected x-value from the distribution is in the given interval.

12. Between 19 and 31

13. At most 22

In Exercises 14–16, consider the points (0, 2), (3, 4), and (4, 5).

14. Find the least squares regression line for the points. Use a graphing utility to verify your answer.

15. Find the sum of the squared differences using the linear model you found in Exercise 14.

16. Find the correlation coefficient r. How well does the linear model you found in Exercise 14 fit the data?

17. Find the least squares regression parabola for the points $(-2, 10)$, $(-1, 4)$, $(0, 2)$, and $(1, 2)$. Use a graphing utility to verify your answer.

18. The guayule plant is one of several plants used as a source of rubber. In a large group of guayule plants, the heights of the plants are normally distributed with a mean of 12 inches and a standard deviation of 2 inches.

    (a) What percent of the plants are taller than 17 inches?

    (b) What percent of the plants are between 7 and 14 inches?

    (c) What percent of the plants are at least 3 inches taller than or at least 3 inches shorter than the mean height?

Proofs in Mathematics

> **Alternative Formula for Standard Deviation** *(p. 888)*
>
> The standard deviation of $\{x_1, x_2, \ldots, x_n\}$ is
>
> $$\sigma = \sqrt{\frac{x_1^2 + x_2^2 + \cdots + x_n^2}{n} - \bar{x}^2}.$$

Proof

To prove this formula, begin with the formula for standard deviation given in the definition on page 887.

$$\sigma = \sqrt{v} \qquad \text{Square root of variance}$$

$$= \sqrt{\frac{(x_1 - \bar{x})^2 + (x_2 - \bar{x})^2 + \cdots + (x_n - \bar{x})^2}{n}} \qquad \text{Substitute definition of variance.}$$

$$= \sqrt{\frac{(x_1^2 - 2x_1\bar{x} + \bar{x}^2) + (x_2^2 - 2x_2\bar{x} + \bar{x}^2) + \cdots + (x_n^2 - 2x_n\bar{x} + \bar{x}^2)}{n}} \qquad \text{Square binomials.}$$

$$= \sqrt{\frac{(x_1^2 + x_2^2 + \cdots + x_n^2) - 2(x_1\bar{x} + x_2\bar{x} + \cdots + x_n\bar{x}) + \underbrace{(\bar{x}^2 + \bar{x}^2 + \cdots + \bar{x}^2)}_{n \text{ terms}}}{n}} \qquad \text{Regroup terms in numerator.}$$

$$= \sqrt{\frac{(x_1^2 + x_2^2 + \cdots + x_n^2)}{n} - \frac{2(x_1\bar{x} + x_2\bar{x} + \cdots + x_n\bar{x})}{n} + \frac{n\bar{x}^2}{n}} \qquad \text{Rewrite radicand as three fractions.}$$

$$= \sqrt{\frac{(x_1^2 + x_2^2 + \cdots + x_n^2)}{n} - \frac{2\bar{x}(x_1 + x_2 + \cdots + x_n)}{n} + \frac{n\bar{x}^2}{n}} \qquad \text{Factor } \bar{x} \text{ out of numerator of second fraction.}$$

$$= \sqrt{\frac{(x_1^2 + x_2^2 + \cdots + x_n^2)}{n} - 2\bar{x}^2 + \bar{x}^2} \qquad \text{Definition of } \bar{x}; \text{ divide out } n \text{ in third fraction.}$$

$$= \sqrt{\frac{(x_1^2 + x_2^2 + \cdots + x_n^2)}{n} - \bar{x}^2} \qquad \text{Simplify.}$$

P.S. Problem Solving ▪ ▪ ▪ ▪ ▪ ▪ ▪ ▪ ▪ ▪ ▪ ▪ ▪ ▪ ▪ ▪

1. **U.S. History** The first 20 states were admitted in the Union in the following years. Construct a stem-and-leaf plot of the data. *(Spreadsheet at LarsonPrecalculus.com)*

   DATA 1788, 1787, 1788, 1816, 1792, 1812, 1788, 1788, 1817, 1788, 1787, 1788, 1789, 1803, 1787, 1790, 1788, 1796, 1791, 1788

2. **Mayflower** The known ages (in years) of adult male passengers on the *Mayflower* at the time of its departure are listed below. *(Spreadsheet at LarsonPrecalculus.com)*

   DATA 21, 34, 29, 38, 30, 54, 39, 20, 35, 64, 37, 45, 21, 25, 55, 45, 40, 38, 38, 21, 21, 34, 38, 52, 41, 48, 18, 27, 32, 49, 30, 42, 30, 25, 38, 25, 20

   (a) Construct a stem-and-leaf plot of the ages.

   (b) Find the median age, and the range of the ages.

   (c) According to one source, the age of passenger Thomas English was unknown at the time of the *Mayflower's* departure. What is the probability that he was 18–29 years old? Explain.

   (d) The first two rows of a *cumulative frequency distribution* for the data are shown below. The *cumulative frequency* for a given interval is the sum of the current frequency and all preceding frequencies. A histogram constructed from a cumulative frequency distribution is called a *cumulative frequency histogram*. Copy and complete the cumulative frequency distribution. Then construct a cumulative frequency histogram for the data.

Interval	Frequency	Cumulative Frequency
[15, 25)	7	7
[25, 35)	11	7 + 11 = 18
?	?	?

3. **Organizing Data** Construct a line plot, frequency distribution, and histogram for the data given in (a) Exercise 1 and (b) Exercise 2.

4. **Constructing a Stem-and-Leaf Plot** Construct a stem-and-leaf plot that has the same distribution of data as the histogram shown below. Describe the process you used.

   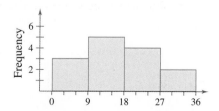

5. **Think About It** Two data sets have the same mean, interquartile range, and range. Is it possible for the box-and-whisker plots of such data sets to be different? If so, give an example.

6. **Singers in a Chorus** The box-and-whisker plots below show the heights (in inches) of singers in a chorus, according to their voice parts. A soprano part has the highest pitch, followed by alto, tenor, and bass, respectively. Draw a conclusion about voice parts and heights. Justify your conclusion.

   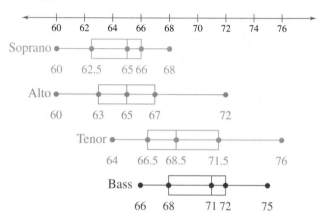

7. **Equation of a Normal Curve** A normal curve is defined by an equation of the form

   $$y = \frac{1}{\sigma\sqrt{2\pi}}e^{-(x-\bar{x})^2/(2\sigma^2)}$$

   where $\bar{x}$ is the mean and σ is the standard deviation of the corresponding normal distribution.

   (a) Use a graphing utility to graph three equations of the given form, where $\bar{x}$ is held constant and σ varies.

   (b) Describe the effect that the standard deviation has on the shape of a normal curve.

8. **Height** According to a survey by the National Center for Health Statistics, the heights of male adults in the United States are normally distributed with a mean of approximately 69 inches and a standard deviation of approximately 3.1 inches.

   (a) What is the probability that 3 randomly selected U.S. male adults are all more than 6 feet tall?

   (b) What is the probability that 5 randomly selected U.S. male adults are all between 65 and 75 inches tall?

9. **Test Scores** The scores of a mathematics exam given to 600 science and engineering students at a college had a mean of 235 and a standard deviation of 28. Use Chebychev's Theorem to determine the intervals containing at least $\frac{3}{4}$ and at least $\frac{8}{9}$ of the scores. How would the intervals change if the standard deviation was 16?

Appendix A   Review of Fundamental Concepts of Algebra

A.1   Real Numbers and Their Properties

Real numbers can represent many real-life quantities. For example, in Exercises 49–52 on page A12, you will use real numbers to represent the federal surplus or deficit.

■ Represent and classify real numbers.
■ Order real numbers and use inequalities.
■ Find the absolute values of real numbers and find the distance between two real numbers.
■ Evaluate algebraic expressions.
■ Use the basic rules and properties of algebra.

Real Numbers

Real numbers can describe quantities in everyday life such as age, miles per gallon, and population. Symbols such as

$$-5, \ 9, \ 0, \ \tfrac{4}{3}, \ 0.666\ldots, \ 28.21, \ \sqrt{2}, \ \pi, \ \text{and} \ \sqrt[3]{-32}$$

represent real numbers. Here are some important **subsets** (each member of a subset B is also a member of a set A) of the real numbers. The three dots, or *ellipsis points*, tell you that the pattern continues indefinitely.

$$\{1, 2, 3, 4, \ldots\} \qquad \text{Set of natural numbers}$$

$$\{0, 1, 2, 3, 4, \ldots\} \qquad \text{Set of whole numbers}$$

$$\{\ldots, -3, -2, -1, 0, 1, 2, 3, \ldots\} \qquad \text{Set of integers}$$

A real number is **rational** when it can be written as the ratio p/q of two integers, where $q \neq 0$. For example, the numbers

$$\tfrac{1}{3} = 0.3333\ldots = 0.\overline{3}, \quad \tfrac{1}{8} = 0.125, \quad \text{and} \quad \tfrac{125}{111} = 1.126126\ldots = 1.\overline{126}$$

are rational. The decimal representation of a rational number either repeats $\left(\text{as in } \tfrac{173}{55} = 3.1\overline{45}\right)$ or terminates $\left(\text{as in } \tfrac{1}{2} = 0.5\right)$. A real number that cannot be written as the ratio of two integers is **irrational.** The decimal representation of an irrational number neither terminates nor repeats. For example, the numbers

$$\sqrt{2} = 1.4142135\ldots \approx 1.41 \quad \text{and} \quad \pi = 3.1415926\ldots \approx 3.14$$

are irrational. (The symbol $\approx$ means "is approximately equal to.") Figure A.1 shows subsets of the real numbers and their relationships to each other.

```
          Real
        numbers
        /       \
Irrational    Rational
numbers       numbers
             /        \
        Integers    Noninteger
                    fractions
                    (positive and
                    negative)
        /        \
Negative      Whole
integers     numbers
            /       \
       Natural      Zero
       numbers
```

Subsets of the real numbers
Figure A.1

EXAMPLE 1   Classifying Real Numbers

Determine which numbers in the set $\left\{-13, -\sqrt{5}, -1, -\tfrac{1}{3}, 0, \tfrac{5}{8}, \sqrt{2}, \pi, 7\right\}$ are (a) natural numbers, (b) whole numbers, (c) integers, (d) rational numbers, and (e) irrational numbers.

Solution

a. Natural numbers: $\{7\}$

b. Whole numbers: $\{0, 7\}$

c. Integers: $\{-13, -1, 0, 7\}$

d. Rational numbers: $\left\{-13, -1, -\tfrac{1}{3}, 0, \tfrac{5}{8}, 7\right\}$

e. Irrational numbers: $\left\{-\sqrt{5}, \sqrt{2}, \pi\right\}$

✓ *Checkpoint* ◀))) *Audio-video solution in English & Spanish at LarsonPrecalculus.com*

Repeat Example 1 for the set $\left\{-\pi, -\tfrac{1}{4}, \tfrac{6}{3}, \tfrac{1}{2}\sqrt{2}, -7.5, -1, 8, -22\right\}$. ■

Real numbers are represented graphically on the **real number line.** When you draw a point on the real number line that corresponds to a real number, you are **plotting** the real number. The point representing 0 on the real number line is the **origin.** Numbers to the right of 0 are positive, and numbers to the left of 0 are negative, as shown in the figure below. The term **nonnegative** describes a number that is either positive or zero.

As the next two number lines illustrate, there is a *one-to-one correspondence* between real numbers and points on the real number line.

Every real number corresponds to exactly one point on the real number line.

Every point on the real number line corresponds to exactly one real number.

EXAMPLE 2   **Plotting Points on the Real Number Line**

Plot the real numbers on the real number line.

a. $-\dfrac{7}{4}$

b. 2.3

c. $\dfrac{2}{3}$

d. -1.8

Solution   The figure below shows all four points.

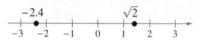

a. The point representing the real number $-\frac{7}{4} = -1.75$ lies between -2 and -1, but closer to -2, on the real number line.

b. The point representing the real number 2.3 lies between 2 and 3, but closer to 2, on the real number line.

c. The point representing the real number $\frac{2}{3} = 0.666\ldots$ lies between 0 and 1, but closer to 1, on the real number line.

d. The point representing the real number -1.8 lies between -2 and -1, but closer to -2, on the real number line. Note that the point representing -1.8 lies slightly to the left of the point representing $-\frac{7}{4}$.

✓ *Checkpoint* ◀))) *Audio-video solution in English & Spanish at LarsonPrecalculus.com*

Plot the real numbers on the real number line.

a. $\dfrac{5}{2}$ b. -1.6

c. $-\dfrac{3}{4}$ d. 0.7

Ordering Real Numbers

One important property of real numbers is that they are *ordered*.

> ### Definition of Order on the Real Number Line
> If a and b are real numbers, then a is *less than b* when $b - a$ is positive. The **inequality** $a < b$ denotes the **order** of a and b. This relationship can also be described by saying that b is *greater than a* and writing $b > a$. The inequality $a \le b$ means that a is *less than or equal to b*, and the inequality $b \ge a$ means that b is *greater than or equal to a*. The symbols $<$, $>$, $\le$, and $\ge$ are *inequality symbols*.

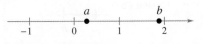

$a < b$ if and only if a lies to the left of b.

Figure A.2

Geometrically, this definition implies that $a < b$ if and only if a lies to the *left* of b on the real number line, as shown in Figure A.2.

EXAMPLE 3 **Ordering Real Numbers**

Place the appropriate inequality symbol ($<$ or $>$) between the pair of real numbers.

a. $-3, 0$ **b.** $-2, -4$ **c.** $\frac{1}{4}, \frac{1}{3}$

Solution

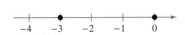

Figure A.3

a. On the real number line, -3 lies to the left of 0, as shown in Figure A.3. So, you can say that -3 is *less than* 0, and write $-3 < 0$.

Figure A.4

b. On the real number line, -2 lies to the right of -4, as shown in Figure A.4. So, you can say that -2 is *greater than* -4, and write $-2 > -4$.

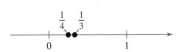

Figure A.5

c. On the real number line, $\frac{1}{4}$ lies to the left of $\frac{1}{3}$, as shown in Figure A.5. So, you can say that $\frac{1}{4}$ is *less than* $\frac{1}{3}$, and write $\frac{1}{4} < \frac{1}{3}$.

✓ **Checkpoint** ◀))) *Audio-video solution in English & Spanish at LarsonPrecalculus.com*

Place the appropriate inequality symbol ($<$ or $>$) between the pair of real numbers.

a. $1, -5$ **b.** $\frac{3}{2}, 7$ **c.** $-\frac{2}{3}, -\frac{3}{4}$

EXAMPLE 4 **Interpreting Inequalities**

See LarsonPrecalculus.com for an interactive version of this type of example.

Describe the subset of real numbers that the inequality represents.

a. $x \le 2$ **b.** $-2 \le x < 3$

Solution

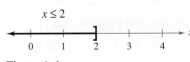

Figure A.6

a. The inequality $x \le 2$ denotes all real numbers less than or equal to 2, as shown in Figure A.6.

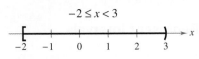

Figure A.7

b. The inequality $-2 \le x < 3$ means that $x \ge -2$ *and* $x < 3$. This "double inequality" denotes all real numbers between -2 and 3, including -2 but not including 3, as shown in Figure A.7.

✓ **Checkpoint** ◀))) *Audio-video solution in English & Spanish at LarsonPrecalculus.com*

Describe the subset of real numbers that the inequality represents.

a. $x > -3$ **b.** $0 < x \le 4$

Inequalities can describe subsets of real numbers called **intervals.** In the bounded intervals below, the real numbers a and b are the **endpoints** of each interval. The endpoints of a closed interval are included in the interval, whereas the endpoints of an open interval are not included in the interval.

$\cdots\cdots\cdots\cdots\cdots\triangleright$

· · REMARK The reason that the four types of intervals at the right are called *bounded* is that each has a finite length. An interval that does not have a finite length is *unbounded* (see below).

Bounded Intervals on the Real Number Line

Notation	Interval Type	Inequality	Graph
$[a, b]$	Closed	$a \le x \le b$	
(a, b)	Open	$a < x < b$	
$[a, b)$		$a \le x < b$	
$(a, b]$		$a < x \le b$	

The symbols ∞, **positive infinity,** and $-\infty$, **negative infinity,** do not represent real numbers. They are convenient symbols used to describe the unboundedness of an interval such as $(1, \infty)$ or $(-\infty, 3]$.

· · REMARK Whenever you write an interval containing ∞ or $-\infty$, always use a parenthesis and never a bracket next to these symbols. This is because ∞ and $-\infty$ are never included in the interval.

$\cdots\cdots\cdots\cdots\cdots\triangleright$

Unbounded Intervals on the Real Number Line

Notation	Interval Type	Inequality	Graph
$[a, \infty)$		$x \ge a$	
(a, ∞)	Open	$x > a$	
$(-\infty, b]$		$x \le b$	
$(-\infty, b)$	Open	$x < b$	
$(-\infty, \infty)$	Entire real line	$-\infty < x < \infty$	

EXAMPLE 5 **Interpreting Intervals**

a. The interval $(-1, 0)$ consists of all real numbers greater than -1 and less than 0.

b. The interval $[2, \infty)$ consists of all real numbers greater than or equal to 2.

✓ **Checkpoint** 🔊))) *Audio-video solution in English & Spanish at LarsonPrecalculus.com*

Give a verbal description of the interval $[-2, 5)$.

EXAMPLE 6 **Using Inequalities to Represent Intervals**

a. The inequality $c \le 2$ can represent the statement "c is at most 2."

b. The inequality $-3 < x \le 5$ can represent "all x in the interval $(-3, 5]$."

✓ **Checkpoint** 🔊))) *Audio-video solution in English & Spanish at LarsonPrecalculus.com*

Use inequality notation to represent the statement "x is less than 4 and at least -2." ∎

Absolute Value and Distance

The **absolute value** of a real number is its *magnitude,* or the distance between the origin and the point representing the real number on the real number line.

Definition of Absolute Value

If a is a real number, then the **absolute value** of a is

$$|a| = \begin{cases} a, & a \geq 0 \\ -a, & a < 0 \end{cases}.$$

Notice in this definition that the absolute value of a real number is never negative. For example, if $a = -5$, then $|-5| = -(-5) = 5$. The absolute value of a real number is either positive or zero. Moreover, 0 is the only real number whose absolute value is 0. So, $|0| = 0$.

Properties of Absolute Values

1. $|a| \geq 0$ $\qquad\qquad$ **2.** $|-a| = |a|$

3. $|ab| = |a||b|$ $\qquad\qquad$ **4.** $\left|\dfrac{a}{b}\right| = \dfrac{|a|}{|b|}, \quad b \neq 0$

EXAMPLE 7   Finding Absolute Values

a. $|-15| = 15$ $\qquad$ **b.** $\left|\dfrac{2}{3}\right| = \dfrac{2}{3}$

c. $|-4.3| = 4.3$ $\qquad$ **d.** $-|-6| = -(6) = -6$

✓ *Checkpoint* 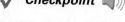 Audio-video solution in English & Spanish at LarsonPrecalculus.com

Evaluate each expression.

a. $|1|$ $\qquad$ **b.** $-\left|\dfrac{3}{4}\right|$ $\qquad$ **c.** $\dfrac{2}{|-3|}$ $\qquad$ **d.** $-|0.7|$

EXAMPLE 8   Evaluating an Absolute Value Expression

Evaluate $\dfrac{|x|}{x}$ for (a) $x > 0$ and (b) $x < 0$.

Solution

a. If $x > 0$, then x is positive and $|x| = x$. So, $\dfrac{|x|}{x} = \dfrac{x}{x} = 1$.

b. If $x < 0$, then x is negative and $|x| = -x$. So, $\dfrac{|x|}{x} = \dfrac{-x}{x} = -1$.

✓ *Checkpoint* Audio-video solution in English & Spanish at LarsonPrecalculus.com

Evaluate $\dfrac{|x + 3|}{x + 3}$ for (a) $x > -3$ and (b) $x < -3$.

The **Law of Trichotomy** states that for any two real numbers a and b, *precisely* one of three relationships is possible:

$$a = b, \quad a < b, \quad \text{or} \quad a > b. \qquad \text{Law of Trichotomy}$$

EXAMPLE 9 Comparing Real Numbers

Place the appropriate symbol ($<$, $>$, or $=$) between the pair of real numbers.

a. $|-4| \quad\blacksquare\quad |3|$ **b.** $|-10| \quad\blacksquare\quad |10|$ **c.** $-|-7| \quad\blacksquare\quad |-7|$

Solution

a. $|-4| > |3|$ because $|-4| = 4$ and $|3| = 3$, and 4 is greater than 3.
b. $|-10| = |10|$ because $|-10| = 10$ and $|10| = 10$.
c. $-|-7| < |-7|$ because $-|-7| = -7$ and $|-7| = 7$, and -7 is less than 7.

✓ *Checkpoint* 🔊))) Audio-video solution in English & Spanish at LarsonPrecalculus.com

Place the appropriate symbol ($<$, $>$, or $=$) between the pair of real numbers.

a. $|-3| \quad\blacksquare\quad |4|$
b. $-|-4| \quad\blacksquare\quad -|4|$
c. $|-3| \quad\blacksquare\quad -|-3|$

Absolute value can be used to find the distance between two points on the real number line. For example, the distance between -3 and 4 is

$$|-3 - 4| = |-7|$$
$$= 7$$

as shown in Figure A.8.

The distance between -3 and 4 is 7.
Figure A.8

Distance Between Two Points on the Real Number Line

Let a and b be real numbers. The **distance between a and b** is

$$d(a, b) = |b - a| = |a - b|.$$

EXAMPLE 10 Finding a Distance

Find the distance between -25 and 13.

Solution

The distance between -25 and 13 is

$$|-25 - 13| = |-38| = 38. \qquad \text{Distance between } -25 \text{ and } 13$$

The distance can also be found as follows.

$$|13 - (-25)| = |38| = 38 \qquad \text{Distance between } -25 \text{ and } 13$$

✓ *Checkpoint* 🔊))) Audio-video solution in English & Spanish at LarsonPrecalculus.com

a. Find the distance between 35 and -23.
b. Find the distance between -35 and -23.
c. Find the distance between 35 and 23.

Algebraic Expressions

One characteristic of algebra is the use of letters to represent numbers. The letters are **variables,** and combinations of letters and numbers are **algebraic expressions.** Here are a few examples of algebraic expressions.

$$5x, \qquad 2x - 3, \qquad \frac{4}{x^2 + 2}, \qquad 7x + y$$

Definition of an Algebraic Expression

An **algebraic expression** is a collection of letters (**variables**) and real numbers (**constants**) combined using the operations of addition, subtraction, multiplication, division, and exponentiation.

The **terms** of an algebraic expression are those parts that are separated by *addition.* For example, $x^2 - 5x + 8 = x^2 + (-5x) + 8$ has three terms: x^2 and $-5x$ are the **variable terms** and 8 is the **constant term.** For terms such as x^2, $-5x$, and 8, the numerical factor is the **coefficient.** Here, the coefficients are 1, -5, and 8.

EXAMPLE 11   **Identifying Terms and Coefficients**

Algebraic Expression	Terms	Coefficients
a. $5x - \dfrac{1}{7}$	$5x, -\dfrac{1}{7}$	$5, -\dfrac{1}{7}$
b. $2x^2 - 6x + 9$	$2x^2, -6x, 9$	$2, -6, 9$
c. $\dfrac{3}{x} + \dfrac{1}{2}x^4 - y$	$\dfrac{3}{x}, \dfrac{1}{2}x^4, -y$	$3, \dfrac{1}{2}, -1$

✓ *Checkpoint*  *Audio-video solution in English & Spanish at LarsonPrecalculus.com*

Identify the terms and coefficients of $-2x + 4$.   ◼

The **Substitution Principle** states, "If $a = b$, then b can replace a in any expression involving a." Use the Substitution Principle to **evaluate** an algebraic expression by substituting numerical values for each of the variables in the expression. The next example illustrates this.

EXAMPLE 12   **Evaluating Algebraic Expressions**

Expression	Value of Variable	Substitute.	Value of Expression
a. $-3x + 5$	$x = 3$	$-3(3) + 5$	$-9 + 5 = -4$
b. $3x^2 + 2x - 1$	$x = -1$	$3(-1)^2 + 2(-1) - 1$	$3 - 2 - 1 = 0$
c. $\dfrac{2x}{x + 1}$	$x = -3$	$\dfrac{2(-3)}{-3 + 1}$	$\dfrac{-6}{-2} = 3$

Note that you must substitute the value for *each* occurrence of the variable.

✓ *Checkpoint*  *Audio-video solution in English & Spanish at LarsonPrecalculus.com*

Evaluate $4x - 5$ when $x = 0$.   ◼

Basic Rules of Algebra

There are four arithmetic operations with real numbers: *addition, multiplication, subtraction,* and *division,* denoted by the symbols $+$, $\times$ or $\cdot$, $-$, and $\div$ or $/$, respectively. Of these, addition and multiplication are the two primary operations. Subtraction and division are the inverse operations of addition and multiplication, respectively.

Definitions of Subtraction and Division

Subtraction: Add the opposite.   **Division:** Multiply by the reciprocal.

$$a - b = a + (-b)$$ $$\text{If } b \neq 0, \text{ then } a/b = a\left(\frac{1}{b}\right) = \frac{a}{b}.$$

In these definitions, $-b$ is the **additive inverse** (or opposite) of b, and $1/b$ is the **multiplicative inverse** (or reciprocal) of b. In the fractional form a/b, a is the **numerator** of the fraction and b is the **denominator.**

The properties of real numbers below are true for variables and algebraic expressions as well as for real numbers, so they are often called the **Basic Rules of Algebra.** Formulate a verbal description of each of these properties. For example, the first property states that *the order in which two real numbers are added does not affect their sum.*

Basic Rules of Algebra

Let a, b, and c be real numbers, variables, or algebraic expressions.

Property		Example
Commutative Property of Addition:	$a + b = b + a$	$4x + x^2 = x^2 + 4x$
Commutative Property of Multiplication:	$ab = ba$	$(4 - x)x^2 = x^2(4 - x)$
Associative Property of Addition:	$(a + b) + c = a + (b + c)$	$(x + 5) + x^2 = x + (5 + x^2)$
Associative Property of Multiplication:	$(ab)c = a(bc)$	$(2x \cdot 3y)(8) = (2x)(3y \cdot 8)$
Distributive Properties:	$a(b + c) = ab + ac$	$3x(5 + 2x) = 3x \cdot 5 + 3x \cdot 2x$
	$(a + b)c = ac + bc$	$(y + 8)y = y \cdot y + 8 \cdot y$
Additive Identity Property:	$a + 0 = a$	$5y^2 + 0 = 5y^2$
Multiplicative Identity Property:	$a \cdot 1 = a$	$(4x^2)(1) = 4x^2$
Additive Inverse Property:	$a + (-a) = 0$	$5x^3 + (-5x^3) = 0$
Multiplicative Inverse Property:	$a \cdot \dfrac{1}{a} = 1, \quad a \neq 0$	$(x^2 + 4)\left(\dfrac{1}{x^2 + 4}\right) = 1$

Subtraction is defined as "adding the opposite," so the Distributive Properties are also true for subtraction. For example, the "subtraction form" of $a(b + c) = ab + ac$ is $a(b - c) = ab - ac$. Note that the operations of subtraction and division are neither commutative nor associative. The examples

$$7 - 3 \neq 3 - 7 \quad \text{and} \quad 20 \div 4 \neq 4 \div 20$$

show that subtraction and division are not commutative. Similarly

$$5 - (3 - 2) \neq (5 - 3) - 2 \quad \text{and} \quad 16 \div (4 \div 2) \neq (16 \div 4) \div 2$$

demonstrate that subtraction and division are not associative.

EXAMPLE 13   **Identifying Rules of Algebra**

Identify the rule of algebra illustrated by the statement.

a. $(5x^3)2 = 2(5x^3)$ **b.** $(4x + 3) - (4x + 3) = 0$

c. $7x \cdot \dfrac{1}{7x} = 1, \quad x \neq 0$ **d.** $(2 + 5x^2) + x^2 = 2 + (5x^2 + x^2)$

Solution

a. This statement illustrates the Commutative Property of Multiplication. In other words, you obtain the same result whether you multiply $5x^3$ by 2, or 2 by $5x^3$.

b. This statement illustrates the Additive Inverse Property. In terms of subtraction, this property states that when any expression is subtracted from itself, the result is 0.

c. This statement illustrates the Multiplicative Inverse Property. Note that x must be a nonzero number. The reciprocal of x is undefined when x is 0.

d. This statement illustrates the Associative Property of Addition. In other words, to form the sum $2 + 5x^2 + x^2$, it does not matter whether 2 and $5x^2$, or $5x^2$ and x^2 are added first.

✓ **Checkpoint**  Audio-video solution in English & Spanish at LarsonPrecalculus.com

Identify the rule of algebra illustrated by the statement.

a. $x + 9 = 9 + x$ **b.** $5(x^3 \cdot 2) = (5x^3)2$ **c.** $(2 + 5x^2)y^2 = 2 \cdot y^2 + 5x^2 \cdot y^2$

> **REMARK** Notice the difference between the *opposite of a number* and a *negative number*. If a is already negative, then its opposite, $-a$, is positive. For example, if $a = -5$, then
>
> $-a = -(-5) = 5.$

Properties of Negation and Equality

Let a, b, and c be real numbers, variables, or algebraic expressions.

Property	**Example**
1. $(-1)a = -a$	$(-1)7 = -7$
2. $-(-a) = a$	$-(-6) = 6$
3. $(-a)b = -(ab) = a(-b)$	$(-5)3 = -(5 \cdot 3) = 5(-3)$
4. $(-a)(-b) = ab$	$(-2)(-x) = 2x$
5. $-(a + b) = (-a) + (-b)$	$-(x + 8) = (-x) + (-8)$
	$\qquad = -x - 8$
6. If $a = b$, then $a \pm c = b \pm c$.	$\frac{1}{2} + 3 = 0.5 + 3$
7. If $a = b$, then $ac = bc$.	$4^2 \cdot 2 = 16 \cdot 2$
8. If $a \pm c = b \pm c$, then $a = b$.	$1.4 - 1 = \frac{7}{5} - 1 \implies 1.4 = \frac{7}{5}$
9. If $ac = bc$ and $c \neq 0$, then $a = b$.	$3x = 3 \cdot 4 \implies x = 4$

> **REMARK** The "or" in the Zero-Factor Property includes the possibility that either or both factors may be zero. This is an *inclusive or,* and it is generally the way the word "or" is used in mathematics.

Properties of Zero

Let a and b be real numbers, variables, or algebraic expressions.

1. $a + 0 = a$ and $a - 0 = a$ **2.** $a \cdot 0 = 0$

3. $\dfrac{0}{a} = 0, \quad a \neq 0$ **4.** $\dfrac{a}{0}$ is undefined.

5. **Zero-Factor Property:** If $ab = 0$, then $a = 0$ or $b = 0$.

▷

REMARK In Property 1, the phrase "if and only if" implies two statements. One statement is: If $a/b = c/d$, then $ad = bc$. The other statement is: If $ad = bc$, where $b \neq 0$ and $d \neq 0$, then $a/b = c/d$.

Properties and Operations of Fractions

Let a, b, c, and d be real numbers, variables, or algebraic expressions such that $b \neq 0$ and $d \neq 0$.

1. **Equivalent Fractions:** $\dfrac{a}{b} = \dfrac{c}{d}$ if and only if $ad = bc$.

2. **Rules of Signs:** $-\dfrac{a}{b} = \dfrac{-a}{b} = \dfrac{a}{-b}$ and $\dfrac{-a}{-b} = \dfrac{a}{b}$

3. **Generate Equivalent Fractions:** $\dfrac{a}{b} = \dfrac{ac}{bc}, \quad c \neq 0$

4. **Add or Subtract with Like Denominators:** $\dfrac{a}{b} \pm \dfrac{c}{b} = \dfrac{a \pm c}{b}$

5. **Add or Subtract with Unlike Denominators:** $\dfrac{a}{b} \pm \dfrac{c}{d} = \dfrac{ad \pm bc}{bd}$

6. **Multiply Fractions:** $\dfrac{a}{b} \cdot \dfrac{c}{d} = \dfrac{ac}{bd}$

7. **Divide Fractions:** $\dfrac{a}{b} \div \dfrac{c}{d} = \dfrac{a}{b} \cdot \dfrac{d}{c} = \dfrac{ad}{bc}, \quad c \neq 0$

EXAMPLE 14 **Properties and Operations of Fractions**

a. $\dfrac{x}{5} = \dfrac{3 \cdot x}{3 \cdot 5} = \dfrac{3x}{15}$

b. $\dfrac{7}{x} \div \dfrac{3}{2} = \dfrac{7}{x} \cdot \dfrac{2}{3} = \dfrac{14}{3x}$

✓ **Checkpoint**  *Audio-video solution in English & Spanish at LarsonPrecalculus.com*

a. Multiply fractions: $\dfrac{3}{5} \cdot \dfrac{x}{6}$.

b. Add fractions: $\dfrac{x}{10} + \dfrac{2x}{5}$. ■

REMARK The number 1 is neither prime nor composite.

▷

If a, b, and c are integers such that $ab = c$, then a and b are **factors** or **divisors** of c. A **prime number** is an integer that has exactly two positive factors—itself and 1—such as 2, 3, 5, 7, and 11. The numbers 4, 6, 8, 9, and 10 are **composite** because each can be written as the product of two or more prime numbers. The **Fundamental Theorem of Arithmetic** states that every positive integer greater than 1 is a prime number or can be written as the product of prime numbers in precisely one way (disregarding order). For example, the *prime factorization* of 24 is $24 = 2 \cdot 2 \cdot 2 \cdot 3$.

Summarize (Appendix A.1)

1. Explain how to represent and classify real numbers *(pages A1 and A2)*. For examples of representing and classifying real numbers, see Examples 1 and 2.

2. Explain how to order real numbers and use inequalities *(pages A3 and A4)*. For examples of ordering real numbers and using inequalities, see Examples 3–6.

3. State the definition of the absolute value of a real number *(page A5)*. For examples of using absolute value, see Examples 7–10.

4. Explain how to evaluate an algebraic expression *(page A7)*. For examples involving algebraic expressions, see Examples 11 and 12.

5. State the basic rules and properties of algebra *(pages A8–A10)*. For examples involving the basic rules and properties of algebra, see Examples 13 and 14.

A.1   Exercises

See **CalcChat.com** for tutorial help and worked-out solutions to odd-numbered exercises.

Vocabulary: Fill in the blanks.

1. The decimal representation of an _____ number neither terminates nor repeats.
2. The point representing 0 on the real number line is the _____.
3. The distance between the origin and a point representing a real number on the real number line is the _____ _____ of the real number.
4. A number that can be written as the product of two or more prime numbers is a _____ number.
5. The _____ of an algebraic expression are those parts that are separated by addition.
6. The _____ _____ states that if $ab = 0$, then $a = 0$ or $b = 0$.

Skills and Applications

 Classifying Real Numbers   In Exercises 7–10, determine which numbers in the set are (a) natural numbers, (b) whole numbers, (c) integers, (d) rational numbers, and (e) irrational numbers.

7. $\left\{ -9, -\frac{7}{2}, 5, \frac{2}{3}, \sqrt{2}, 0, 1, -4, 2, -11 \right\}$
8. $\left\{ \sqrt{5}, -7, -\frac{7}{3}, 0, 3.14, \frac{5}{4}, -3, 12, 5 \right\}$
9. $\{ 2.01, 0.\overline{6}, -13, 0.010110111 \ldots, 1, -6 \}$
10. $\left\{ 25, -17, -\frac{12}{5}, \sqrt{9}, 3.12, \frac{1}{2}\pi, 7, -11.1, 13 \right\}$

Plotting Points on the Real Number Line   In Exercises 11 and 12, plot the real numbers on the real number line.

11. (a) 3   (b) $\frac{7}{2}$ (c) $-\frac{5}{2}$ (d) -5.2
12. (a) 8.5   (b) $\frac{4}{3}$ (c) -4.75 (d) $-\frac{8}{3}$

 Plotting and Ordering Real Numbers   In Exercises 13–16, plot the two real numbers on the real number line. Then place the appropriate inequality symbol (< or >) between them.

13. $-4, -8$
14. $1, \frac{16}{3}$
15. $\frac{5}{6}, \frac{2}{3}$
16. $-\frac{8}{7}, -\frac{3}{7}$

Interpreting an Inequality or an Interval   In Exercises 17–24, (a) give a verbal description of the subset of real numbers represented by the inequality or the interval, (b) sketch the subset on the real number line, and (c) state whether the subset is bounded or unbounded.

17. $x \le 5$
18. $x < 0$
19. $-2 < x < 2$
20. $0 < x \le 6$
21. $[4, \infty)$
22. $(-\infty, 2)$
23. $[-5, 2)$
24. $(-1, 2]$

Using Inequality and Interval Notation   In Exercises 25–28, use inequality notation and interval notation to describe the set.

25. y is nonnegative.
26. y is no more than 25.
27. t is at least 10 and at most 22.
28. k is less than 5 but no less than -3.

 Evaluating an Absolute Value Expression   In Exercises 29–38, evaluate the expression.

29. $|-10|$
30. $|0|$
31. $|3 - 8|$
32. $|6 - 2|$
33. $|-1| - |-2|$
34. $-3 - |-3|$
35. $5|-5|$
36. $-4|-4|$
37. $\dfrac{|x + 2|}{x + 2}, \quad x < -2$
38. $\dfrac{|x - 1|}{x - 1}, \quad x > 1$

 Comparing Real Numbers   In Exercises 39–42, place the appropriate symbol (<, >, or =) between the pair of real numbers.

39. $|-4|$ ▢ $|4|$
40. -5 ▢ $-|5|$
41. $-|-6|$ ▢ $|-6|$
42. $-|-2|$ ▢ $-|2|$

 Finding a Distance   In Exercises 43–46, find the distance between a and b.

43. $a = 126, b = 75$
44. $a = -20, b = 30$
45. $a = -\frac{5}{2}, b = 0$
46. $a = -\frac{1}{4}, b = -\frac{11}{4}$

Using Absolute Value Notation   In Exercises 47 and 48, use absolute value notation to represent the situation.

47. The distance between x and 5 is no more than 3.
48. The distance between x and -10 is at least 6.

•• Federal Deficit ••••••••••••••

In Exercises 49–52, use the bar graph, which shows the receipts of the federal government (in billions of dollars) for selected years from 2008 through 2014. In each exercise, you are given the expenditures of the federal government. Find the magnitude of the surplus or deficit for the year. (*Source: U.S. Office of Management and Budget*)

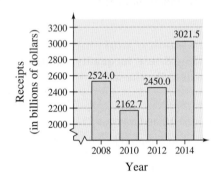

Year

Year	Receipts, R	Expenditures, E	$|R - E|$
49. 2008		\$2982.5 billion	
50. 2010		\$3457.1 billion	
51. 2012		\$3537.0 billion	
52. 2014		\$3506.1 billion	

 Identifying Terms and Coefficients In Exercises 53–58, identify the terms. Then identify the coefficients of the variable terms of the expression.

53. $7x + 4$

54. $2x - 3$

55. $6x^3 - 5x$

56. $4x^3 + 0.5x - 5$

57. $3\sqrt{3}x^2 + 1$

58. $2\sqrt{2}x^2 - 3$

 Evaluating an Algebraic Expression In Exercises 59–64, evaluate the expression for each value of x. (If not possible, state the reason.)

59. $4x - 6$ (a) $x = -1$ (b) $x = 0$

60. $9 - 7x$ (a) $x = -3$ (b) $x = 3$

61. $x^2 - 3x + 2$ (a) $x = 0$ (b) $x = -1$

62. $-x^2 + 5x - 4$ (a) $x = -1$ (b) $x = 1$

63. $\dfrac{x + 1}{x - 1}$ (a) $x = 1$ (b) $x = -1$

64. $\dfrac{x - 2}{x + 2}$ (a) $x = 2$ (b) $x = -2$

Identifying Rules of Algebra

In Exercises 65–68, identify the rule(s) of algebra illustrated by the statement.

65. $\dfrac{1}{h + 6}(h + 6) = 1, \quad h \neq -6$

66. $(x + 3) - (x + 3) = 0$

67. $x(3y) = (x \cdot 3)y = (3x)y$

68. $\frac{1}{7}(7 \cdot 12) = \left(\frac{1}{7} \cdot 7\right)12 = 1 \cdot 12 = 12$

Operations with Fractions

In Exercises 69–72, perform the operation. (Write fractional answers in simplest form.)

69. $\dfrac{2x}{3} - \dfrac{x}{4}$

70. $\dfrac{3x}{4} + \dfrac{x}{5}$

71. $\dfrac{3x}{10} \cdot \dfrac{5}{6}$

72. $\dfrac{2x}{3} \div \dfrac{6}{7}$

Exploration

True or False? In Exercises 73–75, determine whether the statement is true or false. Justify your answer.

73. Every nonnegative number is positive.

74. If $a > 0$ and $b < 0$, then $ab > 0$.

75. If $a < 0$ and $b < 0$, then $ab > 0$.

76. **HOW DO YOU SEE IT?** Match each description with its graph. Which types of real numbers shown in Figure A.1 on page A1 may be included in a range of prices? a range of lengths? Explain.

(i) [number line: 1.87 1.88 1.89 1.90 1.91 1.92 1.93]

(ii) [number line with points: 1.87 1.88 1.89 1.90 1.91 1.92 1.93]

(a) The price of an item is within \$0.03 of \$1.90.

(b) The distance between the prongs of an electric plug may not differ from 1.9 centimeters by more than 0.03 centimeter.

77. Conjecture

(a) Use a calculator to complete the table.

n	0.0001	0.01	1	100	10,000
$\dfrac{5}{n}$					

(b) Use the result from part (a) to make a conjecture about the value of $5/n$ as n (i) approaches 0, and (ii) increases without bound.

A.2 Exponents and Radicals

Real numbers and algebraic expressions are often written with exponents and radicals. For example, in Exercise 69 on page A24, you will use an expression involving rational exponents to find the times required for a funnel to empty for different water heights.

■ Use properties of exponents.
■ Use scientific notation to represent real numbers.
■ Use properties of radicals.
■ Simplify and combine radical expressions.
■ Rationalize denominators and numerators.
■ Use properties of rational exponents.

Integer Exponents and Their Properties

Repeated *multiplication* can be written in **exponential form.**

Repeated Multiplication	Exponential Form
$a \cdot a \cdot a \cdot a \cdot a$	a^5
$(-4)(-4)(-4)$	$(-4)^3$
$(2x)(2x)(2x)(2x)$	$(2x)^4$

Exponential Notation

If a is a real number and n is a positive integer, then

$$a^n = \underbrace{a \cdot a \cdot a \cdots a}_{n \text{ factors}}$$

where n is the **exponent** and a is the **base.** You read a^n as "a to the nth **power.**"

An exponent can also be negative or zero. Properties 3 and 4 below show how to use negative and zero exponents.

Properties of Exponents

Let a and b be real numbers, variables, or algebraic expressions, and let m and n be integers. (All denominators and bases are nonzero.)

Property	Example
1. $a^m a^n = a^{m+n}$	$3^2 \cdot 3^4 = 3^{2+4} = 3^6 = 729$
2. $\dfrac{a^m}{a^n} = a^{m-n}$	$\dfrac{x^7}{x^4} = x^{7-4} = x^3$
3. $a^{-n} = \dfrac{1}{a^n} = \left(\dfrac{1}{a}\right)^n$	$y^{-4} = \dfrac{1}{y^4} = \left(\dfrac{1}{y}\right)^4$
4. $a^0 = 1$	$(x^2 + 1)^0 = 1$
5. $(ab)^m = a^m b^m$	$(5x)^3 = 5^3 x^3 = 125x^3$
6. $(a^m)^n = a^{mn}$	$(y^3)^{-4} = y^{3(-4)} = y^{-12} = \dfrac{1}{y^{12}}$
7. $\left(\dfrac{a}{b}\right)^m = \dfrac{a^m}{b^m}$	$\left(\dfrac{2}{x}\right)^3 = \dfrac{2^3}{x^3} = \dfrac{8}{x^3}$
8. $|a^2| = |a|^2 = a^2$	$|(-2)^2| = |-2|^2 = 2^2 = 4 = (-2)^2$

Scientific Notation

Exponents provide an efficient way of writing and computing with very large (or very small) numbers. For example, there are about 366 billion billion gallons of water on Earth—that is, 366 followed by 18 zeros.

$$366,000,000,000,000,000,000$$

It is convenient to write such numbers in **scientific notation.** This notation has the form $\pm c \times 10^n$, where $1 \le c < 10$ and n is an integer. So, the number of gallons of water on Earth, written in scientific notation, is

$$3.66 \times 100,000,000,000,000,000,000 = 3.66 \times 10^{20}.$$

The *positive* exponent 20 tells you that the number is *large* (10 or more) and that the decimal point has been moved 20 places. A *negative* exponent tells you that the number is *small* (less than 1). For example, the mass (in grams) of one electron is approximately

$$9.1 \times 10^{-28} = 0.00000000000000000000000000091.$$

28 decimal places

EXAMPLE 5   Scientific Notation

a. $0.0000782 = 7.82 \times 10^{-5}$

b. $836,100,000 = 8.361 \times 10^8$

✓ *Checkpoint* 🔊)) *Audio-video solution in English & Spanish at LarsonPrecalculus.com*

Write 45,850 in scientific notation.

EXAMPLE 6   Decimal Notation

a. $-9.36 \times 10^{-6} = -0.00000936$

b. $1.345 \times 10^2 = 134.5$

✓ *Checkpoint* 🔊)) *Audio-video solution in English & Spanish at LarsonPrecalculus.com*

Write -2.718×10^{-3} in decimal notation.

▷ **TECHNOLOGY** Most calculators automatically switch to scientific notation when showing large (or small) numbers that exceed the display range.

To enter numbers in scientific notation, your calculator should have an exponential entry key labeled

EE   or   EXP.

Consult the user's guide for instructions on keystrokes and how your calculator displays numbers in scientific notation.

EXAMPLE 7   Using Scientific Notation

Evaluate $\dfrac{(2,400,000,000)(0.0000045)}{(0.00003)(1500)}$.

Solution   Begin by rewriting each number in scientific notation. Then simplify.

$$\frac{(2,400,000,000)(0.0000045)}{(0.00003)(1500)} = \frac{(2.4 \times 10^9)(4.5 \times 10^{-6})}{(3.0 \times 10^{-5})(1.5 \times 10^3)}$$

$$= \frac{(2.4)(4.5)(10^3)}{(4.5)(10^{-2})}$$

$$= (2.4)(10^5)$$

$$= 240,000$$

✓ *Checkpoint* 🔊)) *Audio-video solution in English & Spanish at LarsonPrecalculus.com*

Evaluate $(24,000,000,000)(0.00000012)(300,000)$.

Radicals and Their Properties

A **square root** of a number is one of its two equal factors. For example, 5 is a square root of 25 because 5 is one of the two equal factors of 25. In a similar way, a **cube root** of a number is one of its three equal factors, as in $125 = 5^3$.

Definition of *n*th Root of a Number

Let a and b be real numbers and let $n \geq 2$ be a positive integer. If

$$a = b^n$$

then b is an ***n*th root of *a*.** If $n = 2$, then the root is a **square root.** If $n = 3$, then the root is a **cube root.**

Some numbers have more than one *n*th root. For example, both 5 and -5 are square roots of 25. The *principal square* root of 25, written as $\sqrt{25}$, is the positive root, 5.

Principal *n*th Root of a Number

Let a be a real number that has at least one *n*th root. The **principal *n*th root of *a*** is the *n*th root that has the same sign as a. It is denoted by a **radical symbol**

$$\sqrt[n]{a}. \qquad \text{Principal } n\text{th root}$$

The positive integer $n \geq 2$ is the **index** of the radical, and the number a is the **radicand.** When $n = 2$, omit the index and write $\sqrt{a}$ rather than $\sqrt[2]{a}$. (The plural of index is *indices*.)

A common misunderstanding is that the square root sign implies both negative and positive roots. This is not correct. The square root sign implies only a positive root. When a negative root is needed, you must use the negative sign with the square root sign.

Incorrect: $\sqrt{4} = \pm 2$ ✗   *Correct:* $-\sqrt{4} = -2$ *and* $\sqrt{4} = 2$

EXAMPLE 8   **Evaluating Radical Expressions**

a. $\sqrt{36} = 6$ because $6^2 = 36$.

b. $-\sqrt{36} = -6$ because $-\left(\sqrt{36}\right) = -\left(\sqrt{6^2}\right) = -(6) = -6$.

c. $\sqrt[3]{\dfrac{125}{64}} = \dfrac{5}{4}$ because $\left(\dfrac{5}{4}\right)^3 = \dfrac{5^3}{4^3} = \dfrac{125}{64}$.

d. $\sqrt[5]{-32} = -2$ because $(-2)^5 = -32$.

e. $\sqrt[4]{-81}$ is not a real number because no real number raised to the fourth power produces -81.

✓ **Checkpoint**  *Audio-video solution in English & Spanish at LarsonPrecalculus.com*

Evaluate each expression, if possible.

a. $-\sqrt{144}$ **b.** $\sqrt{-144}$

c. $\sqrt{\dfrac{25}{64}}$ **d.** $-\sqrt[3]{\dfrac{8}{27}}$

Here are some generalizations about the nth roots of real numbers.

Generalizations About nth Roots of Real Numbers			
Real Number a	Integer $n > 0$	Root(s) of a	Example
$a > 0$	n is even.	$\sqrt[n]{a}, \; -\sqrt[n]{a}$	$\sqrt[4]{81} = 3, \; -\sqrt[4]{81} = -3$
$a > 0$ or $a < 0$	n is odd.	$\sqrt[n]{a}$	$\sqrt[3]{-8} = -2$
$a < 0$	n is even.	No real roots	$\sqrt{-4}$ is not a real number.
$a = 0$	n is even or odd.	$\sqrt[n]{0} = 0$	$\sqrt[5]{0} = 0$

Integers such as 1, 4, 9, 16, 25, and 36 are **perfect squares** because they have integer square roots. Similarly, integers such as 1, 8, 27, 64, and 125 are **perfect cubes** because they have integer cube roots.

Properties of Radicals

Let a and b be real numbers, variables, or algebraic expressions such that the roots below are real numbers, and let m and n be positive integers.

Property

1. $\sqrt[n]{a^m} = \left(\sqrt[n]{a}\right)^m$

2. $\sqrt[n]{a} \cdot \sqrt[n]{b} = \sqrt[n]{ab}$

3. $\dfrac{\sqrt[n]{a}}{\sqrt[n]{b}} = \sqrt[n]{\dfrac{a}{b}}, \quad b \neq 0$

4. $\sqrt[m]{\sqrt[n]{a}} = \sqrt[mn]{a}$

5. $\left(\sqrt[n]{a}\right)^n = a$

6. For n even, $\sqrt[n]{a^n} = |a|$.

For n odd, $\sqrt[n]{a^n} = a$.

Example

$\sqrt[3]{8^2} = \left(\sqrt[3]{8}\right)^2 = (2)^2 = 4$

$\sqrt{5} \cdot \sqrt{7} = \sqrt{5 \cdot 7} = \sqrt{35}$

$\dfrac{\sqrt[4]{27}}{\sqrt[4]{9}} = \sqrt[4]{\dfrac{27}{9}} = \sqrt[4]{3}$

$\sqrt[3]{\sqrt{10}} = \sqrt[6]{10}$

$\left(\sqrt{3}\right)^2 = 3$

$\sqrt{(-12)^2} = |-12| = 12$

$\sqrt[3]{(-12)^3} = -12$

A common use of Property 6 is $\sqrt{a^2} = |a|$.

EXAMPLE 9 **Using Properties of Radicals**

Use the properties of radicals to simplify each expression.

a. $\sqrt{8} \cdot \sqrt{2}$ **b.** $\left(\sqrt[3]{5}\right)^3$

c. $\sqrt[3]{x^3}$ **d.** $\sqrt[6]{y^6}$

Solution

a. $\sqrt{8} \cdot \sqrt{2} = \sqrt{8 \cdot 2} = \sqrt{16} = 4$ **b.** $\left(\sqrt[3]{5}\right)^3 = 5$

c. $\sqrt[3]{x^3} = x$ **d.** $\sqrt[6]{y^6} = |y|$

✓ *Checkpoint* ◀))) *Audio-video solution in English & Spanish at LarsonPrecalculus.com*

Use the properties of radicals to simplify each expression.

a. $\dfrac{\sqrt{125}}{\sqrt{5}}$ **b.** $\sqrt[3]{125^2}$

c. $\sqrt[3]{x^2} \cdot \sqrt[3]{x}$ **d.** $\sqrt{\sqrt{x}}$

Simplifying Radical Expressions

An expression involving radicals is in **simplest form** when the three conditions below are satisfied.

1. All possible factors are removed from the radical.
2. All fractions have radical-free denominators (a process called *rationalizing the denominator* accomplishes this).
3. The index of the radical is reduced.

To simplify a radical, factor the radicand into factors whose exponents are multiples of the index. Write the roots of these factors outside the radical. The "leftover" factors make up the new radicand.

· · REMARK   When you simplify a radical, it is important that both the original and the simplified expressions are defined for the same values of the variable. For instance, in Example 10(c), $\sqrt{75x^3}$ and $5x\sqrt{3x}$ are both defined only for nonnegative values of x. Similarly, in Example 10(e), $\sqrt[4]{(5x)^4}$ and $5|x|$ are both defined for all real values of x.

EXAMPLE 10   Simplifying Radical Expressions

Perfect cube   Leftover factor

a. $\sqrt[3]{24} = \sqrt[3]{8 \cdot 3} = \sqrt[3]{2^3 \cdot 3} = 2\sqrt[3]{3}$

Perfect 4th power   Leftover factor

b. $\sqrt[4]{48} = \sqrt[4]{16 \cdot 3} = \sqrt[4]{2^4 \cdot 3} = 2\sqrt[4]{3}$

c. $\sqrt{75x^3} = \sqrt{25x^2 \cdot 3x} = \sqrt{(5x)^2 \cdot 3x} = 5x\sqrt{3x}$

d. $\sqrt[3]{24a^4} = \sqrt[3]{8a^3 \cdot 3a} = \sqrt[3]{(2a)^3 \cdot 3a} = 2a\sqrt[3]{3a}$

e. $\sqrt[4]{(5x)^4} = |5x| = 5|x|$

✓ **Checkpoint**))) *Audio-video solution in English & Spanish at LarsonPrecalculus.com*

Simplify each radical expression.

a. $\sqrt{32}$ **b.** $\sqrt[3]{250}$ **c.** $\sqrt{24a^5}$ **d.** $\sqrt[3]{-135x^3}$ ■

Radical expressions can be combined (added or subtracted) when they are **like radicals**—that is, when they have the same index and radicand. For example, $\sqrt{2}$, $3\sqrt{2}$, and $\frac{1}{2}\sqrt{2}$ are like radicals, but $\sqrt{3}$ and $\sqrt{2}$ are unlike radicals. To determine whether two radicals can be combined, first simplify each radical.

EXAMPLE 11   Combining Radical Expressions

a. $2\sqrt{48} - 3\sqrt{27} = 2\sqrt{16 \cdot 3} - 3\sqrt{9 \cdot 3}$ Find square factors.

$= 8\sqrt{3} - 9\sqrt{3}$ Find square roots and multiply by coefficients.

$= (8 - 9)\sqrt{3}$ Combine like radicals.

$= -\sqrt{3}$ Simplify.

b. $\sqrt[3]{16x} - \sqrt[3]{54x^4} = \sqrt[3]{8 \cdot 2x} - \sqrt[3]{27x^3 \cdot 2x}$ Find cube factors.

$= 2\sqrt[3]{2x} - 3x\sqrt[3]{2x}$ Find cube roots.

$= (2 - 3x)\sqrt[3]{2x}$ Combine like radicals.

✓ **Checkpoint**))) *Audio-video solution in English & Spanish at LarsonPrecalculus.com*

Simplify each radical expression.

a. $3\sqrt{8} + \sqrt{18}$ **b.** $\sqrt[3]{81x^5} - \sqrt[3]{24x^2}$ ■

Rationalizing Denominators and Numerators

To rationalize a denominator or numerator of the form $a - b\sqrt{m}$ or $a + b\sqrt{m}$, multiply both numerator and denominator by a **conjugate**: $a + b\sqrt{m}$ and $a - b\sqrt{m}$ are conjugates of each other. If $a = 0$, then the rationalizing factor for $\sqrt{m}$ is itself, $\sqrt{m}$. For cube roots, choose a rationalizing factor that produces a perfect cube radicand.

EXAMPLE 12 Rationalizing Single-Term Denominators

a. $\dfrac{5}{2\sqrt{3}} = \dfrac{5}{2\sqrt{3}} \cdot \dfrac{\sqrt{3}}{\sqrt{3}}$ $\sqrt{3}$ is rationalizing factor.

$= \dfrac{5\sqrt{3}}{2(3)}$ Multiply.

$= \dfrac{5\sqrt{3}}{6}$ Simplify.

b. $\dfrac{2}{\sqrt[3]{5}} = \dfrac{2}{\sqrt[3]{5}} \cdot \dfrac{\sqrt[3]{5^2}}{\sqrt[3]{5^2}}$ $\sqrt[3]{5^2}$ is rationalizing factor.

$= \dfrac{2\sqrt[3]{5^2}}{\sqrt[3]{5^3}}$ Multiply.

$= \dfrac{2\sqrt[3]{25}}{5}$ Simplify.

✓ **Checkpoint** ◀))) *Audio-video solution in English & Spanish at LarsonPrecalculus.com*

Rationalize the denominator of each expression.

a. $\dfrac{5}{3\sqrt{2}}$ **b.** $\dfrac{1}{\sqrt[3]{25}}$

EXAMPLE 13 Rationalizing a Denominator with Two Terms

$\dfrac{2}{3 + \sqrt{7}} = \dfrac{2}{3 + \sqrt{7}} \cdot \dfrac{3 - \sqrt{7}}{3 - \sqrt{7}}$ Multiply numerator and denominator by conjugate of denominator.

$= \dfrac{2(3 - \sqrt{7})}{3(3 - \sqrt{7}) + \sqrt{7}(3 - \sqrt{7})}$ Distributive Property

$= \dfrac{2(3 - \sqrt{7})}{3(3) - 3(\sqrt{7}) + \sqrt{7}(3) - \sqrt{7}(\sqrt{7})}$ Distributive Property

$= \dfrac{2(3 - \sqrt{7})}{(3)^2 - (\sqrt{7})^2}$ Simplify.

$= \dfrac{2(3 - \sqrt{7})}{2}$ Simplify.

$= 3 - \sqrt{7}$ Divide out common factor.

✓ **Checkpoint** ◀))) *Audio-video solution in English & Spanish at LarsonPrecalculus.com*

Rationalize the denominator: $\dfrac{8}{\sqrt{6} - \sqrt{2}}$.

Sometimes it is necessary to rationalize the numerator of an expression. For instance, in Appendix A.4 you will use the technique shown in Example 14 on the next page to rationalize the numerator of an expression from calculus.

EXAMPLE 14   **Rationalizing a Numerator**

$$\frac{\sqrt{5} - \sqrt{7}}{2} = \frac{\sqrt{5} - \sqrt{7}}{2} \cdot \frac{\sqrt{5} + \sqrt{7}}{\sqrt{5} + \sqrt{7}}$$ Multiply numerator and denominator by conjugate of numerator.

$$= \frac{(\sqrt{5})^2 - (\sqrt{7})^2}{2(\sqrt{5} + \sqrt{7})}$$ Simplify.

$$= \frac{5 - 7}{2(\sqrt{5} + \sqrt{7})}$$ Property 5 of radicals

$$= \frac{-2}{2(\sqrt{5} + \sqrt{7})}$$ Simplify.

$$= \frac{-1}{\sqrt{5} + \sqrt{7}}$$ Divide out common factor.

✓ *Checkpoint* ◀))) Audio-video solution in English & Spanish at LarsonPrecalculus.com

Rationalize the numerator: $\dfrac{2 - \sqrt{2}}{3}$.

• • **REMARK** Do not confuse the expression $\sqrt{5} + \sqrt{7}$ with the expression $\sqrt{5 + 7}$. In general, $\sqrt{x + y}$ does not equal $\sqrt{x} + \sqrt{y}$. Similarly, $\sqrt{x^2 + y^2}$ does not equal $x + y$.

Rational Exponents and Their Properties

Definition of Rational Exponents

If a is a real number and n is a positive integer such that the principal nth root of a exists, then $a^{1/n}$ is defined as

$$a^{1/n} = \sqrt[n]{a}.$$

Moreover, if m is a positive integer, then

$$a^{m/n} = (a^{1/n})^m.$$

$1/n$ and m/n are called **rational exponents** of a.

• • **REMARK**   If m and n have no common factors, then it is also true that $a^{m/n} = (a^m)^{1/n}$.

The numerator of a rational exponent denotes the *power* to which the base is raised, and the denominator denotes the *index* or the *root* to be taken.

$$b^{m/n} = \left(\sqrt[n]{b}\right)^m = \sqrt[n]{b^m}$$

where the numerator is the Power and the index denotes the Index.

When you are working with rational exponents, the properties of integer exponents still apply. For example, $2^{1/2}2^{1/3} = 2^{(1/2)+(1/3)} = 2^{5/6}$.

EXAMPLE 15   **Changing From Radical to Exponential Form**

a. $\sqrt{3} = 3^{1/2}$

b. $\sqrt{(3xy)^5} = \sqrt[2]{(3xy)^5} = (3xy)^{5/2}$

c. $2x\sqrt[4]{x^3} = (2x)(x^{3/4}) = 2x^{1+(3/4)} = 2x^{7/4}$

✓ *Checkpoint* ◀))) Audio-video solution in English & Spanish at LarsonPrecalculus.com

Write (a) $\sqrt[3]{27}$, (b) $\sqrt{x^3y^5z}$, and (c) $3x\sqrt[3]{x^2}$ in exponential form.

▷ TECHNOLOGY There are four methods of evaluating radicals on most graphing utilities. For square roots, you can use the *square root key* $\boxed{\sqrt{}}$. For cube roots, you can use the *cube root key* $\boxed{\sqrt[3]{}}$. For other roots, first convert the radical to exponential form and then use the *exponential key* $\boxed{\wedge}$, or use the *xth root key* $\boxed{\sqrt[x]{}}$ (or menu choice). Consult the user's guide for your graphing utility for specific keystrokes.

EXAMPLE 16 **Changing From Exponential to Radical Form**

See LarsonPrecalculus.com for an interactive version of this type of example.

a. $(x^2 + y^2)^{3/2} = \left(\sqrt{x^2 + y^2}\right)^3 = \sqrt{(x^2 + y^2)^3}$

b. $2y^{3/4}z^{1/4} = 2(y^3z)^{1/4} = 2\sqrt[4]{y^3z}$

c. $a^{-3/2} = \dfrac{1}{a^{3/2}} = \dfrac{1}{\sqrt{a^3}}$

d. $x^{0.2} = x^{1/5} = \sqrt[5]{x}$

✓ **Checkpoint** ◀))) Audio-video solution in English & Spanish at LarsonPrecalculus.com

Write each expression in radical form.

a. $(x^2 - 7)^{-1/2}$ **b.** $-3b^{1/3}c^{2/3}$

c. $a^{0.75}$ **d.** $(x^2)^{2/5}$ ◼

Rational exponents are useful for evaluating roots of numbers on a calculator, for reducing the index of a radical, and for simplifying expressions in calculus.

EXAMPLE 17 **Simplifying with Rational Exponents**

a. $(-32)^{-4/5} = \left(\sqrt[5]{-32}\right)^{-4} = (-2)^{-4} = \dfrac{1}{(-2)^4} = \dfrac{1}{16}$

b. $(-5x^{5/3})(3x^{-3/4}) = -15x^{(5/3)-(3/4)} = -15x^{11/12}, \quad x \neq 0$

c. $\sqrt[9]{a^3} = a^{3/9} = a^{1/3} = \sqrt[3]{a}$ Reduce index.

d. $\sqrt[3]{\sqrt{125}} = \sqrt[6]{125} = \sqrt[6]{(5)^3} = 5^{3/6} = 5^{1/2} = \sqrt{5}$

e. $(2x - 1)^{4/3}(2x - 1)^{-1/3} = (2x - 1)^{(4/3)-(1/3)} = 2x - 1, \quad x \neq \tfrac{1}{2}$

✓ **Checkpoint** ◀))) Audio-video solution in English & Spanish at LarsonPrecalculus.com

••REMARK The expression in Example 17(b) is not defined when $x = 0$ because $0^{-3/4}$ is not a real number. Similarly, the expression in Example 17(e) is not defined when $x = \tfrac{1}{2}$ because

$$\left(2 \cdot \tfrac{1}{2} - 1\right)^{-1/3} = (0)^{-1/3}$$

is not a real number.

Simplify each expression.

a. $(-125)^{-2/3}$ **b.** $(4x^2y^{3/2})(-3x^{-1/3}y^{-3/5})$

c. $\sqrt[3]{\sqrt[4]{27}}$ **d.** $(3x + 2)^{5/2}(3x + 2)^{-1/2}$ ◼

Summarize (Appendix A.2)

1. Make a list of the properties of exponents *(page A13)*. For examples that use these properties, see Examples 1–4.

2. State the definition of scientific notation *(page A16)*. For examples involving scientific notation, see Examples 5–7.

3. Make a list of the properties of radicals *(page A18)*. For examples involving radicals, see Examples 8 and 9.

4. Explain how to simplify a radical expression *(page A19)*. For examples of simplifying radical expressions, see Examples 10 and 11.

5. Explain how to rationalize a denominator or a numerator *(page A20)*. For examples of rationalizing denominators and numerators, see Examples 12–14.

6. State the definition of a rational exponent *(page A21)*. For examples involving rational exponents, see Examples 15–17.

A.2 Exercises

See **CalcChat.com** for tutorial help and worked-out solutions to odd-numbered exercises.

Vocabulary: Fill in the blanks.

1. In the exponential form a^n, n is the _____ and a is the _____.
2. A convenient way of writing very large or very small numbers is _____ _____.
3. One of the two equal factors of a number is a _____ _____ of the number.
4. In the radical form $\sqrt[n]{a}$, the positive integer n is the _____ of the radical and the number a is the _____.
5. Radical expressions can be combined (added or subtracted) when they are _____ _____.
6. The expressions $a + b\sqrt{m}$ and $a - b\sqrt{m}$ are _____ of each other.
7. The process used to create a radical-free denominator is known as _____ the denominator.
8. In the expression $b^{m/n}$, m denotes the _____ to which the base is raised and n denotes the _____ or root to be taken.

Skills and Applications

 Evaluating Exponential Expressions In Exercises 9–14, evaluate each expression.

9. (a) $5 \cdot 5^3$ (b) $\dfrac{5^2}{5^4}$

10. (a) $(3^3)^0$ (b) -3^2

11. (a) $(2^3 \cdot 3^2)^2$ (b) $\left(-\dfrac{3}{5}\right)^3\left(\dfrac{5}{3}\right)^2$

12. (a) $\dfrac{3}{3^{-4}}$ (b) $48(-4)^{-3}$

13. (a) $\dfrac{4 \cdot 3^{-2}}{2^{-2} \cdot 3^{-1}}$ (b) $(-2)^0$

14. (a) $3^{-1} + 2^{-2}$ (b) $(3^{-2})^2$

Evaluating an Algebraic Expression In Exercises 15–20, evaluate the expression for the given value of x.

15. $-3x^3$, $x = 2$
16. $7x^{-2}$, $x = 4$
17. $6x^0$, $x = 10$
18. $2x^3$, $x = -3$
19. $-3x^4$, $x = -2$
20. $12(-x)^3$, $x = -\frac{1}{3}$

 Using Properties of Exponents In Exercises 21–26, simplify each expression.

21. (a) $(5z)^3$ (b) $5x^4(x^2)$
22. (a) $(-2x)^2$ (b) $(4x^3)^0$
23. (a) $6y^2(2y^0)^2$ (b) $(-z)^3(3z^4)$
24. (a) $\dfrac{7x^2}{x^3}$ (b) $\dfrac{12(x+y)^3}{9(x+y)}$
25. (a) $\left(\dfrac{4}{y}\right)^3\left(\dfrac{3}{y}\right)^4$ (b) $\left(\dfrac{b^{-2}}{a^{-2}}\right)\left(\dfrac{b}{a}\right)^2$
26. (a) $[(x^2y^{-2})^{-1}]^{-1}$ (b) $(5x^2z^6)^3(5x^2z^6)^{-3}$

 Rewriting with Positive Exponents In Exercises 27–30, rewrite each expression with positive exponents. Simplify, if possible.

27. (a) $(x + 5)^0$ (b) $(2x^2)^{-2}$
28. (a) $(4y^{-2})(8y^4)$ (b) $(z + 2)^{-3}(z + 2)^{-1}$
29. (a) $\left(\dfrac{x^{-3}y^4}{5}\right)^{-3}$ (b) $\left(\dfrac{a^{-2}}{b^{-2}}\right)\left(\dfrac{b}{a}\right)^3$
30. (a) $\dfrac{3^n \cdot 3^{2n}}{3^{3n} \cdot 3^2}$ (b) $\dfrac{x^2 \cdot x^n}{x^3 \cdot x^n}$

 Scientific Notation In Exercises 31 and 32, write the number in scientific notation.

31. $10{,}250.4$ 32. -0.000125

Decimal Notation In Exercises 33–36, write the number in decimal notation.

33. 3.14×10^{-4} 34. -2.058×10^6
35. Light year: 9.46×10^{12} kilometers
36. Diameter of a human hair: 9.0×10^{-6} meter

Using Scientific Notation In Exercises 37 and 38, evaluate each expression without using a calculator.

37. (a) $(2.0 \times 10^9)(3.4 \times 10^{-4})$
    (b) $(1.2 \times 10^7)(5.0 \times 10^{-3})$

38. (a) $\dfrac{6.0 \times 10^8}{3.0 \times 10^{-3}}$ (b) $\dfrac{2.5 \times 10^{-3}}{5.0 \times 10^2}$

Evaluating Radical Expressions In Exercises 39 and 40, evaluate each expression without using a calculator.

39. (a) $\sqrt{9}$ (b) $\sqrt[3]{\frac{27}{8}}$ 40. (a) $\sqrt[3]{27}$ (b) $\left(\sqrt{36}\right)^3$

Using Properties of Radicals In Exercises 41 and 42, use the properties of radicals to simplify each expression.

41. (a) $\left(\sqrt[5]{2}\right)^5$ 　　　(b) $\sqrt[5]{32x^5}$

42. (a) $\sqrt{12} \cdot \sqrt{3}$ 　　　(b) $\sqrt[4]{(3x^2)^4}$

 Simplifying Radical Expressions In Exercises 43–50, simplify each radical expression.

43. (a) $\sqrt{20}$ 　　　(b) $\sqrt[3]{128}$

44. (a) $\sqrt[3]{\frac{16}{27}}$ 　　　(b) $\sqrt{\frac{75}{4}}$

45. (a) $\sqrt{72x^3}$ 　　　(b) $\sqrt{54xy^4}$

46. (a) $\sqrt{\frac{18^2}{z^3}}$ 　　　(b) $\sqrt{\frac{32a^4}{b^2}}$

47. (a) $\sqrt[3]{16x^5}$ 　　　(b) $\sqrt{75x^2y^{-4}}$

48. (a) $\sqrt[4]{3x^4y^2}$ 　　　(b) $\sqrt[5]{160x^8z^4}$

49. (a) $2\sqrt{20x^2} + 5\sqrt{125x^2}$
　　　(b) $8\sqrt{147x} - 3\sqrt{48x}$

50. (a) $3\sqrt[3]{54x^3} + \sqrt[3]{16x^3}$
　　　(b) $\sqrt[3]{64x} - \sqrt[3]{27x^4}$

 Rationalizing a Denominator In Exercises 51–54, rationalize the denominator of the expression. Then simplify your answer.

51. $\dfrac{1}{\sqrt{3}}$ 　　　　**52.** $\dfrac{8}{\sqrt[3]{2}}$

53. $\dfrac{5}{\sqrt{14} - 2}$ 　　　　**54.** $\dfrac{3}{\sqrt{5} + \sqrt{6}}$

 Rationalizing a Numerator In Exercises 55 and 56, rationalize the numerator of the expression. Then simplify your answer.

55. $\dfrac{\sqrt{5} + \sqrt{3}}{3}$ 　　　　**56.** $\dfrac{\sqrt{7} - 3}{4}$

 Writing Exponential and Radical Forms In Exercises 57–60, fill in the missing form of the expression.

Radical Form	Rational Exponent Form
57. $\sqrt[3]{64}$	
58. $x^2\sqrt{x}$	
59.	$3x^{-2/3}$
60.	$a^{0.4}$

 Simplifying Expressions In Exercises 61–68, simplify each expression.

61. (a) $32^{-3/5}$ 　　　(b) $\left(\frac{16}{81}\right)^{-3/4}$

62. (a) $100^{-3/2}$ 　　　(b) $\left(\frac{9}{4}\right)^{-1/2}$

63. (a) $\sqrt[4]{3^2}$ 　　　(b) $\sqrt[6]{(x+1)^4}$

64. (a) $\sqrt[6]{x^3}$ 　　　(b) $\sqrt[4]{(3x^2)^4}$

65. (a) $\sqrt{\sqrt{32}}$ 　　　(b) $\sqrt{\sqrt[4]{2x}}$

66. (a) $\sqrt{\sqrt{243(x+1)}}$
　　　(b) $\sqrt{\sqrt[3]{10a^7b}}$

67. (a) $(x-1)^{1/3}(x-1)^{2/3}$
　　　(b) $(x-1)^{1/3}(x-1)^{-4/3}$

68. (a) $(4x+3)^{5/2}(4x+3)^{-5/3}$
　　　(b) $(4x+3)^{-5/2}(4x+3)^{2/3}$

• • 69. Mathematical Modeling • • • • • • • •

A funnel is filled with water to a height of h centimeters. The formula

$$t = 0.03\left[12^{5/2} - (12 - h)^{5/2}\right], \quad 0 \le h \le 12$$

represents the amount of time t (in seconds) that it will take for the funnel to empty. Use the *table* feature of a graphing utility to find the times required for the funnel to empty for water heights of $h = 0$, $h = 1$, $h = 2$, . . . , $h = 12$ centimeters.

70. HOW DO YOU SEE IT? Package A is a cube with a volume of 500 cubic inches. Package B is a cube with a volume of 250 cubic inches. Is the length x of a side of package A greater than, less than, or equal to twice the length of a side of package B? Explain.

Exploration

True or False? In Exercises 71–74, determine whether the statement is true or false. Justify your answer.

71. $\dfrac{x^{k+1}}{x} = x^k$ 　　　　**72.** $(a^n)^k = a^{n^k}$

73. $(a+b)^2 = a^2 + b^2$

74. $\dfrac{a}{\sqrt{b}} = \dfrac{a^2}{(\sqrt{b})^2} = \dfrac{a^2}{b}$

A.3   Polynomials and Factoring

Polynomial factoring has many real-life applications. For example, in Exercise 84 on page A34, you will use polynomial factoring to write an alternative form of an expression that models the rate of change of an autocatalytic chemical reaction.

· · · · · · · · · · · · · · · · ▷

· **REMARK**   Expressions are not polynomials when a variable is underneath a radical or when a polynomial expression (with degree greater than 0) is in the denominator of a term. For example, the expressions $x^3 - \sqrt{3x} = x^3 - (3x)^{1/2}$ and $x^2 + (5/x) = x^2 + 5x^{-1}$ are not polynomials.

- ■ Write polynomials in standard form.
- ■ Add, subtract, and multiply polynomials.
- ■ Use special products to multiply polynomials.
- ■ Factor out common factors from polynomials.
- ■ Factor special polynomial forms.
- ■ Factor trinomials as the product of two binomials.
- ■ Factor polynomials by grouping.

Polynomials

One of the most common types of algebraic expressions is the **polynomial.** Some examples are $2x + 5$, $3x^4 - 7x^2 + 2x + 4$, and $5x^2y^2 - xy + 3$. The first two are *polynomials in x* and the third is a *polynomial in x and y.* The terms of a polynomial in x have the form ax^k, where a is the **coefficient** and k is the **degree** of the term. For example, the polynomial $2x^3 - 5x^2 + 1 = 2x^3 + (-5)x^2 + (0)x + 1$ has coefficients 2, -5, 0, and 1.

Definition of a Polynomial in x

Let $a_0, a_1, a_2, \ldots, a_n$ be real numbers and let n be a nonnegative integer. A polynomial in x is an expression of the form

$$a_n x^n + a_{n-1} x^{n-1} + \cdots + a_1 x + a_0$$

where $a_n \neq 0$. The polynomial is of **degree** n, a_n is the **leading coefficient,** and a_0 is the **constant term.**

Polynomials with one, two, and three terms are **monomials, binomials,** and **trinomials,** respectively. A polynomial written with descending powers of x is in **standard form.**

EXAMPLE 1   Writing Polynomials in Standard Form

Polynomial	Standard Form	Degree	Leading Coefficient
a. $4x^2 - 5x^7 - 2 + 3x$	$-5x^7 + 4x^2 + 3x - 2$	7	-5
b. $4 - 9x^2$	$-9x^2 + 4$	2	-9
c. 8	8 or $8x^0$	0	8

✓ *Checkpoint*  *Audio-video solution in English & Spanish at LarsonPrecalculus.com*

Write the polynomial $6 - 7x^3 + 2x$ in standard form. Then identify the degree and leading coefficient of the polynomial. ■

A polynomial that has all zero coefficients is called the **zero polynomial,** denoted by 0. No degree is assigned to the zero polynomial. For polynomials in more than one variable, the degree of a *term* is the sum of the exponents of the variables in the term. The degree of the *polynomial* is the highest degree of its terms. For example, the degree of the polynomial $-2x^3y^6 + 4xy - x^7y^4$ is 11 because the sum of the exponents in the last term is the greatest. The leading coefficient of the polynomial is the coefficient of the highest-degree term.

Operations with Polynomials

You can add and subtract polynomials in much the same way you add and subtract real numbers. Add or subtract the *like terms* (terms having the same variables to the same powers) by adding or subtracting their coefficients. For example, $-3xy^2$ and $5xy^2$ are like terms and their sum is

$$-3xy^2 + 5xy^2 = (-3 + 5)xy^2 = 2xy^2.$$

EXAMPLE 2   **Adding or Subtracting Polynomials**

a. $(5x^3 - 7x^2 - 3) + (x^3 + 2x^2 - x + 8)$

$\quad\quad = (5x^3 + x^3) + (-7x^2 + 2x^2) + (-x) + (-3 + 8)$ Group like terms.

$\quad\quad = 6x^3 - 5x^2 - x + 5$ Combine like terms.

b. $(7x^4 - x^2 - 4x + 2) - (3x^4 - 4x^2 + 3x)$

$\quad\quad = 7x^4 - x^2 - 4x + 2 - 3x^4 + 4x^2 - 3x$ Distributive Property

$\quad\quad = (7x^4 - 3x^4) + (-x^2 + 4x^2) + (-4x - 3x) + 2$ Group like terms.

$\quad\quad = 4x^4 + 3x^2 - 7x + 2$ Combine like terms.

✓ **Checkpoint** 🔊))) *Audio-video solution in English & Spanish at LarsonPrecalculus.com*

Find the difference $(2x^3 - x + 3) - (x^2 - 2x - 3)$ and write the resulting polynomial in standard form. ◾

> • **REMARK** When a negative sign precedes an expression inside parentheses, remember to distribute the negative sign to each term inside the parentheses. In other words, multiply each term by -1.
>
> $-(3x^4 - 4x^2 + 3x)$
> $\quad = -3x^4 + 4x^2 - 3x$

To find the *product* of two polynomials, use the right and left Distributive Properties. For example, you can find the product of $3x - 2$ and $5x + 7$ by first treating $5x + 7$ as a single quantity.

$$(3x - 2)(5x + 7) = 3x(5x + 7) - 2(5x + 7)$$

$$= (3x)(5x) + (3x)(7) - (2)(5x) - (2)(7)$$

$$= 15x^2 + 21x - 10x - 14$$

Product of First terms	Product of Outer terms	Product of Inner terms	Product of Last terms

$$= 15x^2 + 11x - 14$$

Note that when using the **FOIL Method** above (which can be used only to multiply two binomials), some of the terms in the product may be like terms that can be combined into one term.

EXAMPLE 3   **Finding a Product by the FOIL Method**

Use the FOIL Method to find the product of $2x - 4$ and $x + 5$.

Solution

$$\overset{\text{F} \quad\quad \text{O} \quad\quad \text{I} \quad\quad \text{L}}{(2x - 4)(x + 5) = 2x^2 + 10x - 4x - 20} = 2x^2 + 6x - 20$$

✓ **Checkpoint** 🔊))) *Audio-video solution in English & Spanish at LarsonPrecalculus.com*

Use the FOIL Method to find the product of $3x - 1$ and $x - 5$. ◾

Special Products

Some binomial products have special forms that occur frequently in algebra. You do not need to memorize these formulas because you can use the Distributive Property to multiply. However, becoming familiar with these formulas will enable you to manipulate the algebra more quickly.

Special Products

Let u and v be real numbers, variables, or algebraic expressions.

Special Product   **Example**

Sum and Difference of Same Terms

$$(u + v)(u - v) = u^2 - v^2$$ $(x + 4)(x - 4) = x^2 - 4^2$
$$= x^2 - 16$$

Square of a Binomial

$$(u + v)^2 = u^2 + 2uv + v^2$$ $(x + 3)^2 = x^2 + 2(x)(3) + 3^2$
$$= x^2 + 6x + 9$$

$$(u - v)^2 = u^2 - 2uv + v^2$$ $(3x - 2)^2 = (3x)^2 - 2(3x)(2) + 2^2$
$$= 9x^2 - 12x + 4$$

Cube of a Binomial

$$(u + v)^3 = u^3 + 3u^2v + 3uv^2 + v^3$$ $(x + 2)^3 = x^3 + 3x^2(2) + 3x(2^2) + 2^3$
$$= x^3 + 6x^2 + 12x + 8$$

$$(u - v)^3 = u^3 - 3u^2v + 3uv^2 - v^3$$ $(x - 1)^3 = x^3 - 3x^2(1) + 3x(1^2) - 1^3$
$$= x^3 - 3x^2 + 3x - 1$$

EXAMPLE 4   **Sum and Difference of Same Terms**

Find each product.

a. $(5x + 9)(5x - 9)$ **b.** $(x + y - 2)(x + y + 2)$

Solution

a. The product of a sum and a difference of the *same* two terms has no middle term and takes the form $(u + v)(u - v) = u^2 - v^2$.

$$(5x + 9)(5x - 9) = (5x)^2 - 9^2 = 25x^2 - 81$$

b. One way to find this product is to group $x + y$ and form a special product.

Difference   Sum

$$(x + y - 2)(x + y + 2) = [(x + y) - 2][(x + y) + 2]$$
$$= (x + y)^2 - 2^2 \qquad \text{Sum and difference of same terms}$$
$$= x^2 + 2xy + y^2 - 4 \qquad \text{Square of a binomial.}$$

✓ **Checkpoint** ◀))) *Audio-video solution in English & Spanish at LarsonPrecalculus.com*

Find each product.

a. $(3x - 2)(3x + 2)$ **b.** $(x - 2 + 3y)(x - 2 - 3y)$ ■

Polynomials with Common Factors

The process of writing a polynomial as a product is called **factoring.** It is an important tool for solving equations and for simplifying rational expressions.

Unless noted otherwise, when you are asked to factor a polynomial, assume that you are looking for factors that have integer coefficients. If a polynomial does not factor using integer coefficients, then it is **prime** or **irreducible over the integers.** For example, the polynomial

$$x^2 - 3$$

is irreducible over the integers. Over the *real numbers,* this polynomial factors as

$$x^2 - 3 = \left(x + \sqrt{3}\right)\left(x - \sqrt{3}\right).$$

A polynomial is **completely factored** when each of its factors is prime. For example,

$$x^3 - x^2 + 4x - 4 = (x - 1)(x^2 + 4)$$ Completely factored

is completely factored, but

$$x^3 - x^2 - 4x + 4 = (x - 1)(x^2 - 4)$$ Not completely factored

is not completely factored. Its complete factorization is

$$x^3 - x^2 - 4x + 4 = (x - 1)(x + 2)(x - 2).$$

The simplest type of factoring involves a polynomial that can be written as the product of a monomial and another polynomial. The technique used here is the Distributive Property, $a(b + c) = ab + ac$, in the *reverse* direction.

$$ab + ac = a(b + c)$$ a is a common factor.

Factoring out any common factors is the first step in completely factoring a polynomial.

EXAMPLE 5   Factoring Out Common Factors

Factor each expression.

a. $6x^3 - 4x$

b. $-4x^2 + 12x - 16$

c. $(x - 2)(2x) + (x - 2)(3)$

Solution

a. $6x^3 - 4x = 2x(3x^2) - 2x(2)$ $2x$ is a common factor.

$$= 2x(3x^2 - 2)$$

b. $-4x^2 + 12x - 16 = -4(x^2) + (-4)(-3x) + (-4)4$ -4 is a common factor.

$$= -4(x^2 - 3x + 4)$$

c. $(x - 2)(2x) + (x - 2)(3) = (x - 2)(2x + 3)$ $(x - 2)$ is a common factor.

✓ **Checkpoint** ◀))) *Audio-video solution in English & Spanish at LarsonPrecalculus.com*

Factor each expression.

a. $5x^3 - 15x^2$

b. $-3 + 6x - 12x^3$

c. $(x + 1)(x^2) - (x + 1)(2)$

Factoring Special Polynomial Forms

Some polynomials have special forms that arise from the special product forms on page A27. You should learn to recognize these forms.

Factoring Special Polynomial Forms

Factored Form	**Example**
Difference of Two Squares	
$u^2 - v^2 = (u + v)(u - v)$	$9x^2 - 4 = (3x)^2 - 2^2 = (3x + 2)(3x - 2)$
Perfect Square Trinomial	
$u^2 + 2uv + v^2 = (u + v)^2$	$x^2 + 6x + 9 = x^2 + 2(x)(3) + 3^2 = (x + 3)^2$
$u^2 - 2uv + v^2 = (u - v)^2$	$x^2 - 6x + 9 = x^2 - 2(x)(3) + 3^2 = (x - 3)^2$
Sum or Difference of Two Cubes	
$u^3 + v^3 = (u + v)(u^2 - uv + v^2)$	$x^3 + 8 = x^3 + 2^3 = (x + 2)(x^2 - 2x + 4)$
$u^3 - v^3 = (u - v)(u^2 + uv + v^2)$	$27x^3 - 1 = (3x)^3 - 1^3 = (3x - 1)(9x^2 + 3x + 1)$

The factored form of a difference of two squares is always a set of **conjugate pairs**.

$$u^2 - v^2 = (u + v)(u - v) \qquad \text{Conjugate pairs}$$

Difference   Opposite signs

To recognize perfect square terms, look for coefficients that are squares of integers and variables raised to *even powers*.

EXAMPLE 6 **Factoring Out a Common Factor First**

$$3 - 12x^2 = 3(1 - 4x^2) \qquad \text{3 is a common factor.}$$
$$= 3[1^2 - (2x)^2] \qquad \text{Rewrite } 1 - 4x^2 \text{ as the difference of two squares.}$$
$$= 3(1 + 2x)(1 - 2x) \qquad \text{Factor.}$$

✓ **Checkpoint** ◀))) *Audio-video solution in English & Spanish at LarsonPrecalculus.com*

Factor $100 - 4y^2$.

> •• **REMARK** In Example 6, note that the first step in factoring a polynomial is to check for any common factors. Once you have removed any common factors, it is often possible to recognize patterns that were not immediately obvious.

EXAMPLE 7 **Factoring the Difference of Two Squares**

a. $(x + 2)^2 - y^2 = [(x + 2) + y][(x + 2) - y] = (x + 2 + y)(x + 2 - y)$

b. $16x^4 - 81 = (4x^2)^2 - 9^2 \qquad \text{Rewrite as the difference of two squares.}$
$$= (4x^2 + 9)(4x^2 - 9) \qquad \text{Factor.}$$
$$= (4x^2 + 9)[(2x)^2 - 3^2] \qquad \text{Rewrite } 4x^2 - 9 \text{ as the difference of two squares.}$$
$$= (4x^2 + 9)(2x + 3)(2x - 3) \qquad \text{Factor.}$$

✓ **Checkpoint** ◀))) *Audio-video solution in English & Spanish at LarsonPrecalculus.com*

Factor $(x - 1)^2 - 9y^4$.

A **perfect square trinomial** is the square of a binomial, and it has the form

$$u^2 + 2uv + v^2 = (u + v)^2 \quad \text{or} \quad u^2 - 2uv + v^2 = (u - v)^2.$$

Like signs   Like signs

Note that the first and last terms are squares and the middle term is twice the product of u and v.

EXAMPLE 8   Factoring Perfect Square Trinomials

Factor each trinomial.

a. $x^2 - 10x + 25$ **b.** $16x^2 + 24x + 9$

Solution

a. $x^2 - 10x + 25 = x^2 - 2(x)(5) + 5^2 = (x - 5)^2$

b. $16x^2 + 24x + 9 = (4x)^2 + 2(4x)(3) + 3^2 = (4x + 3)^2$

✓ *Checkpoint* ◀))) *Audio-video solution in English & Spanish at LarsonPrecalculus.com*

Factor $9x^2 - 30x + 25$.

The next two formulas show the sum and difference of two cubes. Pay special attention to the signs of the terms.

Like signs   Like signs

$$u^3 + v^3 = (u + v)(u^2 - uv + v^2) \quad u^3 - v^3 = (u - v)(u^2 + uv + v^2)$$

Unlike signs   Unlike signs

EXAMPLE 9   Factoring the Difference of Two Cubes

$$x^3 - 27 = x^3 - 3^3 \qquad \text{Rewrite 27 as } 3^3.$$

$$= (x - 3)(x^2 + 3x + 9) \qquad \text{Factor.}$$

✓ *Checkpoint* ◀))) *Audio-video solution in English & Spanish at LarsonPrecalculus.com*

Factor $64x^3 - 1$.

EXAMPLE 10   Factoring the Sum of Two Cubes

a. $y^3 + 8 = y^3 + 2^3 \qquad \text{Rewrite 8 as } 2^3.$

$$= (y + 2)(y^2 - 2y + 4) \qquad \text{Factor.}$$

b. $3x^3 + 192 = 3(x^3 + 64) \qquad \text{3 is a common factor.}$

$$= 3(x^3 + 4^3) \qquad \text{Rewrite 64 as } 4^3.$$

$$= 3(x + 4)(x^2 - 4x + 16) \qquad \text{Factor.}$$

✓ *Checkpoint* ◀))) *Audio-video solution in English & Spanish at LarsonPrecalculus.com*

Factor each expression.

a. $x^3 + 216$ **b.** $5y^3 + 135$

Trinomials with Binomial Factors

To factor a trinomial of the form $ax^2 + bx + c$, use the pattern below.

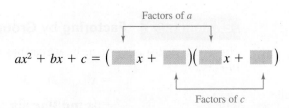

The goal is to find a combination of factors of a and c such that the sum of the outer and inner products is the middle term bx. For example, for the trinomial $6x^2 + 17x + 5$, you can write all possible factorizations and determine which one has outer and inner products whose sum is $17x$.

$$(6x + 5)(x + 1), \ (6x + 1)(x + 5), \ (2x + 1)(3x + 5), \ (2x + 5)(3x + 1)$$

The correct factorization is $(2x + 5)(3x + 1)$ because the sum of the outer (O) and inner (I) products is $17x$.

$$\begin{array}{ccccc} \text{F} & \text{O} & \text{I} & \text{L} & \text{O} + \text{I} \\ \downarrow & \downarrow & \downarrow & \downarrow & \downarrow \end{array}$$

$$(2x + 5)(3x + 1) = 6x^2 + 2x + 15x + 5 = 6x^2 + 17x + 5$$

EXAMPLE 11   **Factoring a Trinomial: Leading Coefficient Is 1**

Factor $x^2 - 7x + 12$.

Solution   For this trinomial, $a = 1$, $b = -7$, and $c = 12$. Because b is negative and c is positive, both factors of 12 must be negative. So, the possible factorizations of $x^2 - 7x + 12$ are

$$(x - 1)(x - 12), \quad (x - 2)(x - 6), \quad \text{and} \quad (x - 3)(x - 4).$$

Testing the middle term, you will find the correct factorization to be

$$x^2 - 7x + 12 = (x - 3)(x - 4). \qquad \text{O} + \text{I} = -4x - 3x = -7x$$

✓ **Checkpoint** 🔊))) *Audio-video solution in English & Spanish at LarsonPrecalculus.com*

Factor $x^2 + x - 6$.

EXAMPLE 12   **Factoring a Trinomial: Leading Coefficient Is Not 1**

See LarsonPrecalculus.com for an interactive version of this type of example.

Factor $2x^2 + x - 15$.

Solution   For this trinomial, $a = 2$, $b = 1$, and $c = -15$. Because c is negative, its factors must have unlike signs. The eight possible factorizations are below.

$$(2x - 1)(x + 15) \quad (2x + 1)(x - 15) \quad (2x - 3)(x + 5) \quad (2x + 3)(x - 5)$$

$$(2x - 5)(x + 3) \quad (2x + 5)(x - 3) \quad (2x - 15)(x + 1) \quad (2x + 15)(x - 1)$$

Testing the middle term, you will find the correct factorization to be

$$2x^2 + x - 15 = (2x - 5)(x + 3). \qquad \text{O} + \text{I} = 6x - 5x = x$$

✓ **Checkpoint** 🔊))) *Audio-video solution in English & Spanish at LarsonPrecalculus.com*

Factor $2x^2 - 5x + 3$.

• **REMARK** Factoring a trinomial can involve trial and error. However, it is relatively easy to check your answer by multiplying the factors. The product should be the original trinomial. For instance, in Example 11, verify that $(x - 3)(x - 4) = x^2 - 7x + 12$.

Factoring by Grouping

Sometimes, polynomials with more than three terms can be **factored by grouping.**

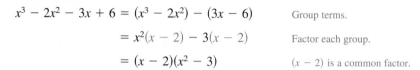

EXAMPLE 13 **Factoring by Grouping**

$$x^3 - 2x^2 - 3x + 6 = (x^3 - 2x^2) - (3x - 6) \qquad \text{Group terms.}$$

$$= x^2(x - 2) - 3(x - 2) \qquad \text{Factor each group.}$$

$$= (x - 2)(x^2 - 3) \qquad (x - 2) \text{ is a common factor.}$$

✓ **Checkpoint** 🔊))) *Audio-video solution in English & Spanish at LarsonPrecalculus.com*

Factor $x^3 + x^2 - 5x - 5$. ■

- - - - - - - - - - - - - - - - ▷

REMARK Sometimes, more than one grouping will work. For instance, another way to factor the polynomial in Example 13 is

$$x^3 - 2x^2 - 3x + 6$$
$$= (x^3 - 3x) - (2x^2 - 6)$$
$$= x(x^2 - 3) - 2(x^2 - 3)$$
$$= (x^2 - 3)(x - 2).$$

Notice that this is the same result as in Example 13.

Factoring by grouping can eliminate some of the trial and error involved in factoring a trinomial. To factor a trinomial of the form $ax^2 + bx + c$ by grouping, choose factors of the product ac that sum to b and use these factors to rewrite the middle term. Example 14 illustrates this technique.

EXAMPLE 14 **Factoring a Trinomial by Grouping**

In the trinomial $2x^2 + 5x - 3$, $a = 2$ and $c = -3$, so the product ac is -6. Now, -6 factors as $(6)(-1)$ and $6 + (-1) = 5 = b$. So, rewrite the middle term as $5x = 6x - x$ and factor by grouping.

$$2x^2 + 5x - 3 = 2x^2 + 6x - x - 3 \qquad \text{Rewrite middle term.}$$

$$= (2x^2 + 6x) - (x + 3) \qquad \text{Group terms.}$$

$$= 2x(x + 3) - (x + 3) \qquad \text{Factor } 2x^2 + 6x.$$

$$= (x + 3)(2x - 1) \qquad (x + 3) \text{ is a common factor.}$$

✓ **Checkpoint** 🔊))) *Audio-video solution in English & Spanish at LarsonPrecalculus.com*

Use factoring by grouping to factor $2x^2 + 5x - 12$. ■

Summarize (Appendix A.3)

1. State the definition of a polynomial in x and explain what is meant by the standard form of a polynomial *(page A25)*. For an example of writing polynomials in standard form, see Example 1.

2. Explain how to add and subtract polynomials *(page A26)*. For an example of adding and subtracting polynomials, see Example 2.

3. Explain the FOIL Method *(page A26)*. For an example of finding a product using the FOIL Method, see Example 3.

4. Explain how to find binomial products that have special forms *(page A27)*. For an example of binomial products that have special forms, see Example 4.

5. Explain what it means to completely factor a polynomial *(page A28)*. For an example of factoring out common factors, see Example 5.

6. Make a list of the special polynomial forms of factoring *(page A29)*. For examples of factoring these special forms, see Examples 6–10.

7. Explain how to factor a trinomial of the form $ax^2 + bx + c$ *(page A31)*. For examples of factoring trinomials of this form, see Examples 11 and 12.

8. Explain how to factor a polynomial by grouping *(page A32)*. For examples of factoring by grouping, see Examples 13 and 14.

A.3   Exercises   See **CalcChat.com** for tutorial help and worked-out solutions to odd-numbered exercises.

Vocabulary: Fill in the blanks.

1. For the polynomial $a_n x^n + a_{n-1} x^{n-1} + \cdots + a_1 x + a_0$, $a_n \neq 0$, the degree is _____, the leading coefficient is _____, and the constant term is _____.
2. A polynomial with one term is a _____, while a polynomial with two terms is a _____ and a polynomial with three terms is a _____.
3. To add or subtract polynomials, add or subtract the _____ _____ by adding or subtracting their coefficients.
4. The letters in "FOIL" stand for F _____, O _____, I _____, and L _____.
5. The process of writing a polynomial as a product is called _____.
6. A polynomial is _____ _____ when each of its factors is prime.
7. A _____ _____ _____ is the square of a binomial, and it has the form $u^2 + 2uv + v^2$ or $u^2 - 2uv + v^2$.
8. Sometimes, polynomials with more than three terms can be factored by _____.

Skills and Applications

 Writing Polynomials in Standard Form
In Exercises 9–14, (a) write the polynomial in standard form, (b) identify the degree and leading coefficient of the polynomial, and (c) state whether the polynomial is a monomial, a binomial, or a trinomial.

9. $7x$
10. 3
11. $14x - \frac{1}{2}x^5$
12. $3 + 2x$
13. $1 + 6x^4 - 4x^5$
14. $-y + 25y^2 + 1$

 Adding or Subtracting Polynomials In Exercises 15–18, add or subtract and write the result in standard form.

15. $(6x + 5) - (8x + 15)$
16. $(2x^2 + 1) - (x^2 - 2x + 1)$
17. $(15x^2 - 6) + (-8.3x^3 - 14.7x^2 - 17)$
18. $(15.6w^4 - 14w - 17.4) + (16.9w^4 - 9.2w + 13)$

 Multiplying Polynomials In Exercises 19–36, multiply the polynomials.

19. $3x(x^2 - 2x + 1)$
20. $y^2(4y^2 + 2y - 3)$
21. $-5z(3z - 1)$
22. $-3x(5x + 2)$
23. $(3x - 5)(2x + 1)$
24. $(7x - 2)(4x - 3)$
25. $(x^2 - x + 2)(x^2 + x + 1)$
26. $(2x^2 - x + 4)(x^2 + 3x + 2)$
27. $(x + 10)(x - 10)$
28. $(4a + 5b)(4a - 5b)$
29. $(2x + 3)^2$
30. $(8x + 3)^2$
31. $(x + 3)^3$
32. $(3x + 2y)^3$

33. $[(x - 3) + y]^2$
34. $[(x + 1) - y]^2$
35. $[(m - 3) + n][(m - 3) - n]$
36. $[(x - 3y) + z][(x - 3y) - z]$

 Factoring Out a Common Factor In Exercises 37–40, factor out the common factor.

37. $2x^3 - 6x$
38. $3z^3 - 6z^2 + 9z$
39. $3x(x - 5) + 8(x - 5)$
40. $(x + 3)^2 - 4(x + 3)$

 Factoring the Difference of Two Squares In Exercises 41–44, completely factor the difference of two squares.

41. $25y^2 - 4$
42. $81 - 36z^2$
43. $(x - 1)^2 - 4$
44. $25 - (z + 5)^2$

 Factoring a Perfect Square Trinomial In Exercises 45–50, factor the perfect square trinomial.

45. $x^2 - 4x + 4$
46. $4t^2 + 4t + 1$
47. $25z^2 - 30z + 9$
48. $36y^2 + 84y + 49$
49. $4y^2 - 12y + 9$
50. $9u^2 + 24uv + 16v^2$

 **Factoring the Sum or Difference of Two Cubes** In Exercises 51–54, factor the sum or difference of two cubes.

51. $x^3 + 125$
52. $x^3 - 8$
53. $8t^3 - 1$
54. $27t^3 + 8$

Factoring a Trinomial In Exercises 55–62, factor the trinomial.

55. $x^2 + x - 2$ 56. $s^2 - 5s + 6$
57. $3x^2 + 10x - 8$ 58. $2x^2 - 3x - 27$
59. $5x^2 + 31x + 6$ 60. $8x^2 + 51x + 18$
61. $-5y^2 - 8y + 4$ 62. $-6z^2 + 17z + 3$

Factoring by Grouping In Exercises 63–68, factor by grouping.

63. $x^3 - x^2 + 2x - 2$ 64. $x^3 + 5x^2 - 5x - 25$
65. $2x^3 - x^2 - 6x + 3$ 66. $3x^3 + x^2 - 15x - 5$
67. $3x^5 + 6x^3 - 2x^2 - 4$ 68. $8x^5 - 6x^2 + 12x^3 - 9$

Factoring a Trinomial by Grouping In Exercises 69–72, factor the trinomial by grouping.

69. $2x^2 + 9x + 9$ 70. $6x^2 + x - 2$
71. $6x^2 - x - 15$ 72. $12x^2 - 13x + 1$

Factoring Completely In Exercises 73–82, completely factor the expression.

73. $6x^2 - 54$ 74. $12x^2 - 48$
75. $x^3 - x^2$ 76. $x^3 - 16x$
77. $2x^2 + 4x - 2x^3$ 78. $9x^2 + 12x - 3x^3$
79. $5 - x + 5x^2 - x^3$ 80. $3u - 2u^2 + 6 - u^3$
81. $2(x - 2)(x + 1)^2 - 3(x - 2)^2(x + 1)$
82. $2(x + 1)(x - 3)^2 - 3(x + 1)^2(x - 3)$

83. **Geometry** The cylindrical shell shown in the figure has a volume of

$$V = \pi R^2 h - \pi r^2 h.$$

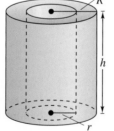

   (a) Factor the expression for the volume.

   (b) From the result of part (a), show that the volume is 2π(average radius)(thickness of the shell)h.

84. **Chemistry**

The rate of change of an autocatalytic chemical reaction is $kQx - kx^2$, where Q is the amount of the original substance, x is the amount of substance formed, and k is a constant of proportionality. Factor the expression.

Exploration

True or False? In Exercises 85–87, determine whether the statement is true or false. Justify your answer.

85. The product of two binomials is always a second-degree polynomial.

86. The sum of two binomials is always a binomial.

87. The difference of two perfect squares can be factored as the product of conjugate pairs.

88. **Error Analysis** Describe the error.
$$9x^2 - 9x - 54 = (3x + 6)(3x - 9)$$
$$= 3(x + 2)(x - 3)$$

89. **Degree of a Product** Find the degree of the product of two polynomials of degrees m and n.

90. **Degree of a Sum** Find the degree of the sum of two polynomials of degrees m and n, where $m < n$.

91. **Think About It** When the polynomial
$$-x^3 + 3x^2 + 2x - 1$$
is subtracted from an unknown polynomial, the difference is $5x^2 + 8$. Find the unknown polynomial.

92. **Logical Reasoning** Verify that $(x + y)^2$ is not equal to $x^2 + y^2$ by letting $x = 3$ and $y = 4$ and evaluating both expressions. Are there any values of x and y for which $(x + y)^2 = x^2 + y^2$? Explain.

93. **Think About It** Give an example of a polynomial that is prime.

94. **HOW DO YOU SEE IT?** The figure shows a large square with an area of a^2 that contains a smaller square with an area of b^2.

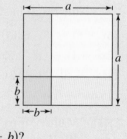

   (a) Describe the regions that represent $a^2 - b^2$. How can you rearrange these regions to show that $a^2 - b^2 = (a - b)(a + b)$?

   (b) How can you use the figure to show that $(a - b)^2 = a^2 - 2ab + b^2$?

   (c) Draw another figure to show that $(a + b)^2 = a^2 + 2ab + b^2$. Explain how the figure shows this.

Factoring with Variables in the Exponents In Exercises 95 and 96, factor the expression as completely as possible.

95. $x^{2n} - y^{2n}$ 96. $x^{3n} + y^{3n}$

A.4 Rational Expressions

Rational expressions have many real-life applications. For example, in Exercise 71 on page A43, you will work with a rational expression that models the temperature of food in a refrigerator.

■ Find domains of algebraic expressions.
■ Simplify rational expressions.
■ Add, subtract, multiply, and divide rational expressions.
■ Simplify complex fractions and rewrite difference quotients.

Domain of an Algebraic Expression

The set of real numbers for which an algebraic expression is defined is the **domain** of the expression. Two algebraic expressions are **equivalent** when they have the same domain and yield the same values for all numbers in their domain. For example,

$$(x + 1) + (x + 2) \quad \text{and} \quad 2x + 3$$

are equivalent because

$$(x + 1) + (x + 2) = x + 1 + x + 2$$
$$= x + x + 1 + 2$$
$$= 2x + 3.$$

EXAMPLE 1 **Finding Domains of Algebraic Expressions**

a. The domain of the polynomial

$$2x^3 + 3x + 4$$

is the set of all real numbers. In fact, the domain of any polynomial is the set of all real numbers, unless the domain is specifically restricted.

b. The domain of the radical expression

$$\sqrt{x - 2}$$

is the set of real numbers greater than or equal to 2, because the square root of a negative number is not a real number.

c. The domain of the expression

$$\frac{x + 2}{x - 3}$$

is the set of all real numbers except $x = 3$, which would result in division by zero, which is undefined.

✓ **Checkpoint** ◀))) *Audio-video solution in English & Spanish at LarsonPrecalculus.com*

Find the domain of each expression.

a. $4x^3 + 3, \quad x \geq 0$ **b.** $\sqrt{x + 7}$ **c.** $\dfrac{1 - x}{x}$ ■

The quotient of two algebraic expressions is a *fractional expression*. Moreover, the quotient of two *polynomials* such as

$$\frac{1}{x}, \quad \frac{2x - 1}{x + 1}, \quad \text{or} \quad \frac{x^2 - 1}{x^2 + 1}$$

is a **rational expression.**

Simplifying Rational Expressions

Recall that a fraction is in simplest form when its numerator and denominator have no factors in common other than ± 1. To write a fraction in simplest form, divide out common factors.

$$\frac{a \cdot \cancel{c}}{b \cdot \cancel{c}} = \frac{a}{b}, \quad c \neq 0$$

The key to success in simplifying rational expressions lies in your ability to *factor* polynomials. When simplifying rational expressions, factor each polynomial completely to determine whether the numerator and denominator have factors in common.

EXAMPLE 2 Simplifying a Rational Expression

$$\frac{x^2 + 4x - 12}{3x - 6} = \frac{(x + 6)\cancel{(x - 2)}}{3\cancel{(x - 2)}} \qquad \text{Factor completely.}$$

$$= \frac{x + 6}{3}, \quad x \neq 2 \qquad \text{Divide out common factor.}$$

Note that the original expression is undefined when $x = 2$ (because division by zero is undefined). To make the simplified expression *equivalent* to the original expression, you must restrict the domain of the simplified expression by excluding the value of $x = 2$.

✓ **Checkpoint** 🔊))) *Audio-video solution in English & Spanish at LarsonPrecalculus.com*

Write $\dfrac{4x + 12}{x^2 - 3x - 18}$ in simplest form. ∎

Sometimes it may be necessary to change the sign of a factor by factoring out (-1) to simplify a rational expression, as shown in Example 3.

EXAMPLE 3 Simplifying a Rational Expression

$$\frac{12 + x - x^2}{2x^2 - 9x + 4} = \frac{(4 - x)(3 + x)}{(2x - 1)(x - 4)} \qquad \text{Factor completely.}$$

$$= \frac{-\cancel{(x - 4)}(3 + x)}{(2x - 1)\cancel{(x - 4)}} \qquad (4 - x) = -(x - 4)$$

$$= -\frac{3 + x}{2x - 1}, \quad x \neq 4 \qquad \text{Divide out common factor.}$$

✓ **Checkpoint** 🔊))) *Audio-video solution in English & Spanish at LarsonPrecalculus.com*

Write $\dfrac{3x^2 - x - 2}{5 - 4x - x^2}$ in simplest form. ∎

In this text, the domain is usually not listed with a rational expression. It is *implied* that the real numbers that make the denominator zero are excluded from the domain. Also, when performing operations with rational expressions, this text follows the convention of listing *by the simplified expression* all values of x that must be specifically excluded from the domain to make the domains of the simplified and original expressions agree. Example 3, for instance, lists the restriction $x \neq 4$ with the simplified expression to make the two domains agree. Note that the value $x = \frac{1}{2}$ is excluded from *both* domains, so it is not necessary to list this value.

⋯⋯⋯⋯⋯⋯⋯⋯▷

⋯REMARK In Example 2, do not make the mistake of trying to simplify further by dividing out terms.

$$\frac{x + 6}{3} = \frac{x + \cancel{6}}{\cancel{3}}$$

$$= x + 2$$

To simplify fractions, divide out common *factors,* not terms.

Operations with Rational Expressions

To multiply or divide rational expressions, use the properties of fractions discussed in Appendix A.1. Recall that to divide fractions, you invert the divisor and multiply.

EXAMPLE 4 **Multiplying Rational Expressions**

$$\frac{2x^2 + x - 6}{x^2 + 4x - 5} \cdot \frac{x^3 - 3x^2 + 2x}{4x^2 - 6x} = \frac{(2x - 3)(x + 2)}{(x + 5)(x - 1)} \cdot \frac{x(x - 2)(x - 1)}{2x(2x - 3)}$$

$$= \frac{(x + 2)(x - 2)}{2(x + 5)}, \quad x \neq 0, x \neq 1, x \neq \tfrac{3}{2}$$

✓ **Checkpoint** ◀))) *Audio-video solution in English & Spanish at LarsonPrecalculus.com*

Multiply and simplify: $\dfrac{15x^2 + 5x}{x^3 - 3x^2 - 18x} \cdot \dfrac{x^2 - 2x - 15}{3x^2 - 8x - 3}$.

▷ **REMARK** Note that Example 4 lists the restrictions $x \neq 0$, $x \neq 1$, and $x \neq \tfrac{3}{2}$ with the simplified expression to make the two domains agree. Also note that the value $x = -5$ is excluded from both domains, so it is not necessary to list this value.

EXAMPLE 5 **Dividing Rational Expressions**

$$\frac{x^3 - 8}{x^2 - 4} \div \frac{x^2 + 2x + 4}{x^3 + 8} = \frac{x^3 - 8}{x^2 - 4} \cdot \frac{x^3 + 8}{x^2 + 2x + 4} \qquad \text{Invert and multiply.}$$

$$= \frac{(x - 2)(x^2 + 2x + 4)}{(x + 2)(x - 2)} \cdot \frac{(x + 2)(x^2 - 2x + 4)}{(x^2 + 2x + 4)}$$

$$= x^2 - 2x + 4, \quad x \neq \pm 2 \qquad \text{Divide out common factors.}$$

✓ **Checkpoint** ◀))) *Audio-video solution in English & Spanish at LarsonPrecalculus.com*

Divide and simplify: $\dfrac{x^3 - 1}{x^2 - 1} \div \dfrac{x^2 + x + 1}{x^2 + 2x + 1}$. ◾

To add or subtract rational expressions, use the LCD (least common denominator) method or the *basic definition*

$$\frac{a}{b} \pm \frac{c}{d} = \frac{ad \pm bc}{bd}, \quad b \neq 0, d \neq 0. \qquad \text{Basic definition}$$

This definition provides an efficient way of adding or subtracting two fractions that have no common factors in their denominators.

EXAMPLE 6 **Subtracting Rational Expressions**

$$\frac{x}{x - 3} - \frac{2}{3x + 4} = \frac{x(3x + 4) - 2(x - 3)}{(x - 3)(3x + 4)} \qquad \text{Basic definition}$$

$$= \frac{3x^2 + 4x - 2x + 6}{(x - 3)(3x + 4)} \qquad \text{Distributive Property}$$

$$= \frac{3x^2 + 2x + 6}{(x - 3)(3x + 4)} \qquad \text{Combine like terms.}$$

▷ **REMARK** When subtracting rational expressions, remember to distribute the negative sign to *all* the terms in the quantity that is being subtracted.

✓ **Checkpoint** ◀))) *Audio-video solution in English & Spanish at LarsonPrecalculus.com*

Subtract and simplify: $\dfrac{x}{2x - 1} - \dfrac{1}{x + 2}$. ◾

A.4 Exercises

See **CalcChat.com** for tutorial help and worked-out solutions to odd-numbered exercises.

Vocabulary: Fill in the blanks.

1. The set of real numbers for which an algebraic expression is defined is the _____ of the expression.
2. The quotient of two algebraic expressions is a fractional expression, and the quotient of two polynomials is a _____ _____.
3. Fractional expressions with separate fractions in the numerator, denominator, or both are _____ fractions.
4. Two algebraic expressions that have the same domain and yield the same values for all numbers in their domains are _____.

Skills and Applications

 Finding the Domain of an Algebraic Expression In Exercises 5–16, find the domain of the expression.

5. $3x^2 - 4x + 7$
6. $6x^2 - 9, \quad x > 0$

7. $\dfrac{1}{3 - x}$
8. $\dfrac{1}{x + 5}$

9. $\dfrac{x + 6}{3x + 2}$
10. $\dfrac{x - 4}{1 - 2x}$

11. $\dfrac{x^2 - 5x + 6}{x^2 + 6x + 8}$
12. $\dfrac{x^2 - 1}{x^2 + 3x - 10}$

13. $\sqrt{x - 7}$
14. $\sqrt{2x - 5}$

15. $\dfrac{1}{\sqrt{x - 3}}$
16. $\dfrac{1}{\sqrt{x + 2}}$

 Simplifying a Rational Expression In Exercises 17–30, write the rational expression in simplest form.

17. $\dfrac{15x^2}{10x}$
18. $\dfrac{18y^2}{60y^5}$

19. $\dfrac{x - 5}{10 - 2x}$
20. $\dfrac{12 - 4x}{x - 3}$

21. $\dfrac{y^2 - 16}{y + 4}$
22. $\dfrac{x^2 - 25}{5 - x}$

23. $\dfrac{6y + 9y^2}{12y + 8}$
24. $\dfrac{4y - 8y^2}{10y - 5}$

25. $\dfrac{x^2 + 4x - 5}{x^2 + 8x + 15}$
26. $\dfrac{x^2 + 8x - 20}{x^2 + 11x + 10}$

27. $\dfrac{x^2 - x - 2}{10 - 3x - x^2}$
28. $\dfrac{4 + 3x - x^2}{2x^2 - 7x - 4}$

29. $\dfrac{x^2 - 16}{x^3 + x^2 - 16x - 16}$
30. $\dfrac{x^2 - 1}{x^3 + x^2 + 9x + 9}$

31. **Error Analysis** Describe the error.
$$\frac{5x^3}{2x^3 + 4} = \frac{5\cancel{x^3}}{2\cancel{x^3} + 4} = \frac{5}{2 + 4} = \frac{5}{6} \quad \times$$

32. **Evaluating a Rational Expression** Complete the table. What can you conclude?

x	0	1	2	3	4	5	6
$\dfrac{x - 3}{x^2 - x - 6}$							
$\dfrac{1}{x + 2}$							

 Multiplying or Dividing Rational Expressions In Exercises 33–38, perform the multiplication or division and simplify.

33. $\dfrac{5}{x - 1} \cdot \dfrac{x - 1}{25(x - 2)}$
34. $\dfrac{r}{r - 1} \div \dfrac{r^2}{r^2 - 1}$

35. $\dfrac{x^2 - 4}{12} \div \dfrac{2 - x}{2x + 4}$
36. $\dfrac{t^2 - t - 6}{t^2 + 6t + 9} \cdot \dfrac{t + 3}{t^2 - 4}$

37. $\dfrac{x^2 + xy - 2y^2}{x^3 + x^2y} \cdot \dfrac{x}{x^2 + 3xy + 2y^2}$

38. $\dfrac{x^2 - 14x + 49}{x^2 - 49} \div \dfrac{3x - 21}{x + 7}$

 Adding or Subtracting Rational Expressions In Exercises 39–46, perform the addition or subtraction and simplify.

39. $\dfrac{x - 1}{x + 2} - \dfrac{x - 4}{x + 2}$
40. $\dfrac{2x - 1}{x + 3} + \dfrac{1 - x}{x + 3}$

41. $\dfrac{1}{3x + 2} + \dfrac{x}{x + 1}$
42. $\dfrac{x}{x + 4} - \dfrac{6}{x - 1}$

43. $\dfrac{3}{2x + 4} - \dfrac{x}{x + 2}$
44. $\dfrac{2}{x^2 - 9} + \dfrac{4}{x + 3}$

45. $-\dfrac{1}{x} + \dfrac{2}{x^2 + 1} + \dfrac{1}{x^3 + x}$

46. $\dfrac{2}{x + 1} + \dfrac{2}{x - 1} + \dfrac{1}{x^2 - 1}$

Error Analysis In Exercises 47 and 48, describe the error.

47.
$$\frac{x+4}{x+2} - \frac{3x-8}{x+2} = \frac{x+4-3x-8}{x+2}$$
$$= \frac{-2x-4}{x+2}$$
$$= \frac{-2(x+2)}{x+2}$$
$$= -2, \quad x \neq -2$$

48.
$$\frac{6-x}{x(x+2)} + \frac{x+2}{x^2} + \frac{8}{x^2(x+2)}$$
$$= \frac{6-x+(x+2)^2+8}{x^2(x+2)}$$
$$= \frac{6-x+x^2+4x+4+8}{x^2(x+2)}$$
$$= \frac{x^2+3x+18}{x^2(x+2)}$$

 Simplifying a Complex Fraction In Exercises 49–54, simplify the complex fraction.

49.
$$\frac{\left(\dfrac{x}{2}-1\right)}{x-2}$$

50.
$$\frac{x+5}{\left(\dfrac{x}{5}-5\right)}$$

51.
$$\frac{\left[\dfrac{x^2}{(x+1)^2}\right]}{\left[\dfrac{x}{(x+1)^3}\right]}$$

52.
$$\frac{\left(\dfrac{x^2-1}{x}\right)}{\left[\dfrac{(x-1)^2}{x}\right]}$$

53.
$$\frac{\left(\sqrt{x}-\dfrac{1}{2\sqrt{x}}\right)}{\sqrt{x}}$$

54.
$$\frac{\left(\dfrac{t^2}{\sqrt{t^2+1}}-\sqrt{t^2+1}\right)}{t^2}$$

Factoring an Expression In Exercises 55–58, factor the expression by factoring out the common factor with the lesser exponent.

55. $x^2(x^2+3)^{-4} + (x^2+3)^3$

56. $2x(x-5)^{-3} - 4x^2(x-5)^{-4}$

57. $2x^2(x-1)^{1/2} - 5(x-1)^{-1/2}$

58. $4x^3(x+1)^{-3/2} - x(x+1)^{-1/2}$

 Simplifying an Expression In Exercises 59 and 60, simplify the expression.

59.
$$\frac{3x^{1/3}-x^{-2/3}}{3x^{-2/3}}$$

60.
$$\frac{-x^3(1-x^2)^{-1/2}-2x(1-x^2)^{1/2}}{x^4}$$

Simplifying a Difference Quotient In Exercises 61–64, simplify the difference quotient.

61.
$$\frac{\left(\dfrac{1}{x+h}-\dfrac{1}{x}\right)}{h}$$

62.
$$\frac{\left[\dfrac{1}{(x+h)^2}-\dfrac{1}{x^2}\right]}{h}$$

63.
$$\frac{\left(\dfrac{1}{x+h-4}-\dfrac{1}{x-4}\right)}{h}$$

64.
$$\frac{\left(\dfrac{x+h}{x+h+1}-\dfrac{x}{x+1}\right)}{h}$$

 Rewriting a Difference Quotient In Exercises 65–70, rewrite the difference quotient by rationalizing the numerator.

65.
$$\frac{\sqrt{x+2}-\sqrt{x}}{2}$$

66.
$$\frac{\sqrt{z-3}-\sqrt{z}}{-3}$$

67.
$$\frac{\sqrt{t+3}-\sqrt{3}}{t}$$

68.
$$\frac{\sqrt{x+5}-\sqrt{5}}{x}$$

69.
$$\frac{\sqrt{x+h+1}-\sqrt{x+1}}{h}$$

70.
$$\frac{\sqrt{x+h-2}-\sqrt{x-2}}{h}$$

71. Refrigeration

After placing food (at room temperature) in a refrigerator, the time required for the food to cool depends on the amount of food, the air circulation in the refrigerator, the original temperature of the food, and the temperature of the refrigerator. The model that gives the temperature of food that has an original temperature of 75°F and is placed in a 40°F refrigerator is

$$T = 10\left(\frac{4t^2+16t+75}{t^2+4t+10}\right)$$

where T is the temperature (in degrees Fahrenheit) and t is the time (in hours).

(a) Complete the table.

t	0	2	4	6	8	10	12
T							

t	14	16	18	20	22
T					

(b) What value of T does the mathematical model appear to be approaching?

72. Rate A copier copies at a rate of 50 pages per minute.

  (a) Find the time required to copy one page.

  (b) Find the time required to copy x pages.

  (c) Find the time required to copy 120 pages.

Probability **In Exercises 73 and 74, consider an experiment in which a marble is tossed into a box whose base is shown in the figure. The probability that the marble will come to rest in the shaded portion of the base is equal to the ratio of the shaded area to the total area of the figure. Find the probability.**

73.

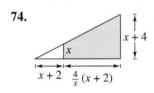

74.

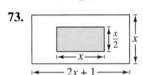

75. Interactive Money Management The table shows the numbers of U.S. households (in millions) using online banking and mobile banking from 2011 through 2014. (*Source: Fiserv, Inc.*)

Year	Online Banking	Mobile Banking
2011	79	18
2012	81	24
2013	83	30
2014	86	35

Mathematical models for the data are

$$\text{Number using online banking} = \frac{-2.9709t + 70.517}{-0.0474t + 1}$$

$$\text{Number using mobile banking} = \frac{0.661t^2 - 47}{0.007t^2 + 1}$$

where t represents the year, with $t = 11$ corresponding to 2011.

  (a) Using the models, create a table showing the numbers of households using online banking and the numbers of households using mobile banking for the given years.

  (b) Compare the values from the models with the actual data.

  (c) Determine a model for the ratio of the number of households using mobile banking to the number of households using online banking.

  (d) Use the model from part (c) to find the ratios for the given years. Interpret your results.

76. Finance The formula that approximates the annual interest rate r of a monthly installment loan is

$$r = \frac{24(NM - P)}{N} \div \left(P + \frac{NM}{12} \right)$$

where N is the total number of payments, M is the monthly payment, and P is the amount financed.

  (a) Approximate the annual interest rate for a five-year car loan of $28,000 that has monthly payments of $525.

  (b) Simplify the expression for the annual interest rate r, and then rework part (a).

77. Electrical Engineering The formula for the total resistance R_T (in ohms) of two resistors connected in parallel is

$$R_T = \frac{1}{\left(\dfrac{1}{R_1} + \dfrac{1}{R_2} \right)}$$

where R_1 and R_2 are the resistance values of the first and second resistors, respectively. Simplify the expression for the total resistance R_T.

78. **HOW DO YOU SEE IT?** The mathematical model

$$P = 100\left(\frac{t^2 - t + 1}{t^2 + 1} \right), \quad t \geq 0$$

gives the percent P of the normal level of oxygen in a pond, where t is the time (in weeks) after organic waste is dumped into the pond. The bar graph shows the situation. What conclusions can you draw from the bar graph?

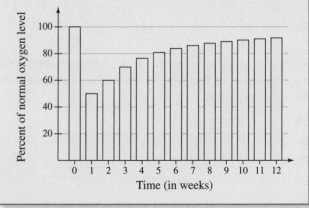

Exploration

True or False? **In Exercises 79 and 80, determine whether the statement is true or false. Justify your answer.**

79. $\dfrac{x^{2n} - 1^{2n}}{x^n - 1^n} = x^n + 1^n$

80. $\dfrac{x^2 - 3x + 2}{x - 1} = x - 2$, for all values of x

A.5   Solving Equations

- ■ Identify different types of equations.
- ■ Solve linear equations in one variable and rational equations that lead to linear equations.
- ■ Solve quadratic equations by factoring, extracting square roots, completing the square, and using the Quadratic Formula.
- ■ Solve polynomial equations of degree three or greater.
- ■ Solve radical equations.
- ■ Solve absolute value equations.
- ■ Use common formulas to solve real-life problems.

Linear equations have many real-life applications, such as in forensics. For example, in Exercises 95 and 96 on page A57, you will use linear equations to determine height from femur length.

Equations and Solutions of Equations

An **equation** in x is a statement that two algebraic expressions are equal. For example, $3x - 5 = 7$, $x^2 - x - 6 = 0$, and $\sqrt{2x} = 4$ are equations. To **solve** an equation in x means to find all values of x for which the equation is true. Such values are **solutions.** For example, $x = 4$ is a solution of the equation $3x - 5 = 7$ because $3(4) - 5 = 7$ is a true statement.

The solutions of an equation depend on the kinds of numbers being considered. For example, in the set of rational numbers, $x^2 = 10$ has no solution because there is no rational number whose square is 10. However, in the set of real numbers, the equation has the two solutions $x = \sqrt{10}$ and $x = -\sqrt{10}$.

An equation that is true for *every* real number in the domain of the variable is an **identity.** For example,

$$x^2 - 9 = (x + 3)(x - 3) \qquad \text{Identity}$$

is an identity because it is a true statement for any real value of x. The equation

$$\frac{x}{3x^2} = \frac{1}{3x} \qquad \text{Identity}$$

is an identity because it is true for any nonzero real value of x.

An equation that is true for just *some* (but not all) of the real numbers in the domain of the variable is a **conditional equation.** For example, the equation

$$x^2 - 9 = 0 \qquad \text{Conditional equation}$$

is conditional because $x = 3$ and $x = -3$ are the only values in the domain that satisfy the equation.

A **contradiction** is an equation that is *false* for *every* real number in the domain of the variable. For example, the equation

$$2x - 4 = 2x + 1 \qquad \text{Contradiction}$$

is a contradiction because there are no real values of x for which the equation is true.

Linear and Rational Equations

Definition of a Linear Equation in One Variable

A **linear equation in one variable** x is an equation that can be written in the standard form

$$ax + b = 0$$

where a and b are real numbers with $a \neq 0$.

A linear equation in one variable has exactly one solution. To see this, consider the steps below. (Remember that $a \neq 0$.)

$$ax + b = 0 \qquad \text{Write original equation.}$$

$$ax = -b \qquad \text{Subtract } b \text{ from each side.}$$

$$x = -\frac{b}{a} \qquad \text{Divide each side by } a.$$

The above suggests that to solve a conditional equation in x, you isolate x on one side of the equation using a sequence of **equivalent equations,** each having the same solution as the original equation. The operations that yield equivalent equations come from the properties of equality reviewed in Appendix A.1.

Generating Equivalent Equations

An equation can be transformed into an *equivalent equation* by one or more of the steps listed below.

	Given Equation	Equivalent Equation
1. Remove symbols of grouping, combine like terms, or simplify fractions on one or both sides of the equation.	$2x - x = 4$	$x = 4$
2. Add (or subtract) the same quantity to (from) *each* side of the equation.	$x + 1 = 6$	$x = 5$
3. Multiply (or divide) *each* side of the equation by the same *nonzero* quantity.	$2x = 6$	$x = 3$
4. Interchange the two sides of the equation.	$2 = x$	$x = 2$

In Example 1, you will use these steps to solve linear equations in one variable x.

EXAMPLE 1 **Solving Linear Equations**

a. $3x - 6 = 0 \qquad$ Original equation

$3x = 6 \qquad$ Add 6 to each side.

$x = 2 \qquad$ Divide each side by 3.

b. $5x + 4 = 3x - 8 \qquad$ Original equation

$2x + 4 = -8 \qquad$ Subtract $3x$ from each side.

$2x = -12 \qquad$ Subtract 4 from each side.

$x = -6 \qquad$ Divide each side by 2.

✓ *Checkpoint* ◀)) *Audio-video solution in English & Spanish at LarsonPrecalculus.com*

Solve each equation.

a. $7 - 2x = 15$

b. $7x - 9 = 5x + 7$

> **• REMARK** After solving an equation, you should check each solution in the original equation. For instance, here is a check of the solution in Example 1(a).
>
> $3x - 6 = 0 \qquad$ Write original equation.
>
> $3(2) - 6 \overset{?}{=} 0 \qquad$ Substitute 2 for x.
>
> $0 = 0 \qquad$ Solution checks. ✓
>
> Check the solution in Example 1(b) on your own.

REMARK An equation with a single fraction on each side can be cleared of denominators by *cross multiplying.* To do this, multiply the left numerator by the right denominator and the right numerator by the left denominator.

$$\frac{a}{b} = \frac{c}{d} \qquad \text{Original equation}$$

$$ad = cb \qquad \text{Cross multiply.}$$

A **rational equation** involves one or more rational expressions. To solve a rational equation, multiply every term by the least common denominator (LCD) of all the terms. This clears the original equation of fractions and produces a simpler equation.

EXAMPLE 2 **Solving a Rational Equation**

Solve $\dfrac{x}{3} + \dfrac{3x}{4} = 2$.

Solution

$$\frac{x}{3} + \frac{3x}{4} = 2 \qquad \text{Write original equation.}$$

$$(12)\frac{x}{3} + (12)\frac{3x}{4} = (12)2 \qquad \text{Multiply each term by the LCD.}$$

$$4x + 9x = 24 \qquad \text{Simplify.}$$

$$13x = 24 \qquad \text{Combine like terms.}$$

$$x = \frac{24}{13} \qquad \text{Divide each side by 13.}$$

The solution is $x = \frac{24}{13}$. Check this in the original equation.

✓ **Checkpoint** ◀))) *Audio-video solution in English & Spanish at LarsonPrecalculus.com*

Solve $\dfrac{4x}{9} - \dfrac{1}{3} = x + \dfrac{5}{3}$.

When multiplying or dividing an equation by a *variable expression,* it is possible to introduce an **extraneous solution,** which is a solution that does not satisfy the original equation.

EXAMPLE 3 **An Equation with an Extraneous Solution**

See LarsonPrecalculus.com for an interactive version of this type of example.

Solve $\dfrac{1}{x - 2} = \dfrac{3}{x + 2} - \dfrac{6x}{x^2 - 4}$.

Solution The LCD is $x^2 - 4 = (x + 2)(x - 2)$. Multiply each term by the LCD.

$$\frac{1}{x - 2}(x + 2)(x - 2) = \frac{3}{x + 2}(x + 2)(x - 2) - \frac{6x}{x^2 - 4}(x + 2)(x - 2)$$

$$x + 2 = 3(x - 2) - 6x, \quad x \neq \pm 2$$

$$x + 2 = 3x - 6 - 6x$$

$$x + 2 = -3x - 6$$

$$4x = -8$$

$$x = -2 \qquad \text{Extraneous solution}$$

In the original equation, $x = -2$ yields a denominator of zero. So, $x = -2$ is an extraneous solution, and the original equation has *no solution.*

✓ **Checkpoint** ◀))) *Audio-video solution in English & Spanish at LarsonPrecalculus.com*

Solve $\dfrac{3x}{x - 4} = 5 + \dfrac{12}{x - 4}$.

Quadratic Equations

A **quadratic equation** in x is an equation that can be written in the general form

$$ax^2 + bx + c = 0$$

where a, b, and c are real numbers with $a \neq 0$. A quadratic equation in x is also called a **second-degree polynomial equation** in x.

You should be familiar with the four methods for solving quadratic equations listed below.

Solving a Quadratic Equation

Factoring

If $ab = 0$, then $a = 0$ or $b = 0$.   Zero-Factor Property

Example: $x^2 - x - 6 = 0$

$$(x - 3)(x + 2) = 0$$

$$x - 3 = 0 \implies x = 3$$

$$x + 2 = 0 \implies x = -2$$

Extracting Square Roots

If $u^2 = c$, where $c > 0$, then $u = \pm\sqrt{c}$.   Square Root Principle

Example: $(x + 3)^2 = 16$

$$x + 3 = \pm 4$$

$$x = -3 \pm 4$$

$$x = 1 \quad \text{or} \quad x = -7$$

Completing the Square

If $x^2 + bx = c$, then

$$x^2 + bx + \left(\frac{b}{2}\right)^2 = c + \left(\frac{b}{2}\right)^2 \qquad \text{Add } \left(\frac{b}{2}\right)^2 \text{ to each side.}$$

$$\left(x + \frac{b}{2}\right)^2 = c + \frac{b^2}{4}.$$

Example: $x^2 + 6x = 5$

$$x^2 + 6x + 3^2 = 5 + 3^2 \qquad \text{Add } \left(\frac{6}{2}\right)^2 \text{ to each side.}$$

$$(x + 3)^2 = 14$$

$$x + 3 = \pm\sqrt{14}$$

$$x = -3 \pm \sqrt{14}$$

Quadratic Formula

If $ax^2 + bx + c = 0$, then $x = \dfrac{-b \pm \sqrt{b^2 - 4ac}}{2a}$.

Example: $2x^2 + 3x - 1 = 0$

$$x = \frac{-3 \pm \sqrt{3^2 - 4(2)(-1)}}{2(2)}$$

$$= \frac{-3 \pm \sqrt{17}}{4}$$

• • • • • • • • • • • • • • • • ▷

•• **REMARK**   It is possible to solve every quadratic equation by completing the square or using the Quadratic Formula.

EXAMPLE 4   Solving Quadratic Equations by Factoring

a. $2x^2 + 9x + 7 = 3$ Original equation

$2x^2 + 9x + 4 = 0$ Write in general form.

$(2x + 1)(x + 4) = 0$ Factor.

$2x + 1 = 0$ ⟹ $x = -\frac{1}{2}$ Set 1st factor equal to 0 and solve.

$x + 4 = 0$ ⟹ $x = -4$ Set 2nd factor equal to 0 and solve.

The solutions are $x = -\frac{1}{2}$ and $x = -4$. Check these in the original equation.

b. $6x^2 - 3x = 0$ Original equation

$3x(2x - 1) = 0$ Factor.

$3x = 0$ ⟹ $x = 0$ Set 1st factor equal to 0 and solve.

$2x - 1 = 0$ ⟹ $x = \frac{1}{2}$ Set 2nd factor equal to 0 and solve.

The solutions are $x = 0$ and $x = \frac{1}{2}$. Check these in the original equation.

✓ *Checkpoint* ◄))) *Audio-video solution in English & Spanish at LarsonPrecalculus.com*

Solve $2x^2 - 3x + 1 = 6$ by factoring.

Note that the method of solution in Example 4 is based on the Zero-Factor Property from Appendix A.1. This property applies only to equations written in general form (in which the right side of the equation is zero). So, collect all terms on one side *before* factoring. For example, in the equation $(x - 5)(x + 2) = 8$, it is *incorrect* to set each factor equal to 8. Solve this equation correctly on your own. Then check the solutions in the original equation.

EXAMPLE 5   Extracting Square Roots

Solve each equation by extracting square roots.

a. $4x^2 = 12$

b. $(x - 3)^2 = 7$

Solution

a. $4x^2 = 12$ Write original equation.

$x^2 = 3$ Divide each side by 4.

$x = \pm\sqrt{3}$ Extract square roots.

The solutions are $x = \sqrt{3}$ and $x = -\sqrt{3}$. Check these in the original equation.

b. $(x - 3)^2 = 7$ Write original equation.

$x - 3 = \pm\sqrt{7}$ Extract square roots.

$x = 3 \pm \sqrt{7}$ Add 3 to each side.

The solutions are $x = 3 \pm \sqrt{7}$. Check these in the original equation.

✓ *Checkpoint* ◄))) *Audio-video solution in English & Spanish at LarsonPrecalculus.com*

Solve each equation by extracting square roots.

a. $3x^2 = 36$

b. $(x - 1)^2 = 10$

When solving quadratic equations by completing the square, you must add $(b/2)^2$ to *each side* in order to maintain equality. When the leading coefficient is *not* 1, divide each side of the equation by the leading coefficient *before* completing the square, as shown in Example 7.

EXAMPLE 6   Completing the Square: Leading Coefficient Is 1

Solve $x^2 + 2x - 6 = 0$ by completing the square.

Solution

$x^2 + 2x - 6 = 0$	Write original equation.
$x^2 + 2x = 6$	Add 6 to each side.
$x^2 + 2x + 1^2 = 6 + 1^2$	Add 1^2 to each side.

(Half of 2)²

$(x + 1)^2 = 7$	Simplify.
$x + 1 = \pm\sqrt{7}$	Extract square roots.
$x = -1 \pm \sqrt{7}$	Subtract 1 from each side.

The solutions are

$$x = -1 \pm \sqrt{7}.$$

Check these in the original equation.

✓ **Checkpoint** 🔊))) *Audio-video solution in English & Spanish at LarsonPrecalculus.com*

Solve $x^2 - 4x - 1 = 0$ by completing the square.

EXAMPLE 7   Completing the Square: Leading Coefficient Is Not 1

Solve $3x^2 - 4x - 5 = 0$ by completing the square.

Solution

$3x^2 - 4x - 5 = 0$	Write original equation.
$3x^2 - 4x = 5$	Add 5 to each side.
$x^2 - \dfrac{4}{3}x = \dfrac{5}{3}$	Divide each side by 3.
$x^2 - \dfrac{4}{3}x + \left(-\dfrac{2}{3}\right)^2 = \dfrac{5}{3} + \left(-\dfrac{2}{3}\right)^2$	Add $\left(-\dfrac{2}{3}\right)^2$ to each side.

$\left(\text{Half of } -\dfrac{4}{3}\right)^2$

$\left(x - \dfrac{2}{3}\right)^2 = \dfrac{19}{9}$	Simplify.
$x - \dfrac{2}{3} = \pm\dfrac{\sqrt{19}}{3}$	Extract square roots.
$x = \dfrac{2}{3} \pm \dfrac{\sqrt{19}}{3}$	Add $\dfrac{2}{3}$ to each side.

✓ **Checkpoint** 🔊))) *Audio-video solution in English & Spanish at LarsonPrecalculus.com*

Solve $3x^2 - 10x - 2 = 0$ by completing the square.

EXAMPLE 8 **The Quadratic Formula: Two Distinct Solutions**

Use the Quadratic Formula to solve $x^2 + 3x = 9$.

Solution

$$x^2 + 3x = 9$$

Write original equation.

$$x^2 + 3x - 9 = 0$$

Write in general form.

$$x = \frac{-b \pm \sqrt{b^2 - 4ac}}{2a}$$

Quadratic Formula

$$x = \frac{-3 \pm \sqrt{(3)^2 - 4(1)(-9)}}{2(1)}$$

Substitute $a = 1$, $b = 3$, and $c = -9$.

$$x = \frac{-3 \pm \sqrt{45}}{2}$$

Simplify.

$$x = \frac{-3 \pm 3\sqrt{5}}{2}$$

Simplify.

The two solutions are

$$x = \frac{-3 + 3\sqrt{5}}{2} \quad \text{and} \quad x = \frac{-3 - 3\sqrt{5}}{2}.$$

Check these in the original equation.

✓ *Checkpoint* ◄))) *Audio-video solution in English & Spanish at LarsonPrecalculus.com*

Use the Quadratic Formula to solve $3x^2 + 2x = 10$.

• **REMARK** When you use the Quadratic Formula, remember that *before* applying the formula, you must first write the quadratic equation in general form.

EXAMPLE 9 **The Quadratic Formula: One Solution**

Use the Quadratic Formula to solve $8x^2 - 24x + 18 = 0$.

Solution

$$8x^2 - 24x + 18 = 0$$

Write original equation.

$$4x^2 - 12x + 9 = 0$$

Divide out common factor of 2.

$$x = \frac{-b \pm \sqrt{b^2 - 4ac}}{2a}$$

Quadratic Formula

$$x = \frac{-(-12) \pm \sqrt{(-12)^2 - 4(4)(9)}}{2(4)}$$

Substitute $a = 4$, $b = -12$, and $c = 9$.

$$x = \frac{12 \pm \sqrt{0}}{8}$$

Simplify.

$$x = \frac{3}{2}$$

Simplify.

This quadratic equation has only one solution: $x = \frac{3}{2}$. Check this in the original equation.

✓ *Checkpoint* ◄))) *Audio-video solution in English & Spanish at LarsonPrecalculus.com*

Use the Quadratic Formula to solve $18x^2 - 48x + 32 = 0$.

Note that you could have solved Example 9 without first dividing out a common factor of 2. Substituting $a = 8$, $b = -24$, and $c = 18$ into the Quadratic Formula produces the same result.

Polynomial Equations of Higher Degree

Sometimes, the methods used to solve quadratic equations can be extended to solve polynomial equations of higher degrees.

••••••••••••••••••▷

REMARK   A common mistake when solving an equation such as that in Example 10 is to divide each side of the equation by the variable factor x^2. This loses the solution $x = 0$. When solving a polynomial equation, always write the equation in general form, then factor the polynomial and set each factor equal to zero. Do not divide each side of an equation by a variable factor in an attempt to simplify the equation.

EXAMPLE 10   **Solving a Polynomial Equation by Factoring**

Solve $3x^4 = 48x^2$ and check your solution(s).

Solution   First write the polynomial equation in general form. Then factor the polynomial, set each factor equal to zero, and solve.

$$3x^4 = 48x^2 \qquad \text{Write original equation.}$$
$$3x^4 - 48x^2 = 0 \qquad \text{Write in general form.}$$
$$3x^2(x^2 - 16) = 0 \qquad \text{Factor out common factor.}$$
$$3x^2(x + 4)(x - 4) = 0 \qquad \text{Factor completely.}$$
$$3x^2 = 0 \implies x = 0 \qquad \text{Set 1st factor equal to 0 and solve.}$$
$$x + 4 = 0 \implies x = -4 \qquad \text{Set 2nd factor equal to 0 and solve.}$$
$$x - 4 = 0 \implies x = 4 \qquad \text{Set 3rd factor equal to 0 and solve.}$$

Check these solutions by substituting in the original equation.

Check

$$3(0)^4 \overset{?}{=} 48(0)^2 \implies 0 = 0 \qquad \text{0 checks. } \checkmark$$
$$3(-4)^4 \overset{?}{=} 48(-4)^2 \implies 768 = 768 \qquad -4 \text{ checks. } \checkmark$$
$$3(4)^4 \overset{?}{=} 48(4)^2 \implies 768 = 768 \qquad 4 \text{ checks. } \checkmark$$

So, the solutions are

$$x = 0, \quad x = -4, \quad \text{and} \quad x = 4.$$

✓ **Checkpoint** ◀))) *Audio-video solution in English & Spanish at LarsonPrecalculus.com*

Solve $9x^4 - 12x^2 = 0$ and check your solution(s).

EXAMPLE 11   **Solving a Polynomial Equation by Factoring**

Solve $x^3 - 3x^2 - 3x + 9 = 0$.

Solution

$$x^3 - 3x^2 - 3x + 9 = 0 \qquad \text{Write original equation.}$$
$$x^2(x - 3) - 3(x - 3) = 0 \qquad \text{Group terms and factor.}$$
$$(x - 3)(x^2 - 3) = 0 \qquad (x - 3) \text{ is a common factor.}$$
$$x - 3 = 0 \implies x = 3 \qquad \text{Set 1st factor equal to 0 and solve.}$$
$$x^2 - 3 = 0 \implies x = \pm\sqrt{3} \qquad \text{Set 2nd factor equal to 0 and solve.}$$

The solutions are $x = 3$, $x = \sqrt{3}$, and $x = -\sqrt{3}$. Check these in the original equation.

✓ **Checkpoint** ◀))) *Audio-video solution in English & Spanish at LarsonPrecalculus.com*

Solve each equation.

a. $x^3 - 5x^2 - 2x + 10 = 0$

b. $6x^3 - 27x^2 - 54x = 0$

Radical Equations

A **radical equation** is an equation that involves one or more radical expressions. Examples 12 and 13 demonstrate how to solve radical equations.

EXAMPLE 12   **Solving Radical Equations**

a.

$\sqrt{2x + 7} - x = 2$	Original equation
$\sqrt{2x + 7} = x + 2$	Isolate radical.
$2x + 7 = x^2 + 4x + 4$	Square each side.
$0 = x^2 + 2x - 3$	Write in general form.
$0 = (x + 3)(x - 1)$	Factor.
$x + 3 = 0 \implies x = -3$	Set 1st factor equal to 0 and solve.
$x - 1 = 0 \implies x = 1$	Set 2nd factor equal to 0 and solve.

Checking these values shows that the only solution is $x = 1$.

b.

$\sqrt{2x - 5} - \sqrt{x - 3} = 1$	Original equation
$\sqrt{2x - 5} = \sqrt{x - 3} + 1$	Isolate $\sqrt{2x - 5}$.
$2x - 5 = x - 3 + 2\sqrt{x - 3} + 1$	Square each side.
$x - 3 = 2\sqrt{x - 3}$	Isolate $2\sqrt{x - 3}$.
$x^2 - 6x + 9 = 4(x - 3)$	Square each side.
$x^2 - 10x + 21 = 0$	Write in general form.
$(x - 3)(x - 7) = 0$	Factor.
$x - 3 = 0 \implies x = 3$	Set 1st factor equal to 0 and solve.
$x - 7 = 0 \implies x = 7$	Set 2nd factor equal to 0 and solve.

The solutions are $x = 3$ and $x = 7$. Check these in the original equation.

✓ *Checkpoint*  *Audio-video solution in English & Spanish at LarsonPrecalculus.com*

Solve $-\sqrt{40 - 9x} + 2 = x$.

EXAMPLE 13   **Solving an Equation Involving a Rational Exponent**

Solve $(x - 4)^{2/3} = 25$.

Solution

$(x - 4)^{2/3} = 25$	Write original equation.
$\sqrt[3]{(x - 4)^2} = 25$	Rewrite in radical form.
$(x - 4)^2 = 15{,}625$	Cube each side.
$x - 4 = \pm 125$	Extract square roots.
$x = 129, \ x = -121$	Add 4 to each side.

The solutions are $x = 129$ and $x = -121$. Check these in the original equation.

✓ *Checkpoint*  *Audio-video solution in English & Spanish at LarsonPrecalculus.com*

Solve $(x - 5)^{2/3} = 16$.

Absolute Value Equations

An **absolute value equation** is an equation that involves one or more absolute value expressions. To solve an absolute value equation, remember that the expression inside the absolute value bars can be positive or negative. This results in *two* separate equations, each of which must be solved. For example, the equation

$$|x - 2| = 3$$

results in the two equations

$$x - 2 = 3$$

and

$$-(x - 2) = 3$$

which implies that the original equation has two solutions: $x = 5$ and $x = -1$.

EXAMPLE 14   **Solving an Absolute Value Equation**

Solve $|x^2 - 3x| = -4x + 6$ and check your solution(s).

Solution   Solve the two equations below.

First Equation

$x^2 - 3x = -4x + 6$	Use positive expression.
$x^2 + x - 6 = 0$	Write in general form.
$(x + 3)(x - 2) = 0$	Factor.
$x + 3 = 0 \implies x = -3$	Set 1st factor equal to 0 and solve.
$x - 2 = 0 \implies x = 2$	Set 2nd factor equal to 0 and solve.

Second Equation

$-(x^2 - 3x) = -4x + 6$	Use negative expression.
$x^2 - 7x + 6 = 0$	Write in general form.
$(x - 1)(x - 6) = 0$	Factor.
$x - 1 = 0 \implies x = 1$	Set 1st factor equal to 0 and solve.
$x - 6 = 0 \implies x = 6$	Set 2nd factor equal to 0 and solve.

Check

$|(-3)^2 - 3(-3)| \overset{?}{=} -4(-3) + 6$	Substitute -3 for x.
$18 = 18$	-3 checks. ✔
$|(2)^2 - 3(2)| \overset{?}{=} -4(2) + 6$	Substitute 2 for x.
$2 \neq -2$	2 does not check.
$|(1)^2 - 3(1)| \overset{?}{=} -4(1) + 6$	Substitute 1 for x.
$2 = 2$	1 checks. ✔
$|(6)^2 - 3(6)| \overset{?}{=} -4(6) + 6$	Substitute 6 for x.
$18 \neq -18$	6 does not check.

The solutions are $x = -3$ and $x = 1$.

✔ *Checkpoint* ◀))) *Audio-video solution in English & Spanish at LarsonPrecalculus.com*

Solve $|x^2 + 4x| = 7x + 18$ and check your solution(s). ■

Common Formulas

You will use the geometric formulas listed below at various times throughout this course. For your convenience, some of these formulas along with several others are also given on the inside cover of this text.

Common Formulas for Area *A*, Perimeter *P*, Circumference *C*, and Volume *V*

Rectangle

$A = lw$

$P = 2l + 2w$

Circle

$A = \pi r^2$

$C = 2\pi r$

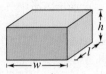

Rectangular Solid

$V = lwh$

Circular Cylinder

$V = \pi r^2 h$

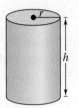

Sphere

$V = \frac{4}{3}\pi r^3$

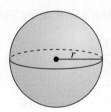

Figure A.9

EXAMPLE 15 Using a Geometric Formula

The cylindrical can shown in Figure A.9 has a volume of 200 cubic centimeters (cm³). Find the height of the can.

Solution The formula for the *volume of a cylinder* is $V = \pi r^2 h$. To find the height of the can, solve for *h*. Then, using $V = 200$ and $r = 4$, find the height.

$$V = \pi r^2 h \implies h = \frac{V}{\pi r^2} = \frac{200}{\pi(4)^2} = \frac{200}{16\pi} \approx 3.98$$

So, the height of the can is about 3.98 centimeters.

✓ **Checkpoint** 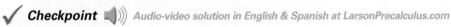 *Audio-video solution in English & Spanish at LarsonPrecalculus.com*

A cylindrical container has a volume of 84 cubic inches and a radius of 3 inches. Find the height of the container.

··REMARK To check that the answer in Example 15 is reasonable, substitute $h = 3.98$ into the formula for the volume of a cylinder and simplify.

$$V = \pi r^2 h$$
$$\approx \pi(4)^2(3.98)$$
$$\approx 200$$

Summarize (Appendix A.5)

1. State the definitions of an identity, a conditional equation, and a contradiction *(page A45)*.

2. State the definition of a linear equation in one variable *(page A45)*. For examples of solving linear equations and rational equations that lead to linear equations, see Examples 1–3.

3. List the four methods for solving quadratic equations discussed in this section *(page A48)*. For examples of solving quadratic equations, see Examples 4–9.

4. Explain how to solve a polynomial equation of degree three or greater by factoring *(page A52)*. For examples of solving polynomial equations by factoring, see Examples 10 and 11.

5. Explain how to solve a radical equation *(page A53)*. For an example of solving radical equations, see Example 12.

6. Explain how to solve an absolute value equation *(page A54)*. For an example of solving an absolute value equation, see Example 14.

7. State the common geometric formulas listed in this section *(page A55)*. For an example that uses a volume formula, see Example 15.

A.5   Exercises

See **CalcChat.com** for tutorial help and worked-out solutions to odd-numbered exercises.

Vocabulary: Fill in the blanks.

1. An _____ is a statement that equates two algebraic expressions.
2. A linear equation in one variable x is an equation that can be written in the standard form _____.
3. An _____ solution is a solution that does not satisfy the original equation.
4. Four methods for solving quadratic equations are _____, extracting _____ _____, _____ the _____, and the _____ _____.

Skills and Applications

 Solving a Linear Equation In Exercises 5–12, solve the equation and check your solution. (If not possible, explain why.)

5. $x + 11 = 15$ 6. $7 - x = 19$

7. $7 - 2x = 25$

8. $7x + 2 = 23$

9. $3x - 5 = 2x + 7$

10. $4y + 2 - 5y = 7 - 6y$

11. $x - 3(2x + 3) = 8 - 5x$

12. $9x - 10 = 5x + 2(2x - 5)$

 Solving a Rational Equation In Exercises 13–24, solve the equation and check your solution. (If not possible, explain why.)

13. $\dfrac{3x}{8} - \dfrac{4x}{3} = 4$ 14. $\dfrac{5x}{4} + \dfrac{1}{2} = x - \dfrac{1}{2}$

15. $\dfrac{5x - 4}{5x + 4} = \dfrac{2}{3}$

16. $\dfrac{10x + 3}{5x + 6} = \dfrac{1}{2}$

17. $10 - \dfrac{13}{x} = 4 + \dfrac{5}{x}$

18. $\dfrac{1}{x} + \dfrac{2}{x - 5} = 0$

19. $\dfrac{x}{x + 4} + \dfrac{4}{x + 4} + 2 = 0$

20. $\dfrac{7}{2x + 1} - \dfrac{8x}{2x - 1} = -4$

21. $\dfrac{2}{(x - 4)(x - 2)} = \dfrac{1}{x - 4} + \dfrac{2}{x - 2}$

22. $\dfrac{12}{(x - 1)(x + 3)} = \dfrac{3}{x - 1} + \dfrac{2}{x + 3}$

23. $\dfrac{1}{x - 3} + \dfrac{1}{x + 3} = \dfrac{10}{x^2 - 9}$

24. $\dfrac{1}{x - 2} + \dfrac{3}{x + 3} = \dfrac{4}{x^2 + x - 6}$

 Solving a Quadratic Equation by Factoring In Exercises 25–34, solve the quadratic equation by factoring.

25. $6x^2 + 3x = 0$ 26. $8x^2 - 2x = 0$

27. $x^2 + 10x + 25 = 0$ 28. $x^2 - 2x - 8 = 0$

29. $3 + 5x - 2x^2 = 0$ 30. $4x^2 + 12x + 9 = 0$

31. $16x^2 - 9 = 0$ 32. $-x^2 + 8x = 12$

33. $\frac{3}{4}x^2 + 8x + 20 = 0$ 34. $\frac{1}{8}x^2 - x - 16 = 0$

 Extracting Square Roots In Exercises 35–42, solve the equation by extracting square roots. When a solution is irrational, list both the exact solution *and* its approximation rounded to two decimal places.

35. $x^2 = 49$ 36. $x^2 = 43$

37. $3x^2 = 81$ 38. $9x^2 = 36$

39. $(x - 4)^2 = 49$ 40. $(x + 9)^2 = 24$

41. $(2x - 1)^2 = 18$ 42. $(x - 7)^2 = (x + 3)^2$

 Completing the Square In Exercises 43–50, solve the quadratic equation by completing the square.

43. $x^2 + 4x - 32 = 0$ 44. $x^2 - 2x - 3 = 0$

45. $x^2 + 4x + 2 = 0$ 46. $x^2 + 8x + 14 = 0$

47. $6x^2 - 12x = -3$ 48. $4x^2 - 4x = 1$

49. $2x^2 + 5x - 8 = 0$ 50. $3x^2 - 4x - 7 = 0$

Using the Quadratic Formula In Exercises 51–64, use the Quadratic Formula to solve the equation.

51. $2x^2 + x - 1 = 0$ 52. $2x^2 - x - 1 = 0$

53. $9x^2 + 30x + 25 = 0$ 54. $28x - 49x^2 = 4$

55. $2x^2 - 7x + 1 = 0$ 56. $3x + x^2 - 1 = 0$

57. $12x - 9x^2 = -3$ 58. $9x^2 - 37 = 6x$

59. $2 + 2x - x^2 = 0$ 60. $x^2 + 10 + 8x = 0$

61. $8t = 5 + 2t^2$ 62. $25h^2 + 80h = -61$

63. $(y - 5)^2 = 2y$ 64. $(z + 6)^2 = -2z$

Choosing a Method In Exercises 65–72, solve the equation using any convenient method.

65. $x^2 - 2x - 1 = 0$

66. $14x^2 + 42x = 0$

67. $(x + 2)^2 = 64$

68. $x^2 - 14x + 49 = 0$

69. $x^2 - x - \frac{11}{4} = 0$

70. $x^2 + 3x - \frac{3}{4} = 0$

71. $3x + 4 = 2x^2 - 7$

72. $(x + 1)^2 = x^2$

 Solving a Polynomial Equation In Exercises 73–76, solve the equation. Check your solutions.

73. $6x^4 - 54x^2 = 0$

74. $5x^3 + 30x^2 + 45x = 0$

75. $x^3 + 2x^2 - 8x = 16$

76. $x^3 - 3x^2 - x = -3$

 Solving a Radical Equation In Exercises 77–84, solve the equation. Check your solutions.

77. $\sqrt{5x} - 10 = 0$

78. $\sqrt{x + 8} - 5 = 0$

79. $4 + \sqrt[3]{2x - 9} = 0$

80. $\sqrt[3]{12 - x} - 3 = 0$

81. $\sqrt{x + 8} = 2 + x$

82. $2x = \sqrt{-5x + 24} - 3$

83. $\sqrt{x - 3} + 1 = \sqrt{x}$

84. $2\sqrt{x + 1} - \sqrt{2x + 3} = 1$

 Solving an Equation Involving a Rational Exponent In Exercises 85–88, solve the equation. Check your solutions.

85. $(x - 5)^{3/2} = 8$

86. $(x^2 - x - 22)^{3/2} = 27$

87. $3x(x - 1)^{1/2} + 2(x - 1)^{3/2} = 0$

88. $4x^2(x - 1)^{1/3} + 6x(x - 1)^{4/3} = 0$

 Solving an Absolute Value Function In Exercises 89–92, solve the equation. Check your solutions.

89. $|2x - 5| = 11$

90. $|3x + 2| = 7$

91. $|x + 1| = x^2 - 5$

92. $|x^2 + 6x| = 3x + 18$

93. Volume of a Billiard Ball A billiard ball has a volume of 5.96 cubic inches. Find the radius of the billiard ball.

94. Length of a Tank The diameter of a cylindrical propane gas tank is 4 feet. The total volume of the tank is 603.2 cubic feet. Find the length of the tank.

• • **Forensics** •

In Exercises 95 and 96, use the following information. The relationship between the length of an adult's femur (thigh bone) and the height of the adult can be approximated by the linear equations

$y = 0.514x - 14.75$ Female

$y = 0.532x - 17.03$ Male

where y is the length of the femur in inches and x is the height of the adult in inches (see figure).

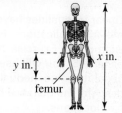

95. A crime scene investigator discovers a femur belonging to an adult human female. The bone is 18 inches long. Estimate the height of the female.

96. Officials search a forest for a missing man who is 6 feet 2 inches tall. They find an adult male femur that is 23 inches long. Is it possible that the femur belongs to the missing man?

• •

Exploration

True or False?   In Exercises 97–99, determine whether the statement is true or false. Justify your answer.

97. An equation can never have more than one extraneous solution.

98. The equation $2(x - 3) + 1 = 2x - 5$ has no solution.

99. The equation

$$\sqrt{x + 10} - \sqrt{x - 10} = 0$$

has no solution.

 100. HOW DO YOU SEE IT?   The figure shows a glass cube partially filled with water.

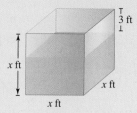

(a) What does the expression $x^2(x - 3)$ represent?

(b) Given $x^2(x - 3) = 320$, explain how to find the capacity of the cube.

A.6   Linear Inequalities in One Variable

Linear inequalities have many real-life applications. For example, in Exercise 104 on page A66, you will use an absolute value inequality to describe the distance between two locations.

- ◾ **Represent solutions of linear inequalities in one variable.**
- ◾ **Use properties of inequalities to write equivalent inequalities.**
- ◾ **Solve linear inequalities in one variable.**
- ◾ **Solve absolute value inequalities.**
- ◾ **Use linear inequalities to model and solve real-life problems.**

Introduction

Simple inequalities were discussed in Appendix A.1. There, the inequality symbols $<$, $\leq$, $>$, and $\geq$ were used to compare two numbers and to denote subsets of real numbers. For example, the simple inequality

$$x \geq 3$$

denotes all real numbers x that are greater than or equal to 3.

Now, you will expand your work with inequalities to include more involved statements such as

$$5x - 7 < 3x + 9 \quad \text{and} \quad -3 \leq 6x - 1 < 3.$$

As with an equation, you **solve an inequality** in the variable x by finding all values of x for which the inequality is true. Such values are **solutions** that **satisfy** the inequality. The set of all real numbers that are solutions of an inequality is the **solution set** of the inequality. For example, the solution set of

$$x + 1 < 4$$

is all real numbers that are less than 3.

The set of all points on the real number line that represents the solution set is the **graph of the inequality.** Graphs of many types of inequalities consist of intervals on the real number line. See Appendix A.1 to review the nine basic types of intervals on the real number line. Note that each type of interval can be classified as *bounded* or *unbounded*.

EXAMPLE 1   Intervals and Inequalities

Write an inequality that represents each interval. Then state whether the interval is bounded or unbounded.

a. $(-3, 5]$ **b.** $(-3, \infty)$

c. $[0, 2]$ **d.** $(-\infty, \infty)$

Solution

a. $(-3, 5]$ corresponds to $-3 < x \leq 5$.   Bounded

b. $(-3, \infty)$ corresponds to $x > -3$.   Unbounded

c. $[0, 2]$ corresponds to $0 \leq x \leq 2$.   Bounded

d. $(-\infty, \infty)$ corresponds to $-\infty < x < \infty$.   Unbounded

✓ *Checkpoint* ◀))) *Audio-video solution in English & Spanish at LarsonPrecalculus.com*

Write an inequality that represents each interval. Then state whether the interval is bounded or unbounded.

a. $[-1, 3]$ **b.** $(-1, 6)$ **c.** $(-\infty, 4)$ **d.** $[0, \infty)$

Properties of Inequalities

The procedures for solving linear inequalities in one variable are similar to those for solving linear equations. To isolate the variable, use the **properties of inequalities.** These properties are similar to the properties of equality, but there are two important exceptions. When you multiply or divide each side of an inequality by a negative number, you must reverse the direction of the inequality symbol. Here is an example.

$$-2 < 5 \qquad \text{Original inequality}$$

$$(-3)(-2) > (-3)(5) \qquad \text{Multiply each side by } -3 \text{ and reverse inequality symbol.}$$

$$6 > -15 \qquad \text{Simplify.}$$

Notice that when you do not reverse the inequality symbol in the example above, you obtain the false statement

$$6 < -15. \qquad \text{False statement}$$

Two inequalities that have the same solution set are **equivalent.** For example, the inequalities

$$x + 2 < 5$$

and

$$x < 3$$

are equivalent. To obtain the second inequality from the first, subtract 2 from each side of the inequality. The list below describes operations used to write equivalent inequalities.

Properties of Inequalities

Let a, b, c, and d be real numbers.

1. Transitive Property

$$a < b \text{ and } b < c \implies a < c$$

2. Addition of Inequalities

$$a < b \text{ and } c < d \implies a + c < b + d$$

3. Addition of a Constant

$$a < b \implies a + c < b + c$$

4. Multiplication by a Constant

$$\text{For } c > 0, a < b \implies ac < bc$$

$$\text{For } c < 0, a < b \implies ac > bc \qquad \text{Reverse the inequality symbol.}$$

Each of the properties above is true when you replace the symbol $<$ with $\leq$ and you replace the symbol $>$ with $\geq$. For example, another form of the multiplication property is shown below.

$$\text{For } c > 0, a \leq b \implies ac \leq bc$$

$$\text{For } c < 0, a \leq b \implies ac \geq bc$$

On your own, verify each of the properties of inequalities by using several examples with real numbers.

Solving Linear Inequalities in One Variable

The simplest type of inequality to solve is a **linear inequality** in one variable. For example, $2x + 3 > 4$ is a linear inequality in x.

EXAMPLE 2   **Solving a Linear Inequality**

Solve $5x - 7 > 3x + 9$. Then graph the solution set.

Solution

$$5x - 7 > 3x + 9 \qquad \text{Write original inequality.}$$
$$2x - 7 > 9 \qquad \text{Subtract } 3x \text{ from each side.}$$
$$2x > 16 \qquad \text{Add 7 to each side.}$$
$$x > 8 \qquad \text{Divide each side by 2.}$$

The solution set is all real numbers that are greater than 8, denoted by $(8, \infty)$. The graph of this solution set is shown below. Note that a parenthesis at 8 on the real number line indicates that 8 *is not* part of the solution set.

Solution interval: $(8, \infty)$

✓ *Checkpoint* 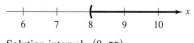 *Audio-video solution in English & Spanish at LarsonPrecalculus.com*

Solve $7x - 3 \le 2x + 7$. Then graph the solution set.

> **REMARK** Checking the solution set of an inequality is not as simple as checking the solution(s) of an equation. However, to get an indication of the validity of a solution set, substitute a few convenient values of x. For instance, in Example 2, substitute $x = 5$ and $x = 10$ into the original inequality.

EXAMPLE 3   **Solving a Linear Inequality**

See LarsonPrecalculus.com for an interactive version of this type of example.

Solve $1 - \frac{3}{2}x \ge x - 4$.

Algebraic Solution

$$1 - \tfrac{3}{2}x \ge x - 4 \qquad \text{Write original inequality.}$$
$$2 - 3x \ge 2x - 8 \qquad \text{Multiply each side by 2.}$$
$$2 - 5x \ge -8 \qquad \text{Subtract } 2x \text{ from each side.}$$
$$-5x \ge -10 \qquad \text{Subtract 2 from each side.}$$
$$x \le 2 \qquad \begin{array}{l}\text{Divide each side by } -5 \text{ and} \\ \text{reverse the inequality symbol.}\end{array}$$

The solution set is all real numbers that are less than or equal to 2, denoted by $(-\infty, 2]$. The graph of this solution set is shown below. Note that a bracket at 2 on the real number line indicates that 2 *is* part of the solution set.

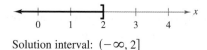

Solution interval: $(-\infty, 2]$

Graphical Solution

Use a graphing utility to graph $y_1 = 1 - \frac{3}{2}x$ and $y_2 = x - 4$ in the same viewing window.

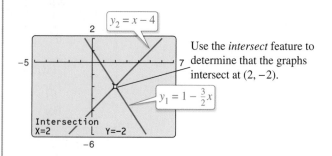

Use the *intersect* feature to determine that the graphs intersect at $(2, -2)$.

The graph of y_1 lies above the graph of y_2 to the left of their point of intersection, which implies that $y_1 \ge y_2$ for all $x \le 2$.

✓ *Checkpoint* 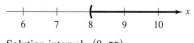 *Audio-video solution in English & Spanish at LarsonPrecalculus.com*

Solve $2 - \frac{5}{3}x > x - 6$ (a) algebraically and (b) graphically.

Sometimes it is possible to write two inequalities as a **double inequality.** For example, you can write the two inequalities

$$-4 \leq 5x - 2$$

and

$$5x - 2 < 7$$

as

$$-4 \leq 5x - 2 < 7. \qquad \text{Double inequality}$$

This form allows you to solve the two inequalities together, as demonstrated in Example 4.

EXAMPLE 4   Solving a Double Inequality

Solve $-3 \leq 6x - 1 < 3$. Then graph the solution set.

Solution   One way to solve this double inequality is to isolate x as the middle term.

$$-3 \leq 6x - 1 < 3 \qquad \text{Write original inequality.}$$

$$-3 + 1 \leq 6x - 1 + 1 < 3 + 1 \qquad \text{Add 1 to each part.}$$

$$-2 \leq 6x < 4 \qquad \text{Simplify.}$$

$$\frac{-2}{6} \leq \frac{6x}{6} < \frac{4}{6} \qquad \text{Divide each part by 6.}$$

$$-\frac{1}{3} \leq x < \frac{2}{3} \qquad \text{Simplify.}$$

The solution set is all real numbers that are greater than or equal to $-\frac{1}{3}$ and less than $\frac{2}{3}$, denoted by $\left[-\frac{1}{3}, \frac{2}{3}\right)$. The graph of this solution set is shown below.

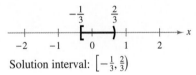

Solution interval: $\left[-\frac{1}{3}, \frac{2}{3}\right)$

✓ **Checkpoint**))) *Audio-video solution in English & Spanish at LarsonPrecalculus.com*

Solve $1 < 2x + 7 < 11$. Then graph the solution set.

Another way to solve the double inequality in Example 4 is to solve it in two parts.

$$-3 \leq 6x - 1 \quad \text{and} \quad 6x - 1 < 3$$

$$-2 \leq 6x \qquad\qquad\quad 6x < 4$$

$$-\frac{1}{3} \leq x \qquad\qquad\quad x < \frac{2}{3}$$

The solution set consists of all real numbers that satisfy *both* inequalities. In other words, the solution set is the set of all values of x for which

$$-\tfrac{1}{3} \leq x < \tfrac{2}{3}.$$

When combining two inequalities to form a double inequality, be sure that the inequalities satisfy the Transitive Property. For example, it is *incorrect* to combine the inequalities $3 < x$ and $x \leq -1$ as $3 < x \leq -1$. This "inequality" is wrong because 3 is not less than -1.

Absolute Value Inequalities

▷ **TECHNOLOGY** A graphing utility can help you identify the solution set of an inequality. For instance, to find the solution set of $|x - 5| < 2$ (see Example 5a), rewrite the inequality as $|x - 5| - 2 < 0$, enter

$$Y1 = abs(X - 5) - 2$$

and press the *graph* key. The graph should resemble the one shown below.

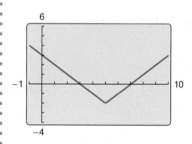

Notice that the graph is below the *x*-axis on the interval (3, 7).

Solving an Absolute Value Inequality

Let u be an algebraic expression and let a be a real number such that $a > 0$.

1. $|u| < a$ if and only if $-a < u < a$.

2. $|u| \le a$ if and only if $-a \le u \le a$.

3. $|u| > a$ if and only if $u < -a$ or $u > a$.

4. $|u| \ge a$ if and only if $u \le -a$ or $u \ge a$.

EXAMPLE 5   **Solving Absolute Value Inequalities**

Solve each inequality. Then graph the solution set.

a. $|x - 5| < 2$ **b.** $|x + 3| \ge 7$

Solution

a. $|x - 5| < 2$ Write original inequality.

$\qquad -2 < x - 5 < 2$ Write related double inequality.

$\quad -2 + 5 < x - 5 + 5 < 2 + 5$ Add 5 to each part.

$\qquad\quad 3 < x < 7$ Simplify.

The solution set is all real numbers that are greater than 3 and less than 7, denoted by (3, 7). The graph of this solution set is shown below. Note that the graph of the inequality can be described as all real numbers less than two units from 5.

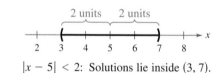

$|x - 5| < 2$: Solutions lie inside (3, 7).

b. $\quad |x + 3| \ge 7$ Write original inequality.

$\qquad x + 3 \le -7 \quad$ or $\quad x + 3 \ge 7$ Write related inequalities.

$x + 3 - 3 \le -7 - 3 \qquad x + 3 - 3 \ge 7 - 3$ Subtract 3 from each side.

$\qquad x \le -10 \qquad\qquad x \ge 4$ Simplify.

The solution set is all real numbers that are less than or equal to -10 *or* greater than or equal to 4, denoted by $(-\infty, -10] \cup [4, \infty)$. The symbol $\cup$ is the *union* symbol, which denotes the combining of two sets. The graph of this solution set is shown below. Note that the graph of the inequality can be described as all real numbers at least seven units from -3.

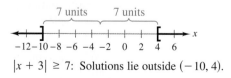

$|x + 3| \ge 7$: Solutions lie outside $(-10, 4)$.

✓ *Checkpoint* ◀))) *Audio-video solution in English & Spanish at LarsonPrecalculus.com*

Solve $|x - 20| \le 4$. Then graph the solution set.

Applications

EXAMPLE 6   **Comparative Shopping**

A car sharing company offers two plans, as shown in Figure A.10. How many hours must you use a car in one month for plan B to cost more than plan A?

Solution   Let h represent the number of hours you use the car. Write and solve an inequality.

$$10.25h + 8 > 8.75h + 50$$
$$1.5h > 42$$
$$h > 28$$

Plan B costs more when you use the car for more than 28 hours in one month.

✓ **Checkpoint** 🔊))) *Audio-video solution in English & Spanish at LarsonPrecalculus.com*

Rework Example 6 when plan A costs $25 per month plus $10 per hour.

Car Sharing Company

Plan A:
$50.00 per month
plus $8.75 per hour

Plan B:
$8.00 per month
plus $10.25 per hour

Figure A.10

EXAMPLE 7   **Accuracy of a Measurement**

You buy a bag of chocolates that cost $9.89 per pound. The scale used to weigh the bag is accurate to within $\frac{1}{32}$ pound. According to the scale, the bag weighs $\frac{1}{2}$ pound and costs $4.95. How much might you have been undercharged or overcharged?

Solution   Let x represent the actual weight of the bag. The difference of the actual weight and the weight shown on the scale is at most $\frac{1}{32}$ pound. That is, $\left| x - \frac{1}{2} \right| \le \frac{1}{32}$. Solve this inequality.

$$-\frac{1}{32} \le x - \frac{1}{2} \le \frac{1}{32}$$
$$\frac{15}{32} \le x \le \frac{17}{32}$$

The least the bag can weigh is $\frac{15}{32}$ pound, which would have cost $4.64. The most the bag can weigh is $\frac{17}{32}$ pound, which would have cost $5.25. So, you might have been overcharged by as much as $0.31 or undercharged by as much as $0.30.

✓ **Checkpoint** 🔊))) *Audio-video solution in English & Spanish at LarsonPrecalculus.com*

Rework Example 7 when the scale is accurate to within $\frac{1}{64}$ pound.

Summarize   (Appendix A.6)

1. Explain how to use inequalities to represent intervals *(page A58)*. For an example of writing inequalities that represent intervals, see Example 1.

2. State the properties of inequalities *(page A59)*.

3. Explain how to solve a linear inequality in one variable *(page A60)*. For examples of solving linear inequalities in one variable, see Examples 2–4.

4. Explain how to solve an absolute value inequality *(page A62)*. For an example of solving absolute value inequalities, see Example 5.

5. Describe real-life applications of linear inequalities in one variable *(page A63, Examples 6 and 7)*.

A.6   Exercises

See **CalcChat.com** for tutorial help and worked-out solutions to odd-numbered exercises.

Vocabulary: Fill in the blanks.

1. The set of all real numbers that are solutions of an inequality is the _____ _____ of the inequality.
2. The set of all points on the real number line that represents the solution set of an inequality is the _____ of the inequality.
3. It is sometimes possible to write two inequalities as a _____ inequality.
4. The symbol ∪ is the _____ symbol, which denotes the combining of two sets.

Skills and Applications

 Intervals and Inequalities In Exercises 5–12, write an inequality that represents the interval. Then state whether the interval is bounded or unbounded.

5. $[-2, 6)$
6. $(-7, 4)$
7. $[-1, 5]$
8. $(2, 10]$
9. $(11, \infty)$
10. $[-5, \infty)$
11. $(-\infty, -2)$
12. $(-\infty, 7]$

 Solving a Linear Inequality In Exercises 13–30, solve the inequality. Then graph the solution set.

13. $4x < 12$
14. $10x < -40$
15. $-2x > -3$
16. $-6x > 15$
17. $x - 5 \geq 7$
18. $x + 7 \leq 12$
19. $2x + 7 < 3 + 4x$
20. $3x + 1 \geq 2 + x$
21. $3x - 4 \geq 4 - 5x$
22. $6x - 4 \leq 2 + 8x$
23. $4 - 2x < 3(3 - x)$
24. $4(x + 1) < 2x + 3$
25. $\frac{3}{4}x - 6 \leq x - 7$
26. $3 + \frac{2}{7}x > x - 2$
27. $\frac{1}{2}(8x + 1) \geq 3x + \frac{5}{2}$
28. $9x - 1 < \frac{3}{4}(16x - 2)$
29. $3.6x + 11 \geq -3.4$
30. $15.6 - 1.3x < -5.2$

 Solving a Double Inequality In Exercises 31–42, solve the inequality. Then graph the solution set.

31. $1 < 2x + 3 < 9$
32. $-9 \leq -2x - 7 < 5$
33. $0 < 3(x + 7) \leq 20$
34. $-1 \leq -(x - 4) < 7$
35. $-4 < \dfrac{2x - 3}{3} < 4$
36. $0 \leq \dfrac{x + 3}{2} < 5$
37. $-1 < \dfrac{-x - 2}{3} \leq 1$
38. $-1 \leq \dfrac{-3x + 5}{7} \leq 2$
39. $\dfrac{3}{4} > x + 1 > \dfrac{1}{4}$
40. $-1 < 2 - \dfrac{x}{3} < 1$
41. $3.2 \leq 0.4x - 1 \leq 4.4$
42. $1.6 < 0.3x + 1 < 2.8$

 Solving an Absolute Value Inequality In Exercises 43–58, solve the inequality. Then graph the solution set. (Some inequalities have no solution.)

43. $|x| < 5$
44. $|x| \geq 8$
45. $\left|\dfrac{x}{2}\right| > 1$
46. $\left|\dfrac{x}{3}\right| < 2$
47. $|x - 5| < -1$
48. $|x - 7| < -5$
49. $|x - 20| \leq 6$
50. $|x - 8| \geq 0$
51. $|7 - 2x| \geq 9$
52. $|1 - 2x| < 5$
53. $\left|\dfrac{x - 3}{2}\right| \geq 4$
54. $\left|1 - \dfrac{2x}{3}\right| < 1$
55. $|9 - 2x| - 2 < -1$
56. $|x + 14| + 3 > 17$
57. $2|x + 10| \geq 9$
58. $3|4 - 5x| \leq 9$

Using Technology In Exercises 59–68, use a graphing utility to graph the inequality and identify the solution set.

59. $7x > 21$
60. $-4x \leq 9$
61. $8 - 3x \geq 2$
62. $20 < 6x - 1$
63. $4(x - 3) \leq 8 - x$
64. $3(x + 1) < x + 7$
65. $|x - 8| \leq 14$
66. $|2x + 9| > 13$
67. $2|x + 7| \geq 13$
68. $\frac{1}{2}|x + 1| \leq 3$

Using Technology In Exercises 69–74, use a graphing utility to graph the equation. Use the graph to approximate the values of x that satisfy each inequality.

Equation	Inequalities	
69. $y = 3x - 1$	(a) $y \geq 2$	(b) $y \leq 0$
70. $y = \frac{2}{3}x + 1$	(a) $y \leq 5$	(b) $y \geq 0$
71. $y = -\frac{1}{2}x + 2$	(a) $0 \leq y \leq 3$	(b) $y \geq 0$
72. $y = -3x + 8$	(a) $-1 \leq y \leq 3$	(b) $y \leq 0$
73. $y = |x - 3|$	(a) $y \leq 2$	(b) $y \geq 4$
74. $y = \left|\frac{1}{2}x + 1\right|$	(a) $y \leq 4$	(b) $y \geq 1$

75. **Think About It** The graph of $|x - 5| < 3$ can be described as all real numbers less than three units from 5. Give a similar description of $|x - 10| < 8$.

76. Think About It The graph of $|x - 2| > 5$ can be described as all real numbers more than five units from 2. Give a similar description of $|x - 8| > 4$.

Using Absolute Value In Exercises 77–84, use absolute value notation to define the interval (or pair of intervals) on the real number line.

77.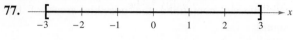

78. ![number line 78]

79. ![number line 79]

80. ![number line 80]

81. All real numbers at least three units from 7

82. All real numbers more than five units from 8

83. All real numbers less than four units from -3

84. All real numbers no more than seven units from -6

Writing an Inequality In Exercises 85–88, write an inequality to describe the situation.

85. During a trading day, the price P of a stock is no less than $7.25 and no more than $7.75.

86. During a month, a person's weight w is greater than 180 pounds but less than 185.5 pounds.

87. The expected return r on an investment is no more than 8%.

88. The expected net income I of a company is no less than $239 million.

Physiology One formula that relates a person's maximum heart rate r (in beats per minute) to the person's age A (in years) is

$r = 220 - A.$

In Exercises 89 and 90, determine the interval in which the person's heart rate is from 50% to 85% of the maximum heart rate. *(Source: American Heart Association)*

89. a 20-year-old   **90.** a 40-year-old

91. Job Offers You are considering two job offers. The first job pays $13.50 per hour. The second job pays $9.00 per hour plus $0.75 per unit produced per hour. How many units must you produce per hour for the second job to pay more per hour than the first job?

92. Job Offers You are considering two job offers. The first job pays $13.75 per hour. The second job pays $10.00 per hour plus $1.25 per unit produced per hour. How many units must you produce per hour for the second job to pay more than the first job?

93. Investment What annual interest rates yield a balance of more than $2000 on a 10-year investment of $1000? $[A = P(1 + rt)]$

94. Investment What annual interest rates yield a balance of more than $750 on a 5-year investment of $500? $[A = P(1 + rt)]$

95. Cost, Revenue, and Profit The revenue from selling x units of a product is $R = 115.95x$. The cost of producing x units is $C = 95x + 750$. To obtain a profit, the revenue must be greater than the cost. For what values of x does this product return a profit?

96. Cost, Revenue, and Profit The revenue from selling x units of a product is $R = 24.55x$. The cost of producing x units is $C = 15.4x + 150,000$. To obtain a profit, the revenue must be greater than the cost. For what values of x does this product return a profit?

97. Daily Sales A doughnut shop sells a dozen doughnuts for $7.95. Beyond the fixed costs (rent, utilities, and insurance) of $165 per day, it costs $1.45 for enough materials and labor to produce a dozen doughnuts. The daily profit from doughnut sales varies between $400 and $1200. Between what levels (in dozens of doughnuts) do the daily sales vary?

98. Weight Loss Program A person enrolls in a diet and exercise program that guarantees a loss of at least $1\frac{1}{2}$ pounds per week. The person's weight at the beginning of the program is 164 pounds. Find the maximum number of weeks before the person attains a goal weight of 128 pounds.

99. GPA An equation that relates the college grade-point averages y and high school grade-point averages x of the students at a college is $y = 0.692x + 0.988$.

(a) Use a graphing utility to graph the model.

(b) Use the graph to estimate the values of x that predict a college grade-point average of at least 3.0.

(c) Verify your estimate from part (b) algebraically.

(d) List other factors that may influence college GPA.

100. Weightlifting The 6RM load for a weightlifting exercise is the maximum weight at which a person can perform six repetitions. An equation that relates an athlete's 6RM bench press load x (in kilograms) and the athlete's 6RM barbell curl load y (in kilograms) is $y = 0.33x + 6.20$. *(Source: Journal of Sports Science & Medicine)*

(a) Use a graphing utility to graph the model.

(b) Use the graph to estimate the values of x that predict a 6RM barbell curl load of no more than 80 kilograms.

(c) Verify your estimate from part (b) algebraically.

(d) List other factors that may influence an athlete's 6RM barbell curl load.

101. Chemists' Wages The mean hourly wage W (in dollars) of chemists in the United States from 2000 through 2014 can be modeled by

$$W = 0.903t + 26.08, \quad 0 \le t \le 14$$

where t represents the year, with $t = 0$ corresponding to 2000. *(Source: U.S. Bureau of Labor Statistics)*

(a) According to the model, when was the mean hourly wage at least $30, but no more than $32?

(b) Use the model to predict when the mean hourly wage will exceed $45.

102. Milk Production Milk production M (in billions of pounds) in the United States from 2000 through 2014 can be modeled by

$$M = 3.00t + 163.3, \quad 0 \le t \le 14$$

where t represents the year, with $t = 0$ corresponding to 2000. *(Source: U.S. Department of Agriculture)*

(a) According to the model, when was the annual milk production greater than 180 billion pounds, but no more than 190 billion pounds?

(b) Use the model to predict when milk production will exceed 230 billion pounds.

103. Time Study The times required to perform a task in a manufacturing process by approximately two-thirds of the workers in a study satisfy the inequality

$$|t - 15.6| \le 1.9$$

where t is time in minutes. Determine the interval in which these times lie.

• •104. Geography• • • • • • • • • • • • • • • •

A geographic information system reports that the distance between two locations is 206 meters. The system is accurate to within 3 meters.

(a) Write an absolute value inequality for the possible distances between the locations.

(b) Graph the solution set.

105. Accuracy of Measurement You buy 6 T-bone steaks that cost $8.99 per pound. The weight that is listed on the package is 5.72 pounds. The scale that weighed the package is accurate to within $\frac{1}{32}$ pound. How much might you be undercharged or overcharged?

106. Accuracy of Measurement You stop at a self-service gas station to buy 15 gallons of 87-octane gasoline at $2.22 per gallon. The gas pump is accurate to within $\frac{1}{10}$ gallon. How much might you be undercharged or overcharged?

107. Geometry The side length of a square is 10.4 inches with a possible error of $\frac{1}{16}$ inch. Determine the interval containing the possible areas of the square.

108. Geometry The side length of a square is 24.2 centimeters with a possible error of 0.25 centimeter. Determine the interval containing the possible areas of the square.

Exploration

True or False? In Exercises 109–112, determine whether the statement is true or false. Justify your answer.

109. If a, b, and c are real numbers, and $a < b$, then $a + c < b + c$.

110. If $a, b,$ and c are real numbers, and $a \le b$, then $ac \le bc$.

111. If $-10 \le x \le 8$, then $-10 \ge -x$ and $-x \ge -8$.

112. If $-2 < x < -1$, then $1 < -x < 2$.

113. Think About It Give an example of an inequality whose solution set is $(-\infty, \infty)$.

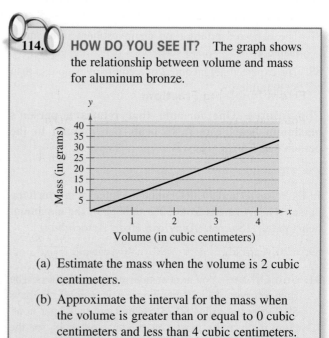

114. **HOW DO YOU SEE IT?** The graph shows the relationship between volume and mass for aluminum bronze.

(a) Estimate the mass when the volume is 2 cubic centimeters.

(b) Approximate the interval for the mass when the volume is greater than or equal to 0 cubic centimeters and less than 4 cubic centimeters.

115. Think About It Find sets of values of a, b, and c such that $0 \le x \le 10$ is a solution of the inequality $|ax - b| \le c$.

A.7 Errors and the Algebra of Calculus

■ Avoid common algebraic errors.
■ Recognize and use algebraic techniques that are common in calculus.

Algebraic Errors to Avoid

This section contains five lists of common algebraic errors: errors involving parentheses, errors involving fractions, errors involving exponents, errors involving radicals, and errors involving dividing out common factors. Many of these errors occur because they seem to be the *easiest* things to do. For example, students often believe that the operations of subtraction and division are commutative and associative. The examples below illustrate the fact that subtraction and division are neither commutative nor associative.

Not commutative	Not associative
$4 - 3 \neq 3 - 4$	$8 - (6 - 2) \neq (8 - 6) - 2$
$15 \div 5 \neq 5 \div 15$	$20 \div (4 \div 2) \neq (20 \div 4) \div 2$

Errors Involving Parentheses

Potential Error	Correct Form	Comment
$a - (x - b) = a - x - b$	$a - (x - b) = a - x + b$	Distribute negative sign to each term in parentheses.
$(a + b)^2 = a^2 + b^2$	$(a + b)^2 = a^2 + 2ab + b^2$	Remember the middle term when squaring binomials.
$\left(\dfrac{1}{2}a\right)\left(\dfrac{1}{2}b\right) = \dfrac{1}{2}(ab)$	$\left(\dfrac{1}{2}a\right)\left(\dfrac{1}{2}b\right) = \dfrac{1}{4}(ab) = \dfrac{ab}{4}$	$\frac{1}{2}$ occurs twice as a factor.
$(3x + 6)^2 = 3(x + 2)^2$	$(3x + 6)^2 = [3(x + 2)]^2$ $= 3^2(x + 2)^2$	When factoring, raise all factors to the power.

Errors Involving Fractions

Potential Error	Correct Form	Comment
$\dfrac{2}{x + 4} = \dfrac{2}{x} + \dfrac{2}{4}$	Leave as $\dfrac{2}{x + 4}$.	The fraction is already in simplest form.
$\dfrac{\left(\dfrac{x}{a}\right)}{b} = \dfrac{bx}{a}$	$\dfrac{\left(\dfrac{x}{a}\right)}{b} = \left(\dfrac{x}{a}\right)\left(\dfrac{1}{b}\right) = \dfrac{x}{ab}$	Multiply by the reciprocal when dividing fractions.
$\dfrac{1}{a} + \dfrac{1}{b} = \dfrac{1}{a + b}$	$\dfrac{1}{a} + \dfrac{1}{b} = \dfrac{b + a}{ab}$	Use the property for adding fractions with unlike denominators.
$\dfrac{1}{3x} = \dfrac{1}{3}x$	$\dfrac{1}{3x} = \dfrac{1}{3} \cdot \dfrac{1}{x}$	Use the property for multiplying fractions.
$(1/3)x = \dfrac{1}{3x}$	$(1/3)x = \dfrac{1}{3} \cdot x = \dfrac{x}{3}$	Be careful when expressing fractions in the form $1/a$.
$(1/x) + 2 = \dfrac{1}{x + 2}$	$(1/x) + 2 = \dfrac{1}{x} + 2 = \dfrac{1 + 2x}{x}$	Be careful when expressing fractions in the form $1/a$. Be sure to find a common denominator before adding fractions.

Errors Involving Exponents

Potential Error	Correct Form	Comment
$(x^2)^3 = x^5$ ✗	$(x^2)^3 = x^{2 \cdot 3} = x^6$	Multiply exponents when raising a power to a power.
$x^2 \cdot x^3 = x^6$ ✗	$x^2 \cdot x^3 = x^{2+3} = x^5$	Add exponents when multiplying powers with like bases.
$(2x)^3 = 2x^3$ ✗	$(2x)^3 = 2^3 x^3 = 8x^3$	Raise each factor to the power.
$\dfrac{1}{x^2 - x^3} = x^{-2} - x^{-3}$ ✗	Leave as $\dfrac{1}{x^2 - x^3}$.	Do not move term-by-term from denominator to numerator.

Errors Involving Radicals

Potential Error	Correct Form	Comment
$\sqrt{5x} = 5\sqrt{x}$ ✗	$\sqrt{5x} = \sqrt{5}\sqrt{x}$	Radicals apply to every factor inside the radical.
$\sqrt{x^2 + a^2} = x + a$ ✗	Leave as $\sqrt{x^2 + a^2}$.	Do not apply radicals term-by-term when adding or subtracting terms.
$\sqrt{-x + a} = -\sqrt{x - a}$ ✗	Leave as $\sqrt{-x + a}$.	Do not factor negative signs out of square roots.

Errors Involving Dividing Out

Potential Error	Correct Form	Comment
$\dfrac{a + bx}{a} = 1 + bx$ ✗	$\dfrac{a + bx}{a} = \dfrac{a}{a} + \dfrac{bx}{a} = 1 + \dfrac{b}{a}x$	Divide out common factors, not common terms.
$\dfrac{a + ax}{a} = a + x$ ✗	$\dfrac{a + ax}{a} = \dfrac{a(1 + x)}{a} = 1 + x$	Factor before dividing out common factors.
$1 + \dfrac{x}{2x} = 1 + \dfrac{1}{x}$ ✗	$1 + \dfrac{x}{2x} = 1 + \dfrac{1}{2} = \dfrac{3}{2}$	Divide out common factors.

A good way to avoid errors is to *work slowly, write neatly,* and *think about each step.* Each time you write a step, ask yourself why the step is algebraically legitimate. For example, the step below is legitimate because *dividing the numerator and denominator by the same nonzero number produces an equivalent fraction.*

$$\frac{2x}{6} = \frac{2 \cdot x}{2 \cdot 3} = \frac{x}{3}$$

EXAMPLE 1   **Describing and Correcting an Error**

Describe and correct the error. $\dfrac{1}{2x} + \dfrac{1}{3x} = \dfrac{1}{5x}$ ✗

Solution   Use the property for adding fractions with unlike denominators.

$$\frac{1}{2x} + \frac{1}{3x} = \frac{3x + 2x}{6x^2} = \frac{5x}{6x^2} = \frac{5}{6x}$$

✓ *Checkpoint* ◀))) *Audio-video solution in English & Spanish at LarsonPrecalculus.com*

Describe and correct the error. $\sqrt{x^2 + 4} = x + 2$ ✗

Some Algebra of Calculus

In calculus it is often necessary to rewrite a simplified algebraic expression. See the following lists, which are from a standard calculus text.

Unusual Factoring

Expression	Useful Calculus Form	Comment
$\dfrac{5x^4}{8}$	$\dfrac{5}{8}x^4$	Write with fractional coefficient.
$\dfrac{x^2 + 3x}{-6}$	$-\dfrac{1}{6}(x^2 + 3x)$	Write with fractional coefficient.
$2x^2 - x - 3$	$2\left(x^2 - \dfrac{x}{2} - \dfrac{3}{2}\right)$	Factor out the leading coefficient.
$\dfrac{x}{2}(x + 1)^{-1/2} + (x + 1)^{1/2}$	$\dfrac{(x + 1)^{-1/2}}{2}[x + 2(x + 1)]$	Factor out the fractional coefficient and the variable expression with the lesser exponent.

Writing with Negative Exponents

Expression	Useful Calculus Form	Comment
$\dfrac{9}{5x^3}$	$\dfrac{9}{5}x^{-3}$	Move the factor to the numerator and change the sign of the exponent.
$\dfrac{7}{\sqrt{2x - 3}}$	$7(2x - 3)^{-1/2}$	Move the factor to the numerator and change the sign of the exponent.

Writing a Fraction as a Sum

Expression	Useful Calculus Form	Comment
$\dfrac{x + 2x^2 + 1}{\sqrt{x}}$	$x^{1/2} + 2x^{3/2} + x^{-1/2}$	Divide each term of the numerator by $x^{1/2}$.
$\dfrac{1 + x}{x^2 + 1}$	$\dfrac{1}{x^2 + 1} + \dfrac{x}{x^2 + 1}$	Rewrite the fraction as a sum of fractions.
$\dfrac{2x}{x^2 + 2x + 1}$	$\dfrac{2x + 2 - 2}{x^2 + 2x + 1}$	Add and subtract the same term.
	$= \dfrac{2x + 2}{x^2 + 2x + 1} - \dfrac{2}{(x + 1)^2}$	Rewrite the fraction as a difference of fractions.
$\dfrac{x^2 - 2}{x + 1}$	$x - 1 - \dfrac{1}{x + 1}$	Use polynomial long division. (See Section 2.3.)
$\dfrac{x + 7}{x^2 - x - 6}$	$\dfrac{2}{x - 3} - \dfrac{1}{x + 2}$	Use the method of partial fractions. (See Section 7.4.)

Inserting Factors and Terms

Expression	Useful Calculus Form	Comment
$(2x - 1)^3$	$\dfrac{1}{2}(2x - 1)^3(2)$	Multiply and divide by 2.
$7x^2(4x^3 - 5)^{1/2}$	$\dfrac{7}{12}(4x^3 - 5)^{1/2}(12x^2)$	Multiply and divide by 12.
$\dfrac{4x^2}{9} - 4y^2 = 1$	$\dfrac{x^2}{9/4} - \dfrac{y^2}{1/4} = 1$	Write with fractional denominators.
$\dfrac{x}{x + 1}$	$\dfrac{x + 1 - 1}{x + 1} = 1 - \dfrac{1}{x + 1}$	Add and subtract the same term.

The next five examples demonstrate many of the steps in the preceding lists.

EXAMPLE 2 **Factors Involving Negative Exponents**

Factor $x(x + 1)^{-1/2} + (x + 1)^{1/2}$.

Solution When multiplying powers with like bases, you add exponents. When factoring, you are undoing multiplication, and so you *subtract* exponents.

$$x(x + 1)^{-1/2} + (x + 1)^{1/2} = (x + 1)^{-1/2}[x(x + 1)^0 + (x + 1)^1]$$
$$= (x + 1)^{-1/2}[x + (x + 1)]$$
$$= (x + 1)^{-1/2}(2x + 1)$$

✓ *Checkpoint*  Audio-video solution in English & Spanish at LarsonPrecalculus.com

Factor $x(x - 2)^{-1/2} + 6(x - 2)^{1/2}$.

Another way to simplify the expression in Example 2 is to multiply the expression by a fractional form of 1 and then use the Distributive Property.

$$[x(x + 1)^{-1/2} + (x + 1)^{1/2}] \cdot \frac{(x + 1)^{1/2}}{(x + 1)^{1/2}} = \frac{x(x + 1)^0 + (x + 1)^1}{(x + 1)^{1/2}}$$
$$= \frac{2x + 1}{\sqrt{x + 1}}$$

EXAMPLE 3 **Inserting Factors in an Expression**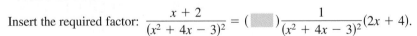

Insert the required factor: $\dfrac{x + 2}{(x^2 + 4x - 3)^2} = ()\dfrac{1}{(x^2 + 4x - 3)^2}(2x + 4)$.

Solution The expression on the right side of the equation is twice the expression on the left side. To make both sides equal, insert a factor of $\frac{1}{2}$.

$$\frac{x + 2}{(x^2 + 4x - 3)^2} = \left(\frac{1}{2}\right)\frac{1}{(x^2 + 4x - 3)^2}(2x + 4)$$

✓ *Checkpoint*  Audio-video solution in English & Spanish at LarsonPrecalculus.com

Insert the required factor: $\dfrac{6x - 3}{(x^2 - x + 4)^2} = ()\dfrac{1}{(x^2 - x + 4)^2}(2x - 1)$.

EXAMPLE 4 **Rewriting Fractions**

Show that the two expressions are equivalent.

$$\frac{16x^2}{25} - 9y^2 = \frac{x^2}{25/16} - \frac{y^2}{1/9}$$

Solution To write the expression on the left side of the equation in the form given on the right side, multiply the numerator and denominator of the first term by $1/16$ and multiply the numerator and denominator of the second term by $1/9$.

$$\frac{16x^2}{25} - 9y^2 = \frac{16x^2}{25}\left(\frac{1/16}{1/16}\right) - 9y^2\left(\frac{1/9}{1/9}\right) = \frac{x^2}{25/16} - \frac{y^2}{1/9}$$

✓ *Checkpoint*  Audio-video solution in English & Spanish at LarsonPrecalculus.com

Show that the two expressions are equivalent.

$$\frac{9x^2}{16} + 25y^2 = \frac{x^2}{16/9} + \frac{y^2}{1/25}$$

EXAMPLE 5 **Rewriting with Negative Exponents**

Rewrite each expression using negative exponents.

a. $\dfrac{-4x}{(1 - 2x^2)^2}$ **b.** $\dfrac{2}{5x^3} - \dfrac{1}{\sqrt{x}} + \dfrac{3}{5(4x)^2}$

Solution

a. $\dfrac{-4x}{(1 - 2x^2)^2} = -4x(1 - 2x^2)^{-2}$

b. $\dfrac{2}{5x^3} - \dfrac{1}{\sqrt{x}} + \dfrac{3}{5(4x)^2} = \dfrac{2}{5x^3} - \dfrac{1}{x^{1/2}} + \dfrac{3}{5(4x)^2} = \dfrac{2}{5}x^{-3} - x^{-1/2} + \dfrac{3}{5}(4x)^{-2}$

✓ *Checkpoint*  Audio-video solution in English & Spanish at LarsonPrecalculus.com

Rewrite $\dfrac{-6x}{(1 - 3x^2)^2} + \dfrac{1}{\sqrt[3]{x}}$ using negative exponents.

EXAMPLE 6 **Rewriting Fractions as Sums of Terms**

Rewrite each fraction as the sum of three terms.

a. $\dfrac{x^2 - 4x + 8}{2x}$ **b.** $\dfrac{x + 2x^2 + 1}{\sqrt{x}}$

Solution

a. $\dfrac{x^2 - 4x + 8}{2x} = \dfrac{x^2}{2x} - \dfrac{4x}{2x} + \dfrac{8}{2x} = \dfrac{x}{2} - 2 + \dfrac{4}{x}$

b. $\dfrac{x + 2x^2 + 1}{\sqrt{x}} = \dfrac{x}{x^{1/2}} + \dfrac{2x^2}{x^{1/2}} + \dfrac{1}{x^{1/2}} = x^{1/2} + 2x^{3/2} + x^{-1/2}$

✓ *Checkpoint* 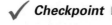 Audio-video solution in English & Spanish at LarsonPrecalculus.com

Rewrite each fraction as the sum of three terms.

a. $\dfrac{x^4 - 2x^3 + 5}{x^3}$ **b.** $\dfrac{x^2 - x + 5}{\sqrt{x}}$

A.7 Exercises

Vocabulary: Fill in the blanks.

1. To rewrite the expression $\dfrac{3}{x^5}$ using negative exponents, move x^5 to the _____ and change the sign of the exponent.

2. When dividing fractions, multiply by the _____.

Skills and Applications

Describing and Correcting an Error In Exercises 3–12, describe and correct the error.

3. $(x + 3)^2 = x^2 + 9$ ✗

4. $5z + 3(x - 2) = 5z + 3x - 2$ ✗

5. $\sqrt{x + 9} = \sqrt{x} + 3$ ✗ 6. $\sqrt{25 - x^2} = 5 - x$ ✗

7. $\dfrac{2x^2 + 1}{5x} = \dfrac{2x + 1}{5}$ ✗ 8. $\dfrac{6x + y}{6x - y} = \dfrac{x + y}{x - y}$ ✗

9. $(4x)^2 = 4x^2$ ✗

10. $\dfrac{1}{a^{-1} + b^{-1}} = \left(\dfrac{1}{a + b}\right)^{-1}$ ✗

11. $\dfrac{3}{x} + \dfrac{4}{y} = \dfrac{7}{x + y}$ ✗ 12. $5 + (1/y) = \dfrac{1}{5 + y}$ ✗

 Factors Involving Negative Exponents In Exercises 13–16, factor the expression.

13. $2x(x + 2)^{-1/2} + (x + 2)^{1/2}$

14. $x^2(x^2 + 1)^{-5} - (x^2 + 1)^{-4}$

15. $4x^3(2x - 1)^{3/2} - 2x(2x - 1)^{-1/2}$

16. $x(x + 1)^{-4/3} + (x + 1)^{2/3}$

Unusual Factoring In Exercises 17–24, complete the factored form of the expression.

17. $\dfrac{5x + 3}{4} = \dfrac{1}{4}(\boxed{})$ 18. $\dfrac{7x^2}{10} = \dfrac{7}{10}(\boxed{})$

19. $\dfrac{2}{3}x^2 + \dfrac{1}{3}x + 5 = \dfrac{1}{3}(\boxed{})$ 20. $\dfrac{3}{4}x + \dfrac{1}{2} = \dfrac{1}{4}(\boxed{})$

21. $x^{1/3} - 5x^{4/3} = x^{1/3}(\boxed{})$

22. $3(2x + 1)x^{1/2} + 4x^{3/2} = x^{1/2}(\boxed{})$

23. $\dfrac{1}{10}(2x + 1)^{5/2} - \dfrac{1}{6}(2x + 1)^{3/2} = \dfrac{(2x + 1)^{3/2}}{15}(\boxed{})$

24. $\dfrac{3}{7}(t + 1)^{7/3} - \dfrac{3}{4}(t + 1)^{4/3} = \dfrac{3(t + 1)^{4/3}}{28}(\boxed{})$

 Inserting Factors in an Expression In Exercises 25–28, insert the required factor in the parentheses.

25. $x^2(x^3 - 1)^4 = (\boxed{})(x^3 - 1)^4(3x^2)$

26. $x(1 - 2x^2)^3 = (\boxed{})(1 - 2x^2)^3(-4x)$

27. $\dfrac{4x + 6}{(x^2 + 3x + 7)^3} = (\boxed{})\dfrac{1}{(x^2 + 3x + 7)^3}(2x + 3)$

28. $\dfrac{x + 1}{(x^2 + 2x - 3)^2} = (\boxed{})\dfrac{1}{(x^2 + 2x - 3)^2}(2x + 2)$

 Rewriting Fractions In Exercises 29–34, show that the two expressions are equivalent.

29. $4x^2 + \dfrac{6y^2}{10} = \dfrac{x^2}{1/4} + \dfrac{3y^2}{5}$

30. $\dfrac{4x^2}{14} - 2y^2 = \dfrac{2x^2}{7} - \dfrac{y^2}{1/2}$

31. $\dfrac{25x^2}{36} + \dfrac{4y^2}{9} = \dfrac{x^2}{36/25} + \dfrac{y^2}{9/4}$

32. $\dfrac{5x^2}{9} - \dfrac{16y^2}{49} = \dfrac{x^2}{9/5} - \dfrac{y^2}{49/16}$

33. $\dfrac{x^2}{3/10} - \dfrac{y^2}{4/5} = \dfrac{10x^2}{3} - \dfrac{5y^2}{4}$

34. $\dfrac{x^2}{5/8} + \dfrac{y^2}{6/11} = \dfrac{8x^2}{5} + \dfrac{11y^2}{6}$

Rewriting with Negative Exponents In Exercises 35–40, rewrite the expression using negative exponents.

35. $\dfrac{7}{(x + 3)^5}$ 36. $\dfrac{2 - x}{(x + 1)^{3/2}}$

37. $\dfrac{2x^5}{(3x + 5)^4}$ 38. $\dfrac{x + 1}{x(6 - x)^{1/2}}$

39. $\dfrac{4}{3x} + \dfrac{4}{x^4} - \dfrac{7x}{\sqrt[3]{2x}}$ 40. $\dfrac{x}{x - 2} + \dfrac{1}{x^2} + \dfrac{8}{3(9x)^3}$

Rewriting a Fraction as a Sum of Terms In Exercises 41–46, rewrite the fraction as a sum of two or more terms.

41. $\dfrac{x^2 + 6x + 12}{3x}$ 42. $\dfrac{x^3 - 5x^2 + 4}{x^2}$

43. $\dfrac{4x^3 - 7x^2 + 1}{x^{1/3}}$ 44. $\dfrac{2x^5 - 3x^3 + 5x - 1}{x^{3/2}}$

45. $\dfrac{3 - 5x^2 - x^4}{\sqrt{x}}$ 46. $\dfrac{x^3 - 5x^4}{3x^2}$

Simplifying an Expression   **In Exercises 47–58, simplify the expression.**

47. $\dfrac{-2(x^2-3)^{-3}(2x)(x+1)^3 - 3(x+1)^2(x^2-3)^{-2}}{[(x+1)^3]^2}$

48. $\dfrac{x^5(-3)(x^2+1)^{-4}(2x) - (x^2+1)^{-3}(5)x^4}{(x^5)^2}$

49. $\dfrac{(6x+1)^3(27x^2+2) - (9x^3+2x)(3)(6x+1)^2(6)}{[(6x+1)^3]^2}$

50. $\dfrac{(4x^2+9)^{1/2}(2) - (2x+3)\left(\dfrac{1}{2}\right)(4x^2+9)^{-1/2}(8x)}{[(4x^2+9)^{1/2}]^2}$

51. $\dfrac{(x+2)^{3/4}(x+3)^{-2/3} - (x+3)^{1/3}(x+2)^{-1/4}}{[(x+2)^{3/4}]^2}$

52. $(2x-1)^{1/2} - (x+2)(2x-1)^{-1/2}$

53. $\dfrac{2(3x-1)^{1/3} - (2x+1)\left(\dfrac{1}{3}\right)(3x-1)^{-2/3}(3)}{(3x-1)^{2/3}}$

54. $\dfrac{(x+1)\left(\dfrac{1}{2}\right)(2x-3x^2)^{-1/2}(2-6x) - (2x-3x^2)^{1/2}}{(x+1)^2}$

55. $\dfrac{1}{(x^2+4)^{1/2}} \cdot \dfrac{1}{2}(x^2+4)^{-1/2}(2x)$

56. $\dfrac{1}{x^2-6}(2x) + \dfrac{1}{2x+5}(2)$

57. $(x^2+5)^{1/2}\left(\dfrac{3}{2}\right)(3x-2)^{1/2}(3)$
$+ (3x-2)^{3/2}\left(\dfrac{1}{2}\right)(x^2+5)^{-1/2}(2x)$

58. $(3x+2)^{-1/2}(3)(x-6)^{1/2}(1)$
$+ (x-6)^3\left(-\dfrac{1}{2}\right)(3x+2)^{-3/2}(3)$

59. **Verifying an Equation**

(a) Verify that $y_1 = y_2$ analytically.

$y_1 = x^2\left(\dfrac{1}{3}\right)(x^2+1)^{-2/3}(2x) + (x^2+1)^{1/3}(2x)$

$y_2 = \dfrac{2x(4x^2+3)}{3(x^2+1)^{2/3}}$

(b) Complete the table and demonstrate the equality in part (a) numerically.

x	-2	-1	$-\frac{1}{2}$	0	1	2	$\frac{5}{2}$
y_1							
y_2							

(c) Use a graphing utility to verify the equality in part (a) graphically.

60. **Athletics**   An athlete has set up a course in which she is dropped off by a boat 2 miles from the nearest point on shore. Once she reaches the shore, she must run to a point 4 miles down the coast and 2 miles inland (see figure). She can swim 2 miles per hour and run 6 miles per hour. The time t (in hours) required for her to complete the course can be approximated by the model

$$t = \dfrac{\sqrt{x^2+4}}{2} + \dfrac{\sqrt{(4-x)^2+4}}{6}$$

where x is the distance (in miles) down the coast from her starting point to the point at which she leaves the water to start her run.

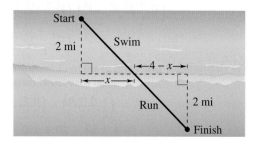

(a) Use a table to approximate the distance down the coast that will yield the minimum amount of time required for the athlete to complete the course.

(b) The expression below was obtained using calculus. It can be used to find the minimum amount of time required for the triathlete to reach the finish line. Simplify the expression.

$$\dfrac{1}{2}x(x^2+4)^{-1/2} + \dfrac{1}{6}(x-4)(x^2-8x+20)^{-1/2}$$

Exploration

61. **Writing**   Write a paragraph explaining to a classmate why

$$\dfrac{1}{(x-2)^{1/2}+x^4} \neq (x-2)^{-1/2} + x^{-4}.$$

62. **Think About It**   You are taking a course in calculus, and for one of the homework problems you obtain the following answer.

$$\dfrac{2}{3}x(2x-3)^{3/2} - \dfrac{2}{15}(2x-3)^{5/2}$$

The answer in the back of the book is

$$\dfrac{2}{5}(2x-3)^{3/2}(x+1).$$

Show how the second answer can be obtained from the first. Then use the same technique to simplify the expression

$$\dfrac{2}{3}x(4+x)^{3/2} - \dfrac{2}{15}(4+x)^{5/2}.$$

15.

x	-1	0	1	2	$\frac{5}{2}$
y	7	5	3	1	0
(x, y)	$(-1, 7)$	$(0, 5)$	$(1, 3)$	$(2, 1)$	$\left(\frac{5}{2}, 0\right)$

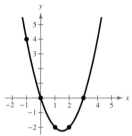

17.

x	-1	0	1	2	3
y	4	0	-2	-2	0
(x, y)	$(-1, 4)$	$(0, 0)$	$(1, -2)$	$(2, -2)$	$(3, 0)$

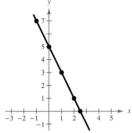

19. x-intercept: $(3, 0)$
y-intercept: $(0, 9)$

21. x-intercept: $(-2, 0)$
y-intercept: $(0, 2)$

23. x-intercept: $\left(\frac{6}{5}, 0\right)$
y-intercept: $(0, -6)$

25. x-intercept: $(-4, 0)$
y-intercept: $(0, 2)$

27. x-intercept: $\left(\frac{7}{3}, 0\right)$
y-intercept: $(0, 7)$

29. x-intercepts: $(0, 0)$, $(2, 0)$
y-intercept: $(0, 0)$

31. x-intercept: $(6, 0)$
y-intercepts: $\left(0, \pm\sqrt{6}\right)$

33. y-axis symmetry
35. Origin symmetry
37. Origin symmetry
39. x-axis symmetry

41.

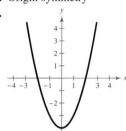

43.

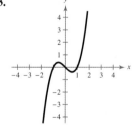

45. x-intercept: $\left(\frac{1}{3}, 0\right)$
y-intercept: $(0, 1)$
No symmetry

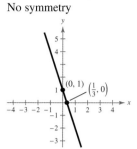

47. x-intercepts: $(0, 0)$, $(2, 0)$
y-intercept: $(0, 0)$
No symmetry

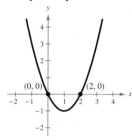

49. x-intercept: $\left(\sqrt[3]{-3}, 0\right)$
y-intercept: $(0, 3)$
No symmetry

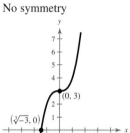

51. x-intercept: $(3, 0)$
y-intercept: None
No symmetry

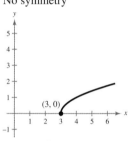

53. x-intercept: $(6, 0)$
y-intercept: $(0, 6)$
No symmetry

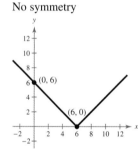

55. x-intercept: $(-1, 0)$
y-intercepts: $(0, \pm 1)$
x-axis symmetry

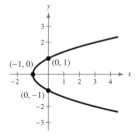

57.

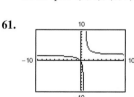

Intercepts: $(6, 0)$, $(0, 3)$

59.

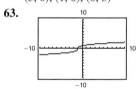

Intercepts:
$(3, 0)$, $(1, 0)$, $(0, 3)$

61.

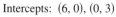

Intercept: $(0, 0)$

63.

Intercepts: $(-1, 0)$, $(0, 1)$

65.

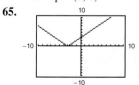

Intercepts: $(-3, 0)$, $(0, 3)$

67. $x^2 + y^2 = 9$ **69.** $(x + 4)^2 + (y - 5)^2 = 4$

71. $(x - 3)^2 + (y - 8)^2 = 169$

73. $(x + 3)^2 + (y + 3)^2 = 61$

75. Center: $(0, 0)$; Radius: 5   **77.** Center: $(1, -3)$; Radius: 3

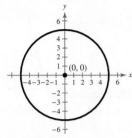   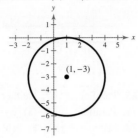

79. Center: $\left(\frac{1}{2}, \frac{1}{2}\right)$; Radius: $\frac{3}{2}$

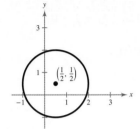

81.

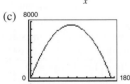

83. (a)   (b) Answers will vary.

(c) 8000   (d) $x = 86\frac{2}{3}$, $y = 86\frac{2}{3}$

(e) A regulation NFL playing field is 120 yards long and $53\frac{1}{3}$ yards wide. The actual area is 6400 square yards.

85. (a) 100   The model fits the data well; Each data value is close to the graph of the model.

(b) 74.7 yr   (c) 1964

(d) $(0, 63.6)$; In 1940, the life expectancy of a child (at birth) was 63.6 years.

(e) Answers will vary.

87. False. $y = x$ is symmetric with respect to the origin.

89. True. *Sample answer*: Depending on the center and radius, the graph could intersect one, both, or neither axis.

91. (a) $a = 1, b = 0$ (b) $a = 0, b = 1$

Section 1.3   *(page 31)*

1. linear   **3.** point-slope   **5.** perpendicular

7. Linear extrapolation   **9.** (a) L_2 (b) L_3 (c) L_1

11.   **13.** $\frac{3}{2}$

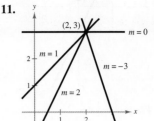

15. $m = 5$ **17.** $m = -\frac{3}{4}$

y-intercept: $(0, 3)$ y-intercept: $(0, -1)$

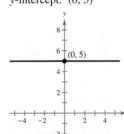

19. $m = 0$ **21.** m is undefined.

y-intercept: $(0, 5)$ y-intercept: none

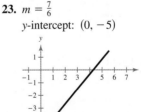

23. $m = \frac{7}{6}$

y-intercept: $(0, -5)$

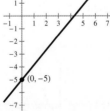

25. $m = -\frac{3}{2}$ **27.** $m = 2$ **29.** $m = 0$

31. m is undefined.   **33.** $m = 0.15$

35. $(-1, 7), (0, 7), (4, 7)$ **37.** $(-4, 6), (-3, 8), (-2, 10)$

39. $(-2, 7), \left(0, \frac{19}{3}\right), (1, 6)$ **41.** $(-4, -5), (-4, 0), (-4, 2)$

43. $y = 3x - 2$

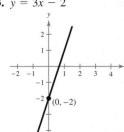

45. $y = -2x$

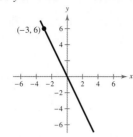

47. $y = -\frac{1}{3}x + \frac{4}{3}$

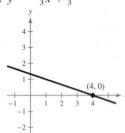

49. $y = -\frac{1}{2}x - 2$

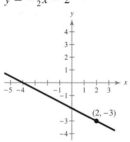

51. $y = \frac{5}{2}$

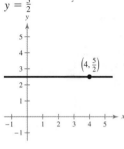

53. $y = 5x + 27.3$

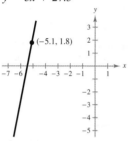

55. $y = -\frac{3}{5}x + 2$

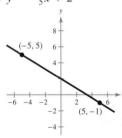

57. $x = -7$

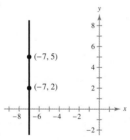

59. $y = -\frac{1}{2}x + \frac{3}{2}$

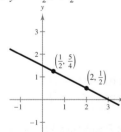

61. $y = 0.4x + 0.2$

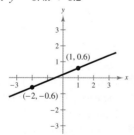

63. $y = -1$

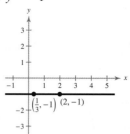

65. Parallel   **67.** Neither   **69.** Perpendicular
71. Parallel   **73.** (a) $y = 2x - 3$ (b) $y = -\frac{1}{2}x + 2$
75. (a) $y = -\frac{3}{4}x + \frac{3}{8}$ (b) $y = \frac{4}{3}x + \frac{127}{72}$
77. (a) $y = 4$ (b) $x = -2$
79. (a) $y = x + 4.3$ (b) $y = -x + 9.3$
81. $5x + 3y - 15 = 0$ **83.** $12x + 3y + 2 = 0$
85. $x + y - 3 = 0$
87. (a) Sales increasing 135 units/yr
    (b) No change in sales
    (c) Sales decreasing 40 units/yr
89. 12 ft
91. $V(t) = -150t + 5400,\quad 16 \le t \le 21$
93. C-intercept: fixed initial cost; Slope: cost of producing an additional laptop bag
95. $V(t) = -175t + 875,\quad 0 \le t \le 5$
97. $F = 1.8C + 32$ or $C = \frac{5}{9}F - \frac{160}{9}$
99. (a) $C = 21t + 42,000$ (b) $R = 45t$
    (c) $P = 24t - 42,000$ (d) 1750 h
101. False. The slope with the greatest magnitude corresponds to the steepest line.
103. Find the slopes of the lines containing each two points and use the relationship $m_1 = -\dfrac{1}{m_2}$.
105. The scale on the y-axis is unknown, so the slopes of the lines cannot be determined.
107. No. The slopes of two perpendicular lines have opposite signs (assume that neither line is vertical or horizontal).
109. The line $y = 4x$ rises most quickly, and the line $y = -4x$ falls most quickly. The greater the magnitude of the slope (the absolute value of the slope), the faster the line rises or falls.
111. $3x - 2y - 1 = 0$ **113.** $80x + 12y + 139 = 0$

Section 1.4   *(page 44)*

1. domain; range; function   **3.** implied domain
5. Function   **7.** Not a function
9. (a) Function
    (b) Not a function, because the element 1 in A corresponds to two elements, -2 and 1, in B.
    (c) Function
    (d) Not a function, because not every element in A is matched with an element in B.
11. Not a function   **13.** Function   **15.** Function
17. Function   **19.** (a) -2 (b) -14 (c) $3x + 1$
21. (a) 15   (b) $4t^2 - 19t + 27$ (c) $4t^2 - 3t - 10$
23. (a) 1   (b) 2.5   (c) $3 - 2|x|$
25. (a) $-\dfrac{1}{9}$ (b) Undefined   (c) $\dfrac{1}{y^2 + 6y}$

27. (a) 1　(b) −1　(c) $\dfrac{|x-1|}{x-1}$

29. (a) −1　(b) 2　(c) 6

31.

x	−2	−1	0	1	2
$f(x)$	1	4	5	4	1

33.

x	−2	−1	0	1	2
$f(x)$	5	$\frac{9}{2}$	4	1	0

35. 5　　**37.** $\frac{4}{3}$　　**39.** ±9　　**41.** 0, ±1　　**43.** −1, 2

45. 0, ±2　　**47.** All real numbers x

49. All real numbers y such that $y \geq -6$

51. All real numbers x except $x = 0, -2$

53. All real numbers s such that $s \geq 1$ except $s = 4$

55. All real numbers x such that $x > 0$

57. (a) The maximum volume is 1024 cubic centimeters.

(b)

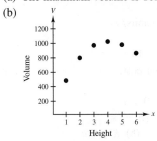

Yes, V is a function of x.

(c) $V = x(24 - 2x)^2$,　$0 < x < 12$

59. $A = \dfrac{P^2}{16}$　　**61.** No, the ball will be at a height of 18.5 feet.

63. $A = \dfrac{x^2}{2(x-2)}$,　$x > 2$

65. 2008: 67.36%
2009: 70.13%
2010: 72.90%
2011: 75.67%
2012: 79.30%
2013: 81.25%
2014: 83.20%

67. (a) $C = 12.30x + 98{,}000$
(b) $R = 17.98x$
(c) $P = 5.68x - 98{,}000$

69. (a)

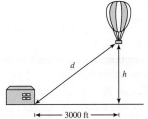

(b) $h = \sqrt{d^2 - 3000^2}$,　$d \geq 3000$

71. (a) $R = \dfrac{240n - n^2}{20}$,　$n \geq 80$

(b)

n	90	100	110	120	130	140	150
$R(n)$	\$675	\$700	\$715	\$720	\$715	\$700	\$675

The revenue is maximum when 120 people take the trip.

73. $2 + h$,　$h \neq 0$　　**75.** $3x^2 + 3xh + h^2 + 3$,　$h \neq 0$

77. $-\dfrac{x+3}{9x^2}$,　$x \neq 3$　　**79.** $\dfrac{\sqrt{5x}-5}{x-5}$

81. $g(x) = cx^2$; $c = -2$　　**83.** $r(x) = \dfrac{c}{x}$; $c = 32$

85. False. A function is a special type of relation.

87. False. The range is $[-1, \infty)$.

89. The domain of $f(x)$ includes $x = 1$ and the domain of $g(x)$ does not because you cannot divide by 0. So, the functions do not have the same domain.

91. No; x is the independent variable, f is the name of the function.

93. (a) Yes. The amount you pay in sales tax will increase as the price of the item purchased increases.

(b) No. The length of time that you study will not necessarily determine how well you do on an exam.

Section 1.5　*(page 56)*

1. Vertical Line Test　　**3.** decreasing

5. average rate of change; secant

7. Domain: $(-2, 2]$; range: $[-1, 8]$
(a) −1　(b) 0　(c) −1　(d) 8

9. Domain: $(-\infty, \infty)$; range: $(-2, \infty)$
(a) 0　(b) 1　(c) 2　(d) 3

11. Function　　**13.** Not a function　　**15.** −6　　**17.** $-\frac{5}{2}$, 6

19. −3　　**21.** $0, \pm\sqrt{6}$　　**23.** ±3, 4　　**25.** $\frac{1}{2}$

27. (a)

0, 6

(b) 0, 6

29. (a)

−5.5

(b) $-\frac{11}{2}$

31. (a)

0.3333

(b) $\frac{1}{3}$

33. Decreasing on $(-\infty, \infty)$

35. Increasing on $(1, \infty)$; Decreasing on $(-\infty, -1)$

37. Increasing on $(1, \infty)$; Decreasing on $(-\infty, -1)$
Constant on $(-1, 1)$

39. Increasing on $(-\infty, -1)$, $(0, \infty)$; Decreasing on $(-1, 0)$

41.

Constant on $(-\infty, \infty)$

43.

Decreasing on $(-\infty, 0)$
Increasing on $(0, \infty)$

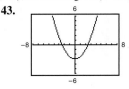

CHAPTER 1

45. 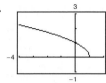   **47.**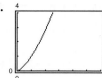

Decreasing on $(-\infty, 1)$ Increasing on $(0, \infty)$

49. Relative minimum: $(-1.5, -2.25)$
51. Relative maximum: $(0, 15)$
  Relative minimum: $(4, -17)$
53. Relative minimum: $(0.33, -0.38)$
55. 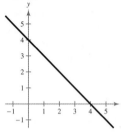   **57.**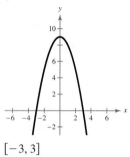

$(-\infty, 4]$ $[-3, 3]$

59.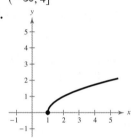

$[1, \infty)$

61. -2 **63.** -1
65. (a)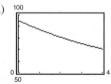

(b) About -6.14; The amount the U.S. federal government spent on research and development for defense decreased by about $6.14 billion each year from 2010 to 2014.

67. (a) $s = -16t^2 + 64t + 6$

(b)   (c) 16 ft/sec

(d) The slope of the secant line is positive.
(e) $y = 16t + 6$
(f)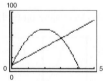

69. (a) $s = -16t^2 + 120t$

(b)   (c) -8 ft/sec

(d) The slope of the secant line is negative.
(e) $y = -8t + 240$
(f)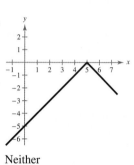

71. Even; y-axis symmetry   **73.** Neither; no symmetry
75. Neither; no symmetry
77. 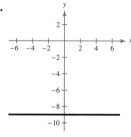   **79.**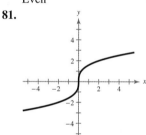

Even   Neither

81.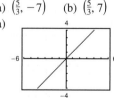

Odd

83. $h = 3 - 4x + x^2$ **85.** $L = 2 - \sqrt[3]{2y}$
87. The negative symbol should be divided out of each term, which yields $f(-x) = -(2x^3 + 5)$. So, the function is neither even nor odd.
89. (a) Ten thousands   (b) Ten millions   (c) Tens
  (d) Ones
91. False. The function $f(x) = \sqrt{x^2 + 1}$ has a domain of all real numbers.
93. True. A graph that is symmetric with respect to the y-axis cannot be increasing on its entire domain.
95. (a) $\left(\frac{5}{3}, -7\right)$ (b) $\left(\frac{5}{3}, 7\right)$
97. (a) 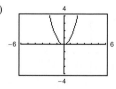   (b)

(c)
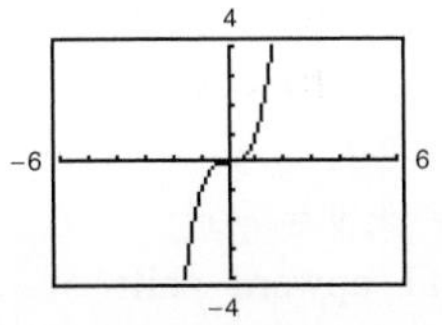
(d)
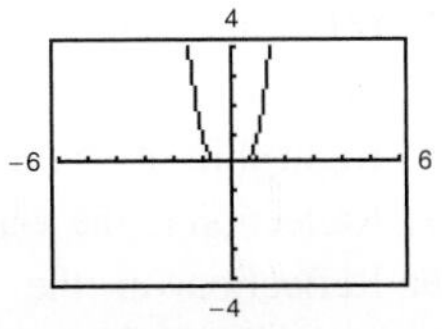
(e)
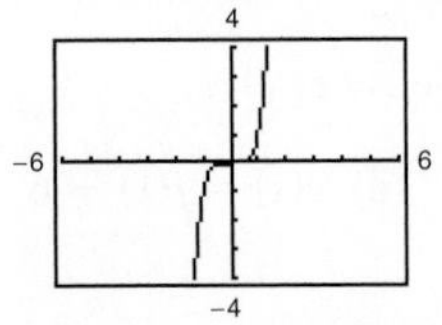
(f)
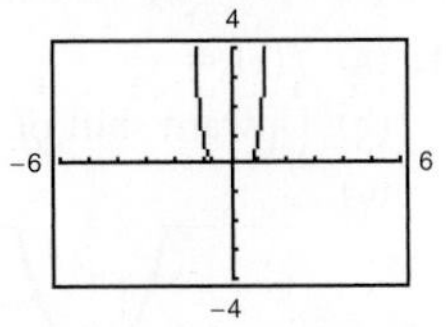

All the graphs pass through the origin. The graphs of the odd powers of x are symmetric with respect to the origin, and the graphs of the even powers are symmetric with respect to the y-axis. As the powers increase, the graphs become flatter in the interval $-1 < x < 1$.

99. (a) Even. The graph is a reflection in the x-axis.
(b) Even. The graph is a reflection in the y-axis.
(c) Even. The graph is a downward shift of f.
(d) Neither. The graph is a right shift of f.

Section 1.6 *(page 65)*

1. Greatest integer function **3.** Reciprocal function
5. Square root function **7.** Absolute value function
9. Linear function
11. (a) $f(x) = -2x + 6$
(b)
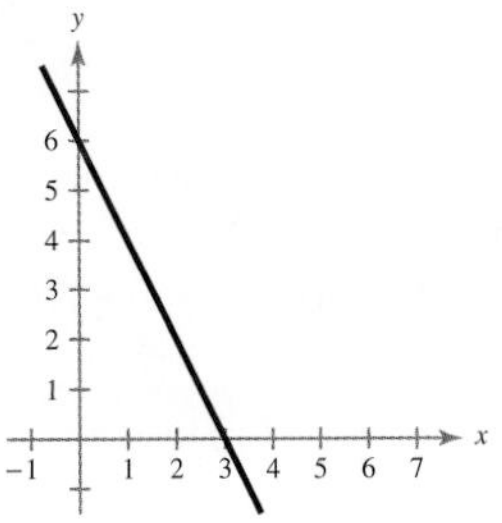

13. (a) $f(x) = \frac{2}{3}x - 2$
(b)
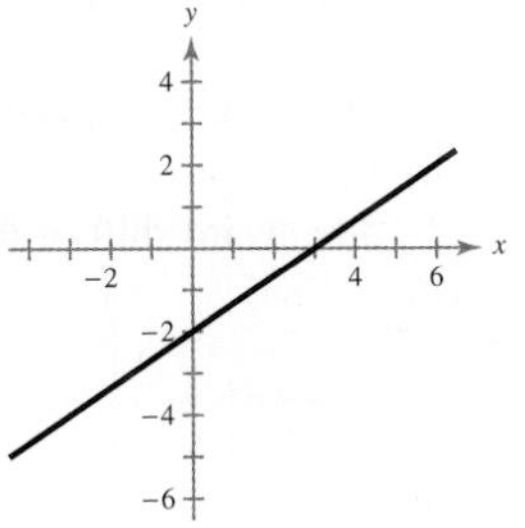

15.
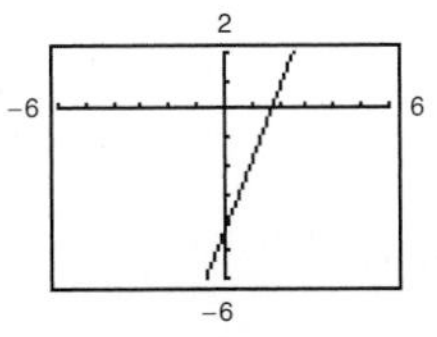

17.
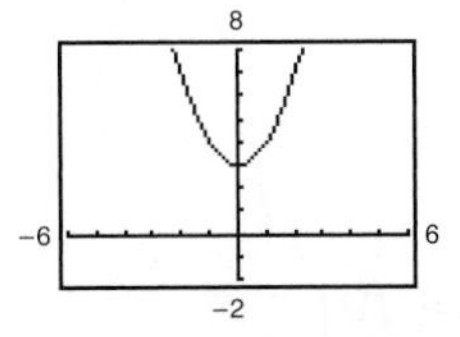

19.
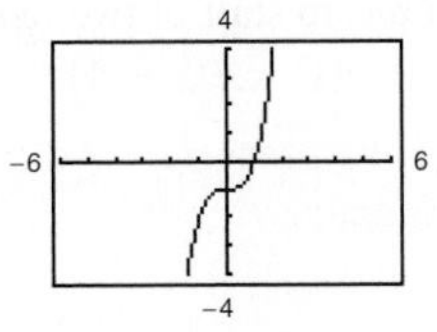

21.
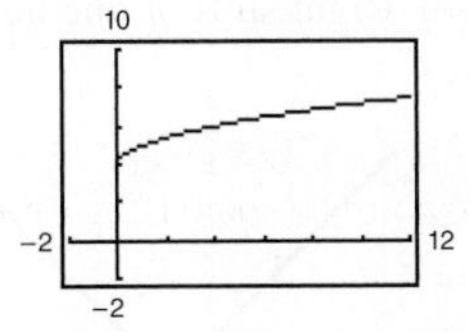

23.
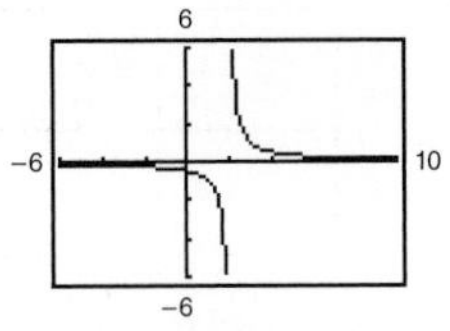

25.
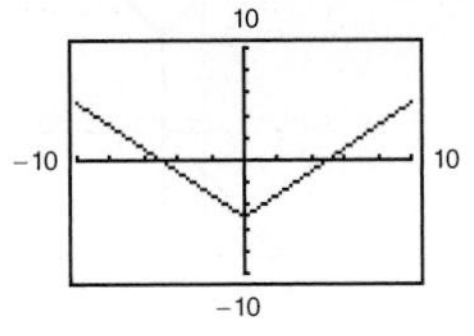

27. (a) 2 (b) 2 (c) -4 (d) 3
29. (a) 1 (b) -4 (c) 3 (d) 2

31.
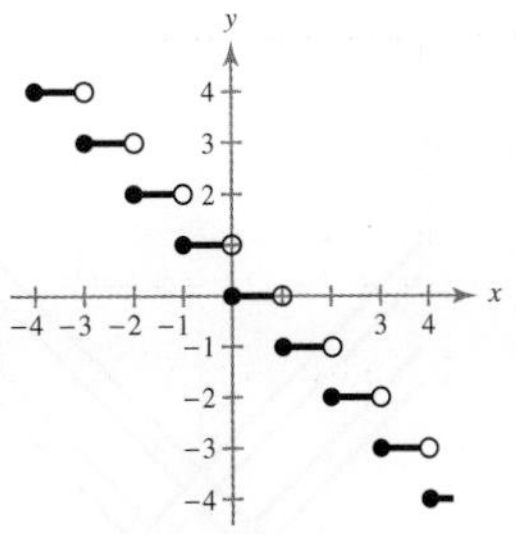

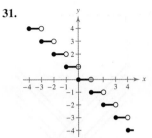

33.
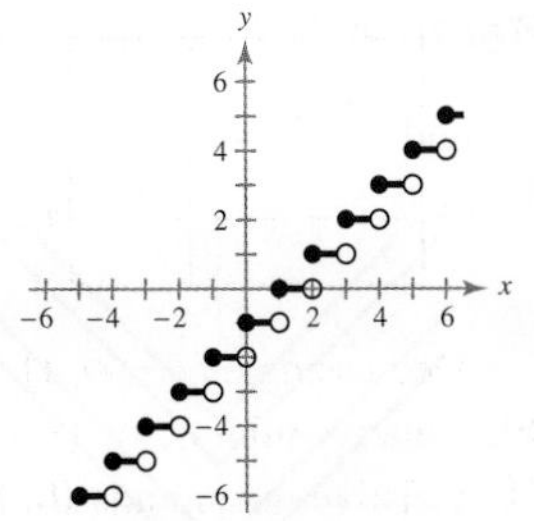

35.
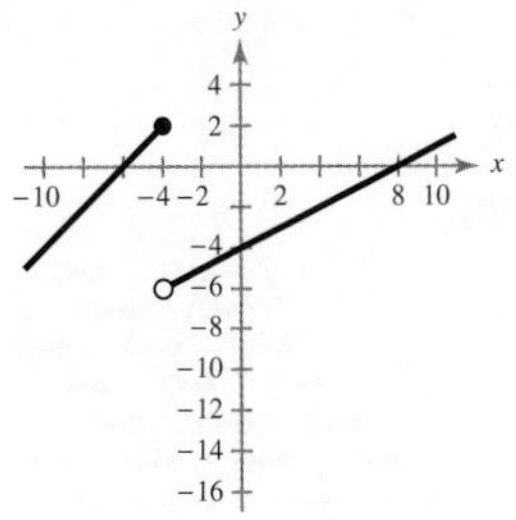

37.
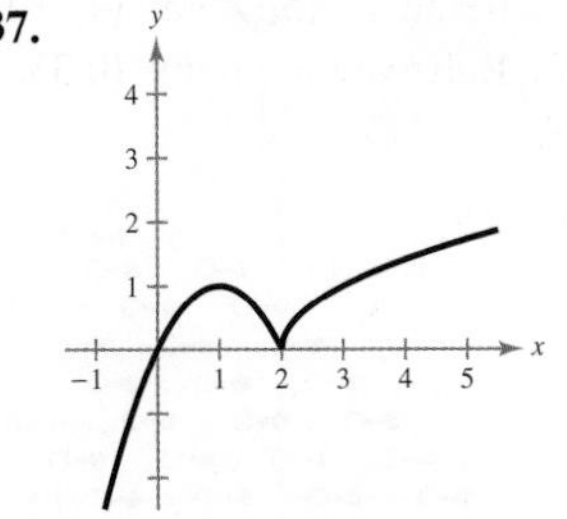

39.
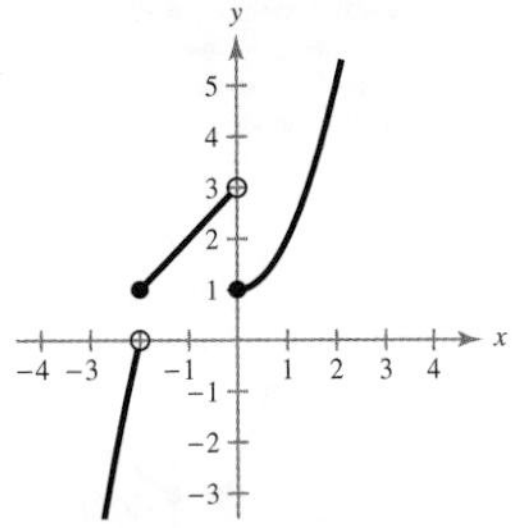

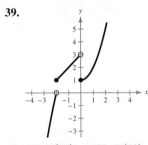

41. (a)
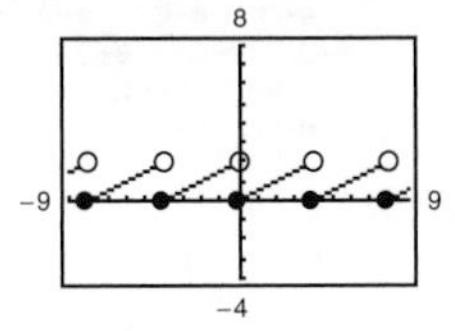

(b) Domain: $(-\infty, \infty)$
Range: $[0, 2)$

43. (a) $W(30) = 420$; $W(40) = 560$; $W(45) = 665$; $W(50) = 770$

(b) $W(h) = \begin{cases} 14h, & 0 < h \leq 36 \\ 21(h - 36) + 504, & h > 36 \end{cases}$

(c) $W(h) = \begin{cases} 16h, & 0 < h \leq 40 \\ 24(h - 40) + 640, & h > 40 \end{cases}$

45.

Interval	Input Pipe	Drain Pipe 1	Drain Pipe 2
[0, 5]	Open	Closed	Closed
[5, 10]	Open	Open	Closed
[10, 20]	Closed	Closed	Closed
[20, 30]	Closed	Closed	Open
[30, 40]	Open	Open	Open
[40, 45]	Open	Closed	Open
[45, 50]	Open	Open	Open
[50, 60]	Open	Open	Closed

47. $f(t) = \begin{cases} t, & 0 \leq t \leq 2 \\ 2t - 2, & 2 < t \leq 8 \\ \frac{1}{2}t + 10, & 8 < t \leq 9 \end{cases}$

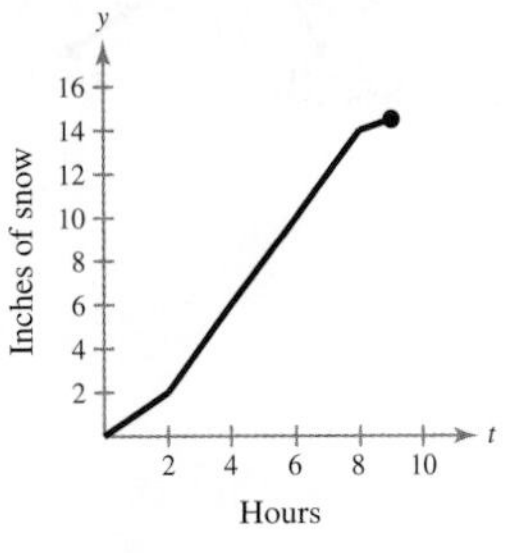

Total accumulation = 14.5 in.

49. False. A piecewise-defined function is a function that is defined by two or more equations over a specified domain. That domain may or may not include x- and y-intercepts.

Section 1.7   *(page 72)*

1. rigid   **3.** vertical stretch; vertical shrink

5. (a)   (b)

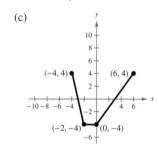

7. (a)   (b)

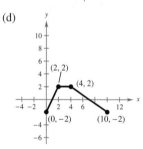

9. (a)   (b)

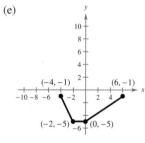

(c)   (d)

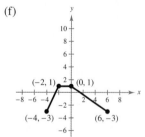

(e)   (f)

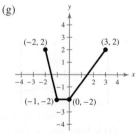

(g)

11. (a) $y = x^2 - 1$ (b) $y = -(x + 1)^2 + 1$

13. (a) $y = -|x + 3|$ (b) $y = |x - 2| - 4$

15. Right shift of $y = x^3$; $y = (x - 2)^3$

17. Reflection in the x-axis of $y = x^2$; $y = -x^2$

19. Reflection in the x-axis and upward shift of $y = \sqrt{x}$;
$y = 1 - \sqrt{x}$

21. (a) $f(x) = x^2$
(b) Upward shift of six units
(c)   (d) $g(x) = f(x) + 6$

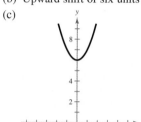

23. (a) $f(x) = x^3$
(b) Reflection in the x-axis and a right shift of two units
(c)   (d) $g(x) = -f(x - 2)$

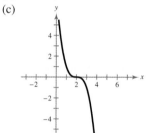

25. (a) $f(x) = x^2$
(b) Reflection in the x-axis, a left shift of one unit, and a downward shift of three units
(c)   (d) $g(x) = -3 - f(x + 1)$

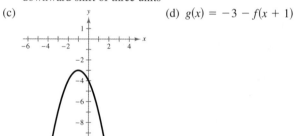

27. (a) $f(x) = |x|$
(b) Right shift of one unit and an upward shift of two units
(c)   (d) $g(x) = f(x - 1) + 2$

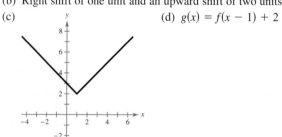

29. (a) $f(x) = \sqrt{x}$ (b) Vertical stretch

(c)   (d) $g(x) = 2f(x)$

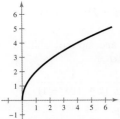

31. (a) $f(x) = [\![x]\!]$

(b) Vertical stretch and a downward shift of one unit

(c)   (d) $g(x) = 2f(x) - 1$

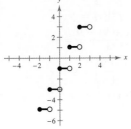

33. (a) $f(x) = |x|$

(b) Horizontal shrink

(c)   (d) $g(x) = f(2x)$

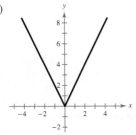

35. (a) $f(x) = x^2$

(b) Reflection in the x-axis, a vertical stretch, and an upward shift of one unit

(c)   (d) $g(x) = -2f(x) + 1$

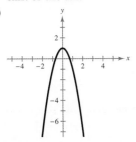

37. (a) $f(x) = |x|$

(b) Vertical stretch, a right shift of one unit, and an upward shift of two units

(c)   (d) $g(x) = 3f(x - 1) + 2$

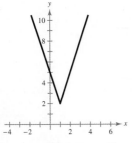

39. $g(x) = (x - 3)^2 - 7$ **41.** $g(x) = (x - 13)^3$

43. $g(x) = -|x| - 12$ **45.** $g(x) = -\sqrt{-x + 6}$

47. (a) $y = -3x^2$ (b) $y = 4x^2 + 3$

49. (a) $y = -\frac{1}{2}|x|$ (b) $y = 3|x| - 3$

51. Vertical stretch of $y = x^3$; $y = 2x^3$

53. Reflection in the x-axis and vertical shrink of $y = x^2$; $y = -\frac{1}{2}x^2$

55. Reflection in the y-axis and vertical shrink of $y = \sqrt{x}$; $y = \frac{1}{2}\sqrt{-x}$

57. $y = -(x - 2)^3 + 2$ **59.** $y = -\sqrt{x} - 3$

61. (a)

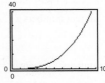

(b) $H\left(\dfrac{x}{1.6}\right) = 0.00001132x^3$; Horizontal stretch

63. False. The graph of $y = f(-x)$ is a reflection of the graph of $f(x)$ in the y-axis.

65. True. $|-x| = |x|$ **67.** $(-2, 0), (-1, 1), (0, 2)$

69. The equation should be $g(x) = (x - 1)^3$.

71. (a) $g(t) = \frac{3}{4}f(t)$ (b) $g(t) = f(t) + 10{,}000$

(c) $g(t) = f(t - 2)$

Section 1.8   *(page 81)*

1. addition; subtraction; multiplication; division

3.

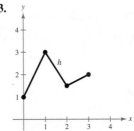

5. (a) $2x$ (b) 4 (c) $x^2 - 4$

(d) $\dfrac{x + 2}{x - 2}$; all real numbers x except $x = 2$

7. (a) $x^2 + 4x - 5$ (b) $x^2 - 4x + 5$ (c) $4x^3 - 5x^2$

(d) $\dfrac{x^2}{4x - 5}$; all real numbers x except $x = \dfrac{5}{4}$

9. (a) $x^2 + 6 + \sqrt{1 - x}$ (b) $x^2 + 6 - \sqrt{1 - x}$

(c) $(x^2 + 6)\sqrt{1 - x}$

(d) $\dfrac{(x^2 + 6)\sqrt{1 - x}}{1 - x}$; all real numbers x such that $x < 1$

11. (a) $\dfrac{x^4 + x^3 + x}{x + 1}$ (b) $\dfrac{-x^4 - x^3 + x}{x + 1}$ (c) $\dfrac{x^4}{x + 1}$

(d) $\dfrac{1}{x^2(x + 1)}$; all real numbers x except $x = 0, -1$

13. 7 **15.** 5 **17.** $-9t^2 + 3t + 5$ **19.** 306

21. $\frac{8}{23}$ **23.** -9

25.   **27.**

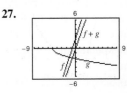

$f(x), g(x)$ $f(x), f(x)$

29. (a) $x + 5$ (b) $x + 5$ (c) $x - 6$
31. (a) $(x - 1)^2$ (b) $x^2 - 1$ (c) $x - 2$
33. (a) x (b) x (c) $x^9 + 3x^6 + 3x^3 + 2$
35. (a) $\sqrt{x^2 + 4}$ (b) $x + 4$
Domains of f and $g \circ f$: all real numbers x such that $x \geq -4$
Domains of g and $f \circ g$: all real numbers x
37. (a) x^2 (b) x^2
Domains of $f, g, f \circ g$, and $g \circ f$: all real numbers x
39. (a) $|x + 6|$ (b) $|x| + 6$
Domains of $f, g, f \circ g$, and $g \circ f$: all real numbers x
41. (a) $\dfrac{1}{x + 3}$ (b) $\dfrac{1}{x} + 3$
Domains of f and $g \circ f$: all real numbers x except $x = 0$
Domain of g: all real numbers x
Domain of $f \circ g$: all real numbers x except $x = -3$
43. (a)   (b)

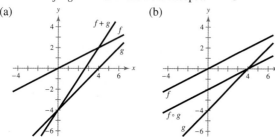

45. (a) 3   (b) 0   **47.** (a) 0   (b) 4
49. $f(x) = x^2,\ g(x) = 2x + 1$ **51.** $f(x) = \sqrt[3]{x},\ g(x) = x^2 - 4$
53. $f(x) = \dfrac{1}{x},\ g(x) = x + 2$ **55.** $f(x) = \dfrac{x + 3}{4 + x},\ g(x) = -x^2$
57. (a) $T = \dfrac{3}{4}x + \dfrac{1}{15}x^2$
(b)

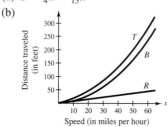

(c) The braking function $B(x)$; As x increases, $B(x)$ increases at a faster rate than $R(x)$.
59. (a) $c(t) = \dfrac{b(t) - d(t)}{p(t)} \times 100$
(b) $c(16)$ is the percent change in the population due to births and deaths in the year 2016.
61. (a) $r(x) = \dfrac{x}{2}$ (b) $A(r) = \pi r^2$
(c) $(A \circ r)(x) = \pi\left(\dfrac{x}{2}\right)^2$;
$(A \circ r)(x)$ represents the area of the circular base of the tank on the square foundation with side length x.
63. $g(f(x))$ represents 3 percent of an amount over \$500,000.
65. False. $(f \circ g)(x) = 6x + 1$ and $(g \circ f)(x) = 6x + 6$
67. (a) $O(M(Y)) = 2\left(6 + \dfrac{1}{2}Y\right) = 12 + Y$
(b) Middle child is 8 years old; youngest child is 4 years old.
69. Proof
71. (a) *Sample answer:* $f(x) = x + 1,\ g(x) = x + 3$
(b) *Sample answer:* $f(x) = x^2,\ g(x) = x^3$

73. (a) Proof
(b) $\dfrac{1}{2}[f(x) + f(-x)] + \dfrac{1}{2}[f(x) - f(-x)]$
$= \dfrac{1}{2}[f(x) + f(-x) + f(x) - f(-x)]$
$= \dfrac{1}{2}[2f(x)]$
$= f(x)$
(c) $f(x) = (x^2 + 1) + (-2x)$
$k(x) = \dfrac{-1}{(x + 1)(x - 1)} + \dfrac{x}{(x + 1)(x - 1)}$

Section 1.9   *(page 90)*

1. inverse   **3.** range; domain   **5.** one-to-one
7. $f^{-1}(x) = \dfrac{1}{6}x$ **9.** $f^{-1}(x) = \dfrac{x - 1}{3}$
11. $f^{-1}(x) = \sqrt{x + 4}$ **13.** $f^{-1}(x) = \sqrt[3]{x - 1}$
15. $f(g(x)) = f(4x + 9) = \dfrac{(4x + 9) - 9}{4} = x$
$g(f(x)) = g\left(\dfrac{x - 9}{4}\right) = 4\left(\dfrac{x - 9}{4}\right) + 9 = x$
17. $f(g(x)) = f(\sqrt[3]{4x}) = \dfrac{(\sqrt[3]{4x})^3}{4} = x$
$g(f(x)) = g\left(\dfrac{x^3}{4}\right) = \sqrt[3]{4\left(\dfrac{x^3}{4}\right)} = x$
19.

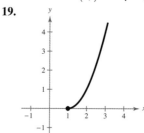

21. (a) $f(g(x)) = f(x + 5) = (x + 5) - 5 = x$
$g(f(x)) = g(x - 5) = (x - 5) + 5 = x$
(b)

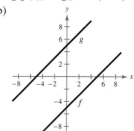

23. (a) $f(g(x)) = f\left(\dfrac{x - 1}{7}\right) = 7\left(\dfrac{x - 1}{7}\right) + 1 = x$
$g(f(x)) = g(7x + 1) = \dfrac{(7x + 1) - 1}{7} = x$
(b)

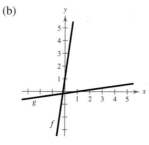

25. (a) $f(g(x)) = f(\sqrt[3]{x}) = (\sqrt[3]{x})^3 = x$
$g(f(x)) = g(x^3) = \sqrt[3]{(x^3)} = x$

(b)

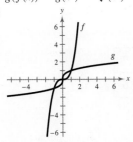

27. (a) $f(g(x)) = f(x^2 - 5) = \sqrt{(x^2 - 5) + 5} = x, \ x \geq 0$
$g(f(x)) = g(\sqrt{x + 5}) = (\sqrt{x + 5})^2 - 5 = x$

(b)

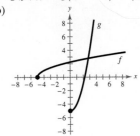

29. (a) $f(g(x)) = f\left(\dfrac{1}{x}\right) = \dfrac{1}{(1/x)} = x$

$g(f(x)) = g\left(\dfrac{1}{x}\right) = \dfrac{1}{(1/x)} = x$

(b)

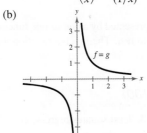

31. (a) $f(g(x)) = f\left(-\dfrac{5x + 1}{x - 1}\right) = \dfrac{-\left(\dfrac{5x + 1}{x - 1}\right) - 1}{-\left(\dfrac{5x + 1}{x - 1}\right) + 5}$

$= \dfrac{-5x - 1 - x + 1}{-5x - 1 + 5x - 5} = x$

$g(f(x)) = g\left(\dfrac{x - 1}{x + 5}\right) = \dfrac{-5\left(\dfrac{x - 1}{x + 5}\right) - 1}{\dfrac{x - 1}{x + 5} - 1}$

$= \dfrac{-5x + 5 - x - 5}{x - 1 - x - 5} = x$

(b)

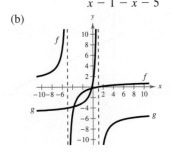

33. No

35.

x	3	5	7	9	11	13
$f^{-1}(x)$	−1	0	1	2	3	4

37. Yes   **39.** No

41.

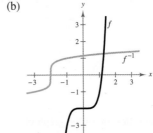

The function does not have an inverse function.

43.

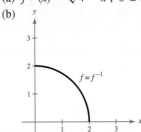

The function does not have an inverse function.

45. (a) $f^{-1}(x) = \sqrt[5]{x + 2}$

(b)

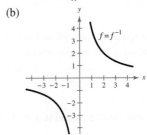

(c) The graph of f^{-1} is the reflection of the graph of f in the line $y = x$.

(d) The domains and ranges of f and f^{-1} are all real numbers x.

47. (a) $f^{-1}(x) = \sqrt{4 - x^2}, \ 0 \leq x \leq 2$

(b)

(c) The graph of f^{-1} is the same as the graph of f.

(d) The domains and ranges of f and f^{-1} are all real numbers x such that $0 \leq x \leq 2$.

49. (a) $f^{-1}(x) = \dfrac{4}{x}$

(b)

(c) The graph of f^{-1} is the same as the graph of f.

(d) The domains and ranges of f and f^{-1} are all real numbers x except $x = 0$.

CHAPTER 1

51. (a) $f^{-1}(x) = \dfrac{2x + 1}{x - 1}$

(b)
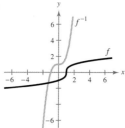

(c) The graph of f^{-1} is the reflection of the graph of f in the line $y = x$.

(d) The domain of f and the range of f^{-1} are all real numbers x except $x = 2$. The domain of f^{-1} and the range of f are all real numbers x except $x = 1$.

53. (a) $f^{-1}(x) = x^3 + 1$

(b)
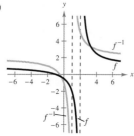

(c) The graph of f^{-1} is the reflection of the graph of f in the line $y = x$.

(d) The domains and ranges of f and f^{-1} are all real numbers x.

55. No inverse function   **57.** $g^{-1}(x) = 6x - 1$

59. No inverse function   **61.** $f^{-1}(x) = \sqrt{x} - 3$

63. No inverse function   **65.** No inverse function

67. $f^{-1}(x) = \dfrac{x^2 - 3}{2}$, $x \geq 0$ **69.** $f^{-1}(x) = \dfrac{5x - 4}{6 - 4x}$

71. $f^{-1}(x) = x - 2$

The domain of f and the range of f^{-1} are all real numbers x such that $x \geq -2$. The domain of f^{-1} and the range of f are all real numbers x such that $x \geq 0$.

73. $f^{-1}(x) = \sqrt{x} - 6$

The domain of f and the range of f^{-1} are all real numbers x such that $x \geq -6$. The domain of f^{-1} and the range of f are all real numbers x such that $x \geq 0$.

75. $f^{-1}(x) = \dfrac{\sqrt{-2(x - 5)}}{2}$

The domain of f and the range of f^{-1} are all real numbers x such that $x \geq 0$. The domain of f^{-1} and the range of f are all real numbers x such that $x \leq 5$.

77. $f^{-1}(x) = x + 3$

The domain of f and the range of f^{-1} are all real numbers x such that $x \geq 4$. The domain of f^{-1} and the range of f are all real numbers x such that $x \geq 1$.

79. 32   **81.** 472   **83.** $2\sqrt[3]{x + 3}$

85. $\dfrac{x + 1}{2}$ **87.** $\dfrac{x + 1}{2}$

89. (a) $y = \dfrac{x - 10}{0.75}$

$x = $ hourly wage; $y = $ number of units produced

(b) 19 units

91. False. $f(x) = x^2$ has no inverse function.

93.

x	1	3	4	6
y	1	2	6	7

x	1	2	6	7
$f^{-1}(x)$	1	3	4	6

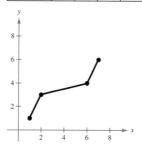

95. Proof   **97.** $k = \frac{1}{4}$

99.
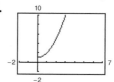

There is an inverse function $f^{-1}(x) = \sqrt{x - 1}$ because the domain of f is equal to the range of f^{-1} and the range of f is equal to the domain of f^{-1}.

101. This situation could be represented by a one-to-one function if the runner does not stop to rest. The inverse function would represent the time in hours for a given number of miles completed.

Section 1.10   *(page 100)*

1. variation; regression   **3.** least squares regression

5. directly proportional   **7.** combined

9.
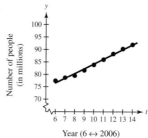

Year (6 ↔ 2006)

The model is a good fit for the data.

11.
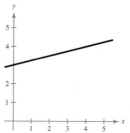

$y = \frac{1}{4}x + 3$

13.
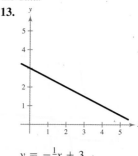

$y = -\frac{1}{2}x + 3$

15.

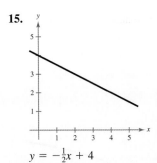

$y = -\frac{1}{2}x + 4$

17. (a) and (b)

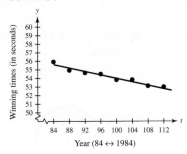

Year (84 ↔ 1984)

Sample answer: $y = -0.1t + 64$

(c) $y = -0.097t + 63.72$ (d) The models are similar.

19. $y = 7x$ **21.** $y = \frac{1}{5}x$ **23.** $y = 2\pi x$

25.

x	2	4	6	8	10
$y = x^2$	4	16	36	64	100

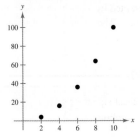

27.

x	2	4	6	8	10
$y = \frac{1}{2}x^3$	4	32	108	256	500

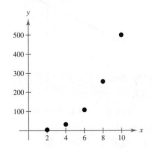

29.

x	2	4	6	8	10
$y = \dfrac{2}{x}$	1	$\frac{1}{2}$	$\frac{1}{3}$	$\frac{1}{4}$	$\frac{1}{5}$

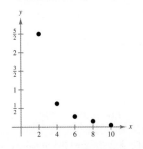

31.

x	2	4	6	8	10
$y = \dfrac{10}{x^2}$	$\frac{5}{2}$	$\frac{5}{8}$	$\frac{5}{18}$	$\frac{5}{32}$	$\frac{1}{10}$

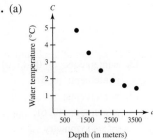

33. Inversely; Answers will vary.   **35.** $y = \dfrac{5}{x}$

37. $y = -\dfrac{7}{10}x$ **39.** $A = kr^2$ **41.** $y = \dfrac{k}{x^2}$

43. $F = \dfrac{kg}{r^2}$ **45.** $R = k(T - T_e)$ **47.** $P = kVI$

49. y is directly proportional to the square of x.

51. A is jointly proportional to b and h.

53. $y = 18x$ **55.** $y = \dfrac{75}{x}$ **57.** $z = 2xy$ **59.** $P = \dfrac{18x}{y^2}$

61. $I = 0.035P$ **63.** Model: $y = \frac{33}{13}x$; 25.4 cm, 50.8 cm

65. $293\frac{1}{3}$N   **67.** About 39.47 lb   **69.** About 0.61 mi/h

71. (a)

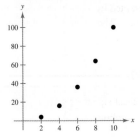

Depth (in meters)

(b) Inverse variation

(c) About 4919.9

(d) 1640 m

73. (a) 200 Hz   (b) 50 Hz   (c) 100 Hz

75. True. If $y = k_1x$ and $x = k_2z$, then $y = k_1(k_2z) = (k_1k_2)z$.

77. π is a constant, not a variable.   **79.** Direct; $y = 2t$

CHAPTER 1

Review Exercises   *(page 106)*

1.

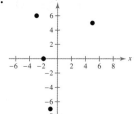

3. Quadrant IV

5.

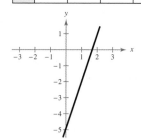

Distance: $3\sqrt{13}$
Midpoint: $\left(1, \frac{3}{2}\right)$

7.

x	-2	-1	0	1	2
y	-11	-8	-5	-2	1

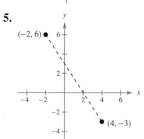

9.

x	-1	0	1	2	3	4
y	4	0	-2	-2	0	4

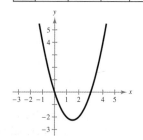

11. x-intercept: $\left(-\frac{7}{2}, 0\right)$
y-intercept: $(0, 7)$

13. x-intercepts: $(1, 0), (5, 0)$
y-intercept: $(0, 5)$

15. x-intercept: $\left(\frac{1}{4}, 0\right)$
y-intercept: $(0, 1)$
No symmetry

17. x-intercepts: $\left(\pm\sqrt{6}, 0\right)$
y-intercept: $(0, 6)$
y-axis symmetry

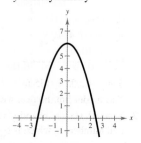

19. x-intercept: $\left(\sqrt[3]{-5}, 0\right)$
y-intercept: $(0, 5)$
No symmetry

21. x-intercept: $(-5, 0)$
y-intercept: $\left(0, \sqrt{5}\right)$
No symmetry

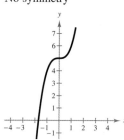

23. Center: $(0, 0)$
Radius: 3

25. Center: $(-2, 0)$
Radius: 4

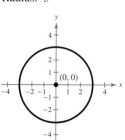

27. $(x - 2)^2 + (y + 3)^2 = 13$

29. $m = -\frac{1}{2}$
y-intercept: $(0, 1)$

31. $m = 0$
y-intercept: $(0, 1)$

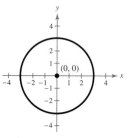

33. $m = -1$

35. $y = \frac{1}{3}x - 7$

37. $y = \frac{1}{2}x + 7$

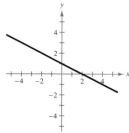

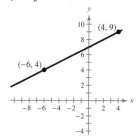

39. (a) $y = \frac{5}{4}x - \frac{23}{4}$ (b) $y = -\frac{4}{5}x + \frac{2}{5}$ **41.** $S = 0.80L$

43. Not a function   **45.** Function

47. (a) 5 (b) 17 (c) $t^4 + 1$ (d) $t^2 + 2t + 2$

49. All real numbers x such that $-5 \le x \le 5$

51. 16 ft/sec   **53.** $4x + 2h + 3$, $h \ne 0$

55. Function   **57.** $\frac{7}{3}, 3$

59.

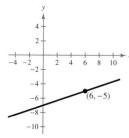

Increasing on $(0, \infty)$
Decreasing on $(-\infty, -1)$
Constant on $(-1, 0)$

61. Relative maximum: $(1, 2)$ **63.** 4

65. Even; y-axis symmetry

67. (a) $f(x) = -3x$ **69.**

(b)

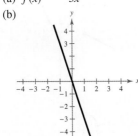

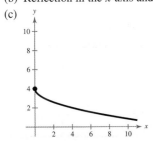

71. (a) $f(x) = x^2$ (b) Downward shift of nine units

(c) (d) $h(x) = f(x) - 9$

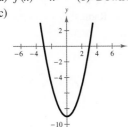

73. (a) $f(x) = \sqrt{x}$

(b) Reflection in the x-axis and an upward shift of four units

(c) (d) $h(x) = -f(x) + 4$

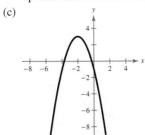

75. (a) $f(x) = x^2$

(b) Reflection in the x-axis, a left shift of two units, and an upward shift of three units

(c) (d) $h(x) = -f(x + 2) + 3$

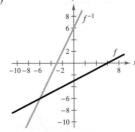

77. (a) $f(x) = [\![x]\!]$

(b) Reflection in the x-axis and an upward shift of six units

(c) (d) $h(x) = -f(x) + 6$

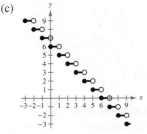

79. (a) $f(x) = [\![x]\!]$

(b) Right shift of nine units and a vertical stretch

(c) (d) $h(x) = 5f(x - 9)$

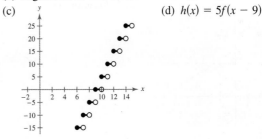

81. (a) $x^2 + 2x + 2$ (b) $x^2 - 2x + 4$

(c) $2x^3 - x^2 + 6x - 3$

(d) $\dfrac{x^2 + 3}{2x - 1}$; all real numbers x except $x = \dfrac{1}{2}$

83. (a) $x - \frac{8}{3}$ (b) $x - 8$

Domains of f, g, $f \circ g$, and $g \circ f$: all real numbers x

85. $f(g(x)) = 0.95x - 100$; $(f \circ g)(x)$ represents the 5% discount before the \$100 rebate.

87. $f^{-1}(x) = \dfrac{x - 8}{3}$

$$f(f^{-1}(x)) = 3\left(\dfrac{x - 8}{3}\right) + 8 = x$$

$$f^{-1}(f(x)) = \dfrac{3x + 8 - 8}{3} = x$$

89. The function does not have an inverse function.

91. (a) $f^{-1}(x) = 2x + 6$

(b)

(c) The graphs are reflections of each other in the line $y = x$.

(d) Both f and f^{-1} have domains and ranges that are all real number x.

93. $x > 4$; $f^{-1}(x) = \sqrt{\dfrac{x}{2} + 4}$, $x \neq 0$

95. (a) and (b) **97.** \$44.80

$B = -5.02t + 135.6$

The model fits the data well.

99. True. If $f(x) = x^3$ and $g(x) = \sqrt[3]{x}$, then the domain of g is all real numbers x, which is equal to the range of f, and vice versa.

CHAPTER 1

Chapter Test   *(page 109)*

1.

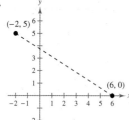

Midpoint: $\left(2, \frac{5}{2}\right)$; Distance: $\sqrt{89}$

2. $V(h) = 16\pi h$

3. x-intercept: $\left(\frac{3}{5}, 0\right)$
y-intercept: $(0, 3)$
No symmetry

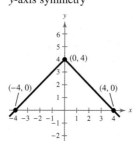

4. x-intercepts: $(\pm 4, 0)$
y-intercept: $(0, 4)$
y-axis symmetry

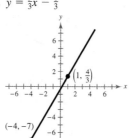

5. x-intercepts: $(\pm 1, 0)$
y-intercept: $(0, -1)$
y-axis symmetry

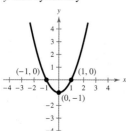

6. $(x - 1)^2 + (y - 3)^2 = 16$

7. $y = -4x - 3$

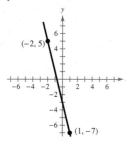

8. $y = \frac{5}{3}x - \frac{1}{3}$

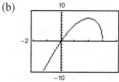

9. (a) $y = -\frac{5}{2}x + 4$ (b) $y = \frac{2}{5}x + 4$

10. (a) $-\dfrac{1}{8}$ (b) $-\dfrac{1}{28}$ (c) $\dfrac{\sqrt{x}}{x^2 - 18x}$ **11.** $x \leq 3$

12. (a) -5
(b)

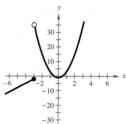

(c) Increasing on $(-5, \infty)$
Decreasing on $(-\infty, -5)$
(d) Neither

13. (a) $0, 3$
(b)

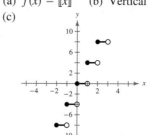

(c) Increasing on $(-\infty, 2)$
Decreasing on $(2, 3)$
(d) Neither

14. (a) $0, \pm 0.4314$
(b)

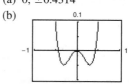

(c) Increasing on $(-0.31, 0)$,
$(0.31, \infty)$
Decreasing on $(-\infty, -0.31)$,
$(0, 0.31)$
(d) Even

15.

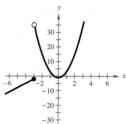

16. (a) $f(x) = [\![x]\!]$ (b) Vertical stretch
(c)

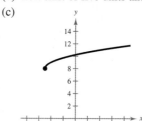

17. (a) $y = \sqrt{x}$
(b) Left shift of five units and an upward shift of eight units
(c)

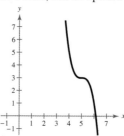

18. (a) $y = x^3$
(b) Reflection in the x-axis, a vertical stretch, a right shift of five units, and an upward shift of three units

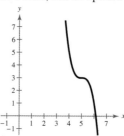

19. (a) $2x^2 - 4x - 2$ (b) $4x^2 + 4x - 12$
(c) $-3x^4 - 12x^3 + 22x^2 + 28x - 35$
(d) $\dfrac{3x^2 - 7}{-x^2 - 4x + 5}$
(e) $3x^4 + 24x^3 + 18x^2 - 120x + 68$
(f) $-9x^4 + 30x^2 - 16$

20. (a) $\dfrac{1 + 2x^{3/2}}{x}$ (b) $\dfrac{1 - 2x^{3/2}}{x}$
(c) $\dfrac{2\sqrt{x}}{x}$ (d) $\dfrac{1}{2x^{3/2}}$
(e) $\dfrac{\sqrt{x}}{2x}$ (f) $\dfrac{2\sqrt{x}}{x}$

21. $f^{-1}(x) = \sqrt[3]{x - 8}$ **22.** No inverse

23. $f^{-1}(x) = \left(\frac{1}{3}x\right)^{2/3}$, $x \geq 0$ **24.** $v = 6\sqrt{s}$

25. $A = \dfrac{25}{6}xy$ **26.** $b = \dfrac{48}{a}$

Problem Solving *(page 111)*

1. (a) $W_1 = 2000 + 0.07S$ (b) $W_2 = 2300 + 0.05S$

(c)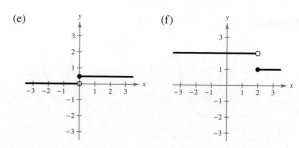

Both jobs pay the same monthly salary when sales equal $15,000.

(d) No. Job 1 would pay $3400 and job 2 would pay $3300.

3. (a) The function will be even.

(b) The function will be odd.

(c) The function will be neither even nor odd.

5. $f(x) = a_{2n}x^{2n} + a_{2n-2}x^{2n-2} + \cdots + a_2x^2 + a_0$
$f(-x) = a_{2n}(-x)^{2n} + a_{2n-2}(-x)^{2n-2}$
$\qquad\qquad\qquad + \cdots + a_2(-x)^2 + a_0$
$\quad = f(x)$

7. (a) $81\frac{2}{3}$ h

(b) $25\frac{5}{7}$ mi/h

(c) $y = \dfrac{-180}{7}x + 3400$

Domain: $0 \leq x \leq \dfrac{1190}{9}$

Range: $0 \leq y \leq 3400$

(d)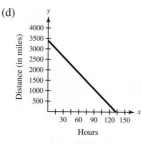

9. (a) $(f \circ g)(x) = 4x + 24$ (b) $(f \circ g)^{-1}(x) = \frac{1}{4}x - 6$

(c) $f^{-1}(x) = \frac{1}{4}x$; $g^{-1}(x) = x - 6$

(d) $(g^{-1} \circ f^{-1})(x) = \frac{1}{4}x - 6$; They are the same.

(e) $(f \circ g)(x) = 8x^3 + 1$; $(f \circ g)^{-1}(x) = \frac{1}{2}\sqrt[3]{x - 1}$;
$f^{-1}(x) = \sqrt[3]{x - 1}$; $g^{-1}(x) = \frac{1}{2}x$;
$(g^{-1} \circ f^{-1})(x) = \frac{1}{2}\sqrt[3]{x - 1}$

(f) Answers will vary.

(g) $(f \circ g)^{-1}(x) = (g^{-1} \circ f^{-1})(x)$

11. (a) 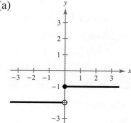   (b)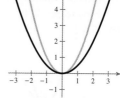

(c) 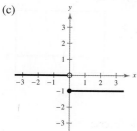   (d)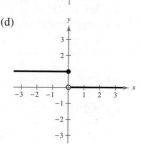

(e) 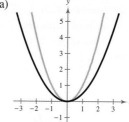   (f)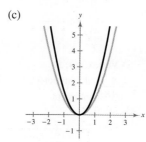

13. Proof

15. (a)

x	-4	-2	0	4
$f(f^{-1}(x))$	-4	-2	0	4

(b)

x	-3	-2	0	1
$(f + f^{-1})(x)$	5	1	-3	-5

(c)

x	-3	-2	0	1
$(f \cdot f^{-1})(x)$	4	0	2	6

(d)

x	-4	-3	0	4
$\lvert f^{-1}(x)\rvert$	2	1	1	3

Chapter 2

Section 2.1 *(page 120)*

1. polynomial   **3.** quadratic; parabola

5. b   **6.** a   **7.** c   **8.** d

9. (a) 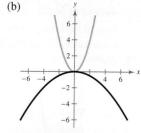   (b)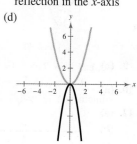

Vertical shrink

Vertical shrink and a reflection in the x-axis

(c)

Vertical stretch

(d)

Vertical stretch and a reflection in the x-axis

CHAPTER 2

11. (a)

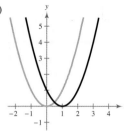

Right shift of one unit

(b)

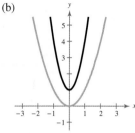

Horizontal shrink and an
upward shift of one unit

(c)

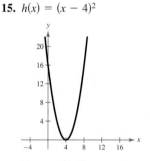

Horizontal stretch and a
downward shift of three
units

(d)

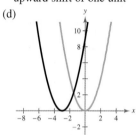

Left shift of three units

13. $f(x) = (x - 3)^2 - 9$

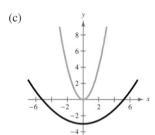

Vertex: $(3, -9)$
Axis of symmetry: $x = 3$
x-intercepts: $(0, 0), (6, 0)$

15. $h(x) = (x - 4)^2$

Vertex: $(4, 0)$
Axis of symmetry: $x = 4$
x-intercept: $(4, 0)$

17. $f(x) = (x - 3)^2 - 7$

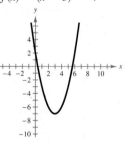

Vertex: $(3, -7)$
Axis of symmetry: $x = 3$
x-intercepts: $\left(3 \pm \sqrt{7}, 0\right)$

19. $f(x) = (x - 4)^2 + 5$

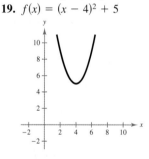

Vertex: $(4, 5)$
Axis of symmetry: $x = 4$
No x-intercept

21. $f(x) = \left(x - \frac{1}{2}\right)^2 + 1$

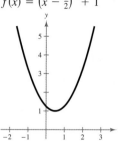

Vertex: $\left(\frac{1}{2}, 1\right)$
Axis of symmetry: $x = \frac{1}{2}$
No x-intercept

23. $f(x) = -(x - 1)^2 + 6$

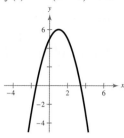

Vertex: $(1, 6)$
Axis of symmetry: $x = 1$
x-intercepts: $\left(1 \pm \sqrt{6}, 0\right)$

25. $h(x) = 4\left(x - \frac{1}{2}\right)^2 + 20$

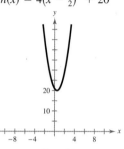

Vertex: $\left(\frac{1}{2}, 20\right)$
Axis of symmetry: $x = \frac{1}{2}$
No x-intercept

27.

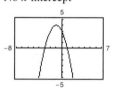

Vertex: $(-1, 4)$
Axis of symmetry: $x = -1$
x-intercepts: $(1, 0), (-3, 0)$

29.

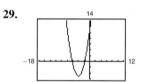

Vertex: $(-4, -5)$
Axis of symmetry: $x = -4$
x-intercepts: $\left(-4 \pm \sqrt{5}, 0\right)$

31.

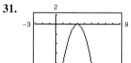

Vertex: $(3, 0)$
Axis of symmetry: $x = 3$
x-intercept: $(3, 0)$

33.

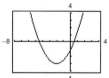

Vertex: $(-2, -3)$
Axis of symmetry: $x = -2$
x-intercepts: $\left(-2 \pm \sqrt{6}, 0\right)$

35. $f(x) = (x + 2)^2 - 1$ **37.** $f(x) = (x + 2)^2 + 5$

39. $f(x) = 4(x - 1)^2 - 2$ **41.** $f(x) = \frac{3}{4}(x - 5)^2 + 12$

43. $f(x) = -\frac{24}{49}\left(x + \frac{1}{4}\right)^2 + \frac{3}{2}$ **45.** $f(x) = -\frac{16}{3}\left(x + \frac{5}{2}\right)^2$

47. $(-1, 0), (3, 0)$ **49.** $(-3, 0), \left(\frac{1}{2}, 0\right)$

51.

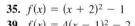

$(0, 0), (4, 0)$

53.

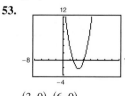

$(3, 0), (6, 0)$

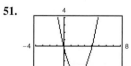

55.

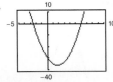

$\left(-\frac{5}{2}, 0\right), (6, 0)$

57. $f(x) = x^2 - 9$
$g(x) = -x^2 + 9$

59. $f(x) = x^2 - 3x - 4$
$g(x) = -x^2 + 3x + 4$

61. $f(x) = 2x^2 + 7x + 3$
$g(x) = -2x^2 - 7x - 3$

63. 55, 55   **65.** 12, 6   **67.** 16 ft   **69.** 20 fixtures

71. (a) $14,000,000; $14,375,000; $13,500,000
(b) $24; $14,400,000; Answers will vary.

73. (a) $A = \dfrac{8x(50 - x)}{3}$ (b) $x = 25$ ft, $y = 33\frac{1}{3}$ ft

75. True. The equation has no real solutions, so the graph has no x-intercepts.

77. $b = \pm 20$ **79.** $f(x) = a\left(x + \dfrac{b}{2a}\right)^2 + \dfrac{4ac - b^2}{4a}$

81. Proof

Section 2.2   *(page 132)*

1. continuous   **3.** $n; n - 1$ **5.** touches; crosses
7. standard   **9.** c   **10.** f   **11.** a   **12.** e
13. d   **14.** b
15. (a)

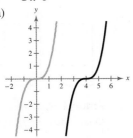

(b)

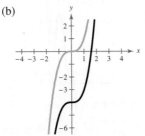

(c)

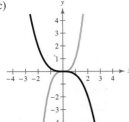

(d)

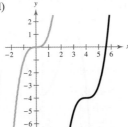

17. (a)

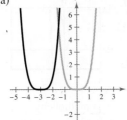

(b)

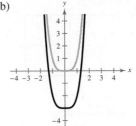

(c)

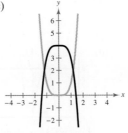

(d)

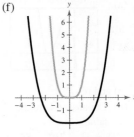

(e)

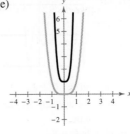

(f)

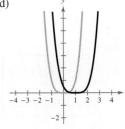

19. Falls to the left, rises to the right
21. Falls to the left and to the right
23. Rises to the left, falls to the right
25. Rises to the left and to the right
27. Rises to the left, falls to the right
29.

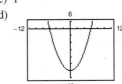

31.

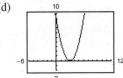

33. (a) ± 6
(b) Odd multiplicity
(c) 1
(d)

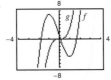

35. (a) 3
(b) Even multiplicity
(c) 1
(d)

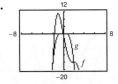

37. (a) $-2, 1$
(b) Odd multiplicity
(c) 1
(d)

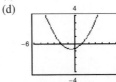

39. (a) $0, 1 \pm \sqrt{2}$
(b) Odd multiplicity
(c) 2
(d)

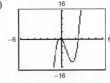

41. (a) $0, 2 \pm \sqrt{3}$
(b) Odd multiplicity
(c) 2
(d)

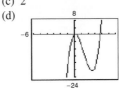

43. (a) $0, \pm\sqrt{3}$
(b) 0, odd multiplicity;
$\pm\sqrt{3}$, even multiplicity
(c) 4
(d)

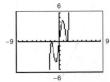

CHAPTER 2

45. (a) No real zero
(b) No multiplicity
(c) 1
(d)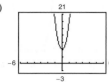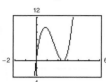

47. (a) $\pm 2, -3$
(b) Odd multiplicity
(c) 2
(d)

49. (a)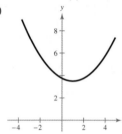
(b) and (c) $0, \frac{5}{2}$
(d) The answers are the same.

51. (a)
(b) and (c) $0, \pm 1, \pm 2$
(d) The answers are the same.

53. $f(x) = x^2 - 7x$ **55.** $f(x) = x^3 + 6x^2 + 8x$
57. $f(x) = x^4 - 4x^3 - 9x^2 + 36x$ **59.** $f(x) = x^2 - 2x - 1$
61. $f(x) = x^3 - 6x^2 + 7x + 2$ **63.** $f(x) = x^2 + 6x + 9$
65. $f(x) = x^3 + 4x^2 - 5x$
67. $f(x) = x^4 + x^3 - 15x^2 + 23x - 10$ **69.** $f(x) = x^5 - 3x^3$
71. (a) Rises to the left and to the right
(b) No zeros   (c) Answers will vary.
(d)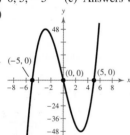

73. (a) Falls to the left, rises to the right
(b) $0, 5, -5$ (c) Answers will vary.
(d)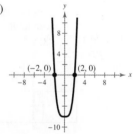

75. (a) Rises to the left and to the right
(b) $-2, 2$ (c) Answers will vary.
(d)

77. (a) Falls to the left, rises to the right
(b) $0, 2, 3$ (c) Answers will vary.
(d)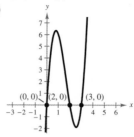

79. (a) Rises to the left, falls to the right
(b) $-5, 0$ (c) Answers will vary.
(d)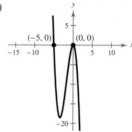

81. (a) Falls to the left, rises to the right
(b) $-2, 0$ (c) Answers will vary.
(d)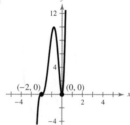

83. (a) Falls to the left and to the right
(b) ± 2 (c) Answers will vary.
(d)

85.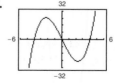
Zeros: $0, \pm 4$,
odd multiplicity

87.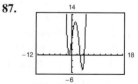
Zeros: -1,
even multiplicity;
$3, \frac{9}{2}$, odd multiplicity

89. (a) $[-1, 0], [1, 2], [2, 3]$ (b) $-0.879, 1.347, 2.532$
91. (a) $[-2, -1], [0, 1]$ (b) $-1.585, 0.779$

93. (a) $V(x) = x(36 - 2x)^2$ (b) Domain: $0 < x < 18$
(c)

6 in. $\times$ 24 in. $\times$ 24 in.

(d)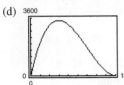

$x = 6$; The results are the same.

95. (a) Relative maximum: $(4.44, 1512.60)$
Relative minimum: $(11.97, 189.37)$
(b) Increasing: $(3, 4.44)$, $(11.97, 16)$
Decreasing: $(4.44, 11.97)$
(c) Answers will vary.

97. $x \approx 200$

99. True. A polynomial function falls to the right only when the leading coefficient is negative.

101. False. The graph falls to the left and to the right or the graph rises to the left and to the right.

103. Answers will vary. *Sample answers:*

$a_4 < 0$ $a_4 > 0$

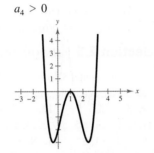

105.

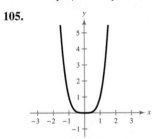

(a) Upward shift of two units; Even
(b) Left shift of two units; Neither
(c) Reflection in the y-axis; Even
(d) Reflection in the x-axis; Even
(e) Horizontal stretch; Even   (f) Vertical shrink; Even
(g) $g(x) = x^3$, $x \geq 0$; Neither   (h) $g(x) = x^{16}$; Even

107.

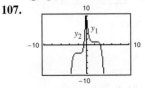

(a) y_1 is decreasing, y_2 is increasing.
(b) Yes; a; If $a > 0$, then the graph is increasing, and if $a < 0$, then the graph is decreasing.

(c)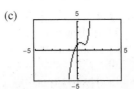

No; f is not strictly increasing or strictly decreasing, so f cannot be written in the form $f(x) = a(x - h)^5 + k$.

Section 2.3 *(page 142)*

1. $f(x)$: dividend; $d(x)$: divisor;
$q(x)$: quotient; $r(x)$: remainder
3. improper   **5.** Factor   **7.** Answers will vary.
9. (a) and (b)

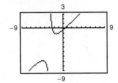

(c) Answers will vary.
11. $2x + 4$, $x \neq -3$ **13.** $x^2 - 3x + 1$, $x \neq -\frac{5}{4}$
15. $x^3 + 3x^2 - 1$, $x \neq -2$ **17.** $6 - \dfrac{1}{x + 1}$
19. $x - \dfrac{x + 9}{x^2 + 1}$ **21.** $2x - 8 + \dfrac{x - 1}{x^2 + 1}$
23. $x + 3 + \dfrac{6x^2 - 8x + 3}{(x - 1)^3}$ **25.** $2x^2 - 2x + 6$, $x \neq 4$
27. $6x^2 + 25x + 74 + \dfrac{248}{x - 3}$ **29.** $4x^2 - 9$, $x \neq -2$
31. $-x^2 + 10x - 25$, $x \neq -10$ **33.** $x^2 + x + 4 + \dfrac{21}{x - 4}$
35. $10x^3 + 10x^2 + 60x + 360 + \dfrac{1360}{x - 6}$
37. $x^2 - 8x + 64$, $x \neq -8$
39. $-3x^3 - 6x^2 - 12x - 24 - \dfrac{48}{x - 2}$
41. $-x^3 - 6x^2 - 36x - 36 - \dfrac{216}{x - 6}$
43. $4x^2 + 14x - 30$, $x \neq -\frac{1}{2}$
45. $f(x) = (x - 3)(x^2 + 2x - 4) - 5$, $f(3) = -5$
47. $f(x) = \left(x + \frac{2}{3}\right)(15x^3 - 6x + 4) + \frac{34}{3}$, $f\left(-\frac{2}{3}\right) = \frac{34}{3}$
49. $f(x) = \left(x - 1 + \sqrt{3}\right)\left[-4x^2 + \left(2 + 4\sqrt{3}\right)x + \left(2 + 2\sqrt{3}\right)\right]$,
$f\left(1 - \sqrt{3}\right) = 0$
51. (a) -2 (b) 1 (c) 36 (d) 5
53. (a) -35 (b) $-\frac{5}{8}$ (c) -10 (d) -211
55. $(x + 3)(x + 2)(x + 1)$; Solutions: $-3, -2, -1$
57. $(2x - 1)(x - 5)(x - 2)$; Solutions: $\frac{1}{2}, 5, 2$
59. $\left(x + \sqrt{3}\right)\left(x - \sqrt{3}\right)(x + 2)$; Solutions: $-\sqrt{3}, \sqrt{3}, -2$
61. $(x - 1)\left(x - 1 - \sqrt{3}\right)\left(x - 1 + \sqrt{3}\right)$;
Solutions: $1, 1 + \sqrt{3}, 1 - \sqrt{3}$

CHAPTER 2

63. (a) Answers will vary.　(b) $2x - 1$
(c) $f(x) = (2x - 1)(x + 2)(x - 1)$　(d) $\frac{1}{2}, -2, 1$
(e)

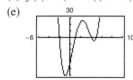

65. (a) Answers will vary.　(b) $(x - 4)(x - 1)$
(c) $f(x) = (x - 5)(x + 2)(x - 4)(x - 1)$　(d) $5, -2, 4, 1$
(e)

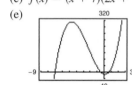

67. (a) Answers will vary.　(b) $x + 7$
(c) $f(x) = (x + 7)(2x + 1)(3x - 2)$　(d) $-7, -\frac{1}{2}, \frac{2}{3}$
(e)

69. (a) Answers will vary.　(b) $x - \sqrt{5}$
(c) $f(x) = (x - \sqrt{5})(x + \sqrt{5})(2x - 1)$　(d) $\pm\sqrt{5}, \frac{1}{2}$
(e)

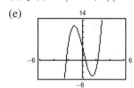

71. (a) $2, \pm 2.236$　(b) 2
(c) $f(x) = (x - 2)(x - \sqrt{5})(x + \sqrt{5})$
73. (a) $-2, 0.268, 3.732$　(b) -2
(c) $h(t) = (t + 2)\left[t - (2 + \sqrt{3})\right]\left[t - (2 - \sqrt{3})\right]$
75. (a) $0, 3, 4, \pm 1.414$　(b) 0
(c) $h(x) = x(x - 4)(x - 3)(x + \sqrt{2})(x - \sqrt{2})$
77. $x^2 - 7x - 8, \; x \neq -8$　**79.** $x^2 + 3x, \; x \neq -2, -1$
81. (a)

3,200,000

0 ──────── 45

−400,000

(b) $250,366$
(c) Answers will vary.
83. False. $-\frac{4}{7}$ is a zero of f.
85. True. The degree of the numerator is greater than the degree of the denominator.
87. $x^{2n} + 6x^n + 9, \; x^n \neq -3$　**89.** $k = -1$, not 1.
91. $c = -210$　**93.** $k = 7$

Section 2.4　(page 150)

1. real　**3.** pure imaginary　**5.** principal square
7. $a = 9, b = 8$　**9.** $a = 8, b = 4$　**11.** $2 + 5i$
13. $1 - 2\sqrt{3}i$　**15.** $2\sqrt{10}i$　**17.** 23　**19.** $-1 - 6i$
21. $0.2i$　**23.** $7 + 4i$　**25.** 1　**27.** $3 - 3\sqrt{2}i$
29. $-14 + 20i$　**31.** $5 + i$　**33.** $108 + 12i$　**35.** 11

37. $-13 + 84i$　**39.** $9 - 2i, 85$　**41.** $-1 + \sqrt{5}i, 6$
43. $-2\sqrt{5}i, 20$　**45.** $\sqrt{6}, 6$　**47.** $\frac{8}{41} + \frac{10}{41}i$
49. $\frac{12}{13} + \frac{5}{13}i$　**51.** $-4 - 9i$　**53.** $-\frac{120}{1681} - \frac{27}{1681}i$
55. $-\frac{1}{2} - \frac{5}{2}i$　**57.** $\frac{62}{949} + \frac{297}{949}i$　**59.** $-2\sqrt{3}$　**61.** -15
63. $7\sqrt{2}i$　**65.** $(21 + 5\sqrt{2}) + (7\sqrt{5} - 3\sqrt{10})i$
67. $1 \pm i$　**69.** $-2 \pm \frac{1}{2}i$　**71.** $-2 \pm \frac{\sqrt{5}}{2}i$
73. $2 \pm \sqrt{2}i$　**75.** $\frac{5}{7} \pm \frac{5\sqrt{13}}{7}i$　**77.** $-1 + 6i$
79. $-14i$　**81.** $-432\sqrt{2}i$　**83.** i　**85.** 81
87. (a) $z_1 = 9 + 16i, z_2 = 20 - 10i$
(b) $z = \dfrac{11{,}240}{877} + \dfrac{4630}{877}i$
89. False. *Sample answer:* $(1 + i) + (3 + i) = 4 + 2i$
91. True.　　　　　$x^4 - x^2 + 14 = 56$
$$(-i\sqrt{6})^4 - (-i\sqrt{6})^2 + 14 \overset{?}{=} 56$$
$$36 + 6 + 14 \overset{?}{=} 56$$
$$56 = 56$$
93. $i, -1, -i, 1, i, -1, -i, 1$; The pattern repeats the first four results. Divide the exponent by 4.
When the remainder is 1, the result is i.
When the remainder is 2, the result is -1.
When the remainder is 3, the result is $-i$.
When the remainder is 0, the result is 1.
95. $\sqrt{-6}\sqrt{-6} = \sqrt{6}i\sqrt{6}i = 6i^2 = -6$　**97.** Proof

Section 2.5　(page 162)

1. Fundamental Theorem of Algebra　**3.** Rational Zero
5. linear; quadratic; quadratic　**7.** Descartes's Rule of Signs
9. 3　**11.** 5　**13.** 2　**15.** $\pm 1, \pm 2$
17. $\pm 1, \pm 3, \pm 5, \pm 9, \pm 15, \pm 45, \pm \frac{1}{2}, \pm \frac{3}{2}, \pm \frac{5}{2}, \pm \frac{9}{2}, \pm \frac{15}{2}, \pm \frac{45}{2}$
19. $-2, -1, 3$　**21.** No rational zeros　**23.** $-6, -1$
25. $-1, \frac{1}{2}$　**27.** $-2, 3, \pm \frac{2}{3}$　**29.** $1, \frac{3}{5} \pm \frac{\sqrt{19}}{5}$
31. $-3, 1, -2 \pm \sqrt{6}$
33. (a) $\pm 1, \pm 2, \pm 4$
(b)　　　　　　　　　　(c) $-2, -1, 2$

35. (a) $\pm 1, \pm 3, \pm \frac{1}{2}, \pm \frac{3}{2}, \pm \frac{1}{4}, \pm \frac{3}{4}$
(b)　　　　　　　　　　(c) $-\frac{1}{4}, 1, 3$

37. (a) $\pm 1, \pm 2, \pm 4, \pm 8, \pm \frac{1}{2}$

(b)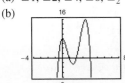

(c) $-\frac{1}{2}, 1, 2, 4$

39. (a) $\pm 1, \pm 3, \pm \frac{1}{2}, \pm \frac{3}{2}, \pm \frac{1}{4}, \pm \frac{3}{4}, \pm \frac{1}{8}, \pm \frac{3}{8}, \pm \frac{1}{16}, \pm \frac{3}{16}, \pm \frac{1}{32}, \pm \frac{3}{32}$

(b)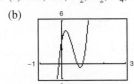

(c) $1, \frac{3}{4}, -\frac{1}{8}$

41. $f(x) = x^3 - x^2 + 25x - 25$

43. $f(x) = x^4 - 6x^3 + 14x^2 - 16x + 8$

45. $f(x) = 3x^4 - 17x^3 + 25x^2 + 23x - 22$

47. $f(x) = 2x^4 + 2x^3 - 2x^2 + 2x - 4$

49. $f(x) = x^3 + x^2 - 2x + 12$

51. (a) $(x^2 + 4)(x^2 - 2)$ (b) $(x^2 + 4)\left(x + \sqrt{2}\right)\left(x - \sqrt{2}\right)$

(c) $(x + 2i)(x - 2i)\left(x + \sqrt{2}\right)\left(x - \sqrt{2}\right)$

53. (a) $(x^2 - 6)(x^2 - 2x + 3)$

(b) $\left(x + \sqrt{6}\right)\left(x - \sqrt{6}\right)(x^2 - 2x + 3)$

(c) $\left(x + \sqrt{6}\right)\left(x - \sqrt{6}\right)\left(x - 1 - \sqrt{2}i\right)\left(x - 1 + \sqrt{2}i\right)$

55. $\pm 2i, 1$ **57.** $2, 3 \pm 2i$ **59.** $1, 3, 1 \pm \sqrt{2}i$

61. $(x + 6i)(x - 6i); \pm 6i$

63. $(x - 1 - 4i)(x - 1 + 4i); 1 \pm 4i$

65. $(x - 2)(x + 2)(x - 2i)(x + 2i); \pm 2, \pm 2i$

67. $(z - 1 + i)(z - 1 - i); 1 \pm i$

69. $(x + 1)(x - 2 + i)(x - 2 - i); -1, 2 \pm i$

71. $(x - 2)^2(x + 2i)(x - 2i); 2, \pm 2i$

73. $-10, -7 \pm 5i$ **75.** $-\frac{3}{4}, 1 \pm \frac{1}{2}i$ **77.** $-2, -\frac{1}{2}, \pm i$

79. One positive real zero, no negative real zeros

81. No positive real zeros, one negative real zero

83. Two or no positive real zeros, two or no negative real zeros

85. Two or no positive real zeros, one negative real zero

87–89. Answers will vary.   **91.** $\frac{3}{4}, \pm \frac{1}{2}$ **93.** $-\frac{3}{4}$

95. $\pm 2, \pm \frac{3}{2}$ **97.** $\pm 1, \frac{1}{4}$

99. d   **100.** a   **101.** b   **102.** c

103. (a)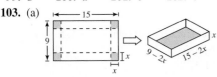

(b) $V(x) = x(9 - 2x)(15 - 2x)$

Domain: $0 < x < \frac{9}{2}$

(c)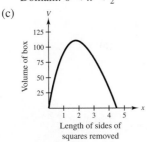

1.82 cm $\times$ 5.36 cm $\times$ 11.36 cm

(d) $\frac{1}{2}, \frac{7}{2}, 8$; 8 is not in the domain of V.

105. (a) $V(x) = x^3 + 9x^2 + 26x + 24 = 120$

(b) 4 ft $\times$ 5 ft $\times$ 6 ft

107. False. The most complex zeros it can have is two, and the Linear Factorization Theorem guarantees that there are three linear factors, so one zero must be real.

109. r_1, r_2, r_3 **111.** $5 + r_1, 5 + r_2, 5 + r_3$

113. The zeros cannot be determined.

115. Answers will vary. There are infinitely many possible functions for f. *Sample equation and graph:*

$f(x) = -2x^3 + 3x^2 + 11x - 6$

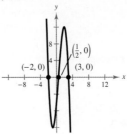

117. $f(x) = x^3 - 3x^2 + 4x - 2$

119. The function should be

$f(x) = (x + 2)(x - 3.5)(x + i)(x - i)$.

121. $f(x) = x^4 + 5x^2 + 4$

123. (a) $x^2 + b$ (b) $x^2 - 2ax + a^2 + b^2$

Section 2.6   *(page 175)*

1. rational functions   **3.** horizontal asymptote

5. Domain: all real numbers x except $x = 1$

$f(x) \to -\infty$ as $x \to 1^-$, $f(x) \to \infty$ as $x \to 1^+$

7. Domain: all real numbers x except $x = \pm 1$

$f(x) \to \infty$ as $x \to -1^-$ and as $x \to 1^+$,

$f(x) \to -\infty$ as $x \to -1^+$ and as $x \to 1^-$

9. Vertical asymptote: $x = 0$

Horizontal asymptote: $y = 0$

11. Vertical asymptote: $x = 5$

Horizontal asymptote: $y = -1$

13. Vertical asymptote: $x = 1$

15. Vertical asymptote: $x = \frac{1}{2}$

Horizontal asymptote: $y = \frac{1}{2}$

17. (a) Domain: all real numbers x except $x = -1$

(b) y-intercept: $(0, 1)$

(c) Vertical asymptote: $x = -1$

Horizontal asymptote: $y = 0$

(d)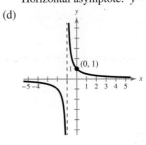

19. (a) Domain: all real numbers x except $x = -4$
  (b) y-intercept: $\left(0, -\frac{1}{4}\right)$
  (c) Vertical asymptote: $x = -4$
   Horizontal asymptote: $y = 0$
  (d)

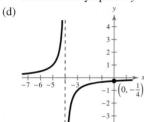

21. (a) Domain: all real numbers x except $x = -2$
  (b) x-intercept: $\left(-\frac{3}{2}, 0\right)$
   y-intercept: $\left(0, \frac{3}{2}\right)$
  (c) Vertical asymptote: $x = -2$
   Horizontal asymptote: $y = 2$
  (d)

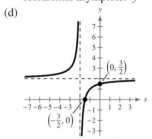

23. (a) Domain: all real numbers x (b) Intercept: $(0, 0)$
  (c) Horizontal asymptote: $y = 1$
  (d)

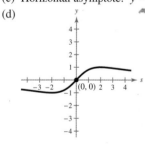

25. (a) Domain: all real numbers s (b) Intercept: $(0, 0)$
  (c) Horizontal asymptote: $y = 0$
  (d)

27. (a) Domain: all real numbers x except $x = 4, -1$
  (b) Intercept: $(0, 0)$
  (c) Vertical asymptotes: $x = -1, x = 4$
   Horizontal asymptote: $y = 0$
  (d)

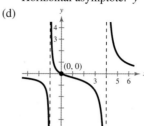

29. (a) Domain: all real numbers x except $x = \pm 4$
  (b) y-intercept: $\left(0, \frac{1}{4}\right)$
  (c) Vertical asymptote: $x = -4$
   Horizontal asymptote: $y = 0$
  (d)

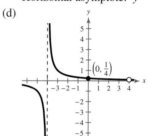

31. (a) Domain: all real numbers t except $t = 1$
  (b) t-intercept: $(-1, 0)$
   y-intercept: $(0, 1)$
  (c) Vertical asymptote: None
   Horizontal asymptote: None
  (d)

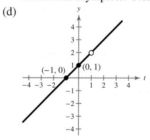

33. (a) Domain: all real numbers x except $x = -1, 5$
  (b) x-intercept: $(-5, 0)$
   y-intercept: $(0, 5)$
  (c) Vertical asymptote: $x = -1$
   Horizontal asymptote: $y = 1$
  (d)

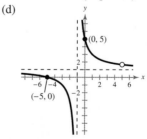

35. (a) Domain: all real numbers x except $x = 2, -3$

(b) Intercept: $(0, 0)$

(c) Vertical asymptote: $x = 2$

Horizontal asymptote: $y = 1$

(d)

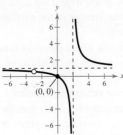

37. (a) Domain: all real numbers x except $x = \pm 1, 2$

(b) x-intercepts: $(3, 0), \left(-\frac{1}{2}, 0\right)$

y-intercept: $\left(0, -\frac{3}{2}\right)$

(c) Vertical asymptotes: $x = 2, x = \pm 1$

Horizontal asymptote: $y = 0$

(d)

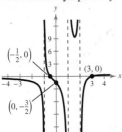

39. d   **40.** a   **41.** c   **42.** b

43. (a) Domain of f: all real numbers x except $x = -1$

Domain of g: all real numbers x

(b)

(c) Because there are only a finite number of pixels, the graphing utility may not attempt to evaluate the function where it does not exist.

45. (a) Domain of f: all real numbers x except $x = 0, 2$

Domain of g: all real numbers x except $x = 0$

(b)

(c) Because there are only a finite number of pixels, the graphing utility may not attempt to evaluate the function where it does not exist.

47. (a) Domain: all real numbers x except $x = 0$

(b) x-intercepts: $(\pm 2, 0)$

(c) Vertical asymptote: $x = 0$

Slant asymptote: $y = x$

(d)

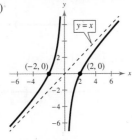

49. (a) Domain: all real numbers x except $x = 0$

(b) No intercepts

(c) Vertical asymptote: $x = 0$

Slant asymptote: $y = 2x$

(d)

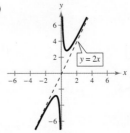

51. (a) Domain: all real numbers x except $x = 0$

(b) No intercepts

(c) Vertical asymptote: $x = 0$

Slant asymptote: $y = x$

(d)

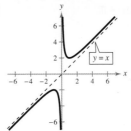

53. (a) Domain: all real numbers t except $t = -5$

(b) y-intercept: $\left(0, -\frac{1}{5}\right)$

(c) Vertical asymptote: $t = -5$

Slant asymptote: $y = -t + 5$

(d)

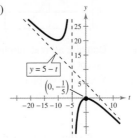

55. (a) Domain: all real numbers x except $x = \pm 2$

(b) Intercept: $(0, 0)$

(c) Vertical asymptotes: $x = \pm 2$

Slant asymptote: $y = x$

(d)

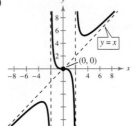

CHAPTER 2

57. (a) Domain: all real numbers x except $x = 1$

(b) y-intercept: $(0, -1)$

(c) Vertical asymptote: $x = 1$

Slant asymptote: $y = x$

(d)

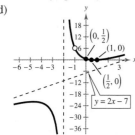

59. (a) Domain: all real numbers x except $x = -1, -2$

(b) y-intercept: $\left(0, \frac{1}{2}\right)$

x-intercepts: $\left(\frac{1}{2}, 0\right), (1, 0)$

(c) Vertical asymptote: $x = -2$

Slant asymptote: $y = 2x - 7$

(d)

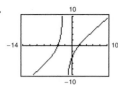

61.

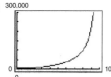

63.

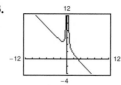

Domain: all real numbers x except $x = -2$

Vertical asymptote: $x = -2$

Slant asymptote: $y = x$

$y = x$

Domain: all real numbers x except $x = 0$

Vertical asymptote: $x = 0$

Slant asymptote: $y = -x + 3$

$y = -x + 3$

65. (a) $(-1, 0)$ (b) -1 **67.** (a) $(1, 0), (-1, 0)$ (b) ± 1

69. (a)

(b) \$4411.76; \$25,000; \$225,000

(c) No. The function is undefined at $p = 100$.

71. 12.8 in. × 8.5 in.

73. (a) Answers will vary.

(b) Vertical asymptote: $x = 25$

Horizontal asymptote: $y = 25$

(c)

(d)

x	30	35	40	45	50	55	60
y	150	87.5	66.7	56.3	50	45.8	42.9

(e) *Sample answer:* No. You might expect the average speed for the round trip to be the average of the average speeds for the two parts of the trip.

(f) No. At 20 miles per hour you would use more time in one direction than is required for the round trip at an average speed of 50 miles per hour.

75. False. Polynomials do not have vertical asymptotes.

77. False. If the degree of the numerator is greater than the degree of the denominator, then no horizontal asymptote exists. However, a slant asymptote exists only if the degree of the numerator is one greater than the degree of the denominator.

79. Yes; No; Every rational function is the ratio of two polynomial functions of the form $f(x) = \dfrac{N(x)}{D(x)}$.

81. $f(x) = \dfrac{x^3}{(x + 2)(x - 1)}$

Section 2.7   *(page 185)*

1. positive; negative   **3.** zeros; undefined values

5. (a) No   (b) Yes   (c) Yes   (d) No

7. (a) Yes   (b) No   (c) No   (d) Yes

9. $-3, 6$ **11.** $4, 5$

13. $(-2, 0)$ **15.** $(-3, 3)$

17. $[-7, 3]$ **19.** $(-\infty, -4] \cup [-2, \infty)$

21. $(-3, 2)$ **23.** $(-3, 1)$

25. $\left(-\infty, -\frac{4}{3}\right) \cup (5, \infty)$ **27.** $(-1, 1) \cup (3, \infty)$

29. $(-\infty, -3) \cup (3, 7)$ **31.** $(-\infty, 0) \cup \left(0, \frac{3}{2}\right)$

33. $[-2, 0] \cup [2, \infty)$ **35.** $[-2, \infty)$

37. The solution set consists of the single real number $\frac{1}{2}$.

39. The solution set is empty.

41. $(-\infty, 0) \cup \left(\frac{1}{4}, \infty\right)$ **43.** $(-7, 1)$

45. $(-5, 3) \cup (11, \infty)$ **47.** $\left(-\frac{3}{4}, 3\right) \cup [6, \infty)$

49. $(-3, -2] \cup [0, 3)$

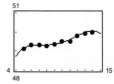

51. $(-\infty, -1) \cup (1, \infty)$

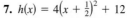

53.

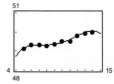

(a) $x \le -1, x \ge 3$

(b) $0 \le x \le 2$

55.

(a) $-2 \le x \le 0$,

$2 \le x \le \infty$

(b) $x \le 4$

57.

(a) $0 \le x < 2$

(b) $2 < x \le 4$

59.

(a) $|x| \ge 2$

(b) $-\infty < x < \infty$

61. $(-3.89, 3.89)$ **63.** $(-0.13, 25.13)$ **65.** $(2.26, 2.39)$

67. (a) $t = 10$ sec (b) 4 sec $< t < 6$ sec

69. $40,000 \le x \le 50,000; \$50.00 \le p \le \$55.00$

71. $[-2, 2]$ **73.** $(-\infty, 4] \cup [5, \infty)$ **75.** $(-5, 0] \cup (7, \infty)$

77. (a) and (c)

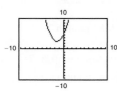

(b) $N = -0.001231t^4 + 0.04723t^3 - 0.6452t^2$
$+ 3.783t + 41.21$

(d) 2017

(e) *Sample answer:* No. For $t > 15$, the model rapidly decreases.

79. 13.8 m $\le L \le 36.2$ m **81.** $R_1 \ge 2$ ohms

83. False. There are four test intervals.

85.

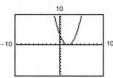

For part (b), the y-values that are less than or equal to 0 occur only at $x = -1$.

For part (c), there are no y-values that are less than 0.

For part (d), the y-values that are greater than 0 occur for all values of x except 2.

87. (a) $(-\infty, -6] \cup [6, \infty)$

(b) When $a > 0$ and $c > 0$, $b \le -2\sqrt{ac}$ or $b \ge 2\sqrt{ac}$.

89. (a) $\left(-\infty, -2\sqrt{30}\right] \cup \left[2\sqrt{30}, \infty\right)$

(b) When $a > 0$ and $c > 0$, $b \le -2\sqrt{ac}$ or $b \ge 2\sqrt{ac}$.

Review Exercises *(page 190)*

1. (a)

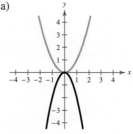

(b)

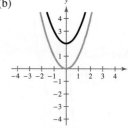

Vertical stretch and a reflection in the x-axis

Upward shift of two units

3. $g(x) = (x - 1)^2 - 1$

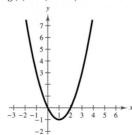

Vertex: $(1, -1)$

Axis of symmetry: $x = 1$

x-intercepts: $(0, 0), (2, 0)$

5. $h(x) = -(x - 2)^2 + 7$

Vertex: $(2, 7)$

Axis of symmetry: $x = 2$

x-intercepts: $\left(2 \pm \sqrt{7}, 0\right)$

7. $h(x) = 4\left(x + \frac{1}{2}\right)^2 + 12$

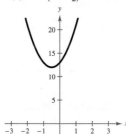

Vertex: $\left(-\frac{1}{2}, 12\right)$

Axis of symmetry: $x = -\frac{1}{2}$

No x-intercept

9. (a) $y = 500 - x$

$A(x) = 500x - x^2$

(b) $x = 250, y = 250$

11.

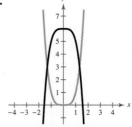

13. Falls to the left and to the right

15. Falls to the left, rises to the right

CHAPTER 2

17. (a) Falls to the left, rises to the right
(b) $-2, 0$ (c) Answers will vary.
(d)

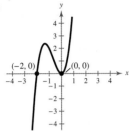

19. (a) Rises to the left, falls to the right
(b) -1 (c) Answers will vary.
(d)

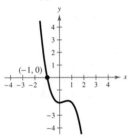

21. (a) $[-1, 0]$ (b) -0.900 **23.** $6x + 3 + \dfrac{17}{5x - 3}$

25. $2x^2 - 9x - 6,\ x \neq 8$

27. (a) Answers will vary.   (b) $(2x + 5), (x - 3)$
(c) $f(x) = (2x + 5)(x - 3)(x + 6)$ (d) $-\frac{5}{2}, 3, -6$
(e)

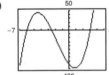

29. $4 + 3i$ **31.** $-3 - 3i$ **33.** $15 + 6i$ **35.** $\frac{4}{5} + \frac{8}{5}i$

37. $\frac{21}{13} - \frac{1}{13}i$ **39.** $1 \pm 3i$ **41.** 2 **43.** $-1, \frac{7}{4}, 6$

45. $(x + 2)(x - 3)(x - 6);\ -2, 3, 6$

47. One or three positive real zeros, two or no negative real zeros

49. Domain: all real numbers x except $x = -10$
Vertical asymptote: $x = -10$
Horizontal asymptote: $y = 3$

51. (a) Domain: all real numbers x except $x = 0$
(b) No intercepts
(c) Vertical asymptote: $x = 0$
    Horizontal asymptote: $y = 0$
(d)

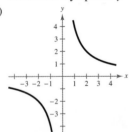

53. (a) Domain: all real numbers x except $x = \pm 4$
(b) Intercept: $(0, 0)$
(c) Vertical asymptotes: $x = \pm 4$
    Horizontal asymptote: $y = 0$
(d)

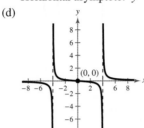

55. (a) Domain: all real numbers x except $x = 0, \frac{1}{3}$
(b) x-intercept: $\left(\frac{3}{2}, 0\right)$
(c) Vertical asymptote: $x = 0$
    Horizontal asymptote: $y = 2$
(d)

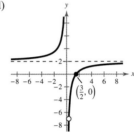

57. (a) Domain: all real numbers x
(b) Intercept: $(0, 0)$ (c) Slant asymptote: $y = 2x$
(d)

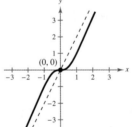

59. (a)

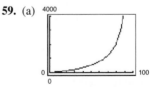

(b) $176 million; $528 million; $1584 million
(or $1.584 billion)
(c) No; The function is undefined when $p = 100$.

61. $\left(-\frac{2}{3}, \frac{1}{4}\right)$ **63.** $[-5, -1) \cup (1, \infty)$

65. 9 days

67. False. The domain of $f(x) = \dfrac{1}{x^2 + 1}$ is the set of all real numbers.

Chapter Test   *(page 192)*

1. (a)

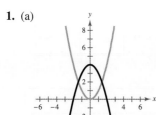

(b)

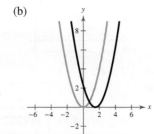

Reflection in the *x*-axis and an upward shift of four units

Right shift of $\frac{3}{2}$ units

2. $y = (x - 3)^2 - 6$

3. (a) 50 ft

(b) 5. Yes, changing the constant term results in a vertical shift of the graph, so the maximum height changes.

4. Rises to the left, falls to the right

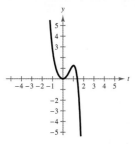

5. $3x + \dfrac{x - 1}{x^2 + 1}$ **6.** $2x^3 - 4x^2 + 5x - 6 + \dfrac{11}{x + 2}$

7. $(2x - 5)\left(x + \sqrt{3}\right)\left(x - \sqrt{3}\right)$; Zeros: $\frac{5}{2}, \pm\sqrt{3}$

8. (a) -14 (b) $19 + 17i$ **9.** $\frac{8}{5} - \frac{16}{5}i$

10. $f(x) = x^4 - 2x^3 + 9x^2 - 18x$

11. $f(x) = x^4 - 6x^3 + 16x^2 - 18x + 7$

12. $-5, -\frac{2}{3}, 1$ **13.** $-2, 4, -1 \pm \sqrt{2}i$

14. *x*-intercepts: $\left(\pm\sqrt{3}, 0\right)$

Vertical asymptote: $x = 0$

Horizontal asymptote: $y = -1$

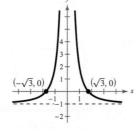

15. *x*-intercept: $\left(-\frac{3}{2}, 0\right)$

y-intercept: $\left(0, \frac{3}{4}\right)$

Vertical asymptote: $x = -4$

Horizontal asymptote: $y = 2$

16. *y*-intercept: $(0, -2)$

Vertical asymptote: $x = 1$

Slant asymptote: $y = x + 1$

17. $x < -4$ or $x > \frac{3}{2}$

18. $x \le -12$ or $-6 < x < 0$

Problem Solving   *(page 195)*

1. Answers will vary.

3. 2 in. $\times$ 2 in. $\times$ 5 in.

5. (a) and (b) $y = -x^2 + 5x - 4$

7. (a) $f(x) = (x - 2)x^2 + 5 = x^3 - 2x^2 + 5$

(b) $f(x) = -(x + 3)x^2 + 1 = -x^3 - 3x^2 + 1$

9. $(a + bi)(a - bi) = a^2 + abi - abi - b^2i^2$

$= a^2 + b^2$

11. (a) As $|a|$ increases, the graph stretches vertically. For $a < 0$, the graph is reflected in the *x*-axis.

(b) As $|b|$ increases, the vertical asymptote is translated. For $b > 0$, the graph is translated to the right. For $b < 0$, the graph is reflected in the *x*-axis and is translated to the left.

13. No. Complex zeros always occur in conjugate pairs.

Chapter 3

Section 3.1   *(page 206)*

1. algebraic   **3.** One-to-One   **5.** $A = P\left(1 + \dfrac{r}{n}\right)^{nt}$

7. 0.863   **9.** 1.552   **11.** 1767.767

13. d   **14.** c   **15.** a   **16.** b

17.

x	-2	-1	0	1	2
$f(x)$	0.020	0.143	1	7	49

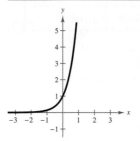

19.

x	-2	-1	0	1	2
$f(x)$	0.063	0.25	1	4	16

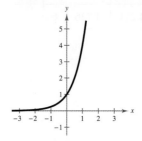

21.

x	−2	−1	0	1	2
f(x)	0.016	0.063	0.25	1	4

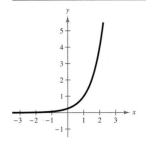

23.

x	−3	−2	−1	0	1
f(x)	3.25	3.5	4	5	7

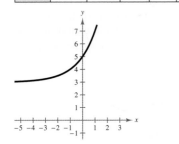

25. 2   **27.** −5   **29.** Shift the graph of f one unit up.

31. Reflect the graph of f in the y-axis and shift three units to the right.

33. 6.686   **35.** 7166.647

37.

x	−8	−7	−6	−5	−4
f(x)	0.055	0.149	0.406	1.104	3

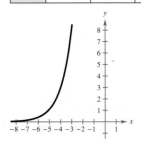

39.

x	−2	−1	0	1	2
f(x)	4.037	4.100	4.271	4.736	6

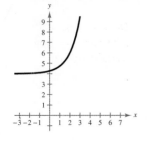

41. 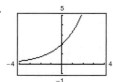   **43.**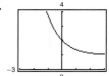

45. $\frac{1}{3}$ **47.** 3, −1

49.

n	1	2	4	12
A	$1828.49	$1830.29	$1831.19	$1831.80

n	365	Continuous
A	$1832.09	$1832.10

51.

n	1	2	4	12
A	$5477.81	$5520.10	$5541.79	$5556.46

n	365	Continuous
A	$5563.61	$5563.85

53.

t	10	20	30
A	$17,901.90	$26,706.49	$39,841.40

t	40	50
A	$59,436.39	$88,668.67

55.

t	10	20	30
A	$22,986.49	$44,031.56	$84,344.25

t	40	50
A	$161,564.86	$309,484.08

57. $104,710.29   **59.** $44.23

61. (a)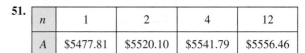

(b)

t	25	26	27	28
P (in millions)	350.281	352.107	353.943	355.788

t	29	30	31	32
P (in millions)	357.643	359.508	361.382	363.266

t	33	34	35	36
P (in millions)	365.160	367.064	368.977	370.901

t	37	38	39	40
P (in millions)	372.835	374.779	376.732	378.697

t	41	42	43	44
P (in millions)	380.671	382.656	384.651	386.656

t	45	46	47	48
P (in millions)	388.672	390.698	392.735	394.783

t	49	50	51	52
P (in millions)	396.841	398.910	400.989	403.080

t	53	54	55
P (in millions)	405.182	407.294	409.417

(c) 2064

63. (a) 16 g (b) 1.85 g

(c)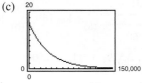

65. (a) $V(t) = 49,810\left(\frac{7}{8}\right)^t$ (b) \$29,197.71

67. True. As $x \to -\infty$, $f(x) \to -2$ but never reaches -2.

69. $f(x) = h(x)$ **71.** $f(x) = g(x) = h(x)$

73.

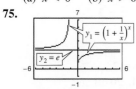

(a) $x < 0$ (b) $x > 0$

75.

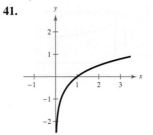

As the x-value increases, y_1 approaches the value of e.

77. (a) 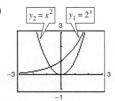 (b)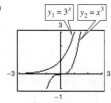

In both viewing windows, the constant raised to a variable power increases more rapidly than the variable raised to a constant power.

79. c, d

Section 3.2   *(page 216)*

1. logarithmic   **3.** natural; e **5.** $x = y$ **7.** $4^2 = 16$

9. $12^1 = 12$ **11.** $\log_5 125 = 3$ **13.** $\log_4 \frac{1}{64} = -3$

15. 6   **17.** 0   **19.** -2 **21.** -0.058 **23.** 1.097

25. 1   **27.** 0   **29.** 5   **31.** ± 2

33.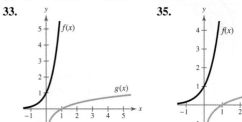

35.

37. a; Upward shift of two units

38. d; Right shift of one unit

39. b; Reflection in the y-axis and a right shift of one unit

40. c; Reflection in the x-axis

41.
Domain: $(0, \infty)$
x-intercept: $(1, 0)$
Vertical asymptote: $x = 0$

43.
Domain: $(0, \infty)$
x-intercept: $\left(\frac{1}{3}, 0\right)$
Vertical asymptote: $x = 0$

45.
Domain: $(-2, \infty)$
x-intercept: $(-1, 0)$
Vertical asymptote: $x = -2$

CHAPTER 3

47.

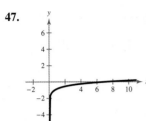

Domain: $(0, \infty)$
x-intercept: $(7, 0)$
Vertical asymptote: $x = 0$

49. $e^{-0.0693\cdots} = \frac{1}{2}$ **51.** $e^{5.521\cdots} = 250$
53. $\ln 7.3890\ldots = 2$ **55.** $\ln \frac{1}{2} = -4x$ **57.** 2.913
59. 6.438   **61.** 4   **63.** 0   **65.** 1
67.

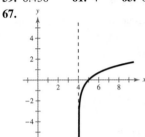

Domain: $(4, \infty)$
x-intercept: $(5, 0)$
Vertical asymptote: $x = 4$

69.

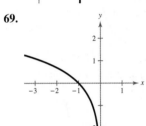

Domain: $(-\infty, 0)$
x-intercept: $(-1, 0)$
Vertical asymptote: $x = 0$

71.

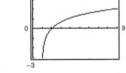

73.

75. 8   **77.** $-2, 3$
79. (a) 30 yr; 10 yr
(b) \$323,179; \$199,109; \$173,179; \$49,109
(c) $x = 750$; The monthly payment must be greater than \$750.
81. (a)

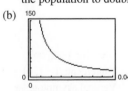

r	0.005	0.010	0.015	0.020	0.025	0.030
t	138.6	69.3	46.2	34.7	27.7	23.1

As the rate of increase r increases, the time t in years for the population to double decreases.
(b)

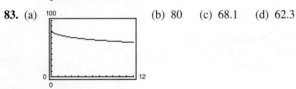

83. (a)

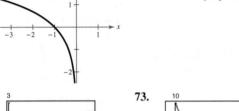

(b) 80   (c) 68.1   (d) 62.3

85. False. Reflecting $g(x)$ in the line $y = x$ will determine the graph of $f(x)$.

87. (a)

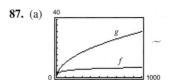

$g(x)$; The natural log function grows at a slower rate than the square root function.
(b)

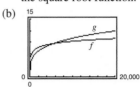

$g(x)$; The natural log function grows at a slower rate than the fourth root function.
89. $y = \log_2 x$, so y is a logarithmic function of x.
91. (a)

x	1	5	10	10^2
$f(x)$	0	0.322	0.230	0.046

x	10^4	10^6
$f(x)$	0.00092	0.0000138

(b) 0
(c)

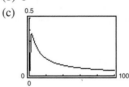

Section 3.3   *(page 223)*

1. change-of-base   **3.** $\dfrac{1}{\log_b a}$ **5.** (a) $\dfrac{\log 16}{\log 5}$ (b) $\dfrac{\ln 16}{\ln 5}$

7. (a) $\dfrac{\log \frac{3}{10}}{\log x}$ (b) $\dfrac{\ln \frac{3}{10}}{\ln x}$ **9.** 2.579   **11.** -0.606
13. $\log_3 5 + \log_3 7$ **15.** $\log_3 7 - 2 \log_3 5$
17. $1 + \log_3 7 - \log_3 5$ **19.** 2   **21.** $-\frac{1}{3}$
23. -2 is not in the domain of $\log_2 x$.   **25.** $\frac{3}{4}$ **27.** 7
29. 2   **31.** $\frac{3}{2}$ **33.** 1.1833   **35.** -1.6542 **37.** 1.9563
39. -2.7124 **41.** $\ln 7 + \ln x$ **43.** $4 \log_8 x$
45. $1 - \log_5 x$ **47.** $\frac{1}{2} \ln z$ **49.** $\ln x + \ln y + 2 \ln z$
51. $\ln z + 2 \ln(z - 1)$
53. $\frac{1}{2} \log_2 (a + 2) + \frac{1}{2} \log_2 (a - 2) - \log_2 7$
55. $2 \log_5 x - 2 \log_5 y - 3 \log_5 z$ **57.** $\frac{1}{3} \ln y + \frac{1}{3} \ln z - \frac{2}{3} \ln x$
59. $\frac{3}{4} \ln x + \frac{1}{4} \ln(x^2 + 3)$ **61.** $\ln 3x$ **63.** $\log_7 (z - 2)^{2/3}$
65. $\log_3 \dfrac{5}{x^3}$ **67.** $\log x(x + 1)^2$ **69.** $\log \dfrac{xz^3}{y^2}$
71. $\ln \dfrac{x}{(x + 1)(x - 1)}$ **73.** $\ln \sqrt{\dfrac{x(x + 3)^2}{x^2 - 1}}$
75. $\log_8 \dfrac{\sqrt[3]{y(y + 4)^2}}{y - 1}$
77. $\log_2 \frac{32}{4} = \log_2 32 - \log_2 4$; Property 2
79. $\beta = 10(\log I + 12)$; 60 dB   **81.** 70 dB
83. $\ln y = \frac{1}{4} \ln x$ **85.** $\ln y = -\frac{1}{4} \ln x + \ln \frac{5}{2}$
87. $\ln y = -0.14 \ln x + 5.7$

89. (a) and (b)

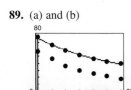

(c)

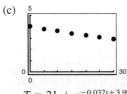

$T = 21 + e^{-0.037t + 3.997}$

The results are similar.

(d)

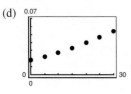

$$T = 21 + \frac{1}{0.001t + 0.016}$$

91. False; $\ln 1 = 0$ **93.** False; $\ln(x - 2) \neq \ln x - \ln 2$
95. False; $u = v^2$

97. $f(x) = \dfrac{\log x}{\log 2} = \dfrac{\ln x}{\ln 2}$ **99.** $f(x) = \dfrac{\log x}{\log \frac{1}{4}} = \dfrac{\ln x}{\ln \frac{1}{4}}$

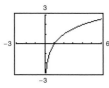   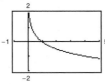

101. The Power Property cannot be used because $\ln e$ is raised to the second power, not just e.

103.

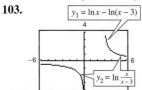

No; $\dfrac{x}{x - 3} > 0$ when $x < 0$.

105. $\ln 1 = 0$ $\ln 9 \approx 2.1972$
$\ln 2 \approx 0.6931$ $\ln 10 \approx 2.3025$
$\ln 3 \approx 1.0986$ $\ln 12 \approx 2.4848$
$\ln 4 \approx 1.3862$ $\ln 15 \approx 2.7080$
$\ln 5 \approx 1.6094$ $\ln 16 \approx 2.7724$
$\ln 6 \approx 1.7917$ $\ln 18 \approx 2.8903$
$\ln 8 \approx 2.0793$ $\ln 20 \approx 2.9956$

Section 3.4   *(page 233)*

1. (a) $x = y$ (b) $x = y$ (c) x (d) x
3. (a) Yes   (b) No   (c) Yes
5. (a) Yes   (b) No   (c) No   **7.** 2
9. 2   **11.** $\ln 2 \approx 0.693$ **13.** $e^{-1} \approx 0.368$
15. 64   **17.** $(3, 8)$ **19.** $2, -1$

21. $\dfrac{\ln 5}{\ln 3} \approx 1.465$ **23.** $\ln 39 \approx 3.664$ **25.** $\dfrac{\ln 80}{2 \ln 3} \approx 1.994$

27. $2 - \dfrac{\ln 400}{\ln 3} \approx -3.454$ **29.** $\dfrac{1}{3} \log \dfrac{3}{2} \approx 0.059$

31. $\dfrac{\ln 12}{3} \approx 0.828$ **33.** 0   **35.** $\dfrac{\ln \frac{8}{3}}{3 \ln 2} + \dfrac{1}{3} \approx 0.805$

37. $-\dfrac{\ln 2}{\ln 3 - \ln 2} \approx -1.710$ **39.** $0, \dfrac{\ln 4}{\ln 5} \approx 0.861$

41. $\ln 5 \approx 1.609$ **43.** $\ln \frac{4}{5} \approx -0.223$

45. $\dfrac{\ln 4}{365 \ln\left(1 + \dfrac{0.065}{365}\right)} \approx 21.330$ **47.** $e^{-3} \approx 0.050$

49. $\dfrac{e^{2.1}}{6} \approx 1.361$ **51.** $e^{-2} \approx 0.135$ **53.** $2(3^{11/6}) \approx 14.988$

55. No solution   **57.** No solution   **59.** No solution
61. 2   **63.** 3.328   **65.** -0.478 **67.** 20.086
69. 1.482   **71.** (a) 27.73 yr   (b) 43.94 yr   **73.** $-1, 0$
75. 1   **77.** $e^{-1} \approx 0.368$ **79.** $e^{-1/2} \approx 0.607$
81. (a) $y = 100$ and $y = 0$; The range falls between 0% and 100%.
   (b) Males: 69.51 in.   Females: 64.49 in.
83. 5 years   **85.** 2011   **87.** About 3.039 min
89. $\log_b uv = \log_b u + \log_b v$
   True by Property 1 in Section 3.3.
91. $\log_b(u - v) = \log_b u - \log_b v$
   False.
   $1.95 \approx \log(100 - 10) \neq \log 100 - \log 10 = 1$
93. Yes. See Exercise 57.
95. For $rt < \ln 2$ years, double the amount you invest. For $rt > \ln 2$ years, double your interest rate or double the number of years, because either of these will double the exponent in the exponential function.
97. (a) 7%   (b) 7.25%   (c) 7.19%   (d) 7.45%
   The investment plan with the greatest effective yield and the highest balance after 5 years is plan (d).

Section 3.5   *(page 243)*

1. $y = ae^{bx}$; $y = ae^{-bx}$ **3.** normally distributed

5. (a) $P = \dfrac{A}{e^{rt}}$ (b) $t = \dfrac{\ln\left(\dfrac{A}{P}\right)}{r}$

7. 19.8 yr; $1419.07   **9.** 8.9438%; $1834.37
11. $6376.28; 15.4 yr   **13.** $303,580.52
15. (a) 7.27 yr   (b) 6.96 yr   (c) 6.93 yr   (d) 6.93 yr
17. (a)

r	2%	4%	6%	8%	10%	12%
t	54.93	27.47	18.31	13.73	10.99	9.16

(b)

r	2%	4%	6%	8%	10%	12%
t	55.48	28.01	18.85	14.27	11.53	9.69

19.

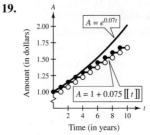

Continuous compounding

21. 6.48 g   **23.** 2.26 g   **25.** $y = e^{0.7675x}$
27. $y = 5e^{-0.4024x}$

29. (a)

Year	Population
1980	104,752
1990	143,251
2000	195,899
2010	267,896

(b) 2019

(c) *Sample answer:* No; As t increases, the population increases rapidly.

31. $k = 0.2988$; About 5,309,734 hits   **33.** About 800 bacteria

35. (a) $V = -150t + 575$ (b) $V = 575e^{-0.3688t}$

(c)

The exponential model depreciates faster.

(d) Linear model: $425; $125
Exponential model: $397.65; $190.18

(e) Answers will vary.

37. About 12,180 yr old

39. (a)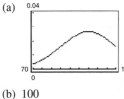

(b) 100

41. (a) 1998: 63,922 sites
2003: 149,805 sites
2006: 208,705 sites

(b)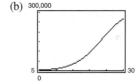

(c) and (d) 2010

43. (a) 203 animals   (b) 13 mo

(c)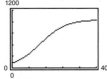

Horizontal asymptotes: $p = 0, p = 1000$. The population size will approach 1000 as time increases.

45. (a) $10^{7.6} \approx 39,810,717$

(b) $10^{5.6} \approx 398,107$ (c) $10^{6.6} \approx 3,981,072$

47. (a) 20 dB   (b) 70 dB   (c) 40 dB   (d) 90 dB

49. 95%   **51.** 4.64   **53.** 1.58×10^{-6} moles/L

55. $10^{5.1}$ **57.** 3:00 A.M.

59. (a) 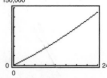   (b) $t \approx 21$ yr; Yes

61. False. The domain can be the set of real numbers for a logistic growth function.

63. False. The graph of $f(x)$ is the graph of $g(x)$ shifted five units up.

65. Answers will vary.

Review Exercises   *(page 250)*

1. 0.164   **3.** 1.587   **5.** 1456.529

7.

x	-1	0	1	2	3
$f(x)$	8	5	4.25	4.063	4.016

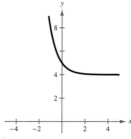

9.

x	-1	0	1	2	3
$f(x)$	4.008	4.04	4.2	5	9

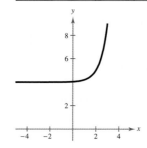

11.

x	-2	-1	0	1	2
$f(x)$	3.25	3.5	4	5	7

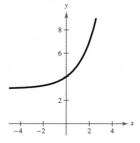

13. 1   **15.** 4   **17.** Shift the graph of f one unit up.

19. Reflect f in the x-axis and shift one unit up.

21. 29.964   **23.** 1.822

25.

x	-2	-1	0	1	2
$h(x)$	2.72	1.65	1	0.61	0.37

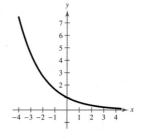

27.

x	-3	-2	-1	0	1
$f(x)$	0.37	1	2.72	7.39	20.09

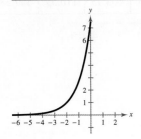

29. (a) 0.283   (b) 0.487   (c) 0.811

31.

n	1	2	4	12
A	\$6719.58	\$6734.28	\$6741.74	\$6746.77

n	365	Continuous
A	\$6749.21	\$6749.29

33. $\log_3 27 = 3$ **35.** $\ln 2.2255 \ldots = 0.8$ **37.** 3
39. -2 **41.** 7   **43.** -5
45. Domain: $(0, \infty)$ **47.** Domain: $(-5, \infty)$
  x-intercept: $(1, 0)$ x-intercept: $(9995, 0)$
  Vertical asymptote: $x = 0$ Vertical asymptote: $x = -5$

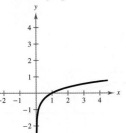   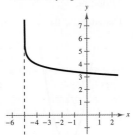

49. 3.118   **51.** 0.25
53. Domain: $(0, \infty)$ **55.** Domain: $(6, \infty)$
  x-intercept: $(e^{-6}, 0)$ x-intercept: $(7, 0)$
  Vertical asymptote: $x = 0$ Vertical asymptote: $x = 6$

57. About 14.32 parsecs   **59.** (a) and (b) 2.585
61. (a) and (b) -2.322 **63.** $\log_2 5 - \log_2 3$
65. $2 \log_2 3 - \log_2 5$ **67.** $\log 7 + 2 \log x$
69. $2 - \frac{1}{2} \log_3 x$ **71.** $2 \ln x + 2 \ln y + \ln z$
73. $\ln 7x$ **75.** $\log \dfrac{x}{\sqrt{y}}$ **77.** $\log_3 \dfrac{\sqrt{x}}{(y + 8)^2}$

79. (a) $0 \le h < 18{,}000$
  (b)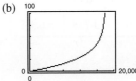
  Vertical asymptote: $h = 18{,}000$
  (c) The plane is climbing at a slower rate, so the time required increases.
  (d) 5.46 min
81. 3   **83.** $\ln 3 \approx 1.099$ **85.** $e^4 \approx 54.598$ **87.** 1, 3
89. $\dfrac{\ln 32}{\ln 2} = 5$ **91.** $\frac{1}{3} e^{8.2} \approx 1213.650$
93. $\dfrac{3}{2} + \dfrac{\sqrt{9 + 4e}}{2} \approx 3.729$ **95.** No solution   **97.** 0.900
99. 2.447   **101.** 1.482   **103.** 73.2 yr
105. e   **106.** b   **107.** f   **108.** d   **109.** a   **110.** c
111. $y = 2e^{0.1014x}$
113.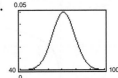
71
115. (a) 10^{-6} W/m^2 (b) $10\sqrt{10}$ W/m^2
  (c) 1.259×10^{-12} W/m^2
117. True by the inverse properties.

Chapter Test   *(page 253)*

1. 0.410   **2.** 0.032   **3.** 0.497   **4.** 22.198
5.

x	-1	$-\frac{1}{2}$	0	$\frac{1}{2}$	1
$f(x)$	10	3.162	1	0.316	0.1

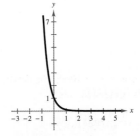

6.

x	-1	0	1	2	3
$f(x)$	-0.005	-0.028	-0.167	-1	-6

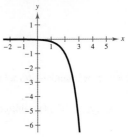

7.

x	-1	$-\frac{1}{2}$	0	$\frac{1}{2}$	1
$f(x)$	0.865	0.632	0	-1.718	-6.389

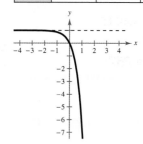

8. (a) -0.89 (b) 9.2

9. Domain: $(0, \infty)$
x-intercept: $(10^{-4}, 0)$
Vertical asymptote: $x = 0$

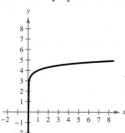

10. Domain: $(4, \infty)$
x-intercept: $(5, 0)$
Vertical asymptote: $x = 4$

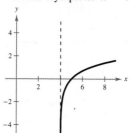

11. Domain: $(-6, \infty)$
x-intercept: $(e^{-1} - 6, 0)$
Vertical asymptote: $x = -6$

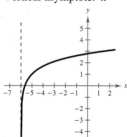

12. 2.209   **13.** -0.167 **14.** -11.047
15. $\log_2 3 + 4 \log_2 a$ **16.** $\frac{1}{2} \ln x - \ln 7$
17. $1 + 2 \log x - 3 \log y$ **18.** $\log_3 13y$ **19.** $\ln \frac{x^4}{y^4}$
20. $\ln \frac{x^3 y^2}{x + 3}$ **21.** -2 **22.** $\frac{\ln 44}{-5} \approx -0.757$
23. $\frac{\ln 197}{4} \approx 1.321$ **24.** $e^{1/2} \approx 1.649$
25. $e^{-11/4} \approx 0.0639$ **26.** 20
27. $y = 2745 e^{0.1570t}$ **28.** 55%

29. (a)

x	$\frac{1}{4}$	1	2	4	5	6
H	58.720	75.332	86.828	103.43	110.59	117.38

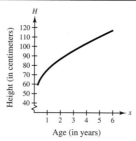

(b) 103 cm; 103.43 cm

Cumulative Test for Chapters 1–3   (page 254)

1.

2.

Midpoint: $\left(\frac{1}{2}, 2\right)$
Distance: $\sqrt{61}$

3.

4.

5. $y = 2x + 2$
6. For some values of x there correspond two values of y.
7. (a) $\frac{3}{2}$ (b) Division by 0 is undefined.   (c) $\frac{s + 2}{s}$
8. (a) Vertical shrink   (b) Upward shift of two units
(c) Left shift of two units
9. (a) $4x - 3$ (b) $-2x - 5$ (c) $3x^2 - 11x - 4$
(d) $\frac{x - 4}{3x + 1}$; Domain: all real numbers x except $x = -\frac{1}{3}$
10. (a) $\sqrt{x - 1} + x^2 + 1$ (b) $\sqrt{x - 1} - x^2 - 1$
(c) $x^2 \sqrt{x - 1} + \sqrt{x - 1}$
(d) $\frac{\sqrt{x - 1}}{x^2 + 1}$; Domain: all real numbers x such that $x \geq 1$
11. (a) $2x + 12$ (b) $\sqrt{2x^2 + 6}$
Domain of $f \circ g$: all real numbers x such that $x \geq -6$
Domain of $g \circ f$: all real numbers x
12. (a) $|x| - 2$ (b) $|x - 2|$
Domains of $f \circ g$ and $g \circ f$: all real numbers x
13. $h(x)^{-1} = \frac{1}{3}(x + 4)$ **14.** 2438.64 kW
15. $y = -\frac{3}{4}(x + 8)^2 + 5$

16.

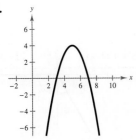

17.

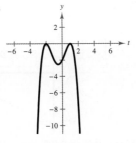

18.

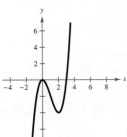

19. $-2, \pm 2i$ **20.** $-7, 0, 3$ **21.** $4, -\frac{1}{2}, 1 \pm 3i$

22. $3x - 2 - \dfrac{3x - 2}{2x^2 + 1}$ **23.** $3x^3 + 6x^2 + 14x + 23 + \dfrac{49}{x - 2}$

24. $[1, 2]$; 1.196

25. Intercept: $(0, 0)$
Vertical asymptotes:
$x = 1, x = -3$
Horizontal asymptote:
$y = 0$

26. y-intercept: $(0, 2)$
x-intercept: $(2, 0)$
Vertical asymptote: $x = 1$
Horizontal asymptote:
$y = 1$

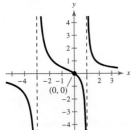

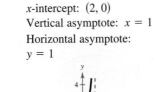

27. y-intercept: $(0, 6)$
x-intercepts: $(2, 0), (3, 0)$
Vertical asymptote: $x = -1$
Slant asymptote: $y = x - 6$

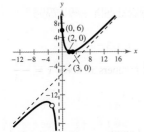

28. $x \le -3$ or $0 \le x \le 3$

29. All real numbers x such that
$x < -5$ or $x > -1$

30. Reflect f in the x-axis and y-axis, and shift three units to the right.

31. Reflect f in the x-axis and shift four units up.

32. 1.991   **33.** -0.067 **34.** 1.717   **35.** 0.390

36. $\ln(x + 5) + \ln(x - 5) - 4 \ln x$ **37.** $\ln \dfrac{x^2}{\sqrt{x + 5}}$, $x > 0$

38. $\dfrac{\ln 12}{2} \approx 1.242$ **39.** $\ln 6 \approx 1.792$ or $\ln 7 \approx 1.946$

40. $e^6 - 2 \approx 401.429$ **41.** \$16,302.05

42. 6.3 h   **43.** 2023

Problem Solving   *(page 257)*

1.

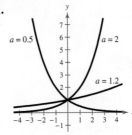

$y = 0.5^x$ and $y = 1.2^x$
$0 < a \le e^{1/e}$

3. As $x \to \infty$, the graph of e^x increases at a greater rate than the graph of x^n.

5. Answers will vary.

7. (a)

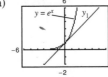

(b)

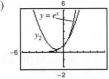

(c)

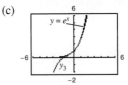

9.

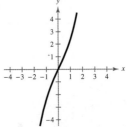

$f^{-1}(x) = \ln\left(\dfrac{x + \sqrt{x^2 + 4}}{2}\right)$

11. c

13. $t = \dfrac{\ln c_1 - \ln c_2}{\left(\dfrac{1}{k_2} - \dfrac{1}{k_1}\right)\ln \dfrac{1}{2}}$

15. (a) $y_1 = 252{,}606(1.0310)^t$
(b) $y_2 = 400.88t^2 - 1464.6t + 291{,}782$
(c)

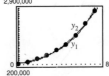

(d) The exponential model is a better fit. No, because the model is rapidly approaching infinity.

CHAPTER 3

17. $1, e^2$

19. $y_4 = (x - 1) - \frac{1}{2}(x - 1)^2 + \frac{1}{3}(x - 1)^3 - \frac{1}{4}(x - 1)^4$

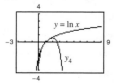

The pattern implies that
$\ln x = (x - 1) - \frac{1}{2}(x - 1)^2 + \frac{1}{3}(x - 1)^3 - \cdots.$

21.

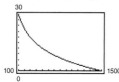

17.7 ft³/min

23. (a)

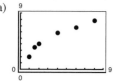

25. (a)

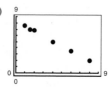

(b)–(e) Answers will vary.   (b)–(e) Answers will vary.

Chapter 4

Section 4.1   *(page 267)*

1. coterminal   **3.** complementary; supplementary
5. linear; angular   **7.** 1 rad   **9.** −3 rad
11. (a) Quadrant I   (b) Quadrant II
13. (a)   (b)

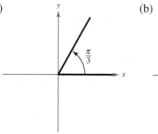

15. *Sample answers:* (a) $\dfrac{13\pi}{6}, -\dfrac{11\pi}{6}$ (b) $\dfrac{7\pi}{6}, -\dfrac{17\pi}{6}$

17. (a) Complement: $\dfrac{5\pi}{12}$; Supplement: $\dfrac{11\pi}{12}$

(b) Complement: none; Supplement: $\dfrac{\pi}{12}$

19. (a) Complement: $\dfrac{\pi}{2} - 1 \approx 0.57$;

Supplement: $\pi - 1 \approx 2.14$

(b) Complement: none; Supplement: $\pi - 2 \approx 1.14$

21. 210°   **23.** −60°

25. (a) Quadrant II   (b) Quadrant IV

27. (a)   (b)

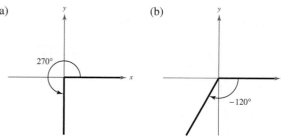

29. (a) 480°, −240°   (b) 150°, −570°
31. (a) Complement: 72°; Supplement: 162°
(b) Complement: 5°; Supplement: 95°
33. (a) Complement: 66°; Supplement: 156°
(b) Complement: none; Supplement: 54°
35. (a) $\dfrac{2\pi}{3}$ (b) $-\dfrac{\pi}{9}$ **37.** (a) 270°   (b) −210°
39. 0.785   **41.** −0.009   **43.** 81.818°   **45.** −756.000°
47. (a) 54.75°   (b) −128.5°
49. (a) 240° 36′   (b) −145° 48′   **51.** 10π in. ≈ 31.42 in.
53. $\dfrac{15}{8}$ rad   **55.** 4 rad   **57.** About 18.85 in.²
59. 20° should be multiplied by $\dfrac{\pi \text{ rad}}{180 \text{ deg}}$.
61. About 592 mi   **63.** About 23.87°
65. (a) 8π rad/min ≈ 25.13 rad/min
(b) 200π ft/min ≈ 628.3 ft/min
67. (a) About 35.70 mi/h   (b) About 739.50 revolutions/min
69.

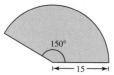

$A = 93.75\pi$ m² ≈ 294.52 m²

71. False. $\dfrac{180°}{\pi}$ is in degree measure.

73. True. Let α and β represent coterminal angles, and let n represent an integer.
$$\alpha = \beta + n(360°)$$
$$\alpha - \beta = n(360°)$$

75. When θ is constant, the length of the arc is proportional to the radius ($s = r\theta$).

77. The speed increases. The linear velocity is proportional to the radius.

79. Proof

Section 4.2   *(page 275)*

1. unit circle   **3.** period
5. $\sin t = \frac{5}{13}$ $\csc t = \frac{13}{5}$ **7.** $\sin t = -\frac{3}{5}$ $\csc t = -\frac{5}{3}$
$\cos t = \frac{12}{13}$ $\sec t = \frac{13}{12}$ $\cos t = -\frac{4}{5}$ $\sec t = -\frac{5}{4}$
$\tan t = \frac{5}{12}$ $\cot t = \frac{12}{5}$ $\tan t = \frac{3}{4}$ $\cot t = \frac{4}{3}$

9. $(0, 1)$ **11.** $\left(-\dfrac{\sqrt{3}}{2}, \dfrac{1}{2}\right)$

13. $\sin \dfrac{\pi}{4} = \dfrac{\sqrt{2}}{2}$ **15.** $\sin\left(-\dfrac{\pi}{6}\right) = -\dfrac{1}{2}$

$\cos \dfrac{\pi}{4} = \dfrac{\sqrt{2}}{2}$ $\cos\left(-\dfrac{\pi}{6}\right) = \dfrac{\sqrt{3}}{2}$

$\tan \dfrac{\pi}{4} = 1$ $\tan\left(-\dfrac{\pi}{6}\right) = -\dfrac{\sqrt{3}}{3}$

17. $\sin\left(-\dfrac{7\pi}{4}\right) = \dfrac{\sqrt{2}}{2}$

$\cos\left(-\dfrac{7\pi}{4}\right) = \dfrac{\sqrt{2}}{2}$

$\tan\left(-\dfrac{7\pi}{4}\right) = 1$

19. $\sin\dfrac{11\pi}{6} = -\dfrac{1}{2}$

$\cos\dfrac{11\pi}{6} = \dfrac{\sqrt{3}}{2}$

$\tan\dfrac{11\pi}{6} = -\dfrac{\sqrt{3}}{3}$

21. $\sin\left(-\dfrac{3\pi}{2}\right) = 1$

$\cos\left(-\dfrac{3\pi}{2}\right) = 0$

$\tan\left(-\dfrac{3\pi}{2}\right)$ is undefined.

23. $\sin\dfrac{2\pi}{3} = \dfrac{\sqrt{3}}{2}$ $\csc\dfrac{2\pi}{3} = \dfrac{2\sqrt{3}}{3}$

$\cos\dfrac{2\pi}{3} = -\dfrac{1}{2}$ $\sec\dfrac{2\pi}{3} = -2$

$\tan\dfrac{2\pi}{3} = -\sqrt{3}$ $\cot\dfrac{2\pi}{3} = -\dfrac{\sqrt{3}}{3}$

25. $\sin\dfrac{4\pi}{3} = -\dfrac{\sqrt{3}}{2}$ $\csc\dfrac{4\pi}{3} = -\dfrac{2\sqrt{3}}{3}$

$\cos\dfrac{4\pi}{3} = -\dfrac{1}{2}$ $\sec\dfrac{4\pi}{3} = -2$

$\tan\dfrac{4\pi}{3} = \sqrt{3}$ $\cot\dfrac{4\pi}{3} = \dfrac{\sqrt{3}}{3}$

27. $\sin\left(-\dfrac{5\pi}{3}\right) = \dfrac{\sqrt{3}}{2}$ $\csc\left(-\dfrac{5\pi}{3}\right) = \dfrac{2\sqrt{3}}{3}$

$\cos\left(-\dfrac{5\pi}{3}\right) = \dfrac{1}{2}$ $\sec\left(-\dfrac{5\pi}{3}\right) = 2$

$\tan\left(-\dfrac{5\pi}{3}\right) = \sqrt{3}$ $\cot\left(-\dfrac{5\pi}{3}\right) = \dfrac{\sqrt{3}}{3}$

29. $\sin\left(-\dfrac{\pi}{2}\right) = -1$ $\csc\left(-\dfrac{\pi}{2}\right) = -1$

$\cos\left(-\dfrac{\pi}{2}\right) = 0$ $\sec\left(-\dfrac{\pi}{2}\right)$ is undefined.

$\tan\left(-\dfrac{\pi}{2}\right)$ is undefined. $\cot\left(-\dfrac{\pi}{2}\right) = 0$

31. 0   **33.** $\frac{1}{2}$ **35.** $-\frac{1}{2}$ **37.** (a) $-\frac{1}{2}$ (b) -2

39. (a) $-\frac{1}{5}$ (b) -5 **41.** (a) $\frac{4}{5}$ (b) $-\frac{4}{5}$

43. 0.5646   **45.** 0.4142   **47.** -1.0009

49. (a) 0.50 ft   (b) About 0.04 ft   (c) About -0.49 ft

51. False. $\sin(-t) = -\sin(t)$ means that the function is odd, not that the sine of a negative angle is a negative number.

53. True. The tangent function has a period of π.

55. (a) y-axis symmetry   (b) $\sin t_1 = \sin(\pi - t_1)$
    (c) $\cos(\pi - t_1) = -\cos t_1$

57. The calculator was in degree mode instead of radian mode.

59. (a)

Circle of radius 1 centered at $(0, 0)$

(b) The t-values represent the central angle in radians. The x- and y-values represent the location in the coordinate plane.

(c) $-1 \le x \le 1,\ -1 \le y \le 1$

61. It is an odd function.

Section 4.3   *(page 284)*

1. (a) v   (b) iv   (c) vi   (d) iii   (e) i   (f) ii

3. complementary

5. $\sin\theta = \frac{3}{5}$ $\csc\theta = \frac{5}{3}$

$\cos\theta = \frac{4}{5}$ $\sec\theta = \frac{5}{4}$

$\tan\theta = \frac{3}{4}$ $\cot\theta = \frac{4}{3}$

7. $\sin\theta = \frac{9}{41}$ $\csc\theta = \frac{41}{9}$

$\cos\theta = \frac{40}{41}$ $\sec\theta = \frac{41}{40}$

$\tan\theta = \frac{9}{40}$ $\cot\theta = \frac{40}{9}$

9. $\sin\theta = \dfrac{\sqrt{2}}{2}$ $\csc\theta = \sqrt{2}$

$\cos\theta = \dfrac{\sqrt{2}}{2}$ $\sec\theta = \sqrt{2}$

$\tan\theta = 1$ $\cot\theta = 1$

11. $\sin\theta = \frac{8}{17}$ $\csc\theta = \frac{17}{8}$

$\cos\theta = \frac{15}{17}$ $\sec\theta = \frac{17}{15}$

$\tan\theta = \frac{8}{15}$ $\cot\theta = \frac{15}{8}$

The triangles are similar, and corresponding sides are proportional.

13. $\sin\theta = \dfrac{1}{3}$ $\csc\theta = 3$

$\cos\theta = \dfrac{2\sqrt{2}}{3}$ $\sec\theta = \dfrac{3\sqrt{2}}{4}$

$\tan\theta = \dfrac{\sqrt{2}}{4}$ $\cot\theta = 2\sqrt{2}$

The triangles are similar, and corresponding sides are proportional.

15.

$\sin\theta = \frac{8}{17}$ $\csc\theta = \frac{17}{8}$

$\sec\theta = \frac{17}{15}$

$\tan\theta = \frac{8}{15}$ $\cot\theta = \frac{15}{8}$

17.

$\sin\theta = \dfrac{\sqrt{11}}{6}$ $\csc\theta = \dfrac{6\sqrt{11}}{11}$

$\cos\theta = \dfrac{5}{6}$

$\tan\theta = \dfrac{\sqrt{11}}{5}$ $\cot\theta = \dfrac{5\sqrt{11}}{11}$

19.

$\csc\theta = 5$

$\cos\theta = \dfrac{2\sqrt{6}}{5}$ $\sec\theta = \dfrac{5\sqrt{6}}{12}$

$\tan\theta = \dfrac{\sqrt{6}}{12}$ $\cot\theta = 2\sqrt{6}$

21.

$\sin\theta = \dfrac{\sqrt{10}}{10}$ $\csc\theta = \sqrt{10}$

$\cos\theta = \dfrac{3\sqrt{10}}{10}$ $\sec\theta = \dfrac{\sqrt{10}}{3}$

$\tan\theta = \dfrac{1}{3}$

23. $\dfrac{\pi}{6};\ \dfrac{\sqrt{3}}{3}$ **25.** $45°;\ \dfrac{\sqrt{2}}{2}$ **27.** $45°;\ \sqrt{2}$

29. (a) 0.3420   (b) 0.3420   **31.** (a) 0.2455   (b) 4.0737

33. (a) 0.9964   (b) 1.0036   **35.** (a) 3.2205   (b) 0.3105

37. (a) $\dfrac{1}{2}$ (b) $\dfrac{\sqrt{3}}{2}$ (c) $\sqrt{3}$ (d) $\dfrac{\sqrt{3}}{3}$

39. (a) $\dfrac{2\sqrt{2}}{3}$ (b) $2\sqrt{2}$ (c) 3   (d) 3

41. (a) $\dfrac{1}{3}$ (b) $\sqrt{10}$ (c) $\dfrac{1}{3}$ (d) $\dfrac{\sqrt{10}}{10}$

43–51. Answers will vary.   **53.** (a) $30° = \dfrac{\pi}{6}$ (b) $30° = \dfrac{\pi}{6}$

55. (a) $60° = \dfrac{\pi}{3}$ (b) $45° = \dfrac{\pi}{4}$

57. (a) $60° = \dfrac{\pi}{3}$ (b) $45° = \dfrac{\pi}{4}$ **59.** $x = 9, y = 9\sqrt{3}$

61. $x = \dfrac{32\sqrt{3}}{3}, r = \dfrac{64\sqrt{3}}{3}$

63. About 443.2 m; about 323.3 m   **65.** $30° = \dfrac{\pi}{6}$

67. (a) About 219.9 ft   (b) About 160.9 ft

69. $(x_1, y_1) = \left(28\sqrt{3}, 28\right)$
$(x_2, y_2) = \left(28, 28\sqrt{3}\right)$

71. $\sin 20° \approx 0.34$, $\cos 20° \approx 0.94$, $\tan 20° \approx 0.36$,
$\csc 20° \approx 2.92$, $\sec 20° \approx 1.06$, $\cot 20° \approx 2.75$

73. (a) About 519.33 ft
(b) About 1174.17 ft
(c) About 173.11 ft/min

75. True. $\csc x = \dfrac{1}{\sin x}$ **77.** False. $\dfrac{\sqrt{2}}{2} + \dfrac{\sqrt{2}}{2} \neq 1$

79. False. $1.7321 \neq 0.0349$

81. Yes, $\tan \theta$ is equal to opp/adj. You can find the value of the hypotenuse by the Pythagorean Theorem. Then you can find $\sec \theta$, which is equal to hyp/adj.

83.

θ	0.1	0.2	0.3	0.4	0.5
$\sin \theta$	0.0998	0.1987	0.2955	0.3894	0.4794

(a) θ is greater.

(b) As $\theta \to 0$, $\sin \theta \to 0$ and $\dfrac{\theta}{\sin \theta} \to 1$.

Section 4.4   *(page 294)*

1. $\dfrac{y}{r}$ **3.** $\dfrac{y}{x}$ **5.** $\cos \theta$ **7.** zero; defined

9. (a) $\sin \theta = \dfrac{3}{5}$ $\csc \theta = \dfrac{5}{3}$
$\cos \theta = \dfrac{4}{5}$ $\sec \theta = \dfrac{5}{4}$
$\tan \theta = \dfrac{3}{4}$ $\cot \theta = \dfrac{4}{3}$

(b) $\sin \theta = \dfrac{15}{17}$ $\csc \theta = \dfrac{17}{15}$
$\cos \theta = -\dfrac{8}{17}$ $\sec \theta = -\dfrac{17}{8}$
$\tan \theta = -\dfrac{15}{8}$ $\cot \theta = -\dfrac{8}{15}$

11. (a) $\sin \theta = -\dfrac{1}{2}$ $\csc \theta = -2$

$\cos \theta = -\dfrac{\sqrt{3}}{2}$ $\sec \theta = -\dfrac{2\sqrt{3}}{3}$

$\tan \theta = \dfrac{\sqrt{3}}{3}$ $\cot \theta = \sqrt{3}$

(b) $\sin \theta = -\dfrac{\sqrt{17}}{17}$ $\csc \theta = -\sqrt{17}$

$\cos \theta = \dfrac{4\sqrt{17}}{17}$ $\sec \theta = \dfrac{\sqrt{17}}{4}$

$\tan \theta = -\dfrac{1}{4}$ $\cot \theta = -4$

13. $\sin \theta = \dfrac{12}{13}$ $\csc \theta = \dfrac{13}{12}$
$\cos \theta = \dfrac{5}{13}$ $\sec \theta = \dfrac{13}{5}$
$\tan \theta = \dfrac{12}{5}$ $\cot \theta = \dfrac{5}{12}$

15. $\sin \theta = -\dfrac{2\sqrt{29}}{29}$ $\csc \theta = -\dfrac{\sqrt{29}}{2}$

$\cos \theta = -\dfrac{5\sqrt{29}}{29}$ $\sec \theta = -\dfrac{\sqrt{29}}{5}$

$\tan \theta = \dfrac{2}{5}$ $\cot \theta = \dfrac{5}{2}$

17. $\sin \theta = \dfrac{4}{5}$ $\csc \theta = \dfrac{5}{4}$
$\cos \theta = -\dfrac{3}{5}$ $\sec \theta = -\dfrac{5}{3}$
$\tan \theta = -\dfrac{4}{3}$ $\cot \theta = -\dfrac{3}{4}$

19. Quadrant I   **21.** Quadrant II

23. $\sin \theta = \dfrac{15}{17}$ $\csc \theta = \dfrac{17}{15}$
$\cos \theta = \dfrac{8}{17}$ $\sec \theta = \dfrac{17}{8}$
    $\cot \theta = \dfrac{8}{15}$

25. $\csc \theta = \dfrac{5}{3}$
$\cos \theta = -\dfrac{4}{5}$ $\sec \theta = -\dfrac{5}{4}$
$\tan \theta = -\dfrac{3}{4}$ $\cot \theta = -\dfrac{4}{3}$

27. $\sin \theta = -\dfrac{\sqrt{10}}{10}$ $\csc \theta = -\sqrt{10}$

$\cos \theta = \dfrac{3\sqrt{10}}{10}$ $\sec \theta = \dfrac{\sqrt{10}}{3}$

$\tan \theta = -\dfrac{1}{3}$

29. $\sin \theta = 1$
    $\sec \theta$ is undefined.
$\tan \theta$ is undefined. $\cot \theta = 0$

31. $\sin \theta = 0$ $\csc \theta$ is undefined.
$\cos \theta = -1$ $\sec \theta = -1$
$\tan \theta = 0$ $\cot \theta$ is undefined.

33. $\sin \theta = \dfrac{\sqrt{2}}{2}$ $\csc \theta = \sqrt{2}$

$\cos \theta = -\dfrac{\sqrt{2}}{2}$ $\sec \theta = -\sqrt{2}$

$\tan \theta = -1$ $\cot \theta = -1$

35. $\sin \theta = \dfrac{2\sqrt{5}}{5}$ $\csc \theta = \dfrac{\sqrt{5}}{2}$

$\cos \theta = \dfrac{\sqrt{5}}{5}$ $\sec \theta = \sqrt{5}$

$\tan \theta = 2$ $\cot \theta = \dfrac{1}{2}$

37. 0   **39.** Undefined   **41.** 1

43. Undefined   **45.** 0

47. $\theta' = 20°$ **49.** $\theta' = 55°$

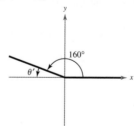

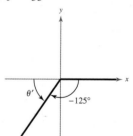

51. $\theta' = \dfrac{\pi}{3}$

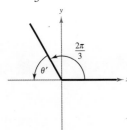

53. $\theta' = 2\pi - 4.8$

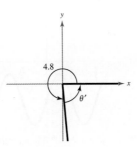

55. $\sin 225° = -\dfrac{\sqrt{2}}{2}$

$\cos 225° = -\dfrac{\sqrt{2}}{2}$

$\tan 225° = 1$

57. $\sin 750° = \dfrac{1}{2}$

$\cos 750° = \dfrac{\sqrt{3}}{2}$

$\tan 750° = \dfrac{\sqrt{3}}{3}$

59. $\sin(-120°) = -\dfrac{\sqrt{3}}{2}$

$\cos(-120°) = -\dfrac{1}{2}$

$\tan(-120°) = \sqrt{3}$

61. $\sin\dfrac{2\pi}{3} = \dfrac{\sqrt{3}}{2}$

$\cos\dfrac{2\pi}{3} = -\dfrac{1}{2}$

$\tan\dfrac{2\pi}{3} = -\sqrt{3}$

63. $\sin\left(-\dfrac{\pi}{6}\right) = -\dfrac{1}{2}$

$\cos\left(-\dfrac{\pi}{6}\right) = \dfrac{\sqrt{3}}{2}$

$\tan\left(-\dfrac{\pi}{6}\right) = -\dfrac{\sqrt{3}}{3}$

65. $\sin\dfrac{11\pi}{4} = \dfrac{\sqrt{2}}{2}$

$\cos\dfrac{11\pi}{4} = -\dfrac{\sqrt{2}}{2}$

$\tan\dfrac{11\pi}{4} = -1$

67. $\sin\left(-\dfrac{17\pi}{6}\right) = -\dfrac{1}{2}$

$\cos\left(-\dfrac{17\pi}{6}\right) = -\dfrac{\sqrt{3}}{2}$

$\tan\left(-\dfrac{17\pi}{6}\right) = \dfrac{\sqrt{3}}{3}$

69. $\dfrac{4}{5}$ **71.** $-\dfrac{\sqrt{13}}{12}$ **73.** $\dfrac{8\sqrt{39}}{39}$ **75.** 0.1736

77. -0.3420 **79.** -28.6363 **81.** 1.4142   **83.** 0.3640

85. -2.6131 **87.** -0.6052 **89.** 1.8382

91. (a) $30° = \dfrac{\pi}{6}, 150° = \dfrac{5\pi}{6}$ (b) $210° = \dfrac{7\pi}{6}, 330° = \dfrac{11\pi}{6}$

93. (a) $60° = \dfrac{\pi}{3}, 300° = \dfrac{5\pi}{3}$ (b) $60° = \dfrac{\pi}{3}, 300° = \dfrac{5\pi}{3}$

95. (a) $45° = \dfrac{\pi}{4}, 225° = \dfrac{5\pi}{4}$ (b) $150° = \dfrac{5\pi}{6}, 330° = \dfrac{11\pi}{6}$

97. (a) 12 mi   (b) 6 mi   (c) About 6.9 mi

99. (a) $B = 24.593 \sin(0.495t - 2.262) + 57.387$

$F = 39.071 \sin(0.448t - 1.366) + 32.204$

(b) February: $B \approx 33.9°, F \approx 14.5°$

April: $B \approx 50.5°, F \approx 48.3°$

May: $B \approx 62.6°, F \approx 62.2°$

July: $B \approx 80.3°, F \approx 70.5°$

September: $B \approx 77.4°, F \approx 50.1°$

October: $B \approx 68.2°, F \approx 33.3°$

December: $B \approx 44.8°, F \approx 2.4°$

(c)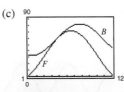

Answers will vary.

101. About 0.79 amp

103. False. In each of the four quadrants, the signs of the secant function and the cosine function are the same because these functions are reciprocals of each other.

105. Answers will vary.

Section 4.5   *(page 304)*

1. cycle   **3.** phase shift   **5.** Period: $\dfrac{2\pi}{5}$; Amplitude: 2

7. Period: 4; Amplitude: $\dfrac{3}{4}$ **9.** Period: $\dfrac{8\pi}{5}$; Amplitude: $\dfrac{1}{2}$

11. Period: 24; Amplitude: $\dfrac{5}{3}$

13. The period of g is one-fifth the period of f.

15. g is a reflection of f in the x-axis.

17. g is a shift of f π units to the right.

19. g is a shift of f three units up.

21. The graph of g has twice the amplitude of the graph of f.

23. The graph of g is a horizontal shift of the graph of f π units to the right.

25.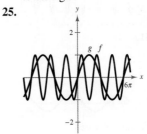

27.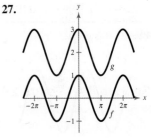

29.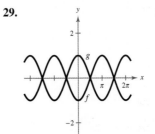

31.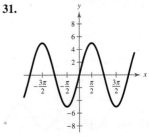

33.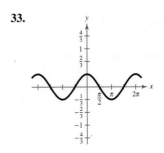

35.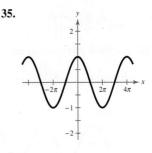

37.

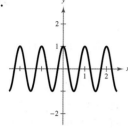

39.

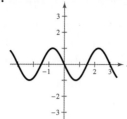

41.

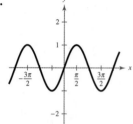

43.

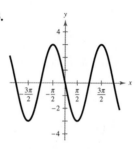

45.

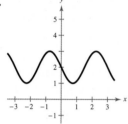

47.

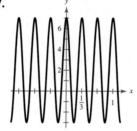

49.

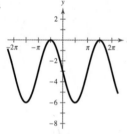

51.

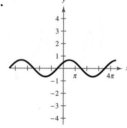

53. (a) Horizontal shrink and a phase shift $\pi/4$ unit right

   (b)   (c) $g(x) = f(4x - \pi)$

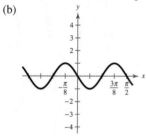

55. (a) Shift two units up and a phase shift $\pi/2$ units right

   (b)   (c) $g(x) = f\left(x - \dfrac{\pi}{2}\right) + 2$

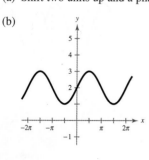

57. (a) Horizontal shrink, a vertical stretch, a shift three units down, and a phase shift $\pi/4$ unit right

   (b)

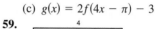

   (c) $g(x) = 2f(4x - \pi) - 3$

59.   **61.**

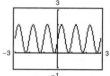

63.

65. $a = 2, d = 1$ **67.** $a = -4, d = 4$

69. $a = -3, b = 2, c = 0$ **71.** $a = 2, b = 1, c = -\dfrac{\pi}{4}$

73.

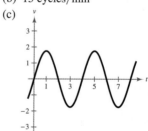

$x = -\dfrac{\pi}{6}, -\dfrac{5\pi}{6}, \dfrac{7\pi}{6}, \dfrac{11\pi}{6}$

75. $y = 1 + 2\sin(2x - \pi)$ **77.** $y = \cos(2x + 2\pi) - \dfrac{3}{2}$

79. (a) 4 sec

   (b) 15 cycles/min

   (c)

81. (a) $\dfrac{6}{5}$ sec   (b) 50 heartbeats/min

83. (a) and (c)

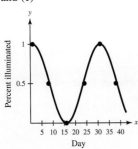

The model fits the data well.

(b) $y = 0.5 \cos\left(\dfrac{\pi x}{15} - \dfrac{\pi}{15}\right) + 0.5$

(d) 30 days   (e) 25%

85. (a) 20 sec; It takes 20 seconds to complete one revolution on the Ferris wheel.

(b) 50 ft; The diameter of the Ferris wheel is 100 feet.

(c)

87. False. The graph of g is shifted one period to the left.

89. True. $-\sin\left(x + \dfrac{\pi}{2}\right) = -\cos x$

91.

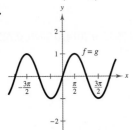

Conjecture:

$\sin x = \cos\left(x - \dfrac{\pi}{2}\right)$

93.

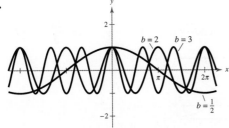

The value of b affects the period of the graph.

$b = \frac{1}{2} \rightarrow \frac{1}{2}$ cycle

$b = 2 \rightarrow 2$ cycles

$b = 3 \rightarrow 3$ cycles

95. (a) 0.4794, 0.4794   (b) 0.8417, 0.8415   (c) 0.5, 0.5

(d) 0.8776, 0.8776   (e) 0.5417, 0.5403

(f) 0.7074, 0.7071

The error increases as x moves farther away from 0.

Section 4.6   *(page 315)*

1. odd; origin   **3.** reciprocal   **5.** π

7. $(-\infty, -1] \cup [1, \infty)$ **9.** e, π **10.** c, 2π **11.** a, 1

12. d, 2π **13.** f, 4   **14.** b, 4

15.

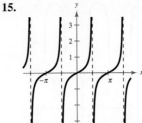

17.

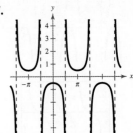

19.

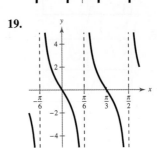

21.

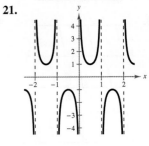

23.

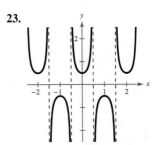

25.

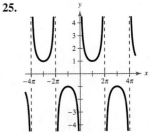

27.

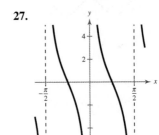

29.

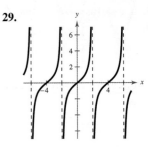

31.

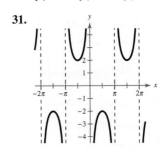

33.

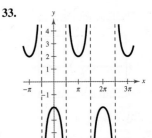

35.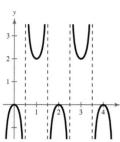

37.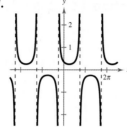

39.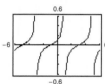

41.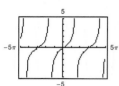

43.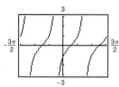

45.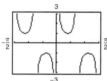

47.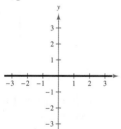

49. $-\dfrac{7\pi}{4}, -\dfrac{3\pi}{4}, \dfrac{\pi}{4}, \dfrac{5\pi}{4}$ **51.** $-\dfrac{7\pi}{6}, -\dfrac{\pi}{6}, \dfrac{5\pi}{6}, \dfrac{11\pi}{6}$

53. $-\dfrac{4\pi}{3}, -\dfrac{2\pi}{3}, \dfrac{2\pi}{3}, \dfrac{4\pi}{3}$ **55.** $-\dfrac{7\pi}{4}, -\dfrac{5\pi}{4}, \dfrac{\pi}{4}, \dfrac{3\pi}{4}$

57. Even   **59.** Odd   **61.** Odd   **63.** Even
65. d, $f \to 0$ as $x \to 0$.   **66.** a, $f \to 0$ as $x \to 0$.
67. b, $g \to 0$ as $x \to 0$.   **68.** c, $g \to 0$ as $x \to 0$.

69.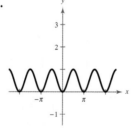

71.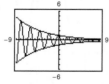

The functions are equal.   The functions are equal.

73.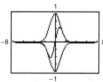

75.

As $x \to \infty$, $g(x) \to 0$.   As $x \to \infty$, $f(x) \to 0$.

77.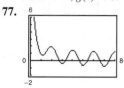

79.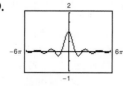

As $x \to 0$, $y \to \infty$.   As $x \to 0$, $g(x) \to 1$.

81.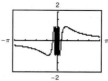

As $x \to 0$, $f(x)$ oscillates between 1 and -1.
83. (a) Period of $H(t)$: 12 mo
      Period of $L(t)$: 12 mo
  (b) Summer; winter   (c) About 0.5 mo
85. $d = 27 \sec x$

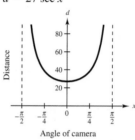

Angle of camera

87. True. For a given value of x, the y-coordinate of $\csc x$ is the reciprocal of the y-coordinate of $\sin x$.
89. (a)

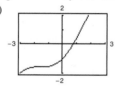

0.7391
  (b) 1, 0.5403, 0.8576, 0.6543, 0.7935, 0.7014, 0.7640, 0.7221, 0.7504, 0.7314, . . . ; 0.7391
91. (a) $f(x) \to \infty$ (b) $f(x) \to -\infty$
  (c) $f(x) \to \infty$ (d) $f(x) \to -\infty$
93. (a) $f(x) \to -\infty$ (b) $f(x) \to \infty$
  (c) $f(x) \to -\infty$ (d) $f(x) \to \infty$

Section 4.7 *(page 324)*

1. $y = \sin^{-1} x$; $-1 \le x \le 1$

3. $y = \tan^{-1} x$; $-\infty < x < \infty$; $-\dfrac{\pi}{2} < y < \dfrac{\pi}{2}$ **5.** $\dfrac{\pi}{6}$ **7.** $\dfrac{\pi}{3}$

9. $\dfrac{\pi}{6}$ **11.** Not possible   **13.** $-\dfrac{\pi}{3}$ **15.** $\dfrac{2\pi}{3}$ **17.** $-\dfrac{\pi}{3}$

19.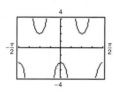

21. 1.19   **23.** -0.85 **25.** -1.25 **27.** Not possible
29. 1.99   **31.** 0.74   **33.** 1.07   **35.** -1.50

37. $-\dfrac{\pi}{3}, -\dfrac{\sqrt{3}}{3}, 1$ **39.** $\theta = \arctan \dfrac{x}{4}$ **41.** $\theta = \arcsin \dfrac{x+2}{5}$

43. $\theta = \arccos \dfrac{x+3}{2x}$ **45.** 0.3   **47.** Not possible

49. $\dfrac{\pi}{4}$ **51.** $\dfrac{3}{5}$ **53.** $\dfrac{\sqrt{5}}{5}$ **55.** $\dfrac{13}{12}$

57. $-\dfrac{5}{3}$ **59.** $-\dfrac{\sqrt{5}}{2}$ **61.** 2   **63.** $\sqrt{1-4x^2}$

65. $\dfrac{1}{x}$ **67.** $\sqrt{1-x^2}$ **69.** $\dfrac{\sqrt{9-x^2}}{x}$ **71.** $\dfrac{\sqrt{x^2+a^2}}{x}$

73.

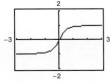

Asymptotes: $y = \pm 1$

75. $\dfrac{9}{\sqrt{x^2+81}}$ **77.** $\dfrac{|x-1|}{\sqrt{x^2-2x+10}}$

79.

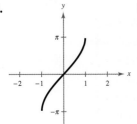

Vertical stretch

81.

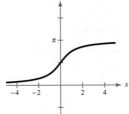

Shift $\dfrac{\pi}{2}$ units up

83.

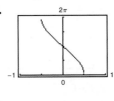

Horizontal stretch

85.

87.

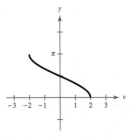

89.

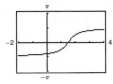

91. $3\sqrt{2}\,\sin\!\left(2t+\dfrac{\pi}{4}\right)$

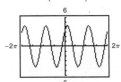

The graph implies that the identity is true.

93. $\dfrac{\pi}{2}$ **95.** $\dfrac{\pi}{2}$ **97.** π

99. (a) $\theta = \arcsin\dfrac{5}{s}$ (b) About 0.13, about 0.25

101. (a) About 32.9°   (b) About 6.5 m

103. (a)

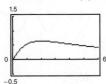

(b) 2 ft   (c) $\beta = 0$; As x increases, β approaches 0.

105. (a) $\theta = \arctan\dfrac{x}{20}$ (b) About 14.0°, about 31.0°

107. True. $-\dfrac{\pi}{4}$ is in the range of the arctangent function.

109. False. $\sin^{-1} x$ is the inverse of $\sin x$, not the reciprocal.

111. Domain: $(-\infty, \infty)$
Range: $(0, \pi)$

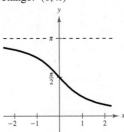

113. Domain: $(-\infty, -1] \cup [1, \infty)$
Range: $[-\pi/2, 0) \cup (0, \pi/2]$

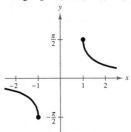

115. $\dfrac{\pi}{4}$ **117.** $\dfrac{3\pi}{4}$ **119.** $-\dfrac{\pi}{2}$ **121.** 1.17

123. -0.12 **125.** 0.19

127. (a) $\dfrac{\pi}{4}$ (b) $\dfrac{\pi}{2}$ (c) About 1.25   (d) About 2.03

129. (a) $f \circ f^{-1}$ $f^{-1} \circ f$

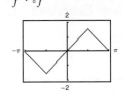

(b) The domains and ranges of the functions are restricted. The graphs of $f \circ f^{-1}$ and $f^{-1} \circ f$ differ because of the domains and ranges of f and f^{-1}.

Section 4.8   *(page 334)*

1. bearing   **3.** period
5. $a \approx 10.39$ **7.** $b \approx 14.21$
    $b = 6$ $c \approx 14.88$
    $B = 30°$ $A = 17.2°$
9. $c = 5$ **11.** $a \approx 52.88$
    $A \approx 36.87°$ $A \approx 73.46°$
    $B \approx 53.13°$ $B \approx 16.54°$
13. 3.00   **15.** 2.50   **17.** About 214.45 ft
19. About 19.7 ft   **21.** About 20.5 ft   **23.** About 11.8 km
25. About 56.3°   **27.** About 75.97°
29. About 3.23 mi or about 17,054 ft
31. (a) $l = 250$ ft, $A \approx 36.87°$, $B \approx 53.13°$
    (b) About 4.87 sec
33. About 508 mi north, about 650 mi east

35. (a) About 104.95 nmi south, about 58.18 nmi west

(b) S 36.7° W; about 130.9 nmi

37. N 56.31° W   **39.** (a) N 58° E   (b) About 68.82 m

41. About 35.3°   **43.** About 29.4 in.   **45.** $d = 4 \sin \pi t$

47. $d = 3 \cos \dfrac{4\pi t}{3}$ **49.** $\omega = 524\pi$

51. (a) 9   (b) $\frac{3}{5}$ (c) 9   (d) $\frac{5}{12}$

53. (a) $\frac{1}{4}$ (b) 3   (c) 0   (d) $\frac{1}{6}$

55. (a)   (b) $\dfrac{\pi}{8}$ (c) $\dfrac{\pi}{32}$

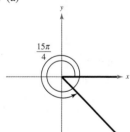

57. (a)

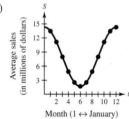

Month (1 ↔ January)

(b) $S = 8 + 6.3 \cos \dfrac{\pi}{6}t$ or $S = 8 + 6.3 \sin\left(\dfrac{\pi}{6}t + \dfrac{\pi}{2}\right)$

The model is a good fit.

(c) 12. Yes, sales of outerwear are seasonal.

(d) Maximum displacement from average sales of $8 million

59. False. The scenario does not create a right triangle because the tower is not vertical.

Review Exercises *(page 340)*

1. (a)   **3.** (a)

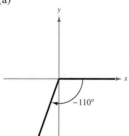

(b) Quadrant IV   (b) Quadrant III

(c) $\dfrac{23\pi}{4}, -\dfrac{\pi}{4}$ (c) 250°, −470°

5. 7.854   **7.** −0.279   **9.** 54°   **11.** −200.535°

13. 198° 24′   **15.** About 48.17 in.   **17.** About 523.60 in.²

19. $\left(-\dfrac{1}{2}, \dfrac{\sqrt{3}}{2}\right)$ **21.** $\left(-\dfrac{\sqrt{3}}{2}, -\dfrac{1}{2}\right)$

23. $\sin \dfrac{3\pi}{4} = \dfrac{\sqrt{2}}{2}$ $\csc \dfrac{3\pi}{4} = \sqrt{2}$

$\cos \dfrac{3\pi}{4} = -\dfrac{\sqrt{2}}{2}$ $\sec \dfrac{3\pi}{4} = -\sqrt{2}$

$\tan \dfrac{3\pi}{4} = -1$ $\cot \dfrac{3\pi}{4} = -1$

25. $\dfrac{\sqrt{2}}{2}$ **27.** $-\dfrac{\sqrt{3}}{2}$ **29.** 3.2361   **31.** −75.3130

33. $\sin \theta = \dfrac{4\sqrt{41}}{41}$ $\csc \theta = \dfrac{\sqrt{41}}{4}$

$\cos \theta = \dfrac{5\sqrt{41}}{41}$ $\sec \theta = \dfrac{\sqrt{41}}{5}$

$\tan \theta = \dfrac{4}{5}$ $\cot \theta = \dfrac{5}{4}$

35. 0.6494   **37.** 3.6722

39. (a) 3   (b) $\dfrac{2\sqrt{2}}{3}$ (c) $\dfrac{3\sqrt{2}}{4}$ (d) $\dfrac{\sqrt{2}}{4}$

41. About 73.3 m

43. $\sin \theta = \frac{4}{5}$ $\csc \theta = \frac{5}{4}$

$\cos \theta = \frac{3}{5}$ $\sec \theta = \frac{5}{3}$

$\tan \theta = \frac{4}{3}$ $\cot \theta = \frac{3}{4}$

45. $\sin \theta = \frac{4}{5}$ $\csc \theta = \frac{5}{4}$

$\cos \theta = \frac{3}{5}$ $\sec \theta = \frac{5}{3}$

$\tan \theta = \frac{4}{3}$ $\cot \theta = \frac{3}{4}$

47. $\sin \theta = -\dfrac{\sqrt{11}}{6}$ $\csc \theta = -\dfrac{6\sqrt{11}}{11}$

$\cos \theta = \dfrac{5}{6}$

$\tan \theta = -\dfrac{\sqrt{11}}{5}$ $\cot \theta = -\dfrac{5\sqrt{11}}{11}$

49. $\sin \theta = \dfrac{\sqrt{21}}{5}$ $\csc \theta = \dfrac{5\sqrt{21}}{21}$

$\sec \theta = -\dfrac{5}{2}$

$\tan \theta = -\dfrac{\sqrt{21}}{2}$ $\cot \theta = -\dfrac{2\sqrt{21}}{21}$

51. $\theta' = 84°$ **53.** $\theta' = \dfrac{\pi}{5}$

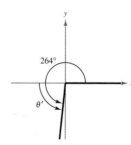

55. $\sin(-150°) = -\dfrac{1}{2}$; $\cos(-150°) = -\dfrac{\sqrt{3}}{2}$; $\tan(-150°) = \dfrac{\sqrt{3}}{3}$

57. $\sin \dfrac{\pi}{3} = \dfrac{\sqrt{3}}{2}$; $\cos \dfrac{\pi}{3} = \dfrac{1}{2}$; $\tan \dfrac{\pi}{3} = \sqrt{3}$

59. 0.9613   **61.** −0.4452

63.   **65.**

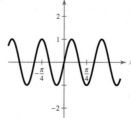
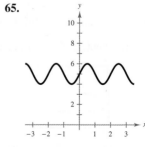

67.

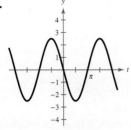

69. (a) $y = 2 \sin 528\pi x$ (b) 264 cycles/sec

71.

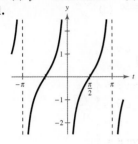

73.

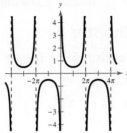

75.

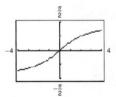

As $x \to +\infty, f(x)$ oscillates.

77. $-\dfrac{\pi}{2}$ **79.** $\dfrac{\pi}{6}$ **81.** -0.92 **83.** 0.06

85.

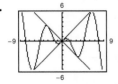

87. $\dfrac{4}{5}$ **89.** $\dfrac{13}{5}$ **91.** $\dfrac{\sqrt{4-x^2}}{x}$

93.

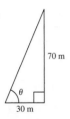

70 m

θ

30 m

$\theta \approx 66.8°$

95. About 1221 mi, about $85.6°$

97. False. For each θ there corresponds exactly one value of y.

99. The function is undefined because $\sec \theta = 1/\cos \theta$.

101. The ranges of the other four trigonometric functions are $(-\infty, \infty)$ or $(-\infty, -1] \cup [1, \infty)$.

Chapter Test   (page 343)

1. (a)

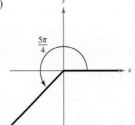

(b) $\dfrac{13\pi}{4}, -\dfrac{3\pi}{4}$

(c) $225°$

2. 3500 rad/min   **3.** About 709.04 ft^2

4. $\sin \theta = \dfrac{3\sqrt{13}}{13}$ $\csc \theta = \dfrac{\sqrt{13}}{3}$

$\cos \theta = \dfrac{2\sqrt{13}}{13}$ $\sec \theta = \dfrac{\sqrt{13}}{2}$

$\cot \theta = \dfrac{2}{3}$

5. $\sin \theta = \dfrac{3\sqrt{10}}{10}$ $\csc \theta = \dfrac{\sqrt{10}}{3}$

$\cos \theta = -\dfrac{\sqrt{10}}{10}$ $\sec \theta = -\sqrt{10}$

$\tan \theta = -3$ $\cot \theta = -\dfrac{1}{3}$

6. $\theta' = 25°$

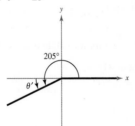

7. Quadrant III   **8.** $150°, 210°$

9. $\sin \theta = -\dfrac{4}{5}$ **10.** $\sin \theta = \dfrac{21}{29}$

$\tan \theta = -\dfrac{4}{3}$ $\cos \theta = -\dfrac{20}{29}$

$\csc \theta = -\dfrac{5}{4}$ $\tan \theta = -\dfrac{21}{20}$

$\sec \theta = \dfrac{5}{3}$ $\csc \theta = \dfrac{29}{21}$

$\cot \theta = -\dfrac{3}{4}$ $\cot \theta = -\dfrac{20}{21}$

11.

12.

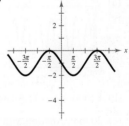

13.

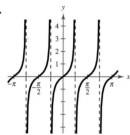

14.

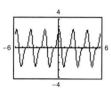

Period: 2

15.

Not periodic;
As $x \to \infty$, $y \to 0$.

16. $a = -2$, $b = \dfrac{1}{2}$, $c = -\dfrac{\pi}{4}$ **17.** $\dfrac{\sqrt{55}}{3}$

18.

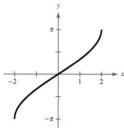

19. About 309.3°   **20.** $d = -6 \cos \pi t$

Problem Solving   *(page 345)*

1. (a) $\dfrac{11\pi}{2}$ rad or 990°   (b) About 816.42 ft

3. $h = 51 - 50 \sin\left(8\pi t + \dfrac{\pi}{2}\right)$

5. (a) About 4767 ft   (b) About 3705 ft

   (c) $w \approx 2183$ ft, $\tan 63° = \dfrac{w + 3705}{3000}$

7. (a)

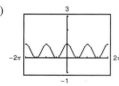

(b)

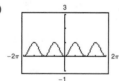

Even   Even

9. (a)

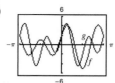 — *(Note: the graphed calculator image for 9(a))*

(b) Period $= \dfrac{3}{4}$ sec; Answers will vary.

(c) 20 mm; Answers will vary.   (d) 80 beats/min

(e) Period $= \dfrac{15}{16}$ sec; $\dfrac{32\pi}{15}$

11. (a)

(b) Period of f: 2π
   Period of g: π
(c) Yes, because the sine
   and cosine functions are
   periodic.

13. (a) About 40.5°   (b) $x \approx 1.71$ ft; $y \approx 3.46$ ft
   (c) About 1.75 ft
   (d) As you move closer to the rock, d must get smaller and
      smaller. The angles θ_1 and θ_2 will decrease along with the
      distance y, so d will decrease.

Chapter 5

Section 5.1   *(page 353)*

1. $\tan u$ **3.** $\cot u$ **5.** 1

7. $\sin x = \dfrac{\sqrt{21}}{5}$ $\csc x = \dfrac{5\sqrt{21}}{21}$

   $\cos x = -\dfrac{2}{5}$ $\sec x = -\dfrac{5}{2}$

   $\tan x = -\dfrac{\sqrt{21}}{2}$ $\cot x = -\dfrac{2\sqrt{21}}{21}$

9. $\sin \theta = -\dfrac{3}{4}$ $\csc \theta = -\dfrac{4}{3}$

   $\cos \theta = \dfrac{\sqrt{7}}{4}$ $\sec \theta = \dfrac{4\sqrt{7}}{7}$

   $\tan \theta = -\dfrac{3\sqrt{7}}{7}$ $\cot \theta = -\dfrac{\sqrt{7}}{3}$

11. $\sin x = \dfrac{2\sqrt{13}}{13}$ $\csc x = \dfrac{\sqrt{13}}{2}$

   $\cos x = \dfrac{3\sqrt{13}}{13}$ $\sec x = \dfrac{\sqrt{13}}{3}$

   $\tan x = \dfrac{2}{3}$ $\cot x = \dfrac{3}{2}$

13. c   **14.** b   **15.** f   **16.** a   **17.** e   **18.** d

19. $\cos \theta$ **21.** $\sin^2 x$ **23.** $\sec x + 1$ **25.** $\sin^4 x$

27. $\csc^2 x (\cot x + 1)$ **29.** $(3 \sin x + 1)(\sin x - 2)$

31. $(\csc x - 1)(\csc x + 2)$ **33.** $\sec \theta$ **35.** $\cos^2 \phi$

37. $\sec \beta$ **39.** $\sin^2 x$ **41.** $1 + 2 \sin x \cos x$

43. $2 \csc^2 x$ **45.** $-2 \tan x$ **47.** $-\cot x$ **49.** $1 + \cos y$

51. $\sin x$ **53.** $3 \sin \theta$ **55.** $2 \tan \theta$

57. $2 \cos \theta = \sqrt{2}$; $\sin \theta = \pm\dfrac{\sqrt{2}}{2}$; $\cos \theta = \dfrac{\sqrt{2}}{2}$

59. $0 \le \theta \le \pi$ **61.** $\ln|\cos x|$

63. $\ln|\csc t \sec t|$ **65.** $\mu = \tan \theta$

67. True. $\csc x = \dfrac{1}{\sin x}$, $\sec x = \dfrac{1}{\cos x}$, $\tan x = \dfrac{\sin x}{\cos x}$, $\cot x = \dfrac{\cos x}{\sin x}$

69. $\infty, 0$ **71.** $\cos(-\theta) = \cos \theta$

73. $\cos \theta = \pm\sqrt{1 - \sin^2 \theta}$ **75.** $\dfrac{\sin \theta}{\cos \theta}$

   $\tan \theta = \pm\dfrac{\sin \theta}{\sqrt{1 - \sin^2 \theta}}$

   $\cot \theta = \pm\dfrac{\sqrt{1 - \sin^2 \theta}}{\sin \theta}$

   $\sec \theta = \pm\dfrac{1}{\sqrt{1 - \sin^2 \theta}}$

   $\csc \theta = \dfrac{1}{\sin \theta}$

Section 5.2 *(page 360)*

1. identity   **3.** tan u **5.** sin u **7.** $-\csc u$

9–41. Answers will vary.

43. $\cot(-x) = -\cot x$

45. (a)

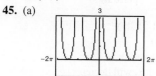

    (b)

    Identity

    (c) Answers will vary.

47. (a)

    (b)

    Not an identity

    (c) Answers will vary.

49. (a)

    (b)

    Identity

    (c) Answers will vary.

51–53. Answers will vary.   **55.** 1

57–61. Answers will vary.

63. False. $\tan x^2 = \tan(x \cdot x) \neq \tan^2 x = (\tan x)(\tan x)$

65. False. An identity is an equation that is true for all real values of θ.

67. The equation is not an identity because $\sin \theta = \pm\sqrt{1 - \cos^2 \theta}$.

    Sample answer: $\dfrac{7\pi}{4}$

69. The equation is not an identity because $1 - \cos^2 \theta = \sin^2 \theta$.

    Sample answer: $-\dfrac{\pi}{2}$

Section 5.3 *(page 370)*

1. isolate   **3.** quadratic   **5–9.** Answers will vary.

11. $\dfrac{\pi}{3} + 2n\pi, \dfrac{2\pi}{3} + 2n\pi$ **13.** $\dfrac{2\pi}{3} + 2n\pi, \dfrac{4\pi}{3} + 2n\pi$

15. $\dfrac{\pi}{6} + n\pi, \dfrac{5\pi}{6} + n\pi$ **17.** $\dfrac{\pi}{3} + n\pi, \dfrac{2\pi}{3} + n\pi$

19. $n\pi, \dfrac{3\pi}{2} + 2n\pi$ **21.** $\dfrac{n\pi}{2}$ **23.** $n\pi, \dfrac{\pi}{6} + n\pi, \dfrac{5\pi}{6} + n\pi$

25. $\pi + 2n\pi, \dfrac{\pi}{3} + 2n\pi, \dfrac{5\pi}{3} + 2n\pi$

27. $\dfrac{\pi}{3} + 2n\pi, \pi + 2n\pi, \dfrac{5\pi}{3} + 2n\pi$ **29.** $\dfrac{\pi}{4}, \dfrac{5\pi}{4}$

31. $\dfrac{\pi}{2}, \dfrac{3\pi}{2}, \dfrac{2\pi}{3}, \dfrac{4\pi}{3}$ **33.** $\dfrac{\pi}{3}, \dfrac{2\pi}{3}, \dfrac{4\pi}{3}, \dfrac{5\pi}{3}$ **35.** No solution

37. $\dfrac{\pi}{2}$ **39.** $\dfrac{\pi}{6} + n\pi, \dfrac{5\pi}{6} + n\pi$ **41.** $\dfrac{\pi}{12} + \dfrac{n\pi}{3}$

43. $\dfrac{\pi}{2} + 4n\pi, \dfrac{7\pi}{2} + 4n\pi$ **45.** $\dfrac{\pi}{3} + 2n\pi$ **47.** $3 + 4n$

49. 3.553, 5.872   **51.** 1.249, 4.391   **53.** 0.739

55. 0.955, 2.186, 4.097, 5.328   **57.** 1.221, 1.921, 4.362, 5.062

59. $\arctan(-4) + n\pi, \arctan 3 + n\pi$

61. $\dfrac{\pi}{4} + n\pi, \arctan 5 + n\pi$ **63.** $\dfrac{\pi}{3} + 2n\pi, \dfrac{5\pi}{3} + 2n\pi$

65. $\arctan \dfrac{1}{3} + n\pi, \arctan\left(-\dfrac{1}{3}\right) + n\pi$

67. $\arccos \dfrac{1}{4} + 2n\pi, -\arccos \dfrac{1}{4} + 2n\pi$

69. $\dfrac{\pi}{2} + 2n\pi, \arcsin\left(-\dfrac{1}{4}\right) + 2n\pi, \arcsin \dfrac{1}{4} + 2n\pi$

71. 0.3398, 0.8481, 2.2935, 2.8018

73. 1.9357, 2.7767, 5.0773, 5.9183

75. $-1.154, 0.534$ **77.** 1.110

79. (a)

    (b) $\dfrac{\pi}{3} \approx 1.0472$

    $\dfrac{5\pi}{3} \approx 5.2360$

    0

    $\pi \approx 3.1416$

    Maximum: $(1.0472, 1.25)$

    Maximum: $(5.2360, 1.25)$

    Minimum: $(0, 1)$

    Minimum: $(3.1416, -1)$

81. (a)

    (b) $\dfrac{\pi}{4} \approx 0.7854$

    $\dfrac{5\pi}{4} \approx 3.9270$

    Maximum: $(0.7854, 1.4142)$

    Minimum: $(3.9270, -1.4142)$

83. (a)

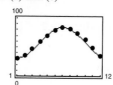

    (b) $\dfrac{\pi}{4} \approx 0.7854$

    $\dfrac{5\pi}{4} \approx 3.9270$

    $\dfrac{3\pi}{4} \approx 2.3562$

    $\dfrac{7\pi}{4} \approx 5.4978$

    Maximum: $(0.7854, 0.5)$

    Maximum: $(3.9270, 0.5)$

    Minimum: $(2.3562, -0.5)$

    Minimum: $(5.4978, -0.5)$

85. 1

87. (a) All real numbers x except $x = 0$

    (b) y-axis symmetry; Horizontal asymptote: $y = 0$

    (c) y approaches 1.

    (d) Four solutions: $\pm\pi, \pm 2\pi$

89. 0.04 sec, 0.43 sec, 0.83 sec

91. January, November, December

93. (a) and (c)

    The model fits the data well.

    (b) $C = 26.55 \cos\left(\dfrac{\pi t}{6} - \dfrac{7\pi}{6}\right) + 57.55$ (d) $57.55°$F

    (e) Above $72°$F: June through September

         Below $72°$F: October through May

CHAPTER 5

95. (a)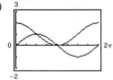

$A \approx 1.12$

(b) $0.6 < x < 1.1$

97. 1

99. True. The first equation has a smaller period than the second equation, so it will have more solutions in the interval $[0, 2\pi)$.

101. The equation would become $\cos^2 x = 2$; this is not the correct method to use when solving equations.

103. (a)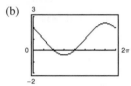

Graphs intersect when $x = \dfrac{\pi}{2}$ and $x = \pi$.

(b)

x-intercepts: $\left(\dfrac{\pi}{2}, 0\right), (\pi, 0)$

(c) Yes; Answers will vary.

Section 5.4   *(page 378)*

1. $\sin u \cos v - \cos u \sin v$

3. $\dfrac{\tan u + \tan v}{1 - \tan u \tan v}$

5. $\cos u \cos v + \sin u \sin v$

7. (a) $\dfrac{\sqrt{2} - \sqrt{6}}{4}$ (b) $\dfrac{\sqrt{2} + 1}{2}$

9. (a) $\dfrac{\sqrt{6} + \sqrt{2}}{4}$ (b) $\dfrac{\sqrt{2} - \sqrt{3}}{2}$

11. $\sin \dfrac{11\pi}{12} = \dfrac{\sqrt{2}}{4}\left(\sqrt{3} - 1\right)$

$\cos \dfrac{11\pi}{12} = -\dfrac{\sqrt{2}}{4}\left(\sqrt{3} + 1\right)$

$\tan \dfrac{11\pi}{12} = -2 + \sqrt{3}$

13. $\sin \dfrac{17\pi}{12} = -\dfrac{\sqrt{2}}{4}\left(\sqrt{3} + 1\right)$

$\cos \dfrac{17\pi}{12} = \dfrac{\sqrt{2}}{4}\left(1 - \sqrt{3}\right)$

$\tan \dfrac{17\pi}{12} = 2 + \sqrt{3}$

15. $\sin 105° = \dfrac{\sqrt{2}}{4}\left(\sqrt{3} + 1\right)$

$\cos 105° = \dfrac{\sqrt{2}}{4}\left(1 - \sqrt{3}\right)$

$\tan 105° = -2 - \sqrt{3}$

17. $\sin(-195°) = \dfrac{\sqrt{2}}{4}\left(\sqrt{3} - 1\right)$

$\cos(-195°) = -\dfrac{\sqrt{2}}{4}\left(\sqrt{3} + 1\right)$

$\tan(-195°) = -2 + \sqrt{3}$

19. $\sin \dfrac{13\pi}{12} = \dfrac{\sqrt{2}}{4}\left(1 - \sqrt{3}\right)$

$\cos \dfrac{13\pi}{12} = -\dfrac{\sqrt{2}}{4}\left(1 + \sqrt{3}\right)$

$\tan \dfrac{13\pi}{12} = 2 - \sqrt{3}$

21. $\sin\left(-\dfrac{5\pi}{12}\right) = -\dfrac{\sqrt{2}}{4}\left(1 + \sqrt{3}\right)$

$\cos\left(-\dfrac{5\pi}{12}\right) = \dfrac{\sqrt{2}}{4}\left(\sqrt{3} - 1\right)$

$\tan\left(-\dfrac{5\pi}{12}\right) = -2 - \sqrt{3}$

23. $\sin 285° = -\dfrac{\sqrt{2}}{4}\left(\sqrt{3} + 1\right)$

$\cos 285° = \dfrac{\sqrt{2}}{4}\left(\sqrt{3} - 1\right)$

$\tan 285° = -\left(2 + \sqrt{3}\right)$

25. $\sin(-165°) = -\dfrac{\sqrt{2}}{4}\left(\sqrt{3} - 1\right)$

$\cos(-165°) = -\dfrac{\sqrt{2}}{4}\left(1 + \sqrt{3}\right)$

$\tan(-165°) = 2 - \sqrt{3}$

27. $\sin 1.8$ **29.** $\sin 75°$ **31.** $\tan \dfrac{7\pi}{15}$ **33.** $\cos(3x - 2y)$

35. $\dfrac{\sqrt{3}}{2}$ **37.** $-\dfrac{1}{2}$ **39.** 0   **41.** $-\dfrac{13}{85}$ **43.** $-\dfrac{13}{84}$

45. $\dfrac{85}{36}$ **47.** $\dfrac{3}{5}$ **49.** $-\dfrac{44}{117}$ **51.** $-\dfrac{125}{44}$ **53.** 1   **55.** 0

57–63. Answers will vary.   **65.** $-\sin \theta$ **67.** $-\sec \theta$

69. $\dfrac{\pi}{6}, \dfrac{5\pi}{6}$ **71.** $\dfrac{5\pi}{4}, \dfrac{7\pi}{4}$ **73.** $0, \dfrac{\pi}{3}, \pi, \dfrac{5\pi}{3}$ **75.** $\dfrac{\pi}{4}, \dfrac{7\pi}{4}$

77. $\dfrac{\pi}{2}, \pi, \dfrac{3\pi}{2}$

79. (a) $y = \dfrac{5}{12} \sin(2t + 0.6435)$

(b) $\dfrac{5}{12}$ ft   (c) $\dfrac{1}{\pi}$ cycle/sec

81. True. $\sin(u \pm v) = \sin u \cos v \pm \cos u \sin v$

83. False. $\sin \alpha \cos \beta = -\cos \alpha \sin \beta$

85. The denominator should be $1 + \tan x \tan \dfrac{\pi}{4} = 1 + \tan x.$

87–89. Answers will vary.

91. (a) $\sqrt{2} \sin\left(\theta + \dfrac{\pi}{4}\right)$ (b) $\sqrt{2} \cos\left(\theta - \dfrac{\pi}{4}\right)$

93. (a) $13 \sin(3\theta + 0.3948)$ (b) $13 \cos(3\theta - 1.1760)$

95. $\sqrt{2} \sin \theta + \sqrt{2} \cos \theta$ **97.** $15°$

99. 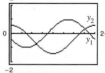   **101.** Proof

No, $y_1 \neq y_2$ because their graphs are different.

Section 5.5   *(page 388)*

1. $2 \sin u \cos u$ **3.** $\frac{1}{2}[\sin(u + v) + \sin(u - v)]$

5. $\pm \sqrt{\dfrac{1 - \cos u}{2}}$ **7.** $n\pi, \dfrac{\pi}{3} + 2n\pi, \dfrac{5\pi}{3} + 2n\pi$ **9.** $\dfrac{2n\pi}{3}$

11. $\dfrac{n\pi}{2}$ **13.** $\dfrac{\pi}{6} + n\pi, \dfrac{\pi}{2} + n\pi, \dfrac{5\pi}{6} + n\pi$ **15.** $3 \sin 2x$

17. $3 \cos 2x$ **19.** $4 \cos 2x$

21. $\sin 2u = -\frac{24}{25}, \cos 2u = \frac{7}{25}, \tan 2u = -\frac{24}{7}$

23. $\sin 2u = \frac{15}{17}, \cos 2u = \frac{8}{17}, \tan 2u = \frac{15}{8}$

25. $8 \cos^4 x - 8 \cos^2 x + 1$ **27.** $\frac{1}{8}(3 + 4 \cos 2x + \cos 4x)$

29. $\dfrac{1}{8}(3 - 4 \cos 4x + \cos 8x)$ **31.** $\dfrac{(3 - 4 \cos 4x + \cos 8x)}{(3 + 4 \cos 4x + \cos 8x)}$

33. $\frac{1}{8}(1 - \cos 8x)$

35. $\sin 75° = \frac{1}{2}\sqrt{2 + \sqrt{3}}$

$\cos 75° = \frac{1}{2}\sqrt{2 - \sqrt{3}}$

$\tan 75° = 2 + \sqrt{3}$

37. $\sin 112° \, 30' = \frac{1}{2}\sqrt{2 + \sqrt{2}}$

$\cos 112° \, 30' = -\frac{1}{2}\sqrt{2 - \sqrt{2}}$

$\tan 112° \, 30' = -1 - \sqrt{2}$

39. $\sin \dfrac{\pi}{8} = \frac{1}{2}\sqrt{2 - \sqrt{2}}$

$\cos \dfrac{\pi}{8} = \frac{1}{2}\sqrt{2 + \sqrt{2}}$

$\tan \dfrac{\pi}{8} = \sqrt{2} - 1$

41. (a) Quadrant I

(b) $\sin \dfrac{u}{2} = \dfrac{3}{5}, \cos \dfrac{u}{2} = \dfrac{4}{5}, \tan \dfrac{u}{2} = \dfrac{3}{4}$

43. (a) Quadrant II

(b) $\sin \dfrac{u}{2} = \dfrac{\sqrt{26}}{26}, \cos \dfrac{u}{2} = -\dfrac{5\sqrt{26}}{26}, \tan \dfrac{u}{2} = -\dfrac{1}{5}$

45. π **47.** $\dfrac{\pi}{3}, \pi, \dfrac{5\pi}{3}$ **49.** $\frac{1}{2}(\cos 2\theta - \cos 8\theta)$

51. $\frac{1}{2}(\cos(-2\theta) + \cos 6\theta)$ **53.** $2 \cos 4\theta \sin \theta$

55. $2 \cos 4x \cos 2x$ **57.** $\dfrac{\sqrt{6}}{2}$ **59.** $-\sqrt{2}$

61. $0, \dfrac{\pi}{4}, \dfrac{\pi}{2}, \dfrac{3\pi}{4}, \pi, \dfrac{5\pi}{4}, \dfrac{3\pi}{2}, \dfrac{7\pi}{4}$ **63.** $\dfrac{\pi}{6}, \dfrac{5\pi}{6}$

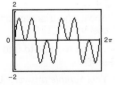   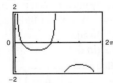

65–69. Answers will vary.

71. (a) $\cos \theta = \dfrac{M^2 - 2}{M^2}$ (b) $\dfrac{\pi}{3}$ (c) 0.4482

(d) 1520 mi/h; 3420 mi/h

73. $x = 2r(1 - \cos \theta)$

75. True. $\sin(-2x) = 2 \sin(-x) \cos(-x) = -2 \sin x \cos x.$

77. Answers will vary.

Review Exercises   *(page 392)*

1. $\cot x$ **3.** $\cos x$

5. $\sin \theta = -\dfrac{\sqrt{21}}{5}$ $\csc \theta = -\dfrac{5\sqrt{21}}{21}$

$\cos \theta = -\dfrac{2}{5}$ $\sec \theta = -\dfrac{5}{2}$

$\tan \theta = \dfrac{\sqrt{21}}{2}$ $\cot \theta = \dfrac{2\sqrt{21}}{21}$

7. $\sin^2 x$ **9.** 1   **11.** $\tan u \sec u$ **13.** $\cot^2 x$

15. $-2 \tan^2 \theta$ **17.** $5 \cos \theta$ **19–25.** Answers will vary.

27. $\dfrac{\pi}{3} + 2n\pi, \dfrac{2\pi}{3} + 2n\pi$ **29.** $\dfrac{\pi}{6} + n\pi$

31. $\dfrac{\pi}{3} + n\pi, \dfrac{2\pi}{3} + n\pi$ **33.** $0, \dfrac{\pi}{2}, \pi, \dfrac{3\pi}{2}$ **35.** $0, \dfrac{\pi}{2}, \pi$

37. $\dfrac{\pi}{8}, \dfrac{3\pi}{8}, \dfrac{9\pi}{8}, \dfrac{11\pi}{8}$ **39.** $\dfrac{\pi}{2}$

41. $0, \dfrac{\pi}{8}, \dfrac{3\pi}{8}, \dfrac{5\pi}{8}, \dfrac{7\pi}{8}, \dfrac{9\pi}{8}, \dfrac{11\pi}{8}, \dfrac{13\pi}{8}, \dfrac{15\pi}{8}$

43. $n\pi, \arctan 2 + n\pi$ **45.** $\arctan(-3) + n\pi, \arctan 2 + n\pi$

47. $\sin 75° = \dfrac{\sqrt{2}}{4}\left(1 + \sqrt{3}\right)$ **49.** $\sin \dfrac{25\pi}{12} = \dfrac{\sqrt{2}}{4}\left(\sqrt{3} - 1\right)$

$\cos 75° = \dfrac{\sqrt{2}}{4}\left(\sqrt{3} - 1\right)$ $\cos \dfrac{25\pi}{12} = \dfrac{\sqrt{2}}{4}\left(\sqrt{3} + 1\right)$

$\tan 75° = 2 + \sqrt{3}$ $\tan \dfrac{25\pi}{12} = 2 - \sqrt{3}$

51. $\sin 15°$ **53.** $-\frac{24}{25}$ **55.** -1

57–59. Answers will vary.   **61.** $\dfrac{\pi}{4}, \dfrac{7\pi}{4}$

63. $\sin 2u = \frac{24}{25}$

$\cos 2u = -\frac{7}{25}$

$\tan 2u = -\frac{24}{7}$

65. Answers will vary.   **67.** $\dfrac{1 - \cos 6x}{1 + \cos 6x}$

69. $\sin(-75°) = -\frac{1}{2}\sqrt{2 + \sqrt{3}}$

$\cos(-75°) = \frac{1}{2}\sqrt{2 - \sqrt{3}}$

$\tan(-75°) = -2 - \sqrt{3}$

71. (a) Quadrant II

(b) $\sin \dfrac{u}{2} = \dfrac{2\sqrt{5}}{5}, \cos \dfrac{u}{2} = -\dfrac{\sqrt{5}}{5}, \tan \dfrac{u}{2} = -2$

73. (a) Quadrant I

(b) $\sin \dfrac{u}{2} = \dfrac{3\sqrt{14}}{14}, \cos \dfrac{u}{2} = \dfrac{\sqrt{70}}{14}, \tan \dfrac{u}{2} = \dfrac{3\sqrt{5}}{5}$

75. $\frac{1}{2}[\sin 10\theta - \sin(-2\theta)]$

77. $2 \cos \dfrac{11\theta}{2} \cos \dfrac{\theta}{2}$ **79.** $\theta = 15°$ or $\dfrac{\pi}{12}$

81. False. If $(\pi/2) < \theta < \pi$, then $\theta/2$ lies in Quadrant I.

83. True. $4 \sin(-x) \cos(-x) = 4(-\sin x) \cos x$

$= -4 \sin x \cos x$

$= -2(2 \sin x \cos x)$

$= -2 \sin 2x$

85. Yes. *Sample answer:* $\sin x = \frac{1}{2}$ has an infinite number of solutions.

CHAPTER 5

Chapter Test *(page 394)*

1. $\sin \theta = \dfrac{2}{5}$ $\qquad\qquad$ $\csc \theta = \dfrac{5}{2}$

$\cos \theta = -\dfrac{\sqrt{21}}{5}$ $\qquad$ $\sec \theta = -\dfrac{5\sqrt{21}}{21}$

$\tan \theta = -\dfrac{2\sqrt{21}}{21}$ $\qquad$ $\cot \theta = -\dfrac{\sqrt{21}}{2}$

2. 1   3. 1   4. $\csc \theta \sec \theta$ 5–10. Answers will vary.

11. $2(\sin 5\theta + \sin \theta)$ 12. $-2 \sin \theta$

13. $0, \dfrac{3\pi}{4}, \pi, \dfrac{7\pi}{4}$ 14. $\dfrac{\pi}{6}, \dfrac{\pi}{2}, \dfrac{5\pi}{6}, \dfrac{3\pi}{2}$ 15. $\dfrac{\pi}{6}, \dfrac{5\pi}{6}, \dfrac{7\pi}{6}, \dfrac{11\pi}{6}$

16. $\dfrac{\pi}{6}, \dfrac{5\pi}{6}, \dfrac{3\pi}{2}$ 17. $0, 2.596$ 18. $\dfrac{\sqrt{2} - \sqrt{6}}{4}$

19. $\sin 2u = -\dfrac{20}{29}, \cos 2u = -\dfrac{21}{29}, \tan 2u = \dfrac{20}{21}$

20. Day 30 to day 310

21. 0.26 min, 0.58 min, 0.89 min, 1.20 min, 1.52 min, 1.83 min

Problem Solving *(page 397)*

1. $\sin \theta = \pm\sqrt{1 - \cos^2 \theta}$

$\tan \theta = \pm\dfrac{\sqrt{1 - \cos^2 \theta}}{\cos \theta}$

$\csc \theta = \pm\dfrac{1}{\sqrt{1 - \cos^2 \theta}}$

$\sec \theta = \dfrac{1}{\cos \theta}$

$\cot \theta = \pm\dfrac{\cos \theta}{\sqrt{1 - \cos^2 \theta}}$

3. Answers will vary.

5. $u + v = w$; Proof

7. (a) $A = 100 \sin \dfrac{\theta}{2} \cos \dfrac{\theta}{2}$ (b) $A = 50 \sin \theta; \theta = \dfrac{\pi}{2}$

9. (a) $F = \dfrac{0.6W \cos \theta}{\sin 12°}$

(b) 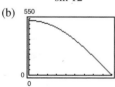   (c) Maximum: $\theta = 0°$
$\qquad\qquad\qquad$ Minimum: $\theta = 90°$

11. (a) High tides: 6:12 A.M., 6:36 P.M.
Low tides: 12:00 A.M., 12:24 P.M.

(b) The water depth never falls below 7 feet.

(c)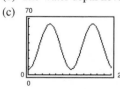

13. (a) $n = \dfrac{1}{2}\left(\cot \dfrac{\theta}{2} + \sqrt{3}\right)$ (b) $\theta \approx 76.5°$

15. (a) $\dfrac{\pi}{6} \le x \le \dfrac{5\pi}{6}$ (b) $\dfrac{2\pi}{3} \le x \le \dfrac{4\pi}{3}$

(c) $\dfrac{\pi}{2} < x < \pi, \dfrac{3\pi}{2} < x < 2\pi$

(d) $0 \le x \le \dfrac{\pi}{4}, \dfrac{5\pi}{4} \le x < 2\pi$

Chapter 6

Section 6.1 *(page 406)*

1. oblique   3. angles; side

5. $A = 30°, a \approx 14.14, c \approx 27.32$

7. $C = 105°, a \approx 5.94, b \approx 6.65$

9. $B = 60.9°, b \approx 19.32, c \approx 6.36$

11. $B = 42° 4', a \approx 22.05, b \approx 14.88$

13. $C = 80°, a \approx 5.82, b \approx 9.20$

15. $C = 83°, a \approx 0.62, b \approx 0.51$

17. $B \approx 21.55°, C \approx 122.45°, c \approx 11.49$

19. $B \approx 9.43°, C \approx 25.57°, c \approx 10.53$

21. $A \approx 10° 11', C \approx 154° 19', c \approx 11.03$

23. $B \approx 48.74°, C \approx 21.26°, c \approx 48.23$

25. No solution

27. Two solutions:
$B \approx 72.21°, C \approx 49.79°, c \approx 10.27$
$B \approx 107.79°, C \approx 14.21°, c \approx 3.30$

29. No solution   31. $B = 45°, C = 90°, c \approx 1.41$

33. (a) $b \le 5, b = \dfrac{5}{\sin 36°}$ (b) $5 < b < \dfrac{5}{\sin 36°}$

(c) $b > \dfrac{5}{\sin 36°}$

35. (a) $b < 80$ (b) Not possible   (c) $b \ge 80$

37. 22.1   39. 94.4   41. 218.0   43. 22.3

45. (a) $\dfrac{h}{\sin 30°} = \dfrac{40}{\sin 56°}$ (b) About 24.1 m

47. From Pine Knob: about 42.4 km
From Colt Station: about 15.5 km

49. About 16.1°

51. (a)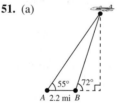
(b) About 6.16 mi
(c) About 5.86 mi
(d) About 4.10 mi

53. $d = \dfrac{2 \sin \theta}{\sin(\phi - \theta)}$

55. True. If an angle of a triangle is obtuse, then the other two angles must be acute.

57. False. When just three angles are known, the triangle cannot be solved.

59. The formula is $A = \dfrac{1}{2}ab \sin C$, so solve the triangle to find the value of a.

61. Yes. $A = 40°, b \approx 11.9, c \approx 15.6$; Yes; *Sample answer:* Use the definitions of cosine and tangent.

Section 6.2 *(page 413)*

1. $b^2 = a^2 + c^2 - 2ac \cos B$ 3. standard

5. $A \approx 38.62°, B \approx 48.51°, C \approx 92.87°$

7. $A \approx 26.38°, B \approx 36.34°, C \approx 117.28°$

9. $B \approx 23.79°, C \approx 126.21°, a \approx 18.59$

11. $B \approx 29.44°, C \approx 100.56°, a \approx 23.38$

13. $A \approx 30.11°, B \approx 43.16°, C \approx 106.73°$

15. $A \approx 132.77°, B \approx 31.91°, C \approx 15.32°$

17. $B \approx 27.46°, C \approx 32.54°, a \approx 11.27$

19. $A \approx 141° \ 45', C \approx 27° \ 40', b \approx 11.87$

21. $A = 27° \ 10', C = 27° \ 10', b \approx 65.84$

23. $A \approx 33.80°, B \approx 103.20°, c \approx 0.54$ **25.** $12.07; 5.69; 135°$

27. $13.86; 68.2°; 111.8°$ **29.** $16.96; 77.2°; 102.8°$

31. Yes; $A \approx 102.44°, C \approx 37.56°, b \approx 5.26$

33. No; No solution   **35.** No; $C = 103°, a \approx 0.82, b \approx 0.71$

37. 23.53 **39.** 5.35 **41.** 0.24 **43.** 1514.14

45. About 373.3 m   **47.** About 103.9 ft

49. About 131.1 ft, about 118.6 ft

51.

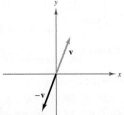

N 37.1° E, S 63.1° E

53. $41.2°, 52.9°$

55. (a) $d = \sqrt{9s^2 - 108s + 1296}$ (b) About 15.87 mi/h

57. About 46,837.5 ft^2 **59.** \$83,336.37

61. False. For s to be the average of the lengths of the three sides of the triangle, s would be equal to $(a + b + c)/3$.

63. $c^2 = a^2 + b^2$; The Pythagorean Theorem is a special case of the Law of Cosines.

65. The Law of Cosines can be used to solve the single-solution case of SSA. There is no method that can solve the no-solution case of SSA.

67. Proof

Section 6.3   *(page 425)*

1. directed line segment   **3.** vector   **5.** standard position

7. multiplication; addition

9. Equivalent; **u** and **v** have the same magnitude and direction.

11. Not equivalent; **u** and **v** do not have the same direction.

13. Equivalent; **u** and **v** have the same magnitude and direction.

15. $\mathbf{v} = \langle 1, 3 \rangle; \|\mathbf{v}\| = \sqrt{10}$ **17.** $\mathbf{v} = \langle 0, 5 \rangle; \|\mathbf{v}\| = 5$

19. $\mathbf{v} = \langle -8, 6 \rangle; \|\mathbf{v}\| = 10$ **21.** $\mathbf{v} = \langle -9, -12 \rangle; \|\mathbf{v}\| = 15$

23. $\mathbf{v} = \langle 16, -26 \rangle; \|\mathbf{v}\| = 2\sqrt{233}$

25.

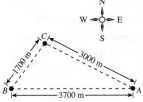

27.

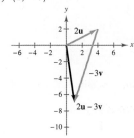

29.

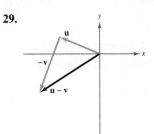

31. (a) $\langle 3, 4 \rangle$ (b) $\langle 1, -2 \rangle$

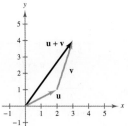
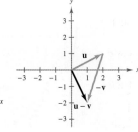

(c) $\langle 1, -7 \rangle$

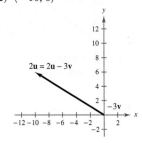

33. (a) $\langle -5, 3 \rangle$ (b) $\langle -5, 3 \rangle$

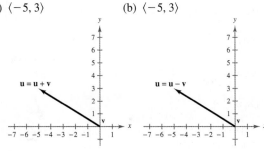

(c) $\langle -10, 6 \rangle$

35. (a) $\langle 1, -9 \rangle$ (b) $\langle -1, -5 \rangle$

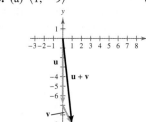

(c) $\langle -3, -8 \rangle$

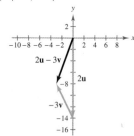

37. 10   **39.** $9\sqrt{5}$ **41.** $\langle 1, 0 \rangle$ **43.** $\left\langle -\dfrac{\sqrt{2}}{2}, \dfrac{\sqrt{2}}{2} \right\rangle$

45. $\left\langle \dfrac{\sqrt{37}}{37}, -\dfrac{6\sqrt{37}}{37} \right\rangle$ **47.** $\mathbf{v} = \langle -6, 8 \rangle$

49. $\mathbf{v} = \left\langle \dfrac{18\sqrt{29}}{29}, \dfrac{45\sqrt{29}}{29} \right\rangle$ **51.** $5\mathbf{i} - 3\mathbf{j}$ **53.** $-6\mathbf{i} + 3\mathbf{j}$

55. $\mathbf{v} = \left\langle 3, -\dfrac{3}{2} \right\rangle$ **57.** $\mathbf{v} = \langle 4, 3 \rangle$

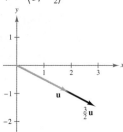   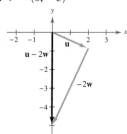

59. $\mathbf{v} = \langle 0, -5 \rangle$

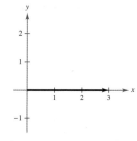

61. $\|\mathbf{v}\| = 6\sqrt{2};\ \theta = 315°$ **63.** $\|\mathbf{v}\| = 3;\ \theta = 60°$

65. $\mathbf{v} = \langle 3, 0 \rangle$ **67.** $\mathbf{v} = \left\langle -\dfrac{7\sqrt{3}}{4}, \dfrac{7}{4} \right\rangle$

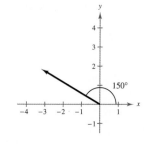

69. $\mathbf{v} = \left\langle \dfrac{9}{5}, \dfrac{12}{5} \right\rangle$

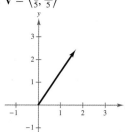

71. $\langle 2, 4 + 2\sqrt{3} \rangle$ **73.** $90°$ **75.** About $62.7°$
77. Vertical ≈ 125.4 ft/sec, horizontal ≈ 1193.4 ft/sec
79. About $12.8°$; about 398.32 N
81. About $71.3°$; about 228.5 lb
83. $T_L \approx 15{,}484$ lb
    $T_R \approx 19{,}786$ lb
85. $\sqrt{2}$ lb; 1 lb   **87.** About 3154.4 lb   **89.** About 20.8 lb
91. About $19.5°$ **93.** N $21.4°$ E; about 138.7 km/h
95. True. See Example 1.   **97.** True. $a = b = 0$
99. $u_1 = 6 - (-3) = 9$ and $u_2 = -1 - 4 = -5$, so $\mathbf{u} = \langle 9, -5 \rangle$.
101. Proof   **103.** $\langle 1, 3 \rangle$ or $\langle -1, -3 \rangle$
105. (a) $5\sqrt{5 + 4\cos\theta}$

(b)

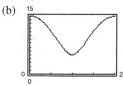

(c) Range: $[5, 15]$
    Maximum is 15 when $\theta = 0$.
    Minimum is 5 when $\theta = \pi$.
(d) The magnitudes of $\mathbf{F}_1$ and $\mathbf{F}_2$ are not the same.
107. (a) and (b) Answers will vary.

Section 6.4   *(page 435)*

1. dot product   **3.** $\dfrac{\mathbf{u} \cdot \mathbf{v}}{\|\mathbf{u}\|\|\mathbf{v}\|}$ **5.** $\left(\dfrac{\mathbf{u} \cdot \mathbf{v}}{\|\mathbf{v}\|^2} \right)\mathbf{v}$ **7.** -19
9. 0   **11.** 6   **13.** 18; scalar   **15.** $\langle 24, -12 \rangle$; vector
17. $\mathbf{0}$; vector   **19.** $\sqrt{10} - 1$; scalar   **21.** -12; scalar
23. 17   **25.** $5\sqrt{41}$ **27.** 6   **29.** $\dfrac{\pi}{2}$ **31.** About 2.50
33. 0   **35.** About 0.93   **37.** $\dfrac{5\pi}{12}$ **39.** About $91.33°$
41. $90°$ **43.** $26.57°, 63.43°, 90°$
45. $41.63°, 53.13°, 85.24°$ **47.** -20 **49.** $12{,}500\sqrt{3}$
51. Not orthogonal   **53.** Orthogonal   **55.** Not orthogonal
57. $\frac{1}{37}\langle 84, 14 \rangle, \frac{1}{37}\langle -10, 60 \rangle$ **59.** $\langle 0, 0 \rangle, \langle 4, 2 \rangle$ **61.** $\langle 3, 2 \rangle$
63. $\langle 0, 0 \rangle$ **65.** $\langle -5, 3 \rangle, \langle 5, -3 \rangle$
67. $\frac{2}{3}\mathbf{i} + \frac{1}{2}\mathbf{j}, -\frac{2}{3}\mathbf{i} - \frac{1}{2}\mathbf{j}$ **69.** 32
71. (a) \$35,727.50
        This value gives the total amount paid to the employees.
    (b) Multiply $\mathbf{v}$ by 1.02.

73. (a) Force $= 30{,}000 \sin d$

(b)

d	0°	1°	2°	3°	4°	5°
Force	0	523.6	1047.0	1570.1	2092.7	2614.7

d	6°	7°	8°	9°	10°
Force	3135.9	3656.1	4175.2	4693.0	5209.4

(c) About 29,885.8 lb

75. 735 N-m   **77.** About 779.4 ft-lb

79. About 10,282,651.78 N-m   **81.** About 1174.62 ft-lb

83. False. Work is represented by a scalar.

85. The dot product is the scalar 0.   **87.** 4

89. $1; \mathbf{u} \cdot \mathbf{u} = \|\mathbf{u}\|^2$

91. (a) $\mathbf{u}$ and $\mathbf{v}$ are parallel.   (b) $\mathbf{u}$ and $\mathbf{v}$ are orthogonal.

93. Proof

Section 6.5 *(page 443)*

1. real   **3.** absolute value   **5.** reflections   **7.** c   **8.** f

9. h   **10.** a   **11.** b   **12.** g   **13.** e   **14.** d

15.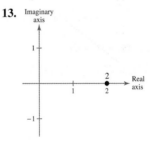

17.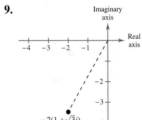

7   10

19.

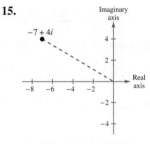

$2\sqrt{13}$

21. $5 + 6i$ **23.** $10 + 4i$ **25.** $6 + 5i$ **27.** $-5 + 7i$

29. $-2 - 2i$ **31.** $10 - 3i$ **33.** $-6i$ **35.** $-3 + 3i$

37. 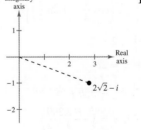   **39.**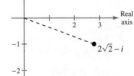

$2 - 3i$ $-1 + 2i$

41. $2\sqrt{2} \approx 2.83$ **43.** $\sqrt{109} \approx 10.44$

45. $(4, 3)$ **47.** $\left(\frac{9}{2}, -\frac{3}{2}\right)$

49. (a) Ship A: $3 + 4i$, Ship B: $-5 + 2i$

(b) *Sample answer:* Find the modulus of the difference of the complex numbers.

51. False. The modulus is always real.

53. False. $|1 + i| + |1 - i| = 2\sqrt{2}$ and $|(1 + i) + (1 - i)| = 2$

55. A circle; The modulus represents the distance from the origin.

57. Isosceles triangle; The moduli are equal.

Section 6.6 *(page 452)*

1. trigonometric form; modulus; argument   **3.** nth root

5.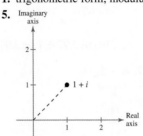

7.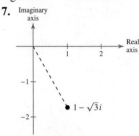

$\sqrt{2}\left(\cos \dfrac{\pi}{4} + i \sin \dfrac{\pi}{4}\right)$ $2\left(\cos \dfrac{5\pi}{3} + i \sin \dfrac{5\pi}{3}\right)$

9. 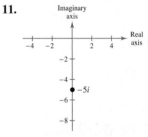   **11.**

$4\left(\cos \dfrac{4\pi}{3} + i \sin \dfrac{4\pi}{3}\right)$ $5\left(\cos \dfrac{3\pi}{2} + i \sin \dfrac{3\pi}{2}\right)$

13. 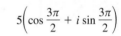   **15.**

$2(\cos 0 + i \sin 0)$ $\sqrt{65}(\cos 2.62 + i \sin 2.62)$

17. 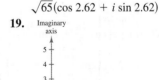   **19.**

$3(\cos 5.94 + i \sin 5.94)$ $\sqrt{29}(\cos 0.38 + i \sin 0.38)$

21.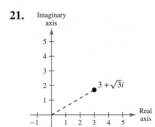

$$2\sqrt{3}\left(\cos \frac{\pi}{6} + i \sin \frac{\pi}{6}\right)$$

23.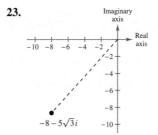

$$\sqrt{139}(\cos 3.97 + i \sin 3.97)$$

25. $1 + \sqrt{3}i$

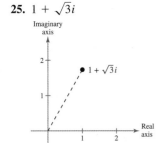

27. $6 - 2\sqrt{3}i$

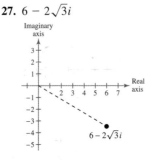

29. $-\dfrac{9\sqrt{2}}{8} + \dfrac{9\sqrt{2}}{8}i$

31. $-4.7347 - 1.6072i$

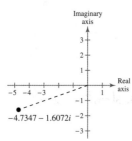

33. $4.6985 + 1.7101i$ **35.** $-1.8126 + 0.8452i$

37. $12\left(\cos \dfrac{\pi}{3} + i \sin \dfrac{\pi}{3}\right)$ **39.** $\dfrac{10}{9}(\cos 150° + i \sin 150°)$

41. $\dfrac{1}{3}(\cos 30° + i \sin 30°)$ **43.** $\cos \dfrac{2\pi}{3} + i \sin \dfrac{2\pi}{3}$

45. (a) $\left[2\sqrt{2}\left(\cos \dfrac{\pi}{4} + i \sin \dfrac{\pi}{4}\right)\right]\left[\sqrt{2}\left(\cos \dfrac{7\pi}{4} + i \sin \dfrac{7\pi}{4}\right)\right]$
   (b) $4(\cos 0 + i \sin 0) = 4$ (c) 4

47. (a) $\left[2\left(\cos \dfrac{3\pi}{2} + i \sin \dfrac{3\pi}{2}\right)\right]\left[\sqrt{2}\left(\cos \dfrac{\pi}{4} + i \sin \dfrac{\pi}{4}\right)\right]$
   (b) $2\sqrt{2}\left(\cos \dfrac{7\pi}{4} + i \sin \dfrac{7\pi}{4}\right) = 2 - 2i$
   (c) $-2i - 2i^2 = -2i + 2 = 2 - 2i$

49. (a) $[5(\cos 0.93 + i \sin 0.93)] \div \left[2\left(\cos \dfrac{5\pi}{3} + i \sin \dfrac{5\pi}{3}\right)\right]$
   (b) $\dfrac{5}{2}(\cos 1.97 + i \sin 1.97) \approx -0.982 + 2.299i$
   (c) About $-0.982 + 2.299i$

51. -1 **53.** $\dfrac{125}{2} + \dfrac{125\sqrt{3}}{2}i$ **55.** -1

57. $608.0 + 144.7i$ **59.** $\dfrac{81}{2} + \dfrac{81\sqrt{3}}{2}i$ **61.** $-4 - 4i$

63. $8i$ **65.** $1024 - 1024\sqrt{3}i$ **67.** $-597 - 122i$

69.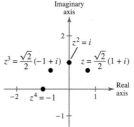

The absolute value of each is 1, and the consecutive powers of z are each 45° apart.

71. (a) $\sqrt{5}(\cos 60° + i \sin 60°)$
   $\sqrt{5}(\cos 240° + i \sin 240°)$
   (b) $\dfrac{\sqrt{5}}{2} + \dfrac{\sqrt{15}}{2}i, -\dfrac{\sqrt{5}}{2} - \dfrac{\sqrt{15}}{2}i$
   (c)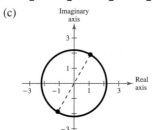

73. (a) $2\left(\cos \dfrac{2\pi}{9} + i \sin \dfrac{2\pi}{9}\right)$ (b) $1.5321 + 1.2856i,$
   $2\left(\cos \dfrac{8\pi}{9} + i \sin \dfrac{8\pi}{9}\right)$ $-1.8794 + 0.6840i,$
                                 $0.3473 - 1.9696i$
   $2\left(\cos \dfrac{14\pi}{9} + i \sin \dfrac{14\pi}{9}\right)$
   (c)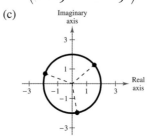

75. (a) $5\left(\cos \dfrac{4\pi}{9} + i \sin \dfrac{4\pi}{9}\right)$ (b) $0.8682 + 4.9240i,$
   $5\left(\cos \dfrac{10\pi}{9} + i \sin \dfrac{10\pi}{9}\right)$ $-4.6985 - 1.7101i,$
                                 $3.8302 - 3.2139i$
   $5\left(\cos \dfrac{16\pi}{9} + i \sin \dfrac{16\pi}{9}\right)$
   (c)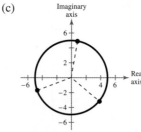

77. (a) $5\left(\cos \dfrac{3\pi}{4} + i \sin \dfrac{3\pi}{4}\right)$ (b) $-\dfrac{5\sqrt{2}}{2} + \dfrac{5\sqrt{2}}{2}i,$
   $5\left(\cos \dfrac{7\pi}{4} + i \sin \dfrac{7\pi}{4}\right)$ $\dfrac{5\sqrt{2}}{2} - \dfrac{5\sqrt{2}}{2}i$

(c)

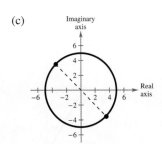

79. (a) $2(\cos 0 + i \sin 0)$

$2\left(\cos \dfrac{\pi}{2} + i \sin \dfrac{\pi}{2}\right)$

$2(\cos \pi + i \sin \pi)$

$2\left(\cos \dfrac{3\pi}{2} + i \sin \dfrac{3\pi}{2}\right)$

(b) $2, 2i, -2, -2i$

(c)

81. (a) $\cos 0 + i \sin 0$

$\cos \dfrac{2\pi}{5} + i \sin \dfrac{2\pi}{5}$

$\cos \dfrac{4\pi}{5} + i \sin \dfrac{4\pi}{5}$

$\cos \dfrac{6\pi}{5} + i \sin \dfrac{6\pi}{5}$

$\cos \dfrac{8\pi}{5} + i \sin \dfrac{8\pi}{5}$

(b) $1, 0.3090 + 0.9511i,$

$-0.8090 + 0.5878i,$

$-0.8090 - 0.5878i,$

$0.3090 - 0.9511i$

(c)

83. (a) $5\left(\cos \dfrac{\pi}{3} + i \sin \dfrac{\pi}{3}\right)$

$5(\cos \pi + i \sin \pi)$

$5\left(\cos \dfrac{5\pi}{3} + i \sin \dfrac{5\pi}{3}\right)$

(b) $\dfrac{5}{2} + \dfrac{5\sqrt{3}}{2}i, -5,$

$\dfrac{5}{2} - \dfrac{5\sqrt{3}}{2}i$

(c)

85. (a) $\sqrt{2}\left(\cos \dfrac{7\pi}{20} + i \sin \dfrac{7\pi}{20}\right)$

$\sqrt{2}\left(\cos \dfrac{3\pi}{4} + i \sin \dfrac{3\pi}{4}\right)$

$\sqrt{2}\left(\cos \dfrac{23\pi}{20} + i \sin \dfrac{23\pi}{20}\right)$

$\sqrt{2}\left(\cos \dfrac{31\pi}{20} + i \sin \dfrac{31\pi}{20}\right)$

$\sqrt{2}\left(\cos \dfrac{39\pi}{20} + i \sin \dfrac{39\pi}{20}\right)$

(b) $0.6420 + 1.2601i,$

$-1 + i,$

$-1.2601 - 0.6420i,$

$0.2212 - 1.3968i,$

$1.3968 - 0.2212i$

(c)

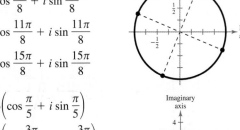

87. $\cos \dfrac{3\pi}{8} + i \sin \dfrac{3\pi}{8}$

$\cos \dfrac{7\pi}{8} + i \sin \dfrac{7\pi}{8}$

$\cos \dfrac{11\pi}{8} + i \sin \dfrac{11\pi}{8}$

$\cos \dfrac{15\pi}{8} + i \sin \dfrac{15\pi}{8}$

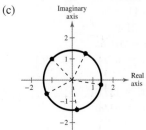

89. $3\left(\cos \dfrac{\pi}{5} + i \sin \dfrac{\pi}{5}\right)$

$3\left(\cos \dfrac{3\pi}{5} + i \sin \dfrac{3\pi}{5}\right)$

$3(\cos \pi + i \sin \pi)$

$3\left(\cos \dfrac{7\pi}{5} + i \sin \dfrac{7\pi}{5}\right)$

$3\left(\cos \dfrac{9\pi}{5} + i \sin \dfrac{9\pi}{5}\right)$

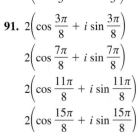

91. $2\left(\cos \dfrac{3\pi}{8} + i \sin \dfrac{3\pi}{8}\right)$

$2\left(\cos \dfrac{7\pi}{8} + i \sin \dfrac{7\pi}{8}\right)$

$2\left(\cos \dfrac{11\pi}{8} + i \sin \dfrac{11\pi}{8}\right)$

$2\left(\cos \dfrac{15\pi}{8} + i \sin \dfrac{15\pi}{8}\right)$

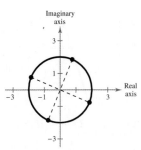

93. $\sqrt[6]{2}\left(\cos \dfrac{7\pi}{12} + i \sin \dfrac{7\pi}{12}\right)$

$\sqrt[6]{2}\left(\cos \dfrac{5\pi}{4} + i \sin \dfrac{5\pi}{4}\right)$

$\sqrt[6]{2}\left(\cos \dfrac{23\pi}{12} + i \sin \dfrac{23\pi}{12}\right)$

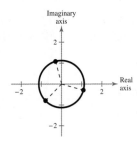

CHAPTER 6

95. (a) $E = 24(\cos 30° + i \sin 30°)$ volts

(b) $E = 12\sqrt{3} + 12i$ volts   (c) $|E| = 24$ volts

97. False. They are equally spaced around the circle centered at the origin with radius $\sqrt[n]{r}$.

99. Answers will vary.

101. Answers will vary; (a) r^2 (b) $\cos 2\theta + i \sin 2\theta$

Review Exercises   *(page 456)*

1. $C = 72°, b \approx 12.21, c \approx 12.36$

3. $A = 26°, a \approx 24.89, c \approx 56.23$

5. $C = 66°, a \approx 2.53, b \approx 9.11$

7. $B = 108°, a \approx 11.76, c \approx 21.49$

9. $A \approx 20.41°, C \approx 9.59°, a \approx 20.92$

11. $B \approx 39.48°, C \approx 65.52°, c \approx 48.24$ **13.** 19.1

15. 47.2   **17.** About 31.1 m

19. $A \approx 16.99°, B \approx 26.00°, C \approx 137.01°$

21. $A \approx 29.92°, B \approx 86.18°, C \approx 63.90°$

23. $A = 36°, C = 36°, b \approx 17.80$

25. $A \approx 45.76°, B \approx 91.24°, c \approx 21.42$

27. No; $A \approx 77.52°, B \approx 38.48°, a \approx 14.12$

29. Yes; $A \approx 28.62°, B \approx 33.56°, C \approx 117.82°$

31. About 4.3 ft, about 12.6 ft   **33.** 7.64   **35.** 8.36

37. Equivalent; **u** and **v** have the same magnitude and direction.

39. $\langle 7, -7 \rangle; 7\sqrt{2}$

41. (a) $\langle -4, 3 \rangle$ (b) $\langle 2, -9 \rangle$

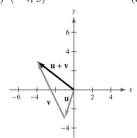

(c) $\langle -4, -12 \rangle$ (d) $\langle -14, 3 \rangle$

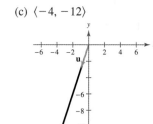

43. (a) $\langle -1, 6 \rangle$ (b) $\langle -9, -2 \rangle$

(c) $\langle -20, 8 \rangle$ (d) $\langle -13, 22 \rangle$

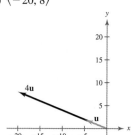
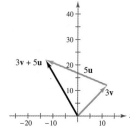

45. (a) $7\mathbf{i} + 2\mathbf{j}$ (b) $-3\mathbf{i} - 4\mathbf{j}$

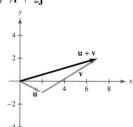
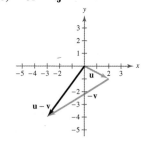

(c) $8\mathbf{i} - 4\mathbf{j}$ (d) $25\mathbf{i} + 4\mathbf{j}$

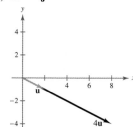
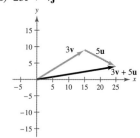

47. (a) $3\mathbf{i} + 6\mathbf{j}$ (b) $5\mathbf{i} - 6\mathbf{j}$

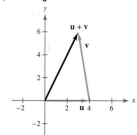
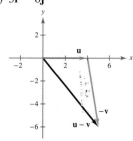

(c) $16\mathbf{i}$ (d) $17\mathbf{i} + 18\mathbf{j}$

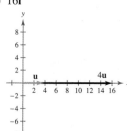
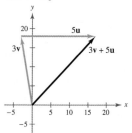

49. $-\mathbf{i} + 5\mathbf{j}$ **51.** $6\mathbf{i} + 4\mathbf{j}$

53. $\langle 30, 9 \rangle$

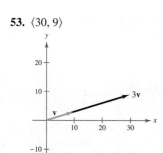

55. $\langle 22, -7 \rangle$

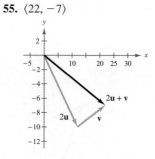

57. $\langle -10, -37 \rangle$

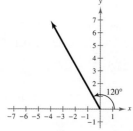

59. $\|\mathbf{v}\| = \sqrt{41}$; $\theta \approx 38.7°$ **61.** $\|\mathbf{v}\| = 3\sqrt{2}$; $\theta = 225°$

63. $\|\mathbf{v}\| = 7$; $\theta = 60°$

65. $\mathbf{v} = \langle -4, 4\sqrt{3} \rangle$

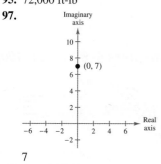

67. About $50.5°$; about 133.92 lb   **69.** 45 **71.** -2

73. 40; scalar   **75.** $4 - 2\sqrt{5}$; scalar

77. $\langle 72, -36 \rangle$; vector   **79.** 38; scalar   **81.** About $160.5°$

83. $165°$ **85.** Orthogonal   **87.** Not orthogonal

89. $-\frac{13}{17}\langle 4, 1 \rangle, \frac{16}{17}\langle -1, 4 \rangle$ **91.** $\frac{5}{2}\langle -1, 1 \rangle, \frac{9}{2}\langle 1, 1 \rangle$ **93.** 48

95. $72{,}000$ ft-lb

97.

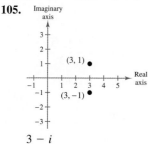

7

99.

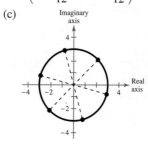

$\sqrt{34}$

101. $3 + i$ **103.** $-2 + i$

105.

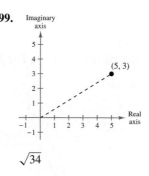

$3 - i$

107. $\sqrt{10}$ **109.** $\left(\frac{5}{2}, 2 \right)$

111.

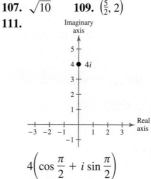

$4\left(\cos \dfrac{\pi}{2} + i \sin \dfrac{\pi}{2} \right)$

113.

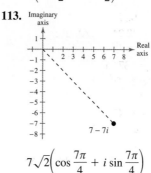

$7\sqrt{2}\left(\cos \dfrac{7\pi}{4} + i \sin \dfrac{7\pi}{4} \right)$

115.

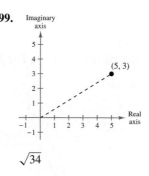

$13(\cos 4.32 + i \sin 4.32)$

117. $4\left(\cos \dfrac{7\pi}{12} + i \sin \dfrac{7\pi}{12} \right)$ **119.** $\dfrac{2}{3}(\cos 45° + i \sin 45°)$

121. $\dfrac{625}{2} + \dfrac{625\sqrt{3}}{2}i$ **123.** $2035 - 828i$

125. (a) $3\left(\cos \dfrac{\pi}{4} + i \sin \dfrac{\pi}{4} \right)$

$3\left(\cos \dfrac{7\pi}{12} + i \sin \dfrac{7\pi}{12} \right)$

$3\left(\cos \dfrac{11\pi}{12} + i \sin \dfrac{11\pi}{12} \right)$

$3\left(\cos \dfrac{5\pi}{4} + i \sin \dfrac{5\pi}{4} \right)$

$3\left(\cos \dfrac{19\pi}{12} + i \sin \dfrac{19\pi}{12} \right)$

$3\left(\cos \dfrac{23\pi}{12} + i \sin \dfrac{23\pi}{12} \right)$

(b) $\dfrac{3\sqrt{2}}{2} + \dfrac{3\sqrt{2}}{2}i$,

$-0.7765 + 2.8978i$,

$-2.8978 + 0.7765i$,

$-\dfrac{3\sqrt{2}}{2} - \dfrac{3\sqrt{2}}{2}i$,

$0.7765 - 2.8978i$,

$2.8978 - 0.7765i$

(c)

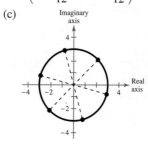

127. (a) $2(\cos 0 + i \sin 0)$ (b) $2, -1 + \sqrt{3}\,i,$
$$2\left(\cos \frac{2\pi}{3} + i \sin \frac{2\pi}{3}\right)$$ $-1 - \sqrt{3}\,i$
$$2\left(\cos \frac{4\pi}{3} + i \sin \frac{4\pi}{3}\right)$$

(c)

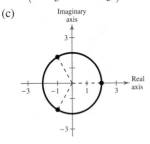

129. $3\left(\cos \dfrac{\pi}{4} + i \sin \dfrac{\pi}{4}\right) = \dfrac{3\sqrt{2}}{2} + \dfrac{3\sqrt{2}}{2}i$

$3\left(\cos \dfrac{3\pi}{4} + i \sin \dfrac{3\pi}{4}\right) = -\dfrac{3\sqrt{2}}{2} + \dfrac{3\sqrt{2}}{2}i$

$3\left(\cos \dfrac{5\pi}{4} + i \sin \dfrac{5\pi}{4}\right) = -\dfrac{3\sqrt{2}}{2} - \dfrac{3\sqrt{2}}{2}i$

$3\left(\cos \dfrac{7\pi}{4} + i \sin \dfrac{7\pi}{4}\right) = \dfrac{3\sqrt{2}}{2} - \dfrac{3\sqrt{2}}{2}i$

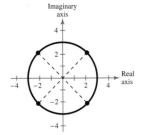

131. $2\left(\cos \dfrac{\pi}{2} + i \sin \dfrac{\pi}{2}\right) = 2i$

$2\left(\cos \dfrac{7\pi}{6} + i \sin \dfrac{7\pi}{6}\right) = -\sqrt{3} - i$

$2\left(\cos \dfrac{11\pi}{6} + i \sin \dfrac{11\pi}{6}\right) = \sqrt{3} - i$

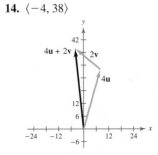

133. True. $\sin 90°$ is defined in the Law of Sines.
135. Direction and magnitude

Chapter Test *(page 459)*

1. No; $C = 88°$, $b \approx 27.81$, $c \approx 29.98$
2. No; $A = 42°$, $b \approx 21.91$, $c \approx 10.95$
3. No; Two solutions:
    $B \approx 29.12°$, $C \approx 126.88°$, $c \approx 22.03$
    $B \approx 150.88°$, $C \approx 5.12°$, $c \approx 2.46$
4. Yes; $A \approx 19.12°$, $B \approx 23.49°$, $C \approx 137.39°$
5. No; No solution

6. Yes; $A \approx 21.90°$, $B \approx 37.10°$, $c \approx 78.15$
7. 2052.5 m^2 **8.** 606.3 mi; $29.1°$ **9.** $\langle 14, -23 \rangle$
10. $\left\langle \dfrac{18\sqrt{34}}{17}, -\dfrac{30\sqrt{34}}{17} \right\rangle$
11. $\langle -4, 12 \rangle$ **12.** $\langle 8, 2 \rangle$

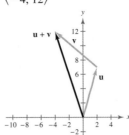 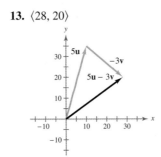

13. $\langle 28, 20 \rangle$ **14.** $\langle -4, 38 \rangle$

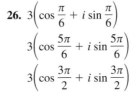 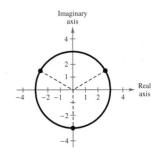

15. 5   **16.** About 14.9°; about 250.15 lb   **17.** 135°
18. Orthogonal   **19.** $\frac{37}{26}\langle 5, 1 \rangle$; $\frac{29}{26}\langle -1, 5 \rangle$ **20.** About 104 lb
21. $4\sqrt{2}\left(\cos \dfrac{7\pi}{4} + i \sin \dfrac{7\pi}{4}\right)$ **22.** $-3 + 3\sqrt{3}\,i$
23. $-\dfrac{6561}{2} - \dfrac{6561\sqrt{3}}{2}i$ **24.** $5832i$ **25.** $4, -4, 4i, -4i$

26. $3\left(\cos \dfrac{\pi}{6} + i \sin \dfrac{\pi}{6}\right)$

$3\left(\cos \dfrac{5\pi}{6} + i \sin \dfrac{5\pi}{6}\right)$

$3\left(\cos \dfrac{3\pi}{2} + i \sin \dfrac{3\pi}{2}\right)$

Cumulative Test for Chapters 4–6 *(page 460)*

1. (a)

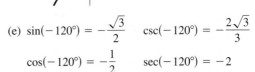

(b) $240°$
(c) $-\dfrac{2\pi}{3}$
(d) $60°$

(e) $\sin(-120°) = -\dfrac{\sqrt{3}}{2}$ $\csc(-120°) = -\dfrac{2\sqrt{3}}{3}$

$\cos(-120°) = -\dfrac{1}{2}$ $\sec(-120°) = -2$

$\tan(-120°) = \sqrt{3}$ $\cot(-120°) = \dfrac{\sqrt{3}}{3}$

2. $-83.079°$ **3.** $\frac{20}{29}$

4.

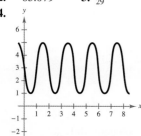

5.

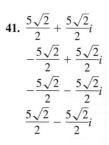

6.

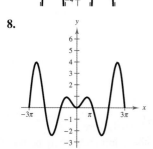

7. $a = -3, b = \pi, c = 0$

8.

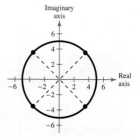

9. 4.9

10. $\frac{3}{4}$

11. $\sqrt{1 - 4x^2}$

12. 1

13. $2 \tan \theta$

14–16. Answers will vary.

17. $\frac{\pi}{3}, \frac{\pi}{2}, \frac{3\pi}{2}, \frac{5\pi}{3}$

18. $\frac{\pi}{6}, \frac{5\pi}{6}, \frac{7\pi}{6}, \frac{11\pi}{6}$ **19.** $\frac{3\pi}{2}$ **20.** $\frac{16}{63}$ **21.** $\frac{4}{3}$

22. $\frac{\sqrt{5}}{5}$ **23.** $\frac{5}{2}\left(\sin \frac{5\pi}{2} - \sin \pi\right)$ **24.** $-2 \sin 8x \sin x$

25. No; $B \approx 26.39°, C \approx 123.61°, c \approx 14.99$

26. Yes; $B \approx 52.48°, C \approx 97.52°, a \approx 5.04$

27. No; $B = 60°, a \approx 5.77, c \approx 11.55$

28. Yes; $A \approx 26.28°, B \approx 49.74°, C \approx 103.98°$

29. No; $C = 109°, a \approx 14.96, b \approx 9.27$

30. Yes; $A \approx 6.88°, B \approx 93.12°, c \approx 9.86$

31. 41.48 in.^{2} **32.** 599.09 m^2 **33.** $7\mathbf{i} + 8\mathbf{j}$

34. $\left\langle \frac{\sqrt{2}}{2}, \frac{\sqrt{2}}{2} \right\rangle$ **35.** -5 **36.** $-\frac{1}{13}\langle 1, 5 \rangle; \frac{21}{13}\langle 5, -1 \rangle$

37.

Imaginary axis

$\bullet$ (3, 2)

Real axis

$\bullet$ (3, −2)

$3 + 2i$

38. $2\sqrt{2}\left(\cos \frac{3\pi}{4} + i \sin \frac{3\pi}{4}\right)$ **39.** $24(\cos 150° + i \sin 150°)$

40. $\cos 0 + i \sin 0 = 1$

$\cos \frac{2\pi}{3} + i \sin \frac{2\pi}{3} = -\frac{1}{2} + \frac{\sqrt{3}}{2}i$

$\cos \frac{4\pi}{3} + i \sin \frac{4\pi}{3} = -\frac{1}{2} - \frac{\sqrt{3}}{2}i$

41. $\frac{5\sqrt{2}}{2} + \frac{5\sqrt{2}}{2}i$

$-\frac{5\sqrt{2}}{2} + \frac{5\sqrt{2}}{2}i$

$-\frac{5\sqrt{2}}{2} - \frac{5\sqrt{2}}{2}i$

$\frac{5\sqrt{2}}{2} - \frac{5\sqrt{2}}{2}i$

Imaginary axis

Real axis

42. About 395.8 rad/min; about 8312.7 in./min

43. 42π yd$^2 \approx 131.95$ yd^2 **44.** 5 ft **45.** About 22.6°

46. $d = 4 \cos \frac{\pi}{4}t$ **47.** About 32.6°; about 543.9 km/h

48. 425 ft-lb

Problem Solving *(page 465)*

1. About 2.01 ft

3. (a)

A 75 mi B
30° 15°
135° 75°
x y
60° Lost party

(b) Station A: about 27.45 mi; Station B: about 53.03 mi

(c) About 11.03 mi; S 21.7° E

5. (a) (i) $\sqrt{2}$ (ii) $\sqrt{5}$ (iii) 1

(iv) 1 (v) 1 (vi) 1

(b) (i) 1 (ii) $3\sqrt{2}$ (iii) $\sqrt{13}$

(iv) 1 (v) 1 (vi) 1

(c) (i) $\frac{\sqrt{5}}{2}$ (ii) $\sqrt{13}$ (iii) $\frac{\sqrt{85}}{2}$

(iv) 1 (v) 1 (vi) 1

(d) (i) $2\sqrt{5}$ (ii) $5\sqrt{2}$ (iii) $5\sqrt{2}$

(iv) 1 (v) 1 (vi) 1

7. Proof

9. (a) $2(\cos 30° + i \sin 30°)$ (b) $3(\cos 45° + i \sin 45°)$

$2(\cos 150° + i \sin 150°)$ $3(\cos 135° + i \sin 135°)$

$2(\cos 270° + i \sin 270°)$ $3(\cos 225° + i \sin 225°)$

$3(\cos 315° + i \sin 315°)$

11. a; The angle between the vectors is acute.

Chapter 7

Section 7.1 *(page 475)*

1. solution **3.** points; intersection

5. (a) No (b) No (c) No (d) Yes **7.** (2, 2)

9. (2, 6), (−1, 3) **11.** (0, 0), (2, −4)

13. (0, 1), (1, −1), (3, 1) **15.** (6, 4) **17.** $\left(\frac{1}{2}, 3\right)$

19. (1, 1) **21.** $\left(\frac{20}{3}, \frac{40}{3}\right)$ **23.** No solution

25. $5500 at 2%; $6500 at 6%

27. $6000 at 2.8%; $6000 at 3.8%

29. (−2, 4), (0, 0) **31.** No solution **33.** (6, 2)

35. $\left(-\frac{3}{2}, \frac{1}{2}\right)$ **37.** (2, 2), (4, 0) **39.** No solution

41. (4, 3), (−4, 3) **43.** (0, 1) **45.** (5.31, −0.54)

47. (1, 2) **49.** No solution **51.** (0.287, 1.751)

53. $\left(\frac{1}{2}, 2\right), \left(-4, -\frac{1}{4}\right)$ **55.** 293 units

57. (a) 344 units (b) 2495 units

CHAPTER 7

59. (a) 8 weeks

(b)

	1	2	3	4
$360 - 24x$	336	312	288	264
$24 + 18x$	42	60	78	96

	5	6	7	8
$360 - 24x$	240	216	192	168
$24 + 18x$	114	132	150	168

61. $y = 2x - 2$, not $-2x + 2$.   **63.** 12 m × 16 m

65. 10 km × 12 km

67. False. You can solve for either variable in either equation and then back-substitute.

69. *Sample answer:* After substituting, the resulting equation may be a contradiction or have imaginary solutions.

71. (a)–(c) Answers will vary.

Section 7.2   *(page 486)*

1. elimination   **3.** consistent; inconsistent

5. (1, 5)   **7.** $(1, -1)$

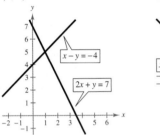 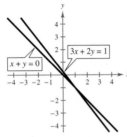

9. No solution   **11.** $\left(a, \frac{3}{2}a - \frac{5}{2}\right)$

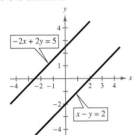 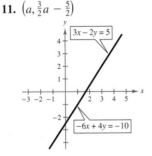

13. (4, 1)   **15.** $\left(\frac{3}{2}, -\frac{1}{2}\right)$ **17.** $\left(-3, \frac{5}{3}\right)$ **19.** $(4, -1)$

21. $\left(-\frac{6}{35}, \frac{43}{35}\right)$ **23.** (101, 96)   **25.** No solution

27. Infinitely many solutions: $\left(a, -\frac{1}{2} + \frac{5}{6}a\right)$ **29.** $(5, -2)$

31. a; infinitely many solutions; consistent

32. c; one solution; consistent

33. d; no solutions; inconsistent

34. b; one solution; consistent   **35.** (4, 1)   **37.** (10, 5)

39. $(19, -55)$ **41.** 660 mi/h; 60 mi/h

43. Cheeseburger: 550 calories; fries: 320 calories

45. (240, 404)   **47.** (2,000,000, 100)

49. (a) $\begin{cases} x + y = 30 \\ 0.25x + 0.5y = 12 \end{cases}$

(b)

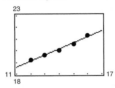

Decreases

(c) 25% solution: 12 L; 50% solution: 18 L

51. $18,000

53. (a) Pharmacy A: $P = 0.52t + 12.9$

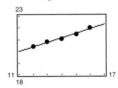

Pharmacy B: $P = 0.39t + 15.7$

(b) Yes. 2021

55. $y = 0.97x + 2.1$

57. (a) $y = 14x + 19$ (b) 41.4 bushels/acre

59. False. Two lines that coincide have infinitely many points of intersection.

61. $k = -4$

63. No. Two lines will intersect only once or will coincide, and if they coincide the system will have infinitely many solutions.

65. Answers will vary.

67. (39,600, 398). It is necessary to change the scale on the axes to see the point of intersection.

69. $u = 1, v = -\tan x$

Section 7.3   *(page 498)*

1. row-echelon   **3.** Gaussian   **5.** nonsquare

7. (a) No   (b) No   (c) No   (d) Yes

9. (a) No   (b) No   (c) Yes   (d) No

11. $(-13, -10, 8)$ **13.** $(3, 10, 2)$ **15.** $\left(\frac{11}{4}, 7, 11\right)$

17. $\begin{cases} x - 2y + 3z = 5 \\ y - 2z = 9 \\ 2x - 3z = 0 \end{cases}$

First step in putting the system in row-echelon form.

19. $(-2, 2)$ **21.** (4, 3)   **23.** (4, 1, 2)   **25.** $\left(1, \frac{1}{2}, -3\right)$

27. No solution   **29.** $\left(\frac{1}{8}, -\frac{5}{8}, -\frac{1}{2}\right)$ **31.** (0, 0, 0)

33. No solution   **35.** $(-a + 3, a + 1, a)$

37. $(-3a + 10, 5a - 7, a)$ **39.** (1, 1, 1, 1)

41. $(2a, 21a - 1, 8a)$ **43.** $\left(-\frac{3}{2}a + \frac{1}{2}, -\frac{2}{3}a + 1, a\right)$

45. $s = -16t^2 + 144$ **47.** $y = \frac{1}{2}x^2 - 2x$

49. $y = x^2 - 6x + 8$ **51.** $y = 4x^2 - 2x + 1$

53. $x^2 + y^2 - 10x = 0$ **55.** $x^2 + y^2 + 6x - 8y = 0$

57. The leading coefficient of the third equation is not 1, so the system is not in row-echelon form.

59. $300,000 at 8%
$400,000 at 9%
$75,000 at 10%

61. $x = 60°, y = 67°, z = 53°$ **63.** 75 ft, 63 ft, 42 ft

65. $I_1 = 1, I_2 = 2, I_3 = 1$ **67.** $y = x^2 - x$

69. (a) $y = 0.0514x^2 + 0.8771x + 1.8857$

(b)

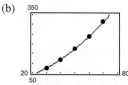

The model fits the data well.

(c) About 356 ft

71. $x = \pm\dfrac{\sqrt{2}}{2}, y = \dfrac{1}{2}, \lambda = 1$ or $x = 0, y = 0, \lambda = 0$

73. False. See Example 6 on page 495.

75. No. Answers will vary.   **77–79.** Answers will vary.

Section 7.4   *(page 508)*

1. partial fraction decomposition   **3.** partial fraction

5. b   **6.** c   **7.** d   **8.** a   **9.** $\dfrac{A}{x} + \dfrac{B}{x-2}$

11. $\dfrac{A}{x+2} + \dfrac{B}{(x+2)^2} + \dfrac{C}{(x+2)^3} + \dfrac{D}{(x+2)^4}$

13. $\dfrac{A}{x} + \dfrac{Bx+C}{x^2+10}$ **15.** $\dfrac{A}{x} + \dfrac{B}{x^2} + \dfrac{Cx+D}{x^2+3} + \dfrac{Ex+F}{(x^2+3)^2}$

17. $\dfrac{1}{x} - \dfrac{1}{x+1}$ **19.** $\dfrac{1}{x-1} - \dfrac{1}{x+2}$

21. $\dfrac{1}{2}\left(\dfrac{1}{x-1} - \dfrac{1}{x+1}\right)$ **23.** $-\dfrac{3}{x} - \dfrac{1}{x+2} + \dfrac{5}{x-2}$

25. $\dfrac{3}{x-3} + \dfrac{9}{(x-3)^2}$ **27.** $\dfrac{3}{x} - \dfrac{1}{x^2} + \dfrac{1}{x+1}$

29. $\dfrac{3}{x} - \dfrac{2x-2}{x^2+1}$ **31.** $-\dfrac{1}{x-1} + \dfrac{x+2}{x^2-2}$

33. $\dfrac{1}{8}\left(\dfrac{1}{2x+1} + \dfrac{1}{2x-1} - \dfrac{4x}{4x^2+1}\right)$

35. $\dfrac{1}{x+1} + \dfrac{2}{x^2-2x+3}$ **37.** $\dfrac{2}{x^2+4} + \dfrac{x}{(x^2+4)^2}$

39. $\dfrac{5}{(x^2+3)^2} - \dfrac{17}{(x^2+3)^3}$ **41.** $\dfrac{2}{x} - \dfrac{3}{x^2} - \dfrac{2x-3}{x^2+2} - \dfrac{4x-6}{(x^2+2)^2}$

43. $1 - \dfrac{2x+1}{x^2+x+1}$ **45.** $2x - 7 + \dfrac{17}{x+2} + \dfrac{1}{x+1}$

47. $x + 3 + \dfrac{6}{x-1} + \dfrac{4}{(x-1)^2} + \dfrac{1}{(x-1)^3}$

49. $x + \dfrac{2}{x} + \dfrac{1}{x+1} + \dfrac{3}{(x+1)^2}$ **51.** $\dfrac{3}{2x-1} - \dfrac{2}{x+1}$

53. $\dfrac{2}{x} + \dfrac{4}{x+1} - \dfrac{3}{x-1}$ **55.** $\dfrac{1}{x^2+2} + \dfrac{x}{(x^2+2)^2}$

57. $2x + \dfrac{1}{2}\left(\dfrac{3}{x-4} - \dfrac{1}{x+2}\right)$ **59.** $\dfrac{60}{100-p} - \dfrac{60}{100+p}$

61. False. The partial fraction decomposition is

$$\dfrac{A}{x+10} + \dfrac{B}{x-10} + \dfrac{C}{(x-10)^2}.$$

63. False. The degrees could be equal.

65. The expression is improper, so first divide the denominator

into the numerator to obtain $1 + \dfrac{x+1}{x^2-x}$.

Section 7.5   *(page 517)*

1. solution   **3.** solution

5.

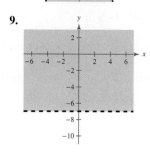

7.

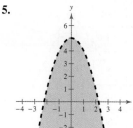

9.

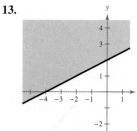

11.

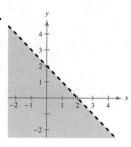

13.

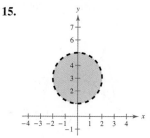

15.

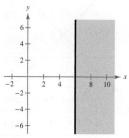

17.

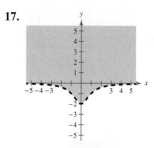

19.

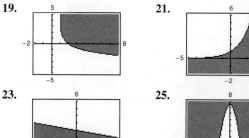

21.

23.

25.

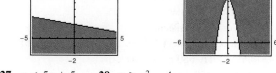

27. $y < 5x + 5$ **29.** $y \geq x^2 - 4$

CHAPTER 7

31.

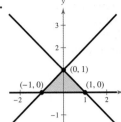

33.

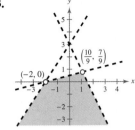

35.

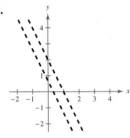

No solution

37.

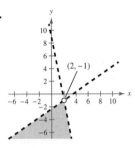

39.

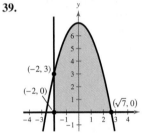

41.

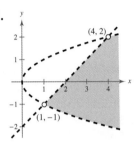

43.

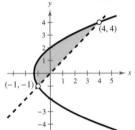

45.

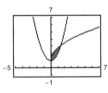

47.

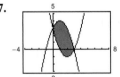

49.

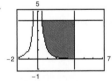

51. $\begin{cases} x \geq 0 \\ y \geq 0 \\ y \leq 6 - x \end{cases}$ **53.** $\begin{cases} x \geq 0 \\ y \geq 0 \\ x^2 + y^2 < 64 \end{cases}$

55. $\begin{cases} x \geq 4 \\ x \leq 9 \\ y \geq 3 \\ y \leq 9 \end{cases}$ **57.** $\begin{cases} y \geq 0 \\ y \leq 5x \\ y \leq -x + 6 \end{cases}$

59. (a)

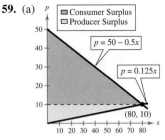

(b) Consumer surplus: $1600
Producer surplus: $400

61. (a)

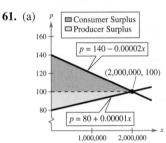

(b) Consumer surplus:
$40,000,000
Producer surplus:
$20,000,000

63. $\begin{cases} x + y \leq 20,000 \\ y \geq 2x \\ x \geq 5,000 \\ y \geq 5,000 \end{cases}$

65. $\begin{cases} x + \frac{3}{2}y \leq 12 \\ \frac{4}{3}x + \frac{3}{2}y \leq 15 \\ x \geq 0 \\ y \geq 0 \end{cases}$

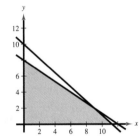

67. (a) $\begin{cases} 180x + 100y \geq 1000 \\ 6x + y \geq 18 \\ 220x + 40y \geq 400 \\ x \geq 0 \\ y \geq 0 \end{cases}$

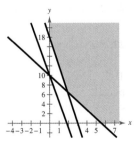

(b) Answers will vary.

69. (a) $\begin{cases} x \geq 50 \\ y \geq 40 \\ 55x + 70y \leq 7500 \end{cases}$

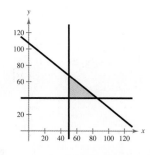

(b) Answers will vary.

71. True. The figure is a rectangle with a length of 9 units and a width of 11 units.

73. Test a point on each side of the line.

75. (a) iv   (b) ii   (c) iii   (d) i

Section 7.6   *(page 526)*

1. optimization   **3.** objective   **5.** inside; on

7. Minimum at $(0, 0)$: 0 **9.** Minimum at $(1, 0)$: 2
Maximum at $(5, 0)$: 20 Maximum at $(3, 4)$: 26

11. Minimum at $(0, 20)$: 140
Maximum at $(60, 20)$: 740

13.

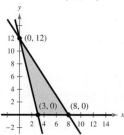

Minimum at $(3, 0)$: 9
Maximum at any point on
the line segment connecting
$(0, 12)$ and $(8, 0)$: 24

15.

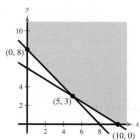

Minimum at $(5, 3)$: 35
No maximum

17.

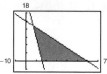

Minimum at $(7.2, 13.2)$: 34.8
Maximum at $(60, 0)$: 180

19.

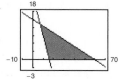

Minimum at $(7.2, 13.2)$: 7.2
Maximum at $(60, 0)$: 60

21. Minimum at $(0, 0)$: 0 **23.** Minimum at $(0, 0)$: 0
Maximum at $(0, 5)$: 25 Maximum at $\left(\frac{22}{3}, \frac{19}{6}\right)$: $\frac{271}{6}$

25. Minimum at $(4, 3)$: 10 **27.** No minimum
No maximum   Maximum at $(12, 5)$: 7

29.

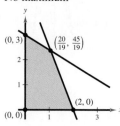

The maximum, 5, occurs at any
point on the line segment
connecting $(2, 0)$ and $\left(\frac{20}{19}, \frac{45}{19}\right)$.
Minimum at $(0, 0)$: 0

31.

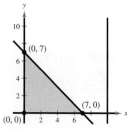

The constraint $x \leq 10$
is extraneous.
Minimum at $(7, 0)$: -7
Maximum at $(0, 7)$: 14

33.

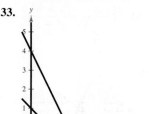

The feasible set is empty.

35.

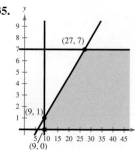

The solution region is
unbounded.
Minimum at $(9, 0)$: 9
No maximum

37. 230 units of the $225 model
45 units of the $250 model
Optimal profit: $8295

39. 2 bottles of brand X   **41.** 13 audits
5 bottles of brand Y   0 tax returns
                                   Optimal revenue: $20,800

43. 60 acres for crop A
90 acres for crop B
Optimal yield: 63,000 bushels

45. $0 on TV ads
$1,000,000 on newspaper ads
Optimal audience: 250 million people

47. True. The objective function has a maximum value at any point on the line segment connecting the two vertices.

49. False. See Exercise 27.

Review Exercises   *(page 531)*

1. $(1, 1)$ **3.** $\left(\frac{3}{2}, 5\right)$ **5.** $(0.25, 0.625)$ **7.** $(5, 4)$

9. $(0, 0)$, $(2, 8)$, $(-2, 8)$ **11.** $(4, -2)$

13. $(1.41, -0.66)$, $(-1.41, 10.66)$ **15.** $(0, -2)$

17. No solution

19.

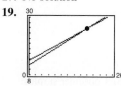

The BMI for males exceeds the BMI for females after age 18.

21. 16 ft × 18 ft   **23.** $\left(\frac{5}{2}, 3\right)$ **25.** $(0, 0)$ **27.** $\left(\frac{8}{5}a + \frac{14}{5}, a\right)$

29. d, one solution, consistent

30. c, infinitely many solutions, consistent

31. b, no solution, inconsistent

32. a, one solution, consistent   **33.** $(100,000, 23)$

35. $(2, -4, -5)$ **37.** $(-6, 7, 10)$ **39.** $\left(\frac{24}{5}, \frac{22}{5}, -\frac{8}{5}\right)$

41. $\left(-\frac{3}{4}, 0, -\frac{5}{4}\right)$ **43.** $(a - 4, a - 3, a)$

45. $y = 2x^2 + x - 5$ **47.** $x^2 + y^2 - 4x + 4y - 1 = 0$

49. 10 gal of spray X   **51.** $16,000 at 7%
    5 gal of spray Y $13,000 at 9%
    12 gal of spray Z $11,000 at 11%

53. $s = -16t^2 + 150$ **55.** $\dfrac{A}{x} + \dfrac{B}{x + 20}$

57. $\dfrac{A}{x} + \dfrac{B}{x^2} + \dfrac{C}{x - 5}$ **59.** $\dfrac{3}{x + 2} - \dfrac{4}{x + 4}$

61. $1 - \dfrac{25}{8(x + 5)} + \dfrac{9}{8(x - 3)}$ **63.** $\dfrac{1}{2}\left(\dfrac{3}{x - 1} - \dfrac{x - 3}{x^2 + 1}\right)$

65. $\dfrac{3}{x^2 + 1} + \dfrac{4x - 3}{(x^2 + 1)^2}$

67. 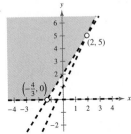   **69.**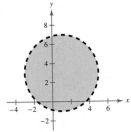

71. 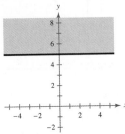   **73.**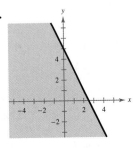

75. 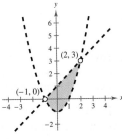   **77.**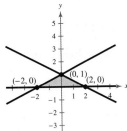

79. 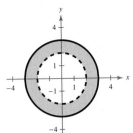   **81.** $\begin{cases} x \geq 3 \\ x \leq 7 \\ y \geq 1 \\ y \leq 10 \end{cases}$

83. (a) 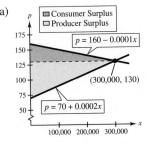   (b) Consumer surplus:
    $4,500,000
    Producer surplus:
    $9,000,000

85. $\begin{cases} 20x + 30y \leq 24{,}000 \\ 12x + 8y \leq 12{,}400 \\ x \geq 0 \\ y \geq 0 \end{cases}$
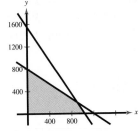
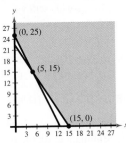

87. 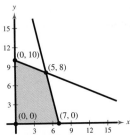   **89.**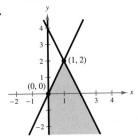

Minimum at $(0, 0)$: 0   Minimum at $(15, 0)$: 26.25
Maximum at $(5, 8)$: 47   No maximum

91. 72 haircuts
    0 permanents
    Optimal revenue: $1800

93. True. The nonparallel sides of the trapezoid are equal in length.

95. $\begin{cases} 4x + y = -22 \\ \frac{1}{2}x + y = 6 \end{cases}$ **97.** $\begin{cases} 3x + y = 7 \\ -6x + 3y = 1 \end{cases}$

99. $\begin{cases} x + y + z = 6 \\ x + y - z = 0 \\ x - y - z = 2 \end{cases}$ **101.** $\begin{cases} 2x + 2y - 3z = 7 \\ x - 2y + z = 4 \\ -x + 4y - z = -1 \end{cases}$

103. An inconsistent system of linear equations has no solution.

Chapter Test   *(page 535)*

1. $(-4, -5)$ **2.** $(0, -1), (1, 0), (2, 1)$ **3.** $(8, 4), (2, -2)$

4. $(4, 2)$ **5.** $(-3, 0), (2, 5)$ **6.** $(1, 4), (0.034, 0.619)$

7. $(-2, -5)$ **8.** $(10, -3)$ **9.** $(2, -3, 1)$

10. No solution   **11.** $-\dfrac{1}{x + 1} + \dfrac{3}{x - 2}$ **12.** $\dfrac{2}{x^2} + \dfrac{3}{2 - x}$

13. $x - \dfrac{5}{x} + \dfrac{3}{x + 1} + \dfrac{3}{x - 1}$ **14.** $-\dfrac{2}{x} + \dfrac{3x}{x^2 + 2}$

15.   **16.**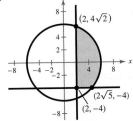

17.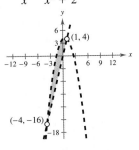

18. Minimum at $(0, 0)$: 0
    Maximum at $(12, 0)$: 240

19. \$24,000 in 4% fund
    \$26,000 in 5.5% fund

20. $y = -\frac{1}{2}x^2 + x + 6$

21. 900 units of model I
    4400 units of model II
    Optimal profit: \$203,000

15. $\begin{cases} a + \quad t \le \quad 32 \\ 0.15a \quad\quad \ge \quad 1.9 \\ 193a + 772t \ge 11{,}000 \end{cases}$

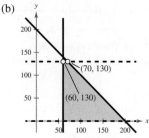

Problem Solving   *(page 537)*

1.

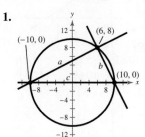

$a = 8\sqrt{5}, b = 4\sqrt{5}, c = 20$
$\left(8\sqrt{5}\right)^2 + \left(4\sqrt{5}\right)^2 = 20^2$
Therefore, the triangle is a right triangle.

17. (a) $\begin{cases} 0 < y < 130 \\ x \ge 60 \\ x + y \le 200 \end{cases}$ **(b)**

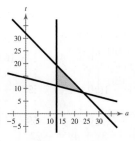

(c) No. The point $(90, 120)$ is not in the solution region.

(d) *Sample answer:* LDL/VLDL: 135 mg/dL;
     HDL: 65 mg/dL

(e) *Sample answer:* $(75, 90)$; $\frac{165}{75} = 2.2 < 4$

3. $ad \ne bc$

5. (a)

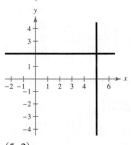

Chapter 8

Section 8.1   *(page 549)*

1. square   **3.** augmented   **5.** row-equivalent

7. 1×2 **9.** 3×1 **11.** 2×2 **13.** 3×3

15. $\begin{bmatrix} 2 & -1 & \vdots & 7 \\ 1 & 1 & \vdots & 2 \end{bmatrix}$ **17.** $\begin{bmatrix} 1 & -1 & 2 & \vdots & 2 \\ 4 & -3 & 1 & \vdots & -1 \\ 2 & 1 & 0 & \vdots & 0 \end{bmatrix}$

19. $\begin{bmatrix} 3 & -5 & 2 & \vdots & 12 \\ 12 & 0 & -7 & \vdots & 10 \end{bmatrix}$ **21.** $\begin{cases} x + y = 3 \\ 5x - 3y = -1 \end{cases}$

23. $\begin{cases} 2x \quad\quad + 5z = -12 \\ \quad y - 2z = 7 \\ 6x + 3y \quad\quad = 2 \end{cases}$

25. $\begin{cases} 9x + 12y + 3z \quad\quad = 0 \\ -2x + 18y + 5z + 2w = 10 \\ x + 7y - 8z \quad\quad = -4 \\ 3x \quad\quad + 2z \quad\quad = -10 \end{cases}$

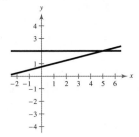

$(5, 2)$
Answers will vary.

(b)

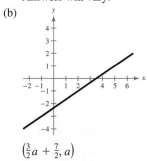

$\left(\frac{3}{2}a + \frac{7}{2}, a\right)$
Answers will vary.

7. 10.1 ft; About 252.7 ft   **9.** \$12.00

11. (a) $(3, -4)$ **(b)** $\left(\dfrac{2}{-a+5}, \dfrac{1}{4a-1}, \dfrac{1}{a}\right)$

13. (a) $\left(\dfrac{-5a+16}{6}, \dfrac{5a-16}{6}, a\right)$

   (b) $\left(\dfrac{-11a+36}{14}, \dfrac{13a-40}{14}, a\right)$

   (c) $(-a+3, a-3, a)$ **(d)** Infinitely many

27. Add 5 times Row 2 to Row 1.

29. Interchange Row 1 and Row 2.
    Add 4 times new Row 1 to Row 3.

31. $\begin{bmatrix} 1 & 2 & \frac{8}{3} \\ 4 & -3 & 6 \end{bmatrix}$ **33.** $\begin{bmatrix} 1 & 1 & 1 \\ 0 & -7 & -1 \end{bmatrix}$

35. $\begin{bmatrix} 1 & 0 & 14 & -11 \\ 0 & 1 & -2 & 2 \\ 0 & 0 & 1 & -7 \end{bmatrix}$

37. $\begin{bmatrix} 1 & 1 & 4 & -1 \\ 0 & 5 & -2 & 6 \\ 0 & 3 & 20 & 4 \end{bmatrix}; \begin{bmatrix} 1 & 1 & 4 & -1 \\ 0 & 1 & -\frac{2}{5} & \frac{6}{5} \\ 0 & 3 & 20 & 4 \end{bmatrix}$

39. (a) (i) $\begin{bmatrix} 3 & 0 & \vdots & -6 \\ 6 & -4 & \vdots & -28 \end{bmatrix}$

(ii) $\begin{bmatrix} 3 & 0 & \vdots & -6 \\ 0 & -4 & \vdots & -16 \end{bmatrix}$

(iii) $\begin{bmatrix} 3 & 0 & \vdots & -6 \\ 0 & 1 & \vdots & 4 \end{bmatrix}$

(iv) $\begin{bmatrix} 1 & 0 & \vdots & -2 \\ 0 & 1 & \vdots & 4 \end{bmatrix}$

(b) $\begin{cases} -3x + 4y = 22 \\ 6x - 4y = -28 \end{cases}$

Solution: $(-2, 4)$

(c) Answers will vary.

41. Reduced row-echelon form   **43.** Not in row-echelon form

45. $\begin{bmatrix} 1 & 1 & 0 & 5 \\ 0 & 1 & 2 & 0 \\ 0 & 0 & 1 & -1 \end{bmatrix}$ **47.** $\begin{bmatrix} 1 & -1 & -1 & 1 \\ 0 & 1 & 6 & 3 \\ 0 & 0 & 0 & 0 \end{bmatrix}$

49. $\begin{bmatrix} 1 & 0 & 0 \\ 0 & 1 & 0 \\ 0 & 0 & 1 \end{bmatrix}$ **51.** $\begin{bmatrix} 1 & 2 & 0 & 0 \\ 0 & 0 & 1 & 0 \\ 0 & 0 & 0 & 1 \\ 0 & 0 & 0 & 0 \end{bmatrix}$

53. $\begin{bmatrix} 1 & 0 & 3 & 16 \\ 0 & 1 & 2 & 12 \end{bmatrix}$ **55.** $\begin{cases} x - 2y = 4 \\ y = -1 \end{cases}$

$(2, -1)$

57. $\begin{cases} x - y + 2z = 4 \\ y - z = 2 \\ z = -2 \end{cases}$

$(8, 0, -2)$

59. $(-3, 5)$ **61.** $(-5, 6)$ **63.** $(-4, -3, 6)$

65. No solution   **67.** $(3, -2, 5, 0)$ **69.** $(3, -4)$

71. $(-1, -4)$ **73.** $(5a + 4, -3a + 2, a)$ **75.** $(4, -3, 2)$

77. $(7, -3, 4)$ **79.** $(0, 2 - 4a, a)$ **81.** $(1, 0, 4, -2)$

83. $(-2a, a, a, 0)$ **85.** The dimension is 4×1.

87. $f(x) = -x^2 + x + 1$ **89.** $f(x) = -9x^2 - 5x + 11$

91. $f(x) = x^2 + 2x + 5$

93. $y = 7.5t + 28$; About 141 cases; Yes, because the data values increase in a linear pattern.

95. $1,200,000 at 8%

$200,000 at 9%

$600,000 at 12%

97. False. It is a 2×4 matrix.   **99.** They are the same.

Section 8.2   *(page 564)*

1. equal   **3.** zero; O **5.** $x = -4, y = 23$

7. $x = -1, y = 3$

9. (a) $\begin{bmatrix} 3 & -2 \\ 1 & 7 \end{bmatrix}$ (b) $\begin{bmatrix} -1 & 0 \\ 3 & -9 \end{bmatrix}$ (c) $\begin{bmatrix} 3 & -3 \\ 6 & -3 \end{bmatrix}$

(d) $\begin{bmatrix} -1 & -1 \\ 8 & -19 \end{bmatrix}$

11. (a), (b), and (d) Not possible   (c) $\begin{bmatrix} 18 & 0 & 9 \\ -3 & -12 & 0 \end{bmatrix}$

13. (a) $\begin{bmatrix} 9 & 5 \\ 1 & -2 \\ -3 & 15 \end{bmatrix}$ (b) $\begin{bmatrix} 7 & -7 \\ 3 & 8 \\ -5 & -5 \end{bmatrix}$ (c) $\begin{bmatrix} 24 & -3 \\ 6 & 9 \\ -12 & 15 \end{bmatrix}$

(d) $\begin{bmatrix} 22 & -15 \\ 8 & 19 \\ -14 & -5 \end{bmatrix}$

15. (a) $\begin{bmatrix} 5 & 5 & -2 & 4 & 4 \\ -5 & 10 & 0 & -4 & -7 \end{bmatrix}$

(b) $\begin{bmatrix} 3 & 5 & 0 & 2 & 4 \\ 7 & -6 & -4 & 2 & 7 \end{bmatrix}$

(c) $\begin{bmatrix} 12 & 15 & -3 & 9 & 12 \\ 3 & 6 & -6 & -3 & 0 \end{bmatrix}$

(d) $\begin{bmatrix} 10 & 15 & -1 & 7 & 12 \\ 15 & -10 & -10 & 3 & 14 \end{bmatrix}$

17. $\begin{bmatrix} -8 & -7 \\ 15 & -1 \end{bmatrix}$ **19.** $\begin{bmatrix} -24 & -4 & 12 \\ -12 & 32 & 12 \end{bmatrix}$ **21.** $\begin{bmatrix} 10 & 8 \\ -59 & 9 \end{bmatrix}$

23. $\begin{bmatrix} -17.12 & 2.2 \\ 11.56 & 10.24 \end{bmatrix}$ **25.** $\begin{bmatrix} -10.81 & -5.36 & 0.4 \\ -14.04 & 10.69 & -14.76 \end{bmatrix}$

27. $\begin{bmatrix} -4 & 6 & -2 \\ 4 & 0 & 10 \end{bmatrix}$ **29.** $\begin{bmatrix} -2 & 0 & 5 \\ -\frac{5}{2} & 0 & \frac{7}{2} \end{bmatrix}$

31. $\begin{bmatrix} 3 & -\frac{1}{2} & -\frac{13}{2} \\ 3 & 0 & -\frac{11}{2} \end{bmatrix}$ **33.** $\begin{bmatrix} 2 & -5 & 5 \\ -5 & 0 & -6 \end{bmatrix}$

35. $\begin{bmatrix} -2 & 51 \\ -8 & 33 \\ 0 & 27 \end{bmatrix}$ **37.** Not possible   **39.** $\begin{bmatrix} 1 & 0 & 0 \\ 0 & 1 & 0 \\ 0 & 0 & \frac{7}{2} \end{bmatrix}$

3×2 3×3

41. $\begin{bmatrix} 70 & -17 & 73 \\ 32 & 11 & 6 \\ 16 & -38 & 70 \end{bmatrix}$ **43.** $\begin{bmatrix} 151 & 25 & 48 \\ 516 & 279 & 387 \\ 47 & -20 & 87 \end{bmatrix}$

45. (a) $\begin{bmatrix} 0 & 15 \\ 6 & 12 \end{bmatrix}$ (b) $\begin{bmatrix} -2 & 2 \\ 31 & 14 \end{bmatrix}$ (c) $\begin{bmatrix} 9 & 6 \\ 12 & 12 \end{bmatrix}$

47. (a) $\begin{bmatrix} 5 & -9 & 0 \\ 3 & 0 & -8 \\ -1 & 4 & 11 \end{bmatrix}$ (b) $\begin{bmatrix} 5 & -9 & 0 \\ 3 & 0 & -8 \\ -1 & 4 & 11 \end{bmatrix}$

(c) $\begin{bmatrix} -2 & -45 & 72 \\ 23 & -59 & -88 \\ -4 & 53 & 89 \end{bmatrix}$

49. (a) $\begin{bmatrix} 19 \\ 48 \end{bmatrix}$ (b) Not possible (c) $\begin{bmatrix} 14 & -8 \\ 16 & 142 \end{bmatrix}$

51. (a) $\begin{bmatrix} 7 & 7 & 14 \\ 8 & 8 & 16 \\ -1 & -1 & -2 \end{bmatrix}$ (b) $[13]$ (c) Not possible

53. $\begin{bmatrix} 5 & 8 \\ -4 & -16 \end{bmatrix}$ **55.** $\begin{bmatrix} -4 & 10 \\ 3 & 14 \end{bmatrix}$

57. (a) $\langle 4, 7 \rangle$ (b) $\langle -2, 3 \rangle$ (c) $\langle 8, 1 \rangle$

59. (a) $\langle 3, 6 \rangle$ (b) $\langle -7, -2 \rangle$ (c) $\langle 17, 10 \rangle$

61. $\langle 4, -2 \rangle$; Reflection in the x-axis

63. $\langle 2, 4 \rangle$; Reflection in the line $y = x$

65. $\langle 8, 2 \rangle$; Horizontal stretch

67. (a) $\begin{bmatrix} 2 & 3 \\ 1 & 4 \end{bmatrix}\begin{bmatrix} x_1 \\ x_2 \end{bmatrix} = \begin{bmatrix} 5 \\ 10 \end{bmatrix}$ (b) $\begin{bmatrix} -2 \\ 3 \end{bmatrix}$

69. (a) $\begin{bmatrix} 1 & -2 & 3 \\ -1 & 3 & -1 \\ 2 & -5 & 5 \end{bmatrix}\begin{bmatrix} x_1 \\ x_2 \\ x_3 \end{bmatrix} = \begin{bmatrix} 9 \\ -6 \\ 17 \end{bmatrix}$ (b) $\begin{bmatrix} 1 \\ -1 \\ 2 \end{bmatrix}$

71. (a) $\begin{bmatrix} 1 & -5 & 2 \\ -3 & 1 & -1 \\ 0 & -2 & 5 \end{bmatrix}\begin{bmatrix} x_1 \\ x_2 \\ x_3 \end{bmatrix} = \begin{bmatrix} -20 \\ 8 \\ -16 \end{bmatrix}$ (b) $\begin{bmatrix} -1 \\ 3 \\ -2 \end{bmatrix}$

73. $\begin{bmatrix} 110 & 99 & 77 & 33 \\ 44 & 22 & 66 & 66 \end{bmatrix}$

75. [$1037.50 $1400 $1012.50]

The entries represent the profits from both crops at each of the three outlets.

77. $\begin{bmatrix} \$23.20 & \$20.50 \\ \$38.20 & \$33.80 \\ \$76.90 & \$68.50 \end{bmatrix}$

The entries represent the labor costs at each plant for each size of boat.

79. $\begin{bmatrix} 0.40 & 0.15 & 0.15 \\ 0.28 & 0.53 & 0.17 \\ 0.32 & 0.32 & 0.68 \end{bmatrix}$

P^2 gives the proportions of the voting population that changed parties or remained loyal to their parties from the first election to the third.

81. True. The sum of two matrices of different dimensions is undefined.

83. $\begin{bmatrix} 1 & 0 \\ 2 & 1 \end{bmatrix} \neq \begin{bmatrix} 0 & 0 \\ 3 & 2 \end{bmatrix}$ **85.** $\begin{bmatrix} 3 & -2 \\ 4 & 3 \end{bmatrix} \neq \begin{bmatrix} 2 & -2 \\ 5 & 4 \end{bmatrix}$

87. $AC = BC = \begin{bmatrix} 2 & 3 \\ 2 & 3 \end{bmatrix}$ **89.** Answers will vary.

91. AB is a diagonal matrix whose entries are the products of the corresponding entries of A and B.

Section 8.3   *(page 574)*

1. inverse   **3.** determinant   **5–11.** $AB = I$ and $BA = I$

13. $\begin{bmatrix} 3 & -1 \\ -5 & 2 \end{bmatrix}$ **15.** $\begin{bmatrix} -3 & 2 \\ -2 & 1 \end{bmatrix}$ **17.** $\begin{bmatrix} 1 & -\frac{1}{2} \\ -2 & \frac{3}{2} \end{bmatrix}$

19. $\begin{bmatrix} 1 & 1 & -1 \\ -3 & 2 & -1 \\ 3 & -3 & 2 \end{bmatrix}$ **21.** Not possible

23. $\begin{bmatrix} -\frac{1}{8} & 0 & 0 & 0 \\ 0 & 1 & 0 & 0 \\ 0 & 0 & \frac{1}{4} & 0 \\ 0 & 0 & 0 & -\frac{1}{5} \end{bmatrix}$ **25.** $\begin{bmatrix} -175 & 37 & -13 \\ 95 & -20 & 7 \\ 14 & -3 & 1 \end{bmatrix}$

27. $\begin{bmatrix} -12 & -5 & -9 \\ -4 & -2 & -4 \\ -8 & -4 & -6 \end{bmatrix}$ **29.** $\begin{bmatrix} 0 & -1.\overline{81} & 0.\overline{90} \\ -10 & 5 & 5 \\ 10 & -2.\overline{72} & -3.\overline{63} \end{bmatrix}$

31. $\begin{bmatrix} 1 & 0 & 1 & 0 \\ 0 & 1 & 0 & 1 \\ 2 & 0 & 1 & 0 \\ 0 & 1 & 0 & 2 \end{bmatrix}$ **33.** $\begin{bmatrix} \frac{5}{13} & -\frac{3}{13} \\ \frac{1}{13} & \frac{2}{13} \end{bmatrix}$

35. Not possible   **37.** $\begin{bmatrix} -4 & 2 \\ 10 & -\frac{10}{3} \end{bmatrix}$ **39.** $(5, 0)$

41. $(-8, -6)$ **43.** $(3, 8, -11)$ **45.** $(2, 1, 0, 0)$

47. $(-1, 1)$ **49.** No solution   **51.** $(-4, -8)$

53. $(-1, 3, 2)$ **55.** $\left(\frac{13}{16}, \frac{11}{16}, 0\right)$

57. $3684.21 in AAA-rated bonds
$2105.26 in A-rated bonds
$4210.53 in B-rated bonds

59. $I_1 = 0.5$ amp   **61.** $I_1 = 4$ amps
$I_2 = 3$ amps $I_2 = 1$ amp
$I_3 = 3.5$ amps $I_3 = 5$ amps

63. 100 bags of potting soil for seedlings
100 bags of potting soil for general potting
100 bags of potting soil for hardwood plants

65. (a) $\begin{cases} 2.5r + 4l + 2i = 300 \\ -r + 2l + 2i = 0 \\ r + l + i = 120 \end{cases}$

$\begin{bmatrix} 2.5 & 4 & 2 \\ -1 & 2 & 2 \\ 1 & 1 & 1 \end{bmatrix} \begin{bmatrix} r \\ l \\ i \end{bmatrix} = \begin{bmatrix} 300 \\ 0 \\ 120 \end{bmatrix}$

(b) 80 roses, 10 lilies, 30 irises

67. True. If B is the inverse of A, then $AB = I = BA$.

69. Answers will vary.   **71.** $k \neq -\frac{3}{2}; k = -\frac{3}{2}$

73. (a) Answers will vary.

(b) $A^{-1} = \begin{bmatrix} 1/a_{11} & 0 & 0 & \cdots & 0 \\ 0 & 1/a_{22} & 0 & \cdots & 0 \\ 0 & 0 & 1/a_{33} & \cdots & 0 \\ \vdots & \vdots & \vdots & & \vdots \\ 0 & 0 & 0 & \cdots & 1/a_{nn} \end{bmatrix}$

75. Answers will vary.

Section 8.4   *(page 582)*

1. determinant   **3.** cofactor   **5.** 4   **7.** 16   **9.** -3

11. 0   **13.** 6   **15.** 0   **17.** -23 **19.** -24

21. $\frac{11}{6}$ **23.** 11   **25.** -1924 **27.** 0.08

29. (a) $M_{11} = -6, M_{12} = 3, M_{21} = 5, M_{22} = 4$
(b) $C_{11} = -6, C_{12} = -3, C_{21} = -5, C_{22} = 4$

31. (a) $M_{11} = 3, M_{12} = -4, M_{13} = 1, M_{21} = 2, M_{22} = 2,$
$M_{23} = -4, M_{31} = -4, M_{32} = 10, M_{33} = 8$
(b) $C_{11} = 3, C_{12} = 4, C_{13} = 1, C_{21} = -2, C_{22} = 2,$
$C_{23} = 4, C_{31} = -4, C_{32} = -10, C_{33} = 8$

33. (a) $M_{11} = 10, M_{12} = -43, M_{13} = 2, M_{21} = -30, M_{22} = 17,$
$M_{23} = -6, M_{31} = 54, M_{32} = -53, M_{33} = -34$
(b) $C_{11} = 10, C_{12} = 43, C_{13} = 2, C_{21} = 30, C_{22} = 17,$
$C_{23} = 6, C_{31} = 54, C_{32} = 53, C_{33} = -34$

35. (a) and (b) -36 **37.** (a) and (b) 96

39. (a) and (b) -75 **41.** (a) and (b) 0

43. (a) and (b) 225   **45.** -9 **47.** 0   **49.** 0

51. -58 **53.** 72   **55.** 0   **57.** 412

59. -126 **61.** -336

63. (a) -3 (b) -2 (c) $\begin{bmatrix} -2 & 0 \\ 0 & -3 \end{bmatrix}$ (d) 6

65. (a) -8 (b) 0   (c) $\begin{bmatrix} -4 & 4 \\ 1 & -1 \end{bmatrix}$ (d) 0

67. (a) 2   (b) -6 (c) $\begin{bmatrix} 1 & 4 & 3 \\ -1 & 0 & 3 \\ 0 & 2 & 0 \end{bmatrix}$ (d) -12

69. $A = \begin{bmatrix} 3 & 3 \\ 1 & 2 \end{bmatrix}$ **71.** $A = \begin{bmatrix} 4 & 2 & -1 \\ 2 & 1 & 0 \\ 1 & 1 & 3 \end{bmatrix}$

73. $A = \begin{bmatrix} 2 & 3 \\ 8 & 12 \end{bmatrix}$ **75–79.** Answers will vary.   **81.** ± 2

83. $-2, 1$ **85.** $-1, -4$ **87.** $8uv - 1$ **89.** e^{5x}

91. $1 - \ln x$

93. True. If an entire row is zero, then each cofactor in the expansion is multiplied by zero.

CHAPTER 8

95. Answers will vary.

97. The signs of the cofactors should be $-, +, -$.

99. (a) Columns 2 and 3 of A were interchanged.
$$|A| = -115 = -|B|$$
(b) Rows 1 and 3 of A were interchanged.
$$|A| = -40 = -|B|$$

101. (a) Multiply Row 1 by 5.
(b) Multiply Column 2 by 4 and Column 3 by 3.

103. (a) 28   (b) -10 (c) -12
The determinant of a diagonal matrix is the product of the entries on the main diagonal.

Section 8.5   (page 595)

1. Cramer's Rule   **3.** $A = \pm\dfrac{1}{2}\begin{vmatrix} x_1 & y_1 & 1 \\ x_2 & y_2 & 1 \\ x_3 & y_3 & 1 \end{vmatrix}$

5. uncoded; coded   **7.** $(1, -1)$ **9.** Not possible

11. $(-1, 3, 2)$ **13.** $(-2, 1, -1)$ **15.** 7   **17.** 14

19. $y = \frac{16}{5}$ or $y = 0$ **21.** 250 mi^2 **23.** Collinear

25. Not collinear   **27.** Collinear   **29.** $y = -3$

31. $3x - 5y = 0$ **33.** $x + 3y - 5 = 0$

35. $2x + 3y - 8 = 0$

37. $(0, 0), (0, 3), (6, 0), (6, 3)$

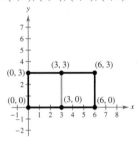

39. $(-4, 3), (-5, 3), (-4, 4), (-5, 4)$

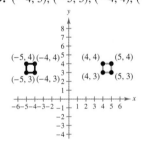

41. 2 square units   **43.** 10 square units

45. (a) Uncoded: $[3 \quad 15], [13 \quad 5], [0 \quad 8], [15 \quad 13], [5 \quad 0],$
$[19 \quad 15], [15 \quad 14]$
(b) Encoded: 48 81 28 51 24 40 54 95 5
10 64 113 57 100

47. (a) Uncoded: $[3 \quad 1 \quad 12], [12 \quad 0 \quad 13], [5 \quad 0 \quad 20],$
$[15 \quad 13 \quad 15], [18 \quad 18 \quad 15], [23 \quad 0 \quad 0]$
(b) Encoded: $-68 \quad 21 \quad 35 \quad -66 \quad 14 \quad 39 \quad -115$
$35 \quad 60 \quad -62 \quad 15 \quad 32 \quad -54 \quad 12 \quad 27 \quad 23 \quad -23 \quad 0$

49. $1 \quad -25 \quad -65 \quad 17 \quad 15 \quad -9 \quad -12 \quad -62 \quad -119$
$27 \quad 51 \quad 48 \quad 43 \quad 67 \quad 48 \quad 57 \quad 111 \quad 117$

51. $-5 \quad -41 \quad -87 \quad 91 \quad 207 \quad 257 \quad 11 \quad -5 \quad -41 \quad 40$
$80 \quad 84 \quad 76 \quad 177 \quad 227$

53. HAPPY NEW YEAR   **55.** CLASS IS CANCELED

57. SEND PLANES   **59.** MEET ME TONIGHT RON

61. $I_1 = -0.5$ amp
$I_2 = 1$ amp
$I_3 = 0.5$ amp

63. False. The denominator is the determinant of the coefficient matrix.

65. The system has either no solutions or infinitely many solutions.

67. 12

Review Exercises   (page 600)

1. 1×2 **3.** 2×5 **5.** $\begin{bmatrix} 3 & -10 & \vdots & 15 \\ 5 & 4 & \vdots & 22 \end{bmatrix}$

7. $\begin{cases} x & + 2z = -8 \\ 2x - 2y + 3z = 12 \\ 4x + 7y + z = 3 \end{cases}$ **9.** $\begin{bmatrix} 1 & 2 & 3 \\ 0 & 1 & 1 \\ 0 & 0 & 1 \end{bmatrix}$

11. $\begin{cases} x + 2y + 3z = 9 \\ y - 2z = 2 \\ z = -1 \end{cases}$ **13.** $\begin{cases} x + 3y + 4z = 1 \\ y + 2z = 3 \\ z = 4 \end{cases}$
$(12, 0, -1)$ $(0, -5, 4)$

15. $(10, -12)$ **17.** $\left(-\frac{1}{5}, \frac{7}{10}\right)$ **19.** No solution

21. $(1, -2, 2)$ **23.** $\left(-2a + \frac{3}{2}, 2a + 1, a\right)$ **25.** $(5, 2, -6)$

27. $(1, 2, 2)$ **29.** $(2, -3, 3)$ **31.** $(2, 6, -10, -3)$

33. $x = 12, y = 11$ **35.** $x = 1, y = 11$

37. (a) $\begin{bmatrix} -1 & 8 \\ 15 & 13 \end{bmatrix}$ (b) $\begin{bmatrix} 5 & -12 \\ -9 & -3 \end{bmatrix}$
(c) $\begin{bmatrix} 8 & -8 \\ 12 & 20 \end{bmatrix}$ (d) $\begin{bmatrix} -2 & 16 \\ 30 & 26 \end{bmatrix}$

39. (a) $\begin{bmatrix} 5 & 7 \\ -3 & 14 \\ 31 & 42 \end{bmatrix}$ (b) $\begin{bmatrix} 5 & 1 \\ -11 & -10 \\ -9 & -38 \end{bmatrix}$
(c) $\begin{bmatrix} 20 & 16 \\ -28 & 8 \\ 44 & 8 \end{bmatrix}$ (d) $\begin{bmatrix} 10 & 14 \\ -6 & 28 \\ 62 & 84 \end{bmatrix}$

41. $\begin{bmatrix} 22 & -17 \\ 14 & 11 \end{bmatrix}$ **43.** $\begin{bmatrix} -16 & -6 \\ -12 & 4 \\ -14 & -8 \end{bmatrix}$ **45.** $\begin{bmatrix} -11 & -6 \\ 8 & -13 \\ -18 & -8 \end{bmatrix}$

47. $\begin{bmatrix} 3 & \frac{2}{3} \\ -\frac{4}{3} & \frac{11}{3} \\ \frac{10}{3} & 0 \end{bmatrix}$ **49.** $\begin{bmatrix} -30 & 4 \\ 51 & 70 \end{bmatrix}; 2 \times 2$

51. $\begin{bmatrix} 100 & 220 \\ 12 & -4 \\ 84 & 212 \end{bmatrix}; 3 \times 2$ **53.** $\begin{bmatrix} 14 & -22 & 22 \\ 19 & -41 & 80 \\ 42 & -66 & 66 \end{bmatrix}$

55. Not possible

57. (a) $\begin{bmatrix} -1 & -1 \\ 18 & -4 \end{bmatrix}$ (b) $\begin{bmatrix} 1 & 14 \\ -2 & -6 \end{bmatrix}$ (c) $\begin{bmatrix} 13 & 6 \\ 8 & 13 \end{bmatrix}$

59. $\langle 2, -5 \rangle$; Reflection in the x-axis

61. $\langle 1, 5 \rangle$; Horizontal shrink   **63.** $\begin{bmatrix} 76 & 114 & 133 \\ 38 & 95 & 76 \end{bmatrix}$

65–67. $AB = I$ and $BA = I$ **69.** $\begin{bmatrix} 4 & -5 \\ 5 & -6 \end{bmatrix}$

71. $\begin{bmatrix} \frac{1}{2} & -1 & -\frac{1}{2} \\ \frac{1}{2} & -\frac{2}{3} & -\frac{5}{6} \\ 0 & \frac{2}{3} & \frac{1}{3} \end{bmatrix}$ **73.** $\begin{bmatrix} 13 & 6 & -4 \\ -12 & -5 & 3 \\ 5 & 2 & -1 \end{bmatrix}$

75. $\begin{bmatrix} 1 & -1 \\ 4 & -\frac{7}{2} \end{bmatrix}$ **77.** Not possible   **79.** $(36, 11)$

81. $(-6, -1)$ **83.** $(2, 3)$ **85.** $(-8, 18)$

87. $(2, -1, -2)$ **89.** $(-3, 1)$ **91.** $\left(\frac{1}{6}, -\frac{7}{4}\right)$

93. 26   **95.** 116

97. (a) $M_{11} = 4, M_{12} = 7, M_{21} = -1, M_{22} = 2$
(b) $C_{11} = 4, C_{12} = -7, C_{21} = 1, C_{22} = 1$

99. (a) $M_{11} = 30, M_{12} = -12, M_{13} = -21, M_{21} = 20,$
$M_{22} = 19, M_{23} = 22, M_{31} = 5, M_{32} = -2, M_{33} = 19$
(b) $C_{11} = 30, C_{12} = 12, C_{13} = -21, C_{21} = -20,$
$C_{22} = 19, C_{23} = -22, C_{31} = 5, C_{32} = 2, C_{33} = 19$

101. -6 **103.** 15   **105.** 130   **107.** $(4, 7)$

109. $(-1, 4, 5)$ **111.** 16   **113.** Collinear

115. $x - 2y + 4 = 0$ **117.** $2x + 6y - 13 = 0$

119. 8 square units   **121.** SEE YOU FRIDAY

123. False. The matrix must be square.

125. If A is a square matrix, then the cofactor C_{ij} of the entry a_{ij} is $(-1)^{i+j}M_{ij}$, where M_{ij} is the determinant obtained by deleting the ith row and jth column of A. The determinant of A is the sum of the entries of any row or column of A multiplied by their respective cofactors.

Chapter Test   *(page 604)*

1. $\begin{bmatrix} 1 & 0 & 0 \\ 0 & 1 & 0 \\ 0 & 0 & 1 \end{bmatrix}$ **2.** $\begin{bmatrix} 1 & 0 & -1 & 2 \\ 0 & 1 & 0 & -1 \\ 0 & 0 & 0 & 0 \\ 0 & 0 & 0 & 0 \end{bmatrix}$

3. $\begin{bmatrix} 4 & 3 & -2 & \vdots & 14 \\ -1 & -1 & 2 & \vdots & -5 \\ 3 & 1 & -4 & \vdots & 8 \end{bmatrix}, \left(1, 3, -\frac{1}{2}\right)$

4. (a) $\begin{bmatrix} 1 & 5 \\ 0 & -4 \end{bmatrix}$

(b) $\begin{bmatrix} 6 & -3 & 12 \\ 0 & 18 & -9 \end{bmatrix}$

(c) $\begin{bmatrix} 8 & 15 \\ -5 & -13 \end{bmatrix}$

(d) $\begin{bmatrix} 10 & -5 & 20 \\ -10 & -1 & -17 \end{bmatrix}$

(e) Not possible

5. $\langle -3, -2 \rangle$; Reflection in the line $y = -x$

6. $\begin{bmatrix} \frac{2}{7} & \frac{3}{7} \\ \frac{5}{7} & \frac{4}{7} \end{bmatrix}$ **7.** $\begin{bmatrix} -\frac{5}{2} & 4 & -3 \\ 5 & -7 & 6 \\ 4 & -6 & 5 \end{bmatrix}$ **8.** $(12, 18)$

9. -112 **10.** 0   **11.** 43   **12.** $(-3, 5)$

13. $(-2, 4, 6)$ **14.** 7

15. Uncoded: $[11 \ 14 \ 15], [3 \ 11 \ 0], [15 \ 14 \ 0], [23 \ 15 \ 15],$
$[4 \ 0 \ 0]$
Encoded: $115 \ -41 \ -59 \ 14 \ -3 \ -11 \ 29 \ -15$
$-14 \ 128 \ -53 \ -60 \ 4 \ -4 \ 0$

16. 75 L of 60% solution, 25 L of 20% solution

Problem Solving   *(page 607)*

1. (a) $AT = \begin{bmatrix} -1 & -4 & -2 \\ 1 & 2 & 3 \end{bmatrix}, AAT = \begin{bmatrix} -1 & -2 & -3 \\ -1 & -4 & -2 \end{bmatrix}$

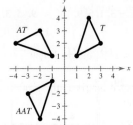

A represents a counterclockwise rotation.

(b) AAT is rotated clockwise 90° to obtain AT. AT is then rotated clockwise 90° to obtain T.

3. (a) Yes   (b) No   (c) No   (d) No   (e) No   (f) No

5. (a) $A^2 - 2A + 5I = \begin{bmatrix} 1 & 2 \\ -2 & 1 \end{bmatrix}^2 - 2\begin{bmatrix} 1 & 2 \\ -2 & 1 \end{bmatrix} + 5\begin{bmatrix} 1 & 0 \\ 0 & 1 \end{bmatrix}$

$= \begin{bmatrix} -3 & 4 \\ -4 & -3 \end{bmatrix} - \begin{bmatrix} 2 & 4 \\ -4 & 2 \end{bmatrix} + \begin{bmatrix} 5 & 0 \\ 0 & 5 \end{bmatrix}$

$= \begin{bmatrix} 0 & 0 \\ 0 & 0 \end{bmatrix}$

(b) $A^{-1} = \frac{1}{5}(2I - A)$

$\begin{bmatrix} 1 & 2 \\ -2 & 1 \end{bmatrix}^{-1} = \frac{1}{5}\left(2\begin{bmatrix} 1 & 0 \\ 0 & 1 \end{bmatrix} - \begin{bmatrix} 1 & 2 \\ -2 & 1 \end{bmatrix}\right)$

$\begin{bmatrix} \frac{1}{5} & -\frac{2}{5} \\ \frac{2}{5} & \frac{1}{5} \end{bmatrix} = \frac{1}{5}\left(\begin{bmatrix} 2 & 0 \\ 0 & 2 \end{bmatrix} - \begin{bmatrix} 1 & 2 \\ -2 & 1 \end{bmatrix}\right)$

$\begin{bmatrix} \frac{1}{5} & -\frac{2}{5} \\ \frac{2}{5} & \frac{1}{5} \end{bmatrix} = \frac{1}{5}\begin{bmatrix} 1 & -2 \\ 2 & 1 \end{bmatrix}$

$\begin{bmatrix} \frac{1}{5} & -\frac{2}{5} \\ \frac{2}{5} & \frac{1}{5} \end{bmatrix} = \begin{bmatrix} \frac{1}{5} & -\frac{2}{5} \\ \frac{2}{5} & \frac{1}{5} \end{bmatrix}$

(c) Answers will vary.

7. $A^T = \begin{bmatrix} -1 & 2 \\ 1 & 0 \\ -2 & 1 \end{bmatrix}$, $B^T = \begin{bmatrix} -3 & 1 & 1 \\ 0 & 2 & -1 \end{bmatrix}$

$(AB)^T = \begin{bmatrix} 2 & -5 \\ 4 & -1 \end{bmatrix} = B^T A^T$

9. $x = 6$ **11.** Answers will vary.

13. $\begin{vmatrix} x & 0 & 0 & d \\ -1 & x & 0 & c \\ 0 & -1 & x & b \\ 0 & 0 & -1 & a \end{vmatrix}$

15. Sulfur: 32 atomic mass units
Nitrogen: 14 atomic mass units
Fluorine: 19 atomic mass units

17. REMEMBER SEPTEMBER THE ELEVENTH

19. $A^{-1} = \begin{bmatrix} 0.0625 & -0.4375 & 0.625 \\ 0.1875 & 0.6875 & -1.125 \\ -0.125 & -0.125 & 0.75 \end{bmatrix}$

$|A^{-1}| = \frac{1}{16}, |A| = 16$

$|A^{-1}| = \frac{1}{|A|}$

CHAPTER 8

Chapter 9

Section 9.1 *(page 617)*

1. infinite sequence   **3.** recursively
5. index; upper; lower   **7.** $-3, 1, 5, 9, 13$ **9.** $5, 3, 5, 3, 5$
11. $-2, 4, -8, 16, -32$ **13.** $\frac{2}{3}, \frac{2}{3}, \frac{2}{3}, \frac{2}{3}, \frac{2}{3}$ **15.** $\frac{1}{3}, \frac{8}{3}, 9, \frac{64}{3}, \frac{125}{3}$
17. $\frac{1}{3}, \frac{1}{2}, \frac{3}{5}, \frac{2}{3}, \frac{5}{7}$ **19.** $0, 0, 6, 24, 60$ **21.** $-\frac{1}{2}, \frac{2}{3}, -\frac{3}{4}, \frac{4}{5}, -\frac{5}{6}$
23. -73 **25.** $\frac{44}{239}$

27. 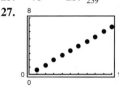   **29.**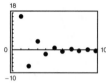

31. 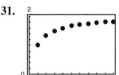   **33.** c   **34.** b   **35.** d
36. a   **37.** $a_n = 4n - 1$
39. $a_n = n^3 + 2$
41. $a_n = (-1)^{n+1}$
43. $a_n = \dfrac{(-1)^n(n+1)}{n+2}$

45. $a_n = \dfrac{n+1}{2n-1}$ **47.** $a_n = \dfrac{1}{n!}$ **49.** $a_n = \dfrac{3^{n-1}}{(n-1)!}$
51. $28, 24, 20, 16, 12$ **53.** $81, 27, 9, 3, 1$ **55.** $1, 2, 2, 3, \frac{7}{2}$
57. $1, 1, 2, 3, 5, 8, 13, 21, 34, 55, 89, 144$
$1, 2, \frac{3}{2}, \frac{5}{3}, \frac{8}{5}, \frac{13}{8}, \frac{21}{13}, \frac{34}{21}, \frac{55}{34}, \frac{89}{55}$
59. $5, 5, \frac{5}{2}, \frac{5}{6}, \frac{5}{24}$ **61.** $6, -24, 60, -120, 210$ **63.** $\frac{1}{30}$
65. $n + 1$ **67.** 90   **69.** $\frac{124}{429}$ **71.** 88   **73.** $\frac{13}{4}$
75. $\dfrac{3}{8}$ **77.** 1.33   **79.** $\displaystyle\sum_{i=1}^{9} \frac{1}{3i}$ **81.** $\displaystyle\sum_{i=1}^{8}\left[2\left(\frac{i}{8}\right)+3\right]$
83. $\displaystyle\sum_{i=1}^{6} (-1)^{i+1}3i$ **85.** $\displaystyle\sum_{i=1}^{7} \frac{i^2}{(i+1)!}$ **87.** $\displaystyle\sum_{i=1}^{5} \frac{2^i - 1}{2^{i+1}}$
89. (a) $\frac{7}{8}$ (b) $\frac{15}{16}$ (c) $\frac{31}{32}$
91. (a) $-\frac{3}{2}$ (b) $-\frac{5}{4}$ (c) $-\frac{11}{8}$ **93.** $\frac{2}{3}$ **95.** $\frac{7}{9}$
97. (a) $A_1 = \$10,087.50, A_2 \approx \$10,175.77, A_3 \approx \$10,264.80,$
$A_4 \approx \$10,354.62, A_5 \approx \$10,445.22, A_6 \approx \$10,536.62,$
$A_7 \approx \$10,628.81, A_8 \approx \$10,721.82$
(b) $\$14,169.09$
(c) No. $A_{80} \approx \$20,076.31 \neq 2A_{40} \approx \$28,338.18$
99. True by the Properties of Sums.   **101.** $\$500.95$
103. Proof   **105.** $\displaystyle\sum_{k=1}^{4} 3 = 3(4) = 12$
107. (a) 0 blue faces: 1
1 blue face: 6
2 blue faces: 12
3 blue faces: 8

(b)

Number of blue faces	0	1	2	3
$4 \times 4 \times 4$	8	24	24	8
$5 \times 5 \times 5$	27	54	36	8
$6 \times 6 \times 6$	64	96	48	8

(c) 0 blue faces: $(n-2)^3$
1 blue face: $6(n-2)^2$
2 blue faces: $12(n-2)$
3 blue faces: 8

Section 9.2 *(page 626)*

1. arithmetic; common   **3.** recursion   **5.** Not arithmetic
7. Arithmetic, $d = -2$ **9.** Arithmetic, $d = \frac{1}{4}$
11. Not arithmetic
13. $8, 11, 14, 17, 20$ **15.** $7, 3, -1, -5, -9$
Arithmetic, $d = 3$ Arithmetic, $d = -4$
17. $-1, 1, -1, 1, -1$ **19.** $2, 8, 24, 64, 160$
Not arithmetic   Not arithmetic
21. $a_n = 3n - 2$ **23.** $a_n = -8n + 108$
25. $a_n = -\frac{5}{2}n + \frac{13}{2}$ **27.** $a_n = \frac{10}{3}n + \frac{5}{3}$ **29.** $a_n = 3n + 85$
31. $5, 11, 17, 23, 29$ **33.** $2, -4, -10, -16, -22$
35. $-2, 2, 6, 10, 14$ **37.** $15, 19, 23, 27, 31$
39. $15, 13, 11, 9, 7$ **41.** -49 **43.** $\frac{31}{8}$ **45.** 110
47. -25 **49.** 10,000   **51.** 15,100   **53.** -7020
55. 1275   **57.** 129,250   **59.** $-28,300$
61. b   **62.** d   **63.** c   **64.** a
65. 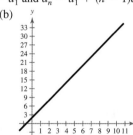   **67.**

69. (a) $\$40,000$ (b) $\$217,500$ **71.** 2430 seats
73. 784 ft   **75.** $\$375,000$; Answers will vary.
77. (a)

(bar graph: Number of new stores vs. Year; years 2011–2015; y-axis 200–450)

(b) $a_n = 229.25 + 36.75n$
(c)

(scatter plot, y-axis 200–450)

(d) $\displaystyle\sum_{n=1}^{5} (229.25 + 36.75n)$; About 1698 stores
79. True. Given a_1 and a_2, $d = a_2 - a_1$ and $a_n = a_1 + (n-1)d$.
81. (a)

(scatter plot a_n, axis to 33)

(b)

(line graph y, axis to 33)

(c) The graph of $y = 3x + 2$ contains all points on the line. The graph of $a_n = 2 + 3n$ contains only points at the positive integers.
(d) The slope of the line and the common difference of the arithmetic sequence are equal.

83. $x, 3x, 5x, 7x, 9x, 11x, 13x, 15x, 17x, 19x$
85. When $n = 50$, $a_n = 2(50) - 1 = 99$.
87. (a) 4, 9, 16, 25, 36   (b) $S_n = n^2$; $S_7 = 49 = 7^2$

(c) $\dfrac{n}{2}[1 + (2n - 1)] = n^2$

Section 9.3   *(page 635)*

1. geometric; common   **3.** $a_1\left(\dfrac{1 - r^n}{1 - r}\right)$
5. Geometric, $r = 2$ **7.** Geometric, $r = 3$
9. Not geometric   **11.** Geometric, $r = -\sqrt{7}$
13. 4, 12, 36, 108, 324   **15.** $1, \frac{1}{2}, \frac{1}{4}, \frac{1}{8}, \frac{1}{16}$ **17.** $1, e, e^2, e^3, e^4$
19. $3, 3\sqrt{5}, 15, 15\sqrt{5}, 75$ **21.** $2, 6x, 18x^2, 54x^3, 162x^4$
23. $a_n = 4\left(\frac{1}{2}\right)^{n-1}$; $\frac{1}{128}$ **25.** $a_n = 6\left(-\frac{1}{3}\right)^{n-1}$; $-\dfrac{2}{59,049}$
27. $a_n = 100e^{x(n-1)}$; $100e^{8x}$ **29.** $a_n = \left(\sqrt{2}\right)^{n-1}$; $32\sqrt{2}$
31. $a_n = 500(1.02)^{n-1}$; About 1082.372   **33.** $a_n = 64\left(\frac{1}{2}\right)^{n-1}$
35. $a_n = 9(2)^{n-1}$ **37.** $a_n = 6\left(-\frac{3}{2}\right)^{n-1}$ **39.** 13,122
41. $\frac{1}{768}$ **43.** $a_3 = 9$ **45.** $a_6 = -2$
47. a   **48.** c   **49.** b   **50.** d
51.

53.

55. 5461   **57.** $-14,706$ **59.** 29,921.311   **61.** 1360.383
63. 1.600   **65.** $\sum\limits_{n=1}^{7} 10(3)^{n-1}$ **67.** $\sum\limits_{n=1}^{6} 0.1(4)^{n-1}$ **69.** 2
71. $\frac{2}{3}$ **73.** 5   **75.** 32   **77.** Undefined   **79.** $\frac{4}{11}$
81.

Horizontal asymptote: $y = 12$
Corresponds to the sum of the series
83. $29,412.25   **85.** Answers will vary.   **87.** $1600
89. $273\frac{8}{9}$ in.^{2} **91.** $5,435,989.84
93. False. A sequence is geometric when the ratios of consecutive terms are the same.
95. (a)

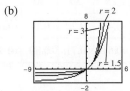

As $x \to \infty$, $y \to \dfrac{1}{1 - r}$.

(b)

As $x \to \infty$, $y \to \infty$.

Section 9.4   *(page 646)*

1. mathematical induction   **3.** arithmetic
5. $\dfrac{5}{(k + 1)(k + 2)}$ **7.** $(k + 1)^2(k + 4)^2$
9. $\dfrac{3}{(k + 3)(k + 4)}$ **11–39.** Proofs
41. $S_n = n(2n - 1)$; Proof   **43.** $S_n = \dfrac{n}{2(n + 1)}$; Proof
45. 120   **47.** 91   **49.** 979   **51.** 70   **53.** -3402
55. Linear; $a_n = 9n - 4$ **57.** Quadratic; $a_n = 2n^2 + 2$
59. Quadratic; $a_n = 4n^2 - 5$
61. 0, 3, 6, 9, 12, 15
   First differences: 3, 3, 3, 3, 3
   Second differences: 0, 0, 0, 0
   Linear
63. 4, 10, 19, 31, 46, 64
   First differences: 6, 9, 12, 15, 18
   Second differences: 3, 3, 3, 3
   Quadratic
65. 3, 7, 16, 32, 57, 93
   First differences: 4, 9, 16, 25, 36
   Second differences: 5, 7, 9, 11
   Neither
67. 5, 3, 9, 7, 13, 11
   First differences: $-2, 6, -2, 6, -2$
   Second differences: $8, -8, 8, -8$
   Neither
69. $a_n = n^2 - n + 3$ **71.** $a_n = \frac{1}{2}n^2 + 2n - 1$
73. $a_n = n^2 + 4n - 5$
75. (a) 16, 15, 15, 15, 13; *Sample answer:* $a_n = 15n + 4636$
   (b) $a_n \approx 14.9n + 4637$; The models are similar.
   (c) Part (a): 4,951,000,
      Part (b): 4,949,900;
      The values are similar.
77. False. P_1 must be proven to be true.

Section 9.5   *(page 653)*

1. expanding   **3.** Binomial Theorem; Pascal's Triangle
5. 10   **7.** 1   **9.** 210   **11.** 4950   **13.** 20   **15.** 5
17. $x^6 + 6x^5 + 15x^4 + 20x^3 + 15x^2 + 6x + 1$
19. $y^3 - 9y^2 + 27y - 27$ **21.** $r^3 + 9r^2s + 27rs^2 + 27s^3$
23. $243a^5 - 1620a^4b + 4320a^3b^2 - 5760a^2b^3$
      $+ 3840ab^4 - 1024b^5$
25. $a^4 + 24a^3 + 216a^2 + 864a + 1296$
27. $y^6 - 6y^5 + 15y^4 - 20y^3 + 15y^2 - 6y + 1$
29. $81 - 216z + 216z^2 - 96z^3 + 16z^4$
31. $x^5 + 10x^4y + 40x^3y^2 + 80x^2y^3 + 80xy^4 + 32y^5$
33. $x^8 + 4x^6y^2 + 6x^4y^4 + 4x^2y^6 + y^8$
35. $\dfrac{1}{x^5} + \dfrac{5y}{x^4} + \dfrac{10y^2}{x^3} + \dfrac{10y^3}{x^2} + \dfrac{5y^4}{x} + y^5$
37. $2x^4 - 24x^3 + 113x^2 - 246x + 207$ **39.** $120x^7y^3$
41. $360x^3y^2$ **43.** $1,259,712x^2y^7$ **45.** $-4,330,260,000y^9x^3$
47. 160   **49.** 720   **51.** $-6,300,000$ **53.** 210
55. $x^{3/2} + 15x + 75x^{1/2} + 125$
57. $x^2 - 3x^{4/3}y^{1/3} + 3x^{2/3}y^{2/3} - y$
59. $81t^2 + 108t^{7/4} + 54t^{3/2} + 12t^{5/4} + t$

61. $3x^2 + 3xh + h^2,\ h \neq 0$

63. $6x^5 + 15x^4h + 20x^3h^2 + 15x^2h^3 + 6xh^4 + h^5,\ h \neq 0$

65. $\dfrac{1}{\sqrt{x+h} + \sqrt{x}},\ h \neq 0$ **67.** -4 **69.** $2035 + 828i$

71. 1   **73.** 1.172   **75.** 510,568.785   **77.** 0.273

79. 0.171

81.

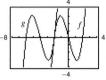

The graph of g is shifted four units to the left of the graph of f.
$g(x) = x^3 + 12x^2 + 44x + 48$

83. Fibonacci sequence

85. (a) $g(t) = -0.056t^2 + 1.06t + 23.1$

(b)

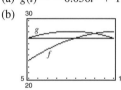

(c) 2010

87. True. The coefficients from the Binomial Theorem can be used to find the numbers in Pascal's Triangle.

89. The first and last numbers in each row are 1. Every other number in each row is formed by adding the two numbers immediately above the number.

91.

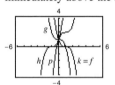

k, f; $k(x)$ is the expansion of $f(x)$.

93–95. Proofs

97.

n	r	$_nC_r$	$_nC_{n-r}$
9	5	126	126
7	1	7	7
12	4	495	495
6	0	1	1
10	7	120	120

This illustrates the symmetry of Pascal's Triangle.

Section 9.6 *(page 663)*

1. Fundamental Counting Principle   **3.** $_nP_r = \dfrac{n!}{(n-r)!}$

5. combinations   **7.** 6   **9.** 5   **11.** 3   **13.** 8

15. 30   **17.** 30   **19.** 64   **21.** 175,760,000

23. (a) 900   (b) 648   (c) 180   (d) 600

25. 64,000   **27.** (a) 40,320   (b) 384   **29.** 120

31. 20   **33.** 132   **35.** 2730   **37.** 5,527,200

39. 504   **41.** 1,816,214,400   **43.** 420   **45.** 2520

47. ABCD, ABDC, ACBD, ACDB, ADBC, ADCB, BACD, BADC, CABD, CADB, DABC, DACB, BCAD, BDAC, CBAD, CDAB, DBAC, DCAB, BCDA, BDCA, CBDA, CDBA, DBCA, DCBA

49. 15   **51.** 1   **53.** 120   **55.** 38,760

57. AB, AC, AD, AE, AF, BC, BD, BE, BF, CD, CE, CF, DE, DF, EF

59. 5,586,853,480   **61.** 324,632

63. (a) 7315   (b) 693   (c) 12,628

65. (a) 3744   (b) 24   **67.** 292,600   **69.** 5   **71.** 20

73. 36   **75.** $n = 2$ **77.** $n = 3$ **79.** $n = 5$ or $n = 6$

81. $n = 10$ **83.** False. It is an example of a combination.

85. $_{10}P_6 > {}_{10}C_6$. Changing the order of any of the six elements selected results in a different permutation but the same combination.

87–89. Proofs

91. No. For some calculators the number is too great.

Section 9.7 *(page 674)*

1. experiment; outcomes   **3.** probability

5. mutually exclusive   **7.** complement

9. $\{(H, 1), (H, 2), (H, 3), (H, 4), (H, 5), (H, 6),$
$(T, 1), (T, 2), (T, 3), (T, 4), (T, 5), (T, 6)\}$

11. $\{ABC, ACB, BAC, BCA, CAB, CBA\}$

13. $\{AB, AC, AD, AE, BC, BD, BE, CD, CE, DE\}$ **15.** $\frac{3}{8}$

17. $\frac{1}{2}$ **19.** $\frac{7}{8}$ **21.** $\frac{3}{13}$ **23.** $\frac{3}{26}$ **25.** $\frac{5}{36}$ **27.** $\frac{11}{12}$

29. $\frac{1}{3}$ **31.** $\frac{1}{5}$ **33.** $\frac{2}{5}$

35. (a) 996,000   (b) $\frac{9}{50}$ (c) $\frac{27}{50}$ (d) $\frac{4}{25}$

37. (a) $\frac{13}{16}$ (b) $\frac{3}{16}$ (c) $\frac{1}{32}$ **39.** 19%

41. (a) $\frac{21}{1292}$ (b) $\frac{225}{646}$ (c) $\frac{49}{323}$ **43.** (a) $\frac{1}{120}$ (b) $\frac{1}{24}$

45. (a) $\frac{5}{13}$ (b) $\frac{1}{2}$ (c) $\frac{4}{13}$ **47.** (a) $\frac{14}{55}$ (b) $\frac{12}{55}$ (c) $\frac{54}{55}$

49. (a) $\frac{1}{4}$ (b) $\frac{1}{2}$ (c) $\frac{841}{1600}$ (d) $\frac{1}{40}$ **51.** 0.27   **53.** $\frac{4}{5}$

55. 0.71   **57.** $\frac{11}{25}$

59. (a) 0.9702   (b) 0.0002   (c) 0.9998

61. (a) $\frac{1}{38}$ (b) $\frac{9}{19}$ (c) $\frac{10}{19}$ (d) $\frac{1}{1444}$ (e) $\frac{729}{6859}$ **63.** $\frac{7}{16}$

65. True. Two events are independent when the occurrence of one has no effect on the occurrence of the other.

67. (a) As you consider successive people with distinct birthdays, the probabilities must decrease to take into account the birth dates already used. Because the birth dates of people are independent events, multiply the respective probabilities of distinct birthdays.

(b) $\frac{365}{365} \cdot \frac{364}{365} \cdot \frac{363}{365} \cdot \frac{362}{365}$ (c) Answers will vary.

(d) Q_n is the probability that the birthdays are *not* distinct, which is equivalent to at least two people having the same birthday.

(e)

n	10	15	20	23	30	40	50
P_n	0.88	0.75	0.59	0.49	0.29	0.11	0.03
Q_n	0.12	0.25	0.41	0.51	0.71	0.89	0.97

(f) 23; $Q_n > 0.5$ for $n \geq 23$.

Review Exercises   *(page 680)*

1. $15, 9, 7, 6, \frac{27}{5}$ **3.** $120, 60, 20, 5, 1$ **5.** $a_n = 2(-1)^n$

7. $a_n = \frac{4}{n}$ **9.** $\frac{1}{20}$ **11.** $\frac{1}{n(n+1)}$ **13.** $\frac{205}{24}$

15. $\displaystyle\sum_{k=1}^{20} \frac{1}{2k}$ **17.** $\frac{4}{9}$

19. (a) $A_1 = \$10,018.75$
    $A_2 \approx \$10,037.54$
    $A_3 \approx \$10,056.36$
    $A_4 \approx \$10,075.21$
    $A_5 \approx \$10,094.10$
    $A_6 \approx \$10,113.03$
    $A_7 \approx \$10,131.99$
    $A_8 \approx \$10,150.99$
    $A_9 \approx \$10,170.02$
    $A_{10} \approx \$10,189.09$
    (b) $\$12,520.59$

21. Arithmetic, $d = -6$ **23.** Not arithmetic

25. $a_n = 12n - 5$ **27.** $a_n = -18n + 150$

29. $4, 21, 38, 55, 72$ **31.** $45,450$ **33.** 80 **35.** 88

37. (a) $\$51,600$ (b) $\$238,500$ **39.** Geometric, $r = 3$

41. Geometric, $r = -3$ **43.** $2, 30, 450, 6750, 101,250$

45. $9, 6, 4, \frac{8}{3}, \frac{16}{9}$ or $9, -6, 4, -\frac{8}{3}, \frac{16}{9}$

47. $a_n = 100(1.05)^{n-1}$; About 155.133

49. $a_n = 18\left(-\frac{1}{2}\right)^{n-1}$; $-\frac{9}{256}$ **51.** 127 **53.** $\frac{15}{16}$ **55.** 31

57. 23.056 **59.** 8 **61.** 12

63. (a) $a_n = 120,000(0.7)^n$ (b) $\$20,168.40$

65–67. Proofs   **69.** $S_n = n(2n + 7)$; Proof

71. $S_n = \frac{5}{2}\left[1 - \left(\frac{3}{5}\right)^n\right]$; Proof   **73.** 2850

75. $5, 10, 15, 20, 25$
    First differences: $5, 5, 5, 5$
    Second differences: $0, 0, 0$
    Linear

77. 15 **79.** 21 **81.** $x^4 + 16x^3 + 96x^2 + 256x + 256$

83. $64 - 240x + 300x^2 - 125x^3$ **85.** 6 **87.** $10,000$

89. 120 **91.** $225,792,840$ **93.** (a) $\frac{1}{5}$ (b) $\frac{3}{5}$

95. (a) 43% (b) 82% **97.** $\frac{1}{1296}$ **99.** $\frac{3}{4}$

101. False. $\dfrac{(n+2)!}{n!} = \dfrac{(n+2)(n+1)n!}{n!} = (n+2)(n+1)$

103. True by the Properties of Sums.

105. The set of positive integers

107. Each term of the sequence is defined in terms of preceding terms.

Chapter Test   *(page 683)*

1. $-\dfrac{1}{5}, \dfrac{1}{8}, -\dfrac{1}{11}, \dfrac{1}{14}, -\dfrac{1}{17}$ **2.** $a_n = \dfrac{n+2}{n!}$

3. $60, 73, 86; 329$ **4.** $a_n = -3n + 60$ **5.** $a_n = \frac{7}{2}(2)^n$

6. $86,100$ **7.** 477 **8.** 4 **9.** $-\frac{1}{4}$ **10.** Proof

11. $x^4 + 24x^3y + 216x^2y^2 + 864xy^3 + 1296y^4$

12. $3x^5 - 30x^4 + 124x^3 - 264x^2 + 288x - 128$

13. $-22,680$ **14.** (a) 72 (b) $328,440$

15. (a) 330 (b) $720,720$ **16.** $26,000$ **17.** 720

18. $\dfrac{1}{15}$ **19.** $\dfrac{1}{27,405}$ **20.** 10%

Cumulative Test for Chapters 7–9   *(page 684)*

1. $(1, 2), \left(-\frac{3}{2}, \frac{3}{4}\right)$ **2.** $(-3, -1)$ **3.** $(5, -2, -2)$

4. $(1, -2, 1)$ **5.** $\$0.75$ mixture: 120 lb; $\$1.25$ mixture: 80 lb

6. $y = \frac{1}{4}x^2 - 2x + 6$ **7.** $-\dfrac{3}{x} + \dfrac{5x - 1}{x^2 + 2}$

8.

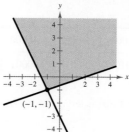

9.

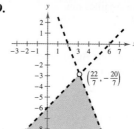

10.

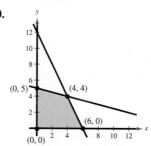

Maximum at $(4, 4)$: 20
Minimum at $(0, 0)$: 0

11. $\begin{bmatrix} -1 & 2 & -1 & \vdots & 9 \\ 2 & -1 & 2 & \vdots & -9 \\ 3 & 3 & -4 & \vdots & 7 \end{bmatrix}$ **12.** $(-2, 3, -1)$

13. $\begin{bmatrix} -3 & 8 \\ 6 & 1 \end{bmatrix}$ **14.** $\begin{bmatrix} 8 & -19 \\ 12 & 9 \end{bmatrix}$ **15.** $\begin{bmatrix} -13 & 6 & -4 \\ 18 & 4 & 4 \end{bmatrix}$

16. Not possible   **17.** $\begin{bmatrix} 19 & 3 \\ 6 & 22 \end{bmatrix}$ **18.** $\begin{bmatrix} 28 & 19 \\ -6 & -3 \end{bmatrix}$

19. $\begin{bmatrix} -175 & 37 & -13 \\ 95 & -20 & 7 \\ 14 & -3 & 1 \end{bmatrix}$ **20.** 203

21. $(0, -2), (3, -5), (0, -5) (3, -2)$

22. Gym shoes: $\$2539$ million
    Jogging shoes: $\$2362$ million
    Walking shoes: $\$4418$ million

23. $(-5, 4)$ **24.** $(-3, 4, 2)$ **25.** 9

26. $\dfrac{1}{5}, -\dfrac{1}{7}, \dfrac{1}{9}, -\dfrac{1}{11}, \dfrac{1}{13}$ **27.** $a_n = \dfrac{(n+1)!}{n+3}$

28. 1536 **29.** (a) 65.4 (b) $a_n = 3.2n + 1.4$

30. $3, 6, 12, 24, 48$ **31.** $\frac{190}{9}$ **32.** Proof

33. $w^4 - 36w^3 + 486w^2 - 2916w + 6561$ **34.** 2184

35. 600 **36.** 70 **37.** 462 **38.** $453,600$

39. $151,200$ **40.** 720 **41.** $\frac{1}{4}$

CHAPTER 9

Problem Solving   *(page 689)*

1. (a) 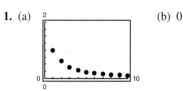   (b) 0

(c)

n	1	10	100	1000	10,000
a_n	1	0.1089	0.0101	0.0010	0.0001

(d) 0

3. $s_d = \dfrac{a_1}{1-r} = \dfrac{20}{1-\frac{1}{2}} = 40$

This represents the total distance Achilles ran.

$s_t = \dfrac{a_1}{1-r} = \dfrac{1}{1-\frac{1}{2}} = 2$

This represents the total amount of time Achilles ran.

5. (a) Arithmetic sequence, difference $= d$
   (b) Arithmetic sequence, difference $= dC$
   (c) Not an arithmetic sequence

7. (a) 7, 22, 11, 34, 17, 52, 26, 13, 40, 20, 10, 5, 16, 8, 4, 2, 1, 4, 2, 1
   (b) $a_1 = 4$: 4, 2, 1, 4, 2, 1, 4, 2, 1, 4
      $a_1 = 5$: 5, 16, 8, 4, 2, 1, 4, 2, 1, 4
      $a_1 = 12$: 12, 6, 3, 10, 5, 16, 8, 4, 2, 1
      Eventually, the terms repeat: 4, 2, 1.

9. Proof   **11.** $S_n = \left(\dfrac{1}{2}\right)^{n-1}$; $A_n = \dfrac{\sqrt{3}}{4}S_n^2$ **13.** $\dfrac{1}{3}$

15. (a) 3 to 7; 7 to 3   (b) 30 marbles

   (c) $P(E) = \dfrac{\text{odds in favor of } E}{\text{odds in favor of } E + 1}$

   (d) Odds in favor of event $E = \dfrac{P(E)}{P(E')}$

Chapter 10

Section 10.1   *(page 696)*

1. inclination   **3.** $\left|\dfrac{m_2 - m_1}{1 + m_1 m_2}\right|$ **5.** $\dfrac{\sqrt{3}}{3}$ **7.** -1

9. $\sqrt{3}$ **11.** 0.4111   **13.** 3.2236   **15.** -4.1005

17. $\dfrac{\pi}{4}$ rad, 45°   **19.** 0.5880 rad, 33.7°   **21.** $\dfrac{3\pi}{4}$ rad, 135°

23. 2.1588 rad, 123.7°   **25.** $\dfrac{\pi}{6}$ rad, 30°   **27.** $\dfrac{5\pi}{6}$ rad, 150°

29. 1.0517 rad, 60.3°   **31.** 2.1112 rad, 121.0°

33. 1.6539 rad, 94.8°   **35.** $\dfrac{3\pi}{4}$ rad, 135°   **37.** $\dfrac{\pi}{4}$ rad, 45°

39. $\dfrac{5\pi}{6}$ rad, 150°   **41.** 1.2490 rad, 71.6°

43. 2.4669 rad, 141.3°   **45.** 1.1071 rad, 63.4°

47. 0.1974 rad, 11.3°   **49.** 1.4289 rad, 81.9°

51. 0.9273 rad, 53.1°   **53.** 0.8187 rad, 46.9°

55. $(1, 5) \leftrightarrow (4, 5)$: slope $= 0$
   $(4, 5) \leftrightarrow (3, 8)$: slope $= -3$
   $(3, 8) \leftrightarrow (1, 5)$: slope $= \frac{3}{2}$
   $(1, 5)$: 56.3°; $(4, 5)$: 71.6°; $(3, 8)$: 52.1°

57. $(-4, -1) \leftrightarrow (3, 2)$: slope $= \frac{3}{7}$
   $(3, 2) \leftrightarrow (1, 0)$: slope $= 1$
   $(1, 0) \leftrightarrow (-4, -1)$: slope $= \frac{1}{5}$
   $(-4, -1)$: 11.9°; $(3, 2)$: 21.8°; $(1, 0)$: 146.3°

59. $\dfrac{\sqrt{2}}{2} \approx 0.7071$ **61.** $\dfrac{2\sqrt{5}}{5} \approx 0.8944$

63. $2\sqrt{2} \approx 2.8284$ **65.** $\dfrac{\sqrt{10}}{10} \approx 0.3162$

67. $\dfrac{4\sqrt{10}}{5} \approx 2.5298$ **69.** 1   **71.** $\dfrac{1}{5}$

73. (a) 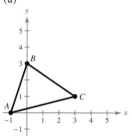   **75.** (a)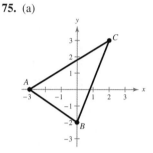

   (b) $\dfrac{11\sqrt{17}}{17}$ (c) $\dfrac{11}{2}$ (b) $\dfrac{19\sqrt{34}}{34}$ (c) $\dfrac{19}{2}$

77. (a)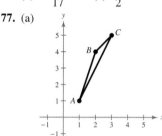

   (b) $\dfrac{\sqrt{5}}{5}$ (c) 1

79. $2\sqrt{2}$ **81.** 0.1003, 1054 ft   **83.** $\theta \approx 31.0°$

85. $\alpha \approx 33.69°$; $\beta \approx 56.31°$ **87.** True. tan 0 = 0

89. False. Substitute $\tan \theta_1$ and $\tan \theta_2$ for m_1 and m_2 in the formula for the angle between two lines.

91. The inclination of a line measures the angle of intersection (measured counterclockwise) of a line and the x-axis. The angle between two lines is the acute angle of their intersection, which must be less than $\pi/2$.

93. (a) $d = \dfrac{4}{\sqrt{m^2 + 1}}$

   (b)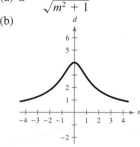

   (c) $m = 0$

   (d) The graph has a horizontal asymptote of $d = 0$. As the slope becomes larger, the distance between the origin and the line, $y = mx + 4$, becomes smaller and approaches 0.

Section 10.2 *(page 704)*

1. conic   **3.** locus   **5.** axis   **7.** focal chord
9. c   **10.** a   **11.** b   **12.** d   **13.** $x^2 = 4y$
15. $x^2 = 2y$ **17.** $y^2 = -8x$ **19.** $x^2 = -8y$
21. $y^2 = 4x$ **23.** $x^2 = \frac{8}{3}y$ **25.** $y^2 = -\frac{25}{2}x$
27. $(x - 2)^2 = -8(y - 6)$ **29.** $(y - 3)^2 = -8(x - 6)$
31. $x^2 = -8(y - 2)$ **33.** $(y - 2)^2 = 8x$
35. $(x - 3)^2 = 3(y + 3)$

37. Vertex: $(0, 0)$
Focus: $\left(0, \frac{1}{2}\right)$
Directrix: $y = -\frac{1}{2}$

39. Vertex: $(0, 0)$
Focus: $\left(-\frac{3}{2}, 0\right)$
Directrix: $x = \frac{3}{2}$

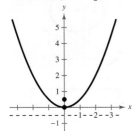

41. Vertex: $(0, 0)$
Focus: $(0, -3)$
Directrix: $y = 3$

43. Vertex: $(1, -2)$
Focus: $(1, -4)$
Directrix: $y = 0$

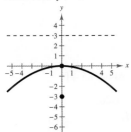

45. Vertex: $\left(\frac{3}{2}, -7\right)$
Focus: $\left(\frac{5}{2}, -7\right)$
Directrix: $x = \frac{1}{2}$

47. Vertex: $(1, 1)$
Focus: $(1, 2)$
Directrix: $y = 0$

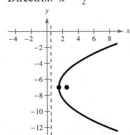

49. Vertex: $(-2, -3)$
Focus: $(-4, -3)$
Directrix: $x = 0$

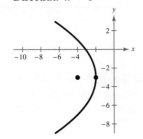

51. Vertex: $(-2, 1)$
Focus: $\left(-2, \frac{5}{2}\right)$
Directrix: $y = -\frac{1}{2}$

53. Vertex: $\left(\frac{1}{4}, -\frac{1}{2}\right)$
Focus: $\left(0, -\frac{1}{2}\right)$
Directrix: $x = \frac{1}{2}$

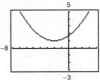

55. $y = \frac{3}{2}x - \frac{9}{2}$ **57.** $y = 4x - 8$ **59.** $y = 4x + 2$
61. $y^2 = 6x$ **63.** $y^2 = 640x$
65. (a) $x^2 = 12{,}288y$ (in feet)   (b) About 22.6 ft
67. $x^2 = -\frac{25}{4}(y - 48)$ **69.** About 19.6 m
71. (a) $(0, 45)$ (b) $y = \frac{1}{180}x^2$
73. (a) $17{,}500\sqrt{2}$ mi/h $\approx 24{,}750$ mi/h
(b) $x^2 = -16{,}400(y - 4100)$
75. (a) $x^2 = -49(y - 100)$ (b) 70 ft
77. False. If the graph crossed the directrix, then there would exist points closer to the directrix than the focus.
79. True. If the axis (line connecting the vertex and focus) is horizontal, then the directrix must be vertical.
81. Both (a) and (b) are parabolas with vertical axes, while (c) is a parabola with a horizontal axis. Equations (a) and (b) are equivalent when $p = 1/(4a)$.
83.

Single point $(0, 0)$; A single point is formed when a plane intersects only the vertex of the cone.

85. (a) $\dfrac{64\sqrt{2}}{3} \approx 30.17$

(b) As p approaches zero, the parabola becomes narrower and narrower, thus the area becomes smaller and smaller.

Section 10.3 *(page 714)*

1. ellipse; foci   **3.** minor axis   **5.** b   **6.** c   **7.** a
8. d   **9.** $\dfrac{x^2}{4} + \dfrac{y^2}{16} = 1$ **11.** $\dfrac{x^2}{49} + \dfrac{y^2}{45} = 1$
13. $\dfrac{x^2}{25} + \dfrac{y^2}{9} = 1$ **15.** $\dfrac{x^2}{9} + \dfrac{y^2}{36} = 1$ **17.** $\dfrac{x^2}{36} + \dfrac{5y^2}{9} = 1$
19. $\dfrac{(x - 2)^2}{1} + \dfrac{(y - 3)^2}{9} = 1$ **21.** $\dfrac{(x - 6)^2}{16} + \dfrac{y^2}{4} = 1$
23. $\dfrac{(x - 2)^2}{9} + \dfrac{y^2}{5} = 1$ **25.** $\dfrac{(x - 1)^2}{9} + \dfrac{(y - 3)^2}{4} = 1$
27. $\dfrac{(x - 1)^2}{12} + \dfrac{(y - 4)^2}{16} = 1$ **29.** $\dfrac{(x - 2)^2}{4} + \dfrac{(y - 2)^2}{1} = 1$

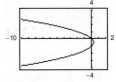

CHAPTER 10

31. Center: $(0, 0)$
Vertices: $(\pm 5, 0)$
Foci: $(\pm 3, 0)$
Eccentricity: $\dfrac{3}{5}$

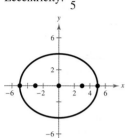

33. Center: $(0, 0)$
Vertices: $(0, \pm 6)$
Foci: $(0, \pm 4\sqrt{2})$
Eccentricity: $\dfrac{2\sqrt{2}}{3}$

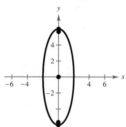

43. Center: $\left(-\dfrac{3}{2}, \dfrac{5}{2}\right)$
Vertices: $\left(-\dfrac{3}{2}, \dfrac{5}{2} \pm 2\sqrt{3}\right)$
Foci: $\left(-\dfrac{3}{2}, \dfrac{5}{2} \pm 2\sqrt{2}\right)$
Eccentricity: $\dfrac{\sqrt{6}}{3}$

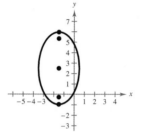

45. Center: $\left(\dfrac{1}{2}, -1\right)$
Vertices: $\left(\dfrac{1}{2} \pm \sqrt{5}, -1\right)$
Foci: $\left(\dfrac{1}{2} \pm \sqrt{2}, -1\right)$
Eccentricity: $\dfrac{\sqrt{10}}{5}$

35. Center: $(4, -1)$
Vertices: $(4, -6), (4, 4)$
Foci: $(4, 2), (4, -4)$
Eccentricity: $\dfrac{3}{5}$

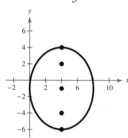

37. Center: $(-5, 1)$
Vertices: $\left(-\dfrac{7}{2}, 1\right), \left(-\dfrac{13}{2}, 1\right)$
Foci: $\left(-5 \pm \dfrac{\sqrt{5}}{2}, 1\right)$
Eccentricity: $\dfrac{\sqrt{5}}{3}$

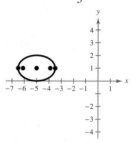

47.

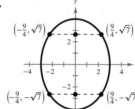

Center: $(0, 0)$
Vertices: $\left(0, \pm \sqrt{5}\right)$
Foci: $\left(0, \pm \sqrt{2}\right)$

49.

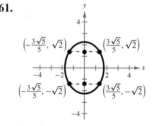

Center: $(5, -2)$
Vertices: $(2, -2), (8, -2)$
Foci: $\left(5 \pm 2\sqrt{2}, -2\right)$

51. $\dfrac{x^2}{25} + \dfrac{y^2}{9} = 1$

53. (a) $\dfrac{x^2}{2352.25} + \dfrac{y^2}{529} = 1$ (b) About 85.4 ft

55. About 229.8 mm   **57.** $e \approx 0.0520$

59.

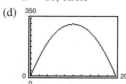

61.

39. Center: $(-2, 3)$
Vertices: $(-2, 6), (-2, 0)$
Foci: $\left(-2, 3 \pm \sqrt{5}\right)$
Eccentricity: $\dfrac{\sqrt{5}}{3}$

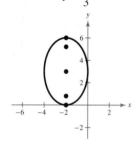

41. Center: $(4, 3)$
Vertices: $(14, 3), (-6, 3)$
Foci: $\left(4 \pm 4\sqrt{5}, 3\right)$
Eccentricity: $\dfrac{2\sqrt{5}}{5}$

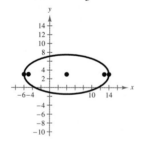

63. False. The graph of $(x^2/4) + y^4 = 1$ is not an ellipse. The degree of y is 4, not 2.

65. $\dfrac{(x - 6)^2}{324} + \dfrac{(y - 2)^2}{308} = 1$

67. (a) $A = \pi a(20 - a)$ (b) $\dfrac{x^2}{196} + \dfrac{y^2}{36} = 1$

(c)

a	8	9	10	11	12	13
A	301.6	311.0	314.2	311.0	301.6	285.9

$a = 10$, circle

(d)

The maximum occurs at $a = 10$.

69. Proof

Section 10.4 *(page 724)*

1. hyperbola; foci   **3.** transverse axis; center

5. b   **6.** d   **7.** c   **8.** a   **9.** $\dfrac{y^2}{4} - \dfrac{x^2}{12} = 1$

11. $\dfrac{(x-4)^2}{4} - \dfrac{y^2}{12} = 1$ **13.** $\dfrac{(y-5)^2}{16} - \dfrac{(x-4)^2}{9} = 1$

15. $\dfrac{y^2}{9} - \dfrac{4(x-2)^2}{9} = 1$ **17.** $\dfrac{(x-2)^2}{4} - \dfrac{(y+3)^2}{8} = 1$

19. Center: $(0, 0)$
Vertices: $(\pm 1, 0)$
Foci: $\left(\pm \sqrt{2}, 0\right)$
Asymptotes: $y = \pm x$

21. Center: $(0, 0)$
Vertices: $(0, \pm 6)$
Foci: $\left(0 \pm 2\sqrt{34}\right)$
Asymptotes: $y = \pm \frac{3}{5}x$

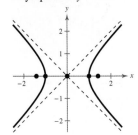
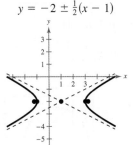

23. Center: $(0, 0)$
Vertices: $(0, \pm 1)$
Foci: $\left(0, \pm \sqrt{5}\right)$
Asymptotes: $y = \pm \frac{1}{2}x$

25. Center: $(1, -2)$
Vertices: $(3, -2), (-1, -2)$
Foci: $\left(1 \pm \sqrt{5}, -2\right)$
Asymptotes:
$y = -2 \pm \frac{1}{2}(x - 1)$

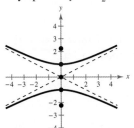
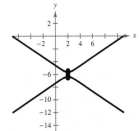

27. Center: $(2, -6)$
Vertices: $\left(2, -\frac{17}{3}\right), \left(2, -\frac{19}{3}\right)$
Foci: $\left(2, -6 \pm \dfrac{\sqrt{13}}{6}\right)$
Asymptotes:
$y = -6 \pm \frac{2}{3}(x - 2)$

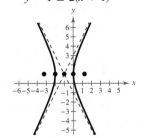

29. Center: $(2, -3)$
Vertices: $(3, -3), (1, -3)$
Foci: $\left(2 \pm \sqrt{10}, -3\right)$
Asymptotes:
$y = -3 \pm 3(x - 2)$

31. Center: $(-1, 1)$
Vertices: $(-2, 1), (0, 1)$
Foci: $\left(-1 \pm \sqrt{5}, 1\right)$
Asymptotes:
$y = 1 \pm 2(x + 1)$

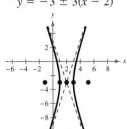

33.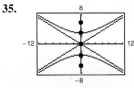
Center: $(0, 0)$
Vertices: $\left(\pm \sqrt{3}, 0\right)$
Foci: $\left(\pm \sqrt{5}, 0\right)$

35.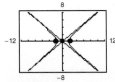
Center: $(0, 0)$
Vertices: $(0, \pm 3)$
Foci: $\left(0, \pm \sqrt{34}\right)$

37.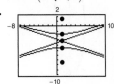

Center: $(1, -3)$
Vertices: $\left(1, -3 \pm \sqrt{2}\right)$
Foci: $\left(1, -3 \pm 2\sqrt{5}\right)$

39. $\dfrac{x^2}{1} - \dfrac{y^2}{25} = 1$ **41.** $\dfrac{17y^2}{1024} - \dfrac{17x^2}{64} = 1$

43. $\dfrac{(x-2)^2}{1} - \dfrac{(y-2)^2}{1} = 1$ **45.** $\dfrac{(y-2)^2}{4} - \dfrac{(x-3)^2}{9} = 1$

47. $\dfrac{(x-4)^2}{16} - \dfrac{(y+1)^2}{9} = 1$

49. (a) $\dfrac{x^2}{1} - \dfrac{y^2}{169/3} = 1$ (b) About 2.403 ft

51. $\dfrac{x^2}{98,010,000} - \dfrac{y^2}{13,503,600} = 1$

53. (a) $x \approx 110.3$ mi   (b) 57.0 mi   (c) $y = \dfrac{27\sqrt{19}}{93}x$

55. Ellipse   **57.** Hyperbola   **59.** Hyperbola
61. Ellipse   **63.** Parabola   **65.** Circle
67. True. For a hyperbola, $c^2 = a^2 + b^2$. The larger the ratio of b to a, the larger the eccentricity of the hyperbola, $e = c/a$.
69. False. The graph is two intersecting lines.
71. Draw a rectangle through the vertices and the endpoints of the conjugate axis. Sketch the asymptotes by drawing lines through the opposite corners of the rectangle.
73. The equations of the asymptotes should be $y = k \pm \dfrac{a}{b}(x - h)$.

75.
(a) $C > 2$ and $C < -\frac{17}{4}$
(b) $C = 2$
(c) $-2 < C < 2, C = -\frac{17}{4}$
(d) $C = -2$
(e) $-\frac{17}{4} < C < -2$

Section 10.5 *(page 733)*

1. rotation; axes   **3.** invariant under rotation   **5.** $(0, -2)$

7. $\left(\dfrac{3 + \sqrt{3}}{2}, \dfrac{3\sqrt{3} - 1}{2}\right)$ **9.** $\left(\dfrac{3\sqrt{2}}{2}, -\dfrac{\sqrt{2}}{2}\right)$

11. $\left(\dfrac{2\sqrt{3} + 1}{2}, \dfrac{2 - \sqrt{3}}{2}\right)$

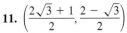

CHAPTER 10

13. $\dfrac{(y')^2}{6} - \dfrac{(x')^2}{6} = 1$

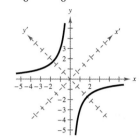

15. $\dfrac{\left(y' + \dfrac{3\sqrt{2}}{2}\right)^2}{12} - \dfrac{\left(x' + \dfrac{\sqrt{2}}{2}\right)^2}{12} = 1$

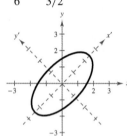

17. $\dfrac{(x')^2}{6} + \dfrac{(y')^2}{3/2} = 1$

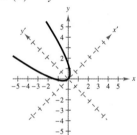

19. $\dfrac{(x')^2}{1} + \dfrac{(y')^2}{4} = 1$

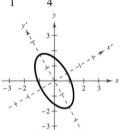

21. $(x')^2 = y'$

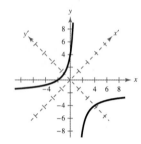

23. $(x' - 1)^2 = 6\left(y' + \dfrac{1}{6}\right)$

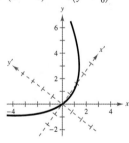

25.

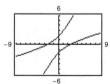

$\theta \approx 37.98°$

27.

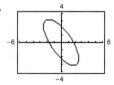

$\theta \approx 36.32°$

29.

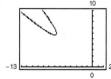

$\theta = 45°$

31. e   **32.** a   **33.** d   **34.** c   **35.** f   **36.** b

37. (a) Parabola

(b) $y = \dfrac{(8x - 5) \pm \sqrt{(8x - 5)^2 - 4(16x^2 - 10x)}}{2}$

(c)

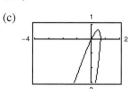

39. (a) Ellipse

(b) $y = \dfrac{6x \pm \sqrt{36x^2 - 28(12x^2 - 45)}}{14}$

(c)

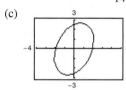

41. (a) Hyperbola

(b) $y = \dfrac{6x \pm \sqrt{36x^2 + 20(x^2 + 4x - 22)}}{-10}$

(c)

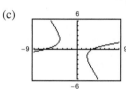

43. (a) Parabola

(b) $y = \dfrac{-(4x - 1) \pm \sqrt{(4x - 1)^2 - 16(x^2 - 5x - 3)}}{8}$

(c)

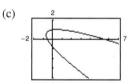

45.

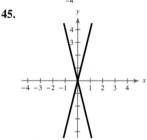

47.

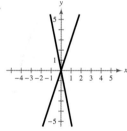

49.

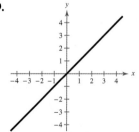

51.

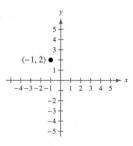

53.

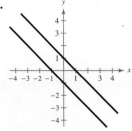

55. $(14, -8), (6, -8)$ **57.** $(1, 0)$ **59.** $\left(1, \sqrt{3}\right), \left(1, -\sqrt{3}\right)$
61. $(-7, 0), (-1, 0)$
63. (a) $(y' + 9)^2 = 9(x' - 12)$ (b) 2.25 ft
65. True. The graph of the equation can be classified by finding the discriminant. For a graph to be a hyperbola, the discriminant must be greater than zero. If $k \geq \frac{1}{4}$, then the discriminant would be less than or equal to zero.
67. Answers will vary.

Section 10.6   *(page 741)*

1. plane curve   **3.** eliminating; parameter
5. (a)

t	0	1	2	3	4
x	0	1	$\sqrt{2}$	$\sqrt{3}$	2
y	3	2	1	0	-1

(b)

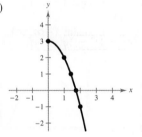

(c)

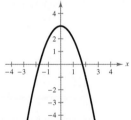

The graph of $y = 3 - x^2$ shows the entire parabola rather than just the right half.

7.

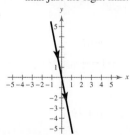

The curve is traced from left to right.

9.

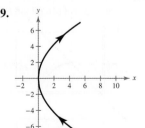

The curve is traced clockwise.

11.

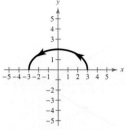

The curve is traced from right to left.

13. (a)

(b) $y = 4x$

15. (a)

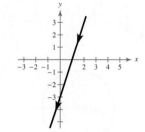

(b) $y = 3x - 3$

17. (a)

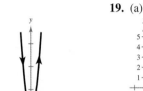

(b) $y = 16x^2$

19. (a)

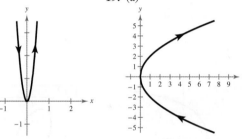

(b) $y = \pm 2\sqrt{x}$

21. (a)

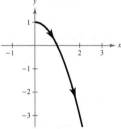

(b) $y = 1 - x^2$, $x \geq 0$

23. (a)

(b) $y = (x + 3)^6$, $x \geq -3$

25. (a)

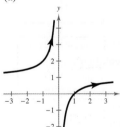

(b) $y = \dfrac{(x - 1)}{x}$

27. (a)

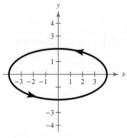

(b) $\dfrac{x^2}{16} + \dfrac{y^2}{4} = 1$

CHAPTER 10

29. (a)

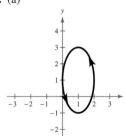

31. (a)

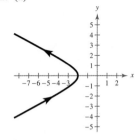

(b) $\dfrac{(x-1)^2}{1} + \dfrac{(y-1)^2}{4} = 1$ (b) $\dfrac{x^2}{4} - y^2 = 1,\ x \le -2$

33. (a)

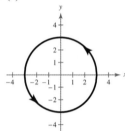

35. (a)

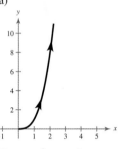

(b) $x^2 + y^2 = 9$ (b) $y = x^3,\ x > 0$

37. (a)

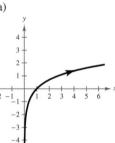

39.

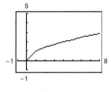

(b) $y = \ln x$

41.

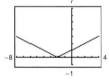

43.

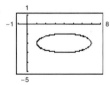

45.

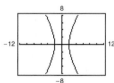

47.

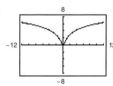

49. Each curve represents a portion of the line $y = 2x + 1$.

Domain	Orientation
(a) $(-\infty, \infty)$	Left to right
(b) $[-1, 1]$	Depends on θ
(c) $(0, \infty)$	Right to left
(d) $(0, \infty)$	Left to right

51. $y - y_1 = m(x - x_1)$ **53.** $\dfrac{(x-h)^2}{a^2} + \dfrac{(y-k)^2}{b^2} = 1$

55. $x = 3t$
       $y = 6t$

57. $x = 3 + 4\cos\theta$
       $y = 2 + 4\sin\theta$

59. $x = 5\cos\theta$
       $y = 3\sin\theta$

61. $x = 5 + 4\sec\theta$
       $y = 3\tan\theta$

63. $x = -5t$
       $y = 2t$
       $0 \le t \le 1$

65. $x = 3\sec\theta$
       $y = 4\tan\theta$
       $\dfrac{\pi}{2} < \theta < \dfrac{3\pi}{2}$

67. (a) $x = t,\ y = 3t - 2$ (b) $x = -t + 2,\ y = -3t + 4$
69. (a) $x = t,\ y = \frac{1}{2}(t - 1)$ (b) $x = -t + 2,\ y = -\frac{1}{2}(t - 1)$
71. (a) $x = t,\ y = t^2 + 1$ (b) $x = -t + 2,\ y = t^2 - 4t + 5$
73. (a) $x = t,\ y = 1 - 2t^2$
       (b) $x = 2 - t,\ y = -2t^2 + 8t - 7$
75. (a) $x = t,\ y = \dfrac{1}{t}$ (b) $x = -t + 2,\ y = -\dfrac{1}{t - 2}$
77. (a) $x = t,\ y = e^t$ (b) $x = -t + 2,\ y = e^{-t+2}$

79.

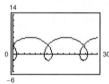

81.

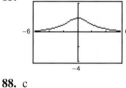

83.

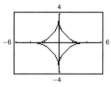

85.

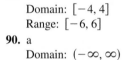

87. b
Domain: $[-2, 2]$
Range: $[-1, 1]$

88. c
Domain: $[-4, 4]$
Range: $[-6, 6]$

89. d
Domain: $(-\infty, \infty)$
Range: $(-\infty, \infty)$

90. a
Domain: $(-\infty, \infty)$
Range: $[-2, 2]$

91. (a)

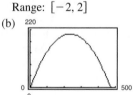

(b)

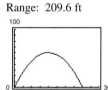

Maximum height: 90.7 ft
Range: 209.6 ft

Maximum height: 204.2 ft
Range: 471.6 ft

(c)

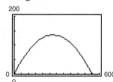

(d)

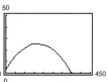

Maximum height: 60.5 ft
Range: 242.0 ft

Maximum height: 136.1 ft
Range: 544.5 ft

93. (a) $x = (146.67\cos\theta)t$
       $y = 3 + (146.67\sin\theta)t - 16t^2$

(b)

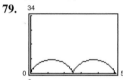

No

(c)

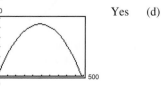

Yes   (d) $19.3°$

95. (a) $x = (\cos 35°)v_0 t$
$y = 7 + (\sin 35°)v_0 t - 16t^2$

(b) About 54.09 ft/sec

(c) 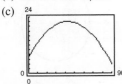   22.04 ft

(d) About 2.03 sec

97. (a) $h = 7$, $v_0 = 40$, $\theta = 45°$
$x = (40 \cos 45°)t$
$y = 7 + (40 \sin 45°)t - 16t^2$

(b)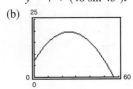

(c) Maximum height: 19.5 ft
Range: 56.2 ft

99. $x = a\theta - b \sin \theta$
$y = a - b \cos \theta$

101. True
$x = t$
$y = t^2 + 1 \implies y = x^2 + 1$
$x = 3t$
$y = 9t^2 + 1 \implies y = x^2 + 1$

103. False. The parametric equations $x = t^2$ and $y = t$ give the rectangular equation $x = y^2$, so y is not a function of x.

105. Parametric equations are useful when graphing two functions simultaneously on the same coordinate system. For example, they are useful when tracking the path of an object so that the position and the time associated with that position can be determined.

107. The parametric equation for x is defined only when $t \geq 1$, so the domain of the rectangular equation is $x \geq 0$.

109. Yes. The orientation would change.

Section 10.7   *(page 749)*

1. pole   **3.** polar

5.   **7.**

$\left(2, -\dfrac{11\pi}{6}\right), \left(-2, -\dfrac{5\pi}{6}\right),$
$\left(-2, \dfrac{7\pi}{6}\right)$

$\left(4, \dfrac{5\pi}{3}\right), \left(-4, \dfrac{2\pi}{3}\right),$
$\left(-4, -\dfrac{4\pi}{3}\right)$

9. 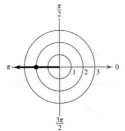   **11.**

$(2, \pi), (-2, 0), (2, -\pi)$ $\left(-2, -\dfrac{4\pi}{3}\right), \left(2, -\dfrac{\pi}{3}\right),$
$\left(2, \dfrac{5\pi}{3}\right)$

13.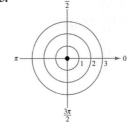

Sample answer: $\left(0, -\dfrac{11\pi}{6}\right), \left(0, -\dfrac{5\pi}{6}\right), \left(0, \dfrac{\pi}{6}\right)$

15.   **17.**

$\left(\sqrt{2}, -3.92\right),$ $(-3, 4.71), (3, 1.57),$
$\left(-\sqrt{2}, -0.78\right),$ $(3, -4.71)$
$\left(-\sqrt{2}, 5.50\right)$

19. $(0, 0)$ **21.** $(0, 3)$ **23.** $\left(-\sqrt{2}, \sqrt{2}\right)$ **25.** $\left(\sqrt{3}, 1\right)$

27. $\left(-\dfrac{3}{2}, \dfrac{3\sqrt{3}}{2}\right)$ **29.** $(-1.85, 0.77)$ **31.** $(0.26, 0.97)$

33. $(-1.13, -2.23)$ **35.** $(-2.43, -0.60)$

37. $(0.20, -3.09)$ **39.** $\left(\sqrt{2}, \dfrac{\pi}{4}\right)$ **41.** $\left(3\sqrt{2}, \dfrac{5\pi}{4}\right)$

43. $(3, 0)$ **45.** $\left(5, \dfrac{3\pi}{2}\right)$ **47.** $\left(\sqrt{6}, \dfrac{5\pi}{4}\right)$

49. $\left(2, \dfrac{11\pi}{6}\right)$ **51.** $(3.61, 5.70)$ **53.** $(5.39, 2.76)$

55. $(4.36, -1.98)$ **57.** $(2.83, 0.49)$ **59.** $r = 3$

61. $\theta = \dfrac{\pi}{4}$ **63.** $r = 10 \sec \theta$ **65.** $r = \dfrac{-2}{3 \cos \theta - \sin \theta}$

67. $r^2 = 16 \sec \theta \csc \theta = 32 \csc 2\theta$ **69.** $r = a \sec \theta$

71. $r = a$ **73.** $r = 2a \cos \theta$ **75.** $r^2 = \cos 2\theta$

77. $r = \cot^2 \theta \csc \theta$ **79.** $x^2 + y^2 = 25$

81. $\sqrt{3}x + y = 0$ **83.** $x = 0$ **85.** $y = 4$

87. $x = -3$ **89.** $x^2 + y^2 + 2x = 0$

91. $x^2 + y^2 - x^{2/3} = 0$ **93.** $(x^2 + y^2)^2 = 2xy$

95. $(x^2 + y^2)^2 = 6x^2 y - 2y^3$ **97.** $x^2 + 4y - 4 = 0$

99. $4x^2 - 5y^2 - 36y - 36 = 0$

101. The graph consists of all points six units from the pole; $x^2 + y^2 = 36$

103. The graph consists of all points on the line that makes an angle of $\pi/6$ with the polar axis; $-\sqrt{3}x + 3y = 0$

105. The graph is a vertical line; $x - 3 = 0$

107. The graph is a circle with center $(0, 1)$ and radius 1; $x^2 + (y - 1)^2 = 1$

109. (a) $r = 30$

(b) $\left(30, \dfrac{5\pi}{6}\right)$; 30 represents the distance of the passenger car from the center, and $\dfrac{5\pi}{6} = 150°$ represents the angle to which the car has rotated.

(c) $(-25.98, 15)$; The car is about 25.98 feet to the left of the center and 15 feet above the center.

111. True. Because r is a directed distance, the point (r, θ) can be represented as $(r, \theta \pm 2\pi n)$.

113. $\left(1, -\sqrt{3}\right)$ is in Quadrant IV, so $\theta = -\dfrac{\pi}{3}$.

115. (a) $(x - h)^2 + (y - k)^2 = h^2 + k^2$; $r = \sqrt{h^2 + k^2}$

(b) $\left(x - \tfrac{1}{2}\right)^2 + \left(y - \tfrac{3}{2}\right)^2 = \tfrac{5}{2}$

Center: $\left(\tfrac{1}{2}, \tfrac{3}{2}\right)$

Radius: $\dfrac{\sqrt{10}}{2}$

Section 10.8   *(page 757)*

1. $\theta = \dfrac{\pi}{2}$ **3.** convex limaçon   **5.** lemniscate

7. Circle   **9.** Limaçon with inner loop

11. Rose curve with 4 petals   **13.** Polar axis

15. $\theta = \dfrac{\pi}{2}$ **17.** $\theta = \dfrac{\pi}{2}$, polar axis, pole

19. Maximum: $|r| = 20$ when $\theta = \dfrac{3\pi}{2}$

Zero: $r = 0$ when $\theta = \dfrac{\pi}{2}$

21. Maximum: $|r| = 4$ when $\theta = 0, \dfrac{\pi}{3}, \dfrac{2\pi}{3}$

Zeros: $r = 0$ when $\theta = \dfrac{\pi}{6}, \dfrac{\pi}{2}, \dfrac{5\pi}{6}$

23. 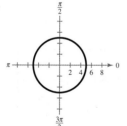   **25.**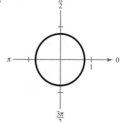

27. 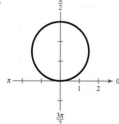   **29.**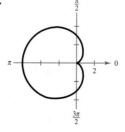

31. 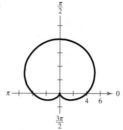   **33.**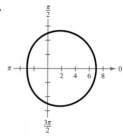

35. 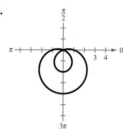   **37.**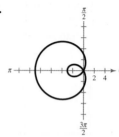

39. 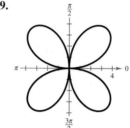   **41.**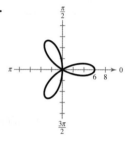

43. 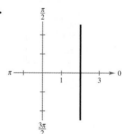   **45.**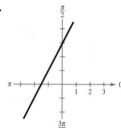

47. 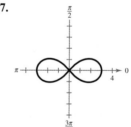   **49.**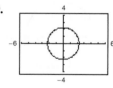

51. 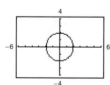   **53.**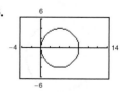

55. 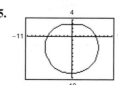   **57.**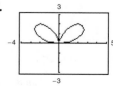

59.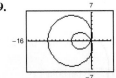

$0 \le \theta < 2\pi$

61.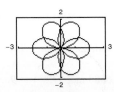

$0 \le \theta < 4\pi$

63.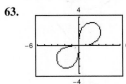

$0 \le \theta < \dfrac{\pi}{2}$

65.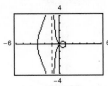

67.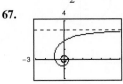

69. (a)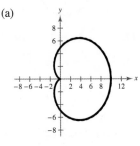

Cardioid

(b) 0 radians

71. True. The equation is of the form $r = a \sin n\theta$, where n is odd.

73. (a) 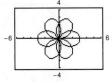   (b)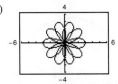

$0 \le \theta < 4\pi$ $0 \le \theta < 4\pi$

(c) Yes. Explanations will vary.

75. (a) 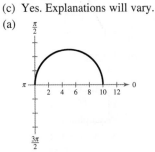   (b)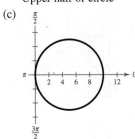

Upper half of circle   Lower half of circle

(c) 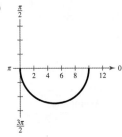   (d)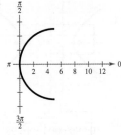

Full circle   Left half of circle

77. Answers will vary.

79. (a) $r = 2 - \dfrac{\sqrt{2}}{2}(\sin \theta - \cos \theta)$ (b) $r = 2 + \cos \theta$

(c) $r = 2 + \sin \theta$ (d) $r = 2 - \cos \theta$

Section 10.9 *(page 763)*

1. conic   **3.** vertical; left

5. (a) $r = \dfrac{2}{1 + \cos \theta}$,   **7.** (a) $r = \dfrac{2}{1 - \sin \theta}$,

parabola   parabola

(b) $r = \dfrac{1}{1 + 0.5 \cos \theta}$,   (b) $r = \dfrac{1}{1 - 0.5 \sin \theta}$,

ellipse   ellipse

(c) $r = \dfrac{3}{1 + 1.5 \cos \theta}$,   (c) $r = \dfrac{3}{1 - 1.5 \sin \theta}$,

hyperbola   hyperbola

9. c   **10.** d   **11.** a   **12.** b

13. Parabola   **15.** Parabola

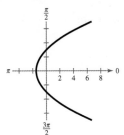   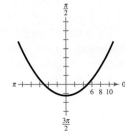

17. Ellipse   **19.** Ellipse

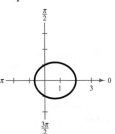   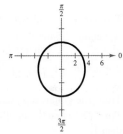

21. Hyperbola   **23.** Hyperbola

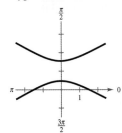   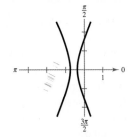

25. 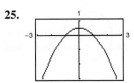   **27.**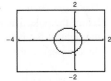

Parabola   Ellipse

29. 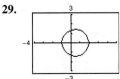   **31.**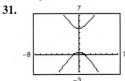

Ellipse   Hyperbola

33. 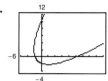   **35.**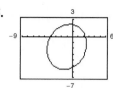

37. $r = \dfrac{1}{1 - \cos \theta}$ **39.** $r = \dfrac{3}{2 + \cos \theta}$

41. $r = \dfrac{2}{1 + 2 \cos \theta}$ **43.** $r = \dfrac{4}{1 + \cos \theta}$

45. $r = \dfrac{10}{1 - \cos \theta}$ **47.** $r = \dfrac{10}{3 + 2 \cos \theta}$

49. $r = \dfrac{20}{3 - 2 \cos \theta}$ **51.** $r = \dfrac{9}{4 - 5 \sin \theta}$

53. Answers will vary.

55. $r = \dfrac{9.2931 \times 10^7}{1 - 0.0167 \cos \theta}$

Perihelion: 9.1405×10^7 mi

Aphelion: 9.4509×10^7 mi

57. $r = \dfrac{1.0821 \times 10^8}{1 - 0.0067 \cos \theta}$

Perihelion: 1.0748×10^8 km

Aphelion: 1.0894×10^8 km

59. $r = \dfrac{1.4038 \times 10^8}{1 - 0.0935 \cos \theta}$

Perihelion: 1.2838×10^8 mi

Aphelion: 1.5486×10^8 mi

61. $\dfrac{3}{2 + \sin \theta} = \dfrac{3/2}{1 + (1/2) \sin \theta}$, so $e = \dfrac{1}{2}$ and the conic is an ellipse.

63. True. The graphs represent the same hyperbola.

65. True. The conic is an ellipse because the eccentricity is less than 1.

67. Answers will vary.   **69.** $r^2 = \dfrac{24{,}336}{169 - 25 \cos^2 \theta}$

71. $r^2 = \dfrac{144}{25 \cos^2 \theta - 9}$ **73.** $r^2 = \dfrac{144}{25 \cos^2 \theta - 16}$

75. The original equation graphs as a parabola that opens upward.

(a) The parabola opens to the right.

(b) The parabola opens downward.

(c) The parabola opens to the left.

(d) The parabola has been rotated.

77. (a) Ellipse

(b) The given polar equation, r, has a vertical directrix to the left of the pole. The equation r_1 has a vertical directrix to the right of the pole, and the equation r_2 has a horizontal directrix below the pole.

(c) $r_2 = \dfrac{4}{1 - 0.4 \sin \theta}$

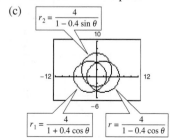

$r_1 = \dfrac{4}{1 + 0.4 \cos \theta}$ $r = \dfrac{4}{1 - 0.4 \cos \theta}$

Review Exercises *(page 768)*

1. $\dfrac{\pi}{4}$ rad, 45°   **3.** 1.9513 rad, 111.8°   **5.** 0.4424 rad, 25.3°

7. 0.6588 rad, 37.7°   **9.** $\dfrac{4\sqrt{5}}{5}$ **11.** Hyperbola

13. $x^2 = 12y$ **15.** $(y - 2)^2 = 12x$

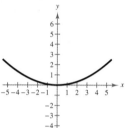   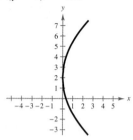

17. $y = -4x - 2$ **19.** $6\sqrt{5}$ m

21. $\dfrac{(x - 2)^2}{9} + \dfrac{(y - 8)^2}{64} = 1$ **23.** $\dfrac{(x - 2)^2}{4} + \dfrac{(y - 1)^2}{1} = 1$

25. 3 feet from the center of the arch at ground level.

27. Center: $(-2, 5)$

Vertices: $(-10, 5), (6, 5)$

Foci: $\left(-2 \pm 2\sqrt{7}, 5\right)$

Eccentricity: $e = \dfrac{\sqrt{7}}{4}$

29. Center: $(1, -4)$

Vertices: $(1, 0), (1, -8)$

Foci: $\left(1, -4 \pm \sqrt{7}\right)$

Eccentricity: $\dfrac{\sqrt{7}}{4}$

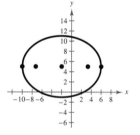   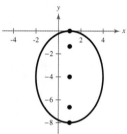

31. $\dfrac{y^2}{36} - \dfrac{x^2}{28} = 1$ **33.** $\dfrac{x^2}{16} - \dfrac{y^2}{9} = 1$

35. Center: $(4, -2)$

Vertices:

$(-3, -2), (11, -2)$

Foci: $\left(4 \pm \sqrt{74}, -2\right)$

Asymptotes:

$y = -2 \pm \tfrac{5}{7}(x - 4)$

37. Center: $(1, -1)$

Vertices:

$(5, -1), (-3, -1)$

Foci: $(6, -1), (-4, -1)$

Asymptotes:

$y = -1 \pm \tfrac{3}{4}(x - 1)$

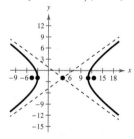   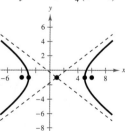

39. $\dfrac{x^2}{10{,}890{,}000} - \dfrac{y^2}{16{,}988{,}400} = 1$

41. Hyperbola   **43.** Ellipse

45. $\dfrac{(y')^2}{10} - \dfrac{(x')^2}{10} = 1$ **47.** $\dfrac{(x')^2}{3} + \dfrac{(y')^2}{2} = 1$

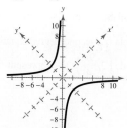   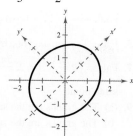

49. (a) Parabola

(b) $y = \dfrac{24x + 40 \pm \sqrt{(24x + 40)^2 - 36(16x^2 - 30x)}}{18}$

(c)

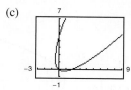

51. (a) Hyperbola   (b) $y = \dfrac{10x \pm \sqrt{100x^2 - 4(x^2 + 1)}}{2}$

(c)

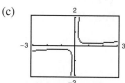

53. (a)

t	-2	-1	0	1	2
x	-8	-5	-2	1	4
y	15	11	7	3	-1

(b)

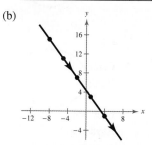

55. (a)

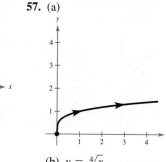

(b) $y = 2x$

57. (a)

(b) $y = \sqrt[4]{x}$

59. (a)

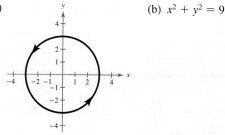

(b) $x^2 + y^2 = 9$

61. (a) $x = t,\ y = 2t + 3$ (b) $x = t - 1,\ y = 2t + 1$
(c) $x = 3 - t,\ y = 9 - 2t$

63. (a) $x = t,\ y = t^2 + 3$ (b) $x = t - 1,\ y = t^2 - 2t + 4$
(c) $x = 3 - t,\ y = t^2 - 6t + 12$

65. (a) $x = t,\ y = 1 - 4t^2$
(b) $x = t - 1,\ y = -4t^2 + 8t - 3$
(c) $x = 3 - t,\ y = -4t^2 + 24t - 35$

67.

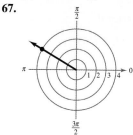

$\left(4, -\dfrac{7\pi}{6}\right), \left(-4, -\dfrac{\pi}{6}\right),$
$\left(-4, \dfrac{11\pi}{6}\right)$

69.

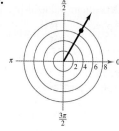

$(7, -5.23),\ (7, 1.05),$
$(-7, -2.09)$

71. $(0, 0)$ **73.** $\left(-\dfrac{1}{2}, -\dfrac{\sqrt{3}}{2}\right)$ **75.** $\left(3\sqrt{2}, \dfrac{\pi}{4}\right)$

77. $\left(\sqrt{10}, \dfrac{3\pi}{4}\right)$ **79.** $r = 9$ **81.** $r = 5 \sec \theta$

83. $r^2 = 10 \csc 2\theta$ **85.** $x^2 + y^2 = 16$ **87.** $x^2 + y^2 = 3x$

89. $x^2 + y^2 = y^{2/3}$

91. Symmetry: $\theta = \dfrac{\pi}{2}$,
polar axis, pole
Maximum value of $|r|$:
$|r| = 6$ for all values of θ
No zeros of r

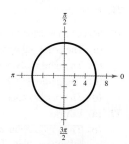

93. Symmetry: polar axis
Maximum value of $|r|$:
$|r| = 4$ when $\theta = 0$
Zeros of r: $r = 0$ when $\theta = \pi$

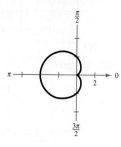

CHAPTER 10

95. Symmetry: $\theta = \dfrac{\pi}{2}$, polar axis, pole

Maximum value of $|r|$: $|r| = 4$ when $\theta = \dfrac{\pi}{4}, \dfrac{3\pi}{4}, \dfrac{5\pi}{4}, \dfrac{7\pi}{4}$

Zeros of r: $r = 0$ when $\theta = 0, \dfrac{\pi}{2}, \pi, \dfrac{3\pi}{2}$

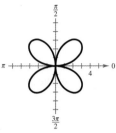

97. Symmetry: $\theta = \dfrac{\pi}{2}$

Maximum value of $|r|$: $|r| = 8$ when $\theta = \dfrac{\pi}{2}$

Zeros of r: $r = 0$ when $\theta = 3.4814, 5.9433$

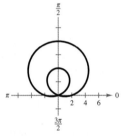

99. Symmetry: $\theta = \dfrac{\pi}{2}$, polar axis, pole

Maximum value of $|r|$: $|r| = 3$ when $\theta = \dfrac{\pi}{2}$

Zeros of r: $r = 0$ when $\theta = 0, \pi$

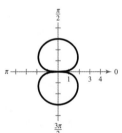

101. Limaçon

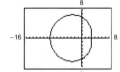

103. Rose curve

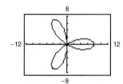

105. Hyperbola

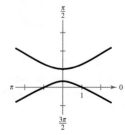

107. Ellipse

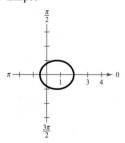

109. $r = \dfrac{4}{1 - \cos \theta}$ **111.** $r = \dfrac{5}{3 - 2 \cos \theta}$

113. $r = \dfrac{7961.93}{1 - 0.937 \cos \theta}$; 10,980.11 mi

115. False. The equation of a hyperbola is a second-degree equation.

117. False. $\left(2, \dfrac{\pi}{4}\right)$, $\left(-2, \dfrac{5\pi}{4}\right)$, and $\left(2, \dfrac{9\pi}{4}\right)$ all represent the same point.

119. (a) The graphs are the same.
    (b) The graphs are the same.

Chapter Test *(page 771)*

1. 0.5191 rad, 29.7°   **2.** 0.7023 rad, 40.2°   **3.** $\dfrac{\sqrt{10}}{10}$

4. Parabola: $y^2 = 2(x - 1)$
Vertex: $(1, 0)$
Focus: $\left(\frac{3}{2}, 0\right)$

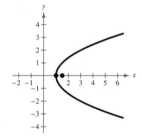

5. Hyperbola: $\dfrac{(x - 2)^2}{4} - y^2 = 1$
Center: $(2, 0)$
Vertices: $(0, 0), (4, 0)$
Foci: $\left(2 \pm \sqrt{5}, 0\right)$
Asymptotes: $y = \pm\dfrac{1}{2}(x - 2)$

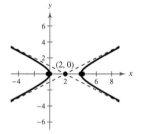

6. Ellipse:
$\dfrac{(x + 3)^2}{16} + \dfrac{(y - 1)^2}{9} = 1$
Center: $(-3, 1)$
Vertices: $(1, 1), (-7, 1)$
Foci: $\left(-3 \pm \sqrt{7}, 1\right)$

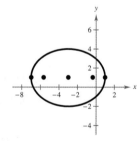

7. Circle: $(x - 2)^2 + (y - 1)^2 = \frac{1}{2}$
Center: $(2, 1)$

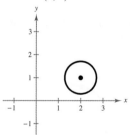

8. $(y + 4)^2 = 12(x - 3)$ **9.** $\dfrac{y^2}{2/41} - \dfrac{x^2}{162/41} = 1$

10. (a) $\dfrac{(y')^2}{2} - \dfrac{(x')^2}{2} = 1$ **11.**

(b)

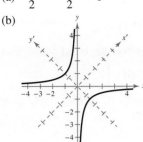

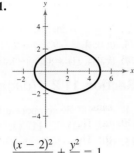

$\dfrac{(x-2)^2}{9} + \dfrac{y^2}{4} = 1$

12. (a) $x = t,\ y = 3 - t^2$ (b) $x = t - 2,\ y = -t^2 + 4t - 1$

13. $\left(\sqrt{3},\ -1\right)$

14. $\left(2\sqrt{2},\ \dfrac{7\pi}{4}\right),\ \left(-2\sqrt{2},\ \dfrac{3\pi}{4}\right),\ \left(2\sqrt{2},\ -\dfrac{\pi}{4}\right),\ \left(-2\sqrt{2},\ -\dfrac{5\pi}{4}\right)$

15. $r = 8$

16. Parabola   **17.** Ellipse

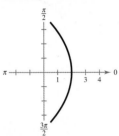   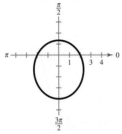

18. Limaçon with inner loop   **19.** Rose curve

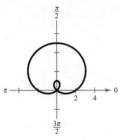   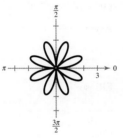

20. Answers will vary. For example: $r = \dfrac{1}{1 + 0.25\sin\theta}$

21. Slope: 0.1511; Change in elevation: 789 ft

22. No; Yes

Problem Solving   *(page 775)*

1. (a) 1.2016 rad   (b) 2420 ft, 5971 ft   **3.** $A = \dfrac{4a^2b^2}{a^2 + b^2}$

5. (a) Because $d_1 + d_2 \le 20$, by definition, the outer bound that the boat can travel is an ellipse. The islands are the foci.

(b) Island 1: $(-6, 0)$; Island 2: $(6, 0)$

(c) 20 mi; Vertex: $(10, 0)$

(d) $\dfrac{x^2}{100} - \dfrac{y^2}{64} = 1$

7–9. Proofs

11. Answers will vary. *Sample answer:*

$x = \cos(-t)$

$y = 2\sin(-t)$

13. (a) $y^2 = x^2\left(\dfrac{2 - x}{2 + x}\right)$

(b) $x = \dfrac{2 - 2t^2}{1 + t^2},\ y = \dfrac{t(2 - 2t^2)}{1 + t^2}$

(c)

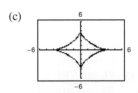

15. (a)

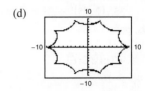

The graph is a line between -2 and 2 on the x-axis.

(b)

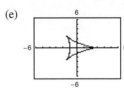

The graph is a three-sided figure with counterclockwise orientation.

(c)

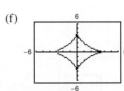

The graph is a four-sided figure with counterclockwise orientation.

(d)

The graph is a 10-sided figure with counterclockwise orientation.

(e)

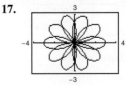

The graph is a three-sided figure with clockwise orientation.

(f)

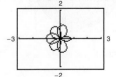

The graph is a four-sided figure with clockwise orientation.

17.

$r = 3\sin\left(\dfrac{5\theta}{2}\right)$ $r = -\cos\left(\sqrt{2}\theta\right),$

$-2\pi \le \theta \le 2\pi$

Sample answer: If n is a rational number, then the curve has a finite number of petals. If n is an irrational number, then the curve has an infinite number of petals.

Chapter 11

Section 11.1 *(page 783)*

1. three-dimensional **3.** octants

5. $\left(\dfrac{x_1 + x_2}{2}, \dfrac{y_1 + y_2}{2}, \dfrac{z_1 + z_2}{2}\right)$ **7.** surface; space

9.
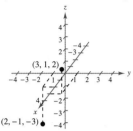

11.
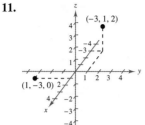

13.
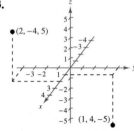

15. $(-2, 4, 3)$ **17.** $(8, 0, 0)$ **19.** $(0, -3, 5)$
21. Octant IV **23.** Octants II, IV, VI, and VIII
25. $\sqrt{33}$ units **27.** $\sqrt{29}$ units **29.** $\sqrt{110}$ units
31. 12 units **33.** $\sqrt{114}$ units **35.** $\left(2\sqrt{5}\right)^2 + 3^2 = \left(\sqrt{29}\right)^2$
37. $6, 6, 2\sqrt{10}$; isosceles triangle
39. $6, 6, 2\sqrt{10}$; isosceles triangle **41.** $\left(\frac{5}{2}, -3, 2\right)$
43. $(0, -1, 7)$ **45.** $\left(\frac{1}{2}, \frac{1}{2}, -1\right)$ **47.** $\left(\frac{5}{2}, 2, 6\right)$
49. $(x - 3)^2 + (y - 2)^2 + (z - 4)^2 = 16$
51. $(x + 3)^2 + (y - 7)^2 + (z - 5)^2 = 25$
53. $\left(x - \frac{3}{2}\right)^2 + y^2 + (z - 3)^2 = \frac{45}{4}$
55. Center: $(3, 0, 0)$; radius: 3
57. Center: $(2, -1, 3)$; radius: 2
59. Center: $\left(1, \frac{1}{3}, 4\right)$; radius: 3
61. Center: $\left(\frac{1}{3}, -1, 0\right)$; radius: 1

63.
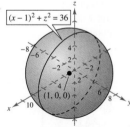

65.
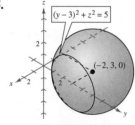

67.
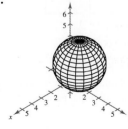

69.
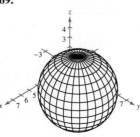

71. $x^2 + y^2 + z^2 = \dfrac{205^2}{4}$

73. False. z is the directed distance from the xy-plane to P.
75. $0; 0; 0$
77. $(x_2, y_2, z_2) = (2x_m - x_1, 2y_m - y_1, 2z_m - z_1)$

Section 11.2 *(page 790)*

1. zero **3.** component form **5.** parallel **7.** $\langle 0, 6, 0 \rangle$
9. $\langle -2, 3, 1 \rangle$ **11.** $\langle 7, -5, 7 \rangle$ **13.** $\langle 3, 1, 0 \rangle$
15. (a)   (b)
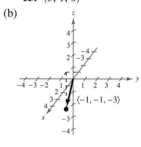

(c)   (d)
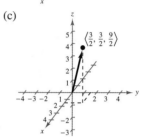

17. (a)   (b)
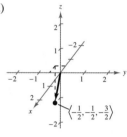

(c)   (d)
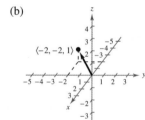

(c)   (d)
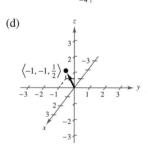

19. $\mathbf{z} = \langle -11, 7, 0 \rangle$ **21.** $\mathbf{z} = \left\langle -15, \frac{23}{2}, -1 \right\rangle$
23. $\mathbf{z} = \left\langle 0, \frac{11}{2}, 2 \right\rangle$ **25.** $3\sqrt{5}$ **27.** $5\sqrt{2}$ **29.** $\sqrt{11}$
31. $\sqrt{74}$ **33.** $\sqrt{34}$
35. (a) $\dfrac{\sqrt{29}}{29}\langle 2, 0, 5 \rangle$ (b) $-\dfrac{\sqrt{29}}{29}\langle 2, 0, 5 \rangle$
37. (a) $\dfrac{\sqrt{74}}{74}(8\mathbf{i} + 3\mathbf{j} - \mathbf{k})$ (b) $-\dfrac{\sqrt{74}}{74}(8\mathbf{i} + 3\mathbf{j} - \mathbf{k})$
39. -4 **41.** 0 **43.** About $124.45°$ **45.** About $109.92°$
47. Parallel **49.** Neither **51.** Orthogonal
53. Orthogonal **55.** Not collinear **57.** Collinear
59. $(3, 1, 7)$ **61.** $\pm\dfrac{3\sqrt{14}}{14}$

63. $\langle 0, 2\sqrt{2}, 2\sqrt{2} \rangle$ or $\langle 0, 2\sqrt{2}, -2\sqrt{2} \rangle$

65. B: 226.52 N, C: 202.92 N, D: 157.91 N

67. True. When $0° \le \theta \le 180°$ and $\cos \theta = 0$, $\theta = 90°$.

69. The angle between **u** and **v** is an obtuse angle.

Section 11.3 *(page 797)*

1. cross product   **3.** $-\mathbf{i} + 7\mathbf{j} + 5\mathbf{k}$ **5.** **0**

7. $-2\mathbf{i} + 14\mathbf{j} + 10\mathbf{k}$ **9.** $\mathbf{i} - 7\mathbf{j} - 5\mathbf{k}$ **11.** 0

13. $\langle 1, 1, 1 \rangle$ **15.** $\langle -2, 3, 3 \rangle$ **17.** $-17\mathbf{i} + \mathbf{j} + 10\mathbf{k}$

19. $-7\mathbf{i} + 13\mathbf{j} + 16\mathbf{k}$ **21.** $-18\mathbf{i} - 6\mathbf{j}$ **23.** $-\mathbf{i} - 2\mathbf{j} - \mathbf{k}$

25. $\dfrac{1}{3}\mathbf{i} - \dfrac{2}{3}\mathbf{j} - \dfrac{2}{3}\mathbf{k}$ **27.** $\dfrac{\sqrt{19}}{19}(\mathbf{i} - 3\mathbf{j} + 9\mathbf{k})$

29. $\dfrac{\sqrt{2}}{2}(\mathbf{i} - \mathbf{j})$ **31.** $\dfrac{\sqrt{5}}{15}(4\mathbf{i} + 5\mathbf{j} + 2\mathbf{k})$ **33.** 6   **35.** 56

37. 1   **39.** $30\sqrt{3}$ **41.** (a) Answers will vary. (b) $6\sqrt{10}$

43. (a) Answers will vary.   (b) $\sqrt{286}$ **45.** $\dfrac{3\sqrt{13}}{2}$

47. $\dfrac{1}{2}\sqrt{4290}$ **49.** 6   **51.** 2   **53.** 2   **55.** 12   **57.** 84

59. (a) $T = \dfrac{p}{2}\cos 40°$

(b)

p	15	20	25	30	35	40	45
T	5.75	7.66	9.58	11.49	13.41	15.32	17.24

61. True. The cross product is not defined for two-dimensional vectors.

63. $2\mathbf{u} \times 2\mathbf{v} = 2(\mathbf{u} \times 2\mathbf{v}) = 4(\mathbf{u} \times \mathbf{v})$ **65.** Proof

Section 11.4 *(page 806)*

1. direction   **3.** symmetric equations

5. (a) $x = 4 + 5t$, $y = 1 + 3t$, $z = 2 + 6t$

(b) $\dfrac{x - 4}{5} = \dfrac{y - 1}{3} = \dfrac{z - 2}{6}$

7. (a) $x = 2 + 4t$, $y = -5 + 3t$, $z = -t$

(b) $\dfrac{x - 2}{4} = \dfrac{y + 5}{3} = -z$

9. (a) $x = 2 + 2t$, $y = -3 - 3t$, $z = 5 + t$

(b) $\dfrac{x - 2}{2} = \dfrac{y + 3}{-3} = z - 5$

11. (a) $x = 2 - t$, $y = 4t$, $z = 2 - 5t$

(b) $\dfrac{x - 2}{-1} = \dfrac{y}{4} = \dfrac{z - 2}{-5}$

13. (a) $x = -3 + 4t$, $y = 8 - 10t$, $z = 15 + t$

(b) $\dfrac{x + 3}{4} = \dfrac{y - 8}{-10} = z - 15$

15. (a) $x = 3 - 4t$, $y = 1$, $z = 2 + 3t$

(b) No symmetric equations

17.

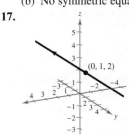

19. $-2x + y - 2z + 10 = 0$ **21.** $-2x + y + 2z - 3 = 0$

23. $-2x - 4y + 3z + 4 = 0$ **25.** $6x - 2y - z - 8 = 0$

27. $y - 5 = 0$ **29.** $y - z + 2 = 0$

31. $7x + y - 11z - 5 = 0$ **33.** Perpendicular

35. Parallel   **37.** Neither; about 61.6°

39. $x = 2$ **41.** $x = 2 + 3t$
$\ y = 3$ $\ y = 3 + 2t$
$\ z = 4 + t$ $\ z = 4 - t$

43. (a) 56.9°   (b) $x = t$, $y = -7t$, $z = -3t$

45. (a) 77.8°   (b) $x = 6t + 1$, $y = t$, $z = 7t + 1$

47.   **49.**

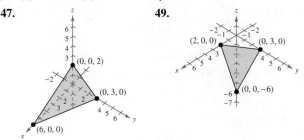

51.

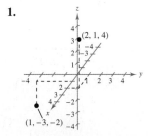

53. $\dfrac{\sqrt{6}}{6}$ **55.** $\dfrac{3\sqrt{5}}{5}$ **57.** $\dfrac{10\sqrt{17}}{17}$ **59.** 89.1°

61. False. Lines that do not intersect and are not in the same plane may not be parallel.

63. The parametric equations should be $x = -2 - 3t$, $y = 5 - 4t$, and $z = 3 + 2t$.

Review Exercises *(page 810)*

1.

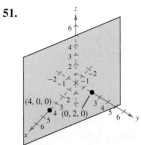

3. $(-5, 4, 0)$ **5.** $\sqrt{41}$ **7.** $\sqrt{61}$

9. $\sqrt{29}$, $\sqrt{38}$, $\sqrt{67}$
$\left(\sqrt{29}\right)^2 + \left(\sqrt{38}\right)^2 = \left(\sqrt{67}\right)^2$

11. $(5, 4, 5)$ **13.** $(-2, 2, 2)$

15. $(x - 1)^2 + (y - 5)^2 + (z - 4)^2 = 4$

17. $(x - 1)^2 + (y + 4)^2 + (z - 2)^2 = 49$

19. $x^2 + y^2 + z^2 = 12$ **21.** Center: $(0, 0, 4)$; radius: 4

23. Center: $(5, -3, 2)$; radius: 2

CHAPTER 11

25. (a) 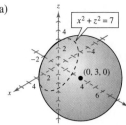   (b)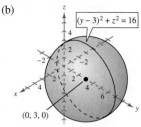

4. $(x - 7)^2 + (y - 1)^2 + (z - 2)^2 = 19$

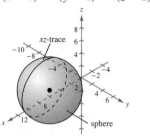

27. (a) $\langle 1, 6, -1 \rangle$ (b) $\sqrt{38}$ (c) $\left\langle \dfrac{\sqrt{38}}{38}, \dfrac{3\sqrt{38}}{19}, -\dfrac{\sqrt{38}}{38} \right\rangle$

29. (a) $\langle -10, 6, 7 \rangle$ (b) $\sqrt{185}$

    (c) $\left\langle \dfrac{-2\sqrt{185}}{37}, \dfrac{6\sqrt{185}}{185}, \dfrac{7\sqrt{185}}{185} \right\rangle$

31. -9 **33.** 1   **35.** About $77.83°$ **37.** Parallel

39. Orthogonal   **41.** Collinear   **43.** Not collinear

45. A: 159.1 lb

    B: 115.6 lb

    C: 115.6 lb

47. $\langle -10, 0, -10 \rangle$ **49.** $4\mathbf{i} + 2\mathbf{j} - 7\mathbf{k}$

51. $\dfrac{\sqrt{11}}{11}\mathbf{i} + \dfrac{3\sqrt{11}}{11}\mathbf{j} + \dfrac{\sqrt{11}}{11}\mathbf{k}$

53. $-\dfrac{71\sqrt{7602}}{7602}\mathbf{i} - \dfrac{22\sqrt{7602}}{3801}\mathbf{j} + \dfrac{25\sqrt{7602}}{7602}\mathbf{k}$

55. (a) Answers will vary.   (b) $2\sqrt{43}$ **57.** 75

59. (a) $x = -2t, y = 5t, z = t$

    (b) $\dfrac{x}{-2} = \dfrac{y}{5} = z$

61. (a) $x = -1 + 4t, y = 3 + 3t, z = 5 - 6t$

    (b) $\dfrac{x + 1}{4} = \dfrac{y - 3}{3} = \dfrac{z - 5}{-6}$

63. $-2x - 12y + 5z = 0$ **65.** $z - 2 = 0$

67.

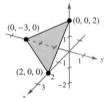

69. $\dfrac{\sqrt{6}}{6}$ **71.** $\dfrac{27\sqrt{17}}{17}$

73. False. $c(\mathbf{u} \times \mathbf{v}) = (c\mathbf{u}) \times \mathbf{v} = \mathbf{u} \times (c\mathbf{v})$

75. Answers will vary.

Chapter Test   *(page 812)*

1.

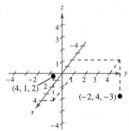

2. No. $\left(\sqrt{76}\right)^2 + \left(\sqrt{102}\right)^2 \neq \left(\sqrt{194}\right)^2$ **3.** $(7, 1, 2)$

5. $\mathbf{u} = \langle -2, 6, -6 \rangle$, $\mathbf{v} = \langle -12, 5, -5 \rangle$

6. (a) 84   (b) $\langle 0, 62, 62 \rangle$

7. (a) $\left\langle -\dfrac{\sqrt{19}}{19}, \dfrac{3\sqrt{19}}{19}, -\dfrac{3\sqrt{19}}{19} \right\rangle$

    (b) $\left\langle -\dfrac{6\sqrt{194}}{97}, \dfrac{5\sqrt{194}}{194}, -\dfrac{5\sqrt{194}}{194} \right\rangle$

8. $46.23°$

9. (a) $x = 8 - 2t, y = -2 + 6t, z = 5 - 6t$

    (b) $\dfrac{x - 8}{-2} = \dfrac{y + 2}{6} = \dfrac{z - 5}{-6}$

10. Parallel   **11.** Orthogonal   **12.** Neither

13. (a) Answers will vary.   (b) $2\sqrt{230}$ **14.** 200

15.

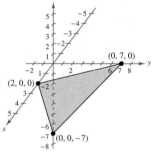

16.

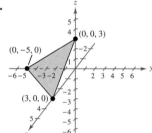

17. $11x + 4y + 16z + 6 = 0$ **18.** $\dfrac{4\sqrt{14}}{7}$ **19.** About $88.5°$

Problem Solving   *(page 815)*

1. (a) 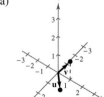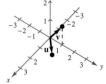   (b) Answers will vary.

    (c) $a = b = 1$

    (d) Answers will vary.

3. (a) Right triangle   (b) Obtuse triangle

    (c) Obtuse triangle   (d) Acute triangle

5. About 860.0 lb   **7.** Answers will vary.   **9–11.** Proofs

13. (a) $1080 \sin \theta$

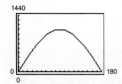

(b) $540\sqrt{2} \approx 763.48$

(c) $\theta = 90°$. This is what should be expected. When $\theta = 90°$, the pipe wrench is horizontal.

15. Proof   **17.** (a) $D = \dfrac{3\sqrt{2}}{2}$ (b) $D = \sqrt{5}$

19. (a) $d = \sqrt{70}$ when $t = 0$.

(b)

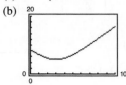

(c) The distance between the two insects appears to lessen in the first 3 seconds but then begins to increase with time.

(d) The insects get within 5 inches of each other.

Chapter 12

Section 12.1   *(page 826)*

1. limit

3. (a)

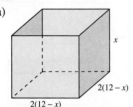

(b) $V(x) = 4x(12 - x)^2$

(c)

x	3	3.5	3.9	4
$V(x)$	972	1011.5	1023.5	1024

x	4.1	4.5	5
$V(x)$	1023.5	1012.5	980

$\displaystyle\lim_{x \to 4} V(x) = 1024$

(d)

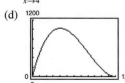

5.

x	0.9	0.99	0.999	1
$f(x)$	9.3	9.93	9.993	?

x	1.001	1.01	1.1
$f(x)$	10.007	10.07	10.7

10; The limit can be reached.

7.

x	-2.1	-2.01	-2.001	-2
$f(x)$	-0.244	-0.249	-0.2499	?

x	-1.999	-1.99	-1.9
$f(x)$	-0.2501	-0.251	-0.256

-0.25; The limit cannot be reached.

9.

x	-0.1	-0.01	-0.001	0
$f(x)$	1.9867	1.9999	1.999999	?

x	0.001	0.01	0.1
$f(x)$	1.999999	1.9999	1.9867

2; The limit cannot be reached.

11.

x	0.9	0.99	0.999	1
$f(x)$	0.2564	0.2506	0.2501	Error

x	1.001	1.01	1.1
$f(x)$	0.2499	0.2494	0.2439

$\dfrac{1}{4}$

13.

x	-0.1	-0.01	-0.001	0
$f(x)$	0.2247	0.2237	0.2236	Error

x	0.001	0.01	0.1
$f(x)$	0.2236	0.2235	0.2225

$\dfrac{\sqrt{5}}{10}$

15.

x	-4.1	-4.01	-4.001	-4
$f(x)$	0.4762	0.4975	0.4998	Error

x	-3.999	-3.99	-3.9
$f(x)$	0.5003	0.5025	0.5263

$\dfrac{1}{2}$

17.

x	-0.1	-0.01	-0.001	0
$f(x)$	-0.1	-0.01	-0.001	Error

x	0.001	0.01	0.1
$f(x)$	0.001	0.01	0.1

0

19.

x	-0.1	-0.01	-0.001	0
$f(x)$	0.9063	0.9901	0.9990	Error

x	0.001	0.01	0.1
$f(x)$	1.0010	1.0101	1.1070

1

21.

x	0.9	0.99	0.999	1
$f(x)$	2.2314	2.0203	2.0020	Error

x	1.001	1.01	1.1
$f(x)$	1.9980	1.9803	1.8232

2

23.

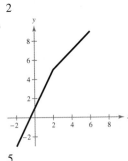

5

25. 2   **27.** Does not exist; Answers will vary.

29. Does not exist; Answers will vary.

31. Does not exist; Answers will vary.

33.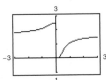

Does not exist
Answers will vary.

35.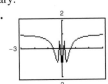

Does not exist
Answers will vary.

37.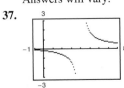

Does not exist
Answers will vary.

39.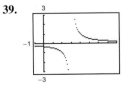

Exists

41. (a) -12 (b) 9 (c) $\frac{1}{2}$ (d) $\sqrt{3}$

43. (a) 8 (b) $\frac{3}{8}$ (c) 3 (d) $-\frac{61}{8}$ **45.** -8

47. 7   **49.** -6 **51.** $-\frac{4}{5}$ **53.** $\frac{1}{10}$ **55.** 3   **57.** $\frac{35}{3}$

59. $e^3 \approx 20.09$ **61.** -1 **63.** $\frac{\pi}{6}$

65. True. The limit of $f(x)$ as $x \to c$ does not exist when $f(x)$ approaches a different number from the right side of c than it approaches from the left side of c.

67. (a) and (b) Answers will vary.

69. As the function's x-value approaches 5 from both the right and left sides, its corresponding output values approach 12.

71. The function may approach different values from the right and left of 2.

73. $\lim\limits_{x \to 0} \tan x = 0$

$\lim\limits_{x \to \pi/4} \tan x = 1$

$\lim\limits_{x \to \pi/2} \tan x$ does not exist.

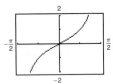

Section 12.2   *(page 837)*

1. indeterminate form   **3.** one-sided limit

5. (a) 2   (b) 0   (c) 0

$g_2(x) = x(x + 1)$

7. 12   **9.** $\frac{1}{5}$ **11.** $\frac{1}{10}$ **13.** $\frac{1}{4}$ **15.** -1

17. Does not exist   **19.** 0

21.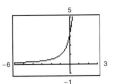

3.000

23.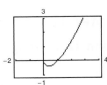

0

25.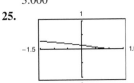

0.135

27.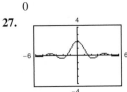

2.000

29.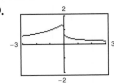

0.333

31. (a)

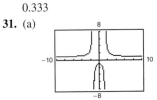

3

(b)

x	1.9	1.99	1.999	2
$f(x)$	3.48	3.041	3.004	Error

x	2.001	2.01	2.1
$f(x)$	2.996	2.96	2.66

3

(c) 3

33. (a)

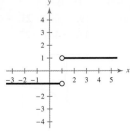

-0.125

(b)

x	16.1	16.01
$f(x)$	-0.12481	-0.12498

x	16.001	16.0001
$f(x)$	-0.124998	-0.1249998

-0.125

(c) $-\frac{1}{8}$

35.

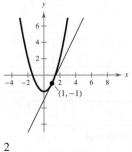

Does not exist

37.

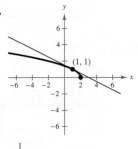

4

39. Limit (a) can be evaluated by direct substitution.

(a) 0   (b) 1

41. -32 ft/sec   **43.** Answers will vary.   **45.** $2x + 2$

47. $\dfrac{1}{2\sqrt{x-2}}$

49.

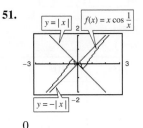

0

51.

0

53. True. See discussion on page 829.

55. Use the dividing out technique to find that the limit is -8.

57. Answers will vary.

Section 12.3   *(page 846)*

1. Calculus   **3.** $\displaystyle\lim_{h\to 0}\dfrac{f(x+h)-f(x)}{h}$ **5.** 0   **7.** $\dfrac{1}{2}$

9.

11.

2

$-\dfrac{1}{2}$

13.

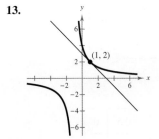

-1

15. -4 **17.** -3 **19.** -1 **21.** $\frac{1}{6}$

23. $m = -2x$; (a) 0   (b) 2

25. $m = -\dfrac{1}{(x+4)^2}$; (a) $-\dfrac{1}{16}$ (b) $-\dfrac{1}{4}$

27. $m = \dfrac{1}{2\sqrt{x-1}}$; (a) $\dfrac{1}{4}$ (b) $\dfrac{1}{6}$ **29.** 0   **31.** 2

33. $4x + 3$ **35.** $-2x^{-3}$ **37.** $\dfrac{1}{2\sqrt{x-7}}$

39. $-\dfrac{1}{(x+1)^2}$ **41.** $-\dfrac{1}{2(x-4)^{3/2}}$

43. (a) 4
(b) $y = 4x - 5$
(c)

45. (a) 1
(b) $y = x - 2$
(c)

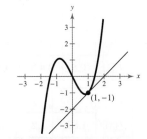

47. (a) $\frac{1}{4}$
(b) $y = \frac{1}{4}x + \frac{5}{4}$
(c)

49. (a) -1
(b) $y = -x - 3$
(c)

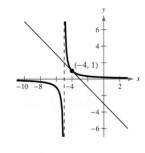

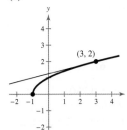

51.

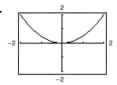

They appear to be the same.

x	-2	-1.5	-1	-0.5	0
$f(x)$	2	1.125	0.5	0.125	0
$f'(x)$	-2	-1.5	-1	-0.5	0

x	0.5	1	1.5	2
$f(x)$	0.125	0.5	1.125	2
$f'(x)$	0.5	1	1.5	2

53.

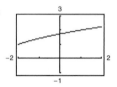

They appear to be the same.

x	-2	-1.5	-1	-0.5	0
$f(x)$	1	1.225	1.414	1.581	1.732
$f'(x)$	0.5	0.408	0.354	0.316	0.289

x	0.5	1	1.5	2
$f(x)$	1.871	2	2.121	2.236
$f'(x)$	0.267	0.25	0.236	0.224

55. $y = -x + 1$ **57.** $y = -6x \pm 8$
59. $f'(x) = 2x - 4$; horizontal tangent at $(2, -1)$
61. $f'(x) = 9x^2 - 9$; horizontal tangents at $(-1, 6)$ and $(1, -6)$
63. $(-1, -1), (0, 0), (1, -1)$
65. $\left(\dfrac{\pi}{6}, \sqrt{3} + \dfrac{\pi}{6}\right), \left(\dfrac{5\pi}{6}, \dfrac{5\pi}{6} - \sqrt{3}\right)$ **67.** $(e^{-1}, -e^{-1})$
69. $(0, 0), (-2, 4e^{-2})$ **71.** Answers will vary.
73. (a) $y = 64.509x^2 - 724.08x + 3045.8$

(b) 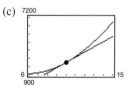 (c)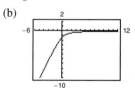

About 566.1; The rate of change in revenue in 2010 was about \$566.1 million.   The slopes are the same.

75. True. The slope is dependent on x.

77. (a) 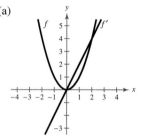 (b)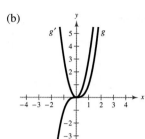

(c) Answers will vary.
$h'(x) = nx^{n-1}$
79–81. Answers will vary.
83. b   **84.** a   **85.** d   **86.** c

Section 12.4   *(page 855)*

1. limit; infinity   **3.** converge   **5.** c   **6.** a   **7.** d
8. b   **9.** 2   **11.** -1 **13.** -3 **15.** 0
17. Does not exist; Answers will vary.   **19.** $-\dfrac{4}{3}$ **21.** -1
23. Does not exist; Answers will vary.   **25.** -4 **27.** -5
29. 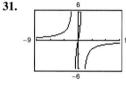   **31.**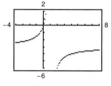

$y = -3$ $y = 0$
$y \to -3$ $y \to 0$

33.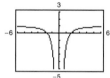

$y = 1$
$y \to 1$

35. (a)

x	10^0	10^1	10^2	10^3
$f(x)$	-0.7321	-0.0995	-0.0100	-0.0010

x	10^4	10^5	10^6
$f(x)$	-1.0×10^{-4}	-1.0×10^{-5}	-1.0×10^{-6}

0

(b)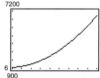

0

37. (a)

x	10^0	10^1	10^2	10^3
$f(x)$	-0.7082	-0.7454	-0.7495	-0.74995

x	10^4	10^5	10^6
$f(x)$	-0.749995	-0.7499995	-0.75

-0.75

(b)

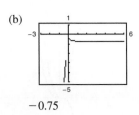

-0.75

39. (a) 1

(b)

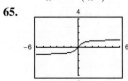

(c) Over a long period of time, the level of the oxygen in the pond returns to the normal level.

41. (a) $\overline{C}(x) = \dfrac{73x + 25{,}000}{x}$ (b) \$98; \$78; \$75.50

(c) \$73; As the number of MP3 players gets very large, the average cost approaches \$73.

43. (a) \$1476.58

(b) \$771.25; As time increases, the average monthly Social Security benefit approaches \$771.25.

(c) Answers will vary.

45. $1, \frac{3}{5}, \frac{2}{5}, \frac{5}{17}, \frac{3}{13}$ **47.** $\frac{1}{3}, \frac{2}{5}, \frac{3}{7}, \frac{4}{9}, \frac{5}{11}$

Limit: 0   Limit: $\frac{1}{2}$

49. $\frac{1}{5}, \frac{4}{7}, 1, \frac{16}{11}, \frac{25}{13}$ **51.** 2, 3, 4, 5, 6

Limit does not exist;   Limit does not exist;

Answers will vary.   Answers will vary.

53. $-1, \frac{1}{2}, -\frac{1}{3}, \frac{1}{4}, -\frac{1}{5}$

Limit: 0

55. $\frac{3}{2}$

n	10^0	10^1	10^2	10^3
a_n	2	1.55	1.505	1.5005

n	10^4	10^5	10^6
a_n	1.50005	1.500005	1.5000005

57. $\frac{16}{3}$

n	10^0	10^1	10^2	10^3
a_n	16	6.16	5.4136	5.3413

n	10^4	10^5	10^6
a_n	5.3341	5.33341	5.333341

59. False. Graph $y = \dfrac{x^2}{x+1}$.

61. True. If the sequence converges, then the limit exists.

63. $5 - \dfrac{2}{x^2} \ne 5\left(\dfrac{-2}{x^2}\right)$

65.

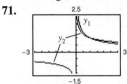

Two; As $x \to \infty$, $y \to 1$, and as $x \to -\infty$, $y \to -1$.

67.

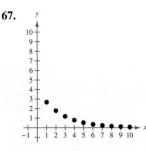

Converges to 0

69.

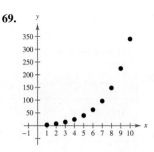

Diverges

71.

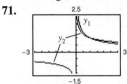

$\sqrt{x}$ is undefined for $x < 0$. Answers will vary.

Section 12.5   *(page 864)*

1. cn **3.** $\dfrac{n^2(n+1)^2}{4}$ **5.** 240   **7.** 44,100

9. 14,430   **11.** 5850

13. (a) $S(n) = \dfrac{n^2 + 2n + 1}{4n^2}$

(b)

n	10^0	10^1	10^2	10^3	10^4
$S(n)$	1	0.3025	0.25503	0.25050	0.25005

(c) $\frac{1}{4}$

15. (a) $S(n) = \dfrac{2n^2 + 3n - 7}{2n^2}$

(b)

n	10^0	10^1	10^2	10^3	10^4
$S(n)$	6	1.185	1.0154	1.0015	1.00015

(c) 1

17. (a) $S(n) = \dfrac{14n^2 + 3n + 1}{6n^3}$

(b)

n	10^0	10^1	10^2	10^3	10^4
$S(n)$	3	0.2385	0.02338	0.002334	0.000233

(c) 0

19. (a) $S(n) = \dfrac{4n^2 - 3n - 1}{6n^2}$

(b)

n	10^0	10^1	10^2	10^3	10^4
$S(n)$	0	0.615	0.66165	0.666167	0.666617

(c) $\frac{2}{3}$

21. 14.25 square units   **23.** $\frac{337}{128}$ square units

25.

n	4	8	20	50
Approximate Area	18	21	22.8	23.52

27.

n	4	8	20	50
Approximate Area	5.156	5.508	5.786	5.912

29.

n	4	8	20	50
Approximate Area	12.516	11.848	11.481	11.341

31.

n	4	8	20	50	100
Approximate Area	46	43	41.2	40.48	40.24

40 square units

33.

n	4	8	20
Approximate Area	14.25	14.8125	15.13

n	50	100
Approximate Area	15.2528	15.2932

$\frac{46}{3}$ square units

35.

n	4	8	20	50	100
Approximate Area	19	18.5	18.2	18.08	18.04

18 square units

37. 4 square units   **39.** 2 square units   **41.** $\frac{10}{3}$ square units
43. $\frac{51}{4}$ square units
45.

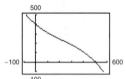

About 105,208.33 ft² ≈ 2.4153 acres
47. Answers will vary.   **49.** True. See Property 2 on page 858.

Review Exercises *(page 868)*

1.

x	2.9	2.99	2.999	3
$f(x)$	16.4	16.94	16.994	?

x	3.001	3.01	3.1
$f(x)$	17.006	17.06	17.6

17; The limit can be reached.
3.

x	2.9	2.99	2.999	3
$f(x)$	0.204	0.2004	0.20004	?

x	3.001	3.01	3.1
$f(x)$	0.19996	0.1996	0.196

0.2; The limit cannot be reached.
5. 2   **7.** 2   **9.** (a) 64   (b) 7   (c) 20   (d) $\frac{4}{5}$
11. 8   **13.** 24   **15.** 1   **17.** $\sqrt{6}$ **19.** $\frac{11}{7}$ **21.** 0
23. 0   **25.** 8   **27.** $\frac{1}{15}$ **29.** $-\frac{1}{3}$ **31.** -1

33. $-\frac{1}{8}$ **35.** $\frac{1}{4}$
37. (a)

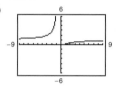

0.1667
(b)

x	2.9	2.99	3	3.01	3.1
$f(x)$	0.1695	0.1669	Error	0.1664	0.1639

0.1667
39. (a)

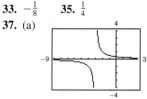

Does not exist
(b)

x	-0.1	-0.01	-0.001	0
$f(x)$	4.85E8	7.2E86	Error	Error

x	0.001	0.01	0.1
$f(x)$	0	1E-87	2.1E-9

Does not exist
41. (a)

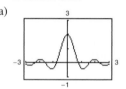

2
(b)

x	-0.1	-0.01	-0.001	0
$f(x)$	1.94709	1.99947	1.999995	Error

x	0.001	0.01	0.1
$f(x)$	1.999995	1.99947	1.94709

2
43. (a)

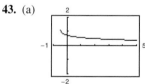

About 0.575
(b)

x	1.1	1.01	1.001	1.0001
$f(x)$	0.5680	0.5764	0.5773	0.5773

About 0.577 $\left(\text{Actual limit is } \dfrac{\sqrt{3}}{3}.\right)$

45. 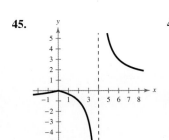   **47.**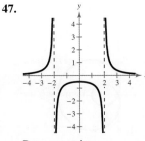

Does not exist   Does not exist

49.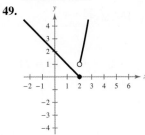

Does not exist

51. -5 **53.** 2

55. 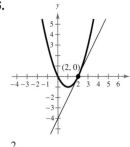   **57.**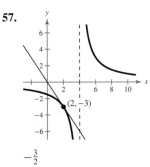

2                      $-\frac{3}{2}$

59. $m = 2x - 4$; (a) -4 (b) 0

61. $m = -\dfrac{4}{(x - 6)^2}$; (a) -4 (b) -1 **63.** $f'(x) = 0$

65. $h'(x) = -\frac{1}{2}$ **67.** $g'(x) = 8x - 1$

69. $f'(t) = \dfrac{1}{2\sqrt{t + 5}}$ **71.** $g'(s) = -\dfrac{4}{(s + 5)^2}$

73. $g'(x) = \dfrac{-1}{2(x + 4)^{3/2}}$

75. (a) 0
   (b) $y = -1$
   (c)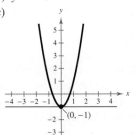

77. 2   **79.** 3   **81.** 0

83. Does not exist; Answers will vary.   **85.** 3

87. $\frac{1}{7}, \frac{1}{4}, \frac{5}{17}, \frac{7}{22}, \frac{1}{3}$ **89.** $-1, \frac{1}{8}, -\frac{1}{27}, \frac{1}{64}, -\frac{1}{125}$

   Limit: $\frac{2}{5}$ Limit: 0

91. $\frac{1}{5}, \frac{1}{2}, \frac{9}{11}, \frac{8}{7}, \frac{25}{17}$

   Limit does not exist;
   Answers will vary.

93. (a) $S(n) = \dfrac{(n + 1)(5n + 4)}{6n^2}$

(b)

n	10^0	10^1	10^2	10^3	10^4
$S(n)$	3	0.99	0.8484	0.8348	0.8335

(c) $\frac{5}{6}$

95. $\frac{27}{4} = 6.75$ square units

97.

n	4	8	20	50
Approximate Area	15.5	14.375	13.74	13.4944

99. 32 square units   **101.** 15 square units

103. $\frac{19}{4}$ square units

105. (a) $y = (-3.376068 \times 10^{-7})x^3 + (3.7529 \times 10^{-4})x^2$
$$- 0.17x + 132$$

(b)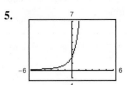

(c) About 87,695.0 ft² (Answers will vary.)

107. True
$$\lim_{x \to c}\left[f(x) + g(x)\right] = \lim_{x \to c} f(x) + \lim_{x \to c} g(x), \text{ if both limits exist.}$$

Chapter Test   *(page 871)*

1. 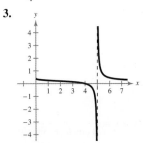   **2.**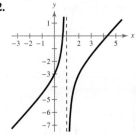

-4 Does not exist

3.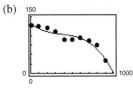

Does not exist

4. 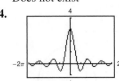   **5.**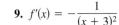

3   2

6. (a) $m = 8x - 1$; 23   (b) $m = 6x^2 + 6$; 12

7. $f'(x) = 0$ **8.** $f'(x) = 4x + 4$ **9.** $f'(x) = -\dfrac{1}{(x + 3)^2}$

10. 0   **11.** -3 **12.** Does not exist; Answers will vary.

13. $0, \frac{3}{4}, \frac{14}{19}, \frac{12}{17}, \frac{36}{53}$

Limit: $\frac{1}{2}$

14. $0, 1, 0, \frac{1}{2}, 0$

Limit: 0

15. $\frac{25}{2}$ square units **16.** 8 square units **17.** $\frac{22}{3}$ square units

18. (a) $y = 8.786x^2 - 6.25x - 0.4$ (b) 81.6 ft/sec

Cumulative Test for Chapters 10–12 *(page 872)*

1. 1.0041 rad; 57.5°

2. Parabola

3. Hyperbola

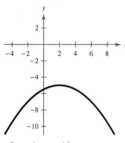

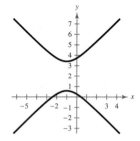

4. $\dfrac{x^2}{1} + \dfrac{(y-2)^2}{4} = 1$

5. $\dfrac{(x')^2}{3/2} - \dfrac{(y')^2}{3} = 1$

6.

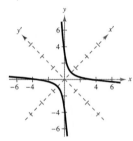

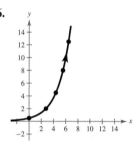

The corresponding rectangular equation is $y = \frac{1}{2}e^{x/2}$.

7.

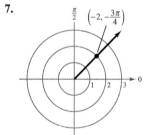

$\left(-2, \dfrac{5\pi}{4}\right), \left(2, -\dfrac{7\pi}{4}\right), \left(2, \dfrac{\pi}{4}\right)$

8. $-8r\cos\theta - 3r\sin\theta + 5 = 0$ or $r = \dfrac{5}{8\cos\theta + 3\sin\theta}$

9. $9x^2 + 20x - 16y^2 + 4 = 0$

10. Circle

11. Ellipse

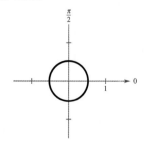

12. Limaçon with inner loop

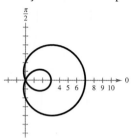

13. $(-6, 1, 3)$ **14.** $(0, -4, 0)$ **15.** $\sqrt{149}$

16. 3, 4, 5

$3^2 + 4^2 = 5^2$

17. $\left(-2, 0, \dfrac{13}{2}\right)$ **18.** $(x-2)^2 + (y-2)^2 + (z-4)^2 = 24$

19.

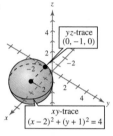

20. $\mathbf{u} \cdot \mathbf{v} = -5$;

$\mathbf{u} \times \mathbf{v} = \langle -10, -20, -15 \rangle$

21. $3\sqrt{6}$ **22.** Neither **23.** Orthogonal **24.** Parallel

25. (a) $x = -1 + 2t$

$y = 2 - 4t$

$z = t$

(b) $\dfrac{x+1}{2} = \dfrac{y-2}{-4} = z$

26. $75x + 50y - 31z = 0$

27.

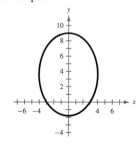

28. $\dfrac{\sqrt{30}}{2} \approx 2.74$ **29.** About 84.26° **30.** $\dfrac{1}{4}$ **31.** -1

32. Does not exist; Answers will vary.

33. $-\dfrac{1}{9}$ **34.** $\dfrac{1}{8}$ **35.** $\dfrac{1}{4}$

36. $m = \dfrac{1}{2}(x+3)^{-1/2}; \dfrac{1}{2}$ **37.** $m = -(x+3)^{-2}; -\dfrac{1}{16}$

38. $m = 2x - 2; 2$ **39.** Does not exist; Answers will vary.

40. -7 **41.** 3 **42.** 0 **43.** 4270 **44.** 672,880

45. 10.5 square units **46.** About 1.566 square units

47. $\dfrac{5}{2}$ square units **48.** $\dfrac{16}{3}$ square units

Problem Solving *(page 875)*

1. (a) i, iv (b) i, iii, iv (c) i, iv

3. $y = 1 + 3\sqrt{x}$

5. (a)

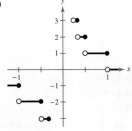

(b) $f\left(\frac{1}{4}\right) = [\![\frac{1}{4}]\!] = 4$

$f(3) = [\![\frac{1}{3}]\!] = 0$

$f(1) = [\![1]\!] = 1$

(c) $\lim\limits_{x \to 1^-} f(x) = 1$

$\lim\limits_{x \to 1^+} f(x) = 0$

$\lim\limits_{x \to (1/2)^-} f(x) = 2$

$\lim\limits_{x \to (1/2)^+} f(x) = 1$

7. (a) Domain: $x \geq -27, \; x \neq 1$

(b)

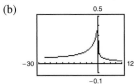

(c) $\frac{1}{14}$

(d) $\frac{1}{12}$

9.

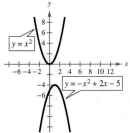

Tangent lines: $y = -2x - 1$ and $y = 4x - 4$

11. (a) $d(m) = \dfrac{|3m + 3|}{\sqrt{m^2 + 1}}$

(b)

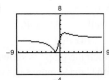

(c) $\lim\limits_{m \to \infty} d(m) = 3$, $\lim\limits_{m \to -\infty} d(m) = 3$. This indicates that the distance between the point and the line approaches 3 as the slope approaches positive or negative infinity.

13. The error was probably due to the calculator being in degree mode rather than radian mode.

15. (a)

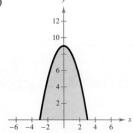

(b) $A = 36$

(c) Base $= 6$, height $= 9$; Area $= \frac{2}{3}bh = 36$

Chapter 13

Section 13.1　*(page 883)*

1. Statistics　**3.** Samples　**5.** Quantitative; qualitative

7. stratified random　**9.** self-selected

11. Population: Students donating at a blood drive

Sample: Students donating who have type O blood

13. (a) "31% (of parents) plan to financially support their children through college graduation and 6% plan to financially support their children through the start of college."

(b) *Sample answer:* More parents plan to support their children through college graduation than through the start of college.

15. Quiz 1:

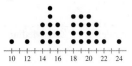

15

Quiz 2:

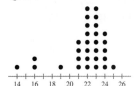

22 and 23

17. (a) 9　(b) $0.19

19.

Stems	Leaves
7	0 5 5 5 7 7 8 8 8
8	1 1 1 1 2 3 4 5 5 5 5 7 8 9 9 9
9	0 2 8
10	0 0

Key: $7|0 = 70$

100, 70

21.

Interval	Tally
[75, 100)	|||
[100, 125)	⠠⠠⠠ ⠠⠠⠠
[125, 150)	⠠⠠⠠ ⠠⠠⠠ ||
[150, 175)	⠠⠠⠠ ||
[175, 200)		
[200, 225)	||

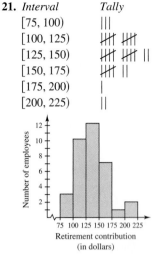

23. *Sample answer:* Use a histogram. It would be easy to group the data into intervals and identify any patterns in the data.

25. True   **27.** Answers will vary.

Section 13.2   *(page 894)*

1. mean   **3.** modes; bimodal   **5.** Quartiles

7. outlier   **9.** standard normal

11. Mean: 9.25; median: 8.5; modes: 7, 12

13. Mean: 10.5; median: 8.5; modes: 7, 12

15. Mean: $8.\overline{1}$; median: 7; mode: 7

17. Mean: $67.14; median: $65.35

19. Mean: about 3.07; median: 3; mode: 3   **21.** Median

23. (a) $\bar{x} = 12$; $\sigma \approx 2.83$ (b) $\bar{x} = 20$; $\sigma \approx 2.83$

25. $\bar{x} = 6$; $v = 10$; $\sigma \approx 3.16$

27. $\bar{x} = 2$; $v = \frac{4}{3}$; $\sigma \approx 1.15$ **29.** $\bar{x} = 54$; $v \approx 76.33$; $\sigma \approx 8.74$

31. About 4.76   **33.** About 107.42   **35.** 1.90

37. Lower quartile: 11; upper quartile: 16

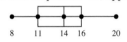

39. Lower quartile: 11.5; upper quartile: 18

41. $\bar{x} = 743.05$, $v \approx 223{,}106.85$, $\sigma \approx 472.34$; 75%

43. 0.16   **45.** 0.68   **47.** 0.68   **49.** 0.975

51. About 0.7257   **53.** About 0.9965

55. About 0.6709   **57.** About 0.16

59. False. Quantitative data can be ordered, so it will always have a median.

61. False. A box-and-whicker plot uses the median, not the mean.

63. The 84th percentile corresponds to a value that is 1 standard deviation about the mean.

65. Mean; Answers will vary.

67. $\bar{x} = 12$ and $|x_1 - 12| = 8$ and all x_i

Section 13.3   *(page 902)*

1. correlation   **3.** squared differences   **5.** 5.96

7. 6.75   **9.** 0.8259   **11.** 0.125   **13.** $y = 1.6x + 7.5$

15. $y = -\frac{100}{29}x + \frac{419}{58}$

17. 0.9923; The model fits the data well.

19. 0.9828; The model fits the data well.

21. $y = 1.301x + 5.099$; 0.01247

23. (a) $y = -1.78x + 127.6$ (b) About 69   (c) About $25

25. $y = -0.75x^2 + 3.5x$ **27.** $y = \frac{10}{11}x^2 - \frac{104}{55}x + \frac{78}{55}$

29. $y = 2x^2 - 3x + 4$ **31.** $y = -\frac{5}{4}x^2 + \frac{9}{20}x + \frac{199}{20}$

33. *Sample answer:* The y-axis scale changes for each unit, which may give the impression that the data points are linear.

35. True. When $|r|$ is close to 0, the data are not linearly correlated.

37. About 67% of elementary school students and 55% of middle school students responded "no."

39. Answers will vary.

Review Exercises   *(page 906)*

1.

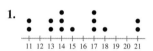

3.

Stems	Leaves
0	2 4 6 6 7 7 9 9
1	5 5 6 7 8 9
2	0 3 9
3	3
4	2 4

Key: $1|5 = 15$

5. Mean: about 9.86; median: 9; mode: 8

7. Mean: 8; median: 7; mode: 6   **9.** Mean or median

11. (a) $\bar{x} = 12$; $\sigma \approx 1.41$ (b) $\bar{x} = 9$; $\sigma \approx 1.41$

13. $\bar{x} = 4$; $v = 4$; $\sigma = 2$ **15.** About 2.72

17. Lower quartile: 13; upper quartile: 21.5

19. 16%   **21.** 0.84   **23.** 0.9735   **25.** 0.9452

27. 0.1974   **29.** (a) 0.975   (b) 0.815

31. 0.9827; The model fits the data well.   **33.** 2.46

35. $y = \frac{11}{2}x + \frac{7}{6}$ **37.** $y = -7x^2 + 14x + 1$

39. *Sample answer:* The angle at which the graph is shown gives the impression that the salesperson C made more sales than the other salespersons.

41. False. Populations are collections of all the outcomes, measurements, counts, or responses that are of interest.

43. True. See Section 13.2, Exercise 65.

45. True. In the graph of $y = 3.29 - 4.17x$, y decreases as x increases.

47. All of the numbers in the data set are equal.

Chapter Test *(page 908)*

1.

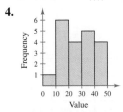

2.

Stems	Leaves
0	7
1	2 4 4 6 6 6
2	0 6 7 7
3	2 2 3 4 7
4	0 3 5 6

Key: $1|2 = 12$

3.

Interval	Tally
[0, 10)	|
[10, 20)	ⵌ |
[20, 30)	||||
[30, 40)	ⵌ
[40, 50)	||||

4.

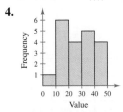

5. Systematic sample

6. Mean: about 6.86; median: 6; mode: 5

7. Mean: 36; median: 36.5; mode: 36

8. $\bar{x} = 3$; $v = 4$; $\sigma = 2$ **9.** $\bar{x} = 4.5$, $v = 4.25$; $\sigma \approx 2.06$

10. Lower quartile: 5; upper quartile: 9

11. Lower quartile: 21.5; upper quartile: 28

12. 0.95   **13.** 0.16   **14.** $y = \frac{19}{26}x + \frac{51}{26}$ **15.** $\frac{1}{26}$

16. 0.9959; The linear model fits the data well.

17. $y = \frac{3}{2}x^2 - \frac{11}{10}x + \frac{17}{10}$

18. (a) About 0.62%

(b) About 83.51%

(c) About 13.36%

Problem Solving *(page 910)*

1.

Stems	Leaves
178	7 7 7 8 8 8 8 8 8 8 9
179	0 1 2 6
180	3
181	2 6 7

Key: $178|7 = 1787$

3. (a)

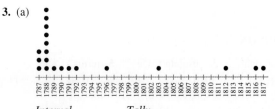
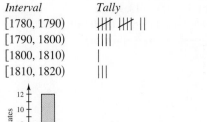

Interval	Tally
[1780, 1790)	ⵌ ⵌ ||
[1790, 1800)	||||
[1800, 1810)	|
[1810, 1820)	|||

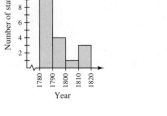

(b)

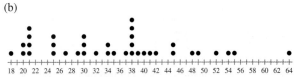

Interval	Tally
[15, 25)	ⵌ ||
[25, 35)	ⵌ ⵌ |
[35, 45)	ⵌ ⵌ |
[45, 55)	ⵌ |
[55, 65)	||

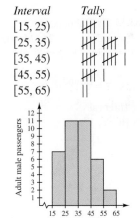

5. Yes. *Sample answer:* 0, 2, 4, 6, 8, and −1, 3, 4, 7, 7

7. *Sample answer:*

(a)

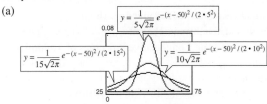

(b) As the standard deviation increases, the graph of the normal curve becomes shorter and wider.

9. [179, 291]; [151, 319]; When $\sigma = 16$, the intervals are [203, 267] and [187, 283].

Appendix A

Appendix A.1 *(page A11)*

1. irrational   **3.** absolute value   **5.** terms
7. (a) 5, 1, 2   (b) 0, 5, 1, 2   (c) $-9, 5, 0, 1, -4, 2, -11$
   (d) $-9, -\frac{7}{2}, 5, \frac{2}{3}, 0, 1, -4, 2, -11$ (e) $\sqrt{2}$
9. (a) 1   (b) 1   (c) $-13, 1, -6$
   (d) $2.01, 0.\overline{6}, -13, 1, -6$ (e) $0.010110111\ldots$
11. (a)   (b)

   (c)   (d)

13.   **15.**

   $-4 > -8$

   $\frac{5}{6} > \frac{2}{3}$

17. (a) $x \le 5$ denotes the set of all real numbers less than or
   equal to 5.
   (b)   (c) Unbounded
19. (a) $-2 < x < 2$ denotes the set of all real numbers greater
   than -2 and less than 2.
   (b)   (c) Bounded
21. (a) $[4, \infty)$ denotes the set of all real numbers greater than or
   equal to 4.
   (b)   (c) Unbounded
23. (a) $[-5, 2)$ denotes the set of all real numbers greater than or
   equal to -5 and less than 2.
   (b)   (c) Bounded

Inequality	Interval
25. $y \ge 0$	$[0, \infty)$
27. $10 \le t \le 22$	$[10, 22]$

29. 10   **31.** 5   **33.** -1 **35.** 25   **37.** -1
39. $|-4| = |4|$ **41.** $-|-6| < |-6|$ **43.** 51   **45.** $\frac{5}{2}$
47. $|x - 5| \le 3$ **49.** 2524.0 billion; 458.5 billion
51. 2450.0 billion; 1087.0 billion
53. $7x$ and 4 are the terms; 7 is the coefficient.
55. $6x^3$ and $-5x$ are the terms; 6 and -5 are the coefficients.
57. $3\sqrt{3}x^2$ and 1 are the terms; $3\sqrt{3}$ is the coefficient.
59. (a) -10 (b) -6 **61.** (a) 2   (b) 6
63. (a) Division by 0 is undefined.   (b) 0
65. Multiplicative Inverse Property
67. Associative and Commutative Properties of Multiplication
69. $\frac{5x}{12}$ **71.** $\frac{x}{4}$
73. False. Zero is nonnegative, but not positive.
75. True. The product of two negative numbers is positive.
77. (a)

n	0.0001	0.01	1	100	10,000
$\dfrac{5}{n}$	50,000	500	5	0.05	0.0005

   (b) (i) The value of $5/n$ approaches infinity as n approaches 0.
      (ii) The value of $5/n$ approaches 0 as n increases without
         bound.

Appendix A.2 *(page A23)*

1. exponent; base   **3.** square root   **5.** like radicals
7. rationalizing   **9.** (a) 625   (b) $\frac{1}{25}$
11. (a) 5184   (b) $-\frac{3}{5}$ **13.** (a) $\frac{16}{3}$ (b) 1
15. -24 **17.** 6   **19.** -48 **21.** (a) $125z^3$ (b) $5x^6$
23. (a) $24y^2$ (b) $-3z^7$ **25.** (a) $\dfrac{5184}{y^7}$ (b) 1
27. (a) 1   (b) $\dfrac{1}{4x^4}$ **29.** (a) $\dfrac{125x^9}{y^{12}}$ (b) $\dfrac{b^5}{a^5}$
31. 1.02504×10^4 **33.** 0.000314
35. $9,460,000,000,000$ km   **37.** (a) 6.8×10^5 (b) 6.0×10^4
39. (a) 3   (b) $\frac{3}{2}$ **41.** (a) 2   (b) $2x$
43. (a) $2\sqrt{5}$ (b) $4\sqrt[3]{2}$ **45.** (a) $6x\sqrt{2x}$ (b) $3y^2\sqrt{6x}$
47. (a) $2x\sqrt[3]{2x^2}$ (b) $\dfrac{5|x|\sqrt{3}}{y^2}$
49. (a) $29|x|\sqrt{5}$ (b) $44\sqrt{3x}$ **51.** $\dfrac{\sqrt{3}}{3}$ **53.** $\dfrac{\sqrt{14}+2}{2}$
55. $\dfrac{2}{3\left(\sqrt{5}-\sqrt{3}\right)}$ **57.** $64^{1/3}$ **59.** $\dfrac{3}{\sqrt[3]{x^2}}$
61. (a) $\frac{1}{8}$ (b) $\frac{27}{8}$ **63.** (a) $\sqrt{3}$ (b) $\sqrt[3]{(x+1)^2}$
65. (a) $2\sqrt[4]{2}$ (b) $\sqrt[8]{2x}$ **67.** (a) $x - 1$ (b) $\dfrac{1}{x-1}$
69.

h	0	1	2	3	4	5	6
t	0	2.93	5.48	7.67	9.53	11.08	12.32

h	7	8	9	10	11	12
t	13.29	14.00	14.50	14.80	14.93	14.96

71. False. When $x = 0$, the expressions are not equal.
73. False. For instance, $(3 + 5)^2 = 8^2 = 64 \ne 34 = 3^2 + 5^2$.

Appendix A.3 *(page A33)*

1. $n; a_n; a_0$ **3.** like terms   **5.** factoring
7. perfect square trinomial
9. (a) $7x$ (b) Degree: 1; Leading coefficient: 7
   (c) Monomial
11. (a) $-\frac{1}{2}x^5 + 14x$ (b) Degree: 5; Leading coefficient: $-\frac{1}{2}$
   (c) Binomial
13. (a) $-4x^5 + 6x^4 + 1$
   (b) Degree: 5; Leading coefficient: -4 (c) Trinomial
15. $-2x - 10$ **17.** $-8.3x^3 + 0.3x^2 - 23$
19. $3x^3 - 6x^2 + 3x$ **21.** $-15z^2 + 5z$ **23.** $6x^2 - 7x - 5$
25. $x^4 + 2x^2 + x + 2$ **27.** $x^2 - 100$ **29.** $4x^2 + 12x + 9$
31. $x^3 + 9x^2 + 27x + 27$ **33.** $x^2 + 2xy + y^2 - 6x - 6y + 9$
35. $m^2 - n^2 - 6m + 9$ **37.** $2x(x^2 - 3)$
39. $(x - 5)(3x + 8)$ **41.** $(5y - 2)(5y + 2)$
43. $(x + 1)(x - 3)$ **45.** $(x - 2)^2$ **47.** $(5z - 3)^2$
49. $(2y - 3)^2$ **51.** $(x + 5)(x^2 - 5x + 25)$
53. $(2t - 1)(4t^2 + 2t + 1)$ **55.** $(x + 2)(x - 1)$
57. $(3x - 2)(x + 4)$ **59.** $(5x + 1)(x + 6)$
61. $-(5y - 2)(y + 2)$ **63.** $(x - 1)(x^2 + 2)$
65. $(2x - 1)(x^2 - 3)$ **67.** $(3x^3 - 2)(x^2 + 2)$
69. $(x + 3)(2x + 3)$ **71.** $(2x + 3)(3x - 5)$
73. $6(x + 3)(x - 3)$ **75.** $x^2(x - 1)$ **77.** $-2x(x + 1)(x - 2)$
79. $(5 - x)(1 + x^2)$ **81.** $-(x - 2)(x + 1)(x - 8)$

83. (a) $\pi h(R + r)(R - r)$ (b) $V = 2\pi\left[\left(\dfrac{R + r}{2}\right)(R - r)\right]h$

85. False. $(4x^2 + 1)(3x + 1) = 12x^3 + 4x^2 + 3x + 1$

87. True. $a^2 - b^2 = (a + b)(a - b)$ **89.** $m + n$

91. $-x^3 + 8x^2 + 2x + 7$

93. Answers will vary. *Sample answer:* $x^2 - 3$

95. $(x^n + y^n)(x^n - y^n)$

Appendix A.4   *(page A42)*

1. domain   **3.** complex   **5.** All real numbers x

7. All real numbers x such that $x \neq 3$

9. All real numbers x such that $x \neq -\frac{2}{3}$

11. All real numbers x such that $x \neq -4, -2$

13. All real numbers x such that $x \geq 7$

15. All real numbers x such that $x > 3$ **17.** $\dfrac{3x}{2}, \; x \neq 0$

19. $-\dfrac{1}{2}, \; x \neq 5$ **21.** $y - 4, \; y \neq -4$ **23.** $\dfrac{3y}{4}, \; y \neq -\dfrac{2}{3}$

25. $\dfrac{x - 1}{x + 3}, \; x \neq -5$ **27.** $-\dfrac{x + 1}{x + 5}, \; x \neq 2$ **29.** $\dfrac{1}{x + 1}, \; x \neq \pm 4$

31. When simplifying fractions, only common factors can be divided out, not terms.

33. $\dfrac{1}{5(x - 2)}, \; x \neq 1$ **35.** $-\dfrac{(x + 2)^2}{6}, \; x \neq \pm 2$

37. $\dfrac{x - y}{x(x + y)^2}, \; x \neq -2y$ **39.** $\dfrac{3}{x + 2}$ **41.** $\dfrac{3x^2 + 3x + 1}{(x + 1)(3x + 2)}$

43. $\dfrac{3 - 2x}{2(x + 2)}$ **45.** $\dfrac{2 - x}{x^2 + 1}, \; x \neq 0$

47. The minus sign should be distributed to each term in the numerator of the second fraction to yield $x + 4 - 3x + 8$.

49. $\dfrac{1}{2}, \; x \neq 2$ **51.** $x(x + 1), \; x \neq -1, 0$ **53.** $\dfrac{2x - 1}{2x}, \; x > 0$

55. $\dfrac{x^2 + (x^2 + 3)^7}{(x^2 + 3)^4}$ **57.** $\dfrac{2x^3 - 2x^2 - 5}{(x - 1)^{1/2}}$ **59.** $\dfrac{3x - 1}{3}, \; x \neq 0$

61. $\dfrac{-1}{x(x + h)}, \; h \neq 0$ **63.** $\dfrac{-1}{(x - 4)(x + h - 4)}, \; h \neq 0$

65. $\dfrac{1}{\sqrt{x + 2} + \sqrt{x}}$ **67.** $\dfrac{1}{\sqrt{t + 3} + \sqrt{3}}, \; t \neq 0$

69. $\dfrac{1}{\sqrt{x + h + 1} + \sqrt{x + 1}}, \; h \neq 0$

71. (a)

t	0	2	4	6	8	10	12
T	75	55.9	48.3	45	43.3	42.3	41.7

t	14	16	18	20	22
T	41.3	41.1	40.9	40.7	40.6

(b) The model is approaching a T-value of 40.

73. $\dfrac{x}{2(2x + 1)}, \; x \neq 0$

75. (a)

Year, t	Online Banking	Mobile Banking
11	79.1	17.9
12	80.9	24.0
13	83.1	29.6
14	86.0	34.8

(b) The values from the models are close to the actual data.

(c) Ratio = $\dfrac{0.0313t^3 - 0.661t^2 - 2.23t + 47}{0.0208t^3 - 0.494t^2 + 2.97t - 70.5}$

(d)

Year	Ratio
2011	0.2267
2012	0.2977
2013	0.3578
2014	0.4061

Answers will vary.

77. $\dfrac{R_1 R_2}{R_2 + R_1}$

79. False. In order for the simplified expression to be equivalent to the original expression, the domain of the simplified expression needs to be restricted. If n is even, $x \neq -1, 1$. If n is odd, $x \neq 1$.

Appendix A.5   *(page A56)*

1. equation   **3.** extraneous   **5.** 4   **7.** -9 **9.** 12

11. No solution   **13.** $-\frac{96}{23}$ **15.** 4   **17.** 3

19. No solution. The variable is divided out.

21. No solution. The solution is extraneous.

23. 5   **25.** $0, -\frac{1}{2}$ **27.** -5 **29.** $-\frac{1}{2}, 3$ **31.** $\pm\frac{3}{4}$

33. $-\frac{20}{3}, -4$ **35.** ± 7 **37.** $\pm 3\sqrt{3} \approx 5.20$ **39.** $-3, 11$

41. $\dfrac{1 \pm 3\sqrt{2}}{2} \approx 2.62, -1.62$ **43.** $4, -8$ **45.** $-2 \pm \sqrt{2}$

47. $-1 \pm \dfrac{\sqrt{2}}{2}$ **49.** $\dfrac{-5 \pm \sqrt{89}}{4}$ **51.** $\dfrac{1}{2}, -1$ **53.** $-\dfrac{5}{3}$

55. $\dfrac{7}{4} \pm \dfrac{\sqrt{41}}{4}$ **57.** $\dfrac{2}{3} \pm \dfrac{\sqrt{7}}{3}$ **59.** $1 \pm \sqrt{3}$ **61.** $2 \pm \dfrac{\sqrt{6}}{2}$

63. $6 \pm \sqrt{11}$ **65.** $1 \pm \sqrt{2}$ **67.** $-10, 6$ **69.** $\frac{1}{2} \pm \sqrt{3}$

71. $\dfrac{3}{4} \pm \dfrac{\sqrt{97}}{4}$ **73.** $0, \pm 3$ **75.** $-2, \pm 2\sqrt{2}$ **77.** 20

79. $-\frac{55}{2}$ **81.** 1   **83.** 4   **85.** 9   **87.** 1   **89.** $8, -3$

91. $-\dfrac{1}{2} - \dfrac{\sqrt{17}}{2}, 3$ **93.** $\sqrt[3]{\dfrac{4.47}{\pi}} \approx 1.12$ in.   **95.** 63.7 in.

97. False. See Example 14 on page A54.

99. True. There is no value that satisfies this equation.

Appendix A.6   *(page A64)*

1. solution set   **3.** double   **5.** $-2 \leq x < 6$; Bounded

7. $-1 \leq x \leq 5$; Bounded   **9.** $x > 11$; Unbounded

11. $x < -2$; Unbounded

13. $x < 3$ **15.** $x < \frac{3}{2}$

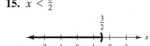

17. $x \geq 12$ **19.** $x > 2$

21. $x \geq 1$ **23.** $x < 5$

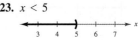

25. $x \geq 4$ **27.** $x \geq 2$

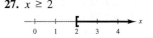

APPENDIX A

29. $x \geq -4$

31. $-1 < x < 3$

33. $-7 < x \leq -\frac{1}{3}$

35. $-\frac{9}{2} < x < \frac{15}{2}$

37. $-5 \leq x < 1$

39. $-\frac{3}{4} < x < -\frac{1}{4}$

41. $10.5 \leq x \leq 13.5$

43. $-5 < x < 5$

45. $x < -2, x > 2$

47. No solution

49. $14 \leq x \leq 26$

51. $x \leq -1, x \geq 8$

53. $x \leq -5, x \geq 11$

55. $4 < x < 5$

57. $x \leq -\frac{29}{2}, x \geq -\frac{11}{2}$

59.

$x > 3$

61.

$x \leq 2$

63.

$x \leq 4$

65.

$-6 \leq x \leq 22$

67.

$x \leq -\frac{27}{2}, x \geq -\frac{1}{2}$

69.

(a) $x \geq 1$　(b) $x \leq \frac{1}{3}$

71.

(a) $-2 \leq x \leq 4$
(b) $x \leq 4$

73.

(a) $1 \leq x \leq 5$
(b) $x \leq -1, x \geq 7$

75. All real numbers less than eight units from 10
77. $|x| \leq 3$　**79.** $|x - 7| \geq 3$　**81.** $|x - 7| \geq 3$
83. $|x + 3| < 4$　**85.** $7.25 \leq P \leq 7.75$　**87.** $r < 0.08$
89. $100 \leq r \leq 170$　**91.** More than 6 units per hour
93. Greater than 10%　**95.** $x \geq 36$　**97.** $87 \leq x \leq 210$
99. (a)

(b) $x \geq 2.9$
(c) $x \geq 2.908$
(d) Answers will vary.

101. (a) $4.34 \leq t \leq 6.56$ (Between 2004 and 2006)
(b) $t > 20.95$ (2020)
103. $13.7 < t < 17.5$　**105.** \$0.28
107. 106.864 in.$^2 \leq$ area ≤ 109.464 in.^{2}
109. True by the Addition of a Constant Property of Inequalities.
111. False. If $-10 \leq x \leq 8$, then $10 \geq -x$ and $-x \geq -8$.
113. *Sample answer:* $x < x + 1$
115. *Sample answer:* $a = 1, b = 5, c = 5$

Appendix A.7　(page A72)

1. numerator
3. The middle term needs to be included.
$(x + 3)^2 = x^2 + 6x + 9$
5. $\sqrt{x + 9}$ cannot be simplified.
7. Divide out common factors, not common terms.
$\dfrac{2x^2 + 1}{5x}$ cannot be simplified.
9. The exponent also applies to the coefficient. $(4x)^2 = 16x^2$
11. To add fractions, first find a common denominator.
$\dfrac{3}{x} + \dfrac{4}{y} = \dfrac{3y + 4x}{xy}$
13. $(x + 2)^{-1/2}(3x + 2)$
15. $2x(2x - 1)^{-1/2}[2x^2(2x - 1)^2 - 1]$　**17.** $5x + 3$
19. $2x^2 + x + 15$　**21.** $1 - 5x$　**23.** $3x - 1$　**25.** $\frac{1}{3}$
27. 2　**29–33.** Answers will vary.　**35.** $7(x + 3)^{-5}$
37. $2x^5(3x + 5)^{-4}$　**39.** $\frac{4}{3}x^{-1} + 4x^{-4} - 7x(2x)^{-1/3}$
41. $\dfrac{x}{3} + 2 + \dfrac{4}{x}$　**43.** $4x^{8/3} - 7x^{5/3} + \dfrac{1}{x^{1/3}}$
45. $\dfrac{3}{x^{1/2}} - 5x^{3/2} - x^{7/2}$　**47.** $\dfrac{-7x^2 - 4x + 9}{(x^2 - 3)^3(x + 1)^4}$
49. $\dfrac{27x^2 - 24x + 2}{(6x + 1)^4}$　**51.** $\dfrac{-1}{(x + 3)^{2/3}(x + 2)^{7/4}}$
53. $\dfrac{4x - 3}{(3x - 1)^{4/3}}$　**55.** $\dfrac{x}{x^2 + 4}$
57. $\dfrac{(3x - 2)^{1/2}(15x^2 - 4x + 45)}{2(x^2 + 5)^{1/2}}$
59. (a) Answers will vary.
(b)

x	-2	-1	$-\frac{1}{2}$	0	1	2	$\frac{5}{2}$
y_1	-8.7	-2.9	-1.1	0	2.9	8.7	12.5
y_2	-8.7	-2.9	-1.1	0	2.9	8.7	12.5

(c) Answers will vary.
61. You cannot move term-by-term from the denominator to the numerator.

Technology

Chapter 1   *(page 3)*

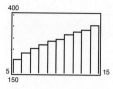

(page 27)

The lines appear perpendicular on the square setting.

(page 40)

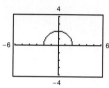   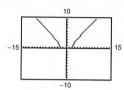

Domain: $[-2, 2]$ Domain: $(-\infty, -2] \cup [2, \infty)$
                         Yes, for -2 and 2.

(page 63)

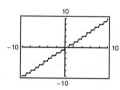   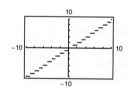

The graph in *dot* mode illustrates that the range is the set of all integers.

Chapter 3   *(page 237)*

$S = 0.00036(2.130)^t$

The exponential regression model has the same coefficient as the model in Example 1. However, the model given in Example 1 contains the natural exponential function.

Chapter 4   *(page 298)*

No graph is visible. $-\pi \le x \le \pi$ and $-0.5 \le y \le 0.5$ displays a good view of the graph.

Chapter 7   *(page 470)*

The point of intersection (7000, 5000) agrees with the solution.

(page 472)

$(2, 0); (0, -1), (2, 1);$ None

Chapter 8   *(page 556)*

$\begin{bmatrix} 1 & 1 \\ 1 & -5 \end{bmatrix}$

Chapter 10   *(page 740)*

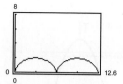

Checkpoints

Chapter 1

Section 1.1

1.

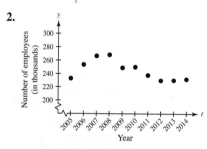

2.

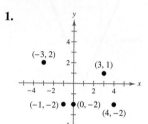

3. $\sqrt{37} \approx 6.08$

4. $d_1 = \sqrt{45}, d_2 = \sqrt{20}, d_3 = \sqrt{65}$
   $\left(\sqrt{45}\right)^2 + \left(\sqrt{20}\right)^2 = \left(\sqrt{65}\right)^2$

5. $(1, -1)$ 6. $\sqrt{709} \approx 27$ yd   7. $4.8 billion

8. $(1, 2), (1, -2), (-1, 0), (-1, -4)$

Section 1.2

1. (a) No   (b) Yes

2. (a)   (b)

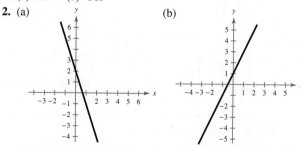

3. (a)

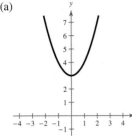

(b)

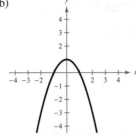

4. *x*-intercepts: $(0, 0), (-5, 0),$
y-intercept: $(0, 0)$

5. *x*-axis symmetry

6.

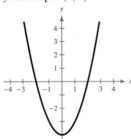

7.

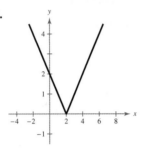

8. $(x + 3)^2 + (y + 5)^2 = 25$ **9.** About 221 lb

Section 1.3

1. (a)

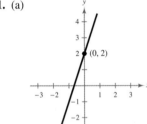

(b)

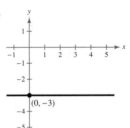

(c)

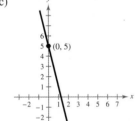

2. (a) 2   **(b)** $-\frac{3}{2}$ **(c)** Undefined   **(d)** 0
3. (a) $y = 2x - 13$ **(b)** $y = -\frac{2}{3}x + \frac{5}{3}$ **(c)** $y = 1$
4. (a) $y = \frac{5}{3}x + \frac{23}{3}$ **(b)** $y = -\frac{3}{5}x - \frac{7}{5}$ **5.** Yes
6. The *y*-intercept, $(0, 1500)$, tells you that the initial value of
a copier at the time it is purchased is $1500. The slope,
$m = -300$, tells you that the value of the copier decreases by
$300 each year after it is purchased.
7. $y = -4125x + 24,750$ **8.** $y = 0.7t + 4.4;$ $9.3 billion

Section 1.4

1. (a) Not a function   **(b)** Function
2. (a) Not a function   **(b)** Function
3. (a) -2 **(b)** -38 **(c)** $-3x^2 + 6x + 7$
4. $f(-2) = 5, f(2) = 1, f(3) = 2$ **5.** ± 4 **6.** $-4, 3$

7. (a) $\{-2, -1, 0, 1, 2\}$
  (b) All real numbers *x* except $x = 3$
  (c) All real numbers *r* such that $r > 0$
  (d) All real numbers *x* such that $x \geq 16$
8. (a) $S(r) = 10\pi r^2$ **(b)** $S(h) = \frac{5}{8}\pi h^2$ **9.** No
10. 2009: 772   2013: 1277
   2010: 841   2014: 1437
   2011: 910   2015: 1597
   2012: 1117
11. $2x + h + 2, h \neq 0$

Section 1.5

1. (a) All real numbers *x* except $x = -3$
  (b) $f(0) = 3; f(3) = -6$ **(c)** $(-\infty, 3]$
2. Function
3. (a) $x = -8, x = \frac{3}{2}$ **(b)** $t = 25$ **(c)** $x = \pm\sqrt{2}$
4.

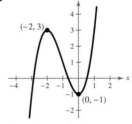

Increasing on $(-\infty, -2)$ and
$(0, \infty)$

Decreasing on $(-2, 0)$

5. $(-0.88, 6.06)$ **6. (a)** -3 **(b)** 0
7. (a) 20 ft/sec   **(b)** $\frac{140}{3}$ ft/sec
8. (a) Neither; No symmetry   **(b)** Even; *y*-axis symmetry
  (c) Odd; Origin symmetry

Section 1.6

1. $f(x) = -\frac{5}{2}x + 1$ **2.** $f\left(-\frac{3}{2}\right) = 0, f(1) = 3, f\left(-\frac{5}{2}\right) = -1$
3.

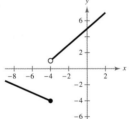

Section 1.7

1. (a)

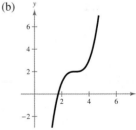

(b)

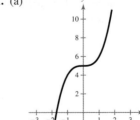

2. $j(x) = -(x + 3)^4$
3. (a) The graph of *g* is a reflection in the *x*-axis of the graph
  of *f*.
  (b) The graph of *h* is a reflection in the *y*-axis of the graph
  of *f*.

4. (a) The graph of g is a vertical stretch of the graph of f.

(b) The graph of h is a vertical shrink of the graph of f.

5. (a) The graph of g is a horizontal shrink of the graph of f.

(b) The graph of h is a horizontal stretch of the graph of f.

Section 1.8

1. $x^2 - x + 1; 3$ **2.** $x^2 + x - 1; 11$ **3.** $x^2 - x^3; -18$

4. $\left(\dfrac{f}{g}\right)(x) = \dfrac{\sqrt{x-3}}{\sqrt{16-x^2}}$; Domain: $[3, 4)$

$\left(\dfrac{g}{f}\right)(x) = \dfrac{\sqrt{16-x^2}}{\sqrt{x-3}}$; Domain: $(3, 4]$

5. (a) $8x^2 + 7$ (b) $16x^2 + 80x + 101$ (c) 9

6. All real numbers x **7.** $f(x) = \frac{1}{5}\sqrt[3]{x}$, $g(x) = 8 - x$

8. (a) $(N \circ T)(t) = 32t^2 + 36t + 204$ (b) About 4.5 h

Section 1.9

1. $f^{-1}(x) = 5x, f(f^{-1}(x)) = \frac{1}{5}(5x) = x, f^{-1}(f(x)) = 5\left(\frac{1}{5}x\right) = x$

2. $g(x) = 7x + 4$

3.

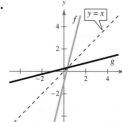

4.

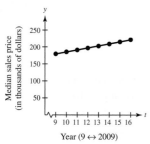

5. (a) Yes   (b) No   **6.** $f^{-1}(x) = \dfrac{5-2x}{x+3}$

7. $f^{-1}(x) = x^3 - 10$

Section 1.10

1.

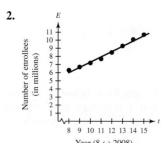

The model is a good fit for the data.

2.

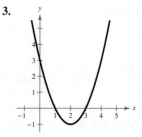

$E = 0.65t + 0.8$

The model is a good fit for the data.

3. $I = 0.075P$ **4.** 576 ft   **5.** 508 units

6. About 1314 ft   **7.** 14,000 joules

Chapter 2

Section 2.1

1. (a)

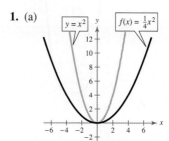

The graph of $f(x) = \frac{1}{4}x^2$ is broader than the graph of $y = x^2$.

(b)

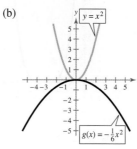

The graph of $g(x) = -\frac{1}{6}x^2$ is a reflection in the x-axis and is broader than the graph of $y = x^2$.

(c)

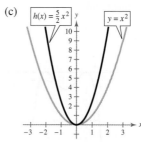

The graph of $h(x) = \frac{5}{2}x^2$ is narrower than the graph of $y = x^2$.

(d)

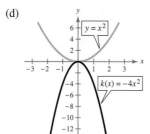

The graph of $k(x) = -4x^2$ is a reflection in the x-axis and is narrower than the graph of $y = x^2$.

2.

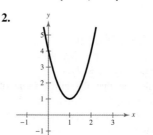

3.

Vertex: $(1, 1)$

Axis: $x = 1$

Vertex: $(2, -1)$

x-intercepts: $(1, 0), (3, 0)$

4. $y = (x + 4)^2 + 11$ **5.** About 39.7 ft

Section 2.2

1. (a) 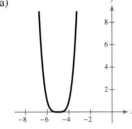   (b)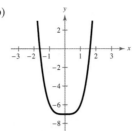

(c) 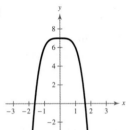   (d)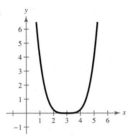

2. (a) Falls to the left, rises to the right
(b) Rises to the left, falls to the right
3. Real zeros: $x = 0, x = 6$; Turning points: 2
4. 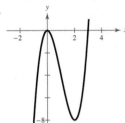   **5.**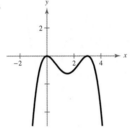

6. $x \approx 3.196$

Section 2.3

1. $(x + 4)(3x + 7)(3x - 7)$ **2.** $x^2 + x + 3, x \neq 3$

3. $2x^3 - x^2 + 3 - \dfrac{6}{3x + 1}$ **4.** $5x^2 - 2x + 3$

5. (a) 1   (b) 396   (c) $-\frac{13}{2}$ (d) -17
6. $f(-3) = 0, f(x) = (x + 3)(x - 5)(x + 2)$

Section 2.4

1. (a) $12 - i$ (b) $-2 + 7i$ (c) i (d) 0
2. (a) $-15 + 10i$ (b) $18 - 6i$ (c) 41   (d) $12 + 16i$
3. (a) 45   (b) 29   **4.** $\frac{3}{5} + \frac{4}{5}i$ **5.** $-2\sqrt{7}$

6. $-\dfrac{7}{8} \pm \dfrac{\sqrt{23}}{8}i$

Section 2.5

1. 4   **2.** No rational zeros   **3.** 5   **4.** $-3, \frac{1}{2}, 2$
5. $-1, \dfrac{-3 - 3\sqrt{17}}{4} \approx -3.8423, \dfrac{-3 + 3\sqrt{17}}{4} \approx 2.3423$
6. $f(x) = x^4 + 45x^2 - 196$
7. $f(x) = -x^4 - x^3 - 2x^2 - 4x + 8$ **8.** $\frac{2}{3}, \pm 4i$
9. $f(x) = (x - 1)(x + 1)(x - 3i)(x + 3i); \pm 1, \pm 3i$
10. No positive real zeros, three or no negative real zeros
11. $\frac{1}{2}$ **12.** 7 in. × 7 in. × 9 in.

Section 2.6

1. Domain: all real numbers x such that $x \neq 1$
$f(x)$ decreases without bound as x approaches 1 from the left and increases without bound as x approaches 1 from the right.
2. Vertical asymptote: $x = -1$
Horizontal asymptote: $y = 3$
3. 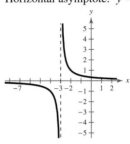   **4.**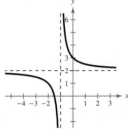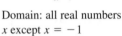

Domain: all real numbers x except $x = -3$

Domain: all real numbers x except $x = -1$

5. 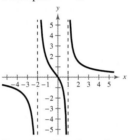   **6.**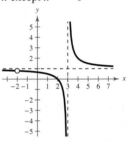

Domain: all real numbers x except $x = -2, 1$

Domain: all real numbers x except $x = -2, 3$

7.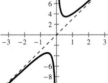

Domain: all real numbers x except $x = 0$
8. (a) \$63.75 million; \$208.64 million; \$1020 million
(b) No. The function is undefined at $p = 100$.
9. 12.9 in. by 6.5 in.

Section 2.7

1. $(-4, 5)$

2. (a) $\left(-\frac{5}{2}, 1\right)$
(b) 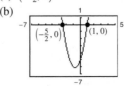   The graph is below the x-axis when x is greater than $-\frac{5}{2}$ and less than 1. So, the solution set is $\left(-\frac{5}{2}, 1\right)$.

3. $\left(-2, \frac{1}{3}\right) \cup (2, \infty)$

4. (a) The solution set is empty.
  (b) The solution set consists of the single real number $\{-2\}$.
  (c) The solution set consists of all real numbers except $x = 3$.
  (d) The solution set is all real numbers.

5. (a) $\left(-\infty, \frac{11}{4}\right] \cup (3, \infty)$ (b) $(-\infty, -17) \cup (6, \infty)$

6. $180{,}000 \le x \le 300{,}000$ **7.** $(-\infty, 2] \cup [5, \infty)$

Chapter 3

Section 3.1

1. 0.0528248

2.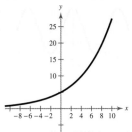

3.

4. (a) 2   (b) 3

5. (a) Shift the graph of f two units to the right.
  (b) Shift the graph of f three units up.
  (c) Reflect the graph of f in the y-axis and shift three units down.

6. (a) 1.3498588   (b) 0.3011942   (c) 492.7490411

7.

8. (a) \$7927.75   (b) \$7935.08   (c) \$7938.78

9. About 9.970 lb; about 0.275 lb

Section 3.2

1. (a) 0   (b) -3 (c) 3

2. (a) 2.4393327   (b) Error or complex number
  (c) -0.3010300

3. (a) 1   (b) 3   (c) 0   **4.** ± 3

5.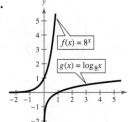

6.

Vertical asymptote: $x = 0$

7. (a)   (b)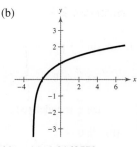

8. (a) -4.6051702 (b) 1.3862944   (c) 1.3169579
  (d) Error or complex number

9. (a) $\frac{1}{3}$ (b) 0   (c) $\frac{3}{4}$ (d) 7   **10.** $(-3, \infty)$

11. (a) 70.84   (b) 61.18   (c) 59.61

Section 3.3

1. 3.5850   **2.** 3.5850

3. (a) $\log 3 + 2 \log 5$ (b) $2 \log 3 - 3 \log 5$ **4.** 4

5. $\log_3 4 + 2 \log_3 x - \dfrac{1}{2} \log_3 y$ **6.** $\log \dfrac{(x + 3)^2}{(x - 2)^4}$

7. $\ln y = \frac{2}{3} \ln x$

Section 3.4

1. (a) 9   (b) 216   (c) $\ln 5$ (d) $-\frac{1}{2}$

2. (a) $4, -2$ (b) 1.723   **3.** 3.401   **4.** -4.778

5. 1.099, 1.386   **6.** (a) $e^{2/3}$ (b) 7   (c) 12

7. 0.513   **8.** $\frac{32}{3}$ **9.** 10

10. About 13.2 years; It takes longer for your money to double at a lower interest rate.

11. 2010

Section 3.5

1. 2018   **2.** 400 bacteria   **3.** About 38,000 yr

4.

495

5. About 7 days   **6.** (a) 1,000,000   (b) About 80,000,000

Chapter 4

Section 4.1

1. (a) $\dfrac{\pi}{4}, -\dfrac{7\pi}{4}$ (b) $\dfrac{5\pi}{3}, -\dfrac{7\pi}{3}$

2. (a) Complement: $\dfrac{\pi}{3}$; Supplement: $\dfrac{5\pi}{6}$

  (b) Complement: none; Supplement: $\dfrac{\pi}{6}$

3. (a) $\dfrac{\pi}{3}$ (b) $\dfrac{16\pi}{9}$ **4.** (a) $30°$ (b) $300°$

5. 24π in. ≈ 75.40 in.   **6.** About 0.84 cm/sec

7. (a) 4800π rad/min   (b) About 60.319 in./min

8. About 1117 ft^2

Section 4.2

1. (a) $\sin \dfrac{\pi}{2} = 1$ $\csc \dfrac{\pi}{2} = 1$

       $\cos \dfrac{\pi}{2} = 0$ $\sec \dfrac{\pi}{2}$ is undefined.

       $\tan \dfrac{\pi}{2}$ is undefined. $\cot \dfrac{\pi}{2} = 0$

  (b) $\sin 0 = 0$ $\csc 0$ is undefined.
       $\cos 0 = 1$ $\sec 0 = 1$
       $\tan 0 = 0$ $\cot 0$ is undefined.

  (c) $\sin\left(-\dfrac{5\pi}{6}\right) = -\dfrac{1}{2}$ $\csc\left(-\dfrac{5\pi}{6}\right) = -2$

       $\cos\left(-\dfrac{5\pi}{6}\right) = -\dfrac{\sqrt{3}}{2}$ $\sec\left(-\dfrac{5\pi}{6}\right) = -\dfrac{2\sqrt{3}}{3}$

       $\tan\left(-\dfrac{5\pi}{6}\right) = \dfrac{\sqrt{3}}{3}$ $\cot\left(-\dfrac{5\pi}{6}\right) = \sqrt{3}$

  (d) $\sin\left(-\dfrac{3\pi}{4}\right) = -\dfrac{\sqrt{2}}{2}$ $\csc\left(-\dfrac{3\pi}{4}\right) = -\sqrt{2}$

       $\cos\left(-\dfrac{3\pi}{4}\right) = -\dfrac{\sqrt{2}}{2}$ $\sec\left(-\dfrac{3\pi}{4}\right) = -\sqrt{2}$

       $\tan\left(-\dfrac{3\pi}{4}\right) = 1$ $\cot\left(-\dfrac{3\pi}{4}\right) = 1$

2. (a) 0   (b) $-\dfrac{\sqrt{3}}{1}$ (c) 0.3

3. (a) 0.78183148   (b) 1.0997502

Section 4.3

1. $\sin \theta = \dfrac{1}{2}$ $\csc \theta = 2$

     $\cos \theta = \dfrac{\sqrt{3}}{2}$ $\sec \theta = \dfrac{2\sqrt{3}}{3}$

     $\tan \theta = \dfrac{\sqrt{3}}{3}$ $\cot \theta = \sqrt{3}$

2. $\cot 45° = 1$, $\sec 45° = \sqrt{2}$, $\csc 45° = \sqrt{2}$

3. $\tan 60° = \sqrt{3}$, $\tan 30° = \dfrac{\sqrt{3}}{3}$ **4.** 1.7650691

5. (a) 0.28   (b) 0.2917   **6.** (a) $\frac{1}{2}$ (b) $\sqrt{5}$

7. Answers will vary.   **8.** About 40 ft

9. 60°   **10.** About 17.6 ft; about 17.2 ft

Section 4.4

1. $\sin \theta = \dfrac{3\sqrt{13}}{13}$, $\cos \theta = -\dfrac{2\sqrt{13}}{13}$, $\tan \theta = -\dfrac{3}{2}$

2. $\cos \theta = -\frac{3}{5}$, $\tan \theta = -\frac{4}{3}$

3. $-1; 0$ **4.** (a) 33°   (b) $\dfrac{4\pi}{9}$ (c) $\dfrac{\pi}{5}$

5. (a) $-\dfrac{\sqrt{2}}{2}$ (b) $-\dfrac{1}{2}$ (c) $-\dfrac{\sqrt{3}}{3}$ **6.** (a) $-\dfrac{3}{5}$ (b) $\dfrac{4}{3}$

7. (a) -1.8040478 (b) -1.0428352 (c) 0.8090170

Section 4.5

1.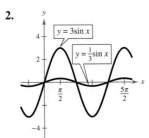

2.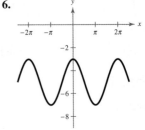

3.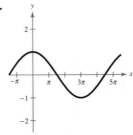

4. Period: 2π
     Amplitude: 2
     $\left[\dfrac{\pi}{2}, \dfrac{5\pi}{2}\right]$ corresponds to one cycle.
     Key points: $\left(\dfrac{\pi}{2}, 2\right)$, $(\pi, 0)$, $\left(\dfrac{3\pi}{2}, -2\right)$, $(2\pi, 0)$, $\left(\dfrac{5\pi}{2}, 2\right)$

5.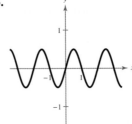

6.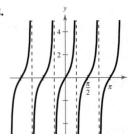

7. $y = 5.6\sin(0.524t - 0.524) + 5.7$

Section 4.6

1.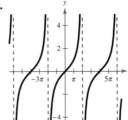

2.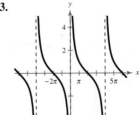

3.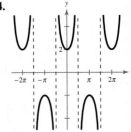

4.

5.

6.

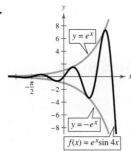

Section 4.7

1. (a) $\dfrac{\pi}{2}$
   (b) Not possible

2.

3. π

4. (a) 1.3670516   (b) Not possible   (c) 1.9273001

5. (a) -14 (b) $-\dfrac{\pi}{4}$ (c) 0.54   6. $\dfrac{4}{5}$ 7. $\sqrt{x^2 + 1}$

Section 4.8

1. $a \approx 5.46$, $c \approx 15.96$, $B = 70°$ 2. About 15.8 ft
3. About 15.1 ft   4. 3.58°
5. Bearing: N 53° W, Distance: about 3.2 nmi
6. $d = 6 \sin \dfrac{2\pi}{3} t$
7. (a) 4   (b) 3 cycles per unit of time   (c) 4   (d) $\frac{1}{12}$

Chapter 5
Section 5.1

1. $\sin x = -\dfrac{\sqrt{10}}{10}$, $\cos x = -\dfrac{3\sqrt{10}}{10}$, $\tan x = \dfrac{1}{3}$, $\csc x = -\sqrt{10}$

   $\sec x = -\dfrac{\sqrt{10}}{3}$, $\cot x = 3$

2. $-\sin x$
3. (a) $(1 + \cos \theta)(1 - \cos \theta)$ (b) $(2 \csc \theta - 3)(\csc \theta - 2)$
4. $(\tan x + 1)(\tan x + 2)$ 5. $\sin x$ 6. $2 \sec^2 \theta$
7. $1 + \sin \theta$ 8. $3 \cos \theta$ 9. $\ln|\tan x|$

Section 5.2

1–7. Answers will vary.

Section 5.3

1. $\dfrac{\pi}{4} + 2n\pi$, $\dfrac{3\pi}{4} + 2n\pi$

2. $\dfrac{\pi}{3} + 2n\pi$, $\dfrac{2\pi}{3} + 2n\pi$, $\dfrac{4\pi}{3} + 2n\pi$, $\dfrac{5\pi}{3} + 2n\pi$ 3. $n\pi$

4. $\dfrac{\pi}{6}, \dfrac{\pi}{2}, \dfrac{5\pi}{6}$ 5. $\dfrac{\pi}{4} + n\pi$, $\dfrac{3\pi}{4} + n\pi$ 6. $0, \dfrac{3\pi}{2}$

7. $\dfrac{\pi}{6} + n\pi$, $\dfrac{\pi}{3} + n\pi$ 8. $\dfrac{\pi}{2} + 2n\pi$

9. $\arctan \frac{3}{4} + n\pi$, $\arctan(-2) + n\pi$ 10. 0.4271, 2.7145

11. $\theta \approx 54.7356°$

Section 5.4

1. $\dfrac{\sqrt{2} + \sqrt{6}}{4}$ 2. $\dfrac{\sqrt{2} + \sqrt{6}}{4}$ 3. $-\dfrac{63}{65}$

4. $\dfrac{x + \sqrt{1 - x^2}}{\sqrt{2}}$ 5. Answers will vary.

6. (a) $-\cos \theta$ (b) $\dfrac{\tan \theta - 1}{\tan \theta + 1}$ 7. $\dfrac{\pi}{3}, \dfrac{5\pi}{3}$

8. Answers will vary.

Section 5.5

1. $\dfrac{\pi}{3} + 2n\pi$, $\pi + 2n\pi$, $\dfrac{5\pi}{3} + 2n\pi$

2. $\sin 2\theta = \frac{24}{25}$, $\cos 2\theta = \frac{7}{25}$, $\tan 2\theta = \frac{24}{7}$

3. $4 \cos^3 x - 3 \cos x$ 4. $\dfrac{\cos 4x - 4 \cos 2x + 3}{\cos 4x + 4 \cos 2x + 3}$

5. $\dfrac{-\sqrt{2 - \sqrt{3}}}{2}$ 6. $\dfrac{\pi}{3}, \pi, \dfrac{5\pi}{3}$ 7. $\dfrac{1}{2} \sin 8x + \dfrac{1}{2} \sin 2x$

8. $\dfrac{\sqrt{2}}{2}$ 9. $\dfrac{\pi}{6} + \dfrac{2n\pi}{3}, \dfrac{\pi}{2} + \dfrac{2n\pi}{3}, n\pi$ 10. 45°

Chapter 6
Section 6.1

1. $C = 105°$, $b \approx 45.25$ cm, $c \approx 61.82$ cm   2. 13.40 m
3. $B \approx 12.39°$, $C \approx 136.61°$, $c \approx 16.01$ in.
4. $\sin B \approx 3.0311 > 1$
5. Two solutions:
   $B \approx 70.4°$, $C \approx 51.6°$, $c \approx 4.16$ ft
   $B \approx 109.6°$, $C \approx 12.4°$, $c \approx 1.14$ ft
6. About 213 yd²   7. About 1856.59 m

Section 6.2

1. $A \approx 26.38°$, $B \approx 36.34°$, $C \approx 117.28°$
2. $B \approx 59.66°$, $C \approx 40.34°$, $a \approx 18.26$ m   3. About 202 ft
4. N 15.37° E   5. About 19.90 in.²

Section 6.3

1. $\|\overrightarrow{PQ}\| = \|\overrightarrow{RS}\| = \sqrt{10}$, $\text{slope}_{\overrightarrow{PQ}} = \text{slope}_{\overrightarrow{RS}} = \frac{1}{3}$
   $\overrightarrow{PQ}$ and $\overrightarrow{RS}$ have the same magnitude and direction, so they are equivalent.
2. $\mathbf{v} = \langle -5, 6 \rangle$, $\|\mathbf{v}\| = \sqrt{61}$
3. (a) $\langle 4, 6 \rangle$ (b) $\langle -2, 2 \rangle$ (c) $\langle -7, 2 \rangle$
4. (a) $3\sqrt{17}$ (b) $2\sqrt{13}$ (c) $5\sqrt{13}$
5. $\left\langle \dfrac{6}{\sqrt{37}}, -\dfrac{1}{\sqrt{37}} \right\rangle$ 6. $-6\mathbf{i} - 3\mathbf{j}$ 7. $11\mathbf{i} - 14\mathbf{j}$
8. (a) 135°   (b) About 209.74°   9. $\langle -96.59, -25.88 \rangle$
10. About 2405 lb   11. $\|\mathbf{v}\| \approx 451.8$ mi/h, $\theta \approx 305.1°$

Section 6.4

1. (a) -6 (b) 37   (c) 0
2. (a) $\langle -36, 108 \rangle$ (b) 43   (c) $2\sqrt{10}$ 3. 45°
4. Yes   5. $\frac{1}{17}\langle 64, 16 \rangle$; $\frac{1}{17}\langle -13, 52 \rangle$ 6. About 38.8 lb
7. About 1212 ft-lb

Section 6.5

1.

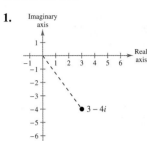

5

2. $4 + 3i$ **3.** $1 - 5i$

4.

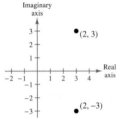

$2 + 3i$

5. $\sqrt{82} \approx 9.06$ units   **6.** $\left(\frac{7}{2}, -2\right)$

Section 6.6

1. $6\sqrt{2}\left(\cos \frac{7\pi}{4} + i \sin \frac{7\pi}{4}\right)$ **2.** $-4 + 4\sqrt{3}i$ **3.** 10

4. $12i$ **5.** $\frac{\sqrt{3}}{2} + \frac{1}{2}i$ **6.** -8 **7.** -4 **8.** $1, i, -1, -i$

9. $\sqrt[3]{3} + \sqrt[3]{3}i, -1.9707 + 0.5279i, 0.5279 - 1.9701i$

Chapter 7

Section 7.1

1. $(3, 3)$ **2.** \$6250 at 6.5%, \$18,750 at 8.5%

3. $(-3, -1), (2, 9)$ **4.** No solution   **5.** $(1, 3)$

6. About 5172 pairs   **7.** 5 weeks

Section 7.2

1. $\left(\frac{3}{4}, \frac{5}{2}\right)$ **2.** $(4, 3)$ **3.** $(3, -1)$ **4.** $(9, 12)$

5.

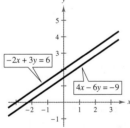

No solution; inconsistent

6. No solution   **7.** Infinitely many solutions: $(a, 4a + 3)$

8. About 471.18 mi/h; about 16.63 mi/h   **9.** $(1,500,000, 537)$

Section 7.3

1. $(4, -3, 3)$ **2.** $(1, 1)$ **3.** $(1, 2, 3)$ **4.** No solution

5. Infinitely many solutions: $(-23a + 22, 15a - 13, a)$

6. Infinitely many solutions: $\left(\frac{1}{4}a, \frac{17}{4}a - 3, a\right)$

7. $s = -16t^2 + 20t + 100$; The object was thrown upward at a velocity of 20 feet per second from a height of 100 feet.

8. $y = \frac{1}{3}x^2 - 2x$

Section 7.4

1. $-\dfrac{3}{2x + 1} + \dfrac{2}{x - 1}$ **2.** $x - \dfrac{3}{x} + \dfrac{4}{x^2} + \dfrac{3}{x + 1}$

3. $-\dfrac{5}{x} + \dfrac{7x}{x^2 + 1}$ **4.** $\dfrac{x + 3}{x^2 + 4} - \dfrac{6x + 5}{(x^2 + 4)^2}$

5. $\dfrac{1}{x} - \dfrac{2}{x^2} + \dfrac{2 - x}{x^2 + 2} + \dfrac{4 - 2x}{(x^2 + 2)^2}$

Section 7.5

1.

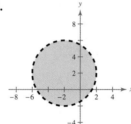

2.

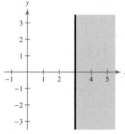

3.

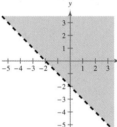

4.

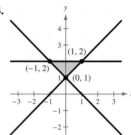

5.

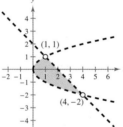

6.

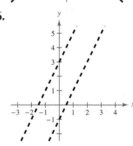

No solution

7.

8. Consumer surplus: \$22,500,000
Producer surplus: \$33,750,000

9. $\begin{cases} 8x + 2y \geq 16 & \text{Nutrient A} \\ x + y \geq 5 & \text{Nutrient B} \\ 2x + 7y \geq 20 & \text{Nutrient C} \\ x \geq 0 \\ y \geq 0 \end{cases}$

Section 7.6

1. Maximum at $(0, 6)$: 30   2. Minimum at $(0, 0)$: 0
3. Maximum at $(60, 20)$: 880   4. Minimum at $(10, 0)$: 30
5. $2925; 1050 boxes of chocolate-covered creams, 150 boxes of chocolate-covered nuts
6. 3 bottles of brand X, 2 bottles of brand Y

Chapter 8

Section 8.1

1. 2×3 2. $\begin{bmatrix} 1 & 1 & 1 & \vdots & 2 \\ 2 & -1 & 3 & \vdots & -1 \\ -1 & 2 & -1 & \vdots & 4 \end{bmatrix}$; 3×4

3. Add -3 times Row 1 to Row 2.
4. Answers will vary. Solution: $(-1, 0, 1)$
5. Reduced row-echelon form   6. $(4, -2, 1)$
7. No solution   8. $(7, 4, -3)$ 9. $(3a + 8, 2a - 5, a)$

Section 8.2

1. $a_{11} = 6, a_{12} = 3, a_{21} = -2, a_{22} = 4$

2. (a) $\begin{bmatrix} 6 & -2 \\ 2 & 3 \end{bmatrix}$ (b) $\begin{bmatrix} 0 & 0 \\ 0 & 0 \\ 0 & 0 \end{bmatrix}$ (c) Not possible (d) $\begin{bmatrix} 0 \\ 0 \\ 2 \end{bmatrix}$

3. (a) $\begin{bmatrix} 4 & -5 \\ 1 & 1 \\ -4 & 1 \end{bmatrix}$ (b) $\begin{bmatrix} 12 & -3 \\ 0 & 12 \\ -9 & 24 \end{bmatrix}$ (c) $\begin{bmatrix} 12 & -11 \\ 2 & 6 \\ -11 & 10 \end{bmatrix}$

4. $\begin{bmatrix} 1 & 2 \\ 10 & -4 \end{bmatrix}$ 5. $\begin{bmatrix} -6 & 6 \\ -10 & 6 \end{bmatrix}$ 6. $\begin{bmatrix} 5 & 0 \\ -1 & 4 \end{bmatrix}$

7. $\begin{bmatrix} -1 & 30 \\ 2 & -4 \\ 1 & 12 \end{bmatrix}$ 8. $\begin{bmatrix} -3 & -22 \\ 3 & 10 \\ -5 & 10 \end{bmatrix}$

9. (a) $\begin{bmatrix} 3 & -1 \\ -9 & 3 \end{bmatrix}$ (b) $[6]$ (c) Not possible

10. $\begin{bmatrix} 7 & 0 \\ 0 & 7 \end{bmatrix}$ 11. (a) $\langle -5, 1 \rangle$ (b) $\langle 17, 23 \rangle$

12. $\langle -3, 1 \rangle$; Reflection in the y-axis

13. (a) $\begin{bmatrix} -2 & -3 \\ 6 & 1 \end{bmatrix}\begin{bmatrix} x_1 \\ x_2 \end{bmatrix} = \begin{bmatrix} -4 \\ -36 \end{bmatrix}$ (b) $\begin{bmatrix} -7 \\ 6 \end{bmatrix}$

14. Total cost for women's team: $2310
    Total cost for men's team: $2719

Section 8.3

1. $AB = I$ and $BA = I$ 2. $\begin{bmatrix} 3 & 2 \\ 1 & 1 \end{bmatrix}$

3. $\begin{bmatrix} -4 & -2 & 5 \\ -2 & -1 & 2 \\ -1 & 0 & 1 \end{bmatrix}$ 4. $\begin{bmatrix} \frac{4}{23} & \frac{1}{23} \\ -\frac{3}{23} & \frac{5}{23} \end{bmatrix}$ 5. $(2, -1, -2)$

Section 8.4

1. (a) -7 (b) 10 (c) 0
2. $M_{11} = -9, M_{12} = -10, M_{13} = 2, M_{21} = 5, M_{22} = -2,$
   $M_{23} = -3, M_{31} = 13, M_{32} = 5, M_{33} = -1$
   $C_{11} = -9, C_{12} = 10, C_{13} = 2, C_{21} = -5, C_{22} = -2,$
   $C_{23} = 3, C_{31} = 13, C_{32} = -5, C_{33} = -1$

3. -31 4. 704

Section 8.5

1. $(3, -2)$ 2. $(2, -3, 1)$ 3. 9 square units
4. Collinear   5. $x - y + 2 = 0$
6. $(0, 0), (2, 0), (0, 4), (2, 4)$ 7. 10 square units
8. $[15 \ \ 23 \ \ 12][19 \ \ 0 \ \ 1][18 \ \ 5 \ \ 0][14 \ \ 15 \ \ 3]$
   $[20 \ \ 21 \ \ 18][14 \ \ 1 \ \ 12]$
9. $110 \ -39 \ -59 \ 25 \ -21 \ -3 \ 23 \ -18 \ -5 \ 47$
   $-20 \ -24 \ 149 \ -56 \ -75 \ 87 \ -38 \ -37$
10. OWLS ARE NOCTURNAL

Chapter 9

Section 9.1

1. $3, 5, 7, 9$ 2. $1, \frac{3}{2}, \frac{1}{3}, \frac{3}{4}$
3. (a) $a_n = 4n - 3$ (b) $a_n = (-1)^{n+1}(2n)$
4. $6, 7, 8, 9, 10$ 5. $1, 3, 4, 7, 11$ 6. $2, 4, 5, \frac{14}{3}, \frac{41}{12}$
7. $4(n + 1)$ 8. 44   9. (a) 0.5555 (b) $\frac{5}{9}$
10. (a) $1000, $1002.50, $1005.01 (b) $1127.33

Section 9.2

1. $2, 5, 8, 11; d = 3$ 2. $a_n = 5n - 6$
3. $-3, 1, 5, 9, 13, 17, 21, 25, 29, 33, 37$ 4. 79   5. 217
6. (a) 630 (b) $N(1 + 2N)$ 7. 43,560   8. 1470
9. $2,500,000

Section 9.3

1. $-12, 24, -48, 96; r = -2$
2. $2, 8, 32, 128, 512$

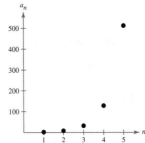

3. 104.02   4. $a_n = 4(5)^{n-1}$; 195,312,500   5. $\frac{2187}{32}$
6. 2.667   7. (a) 10 (b) 6.25   8. $3500.85

Section 9.4

1. (a) $\dfrac{6}{(k + 1)(k + 4)}$ (b) $k + 3 \leq 3k^2$

   (c) $2^{4k+2} + 1 > 5k + 5$
2–5. Proofs   6. $S_k = k(2k + 1)$; Proof
7. (a) 210 (b) 785   8. $a_n = n^2 - n - 2$

Section 9.5

1. (a) 462 (b) 36 (c) 1 (d) 1
2. (a) 21 (b) 21 (c) 14 (d) 14
3. $1, 9, 36, 84, 126, 126, 84, 36, 9, 1$
4. $x^4 + 8x^3 + 24x^2 + 32x + 16$

5. (a) $y^4 - 8y^3 + 24y^2 - 32y + 16$

(b) $32x^5 - 80x^4y + 80x^3y^2 - 40x^2y^3 + 10xy^4 - y^5$

6. $125 + 75y^2 + 15y^4 + y^6$

7. (a) $1120a^4b^4$ (b) $-3{,}421{,}440$

Section 9.6

1. 3 ways **2.** 2 ways **3.** 27,000 combinations

4. 2,600,000 numbers **5.** 24 permutations **6.** 20 ways

7. 1260 ways **8.** 21 ways

9. 22,100 three-card poker hands **10.** 1,051,050 teams

Section 9.7

1. $\{HH1, HH2, HH3, HH4, HH5, HH6, HT1, HT2, HT3, HT4,$
$HT5, HT6, TH1, TH2, TH3, TH4, TH5, TH6, TT1, TT2, TT3$
$TT4, TT5, TT6\}$

2. (a) $\frac{1}{8}$ (b) $\frac{1}{4}$ **3.** $\frac{1}{9}$ **4.** Answers will vary.

5. $\frac{320}{2311} \approx 0.138$ **6.** $\frac{1}{962{,}598}$ **7.** $\frac{4}{13} \approx 0.308$

8. $\frac{66}{529} \approx 0.125$ **9.** $\frac{121}{900} \approx 0.134$ **10.** About 0.116

11. 0.452

Chapter 10

Section 10.1

1. (a) $0.6747 \approx 38.7°$ (b) $\frac{3\pi}{4} \approx 135°$ **2.** $1.4841 \approx 85.0°$

3. $\frac{12}{\sqrt{10}} \approx 3.79$ units **4.** $\frac{3}{\sqrt{34}} \approx 0.51$ unit

5. (a) $\frac{8}{\sqrt{5}} \approx 3.58$ units (b) 12 square units

Section 10.2

1. $x^2 = \frac{3}{2}y$ **2.** $(y + 3)^2 = 8(x - 2)$ **3.** $(2, -3)$

4. $y = 6x - 3$

Section 10.3

1. $\frac{(x - 2)^2}{7} + \frac{(y - 3)^2}{16} = 1$

2.

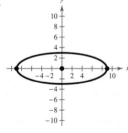

Center: $(0, 0)$

Vertices: $(-9, 0), (9, 0)$

3. Center: $(-2, 1)$

Vertices: $(-2, -2), (-2, 4)$

Foci: $\left(-2, 1 \pm \sqrt{5}\right)$

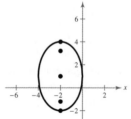

4. Center: $(-1, 3)$

Vertices: $(-4, 3), (2, 3)$

Foci: $(-3, 3), (1, 3)$

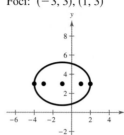

5. Aphelion:

4.080 astronomical units

Perihelion:

0.340 astronomical unit

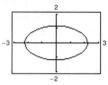

Section 10.4

1. $\frac{(y + 1)^2}{9} - \frac{(x - 2)^2}{7} = 1$

2.

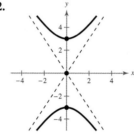

3.

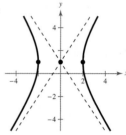

4. $\frac{(x - 6)^2}{9} - \frac{(y - 2)^2}{4} = 1$

5. The explosion occurred somewhere on the right branch of the hyperbola

$$\frac{x^2}{4{,}840{,}000} - \frac{y^2}{2{,}129{,}600} = 1.$$

6. (a) Circle (b) Hyperbola (c) Ellipse (d) Parabola

Section 10.5

1. $\frac{(y')^2}{12} - \frac{(x')^2}{12} = 1$ **2.** $\frac{(x')^2}{1} - \frac{(y')^2}{9} = 1$

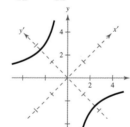 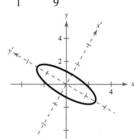

3. $(x')^2 = -2(y' - 3)$

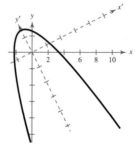

4. Parabola; $y = \dfrac{2x \pm \sqrt{-6x - 10}}{4}$

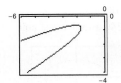

Section 10.6

1.

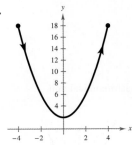

The curve starts at $(-4, 18)$ and ends at $(4, 18)$.

2.

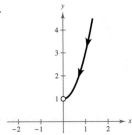

3. (a)

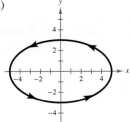

(b)

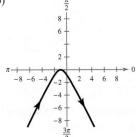

4. (a) $x = t,\ y = t^2 + 2$ (b) $x = 2 - t,\ y = t^2 - 4t + 6$

5. $x = a(\theta + \sin \theta),\ y = a(1 + \cos \theta)$

Section 10.7

1. (a)

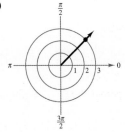

(b)

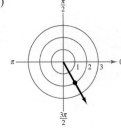

(c)

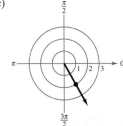

2.

$\left(-1, -\dfrac{5\pi}{4}\right), \left(1, \dfrac{7\pi}{4}\right), \left(1, -\dfrac{\pi}{4}\right)$

3. $(-2, 0)$ **4.** $\left(2, \dfrac{\pi}{2}\right)$

5. (a) The graph consists of all points seven units from the pole; $x^2 + y^2 = 49$

(b) The graph consists of all points on the line that makes an angle of $\pi/4$ with the polar axis and passes through the pole; $y = x$

(c) The graph is a circle with center $(0, 3)$ and radius 3; $x^2 + (y - 3)^2 = 9$

Section 10.8

1.

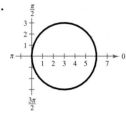

2.

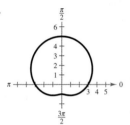

3.

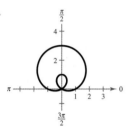

4.

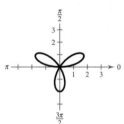

5.

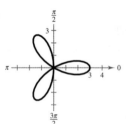

6.

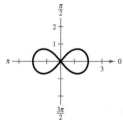

Section 10.9

1. Hyperbola
2. Hyperbola

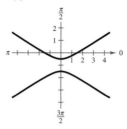

3. $r = \dfrac{2}{1 - \cos \theta}$

4. $r = \dfrac{0.625}{1 + 0.847 \sin \theta}$; about 0.340 astronomical unit

Chapter 11

Section 11.1

1.

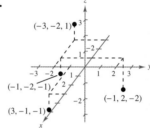

2. $\sqrt{35}$ **3.** $(5, -3, 1)$
4. $(x - 1)^2 + (y - 3)^2 + (z - 2)^2 = 16$
5. Center: $(-3, 2, -4)$; radius: 6
6.

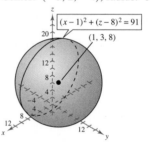

Section 11.2

1. $\langle 1, 6, -4 \rangle$; $\sqrt{53}$; $\left\langle \dfrac{\sqrt{53}}{53}, \dfrac{6\sqrt{53}}{53}, -\dfrac{4\sqrt{53}}{53} \right\rangle$ **2.** -2
3. About $76.5°$ **4.** (a) Not orthogonal   (b) Orthogonal
5. (a) Parallel   (b) Not parallel   **6.** Collinear
7. $Q(3, -5, 1)$ **8.** 140.7 lb, 216.3 lb, 140.7 lb

Section 11.3

1. (a) $-4\mathbf{i} + 2\mathbf{j} + 5\mathbf{k}$ (b) $4\mathbf{i} - 2\mathbf{j} - 5\mathbf{k}$ (c) $\mathbf{0}$
2. $-\frac{1}{3}\mathbf{i} + \frac{2}{3}\mathbf{j} + \frac{2}{3}\mathbf{k}$
3. $\vec{AB} = -\mathbf{i} + \mathbf{j} + 2\mathbf{k}$ is parallel to $\vec{DC} = -\mathbf{i} + \mathbf{j} + 2\mathbf{k}$
   $\vec{AD} = 0\mathbf{i} - 2\mathbf{j} + 4\mathbf{k}$ is parallel to $\vec{BC} = 0\mathbf{i} - 2\mathbf{j} + 4\mathbf{k}$
   Area $= 2\sqrt{21} \approx 9.17$
4. 6.6   **5.** 15

Section 11.4

1. Parametric equations: $x = 2 + 4t,\ y = 1 - 2t,\ z = -3 + 7t$

Symmetric equations: $\dfrac{x - 2}{4} = \dfrac{y - 1}{-2} = \dfrac{z + 3}{7}$

2. Parametric equations: $x = 1 + 2t,\ x = -6 + 11t,\ z = 3 + 5t$

Symmetric equations: $\dfrac{x - 1}{2} = \dfrac{y + 6}{11} = \dfrac{z - 3}{5}$

3. $2x - y + 6z - 13 = 0$ **4.** $3x + y + 2z - 13 = 0$

5. $90°;\ x = t,\ y = 4t,\ z = -2t$ **6.** $\dfrac{6}{\sqrt{30}}$

Chapter 12
Section 12.1

1. 13 in. × 13 in.

2.

x	2.9	2.99	2.999	3
$f(x)$	-2.8	-2.98	-2.998	?

x	3.001	3.01	3.1
$f(x)$	-3.002	-3.02	-3.2

-3

3.

x	0.9	0.99	0.999	1
$f(x)$	0.20408	0.200401	0.200040	?

x	1.001	1.01	1.1
$f(x)$	0.199960	0.199601	0.196078

0.2

4. 7   **5.** -3 **6.** Answers will vary

7. The limit does not exist.   **8.** The limit does not exist.

9. (a) $\dfrac{1}{4}$ (b) 27   (c) $-\dfrac{1}{\pi}$ (d) $2\sqrt{3}$ (e) 0   (f) 4

10. (a) 7   (b) $\dfrac{7}{3}$ (c) $\dfrac{1}{3}$

Section 12.2

1. 1   **2.** $\dfrac{1}{56}$ **3.** $\dfrac{1}{2}$ **4.** 1   **5.** 0

6. $\lim\limits_{x\to 3^-} \dfrac{|x - 3|}{x - 3} = -1$

$\lim\limits_{x\to 3^+} \dfrac{|x - 3|}{x - 3} = 1$

7. 2   **8.** $\lim\limits_{x\to 1^-} f(x) = 24 \neq 28 = \lim\limits_{x\to 1^+} f(x)$

9. 0   **10.** 0

Section 12.3

1. 4

2. 10 degrees per month; The monthly normal temperature in May is about 10 degrees higher than the monthly normal temperature in April.

3. 4   **4.** -3 **5.** $4x;\ -12;\ 8$ **6.** $3x^2 + 2$

7. $\dfrac{1}{2\sqrt{x - 1}};\ \dfrac{1}{2};\ \dfrac{1}{6}$

Section 12.4

1. 0   **2.** (a) 0   (b) -2 (c) Does not exist

3. (a) \$24.00   (b) \$6.00   (c) \$4.20   (d) \$4.00

4. (a) $3, \frac{4}{3}, 1, \frac{6}{7}, \frac{7}{9}$; Limit: $\frac{1}{2}$ (b) $1, \frac{5}{12}, \frac{29}{81}, \frac{11}{32}, \frac{127}{375}$; Limit: $\frac{1}{3}$

(c) $1, \frac{1}{3}, \frac{5}{27}, \frac{1}{8}, \frac{7}{75}$; Limit: 0

5. (a) $\frac{10}{7}$ (b) 1

Section 12.5

1. 385   **2.** 1.85, 1.535, 1.5035, 1.50035   **3.** 0

4. $\frac{9}{2} = 4.5$ square units   **5.** $\frac{3}{2}$ square units   **6.** $\frac{38}{3}$ square units

Chapter 13
Section 13.1

1. (a) "86% (of college graduates) said that college has been a good investment for them personally."

(b) *Sample answer:* Most college graduates consider college a good investment.

2.

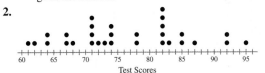

Test Scores

82

3. 20 days   **4.** 9 states

5.

Interval	Tally
60–69	HHH
70–79	HHH HHH |||
80–89	HHH ||||
90–99	

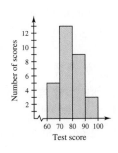

Test score

Section 13.2

1. Mean: 75.8; median: 73.5; mode: 82

2. (a) Mode   (b) Median   (c) Mean or median   **3.** $C;\ B$

4. Data set A: $\sigma = 2$

data set B: $\sigma \approx 2.28$

data set C: $\sigma \approx 1.72$

5. About 1.36   **6.** 100%

7. Lower quartile: 23, upper quartile: 64.5

8.

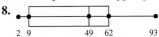

9. 0.815   **10.** 47.5%   **11.** About 0.8849

Section 13.3

1. 9.056775   **2.** $y = \frac{17}{10}x + \frac{2}{5}$

3. $r \approx 0.990$; The linear model fits the data very well.

4. $y = -\frac{1}{2}x^2 - \frac{1}{2}x + 1$

Appendix A

Appendix A.1

1. (a) Natural numbers: $\left\{\frac{6}{3}, 8\right\}$ (b) Whole numbers: $\left\{\frac{6}{3}, 8\right\}$
(c) Integers: $\left\{\frac{6}{3}, -1, 8, -22\right\}$
(d) Rational numbers: $\left\{-\frac{1}{4}, \frac{6}{3}, -7.5, -1, 8, -22\right\}$
(e) Irrational numbers: $\left\{-\pi, \frac{1}{2}\sqrt{2}\right\}$

2.

$$-1.6 \quad -\tfrac{3}{4} \quad 0.7 \quad \tfrac{5}{2}$$

3. (a) $1 > -5$ (b) $\frac{3}{2} < 7$ (c) $-\frac{2}{3} > -\frac{3}{4}$

4. (a) The inequality $x > -3$ denotes all real numbers greater than -3.
(b) The inequality $0 < x \le 4$ denotes all real numbers between 0 and 4, not including 0, but including 4.

5. The interval consists of all real numbers greater than or equal to -2 and less than 5.

6. $-2 \le x < 4$ **7.** (a) 1   (b) $-\frac{3}{4}$ (c) $\frac{2}{3}$ (d) -0.7

8. (a) 1   (b) -1

9. (a) $|-3| < |4|$ (b) $-|-4| = -|4|$
(c) $|-3| > -|-3|$

10. (a) 58   (b) 12   (c) 12

11. Terms: $-2x, 4$; Coefficients: $-2, 4$ **12.** -5

13. (a) Commutative Property of Addition
(b) Associative Property of Multiplication
(c) Distributive Property

14. (a) $\dfrac{x}{10}$ (b) $\dfrac{x}{2}$

Appendix A.2

1. (a) -81 (b) 81   (c) 27   (d) $\frac{1}{27}$
2. (a) $-\frac{1}{16}$ (b) 64
3. (a) $-2x^2y^4$ (b) 1   (c) $-125z^5$ (d) $\dfrac{9x^4}{y^4}$
4. (a) $\dfrac{2}{a^2}$ (b) $\dfrac{b^5}{5a^4}$ (c) $\dfrac{10}{x}$ (d) $-2x^3$
5. 4.585×10^4 **6.** -0.002718 **7.** $864,000,000$
8. (a) -12 (b) Not a real number   (c) $\frac{5}{8}$ (d) $-\frac{2}{3}$
9. (a) 5   (b) 25   (c) x (d) $\sqrt[4]{x}$
10. (a) $4\sqrt{2}$ (b) $5\sqrt[3]{2}$ (c) $2a^2\sqrt{6a}$ (d) $-3x\sqrt[3]{5}$
11. (a) $9\sqrt{2}$ (b) $(3x - 2)\sqrt[3]{3x^2}$
12. (a) $\dfrac{5\sqrt{2}}{6}$ (b) $\dfrac{\sqrt[3]{5}}{5}$ **13.** $2(\sqrt{6} + \sqrt{2})$ **14.** $\dfrac{2}{3(2 + \sqrt{2})}$
15. (a) $27^{1/3}$ (b) $x^{3/2}y^{5/2}z^{1/2}$ (c) $3x^{5/3}$
16. (a) $\dfrac{1}{\sqrt{x^2 - 7}}$ (b) $-3\sqrt[3]{bc^2}$ (c) $\sqrt[4]{a^3}$ (d) $\sqrt[5]{x^4}$
17. (a) $\frac{1}{25}$ (b) $-12x^{5/3}y^{9/10}, x \ne 0, y \ne 0$
(c) $\sqrt[4]{3}$ (d) $(3x + 2)^2, x \ne -\frac{2}{3}$

Appendix A.3

1. Standard form: $-7x^3 + 2x + 6$
Degree: 3; Leading coefficient: -7
2. $2x^3 - x^2 + x + 6$ **3.** $3x^2 - 16x + 5$
4. (a) $9x^2 - 4$ (b) $x^2 - 4x + 4 - 9y^2$
5. (a) $5x^2(x - 3)$ (b) $-3(1 - 2x + 4x^3)$
(c) $(x + 1)(x^2 - 2)$

6. $4(5 + y)(5 - y)$ **7.** $(x - 1 + 3y^2)(x - 1 - 3y^2)$
8. $(3x - 5)^2$ **9.** $(4x - 1)(16x^2 + 4x + 1)$
10. (a) $(x + 6)(x^2 - 6x + 36)$ (b) $5(y + 3)(y^2 - 3y + 9)$
11. $(x + 3)(x - 2)$ **12.** $(2x - 3)(x - 1)$
13. $(x^2 - 5)(x + 1)$ **14.** $(2x - 3)(x + 4)$

Appendix A.4

1. (a) All nonnegative real numbers x
(b) All real numbers x such that $x \ge -7$
(c) All real numbers x such that $x \ne 0$
2. $\dfrac{4}{x - 6}, x \ne -3$ **3.** $-\dfrac{3x + 2}{5 + x}, x \ne 1$
4. $\dfrac{5(x - 5)}{(x - 6)(x - 3)}, x \ne -3, x \ne -\dfrac{1}{3}, x \ne 0$
5. $x + 1, x \ne \pm 1$ **6.** $\dfrac{x^2 + 1}{(2x - 1)(x + 2)}$
7. $\dfrac{7x^2 - 13x - 16}{x(x + 2)(x - 2)}$ **8.** $\dfrac{3(x + 3)}{(x - 3)(x + 2)}$ **9.** $-\dfrac{1}{(x - 1)^{4/3}}$
10. $\dfrac{2(x + 1)(x - 1)}{(x^2 - 2)^{3/2}}$ **11.** $\dfrac{1}{\sqrt{9 + h} + 3}, h \ne 0$

Appendix A.5

1. (a) -4 (b) 8   **2.** $-\frac{18}{5}$ **3.** No solution
4. $-1, \frac{5}{2}$ **5.** (a) $\pm 2\sqrt{3}$ (b) $1 \pm \sqrt{10}$ **6.** $2 \pm \sqrt{5}$
7. $\dfrac{5}{3} \pm \dfrac{\sqrt{31}}{3}$ **8.** $-\dfrac{1}{3} \pm \dfrac{\sqrt{31}}{3}$ **9.** $\dfrac{4}{3}$ **10.** $0, \pm\dfrac{2\sqrt{3}}{3}$
11. (a) $5, \pm\sqrt{2}$ (b) $0, -\frac{3}{2}, 6$ **12.** -9 **13.** $-59, 69$
14. $-2, 6$ **15.** About 2.97 in.

Appendix A.6

1. (a) $1 \le x \le 3$; Bounded   (b) $-1 < x < 6$; Bounded
(c) $x < 4$; Unbounded   (d) $x \ge 0$; Unbounded
2. $x \le 2$

3. (a) $x < 3$
(b)

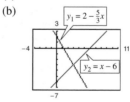

$y_1 > y_2$ for $x < 3$.
4. $(-3, 2)$ **5.** $[16, 24]$

6. More than 68 hours
7. You might have been overcharged by as much as \$0.16 or undercharged by as much as \$0.15.

Appendix A.7

1. Do not apply radicals term-by-term. Leave as $\sqrt{x^2 + 4}$.
2. $(x - 2)^{-1/2}(7x - 12)$ **3.** 3
4. Answers will vary.   **5.** $-6x(1 - 3x^2)^{-2} + x^{-1/3}$
6. (a) $x - 2 + \dfrac{5}{x^3}$ (b) $x^{3/2} - x^{1/2} + 5x^{-1/2}$

Index

Definition of the Six Trigonometric Functions

Right triangle definitions, where $0 < \theta < \pi/2$

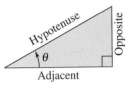

$$\sin \theta = \frac{\text{opp}}{\text{hyp}} \qquad \csc \theta = \frac{\text{hyp}}{\text{opp}}$$

$$\cos \theta = \frac{\text{adj}}{\text{hyp}} \qquad \sec \theta = \frac{\text{hyp}}{\text{adj}}$$

$$\tan \theta = \frac{\text{opp}}{\text{adj}} \qquad \cot \theta = \frac{\text{adj}}{\text{opp}}$$

Circular function definitions, where θ is any angle

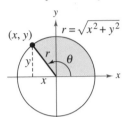

$$\sin \theta = \frac{y}{r} \qquad \csc \theta = \frac{r}{y}$$

$$\cos \theta = \frac{x}{r} \qquad \sec \theta = \frac{r}{x}$$

$$\tan \theta = \frac{y}{x} \qquad \cot \theta = \frac{x}{y}$$

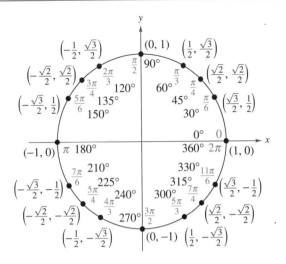

Reciprocal Identities

$$\sin u = \frac{1}{\csc u} \qquad \cos u = \frac{1}{\sec u} \qquad \tan u = \frac{1}{\cot u}$$

$$\csc u = \frac{1}{\sin u} \qquad \sec u = \frac{1}{\cos u} \qquad \cot u = \frac{1}{\tan u}$$

Quotient Identities

$$\tan u = \frac{\sin u}{\cos u} \qquad \cot u = \frac{\cos u}{\sin u}$$

Pythagorean Identities

$$\sin^2 u + \cos^2 u = 1$$
$$1 + \tan^2 u = \sec^2 u \qquad 1 + \cot^2 u = \csc^2 u$$

Cofunction Identities

$$\sin\left(\frac{\pi}{2} - u\right) = \cos u \qquad \cot\left(\frac{\pi}{2} - u\right) = \tan u$$

$$\cos\left(\frac{\pi}{2} - u\right) = \sin u \qquad \sec\left(\frac{\pi}{2} - u\right) = \csc u$$

$$\tan\left(\frac{\pi}{2} - u\right) = \cot u \qquad \csc\left(\frac{\pi}{2} - u\right) = \sec u$$

Even/Odd Identities

$$\sin(-u) = -\sin u \qquad \cot(-u) = -\cot u$$
$$\cos(-u) = \cos u \qquad \sec(-u) = \sec u$$
$$\tan(-u) = -\tan u \qquad \csc(-u) = -\csc u$$

Sum and Difference Formulas

$$\sin(u \pm v) = \sin u \cos v \pm \cos u \sin v$$
$$\cos(u \pm v) = \cos u \cos v \mp \sin u \sin v$$
$$\tan(u \pm v) = \frac{\tan u \pm \tan v}{1 \mp \tan u \tan v}$$

Double-Angle Formulas

$$\sin 2u = 2 \sin u \cos u$$
$$\cos 2u = \cos^2 u - \sin^2 u = 2 \cos^2 u - 1 = 1 - 2 \sin^2 u$$
$$\tan 2u = \frac{2 \tan u}{1 - \tan^2 u}$$

Power-Reducing Formulas

$$\sin^2 u = \frac{1 - \cos 2u}{2}$$

$$\cos^2 u = \frac{1 + \cos 2u}{2}$$

$$\tan^2 u = \frac{1 - \cos 2u}{1 + \cos 2u}$$

Sum-to-Product Formulas

$$\sin u + \sin v = 2 \sin\left(\frac{u + v}{2}\right) \cos\left(\frac{u - v}{2}\right)$$

$$\sin u - \sin v = 2 \cos\left(\frac{u + v}{2}\right) \sin\left(\frac{u - v}{2}\right)$$

$$\cos u + \cos v = 2 \cos\left(\frac{u + v}{2}\right) \cos\left(\frac{u - v}{2}\right)$$

$$\cos u - \cos v = -2 \sin\left(\frac{u + v}{2}\right) \sin\left(\frac{u - v}{2}\right)$$

Product-to-Sum Formulas

$$\sin u \sin v = \frac{1}{2}[\cos(u - v) - \cos(u + v)]$$

$$\cos u \cos v = \frac{1}{2}[\cos(u - v) + \cos(u + v)]$$

$$\sin u \cos v = \frac{1}{2}[\sin(u + v) + \sin(u - v)]$$

$$\cos u \sin v = \frac{1}{2}[\sin(u + v) - \sin(u - v)]$$

FORMULAS FROM GEOMETRY

Triangle:

$h = a \sin \theta$

$\text{Area} = \dfrac{1}{2}bh$

$c^2 = a^2 + b^2 - 2ab \cos \theta$ (Law of Cosines)

Right Triangle:

Pythagorean Theorem
$c^2 = a^2 + b^2$

Equilateral Triangle:

$h = \dfrac{\sqrt{3}s}{2}$

$\text{Area} = \dfrac{\sqrt{3}s^2}{4}$

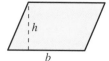

Parallelogram:

$\text{Area} = bh$

Trapezoid:

$\text{Area} = \dfrac{h}{2}(a + b)$

Circle:

$\text{Area} = \pi r^2$

$\text{Circumference} = 2\pi r$

Sector of Circle:

$\text{Area} = \dfrac{\theta r^2}{2}$

$s = r\theta$

θ in radians

Circular Ring:

$\text{Area} = \pi(R^2 - r^2)$

$\quad\quad = 2\pi pw$

$p = $ average radius,

$w = $ width of ring

Sector of Circular Ring:

$\text{Area} = \theta pw$

$p = $ average radius,

$w = $ width of ring,

θ in radians

Ellipse:

$\text{Area} = \pi ab$

$\text{Circumference} \approx 2\pi \sqrt{\dfrac{a^2 + b^2}{2}}$

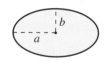

Cone:

$\text{Volume} = \dfrac{Ah}{3}$

$A = $ area of base

Right Circular Cone:

$\text{Volume} = \dfrac{\pi r^2 h}{3}$

$\text{Lateral Surface Area} = \pi r \sqrt{r^2 + h^2}$

Frustum of Right Circular Cone:

$\text{Volume} = \dfrac{\pi(r^2 + rR + R^2)h}{3}$

$\text{Lateral Surface Area} = \pi s(R + r)$

Right Circular Cylinder:

$\text{Volume} = \pi r^2 h$

$\text{Lateral Surface Area} = 2\pi rh$

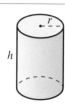

Sphere:

$\text{Volume} = \dfrac{4}{3}\pi r^3$

$\text{Surface Area} = 4\pi r^2$

Wedge:

$A = B \sec \theta$

$A = $ area of upper face,

$B = $ area of base

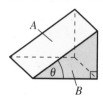